T0292699

CAMBRIDGE LIBRARY COLLECTION

Books of enduring scholarly value

Life Sciences

Until the nineteenth century, the various subjects now known as the life sciences were regarded either as arcane studies which had little impact on ordinary daily life, or as a genteel hobby for the leisured classes. The increasing academic rigour and systematisation brought to the study of botany, zoology and other disciplines, and their adoption in university curricula, are reflected in the books reissued in this series.

The English Flower Garden

An Irish-born gardener and writer, William Robinson (1838–1935) travelled widely to study gardens and gardening in Europe and America. He founded a weekly illustrated periodical, *The Garden*, in 1871, which he owned until 1919, and published numerous books on different aspects of horticulture. Topics included annuals, hardy perennials, alpines and sub-tropical plants, as well as accounts of his travels. This book, his most famous work, was first published in 1883, and fifteen editions were issued in his lifetime. It has been described as 'the most widely read and influential gardening book ever written'. Aimed at both amateurs and experienced gardeners, it sets out clearly the different types of plant suitable for each type of situation, and how to grow them. Robinson advocated a revolution in garden design, rejecting the more formal flower-beds which had long been popular in favour of a more natural and individual style.

Cambridge University Press has long been a pioneer in the reissuing of out-of-print titles from its own backlist, producing digital reprints of books that are still sought after by scholars and students but could not be reprinted economically using traditional technology. The Cambridge Library Collection extends this activity to a wider range of books which are still of importance to researchers and professionals, either for the source material they contain, or as landmarks in the history of their academic discipline.

Drawing from the world-renowned collections in the Cambridge University Library, and guided by the advice of experts in each subject area, Cambridge University Press is using state-of-the-art scanning machines in its own Printing House to capture the content of each book selected for inclusion. The files are processed to give a consistently clear, crisp image, and the books finished to the high quality standard for which the Press is recognised around the world. The latest print-on-demand technology ensures that the books will remain available indefinitely, and that orders for single or multiple copies can quickly be supplied.

The Cambridge Library Collection will bring back to life books of enduring scholarly value (including out-of-copyright works originally issued by other publishers) across a wide range of disciplines in the humanities and social sciences and in science and technology.

The English Flower Garden

Style, Position, and Arrangement

William Robinson

CAMBRIDGE UNIVERSITY PRESS

Cambridge, New York, Melbourne, Madrid, Cape Town,
Singapore, São Paolo, Delhi, Tokyo, Mexico City

Published in the United States of America by Cambridge University Press, New York

www.cambridge.org
Information on this title: www.cambridge.org/9781108037129

This edition first published 1883
This digitally printed version 2011

ISBN 978-1-108-03712-9 Paperback

THE

ENGLISH FLOWER GARDEN.

THE ENGLISH FLOWER GARDEN

STYLE, POSITION, AND ARRANGEMENT;

FOLLOWED BY

A DESCRIPTION, ALPHABETICALLY ARRANGED,
OF ALL THE PLANTS BEST SUITED FOR ITS EMBELLISHMENT;
THEIR CULTURE, AND POSITIONS SUITED FOR EACH.

BY

W. ROBINSON,

FOUNDER OF "THE GARDEN," "FARM AND HOME," AND "GARDENING ILLUSTRATED;"
AUTHOR OF "THE PARKS AND GARDENS OF PARIS," "THE SUBTROPICAL GARDEN,"
"ALPINE FLOWERS FOR ENGLISH GARDENS," "THE WILD GARDEN,"
"HARDY FLOWERS," ETC.

WITH THE CO-OPERATION OF MANY OF THE
BEST FLOWER GARDENERS OF THE DAY.

ILLUSTRATED WITH MANY ENGRAVINGS.

LONDON:
JOHN MURRAY, ALBEMARLE STREET.
NEW YORK: SCRIBNER & WELFORD.
1883.

LONDON:
PRINTED BY WILLIAM CLOWES AND SONS, LIMITED,
STAMFORD STREET AND CHARING CROSS.

PREFACE.

In various books written by the author of this work, and in journals founded by him, a good deal has been done to disturb people's opinions as to the flower garden. The subject has been attacked from many sides, and something has been done to widen men's views on the matter. But while many desire to make changes, little help of a progressive character is to be had from books. This is an attempt, as far as possible within the limits of a volume not beyond the means of every amateur and gardener, to supply the want.

Hitherto I have mostly dealt with the flower garden from special points of view, treating of various subjects that add to the beauty of the garden—rock gardens, plants of fine form, those we may grow without care in any rough or out-of-the-way places, and the many beautiful things included among the hardy flowers of Northern countries. A book on the convenient and simple plan of a Dictionary, embracing all the plants, both hardy and half-hardy, annual and bulbous, suitable in any way for the British flower garden, seemed likely to best meet the wants of the time; and the present work is the result. It is profusely illustrated, with a view to make familiar the aspects of most of the genera, while in the first or general part the illustrations will help as to principles and design. The whole aim of the book is to make the flower garden a reflex, so to speak, of the world of beautiful plant-life, instead of the poor formal array it has long been.

Few know the abounding wealth of flowers fitted for the embellishment of our open-air gardens. Without such knowledge, progress is impossible. It is useless to discuss systems of arrangement if the beauty of the things to be arranged is a sealed book to us. On the other hand, those who are aware of the number of beautiful plants existing will usually contrive to have some of them growing happily near them. No prosaic stereotyped garden of half a dozen kinds of plants will satisfy any one who knows that not six, but hundreds of different and beautiful aspects of vegetation are obtainable in a garden in spring, summer, and autumn. More thought should be given to hardy vegetation in the open air. Every effort should be made to secure stocks of good plants, and some thought should be given to the artistic arrangement as well as to the contents of a garden. At present the rule is, no thought; no arrangement; no bold good grouping; no little pictures made with living colours; no variety; no contrast—repetitions of the poor and ugly patterns we see so often. The choke-muddle shrubbery, in which the poor flowering shrubs dwindle and kill each other, generally supports a few ill-grown and ill-chosen plants, but it is mainly distinguished for wide patches of bare earth in summer, over which, in better hands, pretty green things might crowd. It is disheartening to see how little pleasure men get out of their gardens, and how near to a desert they make them in this country of verdure and fertility. Yet, the smallest garden may be a picture, and a pretty one. Not only may we easily have much more variety in any one garden—and that of the highest beauty, but if men would give up mere imitation, we should be charmed with the contrasts between gardens. Every district should have flower gardens characteristic of itself, and adapted to its soil, climate, and position. Even small suburban gardens might refresh us with their brightness and their pleasant variety. In the larger gardens the opportunities are proportionately great—and seldom used. They are stereotyped at the very seasons when they ought to be full of delightful change.

In the compilation of this book the storehouse of information in THE GARDEN has been taken advantage of, but articles have been specially written where it was thought necessary. In the preparation of the second part of the work valuable help has been given by Mr. W. Goldring, late superintendent of the hardy plant department of the Royal Gardens at Kew, and now for some years on the staff of THE GARDEN. Mr. H. Hyde gave useful aid in the selection of the illustrations, and engraved many of them. The following are the names of the other writers whose contributions are embodied in the book, and frequently marked by their initials:—

J. Allen
J. Atkins
P. Barr
T. C. Boston
F. W. Burbidge
W. Brockbank
J. Birkenhead
J. Britten
Latimer Clarke
Rev. Harpur Crewe
J. Cornhill
A. Dean
R. Dean
J. Douglas
Rev. H. H. Dombrain
Rev. C. Wolley Dod
J. Dundas
Rev. Canon Ellacombe
H. J. Elwes
H. Ewbank
Dr. M. Foster
P. Neill Fraser
O. Froebel
W. Falconer
D. T. Fish
J. Groom
W. E. Gumbleton
P. Grieve
W. B. Hemsley
Rev. Canon Hole
Rev. F. D. Horner
T. Hatfield
E. Harvey
E. Hobday
Miss F. Hope
C. M. Hovey
T. H. Archer-Hind
I. Anderson-Henry
E. Jackson
Miss G. Jekyll
A. Kingsmill
Miss R. Kingsley
E. G. Loder
Max Leichtlin
R. I. Lynch
G. Maw
J. McNab
R. Marnock
J. G. Nelson
J. C. Niven
Miss C. M. Owen
J. T. Poë
A. Perry
R. Potter
A. Rawson
A. Salter
Rev. Canon Swayne
J. Simpson
T. Spanswick
C. W. Shaw
J. Smith
J. Sheppard
J. Stevens
W. Thompson
Rev. F. Tymons
T. Williams
G. F. Wilson
W. Wildsmith
J. Wood
E. H. Woodall

It is hoped the book will be found as complete as possible within the limits laid down. It consists of two parts: First, a general introduction, dealing with the question of "laying out"—the aim here urged for the first time being to make each place at various seasons and in every available situation an epitome of the great flower garden of the world itself. The old and the general plan is to repeat in the garden, and usually in the best position in it, the lifeless and offensive formality of wall-paper or carpet. How to destroy this miserable conventionality is, it is hoped, clearly shown, and that too, not by diminishing the number of flowers, but, on the contrary, by much increasing them.

The second part shows the many beautiful families of plants which grow in the open air in Britain, with their culture, the positions suited for each—a point hitherto little attended to in books. Where kinds are of slight or doubtful value, this is also distinctly stated in the interest of the amateur, who has to select from long lists of strange names.

The illustrations are frequently from THE GARDEN; some from the collection of Messrs. Vilmorin, while a variety of plants and plans have been drawn expressly for the book. The illustrations may well suggest the number of beautiful types of plant-life shut out from our gardens by the few plants used to perpetrate the crudities of "bedding out." During the past dozen years some gardens have been enriched; and even new aspects of garden vegetation have appeared, as, for example, the splendid and numerous Japanese and North American lilies, which look so admirable in the open air. But gardens generally are still poor in variety of flower and form. There is much more in flower gardening than is usually seen. It is an art that in all stages of life might afford men infinite pleasure and work at once innocent, healthy, and refining. But they can only know a few of its charms without a complete change in the narrow and "hard" way in which it is generally practised.

W. R.

Nov. 30th, 1883.

CONTENTS OF PART I.

Illustrations in *Italic* type.

THE

ENGLISH FLOWER GARDEN.

POSITION AND STYLE.

ONE purpose of this book is to help to uproot the common notion that a flower garden is necessarily of set pattern—usually geometrical—placed on one side of the house.

The true idea is that the various wants of flowers can be best met, and their varied loveliness fully shown, in a variety of positions. The very first thing to do is to consider the wisdom of arraying all our flowers in one spot exactly under the same conditions. To suppose that we can ever have a fair representation of a tithe of the beauty of which our gardens are capable by this means is proof that we have no just idea of the world of flowers as it is. The earth will bear flowers anywhere, and to limit its chances of doing so to a narrow floral rug beneath the windows is a mistake for which we have to pay in full. That a flower garden should occupy one spot, and all the flowers be therein shown, is part of the legacy of gardening knowledge which we inherit. We must take no notice of it if our gardens are to be as rich, as varied, and as beautiful and fragrant, with the flowers of all seasons, as every true lover of the garden desires they should be. We are now speaking of the numerous places where there is room for choice in the matter. The settled way has been, and is, to regard one spot with the same soil and aspect —with every condition alike, in fact—as the best and only home for the flowers; yet in the same place, and within a short distance, there may be many different positions each favourable to a different type of flower life. We will, therefore, begin with the question of style, in the hope of showing, by reason and illustration, that the natural one is *right*.

True taste based on law.—So many things are said to be "matters of taste," that to most people it is a surprise when it is said that, even as regards matters said to be "purely of taste," like gardening, all things are matters of law, as much as what are called exact sciences. The shades are so fine, the swift changes so beautiful, the many creatures of wood, grass, air, and wave so many and so fair, as they pass, that we may well forget for a moment that Nature is law and life is law. For all that concerns our lives and our surroundings there are laws which will aid and guide us if we seek for them. The laws meant here are Nature's—not merely landmarks set out oy man for his convenience. We shall never settle the most trifling question by the stupid saying that it is "a matter of taste." I hope in these notes to set down the laws that Nature has written for us, and to apply them to the formation of our gardens, too often wholly disfigured by want of attention to the said laws. Only they are not laws that bind with weary fetters, but full of tenderness like a good mother, and infinite in delightful change as the restless clouds on the hills. No one need fear their slavery, for within them is perfect liberty, and all the sweet wanderings the earth mother has for her children.

There are in books many dissertations on the several styles of laying out gardens; indeed, some have taken us to China and Japan, others to Mexico for illustration. But when all is read and examined, what is the result to anybody who looks from words to things? That there are really two styles: one straitlaced, mechanical, with much wall and stone, or it may be gravel; with much also of such geometry as the designer of wall papers excels in—often poorer than that; with an immoderate supply of spouting water, and with trees in tubs as an accompaniment, and perhaps griffins and endless plaster-work and sculpture of the poorer sort. The other, with right desire, though often awkwardly, accepting Nature as a guide, and endeavouring to illustrate in our gardens, so far as convenience and knowledge will permit, her many treasures of the world of flowers.

We read that, "we are forced, for the sake of accumulating our power and knowledge, to live in cities; but such advantage as we have in association with each other is in great part counterbalanced by our loss of fellowship with Nature. We cannot all have our gardens now, nor our pleasant fields to meditate in at eventide. Then the function of our architecture is, as far as may be, to replace these; to tell us about Nature; to possess us with memories of her quietness; to be solemn and full of tenderness like her, and rich in portraitures of her; full of delicate imagery of the flowers we can no more gather, and of the living creatures now far away from us in their own solitude." What, then, are we to think of those who carry the dead lines and changeless triumphs of the building and the studio into the garden, which, above any other artificial creation, should give us the sweetest and most wholesome "fellowship with Nature?" Simply that their doings result from ignorance of what a true garden ought to be. The worst thing that can be done with it is to introduce any feature which, unlike the materials of our world-designer, never changes.

The two styles : The natural. Lawn view in the gardens at Golder's Hill.

There are positions, it is true, where the intrusion of architecture and embankments into the garden is justifiable; nay, now and then, even necessary. The misfortune is that these accessories are often thought to be necessary when they are not so. The best terrace gardens in Europe are those built where the nature of the ground calls for them, usually where the ground is steep and rugged; it is in positions like these that they are most wanted in this country. There is no code of taste resting on any solid foundation which proves that garden or park should have any extensive stonework or geometrical arrangement. Many instances could be given to prove that the natural, or nearly natural, disposition of the most monotonous ground is preferable to the great majority of expensive geometrical gardens. Let us, then, use as few oil-cloth or carpet patterns and as little stonework as possible in our gardens, and arrange them so that, when our sunny season does come, they may be full of life and change, and of such beauty as is nowhere to be found in the deadly formalism we condemn.

The two styles: The formal. View in the gardens at Trentham.

An example.—In considering the formal kind of gardening we have a capital example in the region of the great fountain basins of the Crystal Palace. Here a most dismal impression is given, though the upper terrace at the Palace illustrates some of the

Example of the formal garden in a position requiring it. View at Shrublands, from Sir Charles Barry's works.

best features of the formal system. But the vast expense for a poor theatrical effect is not the most regrettable matter. This is supplied in the mass of formal, dirty water basins, with their spouting pipes and crumbling margins. There is nothing more melancholy than the walls, fountain basins, clipped trees, and long canals of places like these, not only because they utterly fail to satisfy in themselves, but constantly suggest wasted effort and riches worse than lost. There are, from Versailles to Caserta, a great many ugly gardens in Europe, but it is at Sydenham that the greatest modern example of the waste of enormous means in making hideous a fine piece of garden is to be found. This has been called a work of genius, but it is only the realisation of a misguided ambition to outdo another sad monument of great means prostituted to a base use—Versailles. But Versailles is a relic of a past epoch, and was the expression of such knowledge as men possessed of the gardening art at the date of its creation. Backward as we are now, our means for garden embellishment have increased a hundredfold since Versailles was designed. Therefore, our modern illustration of a barbarous style has none of the excuses which might be urged for Versailles. Instead of a desire to express all that we at present know of pure garden design, and of the wealth of beauty now within our reach, the major idea at the Palace was to out-Versailles Versailles. Mouldering water basins being plentiful in the French garden, it was thought well to make some much larger. Versailles having numerous tall water-spouts, the best way to glorify ourselves at Sydenham was to make some unmistakably taller! Instead of confining the purely geometrical gardening to the upper terrace, by far the greater portion of the ground was devoted to the more antiquated and baser features of a changeless and stony style of garden design, and nearly in the centre the vast fountain basins were placed, with their unclean water and appalling display of pipes. As garden sights these water basins are positively more hideous than the crater of a volcano. The extensive contrivances to enable the water to go downstairs, the temples impudently prominent, the statues, the dead walls, all help to add to the distracting elements of the central region. The special gardens, too, such as the Rose garden, so much better if veiled from the great open central scene, are made as conspicuous as possible.

Evil influences.—When a private individual indulges in expensive fancies, he may not injure anyone but himself; but in the matter of the expenditure on a public garden—which, by the way, is set up as an example of all that is admirable—we have, in addition to the first wasteful outlay, an object hurtful to the public taste which sows the seed of its ugliness all over the country. It may be said that our taste in

England is sufficiently assured against this; but it is not so. Many whose lawns were, or might readily be made, the most beautiful of gardens have ruined them for the mere sake of having a terraced garden. There is a modern castle in Scotland where the embankments are piled one above another, till the whole looks as if Uncle Toby and a whole army of corporals had been carrying out his grandest scheme in fortification. Were such an erection a matter of trifling cost, or one which could be easily removed, or even avoided, it would not be worth notice; but being so expensive that it may curtail for years the legitimate outlay for a garden, and prevent expenditure in living objects of the highest value, it cannot be too strongly denounced. The style is in doubtful taste in climates and positions more suited to it than that of England; but he who

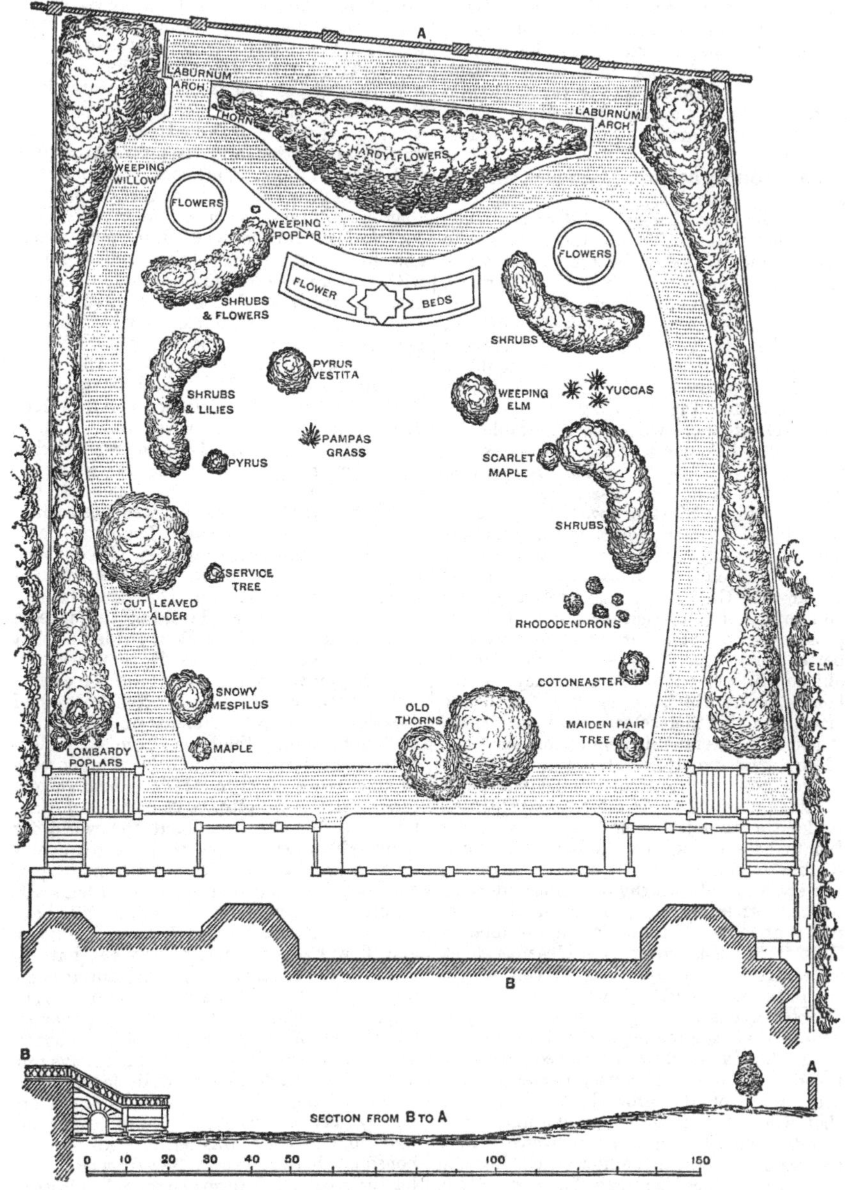

A non-geometrical town garden with open lawn and good trees and shrubs.

would adopt it in the present day, and in the presence of the inexhaustible and magnificent collections of trees and plants now within our reach, is an enemy to every true interest of the garden. The rude stone wall of the hill husbandman supporting a narrow slip of soil for his Olive trees or Vines became, in the ornamental garden of the wealthy Roman, an architectural feature, varied by vases, statues, &c. It is essential to bear in mind that the beauty of an Italian, or geometrical garden of any kind, will depend on the predominance of vegetation over the merely artificial. This may be said to be true of all gardens, and so it is, no doubt; but it applies to the terraced style more than to any other, inasmuch as it is in that style that artificial features are most obtrusive.

Terraced gardens, allowing of an endless variety of architectural work, apart from that of the house, have naturally been much in favour with architects and artists who have taken up the profession of landscape gardener. The landscape gardener proper, so to say, impressed by "orthodox" custom, and not attempting to think for himself, chimes in with the popular notion that every house, no matter what its position, should be fortified by terraces. Accordingly, he busies himself in forming terrace gardens, usually on level ground, which is unsuitable for them. It thus comes that vast sums, ostensibly devoted to the garden, are spent on waterworks, fountains, vases, statues, balustrades, walls, and stucco work. By the extensive use of such materials many a noble lawn is cut up; and sometimes, as at Witley Court, the architectural gardening is pushed so far into the park as absolutely to curtail and injure the prospect—assuming that the prospect of a noble garden of fine trees is better than that of a posing ground for the objects above enumerated. Many of the houses before which we see these formidable arrays would themselves seem to require much further embellishment from the hand of the artist-architect. Indeed, if the cost of the stone and stucco ornament lavished on the garden were, in many cases, spent on its legitimate object—the house, it could not fail to be a change for the better for architecture as well as for gardening. The fact is, the style is only worthy of serious adoption with us when the ground favours it, as—to name an English example—at Shrublands. Here it is used, with a very pleasing effect, to lead from the house down a steep bank to the pleasure grounds below.

Nothing can be more opposed to true gardening than the too popular notion that the right plan is to place a terraced garden on the best front of the house, no matter what the nature of the ground; the fact being that where the ground is level a finer effect results from allowing the turf to sweep up to the walk in front of the house than from an elaborate terrace, as may be seen on the north side of Holland House, and at Cambridge House, and in such gardens as Oak Lodge, where there is a very narrow terrace. In many cases there is need for a formal walk, raised or otherwise, and, it may be, for a small terrace—points which will be governed in each case by the position, and sometimes by the house itself. But where the ground, as in most English gardens, is level, there is no occasion for more than a grassy foreground, which leaves us free to adopt everywhere a purely artistic and natural style. In level town gardens, where the excuse of formal surroundings is used to justify a stony style, it is also a mistake. The highest effect is to be obtained, not by carrying architectural features into the usually small town or city garden, but by securing an absolute contrast between the garden vegetation and its unavoidable formal surroundings. This may be illustrated by a plan of a London town garden, such as that of Montague House, here given. This contrast should be secured in such cases, not by aiming at the sham picturesque, with rocks, cascades, and undulations of the ground, which would be too obviously artificial in such a case, but mainly by the simple majesty of trees and the charm of level turf. Thus, it has been affirmed that none but an Italian garden would have suited South Kensington. Well, we had an elaborate garden designed there, and skilfully designed in its way. The plan was carried out with the greatest care; the planting, &c., was done by experienced men, yet the result, as everybody knows, was unsatisfactory in the highest degree. There are many private gardens in European cities, of a more limited extent and with more formal surroundings, which are beautiful in the highest sense, and as devoid of the aspects that offend many in such a garden as that of South Kensington, as if they were in some happy valley far away from the city.

Terrace walls.—Very often I have had occasion to study the effect of the terrace wall, approaching the house from some pleasant part of the park. This adjunct very often catches the eye where one might expect something better. The parks of England are famous throughout the world for their sylvan beauty, and they are frequently much marred by the somewhat common prevalence of the so-called terrace garden made within the past generation or so. We have often to consider these walls from the house and flower garden points of view, but the most dismal of all is that looking towards the house. If this wall be raised, as it must be in places on the level or nearly level ground, so as to cut off the foreground of the house itself from the park, a bad effect is always produced. A beautiful and well-placed house is one of the things that should not be cut off by any commonplace, hard, and often ugly object such as this terrace wall. Let the house be sheltered and relieved by trees, and let their beautiful forms be brought near and around it. Add as many living graces as you can to the scene without impeding the view from important points, but do not cut off the whole of what ought to be the best considered spot in the place by a terrace wall. In many places, laid out at great cost within the past twenty years or so, the effect from the park beyond the garden is about as cheerful as looking against a piece of the wall of Clerkenwell Prison. The illustration which accompanies this is not what one would call by any means a bad example, or even a typical one, of what is meant, and yet it serves to point out the aspect to which we wish to call attention.

Effect of terrace walls; one of the least objectionable types. The view of the outer wall from the park is worse.

Lawns destroyed.—Another matter of some importance is that these terrace walls very often prevent the formation of a beautiful wide lawn. In many cases they have been made on the site of a beautiful lawn. Now, a simple, large, fairly-kept lawn is one of the best features of a garden. Unhappily, there has been so much cutting up, geometry, and stone work, that it is extremely rare to find a place where a good lawn is left. There is many a site cut up by terrace gardens and other formalities which would be vastly improved by the substitution of a simple, wide, and nobly-fringed lawn. Imagine the effect of a well-built and fine old house, seen from the extremity of a wide lawn, with plenty of trees and shrubs on its outer parts, and nothing to impede the view of the house or its windows but a refreshing carpet of grass; and then, standing in the same position, consider the wisdom of what has so often been done—viz., the facing of such a house with a wide terrace. The truth is, that a very common, poorly-built house with a fine lawn has a better effect than a fine pile with a rectilineal garden and terraces in front of it. If owners of parks were to consider this point fully, and, as they travel about, watch the effect of such lawns as remain to us, and compare them with what has been done by certain landscape gardeners, there would shortly be, at many a country seat, a rapid carting away of the terrace and all its adjuncts.

Exceptions and improvements.—We of course would except cases in which the terrace was really called for by the nature of the ground, and we have no desire to limit the flower department in any way; on the contrary, we would increase it. The few more or less complicated and finnikin beds that are on these terraces, to serve the place of the flower garden, are generally not half sufficient for raising all the beautiful flowers we ought to grow. The attempt to make the varied beauties of our garden flora all conform to the same rules and occupy the same spot will never give a satisfactory result. In removing, in a large place, a needless and unsatisfactory terrace, and in forming a sweet unbroken lawn instead, we should, according to the circumstances of each place, make from two to six times the amount of flower accommodation in various parts of the garden or pleasure ground—on the outer fringes of the lawn, by the sides of the pleasant walks, or anywhere that favourite plants can be grown best and would look best. We should abolish all wormy and pattern beds and adopt simple large forms, oval, circular, or with an irregular outline in the case of very large masses, and in this way would give infinitely more variety and pleasure than with any plan laid out at once like a carpet. The natural features of the garden would in this way relieve the eye and assist the artist; groups or specimens of trees, fine groups of shrubs would come in to produce a good effect between each of the special floral features which are the pride and care of the gardener to develop year by year. The production of these special and distinct features will be dealt with further on and in many parts of this book, and the plants described in it will show how rich is the store to draw upon.

Nature's revenge.—I find the following in one of the recent novels of Victor Cherbuliez concerning a type of garden not uncommon. As an opinion coming from one out of the "speciality," so to say, it is interesting, and takes a just view: "He passed before the open gate of a spacious garden which formerly presented, to the admiration of visitors, beautifully straight alleys, bordered with nicely trimmed Boxwood and globe and cone-shaped Yew trees, shrubs arranged in angles or in the form of a chess-board, or devised to imitate a wall; close-clipped Elm trees and statues everywhere. In the centre, surrounded with flower borders, a large basin in which the water thrown up by dolphins fell back in transparent veils. It was one of those classical gardens, the planners of which prided themselves upon being able to give Nature lessons of good behaviour, to teach her geometry and the fine art of irreproachable lines; but Nature abhors lines; she is for geometers a reluctant pupil, and if she submits to their tyranny she does it with bad grace, and with the firm resolve to take eventually her revenge. Man cannot conquer the wildness of her disposition, and so soon as he is no longer at hand to impose his will, so soon as he relaxes his cares, she destroys his work. The garden in which Lionel had just entered had been very badly kept after Baron Adhémard's death, who cared more for his fields than for his garden. After him decay had degenerated into ruin. The Boxwood was seldom trimmed, the Grass invaded the alleys, the Yew trees stooped through sheer old age. The large basin had no longer any water, and the dolphins which in days gone by supplied it from their throats looked as if they asked each other to what purpose they were in this world. On their faces you could read that melancholy that haunts all beings who ignore or have forgotten the secret of their destiny. But the statues had suffered most; moss and a green dampness had invaded them—their whiteness had disappeared for ever. Some kind of plague or leprosy which stone is heir to had covered them with stains and sores. Pitiless Time had inflicted on them mutilations and insults. One had lost an arm, another a leg; almost all were deprived of the top of their noses. There was in the basin a Neptune whose face was sadly damaged. He had not much more left than his beard and half his trident. Further on could be seen a Jupiter deprived of his head; the rain water sojourned in his hollow neck, and the sparrows came thither to drink, for sparrows have not much respect for aught, and they experience no scruple in turning the neck of a Jupiter into a drinking fountain. In the midst of a thicket stood a little Pan, who, leaning his back against a rock, had blown in his pipes for well nigh two centuries; he had no longer any pipes or hands, or the least breath, and the rock, ignoring the reason of his silence, wondered very much why he no longer heard his song. Elsewhere, a Vertumnus rested his crossed legs and his horn of plenty on a stand which was fast losing its equilibrium;

he had a careworn appearance—he foresaw an accident. Close by, as a contrast, was a pedestal whose statue had disappeared, and

Stone gardening (Caserta) "before decay's effacing fingers!"

which seemed to say, "Where is my god gone to!"

Railway-bank terraces.—A striking example of a style of garden design that for a long time has had an injurious effect on country seats, and above all on the garden and park, is the "railway embankment" phase of landscape gardening madness—one in which we see a series of sharply graded slopes, exactly like well-smoothed railway embankments, more or less relieved by fountains, balustrades, &c. The extraordinary thing is that anybody supposed to have taste should imagine that an arrangement of this kind, intercepting the whole landscape, so to say, should give permanent pleasure to any human being, or do anything but make the home landscape formal and wearisome to the last degree. The man who designs such things may fancy that they look very pretty in his tracery, but when one comes to apply them to the landscape itself, and, looking up a hill towards a house inhabited by English people, sees it thrown into intersections of huge, sharply-graded banks, may, if he has an independent judgment, begin to wonder why people have been led to do so much harm to gardens in England through this fashion. In these days of retrenchment, and, indeed, in all days, the labour of keeping up such places—preserving all their harsh angles, mowing, attending to the "compo" and the brickwork, the fountains, keeping the large areas of gravel clean—must be very great, require much attention, and, as a matter of fact, be very expensive. The cost of formation is so great, that naturally there is little left for the labour to keep it in order. Equally the true gardening, the culture of plants for its own sake, and the annual planting of the finer trees—a thing which never should be neglected by those who have large estates or parks—is injured in consequence of excessive expense devoted to features that must eventually be pronounced worse than useless, and be removed or modified.

EXAMPLES FROM ENGLISH GARDENS.

No plan will help us so well to a clear view of what is best in the way of design or style as a visit to a few typical gardens, the visit being aided by such pictorial views as we can present of them in engravings.

Thoresby is typical of the great houses of the present day in which the elaborate terrace, protecting and hiding the best part of the house, is carried out. To get a pretty view of it, the artist has gone to the riverside, because on nearer approach the hard walls of the terrace cut off the view. In this wide and airy park the views are very fine; and though the vegetation near the house is in many cases young, there is in the outer parts no lack of stately trees. The house is a large pile, as yet not softened by the delicate and beautiful tones that time lays on old buildings, but without any of the "confectionery" with which the modern architect shows his "originality." The question of the position of a mansion in a place where almost any kind of site might be found is a very important one. No doubt there are excellent reasons for choosing the present one, the only question, in the interest of art, seeming to be—are they sufficient? Very few people give an entire attention to the selection of the site of a house. In old times houses were generally put in the worst positions for the sake of securing shelter; and yet some of our old houses—such as Hardwick Hall—do not appear to be any the worse for being fully exposed and on high ground. With all our modern appliances for securing warmer buildings than was possible a few hundred years ago, we should not fear a little exposure if with that we secured the advantage of a commanding and in all ways a satisfactory position. A fine, well-built house facing the warmth and sun to a large extent really makes its own shelter. As to being too much elevated from any point of view, that is scarcely a serious objection, because our forest trees add a good deal towards concealment in time if required. Though it is not always desirable to choose the crest of a hill, yet it is doubtful if building the house partly into the slope, as at Thoresby, is the best way. There is usually some awkwardness in consequence of this selection. At Thoresby this has not been quite escaped, and the effect of the stables above the house, and rather too near it, does not strike one as the best arrangement. The principal front certainly commands a fine site, and rises above a beautiful slope of Grass, but its good effect from the park is somewhat neutralised by the terrace, the lines and other impediments belonging to which preventing the eye from enjoying the view of the building. There can be no question whatever that the beauty of many houses is sadly interfered with by an elaborately planned garden in front. The simplest way to settle this question is to entirely leave aside books and rules, and look at houses with these gardens from the chief points of view, and then look at houses where there is nothing of the sort, and compare the effects. The objection in question particularly applies to where the terrace is a wide and complicated one, pushed out so that it steals away the foreground from the finest front of the house. It shows that the fashion which somehow has come into use of late years of forming a geometrical garden—this is almost a law in connection with big new houses—is extremely unfortunate. There is nothing whatever in gardens of this sort in the way of plants or trees that cannot be shown quite as well in other ways. As a rule, indeed, they may be shown much better, because away from the immediate front of the house we are not controlled by the necessity of conforming everything to one level pattern. Therefore, as a principle, in large places where we are no longer confined, as of old, to bits of ground under the castle walls or within them, it is better that all the choicest flower and other ornamental gardening should be veiled in graceful plantations to the right and left of the house rather *than spread out before it or at the back of it.*

Highclere, occupying a similar situation as to ground, is an example of the better fashion of having a quiet, open vista in such places. There is ample room for every phase of gardening or tree planting in the grounds spreading out on either side and behind the house. I doubt if, to the end of time, any better foreground can be got for the principal front of a noble house than a carpet of turf; and I have never seen an instance in which dead or other walls have been run across it, for any purpose, without marring it. I am speaking of places such as Osborne, Highclere, and Thoresby, and the

Thoresby Park front. The effect of the terrace walls should be seen from the lawn shown in our view, where they cut off the view of the lower part of the house.

Highclere Park front. View of house clear.

great number of lowland English gardens in which the ground is level and does not demand the natural and necessary terraces of the steep hillside or rocky shore. Osborne is perhaps one of the saddest and ugliest examples in England of carrying out the stiff, geometrical idea in a place specially suited for the natural and free style of gardening—not a bit of ground there required any terracing.

Greenlands is an example of a garden in which the principal and river front of the house is a simple sloping lawn, and it comes nearer to our possible English flower garden than any of the examples recently named.

A Thames-side garden (Greenlands, Henley-on-Thames) with wide sloping lawn, and flower gardens partly visible from the lawn seen through the trees.

It is not to the large country seats of great families that one must look for good design in gardens, and Greenlands is one of the places that illustrates the truth of the remark. Originally laid out by Mr. Marnock

View of terrace garden front, Clumber.

for Mr. Majoribanks, it has long been a model of good taste and good keeping. The situation of the house by the river is well shown in the engraving. The various pretty river and other views from the house must be left to the imagination. As will be seen, there are no terrace gardens, no walls for stucco or terra-cotta vases. One passes easily from the house to a pleasant lawn, which slopes gently till it touches the river. From this lawn one approaches the wide and well-planted grounds around, studded with numerous fine trees, among which are beautiful groups of Cedars. It may be noticed that the common fashion of a garden in front of the house is here avoided altogether; but at some little distance there are various flower gardens which, without being exactly under the windows, are within very easy reach, and as richly stored with flowers and flowering shrubs as anybody could desire. It need scarcely be pointed out that this plan keeps the lawn immediately in front of the house fresh, unbroken, and quiet, instead of what it too often is—patched with brown earth or, not always happy, masses of flowers.

Clumber.—Here the drawing (from Sir Charles Barry's works) is so arranged that the house appears not to be cut off from the observer on the near side of the water—a very important position. But it really does cut off the view in this way, and, looked at from various points outside, the garden has a bad effect. So it will appear to those who take the trouble to look at the scene from all-important situations. Such very important positions should not be lightly decided upon, but it is too common to arrange them with reference to a single idea, and thus mar the beauty of a place. While dwelling on the (to me) essential question of the preservation of the main front of the house from the elaborate designs of the geometrical landscape gardener, we may glance at

Goodwood, where a "fine opportunity" to spend a fortune on a garden of stone has not yet been "taken advantage of." Goodwood, broadly considered, not only has great interest as regards landscape gardening and planting, but also as regards position, diversity, and charm of surface of a large portion of its park. As seen in our engraving, no terrace wall stands up to cut off the house from the park with its wide lawns and noble groups of trees, both deciduous and evergreen. Its greatest charms are perhaps the noble groups of evergreen Oaks and the great trees of Cedar of Lebanon standing, as native trees might do in a wide open park, with little semblance of an artificial or pleasure ground aspect. The delightful contrasts between British Oak in fine groups and evergreen Oak in groups equally fine, with Cedars of Lebanon, change completely the character of the landscape from what is usually met with in English parks. Cedars of Lebanon we all see and admire in pleasure gardens, but this is the first place where we have ever seen the majesty and noble variety of this tree truly shown, whether scattered in groves and groups or as single specimens over a large and varied landscape. Assuredly we do not want to be told of the Cedar's beauty and value, and of its great superiority to the majority of the "fashionable" Conifers now being planted everywhere; but we do want to get some idea of its noble uses in park, forest, and hill scenery. No pleasure-ground preparation of deep, well-selected, and gathered soil for these giants of the hills which here compete with the sturdiness of the finest British Oak! Scant soil over chalk and perfect exposure to the hill and sea winds is their fare. One tree is well up the hillside, fully exposed on every side, and, beyond a picturesque leaning of the plumes in one direction, shows quite as little sign of suffering as a group of the common Yew below it. Somewhat larger and more stately in the lower level part of the park, the great Cedars contrast superbly with the great Oaks and Chestnuts. Some wise planter must have been here years ago to give us all this grouping and stateliness of tree life without torturing the sward with dots, or obscuring the view of sea and hill. Oak, and Chestnut, and Lime, fine in stature and in form, are grouped like herds of great creatures. Away on the hills are grove after grove of Fir and Beech, with here and there the dark plumes of the Yew, or the pale green of Box bushes showing themselves prettily outside the masses; while, standing clear out of the nearest downs, are those round tufts of trees so characteristic of the downs country. Now, the very important consideration arises here as to all that would be lost were any contrivance to cut off the view of such a noble lawn and such noble trees from the window of the principal rooms in the house. I could name cases, in places of various sizes, where the view both to and from the best side of the house has been quite shut out by a heavy terrace wall, for which there was no reason whatever.

Pendell Court.—We have just been dealing with great places to which prairie-like parks belong. It will be seen in this illustration that even where it is desired to have the garden, or a portion of it, against the house, it is then by no means necessary that the ground should be made "architectural." It is a great pleasure to see a beautiful old house, made to live in, with nothing to keep one away from the door but the pleasant Grass. From a gardening standpoint there are three distinct views of it which are good; first that of the lawn in front of the house, which, when we saw it, was a flowery meadow yet uncut, and no beds or other impediments between the point of view and the house, with a group of some fine trees on either hand. It was a poem in building and in lawn. Quite on the other side a border of flowers and a wall of climbers ran from the house. Looking along this border to the house, a shower of white climbing Roses is seen falling from the wall, and a quaint gable and a few windows and glistening rich Ivy behind form such a picture, that one regrets to know it is both old and rare. Another view of the house from across the water, showing its west end, is also very beautiful. There is a wild Rose bush on the right and a tuft of Flag leaves on the left; before the beholder, the water and its Lilies; then a smooth, gently rising lawn creeping up to the windows, which on this side are all wreathed with

Goodwood Park front. From a sketch by Alfred Parsons, May 6, 1880.

lovely white climbing Roses. It will be observed that all these different views of the same house, although quite distinct, are all marked by the absence of the impediments which a false art frequently places so very near our houses—that is to say, formal patterns in beds, fountains, statues, and other like objects, which destroy the repose so desirable in such places. Those who care for the beauty of their gardens should not be afraid of a tree, a group of shrubs, a tuft of Ivy on the Grass, a bed or two of Roses on their own roots, or any beautiful object like it. But set patterns of beds, in the very place where the garden should be green and quiet, are a mistake in all cases where there is a more fitting place for them. The view from the house to the left is also free and charming—a wide meadow climbing up the hill through groups of trees, and reminding one a little of alpine pastures in the woody region.

View of lawn-garden at Pendell Court; a terraceless house and garden.

How delightful it is to visit a garden where there is variety as well as beauty in the vegetation, where one meets with old friends not, perhaps, seen for years, or finds them grown in new and better ways, amid fresh associations and groupings!

There is an expression common among persons interested in gardens which is very often used in reference to a place where a few things are grown that one sees everywhere else, and these among the quiet second-rate types of vegetation. The remark is, "There is nothing there." It is not exactly scientific, and yet it does explain the facts sufficiently from our point of view. Quite an opposite phrase must be used for a place like this, in which there is abundance of varied plant life, and great beauty of flower day by day as the seasons pass. There is also the charm of novel combinations of plants which, varying always according to soil, situation, and even slight differences of climate, are capable of being used and shown in ways absolutely without end.

Mr. Marnock on the position of the flower garden.—This may be a fitting place for a few remarks on the position of the flower garden written for this work by Mr. Marnock, so well known as a tasteful designer of gardens. There are with respect to the position and distribution of the floral decoration of the outdoor garden certain canons which apply to this particular department; and although in their application they not only admit, but compel, the use of wide variation in detail, they nevertheless may be accepted as forming a fairly safe basis and guide in determining what ought to be the most suitable position for a flower garden. This is specially so in the case of large mansions where ornamental gardening and the cultivation of flowers are extensively carried on. The flower garden in this case is an appendage to the house, and ought to be easily accessible from it. With this in view, the flower garden should, of course, be placed on the private front adjoining the mansion, where from the windows it can be most effectively seen.

Aspect.—With more especial regard to the position of the flower garden in respect of aspect, it is to be noticed that in the case of a first-class mansion—which these remarks assume to be understood—this particular point would naturally be determined by the position of the house. This usually stands in the open, surrounded by ample park ground, where no special cause interferes with the one only question at issue—the best aspect for the private front. This is generally determined in favour of some point between south and south-east; or in the event of there being two private fronts, which is sometimes the case, the south-east and south-west would probably be the points chosen. Since all past experience has accepted these aspects as the most enjoyable, as well as the most conducive to health, one or both of these private fronts would, therefore, be the proper position for the flower garden.

In cases where an elaborate and extensive garden is deemed an essential appendage to the house, it possesses the advantage of ready access, and is itself necessarily benefited by the shelter of the usual screen walls and adjoining shrubberies. On the other hand, it is fair to remark that there are corresponding disadvantages inseparably connected with an arrangement of this kind, in the constant presence of a little army of workmen daily employed in all the ordinary duties of general cultivation within a few yards of the windows of the house—a state of things to which few families are indifferent. Although the position in question is that which is usually accepted as proper for a flower garden of the first class—consisting, as it would, of a series of beds of elaborate design cut upon the turf, or formed on gravel with Box edging, or it may be a mixture of both, yet this position is equally adapted for a very different form of garden. This latter, though belonging to the same type, represents a simpler style, with fewer and larger flower beds, thus permitting a much larger breadth of open turf. In my somewhat preference for the old English garden, with its terraces and steps, be it known by the name English or Italian, it is by no means intended to commend the propriety of introducing highly artificial decorations where there is nothing in the character of the mansion to warrant their use. Indeed, it is right to admit, and again to repeat, that the excessive use of artificial decorations does very often vulgarise many an otherwise fair garden scene. Though this be true, it is not on that account necessary to be driven to the opposite extreme, and renounce all artificial accompaniments, but to use sparingly and with a delicate hand whatever is good.

Let us now turn to the far more numerous class of smaller gardens of modern times, which are not always connected with houses commanding grounds of large extent. The right use of floral embellishments, whether in the dwelling house or in the garden, must necessarily, to a great extent, depend on the arrangements of the house within and the nature and character of the garden without. This being so, full liberty in the disposition of the flower garden may be freely used, the ever-varying circumstances being kept in view. For the particular class of dwellings at present under consideration it will be obvious that there can be no such thing as a fixed and uniform position for the flower garden. Nor is this any matter of regret; for although the style and greatness of the stately mansion are, in a way, accepted as compensation for the discomfort of having carried on immediately under the windows of the house the busy work of the great and formal garden, on the other hand, the owners of the less pretentious dwelling are under no such obligation, and can enjoy freedom of choice as regards their flower garden, and place it in whatever situation the natural conditions suggest as most convenient and best. In all such cases the flower garden very often, and very properly, finds a site in some quiet and sheltered nook where the flowers can thrive, and where they may or may not be seen from the house.

It may, however, be preferred that these summer flowers should be distributed in isolated beds and scattered groups along the boundary walks, and in a large proportion of cases the adoption of this course will prove the most satisfactory. Unhappily, there is often an aspect of harshness and incongruity in the little knots of severely artificial flower

English cottage garden and lawn, with flowers mostly on the outer fringes of lawn, borders, and around beds and peat shrubs. Gilbert White's garden at Selborne, but sketched in 1880 when occupied by Mr. Bell.

beds dug out in the middle of what would otherwise be a quiet and pleasant lawn. Much may be said on behalf of a gay and bright display of flowers, apart altogether from the question as to the size or shape of the flower beds; still, these little flower plots sometimes form garish spots, and are much out of harmony, not only with all around, but with the general tone of the place itself; whereas, if the same number of plants were distributed more freely over the garden in simpler, bolder masses, the effect of the blending and intermingling of colour throughout would prove both acceptable and pleasing. At the same time if, in addition to the above, the resources of the garden permitted some form of avowed parterre, so much the better; the latter would in that case no longer be an isolated feature.

On small properties it is more frequent than otherwise that the question of aspect is surrounded with conditions which make it often a necessity to accept, not what is best, but the best that surrounding circumstances will permit. It is, besides, very frequently the case, even when the house itself can be properly placed as to aspect, that local obstacles of some kind are pretty sure to arise to prevent the flower garden from being placed on the south front, or even in immediate connection with the house at all. So varied, indeed, are the conditions affecting the aspect of modern buildings—such as the lie of the ground, the presence or absence of towns, buildings, woods, trees, neighbouring properties, and the like—that the selection of a garden site is often a matter of very great difficulty. Even where the surroundings are fairly ample it is not only the aspect of the house, but the stables, the kitchen garden, the farm buildings, the kitchen offices, which have to be provided for, and the necessities of the one very often interfere with the best that can be done for the others.

The whole face of Nature seems to form a standing protest against the prevalent notion that the flower garden must always occupy an unchanging position in relation to the house, wholly independent of all other considerations. This matter is often discussed as if it were a possible thing that every dwelling house, altogether apart from its local surroundings, must stand to a particular aspect; that the kitchen garden, the flower garden, the stables, and offices must all occupy, in every case, exactly the same position in relation to the house and to each other. Few notions could be more futile, and fewer still more impossible. It is perhaps not too much to say that no two houses and gardens, any more than any two human faces, were ever yet found to be, in all particulars, exactly alike. Every plot of ground on which any house, whether mansion or villa, has ever yet stood, or ever can stand, must always present the individual features and character peculiar to itself and to no other.

The exact position, therefore, of each—whether it be flower garden, kitchen garden, or any other department—must be determined, in every case, by the due consideration, not of one, but of all the special features taken as a whole. It is evident that He who provided the flowers of the garden with such a profusion of exquisite beauty intended that we should have our natural craving for novelty and variety gratified to the full. Hence we have, in connection with this ever-varying condition of things, the undulation of hill and valley, and, as a result, the picturesque beauty of the landscape, in place of what would otherwise be dull uniformity. Let us, therefore, seek to profit by the unerring teaching of Nature, and accept and use the implied principles of freedom, whether it be in the application and disposition of plants and flowers with a view to their individual beauty, or in their hardly less pleasing effect when disposed in masses on the open lawn.

It is often very instructive to notice the different views of different persons with respect to the use of plants and flowers. While one prefers a display of red and yellow bedding plants, another, perhaps more artistic, and it may be with more leisure and higher culture, elects to admire and study the beauty of individual plants, while a third cares for none of these, but consecrates all his affections and service to the tending and care of another and more restricted class. Happily, no one with a spark of the love of flowers in his heart would venture to say a word to restrain and hinder this taste for variety, but rather encourage and spread it far and wide.

It may be that, on some points, views have here found expression savouring somewhat of heresy, and apparently making light of generally accepted doctrines of recognised routine. This, however, is not what is aimed at, but rather that what has been said may tend to a freer style and to a more extended use of ornamental plants, thereby adding some additional brightness and increased enjoyment to the homes of both rich and poor.

A lawn and garden.—The reason why a determined stand is made here against formalism in a set position in relation to the house is that it prevents any simple, good use being made of the natural fitness for a garden that the ground may possess. Each garden should have a character of its own—all the better if that character is the best expression of such advantages as the site, soil, and climate allow.

To illustrate the point we will visit a place formed by Mr. Marnock, in which an open lawn, with many beautiful trees on its outer fringes, creeps up to the house placed on a gentle slope. This being so, the designer is perfectly free to treat the ground in the most natural and artistic manner.

The wide belt of pretty and varied positions near the walk present facilities and positions for the culture of flowering plants on as extensive and varied a scale as anyone could desire in and about its groves and borders and tree groups. The great expanse

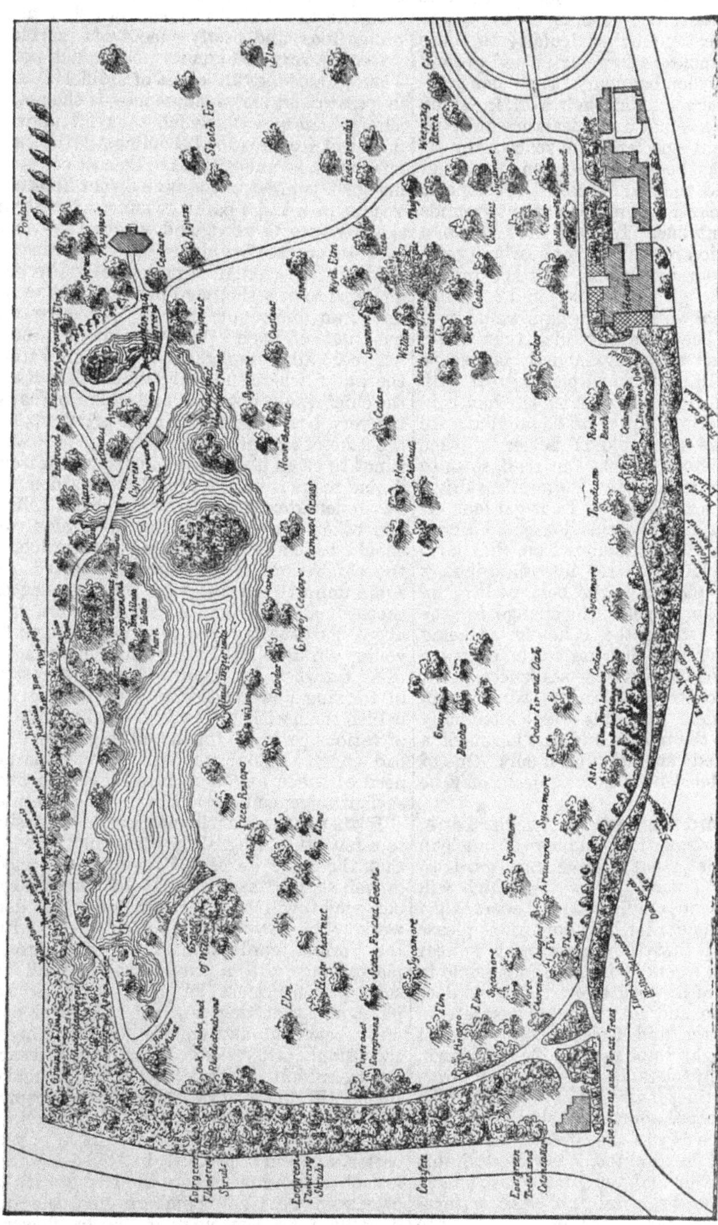

Garden by Mr. Marnock. Here the foreground is a sloping lawn; the flowers bedded are mostly arranged near the kitchen garden, partly shown to right; the hardy ones grouped and scattered in various positions near, or within good view of, the one bold walk which sweeps round the ground.

of Grass is undoubtedly the principal charm of the design; it is a perennial feature, always there, and is pleasant to look on in summer or in winter. What opportunities, too, it presents for gardening, this Grass space! Imagine a spring garden everywhere filling its borders and groves, the plants growing permanently in their situations, and the gardener's mind relieved of the never-ending labour of filling and emptying and wishing his employers out of the way in order that he might root the plants up before they are well out of flower, so as to get in his bedders for an autumn display! Then the design shows no dug borders, or very few of them, and the keeping consists of little else than the cutting and cleaning of the Grass. This is a great point, particularly to those who like to enclose a large extent of ground within the garden boundary. The older landscape gardeners overdid their work in trying to give their grounds an appearance of interminable extent—pottered the space away by winding walks, beds, and groups in a manner that defeated the end in view. The idea of placing the mansion at one end of the grounds is an excellent one. To place the house in a central position has always, or nearly always, been a foregone conclusion with the formal school, and to give it a place in a corner at one end, as shown in the design, would probably have been the last idea that would have occurred to many modern landscape gardeners. In many small places great gain would result from not placing the house in the middle of the space to be laid out. In very many cases it would be better to place the house almost against the road, so as to save the cost and space of a needless drive, and a small, unsatisfactory front garden. If one could easily take the house and all its approaches and surroundings from the centre of many a place frittered into meaningless patches, and place it in the best position on the side of the property, the change for the better would strike the beholder as being truly astonishing. But the waste of space in town or country from useless fences, walls, walks, and obstructions of all sorts is appalling. To deal with this waste effectually will require the temper and the means of a more advanced generation than ours. One of the most offensive of these objects may be dealt with as

Ponds and foul water in gardens.—It is surprising the number of unclean "duck ponds" that deface our gardens. People have a mania for water, which will probably never be eradicated; the effects they see of water under its best conditions please them so well, that they must bring it near their houses, where it is quite impossible to provide any of its good effects. But one does see the scum and filth that accumulate in stagnant water, and feels its deadly smell o'er what ought to be many a pleasant lawn. Water is rarely tolerable, except in such large, bold, and distinct masses, that it reflects light in the landscape. Such bold sheets are best some distance away—the effect is better. In our cold, wet climate, too, it is very desirable that the vapour and more immediate effects of water—even beautiful and clean water—should be kept rather far away, and since this cannot be done in very small places, water is better entirely excluded from such. In one form, however, it is always innocuous, and that is as a streamlet or rill. In this form it can do no harm, and may, in a garden or near it, be always made alive with beauty by bordering it with flowers and trailing shrubs, broken by little bits of smooth Grass here and there. A hundred yards of a chalk stream alive with fish, such as one may see among the Wiltshire Downs, clear as a diamond on the coldest day, is better than many pretentious and costly pieces of "artificial water." A very bad quality of the duck ponds —and, indeed, of all kinds of artificial water in garden and park landscapes—is the astonishing facility with which they fill up with mud and leaves, thus becoming all the more offensive and unhealthful. The cost of cleaning them is great, as many an owner finds out. Where in a large park the effect of water in the distance is good, and where the views over it are really picturesque, cleaning is worth doing, even if it periodically involves an expense of a thousand pounds—not, as we know, an uncommon charge for taking the mud out of such "lakes." But in places where no kind of good effect is obtained from the water, where the scum and the smell are the chief results obtained, and where, perhaps, the very best part of the lawn is cut up for the sake of an ugly water hole, the best way is not to clean it out, but to cut a drain from it, and make it into a snug little garden for Rhododendrons, Azaleas, hardy Ferns, Lilies, and other flowers. Then, at all events, we should have a beautiful thing, into which, if the children fell, they would be picked out again unhurt. By the way, the presence of these duck ponds and fountain basins, &c., about gardens is often a serious danger to young children toddling about; many have been drowned in them. Those interested in forming flower gardens often try to embellish them with water squirts and fountains of various patterns, for the most part with a bad effect. In our wet climate there is no need of them, and it is rare indeed to find such arrangements producing a happy result.

Rhianva.—We will now resume our visits to a few other English gardens, in the hope that the more we see of them, well formed or well stocked as they may be, the clearer our views will be over our flower garden work. We have not only to deal with the formal gardens, made in the wrong places, but with a false idea that all in such gardens must be as "hard" as tin plate, and that they are not suited for our most beautiful and stately flowers. These, accordingly, are "bedded out" with dwarf plants, which are often clipped to make them still more dwarf-like. But, despite the almost invariable clipping and shaving, one may see here and there a little of the better way, and at Rhianva, in Anglesea, there is such an example, where the free growth of evergreen trees and climbers, and the delightful interlacements of hardy flowers, Ferns, and creepers, make a partly formal garden beautiful. Again, I remember a beautiful old garden at Ockham Park in Dr. Lushington's time, which was formal and yet beautiful, through the informality of the vegetation. So again in Italy, where the formal gardening is often a necessity from the nature of the ground, the stiffness of the stone is soon softened by the graceful forms of trees, shrubs, and trailers. Thirty years ago the site of Rhianva, on the banks of the Menai Straits, was a steep field, with the large lumps of grey rock, so characteristic of Anglesea, crossed by a small stream which lost itself in marshy ground by the shore. A couple of old Apple and Thorn trees and a little whitewashed cottage made up its features. The property had long been in the family of the late Sir John Hay Williams, of Bodelwyddan, and he and his wife, Lady Sarah H. Williams, first began to build the house and lay out the garden in 1849.

A terrace garden with free vegetation and informal arrangements. Rhianva, Anglesey (Autumn, 1879).

The garden at Rhianva was laid out by Sir John Hay Williams in 1850 and 1851. The extreme steepness of the ground and the quantity of rock made it very difficult to deal with, and so a number of supporting walls were built to form terraces; and, by the help of a protecting sea wall, the flowers were carried down to the very edge of the water. The climate is mild in winter, and the garden being on a southern slope, the trees and shrubs grew with great rapidity. Hedges of red Fuchsias and of blue and pink Hydrangeas soon hid the stone walls. Myrtles and Camellias, and some Acacias were found to do admirably out-of-doors; and at the present time the only difficulty is to prevent the shrubs from injuring each other through their rapid growth in the limited space. In summer, the luxuriant abundance of the Roses, climbing from bush to bush, the Cypresses, the Tamarisk and the Vines, and the waters of the Menai Straits, with the purple mountains beyond, seem to belong rather to the Lake of Como than to the wilds of Wales. This is an example which enables us to see what a vast gain we make by owning to the absurdity of the orthodox law that a terrace garden, be it right or wrong in design, is only to be properly dealt with as a stiff "bedded-out" garden. At Rhianva we have the terrace garden in a position that called for it—a rocky slope, in which the only possible way of making a garden was by terracing the ground. We have here also a precious example of a terraced garden that shelters every treasure of our garden flora, from the Lily and the Cyclamen to the Tea Rose, Cypress, and Vine. Facing a little to the east of south, the garden is protected from the violence of the westerly gales, while the more tender plants are sheltered by the larger shrubs and trees from the nipping east winds of spring. The foliage of the Cork tree, the Myrtle, the Pomegranate, the Bamboo, and the Yucca gives the garden a southern look which is not often met with in this country. This is one of the most beautiful gardens we have seen. All the borders are mossed over with small green plants; large hardy exotic Ferns, like the Ostrich Fern, are spread into groups and colonies; lacework of Ivy and Vine and creepers is in many parts. A mixed order of planting is the one pursued, but in many cases the shrubs and plants are allowed to spread as they will, and the climbers take advantage of the privilege to become picturesque.

Improvements of the flower garden near the house.—I wish to speak to those who hold that the flower garden, or parterre, is not a proper place for the many lovely things grouped as hardy flowers. It has been said that, however valuable the more beautiful inmates of the flower garden, their place is not the parterre, but that they should be found in some out-of-the-way spot, where they starve and die. Not only may a terrace garden, no matter of what character, be embellished with hardy flowers, but it is one of the best places for them. Thinking over the odd notion that our fairest flowers must not show themselves in the flower garden, one might suppose that the latter was a thing of yesterday, and that there was nothing in the terrace garden before geometrical bedding was invented. But, whatever were the materials they had in old times, we have much more to adorn a true flower garden now. The ugliest and most needless parterre in England may be planted in the most beautiful way with hardy flowers alone. Are we not all wrong in adopting one degree, so to say, of plant life as the only fitting one to lay before the house? Is it well to devote the flower bed to one type of vegetation only—low herbaceous vegetation—be that hardy or tender? What should we say of the gardener who filled his winter garden with low soft-wooded plants only, and omitted from his collection Camellia, and Palm, and Heath, and Azalea? This is what we have long been doing in the flower garden, and by a change we can effect great good at once. We have so long been accustomed to leave flower beds raw, and to put a number of plants out every year, forming flat surfaces of colour, that no one ever thinks of the higher and better way of filling them. But surely it is worth considering whether it would not be right to fill the beds permanently rather than leave them in this naked or flat condition throughout the whole of the year. In Nature, vegetation in its most beautiful aspects is rarely a thing of one effect, but a union or mixing of different types of life, and a succession of different seasons of blooming. So it is in the garden. The most beautiful effects must be obtained by a variety of different forms so combined that they help each other, and give us a succession of pictures and of varied interest, instead of monotony or bareness. If any place asks for permanent planting it is the spot of ground immediately near the house; for no one can wish to see large, grave-like masses of soil frequently dug and disturbed near the windows, and few care for the result of all this, even when the ground is well covered during a good season. But everybody could form beds that would look fairly well at all seasons by the use of our choicer shrubs of many kinds—Spiræa, Rhododendron, Azalea, Dwarf Cypress, Retinospora, Japan Quince, Tree Heath, Comptonia, Coronilla (in mild districts), Viburnum, Clematis, Aralia, Honeysuckle, Weigela, Sweet Brier, Vine, Hydrangea, Arbutus (Croomi), Azara, Skimmia, Rock Rose, Tamarix, Daphne, Yucca, Tree Pæony, Escallonia, new large-flowered Hypericums, choice Hollies, Osmanthus. Why should we not use a beautiful Andromeda, or a Kalmia, or a rare evergreen Barberry in the flower garden in the same way as we do a Camellia, an Acacia, or a Tree Fern in the winter garden?

Secondary results would be improved culture for, and much more beauty from, our many choice hardy shrubs now so often neglected in the shrubbery. The shrubs should be arranged in an open way, the opposite of the crowding now common in our beds of American shrubs. In these all individual character and form are crushed away in the crowd; yet there is scarcely a shrub that has not a peculiarity of form which it will show if allowed room, and the plan must be to allow no possibility of crowding, and to place the *finest and most hardy flowers in groups and colonies between the free untortured shrubs.* Thoroughly prepare the beds; put in the choicest shrubs, which, never growing into low trees or tall shrubs to obscure the view, nevertheless adorn the earth all the winter as well as the summer, and give us a broken as well as a beautiful surface. Between them, in open spaces, and round their feet, so to say, allow for the planting of a variety of the choicer hardy plants, which would come up and flower and pass away, and vary the scene and give us pictures following the seasons. So far from this leading to any kind of monotony, the changes, and combinations, and successions of beautiful things which it would lead to are infinite.

We should not have any definite pattern to weary the eye; but we should have quiet grace, and verdure, and little pictures month by month. The beds, filled with shrubs, and garlanded with Ivy and other evergreens and creepers, would afford everywhere nooks and spaces, where, among the shrubs, some of the many fine hardy Lilies could be grown, with the Tritomas, Gladioli, Phlox, Pentstemon, Iris, tall Anemone, Pæony, and Delphinium. The choice shrubs recommended for such beds are not gross feeders, like trees, and seem to encourage the finer hardy bulbs and flowers. The shrubs also relieve the plants by their bloom or foliage, and when a Lily or Cardinal Flower fades after blooming it is not noticed as it might be in a stiff border. Another important point would be that by this system we should not need the wretched plan of growing a number of low evergreens in pots, or otherwise, to adorn the flower garden in winter.

Crowded beds.—To get artistic effects in a flower garden we must not by any means adopt the close pattern beds usual, because no system ever invented by man can get a good effect from beds crowded one on another, like cakes in a pastrycook's tray. A certain repose and some verdure are needed. Before making the change from the dwarf plants only, be they hardy or tender, it would be well to see that there is ample repose—room for the full expression of the beauty of each group or bed—no complication or crowding, no merely fanciful or angular beds. The contents, and not the outlines of the beds, are what we should see; and in this permanent way of planting with beautiful flowering, summer-leafing, or evergreen shrubs with abundant space to grow, we could have a permanent beauty in our terrace garden beds the hardest day in winter. Between the beautiful bushes or evergreens we could have a variety of dwarf evergreen carpets of alpine plants, and tiny hill shrubs and trailers. Through these the autumn, winter, and spring flowering bulbs could peep to bloom untarnished by the soil-splashing of the ordinary border. Shelter, as well as the best culture, could be thus secured for many a fair favourite, which would there, once well planted, come up year after year. Among our flowering shrubs we have the perpetual flowering Rose on its own roots to help us with our plans. This in many forms is perhaps the greatest treasure of all, and its effect so disposed will be better than in any special arrangement. We must get rid of the old collection of ugly sticks formally arranged as a Rose garden. It should, apart from its use in borders and mixed groups, be planted in a natural manner, without formality, in simple, large, well-prepared beds, the plants trained as bushes or pegged down, so as to let their foliage and shoots spring from the ground. No one can refuse a place to the Rose in our formal garden, however near the house. No Rose beds can be made in this way without offering positions for beautiful hardy plants. We may have many flowers beneath the Roses, or to stand among them or above them, like fair Lilies.

The plan sketched above is only one phase of the culture of the open-air flora which may be possible in the choicest garden, formal or otherwise. What could be more delightful in a formal garden, even right against the house, than a good mixed border? not a stiff collection of staked plants, but a well-stored and well-formed border, filled with Carnations and Picotees in autumn, as well as Stocks, white Lilies, and Christmas Roses, and in spring Daffodils and all the rest of the families that used to adorn these gardens in olden times, as well as those we have now. In the garden at Penshurst Place, for example, there is a raised walk on a terrace above the main garden—a walk alongside of and sheltered by a high wall. Here is the very place for a mixed border, and, accordingly, one has been made there. A noble border of the finer shrubs, with flowers between them, would grace the fairest house or garden yet made.

One of the prettiest garden borders I know of is against a small house. Instead of the walk coming near the windows, a bed of various choice shrubs, varying from 9 feet to 15 feet in width, is against the house. Nothing in this border grows high enough to intercept the view, but it is just high enough to hide the walk beyond it. Looking out of the windows on the ground floor of the house, one sees the foliage and bloom of the border, but not the walk, the eye reaching a green lawn beyond. Among the shrubs are planted tall Evening Primroses, and Lilies, and Meadow Sweets, and tall blue Larkspurs; and these, after the early shrubs have flowered, peer above their leaves. This is an example of the mixed border of choice shrubs which we talked of in connection with beds before. The ground is always furnished, and the effect is good, even in winter. Let no one imagine that this system economy possible with tender plants that is not even more practicable with hardy ones; for in their case we can dispense with the whole of the pot-and-kettle business, glass, coal, &c.—even smoke!

A Somerset garden. Selwood Cottage, near Frome. Rough wall in foreground covered with rock plants.

can be carried out without thought, and taste, and labour. Anyone can change the direction of much of the profitless labour spent over a large area for a temporary aim into work that will last for years. But there is no

Digging in this system is to be abolished entirely. The beds or borders should be thoroughly prepared at first, and the things we want in them carefully planted. We have simply to keep the soil wholly free from weeds, and anything done in the way of cultivation must be in the form of surface dressing in autumn or spring. Only at such times as the ground gets worn out again, or that some portion of it, owing to the excessive growth of a vigorous plant, wants to be replanted and re-enriched, should we take up the whole again, and, thoroughly preparing, replant in the desired way. The surface should be all covered to prevent the effect of bare earth. The cover is afforded by what we call ground plants, delicate things like Stonecrops or Saxifrages, and many others from the hills or rocks, which are allowed to spread all over the surface, and frequently give us a bloom in spring. From these rise the taller plants in such groups as we have mentioned. Part of the common system in mixed arrangements is to repeat a favourite flower at intervals everywhere. The true way is to group enough of any one plant in one or two places, so as to fully enjoy its character, and then be done with it. This would give us different aspects of vegetation as we passed along the border apart from various other important advantages. Primroses, Cowslips, Bluebells, come in tangible, visible masses, which we can enjoy and see. Occasionally they are mixed, as our groups in the border might be mixed, but not in the common way of dotting fifty different things in a few yards' space. We may have artistic and succeeding mixtures without adopting the middle mixture, which spoils the whole border. Treating things in groups in this way, we can see better what each subject is doing. There is less fear of workmen unacquainted with plants destroying them than when a great number of kinds are dotted promiscuously, with or without labels.

This artistic system need not prevent us from growing single specimens if we desire it in a special border, or growing a collection for increase in nursery beds. I might pass on from this to the various beautiful plants which are so full of character, like Yucca and Acanthus, that they might stand by themselves without any aid or arrangement whatever; but enough has been said to show how hollow is the notion that all our fair garden flowers must be grown in obscure borders, where they may be robbed by tree roots, ill-treated, and forgotten, and to show how that, in at least one way, we may fully enjoy them in positions said by many to be unsuitable.

Selwood Cottage. — This pleasant house, the residence of Mr. T. B. Sheppard, is placed on the brow of a hill overlooking a pastoral Somersetshire valley. The trees, by which it is protected from the north winds, form an effective background to the thatched roof and verandah; and from many points of view it forms one of the most pleasing features in the peaceful landscape, with which it is entirely in harmony. The garden has long been well cared for; it is planted with hardy flowers which, coming up year after year in the well-known spot, become as much a part of the home as the house itself. While the collections of hardy perennials had disappeared in most of our nurseries, they were still cultivated by the late Mr. Wheeler, of Warminster, and from his garden they were introduced into many of those in the neighbourhood. In that at Selwood Cottage are several plants which, until the recent revival in favour of hardy flowers, were almost unknown. The small but well-arranged hardy fernery is very good, and a line of stones, overgrown with Ferns and yellow Welsh Poppy (Meconopsis cambrica), makes a very beautiful edging to the shrubbery. The garden is protected by a ha-ha, the wall of which has become beautiful with a luxuriant growth of rock plants: this, backed by a border of Lilies and such like flowers, forms the foreground in the sketch, while the projecting stones and iron handrail make a convenient and picturesque stairway to the fields beyond. When last there I noticed an Araucaria, hideous and stunted as an Araucaria often is, quite glorified by a healthy plant of Canary flower, which had found its way into the branches that surrounded it, and concealed the bare ugliness with a mass of yellow blossoms. This garden shows how much individuality and interest may be displayed in quite a small space, when thought and loving care are bestowed upon it.

Eversley.—The example of Rhianva shows us how many a stony waste may be turned into a garden and how to secure variety for such an area. In the same way, every distinct and good example is a gain in enlarging our views. Thus, by a glance at the late Charles Kingsley's rectory garden at Eversley, we get to see a modest and charming, and simple as charming, type of garden. The walls and borders are full of flowers, while the Grass clothes the central space. When Canon Kingsley became rector of Eversley, in 1844, he found the garden at the rectory in as unsatisfactory a state as was, in other respects, the rest of his parish; but it had capabilities, and these capabilities he used to the utmost. On the sloping lawn between the house and the road stood, and still stands, a noble group of three Scotch Firs, planted at the same time that James I.—who was just then building the grand old house of Brainshill, hard by, as a hunting box for Prince Henry—planted the Scotch Firs in Brainshill Park, and the isolated clumps on Hartford Bridge Flats and Elvetham Mount. A fine Acacia and twin Arbor-vitæ, 30 feet high, completed the trees on the lawn. Most of the garden consisted then of a line of ponds from the glebe fields, past the house, down to the large pond behind the garden and churchyard. The rector at once became his own engineer and landscape gardener. The ponds were drained, with the exception of three in the field, which were in course of time stocked with trout. Plane trees, which threatened in every high gale to fall on the

The garden at Eversley Rectory.

south end of the house, were cut down, and masses of shrubs were planted to keep out the cold draughts of air, which even on summer evenings streamed down from the large bogs a quarter of a mile off, on the edge of Hartford Bridge Flats. What had been a wretched chicken yard in front of the brick-floored room used as a study was laid down in Grass, with a wide border on each side, and the wall between the house and stable was soon a mass of creeping Roses, scarlet Honeysuckles, and Virginian Creeper. Against the south side of the house a Magnolia (M. grandiflora) was trained, filling the air and all the rooms with its fragrance. Lonicera flexuosa, and Clematis montana, Wistaria, Gloire de Dijon and Ayrshire Roses, and variegated Ivy hid the rest of the wall with a veil of verdure and sweetness. In front of the study window, on the lawn, an immense plant of Japanese Honeysuckle grows, trained over an iron umbrella. This was given to the rector by Mr. Standish—when only a small plant with six leaves in a pot—a year or more before he distributed it to the public, and carefully the little treasure was nursed and well it throve, being covered every summer with sweet flowers, and with bright purple berries for autumn decoration. Next to this, the pride of the study garden lay in its double yellow Brier Roses. These grew very freely, and in June the wall of the house and garden was ablaze with the vivid golden blooms, the rooms being always decorated for two or three weeks with dishes of the yellow Roses, mixed with darkest purple Pansies, on a ground of green Ferns. The rector was never able to afford himself a greenhouse, so only a few plants were kept for bedding in a little pit; and, owing to the poor soil and the late and early frosts, which were peculiarly destructive from the low, damp situation of the rectory, none but the hardiest and most common plants could be

View in the late Mr. Hewittson's garden at Weybridge.

grown out-of-doors; but the borders were always gay with such plants as Phloxes, Delphiniums, Alyssums, Saxifrages, Pinks, Pansies, and, above all, Roses and Carnations, which grew, without the least trouble, in the greatest profusion. One bay in front of the house was well covered with Pyracantha, in which a pair of white-throats built undisturbed for many years; the further bay was, up to 1860-61, covered by a single plant of Jaune Desprez Rose, but the severe winter of that year killed it, and its place was taken by hardier climbers. Over the glass porch at the front door Clematis Jackmanni, white cluster Roses, and Pyrus japonica were woven in inextricable confusion. Rhododendrons grew in the greatest luxuriance, and the neighbours always came to see the rector's garden when two beds, on either side of the front, were in blossom. An ancient Yew tree, and a slight hedge of Laburnum, Hollies, Lilac, and Syringa divide the rectory garden from the churchyard, and here, again, the rector turned his mind to making the best of what he had. The church, a plain red brick structure, was gradually covered with Roses, Ivy, Cotoneaster, Pyracantha, &c., and, in order that his parishioners should look on beautiful objects when they assembled in the churchyard for their Sunday gossip before service, the older part of the churchyard was planted with choice trees, flowering shrubs, Junipers, Cypress, Berberis, and Acer Negundo, and the Grass dotted with Crocuses where it was not carpeted with wild white Violets.

A novel garden.—Lastly, we will glance at another distinct type of garden which will serve to show how much we lose by having only one ideal in our heads. True taste here, as in other directions, is unhappily much rarer than many people suppose. No amount of expense, rich collections, good cultivation, large gardens, and expenditure of glass will make up for it. All these and much more it is not difficult to get, but a few acres of garden showing a real love of the beautiful is very rare, and when it is seen it is often rather the result of accident than design.

The garden of which we here give an engraving contains some most delightful bits of garden scenery. Below the house, on the slope over the water of Oatlands Park, and below the usual lawn beds and trees, is a piece of heathy ground, a portion of which is shown. The ground, when we saw it, was partially clad with common Heaths, through which little irregular green paths ran. Abundantly naturalised in the warm sandy soil were the Sun Roses, which are shown in the foreground of the sketch. Here and there among the Heaths, creeping about in a perfectly natural-looking fashion, too, was the Gentian-blue Gromwell, with some varieties of other hardy plants suited to the situation. Among these naturalised groups were the large Evening Primroses and Alstrœmeria aurea, the whole being well relieved by bold bushes of flowering shrubs, so tastefully grouped and arranged as not to have the slightest trace of formality about them in any way. Such plants as these are not set out singly and without preparation, but carefully planted in beds of such naturally irregular outline that, when the plants become established, they seem native children of the soil as much as the Bracken and Heath around. It is remarkable how all this is done without in the least detracting from the most perfect order and keeping. Closely shaven glades and wide Grass belts wind about among such objects as those we allude to, while all trees that require special care and attention show by their health and size that they find all they require. This garden is freer from needless or offensive geometrical twirlings, barren expanse of gravelled surface, and all kinds of puerilities—old-fashioned and new-fangled—than any we have seen for years.

Professor Owen's garden.—The most attractive gardens are by no means the largest. Indeed, the most beautiful in England are comparatively small ones. Professor Owen's garden is one of the simplest and most unpretending, but withal one of the most charming, in the neighbourhood of London. Many a visitor to Richmond Park enjoys the look of his cottage, as it nestles on the margin of the noble sweep of undulating ground near the Sheen Gate, but it is from the other, or the garden, side that the picture is most beautiful. A lawn, quite unbroken, stretches from near the windows to the boundary; it is fringed with numerous hardy trees. Here and there are masses of flowering shrubs and an odd bed of Lilies, while numerous hardy flowers peep from among the Roses and Rhododendrons. Quite near the house stands a noble specimen of Gleditschia triacanthos, graceful in foliage, stately and picturesque in the highest degree. Its long lower arms stretch far out near the turf, and are laden with their Fern-like leaves, and the whole surface of the tree for 80 feet upwards is broken up in the boldest and most picturesque manner. No tree, except, perhaps, old specimens of the Weeping Beech, displays such an uncontrolled variety of picturesque branching. There is in the main part of the garden only one walk, and this takes one round the whole place, and does not needlessly obtrude itself, as it glides behind the outside of the groups which fringe the sweet little lawn. Instead of this walk coming quite close to the house it is cut off from it by a deep border of Rhododendrons, intermingled with Lilies and the finer herbaceous plants. These flowers look into the windows. Instead of looking out, as usual, on a bare gravel walk, the eye is arrested by Rhododendrons or Spiræas, with here and there a Lily, a Foxglove, or a tall Evening Primrose, according to the season. Beyond these, at a distance of 12 feet or so, is a broad, convenient walk. The effect of the border from the other side of the garden is quite charming, the creeper-covered cottage seeming to spring out of a bank of flowers. The placing of a wide border with evergreens near the house is a variation from the ordinary mode of laying out villa shady walks, with a carpet of Grass, are the most enjoyable retreats one can find. Besides, their margins form a capital situation for naturalising many beautiful hardy plants which are seen to great advantage in such

Plan of Professor Owen's garden at Sheen Lodge, Richmond Park. Lawn open, walks few, flowers in various positions.

and cottage gardens which it would be desirable to see adopted. Another agreeable feature of this garden is the Grass walks, which ramble through a thick and shady plantation. Even in our coolest summers there is many a day on which such positions, as, for example, Daffodils, hardy Ferns, Scillas, the Forget-me-not tribe, the Hairbells, Snowdrops, and Snowflakes.

In concluding this chapter I would urge the reader to see as many different types of garden as possible; the practice tends much to enlarge one's views as well as, in itself, to furnish an enlightened pleasure. What we have done imperfectly in the way of showing good gardens should be done well and extensively by all flower gardeners. Many a modest cottage garden has its lessons to give, and the observer will find them in strange places. Let him beware of public gardens and their influences; such are often a long way behind the time. The aim should be never to rest till the garden is a reflex of Nature in her fairest moods; at present it is, as a rule, the very opposite. I heard of a distinguished artist the other day who had complained to a friend of mine of the ugliness of gardens in general. He had been staying at a well-known place on the Thames where the view from the terrace is wide and fine, and the sylvan charms of the place are remarkable. But, said he, "owing to the masses of formal colour stretching far away from the terrace to the river, one cannot paint the scene." The whole of what might be a fine foreground is covered by a "pattern" like a bad carpet. In this men are always pottering about, mowing and carrying on various operations to keep the "geometry" in order, including the working of a kind of bridge, to enable them to get into the middle of the beds. "Generally in gardens," the painter continued, "one can find little that can be painted, except, perhaps, by discovering the tool-house, or some other obscure structure, where trailers have been allowed to have their way." I cannot think it will be always so. When people see that they can have much more beautiful colour than at present without spoiling the home landscape, they will soon make short work of the many garden incongruities they now tolerate

HARDY PLANTS AND THE MODES OF ARRANGING THEM.

HAVING spoken so far of the change and greater variety necessary in the general design of the flower garden, we have next to consider the various ways of arranging the material at our hand to plant it. The extent of this material is shown in the body of this book, and is inexhaustible and rich. The question, then, is how the garden designer is to enjoy as much of this treasure as his tastes and circumstances will allow. As in this country the true flower garden must be formed of the vegetation that will endure the climate and temperature, so precedence will be given to hardy plants. This has not of late been the popular line, but it must become so as soon as it is seen how vastly more beautiful the things are that may be grown in the open air.

Cost and duration.—The question of the expense of growing hardy flowers as compared with tender ones is important. The sacrifice of flower gardens to plants that perish every year has left them so poor of all the nobler plants—has, in fact, caused the expenses of the garden to go to purposes which leaves it at the end of every flowering season almost generally devoid of life. We here take into account the hothouses, the propagation of plants by thousands at certain seasons, the planting out at the busiest and fairest time of the year—in May or June, the no less necessary digging up and storing in autumn, the care in hot and cool structures in the winter, the hardening off, &c. The annual bill must not be forgotten. Now, expenditure should go towards permanent arrangement and planting, and that for the best possible reason—that very often the best things of our garden, a fine old Judas tree, or the Snowdrop tree, or a group of Thorns, or a mass of mountain Clematis, or a Gloire de Dijon against the wall, did not really cost at first a sum that any cottager would grudge. Carry the same idea further, and think of the enormous number of lovely flowering shrubs there are, as well as of hardy plants, and of the many tasteful ways in which we can arrange them. From the contemplation we get glimpses which open vistas of delight in our gardens. The best features of many places are those in which such permanent work has been carried out, notwithstanding the fact that all the so-called "floral decoration" for years past has prevented due attention being paid to such permanent artistic work.

Examples of arrangements not requiring annual renewal.—Referring to the idea that, as opposed to the usual way of planting the beds in May or early June, and digging them up in October or November, there is a much more enduring one, Mr. Thomson says: "The idea, too, that once these hardy plants are planted they will go on satisfactorily for many years without any further cultivation, is one of the greatest delusions possible; for, unless the soil in which they grow is kept in good order, the whole thing is a complete failure, and the vigour and display of bloom ceases to be at its best." This is a statement that admits of proof, and that the idea in question is no "delusion" anyone can satisfy himself of. Perhaps the most astonishing effects from individual plants ever seen in England were Lilies (auratum) grown by Mr. McIntosh, at Weybridge Heath, among his Rhododendrons. So far from frequent culture of such plants being necessary, it would be ruinous to them. The plant mentioned is not alone; it represents scores of others equally beautiful which may be grown in the open air in the same simple way; and not Lilies alone, but many other noble flowers. A few years ago we saw nothing but round monotonous masses of Rhododendrons as soon as the flowers were past; now the idea of growing this bright-flowered evergreen shrub with the nobler bulbous and other hardy plants has spread throughout the world. It means more room for the individuals, greater and more natural beauty of form in consequence, more light, and shade, and grace; mutual encouragement of shrub and plant; no dotting, but colonies and groups of lovely plants among the shrubs. Good preparation and some knowledge are needed here, but no necessity whatever for any but a system that may be called permanent. Overgrowth or accident will in time cause need for attention here and there, but that would be slight, and could be given autumn, winter, or spring. Properly done, such arrangements could be left for at least five years without any radical alteration. In the way above mentioned there are opportunities to grow in many large or medium sized pleasure gardens all the nobler hardy plants introduced; but to show the full security of the position here, it may not be amiss to name a few other modes of arranging flowers which do not require annual planting and digging up.

The rock garden must be named now, because all beautiful rock gardening cannot be of temporary duration. To scatter annuals or bedding plants over such is to degrade it, and to show that the meaning of this beautiful phase of the gardening art is not even understood. In the rock garden we may represent the glorious mountain flora of the northern world; and not the mountain and hill plants only (of which two well known specimens, Gentiana acaulis and Lithospermum fruticosum, may be taken as typical of hundreds equally beautiful), but the hill shrub flora. Of these the alpine Rhododendron and the various hardy Heaths may be taken as the European representatives of a great number of beautiful and hardy shrubs which associate well with the humbler types, both series going to make arrangements lasting and beautiful if tastefully planted and kept free from coarser vegetation. That such would require care and taste need not be said. We have good examples already—at Mr. Backhouse's, Mr. Hammond's, near Canterbury, Mr. Whitehead's, not to mention other places, the number of which is increasing every year. Coming to special arrangements, there are quite a number which, given thorough preparation at first, it would be in no sense wise to interfere with for some years at a time—as, for example, groups or beds of the various Tritomas, Irises, Globe Flowers, Pæonies, Aconites, the free flowering Yuccas (Y. filamentosa and Y. flaccida), Narcissi—these and much more, either grouped with others or in families. When all these exhaust the ground and become too crowded, move them by all means; but this a very different thing from two radical operations each year on the same spot of ground. We know a group of Pæonies that has been in good condition for ten years, and many arrangements can be made to last for half this time without any but the routine attention which all gardens require. Then we have the mixed border, which, as usually seen, is an eyesore, but which well done is beautiful and lasting, too, if the preparation is as good as it ought to be at first. Lastly, we have the various phases of the wild garden, or the more vigorous perennial flowers in various places out of the garden proper, of which we have examples already at Castle Ashby, at Great Tew, and at Crowsley Park; so that, even now, when in one garden out of every hundred no one thinks of seeking to make a given spot permanently beautiful, we have numerous examples of permanent flower gardening of the highest class being not only possible, but easy.

Flower garden plants requiring winter protection.—It is all very well to say that pots may be dispensed with to a great extent in the preparation of tender plants for summer bedding; so they may be in the case of hardy plants also, and so may hothouses, heating, and much other expense. In a climate like ours the need of glass-houses is great for various essential things—we want fruit which cannot be grown out of doors; we want a pleasant winter garden in which we may enjoy a variety of beautiful life at seasons when we cannot do so out of doors; and, in fact, we have sufficient work for our glasshouses without devoting them so much to the production of things that are to adorn the open garden. There are numerous groups of plants that do not require any such attention; so many, indeed, that there cannot be the slightest doubt in the mind of any observant man that ornamental gardens of the most beautiful description are quite possible without any glass at all in this climate. We do not say that it would be desirable for everybody to do as suggested in this case, and we admit, at once, the charm of a fair proportion of plants put out for the summer; but we believe that nothing would have a more healthy tendency than that a certain number of people resolved to see what they could do in their flower gardens without the aid of any glass. The time was when the outdoor gardens—kitchen, fruit, and flower—received much more attention from the good gardener than they do now in proportion to their importance. A good many of us are apt to suppose ourselves made for a conservatory life only, and there are men, indeed, who consider a good deal of glass an essential point in any place to be proud of. To such an extent is this carried, that many regret the absence in the younger men of the complete attention to the more important phases of outdoor gardening that used to be given in olden times. It would in all ways be an improvement if every country were to see what could be done, unaided by the hothouse—so far as the flower garden is concerned; but, pending such a movement, the wise man will reduce the expense of glass, labour, fire, repairs, paint, pipes, and boilers to something like just proportions. In the face of the splendid wealth of our hardy garden flora, the promise of which is now such as men never expected a few years ago, no one need doubt the practicability of making a fair flower garden from hardy plants alone.

The labour and the care required of a gardener in any place where beauty or variety is sought are so great, that it is extremely desirable to reduce to the narrowest limits any frequently recurring work not essential for the full enjoyment of a garden. In all places where a good kitchen and fruit garden exists, the annual labour essential and unavoidable is so great, that any plans which lighten the work in the flower garden, especially in the busy season of spring and early summer, are likely to be an aid to good gardening in all ways, and bring a little needed repose after the laborious preparation of autumn and winter planting.

The true way for all who desire to make their gardens yield a return of beauty for the labour and skill bestowed upon them is the permanent one. Choose some beautiful class of plants and skilfully select a place that will suit them in all ways, not omitting their effect in the garden or the home landscape. Let the garden be as permanently and as well planted as possible, so that beyond the ordinary cleaning there will remain little to be done for years. All plants will not lend themselves to such permanent plan, it is true, but such as do not may be grouped together and treated collectively—for instance, the beautiful beds or masses of Anemones, double and single; the Turban and Persian Ranunculuses, the Clove Carnation, Stocks, Asters, and the finer annuals. All these, which no really good garden should be entirely without, do not lend themselves to such treatment, but preparation for them can be made to a great extent in the autumn, winter, or spring season, and no gardener will begrudge the attention necessary for such fine things if he has not the care of many thousand bedding plants. But a great many delightful plants can be planted permanently, either allowing them to arrange themselves, to group with others, or to loom among peat-loving shrubs which, in hundreds of places, stand bare and unrelieved. Here and there, carrying out this plan, we might have planted tufts of Tritomas and Lilies, Irises and Gladioli, and many other lovely sorts among them.

Large beds.—One of the best reforms will be to keep away from the "rug pattern," and adopt large and simple beds, placed singly or in groups, in positions suited to the plants they are to contain. Then these can be filled permanently, or partially so, with ease, because the planter is free to deal with them in a bolder and more artistic way, and has not to consider the necessity of making them correspond with a number of other things near them. In this way, also, the delight of flowers is much more keenly felt. One sees them relieved, sees them at different times, has to make a little journey to see them when they are not all under his window, stereotyped. Roses—favourites with everybody—grouped in their different classes, and not trained as standards, would lend themselves admirably for culture and grouping with other things. For instance, we might have Moss Roses growing out of a carpet of double Primroses, Tea Roses with the Japanese Irises which require warm soils, or Hybrid Perpetuals and the varied kinds of grand German Irises that are now obtainable. Lilies of the newer and finer kinds do not merely thrive in beds of American plants, but they afford in certain gardens in Surrey the grandest effects I have ever seen, whether in garden, in glasshouse, or in wilderness. Then there are many groupings which could be made by the aid of the finer perennials themselves, such as, say, the splendid Delphiniums and white Phloxes, choosing things that would go well together, where the plants permitted it, finishing with fringes of some other dwarf-like plants to hide the earth and the bare spaces by carpets of beautiful hardy flowers. Other plants, such as Yuccas, of which there are now a good many beautiful kinds, are, perhaps, best by themselves; and noble groups they form, whether in flower or not. The kinds of Yucca that flower very freely, such as Y. recurva and Y. flaccida, lend themselves for grouping with Flame Flowers (Tritoma) and the bolder autumn plants. Year by year the gardener, who is not worried to death with excessive planting in the beginning of summer, by thinking over the matter and visiting extensive collections, could devise some beautiful new feature, or series of groups, and soon a place might be fairly well furnished with such. Then, by way of relief, a few groups of the more suitable tender plants, such as Cannas and Dahlias, mixed would add to the beauty and variety of the whole where the climate was favourable.

Waste of effort.—No plan which involves an expensive yearly effort on the same piece of ground can ever be wholly satisfactory, and mainly because it is great waste. All plants require attention, and then all, as many know, require liberal expenditure to do them justice. But they do not require this annually. The true way is quite a different one—the devotion of the skill, expense, and effort to a new spot or situation each year. The "fresh designs," instead of supplanting those made the previous year for the same spot, should be carefully thought out, and be made to last for half or a whole lifetime, or perhaps generations. The right way does not exclude summer "bedding," but it includes numerous possibilities of lovely and varied aspects of vegetation as to beauty, and even as to colour, far beyond what is attainable in summer "bedding." The plan attempts to make the place generally and permanently beautiful. It also particularly helps to make the skill and labour of the gardener effective for permanent good, and not to be thrown away in annual fireworks. The energy and skill wasted on this "bedding out" during the past dozen years in one small portion of many a large place would, if intelligently devoted to permanent and artistic planting of many flowers, shrubs, and flowering and evergreen trees, make a garden and sylvan paradise of a small estate.

No gardening can be done without care. But I have only to appeal to the common sense of the reader in asking him—Is there not a vast difference between some of the beds and groups just mentioned and those which wholly disappear with the fogs and frosts of October, leaving us nothing but bare earth and nothing in it?

Flowers in their seasons.—The main charm of bedding plants—that of lasting in bloom such a length of time—is really their most serious fault. It is the stereotyped kind of garden which we all have to fight against; we want artistic, beautiful, and gratifying gardens. We should, therefore, have flowers of each season, and the flowers should tell the season. Too short a bloom is always a misfortune; but a bloom may be also too prolonged. Numbers of hardy plants bloom quite as long as could be desired. Some afford a second bloom, as the Delphiniums. Others, like Lilium auratum, bloom one after the other for months; while the short-lived kinds, like Irises, may be well used in combination with those which precede or succeed them.

Beauty.—There is nothing whatever used in bedding out to be compared in any way—colour, scent, size, or bloom—to those specimens belonging to many families of hardy plants now obtainable. Those patronising admissions of "interesting," "pretty," we sometimes hear are ridiculously misplaced. There is no beauty at all among bedding plants comparable with that of Irises, Lilies, Delphiniums, Evening Primroses, Pæonies, Carnations, Narcissi, and a host of others. Are we to put aside all this glorious beauty, or put it into a second place, for the sake of the comparatively few things that merely make beds and lines of colour? Let those who like bedding flowers enjoy them; but no one who knows what the plants of the northern and temperate world are can admit that their place is a secondary one, much less that only this poor phase of gardening should be the leading one in England. It is the simple fact that there is nothing among tender things equal to Windflowers—Anemones in many kinds, flowering in spring, summer, and autumn; Flame flowers (Tritoma), superb in autumn; Columbines; Hairbells (Campanula); Delphinium—no blue or purple flowers equal to these when well grown, some being 8 feet, 9 feet, and 10 feet high; Day Lilies (Hemerocallis), fragrant and showy; Everlasting Peas, several handsome kinds; Evening Primroses (Œnothera), many bold and showy kinds; Pæonies, many both showy and delicate colours, and some fragrant; Phloxes, tall and dwarf in many kinds; Potentillas, double; Pyrethrums, double and single; Ranunculuses, double and single, and the many fine species; Rudbeckia, and all the noble autumn-blooming Compositæ, of which Helenium autumnale grandiceps may be taken as the type; the large blue Scabious and the smaller kinds; the Larkspurs, charming in colour; Spiræas, plumy white and rose coloured; the Globe flowers, fine in form and glowing in colour; Lilies in superb variety, some attaining a height of over 8 feet in the open air; Polyanthuses; coloured Primroses; double Primroses; Auriculas; Wallflowers, double and single; African and Belladonna Lilies in southern counties; Meadow Saffrons, double and single, various; Camassias, several fine hardy kinds; Crocuses, many kinds, both of the spring and autumn; Scillas; Gladioli; Snowflakes; Grape Hyacinths; Narcissi in splendid variety, and quite happy in our coldest springs or heaviest rains; Tulips, fine old florists' kinds, and seedlings from them for border culture; Yucca, free-flowering kinds; Alyssums; Aubrietias; Thrifts in variety; Carnations and Pinks, Dielytras, Veronicas, Cornflowers, Foxgloves, Rhodanthes, Lupines, Stocks, Asters; the Great Scarlet and other Poppies, single and double; Christmas Roses, both of the winter and spring-blooming kinds; Hepaticas; Orobus, spring-flowering kind; and Forget-me-nots.

Blank in spring and summer.—The greatest loss suffered by those who adopt the bedding system is the complete exclusion of the spring and early summer flowers. These, with the exception of a very few, are thoroughly hardy and so beautiful, that the garden may well be allowed to be a little tame later in the year in order to make room for them. I never was more struck with the utter folly of the bedding system, as usually carried out, than when passing through a road of some fifty first-class villas, after a day's ramble amongst the woods and lanes in the neighbourhood of Sevenoaks towards the end of March. The cottage gardens were all ablaze with Primroses of half a dozen colours, Violets, Pansies, Daffodils, Crown Imperials, blue Anemones, purple Aubrietia, and white Arabis, and the woods and lanes were equally bright with Primroses, Violets, Cuckoo flower, and Wood Anemones, as thick as they could find standing room in many places. The villa gardens, on the contrary, were a blank, and showed no more signs of spring than they did at Christmas. Now, Pansies raised from seeds or cuttings in May, and planted so as to be just coming into bloom in November, will bloom from the moment frost is over until the summer gets too hot for them, and will even, in mild winters, yield a nice little nosegay on Christmas morning. Then, beginning with the Snowdrop and the Christmas Rose, there is a regular procession of floral beauty, of Primroses, Daffodils, and

Narcissi, Violets, Anemones, Irises, Pyrethrums, Ranunculus, Fritillaries, Tulips, yielding flowers of every hue, until by the beginning of June we reach the flowering time of the Lilies, the summer-flowering Irises, the Campanulas, the Gladioli, the Columbines, and Delphiniums. If we turn to autumn, what splendid things are the Japanese Anemones, the Phloxes, Senecio pulcher, and the Chrysanthemum, the last thriving so well in towns. The choice is, whether will you have your garden occupied during a third of the year by a few families of plants—not particularly distinguished for beauty, which may bloom well or not, and present as little variety as possible—or, will you have your garden a home for a selection of the most varied and beautiful of Nature's floral productions, presenting a continual succession of lovely and everchanging forms and colours during three-fourths of the year?

Destroying beauty of form.—One motive, which seems to be a very powerful one with the lovers of geometrical beds, is the desire to have everything trimmed, and shaven, and neat; hedges cut to a rectangular outline; creepers nailed to walls with their branches straight and at regular intervals; beds so many feet apart to an inch; Grass turf closely shaven and trimmed at the edges; shrubs and trees such as grow in the most formal shapes only. Now, if there is anything Nature seems always to be striving against, it is formality and an appearance of neatness. Trim a roadside hedge in winter, and by next autumn you will find it decorated with festoons of Briony, Traveller's Joy, and Brambles, and as much in need of trimming as ever; build a house, and immediately Nature begins bending its straight lines into subtle curves, and painting its walls with Lichens; cut a path straight, and immediately the roadside plants begin to break up the formal line. Before any progress can be made in gardening we must get rid of the fondness for the work of the shears, the measuring line, and the edging iron. The root of the whole matter is want of appreciation of grace of line and beauty of form. Love of harmonious colouring is a natural faculty, and can be little altered by cultivation, although it can be increased or decreased in races by cultivation, or the reverse. But appreciation of beauty of form is an enormously more intellectual thing, and can almost be destroyed by the continual contemplation of ugliness; while the more it is cultivated the more do things with formal and lumpish outlines become distasteful. The growing taste for plants, grown for their beautiful forms of foliage alone, is a hopeful sign for the future of gardening, showing as it does that the public are awakening to the fact that there are other beauties in plants besides colour.—J. D.

Various systems and suggestions.—And now to suggest the various plans by which hardy flowers may be best brought near the home. The bounteous storehouse of material lying to hand for this, and how to use it, will be found by means of the alphabetical arrangement of this book. The number is so large of the good things in store, that it is almost hopeless to give even an outline of the numerous ways in which they may be grown. Still, the plan we recommend may help many who, up to the present, have entertained the primitive notion that a flower garden is a thing to be spread out in one spot, the plants mostly of the same height, all growing under the same conditions, their duty being to flower all together and for a long time.

Flower borders.—The usual way, then, in which people generally attempt the cultivation of hardy flowers is in what is called the "mixed border." This sort of garden may be made in a variety of ways, and its success to a great extent will depend upon how it is made, and scarcely less on the position in which it is placed. Frequently it is made on the face of a plantation of trees and bushes which rob it. The roots of the trees and shrubs will, of course, occupy the ground, and there is less for the plants. These plants in their turn require deep digging; the trees and shrubs will be injured by this operation. Therefore, while the effect of a good shrubbery as a background to a mixed border is very good, the result from a cultural point of view is bad, because of the double call on the soil, so to say; yet one of the most charming of mixed borders can be made on the face of a shrubbery by accepting the conditions and meeting them. The face of such a shrubbery should be broken—that is to say, the shrubs should not form a hard line, but the herbaceous plants should begin at that line, and the shrubs should come out to the edge and finish it here and there, thereby breaking the border agreeably. The variety of position and places afforded by the front of a shrubbery is delightful. Even here and there, in a large open space, one might have groups or masses of plants that require good cultivation, but generally it would be best to avoid this attempt, and use things which do not depend for their beauty on high culture,—which, in fact, fight their way among dwarf shrubs—and there are a great many such growths. Many hardy flowers require good culture to be appreciated; certain others take their chance, like an evergreen Candytuft, and the large-leaved Saxifrage, and the Acanthus, and the Day Lily, the Everlasting Pea, and a great many others.

A shrubbery border.—A scattered, dotty, mixed border along the face of a shrubbery produces a miserable effect at all times; whereas a good one may be secured by grouping the plants in the bolder spaces between the shrubs, making a very careful selection of good things, each occupying a sufficient space and carefully studied as regards the shrubs around it. Nothing can be more delightful than a border made in this way; but it wants good taste, a knowledge of plants, and that desire to consider plants in relation to their surroundings which is never shown by those who make a labelled, "dotty," mixed border which is the same all the way along, and in no place looks pretty. The presence of tree and shrub life is a great advantage to those who know how to use it. Here is a group of shrubs over which we can throw a delicate veil of some pretty creeper that would look stiff and wretched against a wall; here is a shady recess beneath a flowering tree. Instead of following it up in the ordinary gardening way and making a shrub wall or a bank of plants, keep it for the sake of its shade. If any more important plants will not grow in it, cover the ground with Woodruff, which will form a pretty carpet and flower very prettily early in the year, and through the Woodruff dot a few common British Ferns. In front of this only use low plants, and thus we shall get a pretty little vista, with shade and a pleasant relief. Next we come to a bare spot of 6 feet or 7 feet or so on the margin, covering it with a strong evergreen Candytuft, and let this form the edge. Then allow a group of Japan Quince to come right into the Grass edge and break the margin; next a carpet of broad-leaved Saxifrage, receding under the near bushes and trees; and so proceed, artistically, making groups and colonies, considering every point, never using a plant of which you do not know and enjoy the effect, and arranging the place, so that with cleaning it may last for years with such slight changes as new additions to your stock may require.

This border plan is capable of considerable variety, depending on whether we are dealing with an established and tall shrubbery, a medium one of flowering trees and ordinary shrubs, or a very choice plantation of flowering Evergreens and American plants. In the last case, owing to the soil and the neat habit of the bushes, we have excellent conditions in which good culture as well as an effective arrangement is possible. One can have the finest things among them—that is to say, if the bushes are not jammed together. The ordinary way of planting shrubs is such that they grow together, and then it is not possible to grow flowers between them, nor can one see the very shape of the bushes, because their forms are lost in one solid, leafy mass. In growing fine things—Lilies or Cardinal Flowers, or tall Evening Primroses—among fairly spaced bushes, we effect a double purpose: we form a delightful kind of garden, we secure sufficient space for the bushes to show their form and habit, and we get some light and shade among them. In such plantations one might in the back parts have "secret" colonies, so to say, of lovely things which it might not be desirable to show on the front of the border, or which were the better of the shade and more perfect shelter that the front did not afford.

The flower border against a wall or building.—In many situations near houses, and especially large and old houses, there are delightful opportunities for a very beautiful kind of flower border. The grey Lichen-stained stone forms an excellent background, and also cuts us off wholly from any exhausting vegetation. Here we have exactly the opposite conditions to those we had in the shrubbery; we are sure there are no hungry giants behind our plants stealing all their food; we have no trenches to cut to get rid of robbing roots. We know that here we really can go in for good cultivation; have our Delphiniums, our Lilies, our Pæonies, our Irises, and all our choice plants perfectly well grown. If the wall happens to be near the house, and we do not wish, for a variety of reasons, to have climbers on it, it is none the less valuable, especially a good stone wall. It may be, if desired, adorned with climbers of more delicate growth—Ivy, or climbing Rose, or Wistaria, or Vine, or Clematis on the walls will help out our beautiful mixed border. Those trees must be to some extent trained, although they may be allowed a certain degree of abandoned grace even on a wall. In this kind of border we have, as a rule, no contrasts from shrubs, and therefore we must get the choicest variety of life we can into the border itself. We should cultivate it well; we may have a more mixed style than was devised for the shrubbery flower border; we must try and secure a constant succession of interest, and this is very easy. There is scarcely a month of the year in which we may not have delightful things opening into bloom. In certain places it may be found that this kind of border will have a bare look in winter if seen from the windows. The variety of our plant resources is, however, so great, that we can make such a border evergreen by employing the evergreen herbaceous plants of various important types. Occasionally it might be desirable to scatter a number of choice evergreen shrubs over the border, which would help to give it a furnished appearance. In large places there are frequently positions which are not seen from the windows—fine massive walls that would form an admirable background to a mixed border, composed wholly of hardy flowers of carefully chosen kinds. On these walls a choice selection of well-grown climbers and evergreens, like the Magnolia and climbing Roses, form an excellent addition to the hardy plants in front. Where the wall is broken with pillars, the effect might be better still by the use of Vines and Wistaria along the top and over the pillars or the buttresses.

The flower border in the fruit or kitchen garden.—In this we have the original and perhaps the commonest form of mixed garden—the borders in the kitchen garden or fruit garden as the case may be. This kind of border is very often badly made, but it may be made the most delightful thing conceivable. The plan is to secure from 6 feet to 12 feet of rich soil on each side of the walk, and cut the borders off from the main garden by a trellis of some kind. This trellis may be of strong iron or galvanised wire, or perhaps, better still, of simple, rough wooden branches—uprights topped by other branches of the same kind. Any kind of rough permanent trellis will do from 6 feet to 9 feet high. On this rough trellis, appropriately used, we have the opportunity of growing the climbing Roses and Clematis, and all the choicer, but not too rampant climbers. Moreover, we can grow them with all their natural grace along the wires or rough branches, or, still better, up and across our rustic wooden trellis, and Rose and Jasmine may show their grace uncontrolled. We fix the main branches to the supports, and leave the rest to the winds. Here, then, we have the best opportunity for the finest type of mixed border, because we have all the graceful climbing plant life we desire in contrast with the plants in the border. There are opportunities for making borders in front of evergreen hedges. In fact, there is scarcely a place in which sites and situations may not be made available. The true art of gardening is to adorn and make the most of the situations we have; the opposite, and the much commoner, way is to suppose that we cannot make much of what we have, and therefore must go to extraordinary expense to create conditions and situations supposed to be necessary for us. We should not so much follow an idea because we have seen it carried out somewhere else, but rather develop features that suit the ground and all the surroundings.

General directions.—Mixed borders may be made in a variety of ways; those interested in them will do well to bear in mind the following points: Select only good plants; throw away weedy and worthless kinds; there is no scarcity of the very best. See good collections, and consult good judges in making your selection. Put, at first, the good kinds selected in lines across 4-ft. nursery beds, so that a stock of young and strong plants may be at hand, and that you may be able to exchange with others as well as form arrangements or groups in any way desired. Place borders where they cannot be robbed by the roots of trees; see that the ground is thoroughly prepared, and rich, and deep enough—never less than 2½ feet of the best friable soil. The soil should be so deep that in a dry season the roots could seek their supplies far below the surface. On the making of the border depends, in fact, whether the vegetation will be noble and graceful or stunted. If limited to one border only, some variety in the soil will be necessary to meet the wants of peat and moisture-loving plants. In planting, plant in groups, and not in the old dotting way. Never repeat the same plant along the border at intervals, as is so often done with favourites. Plant a bold, natural group of it, or two or three groups if you must have so many. Do not be particular to graduate the plant always from the back to the front, as is generally done, but occasionally let a bold and sturdy plant come towards the edge; and, on the other hand, let a little carpet of a dwarf plant pass in here and there to the back, so as to give a broken and beautiful, instead of a monotonous surface. Have no patience with bare ground. Cover the border entirely with dwarf plants; do not put them along the front of the border only, as used to be done. Let Hepaticas and double and other Primroses, and Saxifrages, and Golden Moneywort, and Stonecrops, and Forget-me-nots, and dwarf Phloxes, and many similar plants cover the ground everywhere—the back as well as the front of the border—among the tall plants. Let these little ground plants form broad patches and colonies here and there by themselves occasionally, and let them pass into and under other plants. A white Lily will be none the worse, but all the better, for having a colony of creeping Forget-me-nots about it in the winter or spring. The charming variety that may be thus obtained is infinite. Thoroughly prepared at first, the border should remain for years without any digging in the usual sense. All digging operations should be confined to changes and to the filling up of blanks with good plants, and to the rearrangement of ground plants. If the border is in the kitchen garden, or any other position in which it is desired to cut it off from its surroundings, erect a trellis at its back from 6 feet to 10 feet high, and cover this with climbing plants—Clematises, Roses, Sweet Briers, Honeysuckles, or any beautiful and thoroughly hardy climbing plants, not twined too stiffly, but allowed to grow into free wreaths. Roses of the very hardiest kind only should be employed, so as to guard against gaps in severe winters; the old single Clematis, the mountain and the sweet autumn Clematis (C. Flammula), as well as other single kinds, should have a place here as much as the larger forms. The trellis may be made in the usual way, of wood or iron, or in a simpler and certainly handsomer way, of rough tree posts and branches. In case the soil is not very deep or not very well prepared, and the surface is not covered with green life in the way above advised, it will be well in many cases to mulch the ground by placing a couple of inches of some light sweet dressing on it in summer. When a plant is old and has got rather too thick, never hesitate to move it on a wet day in the middle of August or July as well as in the middle of winter. Take it up and put a fresh bold group in fresh ground; the young plants will have plenty of roots by the winter and will flower much stronger the following spring than if they had been transplanted in spring or in winter. Do not pay over much attention to labelling; if a plant is not worth knowing, it is not worth growing; let each good thing be so boldly and so well grown and placed that it impresses its individuality upon all who see it.

The natural grouping mixed border, in which the plants are placed in rows, each kept to a small, neat specimen, the tall growing ones having their flower-stems tied to neat stakes, rigidly upright, is amongst the worst arrangements possible for hardy flowers; but not so the mixed border, in which spreading plants are allowed to form great patches a yard or two across. Take a mixed lot of Primroses and plant them in a row, and the effect is poor compared with that of the same plants arranged in one irregular clump; the same with Auriculas, Polyanthuses, Daffodils, and most other dwarf growing things. All hardy plants will be found to have the best effect planted in some informal manner, as if in a state of nature. This does not mean that the plants are to be planted in any higgledy-piggledy fashion; this is the very reverse of Nature's arrangements. Plants, when they seed themselves, come up all round the parent plants, forming clumps and masses, but occasional seeds get blown away, or carried by birds, so that approach to a colony of any particular plant is generally signalled by the appearance of stragglers or outliers away from the principal groups. Let one thing be a feature in the several parts of the garden at one season, and all the rest be subordinate. At a particular time, for instance, a corner of the garden might be conspicuous for its Phloxes, at another for its Roses, at another for its Dahlias, again for its Gladioli, for its Japanese Anemones, and so on, always choosing for the conspicuous plants those which remain in bloom for a considerable time, and keeping those subordinate whose blooming period is short. One advantage of the picturesque style of gardening is the great use that can be made of climbing and twining plants. Honeysuckles, Everlasting Peas, Clematis, Passion Flowers, and annual climbers can all be placed almost anywhere—running up poles, over trees and fences, trailing over sunny banks, or trained in rustic arches over the walks. Many beautiful climbing Roses are almost lost to our gardens, because with bedding arrangements there was no place for them.

Mr. F. Miles on border making.—Among the first to see the merits of natural carpeting beds, and who made remarkable a border himself, was Mr. Miles. His own account of his work we give here: "If we are to have mixed borders of herbaceous plants, one thing is quite certain—we can never go back to the borders of our ancestors, in which every plant had a bare space of ground round it. In the spot where once a plant had bloomed there was an end for the year of any flowers. Now, a yard of ground should have bloom on it at least eight months in the year, and this applies to every yard of ground in a really good mixed border. I am certain that once a border is well made it need not be dug up at all. But the question is—what is a well made border? I think a border is not well made, or suitable for growing the most beautiful plants to perfection, unless it is as well made as a Vine border in a vinery. Why we should not take as much trouble with the garden border as the border of a conservatory I cannot imagine, seeing that Lilies will grow 11 feet high in the open air, not less than 10½ inches across the flower, and Irises little less than that. The more I garden, the deeper I get my drainage and the fuller of sand and fibre my soil. I consider, first, that a border must have a bed of broken bricks or other drainage, with ashes over that, to prevent the drainage from filling up; secondly, that that bed of drainage must have 2 feet of light soil over it; thirdly, that that soil must have equal parts of sand, soil, and vegetable matter, Cocoa-nut fibre seemingly being the best. A soil of these constituents and depth is never wet in winter and never dry in summer. During the dry weather I found soil like this, in which quantities of auratum Lilies were growing, to be quite moist an inch below the surface, and I know in winter it always appears dry compared with the natural garden soil. The time may come when a border thus made and cropped as I propose theoretically to crop it will be exhausted, but then is the time to add nourishment on the surface.

"**No bare soil.**—But, for all practical intents and purposes, every 6 inches of ground could contain its plant, so that no 6 inches of bare ground need obtrude on the eye. Almost any kind of bare rock has a certain beauty, but I cannot say bare ground is ever beautiful, seeming so unnatural. A bare rock you do not expect to see verdured over, and the eye allows for that, and takes a delight in its very bareness, but a bit of bare ground seems as though it ought to have flowers and seedling plants growing upon it. We like the aroma of newly ploughed land, but not its bareness, except that some soils are occasionally of a beautiful colour and become an element in an October landscape. Nature covers her bulbs with greenery, and we can do it in our gardens. Well, supposing the back of the border filled with Delphiniums, Phloxes, and Roses pegged down and other summer and autumn blooming plants, and supposing the border to be made as I have described it, I should carpet the ground at the back with winter-blooming flowers, so that when the Roses are bare and the Delphiniums and Phloxes have not pushed above ground, the border should even then be a blaze of beauty. Crocuses, Snowdrops, Aconites, and Primroses are quite enough for that purpose. The whole space under the Roses I should cover with the common Wood Anemone, and the golden Wood Anemone, and early Cyclamens, and the earliest dwarf Daffodils. And among the Roses and Pæonies and other medium tall shrubs I would put all the taller Lilies, such as require continual shade on their roots; and such are pardalinum and the Californian section generally, all the forms of auratum (though the scarlet form does not grow quite so high, and wants to be more in front of the border); Lilies, like excelsum, tigrinum splendens, monadelphum, Martagon album, longiflorum Wilsoni, dalmaticum, Hansoni, and giganteum. Now we come more to the front of the border, and here I would have combinations, such as the Great St. Bruno's Lily (Anthericum Liliastrum majus) and the delicate hybrid Columbines, Primroses planted over hardy autumn Gladioli, so that when the Primroses are at rest the Gladioli should catch the eye; Carnations and Daffodils, planted so that the Carnations form a maze of blue-green for the delicate creams and oranges of the Daffodils. When the Daffodils are gone there are the Carnations in the autumn. A mass of Iberis correæfolia happens to have been the very best thing possible for some Lilium Browni to grow through, for the Iberis flowered early and then made a protection for the young growth of the Browni, and then a lovely dark green setting for the infinite beauty of the Lily flowers. As for saying that this cannot be done, I say that it is nonsense, for the Iberis flowered beautifully under such circumstances, and the Lilies too. If once you get it into your head that no bit of ground ought ever to be seen without flowers or immediate prospect of flowers, heaps of combinations will immediately occur to those conversant with plants and the deep-rooting habits of most bulbs and the surface rooting of most herbaceous plants—for instance, Colchicums and Daffodils, with a surface of Campanula pusilla alba. The big leaves of the Colchicum grow in spring, and there would be nothing but leaves were it not for the masses of Daffodils. By-and-by the leaves of the Colchicums and Daffodils are dry enough to pull away, and then the Campanula, be it pusilla, pusilla alba, or turbinata alba, comes into a sheet of bloom. Before the bloom has passed away the Colchicum blooms begin to push up, and as some of my Colchicums are 5 inches across, of the richest rose colour, I do not exactly feel that this is a colourless kind of gardening, and as I have a hundred different kinds of Daffodils, this little arrangement will not be without interest in spring.

"**The Daffodils** and Colchicums root deeply and grow mostly in winter, requiring water then, and not in summer, when the Campanula carpet is taking it all. There are some, however, which one must be careful about—the common white Lily, for instance, which wants exposing to the sun in the autumn. I do not mind the exquisite French Poppies among these candidum Lilies, because the Poppies die about August, and then the Lilies get their baking and refuse to show the bare earth, soon covering it all with their leaves. For the extreme front of the border hundreds of combinations will occur—Pansies over Daffodils, Portulacas over Central Asiatic bulbs, Christmas Roses and Hellebores over the taller Daffodils, with Gladioli, Tritomas, and giant Daffodils, Hepaticas, and autumn-blooming and spring-blooming Cyclamens, with Scillas and Snowdrops. Mr. Fish says Hepaticas will not do for surface plants, because they lose their leaves. All I can say is, that mine used to lose their leaves before I knew how to grow them, but I determined to grow them as they ought to be grown, and never a leaf do they now lose in the hottest summer, and beautiful their leaves are, and of infinite value. Aubrietias are not very good for carpeting for Croci, as the Crocus leaves obscure the Aubrietia flowers. The Golden Nettle (Lamium maculatum aureum) is good with the blue Camassias and the more delicate Hyacinths peering through it, and I think no one will quarrel with my Golden Crocus coming through a deep crimson-purple Ajuga (reptans atro-purpurea), with a good bunch of the glorious blue autumn Crocus (speciosus) making another combination later in the year. When Anemone japonica is low, up come the taller Tulips, sylvestris for instance, and higher still out of the dark green leaves come the bejewelled Crown Imperials. Veronica rupestris is not a bad surfacer for some bulbs.

"**Advantages**.—As for the cultural advantages, I can imagine this system in the hands of a skilful gardener to be the best of all. In the first place, the plants suffer much less from drought, because there is so much less surface exposed to sun and wind. I daresay this is not what might be imagined, but such is certainly the case. Examine, not right under the root, but under the spreading part of a Mignonette, and see if, on a broiling hot day, the ground is not much cooler and moister than on the bare ground. Irises are almost the only plants I know of that do require the soil bare about their rootstocks, but then Irises are a carpet of green always, and a few clumps of Tiger Lilies or Tiger Irises will not seriously injure their flowering prospects. And what cannot be done with an herbaceous border edge when that edge is the green Grass? Crocuses and Crocuses all the autumn and winter and spring in the Grass. The tiniest Scillas, and Hyacinths, and Daffodils, and Snowdrops are leading into the border without any break. So I believe, and I think many others will believe by-and-by, that every bulbous plant ought to be grown in combination with something else, as Amaryllis Belladonna, for instance, which I plant with Arum italicum pictum. In spring the Arum comes up extremely early, and its leaves protect the far more delicate leaves of the Amaryllis till they are growing freely and the Arum dies down. The ground is surfaced with Violets, so that the Belladonnas are now coming into bloom, not with the bare ground, but with a setting of Violet leaves in beautiful contrast with their pink blossoms. Christmas Roses of all kinds would probably be a more beautiful setting still, but the Belladonnas want a good deal of summer drying up, which the Hellebores could not stand so well.

"**We can never go back** to the mixed border of our ancestors; we have been spoilt for such blank, flowerless spaces as they had by the gorgeousness of bedding out. But we have now a wealth of hardy plants, especially bulbs, which they never had, and this combination of bulbous plants and herbaceous plants will certainly lead to a preparation of the borders which has been hardly dreamt of by people who do not care what they spend on tropical flowers, for it seems to be forgotten that we have Irises as big as plates and Lilies as tall as a tree, all hardy and requiring little attention when once they have been properly planted. The time that used to be spent year after year in digging acres of borders might now be spent in properly making or re-making a few yards of border, till the whole outdoor borders are as exactly suited for the growth of plants to the uttermost perfection—as many as possible being put in the given space—as the borders of a large conservatory. It is in such a border as this that we attain the utmost variety, unceasingly beautiful, every yard different, every week varying, holding on its surface at least three times the value of plant life and successional plant beauty of any ordinary garden. The chief enemy to the system is the slug; but while the Belladonna Delphinium, which is usually half eaten by slugs in most gardens, grows 6 ft. high with me, I am not going to give up my system. And if any one who has read this will send me a hamper of common Wood Anemone roots, or of Lent Lilies (Narcissus Pseudo-Narcissus), I will send them a hamper of Phloxes, and show another year how the back of the border should be as gay in spring as it is in autumn."

Mr. Brockbank's borders.—Mr. Brockbank, a very successful and enthusiastic cultivator of hardy flowers, near Manchester, has borders of another kind, of which the following is a description: "I fancy my experience for the last five years has taught me a good deal; and as at length we are fairly successful here, and have found out how to have gay borders with little trouble, I venture to send you a few remarks upon the subject. I will commence with the most difficult part of it, which is how to have a gay flower garden in place of the usual mosaic work made up entirely of bedding-out plants. We have ended in a compromise by using Pelargoniums in the summer work, but for the rest of the year we manage entirely with hardy flowers. By the help of the annexed diagram I can best explain the plan adopted. The border is a long one, say 150 feet long by 12 feet wide, with a lawn in front. Along the front line are two rows of Crocuses and Daisies, red and white, in succession, and these are trimmed up as they go out of flower. The main line of the border is divided into 8-foot spaces, and in the centre of each is struck a circular bed 3 feet in diameter. These are edged with Saxifrages and Sedums in the early spring, and when these get seedy they are replaced by Myosotis azorica, Lamium maculatum aureum, and other hardy fine-foliaged plants, some of which remain good from year to year, as, for instance, some of the Saxifrages and Sedums. These circular beds are filled with red and white Daisies in early summer, and these are now replaced by scarlet Pelargoniums, chosen for their fine bold trusses rather than

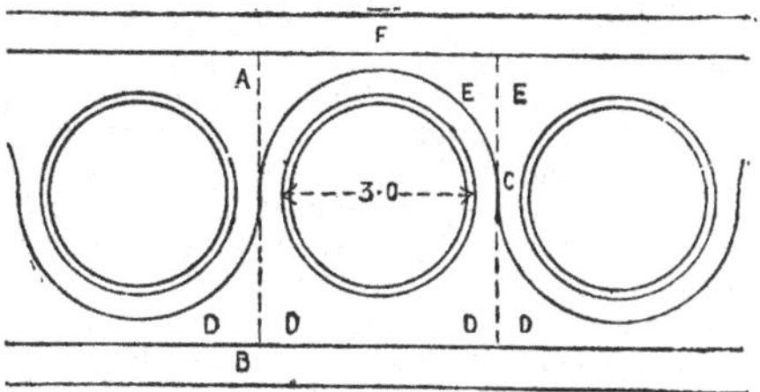

those of dwarf habit. The serpentine line C is also changed. For spring it consists of a wide band of white Daisies, which are removed and replaced by dwarf Calceolarias for the autumn, and these are practically all the changes required from the early summer to the autumn bedding, and a couple of days' work suffices for this alteration. For all the rest of the border the planting is permanent, needing only trimming and replacing as the plants become overgrown or die off, and the addition from time to time of new flowers. By a little system we know where every kind of plant is, so that when they have died down we are sure not to disturb them. Upon the line A B we have at the back groups of Daffodils, then clumps of Anemone Honorine Jobert, and in front Iris anglica. These follow and furnish a succession of bold flowers throughout the season. In the corners E E are double Pyrethrums, Lilium auratum, and Lychnis Viscaria splendens, the space being closely filled up with self Pansies and Violas, including Magpie, Lord Brooke, Admiration, Golden Gem, &c. At this moment these are one mass of flower. The triangular spaces D D on the front line are filled with Aubrietia græca, and we find in practice that this and the Daisies can be left throughout the year. The Aubrietia is always gay, and is just now very beautiful. The Daisies are also late this year, and we shall leave them alone through the autumn. In odd corners also in front we have Campanula carpatica, and on the back line Campanula grandis and alba. Behind all these on the line F we have in early spring Narcissus Horsfieldi and sundry Lilies; and now we have in splendid bloom fine varieties of Delphiniums, especially D. Belladonna, the light blue variety. Behind this again come compact plants of the best Rhododendrons, interspersed with the fine Japanese Hydrangea paniculata plumosa; and again behind these are Lupines, blue and white, and some Hollyhocks. There are hosts of other plants interspersed, but this general description will suffice, and it will be seen that there is everywhere a succession from spring to autumn, so that the border is always full of beauty and interest. A few hours every week suffices to keep it in perfect order by replacing failures and checking overgrowths. This is the only border in which we find it needful to use Pelargoniums and Calceolarias, but I think it would be found the most practical system for general adoption, as it admits of the use of bedding-out plants to any extent. We have another large border around a lawn which is edged in front with yellow Violas and masses of white Pinks, behind which edging are masses of blue and purple Violas, pink Pinks, Pæonies, Phloxes, Aquilegias, Doronicums, Sweet Williams, and other large growing plants, alternating with dwarf Roses and backed with Rhododendrons and Azaleas, and in this bed we have in early spring large clumps of coloured Primroses. The effect of this border, which has no fixed lines, is always charming and always bright. In another we group the larger flowers, such as Pæonies, Delphiniums, Centaureas, blue and white, Hollyhocks, and several sorts of Globe Flower in large clumps, and in front Violas, Primroses, Polyanthuses, Cheiranthus, and Alyssum, &c. This has been a glorious sight during the present early summer, but its beauty is now waning. In another bed we have a fine show of Pansies and Helianthemums, to be followed by Roses, Phloxes, Pentstemons, Gladioli, and Antirrhinums; and by the side walks, which surround this bed we have bold masses of Papaver orientale and bracteatum now in great splendour, to be followed by Lilium candidum and other Lilies, white and orange. It will be seen from this recital that there is no lack of bloom if a garden of hardy plants is properly arranged and stocked; these are only the "bye" beauties of such a garden. The real gems are the smaller flowers, which do not find a place in such beds as these, but are kept in quiet nooks and corners, where they may be sought out and enjoyed. A garden such as described is easily kept up when once established, and affords far greater pleasure than one of a more fashionable description."

Plants for evergreen borders.—The following most carefully made selections will be useful to those who desire to encourage the evergreen type of hardy plants in certain positions. The situations should be sheltered, and the soil free from damp

Hardy evergreen and sub-evergreen plants.

Helleborus ilicifolius
  atro-rubens
  colchicus
  caucasicus
  abchasicus
  orientalis
Papaver orientale & vars.
Ophiopogon japonicus
Fluggea japonica
Heuchera americana
  ribifolia
Tellima grandiflora
Saxifraga ligulata & vars.
  crassifolia and vars.
Saxifraga cordifolia
Sisyrinchium striatum
Epimedium alpinum and vars.
  Perraldianum
  macranthum
  pinnatum
Iberis sempervirens
  semperflorens
  Garrexiana
  Pruiti
  superba
  corifolia
  correæfolia
Sedum populifolium
Statice latifolia
Cynara Scolymus
Nepeta Mussini
Pterocephalus Parnassi
Cardamine trifolia
Armeria cephalotes
Acorus gramineus and fol. variegatus
Morina longifolia
Euphorbia Myrsinites
Ferula gigantea
  persica
  Ferulago
Potentilla
Geum aureum
  sylvaticum
Valeriana Phu and var. aurea
Caltha palustris
  radicans
Galax aphylla
Alyssum saxatile
Vinca minor
Tritoma Uvaria
  caulescens
Rhodea japonica
Meconopsis nepalensis
  Wallichi
Onopordon Acanthium
Yucca filamentosa
  flaccida
Teucrium Chamædrys

It may be added that the Ferulas do not strictly belong to this selection, but as they die down late and come up early, they ought to be included.—W. GOLDRING.

Another selection of the same nature is given by Mr. Burbidge, and is as follows:—

Helleborus niger vars.
  orientalis
  olympicus
  atrorubens
Aubrietia purpurea
Viola tricolor vars.
  odorata vars.
Bellis perennis vars.
Saxifraga coriophylla
  Burseriana
  crassifolia
  Stracheyi
  ciliaris
  cordata
  oppositifolia vars.
Silene acaulis
Cheiranthus Cheiri vars.
Primula acaulis vars.
  Auricula
  marginata
  denticulata
Polemonium cœruleum
Iris germanica vars.
Lilium candidum
Fragaria indica
Cyclamen vernum
Carnations
Arundo conspicua
Omphalodes verna
Phlox setacea
  Nelsoni
Sisyrinchium striatum
Linum arboreum
Androsace lanuginosa
Achillea tomentosa
Alyssum saxatile & others
Arabis albida
  lucida
  lucida fol.-var.
Arenaria balearica
Bellis perennis var.
Campanula pyramidalis
Cerastium Biebersteini
Dryas octopetala
Euphorbia Characias
Gentiana acaulis
Michauxia campanuloides
Hedysarum coronarium
Centaurea argentea
Helleborus fœtidus
  argutifolius
Lamium maculatum
Linnæa borealis
Meconopsis Wallichi
Ramondia pyrenaica
Saxifraga crassifolia
  ligulata
Sempervivums, many kinds
Soldanella alpina
Thymus, various
Tritoma Uvaria
Veronica prostrata
Vinca major
  minor
Eryngium Serra
Epimedium, several sp.
Convolvulus mauritanicus
Cheiranthus Marshalli
Iberis, many kinds
Spiræa Filipendula
Achillea tomentosa and other species
Carex pendula, fine foliage
Myosotis dissitiflora
Papaver bracteatum
Santolina incana
Erinus alpinus
Sedum glaucum
  lividum
  acre
  a. aureum

Rosemary and Lavender in borders.—It may be noted here that the old garden favourites, Rosemary and Lavender, are charming as grouped in borders near or against a house. In summer, Cherry Pie, or Heliotrope, which goes so well with them in colour, may be planted freely with them. A border so formed contrasts to advantage with more richly stored and well arranged borders of brilliant flowers. We should give more space to the old pleasant, sweet-smelling flowers.

Mulching instead of digging borders.—Thus writes "W. H. E.": "I now adopt the plan of mulching my mixed borders and of never digging them, and in some cases I go further. On my peat bed and fernery I do not remove the fallen leaves that may accumulate there, but I cover them over with a slight mulching of Cocoa-nut refuse. In this way I imitate Nature, who, by a thick covering of leaves, provides against the ill effects of plants throwing themselves out of the ground. If I could dare to brave the imputation of untidiness, I would not cut anything down till the spring. You may be sure the flower-stems and decaying root-leaves are an immense protection in winter. I have mixed borders right up to the drawing-room windows. Of course, I am rather particular what plants I put there. There is no formality, and there is plenty of interest. Lilies never did well here, so I made them a special bed for themselves. I chose the *lowest* part of the garden. There I dug out my bed 2 feet deep; I filled 6 inches of this with brick and other rubble, and the remaining 18 inches with coarse river sand and garden soil in equal quantities. In this I planted my Lilies, putting no manure; but when they had been in a year, I mulched the surface with rotten manure, and shall do so again. Hitherto, the result has quite satisfied me. The spaces between the Lilies I intended to fill with Oxalises, but the last winter destroyed many of them. Those that survived evidently like the soil much, and in the front now I have Croci and other small bulbs."

Need of care in choosing plants.—People do not always know what to select. They may give precious space and important positions to plants that, no matter how grown or arranged, would never give a satisfactory result, from poor weedy habit, resemblance to types which are already represented in our fields and hedgerows, shortness of bloom, and various other bad qualities. It is desirable that space should be well and wisely occupied by a variety of beautiful, distinct, and high-class and varied plants.

Scores of plants are under cultivation, taking up much room and eating fine silky loam which should be given to really beautiful and distinct growths. Our botanic gardens are a case in point. Of course, they ought to represent the greatest variety of vegetation possible, but they might do this while helping us to a more beautiful garden flora. We could mention many weedy plants grown in botanic gardens that are not a whit more useful in illustrating the Order or Class to which they belong than the more beautiful plants would be. Whatever excuse there may be for a botanic garden growing weedy types of plants, and perhaps having masses of things that grow in the hedges or fields around, there can be little for a private garden in which we expect to find plants grown for their beauty, and not as representatives of things we see in the hedges or ditches around, or subjects otherwise valueless for cultivation. Sometimes plants are objectionable, not only from the short period during which they flower and the great space they take up, but also from the danger there is of their overrunning precious space and exhausting precious soil. Frequently a plant, not objectionable in itself, may be quite unworthy of a place in consequence of the greater beauty of an allied plant. It would be well to place all new-comers not known to be good on their trial in a nursery plot or border so as to test their value or fitness for the garden, and to increase them if they are worth it. One of the commonest sights is a rock garden overrun with certain rubbishy plants that never ought to have been allowed to grow near it. How often does one see the growth of a Dead Nettle or St. John's Wort or Willow Herb, or even a worse thing, which has driven away the beauty of a true alpine plant from a rock garden. By all means let every man follow his own taste in his own garden. A garden may be a place for intellectual recreation, a place for study, a place for rest, or a place for amusement, as suits the fancy of its owner, and no one has any right to find fault. Nor has anyone any right to recommend this or that as an improvement, except in so far as what is recommended will lead to a wider and higher enjoyment of the beauty of plants. Many plants must, however, be excluded from gardens for reasons totally apart from the question of their beauty. Take the common Dandelion, for instance. There is scarcely a brighter yellow outside the list of florist's flowers, yet, even though it were a rare exotic, instead of one of our commonest weeds, it would have to be excluded from our gardens on account of its free seeding propensities. A well-grown Teazle is a good study for an art metal worker—varied and beautiful in every line and form; but we do not grow Teazles in our gardens, because we can find them in every damp ditch if we want them. That most graceful of the Campanulas, C. rotundifolia, is grown in gardens in the south, but no one would think of growing it in Scotland, where it can be found in nearly every yard of roadside bank. Our native plants can only justify a place in garden borders where great beauty is combined with some amount of rarity, except in the case of spring flowers, which bloom at a season when it is more convenient to enjoy their beauty in one's own garden than in the woods and fields. I have seen the common Meadow Sweet grown as a garden plant, and a very stately and decorative one, too, in districts where it was rare, but who would think of growing it where the plants are as thick as they can stand in every ditch? The greatest difference of opinion will, however, most probably be found about plants of a feathery and light foliaged habit, such as the annual Chrysanthemums and Larkspurs, and Gypsophila paniculata, and plants which carry their flowers at the top of tall stems, and not enough of them to form a mass, such as Geum coccineum fl.-pl. and German Scabious. Place a lot of these plants together and the effect is weedy; but mix them judiciously amongst plants of more decided forms, and they are as useful in contrasting and softening hard outlines as feathery Grasses are in a bouquet. Dwarf growing plants are of course too small to have a weedy effect, and their exclusion or acceptance must be decided on other grounds.

But hand in hand with selection must go the determination to make one's own garden after one's own fashion, the situation, soil, and other circumstances being taken into account. The bane of gardening, in fact, is that everybody imitates everybody else. There must be originality—individuality, if you will—in a garden, or it will not please.

Hardy flower culture.—The "linger-and-die" principle of growing hardy flowers can never give satisfaction. If the vegetation of temperate latitudes is to become a feature in our gardens, the first essential is good culture. Many entertain different ideas as to the meaning of "good culture." Thus, with some the good culture of Grapes, Pines, and Peaches means something tangible—great heaps of turfy loam fresh from upland pastures, well decomposed manure, lime rubbish, tanks filled with manure water, the use of the syringe or hose pipe, and great care and watchfulness during all stages of progress until the crop is secured. So also is it with the good cultivator of vegetables. He never tries to grow his Cauliflowers, Celery, and Peas on the "mixed border" system, sticking a Cabbage here or a Tomato there without either preparation of the earth or thought as regards manure. When fruit and vegetable culture are concerned we generally find that cultivators are in real earnest, and they set themselves to work on some common-sense plan. Nothing is neglected; shelter, good tilth, manurial stimulants, and every attention is given as required; but, let the culture of hardy flowers be the object, and in nine cases out of twelve the "linger-and-die" system is resorted to. How seldom do we see any adequate attention given to hardy flowers. No heaps of fresh turf or manure, no hose pipe or cultural attention is reserved for them, and yet how patent is the fact that all plants, beautiful as well as useful, need great attention if we desire to have them first-rate of their kind! Deep tilth, well enriched and well drained earth, and due rotation of crops are as needful to the full growth of the most common and most effective of hardy flowers as these aids are to the successful culture of cereals or root crops. Then, as to cultural attention, no plants are more grateful for their share of this than hardy herbaceous plants. Plants under glass are syringed, stopped, or pinched, and carefully attended to in many ways, but the hardy flowers are supposed to be able to forego much of this cultural care, and in many cases are expected to take care of themselves after they have been stuck into a poverty-stricken, choked-up border. As a rule, hardy plants are expected to thrive under about one-half of the advantages extended to a bed of spring Cabbages, and the result is poverty and choke-muddle in many gardens where the most orderly display of health and vigorous flower beauty ought to exist. Now, that we are emerging from the dark ages of hardy plant culture, however, we may naturally hope for better things. In all large gardens one or more special men must be charged with the cultural care of the hardy plants, and their labour must be invested with as much importance as is now accorded to workers in the indoor departments. In the hardy flower culture of the future, the hose pipe will prove as useful as the syringe is now indoors. Green fly and other insects must be kept down outside as well as indoors, and root moisture and manure water are as acceptable and beneficial to vigorous outdoor vegetation as they are to hothouse plants. As to the cultural care needed amongst hardy herbaceous plants, we may offer a few examples.

The Chrysanthemum, for instance, is one of the few hardy flowers at all presentable in the more sheltered parts of Britain in November. In most cases, the plants are tall and their lower parts leafless; so tall are they, in fact, that they must perforce either be looped up to walls or fences, or secured to stakes, and the general result of outdoor Chrysanthemum culture is to obtain a few washy flowers on the apex of a broom-like wisp of stems. Here and there, where a little attention has been paid to pinching, results may be a trifle more satisfactory, but, as a rule, outdoor Chrysanthemum culture is a failure. As there is a way out of all difficulties, so is there an improvement on this poor way of growing one of our most showy autumn and winter flowers. When Mr. John Bain was curator of the University Botanic Gardens, Dublin, some years ago, his Chrysanthemums were, during mild winters, very showy and bright; the plants themselves dwarf, bushy, and with green foliage to the ground level. His plan of culture was so simple and so effectual, that it deserves to become widely known. In April or May his plan was to examine all the Chrysanthemums which were planted out in the rich, sheltered sunny borders, in order to prepare them for their summer growth. Each plant was looked over and its shoots thinned out, only ten or twelve of the finest being allowed to remain. Then each plant was lifted, and a hole dug in which to replant it, the plant being lowered so that only the tips of the young growth remained above ground level. In filling in the earth, leaf-mould and well decomposed manure were freely added, and the shoots were arranged as widely apart as possible, so that each had ample space for development. The result of this lowering of the plants was the production of new roots from the buried portion of the shoots, and the new roots found ample nourishment around them in the newly added earth and manure. In due time each shoot was stopped once or twice as growth proceeded, and in very dry weather they were watered freely. Thus treated, the plants were, as described, dwarfy, healthy, and most floriferous. The plan is well worth adoption in those warm and sheltered localities in England where Chrysanthemums generally flower freely in November and December during ordinary seasons.

The stopping system, so beneficial in the case of the Chrysanthemum, may be extended to some other hardy plants, and by its being adopted as a rule, many plants which, as usually grown, become top-heavy and needful of staking, may be rendered self-supporting. As a case in point take Campanula pyramidalis, or the common Hollyhock. Either of these plants, if allowed to grow unrestrained, will need stakes. But if the end of the spiring flower-stems be pinched out when from 15 inches to 18 inches in height, the result is a multitude of lateral branches, which spring from near the ground and need no support whatever; indeed, are far better without support of any kind. Stopping or pinching will be found of the utmost value wherever hardy flowers are grown. Even some Lily stems may be stopped in the same way. A celebrated cultivator tells me that he always stops the flower-stems of a portion of his plants of Lilium longiflorum as soon as they appear in his Rhododendron beds and open borders. The result gained is a secondary growth of spikes, which blooms later, and which escapes the injury that sometimes occurs through the late spring frosts, while the earlier growths which are unstopped sometimes suffer. Again, where the earliest batch of this Lily is grown in pots in a cold frame or greenhouse, it is well to have flowers in succession, and this is another gain in hardy flower culture easily obtained by judiciously stopping some of the early growths. We never take up our Dahlias or Hollyhocks, and in spring when growth commences we cut off the tops as cuttings. The tops root readily in sawdust on a gentle bottom-heat, and when planted out in rich beds and borders in June they give us a few late blossoms, while the old plants are dwarfy and bushy compared with plants allowed to spire up in the usual way. We have here acres of hardy flowers of nearly all kinds, but rarely use a stake in our borders. Aconitums and Delphiniums form charming bushes of bloom when their main spires are stopped sufficiently early. Last season our Lilies of the auratum and speciosum types were greatly improved in freshness of leafage and floral vigour by syringing with the hose pipe on the evenings of hot dry days; and, owing to a liberal allowance of moisture in this way, many bog and swamp plants attained to a respectable state of perfection, as planted out in the ordinary herbaceous borders. As to the moisture side of the question, let us remember the Crocus, Narcissus, and Orchis roots are now deluged with our winter rains, and in some localities actually submerged for weeks together. Drought and poverty of soil are two of the great evils to avoid in hardy plant culture; indeed, in plant culture of all kinds.

Division and digging.—One point in hardy flower culture must always and at all times be insisted on, viz., that division must always be carried on. I make it a practice to collect and sow a portion of all the seeds that ripen with us immediately after they do so, irrespective of season. Newly ripened seeds sown as soon as gathered germinate at once, as a rule, and thus much valuable time is gained. Cuttings are put in during wet days in the summer time, so that we thus have at all times young stock for replanting vacancies, or for giving away in exchange, or otherwise. Another rule here is never to dig borders in which plants and bulbs are growing. We top-dress with burnt garden refuse, leaf-mould, or manure every season, and make it a point to take up our plants (with but few exceptions) every three or four years, and after thoroughly trenching up the border and adding burnt refuse or manure, we replant it again with new stock from the nursery beds, so as to secure a change of soil or rotation of crops as far as is possible. It is only by replanting well and thoroughly in this way that anything like a good and even effect can be relied on. Nothing is so fatal to the well-being of hardy flowers as a system of poking them up every winter with a fork, or slicing their roots with a spade, at a time when root growth is most active.—F. W. B.

An example of good culture.—What may be done in cultivation is shown by the history of the Lily of the Valley, for of all plants that have hitherto been submitted to the gentle art of forcing, none perhaps are more grateful for it, while none is more graceful, than this old inhabitant of our gardens. Long before men knew how to force plants under glass, or had a pane of glass to put them under, this was a famous plant, loved of those who sought it in its native wilds, and those who grew it in gardens. Every garden used to grow it, and ought to have it now, though it was a common thing to see it very badly grown in some odd corner, owing partly to the free, hardy habit of the plant itself. It spreads about so freely, that it soon forms a dense "mat." As the habit often was to leave it in mats and beds, it became attenuated and poor as regards the flower, so that it was rather exceptional to see in gardens a good bloom of the Lily of the Valley. From Nature, in the hill woods it fared a little better, inasmuch as there the struggle for life prevented it from living on itself, so to say. The "mat" of plants was not so dense as in the garden bed, and often a few plants grew alone, as it were, and flowered well. Not very many years ago the forcing of the plant began to be popular, and this gradually led to a better culture. Plants grown in the matted manner, although they flowered when put in pots or boxes in a warm house or frame, never produced a good result; they were starved and weak from overcrowding. Soon it was found out that plants specially prepared for this purpose, or, in other words, well and firmly grown, forced beautifully and gave fine and numerous blooms. The simple fact was that these plants had been fairly well grown, so that the little buds were enabled to fill up, each with a flower in its heart. A very considerable trade was done with the Continent for years, and is done still, in the preparation and growth of the Lily of the Valley for forcing. The despised and too vigorous little plant got, in fact, a chance, with a result greatly to the advantage of the foreign growers. As in many similar cases, this success was put down to the climate; but there is not much in the claim. The same thing is now being done by our own growers. Obviously, the system which enables growers to get good plants for forcing will enable us to get good plants for blooming in the open air.

Hardy flowers in beds.—As separate from the many kinds of beautiful borders and groups and the square or plot garden, we have the possibilities of beds of various degrees of beauty and interest, from the evergreen one of Yuccas and Hellebores, so fine in winter, to the many bright combinations of bulbs in early summer. Here we have to deal, however, with a definite mass or group, which demands a different plan. The combinations of graceful form, beautiful flowers, and fine colour are endless. We can only hope to suggest to beginners a few plans which will not disappoint them; those who know the plants have at command the pleasure of working out new combinations for themselves.

OF HARDY FLOWERS.

Miss Hope's beds.—The following were described by Miss Hope, in her interesting garden near Edinburgh, as among her successful beds. She uses a half-hardy plant or annual, now and then a course to which there can be no sound objection:—

"1. I begin with the latest flowering bed which I have, and which is now (January 20) covered with little white stars, viz., Aster ericoides, the first flowers of which expanded in October. This is a remarkably neat-growing plant, and its Heath-like appearance is attractive, even when not in flower. The pair of small beds which I have of it consist of a centre plant of Golden Queen Holly, the outside edging Euonymus radicans variegata, the rest being filled with alternate plants of the little Aster and dwarf double red Sweet William. Even when done flowering, this Dianthus, owing to its dark reddish foliage, contrasts well with the variegated evergreens and light growing Aster.

"2. Campanula carpatica (blue), with a broad edging of the yellow Œnothera missouriensis. By preventing the Campanula from seeding, this bed will last a mass of flowers nearly four months, beginning with June and July.

"3. Ononis rotundiflora (Rest Harrow), a bright pink Pea-flowered plant, mixed through with Dactylis glomerata variegata. This makes a lovely summer bed.

"4. Common double white Pink, edged with Heuchera lucida. Among the Pinks are small single red Tulips, which come up and flower year after year, and which well contrast with the grey foliage of Dianthus and blue Scilla bifolia, through and outside the brown-leaved Heuchera edging. This bed in summer is a mass of white, and, by dotting in a few plants of crimson Mule Pinks, there are some flowers in autumn. No. 4 is a good winter and spring bed, and a most fragrant one in summer.

"5. Alternate plants of Erica carnea præcox and Heuchera, with room between each of these permanent plants for a tuft of Campanula carpatica or C. turbinata; or, if an annual is preferred, dwarf China Asters, purple, pink, and white; or Tagetes signata pumila, the lively green foliage of which is pleasing even before the plant comes into flower.

"6. (Large bed), centre plant, Golden Yew, the rest of the bed being filled with three shades of red, consisting of the Phlox Drummondi pegged down twice, either mixed, or shaded, from light to dark, or *vice versa*, according to taste. The mass of Phlox is surrounded with a belt of the pure white Œnothera marginata and an outside edging of Heuchera. Dianthus Heddewigi would do equally well with the permanent plants if a change is wished from the Phlox; both come into flower early in July, and are superior to Verbenas, and last until November, when the winter beds require to be planted.

"7. Clematis Jackmani and its varieties, edged with Vinca major elegantissima, minor aurea, or argentea, or Euonymus radicans variegata, or Lonicera reticulata, but, although hardy, this last loses most of its leaves in winter, and V. elegantissima is spoilt by severe frosts; the small Periwinkles and Euonymus are surest. I was surprised to find, on cutting over the Clematis this winter, fine roots where it had been pegged down, as I had always thought it must be increased by grafting. If a high bed is wished for, some suitable branches for the Clematis to crawl over quickly make the bed any height desired. I have seen Tropæolum speciosum grown in this way in Perth and Aberdeenshire with most brilliant and beautiful effect, but it does not succeed about Edinburgh. The Clematis's first flower opened in July, and the last in November.

"8. Erica carnea or præcox, Violas, and Pansies, in alternate lines or rings, and in colours according to taste. It is hardly possible to go wrong with this bed. When the Heaths are too much grown to admit of Violas between them, and you are not inclined to part them, Phlox Drummondi (mixed) will give bright eyes of colour all summer and autumn, and, if thinly planted, will do not the least harm to the Heaths, which, of course, must never be overgrown by other plants, but by clipping over the dead flowers, and trimming in sprawling pieces, they will last for years undivided, if so wished.

"9. Dactylis variegata and Phlox Drummondi intermixed make a light mixed bed.

"10. Anemone Honorine Jobert (white) alternated with Chrysanthemum Bob (dark red), the edging being Heuchera, makes a pretty bed.

"11. The small Fuchsia pumila, with white Vittadenia trilobata, makes a very neat and lasting bed. Some think the Vittadenia too like a Daisy, but I am fond of all rayed flowers, and prize any plant that lasts six months in flower; it sows itself (if one chooses to let it), stands the early frosts, fades into pinkish lilac, so as to produce various shades of colour at the same time on the plants, and requires no pegging or cutting off of dead flowers. As it is not a Daisy, it does not shut up in sunless weather, or in the afternoon.

"12. Groundwork, the little yellow Œnothera prostrata, thinly dotted with Viola Perfection, or any other purple, lilac, or blue Viola or Pansy of compact habit.

"13. Centre plant Yucca gloriosa, set in a groundwork of dark blue Ajuga purpurascens (summer), dotted with Sedum spectabile (autumn), the last having fine large pink tassels of bloom, which stand the first frosts. This bed is edged with variegated Ivy.

"14. Large or small beds of Dianthus Heddewigi are invaluable. This year some beds were the admiration of every one, both nurserymen and amateurs. We never had such beautiful Chinese Pinks, of every shade of red, from delicate pink to dark blood red, lilac to purple, white, and all degrees of blotched, and streaked and spotted in every delicate shade, and decided markings. There were twenty small divisions filled with these bedding Dianthuses, and in each compartment there were different varieties. We regretted having to transplant them, still in flower in November, on account of some necessary alterations that had to be made in the beds. The Dianthuses were edged with Artemisia Stelleriana. They last well in water, and bear close inspection.

"15. Large bed (to look down upon); centre plant, a tuft of best Gardener's Garters (Phalaris arundinacea elegantissima), surrounded with Aster bessarabicus (best blue), then scarlet Pelargoniums (somewhat large plants), a band of Helleborus niger major, the edging being alternate plants of purple-leaved Plantago rubescens and Funkia japonica cordata. As all our beds and borders are edged with hardy bulbs, No. 15 has flowers in winter (Christmas Roses), spring (Crocuses), summer (scarlet Geraniums and lilac Funkia bells), and autumn (Asters); and, owing to the variety of shape and colour of the foliage-plants, beauty is still maintained in the intervals between the times of flowering of the different plants. H. niger major should be grown for a foliage plant, even if it never flowered. In old gardens in the north I have, at a distance, mistaken clumps of this Christmas Rose for those of Pæonies.

"16. Centre plants Spiræa, with one plant of Tritoma grandis on each side. These three plants were surrounded by Delphinium formosum (selected seedlings of good clear blue), the small double yellow Sunflower, then a ring of alternate plants of Saxifraga cordata, and the finest and earliest flowerer of the large-leaved Saxifrages; one we got unnamed from Mr. Niven, of Hull, belonging to the ciliata type. This bed is edged with Heuchera, and between each Saxifrage are dwarf Lilies, L. venustum (red) or L. superbum (yellowish), alternated with the white trumpet L. eximium or longiflorum, and between every plant of Delphinium and Sunflower is L. croceum (orange). This bed has flowers in it in spring (dwarf Narcissus, for bulb edging, and Saxifrages), summer (Spiræa, Lilies, and Delphinium), and autumn (Sunflowers, Lilies, and Tritomas, which last continue to flower long after the Sunflowers).

"17. Foliage-bed—Centre plant, Golden Yew, surrounded with Cineraria acanthifolia or Centaurea ragusina, the last of which stands most winters with us. The Cineraria is hardy with us. Then a broad mass, pegged down, of Red Spinach (seed from Belvoir—very superior to what is usually bought as Red Orache, and does not acquire so much of the rusty colour when old). Next a ring of pegged Artemisia annua, the outside edging by the Grass being Stachys lanata. This bed was so satisfactory that we had it two or three years in succession, and the Spinach came up of itself more than we required, although the beds had been filled for winter and thoroughly dug.

"18. Lobelia fulgens intermixed with Tussilago Farfara variegata. This is a grand bed, and unlike every other; but I must admit that the Lobelia does not succeed with us as I have seen it in Dumfriesshire, where the climate is damp and the soil heavy—conditions which suit both plants. The dark red-pointed foliage of the Lobelia contrasts well with the round, bold Coltsfoot, and it is a good bed before the intense scarlet flower-spikes come into bloom. Our beds are all on Grass, and, being circles and ovals, simple in shape, the grass-cutter runs round them all without loss of time, and in clipping the edges the shears have not to be lifted and reversed. One of the most immediate benefits gained by filling beds with the plants recommended in this paper, and one patent to all, even rigid bedders-out of exotics, is that the extent of glass, always too limited and over-crowded, even in the largest places at times, is set free for its legitimate occupants, and the miserable crammings and makeshifts for storing away the struck and lifted bedding stock through the winter, and still more harrassing time of spring, are done away with. Not merely does the bedding system carried to extremes curtail the variety of plants in our flower gardens, but the interesting and valuable contents of frames, greenhouses, and stoves are all injured and limited on account of this one object; even vineries and Peach houses are not improved by the multifarious and multitudinous collections of bedding plants that are thrust into them."

Long beds along walks, which are usually so tame and uninteresting, can be varied almost infinitely by using hardy plants. The rainbow-like effect of winding lines of scarlet, pink, blue, yellow, and white was very striking when first seen, but gets disgusting when one sees it in every garden repeated every season; and the same may be said of the oval, serpentine, star, and circle and panel beds cut out on the turf. A series of simple rectangular beds, parallel to the walks, varied in planting, soil, and cultivation, would have a far better effect. For these beds there are many plants suitable which look well in a mass, and yet bear close inspection even better than bedding plants. Tender plants need not be excluded, but can be used freely without destroying the effect of the hardy flowers, and their use in summer would allow more latitude in the arrangements. What ought to be aimed at, however, is not sameness, but variety; instead of the beds being as nearly as possible the same throughout the season, the aim should be to produce the variety of Nature. This can be managed and unsightliness avoided by using for tall plants late-flowering ones principally along with those early-flowering ones, which can be cut down immediately their flowering is over without injury. Then, as to cultivation, this is easily managed. Many florists' flowers require rich cultivation; these can be planted in separate beds, which can be dug and manured every year, and still form part of a decorative scheme. Every accident of position can be used to produce variety. Beds facing due south give an opportunity for planting Anemones and Primulas on the north side, afterwards shaded by tall plants. Beds, partially or wholly shaded by trees, may be planted with things that like such positions; while beds, so placed that they get no sun during winter and sunshine nearly the whole day in summer, afford good positions for plants from regions where winter and summer are more decidedly marked in their beginnings and endings than in our climate, and the plants from which are started into growth by unseasonable high temperature early in the year, only to be partially checked by late frosts. In using hardy flowers of tall growth no attempt should be made to form masses of them. I have noticed many beds of Dahlias, Cannas, and sub-tropical plants ruined, in effect, by that treatment. Tall plants should stand alone, or in clumps wide enough apart for their individual beauty to be appreciated, and separated by something wholly distinct.

Single specimens.—The following are suitable, each plant being so distinct, telling, and complete in itself that it loses in value if not seen from head to foot. Dielytra spectabilis: give this a light, well-drained soil and an easterly exposure, so that it will not start growing till late; Campanula pyramidalis, a grand Campanula, 10 feet high when carefully grown, ordinarily 6 feet; Tritomas, the well-known Red-hot Poker Plants; Eryngium amethystinum, a singular plant with blue flowers and flower-stems. Lilium giganteum, a huge Lily with heart-shaped leaves, 12 feet high, with large white flowers, requires a moist, rich, well-drained loam mixed with leaf-mould or peat, and a half-shady situation. Anemone vitifolia (Honorine Jobert) should be planted by itself in a round bed 4 feet across, and allowed to spread until it fills it; a half-shady place suits it best. The Acanthus family are fine plants for planting on turf, as are the Rheums, or true Rhubarbs, and the Veratrum and Verbascum families; these are handsome foliage plants. There are many fine Grasses useful for beds on the lawn, especially in mild, warm districts, besides the well-known Pampas Grass. The best are Eulalia japonica zebrina, Erianthus Ravennæ, Pennisetum caudatum, and Stipa pinnata. The finest Grass, however, after the Pampas is the New Zealand Reed (Arundo conspicua), which is very handsome in the western and southern counties. There, too, the New Zealand flora is often very fine.

The arrangements of flowers herein set forth would not of course please everybody. Those who care for hardy flowers for their own sake will always be satisfied if they get them to grow well and show their individual beauties to advantage. Such people have but little sympathy, as a rule, with the class of mind which regards flowers and plants merely as so much beautiful material to be worked up into decorative effects.

In arranging the plants care must be taken not to get a confused effect; plants which form a plume like a common Fern should rise from a carpet of dwarf plants. Auriculas, Polyanthuses, and Primroses require shading from the hot summer sun; this can be arranged for in planting. The spring flowers are, as a rule, the most dwarf-like so that the beds can be carpeted with these, except over the later flowering plants. Tulips and Daisies, which can be moved after blooming, will leave spaces for half-hardy and tender annuals or bedding plants to add to the late summer and autumn display. Where possible, the foliage of plants should be contrasted as well as the flowers. Very pretty effects could be got in light soils by a judicious use of the broad-leaved Saxifrages; these have leaves like the Water Lily and make fine spreading tufts. The flowers are handsome in spring when the plants have been established a year or two, and the leaves being evergreen contrast nicely with smaller leaved plants, both in summer and winter; they also contrast well with plants with upright tufts of sword-shaped leaves. A pretty bed could be made of Saxifraga cordifolia mixed with Pansies, and having a centre of German Iris and Gladioli, or of Carnations. The large-leaved plants, Acanthuses, Rheums, &c., can be used in the same way as the Saxifrages; massive leaves of simple outline give the eye something to take hold of, and give value to all light feathery plants by contrast. In planting hardy flowers in small beds on Grass, it should be borne in mind that flowering plants present their forms to us in two different ways.

Dwarf plants are, as a rule, designed by Nature to be seen from above as a carpet is seen, and tall plants are designed to be seen from the side. It is possible to arrange dwarf plants to form geometric figures, as the forms are small, and become undistinguishable at a very short distance; but tall plants have distinct forms of their own, which will hide or break up the lines of any geometric arrangement; they cannot, therefore, be used to form decided patterns. In geometric decoration in colour it is best to emphasise and repeat the outlines by bands of more or less decided colour, but the moment we deal with natural forms, or forms borrowed from natural forms, the first object is to judiciously break any straight or rigid outlines.

Tall plants also have generally a straight central line of their own, which gives value to their curved forms, and any strong artificial line brought against them takes from their beauty. For both these reasons, when tall plants are used in small beds, the outlines of the beds should never be emphasised or repeated by any planting in lines parallel to these outlines, but should be broken by three or more tallish plants placed close to the edge of the bed, so as to overhang the Grass and break the outline in an irregular manner. These should not be such very tall plants as to make the bed look like a table with its legs in the air, but should be either round, bushy plants of medium height, or plants which form a tuft of foliage and carry their flowers on tall stalks well above it.

In planting, care should be taken to plant things which will bloom at different seasons interspersed with each other, so that the beds at no time look empty; this can only be done by carefully selecting and placing the plants with due knowledge of their habits.

Mixtures good and natural.—The habits of rooting which different plants have must also be taken into consideration in planting. Gladioli do well with Rhododendrons, because Rhododendrons form a thick mat of roots close to their stems, and leave the soil round them free for other plants; while they also do well with Roses, because these send out a few strong rambling roots and leave plenty of unoccupied soil between them. Some plants root so freely as to keep down weeds, others are so greedy of moisture as to parch everything near them; both these things must be taken into account. If we plant a strong-rooting thing in a rich soil, we must not plant a weaker subject close to it if that weaker subject requires like cultivation; but we may plant something that can live on little with every chance of success; the strong plant will break up the soil with its roots and appropriate the manure and moisture, and so change the soil as to make it suitable for a plant requiring a light, dry soil. Some plants root deep, others root near the surface; these can often be planted together. Then many plants are in active growth only for a short period; these can also be planted closely provided the plants placed together are dormant at different periods. Nature affords many examples of this. Take the hedgerows of a clay country, for instance; in early spring they are bright with Ivy, Celandine, and Arum; a little later the common Squill or Wood Hyacinth comes up; then the hedgerows themselves wake up, and Hawthorn follows Blackthorn and Sloe, to be followed in turn by the wild Roses; following these come Meadow Sweet, several wild Geraniums, Briony, and a host of the Aster family, and last the White Bindweed.

SPRING FLOWERS.

THE meaning and the beauty of a garden in the early year are clear to all who have eyes; and a happy thing it is, the recent revival in hardy and spring flowers, inasmuch as it gives us a whole season of beauty added to our gardens: that is to say, a place full of hardy flowers will have at least three months of clear gain in flowers over such as have no spring flowers, and which depend entirely on bedding out. This, as everybody knows, is usually carried out the first week in June. Being tender plants, the cold rains and storms after that date frequently injure them, and their period of beauty is often much later. As a matter of fact, then, the bedding system reduced the beauty of the English garden by one half.

As yet, the difficulty of getting good plants is great, and therefore those bright Hepaticas and other spring flowers are often seen as poor dots where they ought to be bright sheets or healthy tufts a yard across. The question of obtaining supplies of our hardy spring flowers by those who do not carefully increase them for themselves is a very difficult one. Such is the run on the nurseries that grow these things that people cannot, even at a high price, get a healthy, stout plant. One has to give much more for a little border flower than for a well-established young tree, and then, perhaps, will not get a good plant. But this is a state of things that may be expected to change for the better. What we should counsel those who care for their spring gardens is this: To begin and always to work with a series of nursery beds, while not neglecting their flower beds and borders. Of the two, we should begin with forming and planting nursery beds. If plants when obtained were divided into little bits in lines 1 foot apart in these nursery beds, they would soon increase, so that one could get such a stock as is required. When a plant is at all rare, plant it first in the nursery beds. There are many districts where an observing person may make a fair collection without much trouble. In some parts of the country the cottage gardens were not all cleared of their hardy adornments; this is one source of supply. Then there are the seed catalogues of certain houses, which ought to be examined every spring. Some things are raised as easily as Mustard and Cress—for example, the different Bellflowers. Some hardy flowers are a little difficult, and have to be waited for, but this is partly owing to keeping the seed too long after it is ripe. Hitherto the principle has been to select when the seed catalogues come out in spring and sow afterwards—a good plan, no doubt, if we have not our own fresh seeds to sow when they ripen. There has always been some little difficulty about raising Hepaticas and Christmas Roses from seed, but Mr. Frank Miles, who is very successful with these, says he manages it easily by sowing as soon as the seed is ripe in the open air, and covering the border in which it is sown with slates or bricks. They remain on till the seed begins to germinate. They save the seeds from drought and from the attacks of birds and vermin, and this explains the success of the method. In presence of the difficulty of obtaining good stocks of hardy flowers, all who can should raise them from seed. Mr. Miles's plan might with advantage be applied to other things which do not come easily from seed.

Positions for spring flowers.—The first move in the direction of spring gardening was a kind of bedding-out, and a very attractive thing too—Forget-me-nots, Pansies, Daisies, Catchflies, Violets, Hyacinths in beds and in ribbons; but this is only one way of cultivating spring flowers, and not the best. The easiest and the most artistic thing to do is to scatter about the flowers wherever they will grow—in mixed beds, hedgerows, or plantations. Many of our country seats, like the London parks, are as bare and ugly in their dug borders as a London cemetery. It was quite an exception some years ago to see a beautiful flower in the open air before the time of bedding-out arrived. Now, since we have quite doubled the length of our garden season, so to say, the first mode of growing and enjoying spring flowers is a most important question. Here and there one may see, as at Ribston Hall, in Yorkshire, signs of taste and knowledge in growing spring flowers, and suggestions of what is possible in the future. Every place where there is a pleasure ground, or any open space of Grass with trees on it, may be made delightful with such things as the winter Aconite and Snowdrop, and spring Snowflakes, and the blue Apennine Anemone, and various other flowers dotted in the Grass. It is not a desirable practice in the much-shaven parts, and is best about trees and under the branches of summer-leafing trees. Some little plants that flower and ripen their leaves early find a happy home under Beech or Oak or other deciduous trees; they complete their season's work before the leaves come on the trees, and in spring are seen happy under their branches. Then, again, in any place where wild flowers grow well numerous additions from other countries may be made to them. For instance, if we have a grove where the wood Anemone grows naturally (a common occurrence enough), nothing is easier than to introduce the blue Apennine Anemone along with it. If the soil be chalky, the yellow Anemone (A. ranunculoides) would be a delightful addition to them. Or does the Bluebell or Wood Hyacinth grow with us? Then certainly in the same place, or near it, will also grow its relative the bell-flowered Scilla and S. bifolia—not native plants, but perfectly hardy in our country. Various kinds of Daffodils or Narcissi will grow anywhere the common Daffodil will, as, for example, the different forms of Poet's or Pheasant's-eye Narcissus. The beautiful Wood Forget-me-not may be sown in any wood, copse, or shrubbery, and will give an ample return. Thus it will be seen that, apart from the garden proper, much may be done in adding the glory of spring flowers to any place where there are trees and Grass. The corners in an old orchard are among the places delightful for experiments of this kind.

Coming into the garden, we next look at the many positions in which spring flowers may be grown before we come to their geometrical bedding, which is the most troublesome and the least desirable of all. The fashion of leaving beds of Roses, choice shrubs, &c., bare of all but what might be called their proper contents must now be given up. In many places we know the bare and rich Rose beds alone would furnish a happy home for numerous beautiful spring flowers—Pansies, Violets, choice Daffodils, Scillas; in fact, for many pretty dwarf plants established in colonies between the Roses. Double Primroses are particularly happy in such positions, and flower profusely. The slight shade such plants receive in summer from the other tenants of the bed assists them; they do better than in bare borders. Where the Rhododendron beds are thinly planted, as they often are (and we think that the bushes never ought to be jammed together), a garden of another delightful kind is at our disposal. The peat-loving plants of the world (and there are many fair ones among them) will be quite at home there—much more so than in any bare borders. The white Wood Lily of the American woods, the Virginian Lungwort, the Canadian Bloodroot, and the various Dog's-tooth Violets are plants that enjoy peat beds. Next, we come to borders and beds of favourite flowers, of spring flowers, such as the Polyanthus, the Primroses in their various coloured forms, Cowslips, Auriculas, which in the self-coloured and border kinds are delightful. Special beds of favourite spring flowers are also very desirable. One can "cut and come again" for their flowers; they are also convenient for division and exchange. Then, by some favourite walk or walks in quiet places, a rich border for those glorious Polyanthuses and coloured Primroses and any other favourite free spring flowers may be well worth having. Thus it will be seen that before we come to the formal massing of spring flowers there is a variety of ways of enjoying them, more artistic, more satisfactory, more easily managed than "bedding-out" pure and simple. That may follow the fashion of the hour, and be arranged according to taste, with a considerable variety of material—Forget-me-nots, Daisies (both variegated and green), Silene, Pansy, Violet, Hyacinth, Anemone, Tulip, and so on. If we have a group of beds, and, say, a parterre under a window or any other conspicuous position, a bright and pretty effect may be formed in this way; but, without any such thing as either parterre or formal beds under the windows, fair gardens of spring flowers may be made in every place. If they are so made, the eternal problem of design for the few formal beds of the parterre will not seem so terrible or so necessary a business as is the case at present.

New forms of spring flowers.—Of late a number of beautiful forms of well-known and much-loved flowers have been collected from various countries or raised from seed. For example, it is believed that there now exist over a dozen different forms of the Lily of the Valley, differing in size of bloom, in size of plant, and even in time of flowering. So, again, the Hepaticas, which we know in two or three bright forms, have broken into a much greater number. It needs only a small effort of the imagination to know what we can do with such treasures when they are sufficiently increased to be valuable for general garden decoration. Apart from these new forms of old friends, there are many wholly new species being introduced year by year, thus adding to the already superabundant store from which we may draw to make our spring gardens rich in form, colour, and fragrance.

Belvoir spring gardening. — The massing of spring flowers is so admirably done at Belvoir, the seat of the Duke of Rutland, that we are glad to give, as connected with it, the following list, supplied by Mr. Ingram. This list contains the names of the greater number of the plants used in spring either in beds, borders, or rockwork :—

Alpine Auriculas
Anemone fulgens (largely)
  blanda (largely)
  apennina
  Robinsoniana
  nemorosa fl.-pl. (largely)
Arabis albida (an early and compact variety obtained from seed)
Aubrietia græca
  grandiflora (great variety from seed)
  deltoidea variegata
  seedling (light pink)
Antennaria tomentosa
Alyssum saxatile
  invulanum
Arum italicum
Cardamine trifolia
  rotundifolia (very early)
Cheiranthus Cheiri (very early dwarf yellow variety; grown largely)
Caltha palustris
Corydalis cava
  nobilis
Cowslip (Primula macrocalyx; very early species)
Crocus Imperati
  vernus and others
Daisy (double red, white, crimson, and aucubæfolia; largely)
Doronicum austriacum
Erica carnea
Erythronium Dens-canis
Epimedium pinnatum elegans
  macranthum
  grandiflorum
Eranthus hyemalis
Fritillaria
Fumaria solida purpurea
Gentiana verna
  acaulis
Galanthus Imperati
  nivalis
Grape Hyacinth (Muscari)
Helleborus niger
  n. maximus
  atro-ruber
  orientalis
Hyacinth (large number of single bedding varieties)
Heuchera lucida (for foliage)
Hepatica triloba
  angulosa
Iris bicolor
  reticulata
Limnanthes Douglasi
Lunaria biennis (improved high-coloured variety; raised at Belvoir)
Myosotis dissitiflora (largely)
  sylvatica
Narcissus Pseudo-Narcissus
  obvallaris
Narcissus Empress and Emperor
Narcissus minor and minimus
  poeticus (and large collection of others)
Omphalodes verna
  v. alba
Orobus vernus
  lathyroides
Phlox subulata
  frondosa
  Nelsoni
  The Bride (and others)
Polygala Chamæbuxus
Polyanthus Golden Gem and large number of fine yellow varieties from seed
Primula (double white, yellow, and crimson)
  Magenta King and seedling varieties
  cortusoides amœna or Sieboldi
Pulmonaria azurea
  officinalis
Puschkinia scilloides
Ranunculus montanus and amplexicaulis
Sanguinaria canadensis
Saxifraga Burseriana
  oppositifolia
  hypnoides and a collection of the mossy species
  ligulata (largely)
  crassifolia
  cordifolia
Sedum acre aureum
  glaucum
  Lydium
Scilla sibirica
  bifolia
Silene pendula compacta
Snowdrop
Stachys lanata
Scarlet and yellow Van Thol Tulips
Thymus lanuginosus
  golden and silver
Tussilago fragrans
Viola, Russian
  Queen Victoria

Viola Marie Louise
Veratrum nigrum

Shrubs.

Almond (Amygdalis nana)
Azara microphylla
Berberis Aquifolium
Chimonanthus fragrans
Forsythia suspensa
Jasminum nudiflorum
Lonicera fragrantissima
Magnolia conspicua
Pyrus japonica
Rhododendron præcox
dahuricum atro-virens
Nobleanum
ferrugineum

Spring plants require good cultivation. There is a period in the life of a plant when it produces the best results in blossom; old, worn-out plants often bloom late and poorly, ill-grown young plants produce inferior flowers. The great art is to have plants of the right age, full of vigour, and large enough to produce a good effect; then it is requisite to have clusters or masses where a single plant would be ineffective. It takes many years to secure a stock of plants, and a reserve garden is indispensable. Numerous as are the above plants, a great many more of the northern flora will be found good for the spring garden.

AUTUMNAL FLOWERS.

THE summer garden is usually so full of flower that it may be left to take care of itself. Spring flowers are even more numerous. The day will probably come when the beauty of the spring garden will have a reflex in the almost similar beauty of the autumn garden. I speak first of the coming of many autumn blooming plants similar in character to those which are the glory of the spring. Thus, for example, the autumn-flowering species of the Crocus are probably more numerous than the spring-flowering kinds. I speak of species and natural races rather than hybrids. Taken individually, the autumn flowers are also the boldest and strongest, as, for example, Crocus nudiflorus and Crocus speciosus. Up to the present time these have been very little cultivated, and have not as yet broken into garden varieties. Their beauty in the golden days that often come to us in England in October is quite as delightful in its way as any charm of the spring garden or the rock garden. At present nearly all the autumn Crocuses are scarce, with the exception, perhaps, of two or three kinds. Even the kinds that are common are only seen cultivated in the ordinary way as border or nursery plants, and a very proper way, too. To secure, however, little pictures from such plants we must have them happy in the Grass or Moss, or other dwarf plants which will keep their flowers from being splashed, and in carpets where they may be seen on sunny knolls, or banks, or Grassy corners of the lawn or pleasure garden. So placed, in natural groups, they would not be in the way of other flowers, and their coming in blossom would be worth waiting for. Not less important are the wholly distinct Colchicums, sometimes called Autumn Crocuses, which are in flower at the same time, or earlier, than some of the Crocuses. Here the double kind, white as well as rose, play a distinct part, and perhaps handsomest of all is the new and large Meadow Saffron (Colchicum speciosum) drawn some years ago in *The Garden.* This is a noble plant, a tuft of which, welcoming the sun on a bright October morning, is a brilliant sight—its large open cups glowing with a peculiar rosy light, and its bold unopened buds paler beside them. As regards arrangement, the same remarks nearly apply to those that apply to the Croci. As the meadows of much of Central Europe are bright with the flowers of the common Colchicum in autumn, so should our garden turf be with the various precious garden kinds. Thus, from these two races alone, we could add new and distinct charms to our gardens, and that without much expense or troublesome culture, by first securing some stock of the plants either by purchase or by increase in nursery lines, and then by putting them out, considering the positions well, so that their growth and bloom might go on for years without much trouble on our part.

Omitting stray and premature blooms and plants of slight importance, I saw the following in flower in the open air round London in one day in the first week in October: Autumn Crocuses, Tritomas, Sedum spectabile, Lupinus polyphyllus, Phloxes (spring struck plants freshest), Snapdragons, Lilies (three groups—auratum, tigrinum, and speciosum), Asters (China), Starworts, Chrysanthemum lacustre, Pyrethrum uliginosum, Rudbeckia Newmanni, R. sub-tomentosa, Pentstemons, Aconitum autumnale, A. japonicum, Sunflowers (H. multiflorus fl.-pl. and single), Anemone (Japan in var.), Scammony, Helianthus orgyalis, Coreopsis tenuifolia, C. lanceolata, Cyclamen (hardy), Winter Cherry, Gaillardia grandiflora, Phygelius capensis, Liatris tenuifolia, Silphium laciniatum, Prince's Feather, Evening Primrose, Marigolds (various), Tuberose (in flower in open air), Geum coccineum fl.-pl., Veronica noveboracensis, Boltonia decurrens, Desmodium canadense, Scabiosa suaveolens, Yucca, Agapanthus umbellatus, Arnebia echioides, Senecio pulcher, Diplopappus rigidus, early Chrysanthemums.

There is a quiet and peculiar beauty about the more select Starwort, or Michaelmas Daisy, which is charming in the autumn days. The variety of colour, of form, of height, and of bud and blossom is charming, and for the most part it is quite regardless of autumn cold or rains. The plant is most valuable for cutting and for vases, although it flowers so late and is almost essential to all who desire to realise the full beauty of our garden flora. Less showy than the Chrysanthemum, the Starwort is more refined in colour and form, and when examined will be found full of exquisite grace. Where not introduced into the flower garden, it should always be grown for cutting, and some specimens would thrive admirably in a copse or hedgerow. The essential point is to get the distinct kinds, of which the following are among the best that flower in early October: Aster Amellus, bessarabicus, cassubicus, turbinellus, Chapmani, versicolor, pulchellus, cordifolius, Reevesi, discolor, discolor majus, purpuratus, laxus, horizontalis, ericoides, Shorti, multiflorus, dumosus, Curtisi, lævis, longifolius coccineus, longifolius var. Madame Soynuce, sericeus, and fragilis.

Every year adds to our reserve of autumn blooming hardy plants, good selections of which may be made by visiting, in the autumn, gardens containing collections. In some winters, the autumn flowers bloom late and the spring flowers early, so that one may have a winter garden too. It is only where the growths are various, and where tree and shrub life are duly represented, that the outdoor department yields much interest in midwinter. This can only be expected in favoured districts and by the sea-shore.

HARDY BULBS.

OUR "special culture" garden includes beds for hardy bulbs—a very good and pleasant way of growing them. They are, however, so much used in almost every kind of good gardening with hardy plants, that we can only treat of them separately, and to a very limited extent. It will be remembered that each family and kind is dealt with fully in the body of the book, and its culture there given. Hardy bulbs are among the best plants for affording supplies of flowers for the house.

A curious habit of the Narcissi is to get larger, when cut from the plants early, than they would out of doors. This peculiarity is one that may lead us to greater appreciation and enjoyment of the many lovely flowers included among hardy bulbs. Hitherto, the horror of the gardener has been the cutting of flowers for the house; but if it should prove that cutting prolongs his bloom, strengthens his plants, and gives all who care for flowers a fuller enjoyment of them, the way will be made easy. It is very probable that this is the case with many other bulbs, too. Consider what one may escape in the way of storms, frosts, and other dangers if a flower, cut just on arriving at maturity, lasts longer indoors than out, and actually, as in the case of the Narcissi, gets somewhat larger! Narcissi, by their habit and hardiness and drooping heads, seem calculated to stand our climate better than any other flowers, and yet severe storms will beat them about and destroy what might have been happy for days afterwards in the house. But, when we come to large showy flowers like Tulips and other bulbs exposed to the sky, then we have flowers that must suffer with every heavy shower. So, if we find that these two have the same qualities —which I believe they have, we have here an aid in extending the love of these flowers. I left a bloom of that pretty starry white Allium, which is now getting so much into our markets, on the mantel-piece, and forgot it for a week, and I found it—without water, observe—opening its little starry flowers very contentedly at the end of the week. Bulbous plants, some of them, seem to acquire strength enough in the stem to carry them through. In any case, the fact is, as we say, as regards this pretty little A. ciliatum, and no doubt it is true of the various Alliums allied to it, such as the Naples Allium, which is a pretty plant. There is no need for flowers to do without water, but we mention it as showing the nature of such things as compared with others. The whole of the bulbous flowers of the northern and temperate world are pretty closely allied. Anything which makes the presence of flowers in the house more easy to all is a gain to gardening. The exquisite forms are best seen, and tell their story most eloquently when brought near the eye. A flower or two of our common yellow Tulip opening and closing and showing its exquisite form and yellow satiny texture gives one an idea of the grace and beauty of these bulbs which cannot be gleaned by glancing at a bed of them. Another point worth remembering is, that cutting the flowers and bulbs in the early stage strengthens the plant. If the flower fades on the stem, and is allowed to ripen seed, or even begin to form it, it weakens the plant. A variety of hardy bulbs should, in fact, be grown for their value as cut flowers, apart from their great value in the open garden. The use of bulbs in the wild garden is very important, and is discussed under that heading, and also more fully in my book, the "Wild Garden." In grassy or bushy half-wild places the hardier forms—and they are many—are very much more at home than they are on bare borders; moreover, they do not need care after planting.

Hardy bulbs in beds of American plants.—One of the most marked improvements of the last dozen years in gardening is the one effected through handsome bulbs, the heavy, close masses of Rhododendrons, and similar bushes planted in the best and freest soils. Only when in flower do these look interesting, and even then not always so, owing to the flat and monotonous surface into which the shrubs grow. Seeing the great importance of the precious beds, I have suggested that Lilies and the finer bulbs should be placed among the shrubs, and I urged this view on many occasions. In many cases, where it has been extensively adopted, it has almost changed the entire aspect of gardens, and given us additional types of beautiful life where there was but one before. Many fine and rare bulbs may find a home in such places, which among their other advantages are, as a rule, untortured by the spade. In placing choice peat-loving shrubs, plant more open so that the bushes may fully attain their natural forms, and have the interspaces planted with the finer bulbs. Light and shade, relief and grace, are among the merits of this mode of planting, which, intelligently carried out, presents new features of beauty. Small beds and groups of the smaller shrubs will suit admirably for the smaller and more delicate bulbs. Rose beds serve in the same way the shelter of these and low shrubs generally, being an advantage to many bulbs whose leaves are so liable to be cut by cold winds.

Bulbs in large beds.—Not only in beds in the reserve or special culture garden may we have bulbs produce a very fine effect. They are admirable for the lawn, and also for the quiet corners of the pleasure ground. The showy beds of bulbs which are to be seen in our public and other gardens, and which come so largely into our spring gardens, are familiar to all. The kind of beds suggested here are of a higher and more permanent nature, and are to be placed in positions where they may be let alone. In visiting the very interesting gardens at Moulton Grange last year I saw a bed of Tiger Lilies on a piece of quiet Grass with no other flowers near to mar its beauty. The bed was a large oval one, and the colour of the finely-grown Lilies was brilliant and effective seen through the trees and glades. In point of colour alone, indeed, nothing could be better; the plants were about 6 feet high, and told well in the garden landscape, while the mass of bloom was profuse. The plants had greatly the advantage in habit and form over the usual dwarf type in point of colour, which it is claimed is the strong one in the case of bedding plants. Many hardy flowers of the highest beauty have as good qualities as regards colour if we take the same pains with them. Colour on a 6-foot plant must, in all ordinarily varied gardens, be more effective than on a plant 6 inches or 12 inches high. But this is putting the thing in the lowest way, perhaps, for after all flowers will be judged of for other reasons, and however strict our judgment or rigid our selection, the stronger and finer varieties of the Tiger Lily must find a place with us. The bed, it may be remarked, was within a few yards of a walk, and on one of those little bits of turf which occur by most shrubberies, so that it could be easily examined near at hand. It is, perhaps, better so placed, because other plants of varying height and character were not brought near to confuse or weary the

Bed of Tiger Lilies in pleasure ground at Moulton Grange.

eye. There, in a large circular or oval bed, it can get exactly the culture good for it, and should the plant become tired of the spot, removing it to another home and replacing it with some plant of a wholly different character is easy and simple. Among the most lovely beds it is possible to imagine are those of the nobler Lilies in quiet, sheltered spots. The plan admits of splendid variety, too—of the great hardy kinds alone; of the varieties of one species, say of the Panther Lily, grouped together; or of the finer species mixed. Then, another series of good combinations arise from intermingling Lilies with the finer Irises, surfacing and edging the beds with spring flowers. Among Irises, I. reticulata is, I reiterate, one of the loveliest of spring flowers, with its gold and purple, Violet-scented blossoms as brave as Snowdrops.

Bulbs in the improved shrubbery.—Here the opportunities may be large for the use of the hardier and handsomer bulbs. The present overcrowded mode of forming a shrubbery should be given up by all who care for the beauty of shrubs and low trees. The many fine things among our hardy shrubs, and the good climate we enjoy in which to grow them, should make us care more about these. Grow them better, let each form a specimen, or a group, under the conditions best for it, the plants not mutilated, but well grown and furnished to the ground. Under such conditions, they can never touch each other, because, merely to enjoy the beauty of their form and allow each to assume its natural shape, it would be necessary to have spaces between, such as do not occur in what I call the "choke-muddle shrubbery." Those spaces may be alive with bulbous flowers, carefully chosen, hardy, and beautiful. Any wide belt of shrubbery might be made into the most delightful type of garden, varied, broken, and not dug on the edge; not stiff in any part; full of flowering things as well as beautiful Evergreens; the turf spreading in among them here and there, and shade-loving, or shelter-loving, with many kinds of bulbs happy among them, and all the better for their companionship and mutual occupation of the same soil.

ALPINE, BOG, AND WATER PLANTS AND HARDY FERNS.

AMONG the perversities of the human mind there is nothing more sad than the turn it usually takes in the formation of a "rockery." It is generally constructed in such a way that the plants can neither be seen nor grown upon it. Perhaps the difficulty arises from the fact that so few people really notice what occurs in countries where the rock crops out naturally, and where alpine plants grow. The notion exists that the right thing is to form a structure between a bank and wall of scoria, burnt bricks, or any like rubbish. When the material is better, the same idea is carried out; there is no evidence that anybody remembers the fact that alpine flowers, like other flowers, grow for the most part in the Grass or in the level soil, and that few of them care for a dusty hole between two stones in a lowland country. Frequently, the "rockwork" is made without any space for soil, or, if there be soil below, probably there are dry interstices which prevent the roots getting to the soil. In the Alps, we find alpine flowers by thousands growing on the level ground, where they get months of snow and heavy rains in spring; but in our country, where there may be no snow, but, on the contrary, very often a drying time in spring and autumn as well as in summer, people act as if the plants could live on air and dust. When the masses of artificial cement—rock (!)—work are made which are not uncommon in country seats, the "pockets" for soil are far too small, and, where the ledges project, the plants are often starved. We hear, too, of certain salts or other ingredients being used in the formation of these artificial rocks which destroy the plants. If anyone were to take a number of alpine rock plants and plant them in a field of good free soil anywhere, they would grow better there than in the common rock garden, provided always they were not overrun by coarser plants.

We ought to have fewer rockeries, and when we have, we ought to make them with about one-sixth of the material that is generally used, disposing of it in a wholly different way. The rock or stone should crop out of banks or masses of suitable soil, more rock being supposed to be hidden. This, at all events, is the best way for the beginner. This, also, is the true way for beauty to be seen, inasmuch as, so grown, plants can be seen in little colonies or carpets, and their bright colours enjoyed; whereas, stuck in holes between masses of petrified rubbish or stone, the plants are hidden. Five or six stones of millstone grit, or any other suitable stone, half buried in a mound of earth, make a far better "rockery" than many a pretentious affair costing much money and time in its formation. In our public gardens we do not know an example of what we should call a really good rock garden. The plants do well in the Edinburgh Botanic Gardens, but the square kind of pocket there adopted is very inartistic.

Tree-stumps in rock garden.—As regards tree-stumps of any kind, we never saw any good come of their use in connection with the rock garden, while we have seen much evil. The objections to them are their decaying and unstable character. Everything in such a garden should be firm as a rock. Snow, rain, wind, or the settling of the ground should all tend to improve and moss over the surface with flowers, but in the presence of roots nothing of the kind can occur. Then, as regards vermin, roots are very objectionable; they give them the best protection in the many interstices and in the dry-rot which soon takes place. Roots, too, are wrong artistically. Rock gardens are not suited for herbs that can be grown on any border, but for the choicer alpine flowers from the higher mountains, where no tree has a place. If tree roots have any place at all in gardens, it is for wild Vines or Clematises, or some friendly climbing or creeping plants that quickly cover their nakedness; but we are not sure that they have any fitting use at all. Who would place an alpine flower among these roots with any hope of its finding the kind of conditions which it requires? The right way is to place these flowers, like others, in a deep, moist soil. This question ought to be beyond all discussion, for we have already a good example of what to do in more than one garden. If we had not, the grey rocks on the Downs, the groups of rocks in Kent, or Derbyshire, or Westmoreland should teach us. Some of the more recently formed bits in Mr. James Back-

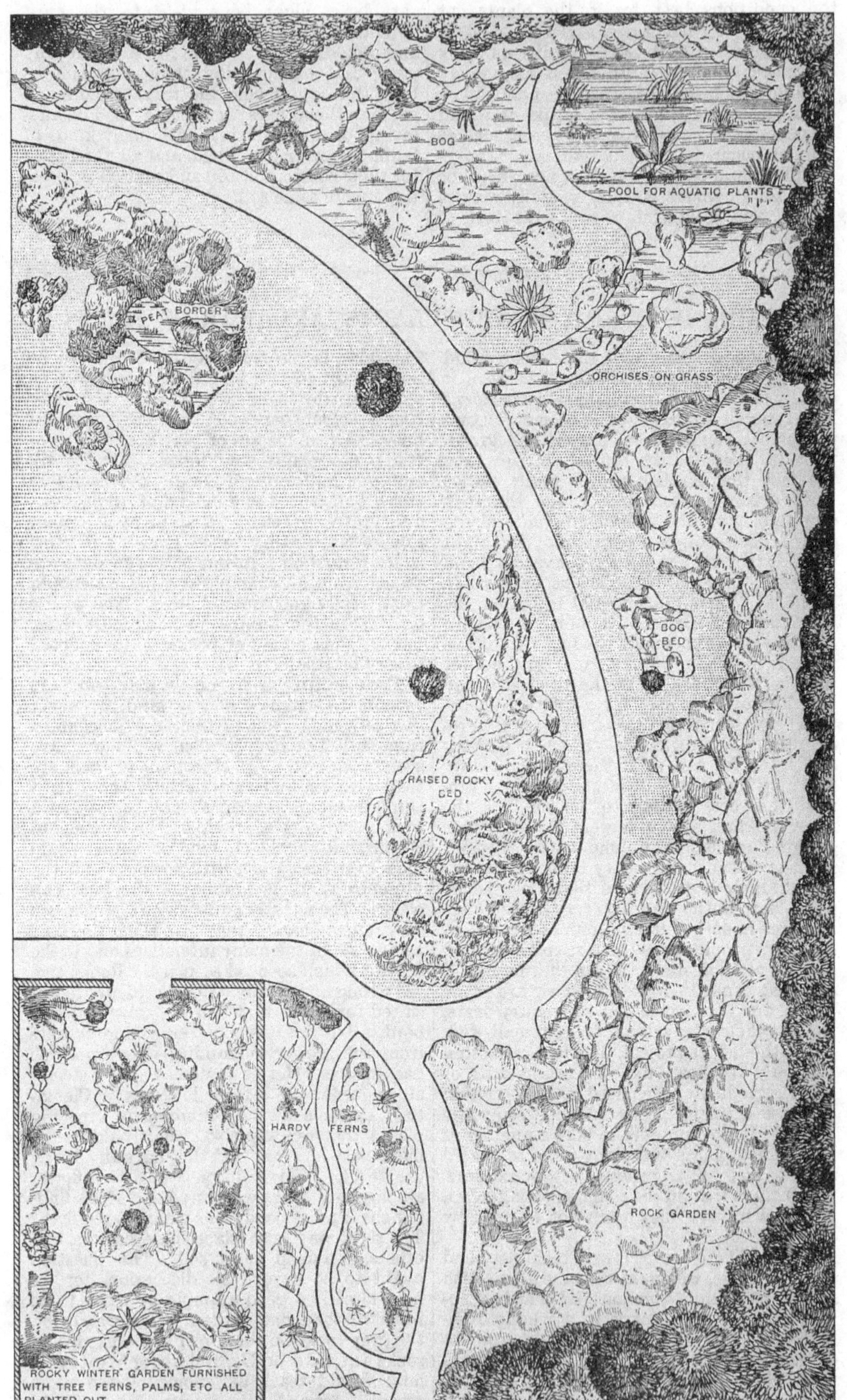

Ground plan of rock garden and its adjuncts at Southwood.

house's garden at York are perfect; Mr. Whitehead's rock garden, quite near London, is also an improvement on the common type, as it accepts the idea of rocky ground rather than the impossible piles described above: still it is too stony in parts.

Grouping and massing alpine flowers.—The rule hitherto has been to place the plants in mixed fashion pretty closely together, and the usual result is, that the strong kinds overrun the dwarf and slow-growing species, which in consequence

Alpine Columbine in rock garden.

soon perish. It is common to see the Wood Strawberry, the Periwinkle, the Tormentil, or still coarser plants, take possession of rock gardens originally planted with choice flowers. But long before fate overtakes the alpine garden, another evil results from the common system, and that is the absence of good or natural arrangement. Dotting each favourite plant all over the place is the way to secure monotony and poverty of effect, as well as the eventual loss of the best plants. In this fashion nothing is gained by having an extensive rock garden, because the same groupings are repeated in all its parts. One never sees anything like the mixture alluded to in the alpine pastures or on the mountain rocks. Often a carpet of one plant charms us there; sometimes the carpet is inlaid with one or two other plants, but generally each bank or mantle of turf, or ledge of rock, has a special character resulting from the individuality of the one or more plants that adorn it. The opposite effect is seen in the garden in which the "general mixture" system gives the same aspect to the whole concern. There is no reason why this should continue. Improvement in artistic treatment, in cultivation, in the preservation of the rarest species and in interest to the observer would follow from the simpler plan of grouping and massing alpine flowers. This may be done with one kind, or with kinds allied in size or character.

If we have, for instance, fifty plants of the common Gentianella (Gentiana acaulis), it is better to make one or more large groups or carpets of them than to scatter them all over the rock garden one by one. Better still would this practice be with the Vernal Gentian, which, being a slower grower and smaller, is more likely to be exterminated by rapid-growing neighbours. Weeds and "interlopers" are seen at a glance in such groups or carpets, and may be promptly dealt with. Good preparation as to soil, stones, position, &c., is more likely to be secured in the case of selecting one or two spots for a favourite plant than in planting it in a score or two different places. Those who know and love such plants may easily acquire the habit of forming groups and masses in a free and natural manner, avoiding every trace of formality, and as a rule only placing one or two kinds on a bank or ledge. In this way, if strong or coarse kinds be placed in the rock garden, the limits of their domain will be more clearly defined than is commonly the case. Often, in consequence of not adopting the natural grouping or massing system, people are tempted to plant Ivies, Periwinkle, Clematis, and like plants, so as to hide the bareness of the ground; but bare ground may be easily covered by following the right system with true alpine flowers. Instead of planting one tuft, the way to cover, and at the same time beautify, the ground is to divide the plant or tuft into as many parts as possible, and thus make the rock garden a nursery, in a sense, as well as a home for the flowers. Some scarce plants will not bear this division, but the majority will. The young roots, emitted by the divided portions, do much more efficient work than the old and broken ones of the single plant or tuft. The chances of success are greater, for if we divide a healthy plant or tuft of Draba aizoides, often planted by itself, we shall have no difficulty in getting a dozen or more young plants, which, carefully "set" among broken stones, even on level borders, will thrive if coarse plants be kept away. Severe winters, excessive wet, or slugs will sometimes destroy a few plants which we do not miss in our carpet or group. Moreover, the bloom is in such cases fairly seen, and its character impressed upon all who see it; whereas a few blooms hidden among numerous other plants may escape even the notice of those who planted them.

Speaking of making the rock garden a nursery bed, it may be added that no rock garden can be successfully and permanently adorned without nursery or reserve beds to fill blanks; it is, however, quite possible by adopting the dividing and grouping system, to do without special nursery beds or with fewer of them. One of the best reasons for wide and natural groups of alpine plants is that they secure distinct aspects of vegetation, as each portion of the garden may be thus easily made to possess a character of its own. Instead of the "general mixture" (or muddle) everywhere, the observer goes from point to point, seeing new aspects of vegetation at every turn. If climbers and vigorous trailers be desired, suitable positions may be selected where they may luxuriate without injury to other subjects. By the adoption of the system here alluded to we may also secure effects of colour which will leave the efforts of the formal "bedding system" behind, as regards brilliancy alone, because we can get more glory of colour without a trace of geometry. The strict limitations of the usual mode of arranging flowers geometrically offend the eye, whereas in the rock garden greater numbers may be grown than anybody could venture to place in formal masses, and the fringes and distant points of the carpets, colonies, or groups fade out of sight in that indefinite way which is so charming in Nature's gardens. It was the wretched effect of the old mixed planting which brought forth the "massing" system, with its original, simple, and telling beds. The great want in our gardens, not only in the rock garden, but in all the departments, is the massing and grouping system, minus geometry.

Rock plants on walls.—The fixing of alpine flowers on old walls, ruins, &c., is one of the most interesting modes of cultivating these plants. They root themselves wherever they get a chance; but when man assists them, then the old wall becomes the most beautiful of gardens. On ruins and old walls near the Alps, particularly in Italy, various alpine flowers and mountain bushes may be seen sown at first by winds or birds. The Colosseum, as everybody knows, was once so rich in flowers, that Dr. Deakin wrote a flora of it; but now these plants have all been cleared off, and the ruin made, in consequence, much less beautiful than when it wore its garlands. On the colossal walls of Caracalla's Baths and many other ruins the Arbutus and the Rock Rose, the Laburnum and the tree Heath, the Fig and the Honeysuckle, the Olive and the Bird Cherry, the Laurustinus and the Ivy, found a home. The Acanthus, model of the column-rearer, carved its own fine leaves far above his capitals; the Maiden-hair Fern came and graced the shady nooks; the Giant Fennel was there in spring with its graceful plume; the Clematis and the Convolvulus, the Cyclamen, the Pink, the Everlasting Pea, the Poet's Narcissus, the Crocus, the Mignonette, the Rosemary, the Violet, the Bee Orchis, and a host of others all grow there, too, just as freely as the Pyrenean Erinus grows on old walls, where it has escaped from cultivation, in Britain. With this fact in mind, it is easy to imagine that the number of plants that could be grown in similar positions is very great. It is not without a useful lesson, apart from that of suggesting how old walls may be embellished. The plants grow on these ruins much better than on the rock gardens so commonly made for their accommodation. And the reason is, that the rock garden, as generally made, is so loose and open in texture that, though the soil may be very suitable, plants perish from drought. On the ruin such moisture as occurs between the stones is prevented from rapidly escaping by the stones of the surface and the firm texture of the whole. Thus, on the ruin we find a garden, and on the so-called rock garden too often nothing but clinkered bricks or bare stones and dust. The best way of establishing plants on walls is by sowing seeds in the chinks and earthy places, the fittest time being early autumn or spring. Small seedling plants may also be inserted carefully in autumn or winter. In parts of the country where Ferns and other plants grow freely on stone walls, good conditions exist for the growth of alpine plants in like places.

Alpine shrubs.—By these are not meant mountain-side or hill shrubs, but the very dwarf woody inhabitants of the higher Alps—microscopic shrubs, indeed, they might be called, for some of them are not over 2 inches high; yet what a world of beauty and interest they present! There are the alpine Andromedas, with numerous white flowers and diminutive stature; the Moss Andromeda, with its Lily of the Valley-like blossoms, borne on a fairy bush of wiry Moss, most delicate and rare; the graceful Linnæa, the small evergreen half-shrubby alpine Speedwells, the very dwarf alpine Willows, the Whortleberry, the Barberries, the little Rhododendrons, the Milkworts, the beautiful dwarf Menziesias, the creeping Azalea, the Partridge-berry, the Mayflower of the woods of North America, the arctic Diapensia, the alpine Daphnes, the small Dogwoods (Cornus), and others, not, perhaps, so decidedly woody as these, but still fitted for association with them. There are, besides, many charming varieties of our own hardy Heaths, attractive from the same point of view, though growing somewhat stronger. All these give materials for the formation of what we may term the "alpine shrubbery." A portion of the rock garden might be devoted to this very interesting group. The plants, for the most part, require pure sand and turfy peat, with plenty of moisture and pure air. Lumps of sandstone or grit-rock on the surface and in the soil will help to prevent evaporation. If the peat gets dry the plants will suffer immediately. Some think that a northern and shaded exposure is necessary to the health of some of the rarer and more fragile alpine shrubs, but this is not so in these islands. It is the drought, not the sun, they suffer from, and with us drought is often severe enough, though it is customary not to take precautions against it; hence, the practice of placing some plants in shade which thrive much better fully exposed to the sun. It is much better to plant so that little or no artificial watering may be required. In the case of these alpine shrubs, by forming a deep, rocky, and well-drained soil, by half burying or inserting pieces of rock in the surface, and, finally, by placing the young plants so that the whole surface where they are planted is covered with their greenery, except the jutting points of rock here and there, it is possible to do without artificial watering except in severe drought. The plants and the surface rocks give that protection to the soil which is given in their native homes by the close turf of the high mountains. To plant such subjects here and there on bare soil, constantly losing its moisture from exposure to the sun, is to run the risk of quickly losing them; though, in favourable soils and moist or elevated districts, some of them will endure even this treatment.

Alpine plants and soil.—Allow me to state that there is a difference between the alpine plants of the primitive and calcareous Alps. Many species prefer the one or the other soil for nutriment; but some cannot be grown except in soil of the one or the other sort. I have tried the experiment by planting forty or fifty plants of the same species in soil containing lime, and the same number in soil containing no lime, side by side. Some of these were found not to thrive, and in a short time to die, while others grew beautifully in soil similar to that of their native place. I therefore came to the conclusion that these plants require their peculiar soil, and will not succeed otherwise. I made this experiment with some plants not very difficult to obtain, or that were sent to me by friends for the purpose. Plants from the primitive rocks are killed in a short time —five or six weeks—in soil containing lime, which is poison for them; but limestone plants are killed only by hunger. It seems that there are many more plants peculiar to the primitive rocks than to limestone. I have also observed that it is of great importance whether the soil is sandy or clayey. Bog plants, such as Drosera and Andromeda polifolia, must be grown without lime. A few species changed their appearance with the change of soil. Some which I raised from seed on alien soil I supposed at first not to have come from seed correctly named. Rhododendron hirsutum without lime appeared to be R. intermedium, Hieracium villosum came like H. alpinum, Papaver Burseri changed to innumerable colours, Leontopodium alpinum changed the colour of the flowers from white to green. It always flowers green unless freshly-powdered limestone sand is frequently scattered over it.

As regards soil, alpine plants may be divided into three sections: 1, chalk-loving plants; 2, plants to which chalk is poison; 3, plants that will succeed in any soil or situation—a very numerous class.—H. Gusmus.

The importance of these views about soil is open to question, though they state the facts as regards the plants in their native country. There are, however, various reasons why a plant should be found on a certain soil in a certain district in its native country, and yet in another country do perfectly well on the soil to which it was supposed to be averse. Many of the plants here described as disliking lime soil may be seen by anybody who observes much in collections thriving perfectly well in the soil they are supposed to dislike. I admit there are certain limitations, and that limy soil is offensive to certain plants, but they are few in number; and there could be no greater mistake than to add a further complication to the culture of alpine flowers. For ages these lovely plants have been kept out of our gardens through an absurd idea that their culture was impossible. When one saw a few alpine flowers they were mostly coddled in a frame, and when a rock garden was made, it was so made that nobody ever saw a true alpine plant upon it. Eventually, Nettles or Briers covered its disgrace.

It is only now that people are beginning to get the true idea that a great number of alpine plants are grown as easily as any other plant, if they escape insect enemies, and are not shaded by coarser plants, or kept dry at the root. Let it not be forgotten that our gardens abound in browsing slugs, wireworms, and other destructive little beasts which are often absent from the high Alps and from poor, heathy places. If anyone really has evidence that a plant will not do with a limy soil, let him say so by all means, but let it not be said because the plant is found on a limy patch or limy district abroad. Let each plant that we suspect to dislike a particular soil be tried—several examples of it—in that soil, and give it a fair chance in other respects, taking care to ascertain that its death, if it should occur, is owing to soil alone. My own experience is, that an enormous majority of alpine plants thrive with the greatest freedom in what we call ordinary soil if fully exposed to the sun and rain, and not over-crowded by coarser plants. Those desiring to go into the culture of alpine flowers will find the subject more fully discussed in "Alpine Flowers for English Gardens" (London: J. Murray).

Hardy Heaths.—These plants may be fittingly spoken of in connection with the rock garden, though they by no means require rock gardens for their culture. A considerable variety of hardy Heaths are now in cultivation, and mixed clumps of them may be arranged so as to have some in flower all the year round or clumps may be formed containing one variety only, which would be preferable. The double-flowering Ling (Calluna vulgaris), which used to form long and much admired edgings, is rarely now to be met with. Numerous beautiful varieties of the single-flowered Calluna vulgaris are admirably adapted for bedding purposes, as are also the varieties of the Cross-leaved Heath (Erica tetralix) and the Bell Heath (Erica cinerea). In addition to these we have the varieties known under the names of Erica ciliaris, E. Mackayana, E. Watsoni, also E. carnea, E. vagans, and E. multiflora. The varieties of the Irish Heath, of the E. mediterranea breed, are also numerous, and flower at various periods. All the hardy Heaths are readily propagated by means of cuttings or layers. Cuttings, which I think make by far the best shaped plants, are readily struck if placed any time during the autumn months in a sand bath in a cold pit, provided they are not exposed to the sun. In such a situation they soon root and start into growth. When rooted and hardened they could be planted out in lines, but I prefer putting them round the edge of pots and placing them in a cold frame partially shaded for a time from the sun. As soon as established the points of the top shoots should all be cut off, when side branches will be freely produced, which will give them a compact habit of growth. After being some months established in pots and sufficiently hardened, they can then be planted out in rows in open-air beds, leaving about 6 inches or 7 inches between plant and plant. When the young shoots have pushed about a couple of inches, cut the tops off again during the following spring with a pair of spring or sheep shears. Thus treated, such plants will soon be ready to place in the bedding compartments intended for their flowering.

In nurseries hardy Heaths are generally propagated by means of layers, an operation easily accomplished, by placing some peat freely mixed with sand in the middle of the plants, which, when rooted, are torn in pieces and planted out in nursery rows. Many species do well in this way, but they flower better when raised from cuttings, besides being better shaped. The single stem of the cutting-made plants seems to favour the shape and the flowering, while layers in many cases have several stems, and scarcely two plants are of exactly the same shape, while with cuttings all the plants are more uniform. If the autumn flowering hardy Heaths are regularly clipped over in spring, they will form nice round patches and produce flowers abundantly, while the spring or summer flowering varieties should always be cut over as soon as possible after the flowering is past. If this is done they will branch out freely and regularly and flower abundantly the following season. When the plants become old they are apt to assume a tall and wiry condition and to flower imperfectly. It is therefore necessary that young plants should always be coming on to take their place. It is not absolutely necessary that hardy Heaths should be grown in peat soil. A mixture of peat and loam, or leaf-mould and sand, will suit them. The above was the late Mr. McNab's practice with the hardy Heaths; he said that the compact beauty of the Heaths on the Scotch hills was partly owing to their being eaten down every year by sheep.

Water and bog plants.—The plants that grow by the waterside, so much admired in natural scenery, are seldom turned to so much advantage in cultivation as they might be; otherwise the bare water edges so often found in connection with lakes and other ornamental water would be less frequent. With the many suitable plants at our service, if appropriately employed, the margins of artificial water might be made to surpass even the examples of natural riverside vegetation. In the majority of cases, if the edges of artificial water are clothed at all, they have a monotonous appearance on account of the continuous fringes of plants of a commonplace type used; whereas if a greater variety of kinds of varied height, habit, and flower were employed and disposed in bold, irregular groups—some close to the margins, others at a distance from them, and some even partly submerged—good effects would be obtained. The principal consideration is a knowledge of the positions in which the plants thrive best, and the degree of moisture in which they will flourish. The grouping of them effectively is easily accomplished. The following enumeration consists of vigorous growing plants that when once planted will take care of themselves. Our native flora supplies numbers of handsome waterside plants in no way inferior to exotic kinds. Amongst the showiest are the

Willow Herbs (Epilobium).—Of these, E. angustifolium is the finest. In rich moist soil it grows 5 feet or 6 feet high, and in summer has many showy purple-red flowers. There is a white variety, a colour uncommon among water plants. The great Willow Herb, or Codlins and Cream (E. hirsutum), is a true water plant, though not so showy as the preceding. The Purple Loosestrife (Lythrum Salicaria) may be said to be the finest of water-side flowers. The Epilobiums are excellent companions for our native Meadow Sweets (Spiræa). Many of the Grasses of the larger type flourish better in moist places near water than elsewhere. One of the handsomest is the Great Reed Grass (Arundo Donax), which grows 10 feet and even 15 feet in height when planted near the margins of water where the roots are continually moist. It requires, however, a rather sheltered position, as it is apt to be injured by severe cold or cutting winds. It requires a warm soil and district to thrive.

The New Zealand Reed Grass (Arundo conspicua), as well as the Pampas Grass (Gynerium argenteum), flourish by water better than in other positions, provided there is not an excessive amount of stagnant moisture about the roots. One or two kinds of Lyme Grass (Elymus) are excellent for planting in wet places where choicer plants would not flourish, the most suitable being E. giganteus, which grows some 4 feet or 5 feet high; E. virginicus and canadensis, both North American species of tall vigorous growth. Some of our bold British Grasses look well if planted in distinct groups, and not allowed to run in a monotonous fringe. The best of these are the common Reed (Arundo Phragmites), which abounds in many parts in wet ditches; the Wood Small Reed (Calamagrostis Epegeios), which grows from 3 feet to 4 feet high and flourishes as well in open wet places as in woods and thickets; Purple Small Reed (C. lanceolata), taller than the last; the Reed Grass (Digraphis arundinacea), from 3 feet to 5 feet, with broad leaves and handsome plumes from 6 inches to 8 inches in length—all interesting when properly planted. There is a variety of this with variegated leaves called the Ribbon Grass or Gardener's Garters.

Bamboos.—There is no other type of hardy plants from which such beautiful effects can be produced by water margins as the various kinds of Bamboos which thrive in our climate. Planted by the side of a running stream, or near the margin of a lake or pool, they succeed, and soon attain a great height. Among the hardiest are Arundinaria falcata, Bambusa arundinacea, Metake, viridis glaucescens, nigra, and Phyllostachys bambusoides. Sedges and Rushes are essentially water plants, and many of them give good effects when planted in bold groups. For this purpose, some of the finest and most suitable are, among Carexes, C. paniculata, a native species, which grows into luxuriant tufts as high as 4 feet if planted in wet boggy places in which little else will grow. Then there is the extremely graceful C. pendula, one of the largest of our native Carexes, with its long catkin-like spikelets, produced in early summer on plants 3 feet high. The Fox Carex, as well as C. acuta, are likewise well adapted for wet places, each attaining 2 feet or 3 feet in height, and of C. acuta there is a handsome variety with variegated foliage. There is also a variegated-leaved variety of C. riparia, which is very handsome and retains its character well, even in water.

One of our handsomest native water-loving plants is the Galingale (Cyperus longus), whose stout stems, terminated by singular tufts of leaves, attain a height of even 4 feet or 5 feet. As it flourishes best when its lower part is wholly submerged, it is a capital subject for planting in shallow water at a little distance from the margin. When disposed in bold groups, and these not repeated too often, it greatly relieves the somewhat monotonous appearance of an even fringe along the water's edge. Another fine Cyperus is vegetus, which has wider leaves than the last and lighter green in colour, but it does not grow so tall. Nearly allied to the Cyperuses are the

Club Rushes (Scirpus).—S. triqueter (3 feet high), S. lacustris (from 4 feet to 8 feet high), and S. Holoschœnus, a stiff Rush-like plant (some 3 feet high), are all excellent water-side plants. Of similar growth is the Prickly Twig Rush (Caladium Mariscus), which is useful for planting in poor and wet soil where little else would thrive.

Irises.—In addition to the common yellow Flags (I. Pseudacorus and fœtidissima) several of the other kinds make good water plants, particularly I. sibirica, a tall-growing kind with glossy foliage and flowers either of a rich purple or white. The beautiful Kæmpfer's Iris, too, though not of a large size, must be included in our list, as it flourishes best in wet places, and if such a position could be allotted to it where the water now and then could be made to flow over the soil for 1 inch or so in depth, it would, if planted in a peaty soil, flourish far better than in an ordinary border. Among plants remarkable for fine foliage few excel the large Water Dock (Rumex Hydrolapathum), the leaves of which grow nearly 3 feet long and nearly 1 foot across, reminding one of a Banana plant in miniature. In some situations it grows as much as 5 feet high, and forms a bold plant close to the water's edge, where the roots would be continually submerged.

The Great Spearwort (Ranunculus Lingua) is another of our bold foliage native plants which grows from 3 feet to 4 feet high, and has long broad leaves of a pale green colour. Its flowers are showy, being of a bright shiny yellow, and more than 1 inch across. A position similar to that recommended for the last suits it best. A flowering branch of this is well shown in our woodcut. The yellow Pond Lily (Nuphar advena), a plant with large, broad, deep green leaves, is one of the noblest of hardy aquatics, and the only kind that sends its leaves erect out of the water to as great a height as 3 feet. The bases of the plants should be submerged to about 1 foot or 1½ feet in depth. They should be planted in bold groups a little way from the margin, and surrounded by Water Lilies and other aquatics with floating leaves. The Butterbur (Tussilago Petasites) is a noble plant when at its largest growth. How pleasing a little colony of it looks by the banks of a stream, where it delights to spread! The Burdocks, too (Arctium), though they naturally affect poor, dry soils, attain enormous dimensions by the side of the water, but they must not be planted so near that their roots are submerged. These plants, however, are best fitted for wild, rough places.

The Sweet Flag (Acorus Calamus).—This is a Reed-like plant growing some 3 feet or more in height. It is very vigorous, and soon spreads itself over a wide area, and will overrun plants of weaker growth if not checked. It is, moreover, a handsome plant, and its highly aromatic leaves make it very desirable. The Great Bulrushes or Cat's-tails (Typha latifolia), which in autumn are furnished with black, club-like flower-spikes, though abundant in many parts of the country, should always be planted where not indigenous, as they are so distinct in aspect from most water plants. T. stenophylla and T. minima are alike graceful plants, growing in tall dense tufts.

Pontederias, of which there are three species, are about 3 feet high. They have arrow-shaped leaves and blue flowers of various tints, produced on stout stalks well above the foliage. The three kinds require to be put in 1 foot or so of water, and are therefore well adapted for planting a little way from the margin. Another noble plant which, unfortunately, is not quite hardy is Thalia dealbata, a Maranta-like plant from South Carolina, growing some 6 feet in height, with large, handsome leaves of a glaucous green hue. The Flowering Rush (Butomus umbellatus), one of our native plants, should adorn the margins of every piece of ornamental water, as it is not only an elegant plant as regards foliage, but its blossoms, which are produced in large umbels, are rosy tinted and beautiful. The Water Plantain (Alisma Plantago) is a bold plant, which often attains 3 feet in height. It grows in watery ditches and edges of streams; the leaves are broad, similar to those of the Great Spearwort.

Caladium virginicum is a bold plant, having large, broad leaves, arrow-shaped and of a deep green. It is excellent for planting in shallow streams or pools, in about 6 inches of water. It rises 2 feet or 3 feet in height, with a habit similar to the Callas or Richardias, which should on no account be omitted. Other highly ornamental plants are the Giant Horsetail (Equisetum Telmateia). This is an extremely fine plant when fully grown, and one which attains several feet in height in moist, shady places, producing graceful plumes of pendulous, thread-like branches in drooping whorls of a cheerful green colour. It is by far the finest of all the Horsetails, but is seldom seen in full growth.

There are many other plants, which though not strictly aquatic, flourish well near water, and have a fine effect, as, for example, the Flame Flowers (Kniphofia). Other plants may be similarly grown, such as the Giant Knotweeds (Polygonum cuspidatum and sachalinense), Astilbe rivularis, Senecio japonicus, North American Lilies, several of the larger Spiræas, Trollius, the Royal Fern (Osmunda regalis), Lysimachias, and many others.

The water plants proper have a beauty specially their own. We mean the various Water Lilies and the plants allied to them, and they are among the most pleasant objects in our natural and artificial waters when not allowed to become too densely matted together. When this occurs no good effect can be expected afterwards. Every kind of Water Lily that is hardy in this country is worth obtaining and growing. The aim should be to so plant them that their effects would be good, and their power of spreading limited. It is possible to do this in artificial waters by confining the rich soil to certain places. It is also desirable, perhaps, to avoid the common British yellow Water Lily, because it is a very vigorous plant, and spreads readily over the muddy bottom of a lake: it is extremely difficult to eradicate.

Artificial bogs.—The artificial bog has not yet received its due share of attention; but I am convinced the day is at hand when it will become as common, and as much esteemed, as the hardy fernery, and will be considered quite as indispensable an adjunct to a perfect garden. Bog plants have many charms of their own, and are so easily managed and so different in aspect to the ordinary class of garden plants, that they cannot fail to please; all that is requisite to form a bog garden is to form a hollow space which will contain water. The simplest way is to buy a large earthenware pan or a wooden tub, bury it 6 inches beneath the surface of the ground, fill it full of broken bricks and stones and water and cover with good peat soil; the margin may be surrounded with clinkers or tiles at discretion, so as to resemble a small bed. In this bed, with occasional watering, all strong-growing bog plants will flourish to perfection; such plants as Osmundas and other Ferns, the Carexes, Cyperuses, &c., will grow to a large size and make a fine display, while the cause of their vigour will not be apparent.

A more perfect bog garden is made by forming a basin of brickwork and Portland cement, about one foot in depth; the bottom may be either concreted or paved with tiles or slates laid in cement, and the whole must be made watertight; an orifice should be made somewhere in the side, at the height of 6 inches, to carry off the surplus water, and another in the bottom at the lowest point, provided with a cork, or, better still, with a brass plug valve to close it. Five or six inches of large stones, bricks, &c., are first laid in, and the whole is filled to the top with good peat soil, the surface being raised into uneven banks and hillocks, with large pieces of clinker or stone imbedded in it, so as to afford drier and wetter spots; the size and form of this garden or bed may be varied at discretion. An oval or circular bed from 5 feet to 6 feet in diameter would look well on a lawn or in any wayside spot, or an irregularly formed corner may be rendered interesting in this way; but it should be in an open and exposed situation; the back may be raised with a rockwork of stones or clinkers imbedded in peat, and the moisture ascending by capillary action will make the position a charming one for Ferns and numberless other peat-loving plants. During the summer the bed should always contain 6 inches of water, but in winter it may be allowed to escape by the bottom plug. It is in every way desirable that a small trickle of water should constantly flow through the bog; ten or twelve gallons per diem will be quite sufficient, but where this cannot be arranged it may be kept filled by hand. The sides of such a bog may be bordered by a very low wall of flints or clinkers, built with mortar, diluted with half its bulk of road-sand and leaf-mould, and with a little earth on the top; the moisture will soon cause this to be covered with Moss, and Ferns and wall plants of all kinds will thrive on it.

Where space will permit, a much larger area may be converted into bog and rockwork intermingled, the surface being raised or depressed at various parts, so as to afford stations for more or less moisture-loving plants. Large stones should be freely used on the surface, so as to form mossy stepping-stones; and many plants will thrive better in the chinks formed by two adjacent stones than on the surface of the peat. In covering such a large area it is not necessary to render the whole of it watertight. A channel of water about 6 inches deep, with drain pipes and bricks at the bottom, may be led to and fro or branched over the surface, the bends or branches being about 3 feet apart. The whole, when covered with peat, will form an admirable bog, the spaces between the channels forming drier portions, in which various plants will thrive vigorously.

Perhaps the best situation of all for a bog is on the side of a hill or on sloping ground. In this case the water flows in at the top, and the surface, whatever its form or inclination, must be rendered watertight with Portland cement or concrete. Contour or level lines should be then traced on the whole surface at distances of about 3 feet, and a ridge, of two bricks in height, should be cemented on the surface along each of the horizontal lines. These ridges, which must be perfectly level, serve to hold the water, the surplus escaping over the top to the next lower level. Two-inch drain tiles, covered with coarse stones, should be laid along each ridge to keep the channel open, and a foot of peat thrown over the whole. Before adding the peat, ridges or knolls of rockwork may be built on the surface, the stones being built together with peat in the interstices. These ridges need not follow the horizontal lines. The positions thus formed are adapted both to grow and to display Ferns and alpine plants to advantage.

In addition to regular bog plants, almost all the choice alpines will luxuriate and thrive better in the drier and more elevated parts than in an ordinary border or in pots. Perhaps the most charming plants to commence with are our own native bog plants—Pinguicula, Drosera, Parnassia, Menyanthes, Viola palustris, Anagallis tenella, Narthecium, Osmunda, Lastrea Oreopteris, Thelypteris spinulosa, and other Ferns; Sibthorpia europæa, Linnæa borealis, Primula farinosa, Campanula hederacea, Chrysosplenium alternifolium and oppositifolium; Saxifraga Hirculus, aizoides, stellaris, &c.; Mimulus luteus, Cyperuses, Carexes, Calthas, Luzulas, Cardamine, Leucojum, Fritillaries, Marsh Orchises, Equisetums. These, and a host of plants from our marshes and from the summits of our higher mountains, will flourish as freely as in their native habitats, and may all be grown in a few square feet of bog; while Rhododendrons, Kalmias, Gunnera scabra, the larger Grasses, Ferns, Carexes, &c., will serve for the bolder features.

I have not space to enumerate the many foreign bog plants of exquisite beauty which abound, and which may be obtained from our nurseries, although many of the best are not yet introduced into this country. In fact, one of the great charms of the bog garden is that everything thrives and multiplies in it, and nothing ever droops or dies, the real difficulty being to prevent the stronger plants from overgrowing, and eventually destroying, the weaker ones. I need scarcely add that a small pool of water filled with water plants forms a charming adjunct to the bog garden. The only caution necessary is to destroy the strong weeds before they gain strength—a single plant of Sheep Rot (Hydrocotyle), for example, would smother and ruin the entire bog in a season.

An irrigated wall.—There is another way in which a minute stream of water may be turned to advantage, and that is by causing it to irrigate the top of a low wall; such a wall should be built 12 inches high, the top course being carefully laid in Portland cement. A course is then formed by bricks projecting over about 2 inches at each side, with a channel left between them along the centre of the wall, which must be carefully cemented. Small drain pipes are laid along this channel and filled in with stones. Large blocks of burr or clinker are then built across the top of the wall, with intervals of 12 inches or 15 inches between them, and these are connected by narrow walls of clinker on each side, so as to form pockets, which are filled with a mixture of peat and sandy loam. The projecting masses of burr stand boldly above the general surface, and, occurring at regular intervals, give a castellated character to the wall, which may be about 2 feet high when finished. Hundreds of elegant wall plants find a choice situation in the pockets, which are kept constantly moist by the percolation of the water beneath them, while Sempervivums and Sedums clothe the projecting burrs. In fact, such a wall forms a garden of blossoms throughout the whole spring and summer.

I cannot too highly commend these low and broad walls to the attention of landscape gardeners, even where irrigation is not attainable. In such case the wall should be built hollow, be 18 inches or 2 feet wide, with bonding bricks at intervals, the interior being filled with old potting soil heaped up high on the top, so as to cover all the wall except the battlement projections. Wallflowers, Sedums of all kinds, Saxifrages, Pinks, Rock Roses, Snapdragons, small Campanulas, Alyssums, and a host of similar gay plants should cover the wall with their greenery and hang down in graceful masses, and as they flower in succession, they keep up a continual interest and beauty. Such very low walls have a nice appearance at the edge of a slope or beside a flight of steps or path, or wherever the object is to mark a separation without forming a boundary.

Wherever a wall can be built against a bank of earth, a charming and striking effect may easily be obtained by building the wall with ordinary mortar, mixed with an equal quantity of good peaty potting mould. In about a month the lime has lost its caustic quality, and it is ready to be planted; it is better, where practicable, to give such a wall a slight slope, and this may, if preferred, be done by setting each course of brick or stone back half an inch, so as to form ledges for the lodgment of seedlings. So far as I have tried, every hardy plant will luxuriate in such a situation; but, by selecting such plants as Saxifrages, Sedums, Aubrietias, Erinus, and others of similar habit, which cover the wall with hanging masses of flowers, an effect is produced which is the admiration of everyone who beholds it, whether fond of flowers or not.—LATIMER CLARK, *Sydenham Hill.*

The garden fernery.—The marriage of the fernery to the flower garden is a very desirable innovation, the many hardy evergreen Ferns we possess being excellent for association with hardy flowers. There are many varieties of our native Polystichums, Hart's-tongues, Blechnums which would be excellent as companions to the evergreen herbaceous plants suited for sheltered, half-shady nooks. There are also many exotic kinds hardy and vigorous. Graceful and new effects may be developed in foregrounds by drives through glades, and in many other positions, by the bold use of hardy Ferns of the larger kind. The Bracken we see everywhere; but some of the others are more graceful in form, and delight in the partial shade of open woods and drives, and do even in the sun. Up to the present time Ferns have, as a rule, been stowed away in obscure corners, and never come into the garden landscape at all. But not only can they give us new and beautiful aspects of vegetation in the garden landscape, but even in parks and ornamental woods of the largest class. The bolder kinds should be selected and multiplied, and the fitness of the position to the Fern must be considered. Very much more may be done with Ferns in the garden than has yet been attempted. Ferns, although their admirers are numerous, have been, and are still, considered by the majority of cultivators merely as plants suitable for planting in and about artificially made mounds of earth, consisting in many cases of the worst possible description of soil, covered with what is called rockwork, and often placed in incongruous positions, where their presence is certainly the reverse of ornamental. Surely, therefore, such exquisitely beautiful plants deserve a better fate; under such circumstances all the charm of contrast which they present when planted in combination with fine-leaved and flowering plants is lost. Such an arrangement is not a natural one, although too

Rock fernery at Danesbury.

popular. People are now, however, beginning to recognise the fact that a garden filled with hardy plants, properly arranged and tended, is much more interesting than one filled with exotics, however gay may be their colours; and, therefore, in the gardens of the future hardy Ferns here and there, in groups and colonies, may be expected to play a leading part. When tastefully made, a good hardy fernery is a pleasant feature. In the home counties there is probably not a better specimen than that of which we give an illustration from the fernery at Danesbury. It is formed on a sloping bank in a rather deep, dell-like valley, overhung with trees and Ivy, in the shade of which the Ferns seem to delight. This charming spot has been further improved by some rockwork artistically made. As regards the planting, the various genera are arranged in distinct and well-defined groups, and each group is assigned a position and provided with soil adapted to its requirements; therefore, all have an equal chance of becoming well developed. "Ah!" says some one, "but these Ferns are mostly indigenous, and therefore do not require any cultural care; simply stick them in the ground, give them one heavy watering, and then let them take care of themselves." Yes; that is how many hardy plants are treated; but not at Danesbury. It should be remembered that the conditions under which such plants are placed are more or less artificial. Ferns in their natural state have, as a rule, both soil and locality exactly suited to their requirements; furthermore, the soil is yearly enriched by the decaying foliage of surrounding trees, which foliage, moreover, forms an invaluable protection to them in winter. Therefore, in order to ensure the best results in hardy Fern culture, as well as in that of other hardy plants when under cultivation, care must be bestowed on them, and that is what is being done at Danesbury. In arranging a fernery, study thoroughly beforehand the habits and requirements of all the species, and allot each such a position as is most likely to produce the best results. At Danesbury the most sheltered, moist spot is given to such plants as the various varieties of the evergreen Blechnums, which delight in a close, damp atmosphere, and the tender forms of Asplenium. Osmunda regalis, which thrives amazingly, is allotted a low, swampy position, which is, however, free from stagnant moisture. The soil used for these Royal Ferns is a mixture of good loam and fibrous peat. The better deciduous kinds of Polypodium, such as P. Phegopteris and P. Dryopteris, are also afforded sheltered positions, and in quiet nooks may be found charming groups of such things as Allosorus crispus, the Parsley Fern, and Cystopteris fragilis, a most delicate and graceful Fern. Lastrea Filix-mas and its varieties occupy the bolder and more exposed positions in company with fine colonies of the evergreen kinds, comprising some unique varieties of the Polystichums, Scolopendriums, Polypodiums, &c. A plentiful supply of water is available for use when requisite in the fernery, and by means of a hose attached to a hydrant abundant soakings of water can be given during long-continued drought—a matter of much importance, seeing that moisture is such an essential element in the well-being of all Ferns. The Ferns here grow superbly, the Royal Ferns really justifying the term.

Carpets for Ferns.—The beauty of every fernery is much enhanced by having the larger kinds of Ferns growing out of some plant of dwarfer habit. The Ferns themselves are much benefited by this plan because there is not an excessive evaporation constantly going on during dry and hot weather. The small Ferns are best without any carpet. They are the choicest, rarest, and most difficult to grow, and as they require a little extra attention, it would be better not to run the risk of their being smothered. As the most appropriate carpet for a fernery is a Fern, I place first on the list the

OAK FERN (Polypodium Dryopteris). It suits the tall Lastreas, Polystichums, Athyriums, Osmundas, Onoclea sensibilis, and Struthiopteris germanica.

SELAGINELLA DENTICULATA.—If I were compelled to adhere to one carpet only, this would be my choice, being well adapted to a moist fernery, and rooting as it runs over the surface. There is an element of tenderness in its constitution, and a severe winter will clear off great patches of it, but some is sure to remain to replace quickly that which was lost. It is a very shallow rooting plant, and looks especially fresh and green during the moist atmosphere of autumn and winter. It loves shade; sunshine soon shrivels it up. It will suit Ferns of large or medium size. Scolopendriums look very well in it.

THE WOOD ANEMONE (Anemone nemorosa) is a glorious carpet in spring when the deciduous Ferns are at rest. It loves shady places.

SAXIFRAGA SARMENTOSA.—The creeping Saxifrage, or our old friend the "Wandering Jew," is well adapted for shady places, where the leaves assume a dark green colour, and the markings become more strongly defined than in full exposure to light. It answers well in rough ferneries where there is plenty of the perpendicular over which it can droop, and in the crevices of which it can root itself. Perfectly hardy; habit that of the Strawberry. I planted it out many years ago, and it has taken care of itself ever since. The foregoing plants have been my groundwork for Ferns for many years past. There are many other plants which would answer equally well. I am sure Campanula hederacea would.—E. JACKSON.

THE SPECIAL CULTURE OF CHOICE AND "FLORISTS'" HARDY FLOWERS.

Muddling things up.—Nothing is more unfortunate in gardens than the way plants of all kinds are huddled together without any relation to their fitness for association in stature, in character, in time of blooming, or needs of culture. This does not show badly in what is called "bedding out," but the moment people begin to deal with the variety of life which may be included under the head of hardy flowers and shrubs, then one sees the want of a greater pleasure in plants and a greater knowledge of them. Hardy plants fall into a variety of groups requiring different treatment and different exposures if we are to have a good result. The common scene of confusion is the shrubbery border, under the wings of which Pinks, Carnations, annuals, alpine flowers, Pæonies are often thrown, there to dwindle and finally to perish. There is no shrubbery border that could not be made more beautiful than any now seen by carpeting it with suitable wood and copse plants of the northern world in broad masses and groups. Thus, one could see and enjoy one thing at a time, and more so, if fittingly placed as regards the shrubs about it or near it. But many of our favourite flowers are not wood plants, and many—for example, Carnations—are not plants that will maintain the struggle with the bushes and trees. Another series of vigorous perennials require isolation if we want their fullest beauty, say the Lilies, and Irises, and Pæonies. The Lilies and Irises will certainly do among bushes if the bushes are well chosen, the best being those called American, which are not coarse feeders, and which thrive best in peat or leafy soil. Their chances in the common shrubbery with its coarse rooting plants and coarse growing bushes is very different. Of all causes which tend to make the garden unsatisfactory, this inconsiderate placing of many things along the fringes of the garden, grove, or shrubbery is the chief. People should think more of the conditions that plants enjoy best, and *divide their hardy collections into two broad series at least—those which will thrive in and near woody growth, and those which can only perish in such places.* The Solomon's Seal and the blue Apennine Anemone are types of plants that one may grow in any shady place, and the Clove, Carnations, Pink, Auricula, represent the large series which must be grown in special beds, away from tree roots, if they are to develop qualities which make them worth having. Among hardy flowers there are, in fact, two sets—one which requires care and good culture, and plenty of sun and air; another which, once started in a fitting place, may be let alone, barring keeping grass weeds out of their way.

One good and certain way.—Aspirations are common enough for a more natural style, but they often end in nothing for want of knowledge. Many seek a more varied garden flora, and many go back to the culture of old favourites pushed out of the way for years, but the result is not always satisfactory. Sometimes one sees a garden where an effort has been made—alpine flowers and coarse herbs mixed up together—a combination hopeless from every point of view. The plants have been badly selected, and there has not been the slightest knowledge shown as to where to place them. Given the best selection possible, nothing can be done unless both the individual habit and peculiarity of the different classes of plants are known. There is at least one plan that all can follow with satisfaction, and it is the growing of various lovely races of plants for their own sakes, without reference to their place in any arrangement or design, and yet not in any kind of "mixed border" or other mixed arrangement.

The common way is to put almost every choice plant in what is called the mixed border, and placing it there very often means losing it. If we are to succeed with our finely bred races of hardy flowers, it cannot be in the mixed border. Each important family of hardy flowers is worthy of special culture, and without it no good result can be obtained. Whether we grow Carnations, Pinks, Pansies, Phloxes, or Lilies, Stocks, double Wallflowers, Cloves, or tall scarlet Lobelias, in every case they ought to have separate attention in fresh soil if we are to reap the full harvest of their beauty. Even an annual, such as the Rhodanthe, or a beautiful ornamental Grass, is not easy to succeed with unless the plant has a fair chance, away from the confusion and weariness of the ordinary flower border.

This special culture for favourite flowers is possible in two ways at least, either in the beds of the flower garden or on the lawn, or in a plot of ground which ought to be set aside for nursery beds of the choicer flowers. Such plants as Carnations, Cloves, Stocks, and Pansies last long in bloom, and may be introduced with good effect in almost any position, except, perhaps, into a set pattern of carpet beds. It is not that they always want the ground wholly to themselves; they may often be grouped with other plants. For instance, Carnations may go with a thin group of standard Roses, or may be interspersed with late-blooming Gladioli rising thinly over the carpet of grey Carnation leaves. So, again, Pansies are admirable for mixtures of this kind; but the mixtures must be of plants that help, not injure each other.

A flower plot.—The plan is to have a piece of ground in or near the kitchen garden or any other open position, sheltered, but not shaded, for the growth of the flowers we are interested in. Such ground should be treated as a good market gardener would treat it—well enriched, open, not encumbered with impedimenta of any kind. It must have a walk through it, but, apart from this, the fewer walks the better. It can be thrown into 4-foot beds, but in this case the little pathways need not be gravelled or edged; they may simply be marked out with the feet. It is better to see ground covered with flowering plants than devoted to edgings and much gravel. If any edging is used, it should be of thin stone sunk in the earth, as natural stone edgings are never offensive, troublesome, or costly. Flints or half-sunken bricks will do as well if the thin stones so easily got in the western counties are not at hand. With the aid of such a division of the garden, the cultivation of many fine hardy plants becomes a pleasure. Be they seedling Verbenas, alpine Auriculas, or any favoured flower much used for cutting, the culture is the most certain that one can adopt. Well furnished, such a garden is a delight. When the things get tired of the soil, or require a change, having no formal plan of beds, it is more easy to establish a rotation among the flowers, making the Carnation bed of the past few years the bulb ones for the next, and so on. It would be easy to change one's favourites from year to year, so that richly feeding plants should follow those of a surface-rooting kind, and thus the freshness and novelty of the garden would be kept up. The abolition of all edgings, beyond one or two main lines through the plot, would tend to more careful culture, as the whole spot could be so readily dug up or otherwise attended to. Such a plot well done would be a paradise for ladies who wish to cut flowers in quantity, and also a great aid in replenishing other arrangements on the margins of shrubberies in the flower garden or on the rock garden. It is also a help to those who wish to exchange with their friends or neighbours, in the generous way all true gardeners do. The space that such an arrangement should occupy will of course depend upon the size and wants of the place in every case; but, wherever the room can be spared, an eighth of an acre of ground might be devoted to the culture in simple beds of favourite flowers, and even the smallest garden should have a small plot of the same kind.

What to grow.—Among the various fair flowers which may be cultivated in this way for their own sakes, each separately and well, are the delightful old Clove Carnations —white, crimson, and scarlet, as well as the various mixed and named races of the same family. Then we have the tall and graceful Phloxes, so fair in the autumn in country gardens; the fine old handsome scarlet Lobelias, splendid in colour, and with erect, sword-like shoots; Pinks of various kinds, white and coloured, and hybrid; the handsome Persian and Turban Ranunculus; the bright old garden Anemones, and the finer species of Anemone, like the scarlet A. fulgens; the many kinds of Lilies, commencing with the beautiful old white Lily, and as many as possible of the splendid species introduced into our gardens from California within the past dozen years; the tall perennial Delphiniums or perennial Larkspurs, with their fountains of lovely blue, surpassing in colour the Gentian; the old double Rockets; the many beautiful Irises, English, Spanish, Japanese, and German; Pansies in great variety, always so faithful and rich in colour, flowering, moreover, nearly throughout the year; the old Tiger Flowers; the beautiful races of Columbine, including the lovely A. cœrulea of the Rocky Mountains and the golden Columbine of the same region; the blue African Lily in various forms, and with it the Belladonna Lily; Pyrethrums, so showy and varied now-a-days; Verbenas, which may now be easily raised from seed during the current year, and are so pretty and varied; Chinese Pinks, rich in colour, large, and finely fringed; the old garden Scabious, with a great variety of delicate and beautiful colour; the blue Cornflower, one of the most precious things we have for cutting, and which should always be sown in autumn, bearing flowers for months in consequence; Sweet Williams; Stocks of many kinds; Wallflowers, double and single; the annual Phlox, which has now broken into a fine series of different colours; Zinnias, which, as grown abroad—that is to say, well and singly grown—are very fine in colour and sometimes as large as Dahlias; China Asters, quilled and others; the Sweet Sultan, in two or three forms, excellent for cutting; the showy tricolored Chrysanthemums; double Daisies, very bright and useful in spring; Grasses of the more useful kinds, suitable for cutting in the winter; Grape Hyacinths; Daffodil or Narcissus in variety—many strong kinds may be grown in the grass or in rough or half-waste places; Meadow Saffrons, pretty in the autumn; Lily of the Valley, of which a variety of kinds are now coming into cultivation, differing in length and size of raceme; Crocuses, the autumnal as well as the vernal kinds; the hardy kinds of Cyclamen, which are at home on the mountains of Europe and perfectly hardy in our own gardens; Forget-me-nots; Dahlias, double and single, and in three or four classes of doubles; Evening Primroses, opening so prettily at night; Pæonies, in splendid variety; Primroses, double and single, many kinds; Pentstemon, graceful autumn flowers; Polyanthus, richly coloured vigorous kinds, for borders; Oxlips the same; Tulips, many early and late kinds; Sweet Violets in great variety, choosing the kinds best liked; American Cowslips; Dog's-tooth Violets and Gladioli, the finest and most stately flowers of autumn; the Christmas Rose and its vernal relations; and, lastly, Everlasting Flowers, which may be grown along with the ornamental Grasses, and, like them, be gathered for house decoration in winter. All these fair

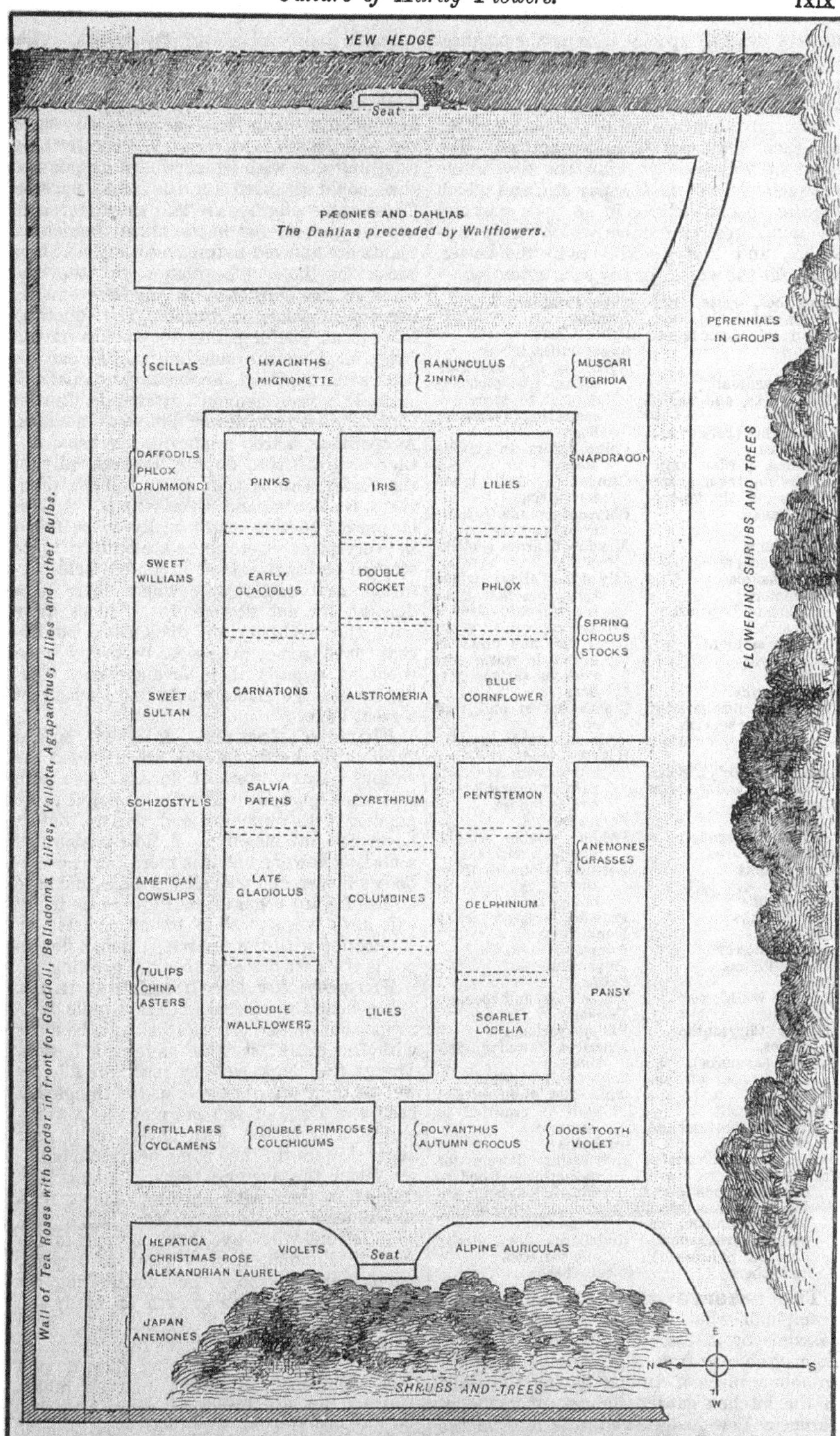

Special culture: reserve garden for the choicer families of hardy plants, grown in beds without reference to general effect, and serving also as a garden for cut flowers and a nursery.

flowers deserve special care in the smallest garden, and should not be trusted to the poor chances of the ill-considered and ill-cultivated slips called "mixed borders."

For convenience, a list of the plants suited for such a garden is here supplied. The amateur will bear in mind the distinction between plants which repay for, and which require, special culture in an open spot and the much greater number which thrive anywhere, and that would be none the better, often far the worse, for any such attention:—

Carnations, white, crimson and scarlet Clove, and mixed and named sorts
Phloxes
Lobelia cardinalis
Clove Pinks, and named sorts
Ranunculus (Persian and Turban)
Anemone, garden varieties and the finer species like the true A. fulgens
Lilies
Delphiniums
Double Rockets
Iris germanica
  Kæmpferi
  English and Spanish
Pansies
Camassia esculenta
Alstrœmeria
Tigridia
Antirrhinums
Columbines, fine varieties and finer species
Agapanthus, the various kinds
Amaryllis (hardy), Belladonna and Ackermanni
Pyrethrums
Schizostylis coccinea
Vallota purpurea
Salvia patens
Verbenas
Anthericums
Chinese Pinks
Scabious
Blue Cornflower
Sweet Williams
Stocks
Double Wallflowers of sorts
Tricolor Chrysanthemums
Thrifts (Armeria), the finer vars. of the common and the Great Thrift
Grape Hyacinth, finer and rarer kinds
Daisies (double in variety)
Grasses, the more graceful kinds, such as the Brizas, Bromus brizæformis, Panicum capillare, Agrostis nebulosa, A. Spica-venti, and others.
Phlox Drummondi
Zinnias
China Asters
Sweet Sultan in var.
Narcissus, finer and rare kinds, till plentiful enough to spare for scattering in Grass or borders
China Asters, in various classes
Campanula, finer kinds for cutting
Chrysanthemums (to flower in open garden)
Meadow Saffrons (Colchicum)
Lily of the Valley; various forms are now coming into cultivation
Crinum capense in vars., a very fine plant as grown in warm corners in various gardens
Crocus, newer and rare species
Cyclamen, hardy species
Hardy Orchids, the finer kinds, such as Cypripedium spectabile and Orchis foliosa
Forget-me-not
Dahlia, double, single, bouquet, and fancy
Evening Primroses (Œnothera), favourite or choice kinds
Pæonies, in great variety
Pentstemon
Primroses
Polyanthus
Oxlips
Tulips, vars. and species
Verbena
Violets (various)
American Cowslips, various
Dog's-tooth Violets
Gaillardias (delicate and slow if crowded in borders)
Gladiolus
Everlasting flowers for gathering — Rhodanthe, Helichrysum, Acroclyne, Gnaphalium orientale
Helleborus, finer species and varieties
Snake's-heads

The reserve garden.—We have an example of what is meant in the preceding passages by a reserve garden at Munstead. As may be seen from the plan, it consists of an oblong piece of ground having the walls of the kitchen garden for two of its boundaries, a Yew hedge sheltering it from the east winds. On the other side it is screened by trees and shrubs, among which are intermingled hardy plants of tall growth. The plants are in beds without reference to the general effect, and all the borders are edged with stone dug on the place, and give no trouble after being once properly set; when old and Moss-grown they look better than anything else with which we are acquainted that could be used for the same purpose. These stone edgings are laid down throughout the garden, and in the mixed border the plants are allowed to run over them, and thus break the lines. The plan shows the contents of the garden as it was one season; every year it may be varied. The following is a list of useful plants for such a garden not placed on the plan, but which may be used when required: Scabious, perennial and annual Chrysanthemums, Marigolds, Chinese Pinks, Sweet Peas, dwarf Phloxes, Camassia, Agapanthus, hardy Amaryllis, Verbena, Anthericum, Thrifts, double Daisies, Lily of the Valley, Orchis (fine hardy kinds), Œnothera, Gaillardia, and Everlastings. A similar garden to this ought really to be found in every place where there are borders to be stocked and maintained in a creditable condition, and particularly where there is a demand for cut flowers, for it does away with the necessity of disfiguring borders that have some attention bestowed upon them as regards their arrangement. Gardeners in large places would find such a plot a great boon.

Florists' flowers.—In these special gardens for hardy flowers are included the various hardy flowers of florists. The term was once applied to flowers supposed to be popular with amateurs and florists, but it never had any meaning. A Rose is above all a florist's flower; but it is more—it is everybody's flower, and we call it a Rose, and have no use for any other term. Flowers are for all who have eyes as well as for specialists. In connection with the reserve garden, a flower-pot is the best place to speak of growing .

Flowers for the breakfast table.—We hear a good deal of dinner-table decoration, but an earnest effort should be made with the breakfast table as regards flowers. Things that look well by night are all very well in their way, but give us the things that bear the light of the morning on a tablecloth. The Briton will never really care about flowers till they come nearer to him—we mean the average being, who has been trained so long with his few yellows, reds, and blues in lines and circles. People who love flowers will have them on their tables, even in London, if they have to buy the samples in the shops; but, considering the opportunities for the growth of hardy and beautiful flowers, and for collecting such things in the hedgerows, the dwellers in the country have, it is surprising they do not more frequently use them on their tables. Getting the fair forms of many beautiful flowers near the eye is the best way to make converts. The enemy in the way hitherto against plenty of cut flowers has, it must be confessed, been the gardener. But the poor fellow was confined in his cutting operations to a greenhouse or conservatory, which he naturally wished to have gay, and cutting flowers has been the bone of contention in many a place. Now-a-days, when we have recognised the beauty of innumerable things that can be grown easily out of doors, there should be no more trouble in this respect. Every gardener should do his best to encourage the cutting of flowers in every way that he can, barring such few rarities or novelties as he wishes to keep for seed. A supply equal to that of twenty plant houses can be got from an open square in the kitchen garden or any other piece of good ground. We noticed a vigorous bed of Carnations last year flowering for nearly three months, while baskets of flowers were taken from it every week. For eight months of the year there is a continual progress of open-air flowers, which can be grown easily in sufficient quantity to allow of plenty being cut for every want. A bed, or a few lines of each favourite in a patch of good soil, would give a great number of flowers, and these, aided by the Roses and other bush and tree flowers about the garden, would yield all the flowers that the largest house would require, besides many for those who have no garden and for hospitals. The flower-cutting department might also be a nursery for some of the finer things we have to give away; but flowers grown for cutting should be carefully selected as regards their odours, forms, and colours. The gardener, we repeat, should be active to encourage an idea which would tend so much to enhance the importance and resources of his own art, and give people a pleasure in their homes. The smallest country place can afford a spot of ground to grow cutting flowers. In large gardens this should be made a feature of, and we venture to say that it would be as much enjoyed for its beauty as the best part of the flower garden. From this and the general open garden collections—the woods, hedgerows, and copses—every charm of flower life might be culled in abundance, from one end of the year to the other, for the embellishment of our tables. Here flowers, the most choice and elegant in form, should be brought near the eye, not in coarse, unconsidered jumbles, but each thoughtfully placed where it would look best, and in such a quantity as best shows its form. The mere dab of mixed flowers in a set place is only the rude dawn of art in this direction; many things have to be placed in half a dozen different positions in a room before we know where they will look best. Many flowers deserve to be seen alone, and even a small bit, or two or three bits, will be best. By trying each beautiful species or form in different positions in the room, in relation to different lights and backgrounds, we find how much we may enhance the beauty of the flower by studying the position or light in which it is best.

Hardy annuals and biennials.—These hardy annuals are amongst the most showy and useful plants that we possess, and they may be raised at a minimum of cost, though, up to the present time, few have successfully used them on a large scale. All that is necessary in the case of hardy annuals is to sow them in the beds or borders, which may be done either in the autumn or any time during February, March, or April, or even later, according to the varieties to be grown and the season at which it is desired to have them come into bloom. The great advantage in sowing in autumn is that they not only flower much sooner, but they are, where the soil and climate suit them, always stronger and finer. In the management and treatment of autumn-sown annuals the best way is to have them in a sheltered situation during winter, and then plant out the finest and strongest plants thinly very early in spring. In doing this, warm, showery weather should be chosen, as then they receive less check and become re-established at once. Why they are so often seen in poor condition is owing to their being sown on thin, hungry soil, and to their being left to grow thick and crowded. The following annuals do well, autumn sown, in the home counties, other kinds would succeed in mild districts:—

Silene of sorts	Larkspur
Sweet Sultan	Sweet Peas
Cyanus	Eschscholtzia
Centaurea	Gilia tricolor
Coreopsis	Nemophila
Godetia	Venus's Looking-glass
Limnanthes	Viscaria
Clarkia	Erysimum
Saponaria	

Autumn-sown annuals.—Concerning these, "Salmoniceps" writes—"I have just measured a plant to-day (October 4) of Nemophila insignis, sown more than a year ago. It has been in flower since May, and measures now 4 feet by 3 feet 10 inches. It would take a long time to count the blossoms, although they are not so large as the earlier ones. The plant grows in a new and rich border. According to the ordinary way of sowing annuals, this single plant occupies the space which is usually allotted to a whole packet of seed. Of course a moist climate favours the growth of most annuals." All the grown kinds are included in the second part of the work, and their cultivation given. It would be interesting to see what a good gardener could do with annuals, each kind considered as to its requirements, and sown at the season to get the best and longest bloom. Those desirous of making the most complete selections of plants from biennials and annuals are referred for directions to the body of this work.

THE GARDEN OF SWEET-SMELLING FLOWERS.

In placing sweet-smelling plants, some attention is due to their habits. Some are lavish of fragrance and give it spontaneously. Of these, in the shrubbery, we have Lilacs, Mock Orange, Azaleas, Sweet Briers, double Gorse, various Brooms and Thorns, Acacias, and Honeysuckles. In the borders, Tulips, Hyacinths, and Daffodils, Triteleias, alpine Auriculas, Musk, double Rockets, Lupines (annual and perennial), Fraxinella, White Lily, Musk Mallow, Phloxes, Mignonette, and Sweet Peas, with several kinds of Scotch and other Brier Roses. For wilder parts, common Gorse, Broom, and Hawthorn, wood Hyacinths, Cowslips, Agrimony, Meadow Sweet and Marsh Marigold. A peculiar and delightful fragrance rises from a sun-baked bank of Heather in late summer, and who does not know the sweetness of a Clover field, and of a warm breeze perfumed with Pine trees, and, better still, though perhaps less commonly known, an April night full of the sweet breath of the young Larch trees? All these are plants and trees that give off their sweetness bountifully, and even from some distance, but the fragrance of many others can only be enjoyed by touching, or, at least, by closely approaching them. Of these the most important are Myrtle, Lavender, Rosemary, Balm of Gilead, Southernwood, Escallonia macrantha, Bay, Bog Myrtle and the Fern-leaved Gale (Comptonia adiantifolia), Juniper, Thyme, Marjoram, and other sweet herbs. A good plan would be to plant these in a "wilderness," with narrow walks or spaces of turf between good groups of each, so that one would brush against the living masses of sweetness, the turf being full of Thyme and the free-smelling shrubs and trees beyond. What a delight it would be to take a blind person into such a garden!

The Gum Cistus in autumn gives off a pungent and agreeable smell, though its flowers have none, and in early winter the foliage of Violets and Woodruff and the dying Strawberry leaves are sweet good-byes of the garden year. There are many of our smaller treasures, to enjoy whose sweetness we must either bend low to, or gather. Linnæa borealis, whose tiny twin-flowers smell like Almonds, the New Zealand Mayflower (Epigæa repens), Polygala Chamæbuxus, also Almond-scented Pyrolas, the sweet-scented Orchis (Gymnadenia conopsea), like white Lilac-blossom, and the Butterfly Orchis, fragrant in the evening; Iris graminea, whose flowers, hiding low among the grassy leaves, have exactly the smell of ripe Plums. The Water Hawthorn (Aponogeton distachyon) is strongly perfumed. The Lily of the Valley need hardly be mentioned. Of other sweet border flowers there are Chinese Pæonies, delicately Tulip-scented; Grape Hyacinths, the Musk Hyacinth, Snapdragons, Salvias, including the variegated-leaved Yuccas; the large white Plantain Lily, as sweet as the white Lily and more delicate, but faintly delicious, smells like those of Crocus, Water Forget-me-not, and Pansy. The Rose family give a whole scale of sweet notes. The wild Roses have a scent as tender as their colouring. The Burnet Rose and its descendants, the whole race of Scotch Briers, have a delicate smell quite peculiar among their kind, as have also the Austrian and Persian Briers respectively, though in their case it is less agreeable. It is distinct again in the Damask Rose and in the sweet old Provence; while in Hybrid Perpetuals we have at least three distinct types of perfume, and as many in the Teas, the most marked type among the latter being that of Gloire de Dijon. The scent differs again in China Roses, and again in the clustered climbing kinds. In Moss Roses the very peculiar and delightful smell seems to come mostly from the viscid matter on the mossy calyx and stalk. This is also the case with some of the garden Brambles, notably Rubus speciosus and with Fraxinella, Escallonia macrantha, and Gum Cistus. Among sweet-smelling plants we must not omit those of the wholesome aromatic character, such as Wormwood, Chrysanthemum, Chamomile, Santolina, and Tansy.—G. J.

There is a list of sweet-smelling plants in "Hardy Flowers," but there are many plants that have a delicate fragrance which are not named there. Much has yet to be learned concerning the fragrance of even garden flowers. All that is certain is that we know enough of them now to satisfy every taste.

THE GARDEN OF BEAUTIFUL FORM.

Our plan is in this part of the work to treat of all the hardy classes before the tender and half-hardy bedding plants, important as these latter are. This phase of gardening sprang out of the bedding plants, and is more recent; but, as many of the plants used in it are hardy, and the best of them are so, we give it place, using it as a connecting chapter between the hardy and the half-hardy growths. Sub-tropical gardening was introduced in order to avoid the eternal round of Geraniums, Calceolarias, and a few other common bedding plants, which, however worthy of culture they may be, can hardly be said to convey any adequate idea of the beauty of the plant world. For some years, however, the very name proved sufficient to deter many from giving the system a trial, from the erroneous notion that only sub-tropical vegetation was admissible. This arose through Musas, Caladiums, and similar tender plants being recommended in the first instance, together with costly preparations for supplying bottom heat, special soils, &c., all of which were at one time considered necessary. But not only have such unsuitable subjects been generally discarded, but more recent events have proved that even better and more lasting results can be attained by the use of plants of even a much hardier character than soft-wooded bedding plants, the sole representatives of out-door decoration some twenty years ago. Now, with the rich store of fine-foliaged plants from temperate climes at command, it is quite possible to have a beautiful garden of hardy plants alone, for such subjects as the Ailantus, Paulownias, and other fine-leaved trees make fine growth if cut down close to the ground every year like herbaceous plants. We have also the hardy Palm (Chamærops), the Yuccas, and graceful Bamboos. As regards the dwarf subjects suitable for edgings, we have such hardy plants as Siebold's Plantain Lily (Funkia), Acanthuses, and plants of a similar character. Amongst plants that are annually raised from seeds, and which merely require the protection of glass to start them, we have much variety between the stately Ricinus or Castor-oil plant, and the dwarf Centaurea or Chamæpeuce. It is by selection from the various sections that the best results are obtained.

A Dracæna in a Wiltshire garden.

One great advantage of using permanent plants in pots for central objects in groups or as isolated specimens on the turf is, that, while they add variety to the arrangement during summer, they can be taken up and utilised for indoor decoration as soon as the beauty of the outdoor garden begins to fade. In this way a maximum of interest and pleasure may be enjoyed from both the outdoor and indoor garden, according to the season. Although tender plants in pots taken out of the houses for the summer are decidedly effective for special occasions in a general way, any plants that are not able to stand out-of-doors from the first week in June until the last week in September can hardly be called fit for summer bedding. Amongst the most suitable are several kinds of Palms, such as Seaforthia elegans, Chamærops excelsa, and C. humilis; Aralias of various sorts; Dracæna australis and D. indivisa; Phormium tenax and its variegated form; Yucca aloifolia variegata, Ficus elastica, and Eucalyptus globulus, the bluish grey tint of which is unique in its way. Erythrinas make fine autumn beds; being very brilliant in colour, they are useful for lighting up sombre masses of foliage. Bamboos have such beautiful feathery foliage, that where the winter is too severe to trust them out, they should be utilised as large pot plants for plunging out-of-doors in summer; they luxuriate on the margins of water, and look particularly well in irregular groups or clumps. Abutilons are particularly well adapted for open-air decoration, either planted or plunged; Abutilon Boule de Neige, Boule d'Or, and Darwini are all good; A. tessellatum and A. Thompsoni are also very effective, the markings of the foliage being rich and varied. The hardiest Tree Fern, Dicksonia antarctica, looks well plunged in shady dells where a good canopy of overhanging foliage gives shelter and shade; and several varieties of dwarf Ferns, such as the Bird's-nest Fern, are admirably adapted as undergrowth to the above. Plants raised from seed will, however, form the majority in most places, from the lack of room under glass for sheltering many large plants. Of these the most generally useful are Cannas. If sown in February in strong heat, they make fine plants for bedding; their foliage is lovely, and the flowers rich in colour. The underground roots are permanent, and increase in size and strength every year. They may be taken up and wintered under glass, or securely protected in the soil by means of

Example of hardy Palm in the open air in English garden (Cornwall).

external coverings. The Cannas one sees in the public parks are generally protected in winter. The tall light green foliaged varieties make excellent centres for groups, as they mostly flower freely, and the dwarf bronze foliaged sorts are good for edging.

Solanums are also very effective. The spiny-leaved S. robustum, the elegant cut-leaved S. laciniatum, and S. Warscewiczi make good single specimens or edgings to groups of taller plants. Wigandias, Ferdinanda eminens, and Melianthus major are likewise all good plants treated as annuals; and among dwarf subjects that can be raised with them I may mention Brazilian Beet, with its richly tinted leaves and midribs; Acacia lophantha, Amarantus in variety, Cineraria maritima, and Centaureas, with their silvery, elegantly cut leaves. Bocconias, with their tall spikes of graceful flowers and noble foliage, make effective and permanent plants for isolated groups. The Pampas Grass and Arundo ought also to find a place in every garden, for they are beautiful at all times of the year, and on the margins of water are quite at home. Several varieties of Rhus or Sumach have very beautiful foliage, Rhus glabra laciniata being especially elegant. Aralia japonica is also well adapted for flower garden decoration, as are the Tritomas, with their fine Rush-like foliage and flame-like spikes of flower, the Funkias, the variegated Grasses, such as Poa trivialis argentea; in fact, the material to select from is unlimited, even if we confine ourselves to fine-foliaged plants.

The Pampas Grass, Yucca, Crinum, and some of the bolder types of hardy plants.

As to arrangement, in all cases beds or sets of beds of the simplest design are the best. Shelter from wind is also of the first importance, and for this reason recesses in shrubberies or banks clothed with foliage form the most fitting background for beds or groups to nestle in. Avoid Musas or Caladiums, the leaves of which tear into shreds if winds cannot be entirely shut out; also plants that look unhappy on the occurrence of a cold night or two; and concentrate effort on subjects that grow and look luxuriant under nearly all conditions. If a dell or garden overhung by trees is at command, where ordinary flowering plants run all to leaf and refuse to flower, take advantage of it for fine-foliaged plants. It will form a charming change from brilliant bedding plants, or severely geometric carpet beds, to come upon a garden where each plant spreads forth its delicate foliage.—J. G.

As an example of fine form from hardy plants, we cannot do better than give this illustration of the New Zealand Reed (Arundo conspicua), furnished by a plant in the gardens of Orchardleigh Park, Somerset. This handsome Grass produces its blossom-spikes earlier than the Pampas, and is decidedly more elegant in its habit, the silky white tufts bending like ostrich plumes at the end of their slender stalks. A sheltered corner suits it best, where it is protected from rough winds, and relieved against a background of darker foliage. The specimen here given produced in 1878 thirty-one spikes, the highest measuring 12 feet from the ground.

As to what may be done with tender plants in the open air, it would be difficult to get a better illustration than is furnished by this stately

Musa Ensete in Berkshire.—In sheltered nooks in the southern counties this plant makes a very fair growth in the summer. In 1877 we were struck with the health and vigour it displayed at Park Place, Henley-on-Thames. Mr. Stanton, the gardener there, raised a batch from seed, and it was surprising what noble plants these became in the course of fifteen months from the time of sowing. They were sown in April, and planted out early in June. Placed indoors. early in October, they remained throughout the winter in a warm greenhouse, and were again planted out in June the following year, when they formed fine young leaves. The plant is quite as effective when seen in a conservatory, or any other large glass house in the winter, as it is when put out-of-doors in warm, sheltered places during summer.

Some arrangements of fine-leaved plants.—In large beds this system gives elevation to the low masses of bright colours, and at the same time relieves and tones down the glare. If only the larger beds in a large design were planted in this way, leaving the smaller beds to be filled as before with dwarf plants, the effect would probably be better than if too much were aimed at. This style of decoration admits of a good deal more freedom of treatment in parterre gardening than the common way of training the plants all to one level or nearly so—especially in the arrangement of the ground and the shape of the beds. Although it may be applied to the formal geometrical patterns, yet it is more in character, and looks better, if introduced in the shape of raised beds and undulating glades of turf. There is no better way of setting out such beds than to take a long waggon rope, and coil it into the desired outlines—securing at the same time the size and position that will look best from

The New Zealand Reed (Arundo conspicua) as grown in Orchardleigh Gardens, Somerset.

the particular point from which it would be viewed most frequently. Where, from the nature of surrounding objects, it is practicable to adopt a picturesque arrange-

Musa Ensete in the open garden at Park Place, Henley-on-Thames.

ment, the elevations and depressions should be easy and graceful, not only for the purpose of securing the best general effect, but also with a view to economy in keeping up good order in mowing, &c., afterwards. Abrupt elevation not only costs a good deal to keep in order when the work has to be done with scythe and shears, but the turf is more liable to become unsightly in hot weather. The following are a few of the best mixed arrangements I noted down in different places as they came under my notice last autumn :—

1. Wigandia caracasana planted 4 feet apart, with a carpet of the Variegated Cocks-foot Grass (Dactylis glomerata variegata)

underneath, broad band of Geranium Beauty of Calderdale, and edged with Variegated Coltsfoot (Tussilago Farfara variegata). The latter is a striking, broad-leaved, hardy plant that dies down and disappears in winter; its creeping underground stems often travel a considerable distance, sometimes coming up in spring several feet away; therefore, the young pale buds must be carefully watched for when digging or hoeing is going on.

2. Ricinus Gibsoni (one of the most effective of the Castor-oil plants) planted 3 feet apart, groundwork variegated Geraniums, broad edging of Viola Perfection.

3. Tall Castor-oil plants 4 feet apart, Prince's Feather pegged down as groundwork, band of Iresine Lindeni, next Golden Feather 1 foot wide, and two rows, nearly flat, of Echeveria secunda glauca outside.

Bed of fine-leaved plants in Hyde Park. Drawn September, 1881.

4. Scarlet Geraniums with Giant Hemp 4 feet apart, and broad band of Gnaphalium lanatum outside. This arrangement is only suitable for a large bed, as the Hemp will in good soil grow to a great height.

5. Giant Fennel with the flowers picked off, 3 feet apart, groundwork Amarantus melancholicus ruber, and edging of Crystal Palace Gem Geranium, two rows.

6. Variegated Maize 2 feet apart, groundwork scarlet Verbenas, with broad edging of Leucophytum Browni.

7. Solanum marginatum 3 feet apart, groundwork Coleus Verschaffelti, edge broad band of Golden Feather. This is a very simple, but an exceedingly rich combination.

8. Solanum robustum, a red-spined variety, 3 feet apart, pink Geranium groundwork, with broad band of Cocksfoot Grass and Lobelias mixed round the edge.

9. Ficus elastica, small plants from 2 feet to 3 feet high, 2 feet apart, golden-leaved Geranium Crystal Palace Gem groundwork, edging Iresine Lindeni.

10. Dracæna indivisa 3 feet to 4 feet apart, according to size; groundwork of dark blue Heliotrope, and broad band of variegated Mesembryanthemum, with single plants of Lobelia intermedia—a very dark foliaged, scarlet-flowered variety—2 feet apart along the centre of the Mesembryanthemum.

11. Canna expansa, 3 feet apart, golden Geranium groundwork, with broad edging of Coleus Verschaffelti.

Here is an illustration of a bold mass that used to be near Hyde Park Corner, which seemed in itself to illustrate some of the best features of the fine-leaved gardening of recent years. It was a very large circle, with a great plant of the Abyssinian Plantain in the middle, fringed by a few sub-tropical plants, and edged by an extraordinary fringe of the fine hardy, and in beauty, long-enduring Siebold's Plantain Lily (Funkia Sieboldi). The reason of the success of this bed is clear enough, and it is always worth ascertaining the reason of any good effect in gardens. The "it-is-all-a-matter-of-taste" men are of no use at all under the new gardening. The bed was right; first, by its form, not a finniking angle or a wormy scrawl, but a bold circle, presenting no confusion to the observer, who simply saw the plants rising in a well-defined group from the roomy turf. The bed was by itself, which allowed it to be seen unopposed—not muddled up with a lot of other beds near and around it—a fault too frequent. In passing, it may not be amiss to note that the eye does not, as a rule, care for more than one thing at a time, and if invited to look at a picture made up of many things, it rests with pleasure in some one spot. Lastly, the plant forms were strong and well selected, and contrasted well with the ordinary tree vegetation near, there being plenty of Grass about the bed to allow of contrast, without confusion, with rival subjects. The way in which the Plantain Lilies began early in the year to adorn the spot, and continued to do so throughout the whole summer and autumn, was a pleasure to see. The drawing was made about the end of September, shortly after some heavy storms had taken place, which tore the Musa a little, but the bed remained excellent in effect till October. Some of the Plantain Lily leaves began to fade at that time, but still produced a very fine effect.

Yuccas in groups.—Wherever space can be afforded, and suitable situations can be found for them, the hardy Yuccas should be grown. Few hardy subjects are so distinct in leafage and manner of growth as these; but to see them to the best advantage they should be arranged in bold groups, and in the immediate vicinity of such trees and shrubs as are best fitted to form a harmonious contrast to them. Perhaps, the best situation for them is a sloping ground fully exposed to the mid-day sun, and backed up by evergreens of some kind. Thus placed, and allowed ample space for development, they will gain in beauty from year to year, and when in bloom will form a fine feature. The handsome spikes of large cream-coloured flowers are extremely effective, especially when relieved by masses of verdure behind them. Yuccas like a well-drained soil, and thrive well where the subsoil is pure chalk. They delight in full exposure to the sun, and enjoy the shelter from rough winds which a taller vegetation affords. Hence, the advisability of planting them in close proximity to trees or shrubs of some kind which may partly screen them. The Yucca is a hardy plant, but the foliage of several of the most ornamental species, such as filamentosa, is apt to get torn when the plants are much exposed.

In grouping Yuccas, a more natural effect is obtained where a proportion of the specimens employed have attained sufficiently large dimensions to raise the head of foliage some 3 feet to 6 feet above the soil. These tall plants should not, however, be placed in a regular manner in a back line, but one here and there should be allowed to advance somewhat into the foreground, with some of the smaller specimens nestling at their feet. The effect of a group thus arranged charms by its irregularity and quaint beauty. I have often thought that we do not sufficiently value these noble hardy plants, for one seldom sees a good use made of them, the owners of gardens generally being content with employing them on the dot system, which cannot be said to convey an idea of their qualities. It should be the aim of all who may have a large extent of pleasure ground to embellish it with plants of an enduring character and to create as much diversity as possible. This can be well carried out by grouping families of plants that have distinctive features in situations most favourable to them, avoiding, as far as possible, all semblance of regularity and formal outline. The hardy Yuccas are children of the sun, and they do not as a rule flower freely unless they get a good deal of it in the summer. With respect to soil, they can scarcely be termed fastidious, but it must be well drained. They appear to be perfectly at home where the subsoil is chalk, attaining a rude vigour and flowering freely when thus planted, but we have not noticed them to fail on any ordinary soil of fair quality.—J. C.

THE WILD GARDEN.

THIS term is especially applied to the placing of perfectly hardy exotic plants in places, and under conditions, where they will become established and take care of themselves. It has nothing to do with the old idea of the "wilderness," though it may be carried out in connection with it. It does not necessarily mean the picturesque garden, for a garden may be highly picturesque, and yet in every part be the result of ceaseless care. What it does mean is explained by the Winter Aconite flowering under a grove of naked trees in February; by the Snowflake growing abundantly in meadows by the Thames side; by the perennial Lupine dyeing an islet with its purple in a Scotch river; and by the Apennine Anemone staining an English wood blue before the blooming of our Bluebells. Multiply these instances a thousandfold, illustrated by many different types of plants and hardy climbers from countries as cold and colder than our own, and one may get a just idea of the wild garden. Some have erroneously represented it as allowing a garden to run wild, or sowing annuals promiscuously; whereas it studiously avoids meddling with the garden proper at all, except at attempting the improvements of bare shrubbery borders in the London parks and elsewhere; but these are waste spaces, not gardens.

I wish the wild garden to be kept distinct

in the mind from the various important phases of hardy plant growth in groups, beds, and borders, in which good culture and good taste may produce many happy effects; distinct from the rock garden or the borders reserved for choice hardy flowers of all kinds; from the best phase of the sub-tropical garden—that of growing hardy plants of fine form; from the ordinary type of spring garden; and from the gardens, so to say, of our beautiful native flowers in our woods and wilds. How far the wild garden may be carried out as an aid to, or in connection with, any of the above in the smaller class of gardens can be best decided on the spot in each case. In the larger gardens, where, on the outer fringe of the lawn, in grove, park, copse, or by woodland walks or drives, there is often ample room, fair gardens and wholly new and beautiful aspects of vegetation may be created by its means.

A mixed colony in the wild garden.

Daffodils in parks.—In passing through Sussex in March I came upon several fields of Daffodils, and was charmed with their beauty. It was at the time of harsh east winds, and yet not a speck or trace of decay of any kind was visible among them. They were not in tufts or groups, as in some woods, but scattered through the Grass, as if they had taken their chance in the struggle for life with the other vegetation, and perhaps been mown down repeatedly. They were somewhat smaller than usual, perhaps owing to this fact, but in point of beauty I have seen nothing to surpass them, as they danced in the sun and realised Wordsworth's poem to the full. Having passed through several parks soon afterwards, and not seeing any of this glorious flower life in its wide breadths of turf, I asked myself why, if this occurs by accident in secluded farms, it should not be made a feature of in our parks here and there? Nothing would be more easy, and no effect of the whole floral year would be more bright or cheering. These Daffodils are fitted to our climate in a peculiar way, and as no other flowers are. Harsh winds or rains may disturb others, but the Daffodil seems regardless of them. The little hardy Hepatica of the mountains of Europe is brave enough, and yet it will be almost destroyed, flowers and all, if much exposed to a long-continued east wind, of which the Daffodils take no notice. Therefore we say to those who have means to enjoy this bold and noble phase of gardening—Consider the Daffodils. It is not only the old and common kind which we may establish, but a variety of other kinds no less beautiful and distinct from the common one. They will prolong the season of bloom, and give a fine variety of form. Of the bolder kinds suited for park scenery I should name Narcissus maximus, N. incomparabilis and its forms, N. poeticus and its varieties (they succeed each other in blossoming), N. odorus (the large Jonquil), and the common kind itself is now obtainable in several forms. Some of its allies, too, are very fine.

Snowdrops and Snowflakes for waste places.—Whoever has seen the Snowdrop at home, in the copses and thickets of many parts of England, must have felt a pleasure totally different from that of seeing it in gardens and shrubberies. About a mile from where I live is a large wood, with a green drive up its centre. A stony-bedded brook crosses a portion of it and runs through a park. In one of the wildest parts of this wood are a few clumps of Snowdrops, and although I have as many of these plants as most people, I always go every year to see these Snowdrops in the wood. Now there are hundreds of miles of woodland walks and drives in the kingdom where one simply walks among trees. How interesting such places might be made by planting them with Snowdrops and similar plants. We have all heard of the Lily of the Valley woods at Woburn, and how jealously they are guarded. But plant such things in profusion, and they need not be guarded. Some imagine that flowers of this or any other class never ought to be out of a garden. They may as well imagine that a bird never looks well except in a cage. If all surplus stock of Snowdrops, Crocuses, and even Hyacinths were distributed in the woods and copses, and by path sides in grounds, a fine feature would be added to many of them. The great fault to be found with most of the places in England is that no matter how great their capabilities may be, gardening only begins at the garden gate. What I have said in reference to the Snowdrop also applies to the summer Snowflake (Leucojum æstivum). This has some claim to be considered indigenous. It is a very hardy, somewhat stately-looking plant (that is, compared with the Snowdrop), and its introduction into damp spots in woods and groves would provide a charming feature.—T. W.

Turk's-cap Lilies in wild garden at Castle Ashby.

Wild gardening on Grass.—Even in the smallest gardens there are places under deciduous trees where the Grass grows thin all the year round, and where it is useless trying to renovate it. These are the very spots for Snowdrops, Aconites, Jonquils, Daffodils, Anemones, Scillas, and similar flowers. Take out a good spade's depth of earth, and, if poor, put in some rich soil, then the roots, and then fill up with the old soil. They will make these bare spots look beautiful in spring, and as the trees put on their verdure in summer, the bulbs will go to rest out of sight, and in the autumn the fallen leaves should be left as a protection for them. The worms will draw most of them into the ground, or, if in very conspicuous places, where on the score of neatness they must be removed, put them in pits, and the following year spread them in a decayed or leaf-mould state over the bulbs. There are, however, plants that require a lighter position to bring them to perfection—that look better springing from the turf than in any other way, such, for instance, as Bocconia cordata, Solomon's Seal, Acanthus latifolius and mollis, the hardy Fuchsias, like Riccartoni, which are cut down annually like herbaceous plants, Heracleum giganteum with its noble foliage, and many others. On mossy banks beautiful effects may be produced by planting various coloured Primroses, single and double Polyanthuses, Oxlips, Hepaticas, Auriculas, Gentians, Cyclamens, Dog's-tooth Violets, Lily of the Valley, Colchicums, and similar plants, all of which will have a better effect so situated than in beds of freshly dug earth, there being no splashing of the blossoms with heavy rains; and there is no need to remove the plants, as is too often the case when used in connection with ordinary spring flower gardens.

A border in the rectory garden at Bingham, with Foxgloves and tall perennials.

The accompanying engraving, from a photograph taken in Mr. Miles' garden at Bingham, shows an interesting aspect of garden vegetation. Though not a large garden, and though what is done is mostly what may be called flower gardening, there is one part of a border surrounding the little lawn where some resemblance to wild gardening may be seen. Its merits consist in the beautifully broken effect of the Foxgloves and like plants, which vary so well the ordinary formal level surfaces usually seen in gardens. Poppies and the bolder Ferns lend their aid, and the ground is quite green with dwarf plants beneath. The effect is very charming and picturesque. There is plenty of colour and much quiet grace.

Exotic plants suitable.—For every beautiful native plant, like the Foxglove, many exotics are equally suitable, because there are many countries with a larger and more varied flora than our own, while quite as hardy. Some of our most charming flowers occur everywhere, and cannot indeed well be left out of the Grass, as in the case of the Primrose and Cowslip. The Snowdrop is, apart from gardens, frequently naturalised in England and Scotland, and supposed to be truly native in the western counties, and it would be strange if it did not occur in gardens too. The Lily of the Valley is a true native plant, said in Hooker to be wild from Moray to Kent and Somerset. Yet, lovely as this plant is, I never saw a deliberate attempt to make it, in a wild state, one of the permanent ornaments, so to say, of a country seat till Mr. George Berry planted out a good deal of it at Longleat some eight years ago. The Vernal Crocus is so freely naturalised in meadows, as to be by many considered a native. The Narcissus is common as a native plant. I have seen meadows of it fairer than any collection in garden or nursery. What I urge is, if so much is done by Nature herself, how much more might be done if Art stepped in and that due attention were given to the many charming things that come from the vast cold regions of the northern world—Europe, Asia, and America. Our own wild flowers are so beautiful and so common about us that their presence is not so important to secure as that of plants equally hardy, equally beautiful, but wholly distinct. For example, wherever the Wood Anemone grows wild, there can be no reason for putting it in the Grass; whereas, the addition of the Apennine Anemone would add a new charm. The subject is a most important one; in this way alone the great question of spring gardening might be for ever settled. Unhappily, the rule is bareness and shaven or naked surfaces everywhere. Not only are there few, except native, flowers in the Grass, but the very borders avowedly made for flowers are bare, and not in winter or spring only, but even at midsummer.

The wild garden in Pine woods.—With care in the selection of positions, a good deal of very charming work might be done in Pine woods with such material as the following: The Twin Flower (Linnæa), Winter Greens (Pyrola), Partridge Berry (Mitchella repens), Oregon and Winter Greens (Gaultheria Shallon and procumbens), Cornish Money-wort (in shady and moist ditches), yellow Peruvian Lily (Alstrœmeria aurantiaca), Lithospermum prostratum, Sun Roses (Cistus), and Rock Roses (Helianthemum). Of course we are assuming there are open sunny places as well as shady ones. The Mayflower (Epigæa repens) does best in a sandy wood under shade; Hellebores will do with goodish soil and the "pan" broken up. No doubt there are many other plants, but almost all of the above have been tried to my own knowledge and found to succeed. Cypripedium spectabile will do in the bogs, and the showy Asclepias tuberosa on the sand hills. A whole world of beauty exists in the varieties of hardy Heaths perfectly suited for this kind of gardening.

Shady and sunny Grass walks.—In our climate, it is well to have some shady walks near the house on sunny days. So, too, it is delightful in taking a walk around the grounds on warm May days to do so on a sunny Grass walk, just a strip of the turf mown without any further preparation going here and there through the shrubbery, and by groups of Lilacs and fragrant bushes. Even where a bold and well-planned walk goes around the garden or pleasure ground it is easy and desirable to have this. In these days when people begin to see the advantage of putting flowers to grow in the Grass, such walks would be all the more delightful. Not less precious is the shady Grass walk, which is of course cool in hot weather, and which enables us to grow Ferns and shade-loving plants, offering a complete relief to the open sunny walks just alluded to.

HARDY CLIMBING PLANTS.

NOVEL and varied combinations from all sources are so much advocated in this book, that we must include something on the climbing plants. With regard to such we can draw no hard lines between the woody and the herbaceous. The numerous hardy climbers which we possess are very rarely seen to advantage, owing to their being stiffly trained against walls. Indeed, the greater number of hardy climbers have gone out of cultivation, owing to there being no just ideas as to their proper use. One of the happiest of all ways of using them is that of training them in a free manner against trees; in this way many good effects may be secured. The trees must not, of course, be those crowded in shrubberies, but standing on the turf. On some low trees the graceful companion may garland their heads; in tall ones the stem only may at first be adorned. But some vigorous climbers could in time ascend the tallest trees, and we know of nothing more beautiful than a veil of such an one as Clematis montana suspended from the branch of a tall tree. Various handsome plants may be seen to great advantage in this way, apart from the well-known and popular climbing plants. There are, for example, many species of Clematis which have never come into cultivation, but which are quite as beautiful as climbers can be, and which may be favourably seen in this way. The same may be said of the Honeysuckles, wild Vines, and various other families of which the names may be found in catalogues. In consequence, however, of the fact that no good plan of growing these plants has been carried out in our gardens, nurseries are by no means rich in them. Much of the northern tree and shrub world is garlanded with creepers which may be grown in the way I suggest and in similar ways, as, for example, on banks and in hedgerows. The naked stems of the trees in our pleasure grounds, however, have the first claim on our attention in planting garlands. There would seldom be need to fear injury to well-grown trees.

Evergreen wall plants.—We now possess a varied and excellent assortment of plants, both deciduous and evergreen, suitable for all situations and for mostly all purposes —for covering walls, open fences, palings, rooteries, trees, and for forming screens for covering unsightly objects. Taking evergreen species first, let us begin with the Ivy, of which there are numerous varieties, some of them exceedingly neat and pretty, and particularly effective in winter and spring, when many other plants are leafless. For covering walls or other objects quickly, none of the varieties surpass the common Ivy. A few plants put in here and there will cover a large space in one or two years. Ivy delights in a good rich soil, and it is worth while to place some well-rotted manure about the roots at planting time if quick and luxuriant growth be an object. When planting Ivy for covering walls, or for forming screens, we have usually taken up good patches of it where it grew on the ground, and planted them in the mass. This beats any other plan for quickness. Ivy does not grow readily unless it has good roots to begin with, and these it never wants when growing on the ground. In many places it is employed for covering bare surfaces under trees where nothing else would succeed, and from such places large patches of it may be cut out with a spade; but they should always be taken from the outside, where the growing points are, cutting in with the spade in a parallel line with the growth as much as possible, till roots are found in quantity, which will be a few feet from the outer edge, and then cutting across so as to secure a large square piece with plenty of growing points. It should be removed with the sod and soil adhering to the roots, and planted against the object to be covered, taking care to bury the roots deep enough, and to support the top with strips of shred till growth begins and the shoots take hold themselves. I have seen unsightly walls completely covered in this way in a single summer. There are a number of green varieties of the Ivy, with large heart-shaped and lobed leaves, which are ornamental in their way and rapid growers, but which can hardly be said to be superior to the common kind. The variegated kind, named Hedera maculata major, may be mentioned, however, as an excellent variety, and nearly as good a grower as the common Ivy; but it is apt to run back to the green state if grown in too rich a soil. There are several neat-habited green sorts that are highly ornamental and suitable for covering limited spaces on walls. Against dark-coloured stone they are not seen to advantage, but a light ground shows off their small, neatly-cut, and closely-placed green leaves effectively. Some of the weaker-growing kinds do not cling readily of their own accord, and must be nailed; but such sorts it is better not to plant. Those who wish to make a selection should first see the different varieties together. The Bramble is not reckoned an evergreen, perhaps, but it retains its leaves so long in some situations as almost to come under that head. The very free-fruiting, Parsley-leaved variety is a tremendous rambler, and grows at a great pace, and when planted against palings or open fences, it soon forms a formidable barrier. It is worth while planting it in good soil at the beginning for the sake of its fruit, which is so excellent and useful; and if the long shoots be trained along the fence for the first year or two, they will take care of themselves afterwards, with a little trimming now and then. The Evergreen Rose (R. sempervirens) is a strong and rapidly growing variety, unsurpassed by any other for covering walls or roofs and fences, and it is very hardy. There are two or three varieties of it, some producing red and some white flowers. It may be planted in almost any situation, and is a sure grower. For covering arbours and covered ways it is an excellent subject.

Clothing walls.—A garden of creepers! Yes, why not? If anyone likes to carry out the idea, a most interesting garden could be made of nothing but creepers, twiners, and climbers. Not a garden of trim formal beds, I grant, as the growth of such plants could not—in fact, should not—be kept within set bounds. What groups and clusters of climbing Roses, Honeysuckles, Jasmines, Clematis, and Ivies one might possess in such a garden! One of the prettiest effects I have seen lately was a Larch with a stout plant of the common sweet-scented Clematis Flammula, or Virgin's Bower, festooning and hanging about its graceful branches. But, apart from the garden of creepers in the many phases of gardening which the future may develop, creepers are destined, I think, to be more used. The common way of clothing buildings or walls is to plant a heterogenous mixture of creepers, and allow them to blend as they will; the strong growers rambling up to the top of the wall or building, and the weakly plants remaining to fill up the base, being occasionally crowded out during the struggle for life which ensues. For old-fashioned houses this mixture is not inappropriate, as there is a degree of freedom about it which seems to accord with the rambling character.

In a village not far from where I am writing there is a house covered with Pyracantha, and it is without exception the most presentable in the place. Every passer-by stops to look at the heavy masses of bright crimson berries nestling among the dark green foliage. The whole front of another house is covered with the small-leaved gold and silver Ivies, and a very pleasing effect is produced with very little labour, as but little pruning is required, the growth being naturally slow and hugging the wall closely. Another interesting wall plant is Veitch's variety of the Virginian Creeper, its metallic-tinted, close-growing foliage being even more effective than the gold and silver Ivies, with which it associates so well.

In thinking over this matter of the creepers and their uses, those who have not made a study of them may be aided in the consideration of their merits if they group them according to the season in which they are most effective. Cotoneaster microphylla and C. Simonsi are desirable wall plants for winter effect, as well as for training up a pole or planting on the top of a mound of rock-work. Escallonia macrantha, Berberis Darwini, and B. stenophylla are useful to cover walls up to 12 feet or 14 feet. Chimonanthus fragrans should be planted for the sake of the delicious spicy scent of its somewhat inconspicuous flowers in winter when the foliage is down. Garrya elliptica produces its long singular catkin-like flowers freely in winter, and will grow from 20 feet to 30 feet high. Ligustrum japonicum (Japanese Privet) is a good wall plant, as is also the Laurustinus. Euonymuses are beautiful, the small-leaved variegated kind (radicans) being specially effective for covering low walls, to which it clings somewhat after the manner of Ivy. It grows slowly, but the effect is very good when the wall is covered with it. Elæagnus pungens variegata, and in warm situations Photinia serrulata, may be so employed. In sheltered places in the south of England Magnolias are grand-looking plants in winter, their large leaves being so noble; but in many places they require to be protected in severe weather, and therefore should not be used for winter effect.

For the spring we have also a long list. Clematis montana grandiflora, Wistaria sinensis, and alba, the white form, are both good for covering high walls, or for training over anywhere where rapid growing creepers are required. Forsythia viridissima and Cydonia japonica are plants of lower growth, but are both very desirable for making out the outlines of panels, or for covering the face of a buttress or pier. The yellow flowering Jasmines, chrysanthemum and revolutum, flower early. All the Vincas are useful for covering low walls, especially elegantissima, and they thrive well in a north aspect; and lastly, the spring offers us a few Roses, for the old pink China or Monthly is the earliest as well as the latest to flower, and Gloire de Dijon generally opens its first flowers early in May.

But it is in the summer when creepers are at their best, for then the Rose, the Clematis, the Honeysuckle, the Magnolias, and the Jasmine are in season; and what visions of sweetness and beauty the mere mention of their names conjure up! Roses alone are capable of transforming the most commonplace buildings we see around us by hiding their defects. Besides the plants named for summer there is the Passion Flower (Passiflora cœrulea), but it is only fit for sheltered places, as it is not quite hardy in our coldest seasons; still, if the base of the plant is sheltered with some dry Fern, it will spring again from the base. The Stauntonia latifolia is a very rapid-growing creeper in the south of England. The Birthworts—Aristolochia Sipho and A. tomentosa—are good and useful climbing plants, the last-named having silvery leaves. Then for warm sheltered places the Ceanothuses are very beautiful, producing freely blue flowers of various shades. And, besides the hardy creepers, a long list might be added of summer or annual creepers, bright and effective, such as the Tropæolum, Lophospermum, Maurandia, Convolvulus, and Ipomæa, in many varieties. In the autumn there is also much beauty of leafage and fruit or berry, if there is less of blossom. The Virginian Creepers are then in all the splendour of crimson and bronze; and the Japanese Honeysuckle is resplendent in its network of gold, with the bright berries of the Pyracantha and the Cotoneaster—most fitting and appropriate autumn decorations. Again, hardy climbers in gardens should for the most part, be what they are in their native places, trailing over trees, or shrubs, or stumps, or banks, and, in addition, over such artificial supports as railings, rustic work, &c. No plant bears repression and continual pruning so ill as a vigorous climber. In that way, moreover, its beauty can rarely be well seen. The evergreen or other shrub that does not climb is often more amenable to training on walls, as, for example, the evergreen Euonymus, the Pyracantha, and certain evergreen Barberries. The value of the hardy native American and other Vines for covering wall-surfaces must not be forgotten. We have seen them clambering up forest trees, spreading into huge masses of fine foliage on the ground, and sending out long arms to find out the nearest trees. They clothe with equal and exquisite grace the cottage wall. The flower gardener who works with creepers and trailers cannot well turn his back upon such a rich resource.

Shrubberies and flowers.—There is one feature in our parks and gardens which requires thorough reform, and this is the shrubberies with their miserable fringe of earth in which nothing but poor bedding plants will grow, and these badly. Parks and gardens are too often designed to look pretty on paper, the shrubberies and beds forming harmonious combinations of flowing curves in the plan. The ground plan of a garden has no business to look pretty, and flowing curves on paper are almost certain to be tame and ugly curves when carried out in walks and beds. Gardens should be designed and staked out on the ground they are to occupy—not drawn on paper and then transferred to the ground. The main difference between real mediæval building and modern imitations of it is that the old work was staked out on the ground from a rough sketch and the details filled in as the work proceeded; whereas, the modern work always fails in picturesque effect, because it always looks like a built drawing. Our gardens have exactly the same fault—the shrubberies never look as if Nature had planted them; there is about them an utter absence of change, of variety, and of sudden effects and surprises. There are grand opportunities now for persons laying out new gardens to boldly go in for a new style of shrubbery, and to do away with the great belts of shrubs which wind along near the principal walks. They might likewise abolish all evergreens of the rounded or bee-hive shaped type; plant instead single trees, groups, and masses of fruit trees which are ornamental in flower as well as in fruit, flowering deciduous trees and shrubs, with here and there a feathery Birch, a graceful Linden, or a Mountain Ash for its scarlet berries. Leave alleys and glades running in amongst the trees—here giving a glimpse of distance; there narrowing and passing out of sight. Let these vary in width from 3 feet to 50 feet. Plant here a great mass of Pansies, at another place some hundreds of Primroses of all colours; here a mass of alpine Auriculas, there a sheet of Gentianella or spring Phloxes, or Daisies, or Daffodils, or Anemones. In smaller groups plant the larger subjects, clumps of Foxgloves, Campanulas, Phloxes, &c., each in its own proper soil and station, and let the tongue or spit of ground where two walks meet be a great bed of mixed plants. Every yard of a garden would in this way have its own distinctive character. The succession of flowering would in this way be along the borders instead of being all mixed up. We have had illustrations of Fern dells; why not institute a glade of Pansies? A little winding walk with some 30 feet of gently sloping and undulating bank on either side filled with Pansies only; the mixed flowers for ground and the selfs in clumps 6 feet or 9 feet across; here a patch of blues, there a patch of russets, here a clump of whites, there a clump of crimsons, and so on through all the other colours. Many other plants could be treated in the same way. Rose walks might be made, or Clematis walks, or both combined. All that is required is that people shall avoid doing as their neighbours do, and strike out new paths for themselves. The besetting sin of the English mind is a tendency to get into ruts. One man starts a way of doing certain things, and all others follow like a flock of sheep. In things involving such a multitude of details and varieties as gardening there is no right way, but a great many right ways and a great many wrong ones; one of the worst ways is that of making all gardens consist of the same features, and every hundred yards of a garden like every other hundred yards. The shrubbery may itself be made a fair garden instead of the museum of crowded half-dead trees and shrubs it usually is.—J. D.

ROSERIES, PAST AND PRESENT.

Up till the present time the Rose has almost always been confined to what is called the "rosery," and has been but sparingly used for furnishing borders and shrubberies in connection with other plants, the curious and erroneous idea prevailing with some that Roses do not associate well with other forms of vegetation. In the "Gardener's Assistant," published only a few years ago, we find the writer on Roses stating that these should be confined to a rosery, because "the stems of standards, &c., contrast unfavourably with shrubs, and, indeed, all other vegetation which is not of a dwarf character"—a strange assertion, when one remembers the unnatural appearance standards present arranged, as they generally are, in roseries in the objectionable dotting system. When the varieties of the Rose were few and poor, there was less objection to the rosery; but it was not always planned quite in the same way as modern roseries, many of which are of the most formal pattern it is possible to conceive. One, for example, we were acquainted with consisted of a parallelogram-shaped piece of ground, nearly half an acre in extent, intersected at right angles by walks, the plots being filled up on the general mixture principle—standards, half standards, and dwarfs alternating in rows set in parallel lines with the walks. There was nothing beautiful nor artistic in such an arrangement. Another example, designed by a modern landscape gardener, consisted of concentric rings laid out on Grass, with a fence of iron pillars at the outside, on which climbing Roses were trained as festoons. This looked better than the straight bed mentioned, but had an artificial and inartistic look also, as all such designs must have. Rose planting and arrangement, indeed, is conducted on much the same lines as it was some sixty or eighty years ago; but Loudon did so far anticipate modern taste in flower gardening that he recommended some of the more distinct species or varieties to be massed by themselves, which was a step in the right direction, had it only been more generally adhered to. Formal roseries are just as objectionable as the formal "bedded-out" garden in its barest and poorest aspect, if not more so; and it is no excuse for their existence that they enable the cultivator to give the bushes special culture, for this can be given equally well in more natural situations. We are not speaking of growing Roses for exhibition purposes, but of their culture for the general decoration of the garden. Hitherto this purpose has been almost quite neglected, because the "rosery" has always absorbed the main supply of plants; and one of the saddest facts connected with our Roses and their culture now is the mortality amongst them annually. Whether due to the severe pruning the plants are subjected to or their poor constitutions, we do not need to determine at present; but it is certain that the mortality has become greater since varieties have so much increased and severer methods of pruning have been adopted.

Places for Roses.—These should be everywhere in the garden where they will grow. We cannot have too many Roses, or Roses in too many places any more than we can have too many other popular flowers which we are now trying to extend so much. But, in order to extend Rose plantations, especially where the cost of procuring and keeping up a stock is a consideration, we must abolish the "rosery" and send its contents to places where they are more wanted. We never saw a garden yet where either places or position needed to be invented for Roses. The walls, fences, and shrubberies, &c., will always absorb as many of the climbing or trailing varieties as can be spared and cultivated to good purposes, and numberless aspects and positions can always be found for Rose bushes. These may be in beds, borders, or in shrubberies; and if it be true that the best and most pleasing effects with garden flowers are produced by planting them in good, visible, striking groups or masses, then that is assuredly the way to use the Rose, and no plant is likely to be found more accommodating and useful in this way. At present we mix summer and autumn flowering sorts and their varieties indiscriminately; but, on the grouping principle, we should have to divide these, or, as occasion required, mix them with a view to a succession of bloom, as far as possible keeping the colours to themselves. If this system were generally adopted, it would tend to shorten our Rose lists sooner than any other plan one can think of, for gardeners would soon discern what varieties suited their purpose best, and they are strongly conservative in making changes. Thus, we should have bold masses of the vivid General Jacqueminot and its race, Gloire de Dijon, John Hopper, La France, Souvenir de la Malmaison, Victor Verdier, Moss Roses, and all the best and most striking kinds, which are comparatively few in number. For the ramblers, where they could be grown unrestrictedly, many suitable positions could be found, where, Brier-like, they could grow at their own sweet will, and produce masses of bloom to be estimated by square yards, as we not unfrequently see on older varieties which have, by chance, established themselves and made headway, aided by a good soil and favourable position. It would be just as easy to cultivate Roses in this way as a formal rosery, and it would enable us to choose the most favourable positions for the more tender and select kinds. The plants could be manured, pruned, and tended just as readily as under any other system, and they would look in their natural and proper place, and afford more pleasure and satisfaction to the owner than they could possibly do if all were planted in one place in the formal, and often inconveniently situated, rosery. This system of planting would, some may think, unduly restrict the number of varieties in a collection. This it would do, if only one sort were planted in each place—a most desirable object under certain circumstances. But, by planting several varieties of the best reds in one place, yellows in another, and so on, the number could be extended as much as desired, and the plan would soon show which were the best in their respective classes, and weeding out would follow as a natural consequence.

Certain varieties of Roses could also be mixed with other suitable flowering subjects in the same way that we plant Gloire de Dijon Rose among the purple Clematis now. Roses would do well among Rhododendrons and Azaleas, and be benefited by their shelter, and would flower when the former's season of bloom was over, as well as with many other things that will suggest themselves to the reader. The Rose patches, wherever they happen to be, could also be carpeted over to better purposes than beds of most other things. They might be literally crammed with spring and summer flowers of dwarf, low habit, including many kinds of bulbs, Saxifrages, and the like, which would just suit the Roses, and afford protection to their roots in winter, for no cultivator doubts that one of the principal causes of mortality among Roses is the absence of a covering to the soil. Even the Brier dies under such circumstances, while in the hedgerow or the copse it braves our hardest winters, and so do many old and good Roses under similar circumstances.

Thick planting.—Another fault to be mentioned is thin planting. The desire to cover much ground has been the cause of this, and Roses are generally planted as far apart as Gooseberry bushes, a great mistake in the case of plants pruned to stumps every winter as Roses are. Both for effect and for shelter, Roses should be planted thickly according to their vigour, and "standards" should only be used to fill up backgrounds of clumps of Roses. Standards have been condemned as in bad taste, and certainly it cannot be said that a mere addition of bare stem can in any way add to the beauty of such a stiffly habited plant as the Rose. But when standards are planted in the formal manner which has been so long the fashion, either as single objects on lawns, or in regular rows in the rosery—heads all on the same level—their objectionable appearance is forced prominently upon the sight. Standards would be useful in clumps placed against shrubs and other objects, where they could be planted according to their size, to fill up the background, the front being occupied by dwarf plants. Standards are not, however, a necessity any way, and they are both dear and more troublesome to grow.

Large bushes of Roses and pruning.—It is a pertinent question to ask whether Rose pruning is not carried a trifle too far in general culture. There is no bush in the garden so closely cut in annually as the Rose, and it is to be feared the advantages of the practice are more apparent than real. Close pruning is advocated to induce vigorous growth, and the weaker varieties are pruned most severely for that reason. Under good management and protection this system answers well enough, and is the best plan for the nurseryman, who has to keep his plants within bounds for commercial reasons; but in private gardens, where a show of Roses is desired, and desired soon, such severe pruning is best avoided. It may be different in the case of Roses grown for exhibition. One fact is pretty well ascertained in regard to pruning, which is, that the most severely pruned plants suffer most in severe winters, and are shorter lived than plants grown as unpruned bushes. It is quite an error to suppose that Hybrid Perpetuals, Bourbons, or Teas, &c., grow weakly and poorly by being left to develop their tops as largely as they will. All the ordinarily vigorous kinds will make quite large bushes if allowed, and will continue to grow and do well for many years. Climbers excepted, the plants never grow above a certain height, but they spread out in circumference, and throw up strong shoots annually, which produce vast quantities of flowers season after season. The cottager's Roses are never pruned, and they are sometimes of remarkable age and size. Ten years since we split up and planted a large number of Hybrid Perpetuals, Teas, and other Roses as a hedge, and ceased pruning them, and they are now all fine, large bushes 4 feet and 5 feet high. The vigour and length of the annual shoots produced is surprising. These are often over 5 feet long and an inch in diameter near the base—just like Brier shoots, in fact. None of them have died, and the plants must now be twenty years of age, some of them probably a good deal older; while in the borders, among the pruned and select Roses, our losses have been great. During the severe winters of the last seven years these old bushes have stood unscathed, and are at present as vigorous as ever. The quantity of Roses they produce is surprising, and drives the productions of the highly cultivated, much-pruned plants quite into the shade. A man has frequently filled for drying several large vegetable baskets with Rose petals from them in the course of an hour without making much difference to the show of flowers. The way the unpruned Rose behaves is this: The plant, as soon as fairly established in a good soil, throws up strong shoots, and plenty of them; and the following year these shoots break their buds freely along the stem, and each branch produces a mass of bloom, which, after a shower, is so heavy as to weigh the branch down almost to the ground. The older branches flower equally well, but any pruning such bushes need consists principally in thinning the oldest branches out and leaving the younger ones. What takes place in the case of dwarf bush Roses happens just the same in climbers, only the mass of bloom is ten times greater. It is the climbing or scrambling Roses that show what the Rose is capable of when cultivated in this natural and free manner. One of our most attractive specimens is an old double white Ayrshire Rose growing in a group of common Laurel in the shrubberies. We cannot tell how old the plant may be, but it has probably been in its present situation for thirty years, struggling the best way it could to keep its place among the tall-growing Laurels, sometimes sending out a shoot of white flowers on this side, and sometimes on that side of the clump of bushes, and sometimes scrambling up to the tops of the tallest limbs, and draping them with its blossoms throughout June and July. Nearly three years ago we had the Laurels headed down to within 6 feet of the ground, leaving the straggling limbs of the Rose which were found amongst them, and since then it has grown and thriven amazingly, and now fairly threatens to gain the mastery. We had the curiosity to measure the plant the other day, and found it rather over 70 feet in circumference. Within this space the plant forms an irregular undulating mound, in all parts so densely covered with Roses, that not so much as a hand's breadth is left vacant anywhere, and the Laurel branches are quite hidden, and, in fact, are now dying, smothered by the Rose. A finer example of luxurious development we never saw. The plant has been a perfect sheet of bloom for a month or more, and there are thousands of buds yet to expand, and hundreds of bunches of buds have been cut just at the opening stage, when they are neater and whiter than a Gardenia, to send away. The tree has never received the least attention or assistance, with the exception of the removal of the Laurel tops before mentioned, to let the light into it. It is growing in a tolerably deep and strong dry loam, and this, together with head room, seems to be all it requires. There are far too few examples of this kind, for our efforts in this culture have not been in the direction of showing what could be done with the Rose as a tree or bush. The common Brier teaches us a good lesson in Rose pruning. No one has ever ventured to say the wild Rose was out of place in the hedgerow; and how does it behave there? It forms a mighty mound of branches, the older stems dying down as the young ones accumulate, till a large bush is formed, covered with flowers in their season. On the roads and lanes we have seen wild Roses which old people say had been there ever since they remembered, and I have taken note of Briers in the woods that, to all appearance, may have been more than fifty years of age; and yet our strongest-growing Roses on Brier stocks in gardens are short-lived. From these facts, the Rose grower may learn that, in order to have Rose bushes and plenty of Roses in his garden, his plan is to plant good kinds of known repute for hardiness and vigour, and confine his use of the knife to thinning out the shoots principally.

Roses on lawns. — Climbing and strong-growing Roses make very ornamental objects in a few years. Climbers want something to support them at first; and for this purpose nothing is better or more convenient than four or five thin flat iron rods

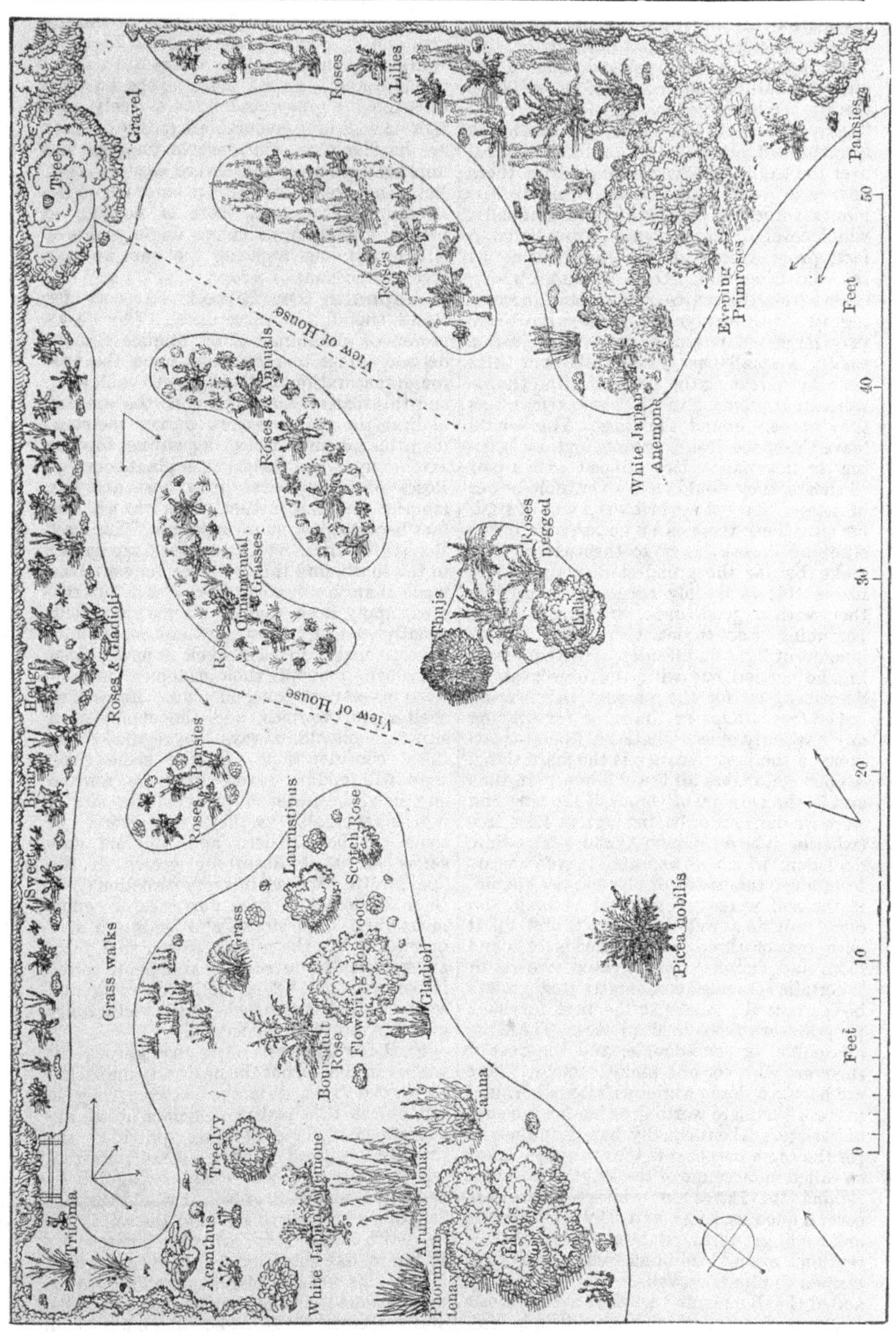

Plan of a rosery, varied with other plants, in a London suburban garden.

bent over each other at right angles, with their ends in the ground, so as to form a circular mound about 4 feet high and 6 feet through. The Rose tree is planted under the hoops, in the centre of the circle, and as it grows a limb is trained each way to form the foundation of the future plant; after the first or second year it is left entirely to itself. The supports are soon covered, and a large and natural-looking mound of Roses is the result. We have seen specimens of this kind twenty years old in which the wood had accumulated about 2 feet or more deep all over the trellis, and yet nowhere was there any dead wood to be seen, owing to the plants throwing out fresh limbs annually, which covered the old ones. The plants, in fact, grow exactly in the same manner as the wild Brier, which keeps sending up long shoots from the centre of the plant, increasing its size every year. Those who desire very large mounds may have them just as easily as small ones, and in about as little time, by increasing the size of the iron framework and planting the Roses—as many as they choose—round the sides. This would leave the space inside vacant, and, by keeping the internal surface clipped with a pair of shears, they would have a veritable bower of Roses. Except against walls and in similar situations, there is no occasion to prune climbing Roses. Left to themselves, they make by far the grandest display, and to insure this it is only necessary to provide them with a good, deep, strong soil at the beginning, and to let them have a fair amount of light on all sides. Whether planting be carried out with the object above described, or for the purpose of covering naked tree stumps or limbs, or for draping any unsightly object whatever, liberal treatment in the first instance is the main thing. A good soil makes all the difference in time and in the permanent vigour of the tree, and were we desirous of having a great Rose tree (whether it be a common Ayrshire or a Gloire de Dijon, which we expected in a few years to produce thousands of blooms), we should, if the soil were not naturally strong and deep, provide a well-drained pit, and fill it with two or three good cartloads of sound loam and manure; thus treated, the result is certain, provided an unrestricted growth be permitted. Amongst the best for such purposes are the Ayrshire Roses. These are all double, or semi-double, and the best of them are white or of a pinkish colour. There are hardly a dozen varieties altogether, and most of them are worth growing for the sake of variety. About equally hardy and useful for the same purposes is the evergreen Rose, so called on account of the length of time it retains its leaves in winter. There are several good varieties, and they are vigorous and rapid growers, and need little or no attention, except thinning when they are trained to pillars or walls. To these may be added the Boursault varieties, a few of the Bourbon kinds, of which Souvenir de la Malmaison is the best example, and Gloire de Dijon and Maréchal Niel among the Tea-scented class. The last two are worthy of all the attention that can be bestowed upon them. Gloire de Dijon comes in early and goes out late, the second growth often producing flowers in abundance late in the season, and in the south the Maréchal Niel does well either on the wall or as a rambler in the open ground. As far north as the Tweed it occasionally grows and flowers freely on a roof or wall in a favourable exposure. Both are hardier than some people imagine, and unpruned plants stand severe winters best. Both, too, are great ramblers, and, where the situation suits them, there is nothing to prevent anyone from having single plants of either quite as large as the one we have given an account of above.

Aspects for Roses.—Aspects for Roses should be warm ones. The Rose's powers of endurance in an English climate depends, next to protection, upon the thorough maturation of their growth each year, and this maturity depends upon the amount of heat the plants receive during the summer, this amount of heat depending, to some extent, upon the position the plants occupy. Roses on sunny sites grow best and live longest, and the farther north we go, this fact becomes the more apparent. The best Roses, like the best Peach trees, are grown in the south, and this is better known to the trade than to private growers. All nurserymen supply Roses, but all do not grow them equally well, and many northern nurserymen have to make up their stock annually from the south, or supply their customers on commission—a very common plan. Roses love shelter and warmth, and the choice of a situation should always be regulated by these considerations. At the same time, they will endure severe exposure, and no one need hesitate to plant if the site is not as favourable as they could desire. In some of the northern and midland nurseries acres of Roses are grown in the open, quite exposed in every direction; but then the losses in such nurseries are enormous, and both stocks and buds are at a premium. For the private grower such facts are sufficiently instructive, and point to the importance of shelter, both for roots and tops, in all situations where the vicissitudes of temperature are uncertain.

Soil for Roses.—Any good garden soil will do for Roses, but the nearer it approaches to a good fresh loam the better. They do not object to a rather tenacious loam, approaching a clayey texture, provided the drainage is good; and in some nurseries, where Roses are successfully propagated by thousands and well grown, it would surprise fastidious growers to see what the soil is like—a stiff, clayey loam, which in summer lies in lumps, like baked bricks, upon the surface. Yet in this soil the plants make fine growth and flowers—on the Brier stock especially. But the best soil for a general plantation of Roses is one neither too light nor too heavy; for Roses in their great variety differ in their rooting power, more particularly when grown on their own roots, the more tender Teas and such like doing best in moderately light and rich composts, and the strong briery climbers

Roses on palings.

in deep, strong loams, in which they grow quickly and live longest. The habit of the variety itself indicates the nature of the compost likely to suit it. Strong growers are most likely to succeed in strong soils, and weak growers in soils that are light and open; and in planting, it is not a difficult matter to give each kind the soil it likes best, at least in starting at the beginning. —S.

Roses on palings.—One of the prettiest and most useful purposes to which the Rose can be turned as a climbing plant is to cover palings or fences. The accompanying illustration represents a portion of a long fence in the Waltham Cross Nurseries, formed of common deal laths, which, otherwise bare and cheerless, is made attractive and beautiful by a covering of Roses, which, when well grown and judiciously encouraged by careful pruning, gracefully entwine themselves in and out the pales. The varieties best adapted for this purpose are Noisettes, Hybrid Chinas, and Tea-scented classes, together with some few of the Bourbons. If the exposure be northerly or the fence high, many of the Boursault, Ayrshire, and evergreen Roses will be found very suitable. The following list embraces all shades of colour, and will give a succession of flowers throughout the summer and autumn:—*Noisettes:* Celine Forestier, yellow; Fellenberg, crimson; Aimée Vibert, white; La Biche, flesh; Rêve d'Or, coppery yellow; Bouquet d'Or, salmon-yellow. ***Hybrid Chinas,*** *&c.:* Blairi No. 2, blush-pink; Charles Lawson, rose; Chenédole, bright crimson; Madame Plantier, white; Paul Ricaut, crimson; Vivid, bright red. *Tea-scented:* Gloire de Dijon, salmon; Belle de Bordeaux, pink; Belle Lyonnaise, canary-yellow; Madame Berard, salmon-yellow; Sombreuil, white; Madame Trifle, salmon. ***Bourbons:*** Souvenir de la Malmaison, pale flesh; Sir J. Paxton, bright rose. When pruned the plants should not be much cut back, but rather have the weak shoots removed altogether, as it is masses of flower, in preference to individual fine specimens, that are to be sought after.

Wall Roses.—There are not many positions in which Roses are grown where they are more effective than against walls when well managed. The protection thus afforded to the Tea-scented varieties is peculiarly favourable to growth. Nevertheless, judging from the condition in which they are often seen, Roses are frequently the least inviting of all wall plants. This arises from several causes, which may be enumerated as follows: First, the selection of unsuitable kinds; second, after a few years the soil in which they are first planted becomes exhausted, and sufficient means are not taken to enrich it; another is unskilful training, which, in the case of strong-growing sorts, permits the lower portion of the wall, if a high one, to become bare of young flower-producing shoots. The prevalent reason, however, of such an unsatisfactory condition of wall-trained Roses is the difficulty of keeping insects in check. Roses, more than the majority of plants, are very susceptible to the attacks of insects when trained to a wall. The pests to which they are more particularly subject are aphides, red spider, and the Rose maggot. The difficulty in freeing the plants from the worst of these is all the greater on account of the walls offering an obstacle to the application of most kinds of liquid, excepting clean water, for their destruction. Most of the insecticides used in a liquid state, including Tobacco water, leave ugly stains on the walls, unless washed off with clean water before they become dry. But in this case, whatever is employed to destroy the insects is also washed off before it has effected the desired object. Since, however, paraffin in a very diluted state is proved to be such an effectual insecticide without injuring the plants, and at the same time is cheap and does not leave traces on the walls, it answers the purpose well, and so the worst drawback connected with the culture of wall-trained Roses is overcome. Within the last year or two I have regularly applied it against a light-coloured wall, where anything that left a mark could not have been used. Prevention is always better than cure, and as soon as the Roses are pruned and nailed or tied to the trellises, they should have a good washing by either syringe or garden engine with paraffin water, for although it may be supposed that severe frosts destroy the eggs of aphides, such is not the case. One recommendation which Roses against walls possess is that, except in the colder parts of the country, they thrive better on east, or even northern, aspects than they do on the southern, which is no small advantage; for on a northern aspect there are few other plants that will flower successfully. As to the sorts suitable for walls, much depends upon the height and space to be covered. Any of the strongest and free-growing Hybrid Perpetual varieties may be grown, as well as the more commonly planted Tea-scented kinds; but there is much to be said in favour of the Teas, such as their more continuous habit of flowering and their handsomer and better foliage. Besides, when well managed, they succeed better against a wall than in the open, which is not so in respect of the Hybrid Perpetual kinds. Many a gloomy-looking and unsightly north wall, instead of being an eyesore, might be made attractive if clothed with well-selected Roses and carefully attended to. But for this purpose, more than in any other position, it is requisite that the plants be either on their own roots or grafted very low, for unless young shoots are produced annually from the base, the essential point to a well-furnished wall is absent.—T. B.

Roses for suburban gardens.—It can scarcely be too often reiterated to the possessors of suburban gardens that standard Roses are about the worst investment they can make. In most suburban gardens near London and other large towns standard Roses will not grow. Where they can exist they do not thrive, and never produce a flower such as can be bought in the streets for a penny. Even in a country garden the standard Rose, as generally seen, is only a vexation and a disfigurement. The reason is not far to seek. To grow the Rose well as a standard requires a good soil, a sheltered, sunny situation, a slightly moist atmosphere, and, above everything, soil open as a Cabbage or Lettuce bed, so that it can be hoed, mulched, and manured as required. The Roses must have the bed all to themselves wherever they are planted. Now, when amateurs go in for standard Roses they want to use them in all sorts of impossible ways. They want to plant them on the Grass, with the turf close up to the stem, or in nice little round beds, one in each, with bedding plants all round them, or rising out of well-kept beds of Rhododendrons; in every way, in fact, in which a standard Rose cannot be grown. The Rose will do very well in many of these positions, but the English Brier or Dog Rose, which forms the root and stem of the standard, will not. Its native place is the hedgerow, where its stem is sheltered alike from the cold of winter and the heat of summer, and where its roots are always cool and moist. Placed in a bare plot, where its stem is bitten by winter frost and scorched by summer heat, it quickly dies, and the Rose budded on it dies with it. The Dog Rose will not stand smoke, and so cannot be made to live in any but country gardens.

Bush Roses, on the contrary, always thrive in a satisfactory manner, and produce good flowers if suitable varieties are chosen; they have also a better effect in the garden than standards. Many kinds thrive fairly where a Dog Rose could not live, and once having got the Roses, it is open to anyone to plant Briers and get them into a healthy state, and then bud the Roses on them, the result being far more healthy standards than can be purchased at a nursery. I saw a shoot of Dog Rose 8 feet high flowering freely in a garden in the outskirts of London, where the planted standards were either dying or dead, showing that the garden was in such a position that Briers might be grown and established, and Roses budded on them, while too unfavourable for made standards to gain sufficient vigour to thrive after the check given by transplanting. The best Roses for suburban gardens are some of the old kinds which have been discarded from the Rose nurseries to make room for the exhibition kinds. The very best are the old white or alba Rose and the Perpetual Damask. The Maiden's Blush is also a good hardy Rose. These cannot now be purchased from the Rose nurseries, but have to be collected from old-fashioned gardens where they still grow and flower—perfect instances of the survival of the fittest. As to the form of plant, there is no doubt that, for general garden purposes, which means for ninety-nine out of every hundred people who wish to grow Roses, those on their own roots are the best. When fairly established they produce almost as good flowers as the budded plants, and a greater average of blooms useful for all purposes except exhibition. They keep up a constant supply of young, healthy wood by means of suckers from the root, and in a severe winter there are no losses when the standards are murdered wholesale. The only difficulty with Roses on their own roots is getting the plants. For some unfathomable reason or other, the nurserymen stick tenaciously to their budding, so that the only way to get own-root Roses is to buy the plants as budded dwarf bushes, and then make own-root Roses from them by means of layers and cuttings, or by inducing the budded plants to make roots of their own.

To those who still believe in standards I would say—Take care what you buy. Very many of the standard Roses which possessors of small gardens buy are made on Briers, which ought never to have been budded with a valuable Rose, and which cannot possibly thrive and produce good flowers. Choose only those standards whose bark looks fresh and young, and of a greenish or brownish colour, and have nothing to do with those in which the bark looks grey, dried up, or cracking, and in which the remains of the stout thorns look old. Purchase no standard which has a club of wood at the lower extremity resembling a hockey stick or golf club, and avoid all which have few root fibres and any thick root-like pieces projecting from the lower part of the stem these will only produce persisting crops of suckers from the root, and will refuse in most cases to send sap up the stem to the Rose budded at the top. The only good way to grow the Rose as a standard is to buy suitable young Briers every autumn and bud them every summer, discarding the old plants as soon as they show signs of failing.—J. D.

ECONOMY AND WASTED EFFORT.

In trying to carry out any of the suggestions in this book owners should remember that, as a rule, there is no merit in mere extent of area. This goes for something in a chain of mountains, in a wide sweep of plain or prairie, or in a broad expanse of ocean, but in a garden mere size is nothing. There is no doubt whatever that great size is an emeny to progress and good gardening; and this is especially true when the reduction of expenses is common. Our gardens, to a great extent, are laid out in an old-fashioned and bad way; the kitchen garden especially. There are so many needless walks and Box edgings and other *impedimenta* in it, that nobody can get to work in any simple way. Half the time is lost in "niggling about," cleaning the feet, seeing that edges are not injured, or in repairing them when they are; in fact, a lot of nonsensical labour is gone through, the time expended in which ought to be bestowed on the growth of good crops. Hitherto, we have had so much cheap labour to employ in such cases, that this did not matter much; but now things are taking a different turn, and we shall probably soon be in the position of our cousins across the sea, who long ago have had to take means to cultivate their gardens more economically. There are many large places in the United Kingdom where as much space is devoted to walks in the kitchen garden as would, if thoroughly cultivated, be quite sufficient to grow vegetables enough for a family. Our gardens being for the most part permanently and expensively laid out, the adoption of the true and simple plan for a kitchen garden may often be difficult, but, in all cases, excessive size may be pruned in. Where the kitchen garden is a maze of complication, clearings may be made, and greater simplicity of plotting secured.

Grass.—Here we have the endless shaving of lawns. No one admires more than we do the soft turf, which is the glory of our gardens, and forms the most delightful playground; but who can say that it is needful or wise to mow, as people have boasted of before now, 40 acres of kept lawn? Carpets we want near our houses, and cannot take too much care of them, but the wide stretches of acres that are mown in many large country seats would be better cut at the proper season. Growing Grass, flowering Grass, is one of the most beautiful things in Nature, and our trees and shrubs, and park or garden lawns would in many positions be quite as lovely with the Grass allowed to grow as when shaven off—we think much lovelier. There are new ideas coming in which this lawn Grass helps us to carry out. Some think that the Grass itself should be a beautiful garden; that we should see in it, as we often see on an alpine meadow, fair bulbous flowers and other plants which will grow in English as well as in alpine turf. By allowing the Grass to grow in spring and till maturity, this phase of the wild garden will be enjoyed and plants will come up year after year to reward us. A squire in Oxfordshire, who carried out the idea of the wild garden in his Grass, was pleased to find a great deal saved in mowing, having left much of his Grass to grow so that he might see the flowers that he had sown and planted bloom in it. Then he cut it down in the usual way. Of course he left some smooth near the house, but we remember seeing a flowering meadow on one side of the house which used before to be shaved once a fortnight.

Walks.—After the Grass comes the needless walks in such places. Our own landscape gardeners are a little more economical in these hideous things than are the French; but we very often have four times too many walks, which torment the poor gardener by needless and painful and stupid labour, always hoeing and weeding and salting—a contemptible and profitless business. The planning of these walks in various elaborate ways has been supposed hitherto to have some relation to good landscape gardening, but the presence of one needless walk often spoils all possibility of good landscape gardening in its vicinity. Walks are essential, but they should be boldly and carefully designed to go with one sweep wherever it is necessary they should go; they should be concealed as much as possible, and, as before remarked, reduced to the most modest dimensions. Generally, all flower beds are best set in Grass, and those who care to see them will approach them in that way quite as soon as if hard walks are brought near them. For the three or four months of our dullest season there is little need of any frequent resort to flower beds, but for the rest of the year the turf is better than any walk. We do not mean that a walk should not lead to the flower garden or any other interesting spot, but that every one not necessary for frequent use should be suppressed. Few have any idea how much they would gain, not merely in saving labour, but in the beauty of their gardens by abolishing walks where possible, and substituting turf instead.

Flower gardens.—The great wide half-cultivated kitchen garden, not green and full of life as a kitchen garden ought to be, but often spotted and starved, has its counterpart in the poor flower garden. In most places there is no real flower gardening at all; a few wretched plants are stuck out every year in the parterre, or whatever answers for it, and a few stunted hardy flowers are perhaps scratched in round poor shrubberies—so little love or labour is bestowed on the growth of flowers. In many places, which we need not name, miles of walks may be seen, bordered in many cases by long naked stretches of earth, as cheerful as the Woking Cemetery in its early years. Mere size is all such places have to boast of. The gardener, with his dozen or twenty men, is impotent to turn such a vast waste into a paradise; his time and his thoughts are eaten up by trivialities—barber's work, shaving Grass and weeding walks eternally. As for the poor flower garden, it is a farce. We went into one the other day in one of the finest places in England, as regards site and associations, and saw six masses of the lugubrious Perilla, as many of yellow Calceolaria, five other well-known plants in equally liberal masses—saw all this at one saddened glance, and then instantly turned aside to find relief in the trees around. This vulgar daubing is not gardening at all, and its continuance is partly owing to the error in the size of gardens. The gardeners say that, in the face of the annual trouble of this system and of their other work, they have not time or help to devote to truer gardening. To form a garden of Roses or groups of choice shrubs, or beds of Lilies or other noble hardy plants properly, as regards the soil and its depth that the bed might fairly nourish its tenants for a dozen years, is real gardening, and to such ends all good gardeners' labour should be directed. Instead of the never-ending scratching of large surfaces in autumn and spring, we ought to have a thoroughly careful and thoughtful preparation of one portion of the garden each year, so that it would yield us quiet beauty for many years to come without much other labour than surface cleaning.

Digging shrubberies.—The absurd digging every year of shrubberies is an appalling spectacle to anyone who looks at any large place treated in this way. The orthodox notion is to dig the shrubbery every winter. This is often carried out without giving a particle of food to the soil in the shape of manure, and harm is actually done to an enormous extent by mutilating the roots of the shrubs which come up to the bit of good soil there is to feed on. Anyone can see the result of this in large places every year, and also in the London parks. The labour and time wasted in this way, if devoted to the true enrichment and proper culture of a portion of the ground each year, would make our gardens delightful indeed. All our fair flowers require culture as well as the inhabitants of the market garden, but they very seldom get it. Few need be told that many shrubs, as fair as any flower requiring the shelter of glass, have been introduced into this country; but for the most part they have been destroyed by the promiscuous muddle and bad cultivation of the shrubbery. Our system is too wild, too "scratchy," and too thoughtless to admit of our taking some beautiful group of hardy shrubs—say the various evergreen Barberries —and giving them as much care as is given to plants in pots. Even the white Lily starves on the fringe of the impoverished shrubbery. In the majority of country seats and gardens a revolution must be carried out as regards size if we are to enjoy our gardens to the full. Doubtful departments must be suppressed and turned to the growth of meadow Grass or some other useful end. Sometimes, fair limitation of gardens might lead to improvement as regards the orchard, which is the rarest thing to find at a country seat. We never saw a really good orchard at a country seat but once, although the cry for more fruit is universal, and that orchard trees with Grass growing under them form one of the most pleasant of sights.—*Field.*

SUMMER BEDDING.

When the bedding system first came into vogue, it was no doubt the extreme brightness, or what we should now call gaudiness, that for a time caused it to hold the position in popular favour that it did; but, like every other hobby that from time to time takes possession of the gardening fraternity, it was soon done to death. Only scarlet Geraniums, yellow Calceolarias, blue Lobelias, or purple Verbenas were used; and, by way of a change, the following year there were Verbenas, Calceolarias, and Geraniums, the constant repetition of this scarlet, yellow, and blue being nauseating even to those who had little taste in gardening matters, whilst others with finer perceptions of the beautiful began to inquire for the Parsley bed, by way of relief. As a matter of course, such an unsatisfactory state of things could not continue; but yet the system could not for several reasons be given up—a very good one being that the great bulk of hardy flowers had been ruthlessly swept out of the garden to make room for bedding plants, and so, gardeners being, as it were, in desperate straits, the developing of the bedding system began, foliage plants of various colours being mixed with the flowers. Then followed standard graceful foliage plants, hardy carpeting plants, and now dwarf-growing shrubs are being used freely in association with the commoner types of bedding plants. Indeed, the improvement both in taste and arrangement, as well as in variety of plants, has been so rapid, that the most relentless opponents of the system have admitted that the system has some redeeming qualities. The writer, however, thinks that the strongest reason of all for its retention is its suitability to formal or geometrical designed parterres. Perhaps, the general re-action now going on in favour of hardy herbaceous perennial plants may quite oust the system; but, whether it does or no, so long as it is practised, the question of *colour* must ever be of paramount importance. Most people have their own notions as to what constitutes perfection of colour in bedding arrangements. This the writer has not attained to, nor is he certain that he has any decided preference for one colour over another; but he has very decided notions as to arrangement of the various colours, which are, that the whole shall be so completely commingled, that one would be puzzled to determine what tint predominates in the entire arrangement. This rule we have followed for years, and have had a fair amount of success in the working of it out. We are even still learning, our latest lesson being that, if any colour at all may predominate, it is what gardeners know as "glaucous," that is, a light grey or whitish green. Of such a colour the eye never tires, perhaps, because it is in such harmony with the predominant tints of the landscape, and particularly of the lawn. To successfully carry out the rule as to arrangement of colours here advanced, there are others which may be termed collateral—rules that must be studied. The first is that high colours, such as scarlet and yellow, must be used much more sparingly, that is, in less proportion than colours of a softer tint, for the obvious reason that these colours overweigh, as it were, all others. Secondly, there must be no violent transition from one colour to another—the contrasting of colours must be avoided as much as possible in favour of their gradual intermingling or harmonising—and, thirdly, the most decided or high colours, being the heaviest, ought to occupy the most central position of beds, or else be distributed in due proportion over the entire garden, to ensure an even than is generally grown in gardens where bedding out is practised to any extent. But, lest the owners of gardens of small dimensions may think it an impossible attainment in their case, it may be well to add that there is *colour* and *colour*, and that, if they cannot, owing to lack of appliances, &c., for raising plants, have elaborate designs and variety in colour, they may have an equivalent in the way of graceful foliage and beautifully tinted shrubs of varying hues and habit, from deep green to bright yellow, tapering, weeping, and feathery. Retinosporas, Cypresses, Yews, Yuccas, and many others not only associate well with all kinds of bedding plants, but with the various kinds

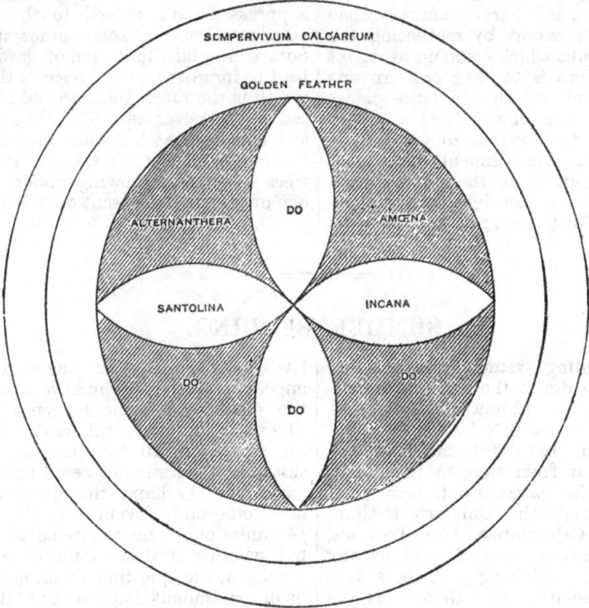

Bed of hardy and half-hardy plants.

balance of weight throughout. Further, when dealing with such colours, just use them in necessary proportion, and no more, and, if you err at all, take care that it be on the side of niggardliness. By close adherence to these rules, we have for years had no difficulty in producing a harmony of colour that has worn so well as to be just as welcome, when viewed at the end, as it was at the beginning of the season; for the quieter the colouring the more lasting is its enjoyment. And it is pleasant to observe the great advance yearly made in favour of what may be called neutral or the quieter tints—gaudiness, as applied to bedding out, being now the exception rather than, as was the case but a few years back, the rule. To fully carry out the ideal of colour here advocated, a great variety of plants is needed, though not more

of hardy Sedums, Saxifrages, and Veronicas; and these are all within the means of most owners of small gardens and might be arranged in bedding-out form, the shrubs for centres and panels, and the dwarf hardy plants for massing and carpeting. Such a garden, though somewhat sombre in colouring, would be infinitely more interesting than if it were planted after the orthodox fashion of bedding out.

Soil and cultivation of bedding plants.—Next to position, soil is the most important element in the formation of a garden, and yet how very rarely it happens, in selecting a site for a mansion, that the nature of the soil and its adaptability to gardening purposes is taken into account. The garden, so to speak, has to take its chance, and if the soil be not suitable, the inevitable has to be accepted. This often ends in the spending of a large amount of money in artificially rendering the soil fit for garden purposes—an item that ofttimes might be saved were the question of garden, as well as site of mansion, taken into consideration. In selecting a suitable soil for flower garden purposes, two things should be kept in view —first, the fact that an open or well-drained soil assists climate (that is, the more porous a soil is, the warmer is the ground, and so plants are better able to withstand extreme cold); and secondly, depth of soil. This second requisite is not of less moment than porosity of soil; for, unless there is depth, permanent large-growing trees will not flourish satisfactorily. Even for less permanent subjects, depth is just as important, as it renders unnecessary frequent manurings or dressings of fresh soil to maintain fertility. Whereever these conditions of soil exist naturally, flower and ornamental gardening become comparatively easy and inexpensive; but in many cases the very opposite conditions have to be dealt with, and though it is a hopeless task to attempt to rival a naturally good or suitable soil, a very near approach can be made to it when men and means are at command. The best soil for all purposes is what in gardening phraseology is termed good loam, that is, soil of a clayey nature, but sufficiently sandy not to be adhesive or plastic. The stiffer it is, so long as the latter is not the case, the more fertile will it be. Of the two conditions, light or heavy, the former is much to be preferred, because it is warmer, can more easily be deeply cultivated, and is more amenable to special treatment in the way of manuring or not, according to the requirements of certain plants. In dealing with heavy soil, the first requisite is plenty of drainage, then deep tilth, and the working-in with it of any material that will render it more porous, such as half-decayed leaves, mortar or brick rubble, charcoal, and ashes. If manure be needed, it should be used in the long straw state as it comes from the stables. One great mistake frequently made with regard to the question of soil is, that sufficient attention is not paid to the kind of plants—more particularly with respect to permanent trees and shrubs—that the soil of a given district is best suited for. Were this always done, we should see but few garden failures, and the gardening in different districts would possess an interest because of its increased variety. At present this is to be observed in very rare instances only, simply because of the apparently inherent disposition there is in the gardening mind to copy. If it were not so—if each possessor of a garden were determined on striking out a line for himself—the question of suitability of soil would soon right itself; for a man would then be too observant to even think of planting a Rhododendron in

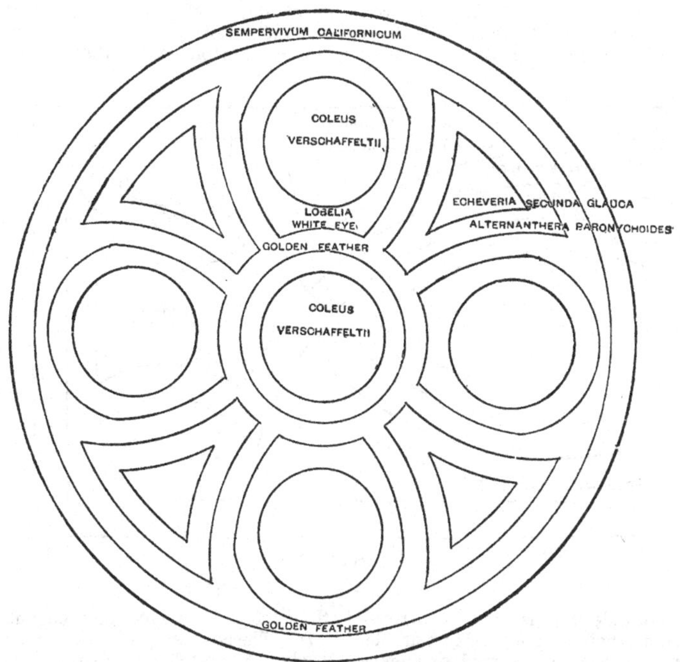

chalky soil because he had admired a friend's Rhododendrons that were growing in peaty or vegetable soil; and surely it is infinitely preferable to have a healthy Yew or Box, that one can behold with pleasure, than a sickly Rhododendron, notwithstanding its flowers. In flower and ornamental gardening, regular cultivation of the soil does not receive that attention that its merits demand. Those parts of it that are the least enduring—the parterre flower beds—are frequently the only portions that come in for yearly doles of manure. Of course, the annual dressing of these is needed to get the best effects; and we would say, by all means continue the same, but do not let it be to the entire neglect of hardy flowers and handsome shrubs and trees. These, though they will do a long time without fresh food, fully appreciate rich top-dressings of good soil or manure; and, indeed, it is only by so treating them that their grandest effects are developed, as may soon be observed by anyone if some are dressed and others left undone. The roots of coniferous trees and similar ornamental subjects are just as easily attracted by fresh dressings of soil or manure as are Grape Vines; and this being so, it follows that, if we value them, we are in duty bound to feed them as often as circumstances permit or their requirements demand. The regular routine of cultivation is not so imperative with respect to permanent plants as it is for annual flowers and bedding plants; hence the neglect.

Deep digging flower beds.—There may be nothing new in the fact that our flower beds require to be deeply dug occasionally, but the operation is so important that it can hardly be too frequently adverted to. Trenching is perhaps the proper term, and yet it scarcely expresses what I mean. The time to do it is, of course, when the beds are empty. We trench up our flower beds once in two years, and I find this sufficient for the purpose. We generally do it in autumn, after the summer bedders are removed, and before we put in our spring-flowering plants. As we have not a sufficient depth of good soil to trench the beds in the ordinary sense of the word—that is, to bring the bottom soil to the surface—we take out a couple of good barrowloads of the surface soil from one end of the bed and place it on one side; then we commence to dig up the bottom, bringing the surface soil from the next trench on to the top of that just dug up. By this means we are able to dig up the bottom and still retain the surface soil on the top; but where there is a sufficient depth of good soil I should prefer to trench in the ordinary way, that is, to bring the bottom soil to the top, as it would afford more change for the occupants of the beds. The advantages of stirring flower beds deeply are two-fold; it creates a wider field of action for the roots, and it also gives the roots an opportunity of getting down out of the reach of drought in a dry season, and it makes a better drainage in a wet one. A clever cultivator of my acquaintance, who has to deal with a thin, light, poor soil, has adopted the plan of deep digging his flower beds for many years past, and his success has been complete. Owing to a short supply of water, he found in a dry season that his flower beds presented anything but a respectable appearance. He adopted the plan of deep digging, and since he has done so his flower beds will compare favourably with any in better soil. He says that the roots are better able to take care of themselves in a deep soil without water in a dry season than they can with liberal supplies in a shallow soil.

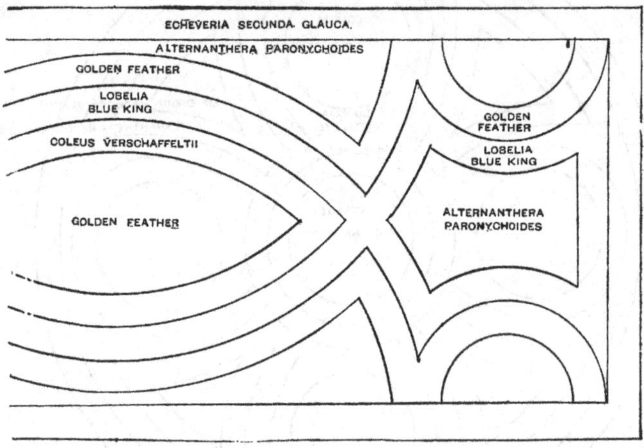

'Carpet' bed.

Coloured foliage.—The use of coloured and other fine-foliaged plants in the flower garden has of recent years greatly increased, the causes for such extended use being, first, the new introduction of such a large number of suitable plants; secondly, the foliage and sub-tropical bedding of the metropolitan parks; and, lastly, the weather, which for the past six or seven years has been so gloomy and wet throughout the summer season, that no sooner have ordinary bedding plants got into full flower than they have been dashed to pieces by the rain. Hence the desire that has arisen to have plants that would withstand such washings, and yet give us brightness appropriate to the season. The sunless and wet character of our summers of late is no doubt the strongest reason why the adoption of coloured foliage is to be commended and practised, at least within reasonable limits. What that limit shall be must of course be left to the taste of owners of gardens. For my own part, as regards coloured foliaged bedding plants in particular, I do not think that if half of the whole of the plants used were what are termed foliage plants, it would be at all out of proportion; of course, in such coloured foliage I would include all the variegated Pelargoniums which, if thought well, could be allowed to flower. Hardy variegated plants would also be included, such as Japanese Honeysuckles, variegated Periwinkles, Ivies, and similar plants; also the hardy Sedums, Saxifrages, and others of the carpeting type. The grand effects that can be had with this class of plants and variegated and coloured-leaved plants of the tender section, with graceful-leaved plants in combination, are infinitely greater than any that can be had with flowering plants alone, not to mention the additional merit of standing all weathers without injury. One of the brightest and most perfect beds as to colouring—planted in geometrical form, for summer effect—that I have ever seen was composed of the following plants, viz.: Sedum acre elegans, creamy white; Sedum glaucum, grey; Herniaria glabra, green; Mesembryanthemum cordifolium variegatum, light yellow; and the bright orange and scarlet Alternantheras, all dwarf plants; the standard or central plants being Grevillea robusta and variegated Abutilons. Thus much with reference to coloured foliage. The question of to what extent it should be used for shrubbery and landscape effect must also depend largely on individual taste and on the selection of plants made. For instance, if the favourite foliaged plants be, say, variegated Hollies, then it is almost impossible to get too large a percentage of coloured-leaved plants, at least of this type, as they harmonise both in colour and habit with shrubs and trees of almost every description, and are alike useful for the shrubbery or for isolation as specimen trees on turf. The variegated Maples, purple Beeches, and Hazels should

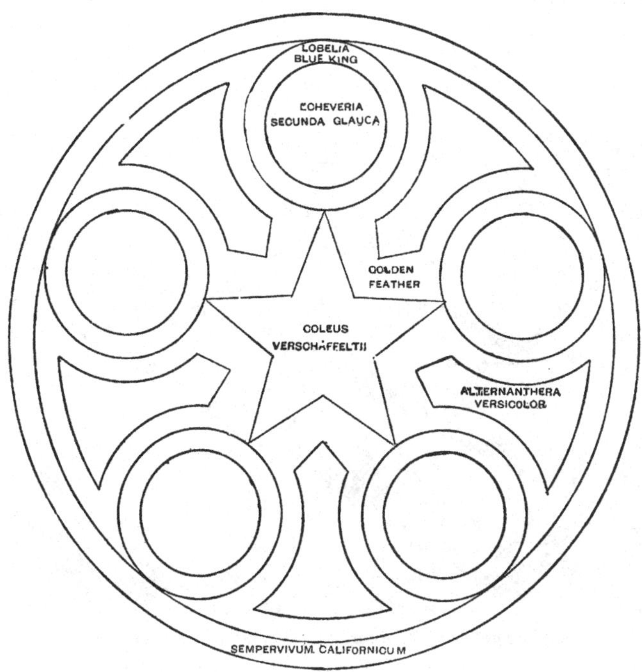

be used much more sparingly, as colours so decided, and so different from the generality of ornamental trees, are so conspicuous that the eye is unwillingly diverted from the general landscape effect to these particular trees. If they *must* be used in large numbers, grouping them together would be the best plan to adopt, the groups forming foregrounds to the general shrubbery or plantation. These remarks only apply to plants of such distinctive colouring as those just named. All the varied tints of greens and greys to be seen in the numerous coniferous family can safely be employed to any extent without the slightest danger of their appearing to overpower any other colour in the landscape, and the nearest approach to perfection in shrubbery and landscape planting is when it is impossible to decide as to what tint of greenery or other colour predominates.

Bedding and foliage plants.—The question of how far to modify what we call bedding-out flowers with foliage plants has been often discussed of late years, and has been illustrated by our best gardeners, both in public and private gardens. There can be no doubt that the use of the freer growing green and graceful fine-leaved plants has done a deal of good in our flower gardens. In the south of England the variety of plants of this kind that one may grow is very great. A number of greenhouse and even of stove plants may be placed in the open air without injury, and even with benefit to themselves. But it wants some discrimination, because some plants which we put out are sickly-looking all the summer and make no good growth. Others always look well, even in the face of storms that injure them. Good judgment, in fact, is required in all such cases, and a very careful selection. Where the climate may be against the tenderer tribes of plants a very good selection may be made from hardy subjects, either of young trees cut down and kept in a single-stemmed state, shrubs, or plants like the Yucca. This illustration (from André's "L'Art des Jardins") shows a smaller and neater type of foliage inserted in the bed. But it is well to confess that there are errors in the system itself from which these things cannot save us. A geometrical bed is none the less formal, none the less geometrical, because we place green-leaved or graceful plants in the middle of it. A more radical alteration still is required, and must be carried out by young and thinking men, and that is the abolition of geometry itself, formalism and straight lines, and all the other hateful gyrations which place the art of gardening on a level so much lower than it deserves to occupy. We can have all the variety, all the grace, all the beauty of form, all the glory of colour of which the world of flowers and plants is capable without any of the pattern business which is now the rule. But we can only make much progress in this direction by suppressing the bed as much as possible, by letting the vegetation tell its own story. The plants we must feed and the soil we must enrich, but finniking beds, reminding one of the art on fire shovels and all other such productions, are not necessary. Let us then, to begin with, adopt a bold,

Summer bedding, with occasional use of fine-foliaged plants.

large, and simple type of bed, from which the flowers will spring and make us think more of them than of the pattern.

Succulents.—By way of variety, succulents are a desirable class of plants to employ in the flower garden, more particularly in dry positions and under the shade of trees, where other kinds of bedding plants do not flourish satisfactorily. Perhaps they may be considered quaint rather than pretty; nevertheless, arranged on a groundwork of dwarf Sedums, Saxifrages, Antennaria, Herniaria, and similar plants, few bedding arrangements elicit more admiration. But, apart from this reason, their power of withstanding storms of wind and rain, or even drought and cold, they are always in good form; they should have a place in summer flower garden arrangements, *i.e.*, in gardens of any extent. Their variety is greater than that of many other classes of bedding plants, and this merit is enhanced by the fact that they harmonise well with many kinds of hardy plants that serve as cushions on which to, as it were, display to the full their quaintness. The term succulent includes all plants of a fleshy, juicy character, the more common types being the Sempervivums, Echeverias, and Mesembryanthemums—Cotyledons, Kleinias, Pachyphytons, Agaves, and Aloes being more rare, but none the less valuable for bedding purposes. The following are the best bedding kinds: Sempervivum montanum, tabulæforme, arachnoideum, soboliferum, canariense, calcareum, arboreum, arboreum variegatum, ciliatum, globiferum, urbicum, and Hendersoni; Echeveria pumila, rosea, secunda glauca, metallica, glauca metallica, farinosa, and Peacocki; Mesembryanthemum cordifolium variegatum, caulescens, tigrinum, inclaudens, and conspicuum; Agave americana and A. variegata; Pachyphyton bracteosum and roseum; Cotyledon pulverulentum and falcatum; Kleinia repens and tomentosa. To these, though strictly speaking not succulents, may be added many kinds of Sedums and Saxifrages, the most suitable varieties being Sedum glaucum, Lydium, corsicum, acre elegans, dasyphyllum, and hispanicum; Saxifraga Aizoon and its small variety hirta, rosularis, pectinata, and hypnoides. When to this list, which is in every way suitable for association with all the foregoing, is added Antennaria tomentosa, Herniaria glabra, Leptinella scariosa, several prostrate growing kinds of the Veronicas and Thymes, it will be seen that, with such range of variety, how very easy it is to make succulent bedding interesting.

Bed of succulents and coloured-leaf plants.

Vases and baskets.—In their proper place, and in due proportion, vases and baskets are indispensable in flower gardens, but not unfrequently they are used out of all proportion to the style of the garden and its surroundings, in which case they become objectionable. Perhaps, the tendency to overdecorate in this way must be credited to the severely geometrical plan of many gardens, when, to square with many meaningless angles, vases are placed on every available pedestal and corner. Happily, this style of gardening is giving place to a less formal one in which vases and baskets can be used or not, according to the taste of the owner, and yet the gardens appear perfect in either case. For a geometrical parterre they are needed to complete the design; at the same time there is no reason whatever why they should be used so largely as to offend good taste. A pair of vases on the pedestal of steps, and others on the turf, at the angles of the most formal divisions of the garden, to break the lines somewhat, are about all that are ever required, even in gardens of the most formal type. When, however, they are present in too large numbers, much may be done to remove this objection by planting in the vases plants of a drooping character; indeed, vases in every position look most natural when planted with a mixture of plants, and when trailers or climbers droop over the sides. Basket-formed beds are well suited for almost any position in pleasure grounds; but the best of all spots is in an isolated recess on the turf, and next, as a central bed in a flower garden, where the surrounding beds are circles or ovals. We have one, the extreme length of which is 16 feet; it is 8 feet wide in the middle, and stands 2 feet 6 inches above the turf, and is made of Portland cement. The principal plants used in it are Marguerites, Pelargoniums, Heliotropes, Fuchsias, Marvel of Peru, Abutilons, Castor-oils, Cannas, Japanese Honeysuckles, and Tropæolums. More rustic forms of baskets than this would be better suited for isolation on the turf and distant parts of the pleasure grounds; and very good ones can be formed with wirework, lined inside with zinc; also with barked Oak boughs instead of wirework. For ferneries, rock gardens, and wild flower dells, the best forms of vases and baskets are those that are carved out of decayed tree stems, which, by a little manipulation in the matter of nailing on Oak boughs, can be made to look just as handsome as the specially made baskets. In baskets and vases of this kind permanent plants should be used, such as the variegated Ivies, Periwinkles, Japanese Honeysuckles, Clematises, and climbing Roses, space being reserved for flowering plants in summer and for small shrubs in winter.

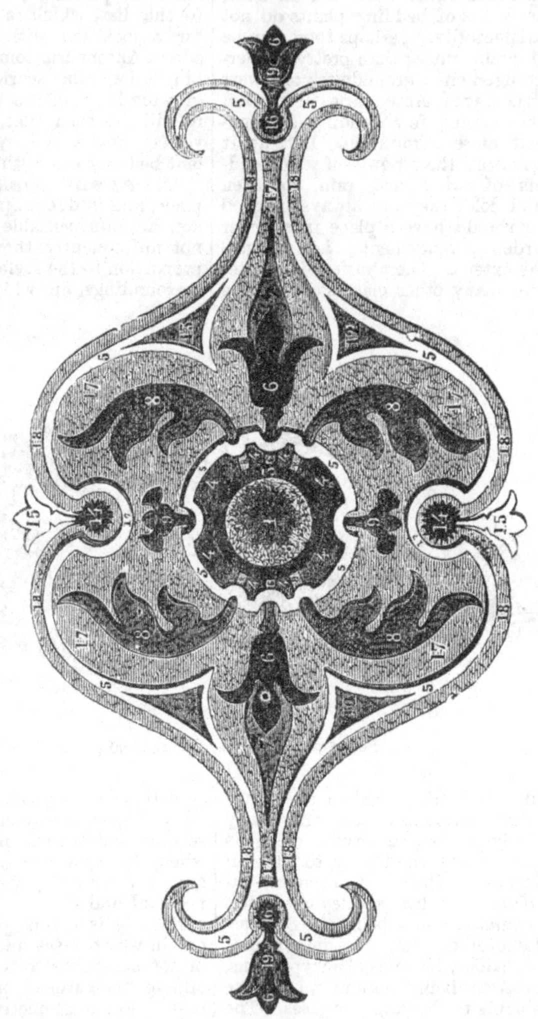

A bed of succulent and dwarf bedding plants: 1, a Palm; 2, Begonias; 3, Coleus; 4, Lobelia; 5, Pyrethrum; 6, Alternanthera versicolor; 7, A. amabilis; 8, A. amœna; 9, Sedum sexangulare; 10, Crassula Cooperi; 11, Echeveria pachyphytum; 12, Kleinia repens; 13, Crassula Bolusi; 14, Yucca quadricolor; 15, Mesembryanthemum cordifolium; 16, Echeveria glauca metallica elegans; 17, Antennaria tomentosa; 18, Echeveria secunda glauca; 19, Alternanthera paronychioides.

Sub-tropical bedding.—There are four types of summer flower gardening. 1, the massing (the oldest); 2, the carpet; 3, the neutral, quiet and low in colour, mainly through use of succulents; and 4, the sub-tropical, in which plants of noble growth and graceful foliage play the chief part. To my mind, a mixture of the four classes, skilfully worked out in conformity with the surrounding architecture, landscape scenery, and ground formation, is the beau ideal of flower gardening. It is possible to plant a formal garden in such a manner that the severest critic could not complain of excessive formality; for, after all, it is the abuse of carpet bedding that has brought it into disrepute. And justly so, for when one sees bed after bed and arrangement after arrangement repeated without end, with no plants to break or relieve the monotony of flat surfaces, one has good reason to protest against the system. I have charge of a terrace garden—certainly not of the severest geometrical type—which has to be planted with the view of obtaining the best display from June to November, and therefore I am compelled to adopt the carpet bedding system; but I supplement it by dotting over the surface, of necessarily formal arrangements, plants of noble or graceful aspect, such as Acacia lophantha, Dracæna indivisa, D. australis, Grevillea robusta, Yucca recurva, and others. For formal succulent arrangements, "dot" plants of similar kinds are best suited, such as Sempervivum arboreum, S. arboreum variegatum, S. phyllioides, Agave americana variegata, and Yucca gloriosa. In such arrangements a judicious blending of beds of flowering plants, principally Pelargoniums, adds brightness to the whole; but on no account should succulents and flowers be planted in the same bed, or even fine foliaged and flowering plants, save under exceptional circumstances. The colour-massing or grouping style of summer gardening is best adapted to a terrace or parterre that is well backed up or surrounded by evergreens, as these afford relief from the glare of brilliant colours, and at the same time set them off to advantage. In arranging such gardens, the point to be kept in view beyond all others is so to dispose the colours that no uneven balance shall be discernable, and also, to guard as much as possible against gaudiness, by using a preponderance of quiet tints such as pink, white, and blue, colours of which the eye never tires. A few plants of noble aspect, distributed at regular distances apart over the entire garden, and especially in beds of glaring colours, will be found to enhance the beauty of the whole. Retinosporas, standard variegated Euonymus, Thuja aurea, and several varieties of Yuccas are all suitable plants for this purpose. My view of sub-tropical gardening is, that it is only suitable for positions where it can be associated with water, or, for sheltered nooks and dells, where the force of the wind is broken before it comes in contact with the plants; otherwise, the first squall will so injure them as to make them unpresentable for the remainder of the season. Where such positions are not at command, the hardier class of noble or handsome-foliaged plants should be selected, many of which may be permanently planted, such as Ailantus glandulosa, Rhus glabra, Phormium tenax, Arundo conspicuà, Salisburia adiantifolia, the Yuccas, and the hardy Palm (Chamærops humilis). Of half-hardy plants that will withstand wind, there are numbers, such as Araucaria excelsa and A. Bidwilli, Acacia lophantha, Ficus elastica, Cycas revoluta, Dracæna indivisa, Aralia japonica, and others. The Abutilons, Ricinus, Solanums, Wigandias, and Ferdinandas are soon torn with the wind; but, nevertheless, they rank amongst the most effective of all plants where sheltered positions can be afforded them. In planting sub-tropical plants, care should be taken to guard against the beds having what may be termed a "bunchy" appearance when fully furnished. To avoid this, plant thinly, and supplement by planting as undergrowth dwarfer plants, of which there are many suitable kinds, some of the best being Gnaphalium lanatum, Cineraria maritima, Pyrethrum Golden Feather, Coleus, Iresine, Mignonette, and Periwinkles. The stronger-growing Sedums and Saxifrages might also be used in this way.

Summer and winter bedding combined.—Now that there is such a wealth of material in the shape of plants suited for the furnishing of flower beds in winter, there can be no excuse for these remaining empty after the summer-bedding plants have been cleared away, *i.e.*, if labour and the plants are forthcoming. Of course, it is useless to suppose that additional labour is not required to carry out both summer and winter bedding; but, where this is allowed, or can be obtained for the asking, then I strongly recommend this kind of decoration being carried out to the fullest extent commensurate with means and demands. After several years' experience, I would not now give up my bit of winter bedding, even if challenged to do so by the withdrawal of the little extra expense which it entails. There are potent reasons why winter bedding should be encouraged. First, there is the season when all around us is bleak, dull, and bare—leaden skies, leafless trees, flowerless meadows, and silent woods, all of which have a somewhat depressing effect on most temperaments. It, therefore, behoves us to endeavour to neutralise this prevailing dulness by making our gardens as cheerful as possible. Another reason, and one which to those fond (as thousands are) of summer bedding should be the great reason for adopting the system, is the short period during which summer bedding continues in perfection, for the thought is continually haunting one that it will fade all too soon. The adoption of winter bedding, however, in my own case has obliterated such thoughts, and now one looks forward to and has a continuation of real pleasure derivable from both systems. Nor has this been the only result; others equally beneficial, perhaps more so, have, as it were, accidentally fallen in our way, and in this wise: It having been a *sine-qua-non* that summer and winter bedding must meet, compulsory ingenuity had to devise ways and means for the accomplishment of this result in the most effective and economical manner possible. This has led to our searching out and using as summer bedders many hardy plants which otherwise we should not have thought of using, but which are just as effective as tender exotics; nay, in some cases, more so; and which when planted in the spring serve till the following spring, when they are taken up, divided, and replanted for another year. It is, doubtless, the endless labour connected with the short duration of summer bedding that has brought it into disrepute in some quarters, and one is obliged to admit very justly. But if we can combine, as I contend we can, summer and winter bedding, then much of the opposition to formal parterre planting will cease. From the above it will be perceived that a sort of compromise, a give and take, a tender and hardy, system of bedding is that which is here advocated. For instance, in summer planting a goodly number of the more handsome and showier growing Conifers and other shrubs should be brought into use for the filling of vases, centres of beds, and as dot plants in the larger beds. Amongst the best plants for this purpose are Retinosporas, Thujas, Junipers, Cypresses, Biotas, Yuccas, standard Golden Yews, standard Cotoneasters, variegated Hollies, and Euonymuses. For lines and groundwork the following are most suitable, viz., Sedums of many kinds, Saxifrages, Sempervivums, Cerastium arvense, Mentha Pulegium gibraltarica, Veronica repens, V. incana, variegated Thyme, variegated Arabis, and others. These hardy plants being worked in with the summer arrangements, and, so far as my own experience goes, with evident improvement to the tenderer subjects, give so much the less labour when the time for the winter arrange-

Panel of dwarf bedding plants.

ment arrives, a few days at the most sufficing for the clearing out of tender plants and replacing them with hardy ones. Amongst the best for taking the place of Pelargoniums, &c., are hardy Heaths, variegated Ivies, variegated Periwinkles, Cotoneasters, Berberis, Aucubas, variegated Box, variegated Euonymus, Lavender, common and variegated Thyme, Pernettyas, &c.

Shrubs for winter bedding.—The furnishing of flower beds in winter is not so generally practised as might be expected, seeing that now there is such an abundance of suitable shrubs, to be had at very cheap rates, and the free use of these, arranged somewhat after the fashion of bedding plants as regards harmony of colour, is just as effective, and, considering the season, as pleasing as are the most tastefully arranged summer bedding plants. Perhaps, for isolated beds or series of beds in distant parts of pleasure grounds that are not much frequented in winter, it may not pay to go to the labour and expense of thus planting them; but for gardens immediately under the windows of a mansion, as most bedded-out gardens are, the winter filling of the beds is of the utmost importance if permanent enjoyment is desired. There can certainly be none in looking out upon bare beds, and that at a time of year which is usually gloomy enough, without having a daily spectre of barrenness directly under our eyes. The fact of naked beds in winter is, no doubt, the strongest point against the bedding-out system, and, whilst acknowledging its justness, it is well to bear in mind that, to a great extent, the same objection is applicable to hardy herbaceous plants, as there are few of these that do real furnishing service during the winter months. Indeed, taking into account position, namely, a garden overlooked from the windows, we should certainly give preference to the bedding-out plan, and supplement that by shrub-planting for the winter, not necessarily of shrubs alone, but of other hardy evergreen plants, particularly the dense-growing Sedums and Saxifrages, as well as other similar plants, many of which are now used with excellent effect in combination with ordinary bedding plants in summer arrangements. Of course, the more plants of this type that it is possible to work in with the summer bedders, so much is the amount of labour lessened when the time arrives for the winter planting. For some years we have thus striven to combine these two opposite types of plants, and with what success may be inferred when it is said that, to fully equip the parterre in winter dress, we do not now require to move more than half the plants that have done duty in summer. For instance, the following may be quoted as a fair example of this matter: A large circular bed edged with Sedum glaucum, raised intersecting lines to form pattern 9 inches wide of Veronica incana, in the centre a pyramidal-shaped plant of Retinospora pisifera aurea in a circle of groundwork of Saxifraga hirta; all these have done duty during the summer, the smaller panels—oblongs and circles—being filled either with Alternantheras, Mesembryanthemums, Lobelias, or tricolour Pelargoniums. These have now given place to, in the small circles, a small Cupressus erecta viridis, carpeted with Sedum Lydium, which is grown in a reserve garden for the purpose, and taken off in flakes with a spade, and simply pressed on the ground, and thus is effective at once; in the oblongs are central plants of variegated Aucuba, the undergrowth here being Sedum glaucum, grown and placed in position in the same way as the one just named. This example will suffice to illustrate what is meant by a combination of the two sets of plants, and to indicate the direction in which to obtain flower-garden cheerfulness in winter. It now only remains to give a list of the best kinds of shrubs for the purpose, the list being strictly composed of kinds that bear removal well, do not grow fast, or change colour after planting. It is as follows: Aucuba japonica, both green and variegated, Cotoneaster microphylla, Cupressus Lawsoniana, C. Lawsoniana erecta viridis, Erica herbacea carnea, E. mediterranea alba, Euonymus aureo-variegatus, E. radicans variegatus, Hollies—variegated kinds, Juniperus macrocarpa, J. tamariscifolia, J. chinensis, J. occidentalis, Laurustinus, Berberis Aquifolium, Osmanthus ilicifolius and aureo-variegatus, Pernettya mucronata, Portugal Laurels, Retinospora obtusa aurea, R. pisifera, R. pisifera aurea, R. plumosa, Rhododendrons, Thujopsis borealis, Vinca elegantissima variegata, Yucca recurva, Yews—green and variegated.—W. W.

Wintering bedding plants.—Pelargoniums may be put in boxes and kept in a very small space. The most tender amongst them are tricolors and other sorts having variegated leaves, all of which should be got up first, and when brought to the shed, be thinned out or cut back, according to the room at disposal. After this is done have any loose leaves picked off and the roots shortened, when they will be ready for boxing or potting. Boxes, however, are preferable, and the handiest sizes are those of about 2 feet long, 9 inches wide, and 4 or 5 inches deep, sizes which hold quite enough soil, and admit of being easily lifted and carried from place to place when it is desired to shift them from one house or frame to another. In putting the plants into the boxes, the best way is to lay about 1 inch of sharp sandy soil in the bottom, and then proceed to pack the plants as thickly as they can be got, covering the roots well as the work proceeds; and to settle the earth and give them a start the next thing is to water well through a coarse rose, and when sufficiently drained they are ready for storing for the winter.

The inner sills of windows or any dry spare room in a house where there is sufficient light answers well, but the best place is the back shelf of a vinery or other cool glass structure, where, if frost is kept out and damp expelled by having a little fire on during wet or foggy weather, they will break well at the joints and become furnished with fresh shoots so as to make fine bushy plants by the spring. These will be found preferable to plants obtained from cuttings that have been made in the autumn, as they not only at once fill a large space in the beds when turned out, but they flower much better on account of the stems being harder and more woody, and the growth which they make more moderate and more firm. This being so, it is advisable to save all that can be stored in any way, and if kept dry it is astonishing how easily this may be managed, as with care they may even be wintered in cellars or lofts, and on several occasions we have kept them successfully in pits covered with straw and earth, ridged over and clamped after the manner in which Potatoes are pitted. The chief thing to be particular about, as regards saving Pelargoniums, is to see that they are safely under cover before being frost-bitten, for if the bark of the shoots is injured they are almost sure to decay.

In regard to Dahlias, the best way of treating them is to cut them down to within 6 inches or so of the ground as soon as their beauty is over and then dig up the roots with a fork, after which the wet soil should be shaken out from among the tubers, when they will be in a fit condition for storing. To hang them up and let them dry and shrivel as some do is a mistake, as they lose much of their vitality and strength, but if buried in dry mould on the floor or shelves of a shed they remain plump, and start vigorously again in the spring, when they may be divided, shoots struck from them, and the stock increased to any extent.

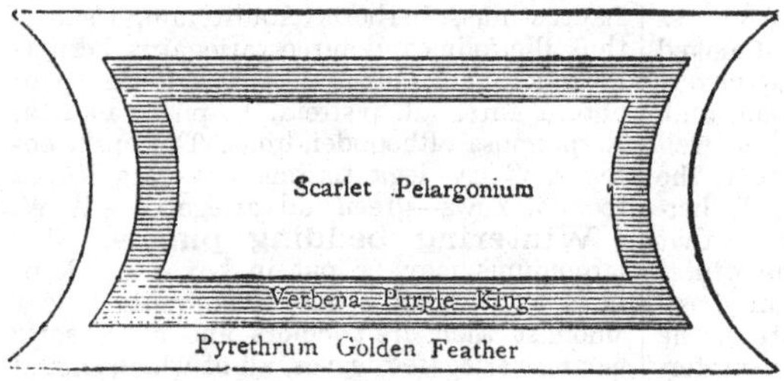

Scarlet Pelargonium

Verbena Purple King

Pyrethrum Golden Feather

The lovely blue Salvia patens, too, winters securely and well in the same way, the tuberous roots being much like those of Dahlias in both their form and character; and then again there are the Begonias, which are so valuable for bedding and for pot culture, that every one should grow them. These have large round fleshy tubers that may be wintered in damp sand or soil, but the corms of Gladioli must be kept somewhat drier, or they will be lost. The way in which we have always been most successful with these is having separate pots for each, in which they are dropped, and the names put with them, after having being covered with sand or dry mould, which keeps the air from them and assists greatly in preserving them in a sound condition.

Cannas, now so much grown as fine foliaged plants, may either be left in the ground where they are and covered deeply with half rotten leaves, tan, ashes, or Cocoa-nut fibre, or lifted and put into a shed. Verbenas, Heliotropes, Ageratums, and hosts of other bedding subjects of a similar character do by far the best propagated in the spring, and all we do with these, although we require great numbers, is to winter one or two large and strong plants of each from which to take cuttings, and these inserted in a little heat strike readily, and are then pricked out in prepared soil in cold frames, from whence they go direct to their summer quarters and are never potted at all. This plan of dealing with them saves a vast amount of labour, and the plants are altogether better, as may be seen by the quick way they start and cover ground.

Bedding plants from seed.—It is not everyone who has the space or means to provide and winter a large number of Pelargoniums and other tender bedding plants. No matter how favourably one may be situated, the keeping of a large stock of such plants involves a good deal of trouble, and takes up space that might be more profitably occupied. Moreover, it is quite practicable to make a garden very gay in summer and autumn with seedlings alone without keeping or purchasing a single plant, and a comparatively small amount will buy the seeds required. There will, of course, in some cases be a difference in habit and some variation in colour, as in those particulars a certain natural freedom, involving some departure from the normal type, is nearly always perceptible in seedlings and must be expected; but to many people this will not be an objection. There are, however, a few plants which come true from seed through many generations; one of these we have in Verbena venosa, one of the best and most pleasing hardy bedding or border plants. We have raised many thousands of this Verbena from seed and never saw any variation in them. It would be a great help to flower gardeners if other shades of colour of this useful old plant could be obtained. Another plant that comes true from seed is the Salvia patens; but both this and the preceding should be sown in a hotbed early in spring, in order to get them into flower early. Lobelias of the speciosa section come fairly true from seed; there may be some variation in strength and compactness of habit, but this feature can be subdued and regulated by cutting or pinching back any plants that seem disposed to outgrow their neighbours. Can anything, again, as a mixed bed be more showy than a mass of seedling Petunias? The colours are not harsh and irritating like the scarlet Geranium, but soft and pleasing. Seedling Verbenas make a handsome bed, little inferior to the Petunia; and for a large bed, where the soil is good and deep, few things are superior to the double Zinnias, which can be had in various colours, separately if desired. Balsams, again, are not half so much used for open-air decoration as they deserve to be. Those who have only seen them starving in small pots cannot form an idea how beautiful they are when planted out in good soil in an open situation away from trees. Among yellow-flowering plants, the small single form of Tagetes signata pumila is useful, but it is not equal to the double variety, which really makes a very handsome bed. For large beds or back positions in borders, the lemon and orange varieties of the African Marigold are very lasting and showy. The dwarf kinds of Ageratum, if selected and saved with care, may be raised in a gentle heat with but little trouble, and, with a little management, in summer very effective masses may easily be had. The tuberous Begonias form a prime feature in sheltered positions; they will grow in all the colours of the Pelargonium without their liability to be washed out by rain, and they require no expensive preparation to keep them through the winter. We have said nothing about the large number of perfectly hardy plants well suited for massing, and that cost nothing to keep. Take, for instance, the large familly of Violas, in almost all shades of purple, yellow, and white. Varieties which formerly took two years to come to perfection may, now that selection and fertilisation have so much improved them, be raised in the early spring for bedding out the same summer. This applies especially to such plants as the Verbena, Viola and Pansy, Geranium, Golden Feather, Salvia patens, S. argentea, Heliotropes, dwarf Antirrhinums, which should be sown in January in heat, much as the Verbena already described; also to Petunias, Phlox Drummondi, Dianthus Heddewigi, Indian Pinks, &c., Ageratums and Lobelias, which should be sown in pans in heat in February, and, if kept growing, will be ready for planting out in May. Begonias, for bedding, may be grown from seed in the same year, but are much more effective if raised the preceding year, and selected according to colour, and stored in the winter ready for bedding out early in the summer. The same system may be employed for indoor decoration, for Gloxinias begin to flower in June if sown in January or February; Begonias in July, and then they last throughout the autumn, when last year's bulbs are over-blown indoors. Fuchsias sown in January flower well in August, and many other plants also. Of fine foliaged plants adapted for bedding which can be raised from seed, there are the useful Amarantus melancholicus ruber and the drooping

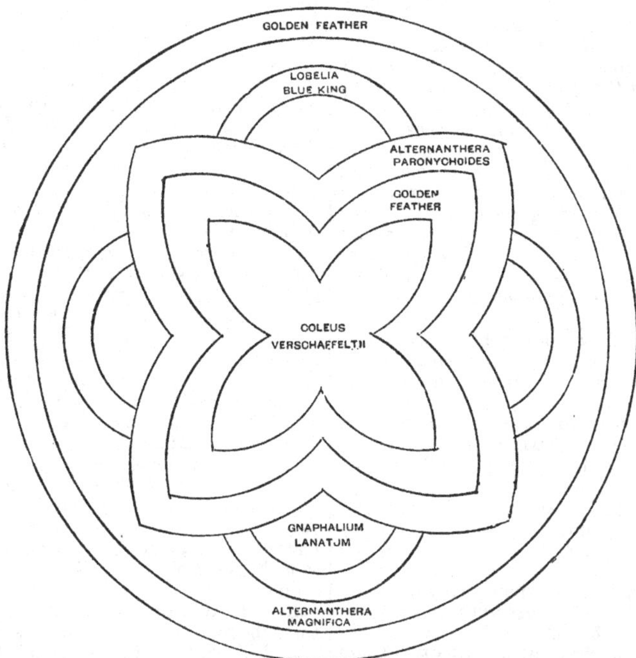

A. salicifolius; Celosia Huttoni, with its fine habit and effective colouring; Centaurea ragusina and its variety compacta; Cineraria maritima and the Humea elegans. To these can be added Dell's crimson-leaved Beet and the variegated broad-leaved Cress. There are many other fine and useful foliage plants, such as Cannas, Chamæpeuce, Nicotiana, Ricinus, Solanum, and Wigandia. In fact, if we were not so much accustomed to depend on cutting plants stored over the winter, we could make a display with seedling plants alone. If Fuchsias may be grown to the flowering stage during the current year there can scarcely be any difficulty in getting a large stock of plants for the open garden in the same way. The cleanliness resulting from this plan would be a gain in itself, because the tendency of old propagated plants is to harbour the eggs of vermin through the winter, which are always ready to eat up the collection if neglected for a week. But, starting with thoroughly clean houses and frames, and seeds in early spring, one could make a better fight against the many small insect enemies of the gardener. As regards the bedding plants and the things one would like to raise in this way, a main point to desire is that the seedsmen select and fix distinct colours of different races of plants. For example, we suppose it would not, with care, be difficult to select a bluish or purple Verbena which one might count on as coming pretty true from seed. In any case we may be assured that a good deal more may be done from seed than has ever hitherto been the case. So much have we relied hitherto upon cuttings and old plants, that the raising of seedlings in proper health has never had a fair amount of attention. We know well that everybody raises seeds, but the early thinning, the perfect exposure to light, the sturdy growth, the unchecked culture that seedlings require to be done really well, are seldom given them, owing to the little space and little thought they usually occupy.

Watering flower gardens.—There is much unnecessary labour expended in watering. Ofttimes if the surface soil only looks dry, it is concluded that watering is needed, and accordingly it is given. We will put the needlessness of watering under such conditions on one side and only look at it in reference to its effects on the plants when thus thoughtlessly administered. First, then, the flower garden in dry weather does require to be watered, but when this is necessary it should be really watered, that is, sufficient should be given to moisten the soil to the full depth of which it was dug or trenched. One such watering, in the driest weather, is sufficient to serve a whole week, and will keep the plants in luxuriant growth; but daily surface driblets, miscalled waterings, have a contrary effect, for, naturally enough, the roots of the plants will go in search of moisture, and this being merely superficial, the roots are formed on the surface only. The consequence is, first, that no matter how deep and good the soil is, being dry, it is useless to the plants; and, secondly, the roots that are formed on the surface have a hard battle in holding their own against drought and hot sunshine; consequently, the growth of the plants is unsatisfactory. A thorough watering, when the plants are first put out, to settle the soil to the roots is all that is necessary till new roots are established, which is generally about a fortnight or so after planting. Then, if the weather continues dry, a soaking once a week is all that is needed to keep the plants in perfect condition. Of course, the heavy waterings here advised render it necessary to stir the soil with a small hoe or pointed stick a day or two after watering; but this operation would be but rarely required were the good old plan of mulching more general than is now the case, except it be in regard to Calceolarias, Violas, Pansies, and Verbenas. Nearly everyone believes in mulching such plants as these, which is quite right; but why these only, when the practice is just as beneficial to every other kind of bedding plant? If applied as soon as the plants are put out in May, it prevents a lot of labour, both as to watering and stirring the soil, to keep it open. The benefits rising from the practice are, by general consent, admitted to be great. Then comes the question as to what are the best materials for the purpose. Cocoa fibre refuse is the neatest; it always looks well, is cleanly to use, and is now so cheap, that cost cannot be considered a valid reason for its exclusion. The next best material is decayed vegetable or leaf soil. This, sifted, also looks very neat, but under bright sunshine soon shrinks up to nothing, and requires renewing. Well-rotted stable manure, broken up very fine, or else sifted, is another good material—too good in a certain sense, for some kinds of plants; and therefore its use should be limited to those kinds only that delight in rich food, amongst which are all the sub-tropicals, Dahlias, Violas, Verbenas, and Calceolarias. But it would be better that all plants requiring a liberal amount of manure should be given it directly in the soil when the beds are being prepared by digging or trenching, and thus the mulching material may be alike throughout.

Edgings for walks and flower-beds.—Let us speak of living edgings first. Of these the most common is Box, though it is far less used now than formerly, owing to the labour entailed in keeping it. Where exactness and formality of design have to be adhered to, there is perhaps nothing to equal Box, provided the often-occurring blanks are made up regularly and it is kept in good order by the shears, for a neglected Box edging sadly mars the appearance of a garden, and those who cannot afford the expense of keeping it as it should be should dispense with it altogether. The most beautiful of evergreen edgings is formed by the Irish Ivy, where these are carefully looked after, as in the public gardens of Paris. A broad, well-kept edging of Irish Ivy has a fine effect in various situations, but the edgings of this sort laid down in the gardens of the Thames Embankment have become ugly from neglect. They do not receive the frequent pinchings and careful attention they get in Paris. Among other edgings of the shrubbery class is the common Bilberry or Bleaberry (Vaccinium Myrtillus). In a tastefully laid out garden, where some of the beds were edged with various kinds of shrubs, the Bilberry was one, and it grew well, although it was clipped like Box. It likes a peaty soil, but grows well enough on a rather dry loam, and leaf-mould suits and promotes its growth. Neat edgings of Oak are also used; though leafless in winter, it has a good effect in summer, and is kept low by being occasionally pruned. The common Heath (Erica vulgaris) forms an edging, but peat or leaf-mould should be added to the soil in which it grows. In some gardens it is used for scroll work, and lasts long in spite of constant clipping. The Yew, of which there are edgings here that have lasted for twenty years, makes a good hardy edging. Though always kept down to a height of 6 inches or thereabouts, they are yet as stiff and dense as a cushion. Such edgings are usually trimmed once or twice during the summer. Other trees and shrubs might be used for edgings, their fitness for such a purpose being judged by their habit. No doubt, the close-growing habit of the Box and Yew first suggested their usefulness in this respect; but these qualities are by no means essential to a good edging, for, by clipping and pruning, it can always be made almost anything one likes. It is only, however, in bold shrub-planting that such strong edgings are needed. Next to shrubs come hardy plants, some of which make good edgings. Among the best are double Daisies, Thrift, Arabis, Phlox procumbens and other dwarf Phloxes, Auriculas, Primroses, Violets, Gentian, Lithospermum prostratum, Forget-me-nots, and others. All these are common, and easily and speedily propagated by division, except the Lithospermum, which strikes freely from cuttings outdoors. They should be planted in thin lines, for they soon spread, and periodically they should be lifted, divided, and replanted. Hardly enough is made of these kinds of plants as edgings. The crimson and white double Daisies, for example, are perfectly charming for a long while in early summer, and they increase so fast and are so easily grown. Let us now advert to dead edgings. Burnt tiles with a bead on the top are often used for the purpose, because they are easily procured and cheap; but they are useless, for the frost breaks them up in a couple of years unless they are very hard burnt, and then they are not often straight. They are always hard and ugly, too, and should never be used by the tasteful gardener, at all events, in the flower garden. A thin kind of tile, made of fireclay and glazed, is also used, but, as a rule, all the burnt clay edgings are to be suspected. Terra-cotta does stand fairly, but it is expensive. Stout blue slate, with sawn edges, is the most enduring; this, too, is expensive at first, but its great lasting properties are a compensation, and it can be had in long lengths, and is easily laid.

Where stone is abundant it is excellent for edgings if tolerably hard. In all quarries there is generally a quantity of refuse which might be used for such purposes. If they are about the same thickness—and they are generally pretty uniform in this respect—pieces of all shapes and sizes may be used, and in laying them down do the same as masons when building a rough-hewn wall: take the pieces as they come, and stick them in close together, keeping the tops just above the ground-line. When finished a rustic edging will be the result, but it will look much better than any made of more costly materials, and be a great deal cheaper. The late Mr. Charles M'Intosh had the edges of the walks in some parts of the extensive kitchen gardens at Dalkeith laid simply with oblong water-worn stones from the river bed. The stones were all selected about the same size, tolerably heavy, and were laid end for end along the walk side; and well they served their purpose, which was simply to keep a clear division between the soil and the road. Moreover, they always looked clean and trim, required no keeping, and the stones were easily replaced if they happened to be dislodged. A single row of stones is sufficient. The charm of edgings made in this way is owing to their getting a good colour through being clothed with fine Mosses—they often have this recommendation, indeed, when laid down. They require no hard, rigid setting, looking much better without it. Unlike the set cast tiles, they are none the worse if they fall out of line, and are easily removed and easily reset. They are, besides, the cheapest of all borders, and give, by far, the best effect.

Dead edgings and living rock plants.—We have condemned the tile edgings which are now so common, and it is not necessary to say any more about them. Edgings are much better made of the natural stone, which abounds in various counties, or even of a line of flints. An edging in a garden may be a beautiful thing by reason of dwarf rock plants being associated with it. If we have an ugly stiff chamfered or "roped" tile edging, one would hardly think of putting an alpine flower near it; but if, on the other hand, we have any good natural stone edging, it forms the most delightful place for a great number of easily grown rock and mountain plants. If the stones are irregular in size and outline, the best way is to sink them partly in the ground, as we do in the case of a flint edging. Among the sinuosities of these stones we can have a whole flora of good things running in and out and spreading into gravel walk or soil. For edgings in a kitchen garden it is not perhaps desirable to bring any plants near them, but in the great number of cases in which stone edgings border choice shrubs or flowers or the ornamental parts of the garden, nothing better can be done than to adorn them with such plants as the choicer silvery and green Saxifrages, the Gentianella, Stonecrops, Houseleeks, Saxifrages, Aubrietias, and other plants which are often starved in pots, or treated no better on a poor, ill-formed rock garden. Different kinds of stone would have to be used in different localities, but there is scarcely a place in which a yard of space need be unadorned. There has been some talk of the necessity of studying the different soils which alpine and rock plants live in in a wild state. There is no doubt "something" in this, to use a common saying, but not very much. The thing that must strike anyone who has observed hardy plants in gardens is the curious freedom with which alpine and rock plants grow on a great variety of different soils, poor or rich. There need be no difficulty whatever in growing numbers of alpine flowers in the position we mention. So few people make rock gardens well, and so few have the means to make them in the expensive manner sometimes adopted, that it is well to state the various simple ways in which they may be made, and none of them are more simple than this. Even the sunk brick edging, which forms such a good margin in the kitchen garden, might be adorned in the way we propose; but it is where the stones are irregularly formed, and where the plants can run in and out among, and occasionally lap over, them and stray into the walk, that one gets the very best effect. And, by the way, in all borders of hardy and alpine plants, an excellent plan is to let tufts of vigorous dwarf plants grow a little into the walk. Make the edging so that it will welcome such outgrowths and not repel them. This is desirable from the good effects which one sees on looking along a border where a large pink Carnation or evergreen Candytuft pushes its growth boldly out. In using these natural stone edgings it would often be desirable to plant different things along the lines instead of using the same plant throughout. In this way a considerable variety of fine dwarf plants might be grown, and grown better than in situations made with care for them. All we have to do is to adapt the vegetation to the soil; but there are so many hardy and vigorous little mountain plants, that little care is wanted even in this respect.

COLOUR IN THE FLOWER GARDEN.

ONE of the most important points in the arrangement of a garden is the placing of the flowers with regard to their colour-effect, and it is one that has been greatly neglected. Too often a garden is an assemblage of plants placed together haphazard, or if any intention is perceptible, as is commonly the case in the bedding system, the object aimed at is as great a number as possible of the most violent contrasts, with the result of a hard, garish vulgarity. Then, in the case of mixed borders, what is usually seen is either lines or evenly distributed spots of colour, wearying and annoying to the eye, in no way interesting, and proving only how poor an effect can be got by the misuse of possibly the best materials. Should it not rather be remembered that in setting a garden we are painting a picture, only it is a picture of hundreds of feet or yards instead of so many inches, painted with living flowers and seen by open daylight, so that to paint it rightly is a debt that we owe to the beauty of the flowers and the light of the sun? Therefore, the colours should be placed with careful forethought and deliberation, as a painter employs them on his picture, and not dropped down in lifeless dabs, as he has them on the palette.

HARMONY RATHER THAN CONTRAST.—Splendid harmonies of rich and brilliant colour, and proper sequences of such combinations, should be the main rule; there should be large effects, each well studied and well placed, varying in different portions of the garden scheme. One of the commonest faults in arrangement is a want of simplicity of intention, or an obvious absence of any definite plan of colouring. A difficulty will arise in the case of many people who have not given any attention to this question of colour-harmony, or who have not by nature the gift of perceiving it. Let them learn it by observing some natural examples of happily related colouring, taking separate families of plants whose members are variously coloured. Some of the best to study would be American Azaleas, Wallflowers, German and Spanish Iris, alpine Auriculas, Polyanthus, and Alstrœmerias, avoiding of course certain abnormal colourings produced in some families of plants by the over-zeal of industrious florists.

BREADTH OF MASS AND INTERGROUPING.—It is important to notice that the mass of each colour should be large enough to have a certain dignity, though it must never be so large as to be wearisome; a certain breadth in the masses is also wanted to counteract the effect of foreshortening when the border is seen in any degree from end to end. When a definite plan of colouring is decided on, it will be found to save trouble if the plants, whose flowers are approximately the same in colour, are grouped together to follow each other in season of blooming. Thus, in a part of the border assigned to red, Oriental Poppies would be planted among or next to Tritomas, with scarlet Gladioli between both, so that in that part there should be a succession of scarlet flowers, the places occupied by the Gladioli being filled beforehand with red Wallflowers.

WARM COLOURS are not difficult to place; scarlet, crimson, pink, orange, yellow, and warm white are easily arranged so as to pass agreeably from one to the other.

PURPLE AND LILAC group well together, but are best kept well away from red and pink; they do well with the colder whites, and are seen at their best when surrounded and carpeted with grey-white foliage, such as Cerastium tomentosum or Cineraria maritima; but if it be desired to pass to purple and lilac from a group of warm colour, it may be done through a good breadth of pale yellow or warm white.

WHITE FLOWERS.—Care must be taken in placing very cold white flowers, such as Iberis correæfolia; this may best be used as quite a high light, led up to by whites of a rather softer character. Frequent repetitions of white patches catch the eye unpleasantly; it will generally be found that one mass or group of white will be enough in any piece of border or garden arrangement that can be comprehended from any one point of view.

BLUE requires rather special treatment, and is best approached by delicate contrasts of warm whites and pale yellows, such as the colours of double Meadow Sweet and Œnothera Lamarckiana, rather avoiding the direct opposition of strong blue and full yellow. Blue flowers are also very beautiful when completely isolated and seen alone among rich dark foliage.

A PROGRESSION OF COLOUR to be recommended in a mixed border might begin with strong blues, light and dark, grouped with white and pale yellow, passing on to pink; then rose colour, crimson, and strongest scarlet, leading to orange and bright yellow. A paler yellow followed by white would distantly connect the warm colours with the lilacs and purples, and a colder white would combine them pleasantly with low-growing plants with cool-coloured leaves.

SILVERY-LEAVED PLANTS are most valuable as edgings and carpets to purple flowers, bearing the same kind of relation to them as the warm-coloured foliage of some plants does to their strong red flowers, as in the case of the Cardinal Flower and double crimson Sweet William. The bright clear blue of Forget-me-not is best with fresh pale green, and pink flowers are beautiful with pale foliage striped with creamy white, such as the variegated forms of Jacob's Ladder or Iris Pseudacorus. A useful carpeting plant, Acæna pulchella, assumes in spring a rich bronze colour between brown and green, valuable with Wallflowers of the brown and orange colours. These few examples, out of many that will come under the notice of any careful observer, will be enough to indicate what should be looked for in the way of accompanying foliage; such foliage, well chosen and well placed, may have the same value to the flowering plant that a worthy and appropriate setting has to a jewel.

IN SUNNY PLACES warm colours should preponderate; the yellow colour of sunlight brings them together and adds to their glowing effect.

A SHADY BORDER, on the other hand, seems best suited for the cooler and more delicate colours. A beautiful scheme of cool colouring alone might be arranged for a retired spot, out of sight of other brightly-coloured flowers, such as a border near the shady side of any such shrubbery or wood as would afford a good background of dark foliage. Here would be the best opportunity for using blue, cool white, palest yellow, and fresh green. A few typical plants for such a place would be the great Larkspurs, Monkshoods, and Columbines, Anemones, such as japonica, sylvestris, apennina, Hepatica, and the single and double forms of nemorosa, white Lilies, Trilliums, Pyrolas, Habenarias, Primroses, white and yellow, double and single, Daffodils, white Cyclamen, Ferns and mossy Saxifrages, Lily of the Valley, and Woodruff. As a background to such flowers the most appropriate would be the shrubs and trees which would give an effect of rich sombre masses of dusky shadow rather than of positive green colour. Among these would be Bay, Phillyrea, Box, Yew, and Evergreen Oak. Such a harmony of cool colouring, in a quiet shady place, could not fail to present a delightful piece of gardening.

BEDDED-OUT PLANTS, in such parts of a garden as may require that system, may be arranged on the same general principle of related, rather than of violently opposed, masses of colour. As an example, a fine effect was obtained by bedding with some of the half-hardy annuals, mostly kinds of Marigold, Chrysanthemum, and Nasturtium, all shades of yellow, orange, and brown. This was in a finely designed garden of the formal kind before the principal front of one of the stateliest of the great houses of England. It was a fine lesson in temperance this employment of a simple scheme of restricted colouring, yet it left nothing to be desired in the way of richness and brilliancy, and served its purpose well as a dignified ornament and worthy accompaniment to the fine old palace.

CONTRASTS — HOW TO BE USED. — The greater effects being secured, some carefully-arranged contrasts may be used with good results to strike the eye when passing; for opposite colours in close companionship are not telling at a distance, and are still less so if interspersed, their tendency then being to neutralise each other. Here and there a bold contrast well placed may have a charming effect, such as a mass of orange Lilies against Delphiniums, or Gentians with alpine Wallflowers; but these violent contrasts had better be used sparingly as brilliant accessories rather than trustworthy principals.

CLIMBING PLANTS ON WALLS.—There is often a question about the suitability of variously coloured creepers on house and garden walls. The same principle of harmonious colouring is the best guide. A warm-coloured wall, such as one of Bath stone or buff bricks, is easily dealt with. On this all the red-flowered, leaved, or berried plants look well. Japan Quince, red and pink Roses, Virginia Creeper, Cratægus Pyracantha, and the more delicate harmonies of Honeysuckle, Banksian Roses, and Clematis montana, and Flammula, while C. Jackmani and other purple and lilac kinds are suitable as occasional contrasts. The large purple and white Clematises harmonise perfectly with the cool grey of Portland stone; also dark-leaved climbers, such as white Jasmine, Passion Flower, and green Ivy. Red brickwork, especially when new, is not a happy ground colour; perhaps it is best treated with large-leaved climbers—Magnolias, Vines, Aristolochia—to counteract the fidgetty look of the bricks and white joints. When brickwork is old and overgrown with grey Lichens, there can be no more beautiful ground for all colours of flowers from the brightest to the tenderest—none seem to come amiss.

Colour in bedding-out.—We must here put out of mind nearly all the higher sense of enjoyment that we have in flowers; the delight in their beauty individually or in natural masses; the pleasure derived from a personal knowledge of their varied characters, appearances, and ways, and that gives so much of human interest and lovableness to them; and must regard them merely as so much colouring matter, to fill such and such spaces for a few months. We are restricted to a kind of gardening that, though better in degree, is in kind not far removed from that in which the spaces of the design are filled in with pounded brick, slate, shells, or some colouring substances other than flowers. The best rule in the arrangement of a bedded garden is to keep the scheme of colouring as simple as possible. The truth of this is easily perceived by an ordinary observer when shown a good example, and is obvious, without any showing, to one who has studied the question of colour effects; and yet the very opposite intention is the thing most commonly seen in gardens, to wit, a garish display of the greatest number of crudely contrasting colours, than which nothing can be worse from the point of view of refined taste. How often do we see combinations of scarlet Geranium, Calceolaria, and blue Lobelia—three subjects that have excellent qualities as bedding plants if used in separate colour schemes, but that in combination can hardly fail to look bad? In this kind of gardening, as in any other, let us by all means have our colours in a brilliant blaze, but never in a discordant glare. One or two colours, used temperately and with careful judgment, will produce nobler and richer results than many colours purposely contrasted or wantonly jumbled. The formally designed garden that is an architectural adjunct to an imposing building demands a dignified unity of colouring rather than the petty and frivolous effects so commonly obtained by the misuse of many colours. As practical examples of simple harmonies, let us take a scheme of red for summer bedding. It may range from palest pink to nearly black, the flowers being Geraniums in many shades of pink, rose, salmon, and scarlet; Verbenas, red and pink; and judicious mixtures of Iresine, Alternanthera, Amaranthus, the dark Ajuga, and red-foliaged Oxalis. Still finer is a colour scheme of yellow and orange, worked out with some eight varieties of Marigold, Zinnias, Calceolarias, and Nasturtiums—a long range of bright rich colour from palest buff and primrose to deepest mahogany. Such examples as these of strong warm colouring are admirably adapted for large spaces of bedded garden. Where a small space has to be dealt with, it would better suit arrangements of blue, with white and palest yellow or of purple and lilac, with grey foliage. A satisfactory example of the latter could be worked out with beds of purple and lilac Clematis, trained over a carpet of Cineraria maritima, or one of the white-foliaged Centaureas, and Heliotropes and purple Verbenas, with silvery foliage of Cerastium, Antennaria, or Stachys lanata. These are some simple examples easily carried out. The principle once seen and understood, and the operator having—or having acquired—a perception of colour harmonies, modifications will suggest themselves, and a correct working with two or more colours will be practicable; but the simpler ways are the best, and will always give the nobler result. There is one peculiar form of harmony to be got even in varied colours by putting together those of nearly the same strength or depth. As an example in spring bedding, Myosotis dissitiflora, Silene pendula (not the deepest shade), and double yellow Primrose or yellow Polyanthus, though distinctly red, blue, and yellow, yet are of such tender and equal depth of colouring, that they work together charmingly, especially if they are further connected with the grey-white foliage of Cerastium.—G. J.

The arrangement of the plants in a garden, with a view to produce a harmonious effect of colour, is one of the most important parts of garden design, and one which has hitherto been little attended to where hardy plants are used. In the old-fashioned styles of gardening the plants were mostly grown for their own sakes; and if any good effect of colour was produced, it was usually the result of chance, and not of design. The introduction of the bedding system changed all this, and showed the possibility of arranging plants so as to produce a preconceived colour effect, and no doubt the great popularity of bedding is wholly owing to that. But, however useful the bedding system may have been in directing attention to a branch of garden design previously neglected, it has many drawbacks, and does not by any means represent finality. Besides the disadvantages which have often been pointed out as belonging to the system, there is another disadvantage which does not belong to it, but is nevertheless generally associated with it—namely, the utterly inharmonious and ill-balanced combinations of colour which are carried out under its name, and pointed to as the main reason for its existence and retention, and its principal advantage as a scheme of garden decoration over borders of hardy plants.

Harmonious arrangement of colours.—If we arrange the pure colours in the fashion of a rainbow, beginning with crimson and passing through red, scarlet orange, yellow, green, blue, purple, and violet to crimson again, with all the intermediate hues, we may say that any two colours different to each other, or taken from distant parts of the above arrangement and placed together will produce a discord—as red and blue, red and yellow, red and green, yellow and blue, or orange and blue. Tints of these colours which are paler than the full hues are, in juxtaposition, still more discordant, as pink and pale blue, or pink and pale yellow. Combinations of two of these pale tints are the worst combinations of bright colour that can be made. The reason of this would seem to be that the beauty of the full colour, to a certain extent, overpowers the discord produced by the combination; while in the pale tint the amount of colour present is swallowed by the discord, and a sickly effect is produced. Two colours distinct, but nearly related to each other, are the only harmonious combination that can be made of two pure colours, such as crimson and scarlet, yellow and orange, green and blue, turquoise-blue, and violet-blue. The only exceptions to this rule are the hues between violet and crimson, which do not harmonise with each other. Many combinations of two discordant hues are harmonised by the introduction of a third, which would produce a discord with either separately; yellow and blue are harmonised in this way by the introduction of red, as seen in some macaws; red and green are harmonised in a lesser degree by the introduction of yellow, as in parrot Tulips. All pure colours are harmonised by neutrals or greys, from white to black, and also by darker or lighter tones of themselves, or of other harmonising colours. All pure colours look well along with mixtures of quiet greys, blues, browns, and bronze-greens, which is the background Nature gives flowers.

PURE COLOURS can be used in combination with impunity if well separated by neutral tints or low tones of mixed colour. The finest combinations that can be made with pure colour without the use of white, black, or grey would consist of various shades of red, blue, and yellow, from light to dark, in the proportion of three red, five blue, and two yellow, separated by green, orange, and purple, mostly in the form of subdued bronze and sage greens, browns and maroons, with just a speck of the pure tints here and there. The use of any great expanse of any pure colour, such as the bright yellow-green of Grass turf, renders a harmonious use of pure bright colour alone impossible. If any colour is used out of the above proportion, white or grey (black of course is not available in gardens) must be used as a counterfoil—that is, if a full harmony of colour is desired. The great evil of the bedding system before the present extensive use of foliage plants was the huge masses of red shades, the horrible discord of which, with the bright Grass turf, completely ruined any attempt at colour harmony, overpowering all delicate effect as much as the salute of a dozen ironclads would overpower a quartet of violins.

BACKGROUNDS FOR FLOWERS.—In arranging the colours in a garden, the nature of the background they will be seen against must be considered. In our English style of garden the colours of the flowers will be either seen against Grass turf or against shrubs. Well-kept Grass turf is always some shade of bright green, and approaches pure yellow in certain lights in sunshine; it should, therefore, never be brought into contact with anything but the colours of spring or autumn foliage, or with white or grey, yellow or orange, or very pale pink, lilac, or blue—the nearer white the better. Buttercups, Daisies, Cowslips, and Cuckoo Flowers are Nature's colouring for a background of green turf; and if we put other colours against it, the colours of both suffer—a sure sign of discord, for discordant colours always look duller when placed together than when separate. Although the colour of most leaves is green when looked at at right angles to their surfaces, there is very little bright green in the general effect of shrubbery seen in ordinary daylight, and still less when seen under the effect of sunshine. The green colouring matter of leaves is in the middle of the leaf, and does not show much when seen sideways. Where the upper surface of the leaf is glossy, the colour mostly seen is made up of a very neutral greenish grey, such as can be matched in painting by mixing a dull blue with a yellowish brown, and this colour is made still more neutral by grey reflections from the sky. The colour of leaves which are hairy or woolly is so much broken by the hairs as to be always more or less of a dull sage-green; the same may be said of the under sides of the leaves. The shadows amongst the stems and foliage of shrubs and trees are various shades of very neutral greens, browns, purples, and black. The only part of foliage which appears green is where the under sides of the leaves are visible, while the light shines through them from above, the number of leaves so seen being generally just sufficient to give a lively bronzy effect to contrast with the more neutral tints. In spring, the foliage is of a

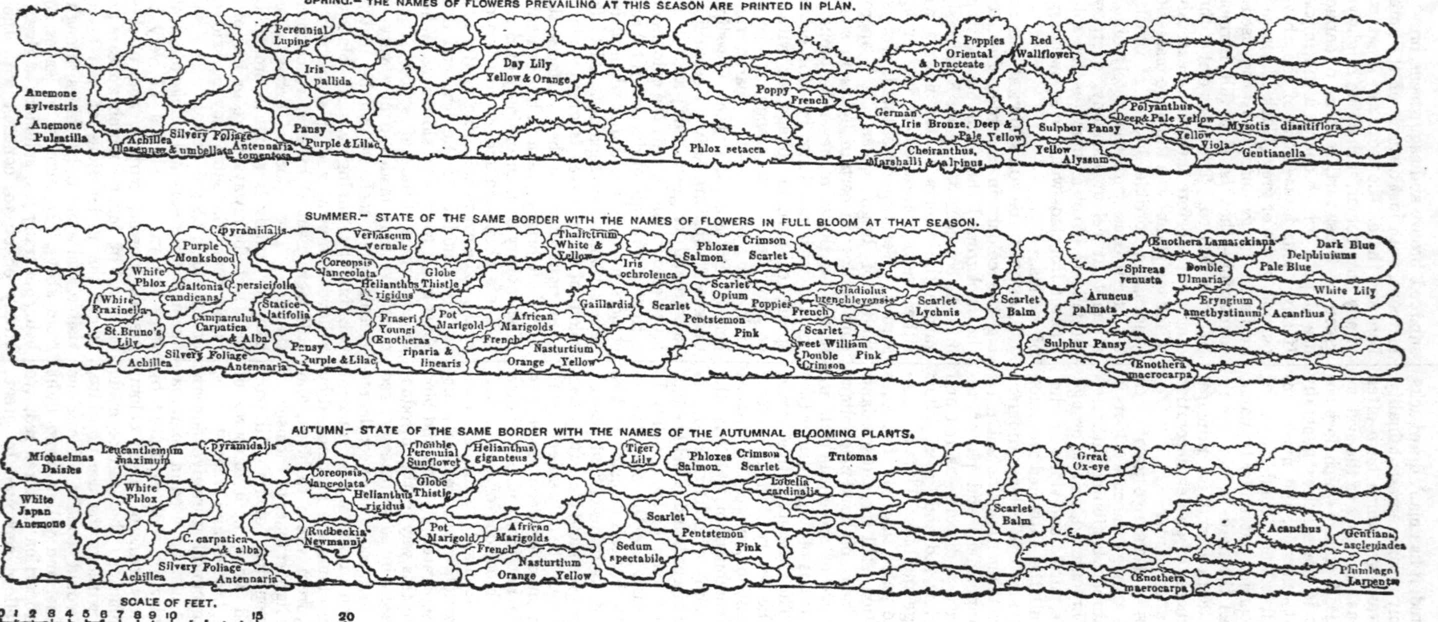

Plan showing the principal groups in a border of hardy flowers; the plants placed to form masses of harmonious colouring, and their progression simply, but carefully, arranged to produce a fine colour-effect. Many groups of small plants and bulbs, that could not be shown on the plan, are planted between and among the larger masses, their colour always agreeing with that of the surrounding flowers.

more decided green, through the cells which compose the leaves being more transparent; but this is modified to a certain extent by reflections and shadows, and by the reddish shades of the tips of shoots and of the bud scales. Foliage being mostly neutral is a far more satisfactory background for flowers than Grass turf, and every colour can be brought against it with advantage; the present system of cutting out beds on lawns and filling them with flowers should, therefore, be abandoned as far as possible, and the beds which are left filled with dwarf shrubs and tall striking plants, such as Tritomas, tall Irises, Hollyhocks, and Pampas and similar Grasses, which would show their foliage only against the turf, and their flowers against the shrubs beyond. The distant edges of lawns nearest shrubberies could be decorated on the same principle with smaller plants. Beds near the house could be filled with brighter-coloured flowers for individual inspection, and not massed for colour effect; such flowers, in fact, as would be useless for massing, through not producing a sufficient number of blooms at once.

MASSING OF PLANTS FOR COLOUR EFFECT.—The principal massing for colour effect should be carried out in beds where the colours of the plants will be seen against a shrubbery, or shrub background, and the groupings should be so arranged as to give their principal effect when seen from important points, although admitting of close inspection. For instance, a standard of the Noisette Rose Aimée Vibert might be so placed as to tell as a mass of white without detracting from its beauty as a single object. White will always rule any combination of colour into which it is introduced in a garden, and will attract the eye to itself, throwing everything else into subordination; but the leading masses of flowers need not be all white—yellows, pinks, and lilacs will all coalesce into one mass with it; but dark and purple or blue shades will not.

Not even spotting or striping of colour should be arranged, and in any one field of view there should always be a great leading mass of light colour rather than one, or at most two, unequal secondary masses, and a few scattered smaller masses. The leading mass should lead in height, as well as in extent, if possible. The deeper and richer colours can be arranged between the light-coloured masses, bearing in mind that the nearer they are to the shrubs in general light and shade, the more they will fall into the same mass with the background. Blue should always stand by itself, and be seen, if possible, against deep shadow. The spires of the bright blue Delphiniums never look better than when breaking up the edge of a mass of deep gloom under trees. Blue and white should never be mixed for distant effect, although a broidery of blue or white, or the converse, looks well in dwarf plants for close inspection. With hardy plants, it is impossible to have a leading mass of flowers at the same spot all the summer, but this is neither necessary nor desirable; the leading mass might be to the left of the picture in spring, to the right of the centre in early summer, to the left of the centre in late summer, and to the right of the picture in autumn. For the sake of clearness, I will describe one planting only.

LEADING MASSES AND GROUPS.—The finest and most telling effect would be on the principle of some parts of a firework display, the subordinate spots coming into bloom first, then the smaller masses, while the principal mass grows in beauty until it finishes with a grand show of the most striking plants of the season. A continuous display of bloom would be kept up amongst the dwarf plants, by their falling in with each succeeding grouping, so that as each arrangement passed its climax, the flowering would begin to lead up to the next. All the masses of spring flowers might lead up to the blooming of clumps of St. Bruno's Lily, and of Lilacs, double Cherries, and Laburnums. The late spring and early summer flowers might lead up to masses of white Lilies, Iris, Campanulas, and such Roses as Harrison's and Persian Yellow, Madame Plantier, and the blush Boursault, which produce conspicuous masses of bloom. The late summer bloom might culminate in the late flowering Lilies and Phloxes and Noisette and Tea Roses, and the autumnal bloom in masses of the hardier Chrysanthemums and Pompones. Flowers, seen from a low point of view, mass themselves in distant effect, although scattered in reality. Everyone must have noticed how the Buttercups in a meadow become a solid mass of yellow at its distant edge, although no thicker there than close at hand, where they grow quite thinly scattered. Advantage can be taken of this to produce masses of colour with plants in level beds, leaving room for several changes in effect in the course of the year, by mixing plants which flower successively.

CLIMBERS AND COLOUR.—A grand and conspicuous variation might be produced by a more plentiful use of climbing and twining plants. Most of these are extremely informal and picturesque, and are capable of being formed into telling groups by planting them behind dwarf plants whose flowers harmonise in colour with them. White Dutch Honeysuckle, for instance, groups beautifully with white and yellow Pansies, Anemones, and Narcissi, and the common Honeysuckle with Auricula-eyed Sweet Williams. The common Canary Creeper tells beautifully behind delicately tinted Asters, and climbing Convolvuli behind Petunias, Pansies, or Verbenas. Roses, which delight to ramble, can be treated in the same way, and a Boursault and an Ayrshire Rose will look better trailing among trees and shrubs than nailed up or trained. While the distant effect of masses of plants, as arranged for smaller groups, can be designed in the same way, the plants which form a distant mass can be arranged in varying groups amongst themselves for closer inspection. Plants which do not produce a solid mass of bloom, and those which have feathery flowers or spikes of bloom, should be placed, if possible, so that their flowers are seen against dark backgrounds. Spiræas, Foxgloves, Campanulas, Lilies, Pyrethrums, &c., will be found to tell best in this way. Lumps or trusses of flower, such as Sweet Williams produce, look best against other flowers or light foliage.

QUANTITIVE PROPORTION OF COLOUR.—Although the proportion of three red, five blue, and two yellow is about the best for a composition of pure colour only, the introduction of such a preponderance of neutral shades as is unavoidable in garden decoration requires a totally different proportion. The quantity of violet-grey in the shadows of plants and shrubs necessitates a great decrease in the quantity of blue used and an increase in the proportion of red; while the bright yellow-green of the turf requires a reduction in the quantity of yellow. The natural proportions of colour among hardy plants are about the best. By natural, I mean omitting those red flowers which are the result of cultivation. This would give a large proportion of white, pink, and lightly tinted flowers, the proportion of yellow being greatest in spring and early autumn and late autumn, and of red greatest from midsummer to mid-autumn, when red shades are absent from trees and shrubs. Red is the most difficult colour to manage in a garden; it is always brightest when seen against white, but a good way to introduce it is, as in its place in the rainbow, after orange, but with white, pink, and lilac preceding the yellow—as, for instance, in the following order: pink, white, lilac, primrose, yellow, orange, scarlet. The same gradation may be continued through crimson, maroon, purple, and violet, and will look well if the scarlet and crimson are omitted, and the orange brought into contact with maroon. Bright scarlet, magenta, or light crimson should be used in small quantities, or they will kill all quieter hues containing red. The best way to use red in a garden is to mass the red shades, as already described for the lighter colours, letting the conspicuous points be scarlet, magenta, or light crimson, but not placing these pure hues together. This allows of the most perfect use of all the rich crimsons, purples, maroons, violets, pinks, and rose colours of which we have so many and such profuse bloomers. Pink acquires a new value against deep maroon or dark crimson, and the same may be said of salmon-rose and salmon-scarlet. A spire of yellow tells well, rising among dwarf deep crimson flowers. Pink and yellow may be used to connect the masses of light flower with those of darker colours by being mixed with both.

Pure yellow, the crux of the bedding system, tells best as a contrast, with maroon or very deep brownish crimson, and should be separated from scarlet, crimson, or blue by three times its quantity of white. Its best place is falling into a mass with white and pale tints, as already indicated. With violet it is heavy; but good, if white or greenish white, such as many foliage plants give, is associated with it. Pale yellow or primrose looks well with lilac and lilac-rose, and groups with such plants as Pinks, which have a glaucous foliage. A great deal may be done in grouping of colours by judicious arrangement of beds. As a simple instance of this, suppose the shrubberies in a garden are brought round the lawn opposite the front of the house in two convenient sweeps. By making the inward ends of these sweeps overlap, and finish—one some distance behind the other—an irregular or oval bed could be planned in a slanting direction between the ends, and continued behind the nearest one, so that the flowers in it would mass with those in front of the nearest bed of shrubs, and be seen against the flowers and shrubs of the furthest mass.

FLOWER BEDS ON THE LAWN.—Very small beds should be either turfed over or filled with specimen shrubs or plants, but large beds might be filled with masses of florists' flowers requiring special rich soil and attention. Such beds might be edged with flowers which would harmonise with Grass turf and with clumps or masses of taller plants in the middle without looking too formal. Carnations, Pinks, Stocks, Pentstemons, Phloxes, and Dahlias would all look well planted in this way, and could be edged with borders of varying width of plants flowering at other seasons. Beds near the edges of lawns could be filled with dwarf plants of one colour, or of varying shades of the same colour, avoiding figures of any kind, and using pale colours or blues and violets, which would agree with the colour of the turf. Sufficient use has never been made, even under the bedding system, of monochrome effects by planting each bed with contrasting shades of one colour only—such as the varying shades of Phlox Drummondi from nearly black to pale pink—in lines or masses, with perhaps just one line of sage-green for a contrast, another bed being all blues, a third all yellows, and so on. All shades of colour look well in this way, and bright scarlets and crimsons can be used sparingly with good effect. The same effect can be got by planting in informal clumps if the beds are not too small. Small beds should be filled with one colour only, avoiding decidedly red hues. Where there is room for two colours only, a combination of two contrasting colours can always be rendered less raw by introducing a spot or two of each colour into the mass of the other. The most satisfactory disposal of colours in a mixed border is, generally speaking, to place the richest and deepest colours in the middle of the border, so that an irregular base line of paler tints may run up here and there to telling clumps of light colours, the heavier colours appearing above this base line and between it and the background. This should be arranged, however, without any appearance of striping or formality; indeed, if a fair selection of plants is chosen, they will mostly fall into this arrangement. Such plants as Pansies can be arranged in mild borders; while others, such as the Star Anemones and Tulips, will throw their flowers against the middle or second zone of a border, when seen from a little distance, although they are planted in the front row.

PLANTS WHICH DO NOT PRODUCE MASSES OF FLOWERS need not be discarded for that reason. To decorate the garden with flowers in masses only would be to adopt the bedding system writ large. Plants can be arranged for effect of foliage as well as of flowers; and plants unsuitable for massing for colour effect can form part of those quieter groupings, or can be cultivated in spots not visible from the reception-room windows, if desired. Such plants as exhibition Roses of moderately free-flowering kinds, and Chrysanthemums, which require a long period of careful cultivation, should generally be so cultivated. The latter, however, have a good effect placed in a double row next a sheltering wall or shrubbery, which they mass with when out of flower, and come out grandly with when in flower, with the front row drooping their shoots, while kept from trailing on the ground by being tied to neat stakes. A good way is to plant showy half-hardy annuals in front of them, such as Balsams, Stocks, Asters, and Zinnias, as the richly manured and well-worked soil which these require will suit the Chrysanthemums. If the annuals are pulled up when over, so as to disturb the roots of the Chrysanthemums as little as possible, these last will quickly appropriate the remaining nourishment of the border and bloom finely. For border display, Chrysanthemums should never be stopped.

BEDDING PLANTS IN THE PICTURESQUE GARDEN.—Bedding plants can be used for picturesque groups exactly in the same way as described for hardy plants. The very same plants now used for carpet beds could be used for decorating a sloping bank in the full sun, and would look much better in the same colour combinations in a picturesque group than in a geometric pattern. Half the effect of geometric beds is not due to the pattern, which cannot be seen except from above, but to chance perspective combinations which are not designed. Were the plants arranged to be seen in the perspective, a better effect would be got. To explain this more clearly, suppose we have a round bed arranged in a repeating pattern of six divisions with an edging: from many points of view we have a mass of three divisions, then two together, while the third appears some little distance off, which is a very satisfactory grouping indeed, but one wholly unintentional. In the same way, rectangular beds appear irregular masses, with a long tail from most points of view, and could be made to look better if an irregular shape and grouping were adopted. There is really no necessity for the introduction of geometrical patterns into garden design, as all that is got by them can be more successfully produced by a less formal arrangement. A spotty arrangement must, however, be avoided; the best for dwarf plants will be found to be to plant them in flakes or cloud forms. The principal thing to be borne in mind in arranging the tall plants is, that the masses are to be produced, not necessarily by plants placed close together, but by plants which will group together from the principal points of view, even though the plants themselves may be yards apart.

COLOUR IN OLD GARDENS.—I have had a long walk through an unfrequented corner of Sussex where the bedding system has never been adopted, and where nearly all the gardens are, consequently, in the old-fashioned style. One border I noticed particularly as showing how nicely the newest introductions fall in with old-fashioned plants. There were spring flowers in plenty, sleeping quietly through the summer; in flower were Roses in standard and bush form, Dahlias, Phloxes, Marigolds (common, French, and African), autumn-flowering Monkshood (with blue and white flowers), Everlasting Peas, Sweet Peas, Coreopsis, Crimson Flax, annual Chrysanthemums, Phlox Drummondi, Indian Pinks, Asters, Stocks, Zinnias, Lupines, early-flowering Chrysanthemums, Carnations, Petunias, double Achillea Ptarmica, Japanese Anemones, Verbenas, scarlet Pelargoniums, yellow Calceolarias, and many annuals I cannot remember. I saw the border from one end, where, of course, no gaps were visible; it was over 80 yards long. The colours were well harmonised, mainly through a predominance of Gloire de Dijon and other light Roses, pale Phloxes, and white and yellow Daisy-shaped flowers, the Geraniums, Stocks, and Verbenas giving a base of heavier colour. The whole effect was as bright as anyone could wish without being in the least gaudy. There were no stripes, patterns, or formality; consequently the blossoms grouped themselves into wreaths and sprays like summer clouds, varying with every change of view.

FOLIAGE *versus* GRASS.—It was wonderful, too, to notice how much the flowers gained in brilliancy and effect from being seen against a background of their own foliage, and of bushes and small fruits, instead of against the green of Grass turf. The same was the case with all the smaller gardens; there was no turf seen along with the flowers. Everything was bright, and pleasant, and natural, and every speck of colour had its proper value. Every decorative artist very quickly learns that there is one colour which must be excluded from all colour combinations, except such as have ample ground of white, black, or neutral tint, and this is bright grass green, just the hue of well-kept Grass turf, from its colour in a sunny day approaching yellow, to the full green it appears on a cloudy day. These hues can only be used in the most minute specks if in combination with scarlets, crimsons, blues, and purples. Flower beds filled with these colours cut out on well-kept turf are enough to set a colourist's teeth on edge; such a combination argues an approach to colour blindness. The quiet neutral greens of most foliage, broken by reflections and shadows, are very different in effect, and are favourable to every colour; therefore, those who enjoy colour and wish to see all their flowers look their best and brightest—every blossom attracting the eye and sparkling like a gem—should avoid planting masses of low flowers in beds on Grass turf. In the long walk already mentioned through miles of villages where the Honeysuckle and Jasmine, Everlasting Pea, and crimson China and old Musk Roses covered the walls, and invaded the roofs of the cottages, I came upon a very ugly house, with a most formal garden of closely shaven turf, ornamented with beds of scarlet, blue, and yellow. The effect, so emphasised by sudden contrast with a perfect style, was simply horrible; it was like nothing so much as the sudden appearance of Mr. Punch's "Arry" with all his unconscious vulgarity and pretence in the middle of a society of gentlemen, honest farmers, and simple country people.

FLOWERS SUITED FOR THE GRASS.—Where a broad expanse of turf wants breaking up in some way, there are many hardy plants with white, yellow, or pale pink or lilac flowers which will answer the purpose admirably, and will not be ruined by being seen against the Grass. There is the Acanthus family, the large-leaved Saxifrages, pale-coloured Phloxes, light-coloured Tea Roses, white, yellow, and pink Lupines, Sweet Rockets, the light-coloured varieties of Iris and Gladioli, the white and pale pink Carnations. Many fine Grasses are also available for the same purpose. Then all our native spring flowers look well on Grass, in evidence of Nature's perfect colouring, for the Grass on the waste lands is green when these are in bloom, and not of a soft neutral tint as it appears in summer. A good way of filling large beds on the Grass would be to carpet the ground with very dwarf evergreen plants, and use tall bright-coloured plants in the centre. Very pretty beds can be made of dwarf and creeping plants, beautiful even when out of flower, and suitable for all soils and situations. The use of these, interspersed with tufts of the smaller growing bulbs, would separate bright colours from the turf. This could only be carried out well, however, in beds from 20 feet to 30 feet across.—J. D.

NATURE A GOOD COLOURIST.—One may have a beautiful garden without paying much technical attention to colour, provided a natural style of treatment is adopted. It is not by any means necessary that a flower should be set off by contrast with some other flower of a different shade of colour near it. We do not need to take our garden plans from the canvas of the artist. For such ends, Nature herself is a good guide, and she does not paint the landscape in a gradation of colours all harmonising one with the other after fixed rules. It is Nature the artist copies, or professes to do, and the gardener has much greater need than the artist to draw his inspiration from the same quarter. I can well remember the time when the bedding-out mania set in, over thirty years ago. At that time the system presented less formality and flatness than it did later on. Many plants, including Phloxes, Dahlias, Hollyhocks, Lobelia cardinalis, Lilies, and numerous other herbaceous plants were used; but the straining after effects in colour led to the banishment of all these until we had nothing left but Geraniums, Lobelias, Calceolarias. For a grand work everything was sacrificed to colour, in the contrasting of which the artist was continually invoked, just as he is now by those who would have us believe that colour effects should be the principal aim in the planting of herbaceous plants—the narrowest conception of the hardy garden that has yet been conceived, and one in direct opposition to the leadings of those who have done most to popularise the use of such flowers. To see one good thing at a time is a good maxim in gardening. For example, we can admire a fine group of Lilies by themselves on the lawn or in the shrubbery, or a mass of Delphiniums or Phloxes, Anemones, or Primroses, and a place can be found for them in that way without encroaching on other subjects. But it is by no means always necessary to show the one off against the other by way of contrast; or, by means of intricate combinations and a laboriousness of detail in arrangement, to make the simple pleasure of gardening such a "high art" affair, as might frighten beginners from attempting it altogether.—J. S.

LABELLING AND STAKING.

ANYTHING which facilitates the acquirement of a knowledge of beautiful and interesting natural objects is valuable. A distinguished botanist once told us that he was first led to the study of the subject by the considerable amount of information and facts about plants placed so simply before the reader in some of Loudon's books on plant nomenclature. But books on botany, horticulture, or any other such subjects are not usually referred to till they are wanted. On the turf of lawn or garden is the most suitable place to make first impressions in this way. For many years past, the discovery of a really useful label has been a great object with our leading horticulturists. Yet it is very rarely that we see the plants well named, even in our best public gardens; and to find them so in a private garden is very rare indeed. Therefore, it may not be unacceptable if we describe the best and simplest system of labelling. Most people seek a permanent label for their plants; but we think it a bad plan, and for these reasons: the contents of a garden are usually in a state of change; we are continually adding to and taking from them; new plants are introduced, and surpass the older kinds or new varieties, and then we throw them away; a severe winter kills a number of Pines or shrubs, which we determine not to plant again. Fashion changes the garden vegetation, too, and then what becomes of the permanent labels that are cast and burnt into the face of hardware, and cemented into cast-iron? They are generally useless and thrown aside. The label which can be used again is the best, and therefore we prefer a cast-iron label of what is usually called **T** shape, or, in other words, a slip of cast-iron with an oblong head slightly thrown back. These are cast very cheaply in the iron districts, and will last for centuries. Of course you will have to paint and write the names of the trees on them when they come to hand; but that can be readily done by

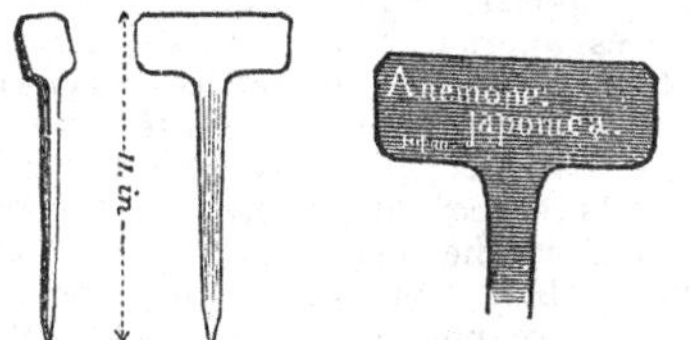

Cast-iron labels: the simplest, neatest, and best form for shrubs, bold herbaceous plants, and all cases in which the label has to be fixed in the ground.

any handy painter, who will probably be glad of such a job in the winter. In a large garden, or a public garden, where much naming is required, the right way is to train a boy or youth who is likely to remain in the place to do it; and we have done that in a few weeks by placing a copy of the desired kind of letters before him. We have found it of great advantage to give the face of the label a coat of copal varnish when the letters are dry, and we usually use white letters on a black ground, giving three coats of black over one of red-lead. These are the best labels for the usual shrub and choice young tree vegetation of a pleasure-ground or flower garden. They will require repainting probably every half dozen years or so, and should we from any cause cease to cultivate the plants to which they belong, the labels may be newly painted and re-used at pleasure. One can get more than one hundred of them for the price of two or three of the permanent labels recommended for choice specimens.

Next we come to old trees, or any trees of respectable elevation and bole, or body, so to speak. When a rare tree attains size and dignity, like many of those at Sion House and in hundreds of English pleasure grounds, it is more desirable to label it than a young specimen, however choice. With such big trees it is always a mistake to use a ground label, which, indeed, we only recommend for the younger and choicer subjects, because another kind could not be affixed to the tree in a satisfactory way. The cheapest, best, and simplest

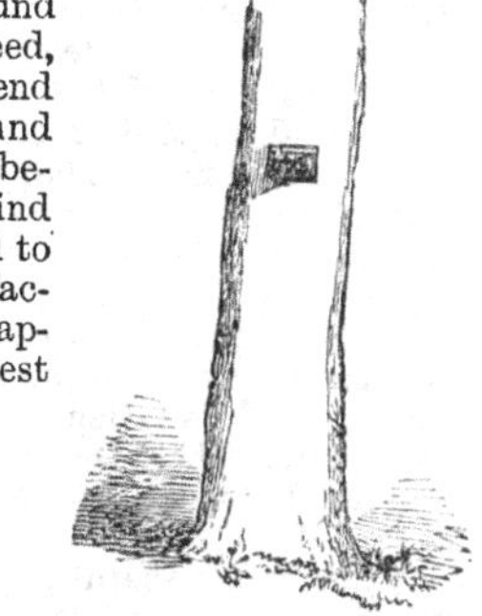

Scarlet Oak.
Quercus coccineus.
N.Amer

The simplest and best label for trees.

Position for tree label.

of all labels for large trees are made of pieces of tin about 4½ inches long by 3½ inches deep. About half an inch of the upper edge should be bent down at a right angle so as to form a little coping for the label, two holes being made, just beneath the little angle, through which is put a strong copper wire, that is firmly nailed to the tree. Place it so that it may be easily read, and at about 5½ feet or 6 feet from the ground. Paint it dark brown or black with white letters, clearly written. This label will last for many years, and is in every way satisfactory. All labels inserted in the Grass in pleasure grounds are liable to be pulled up by mowers or others, and in this way frequently get lost. The labels on the bole are safe from all such mishaps, and are more satisfactory than any other kind.

For Roses, and any kind of tall or open shrub, perhaps the simplest and best kind of label is a common wooden one, 3 inches long or so, and with a hole bored at one end, through which a copper wire may be passed to attach it to the tree, the name being written on while the paint is moist, and the copper rather loosely, though securely, bound round the branch, so as to allow for its expansion. In some places where large collections are kept it is usual to put numbers to the plants, and enter the names with these numbers in a book. In such cases all that has to be done is to provide suitable numbering material, and the best known are little narrow strips of lead, on which the number is impressed with type for the purpose, and then the strip is wrapped round a small branch rather loosely, but securely.

For all bedding plants, ordinary wooden labels are at once the most convenient and the simplest. If required to be of a lasting character, dip the ends in tar or pitch. In most gardens it is the practice in writing these wooden labels to write the name starting from the part that goes in the ground to the top—a bad way, inasmuch as the label always begins to decay at the base, and thus the beginning of the name gets obliterated, while the end of it may be quite legible. *Always begin to write it at the top*, and then if it does decay at the bottom, the commencement of the names will, in most cases, lead to their recognition. This may seem a small matter, but it really is of much importance where there are many plants named with common wooden labels. It becomes as easy to write from the top after a little practice as the other way, and in writing the names always begin as near the top as possible. The wooden labels are most readily made from laths if they be not bought by the bundle, in which way they are frequently sold.

When we have hardy plants it is often desirable to furnish them with more permanent and horizontally written labels. The readiest and best way is as follows: there are zinc labels of several shapes sold by seedsmen, usually written upon with some acid, but they are generally unsatisfactory, and not sufficiently legible. By painting these with a little white paint made without oil, one may write upon the surface with common ink, and by placing over the lettering and general surface of the label when dry a touch of the best copal varnish, a neat and convenient label is provided. The labels can be written by anybody, and are as quickly done as any. It is best to write each letter distinctly—to "print" it after a fashion, in fact. There are others equally suitable, but one may obtain a dozen of these for the price of one of the impressed labels, and they may be used again and again, like the cast-iron labels recommended for trees. We are the more desirous of recommending these economical and efficient articles, from having frequently seen the labelling of a good garden given up in disgust, in consequence of the failure of expensive glazed and other labels. In places where large collections of hard-wooded plants are grown, the small strips of lead, with the number impressed on one end and the strip folded on so that the numbered end shall be presented to the eye, are excellent. They can be used as well for Roses, fruit trees, &c., in the open air.

WAY OF WRITING A LABEL.—But numbering is not a good system for general use. It is generally a good plan to give a place to the "common name." The Columbian Maple, the Dove Plant, the Maiden-hair—these and such as these that are really "common," or have some recognition, or some meaning, or association should be given; but to merely translate the Latin name, to give us something like the "acuminate-leaved Sarcoglottis," or the "long-tubed Brain-bane," is not desirable. Never omit the generic name in full. The native country, in addition to the scientific name, may be given; to put the date of introduction or natural order is unnecessary. One may see labels in some Botanic Gardens so much covered with authorities, and one thing or the other, as to scare away the visitor whom they ought to help. Of course this refers to large and important things; for small ordinary plants the single name only should be used.

After trying every way I am quite satisfied that the best way of all is to have a stock of strong, but neat iron labels, and have some person to re-write additions to them occasionally in winter. The most suitable size for general purposes of labelling permanent plants is about 9 inches or 10 inches high, 1 inch wide in the shank, the head 4 inches or 5 inches across, and 1½ inches or 2 inches deep. In the first instance I made a model in wood of the desired kind of label, and, sending it to Glasgow, had a number cast at a cheap rate per hundredweight—getting somewhat more than a hundred labels per hundredweight. In writing the labels it is best first to write the outlines, and then fill in the white paint rather thickly; the paint to be finely strained, of course. No label deserves general adoption in a large garden which will not permit of being used again and again, if from any cause the plant it was originally used for disappears from the garden, or perhaps receives another name. As to permanence, that question is already settled. Iron labels, painted black and with white lettering, covered with a coat of copal varnish, done ten years ago in the herbaceous department of the Botanic Gardens in the Regent's Park, are now in 1883 legible. This is permanent enough for all ordinary purposes. Grand old Oaks, or other objects no more liable to suffer from vicissitudes than granite rocks, may require as permanent labels as can be devised; but the painted tin tree label before named lasts twenty or thirty years.

Label at Bitton.—The collection of hardy plants at Bitton is the best named I have seen. A small T-shaped, cast-iron label

is used. It is first painted white, then black; when the black has been an hour or so on, the name of the plant is written with a fine, but round-pointed bit of iron, thus exposing the white surface below. In this way effective lettering is produced, and the labels, being black, are not at all so offensive to the eye as white ones. When it is required to read them, moreover, the names are perfectly legible. This mode of writing labels was first published by Mr. Green when gardener to Mr. Wilson Saunders, at Reigate.

Zinc garden labels.—As this question of zinc labels has given rise to a great deal of discussion, we have thought it well to supply the following remarks on the subject, as advanced by experts, in *The Garden*—mostly gentlemen we know who have named their collections satisfactorily.

After trying lead and various other sorts of labels for garden borders, I have found the kind here figured the most satisfactory; the labels are absolutely imperishable, and the numbers cannot be effaced; moreover, they do not bend with any slight pressure, as leaden ones do. They are made of zinc, are cheap, and the time occupied in numbering them is short. I find a very convenient size to be 7 inches long, $2\frac{1}{2}$ inches wide at the upper end. The tools for cutting the numbers are a mallet, a chisel, for the longer strokes, five-eighths of an inch, and for the shorter one half an inch wide. These should be bevelled on both sides to bring them to a very sharp wedge shape. A bradawl, ground down to form a punch for the circular holes, and a block of wood, on which the labels are placed while cutting the numbers, complete the necessary apparatus. The numbers I use, which are the only ones suitable for this plan, are those first published, I believe by the late Mr. Loudon. They are easily learnt by anyone, and soon become as familiar as any other kind of figures; they should be cut quite through the metal.—J. G. N.

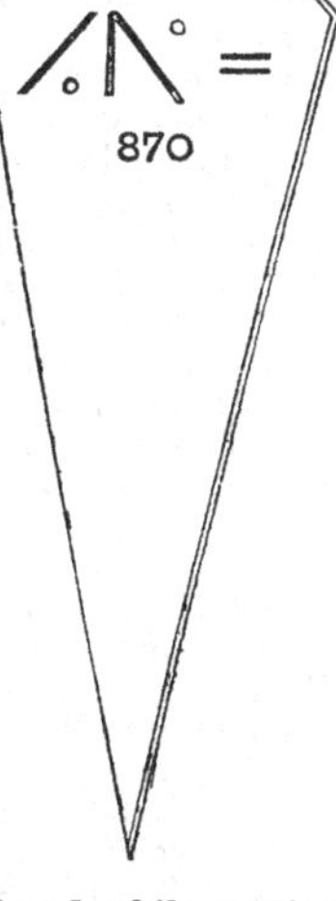

I use some thousands of labels in the course of a year, and find that, for ordinary work, zinc labels with the names bitten in with acid, answer admirably. I make my own labels, and, of course, can manufacture them more cheaply than if I bought them. Obtain a shilling bottle of stopping-out varnish from Buck's or Fenn's, in Newgate Street, a sufficient quantity of labels, some spirits of turpentine, a slate pencil, some pieces of rag or sponge, and some spirits of salts. Just touch the face of the label with the acid, wipe it clean, and cover with a thin coat of varnish. With the pencil, finely pointed, write the name on the surface thus prepared, taking care to remove all the varnish where the acid is to bite, as otherwise no mark would be made. Then pour a few drops of acid on the writing, and let it lie for about five minutes. The acid should be wiped off with a piece of rag, and the varnish removed with a sponge or rag saturated with turpentine, when the operation is complete. Labels thus made cost, after they are written, about 3s. 6d. per 1000, and will last good for nine or ten years.—W. J. MAY.

The ink I use for zinc labels is composed of twelve grains bichloride of platinum to one ounce distilled water. It should be specially made up, as the ink of this description usually kept by chemists for test purposes is much too weak. The zinc must have a bright surface before using, which is easily obtained by slightly rubbing with fine emery paper. The size and shape are matters of taste, but they should be of ample length to prevent their being easily knocked out of the ground or thrown out by frost. Any I have seen in seedsmen's shops are much too thin and short. They can be had from any plumber or tinsmith, who has always plenty of the raw material in stock from which to cut them. The price of the zinc is somewhere about 3d. per lb., to which has to be added the expense of cutting and waste. Some gardeners write from the top or bottom indifferently, so that the neck has to be twisted in one or other direction alternately in reading the labels in a border—much to the trial of one's temper. These labels can be used a dozen times if wanted. Scrape them with an ordinary carpenter's chisel, which is easily done, and they are ready for use.—P. N. FRASER.

Legible labels.—A label is worth nothing unless it gives the name of the plant to which it belongs with unmistakable clearness, while it remains itself in the background hardly seen. If we are to have great, gawky figure heads in our gardens we had better have done with nomenclature altogether, or learn to carry it in

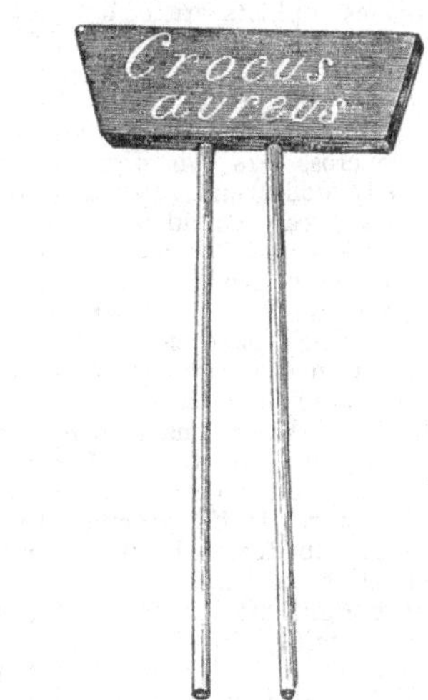

our heads. Let us never forget that the labels belong to the plants, and not the plants to the labels. On the other hand, a label is a deceit and a snare if it does not tell you at once all you want to know. There should be no neck-breaking process required, and no stooping at all. It should be as easy to read the names of our plants as we walk along the garden path as it is to read the pages of a book in the library. The difficulty has been, for those who go along with all this, to unite unobtrusiveness in the label with perfect clearness of the name which is written upon it. But I think it can be done, and the method which I recommend is the following: let the cross-bar of the label be of no greater size than is wanted to carry the name; let this receive two good coats of white paint (the white must be decided, or the letters will suffer for it afterwards) and then when the white paint is dry and you have plants to be named, let one surface of deep black be given to it, and let the words that are needed be scratched in before the black paint is set. The effect of it all is that you have names which are inscribed on a black ground in white letters of remarkable clearness, and I will venture to say that from the point of view of good taste, nothing ever will be found that looks better. The labels assert themselves in the borders in the very slightest possible manner, while the information you want is obtained from them at once. But there are two or three other points to be noticed. If you want to decipher a label with ease which is stuck into the ground, it stands to reason that the inscription on it must be written horizontally, and not from the bottom to the top, or even from the top to the bottom, and this one consideration will govern a good deal. It leads you straight to what I may perhaps call a composite label. I have, it is true, seen the names of plants written horizontally on labels which are all of a piece, but I have never seen such labels which are handy to use, nor could I get on with them. But the whole thing is managed if you give to each of the little cross-bars two wire supports. This is easily done, and after a trial of several years I can confidently say that success is complete. For these little wire supports are very nearly invisible; they scarcely meet the eye at all, and so the main points which have been referred to above are not in the least interfered with. The background of the label still remains of the smallest, you have a name before you, and only a name. And to this I may add some other very clear gains. These little wire legs never rot in the ground. Unlike wood, they are unaffected by damp, and no fungus will grow on them. Next, they remain quite upright after a frost. Wooden labels are certainly lifted out of their places by frost, and zinc labels cut the ground so sharply, that they are even less to be trusted. Thirdly, where cheapness is thought of, iron wire will bear comparison with anything else. I by no means say that the label which I venture to recommend with such confidence is good for all circumstances, all places, and all plants. I wonder that the thought of an ideal label should ever have entered into anyone's head at all. What is suitable for one set of conditions may very likely not be suitable for another. I do not recommend a composite label with white letters on a black ground for the back rows of a large border, or for a shrubbery, or for the wild garden; but I think it is invaluable for a rockery where you have hundreds of small plants which should just be distinguished from each other, and which can bear no advertising at all. I have found it to answer in beds which are crowded with bulbs, and where clearness is especially to be desired. It does well for the front rows of the herbaceous border, or for any border which the eye can command at a glance, or for the glazed pit; but I should never let it be covered over by the huge boughs of some Pinus insignis, or even do battle with a rampant Polygonum. Something more demonstrative is there absolutely required, and chrome yellow would be of use.—HENRY EWBANK, *Ryde*.

A good zinc label.—Having tried many different kinds of labels for flowers, such as alpine and herbaceous plants, Roses, &c., and having very great objections to labels which attract the eye, or such as can be readily seen amongst flowers in the border, or on the rockery, I find no label so suitable as a zinc one, 6 inches in length by 1 inch in breadth, and pointed at one end, having the name of the plant written upon it with indelible ink for zinc labels. I get the plumber to cut me out 500 labels of the size above mentioned at a time, using for the purpose bright untarnished zinc, and when the labels have been cut out they are placed in a dry, close, wooden box for use as may be required. When labels are needed for the flower borders or rockery, I take the requisite quantity out of the box, and while they are bright I write with a sharp quill the names of the plants upon them, commencing the writing as close to the top of the label as possible, and writing towards the end to be inserted in the ground. When the writing becomes dry, which it does in a minute or two, the labels are inserted 4½ inches into the ground near the plants, leaving only 1½ inches exposed above the soil. The result is that the labels always remain where I place them; they never decay; the writing never becomes illegible, and, what to me is the greatest importance, they cannot be seen unless one looks for them. For Roses, I use an oval-headed label, printing the name with the indelible ink, and when dry varnishing the face with spirit varnish. A very suitable size I find to be 6 inches long and 1 inch broad with the oval part at the top 2½ inches long by 1 inch broad. If the names are clearly printed they can be read without the necessity of stooping, and yet the labels are not in any way conspicuous in a bed of Roses on the lawn.—G. M.

Staking plants.—None of the work of the flower garden is of greater importance than staking, and the difficulty of getting it neatly and well done is often too evident. Even where there is time to attend to the staking, the beauty of the garden is often marred by the plants being tied into broom-like bundles. To my mind, herbaceous plants in spring, as they successively rise from the earth with their infinity of form and colour, furnish a sight beautiful in itself. What a

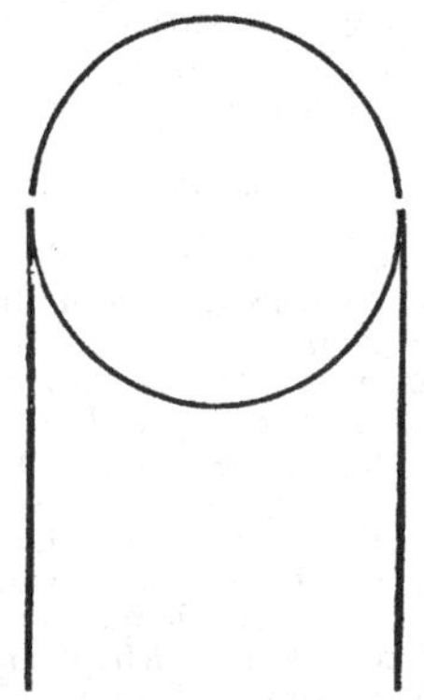

Plant girdle; prongs of upper half not shown.

cheery, vigorous-looking object a well-established clump of Columbine is as it just emerges from the soil! But when the plants throw up their flower-stems and staking commences, transforming the border into a miniature forest of sticks, all beauty disappears, simply from the injudicious manner in which the tying up and staking are too often performed. Those who practice excessive primness with their plants might think the way in which I manage mine careless, and even slovenly. Nevertheless, there is a degree of natural wildness in my borders that (to me at least) is infinitely more satisfactory than the permitting of every plant to be tied like a martyr to a stake. I am acquainted with a garden as well managed and supported as any in England, and one in which hardy herbaceous plants are cherished and cared for; in fact, they are spoiled by being too well cared for. There, every plant, no matter of what character or however graceful naturally, is doomed to be transformed into a besom!

Such plants as Hollyhocks, some Lilies, and tall Larkspurs must be staked. The supports for these should be stout-pointed iron rods (like the round iron rods used in making hurdles). They need not be of solid iron, as stakes can now be bought which are hollow, yet strong, and quite as efficient as the solid ones. These will last for years, and can, without detriment, be thrust into the mass of a plant, or so near it, as to be almost invisible. The tying material should be thin, soft, tarred twine, which should be worked in under the leaves, so that the latter may hide the tie. How are plants generally tied? Let us take a border plant, say a strong, established Carnation with, perhaps, from thirty to forty stems; we generally see this plant with a multitude of stakes (green, perhaps, and from four to six stems tied to each), presenting, when in bloom, handfuls of flowers, or we see plants tied round the middle, a plan by which all individual beauty is destroyed.

To do away with such an unsightly method, I have used for the purpose what I have called girdles, made of different sizes and strength of wire; these are half circles of wire, with the ends turned down to form prongs; two of these are placed round a plant just of sufficient diameter to allow it to stand and spread out in a natural manner, and not as a bundle, the strength and height of such girdles being determined by the plant. A clump of Pinks would merely require a low and slender one, while the double Pæony (the most top-heavy plant I know of) would need one 18 inches high (clear of the ground), and made of very strong wire. When neatly placed round a plant, with the foliage brought over them, these girdles are almost invisible. Cleaned and put by when done with, they will last for years.

Another useful stake for Carnations and similar plants is the one here figured which is intended to provide a stake at once cheap, unobtrusive, everlasting, and, if we may so speak, self-tying. The illustration shows it first by itself, as it is made for Picotees, &c., by taking galvanised wire about 1-12th of an inch thick (hard drawn), and twisting it

Fig. 1. Fig. 2.

in a long spiral way about four times round a stiff straight piece of wire about a quarter of an inch thick (once round in about 3 inches), finishing with a sharper turn round the top about three-fourths of a revolution, and then cutting off, so as to leave half an inch of straight wire at the tip, the straight bottom being about 9 inches in length. Fig. 2 shows it applied to a Picotee, and stuck in so deep that the top bud shall just rise above it. The plant is then wormed round so as to let it fall into a position in the middle of the spiral stick; and when the top is also slipped into its place it will be found as safe as if it were tied ever so well, being supported by the bit of straight wire left for the purpose. A few yards away this stake is scarcely observed, so neat is it compared with ordinary sticks. Another good point is the freedom which it allows to the foliage and long hanging flower-stalks to fall away from it if they wish; and, in addition, if the stem grows after it is applied—as it nearly always does—it slides up without making any ugly hitch in an attempt to push up against ties, as in the old way. A man might make a hundred in a single evening out of half-a-crown's worth of wire, and apply them to plants next day in less than half an hour.

"Brockhurst," in *The Garden*, remarks: "I daresay most gardeners have lots of old fencing wire in old corners, which fetches nothing if sold, and is awkward to stow away. Take this wire, straighten it, and cut it up into lengths of from 1½ feet to 3½ feet; then take a strong pair of pliers, or use an anvil swage to bend over one end into a ring, as here shown. Then dip them into black varnish, kept in an old 2-inch gas tube, stopped at one end, and they are thus quickly and permanently painted. Where a group of flower-stalks is to be tied up, such as a Phlox or Carnation, we take three of the wires and pass a tarred cord or shred of twisted matting through the three eyes around the group. It is not safe to use the wires without bending over the end, as it is apt to lead to accidents, because they are nearly invisible, and your face may unwittingly come in contact with the sharp point when you are looking at or smelling a flower. We have many hundreds of these supports in use, and they appear, as far as one can judge, to be everlasting."

But besides wire and iron stakes there are other kinds which are quite as useful and far more easily obtained. Apple prunings, for example, make excellent flower stakes when dried; and, if properly applied to the plant, are not at all so objectionable as painted stakes. Hazel rods, again, make good flower stakes. They are pliable, and the colour of their bark renders them inconspicuous. Where Michaelmas Daisies are largely grown, they make good stakes, as such plants do not need much support. When the leaves of the Asters fall, the stems may be cut down to the ground, and if tied in bundles and kept dry, they will be ready for use in spring. Bamboo canes, too, are now cheap and exceedingly useful; they last almost as long as iron, and require no painting.

Bad effect of staking hardy plants.

THE

ENGLISH FLOWER GARDEN;

OR,

FLOWERS OF THE OPEN AIR.

Abobra viridiflora, a fragile South American twiner, belonging to the Cucumber family, occasionally grown in gardens, but of little value, easily raised from seed, and trails over a low trellis or against a low wall. It is graceful in habit, but does not seem to succeed in our climate generally.

Abronia *(Sand Verbena)*.—A small genus of Californian plants, numbering some seven species of annual or perennial duration. They are all of a dwarf trailing habit, and bear showy blossoms in dense Verbena-like clusters. Four kinds are known in gardens, viz., A. arenaria, known also as A. latifolia, a perennial having procumbent trailing stems and dense clusters of lemon-yellow flowers, with a honey-like fragrance; A. umbellata, an annual also with succulent trailing stems and dense terminal clusters of rosy-purple and slightly fragrant flowers; A. fragrans, a perennial more or less erect in growth, forming large branching tufts from 1½ ft. to 2 ft. in height, and producing terminal and axillary umbels of pure white flowers which expand late in the afternoon, and then emit a delicate vanilla-like perfume; A. Crux Maltæ, a pretty species with white flowers and sweetly-scented. A. arenaria and A. umbellata succeed best in rather poor, light, and dry soil; in richer and moister ground they are apt to grow weedy, and the flowers become less conspicuous. The position best suited to them is one fully exposed—either an open, flat, but well drained border or rockwork. A. fragrans succeeds best in friable or light soil, but being of larger and taller growth than the others, should not be grown on a rockery, but in a well drained border. The propagation of all the species can only be effected by means of seed, which in favourable seasons may be obtained from A. arenaria and A. umbellata, but as A. fragrans does not ripen seed in this country, imported seeds of it must be procured. The seeds both of this species and of A. fragrans frequently remain dormant some time before vegetating; those of A. umbellata germinate more readily. The plants flower in summer and autumn, and being pretty in bloom, may be used with good effect in the varied flower garden, somewhat as dwarf Verbenas are.

Abutilons.—A genus of plants mostly requiring greenhouse temperature in winter, but which grow freely in ordinary garden soil out-of-doors in summer, and are a graceful aid in the flower garden. A. Darwini and its forms, as well as the varieties related to A. striata, under favourable conditions, grow from 4 ft. to 8 ft. in height. They can be stopped and made bushy according to the height required, and they flower and look better than they do in pots. They are useful among the taller and more graceful plants for the flower garden, and are easily raised from seed and cuttings. These varieties are now of various colours, and others are being raised every day. There are some splendid variegated kinds suitable also for out-door work. A. vitifolia is a very handsome wall plant in mild districts. A. Sellowianum marmoratum is a fine variety. Among the best in cultivation are the following, and new varieties are being continually raised: Admiration, Anna Crozy, Buisson d'Or, Darwini robustum, Darwini majus, Elegantissima, Grandiflorum, Lemoinei, Lady of the Lake, Leo, Orange Perfection, Boule de Neige, Delicata, Pactole, Darwini tesselatum, Thomsoni variegata, and vexillarium variegatum.

Acacia lophantha.—This elegant plant, though not hardy, is one of those which grows freely in the open air in summer. It will prove more useful for the flower garden than it has ever been for the houses. The beauty of its leaves and its quick growth in the open air make it a boon to the flower gardener who wishes for graceful verdure amongst the brighter ornaments of his beds. It has a close, compact, and erect habit, which permits us to closely associate it with flowering plants without shading them or robbing them. By confining it to a single stem and using it in a young state, we get the fullest size and grace of which the leaves are capable. It may be raised from seed as easily as a common bedding plant. By sowing it early in the year it may be had fit for use by June 1; but plants a year old or so, stiff, strong, and well hardened off for planting out at the end of May are the best. It would be desirable to raise an annual stock, as it is almost as useful for room decoration as for the garden. New Holland.

A. Julibrissin.—A native of Persia, with large and elegant much-divided leaves, and flowers somewhat like short tinted brushes from the numerous purple stamens. By confining it to a single stem and using young plants, or plants that have been cut down every year, we get an erect stem covered with leaves more graceful than a Fern, and a pretty object amidst low-growing flowers. The leaves, like those of some other plants of the Pea tribe, are slightly sensitive. On fine sunny days they spread out fully and afford a pleasant shade; on dull ones the leaflets fall down. Seed of A. Julibrissin—or the Silk Rose, as it is called by the Persians, in consequence of its silky stamens—is readily obtained, and it is much better raised from seed, as we then get those single-stemmed and vigorous young plants which are to the flower garden what an elegant Fern is to the greenhouse. It may be protected at the root and cut down every year in spring, or strong young plants may be put out annually, in much the same way as those of A. lophantha.

Acæna.—This genus is wholly confined to the Southern Hemisphere, and though possessing no attractive features as regards the flowers, if we except those intensely crimson spines that give such a charm to the little New Zealand A. microphylla, the species have a neat habit of growth that renders them deserving of culture. When visiting a place in which a very extensive rockwork was formed a couple of years ago, we found an illustration of the adaptability of Acæna pulchella as a rock plant. It was growing on a ledge, above a mass of rock some 4 ft. high, and its branches, which were covered throughout their whole length with pretty bronzy leaves, were suspended in graceful festoons over the face of the stone, even down to the rocky pathway, over which many of them spread in wild profusion, and often measured as much as 7 ft. or 8 ft. in length, the result of but little more than one season's growth

A. microphylla *(Rosy-spined Acæna).*—This spreads into dense tufts no taller than a Moss, and in summer and autumn becomes thickly bestrewn with showy and singular globes of spines. It is quite easily increased by division, is perfectly hardy, grows in ordinary soil, but thrives much the best in that of a fine sandy and somewhat moist character. Its home is on bare level parts of the rockwork, and it is also good as a border or even an edging plant in soils where it thrives. Occasionally it may be used with a singularly good effect to form a carpet beneath larger plants not thickly placed. It would form a good ground or protecting plant for choice hardy Orchids and Trilliums, or like plants in the bog bed. *Syn*—A. Novæ-Zealandiæ.

A. millefolia is much unlike either of the preceding. It has finely divided pale green foliage, and is a graceful plant for hanging down a bank. Its defect is its unsightly fruiting spikes (not heads, as in the others). Its points will adhere to anything they come in contact with, and the result is that seedlings of this species come up in the most unexpected spots. There are many other kinds in cultivation, such as A. ovalifolia and A. sarmentosa, but those that have been mentioned are distinct and characteristic enough, representing as much of the genus as may be admissible in a collection.

A. pulchella, though in floral and fruiting characters, never at any time showy, is nevertheless a plant worthy a place. Any soil or situation will suit it, but the best position we ever saw it in was growing from interstices of stone by the steep side of a sunken rocky path. Its branches, rooting as they went, were covered throughout their entire length with pretty bronzy leaves, and suspended in graceful festoons over the face of the stone, even to the rocky path beneath, in wild profusion, often measuring as much as 7 ft. or 8 ft. in length, the result of little more than one year's growth. Its flowers are inconspicuous.

Acantholimon *(Prickly Thrift).*—An Eastern genus, extending from Syria and the east of Greece to Western Thibet, and having its head-quarters in Persia. The flowers resemble those of the Statice and Armeria, but the habit of the species of Acantholimon at once distinguishes it form either of these. The Acantholimons do, indeed, form branching, cushion-like tufts, somewhat after the style of Thrift; but the leaves, instead of being soft and Grass-like, are rigid and sharply pointed, or even spiny. As many as eighty-four species have been described by Bunge, while Boissier has only about half as many; it is probable that the higher number is capable of considerable reduction. The species much resemble one another.

CULTURE AND POSITION.—As known to us, they are dwarf evergreen rock garden and choice border plants. The following species we have had for years, but have not been very successful in propagating them, except A. glumaceum, which is the freest growing of the set; the others are very slow growing. Cuttings taken off in late summer and kept in a cold frame during winter make nice little plants in two years, but by layering one gets larger plants in less time. All are hardy and seem to prefer warm, sunny situations planted in sandy loam. Few would seem to have any idea of how much is lost by having many of our herbaceous and alpine plants in pots. There are three species in cultivation—A. glumaceum, venustum, and androsaceum, but the introduction of others is much to be desired. A. Kotschyi is a handsome species, with long spikes rising well above the leaves and numerous white flowers; A. melananthum has short, very dense, capitate spikes, the limb of the calyx being bordered with dark violet or black; A. Phrygium is in the way of A. venustum, but

PLATE I.

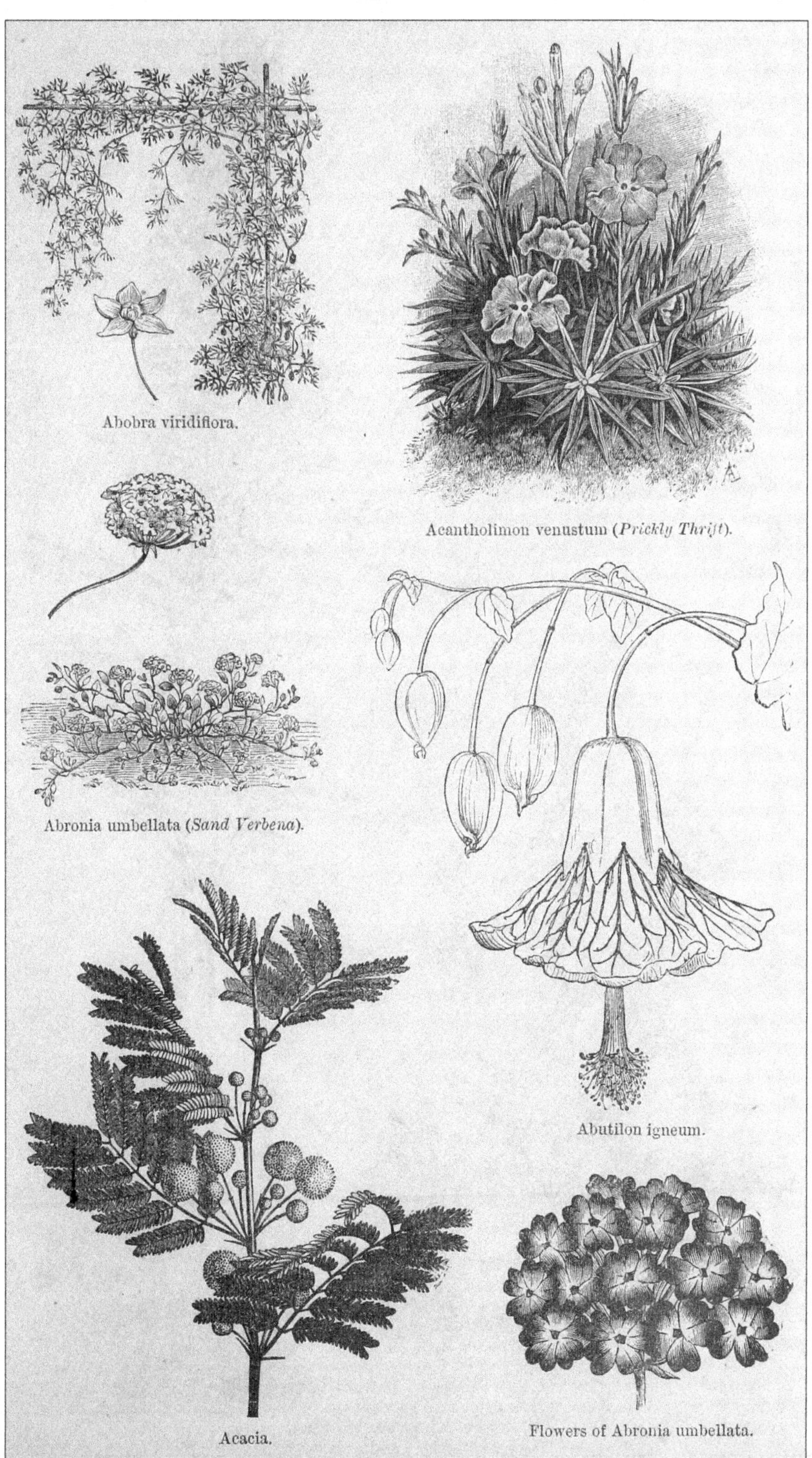

Abobra viridiflora.

Acantholimon venustum (*Prickly Thrift*).

Abronia umbellata (*Sand Verbena*).

Abutilon igneum.

Acacia.

Flowers of Abronia umbellata.

[*To face page* 2.

PLATE II.

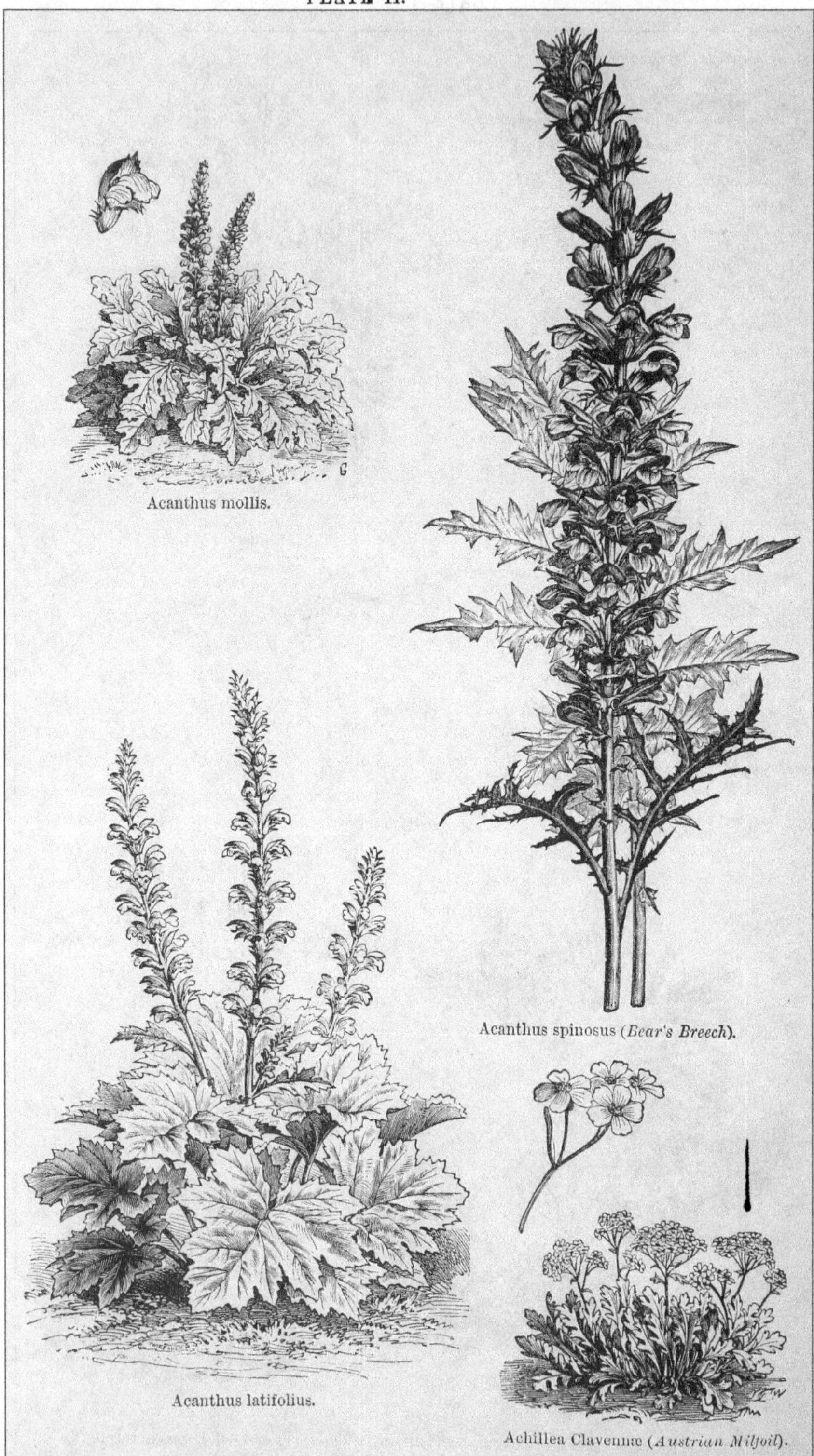

Acanthus mollis.

Acanthus spinosus (*Bear's Breech*).

Acanthus latifolius.

Achillea Clavennæ (*Austrian Milfoil*).

[*To face page* 3.

is a more robust plant, with longer spikes.

Acanthus *(Bear's Breech)*.—A long-neglected group of hardy plants, with singularly fine foliage and stately habit when well grown. The plants mostly come to us from the countries round the Mediterranean, and are hardy. Though the foliage may suffer now and then, we have not remarked that the roots ever perish.

CULTURE AND POSITION.—The plants are not so showy for the mixed border as other perennials, but possess a peculiar grace when they flower, which more than compensates for this; but they are worth growing for the sake of their foliage alone. Their fine habit and foliage make them valuable for planting on the turf, either singly or in groups, best in groups. The grouping should be gracefully done, the plants not dotted about and in a geometrical manner, but irregularly. not too close together. On rocky banks, or about ruins, or on terrace gardens of the less formal kind, they look very well. They will live in shade; to flower well they should have full sun. The fine form of the leaves, and their leathery texture, have made the Acanthus much admired as a house plant, as it is easily grown in a window, in which position we have seen it flower. These plants only require to be planted in any kind of soil in order to grow freely, but attain fair stature and their best bloom on warm, deep soil—a free sandy loam, or any deep and open soil. Clay they do poorly on, not flowering freely or at all—we mean a bad clay like that in the northern suburbs of London. When placed singly or in groups on the lawn, the greatest care should be taken to give them deep, good soil. They are not difficult to increase by careful division of the roots in autumn or winter, and they may be raised from seed in a gentle hot-bed, or out of doors with other perennials.

THE HARDY SPECIES introduced are not numerous: A. hirsutus, S. Europe; A. syriacus, Syria; A. mollis, S. Europe; A. spinosus, Turkey; A. longifolius, Dalmatia; A. spinosissimus, S.E. Europe. The fine bold kind known as A. latifolius (syn., lusitanicus) is thought to be a variety of the oldest cultivated species, A. mollis, but as a garden plant it is wholly distinct from that in size and appearance, and the most valuable kind. The plants vary in height from 1½ ft. to 4 ft. according to the richness of the soil in which they are grown. When in flower in warm districts and on good soils, we have seen them attain to nearly 6 ft. in height, and in this state a good specimen is one of the handsomest of plants.

Aceranthus diphyllus.—Synonymous with Epimedium diphyllum, a Japanese species.

Aceras *(Man Orchis)*.—A small genus of terrestrial Orchids of no garden value.

Achillea *(Milfoil, Yarrow)*.—A numerous family of hardy plants spread through Northern Asia, Italy, Greece, Turkey, Hungary, &c., but more in Southern Europe than in Central or Northern. In the Alps and Pyrenees numerous species are found. A careful selection is very useful for garden culture. The plants vary in height from 2 in. to 4 ft. Their flowers are pale lemon, yellow, and white, rarely pink or rose. A. Millefolium rubra is a fine plant when grown in poor soil; A. tomentosa is charming on dry gravelly soil. A. Ptarmica fl.-pl. (a double variety of the Sneezewort) is a handsome plant with snow-white flowers, but hard to keep within bounds. It should be lifted every year and replanted in clumps. Perhaps twenty of this large family are available as border plants. A. Eupatorium is a noble plant, bearing very large corymbs of yellow flowers. A. ægyptiaca is, perhaps, the most graceful of the family. A. pectinata is a pretty snow-white flowered plant, and one which should be treated as an alpine, and grown almost in gravel; it is a native of Hungary. A. Clavennæ, a silvery Austrian species, is one of the most striking of the family. Its leaves are divided into club-shaped divisions. Planted with Trifolium rubrum or Ajuga purpurea, it has a charming effect.

CULTURE AND POSITION.—Of the easiest possible culture, growing freely and increasing even too freely in any soil, with the exception of the dwarfer mountain species. Some of the large kinds are fine plants for the choicest borders or groups, as A. Eupatorium; the alpine species, such as A. tomentosa and A. Aizoon, are well suited for the rock garden while the coarser, very pretty-growing white kinds, unfit for garden culture, are strikingly effective in flower when naturalised in rough shrubberies and like places.

A. ageratoides.—Originally introduced into the country under the name of Anthemis Aizoon. It is a dwarf, compact-growing, silvery plant, with narrow leaves arranged in dense rosettes, and the margins exquisitely crimped. The flowers are produced singly on stalks about 6 in. or 8 in. high, pure white, and of large size. In many respects the gem of the genus. A rock garden plant. Greece.

A. ægyptiaca may possibly be, as the name would indicate, an Egyptian plant, but we think it is more probably a native of the island of Crete. It has beautifully cut white silvery leaves, and is of compact growth; the flowers are bright yellow in colour, raised on stems about 15 in. high, and would be a desirable plant were it only for its foliage. It is hardy, and a good plant in the warm border.

A. aurea, a native of the Levant, is frequently confused with the foregoing species, but is quite distinct. Its habit of growth is tufted, not creeping. Its leaves are larger, and its flower-stems attain a height of at least 15 in. The flowers are a golden yellow, and are produced in the autumn as well as early summer. It is a somewhat tender plant, and is now rarely met with.

A. Clavennæ.—A fine old plant with hoary leaves, deeply jagged as to the margins, of dwarf tufty habit of growth. Flowers white, in corymbs about 9 in. to 12 in. high. It is a native of Carinthia and the Austrian Alps. Under cultivation it loves dry sandy soil. In strong loam it rarely ever survives a winter without protection. A rock garden or edging plant, and also suited for beds.

A. Eupatorium (sometimes called *A. Filipendula*) is a tall-growing, vigorous, herbaceous plant, somewhat woody in the character of its lower growth. Its flowering corymbs are flat, bright yellow in colour, and elevated on stout stems at a height of 3 ft. to 4 ft.; they retain their beauty and freshness for at least two months. This is admirably adapted for a shrubbery border, where its brilliant yellow flowers and its erect habit of growth show to wonderful advantage amongst the evergreen foliage. It is a native of the shores of the Caspian Sea, and is one of the finest of perennial plants. It is very easily increased by division, and is worthy of good culture and a good position. It would go well with groups of the nobler hardy plants whether deserving the name for their foliage or flowers.

A. Millefolium rosea (*Rosy Yarrow*) is a lovely plant with rose-coloured flowers of so deep a tint as to come near crimson. It is a strong grower, height 2 ft., and blooming freely. It deserves a place in every herbaceous border. There are a number of species so closely related to the Milfoil, that it is needless to specify them. They may have a distinctive specific character, but for cultural purposes they may be taken as one.

A. Ptarmica (*Sneezewort*) is fairly distributed through Britain as one of our upland woodland plants, somewhat meagre and scattered in its native habitat, but when introduced into garden culture becoming a very showy and vigorous border plant, of some 2 ft. in height, with pure white flowers in corymbs. The double variety (A. Ptarmica fl.-pl.) is one of the loveliest white flowers we possess, and as a border plant of free growth and perfect hardiness has few rivals. Worth a place among plants grown for cutting.

A. rupestris (*Rock Yarrow*).—Among the dwarfer species this is one of the best, forming low tufts covered with pretty pure white flowers. Capital for the rock garden or borders, growing well in poor sandy soil.

A. serrata.—A distinct species of dwarfish habit; height about 15 in.; leaves white, with adpressed hairs; flowers in corymbs, a good clear white. Alps of Central Europe. There is a double form, but it is not so good as the double Sneezewort.

A. tomentosa stands first amongst those with yellow flowers. Of creeping habit, its flower-stems scarcely exceed a height of 9 inches, and its flowers are a bright yellow, produced in quantities in the month of June. Its foliage is much divided, and forms a dense carpet of bright green. Rock garden and borders; not in wet places.

A. umbellata is a dwarf compact grower of a tufty habit, scarcely exceeding 8 in. in height, the whole surface of leaf and stem being clothed with a dense covering of short hairs of silvery whiteness. This peculiarity has given it an introduction into the flower garden. A neat and attractive border or rock plant. The flowers are white, but, owing to the silvery character of the plant, they are inconspicuous. Greece.

Achlys (*Oregon May Apple*).—One species, A. triphylla, is the only one in cultivation. It is a North American plant, belonging to the Barberry family, and of doubtful value.

Achyrachæna mollis.—An annual Composite from California; of little value for the garden.

Achyrocline.—The only species, A. Saundersoni, is a small shrubby plant of the Composite family, having small leaves covered with a cottony material; of doubtful hardiness and merit.

Acis.—A small genus of bulbous natives of South Europe, of which some three or four species are in cultivation. The best known and prettiest is A. autumnalis, a very slender-leaved little bulb, with stems rising 3 in. or 4 in. high, and bearing a couple of flowers that may be described as delicate pink Snowdrops, drooping elegantly on short reddish footstalks, of a deep red colour round the seed-vessel, and blooming in autumn before the leaves appear. It is a true gem for the rockwork, where it should be planted in a warm soil and sunny position, sheltered with a few stones, and on which it would look very well springing from a carpet of delicate, feeble-rooting Sedum or other dwarf plant. The other kinds are A. trichophylla, rosea, and hyemalis, all of which will thrive where the soil is of a fine sandy nature, but as yet so rare as to be worthy of the best position and care. Mr. Elwes doubts if any of these plants will thrive in the open air in England. Acis autumnalis used to thrive in the nurseries at Edinburgh in the open air in fine sandy soil.

Aconite (*Aconitum*).

Aconite, Winter (*Eranthis hyemalis*).

Aconitum (*Monkshood*).—An important though dangerous family of plants, from the poisonous nature of their roots. There are too many names—not so many species—and judiciously placed, the best are of much value for our gardens.

CULTURE AND POSITION.—Position here is important, and few would care to risk their being planted anywhere the roots could be by any chance dug up by mistake for edible roots, as they are so poisonous. Nevertheless, some of the kinds are so handsome and stately when in bloom, that they are worthy of a place beside the finest hardy plants; as, for example, the blue and white

PLATE III.

Achillea Ptarmica (*Sneezewort*).

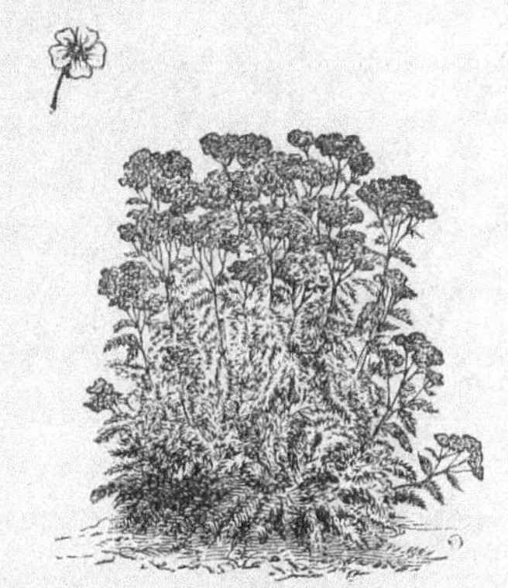

Achillea Millefolium rosea (*Rosy Yarrow*).

Achillea Eupatorium (*Caspian Milfoil*).

Achillea tomentosa (*Creeping Milfoil*).

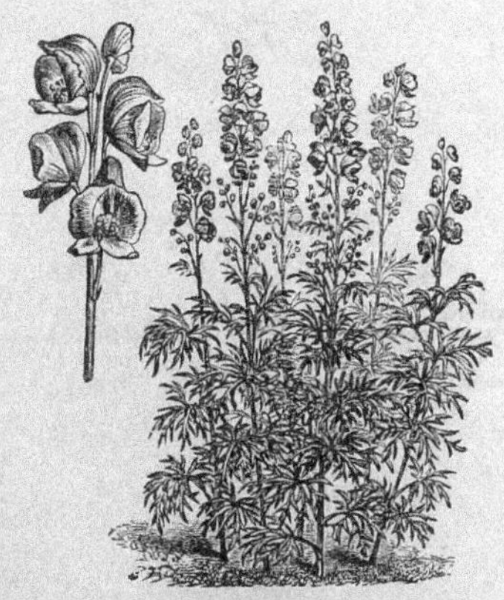

Aconitum versicolor (*Variegated Monkshood*).

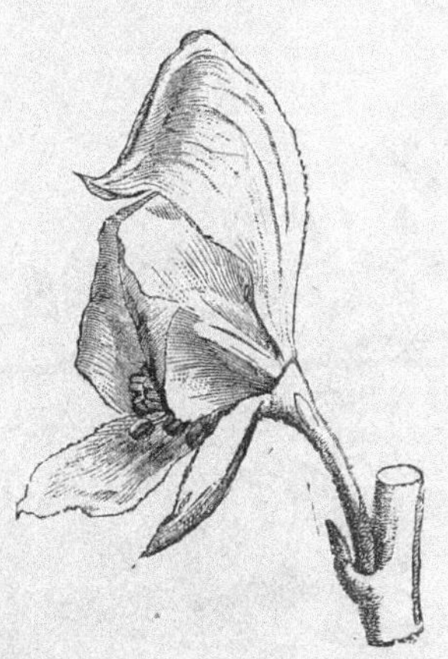

Flower of Aconitum autumnale.

[*To face page* 4.

PLATE IV.

Acorus Calamus (*Sweet Flag*).

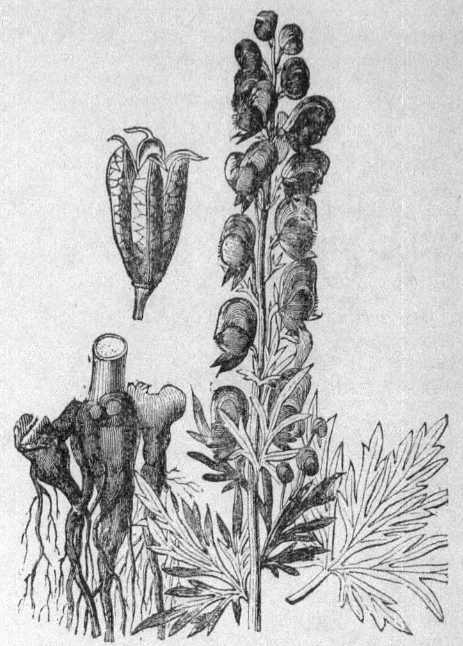

Flower-stem and root of Aconitum Napellus (*Monkshood*).

Acroclinium roseum.

Adenophora liliifolia (*Gland Bellflower*).

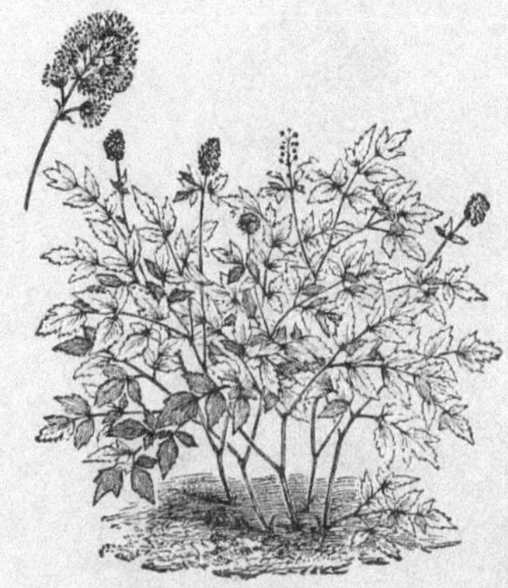

Actæa spicata (*Baneberry*).

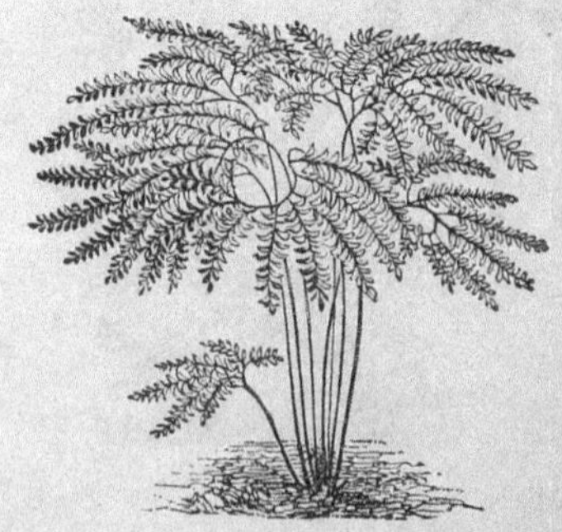

Adiantum pedatum (*American Maiden-hair Fern*).

[*To face page* 5.

A. versicolor, which is a beautiful object in the good soil and partial shade of many a cottage garden. Almost all the kinds may be easily naturalised in copses or shrubberies away from the garden proper, or beside streamlets, or in openings in rich bottoms.

The best kinds are A. Napellus and its forms, versicolor, and others; A. chinense, A. autumnale, A. japonicum, and A. tauricum; A. Lycoctonum is a yellow-flowered and vigorous species. All tall plants, from 3 ft. to 5 ft. high.; flowering from July to September.

Acorus *(Sweet Flag)*.—A small family of water-side or marsh plants, occasionally cultivated, and of wide distribution in nature. The Sweet Flag is fond of rather stiff moist soil, and may be planted either on the margins of pieces of water, or in the water itself. Easily multiplied, like the Iris, by division. Acorus Calamus (Sweet Flag) is a marsh or water-side plant, now naturalised in most parts of Europe. A variety has gold-striped leaves, and has been called A. japonicus folis aureo-striatis. A. gramineus (Grass-leaved Acorus) is a species with a slender creeping rhizome covered with numerous Grass-like leaves, which are from 4 in. to 6 in. in length. A native of China and Japan. A variety with white-streaked leaves (A. g. variegatus). This plant is often seen in the little bronze trays of water plants seen in Japanese gardens and houses. The plants look pretty on the margins of fountain-basins.

Acroclinium.—A. roseum, the only species, is a pretty half-hardy annual from Western Australia. It grows over 1 foot high and bears pretty rosy-pink flowers, which, owing to their chaffiness, are used as "everlasting" flowers. Seeds should be sown in frames in March, and the seedlings planted at the end of April or early in May in a warm border, or the seeds may be sown in the open ground at the end of April in fine rich soil. If the flowers are desired for preservation as everlastings it will be well to gather them when fresh and young, some scarcely out of the bud state. It does best in a warm, sunny border, in good open and well enriched soil. This annual might be made graceful use of in mixed beds. There is a white variety, and the two look well mixed.

Actæa *(Baneberry, Herb Christopher)*.—A small genus and not very important for gardens. Plants rather tall, 3 ft. to 6 ft., thriving in free soil; spikes, white and long, with showy berries where they fruit freely. The white Baneberry has white berries with red footstalks. The var. rubra of A. spicata has showy fruit; the plants are best suited for the wild garden in rich bottoms, as the foliage and habit are good. The flower is somewhat short-lived in the ordinary border, and they are somewhat coarse in habit. A. spicata (common Baneberry), A. racemosa (Black Snakeroot), A. alba (white Baneberry), having white berries with red stalks, and one or two American forms of the common Baneberry are the kinds in cultivation.

Actinella.—A small genus of North American Composite plants of which there are three kinds in gardens. They are all dwarf-growing plants with yellow flowers. The finest is A. grandiflora, a native of Colorado; it is a dwarf alpine plant with flower-heads 3 in. in diameter. The plant is more or less branched, and grows from 6 in. to 9 in. high. The other species, A. Brandige and A. scaposa, are somewhat similar. They are all perennial, and thrive in an open, warm border of light soil.

Actinomeris.—Coarse growing North American plants of the Composite family. A. squarrosa and A. helianthoides are the two kinds known in gardens.

Adam's Needle *(Yucca)*.

Adamsia *(Puschkinia)*.

Adder's Tongue *(Ophioglossum)*.

Adenophora *(Gland Bellflower)*.—Elegant plants of the Campanula Order, not many of which are in cultivation. Mostly from Siberia and Dahuria, and generally blue in colour. Some of the most distinct species are A. communis, A. coronata, A. liliifolia, A. Lamarcki, A. stylosa, and A. pereskiæfolia. In these there occur slight variations in colour and size of flower, and also in the form of the radical leaves. The thick fleshy roots of the Adenophora revel in a strong rich loam, and like a moderately damp sub-soil; they are impatient of removal, and should not be increased by division. Unlike the Platycodons, they produce their seeds freely, and the seedlings reproduce very constantly the specific characteristics of the parent. They vary in height from 18 in. to over 3 ft., and are well suited for the mixed border, flowering in summer.

Adenostylis.—Small growing composite plants of little garden interest. Three kinds are in cultivation—A. Petasites, pyrenaica, and alpina, all natives of Europe.

Adiantum *(Maiden-hair Fern)*.—Some of the species of this lovely genus of Ferns are quite hardy in this country. The soil best adapted for their growth is rough fibry peat, mixed rather liberally with sand and lumps of broken stone or brick. A. pedatum, the fine American kind, might be usefully employed for forming a carpet for other shade-loving plants. It is also excellent for association with the more beautiful wood flowers in the wild garden, such as Trillium, Hepatica, blue Anemone, and the like. The fronds of this Fern rise from a creeping rhizome, therefore care must be taken to plant it in positions in which it is likely to be little disturbed. Like all the Adiantums, it is fond of moisture while growing, care being taken, however, to always provide plenty of drainage, as stagnant moisture around its roots would speedily prove fatal. A. Capillus-veneris, the British Maiden-hair Fern, succeeds best in a very sheltered, warm position, as, for instance, in a little nook at the foot of a shady wall, associated with some equally moisture-loving hardy plants. In such a position it ought to thrive well, and so placed, it would be easy and advisable to protect it with some kind of portable covering during severe winters. Its native habitat is amongst the sheltered rocks of Cornwall, Devon, and Wales, and in various parts of Ireland; therefore some idea may be formed of the sort of climate in which it luxuriates. This Fern has a great predilection for damp, warm walls, which it speedily covers with a carpet of verdure.

There are several varieties or forms of this Maiden-hair, amongst which Adiantum Capillus-veneris incisum is a distinct kind, found in Ireland. In this the pinnules are much more divided than in the type. A. Capillus-veneris rotundatum, found in the Isle of Man, is also a beautiful variety, though very variable. The fronds are narrower and rounder than in the type. A. Capillus-veneris Footi, a large form, which sometimes grows upwards of 1 ft. in height, has fronds beautifully cut and divided. The Cornwall variety (cornubiense) is a fine plant and very distinct from the others, but is as yet somewhat rare. A. C.-v. Luddemannianum is a crested variety. A. C.-v. magnificum is a fine form with an A. farleyense-like port.

Adlumia (*Climbing Fumitory*).—One species only (A. cirrhosa) is known of this genus. It is a rapid grower, and soon covers the object against which it is placed. Its Maiden-hair Fern-like leaves are borne in profusion on the slender, twining stems, and the blossoms, which are white and about ½ in. long, are also borne very freely. There is a variety of it with purple flowers (A. cirrhosa purpurea), which, if grown with the type, forms a pleasing contrast. It is strictly a biennial; that is, it makes growth one year, flowers the next, and then dies, but it bears seed so profusely—which comes up year after year without being sown—that it may well claim to be a perennial. It is a native of North America, and was formerly known under the name of Corydalis fungosa. It requires a warm good soil to make it worth having, and its place is trailing over a shrub or twiggy branch, placed either against a wall or in the open.

Adonis.—Plants belonging to the Ranunculus or Buttercup family. They are chiefly natives of corn-fields in Southern Europe and Western Asia, and are dwarf in stature, with finely-divided leaves, and red, straw-coloured, or yellow flowers. There are about fifteen or sixteen species, most of which are annuals, and not very striking or ornamental in appearance; consequently, with the exception of two or three fine kinds, they are seldom seen in gardens.

A. autumnalis (*Pheasant's-eye*).—A British annual which grows 1 ft. or more high and blooms in the end of summer or early in autumn. The flowers are of a bright scarlet colour. Individually, the plants are not very effective, as they are rather straggling in growth, but when grown in masses in borders or flower beds along with other autumn-flowering annuals, they are pretty, though the plant is not popular in gardens. May be sown in the open ground in autumn or in spring.

A. vernalis forms dense tufts 8 in. to 15 in. high of finely-divided leaves in whorls along the stems. It flowers in spring, when the tufts are covered with numerous large, brilliant yellow, Anemone-like flowers 3 in. in diameter, a single flower being produced at the end of each stem. Of A. vernalis there are several varieties, the chief of which is A. v. sibirica, which differs from the type only in having larger flowers. A. apennina is a later blooming form. A. pyrenaica.—This is a fine and closely allied kind from the Eastern Pyrenees with large deep yellow flowers, resembling those of A. vernalis, but with broader petals. It flowers in April and May, and may be grown in the same way as the last-named species. Free, sandy, moist loam, and not often disturbed, robbed, or shaded by coarser plants. A choice border of rock plants or the rock garden suits the handsome perennial kinds well, and if the soil is poor it may be enriched with leaf-mould or any other decayed manure. They are increased by careful division or by seed sown as soon as gathered.

Ægilops.—A small genus of Grasses allied to Triticum (the Wheat Grasses); of little garden value.

Ægopodium Podagraria (*Gout Weed*).—A troublesome, indigenous weed common in many gardens. It is very difficult to eradicate, and there is no more effectual way than constantly digging out the roots. There is a rather pretty variegated-leaved variety of it, which, however, soon turns green again in gardens, and this circumstance partly accounts for the species being so widely spread in cultivation. No pains should be spared to root it out.

Æthionema.—This is a beautiful genus of the Arabis family, but differs from the greater number of the Crucifers in light, elegant habit and wiry stems, and usually glaucous leaves. They are mostly found on the sunny mountains near the Mediterranean, particularly eastward, and are especially valuable for gardens, forming stronger and more free-flowering tufts in them than in a wild state. These little plants will grow freely enough in borders of well-drained sandy loam, but their true home is in the rock garden. The tall Æ. grandiflorum forms a spreading bush about 1 ft. high, from which spring numerous racemes of pink and lilac flowers. It, too, grows well in borders in ordinary soil; seems to be a true perennial, and when in flower in summer, is among the loveliest of alpine half-shrubby plants. It succeeds perfectly well on the

PLATE V.

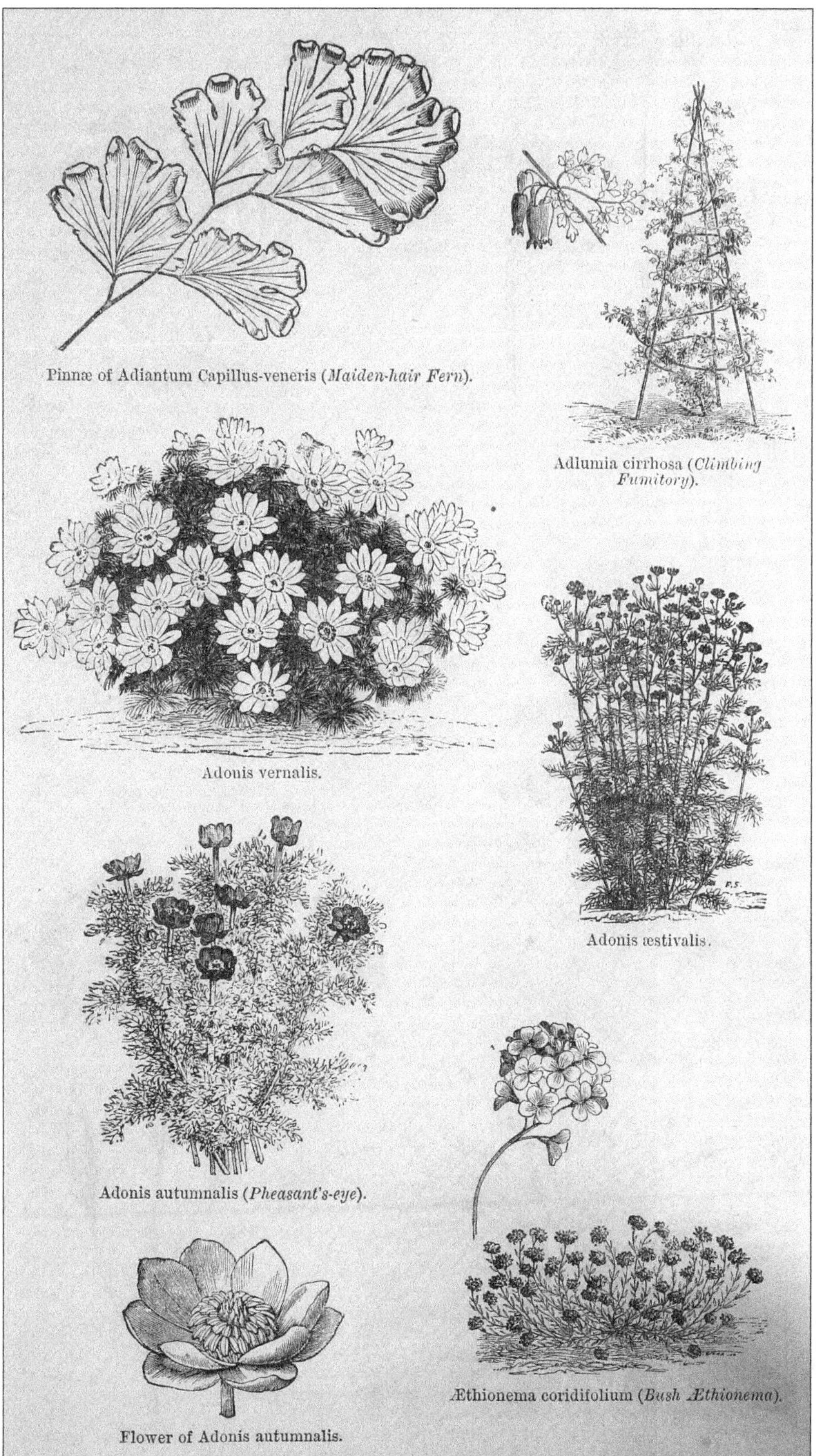

Pinnæ of Adiantum Capillus-veneris (*Maiden-hair Fern*).

Adlumia cirrhosa (*Climbing Fumitory*).

Adonis vernalis.

Adonis æstivalis.

Adonis autumnalis (*Pheasant's-eye*).

Flower of Adonis autumnalis.

Æthionema coridifolium (*Bush Æthionema*).

[*To face page* **6.**

PLATE VI.

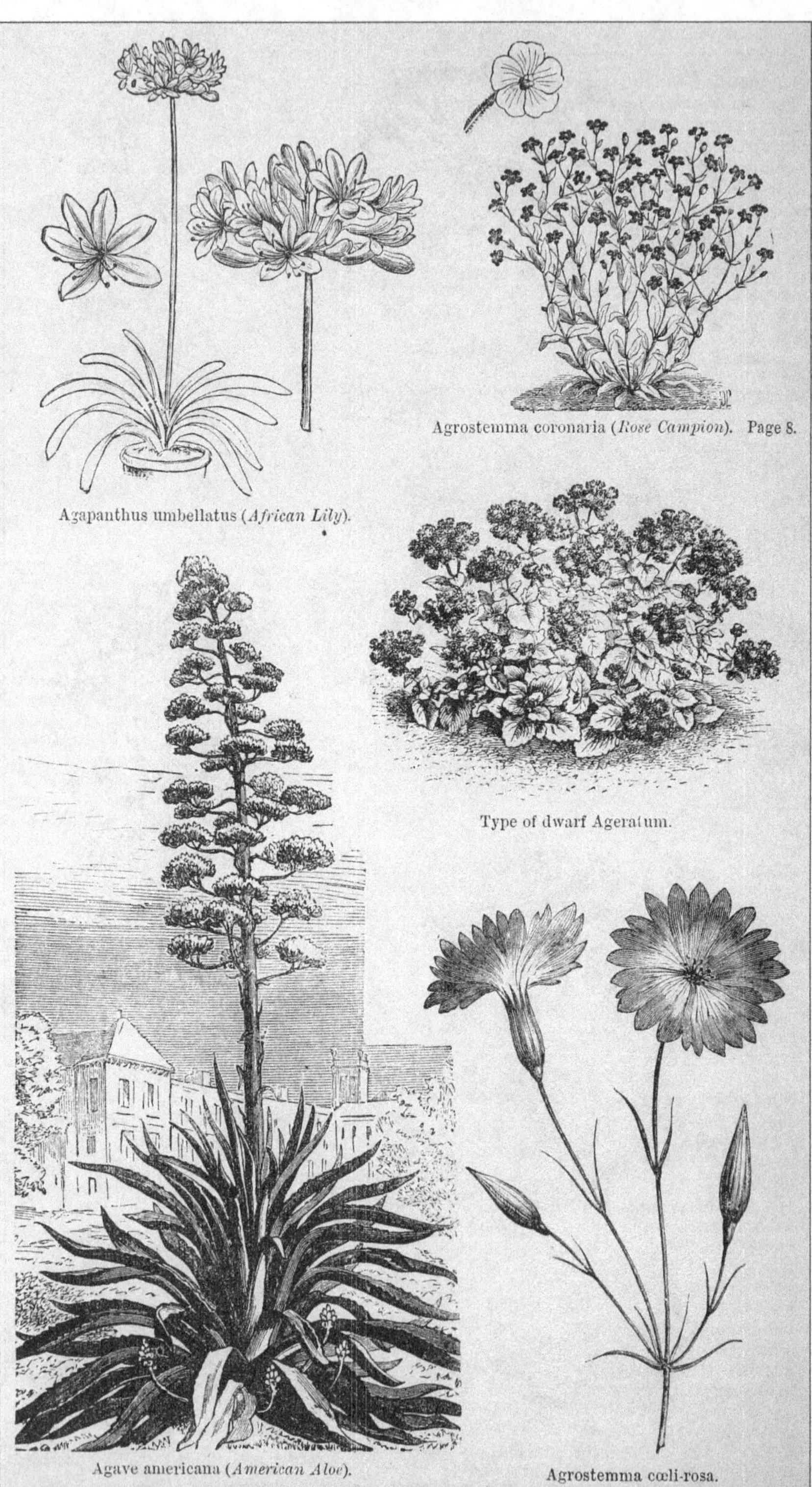
Agapanthus umbellatus (*African Lily*).

Agrostemma coronaria (*Rose Campion*). Page 8.

Type of dwarf Ageratum.

Agave americana (*American Aloe*).

Agrostemma cœli-rosa.

[*To face page* 7.

front margin of the mixed border; and though rockwork is not required, it will certainly be a gain where the highest health of the plants is sought. In consequence of the prostrate spreading habit of the stems, a pleasing effect will be produced by planting these in positions where the roots may descend into deep earth, and the stems fall over the face of rocks at about the level of the eye. The species are easily raised from seed, and thrive well in sandy loam. There are many species, but there are but few in gardens. All the cultivated kinds are dwarf, and may be grouped with alpine or rock plants. The best known kinds are A. coridifolium, pulchellum, and grandiflorum.

African Lily *(Agapanthus).*

Agapanthus *(African Lily).*-Beautiful plants from the Cape, with blue or white flowers in umbels on stems 18 in. to 4 ft. high. A. umbellatus, the old kind, is hardy in some mild seashore districts, and a fine plant in some rich beds or borders, but the better for protection of leaves or Cocoa fibre round the root in winter. Everywhere well worth growing for the flower garden and vases in summer, protecting it in winter by storing under stages, or in sheds or cellars. The fleshy roots may be so stored without potting. Enjoys plenty of water during period of growth out-of-doors, and is easily increased by division. Various new kinds have been introduced, but their value out-of-doors has not been so well tested as the favourite old African Lily. Of the best known kind, A. umbellatus, there are several varieties; major and maximus are both larger than the type, and of the latter there is a white-flowered variety. There is also another with white flowers, but smaller, and one with double flowers (flore-pleno). The variegated-leaved kinds, fol. albo vittatus and fol. aureo vittatus, are likewise desirable for the sake of variety. These are variegated forms of A. umbellatus pallidus. Saundersonianus is a distinct variety with deeper-coloured flowers than the type.

The largest of all is the new A. u. giganteus, the flower-spikes of which attain a height of from 3 ft. to 4 ft., with the umbels of flowers proportionately large, bearing from 150 to 200 flowers. The colour is a pure gentian blue, while the buds are of a still deeper hue. If it proves as hardy as the other species, this will be a very valuable plant. Agapanthus umbellatus candidus, pure white, seems a variety, as the seedlings will not come true from seed, about 70 per cent. only being true. A. u. pallidus is a good pale porcelain blue, a short-leaved variety. A. u. minor is a dwarf variety. Of A. umbellatus there is a true double-flowered variety, a very distinct and good plant. There is, moreover, A. u. atrocœruleus, a dark shaded violet variety. A. u. maximus is a variety with flower-stalks 4 feet long, and bears very full heads of flowers, one set opening while a second is rising to fill up the truss as the first crop fades. In size and colour the flowers are the same as those of A. umbellatus. A. u. Mooreanus is a deciduous and very hardy form; it grows from 12 in. to 18 in. high, has neat narrow leaves, and comes true from seed. There is likewise a pure white kind that is deciduous, the leaves turning yellow in autumn and dying off. It forms a stout root-crown. This is called A. u. albiflorus.

Agathæa cœlestis.—A tender Daisy-like plant, with blue flowers, used for the margins of beds. There is also a pretty golden variegated form. Single plants, or carpets of this are pretty if not very showy. Easily propagated by cuttings in spring or autumn on slight heat, or from seed. *Syn.*—Charieis heterophylla.

Agave americana.—This and its variegated varieties are useful for placing out of doors in summer in vases, tubs, or pots plunged in the ground, and also for the conservatory or large hall in winter. When the plant flowers, which it does only once, and after several years' growth, it sends up a flowering stem, from 26 ft. to nearly 40 ft. high. The flowers are of a yellowish-green colour, and are very numerous on the ends of the chandelier-like branches. It will grow in any moderately dry greenhouse or conservatory in winter, or even in a large hall, and may be placed out of doors at the end of May and brought in in October. Large plants in tubs have a fine effect out of doors in summer. This old plant was so used long before sub-tropical gardening was known in the land. All the varieties are easily increased from suckers. North America.

About four species of this genus, natives of North America, have lately come into cultivation, which are supposed to be hardy in this country, in which case they will be interesting subjects for the rock garden. These are A. Deserti, utahensis, cœrulescens, and Shawi.

Ageratum.—Tender plants, but much used for the flower garden, varying in height from 6 in. to 24 in., with pale blue, lavender, or white blossoms, easily propagated from cuttings in spring or autumn raised on a slight hotbed. The dwarf Ageratums are among the best of summer flower-garden plants, their blue or lavender coloured blossoms being very attractive when seen in contrast with Pelargoniums. The varieties should be increased by means of cuttings, as seedlings usually grow rank and are uneven in height. But the tall old kinds are as well deserving of culture as the dwarfs which have been raised from them. Flowering from June to October.

They are among the best or at all events the most lasting of summer bedding plants, a great point in their favour being their comparative hardiness, as they will withstand a few degrees of frost, consequently may be planted out earlier than most of the bedding plants, and continue longer in bloom. The flowers of all the varieties have plenty of substance in them, and are, therefore, not readily injured by rain, and what is still better, the flowers are lasting and do not fade in colour, but continue the same throughout the flowering season. There are numerous varieties of varying merit, both as to flowering properties and habit, some attaining in good soil a height of 2 ft., others not more than 6 in. The mean between the two heights will be found the best for parterre bedding, and the variety named Cupid has not yet been excelled. Its average height is 9 in.; it has bluish lavender-coloured flowers of great size, well set off with bright green foliage, and never fails to continue in flower till severe frost renders a general clearance of bedding plants desirable. Countess of Stair is nearly equally good, but of taller habit (about 12 in.) and a paler blue colour. Queen is also a very excellent kind, being distinctive in colour, a grey-blue; this grows about 1 ft. in height, and is fine for massing or for boundary lines to scarlet Pelargoniums. Swanley Blue is a very excellent dwarf kind, about 8 in.; the flowers are a dark lavender-blue, very pleasing arranged as a belt to variegated Pelargoniums. The very dwarf kinds are disappointing; they flower so freely, and the growth of the plants is so sparse that they have always the appearance of being stunted. The best of this class is a recently introduced variety named Cannell's Dwarf, a good kind for edgings to small beds, or for panel planting in carpet bedding arrangements. For back lines in long borders, or for grouping in mixed flower borders, there is no variety equal to the oldest of all, viz., mexicanum. Several so-called white-flowered kinds have of late years been introduced, but a single trial has generally shown that the white proved a dingy yellow. All the kinds are easily propagated from cuttings at any season. They strike best when placed on a gentle bottom-heat, and will winter in any position where there is plenty of light and the temperature does not go below 40°.—W. W.

Agrimonia *(Agrimony).*—A genus of the Rose family, most of which are of little garden value.

Agrostemma coronaria *(Rose Campion).*—A beautiful old garden flower, and one of the most precious we have; hardy and free in most places, but perhaps most at home in chalky and dry soils, where it seems to establish itself without care. It is a woolly herb, generally about 2 feet high, covered with rosy crimson flowers and always a bright and welcome object, flowering in summer and autumn, easily raised from seed. Excellent for borders, beds, and naturalisation on banks. It is biennial or frequently perishes on some soils. There is a white and a double red variety; the last uncommon and a good plant. The generic name is sometimes given to the annual Viscarias. A. Githago is a large annual plant occasionally grown in botanic gardens.

Agrostis *(Cloud Grass).*—A large family of Grasses, but few of which are of importance in the garden. The most desirable are the annual kinds so useful when dried and for preservation with "everlasting" flowers. There are some half-a-dozen annual species grown, but the best is A. nebulosa, which forms delicate tufts about 1 ft. or 15 in. in height, terminated when in flower by graceful panicles of spikelets. Valuable for bouquets, vases, baskets, room and table decoration. If cut shortly before the seed ripens, and dried in the shade, it will keep for a long time. Dyed in various colours it is much used by makers of artificial flowers. It may be sown either in September or in April or May. In the former case it will flower from May to July, in the latter from July to September. The seed, being very fine, should be only slightly covered. A. Steveni, multiflora, plumosa, and pulchella require the same treatment. A. Spica-venti is very graceful, especially if it is grown as well as in the cornfields, *i.e.*, from self or autumn-sown plants.

Ailantus *(Tree of Heaven).*—A well-known hardy tree, young plants of which cut down every year give a fine effect in the flower garden, and can be depended upon to do this in all seasons. The Ailantus should be kept in a young state, with a single stem clothed with its superb pinnate leaves; it can readily be kept in this form by cutting down annually, taking care to prevent it from breaking into an irregular head. Vigorous young plants and suckers in good soil will produce handsome arching, elegantly divided leaves 5 feet to 6 feet long, not surpassed by those of any stove plant. Propagated easily by cuttings of the roots. China and Japan.

Aira pulchella.—One of the prettiest Grasses, with numerous hair-like stems, growing in light tufts 6 in. to 8 in. high. It is useful for forming very graceful edgings, or for interspersing amongst plants in borders, or growing in vases or pots for room decoration. Its delicate panicles give a charm to the finest bouquets. May be sown either in September or in April. South Europe. The British A. cæspitosa is a handsome plant, but too common to need culture. Aira cæspitosa vivipara resembles a miniature Pampas Grass, with innumerable panicles of graceful viviparous awns.

Ajuga *(Bugle).*—A small family of dwarf plants, flowering mostly in spring and early summer, and having blue flowers. They are inhabitants of pastures, either mountain or lowland, and are easily cultivated in any position, and readily increased by division. A. genevensis is distinguished from the Common Bugle (A. reptans) by the absence of creeping shoots. The flower-stems are erect, from 6 in. to 9 in. high; the flowers deep blue, arranged in dense whorls along fully half the length of the stem, forming a close pyramidal spike. It is suitable for the front of mixed borders or on the margin of shrubberies, and also for naturalising in rough rocky places, where it will

PLATE VII.

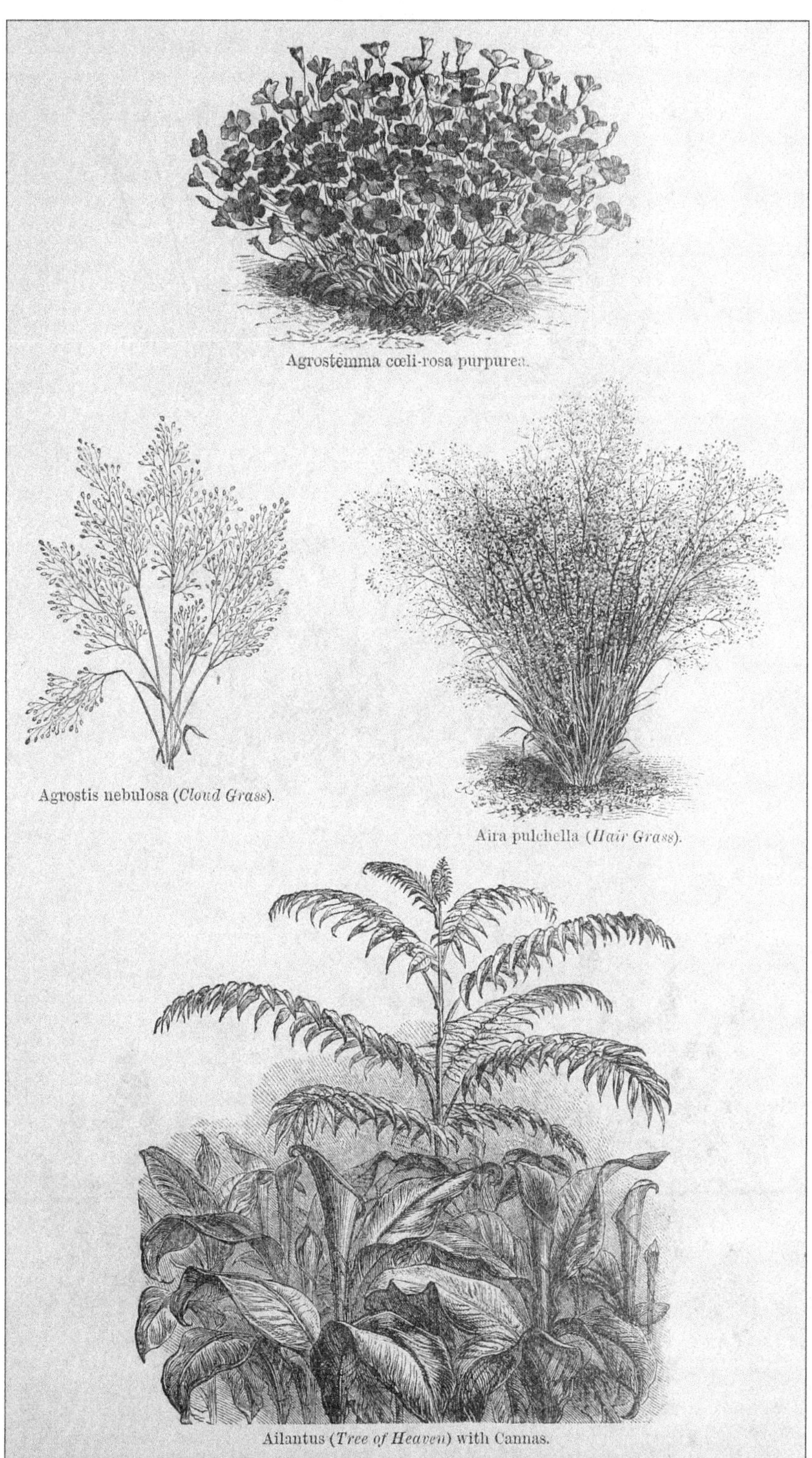

Agrostemma cœli-rosa purpurea.

Agrostis nebulosa (*Cloud Grass*).

Aira pulchella (*Hair Grass*).

Ailantus (*Tree of Heaven*) with Cannas.

[*To face page* 8.

PLATE VIII.

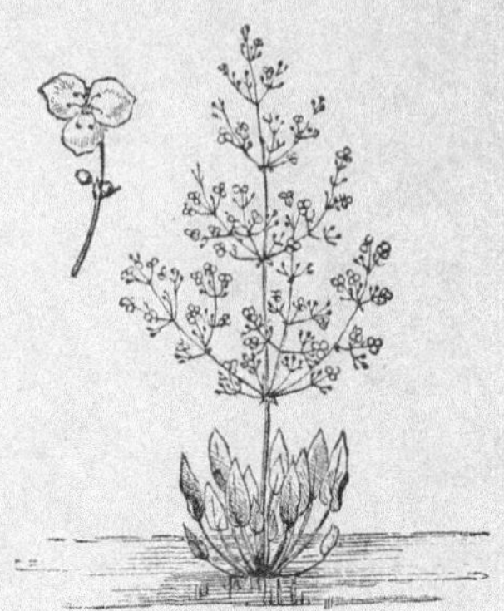

Alisma Plantago (*Water Plantain*)

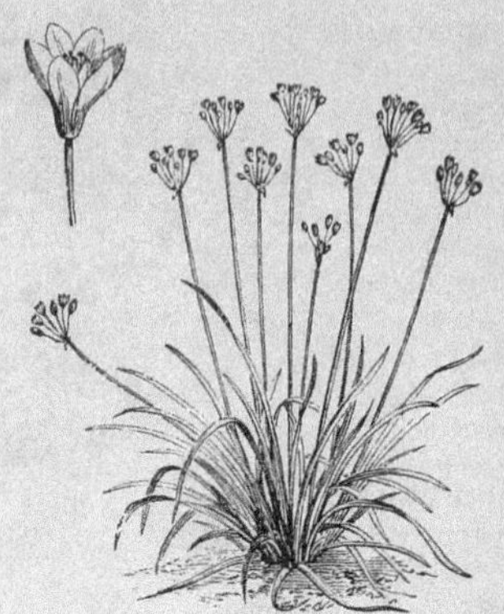

Allium fragrans (*Sweet-scented Allium*).

Allium Moly (*Yellow Allium*).

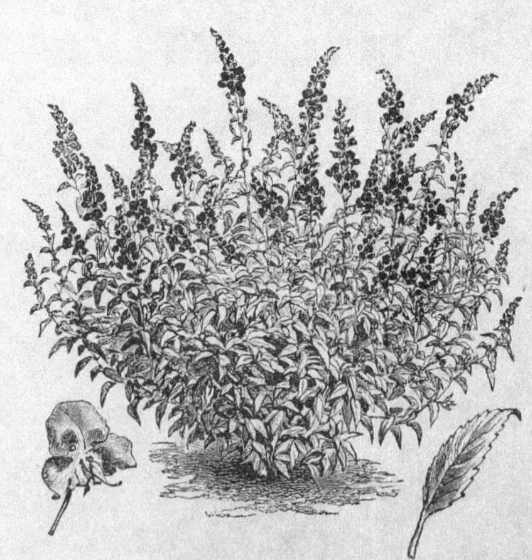

Alonsoa incisifolia.

Alonsoa myrtifolia (*Mask Flower*).

[*To face page* **9.**

establish itself in the fissures if the necessary soil, moisture, and shade can be secured. A. reptans is an indigenous plant, flowering in early summer. The flowers are blue, but there is a white variety of it, and there is also a form with variegated leaves and one with purplish ones, sometimes used in the flower garden. A. Brockbanki is said to be a good kind; a mass of blue flowers in May.

Alecost=**Costmary** *(Balsamita vulgaris).*

Alehoof, an English name for the Ground Ivy *(Nepeta Glechoma).*

Aletris *(Star Grass).*—The only hardy kind A. farinosa is an interesting plant of the Blood Root (Hæmodoracæ) family. It is a dwarf perennial with lance-shaped leaves, forming numerous tufts, from the centre of which spring the flower-stems. The latter are from 15 in. to 18 in. high, on which are densely arranged the pure white, bell-shaped flowers, which are about ½ in. in length. The outside of the blossom is very much wrinkled, which gives it a mealy appearance; hence its specific name. A cool and deep peaty soil with partial shade seems to suit its requirements. Of its garden value little is known.

Alexandrian Laurel *(Ruscus racemosus).*

Alfredia.—A. cernua is the best known kind. It is a Thistle-like perennial herb 4 ft. to 7 ft. high, but only suitable for planting among rampant perennials where those of a Thistle-like and giant character are grown in groups or otherwise in the picturesque garden. In large places a bed or group of such plants fittingly placed in a quiet nook would have a certain charm for those who love bold form in plants. Division or seeds. Siberia.

Alisma *(Water Plantain).*—This genus consists of a few water plants, of which two are desirable for growing with other hardy aquatic plants. A. Plantago, a common and indigenous waterside plant, is rather stately in habit, having broad foliage and tall panicles of pretty pink flowers. When once planted it sows itself freely, and is no further trouble except that it is apt to become too plentiful. As the leaves are poisonous to animals, due regard should be taken as to where it is planted. The other kind is A. ranunculoides, which grows a few inches high and bears in summer an abundance of rosy blossoms. Both adapted for wet ditches, margins of pools and lakes. A. natans is a small floating species, a pretty British plant for collections of hardy water plants. There are one or two Chinese kinds, single and double, as yet little known in our gardens, but which promise to be interesting.

Alkanet *(Anchusa tinctoria).*

Alkekengi *(Physalis Alkekengi).*—The winter Cherry.

Allium.—Not an important garden family, and somewhat objectionable from the odour of the stems and foliage of many species when crushed; but to growers of collections, there are among the great number of known species some interesting kinds, of which a few like neapolitanum, ciliatum, pedemontanum, and the American rose-coloured kinds, have some claims for their beauty. One or two of the above-mentioned kinds are worth growing for cutting, as their white, starry flowers are in great request in spring. The others are mostly for the curious border or bulb garden, and of easy culture in ordinary soil, the bulbs increasing rapidly. Some kinds give off little bulblets, which in certain situations might make them too free in growth. The following are among the known kinds that are worthy of culture: A. neapolitanum, paradoxum, ciliatum, subhirsutum, Clusianum, triquetrum, all with white flowers, azureum and cœruleum, both blue, pedemontanum (mauve), Moly and flavum (yellow), fragrans (sweet-scented), oreophyllum (crimson), descendens (deep crimson), narcissiflorum (purplish), Murrayanum, acuminatum, and Macnabianum (deep rose). These mostly grow from 1 ft. to 18 in. high, some attaining 2 feet or 3 feet, but these are not among the species best worth growing for ornament.

Allosorus crispus *(Parsley Fern).*—A beautiful diminutive Fern, found in some mountainous districts, where it grows out of the crevices of the rocks, its colour being a pleasant shade of green; the fronds grow in dense masses, and from their resemblance to Parsley have obtained for it the name of Parsley Fern. This pretty little Fern will not do well if too much confined. It requires abundance of air and light, and should only be shaded from the hot sun. When planted permanently out-of-doors it should have some care shown to it in the matter of soil, drainage, and light. On rockwork it does well planted between large stones, with broken stones about its roots and just its fronds peeping out of the crevice. Growing in this way out of an interstice of the rockwork it looks very well, and seems to be quite at home in such a situation; but this favourable result need not be looked for if it is deprived of light by other plants overhanging it too closely. It is well suited for planting in chinks on the rock garden, and associates well with choice alpine plants. A British plant.

Alonsoa *(Mask Flower).*—Plants mostly of annual duration. The best species are A. Warscewiczi, which grows over 1 ft. high, and has small bright orange-red flowers. A. linearifolia grows from 1 ft. to 1½ ft. in height, and is bushy and compact. A. acutifolia is a slender growing herb, 1 ft. to 2 ft. in height. Similar to this is A. incisifolia, likewise a very pretty kind. A. myrtifolia (Roezl) is from 2 ft. to 2½ ft.; it is of a very vigorous growth. The individual flowers are larger far than in any other species of this genus, and of a more intense scarlet than those of A. linearifolia. A. patagonica, a pretty species from Patagonia, is an early and free-flowering annual.

It grows about 15 in. in height, and forms densely branched, compact bushes with flowers vermilion-scarlet, of fine form, and disposed in densely set spikes. All are easily grown, and are susceptible of both pot and open ground culture. The seeds should be sown in March, and they will flower early in July. They may also be propagated by cuttings in the spring. A. Warscewiczi is more perennial in character than the rest, and it is more shrubby in growth, but resembles the others in flowers and foliage. As a pot plant it will flower freely from early spring until late autumn without intermission if the roots are kept well nourished. It is rather dwarf in growth, and can be propagated at any time from February to September. The treatment given to the general stock of bedding plants during the winter season will suit this plant. The Alonsoas may be used as "ground plants" among tall fine-leaved or other plants.

Alopecurus pratensis fol. var. is a graceful golden variegated Grass, somewhat ornamental as an edging, but not so much as other variegated forms of native Grasses.

Alpine Sanicle *(Cortusa Matthioli).*

Alsophila excelsa.—A tree Fern, native of Norfolk Island, where it attains a height of 40 ft., with a crest of fine fronds. It stands well in the open air in this country in shady, moist, and thoroughly well sheltered places. It should be put out at the end of May, and taken indoors at the end of September or early in October, and receive warm greenhouse or temperate house treatment in winter. The same remarks apply to A. australis, and probably others of the family will be found to do in the open air where they can be spared that position, for which, however, we do not recommend them, believing that our flower gardens should be adorned for the most part with plants that will brave our climate.

Alstrœmeria *(Peruvian Lily).*—A distinct and fine genus which does not seem to have found a home in our gardens to the extent that might be expected. Probably this arose from trying kinds not really hardy. One or two kinds, however, are hardy and charming as any flowers on warm soil. A. aurantiaca, a handsome plant, is so free that it is quite easily naturalised in any loose sandy soils, or even in those of a heavier nature which have been thrown up into banks. Few things are prettier than a colony of it thus grown. A. Pelegrina comes nearest to this in freedom, so far as our experience goes, enduring for years in favourable soils.

As regards the culture of the hybrid kinds, and the planting in beds of all the hardy or half-hardy sorts, it is useless attempting their cultivation unless the border intended for them is drained, so as to add to the warmth of the soil, and prevent the tuberous roots suffering from an excess of moisture. The best place in which to grow them is a south border, or along the front of a wall having a warm aspect, where, if the soil is not light and dry, it should be made so. Thoroughly prepare a spot for them at the outset. Dig out the soil to the depth of 3 ft., and 6 in. or so of brick rubbish should be spread over the bottom of the border. Shake over the drainage a coating of half-rotten leaves or short littery manure, so as to prevent the soil from running among the interstices of the bricks, and thus stopping up the drainage. If the natural soil is stiff, a portion of it should be wheeled away, and an equal quantity of leaf soil, or other light vegetable mould, substituted; to this a barrow-load of sand should be added together, mixing well. The plants should be procured in pots, as they rarely succeed from divisions; and once planted, they should never be interfered with. In planting they should be placed in rows about 18 in. apart, and 1 ft. from plant to plant. If planted during the winter, they should be placed from 6 in. to 9 in. deep, so as to keep them from frost; and a few inches of half-rotten leaves should be shaken over the surface of the soil. Should there be any difficulty in obtaining established plants in pots to start with, seed may be had; and this should be sown in beds where the plants are to remain or in pots.

The seeds being as large as Peas, they may be sown 2 inches or 3 inches deep; and, in order to ensure a regular plant, three or four seeds should be placed in a patch. If well treated, they will begin to bloom at a year old, and will continue increasing in strength and beauty every season, provided they are not disturbed. When grown in masses in this way they are strikingly beautiful, as every stem furnishes a large number of flowers, and as they vary much in their colour markings they make a gorgeous display. While growing and blooming they should have an occasional watering, as, on account of the liberal drainage required to keep their roots in a healthy state during winter, they would otherwise become too dry, and ripen off prematurely. A good mulching of old Mushroom dung or leaf soil is of great assistance to them while in bloom. When going out of flower the seed-heads should be carefully removed, otherwise the plants are apt to become exhausted, as almost every flower sets, and therefore such a load of seed should not be permitted to ripen. In removing the pods, care should be taken not to shorten the stems or reduce the leaves in any way; all are needed to ripen the tubers and form fresh crowns for the following year. The stems should, therefore, not be cut down, but should die away naturally. Anyone having deep light sandy soil resting on a dry bottom may grow these beautiful flowering plants without any artificial preparation, all that is necessary in that case being to pick out a well sheltered spot, and to give the surface a slight mulching on the approach of severe

PLATE IX.

Alsophila excelsa (*Norfolk Island Tree Fern*).

Flowers of Alstrœmeria Pelegrina (*Peruvian Lily*).

[*To face page* **10.**

PLATE X.

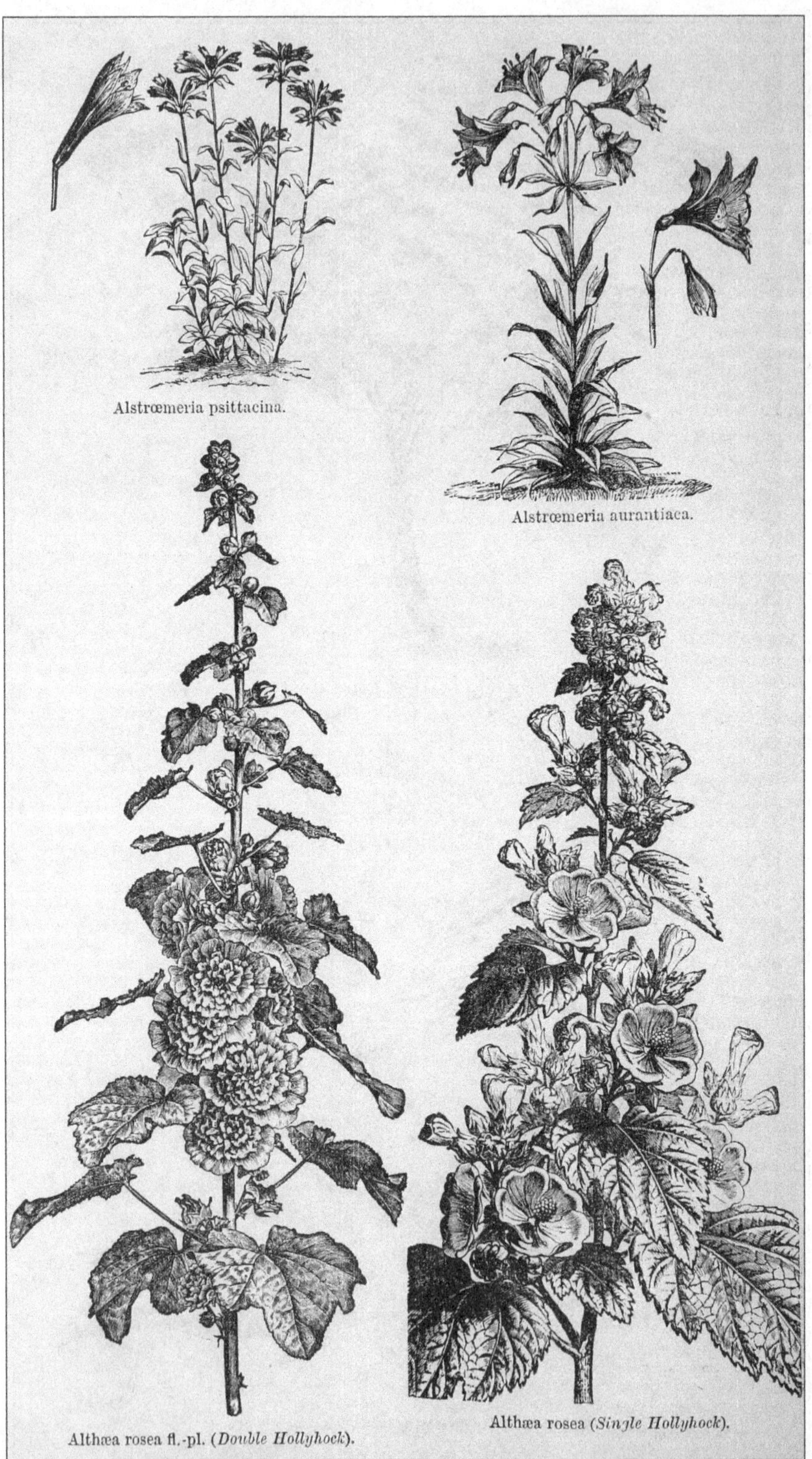

Alstrœmeria psittacina.

Alstrœmeria aurantiaca.

Althæa rosea fl.-pl. (*Double Hollyhock*).

Althæa rosea (*Single Hollyhock*).

[*To face page* 11.

weather. No trouble is involved in staking and tying, as with most plants, for the stems of these are quite strong enough to support themselves, unless in very exposed situations. Independently of the handsome bloom which they make when growing, they are quite worth cultivating for supplying cut flowers, and they last long when cut.

The species in cultivation are

A. aurantiaca (*A. aurea*), a vigorous growing kind, from 2 ft. to 4 ft. high, flowering in summer and autumn. The blossoms are large, orange yellow, streaked with red, produced in umbels of from 10 to 15 blooms terminating the stems. A native of Chili and island of Chiloe.

A. Pelegrina.—Not so tall or robust as the last; the flowers are larger, whitish, beautifully streaked and veined with purple. There are several varieties of this species, including a white variety.

A. psittacina (*A. braziliensis*).—This kind grows about $1\frac{1}{2}$ ft. high, each stem being terminated by an umbel of from seven to nine flowers, which are smaller than either of the preceding, and green and deep red in colour.

"A. pulchra, Sims (Bot. Mag., t. 2421); A. tricolor, Hook; A. Flos Martini, Kev.—These three names were also given to the same plant within the same year, but Herbert says pulchra has priority. It is one of the most beautiful species, and though now very scarce is said to be fairly hardy. From South Chili. There are many other beautiful species or varieties described and figured by Herbert which seem to be lost to our gardens, though they might easily be re-introduced from Chili, The names of any species of Alstrœmeria must be accepted with caution, the genus being much confused."—H. J. ELWES.

Besides these species, the following may be met with in some gardens, though but seldom: A. Ligtu, versicolor, chilensis, oculata, and peruviana.

Alternanthera.—These plants are natives of Brazil, and consequently tender, and can only be used in the more favoured parts of the country. There are few plants that produce that degree of colour and general effectiveness that they possess where they thrive. The varieties range in colour of foliage from deep purple to bright yellow, and all are most effective when used in masses or blocks, surrounded with plants of contrasting colours and similar habit of growth, such as the variegated Mesembryanthemum, Golden Feather Pyrethrum, Veronica incana, and Cerastium tomentosum. The best varieties are A. paronychioides, deep brown tipped with scarlet; paronychioides major, the same colour, but more robust and broader foliage; paronychioides major aurea, the same habit of plant as the last, but deep yellow and red foliage, very fine; A. amabilis latifolia, very broad foliage, deep orange and scarlet colour; magnifica is much the same as the first-named kind, but has wider and deeper scarlet foliage; A. versicolor has dark purple and rose-coloured foliage, and is a strong grower, an excellent contrasting plant to Pyrethrum Golden Feather; A. amœna and amœna spectabilis are the brightest coloured of all, but most tender, and are therefore not suited for any but the most favoured southern districts. There are other varieties, all more or less meritorious, but those named are among the best that have come under our notice. They are all easy of propagation; the smallest particle will strike root in a couple of days in a bottom-heat of 75°. The quickest way to get up a large stock is to make up a hotbed of leaves and stable litter; on this place frames and insert the cuttings firmly in about 4 in. of soil; any kind that is of a light sandy nature will do. Well water them in and keep the frame shaded and close for the first week, afterwards give air as for other soft-wooded bedding plants. This plan of propagation is only intended for the spring season, the plants being transferred in due time from their cutting beds direct to their summer quarters. For stock plants to stand the winter, the cuttings are best inserted in 6-in. pots, plunged in bottom-heat, and as soon as well rooted transferred to shelves in warm houses, there to remain till March, when they may be planted out on hotbeds, where they will quickly produce abundance of cuttings.—W. W.

Althæa (*Hollyhock*).—A genus of the Mallow family consisting chiefly of coarse growing plants, though some, such as A. rosea, from which the Hollyhock has sprung, are showy garden flowers. The original form of the Hollyhock with single flowers is considered by some desirable as well as the double-flowered sorts. There is another species, called A. Irolowiana, a tall plant with arge orange and red flowers, which are very showy. The other species are characterised generally by great vigour of growth, hence not very suitable for the garden. They grow vigorously in almost any situation or soil. Among them A. armenaica, officinalis, narbonnensis, cannabina, and ficifolia are the best — mostly natives of South Europe; flowering in summer and autumn.

A. rosea (*Hollyhock*).—The Hollyhock is one of the noblest of hardy plants, and there are many positions in almost all gardens where Hollyhocks would add finely to the general effect. For breaking up ugly lines of shrubs or walls, and for forming backgrounds, their tall column-like growth is well fitted. So, too, wherever bold and stately effects are desired among or near flower beds, they are valuable. Cottage beekeepers would do well to grow a few Hollyhocks, for bees are fond of working amongst their flowers.

CULTURE.—To obtain fine flowers it is necessary to treat the plants liberally. Deep cultivation, a liberal supply of manure, frequent waterings in dry weather, with occasional soakings of liquid manure, will alone secure fine spikes and well developed flowers. Hollyhocks require good garden soil, well trenched to the depth of 2 ft., and plenty of thoroughly decomposed manure. A wet soil is good in summer, but in winter injurious to them, and to prevent surface wet injuring old plants left in the open ground, we remove the mould round their necks and fill up with about 6 in. of white sand. This preserves the crowns of the plants. It is best, however, to plant young plants every year, as one would Dahlias, *i.e.*, if fine flowers are desired. Plant them not less than 4 ft. from row to row, and 3 ft. apart in the row; if grouped in beds, not nearer than 3 ft. each way. In May or June, when the spikes have grown 1 ft. high, thin them out according to the strength of the plant; if well established and very strong, leave four spikes; if weak, two or three. When they are required for exhibition, only one spike must be left. Stake them before they get too high, tying them securely so as to induce them to grow erect. The most robust among them will not require a stake higher than 4 ft. above the ground level. If the weather is dry they may be watered with a solution of guano or any other liquid manure poured carefully round the roots, but not on or too near the stem. If fine blooms are required cut off the side shoots, thin the flower buds if crowded together, and remove the top of the spike, according to the height desired, taking into consideration the usual height and habit of the plant. By topping, be it observed, you increase the size of the flower, but at the same time shorten its duration, and perhaps disfigure the appearance of the plant.

The best way of showing Hollyhocks is in the form of spikes, and, in judging, the first point should be the individual flowers, the perfection of which consists in the petals being of good substance, while the edges should be smooth and even, and the florets occupying the centre full and compact, closely arranged, rising high in the middle, and of a globular form, with a stiff guard petal extending about ¾ in., or in proportion to the size of the centre ball, so that the different parts of the flower may present a uniform appearance. The next point should be the arrangement of the flowers on the spike. They should be regular, not crowded together in a confused mass, nor hanging loosely with open spaces between each flower, but so disposed that the shape of each, when fully blown, may be distinctly seen, the uppermost covering the top of the spike. A few small green leaves between the flowers also give an improved appearance. The third point is colour; the brightest, strongest, and most distinct should stand first, but as it is desirable to obtain shades of all kinds, anything new or distinct in this way should be encouraged.

PYRAMID HOLLYHOCKS.—I am a great admirer of Hollyhocks, but I dislike the coarse and unsightly appearance which they assume in a short time after the first flowers have faded, or been reduced to a pulp by rain or strong sunshine, and also the inferiority of the extreme terminal flowers of the spike. To obviate this last effect I have been in the habit of cutting out the top while the lower blooms were in perfection. This year, from the middle to the end of June, the tops were removed before the flowers appeared. The result has been to spoil the main spike, which grew up stunted and closely packed; but to make up for the disappointment, from every axil on the main stalk there have sprung out a number of shoots, forming elegant and graceful branches, to the number of more than twenty on some of the plants. These are now covered with perfect flowers, nicely spaced and distinct, alternating with very small leaves, and they look as if they would continue in perfection for a long time. As these shoots grow uniformly round the stem, the general outline is that of a trained pyramidal tree with about 100 flowers expanded at once, and one may conceive what a fine effect the various beautiful shades of colour must produce. I would recommend having only one stalk left on the stool, which can be more easily staked and secured than a greater number, and will form a more elegant object than a number crowded together as usual. One of my plants is 7 ft. high, and 4 ft. across at the lower part, tapering to the top. The side shoots do not exceed ½ in. in diameter.—W.

PROPAGATION.—This is effected from eyes, cuttings, seeds, or careful division. Hollyhocks may be propagated by means of single eyes put in in July and August, and also by cuttings put in the spring, on a slight hotbed. Plants raised in summer are best preserved by potting them in October into 4-in. or 5-in. pots in light, rich, sandy earth, and then placing them in a cold frame or greenhouse, giving them plenty of air on all favourable occasions. Thus treated they will grow a little during the winter. In March or April turn them out into the open ground, and they will bloom as finely and as early as if they had been planted in autumn. Plants even put out in May will flower the same year. If the seeds of Hollyhocks are sown in autumn, as soon as they are ripe, in a box or pan in heat, and potted off and grown on in a pot through the winter, and planted out the following April, they will flower the same summer and autumn. If allowed to remain in the beds or borders where they have flowered, choice Hollyhocks often perish from damp or snow settling round their collars, or from its penetrating through the hollow cavity left by the too close removal of the flower-stems. It is a good plan at the approach of winter, say in October, to carefully lift all which it is desired to save, and lay them close together in a slanting direction, at an angle of about 45°, in a warm mellow soil at the foot of a wall or hedge, where in hard weather shelter can easily be given them. But in wet, heavy soils it is snow and damp that has the most destructive effect. Lifting them thus not only makes them safe, but it gives an opportunity to have the land that is to receive them thoroughly worked in winter, and then, when re-planted in March or April, if a little rotten turf is worked in with them, good spikes and large individual flowers may be expected. Choice and scarce varieties may either be potted up or planted out in a frame. Potting them is the best way, because the plants can be placed in a greenhouse or Vinery on shelves near the glass. Some of the stools will have numerous growths starting from them, and unless the plants have the advantage of a little heat early in the year, many of the cuttings cannot be propagated early enough to flower the same season. Growers in the south of England have an advantage with these spring-struck cuttings over the northern florists. There is quite three weeks difference between the time of flowering in the south and in the northern districts of England and Scotland. Root-grafting gives the propagator a little advantage, and early in the year the plants are propagated more readily in a light frame fixed in a heated propagating house. A hot-bed is uncertain, as there is sometimes too much heat and moisture, and then not enough. Although the young side-shoots produced by old stocks will readily root in a gentle bottom-heat in spring, they may also be propagated in July, just before the plants come into flower. The side-shoots from the flower-spikes, or the smaller flower-spikes, if they can be spared, should be cut up into single joints, and dibbled in thickly in a prepared bed in a frame or pit, where they can be kept close and properly cared for by shading from bright sunshine, and sprinkling occasionally with water that has been warmed by standing in the sun. Thus treated, nearly every cutting will develop a bud from the axil of the leaf, rapidly strike root, and make a good strong plant by the following spring; as a rule, young plants propagated at this season usually produce the best spikes. When cutting down the flowering stems of Hollyhocks after blooming, they should be left a good length, as they are exceedingly impatient of damp about their crowns; in spring their old stems may be removed altogether.

INSECT PESTS AND DISEASES.—Red spider and thrips are both very troublesome enemies to the Hollyhock, but the first named does most injury. It appears on the under sides of the leaves as soon as the hot weather sets in, and is very difficult to dislodge. Before planting out, if there is any trace of red spider, the whole plant, exclusive of the roots, should be dipped in a pail of soft soapy water, to which a pint or so of Tobacco liquid has been added. It will be necessary to syringe the under-sides of the leaves with the mixture if they have been planted out before the pest is perceived. Thrips may be destroyed in the same way. It is well to syringe the plants every day in hot weather.

THE HOLLYHOCK FUNGUS (Puccinia malvacearum) has been and still is quite as destructive to the Hollyhock as the Phylloxera is to the Grape Vine. When once it seizes a collection, probably the best way is to destroy the whole of the plants affected. Those that do not seem to be attacked should as a precaution be washed with soapy water, in which a liberal proportion of flowers of sulphur has been dissolved. The sulphur will settle at the bottom of the vessel, and must be frequently stirred up when the water is being used. Sulphur seems to destroy almost any fungus, and may this in its very earliest stages; but it will not move it when established. —D.

Alyssum *(Madwort)*.—A very hardy family of Crucifers, and very numerous in rocky and alpine districts, but resembling each other too much to make the culture of many kinds desirable. Alyssum saxatile (the Rock Madwort) is one of the most valuable of the yellow flowers of spring, hardy in all parts of these islands. The brilliancy of its masses of bloom and its vigour have made it one of the best known plants. It is often grown in half-shady places; but it, like most rock plants, should be fully exposed. It and its forms are the best of the genus, and well fitted for the spring garden, the mixed border, and rock-work, and also for association with the evergreen Candytufts and Aubrietias for fringing shrubberies. It perishes in winter in heavy, rich clays when on the level ground. Comes from Podolia, in Southern Russia, and flowers with us in April or May. There is a dwarfer variety, distinguished by the name of A. saxatile compactum, but it differs very little from the old plant. Alyssum montanum is a distinct species, spreading into compact tufts, 2 in. or 3 in. high. In April the flowers commence to open, and in May the plants are studded with yellow, alpine Wallflower-like blooms, sweet scented, and produced abundantly on healthy specimens, on the rock garden in good sandy soil, or in some slightly elevated position. Increased by division, cuttings, or seeds, though it does not often seed with us. Alyssum spinosum is a distinct, silvery little bush with showy flowers. Small plants quickly become Lilliputian bushes, 3 in. to 6 in. high; when fully exposed, almost as compact as Moss.

Among other kinds sometimes grown are A. Wiersbecki and A. olympicum, neither of which equal in habit, bloom, or endurance A. saxatile, which from its showy bloom in spring has been called Golden Tuft. This kind is very easily raised from seed, also by cuttings in spring under a cloche in a cool border. The alpine and rock kinds are of easy culture in light or dry soil, as indeed are all the species. A. maritimum is the Sweet Alyssum, a small but hardy and vigorous prostrate plant, with white flowers not very showy, but fragrant. It is useful as a dwarf annual as a carpet plant, and grows easily on the tops of walls in the west country, and also in bare places. In these situations it is sometimes perennial. A native of England or a naturalised plant. Easily raised from seed in spring or autumn, sowing itself freely. There is a variegated form used in the flower garden.

Amarantus *(Prince's Feather, Love-lies-bleeding)*.—Among annuals none are more in want of judicious use and appreciation than these. The few we grow are usually treated as rough, common annuals, and sown so thickly that they never attain half their true development, or never fulfil any of the graceful uses for which they are adapted. The old Love-lies-bleeding (A. caudatus), with its dark red pendent racemes, is a very striking object when well grown, but A. speciosus and some of the more recent varieties are still more so. The more hardy and vigorous species grow from 2 ft. to 5 ft. high. It is advisable to give them plenty of room to spread; otherwise much of their picturesque effect will be lost; and to use them in positions where their fine and peculiar habit may be seen to advantage, as, for example, in large vases, edges of large beds of or dotted among low-growing flowering plants. Easily raised as any annual, they deserve to be properly thinned out, and each plant isolated in rich ground, so that it may attain its full size. The foliage of some varieties is very ornamental, and rivals flowers in the richness of its hues. Planted along with large-leaved subjects, such as Canna, Wigandia, Ricinus, Solanum, their effect is very good. They may also be advantageously employed in borders and flower-beds of all sizes. The varieties of A. tricolor are a little more tender than the other kinds, and a light soil and a warmer position are necessary for them. They do well in gardens by the seaside. They should be sown in April in a hotbed, pricked out in a hotbed, and finally planted permanently about the end of May. The cultivated kinds embrace bicolor, tricolor, atro-purpureus, half-hardy annuals, highly coloured ornamental foliaged plants useful for centres of beds and borders. A. melancholicus ruber, a useful bedding plant with bright crimson leaves; A. Henderi, A. Princess of Wales, and A. salicifolius may be used in the summer garden with good effect. Natives of South America; flowering in summer and autumn.

Amaryllis.—None of the species are perfectly hardy, but the beautiful Belladonna Lily (A. Belladonna) may be grown successfully in the open under certain conditions. It is a noble bulbous plant from the Cape of Good Hope, growing from 1½ ft. to 3 ft. high. It blooms late in summer, the flowers being as large as the white Lily, and of delicate silvery rose tint. They are produced in umbels of from 5 in. to 12 in. in clusters on stout leafless stems, arising from thelarge pear-shaped bulbs. The following are the conditions to be observed for the successful cultivation of this beautiful plant: If the soil be stiff, it should be well drained. Choose a situation such as that on the south side of a house or wall, take out the whole of the soil to the depth of 3 ft. or so, and about 6 in. of broken brick should then be placed in the bottom. Over this some half-rotten manure should be scattered to keep the drainage open, and to form a supply of rich food for the plant to feed on. If the natural soil is not good, some fresh sandy mellow loam should be substituted or added, Should the soil be at all stiff, a few barrow-loads of decomposed leaf soil and one or two of sharp sand should be mixed with it. Having trod this tolerably firm, the bulbs should be planted singly, or in threes if plentiful and it is desired to furnish the border quickly. In the latter case each clump should be about 1 ft. apart, and if the border is of such a width as to require a double row, the plants in the second should be alternate with those in the first. In planting, a handful or so of sharp sand should be placed round the bulbs to keep them from rotting. If planted in autumn, or at any time during the winter, it will be necessary to protect the bulbs from severe weather by applying a good coating of half-rotten leaves, Cocoa-nut fibre, or some other kind of protection. Nothing further will remain to be done till the plants begin to push forth their new leaves, which they do rather early in the spring, and upon the freedom with which they send forth these during the summer the abundance or otherwise of bloom in the autumn in a great measure depends. When once they get fairly into growth they should have plenty of assistance during the dry weather by giving them an occasional soaking with clear water, and from time to time with liquid manure, as it may be required. The object that must be aimed at is a full and free development of leaf growth, and, this accomplished, an abundance of flowers is sure to follow. As soon as the foliage ripens off it should be carefully removed, and the border cleaned and neatly raked over before the blooms begin to protrude through the soil, or they will become injured. Blanda is a variety of the preceding, with much larger bulbs and general development, bearing noble umbels of white blossoms, turning to pale rose, not scented, blooming in summer. There are several other varieties all worthy of cultivation in similar positions.

Amaryllis Ackermanni is found to be hardy in various districts, and to flower well in the open ground. It suggests that others of the scarlet kinds usually grown in a high temperature are worth trial in the open air on warm borders.

Amberboa *(Centaurea)*.

Amblyolepis setigera. — A dwarf growing half hardy annual, producing small heads of orange-yellow flowers which are sweetly scented. It belongs to the composite family and is a native of Texas.

American Cowslip *(Dodecatheon)*.

PLATE XI.

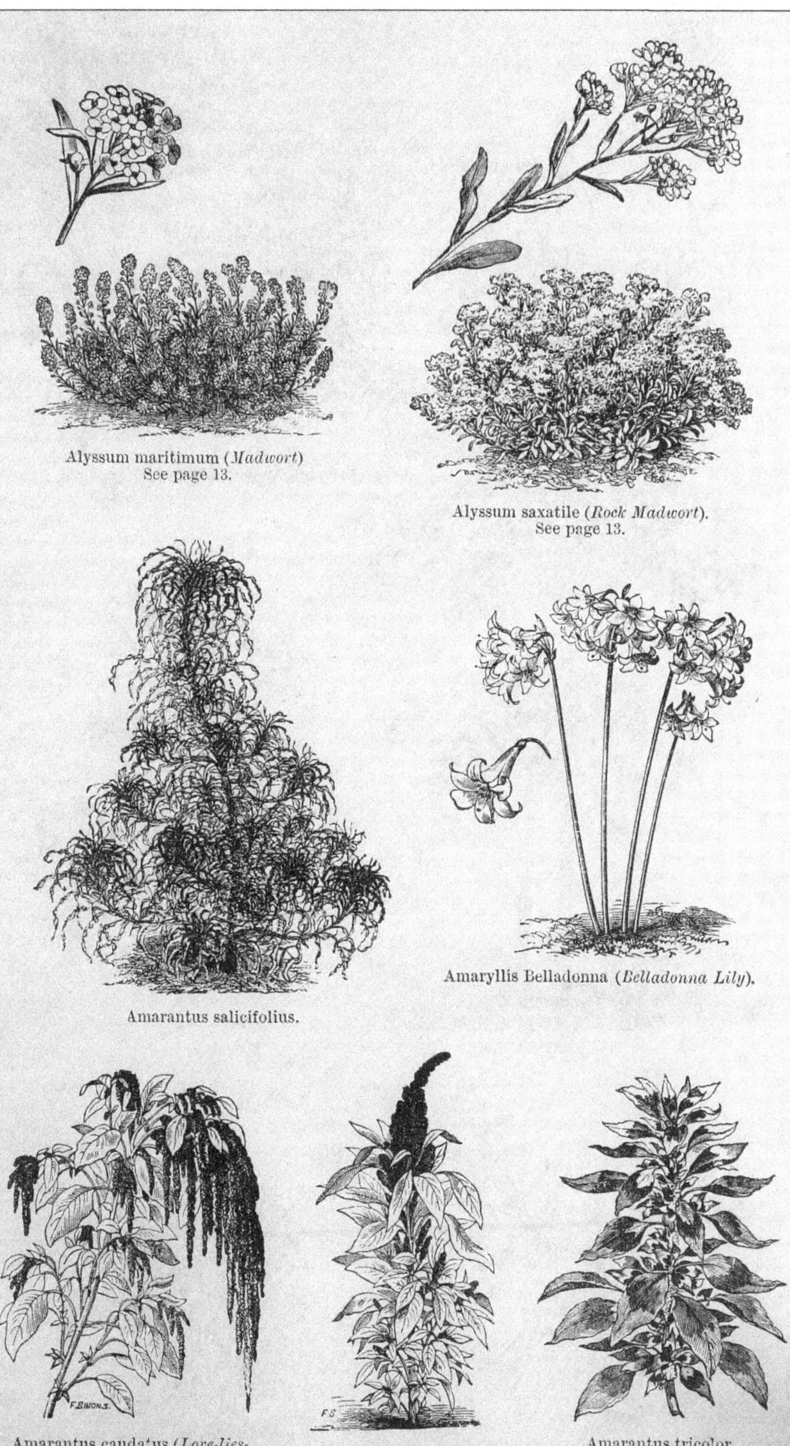

Alyssum maritimum (*Madwort*)
See page 13.

Alyssum saxatile (*Rock Madwort*).
See page 13.

Amarantus salicifolius.

Amaryllis Belladonna (*Belladonna Lily*).

Amarantus caudatus (*Love-lies-bleeding, Prince's Feather*).

Amarantus speciosus.

Amarantus tricolor.

[*To face page* **14.**

PLATE XII.

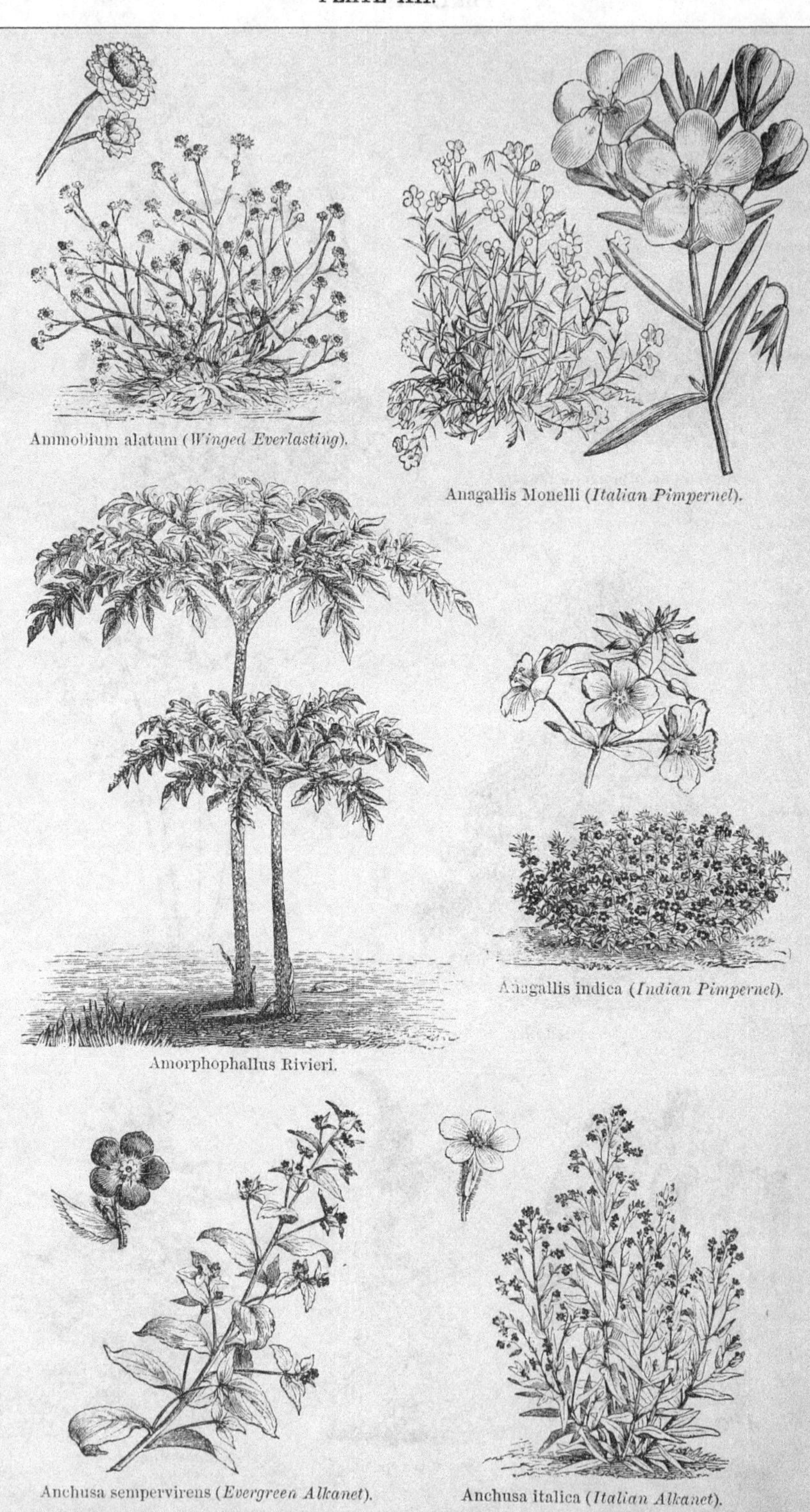

Ammobium alatum (*Winged Everlasting*).

Anagallis Monelli (*Italian Pimpernel*).

Amorphophallus Rivieri.

Anagallis indica (*Indian Pimpernel*).

Anchusa sempervirens (*Evergreen Alkanet*).

Anchusa italica (*Italian Alkanet*).

[*To face page* **15.**

Amianthium muscætoxicum *(Fly Poison).*--A North American Liliaceous plant 1 ft. to 2 ft. high, with broadly, linear leaves, and a dense raceme of white flowers, which turn green with age. Thrives best in a rather moist sandy soil. Not a showy garden plant.

Amicia zygomeris.—A quaint greenhouse plant which has been occasionally used in the sheltered flower garden. Though free in growth, it has few claims for general cultivation. Mr. E. H. Woodall says of it: "I put it on a par (in England) with Melianthus major—good for those who like a bold and distinct growth in a warm situation in summer, and have means to protect or take it up and pot it in winter. I saw it in remarkable beauty last year at Mr. Lane Fox's unrivalled gardens at Bramham, and this is the first year I have grown it, and it has stood the cold, rain, and gales far better than the variegated Maize and big Solanums; indeed, it looks none the worse. The flower, though bright, is not large enough to be effective."

Ammobium *(Winged Everlasting).*—The only kind, A. alatum, is a handsome Everlasting, covered with soft silky hairs. It grows about 1½ ft. to 3 ft. high, and produces its white chaffy flowers with yellow discs from May till September. A native of New Holland. Grown in sandy soil it is generally perennial, but on some heavy and damp soils it must be treated as annual or biennial. Easily raised from seed, and may be treated as an annual plant, and among such plants it is worth a place. Compositæ.

Amorphophallus.—A small genus of Aroidaceous plants, of which one or two are used with good effect in summer in the open air. The stems are stout, and arise erect and are terminated by spreading branches. The species cultivated are A. Rivieri and A. nivosa, which require to be kept over the winter in a warm house, grown on in spring, and put out-of-doors at the end of May in rich soil. Our country is hardly warm enough for them.

Amsonia.—A genus of herbaceous perennials from North America from. 2 ft. to 3 ft. high, and bears small pale blue or purple flowers in terminal clusters in summer. They are easily propagated by division or seeds, but are plants of very little garden value, except for botanic gardens and large and curious collections.

Anacharis Alsinastrum.—An American water plant, now become a most troublesome weed in most lakes, rivers, ponds. Swans eating it down is considered the best remedy. It seems after a time to lose its great vigour.

Anacyclus Pyrethrum *(Pellitory of Spain).*—A Composite plant, native of Barbary, Syria, and Arabia, the roots of which are used for medicinal purposes. For botanic gardens only.

Anagallis *(Pimpernel).*—The species in cultivation are chiefly half hardy annuals, the best known of which is the Italian Pimpernel (A. Monelli), with large blossoms of a deep blue, shaded with rose. There are several varieties of this ornamental plant, the chief of which are rubra grandiflora, Wilmoreana, bright blue purple, yellow eye; lilacina Phillipsi, deep blue, rose-coloured centre; Breweri, intense blue; Impératrice Eugénie, bright blue edged with white; linifolia, fine blue, very dwarf; Napoleon III., maroon: and sanguinea, bright ruby. A packet of mixed seed gives a good variety. These flower from July to September. The Indian Pimpernel (A. indica) is similar to A. Monelli, but has smaller flowers of a bright blue. It is a hardy annual, but the Italian Pimpernel is tender and should receive the treatment accorded to half-hardy annuals. The seed may be sown any time from March till July, the latter sowings to be made in pots and put into a greenhouse or window in autumn. They grow well in any common garden soil and are used with excellent effect planted out in broad masses in borders, and they are also suitable for edgings to beds, and make excellent pot plants as well as being useful in warm borders. The pretty little Pimpernel (A. tenella) is a native plant found in bogs. It is a creeping plant, bearing slender stems with small round leaves, among which are produced myriads of tiny pink flowers. It is excellent for growing in suspended pots or pans, and may be grown easily in the bog or rock garden, or anywhere where the soil is moist and spongy, and the vegetation dwarf and fragile like itself.

Anaphalis triplinervis *(Antennaria).*

Anchusa *(Alkanet).*—A small genus of Borageworts, containing a few plants well deserving of culture. The finest species are A. italica, a vigorous and showy plant, 3 ft. to 4 ft. or more high, with beautiful blue blossoms in panicled racemes. A. hybrida is similar to the foregoing, and probably no species of this genus better merits cultivation. It is dwarf, about 2 ft. high, and produces an abundance of flowers of rich violet colour, scarcely so large as that of A. italica. It is of biennial duration, and, like A. italica, is a native of Southern Europe. A. capensis is a pretty plant with large bright blue flowers, but it is more tender than the preceding; it should be planted in a sheltered, yet well drained border of light rich soil. A. incarnata is a pretty plant, growing about 2 ft. high, yielding an abundance of fresh coloured blossoms. On the whole, they are not important as garden plants. A. sempervirens is a British species, growing from 1½ feet to 2 feet high, bearing spikes of blue flowers. It is scarcely attractive enough for general culture as a border plant, but worthy a position in woods and semi-wild places. Seeds or division. Flowering spring and summer.

Andromeda.—A genus of hardy shrubs, the dwarfer species of which are frequently associated with rock plants and hardy Heaths. Andromeda tetragona is one of the prettiest of all the shrubs introduced to cultivation, seldom growing more than 8 in. high. It is a native of Northern Europe and America, quite hardy, and requiring a moist peat or very fine sandy soil. It is a fitting ornament for planting on the margins of beds of choice dwarf shrubs in sandy peat, loves abundance of moisture in summer, and is easily increased by division wherever it grows vigorously. If on the rock garden it ought to be in a deep bed of soil. Andromeda fastigiata is one of the most rare and beautiful plants that we have obtained from the Himalayas. It should have sandy, moist peat soil. It is most likely to thrive in moist and elevated districts; but, safely planted on rockwork in deep, moist, but well-drained soil, and carefully guarded against drought during the warm season, it may be grown without difficulty. Andromeda hypnoides is a minute, Moss-like shrub, 1 in. to 4 in. high, and one of the most interesting and beautiful of all alpine plants, and one of the most difficult to grow, being very rarely seen in a healthy state. Drought is fatal to it. It is a native both of Europe and America, either far north into the coldest regions of these countries, or on the summits of high mountains. Carefully peg down the slender main branches, and place a few stones round the "neck" of the plant, so as to prevent evaporation. A. polifolia is easily grown in various soils, and is a good plant for the bog garden. The dwarf Andromedas are among the plants which, as regards their propagation, are best left to nurserymen. Some of the kinds may be found in nurseries where hardy Heaths and American plants are grown; others are extremely rare. They thrive well in the Matlock and Edinburgh nurseries. Ericaceæ.

Andropogon.—Tall growing Grasses, suitable for planting singly on lawns or where their noble habit of growth can be fully developed. A. halepensis is very ornamental, forming large tufts 6 ft. in height. A. strictus grows about 4 ft. high, and has graceful silky panicles of bloom. A. furcatus and A. scoparius are also in cultivation, but neither are so desirable as the others. All the kinds named thrive, provided the situation is not too exposed, or the soil too heavy and damp. In order to grow these Grasses finely, it is needful to plant them in a deep soil well enriched and not wet. Our climate is not quite warm enough for them. A few degrees further south they are among the best ornamental Grasses for gardens. In the Paris gardens effective use is made occasionally of them, particularly of A. squarrosus.

Androsace.—The most alpine of alpine plants. Other families send down representatives to the hill pastures or the sea rocks, or sunny heaths, as the Primroses and Hairbells do, but not so these. They are more alpine even than the Gentians, which are as handsome in a hill meadow as on the highest slopes; and as Androsaces are, among flowering plants, those most confined to the snowy region, so, as might be expected, they are the dwarfest of this class. They belong to the Primrose family, and resemble it in the flowers, but even dwarf alpine Primroses are giants to these. Growing at such elevations, where the snow falls very early in autumn, they flower as soon as the snow melts. Sometimes, like some other alpine flowers, they frequent high cliffs with a vertical face, or with portions of the face receding here and there into shallow recesses. Here they must endure intense cold—cold which would destroy all shrub or tree life exposed to it. And here in spring they flower. As yet far from common in our gardens, it is, nevertheless, the aim of every lover of alpine flowers to possess them in good health. This is not difficult where there is a properly-formed rock garden in a pure air. They are among the plants that are almost sure to perish in a smoky atmosphere. Their small evergreen leaves, often downy, retain much more dust and soot than smoother and larger-leaved evergreen alpine plants do. The Androsaces enjoy in cultivation small fissures between rocks or stones, firmly packed with pure sandy peat, or very sandy or gritty loam, not less than 15 in. deep. They should be so placed that no wet can gather or lie about them, and they should be so planted in between rocks or stones that, once well rooted into the deep earth—all the better if mingled with pieces of broken sandstone—they could never suffer from drought. It is easy to arrange rocks and soils so that, once the mass below is thoroughly moistened, an ordinary drought can have little effect in drying it.

Mr. Hatfield finds that all Androsaces are more or less surface rooters. "Some, such as A. Chamæjasme, may send down their roots to a considerable depth, but not necessarily so, providing sufficient moisture be found on or near the surface. Their woody roots (in cases where the roots are woody) would prefer to run horizontally along the face of some stone at a small depth than to sink vertically to a considerable distance. I have tried various species of Androsaces in almost all imaginable positions in the sun, shade being out of the question, and nowhere do I find them succeed better than in a sandy, well-drained peat bed on the level. In this country, and especially on the eastern side, no vertical or horizontal fissures can be ever so carefully made as to ensure the necessary amount of moisture. Pack stones ever so well about them, they are certain to become too dry. I am quite sure that all this (1881) summer the soil, even in our best positions, has been dry to quite 1 ft." There can be no doubt, we think, that these and a great number of other and more precious alpine flowers would thrive better on a fully exposed level bed of sandy soil kept moist than on many a rock garden. They would be safer from drought, and they would be sure of the feeding ground for the roots so often denied them in the dusty rockworks which have

PLATE XIII.

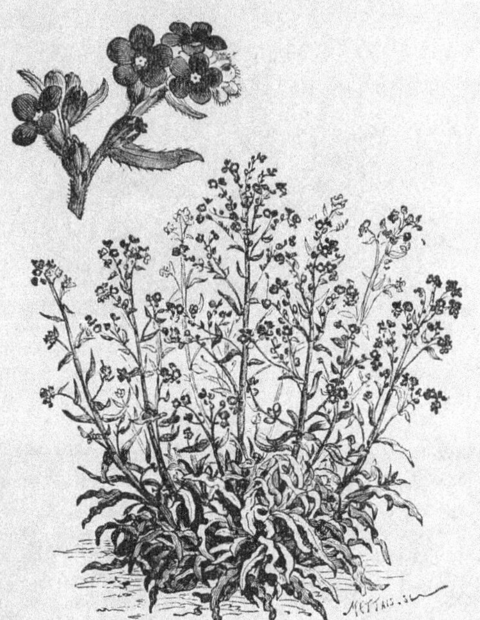

Anchusa capensis. See page 15.

Andropogon halepensis.

Andropogon strictus.

Androsace lanuginosa (*Himalayan Androsace*). See page 17.

[*To face page* 16.

Androsace Laggeri.

Anemone alpina (*Alpine Windflower*).

Anemone apennina (*Apennine Windflower*).

Flower of Anemone angulosa (*Great Hepatica*).

[*To face page* 17.

been, and are still, the rule. Only they must on the level be guarded from the coarse browsing slugs of the lowland garden, and be kept free from the shade of coarse plants which soon take possession of such moist, gritty, sandy, or peaty beds as would suit the Androsaces. It is well to bear in mind, however, that on a properly-formed rock garden there will be level spaces of good soil as well as "rocky" slopes.

A. carnea *(Rose-coloured A.)*.—One of the prettiest and most distinct, coming from the summits of the Alps and Pyrenees; opening in our gardens in the early spring before any of its relatives. Known by its small-pointed leaves, not gathered in tiny rosettes, but regularly clothing a somewhat elongated stem, like a small twig of Juniper, or of the Juniper Saxifrage. The flowers are of a lively pink or rose, with a yellow eye. It is not difficult to cultivate in a mixture of sandy loam and peat on rockwork—the spot to be exposed, and the soil deep and firm. Like most of the species, it may be easily raised from seed, which should be carefully sown in pans of sandy soil as soon as gathered; also by division.

A. carnea var. eximia is a large variety and likely to supersede the type, because it is hardier or more robust. Were it not the counterpart of the type when in flower, except that the flowers are much larger and of greater substance, I should be disposed to think it a distinct species, judging by its appearance at other times. It grows in tufts about 3 in. high very rapidly, and in measuring its leaves this day (August 25) I found them above 1 in. long, and $\frac{1}{8}$ in. broad at the base. Our plants root about 6 in.; a calcareous soil is noxious to it. Increased by division.—H.

A. brigantica in every way resembles the variety of A. carnea, except that its flowers are white. The same position and soil suits it; that is to say, sunny, and in sandy peat free from lime, well drained, but liberally supplied with moisture.

A. Chamæjasme *(Rock Jasmine)*.—This does not nestle into close moss-like cushions, like the Helvetian and other Androsaces, the foliage forming large rosettes of fringed leaves. The blooms are borne on stout little stems frequently not more than 1 in. high, but varying from that to 5 in., according to the vigour of the plants and the position in which they grow. When in good health, it flowers abundantly, is one of the most worthy of culture of all alpine plants, and one of the easiest to grow on an open spot on rockwork, in deep and well-drained rich light loam, the surface nearly covered with small pieces of broken rock, to prevent evaporation and also to preserve the plant from injury. It should get abundance of water in summer, be exposed to the full sun, and be preserved from being overrun by weeds or grazed down by slugs. A native of the Alps of Europe. Mr. Hatfield says he finds it dislikes lime and succeeds in river sand and peat.

A. ciliata *(Fringed A.)* is by some considered a variety of the preceding, with the flower-stems twice as long as the leaves. Androsace cylindrica is another variety with the stems rising to half an inch high, with persistent leaves, which form columns on the stems. It is by some considered a species, bears pure white flowers in spring, and should be treated like A. pubescens.

A. helvetica *(Swiss A.)* forms dense cushions, about $\frac{1}{2}$ in. high, of diminutive ciliated leaves, tightly packed in little rosettes. Each rosette rests on the summit of a little column of old and dead, but hidden half-dried and persistent leaves. A white flower, with a yellowish eye, rises from every tiny rosette, each flower being almost twice as large as the rosette of leaves from which it has arisen. Requires considerable care in cultivation, perfect exposure to sun, and a thoroughly well-drained, but never dry position.

A. imbricata (*Silvery A.*) differs from the Pyrenean and Swiss Androsaces in having the rosettes of a silvery white colour. The pretty white flowers are without stalks, and rest so thickly on the rosettes as often to overlap each other. It will grow freely in rich loamy soil in narrow well-drained fissures of rockwork. A native of the Pyrenees and Alps. Flowers in summer, and is propagated by seeds and division (A. argentea).

A. Laggeri.—This exquisite little gem is one of the most distinct of the family to which it belongs, and is easily recognised by its tiny rosettes of sharp-pointed leaves, which resemble a twig of a Juniper, or more especially the Juniper-leaved Saxifrage. It inhabits the Pyrenean Alps, where it flowers as soon as the snow is melted in summer. Its blossoms are of a lively pink colour, with the centre of a lighter hue.

A. lanuginosa *(Himalayan A.)*.—Distinguished by its spreading and even sometimes, when in vigorous health, long trailing shoots, and bearing umbels of flowers of a delicate rose, the leaves covered with silky hairs. When grown freely it is a lovely plant. Some parts of the country are too cold for this plant, and the southern and western counties, or warm and genial places near the sea, are those in which it may be grown with most success. It is, however, so pretty that in cold places it will be well to preserve it over the winter in dry pits, and plant it out in summer. The most suitable position for it is on the rockwork, planted in sandy peat or very sandy light loam, and so placed that its shoots may fall over the edge of a low rock. Where the soil is free, and not too wet in winter, and the air moist and genial, it may be tried as a border plant. It used to form wide and beautiful tufts on a sandy border in the College Gardens at Dublin. Where increased freely it would prove a beautiful ground or carpet plant. Propagated by cuttings, and flowers from June to October. Himalayas.

A. obtusifolia *(Blunt-leaved A.)*.--This is said to be allied to A. Chamæjasme, but has rather larger rosettes of leaves, and two to five white or rose-coloured flowers with yellow eyes. It seems to grow taller and more vigorously than A. Chamæjasme. Widely distributed over the European Alps. Culture the same as for A. Chamæjasme.

A. pubescens *(Downy A.)*.—Allied to the Swiss and Pyrenean Androsaces in its rather large solitary white flowers, with pale yellow eyes, just rising above the densely packed, slightly hoary leaves, the surface of which is covered with stalked and star-like hairs. The unopened blooms look like small pearls set firmly in a tiny five-cleft cup, and are held on stems barely rising above the dwarf cushion formed by the plant; flowering in July and August in its native state, and in our gardens in spring or early summer. It seems to grow without difficulty on sunny fissures in deep sandy and gritty peat. Alps.

A. pyrenaica *(Pyrenean A.)*.--Like the Swiss Androsace, but the paper-white flowers with yellowish eyes are not quite so well formed as those of that kind, and the flower, instead of being seated or almost seated in the rosettes of leaves, rises on a stem from a quarter to half an inch high. Grows in fissures between rocks, with, however, deep firm rifts of sandy peat and loam in them. It will also grow on a level exposed spot, but in such a position should be surrounded by half-buried stones.

A sarmentosa.—A dwarf-tufted alpine from the Himalayas, and growing in the Ramaon territory at elevations varying from 11,000 ft. to 12,000 ft. It is said to be very nearly related to A. lanuginosa; the latter is, however, a more silvery plant—indeed nearly white. They are borne in trusses of ten to twenty flowers on an erect Primula-like scape, and the whole inflorescence at first sight closely resembles that of a bright, rosy, white-eyed Verbena. Propagated by its runners, which hang over the sides of the pot in profusion. Like many other woolly-leaved alpines, this beautiful Himalayan species is very difficult to keep alive through our cold and damp winters. I have taken the precaution to place a piece of glass in a slanting position about 6 in. above the plant, and I find that this effectually preserves it. Care should also be taken to put finely broken sandstone immediately under the rosettes of leaves and over the surface of the soil to keep every part of the plant, except the roots, from being in contact with the soil. Mr. Hatfield finds a dry calcareous loam best for this.

A. villosa *(Shaggy A.)*.—A very pretty dwarf species, found on many parts of the Alps, with leaves and stems thickly covered with soft white hair or down. It is more inclined to spread than any of the nearly allied sorts, as it throws out runners, and is therefore suitable for planting, so that one side of the specimen may fall down the face of a rock. It should be planted in loam and a mixture of peat, in a properly made fissure between sandstone rocks or large stones, but it may also be grown on level spots. In all cases it should have abundant moisture, is increased by seeds, and in our gardens flowers about the beginning of May.

A. Vitaliana *(Yellow Androsace)* rarely grows above 1 in. high, and produces, scarcely above the leaves, flowers large for so small a plant, and of a rich yellow. It is lovely for association with the freer-growing Androsaces, dwarf Gentians, Primulas, in the rock garden, it may even be grown on a border in a not too dry district where the soil is open. On the rockwork it should be kept moist during the dry months; and when it is tried as a border plant on the level ground, it should be surrounded by stones, half plunged in the ground, to prevent evaporation, as well as to protect it from being trampled upon. It is abundant on the Alps in various parts of Europe, and is increased by careful division or by seeds. Androsace Heeri, bryoides, Charpentieri, Wulfeni, and Haussmanni are among the other finest kinds, and there are one or two annual and biennial kinds not of much value for the garden, except A. coronopifolia.

Androstephium.—A small genus of North American bulbous plants, about which little is known in this country as regards their culture or hardiness. The kinds are breviflorum, flowers violet, borne in umbels of four to seven, in early spring. A stouter plant than the following, and with smaller flowers. A. violaceum, a rare and showy species from Texas, 6 in. to 8 in. high, with a small, coated bulb, which surmounts a depressed globular bulb or corm. Flowers violet, borne in umbels, slightly fragrant.

Andryala.—Small growing plants of the Dandelion order; some with woolly leaves. Two species are in cultivation. A. mogadorensis, shrubby, forms snowy masses on a little islet on the Western Morocco coast, and has not been found elsewhere. It bears flowers as large as a half-crown, of a bright yellow colour, the disc being bright orange. But little is known of its culture and hardiness. A. lanata has white woolly leaves, which make it desirable in some arrangements. It grows well in any soil provided it is not too damp and moist.

Anemiopsis californica.—A North American plant of no garden value.

Anemone *(Windflower)*.—A noble genus of plants to which very much of the beauty of the world in spring and early summer in all northern and temperate countries is due. There are over seventy species known. When in early spring, or what is to us in Northern Europe winter, the valleys of Southern Europe and sunny sheltered spots all round the great rocky basin of the Mediterranean are beginning to glow with colour, we see the earliest Windflowers in all their loveliness. These arid and huge masses of mountain that look so barren and verdureless carry on their sunny sides carpets of Anemones in countless variety of hue. These belong to well-known and very old favourites in our gardens—the common Windflower (A. coronaria) and the Peacock Anemone. Later on the Star Anemone (A. stellata) begins, troops in thousands over the small cultivated terraces, meadows, and fields of the same regions. Climbing the arid-looking mountains in April, the Anemone hepatica soon shows itself nestled in many nooks all over the bushy parts of the hills, and much later on it emerges from the snow in the Swiss Alps, and welcomes the early traveller. Further east, while the common Anemones are aflame along the Riviera valleys and terraces, the lovely blue winter Anemone (A. blanda) is open on the hills of Greece; a little later its congener, the Apennine Anemone (A. apennina), blossoms in Southern Italy. Meanwhile, our Wood Anemone has begun to adorn the woody places throughout the northern world, afterwards ascending to high, treeless places on the mountains, at the same time the open, wind-swept downs beginning to show the purple of the Pasque-flower (A. Pulsatilla) here and there through the brown Grass. The Grass has become tall and richly green before the stately and graceful Alpine Windflower (A. alpina) adorns with its flowers all the natural meadows of the Alps; as soon as its large flowers are succeeded by the long silky heads of fruit, the snow is melting from the high Alpine Windflowers, which soon flower, fruit, and are ready to take their eight or nine months in their snowy bed. These are but few of the many examples of what is done for the northern and temperate world by these plants.

Anemone alpina *(Alpine Windflower)* —On nearly every great mountain range in northern and temperate climes, one of the most frequent and handsome plants, growing to 18 in. and even 2 ft. high. It is a slower plant in gardens than most of the other kinds. Being of a strong rooting and vigorous character, it should, if placed on rockwork, have a level spot with abundance of soil to grow in, and being also tall, it would be the better of close association with neat shrubs, plants of the stature of the vernal Adonis, and other choice perennials. Where the soil is good it grows quite freely as a border plant. Flowers in its native countries as the snow disappears, and in our gardens at the end of April or beginning of May. When plants are well established in good soil they may be taken up and readily divided; it may also be raised from seed. One form has yellow flowers, in which state it is known as A. sulphurea.

A. angulosa *(Great Hepatica)*.—Larger than the common Hepatica, with flowers of sky-blue, as large as a crown-piece, and distinguished from the common kind by its five-lobed and toothed leaves. It is a native of Transylvania, and hardy. It is naturally more an inhabitant of the hill copse than the crest of the Alps, and enjoys partial shade. In sandy soil in shrubberies it attains a height of more than 1 ft. when not in flower, and the shelter and slight shade received from surrounding objects are favourable to it. In all properly formed rock gardens, or near them, it will be possible to give it a suitable position; while in spaces between American plants and choice dwarf shrubs in beds it will succeed. When plentiful enough, it may be used as an edging to beds of choice spring-flowering shrubs, and for planting in wild open spots in shrubberies, or in open, rather bare, and unmown spots along the margins of wood walks. Time will, no doubt, see it sport into several colours like our common Hepatica. Increased by seed and division.

A. apennina *(Apennine Windflower)*.—Has erect flowers of a good sky-blue, the plants growing in dense tufts, so that, though there is but one flower to a stem, they are thickly scattered over the low cushion of soft green leaves. Although figured in most of our works on British plants, and naturalised in various places, it is not a true native of this island, but the hardiest of our native plants take not more kindly to our clime. It is one of the sweetest of spring flowers, and among the many lovely plants that gem the alpine or Apennine pastures there is not one more worthy of being abundantly naturalised in groves and shrubberies. It is welcome in the garden and on the rockwork; but it is when we see it scattered amongst the native Anemones in our woods, or making glorious mixtures of gold and blue with the Buttercup-like Windflower, or running wild among dwarf plants in woods or pleasure-grounds that we see how this Italian plant adds a new charm to the British spring. The Apennine Anemone flowers in March and April, is very readily increased by division, and grows about 4 in. to 6 in. in height. Italy.

A. blanda *(Blue Winter Windflower)*.—A near relative of the Apennine Windflower, and a very lovely plant, deserving to be cultivated in every garden. It is of a deep sky-blue, like A. apennina, and has larger and more finely rayed flowers, dwarfer, harder, and smoother leaves, and blooms in early spring, during mild, open winters, and in warm parts showing as early as Christmas, flowering continuously too, so that it may be seen in flower late in spring with its relative, A. apennina. It is hardy and vigorous, and, from the harder and smoother texture of the leaves, can stand exposure even better than the very hardy Apennine A. It should be grown in every rock garden, and planted on bare banks that catch the early sun in the pleasure ground; should adorn the spring garden, and, when sufficiently plentiful, might be naturalised in bare, half-wild places. It does not grow more than 4 in. high, and is multiplied easily by division.

Botanically, this is chiefly distinguished from A. apennina by its carpels being topped with a black-pointed style, and by the sepals being smooth on the outside. Increased by careful division and by seed. Greece.

A. coronaria *(Poppy Anemone).*—A native of sub-humid pastures in the south of Europe, this plant has been one of the most popular in our gardens from the very earliest times. There are a great number of varieties, both single and double, all worthy of cultivation, and great ornaments of the spring garden. The single sorts may be readily grown from seeds, and they should be thus raised by those wishing a large stock of effective spring flowers. They may be sown in the open air in April. Infinitely varied as they are in colour, and possessing most vigorous constitutions, they deserve to be cultivated, even more than many double varieties annually offered by our seedsmen. The plantation of these double varieties may be made in autumn or in spring, or at intervals all through the winter to secure a continuity of flowers; but the best bloom is secured by September or October planting. The Poppy Anemone does best in a rich deep loam, but is not very fastidious. The roots of the more select kinds may be taken up when the leaves die down, but they are in few cases worth this special attention, as many splendid varieties may be grown from seed as readily as any native herbaceous plant. If the seed be sown in June, and the plants pricked out in autumn, they will flower very well the following spring, so that this fine old plant may be said to be almost as easily raised as an annual. Flowers in April and May, and often through the winter, red, white, and purple in variety. Height, 6 in. to 15 in. Propagated readily by seed or division. Apart from the old florists' or double Anemones and the single ones, there are certain good races of French origin of much value—the Anemones de Caen, for example. These are raised from the same species, but are more vigorous and have larger flowers than the older Dutch kinds. Of the Caen Anemones there are both single and double kinds. The Chrysanthemum-flowered are another fine race, double. The splendid variety of the Poppy Anemones leads to mixed collections being commonly seen. While it is desirable to have occasional mixtures, a better way is for each gardener to select and keep true some of the finer forms in whatever colour may be admired. In each case a fine scarlet, purple, or violet should be grown by itself and for itself, and in that way, aided by judicious mixtures where desirable, the Poppy Anemone will be a greater aid to the garden artist. They all thrive in garden soils of fair quality, and, like most plants, will be benefited by manure.

SEEDLINGS.—The following method will enable anyone to raise seedlings with success, provided the soil be suitable—namely, a moist loam. To save time I always sow as soon as the seed is ripe, taking care only to select it from the very brightest-coloured flowers. First, as to preparing the seed, pains must be taken to separate it thoroughly. Spread a newspaper on the table, pour over it a quart of sand, or dry ashes, or fine earth; sprinkle the seed over this, and rub the seed together till the separation of the seed be complete; and as to the seed bed, it need not be larger than 3 feet by 9 feet. Choose the sunniest part of the garden. Dig and rake till the surface is very fine, tread it down, and give it a good watering. Wait until the surface is dry enough to scratch with a fine rake; sow broadcast, covering the seed about the thickness of a shilling; beat flat with a spade, and give a light sprinkling of water. Now comes the most important point in my method. Never let a ray of sunshine reach the bed; cover it with newspapers, spreading a few Pea sticks or something to retain the covering in its place. Keep the surface of the bed always moist. In about twenty days the young plants will begin to appear; when all seem up remove the covering, and no further care will be needed except watering. This must be strictly attended to, for if the bed becomes once thoroughly dry, the plants are apt, after forming small bulbs about the size of Peas, to stop growing, the foliage to die, and the bulbs to lie dormant for months, but if kept well watered through the summer they will go on growing all through the winter, and begin to blossom the following spring. The seedlings may either be left to blossom where they were sown, or be transplanted in September or October.—J.

CULTURE OF THE CHOICER SORTS.—Although all the beauty of the Poppy Anemone in its brilliant variety of colour may be enjoyed by simple culture, especially in good or warm soils, it is desirable to give here what is considered the best way of growing the finer and named varieties, which have been favourite florists' flowers for a long time. Messrs. Vilmorin and Andrieux, writing to us from Paris (Sept., 1881), say that the growth "of the Anemone as a florist's flower in France is of ancient date. The finest were known to come from Caen and Bayeux, nearly 100 years ago, and it is there that the best in France are now cultivated." There is no better soil for the Anemone than a good yellow, gritty, or sandy and friable loam; but as in most gardens little or no choice is to be had, none need doubt that the choicest Anemones may be well grown in ordinary garden soil of fair quality enriched with decomposed cow manure or decayed leaf-mould. In the more loamy soils decayed old hot-bed manure will do. In some very heavy clayey soils the cultivation of the plant cannot be successful without taking more trouble than it would be wise to give.

PLANTING.—In regard to planting, the florists of forty years ago had two seasons for doing this; one the middle of October, the other at the end of January. "The

PLATE XV.

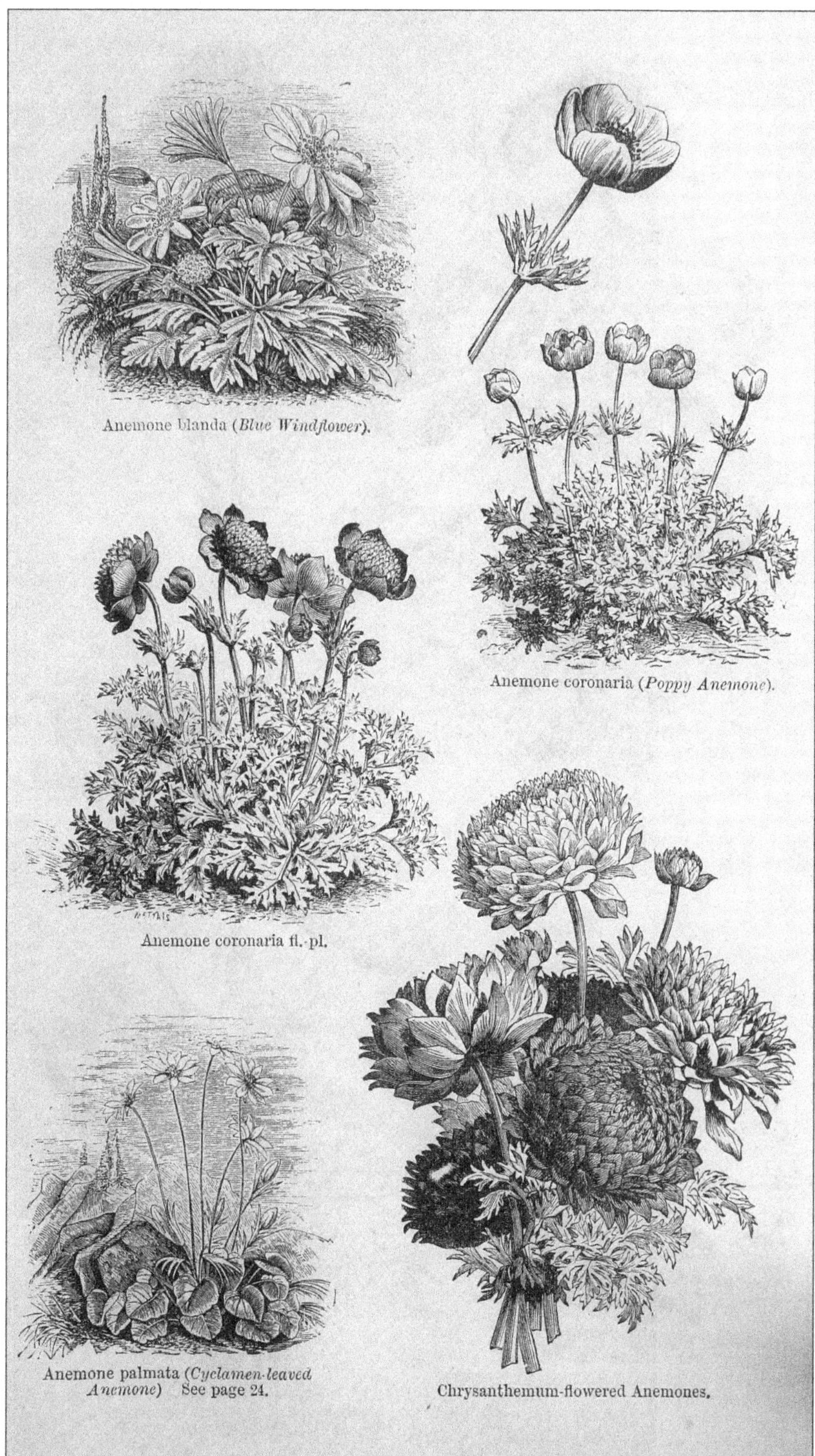

[*To face page* **20.**

PLATE XVI.

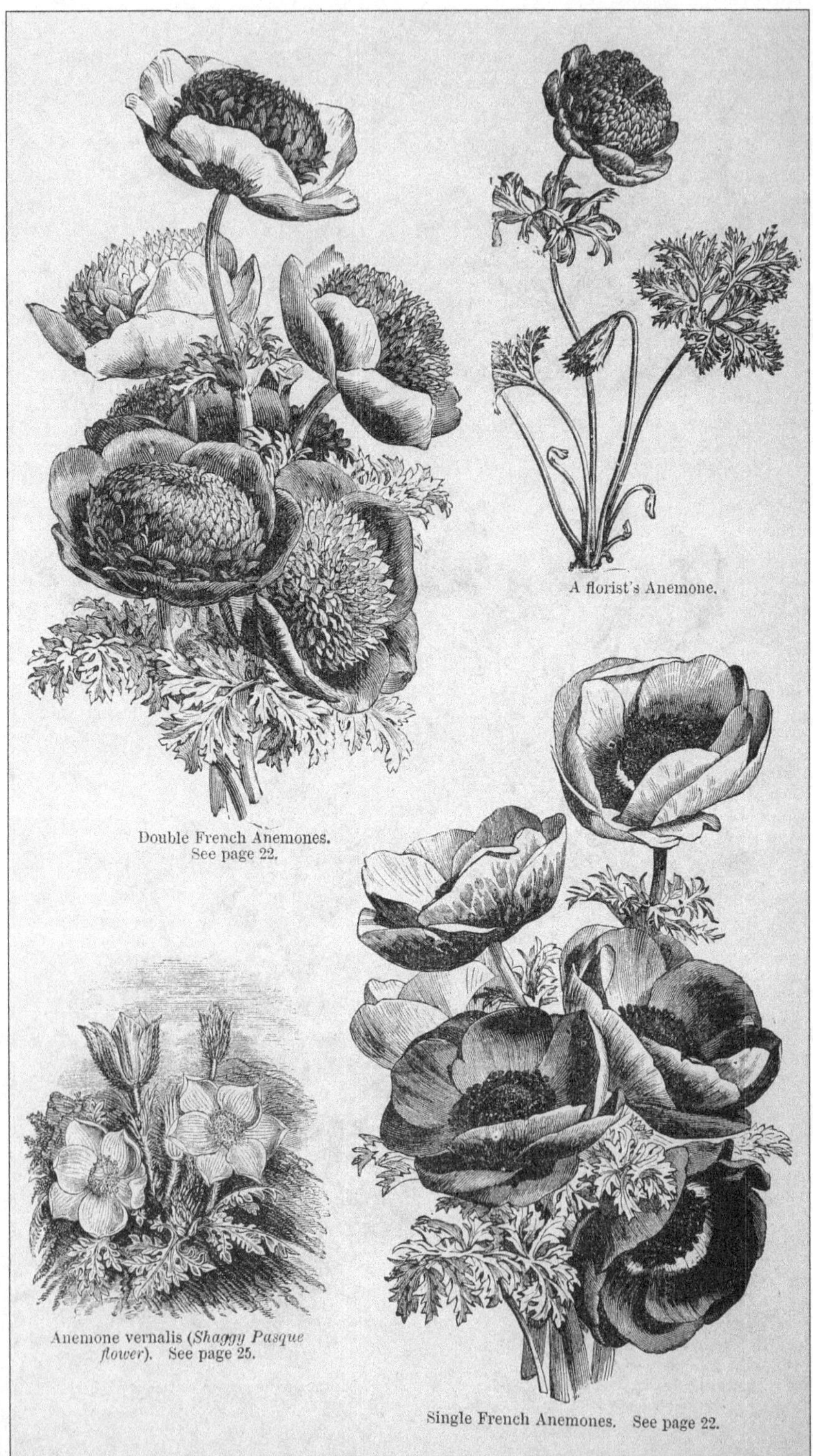

Double French Anemones. See page 22.

A florist's Anemone.

Anemone vernalis (*Shaggy Pasque flower*). See page 25.

Single French Anemones. See page 22.

[*To face page* 21.

early vegetation of such roots as are left in the ground would intimate that the former is the most natural season, and undoubtedly October-planted tubers make stronger plants, throw up more flower buds, bloom earlier, and, when the season is favourable, mature finer blossoms than those that are planted in spring. The main drawback is that the blossoms expand before the frosts have ceased, and hence a larger amount of care and protection is necessary." It was generally held as a part of the florist's experience that a bed planted the first week in October would be in bloom about the second week in May ; and as this is a period when severe frosts often happen the blossoms are injured if not protected. But it is important to remember that the best bloom is always from autumn-planted kinds, and the double ones should be planted in September in a sunny situation if possible, and even early in the month. Where the cultivation of the Anemone is seriously taken up it will be best to plant at several different epochs, particularly in the case of the single and semi-double kinds. We have had very fine blooms from late spring planting, and an autumnal bloom may be had by planting in midsummer. But the glory of the plant is as a brave spring bloomer, and we do not gain by forcing it to deviate much from its natural season.

Correct planting is a matter of much importance. Many persons have purchased Anemone and Ranunculus roots, and by planting them badly, failed. A bed of prepared compost should be made; it matters not where the position is, so that it is a subsoil well drained. A stagnant soil is hurtful to the Anemone, especially in a wet winter. The Anemone roots deeply, and the bed should be some 18 in. deep or more of good soil. When planting, the surface of the bed should be raked level, and the bed marked in cross rows. Mr. Carey Tyso recommends a bed 3 ft. 4 in. in width, and five roots planted in a row, which will allow of each being 6 in. or 7 in. apart. "As the tubers are varied in form and size, the hand or a trowel should be used in making the holes, 2 in. deep and large enough to admit the root to rest evenly on the soil, avoiding much pressure, as the limbs of the tubers are often slenderly attached to the crown and are easily broken off." Anemone tubers vary much in size and shape, according to the variety, and are "formed of irregular fleshy bunches, having a number of small protuberances called crowns. These crowns are distinguishable as tufted pieces, or obtuse points, often a shade darker in colour than the surrounding skin. They are frequently found in clusters near the centre, and sometimes singly at the extremities of the projecting limbs They are easily recognised by the practised eye ; but, as amateurs have been known to plant them upside down, some attention to this matter is needful. The base or lower part of the tuber is known by the remaining fragments of the fibrous root of the former year, unless, indeed, they have been very carefully cleared away. The direction to plant the right side upwards seems trite, but is not superfluous."

In the days when the Anemone was much more of a florist's flower that it is now, it was customary to spread over the surface of the bed 2 in. of half-decayed leaves for a protection against frost. The florist dreaded the effects of frost on his Anemones, Ranunculus and Tulips, and made provision to mitigate its effects. The surface covering of leaves was placed on the bed for this purpose; but it was necessary when the plants commenced to come through the soil to liberate rising foliage as the decaying mulch, coming into combination during wet weather, and becoming matted together when dry, would injure the leaves. The surface over the plants should be broken fine, or pressed as might appear needful. Should the protecting material be thought untidy as the spring advances, it may be carefully drawn off.

Thinning the flowers was done with advantage when the Anemone was an exhibition flower, like the Tulip, as some of them would show signs of defectiveness in their formation; some of them would become blind, or without the complement of centre petals, that give so much symmetry of appearance to the double forms, and these were pinched off to strengthen the remainder. Weeding and watering are two important matters of detail. On the rich soil of the beds weeds will be certain to grow with vigour, and it is obvious they must be plucked out, and the surface of the bed kept stirred and tidy in appearance. The Anemone is a moisture-loving plant, and water should be given freely in dry weather, giving a good soaking when it is administered.

When the roots show signs of ripening and the foliage decays they should not be neglected. If drought prevails water should be given at times, so that the roots be fully prepared by maturation for service another year. At the end of July, or in August, they should be dug up, and put into boxes with some soil attaching to the roots, and put away in a cool, dry place of safety for planting out another season. The strongest can be divided, and in this way a collection is extended. The roots should be looked over occasionally during the winter to see that they are in good condition.

VARIETIES OF DOUBLE AND CHOICE-NAMED POPPY ANEMONES.—The Dutch growers send us every year collections of named varieties, both double and single, and they are very pretty indeed, and well worthy a place in the garden. Those who grow them praise them highly. The old double and single scarlet Anemones are as vivid and striking in colour as they are beautiful, and there are the double and single blue also. Those who are not desirous of purchasing named varieties can have them in mixtures—the double forms by themselves, and the single ones by themselves ; and, when planted in beds or in clumps in the border, they are highly effective. There are now many varieties of Anemone coronaria, and year by year the Dutch growers in particular make additions by means of seedlings raised by them. They have evidently attached more importance to the double than to the single varieties, as is shown by the lists they publish, the list of the double varieties being much more lengthy than that of the single forms, and each raiser has his own list of named flowers. The following list of double flowers is found in Dutch catalogues: *Bleu Aimable,* dark blue; *Cedo Nulli,* purple and carmine scarlet; *Cervantes,* lilac; *Cramoise superbe,* scarlet; *Duchesse de Lorraine,* rosy red; *Fanny,* scarlet; *Feu incomparable,* scarlet and white; *Gloria florum,* red and purple; *Hamlet,* lilac, fine; *King of Scarlets,* rich scarlet; *King of the Blues,* rich deep blue; *l'Eclair,* scarlet; *Lord Nelson,* rosy lilac; *Lord Palmerston,* dark blue; *Marie Stuart,* red and purple; *Mrs. Beecher Stowe,* red and white; *Marie Antoinette,* variegated; *Ornament de la Nature,* dark blue and violet; *Prince of Wales,* dark blue; *Princess Alice,* rosy red and white; *Robert Burns,* violet and lilac; *Sir Walter Scott,* carmine; *Sir Robert Peel,* blue and lilac; *Sophia,* purple and scarlet; *Thalia,* white, rose, and green. It is not customary to give lists of single varieties; they are simply offered in collections.

What are termed French Anemones are considered to be a great improvement on the Dutch varieties, and have very large flowers, brilliant in hue, and varied in colour, as well as of a vigorous growth. At present the varieties are few in number, and consist of—*Gloire de Nantes,* blue; *La Brilliante,* crimson red; *Lilas,* brilliant reddish-lilac; *Mauve Clair,* pale mauve; *Ponceau,* deep scarlet; *Rosine,* peach, shading to scarlet.

Poppy Anemones, double and single. are useful for making edgings and for the decoration of borders either singly or in tufts, of one colour or a well-selected mixture. They are cultivated in beds alone or clumps in borders, and answer well for planting underneath standard Rose trees or other light and thinly planted shrubs. The flowers cut when just open and as buds keep for a length of time in water, which makes them very useful as cut flowers and for the making of bouquets.

A. fulgens *(The Scarlet Windflower).*—The Scarlet Windflower is a native of the south of France, where it occupies but a very limited area, and that for the most part cultivated land, especially in vineyards. Although it is nearly related to Anemone stellata (Lamarck), there appears to be quite sufficient ground for considering it a distinct species, as Gay did when he described it as A. fulgens. In fact, the localities in which A. fulgens and A. stellata are found are far distant from each other, and the seedlings of A. fulgens, although very often distinct from their parent, in no way ever revert to A. stellata. On the other hand, it seems to be perfectly certain that the plant known as A. Pavonina is only the double-flowered form of A. fulgens, as its roots, leaves, and other characters are perfectly identical with those of A. fulgens; and, moreover, it frequently turns up among seedlings of the latter, and is sometimes even intermixed with it in a wild state. As A. Pavonina yields no seed, and is propagated only by roots, the reason is obvious why it never under cultivation reverts to A. fulgens. The Scarlet Windflower may be considered to be perfectly hardy, inasmuch as it has been known to withstand, in the open border, the severest frost of the last ten or twelve years; it is scarcely, indeed, if ever, injured by mere cold, but stagnant moisture is very detrimental to it. No hardy spring flower with which I am acquainted can compete with it as regards brilliancy and colour, which, when lit up by bright sunshine, becomes perfectly dazzling. In good, well-drained soils it will succeed anywhere, but it thrives best in a rich loam on a northern aspect and in a somewhat shaded situation. To insure success, it should have a liberal supply of manure incorporated with the soil, which should be mulched with stable manure before frost sets in. Division of the roots is the surest and most rapid way of propagating it, as it is liable to sport if raised from seeds. Seedlings, as a rule, lack the bright colour and the substance of the parent plant, while some will become double, and resemble, more or less exactly, A. Pavonina as grown in gardens. Roots of this Anemone may be transplanted almost all the year round, although the resting time extends only from June to August; but in order to insure early and good flowers, they should be planted as early as possible in the autumn. Some leaves will make their appearance in September or October with a rounded three to five-lobed outline; these will be succeeded in January by finely and deeply-cut leaves, and soon afterwards by flowers. A good bed of well-grown plants of A. fulgens in full bloom is a gorgeous sight; but it is not only useful for out-door decoration alone, inasmuch as the cut flowers will be found to expand beautifully in water, and last for a week or more if cut when just coming into bloom and kept in a moderately warm room.—H. V.

In the extensive experiments made by M. Henry Vilmorin in the culture and propagation of this plant it was found that many of the seedling plants were not of the brilliant colour so remarkable in the true and finest strain, but of a red with a shade of brick in it. We have seen many of these plants which were carefully separated from the pure stock. They are singularly alike in hue, and manifest no tendency towards A. stellata. On the other hand, plants of the true colour are raised in this way, and sometimes remarkably fine ones, but seed is, nevertheless, not to be depended on for reproducing the plant in its finest form.

Anemone (Hepatica) triloba.

Anemone fulgens (*Scarlet Windflower*).

[*To face page* **22.**

PLATE XVIII.

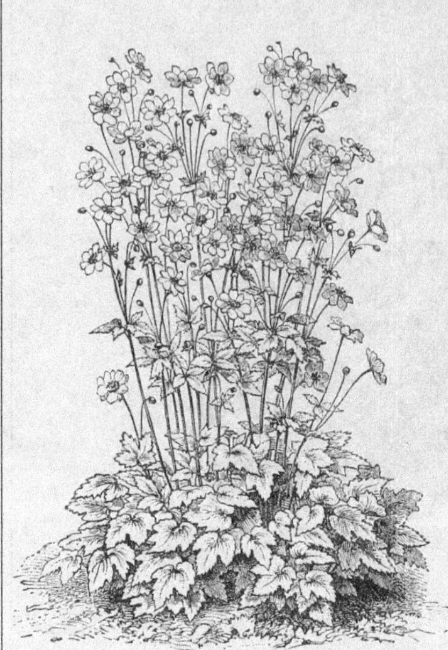

Anemone Honorine Jobert (*Japan Anemone*).

Flower of Anemone japonica (*Japan Anemone*).

Anemone japonica (*Japan Anemone*) naturalised.

[*To face page* **23.**

A. hepatica *(Common Hepatica)*.—An exquisite flower of early spring, hardy everywhere, not fastidious as to soil, though it loves a deep loam. In positions somewhat sheltered, and where the soil is porous, the foliage will usually stand well through the winter. The Hepatica is a deep rooter—indeed, having regard to the shallow nature of its crowns, few plants send their roots deeper into the soil; hence it thrives so well upon made banks and rockwork, and especially likes positions where moisture will not lodge about its crowns. It will do as well as Primroses or Violets in any good garden soil, and where let alone, and not ruthlessly dug about or cut with the hoe, or often lifted and pulled to pieces, will make tufts strong and broad. Clumps of the rich-coloured blues and reds when a mass of bloom are in the month of March exceedingly beautiful. Few plants are more impatient of frequent division. The best plan is to put a young plant into good soil, and let it remain until it has become a strong clump—this would give perhaps twenty or more single crowns—then plant again, allow these to remain a few years, and then re-divide. Thus a good stock may be obtained. The Hepatica is a native of many hilly parts of Europe and also of North America. The best known kinds are the double red and single blue, both amongst the hardiest of the section. Then there are the single white (a charming sort for bouquets), single red, double blue, rich in colour, and always scarce Barlowi, evidently a rich-coloured sport from the single blue; there is a single red known as splendens; lilacina, a pretty mauve kind; and some others. Every variety of the common Hepatica is worthy of care and culture. Is it possible to imagine a more beautiful feature than we may produce by planting a mixed edging of the various colours round say a bed of dwarf American plants? It is but one of many ways in which we may tastefully use them. Usually found in copses and half-shady positions these will be found to suit it best in a cultivated state also. Where plants and space are to spare, there would be little trouble in naturalising this in any copse or among bushes, or in a thin shrubbery—anywhere, in fact, the plants would not be overrun by larger things. Mr. Frank Miles, who has been very successful in raising seedlings, describes his practice as follows:—

HEPATICAS FROM SEED.—Sow the seed directly it is ready to fall, in light sandy loam. If it once gets dry before sowing it is unlikely to germinate. Put flat on the soil slates or bricks. Slates used for tiling are the best; they will not then require either watering or weeding. In October or November the seeds will begin to germinate. Remove the slates and put the boxes under glass without any heat. By spring time every seed will have germinated. During the summer keep the seedlings in a shady place. Some will then make their first leaf and probably bloom the following spring, but from the time of sowing the seed it will take three years before the plants show their free character and blossom well. I adopt exactly the same treatment for Hellebores. I owe this system to Mr. F. Gibbons, of Bramcote, near Nottingham, one of the cleverest herbaceous plant growers in England.

A. japonica *(Japan Anemone)*.—A tall, autumn-blooming species, 2 ft. or 3 ft. high or more, with fine foliage, the flowers large and rose coloured. The variety named Honorine Jobert, with pure white flowers, is a beautiful and effective plant. Both should be cultivated where supplies of cut flowers are required in autumn. The white is a sport from A. japonica, and is a more valuable plant than the type. Both are first-rate plants for the flower garden, for groups, borders, or the wild garden. By having them in various situations, some on a north border, some on a warm one, the bloom may be prolonged. The white form is charming in the shade of a wood. Various hybrids raised between the Japan Anemone and A. vitifolia were obtained at Chiswick, but they appear to have been lost for the most part. Any of them we have seen were not so good as the type, and the best of all, the white. Some suppose that the white is the original form of the plant, and we think with good reason. As to propagation, every bit of the root grows when divided. A rich soil is desirable for these plants.

For greenhouse purposes, "Brockhurst" writes: "We keep a stock of good-sized plants of it in the reserve garden, potting them when the buds are well formed, and then place them in the conservatory or cool greenhouse, where they bloom very freely, and are even more beautiful than those in the open ground. By a little management it is easy to have these Anemones in flower in the conservatory for nearly two months. For dinner-table decoration it is also exceedingly useful, either in pots or in the shape of cut blooms. The latter last fully a week in water if cut when freshly opened. It is perfectly hardy, and needs no skilful cultivation." "Anemone Honorine Jobert," according to M. O. Frœbel, "is not a garden hybrid between A. vitifolia and A. japonica. The variety originated at Verdun-sur-Meuse in the garden of M. Jobert. He obtained it from a large tuft of the old A. japonica, with red flowers, from which plant a root branch flowered with pure white flowers.

The uses of the various forms of the Japan Anemone are various and important—borders, groups, fringes of shrubbery, and here and there in half-shady places in woods and by wood walks. One of the plants best worth growing for affording cut flowers.

A. nemorosa *(Wood Anemone)*.—This native plant adorns our woods in spring, and also those of nearly all Europe and Russian Asia. It is so abundant in the British Isles, that there is little need to plead for its culture. There are double varieties, and the colour of the flower is occasionally lilac, or reddish, or purplish. Flowers from March to May; white, and reddish outside. Height, 6 in. Division.

The sky blue variety of the Wood Anemone (A. Robinsoniana, Hort.) has become a very-much-sought-after plant of recent years. It is a plant of easy culture and of peculiar beauty, especially if seen as the morning or noon-day sun is on the flowers. This plant is fitted to grace a ledge of the rock garden as a colony or wide-spreading tuft. Also for the margins of borders, as a groundplant beneath taller subjects, forming a carpet for beds of choice shrubs with ample space between, in the small beds beneath standard Roses, and also for the wild garden, and dotting through the Grass in the pleasure ground in spots not mown early. In this case some taste should be exercised to get the groups or colonies of easy natural outline. After a time the plant will take care of itself.

The Rev. H. Harpur Crewe writes as follows concerning Anemone Robinsoniana: Of a numerous and very beautiful family, this is, to my mind, the undoubted queen. There is a gorgeous splendour about A. fulgens and Pavonina, and a dazzling beauty in A. stellata and coronaria; there is much delicate grace about A. bracteata, trifoliata, apennina, blanda, sulphurea, alpina, nemorosa, and narcissiflora; we all admire the purple and silk in which A. Pulsatilla and vernalis love to clothe themselves, and seldom tire of gazing at the golden sheen of A. palmata and ranunculoides; A. japonica and vitifolia, and their varieties and hybrids, have much pleasant autumnal brightness; but, to my mind, all fade before the simple and innocent loveliness of A. Robinsoniana. Most botanists seem to agree that, though in many respects very distinct, it is a variety of the Wood Anemone, A. nemorosa. It is a much dwarfer plant, blooms later, and both leaves and flowers possess more strength and substance; but its distinguishing characteristic is the pure, pale, cœrulean blue of the inner surface of its petals. I know of nothing more exquisitely lovely than a fully-expanded patch of this beautiful flower on a bright spring morning. It is a rare British wild plant; I know of its occurrence in Norfolk and Essex, and I believe it has also been found in Kent, Sussex, and Oxford. It was practically unknown to the general public till a few years ago, when Mr. Robinson found it growing wild, and, struck with its marvellous beauty, so frequently spoke of its charms that it became a general favourite, and, in compliment to its champion, took the name of A. Robinsoniana. This name was first given it in a garden in a northern suburb of London.

A. palmata *(Cyclamen-leaved Anemone).*—A very distinct kind, with leathery leaves and large handsome flowers, of a glossy golden yellow, only opening to meet the sun. A native of North Africa, Spain, and other places on the shores of the Mediterranean, this charming flower requires and deserves a little more attention than most of its cultivated sisters. It should be planted in deep turfy peat or light fibrous loam with leaf-mould. It should not be placed in positions on the face of rocks suited for Saxifrages and many other plants that are content with mere crevices, and drape the face of the rocks with the slightest encouragement, but rather on level spots, where it could root deeply and spread into firm tufts. There is a double variety, A. palmata fl.-pl., and a white one, A. palmata alba. Flowers in May and June, 6 in. to 14 in. high, and is propagated by division or seeds.

A. Pulsatilla *(Pasque-flower).*—Though sparsely distributed in Britain, this fine old border plant is a true native, and when it does occur on a bleak chalk down, it is generally freely dotted over the turf. The position is usually such as to suggest the aptness of the name Windflower for the family generally. There are few sights more pleasant to the lover of spring flowers than to see its purple blooms just showing through the hard Grass of a bleak down on an early spring day. The plant is much smaller in a wild than in a cultivated state, usually having a solitary flower. In the garden it forms rich healthy tufts, and flowers more abundantly and vigorously. There are several varieties, including red, lilac, and white kinds, but these are now rare. There is also a double variety. It prefers well-drained and light, but deep soil. Flowers in March, April, May; purplish. Height, 3 in. to 12 in. Propagated by division or by seeds. A border or edging or rock plant.

A. ranunculoides *(Yellow Wood Anemone).*—Not unlike the Apennine and the common Wood Anemone in habit, this species is so very distinct in its clear golden flowers, that it is well worthy of cultivation even by the side of the most admired kinds. It is a South European species, and apparently is not so free on the generality of our soils as the Apennine A., but when grown into well-established tufts on a light or warm and well-drained soil, it blooms in a way of which those who have merely seen isolated plants or figures of the plant can have no idea. I have not found it do well on clay soil, but on chalk soil it seems to grow as freely as the common Crowfoot. It is quite charming for association with tufts of the Apennine or the Wood Anemone, the Pasque-flower, any of the varieties of A. Hepatica, the Aubrietias, and like plants. It is among the naturalised group of British plants, and grows in a semi-wild condition in Herts, in Notts, and it is also reported to occur in several other counties. Flowers in the end of March and beginning of April. Height, 4 in. to 6 in. Propagated readily by division.

A. stellata *(Star Windflower).*—This native of Southern Germany, France, Italy, and Greece, if not so showy, is quite as beautiful as the common Anemone. The star-like flowers, ruby, rosy purple, rosy, or whitish, springing from the much dissected leaves,

PLATE XIX.

Anemone nemorosa var. Robinsoniana.

Anemone sylvestris (*Snowdrop Anemone*) naturalised

[*To face page* 24.

PLATE XX.

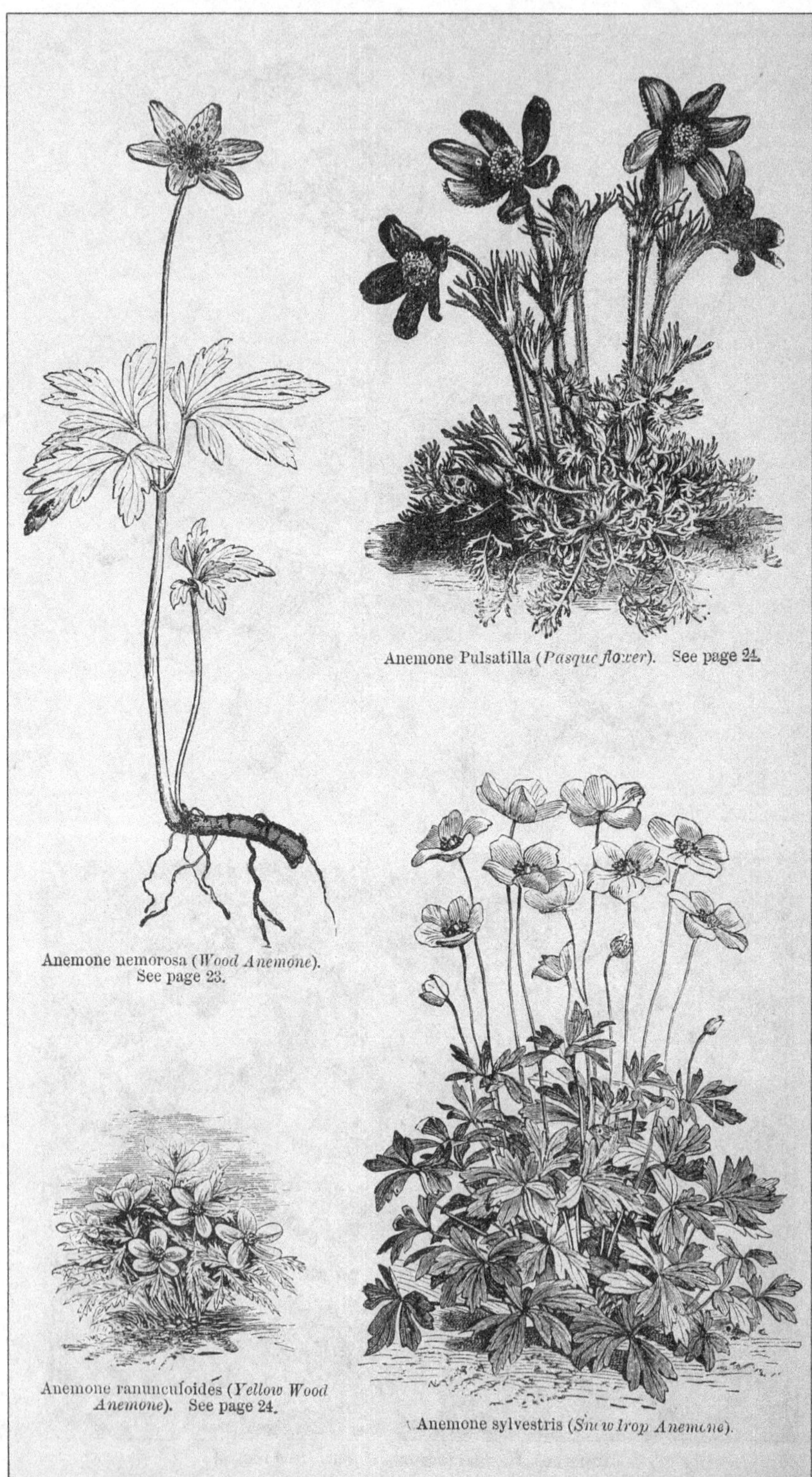

Anemone Pulsatilla (*Pasque flower*). See page 24.

Anemone nemorosa (*Wood Anemone*). See page 23.

Anemone ranunculoides (*Yellow Wood Anemone*). See page 24.

Anemone sylvestris (*Snowdrop Anemone*).

vary in a very charming way, and usually have a large white eye at the base, which contrasts agreeably with the gay or delicate coloration of the rest of the petals, and with the rich brownish violet of the stamens and styles that occupy the centre of the flower. It is not so vigorous in constitution as the Poppy A., and requires a little more care than that does, but this will only make it the more interesting to all who love variety in their collections of hardy plants. It likes a sheltered, yet warm position, a light, sandy, well-drained soil, and seems to make little or no progress on heavy clay soils. It is suitable for association with the choicer kinds of Anemone on the rockwork, the mixed border, and the choice spring garden, and should be grown in every garden where spring flowers are appreciated. Here, as in the case of the finely-coloured Poppy Anemones, the best way will be to select and increase certain fine forms. It is the more desirable in the case of the Star Anemone, because it does not do well on all soils. Where the soil and district suits the plant, it is well to encourage it. Mr. Ellacombe, of Bitton, speaks highly of the white variety: "At first it is rather stained with purple, but when fully out it is a pure white star, with pale purple under petals. This with the black eye and the pretty foliage make it a striking flower, and a very good addition to spring flowers." Flowers in May. Height, 10 in. Propagated by division or by seeds. Sometimes known as A. hortensis.

A. sulphurea is a beautiful soft yellow form of the Alpine Windflower (A. alpina).

A. sylvestris *(Snowdrop Windflower).*—A free and handsome species, growing vigorously on almost any soil, the white flowers, as large as a crown-piece, being freely produced over a mass of fresh green leaves. A native of Siberia and Central Europe, it is perfectly at home in this country, and should be grown wherever first-rate border flowers are appreciated; it will associate well with the Alpine Windflower, and plants of like size, about the lower parts of the rock garden. Being naturally a native of the grove, it will be found perfectly at home along our wood walks and half wild spots, in shrubberies. The aspect of the drooping unopened buds has suggested its English name—the Snowdrop Anemone. Flowers in April and May; pure white. Height, 1 ft. to 15 in. Propagated readily by division of root.

A. thalictroides *(= Thalictrum anemonoides).*

A. vitifolia *(Vine-leaved Anemone).*—As regards mode of growth and flowering, this closely resembles the Japanese Anemone before alluded to, but is much more downy, and it flowers a fortnight earlier than A. japonica. It is a native of the moist shady valleys of Nepaul, where it is said to be one of the commonest and most beautiful of all wild flowers. The name A. vitifolia occurs in many catalogues and on labels, but we have not often seen any plant under the name that differs from what is known as A. japonica or one of its forms. A plant in Mr. Elwes' garden has the leaves simple and Vine-like, as the name implies, whereas the white japonica has the leaves compound with the exception of a few root leaves. A. vitifolia is earlier in flower, and the blossoms not quite so large, the white having a slight tinge of pale purple outside. It really is in no way better than the white Japan Anemone, if so good; they are close allies, and there is some reason to believe the large white kind to be the original type, and the red forms sports from it, the contrary being the usually accepted notion. A. vitifolia was in cultivation before Mr. Gordon raised a number of hybrids between it and A. japonica.

The previously named Anemones are the most beautiful of the family, which, however, contains many other interesting and useful plants. These, from their rarity, slowness of growth, or from various other causes, are only enumerated here: A. acutipetala, A. alba, A. baldensis, A. nemorosa bracteata, A. collina, A. dichotoma, A. Halleri, A. Hudsoniana, A. montana, A. multifida, A. narcissiflora, A. Nuttalliana, A. obtusiloba, A. ochotensis, A. patens, A. pennsylvanica, A. Popeana, A. pratensis, A. rivularis, A. scaposa, A. sibirica, A. thalictroides, A. trifoliata, A. virginiana, and A. vernalis. There are also tender species not included here; most of the above are of easy culture with the exception of the alpine species, like A. vernalis, which are slow and require to be carefully grown in fully exposed spots in moist, gritty soil.

Anemonopsis.—A small genus of Crowfoots, of which one species (A. macrophylla) is in cultivation. It is a beautiful plant similar to the Japanese Windflower (Anemone japonica), but smaller in all its parts. The thick and shining leaves rise to a height of 12 in.; the flower-stems are slender, about 18 in. in height, on which are borne numerous drooping blossoms, about 1½ in. across, of a pale purple colour. The flowers differ from the Anemone in having two rows of petals, one outside and spreading, the other forming a cone in the centre. It is a native of Japan. It thrives in a rich deep soil in a partially shaded border well drained.

Angelica.—Bold growing plants of the Celery Order, which would be of some use for their form had we not so many finer hardy plants in the same family. One is a native of our woods and river banks, and another is a well known plant grown in most kitchen gardens.

Anigosanthus coccineus.—An Australian plant producing numerous crimson flowers with green tips, hardy in sheltered spots, but of no proved value out-of-doors in our climate.

Anisodus luridus.—A hardy perennial of the Solanum family from Nepaul. It has greenish yellow bell-shaped flowers and ample bright green foliage. Of no garden value. *Syn.*, Scopolia lurida.

Anomatheca cruenta.—A pretty diminutive South African bulbous plant growing from 6 inches to 12 inches high. The flowers are about ½ in. across, of a rich carmine crimson colour, three of the lower segments being marked with a dark spot. They are produced in loose clusters on slender stems overtopping the narrow sword-shaped foliage. It is quite hardy on many soils, but in other places it should be planted on warm slopes of rockwork, in very sandy dry soil, or on warm borders among the smaller and choicer bulbous plants; the bulbs should be planted rather deep. In many soils it increases rapidly without attention.

Antennaria *(Cat's-ear)*.—A small genus of Composites, the cultivated species of which are all perennial. A. margaritacea, the Pearly Everlasting, is a North American plant, growing about 2 feet high; the flowers, produced in broad flat clusters, are white and of a chaffy nature, hence are used for winter bouquets in a dry state, and are also dyed in various colours. The Pearly Everlasting, though one of the oldest and commonest plants in our gardens, is not worth a place for any purpose. It does not yield the flowers of the true Immortelle wreath, which are those of a much dwarfer plant (Gnaphalium arenarium). A. plantaginifolia, another North American species, is an unimportant plant for the garden. The Mountain Cat's-ears, A. dioica and A. alpina, and varieties minima and tomentosa, are neat-growing dwarf plants with white downy foliage, hence are largely used as carpeting plants in masses. All are of the simplest culture in any ordinary soil in exposed positions. These are good rock-garden plants, and the pretty little rosy heads of one form of the Mountain Everlasting may be seen in the cottage gardens of Warwickshire. These last kinds only grow a few inches high, and are very easily increased by division. A. tomentosa (Hort.) is a plant of a similar character that has been much used as a dwarf silvery plant in the flower garden. It is hardy and of easy increase and culture in bare spots.

Anthemis *(Camomile)*.—Of the numerous kinds of these in cultivation there are but few worth growing. A. Aizoon is a dwarf silvery rock plant—from 2 in. to 4 in. high, having small white Daisy-like flowers. Its chief beauty is the leaves, which are covered with a white downy substance. It should be grown in the rock garden in exposed places. A. Kitaibeli is a pretty object in the mixed border. It grows rather tall, and requires to be neatly staked, for its large, pale, lemon-coloured, Marguerite-like flowers are very showy. A. tinctoria is very similar, and both are excellent for cutting. The double flowered form of the Corn Camomile (A. arvensis) is sometimes cultivated among annual plants. Lasting, as it does, a length of time in water, its flowers being double and pure white, the flowers are useful for cutting.

Anthericum *(St. Bruno's Lily)*.—A large genus of the Lily family, containing but a few species hardy in this country. These are the European kinds, some of which are among the most beautiful of hardy flowers, and well worth cultivating in every garden.

A. Hookeri *(Chrysobactron Hookeri)* is a showy perennial, growing from 1 foot to 20 in. high. It flowers in early summer, the blooms being bright yellow, nearly ½ inch across, freely produced in racemes, 3 in. to 5 in. long. The leaves form dense tufts in ordinary soil, but the plant grows best in one that is moist and deep, such as an artificial bog. A native of New Zealand.

A. Liliago *(the St. Bernard's Lily)*.—This grows from 1 foot to 2 feet high, producing single, sometimes branched flower-spikes that bear numerous pure white flowers in early summer. It is known also as Phalangium Liliago. A. ramosum has the flower-stems about 2 ft. high, much branched, and bearing small white flowers; it has narrow Grass-like leaves, and the plant soon grows into large tufts. It is sometimes called in gardens and nurseries A. graminifolium.

A. Liliastrum *(St. Bruno's Lily)*.—A most graceful alpine meadow plant. It requires to be planted in deep free sandy soil, and in early summer throws up spikes of snowy white, Lily-like blossoms. In dry soils a good mulching with rotten manure would be a great help to it, and in early spring the plants must be protected from slugs and caterpillars, from attacks of which they are liable to suffer. It is propagated by division of the roots in autumn, which is the best time to plant, or it may be raised from seed. It usually grows about 15 in. high, and is an excellent plant for choice borders. It would look better still as a good colony or group in an open space between dwarf shrubs. Where plentiful it would be an interesting subject to naturalise in a Grassy place. This species is also known under the generic names of Paradisea and Czackia.

The major variety of the St. Bruno's Lily has much larger flowers (2 in. across) than the type, and possesses the peculiarity of sending up large single flowers from the root. These open before the flowers on the spike and are larger, resembling the white blooms of a Pancratium. Generally one solitary bloom of this kind grows on each plant at the root and far below the spike, and it seems as if strange flowers appeared in the bed. This peculiarity of the plant points to it as distinct from the ordinary type of St. Bruno's Lily. It grows 3 ft. high in good soil, and is a fine border plant, but though many think highly of it, the species is more elegant in form and to us more precious.

Antholyza.—A genus of bulbous plants of the Iris family, numbering about a dozen species, all natives of the Cape of Good Hope; of these only four are worth cultivating. These have narrow, erect, Iris-like leaves and flower-spikes that overtop the foliage, bearing numerous flowers of a bright red, though they are not very attractive on account of the flowers not expanding

PLATE XXI.

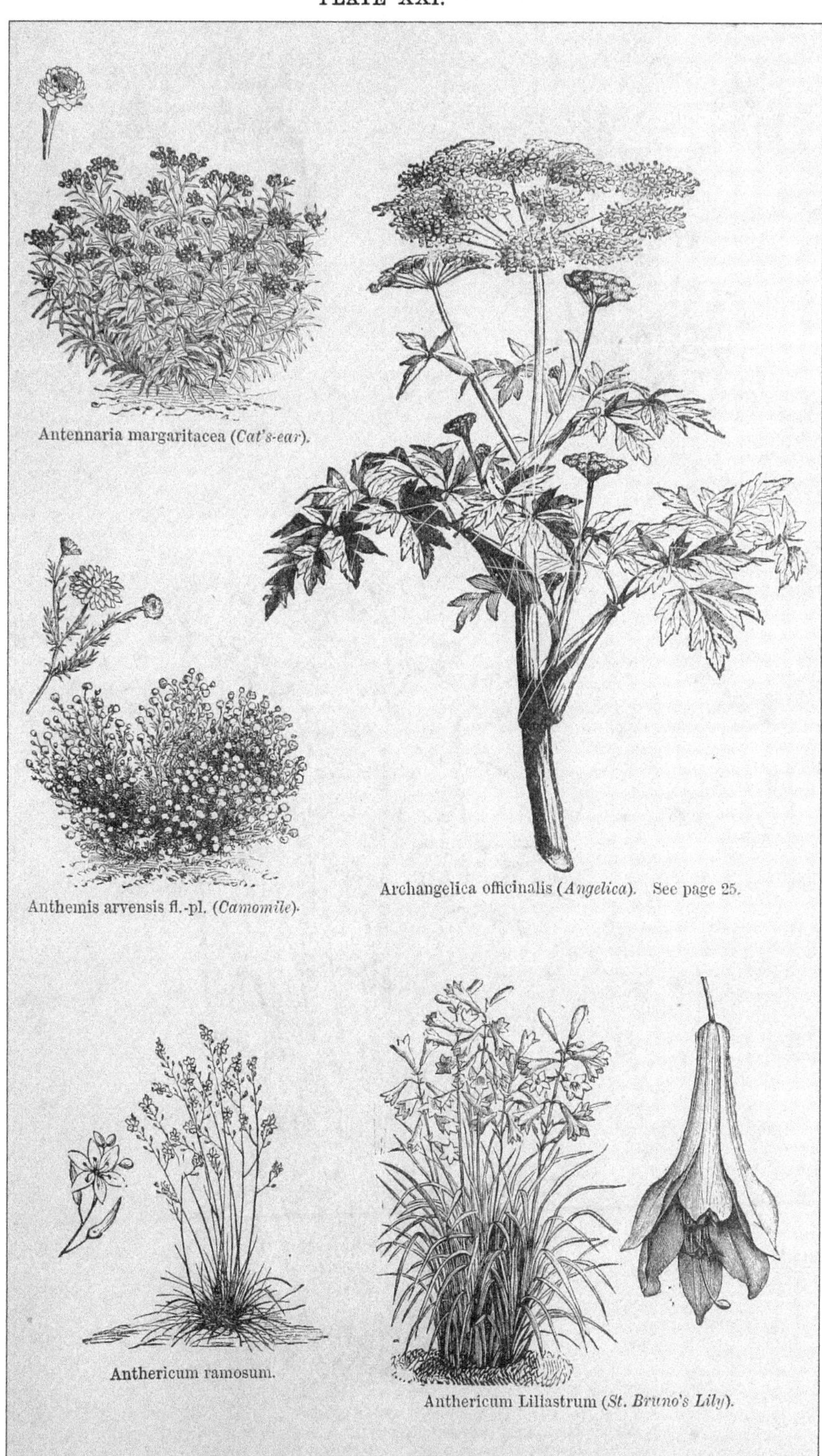

Antennaria margaritacea (*Cat's-ear*).

Archangelica officinalis (*Angelica*). See page 25.

Anthemis arvensis fl.-pl. (*Camomile*).

Anthericum ramosum.

Anthericum Liliastrum (*St. Bruno's Lily*).

[*To face page* **26.**

PLATE XXII.

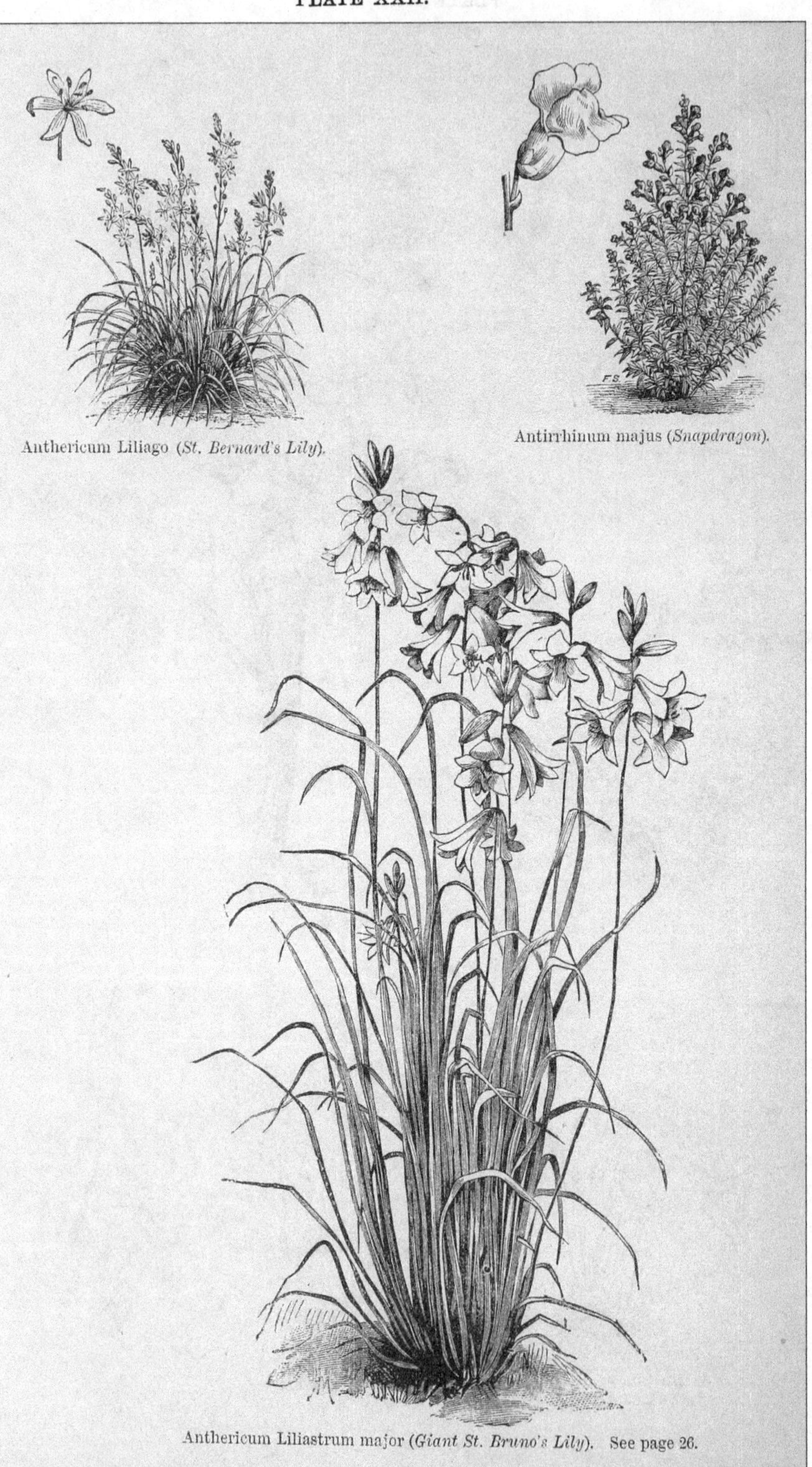

Anthericum Liliago (*St. Bernard's Lily*).

Antirrhinum majus (*Snapdragon*).

Anthericum Liliastrum major (*Giant St. Bruno's Lily*). See page 26.

[*To face page* 27.

sufficiently. The names of these are: A. æthiopica, A. ringens, A. bicolor, and A. Cunonia. The latter is a distinct plant, but the others much resemble each other. They are quite hardy if planted in a warm, sheltered border thoroughly drained. The soil should consist of a good sandy loam enriched by well decayed leaf-mould. The plants may be propagated by the offsets or small bulbs which are annually produced in numbers from the parent bulbs or by seeds, which in favourable seasons are freely produced. These should be sown as soon as ripe in light soil in an ordinary hot-bed, where they will germinate the following spring. In the current summer the seedlings should be planted out and treated in the same manner as mature bulbs. It is advisable to lift the plants every autumn, so as to separate the small bulbs. This should be done about the end of August, and the bulbs should be replanted in October and November, or may be delayed till February. A handful of litter over the bulbs during winter would be the best way to ensure the safety of the plants. Watsonia Meriana and others are known in some gardens as species of Antholyza, but they correctly belong to Watsonia.

Anthyllis *(Kidney Vetch).*—A genus of the Pea family, of which there are some half a dozen species in cultivation; few are worth growing. A. montanus, the Mountain Kidney Vetch, is a plant seldom seen in our gardens, though a very hardy rock plant. It is very dwarf, about 6 in. high, the leaves being pinnate, and nearly white with down. On good light soils it grows larger. The pinkish flowers are produced in dense heads, rising little above the foliage, and forming with the hoary leaves pretty little tufts. This plant is desirable for every kind of rockwork, but chiefly valuable for its power of thriving on stiff, cold, and bad soils. Resisting any cold or moisture, it is peculiarly fitted for a position among the dwarf plants in the front rank of the mixed border. A native of the Alps of Europe. A. erinacea is a singular-looking, much branched, tufted, spiny, almost leafless shrub, about 1 ft. high, with purplish flowers. A. Vulneraria (Woundwort), a common native plant, is pretty and well worth growing in a mixed border or on dry banks. There is a white-flowered variety and a red one.

Anticlea *(Zygadenus).*

Antirrhinum *(Snapdragon).*—A distinct family of northern plants, by far the most popular of which is the handsome A. majus, the common Snapdragon. Like the Wallflower, this claims a place from the facility with which it may be grown on old walls and ruins, or even on the tops of walls far from old. Had we but the common variety, it would be well worthy of our attention from this habit, but when it is considered how many beautiful striped, and self-coloured, and flaked, and mottled, and delicately dotted kinds are now abundant in gardens, and raised from seed as easily as Grass, few will doubt the claims of this plant. The varieties have been much improved of late years, and will produce from seed a charming and profuse variety of brilliant colours, the white-throated and some of the striped varieties being exceedingly handsome. The Tom Thumb varieties are useful for small beds, and even for the rock garden where space is to spare. A good strain only should be grown, as it is waste of time to cultivate inferior forms; and now that a preponderance of striped flowers is certain to occur in a good strain, that is an additional reason for having a plantation in a garden.

But self-coloured kinds should be selected as even more important than striped ones, giving a better effect in groups or masses. Cold wet springs have an injurious effect on old plants grown in a cold, wet soil. On the other hand, plants in a light, drier soil do wonderfully well; and really very fine specimens can often be seen in gardens in full bloom. Those who have a wet, cold, uncongenial soil will do best to raise plants in the autumn, winter them in store boxes in cold frames, and plant them out in March and April, when the weather is favourable. When those who treat the Antirrhinum as a biennial lose their plants through the severity of the winter, they should raise them as annuals. Hard frosts following a wet autumn are destructive to Snapdragons. We saw in a nursery a few days ago some large beds in full bloom of very fine Antirrhinums, all of which had been raised from seed sown in heat in January and February last; the young plants had been grown fast and then planted out in the open ground. The plants in bloom were of large size, and were producing large spikes of very fine flowers, some of the striped varieties being of great beauty. On poor, dry, or high-lying land, no plants give a better return in flowers. They revel in such places.

Among the numerous species, some few are seen in cultivation from time to time, but they do not take a permanent place in gardens. Among the best are A. Asarina and rupestre.

Apera arundinacea.—A dainty slender-growing Grass, and one of the few which can afford to compete with the Feather Grass in graceful beauty. In growth it is almost Rush-like, bearing long weeping plumes of a glossy purplish-brown colour. Being perfectly hardy, it will prove extremely useful in places where graceful spray for indoor decorative uses is in demand. The plumes dry well, and endure for a long time.—F. W. B.

Aphanostephus ramosissimus.—A pretty half-hardy annual of the Composite family from Texas. It grows scarcely more than 4 in. in height, very much branched, as its specific name implies, every shoot producing a flower-head about 1 in. across, with a yellow disc, and violet-blue ray florets. Its peculiarly branching habit conduces to a close carpet-like growth, and ensures an abundance of bloom throughout the summer. It requires the same treatment as other half-hardy annuals.

Aphyllanthes monspeliensis.—A pretty Rush-like plant, forming dense, erect tufts 1 ft. or more high. It flowers in summer, the blossoms being deep blue and about ¾ in. across. The leaves are slender, and the root fibrous. South of France. Borders, in light soil. Propagated by division and seed. A very hardy, though slow growing plant, mainly of botanical interest, though a tuft on the rougher slopes of the rock garden will not be out of place.

Apios tuberosa *(Ground Nut)* is an interesting climbing perennial with a tuberous root. A native of North America. The flowers are a dull brownish purple, sweet scented, produced in summer in axillary racemes. It is desirable for covering arbours or for rambling over shrubs, &c.; the fragrance of its flowers, which much resemble that of Violets, pervades the atmosphere round it, especially after a shower. Belongs to the Pea flower Order, and is increased by division of tubers.

Aplectrum hyemale *(Adam and Eve).*—An interesting terrestrial Orchid, growing from 6 in. to 12 in. high. The thick bulb sends up a large oval leaf in late summer, which lasts until the next summer, when the flower-stalk appears with a raceme of large flowers, which are greenish-brown and speckled with purple. Sandy moist spots in rich leaf-mould. Native of the Eastern United States. Of doubtful value for the garden.

Aplopappus.—A little-known genus of North American Composites. Two species are in cultivation, one an annual the other a perennial. A. ciliatus is a robust growing annual, with stems 3 ft. to 4 ft. high, and bearing flower-heads about 2 in. across, of a bright yellow. It blooms throughout the autumn months, commencing in July. It may be treated either as a biennial, in which case the seed should be sown about August, or as a half-hardy annual, the seeds being sown under glass as early as possible. A. Fremonti is a species growing from 6 in. to 12 inches high, with erect stems and flower-heads, of a bright yellow, 1 in. across. It is a native of Colorado in the sub-alpine districts, and requires the treatment of other plants of its class.

Aplotaxis.—A genus of Compositæ of no garden value.

Apocynum *(Dogbane).*—A small genus of North American plants, of which there are four species in cultivation, but all are of little garden value. A. androsæmifolium (Fly-trap or Spreading Dogbane) is an interesting and curious plant growing from 2 ft. to 3 ft. high, and bearing small rose-coloured blossoms. A. cannabinum, the Indian Hemp, is similar, and so are A. hypericifolium and A. pubescens, probably only varieties of A. cannabinum.

Aponogeton *(Cape Pond Weed).*—A beautiful sweet-scented aquatic plant from the Cape of Good Hope, which, fortunately, is quite hardy in many parts of these islands, and in the north under special circumstances. No one has been more successful with it than the late Jas. McNab, of the Botanic Gardens at Edinburgh, who wrote of it: Aponogeton distachyon has been growing in the pond of the Royal Botanic Gardens at Edinburgh for the last 40 years, and now forms many large patches in various parts of it, the largest being 48 ft. in circumference. The situation where the pond stands was originally a marsh; when it was made, the bottom was causewayed with stones, placed ½ in. apart, in order to allow the numerous springs, peculiar to that portion of land, freedom to rise between them. The pond varies from 2 ft. to 5 ft. in depth, and the bottom is thickly coated with mud, arising from the tree leaves which are annually blown into it. In this decomposed vegetable matter the Aponogeton thrives well, and seeds, which are abundantly produced during the autumn months, germinate freely in the muddy bottom. In consequence of the number of springs which exist, portions of the pond are never coated with ice, even during the most severe winters. The overflow is very large, and is never found to vary at any time throughout the year, not even during very dry summers. To these circumstances I attribute the healthy condition of the beautiful Pond Weed, which flowers abundantly every year, not only during the spring, summer, and autumn, but often during the winter, particularly if the weather is at all mild. Plants of it have been sent from this establishment to many ponds throughout Great Britain, but in few has it been successful, evidently owing to the want of constant springs bubbling up amongst their roots, which causes a continual change of water. Plants of it have, however, succeeded in several mill-ponds where the water is kept warm by the condensed steam constantly thrown into them.

About London during the severe winters there has been no more interesting sight than the profuse bloom of the fragrant Cape Aponogeton. In the open air, in Mr. Parker's nursery at Tooting, myriads of its handsome flowers, with a scent like that of the Hawthorn, floated gracefully on the water, and long and graceful fresh green leaves quite flat on its surface. In the midland and cold districts it is necessary, for the perfect culture of this plant in the open air, to grow it in spring or other water that does not freeze; but in mild districts and in the south of England this is not needed. In Mr. Parker's nursery, where it grows so freely, it is supplied with water from an Artesian well, conducted into the reservoir in which it is grown. Where natural springs keep a pond from freezing it will grow well in any part of the country, with, of course, a little care in starting it. In addition to its blooming

PLATE XXIII.

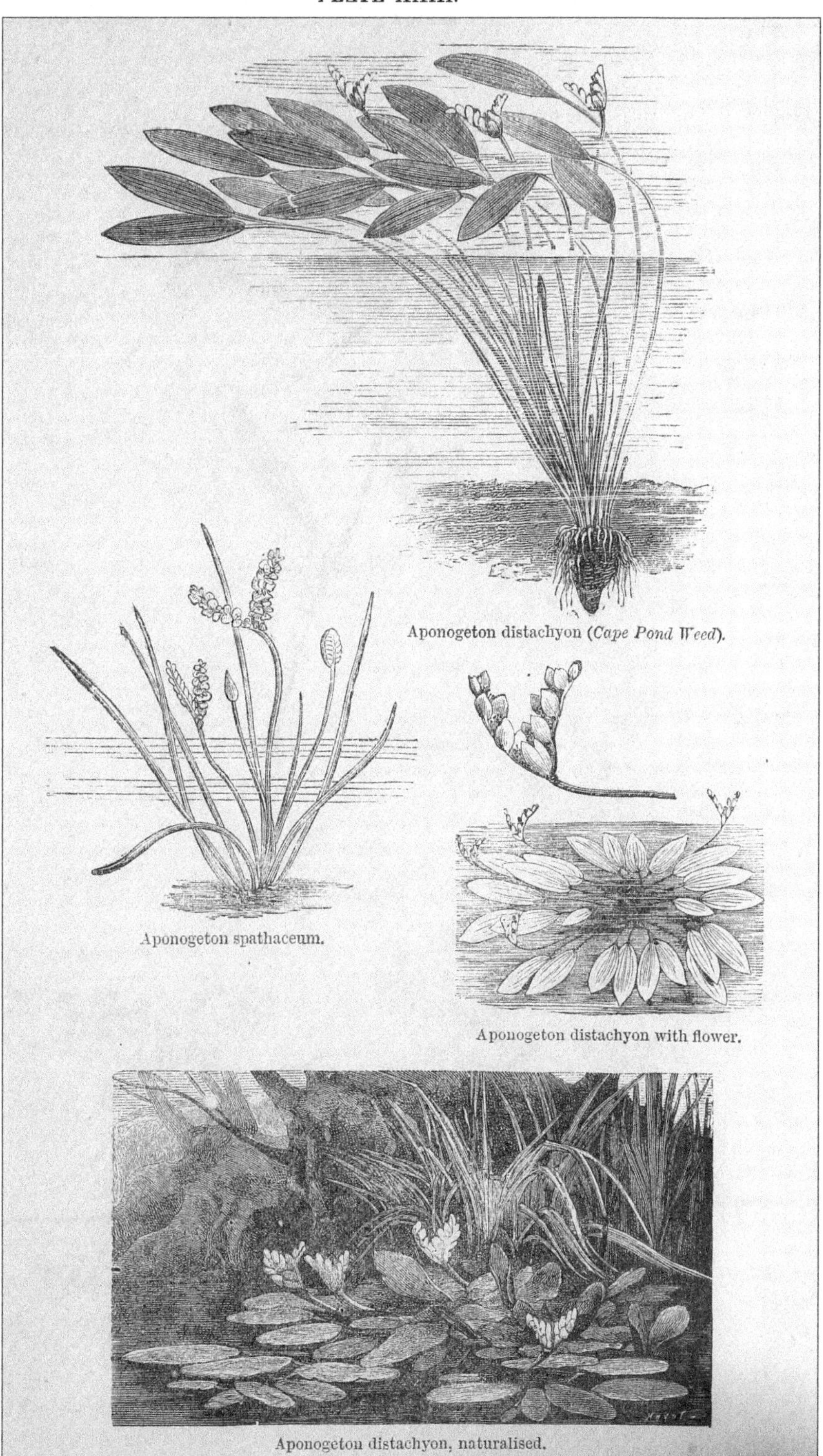

Aponogeton distachyon (*Cape Pond Weed*).

Aponogeton spathaceum.

Aponogeton distachyon with flower.

Aponogeton distachyon, naturalised.

[*To face page* **28.**

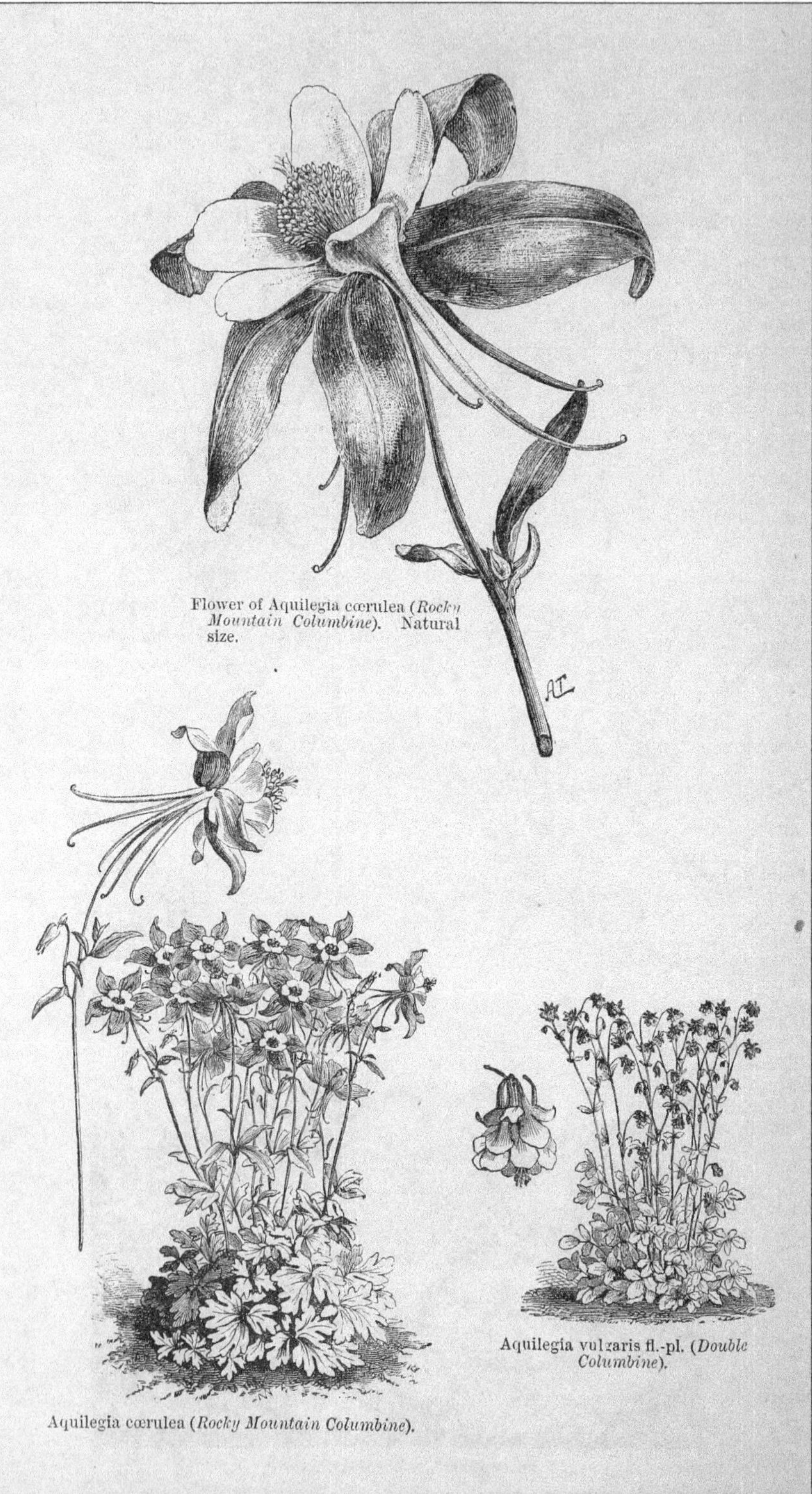

Flower of Aquilegia cœrulea (*Rocky Mountain Columbine*). Natural size.

Aquilegia cœrulea (*Rocky Mountain Columbine*).

Aquilegia vulgaris fl.-pl. (*Double Columbine*).

[*To face page* 29.

throughout the winter in England in the open air, it may also be grown and flowered well as a pot plant. In suitable positions it may be had in bloom all the year round, and it may be well flowered in an inverted bell-glass in a room. In Devonshire it is grown with greater ease and to greater perfection than in the home counties or near London. Failures often result from putting it in too shallow water.

Aponogeton spathaceum.—Unlike its relative A. distachyon, this is of erect habit, and is altogether a smaller growing plant. Its flowers are tinged with rose, but quite devoid of scent, a property so much appreciated in A. distachyon. It is, in fact, a very poor relation of the Cape Pond Weed.

Aquilegia *(Columbine)*.—A valuable family of plants for the garden, often beautiful in habit of plant, colour, and form of flower. It is a northern genus, being widely distributed over the northern and mountain regions of Europe, Asia, and America. The Columbines rank amongst the next successional flowers to those that belong purely to the spring months—their flowering period extending throughout May and June. Among them may be found great variety in the way of colour—white, rose, buff, blue, and purple, and also stripes and intermediate shades even in the same flower. Then amongst the American kinds we have yellow, orange, and scarlet, and most delicate shades of blue. Besides colour, too, there is also considerable variation in the shape of the flowers. In some the petals are reduplicated, and in the very double forms of our common garden Columbine, on removing one of the five petals, which are usually distinguished by their brighter colour and almost invariably by the presence of a spur-like appendage, it will be found that a series of from six to a dozen, or even more, petals are beautifully arranged one inside the other.

The Columbines are frequently of greater stature than most of the plants strictly termed alpine, but are, nevertheless, true alpine plants, and among the most singularly beautiful of the class. Where single plants of the wild form of the common Columbine are met with in the open copses and by the mountain streamlets in Northern England, it looks a queen among the other flowers of the region. The blue, and blue and white alpine kinds, living in the high bushy places in the Alps and Pyrenees, and, indeed, of all European and North-Asian mountain chains, are among the fairest of all flowers. Climbing the sunny hills of the sierras in California one meets with a large scarlet Columbine (Aquilegia eximia) that has the vigour of a Lily and the grace of a Fuchsia; and in the mountains above Salt Lake City, in Utah, and on many others in the Rocky Mountain region, there is the Rocky Mountain Columbine (A. cœrulea), with its long and slender spurs and lovely cool tints in its erect flowers. Indeed, there is no family that has a wider share in adorning the mountains. The finer Columbines are to the smaller alpine flowers what the Birches are to the hill shrubs. Some of the alpine species, however, are much smaller than those commonly grown, as, for example, the Pyrenean Columbine. Although our cottage gardens are alive with Columbines in much beauty of colour in early summer, there is some difficulty experienced in cultivating the rarer alpine varieties. Hence, such highly-valued kinds as the Altaian Columbine (A. glandulosa), the Alpine Columbine (A. alpina), are too rarely seen flowering well in gardens, and frequently disappear where introduced. They require carefully planting in free sandy or gritty, though always moist, ground, and in well-drained ledges in the rock garden, mainly in half shady positions or northern exposures. Most rare Columbines, however, fail to form enduring tufts in our gardens, and where this is the case they must be raised from seed as frequently as good seed can be got. It is the alpine character of the home of many of the Columbines which makes the culture of some of the lovely kinds so uncertain, and which causes them to thrive so well in the north of Scotland when they fail in our ordinary dry garden borders.

To those familiar with the vigour of our common garden Columbine it must appear strange that there should be any difficulty in the cultivation of the various species, and yet no plants are more capricious; take, for instance, the charming A. glandulosa, grown like a weed at Forres, in Scotland, and which is so short-lived and unsatisfactory in most gardens. Nor is this species an exception; it is characteristic of all the mountain species. Let us for a moment examine the conditions under which they naturally grow, and possibly we may get some clue to those conditions essential to success. Their natural habitat is often on the banks of mountain streams and moist slopes or ledges, where, on deposits of gradually accumulated rich alluvial soil, their roots find the special nourishment they require with perfect drainage; and no doubt the shelter of their position, supplemented by the overhanging branches and adjacent vegetation, helps to protect their young spring growth, as on its protection hinges the vigour of the summer bloom. Mr. Whittaker, of Mosely, near Derby, has been very successful with both A. glandulosa and the blue variety of A. leptoceras, and he told Mr. Niven that he grows them in a thoroughly drained, deep, rich, alluvial loam soil; the same were the conditions of Mr. Grigor's success.

Mr. Brockbank speaks hopefully of growing the finer kinds from seed. He says, "I attribute failures to plants sent by nurserymen in very small pots. I believe it will be found that you can never get up a good stock of Aquilegias by purchase. The proper way is to grow your own from seed. Sow in shallow wooden trays, or in pots, and grow the plants on carefully in a cold frame. When the seedlings are sufficiently large, prick them out into the places wherein you wish them to grow—some in pots and some in the garden—and plant them in various situations, here in the shade and there in the open, so as to have as many chances of success with them as possible. We always plant three plants in a triangle, 4 in. apart, so that any group can readily be taken up and potted if we wish it. Once planted, leave them alone ever afterwards, or if you move them, take up a large ball of earth with them, so as not to loosen the soil about the roots more than can be helped. When the plants have flowered and the seed has ripened, my practice is to gather some for future sowing and to scatter the rest around the plant, raking the soil lightly first, and shaking the seed out of the pods every three or four days. From the seed thus scattered young plants come up by hundreds, often as thick as a mat, and may be transplanted, when suitably grown, into proper situations. In this way we have here abundance of Columbines, and amongst these plenty of A. glandulosa self-sown, and as strong and hardy as any." Further details as to culture and position will be found under the various more important kinds.

Mr. J. C. Niven suggests that all the Columbines, except the common one, should be looked upon as biennials rather than good persistent perennials. The seeds should be sown early in spring, and the young plants pricked out into pans or into an old garden frame as soon as they are fit to handle, removing them early in August to their permanent positions; select a cloudy day for the work, and give them a little artificial shading for a few days. Carry out the same process year after year, the old plants being discarded after flowering. Any attempt at dividing the old roots is usually attended with a very small amount of success. There are, however, instances, especially on light soils and hilly districts, where several of them remain good for years.

COLUMBINES IN POTS. — According to a writer in the *Field*, few who have never grown these and other similar hardy plants in pots can realise how fresh and beautiful they are in spring in a cool house. They may be raised from seed when ripe; and although the seedlings may vary a little in colour and habit, yet that scarcely detracts from their value. Sow thinly in light sandy soil, place the seed pots or pans in a close frame, prick off when large enough, and, finally, pot into single pots and shift on during the summer into 6-in. pots, standing the pots on a coal-ash bed—or, better still, plunging them in the ashes to save watering. At the approach of cold weather move them into a cold frame, and if the frost is very severe, lay some litter over the glass to keep the frost from breaking the pots. They are not near so much trouble to grow as some things cultivated in pots for decorating the cool greenhouse and to produce flowers for cutting at this season, and they are very beautiful for both purposes.

A. alpina *(Alpine Columbine)*.—This plant, widely distributed over the higher parts of the Alps of Europe, is a good addition to the choice collection of alpine plants. The stems rise from less than 1 ft. to more than 2 ft. high, bearing showy blue flowers, and leaves deeply divided into linear lobes. There is a lovely variety with a white centre to the flower, which, in consequence of its exquisite tones of colour, is certain to be preferred, and many will say they have not got the "true" plant if they possess only the variety with blue flowers. It does not require any very particular care in culture, but should have a place among the taller ornaments of the rockwork, and be planted in a rather moist and sheltered, but not shady, spot in deep sandy loam or peat. It may be increased by seed or division. In moist districts, and in good free soil, it will prove a first-class border plant. Distinguished from A. vulgaris by the stamens being longer than the petals and by its larger flowers.

A. californica *(Californian Columbine)*.—One of the strongest growers of the American species. The tendency of the plant is to produce one bold woody stem, which under favourable conditions will rise to the height of 3 ft.; the sepals are orange-coloured and blunt-pointed, being closely adpressed to the petals, which are also blunt; they give one the idea that they had been trimmed round with a pair of scissors; hence the appropriateness of one of the specific terms, truncata. The spurs are long, bright orange, more attenuated than in Skinner's Columbine, but to appreciate the full beauty of the flower it must be turned up from its naturally pendent position; then the beautiful shell-like arrangement of the petals becomes at once visible, the bright yellow marginal line gradually shading off into deep orange. The seeds of this species should be carefully looked after, as having once blossomed the old plant is liable to perish. I have never been disappointed with the seedlings diverging from their parent type in character.—J. C. N. *Syn.*—A. eximia. A. truncata.—This plant thrives best on a deep sandy loam and moist.

A. canadensis *(Canadian Columbine)*.—This was once our only New World Columbine, having been introduced from Virginia by the younger Tradescant. It may be taken as the type of the scarlet-orange and yellow group. The flowers are smaller than the Western American kinds; this, however, is amply compensated for by the brilliancy of the scarlet colour of the sepals and the erect somewhat capitate spurs, and the bright yellow of the petals. The true A. canadensis is a slender grower, scarcely exceeding 1 ft. in height, with sharply-notched irregularly-ternate leaves. As seen in cultivation it is often a cross, with an increased vigour of growth nd a decreased brilliancy of colour Easily raised from seed. It is not so valuable since the introduction of the nobler American species, but it is always a free grower. There is a yellow form. It is a plant for borders or the shrubbery, for placing here and there among dwarf shrubs and plants in the rougher parts of the rock garden, but cannot be included among the very best species. Writing of this species, Mr. Falconer says: "To see it at its best you should see it among the rocks. The Canada Columbine grows in abundance in our woods and always in high rocky places; there it springs from the narrowest chink a little bush of leaves and flowers, or maybe in an earthy mat upon a rock you find a colony of Columbines, Virginian Saxifrages, and pale Corydalis; they usually grow together."

A. chrysantha *(Golden Columbine).* This is a very tall, vigorous, and beautiful species, lasting on many soils as a good perennial where the other species perish. This plant was at first by persons who look at herbarium distinctions only erroneously supposed to be a variety of the Rocky Mountain Columbine, and named such by Torrey and Gray. After cultivating the plant, however, for several years, and comparing it in a living state with the Rocky Mountain Columbine (A. cœrulea), Dr. Gray described it as a new species, A. chrysantha. It is one of those cases in which other than purely botanical characters have weight; but it also differs in the flowers being not nearly so much distended as in A. cœrulea, and the plant is far more robust. The plant comes from a different geographical range, grows taller, flowers nearly a month later, and blooms for two months continuously; these peculiarities, added to its full yellow colour, seem to warrant it to rank as a species. Like the Rocky Mountain Columbine, it has a very long and slender spur, often over 2 in. in length. It is perfectly hardy, more so than the Rocky Mountain species. It thrives even on the stiff clay soils north of London and enjoys wet, though it is none the less free in more happy situations. It comes true from seed, which is most safely raised under glass, and pricked out carefully when young. Attaining a height of 4 ft. under good culture, it becomes an important plant for the centre of a bed of the finer perennials or for a group in the properly arranged mixed border. Should seedlings from it prove crossed with inferior kinds, seed must be obtained from wild plants, which cannot be difficult through the American houses. It would be a great pity if such a distinct, beautiful, and hardy plant should degenerate in our gardens.

A. cœrulea *(Rocky Mountain Columbine).*—Beautiful as it is distinct, the spurs of the flower almost as slender as a thread, a couple of inches long, with a tendency to twist round each other, and with green tips. But it is in the blue and white erect flower that the beauty lies, the effect being even better than in the blue and white form of the alpine Columbine. It is a hardy herbaceous plant, flowers rather early in summer, continuing a long time in flower. I have seen it flowering freely on light soil in an exposed spot in Suffolk so late as September. It grows about from 12 in. to 15 in. high, and is worthy of the choicest position on the rock garden, and is suitable for the front margin of the choice mixed border, where the soil is sandy and deep, and not too wet in winter. Unlike the Golden Columbine, it is not a true perennial on many soils, though a better report in this respect comes from the cool hill gardens. To get strong healthy plants that will flower freely, seeds of this kind should be sown annually, and treated after the manner of biennials, as it rarely does well after standing the second year, and in many cases dies out altogether at or before that time. The flowers are, however, so lovely and so useful for cutting, that it is deserving of any amount of trouble and attention to have it in good condition, a result which can only be attained by treating it in the manner just indicated. All the Columbines delight in a deep rich sandy soil where they can find plenty of moisture below for the roots, and as they make their growth early, the friendly shelter of shrubs or rock to keep off cold cutting winds and frosts is of use, if not too near to rob them or restrict their root room.

This is one of the many good plants which deserve a home in the nursery department, so to say. It deserves a choice little bed to itself, from which its lovely flowers could be gathered in abundance for cutting and plants obtained for the rock garden or choice borders. A coating of 2 in. of half rotten leaves or other convenient material in summer would assist the bloom. The seed is best sown as soon as may be after it is ripe, in cool frames near the glass, or in rough boxes in cool frames. With abundance of fresh seed there will be no difficulty in raising it in fine beds of soil in the open air, protecting the beds from birds or slugs, but the seed is usually too precious to risk in the open air.

What is supposed to be a white variety of this plant is sometimes called A. leptoceras, which was indeed the first name.

"M.," writing from Utah, says: "Some plants of this species seen in Utah seem to belong to a distinct variety; their colour is not blue, or blue and white, but pure white or yellowish-white. They were flowering in great quantity 10,000 ft. above the sea wherever any tiny stream trickled down the mountain slopes, and the flowers at a little distance reminded one more of those of Eucharis amazonica than anything else. The plant grows in handsome tufts 2 ft. or 3 ft. high, the flowers large and broad, and the spurs very long (2 in. at least), with a rounded knob at the top. This is probably a distinct plant, and well worth the attention of cultivators. The seed should be easily obtainable in quantity from America. Its treatment, so far as known, is the same as that for the blue form.

A. glandulosa *(Altaian Mountain Columbine).*—This is a very beautiful species with handsome blue and white flowers and a tufted habit. Flowers in early summer—a fine blue, with the tips of the petals creamy-white, the spur curved backwards towards the stalk, the sepals dark blue, large, and nearly oval, with a long footstalk. The upper part of the stem is covered with glandular hairs. A native of the Altai Mountains, and one of the most desirable kinds for the rock garden or the select border, in well-drained deep sandy soil. Increased by seed and by very careful division of the fleshy roots, when the plant is in full leaf. If divided when it is at rest, the roots are almost certain to perish, at least on cold soils.

The Forres Nurseries, in Morayshire, have long been famous for the successful growth of this plant; it has no special care there, and there is no trade secret about the treatment, which is entirely in the open air. The soil is described as "a rich mellow earth, partaking a little of bog or peat earth, and rather cool and moist than otherwise." It flowers the year after sowing, and when full grown is impatient of removal, like most Columbines, but if not transplanted when more than two years old, it continues to flower for at least five or six years, sometimes for more. Those who can get true seed of this fine plant will do well to raise it with care, and plant out when very young into well-prepared beds of moist, deep peaty or sandy soil, putting some of the plants in a northern or cool position. It would be well also to sow some seeds where the plants are to remain, and in various other ways to try and overcome the difficulty which has hitherto attended the culture of this lovely plant. The seeds of other Columbines have a bright perisperm, while those of this species are unburnished, arising from little corrugated markings with which the microscope shows them to be covered.

In many cases an inferior plant bears the name glandulosa. Mr. Brockbank says: I have referred to the original specimen of A. glandulosa, sent by Prof. Regel, of the St. Petersburg Botanic Gardens, from the Altai Mountains. It is a different plant from the A. glandulosa jucunda, being more than twice as tall and in every way more robust. The specimen at Kew is nearly 1½ times the height of the large folio paper in which it is preserved, and the flower measured 4½ in. in diameter. The plants in Kew Gardens are not this variety—the true variety—of A. glandulosa, and, as far as I know, it is not to be found with any of our nurserymen.

A. Skinneri (*Skinner's Columbine*).—This is a distinct and elegant kind. The flowers are on long slender pedicels, the sepals being greenish coloured and lanceolate; the petals are small and yellow; the spurs are nearly 2 in. long, of a bright orange-red, and attenuated into a slightly incurved club-shaped extremity; the leaves are very glaucous, their divisions being sharply incised; the flower-stems 18 in. to 2 ft. high. Though coming from so far south as Guatemala, owing to the fact that it is met with in the higher mountain districts, it is nearly, if not altogether, hardy, and should be more frequently cultivated than it is. Here, again, crossing steps in and too frequently mars its beauty. While the name may be often seen, the true plant is rare, nor are the conditions that insure its perfect development well known if they exist with us. It is a late bloomer.

A. viridiflora.—As a rule green flowers are not very much admired, but this Columbine is an exception; the Sage green of the flower and delicate tint of the leaf form a striking contrast. Out of doors in the border the plant may not be noticed, but if a flowering spray or two is cut and placed in a small glass its great beauty of form and colour, too, are soon recognised. There is a variety of it known as A. atropurpurea. The sepals are green, but the petals a deep chocolate. The plant is a vigorous grower, a native of Siberia, and is synonymous with Fischer's A. dahurica. The green Columbine is a plant for a quiet corner in a bed of shrubs or any other place in which it will not be called upon for a blaze of colour. We also think it well deserves a place in the nursery or cutting beds of hardy flowers. It has a delicate and exquisite fragrance as good in its way as the singular beauty of form of the flower when closely examined and as the novelty of the Sage green of the bloom. It is easily raised from seed.

A. vulgaris (*Common Columbine*).—A familiar occupant of nearly every cottage garden. There are many good and distinct forms in various colours, including various double kinds, flowering from May till towards the end of summer. The common Columbine grows with a vigour, and increases itself by means of seed with a persistency and a power of variation that is quite surprising. Whether it is a true native is doubtful, but, be that as it may, it has become thoroughly at home, and no one who has once seen it wild will readily forget the combination of grace and beauty which it presents. In order to stimulate those who possess extensive wooded estates to its cultivation or establishment along the margins of drives, where shelter is afforded, Mr. Niven states that at Broughton Woods, in North Lincolnshire, this Columbine raises its stately erect stems to a height of 3 ft.; and in the month of May, when the Lily of the Valley, which grows naturally by acres there, adds its delicious perfume, the charm of a walk there, especially after a gentle shower of rain, is very great indeed. But, however valuable for the wild garden, the many forms of the common Columbine are most valuable plants for gardens in which it is worth while now and then raising a batch of them from fresh seed of a good mixed strain. It would be most desirable also to select and fix varieties of the common Columbine of good distinct colours. One may often see a variety of this plant nearly as handsome as any of the choicer kinds. There

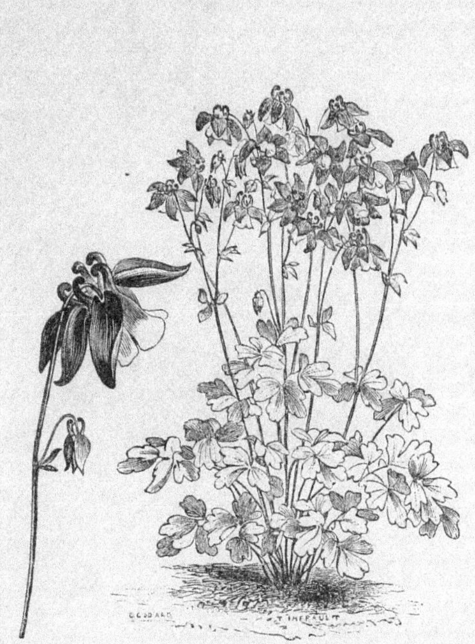

Aquilegia olympica (*Olympic Columbine*).

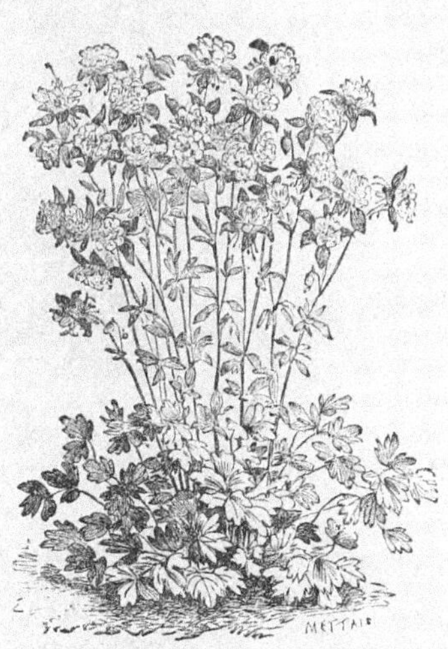

Aquilegia cœrulea fl.-pl. (*Double Rocky Mountain Columbine*).

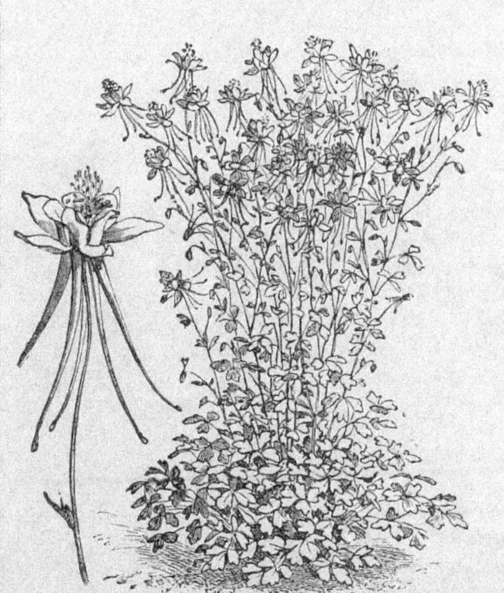

Aquilegia chrysantha (*Golden Columbine*).

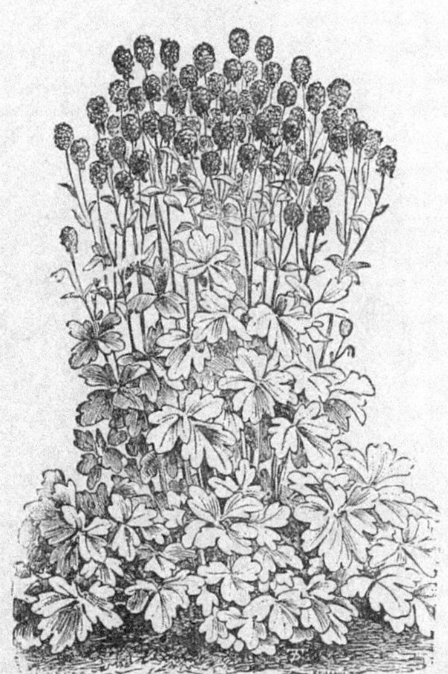

Aquilegia sibirica (*Siberian Columbine*).

[*To face page* **32.**

PLATE XXVI.

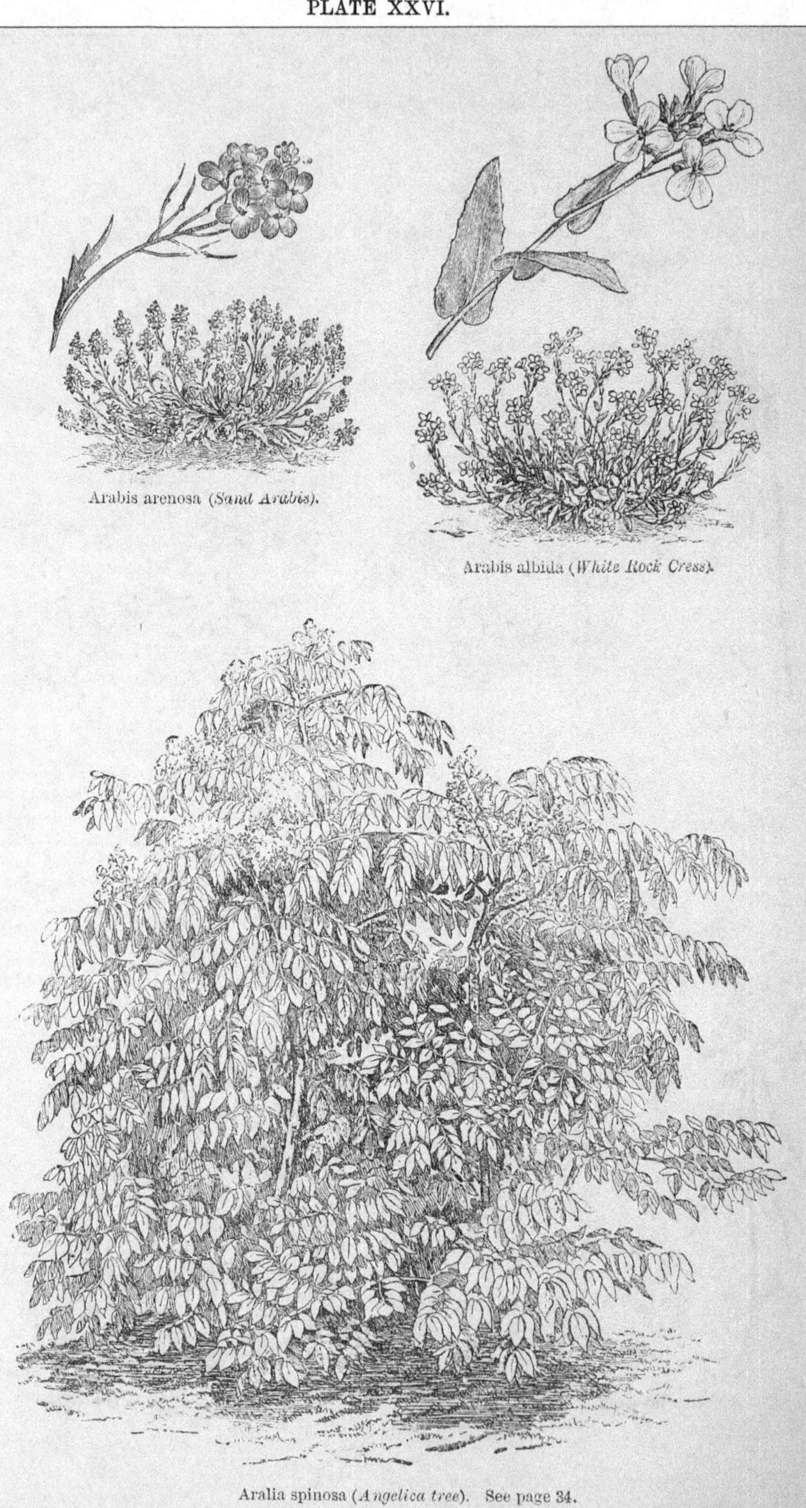

Arabis arenosa (*Sand Arabis*).

Arabis albida (*White Rock Cress*).

Aralia spinosa (*Angelica tree*). See page 34.

[*To face page* 33.

is a bold single white form which is a most distinct and handsome plant. A, Vervæneana is a form with mottled leaves.

The varieties of our common Columbine and some hybrid forms are so free and hardy, that they may well be used in the wilder and more picturesque parts of large pleasure grounds or parks in the long Grass, by streams, or in copses, among Foxgloves, Geraniums, or long Grasses. To establish them the ground should be well dug if the vegetation is dense, and the seed sown on the spot, or raised in beds and the seedlings transplanted. Where bare patches occur from any cause, and where the seedlings have a chance of coming up without being strangled by other plants, seed may be scattered with the hope of establishing some tufts or colonies in wild places. The time to sow the seed is soon after it is ripe.

A. Witmanniana.—A species quite distinct from A. glandulosa, but often sold for it. It is one of the most vigorous in growth and free in flowering of all the Columbines in cultivation. A plant of it has grown in the same spot in my garden for ten years, and produces five or six stems more than 3 ft. high, each bearing twenty to thirty flowers. The flowers have dark purple sepals and a white corolla, but are far inferior in every way to those of A. glandulosa; in fact, it is less ornamental than many seedling varieties of A. vulgaris, which I have raised, but which, unfortunately, I can neither divide nor reproduce true from their own seed.—B.

It were easy to say more of other species, but among the preceding the cultivator will be certain to find the highest beauty of form or colour and the fullest vigour which the genus is known to possess. Collectors will very properly seek other species, but for all gardening and artistic purposes the preceding are the best so far as the family is now known. For the choice rock garden some of the dwarfer and more delicate alpine species may be desirable. A. fragrans is praised for its scent, but the green Columbine is very good in this way.

Arabis *(Rock Cress).*—An extensive family of hill plants, few of which are grown, though some are worth a place.

A. albida *(White Rock Cress).*—Through long years of neglect of nearly all sorts of dwarf hardy plants this, the white Arabis of our gardens, has held its own well—the most popular plant in our gardens, and in the barrow of every London flower-hawker in the spring. It will grow in any soil or situation, flourishing far into our cities as well as in the open country, where its profuse sheets of snowy bloom may expand in early spring. By seed, or division, or cuttings, it is as easily increased as a native weed, and is a valuable ornament of the mixed border, the spring garden, the rockwork, and for naturalisation in wild and bare rocky spots. In the rock garden it is well fitted for falling over the ledges of rocks; it may also be used as an edging to clumps of shrubs, though it is in better taste to associate it in such positions with groups of plants like the Aubrietias, the rock Alyssum, and other easily grown alpine flowers that bloom early in the year. A. albida is closely allied to the Alpine Rock Cress (A. alpina), so widely distributed on the Alps, and by some would be considered a sub-species of that plant, but it is sufficiently distinct, and by far the best kind. There is a variegated variety in cultivation, known by the name of Arabis albida variegata, which is useful as an edging plant both in spring and summer flower gardens. It is the dwarfest and whitest of the variegated Rock Cresses that are grown under the name of A. albida variegata. The yellower and stronger variety, frequently called A. albida variegata, and which is the best for general purposes, is a form of Arabis crispata, of which the ordinary green form is not worthy of cultivation.

A. blepharophylla *(Rosy Rock Cress).*—This is not unlike the white Arabis, but the flowers are of a rosy purple. It varies a good deal, but there is no difficulty in selecting a strain of the deepest rose. It is impossible to have anything more effective than healthy tufts of this plant in the month of April. It is best raised every year from seed, which, like most Cruciferous plants, it yields freely. In mild districts and on light soils plants should be tried out every winter. It does not appear to have answered well after many trials, and as a rock plant its annual and tender character are against it. A native of North America; easily increased by seed.

Among other kinds of Arabis, A. procurrens, a dwarf spreading kind, with shining leaves and small whitish flowers, is often grown, but is not worthy of culture. There is, however, a brilliantly variegated form of it (A. procurrens variegata) which is worthy of a place in a collection of silvery and variegated hardy plants. The prettiest of the variegated Rock Cresses is A. lucida variegata. It forms very neat and effective edgings in winter, spring, and summer flower gardens, from its striking and distinct character is effective on rockwork, and thrives best and is easiest to increase by division in open, sandy, and yet moist soil. The best time to divide it is early in autumn, April, or very early in May. It need scarcely be added that the flowers should be removed when they appear. The green form of this (A. lucida) has a fine effect when associated with such plants as the Echeverias, and, if judiciously used, would give character and, perhaps, add a new feature to this style of bedding. Its habit is good, and it makes a beautiful edging—plant for plant—with Echeveria secunda glauca. A. purpurea, an interesting species for botanical, large, or curious collections, and bearing pale bluish and lilac flowers, is not worthy of general cultivation while we possess such brilliant plants as the purple Aubrietias. A. arenosa, from the south of Europe, is a pretty annual kind that may prove useful in the spring garden, and which might be naturalised on old ruins or dry bare banks. A. petræa (Northern Rock Cress) is a neat, sturdy little plant, with pure white flowers, a native of some of the higher Scotch mountains, and very rarely seen in cultivation, but when well developed on a moist yet well-exposed spot on rockwork is very pretty.

Aralia.—These embrace many plants of very diverse aspects, and few that are fitted for the open air in our climate; but in the case of A. canescens and its relative (A. spinosa), the Angelica tree of North America, we have subjects which thrive perfectly well in our gardens, and which in the size and beauty of their leaves are far before many "foliage-plants" carefully cultivated in hothouses at a perpetual expense. These and even the tender kinds may find a place in the flower garden, and the vigorous herbaceous in the woodland. By cutting the shrubby species down occasionally they may be kept in the condition of fine-leaved plants, like the Ailantus treated in a similar manner.

A. papyrifera (*Chinese Rice-paper plant*).—This, though a native of the hot island of Formosa, flourishes vigorously with us in the summer months, and is one of the most valuable plants in its way, being useful for the greenhouse in winter and the flower garden in summer. It is handsome in leaf and free in growth, though to do well it must, like all the large-leaved things, be protected from cutting winds. If this Aralia be planted in a dwarf and young state, it is likely to give more satisfaction than if planted out when old and tall. The leaves spread widely out near the ground, and then it is very ornamental through the summer. Prefer, therefore, dwarf stocky plants when planting it in early summer. It should have rich, deep soil and plenty of water during the hot summer months. In Battersea Park a bed of A. papyrifera, 13 ft. in diameter, attained a height of 5 ft., from cuttings struck in the spring. The plants were left out all the winter, and, though killed to the ground, sent up many suckers the following spring. It is so fine in form of leaf and habit that no one will be at a loss to place it where it will have a good effect. It is not so useful northwards in the open air as about London and in the south. "Its stem being nearly all pith does not strike readily; therefore a stock of it cannot be got quickly in that way, but if the roots near the stem are examined they will be found to be fleshy. When the plant is taken up, let each of these be cut into pieces about 1 in. long; insert them in light soil, letting the whole be covered, and place them on a shelf near the glass in a temperature of 80°. They will, in due time, throw out shoots, which should be left until they are 3 in. in length, when they may be potted, leaving them in heat, say of 70°, until established. They should then be removed to a colder temperature, and thence to a frame to harden off. In the south of England this plant is almost hardy, and may be wintered in a cold frame, but it should be kept dry."—J.

A. Sieboldi.—A shrubby species, distinct, with fleshy, dark green leaves. It is usually treated as a greenhouse plant, but is hardy and makes a very ornamental plant on soils with a dry porous bottom. It grows remarkably well in the dwelling-house; in fact, it is one of the very few plants of like character that will develop their leaves therein in winter. Not difficult to obtain, it may be used with advantage in the flower garden or pleasure-ground. It would form striking isolated specimens on the turf, and is also fitted for association with fine-leaved plants. *Syn.*—Fatsia japonica.

A. californica, edulis, and racemosa are vigorous herbaceous kinds of very little value, except where a large number of such perennials are desired.

Arctomecon californicum. — A North American plant of the Poppy family, but of which little is known in cultivation.

Arctostaphylos Uva-ursi (*Bearberry*).—A neat evergreen mountain shrub, 1 ft. high, often less, sometimes associated with rock or alpine plants. Has small rose flowers in early summer and red berries in autumn. Abundant in hilly places in Europe and North America. Borders and rough rock-work, especially for hanging over the brows of rocks. It grows in any soil, but prefers a moist one. Increased by division.

Arctotis.—Showy half-hardy perennials of the Daisy Order, natives of S. Africa and Abyssinia. A. breviscapa, orange and black; A. grandiflora, yellow; A. repens, yellow; are useful Gazania-like summer bedding plants. They grow freely from cuttings put in in July. Mr. H. Harpur Crewe gives high praise to Arctotis grandiflora. "I have grown it for some years as a summer bedding plant in a mixed herbaceous border, where it is most effective. I procured it originally from Tresco Abbey, in the Scilly Isles, where it is perfectly hardy, and forms one of the principal ornaments of the beautiful gardens of my friend, Mr. Dorrien Smith. Why does not some enterprising nurseryman re-introduce the other beautiful species of this showy family which our forefathers used to grow? It grows freely from cuttings put in in July; if inserted later they are apt to damp off."

Arenaria (*Sandwort*).—A very numerous genus, of vast distribution over northern and alpine ranges, and also in temperate countries. Comparatively few species are in cultivation compared to the great number known, and these are dwarf plants, easy to grow, and suited mainly for the rock garden.

A. balearica (*Balearic Sandwort*).—A pretty little plant, which coats the face of rocks and stones with the dwarfest verdure, and then scatters over the green mantle countless white starry flowers. Plant firmly in any common soil near the stones or rocks it is desired to cover, and it will soon approach and begin to clothe them. Flowers in spring and continuously, and is readily increased by division or seeds, and quite easy to grow. On cold ones it perishes in winter, but its

PLATE XXVII.

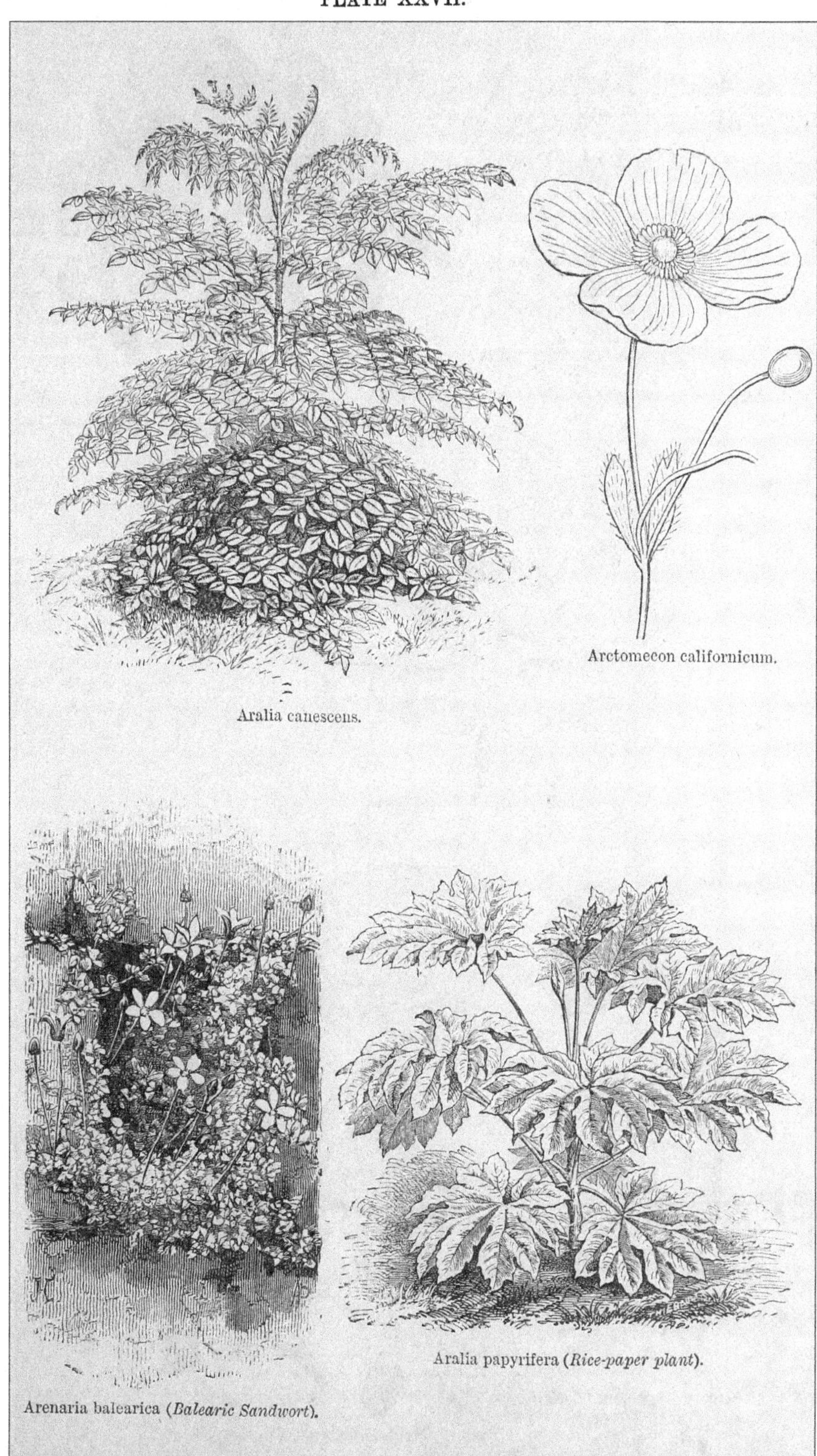

Aralia canescens.

Arctomecon californicum.

Arenaria balearica (*Balearic Sandwort*).

Aralia papyrifera (*Rice-paper plant*).

[*To face page* 34.

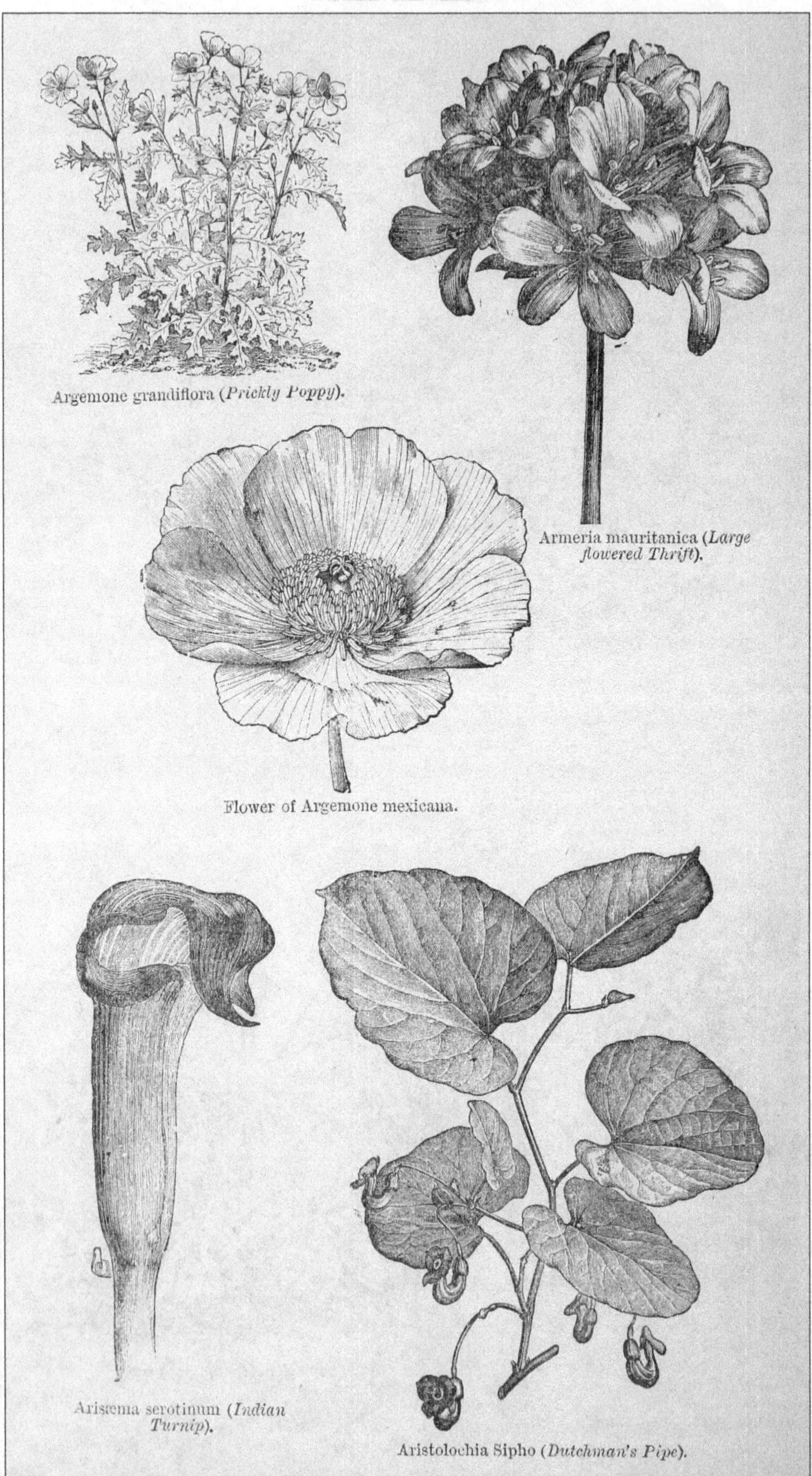

Argemone grandiflora (*Prickly Poppy*).

Armeria mauritanica (*Large flowered Thrift*).

Flower of Argemone mexicana.

Arisæma serotinum (*Indian Turnip*).

Aristolochia Sipho (*Dutchman's Pipe*).

[*To face page* 35.

true home is on the rockwork. It is easily known at any season by its dense tufted cushions of very small leaves. Corsica.

A. montana *(Mountain Sandwort)*.—A very ornamental rock plant, having the habit of a Cerastium, and numerous fine pure white and large flowers. It is the best of the large Sandworts, and should be in every collection of herbaceous or alpine plants. On rockwork it would be well to plant it where its shoots might fall over the face of a rock, giving it any kind of light soil. France. Easily raised from seed.

A. purpurascens *(Purplish Sandwort)*.—Distinguished from other cultivated kinds by its purplish flowers, produced in abundance on a dwarf densely tufted mass of smooth, narrow, oval pointed leaves. It grows plentifully over all the Pyrenean chain, is perfectly hardy, and, like the other kinds, increased by seed or division. It should be associated on the rockwork with the smallest of its brethren, or with dwarf Saxifrages and other slow-growing plants.

Of other Arenarias in cultivation, the best are A. ciliata, a rare British plant; A. verna, A. triflora, a neat species; A. laricifolia, A. graminifolia, and A. tetraquetra. These, however, are scarcely worth growing except in botanical collections.

Arethusa bulbosa.—A beautiful North American hardy Orchid. It grows naturally in wet meadows or bog land, and blossoms in May and June, each plant bearing a bright rose-purple flower that shows conspicuously in its sombre bed of Sphagnum and Cranberry Grass, Sedge, and the like. The little bulbs grow in a mossy mat formed by the roots and decaying herbage of other plants and moss. In cultivation it requires the same treatment as the Pogonia. Care must be taken to get the leaf as well matured as possible. Coming as it does after the flower, the leaf is apt to be neglected. A good outdoor position for growing these Orchids is a shady, moist spot having a northern exposure, using for soil a mixture of well rotted manure and Sphagnum. During winter, the bed in which it grows is better to be protected with some rough mulch, for it appears that when taken from its moist home and placed in higher ground, it loses its extremely hardy nature. It should not be treated as a tender plant, or failure is sure to be the result. It is a plant for the select bog garden or for a moist or irrigated flat bed near the rock garden.—F.

Aretia Vitaliana *(Androsace)*.

Argemone *(Prickly Poppy)*.—Handsome Poppy-like plants, said to be perennial, but usually perishing on moist soils after the first year, and therefore requiring to be frequently raised from seed. Coming from the warmer parts of California and Mexico, even there growing on dry hill-sides and warm valleys, their aptness to perish here may be easily understood. They usually grow about 2 ft. high, and bear large white flowers sometimes 4 in. across with a bunch of yellow stamens in the centre. May be easily raised from seed in a warm frame. They should have a warm sandy loam, and be associated with the choicest annual and biennial plants. The kinds mostly grown are A. mexicana, A. grandiflora, and A. hispida, which are so nearly alike in habit and requirements as to render a separate description for each needless. They might be used with good effect in bold mixtures, and as groups or masses with slenderer and taller plants among them.

Arisæma *(Indian Turnip)*.—A family of N. American plants belonging to the Arum Order, and though curious for the botanical garden, of little merit for the ordinary garden. In A. triphyllum the spathe is curiously marked with purple and white stripes. A. Dracontium is the "Green Dragon," and A. triphyllum is known in America by the common name of Jack in the Pulpit. They are hardy and grow easily on the fringes of beds of shrubs.

Aristolochia Sipho *(Dutchman's Pipe)*.—This well-known large-leaved plant has an excellent effect where large and distinct foliage is desired. Generally it is used as a wall plant, but it is far finer when used to cover bowers or any like structure, or clamber up trees or over stumps. A. tomentosa is smaller, but distinct in tone of green, well worthy of a place, and to be employed in like manner. N. America. These plants grow in ordinary garden soil with freedom. The hardy non-climbing herbaceous plants of the genus have no claim to cultivation. The genus is a large one, mainly tropical, but some of which go into northern countries. Easily propagated from cuttings.

Armeria *(Thrift, Sea Pink)*.—A pretty little family of the Statice Order, most of which are worthy of culture. Best known is the common A. vulgaris (common Thrift). This inhabitant of our sea-shores, and also of the tops of the Scotch mountains and the Alps of Europe, is very pretty with its soft lilac or white flowers springing from dense cushions of grass-like leaves; but it is the deep rosy form of it, which is rarely seen wild, that best deserves cultivation in gardens. It is like the common Thrift in all respects but the colour of the flowers, which are of a showy rose. It is useful for the spring garden, for covering bare banks or borders in shrubberies, for making edgings, and for the rock garden. Easily propagated by division, and as old and large plants do not bloom so long or so continuously as younger ones, occasional replanting is desirable. In addition to the white and the old dark red variety there are several others bearing names—Crimson King, Grandiflora, Pygmæa, and Pink Beauty.

A. cæspitosa.—This is a pretty rose-coloured species from the south of Europe, where it is found growing at an altitude of from 5000 ft. to 8000 ft. above the sea level. Its flower-heads each measure from ¾ in. to 1 in. in diameter, and are borne on slender stems from 1 in. to 2 in. high, from June to September. The leaves are narrow and flat on the upper surface, with a slight depression down the centre. They are from ½ in. to ¾ in. long, in dense tufts, with a branching, woody rootstock. A choice rock garden plant, thriving freely in any well drained, rather poor, sandy loam. In wet weather they are liable to damp off between the foliage and the root in rich soil. They are propagated by seed. There are various other alpine species, but the above are their best representatives.

A. cephalotes *(Great Thrift).*—One of the finest perennials in cultivation. It should be in every select mixed border, and on every rock garden among the taller and stronger plants. It comes from North Africa and Southern Europe, and though hardy on free and well-drained soils, occasionally perishes during a very severe winter, especially on cold soils; it should therefore have well drained, deep, and good sandy loam. It is known under various names—Armeria formosa, A. latifolia, A. mauritanica, A. Pseudo-Armeria, Statice lusitanica, and Statice Pseudo-Armeria. Easily raised from seed; and as it is not so easily increased by division, it is a good plan to sow a little of it every year. Varies a little when raised from seed, but all the forms are worthy of cultivation. This species and its forms have flowers very much larger than the common Thrift.

A. setacea.—A good alpine species. Its little globose heads of pink flowers are produced so plentifully as almost to conceal the other portions of the plant. The flower-stems vary from 1 in. to 3 in. high. This and A. juncea are found on barren, stony mounds, on elevated table-land, in the south of France. They have been cultivated in the nurseries at York, and have proved to be quite hardy and of easy culture. They grow freely in sandy or stony earth, either in the open border, on rockwork, or in pots, and their neatness and compactness of habit fit them for association with the choicest of alpine flowers on the rock garden.

Arnebia echioides *(The Prophet Flower).*—The Borage-worts afford us some of the handsomest of border flowers. This, though not so showy as some, is certainly amongst the most remarkable. It grows 1 ft. to 18 in. high, and the flowers, which are of a bright primrose-yellow, have five black spots on the corolla, which gradually fade to a lighter shade, and finally quite disappear. It is perfectly hardy, and succeeds either on the rockery or in a well drained border, and apparently prefers partial shade. It is a native of the Caucasus, Northern Persia, &c., and, though it has been introduced some considerable time, it is still amongst the rarest of our cultivated hardy flowers. Young plants of it flower long and continuously, which adds to their charms. Increased by seeds, which are not freely produced, and by cuttings. A. Griffithi is an annual, tender and though pretty, not so valuable as A. echioides.

Arnica.—One or two of these are sometimes cultivated, but of little value, considering the great number of ornamental plants of the same composite family that are in cultivation. A. montana with orange yellow flowers is occasionally grown in botanical collections.

Aronicum. — This is a genus of few species, some of which are in cultivation, but though striking when in flower, not of the first merit for gardens. A. glaciale (Glacier A.) grows from 6 to 9 in. high; has large yellow flowers, one to a stem. A. scorpioides (Mountain A.) grows about 1 ft. high, and has large orange flowers, one to a stem. Alps of Europe; seed and division.

Artemisia *(Wormwood).*—A very numerous genus occupying a large part of the earth's surface in northern and various arid regions. Though often poor weeds, some have a use in gardens, though seldom for their flowers. A. anethifolia is one of the most graceful herbaceous perennials, as far as habit is concerned, that we know of. It attains 5 ft. in height. A. annua is an exceedingly graceful kind of Wormwood, with tall stems reaching to a height of 5 ft. or 6 ft. in a season; the foliage is small and fine, and the flowers inconspicuous, but arranged in not inelegant panicles. The hue of the plant is a peculiarly fresh and pleasing green, and it forms an elegant object in the centre of a flower-bed or group with plants of like character. A. gracilis is an exceedingly graceful plant, 3 ft. or 4 ft. high, with leaves cut into very fine hair-like segments, having some resemblance to Fennel or other umbelliferous plants with minutely cut leaves and of a deep grass-green, except in the hearts of the shoots, where the young leaves are unfolding, where there is a slight hoary pubescence. Other kinds, such as A. alpina and A. frigida, belong to an alpine group at home on the rock garden; there are many taller herbaceous and half-woody plants of a silvery hue, such as A. Stelleriana, A. cana, A. maritima, and some with handsome Fern-like foliage like A. tanacetifolia. Other species of Artemisia are for the most part not ornamental, scarcely so much so as the common Wormwood and Southernwood. There is, however, what is called an Indian form of the common A. vulgaris, which is a wonderfully graceful plant in its growth, and very tall, 8 ft. Mostly raised from seed and also from cuttings. The taller and more graceful kinds of Wormwood are very effective and distinct among groups of plants of striking habit or graceful foliage. In districts too cold for hardy plants some of them might be of good service, as they would in gardens where a variety or change are sought.

Arrow-head *(Sagittaria sagittifolia).*

Artichoke, Globe *(Cynara Scolymus).*

Arum Lily *(Calla æthiopica).*

Arum *(Cuckoo Pint).*—A large family of plants, mostly from tropical or warmer countries than ours, but some of which are South European, hardy, and of some interest in our gardens, as, for example, the Italian Arum (A. italica), the foliage of which is

PLATE XXIX.

Flower-head of Armeria cephalotes (*Great Thrift*). Natural size.

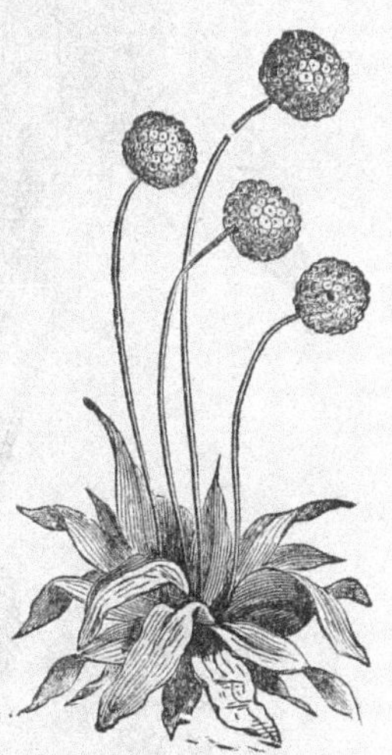

Armeria cephalotes.

Arnebia echioides (*Prophet Flower*).

Arum Dracunculus (*Dragon, or Snake plant*).

[*To face page* 36.

Arum maculatum (*Lords and Ladies, Cuckoo Pint*).

Arum crinitum (*Dragon's Mouth*).

Arum italicum (*Italian Arum*).

Arundo conspicua (*New Zealand Reed*).

[*To face page* 37.

handsome in winter and spring. The old Dragon's Arum (A. Dracunculus) is a curious plant, and still more so is Arum crinitum. They thrive best in warm borders and about the sunny side of garden walls, the Italian Arum thriving more freely. Some nine or ten kinds are found in South Europe, two coming as far north as our own country. Almost contemporaneously with the Primroses, and not unfrequently associated with them, we find the shining green leaves, peering through the Moss or Ivy-clad bank, of the Wild Arum, familiarly known as "Lords and Ladies" and "Cuckoo Pint." Closely related to it comes Arum Arisarum of the South of Europe and A. azoricum of the Azores.

A. crinitum *(Dragon's Mouth).*—The appearance of this plant when in flower is very grotesque from the singular shape of its broad, speckled spathe. The leaves are divided into five or seven deep segments, the central division being much broader than the others, and the leaf-stalks, overlapping each other, form a sort of spurious stem 1 ft. or 14 in. high, marbled and spotted with purplish black. The treatment for this plant is similar to that given for A. Dracunculus; but as it is rather more tender, it will require a little more care and shelter in winter. Warm borders, fringes of shrubberies, or beds of the smaller subtropical plants will suit it best. The appearance of the flower is rather repulsive. In this species the carrion-like smell named occurring in the last becomes strongly pronounced, and doubtless gave rise to Linnæus' old name of A. muscivorum, the smell being sufficiently strong to attract the larger flies in quest of a suitable place wherein to deposit their eggs. It is a strange plant seen in a group of fine-leaved subjects, or holding its awful "blossoms" from out a mass of low shrubs. Increased by division of tubers. S. Europe.

A Dracontium *(Green Dragon Arum)* grows abundantly in the moist and swampy districts of Virginia and New England. The graceful curving of the veins, which is noticeable in all the species belonging to this section, add a special charm to the appearance of the plant; the spathe is greenish coloured. This species is rarely met with in cultivation. A. triphyllum, A. gramineum, A. spirale, A. corsicum, A. tenuifolium, and other species are in cultivation, but not so valuable as the preceding kinds.

A. Dracunculus *(Dragons, Snake Plant)* generally attains a height, when growing vigorously, of from 2 ft. to 3 ft.; the leaves are large; the petioles and stem of a fleshy colour, deeply and irregularly mottled with black, reminding one of the skin of a snake, whence originates its popular name of the Snake Plant; the spathe is of a deep chocolate colour, and large. At certain stages a disagreeable odour is given out by the flower, reminding one of decomposing animal matter; the emission of this odour appears to be quite spasmodic, as at one time it may be sufficiently powerful to cause nausea, and in a few minutes every trace of it will have disappeared. This species is a native of Southern Europe, and forms a handsome border plant, unique in its habit. The Snake Plant loves best a corner to itself at the foot of a south wall in sandy loam. Beyond use in this way it is not important, except as a curious and distinct plant. Many would not care to give place to a plant having so offensive an odour. Division.

A. italicum *(Italian Arum)* is of larger growth than our native Arum; the principal veins are blotched with yellow, giving the leaves a marbled appearance, and as they are produced very early in the season, attaining their full development in the month of March, they form an attractive feature in the flower border. In the autumn, when the leaves have died away, the groups of scarlet berries, supported on foot-stalks 10 in. or 12 in. long, have a very attractive appearance, which they retain for a considerable time. It is a native of Spain, Italy, and Southern Europe. Occasionally we meet in our own woods with a very similar variegated form of A. maculatum, but it is smaller in every respect than its Italian relative, which is a decidedly desirable plant for cultivation. The true use for it in gardens is as a naturalised plant or in the shrubbery or grove, where its handsome leaves will come up bravely in spring. Although it is a very hardy plant, and will thrive almost anywhere in moist soil and a shady position, it will be better to place it in sheltered positions along the sunny fronts of shrubberies, amidst low-spreading evergreens, and in cosy spots about ferneries, to prevent its handsome foliage from being disfigured by wintry winds. One great merit of this is that it may be used to ornament positions in which few other plants will thrive—as, for instance, under trees, groups of shrubs; easily multiplied by division in the end of summer. South of Europe.

Arundinaria *(Bambusa).*

Arundo *(Great Reed).*—An important genus of Grasses, some of them of great value.

A. conspicua *(New Zealand Reed).*—A companion for the Pampas Grass, especially in the western and warmer counties and on good soils. In some very fine deep loams it attains a height of nearly 12 ft., but this is rare. It is well worth growing, even in districts where it does not attain a great development. It comes into flower before the Pampas Grass, and may be considered as a sort of forerunner of that magnificent plant. Suitable for grouping with the finer leaved hardy plants, but making a noble specimen for the lawn, where it grows well. The New Zealand Arundo generally commences blooming early in July, and lasts until the end of October. The Pampas rarely comes out in full flower before November; the Arundo, therefore, while as beautiful when well grown, has a peculiar value in being so early. Grown in tubs in a cool greenhouse or winter garden, this noble Reed forms a striking object, and its silky plumes last in perfection much longer than when produced out of doors. It likes a strong fibrous loamy soil, and a plentiful supply of water nearly all the year round. Seeds or division. It requires careful planting and generally several years' growth before flowering after transplanting.

A. Donax (*Great Reed*).—This great Reed of the south of Europe is a very noble plant on good soils. In the south of England it forms canes 10 ft. high, and has a very distinct and striking aspect. It will grow higher than that if put in a rich deep soil in a favoured locality; and those who so plant clumps of it on the turf in their pleasure-grounds will not be disappointed at the result. It seems much to prefer deep sandy soils to heavy ones. Like all large-leaved plants, it loves shelter. It has suffered much in recent severe winters, and perished in some of our southern gardens where it was finest. But, fine as it is for effect and distinctness, its variegated variety is of more value for the flower garden proper.

A. Donax versicolor.—This will be found pretty hardy in the southern counties, and, considerably north of London, may be saved by a little mound of Cocoa-fibre, sifted coal-ashes, or any like material that may be at hand. In consequence of its effective variegation, it never assumes a large development, like the green or normal form of the species, but keeps dwarf. It is, of course, best suited to warm, free, and good soils, and abhors clay. For a centre to a circular bed nothing can surpass it in the summer and autumn flower garden, while numerous other charming uses may be made of it. Not the least happy of these would be to plant a group of it on the green turf, in a warm spot, near a group of choice shrubs. It is better to leave the plant in the ground, in a permanent position, than to take it up annually. Protect the roots in the winter, whether it be planted in the middle of a flower bed or by itself in a little circle on the Grass. Increased by placing a shoot or a stem in a tank of water, when little plants with roots will soon start from every joint; they should be cut off, potted, and placed in frames, where they will soon become strong enough for planting out.

A. mauritanica.—This is a fine Grass—one might describe it as a stately one were it not for its great relation A. Donax. A. mauritanica is a native of the southern shores of the Mediterranean, hardy in the neighbourhood of Paris, and reaching a height of about 4 ft. Might be grown with a collection of aquatics or Grasses.

A. Phragmites (*Common Reed*).—A native marsh or water plant, 5 ft. or 6 ft. high, bearing when in flower a large, handsome, spreading, purplish panicle. The stems are smooth, simple, very erect, and grow closely together. The plant is only attractive when in flower, as its flat, ribbon-like leaves do not of themselves present any very striking appearance. Useful for the margins of artificial waters, &c., to which it may be brought from its wild haunts. It should, however, if possible, be kept in one spot and not allowed to spread too much. Where it grows wild there is usually no scarcity of it.

Aruncus (*Spiræa Aruncus*).

Asarum (*Asarabacca*).—A small genus of very hardy plants resembling Cyclamens in their foliage, but having little value except as curiosities, and occasionally as wood or shrubbery plants. Asarum canadense is the Canada Snake-root. Bears in spring curious brownish-purple flowers. Root strongly aromatic, like Ginger. Asarum virginicum is the Heart Snake-root. Leaves thick and leathery, with the upper surface mottled with white. Sometimes used as a spice; hence the common name Wild Ginger. They are more or less used in medicine. Asarum caudatum is from Oregon. Much like the others in habit, but the divisions of the flower have long tail-like appendages. Asarum europæum is the Asarabacca. Flowers greenish, about ½ in. long, appearing close to the ground.

Asclepias (*Milk-weed, Silk-weed*).—To this genus belongs a large group of plants confined to the New World, and, with but few exceptions, to the northern part of it. Almost all of them are hardy in this country. They are for the most part ornamental, though not showy, and will be found useful both in the wild garden and shrubbery borders, where they can ramble at pleasure. The following are a few of the most distinct:—

A. Cornuti (*the Common Milk-weed*).—This is a very vigorous kind, 4 ft. in height, bearing nodding umbels of deep purple flowers, which are fragrant. Bees seem to be fond of this plant, and from the fact of its being quite hardy, easily increased, and lasting a considerable time in bloom, it would prove valuable as a bee flower. It is somewhat coarse for the border, but may be worth a place in the wild garden. *Syn.*—A. syriaca.

A. incarnata (*Swamp Milk-weed*).—This attains 3 ft. in height, and has long leaves and leafy stems, bearing umbels of rosy purple flowers in pairs. The variety called pulchra has broader foliage than that of the type. A good water-side plant.

A. purpurascens (*Purple Milk-weed*).—This is a very distinct species, the stems of which are slender, and from 2 ft. to 3 ft. in height, and bearing umbels of bright purple blossoms. (Amœna, Mich.).

A. quadrifolia (*Four-leaved Milk-weed*).—A deliciously fragrant flower, and earliest to blossom of the hardy species of the genus, coming into bloom as it does about June 1. It grows 1 ft. to 2 ft. high, has one or two whorls of four leaves about the middle of the stem, but the other leaves—lower and upper—are in pairs, and terminal heads of lilac-tinged white flowers that are sweet and pretty.—W. F.

PLATE XXXI.

Arundo Donax (*Great Reed*).

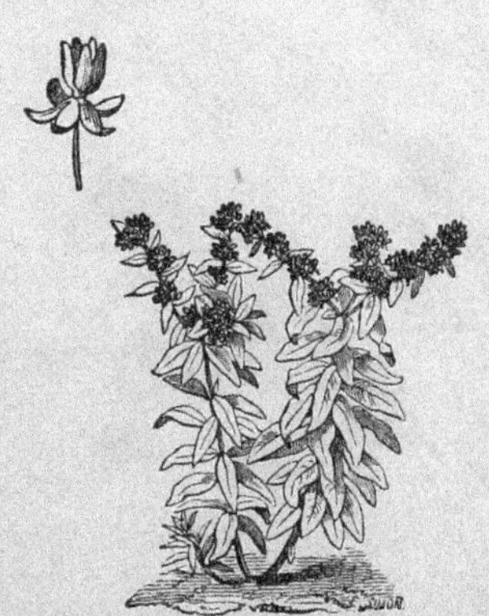

Asclepias tuberosa (*Butterfly Weed*).

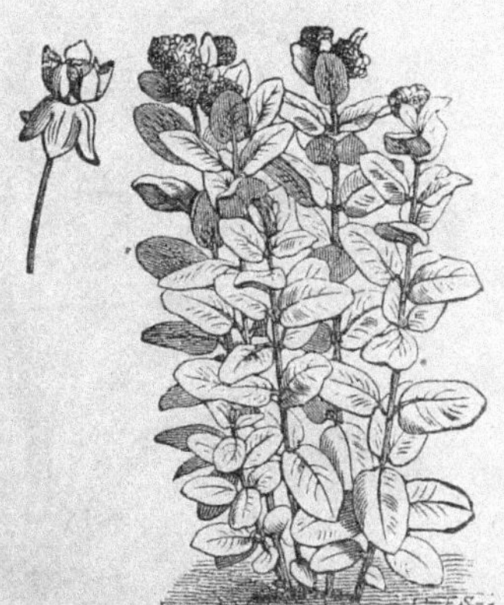

Asclepias Cornuti (*Common Milk Weed*).

[*To face page* 38.

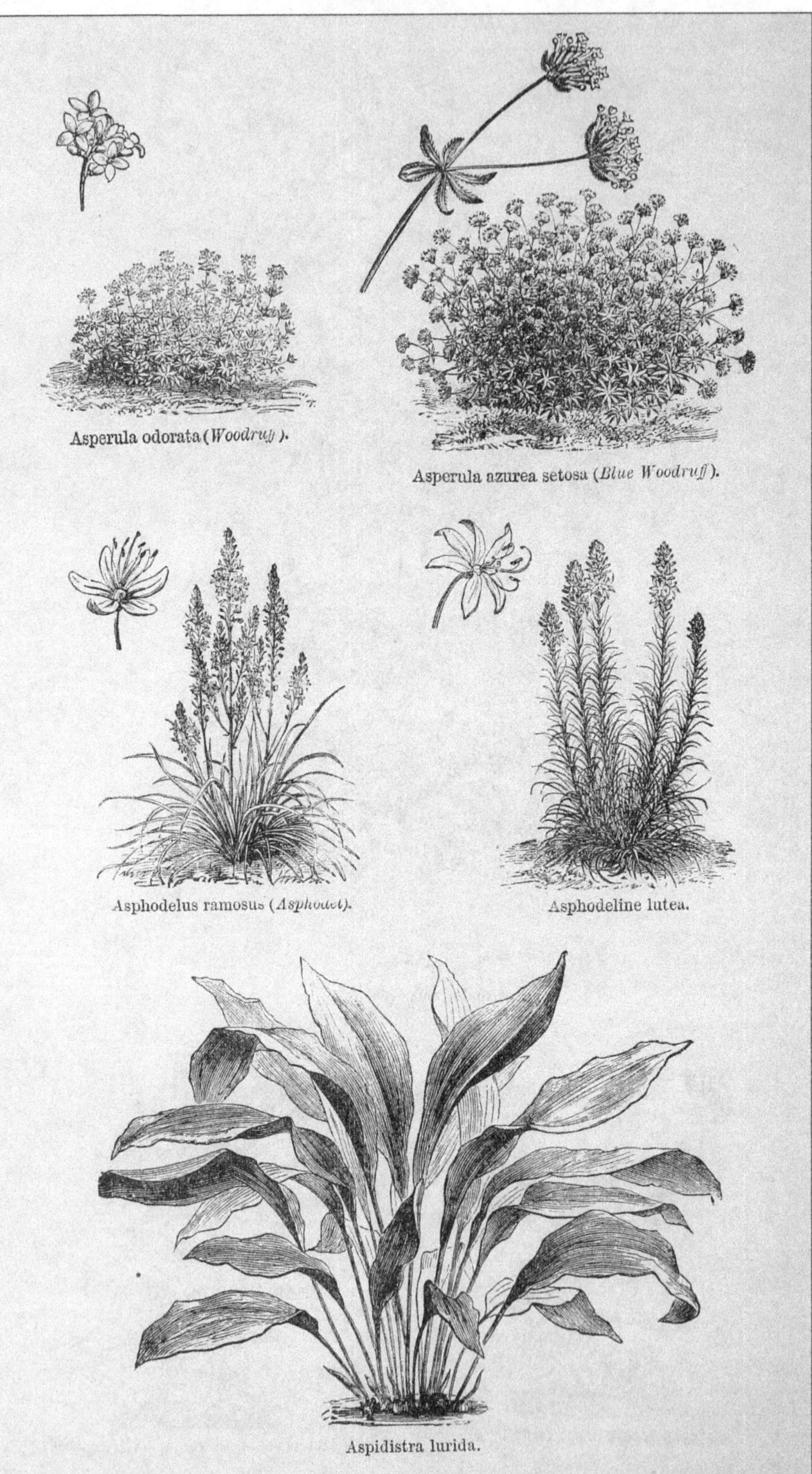

Asperula odorata (*Woodruff*).

Asperula azurea setosa (*Blue Woodruff*).

Asphodelus ramosus (*Asphodel*).

Asphodeline lutea.

Aspidistra lurida.

[*To face page* **39.**

A. rubra *(Red Milk-weed).*—This distinct kind has long, lanceolate, bright green foliage, and stems from 3 ft. to 4 ft. in height; umbels large, deep purplish-red, from two to five on a naked, terminal peduncle. *Syn.*—A. acuminata.

A. Sullivanti is similar to A. Cornuti, but the flowers are larger and deeper in colour, and fragrant.

A. tuberosa *(the Butterfly Weed)* is one of the most beautiful of our autumnal flowers; it is a hardy perennial, having a thick root, and erect leafy stems about 2 ft. in height, crowned with terminal corymbs of bright orange-red flowers. It prefers a warm, sandy soil, and when thoroughly established is a very ornamental plant. It bears seeds occasionally during hot autumns, from which good flowering plants may be obtained in three years, but is mostly increased by division of tubers. A fine border plant.

A. variegata *(Variegated Milk-weed).*—Stems spotted and downy, from 2 ft. to 3 ft. in height; flowers white, with a reddish centre, in large umbels. This is one of the showiest of the family.

A. verticillata *(Whorled Milk-weed).*—Quite unlike the majority of Milk-weeds in its slender habit, narrow leaves, and delicate, small white flowers, produced in abundance in summer. Grows in poor soil to nearly 2 ft.

Asclepiodora decumbens *(Green Milk-weed).*—A low stout herb with umbels of greenish and purplish flowers, which resemble in appearance those of the Milk-weeds, but differ in structure. Texas. Of no garden value.

Ascyrum *(St. Peter's Wort).*—A small genus of sub-shrubby plants of little value in the garden. The best known, A. Crux Andreæ (St. Andrew's Cross), is a small yellow-flowered kind, often classed with the St. John's Wort (Hypericum).

Asparagus.—Some of the species of this family are interesting from their elegant foliage or bold habit, and worth a place for these. A. Broussoneti, a vigorous and tall species, is quite hardy, so are A. tenuifolius and others. The common Asparagus is as good as any of them, and a tuft or group of it is effective and graceful in a border of flowers or bed of fine-leaved hardy plants.

Asperula odorata *(Woodruff).*—This little wood plant, abundant in some parts of Britain, is worthy of a place in the garden or shrubbery, especially in localities where it does not occur wild. Many would like to cut and preserve its stems and leaves for the sake of the fragrant hay-like odour which they give off when dried; and in May the white small flowers, profusely dotted over the tufts of whorled leaves, look very pretty. It is one of the many plants that may be allowed to cover the earth in a shrubbery where the barbarous practice of annually digging and rooting up the borders is not resorted to. It is sometimes used as an edging to the beds in cottage gardens. It is, however, as a wood or shrubbery plant—as a companion to the Wood Hyacinth and the Wood Anemone—that it will be at home. It is largely used in Germany for flavouring summer drinks. It mixes charmingly with Ivy where that is allowed to clothe the ground. It belongs to a considerable family of plants, few, however, of which have come into cultivation.

A. azurea setosa is a pretty early spring flowering hardy blue annual, flowering in April and May. Sow the previous autumn. A. cynanchica is a rosy red perennial, forming a good bank or rough rock plant.

Asphodel *(Asphodelus).*—The name is also applied to Narcissus poeticus in some parts of the country.

Asphodelus *(Asphodel).*—Liliaceous plants of no great value for gardens. There are some half-a-dozen kinds in cultivation, the best known of which is A. ramosus (tall Asphodel), a bold South European species, familiar in most of the old herbaceous plant borders, but better fitted for the shrubbery or wild garden. The stems grow from 3 ft. to 5 ft. high, bearing numerous white flowers. Varieties of this species are albus, cerasiferus, and microcarpus, all similar to the type in general aspect. Other kinds are A. fistulosus and tenuifolius, both with white flowers, and each growing from 1½ ft. to 3 ft. high, and grow well in any border or position. The last named kind has delicate feathery foliage and is a contrast to bold foliage plants. A. acaulis (the stemless Asphodel) is a singularly interesting plant, as it is only an inch or so high, and bears its flowers in a dense cluster, surrounded by a tuft of narrow Grass-like foliage.

Asphodeline.—This genus is nearly allied to the preceding, but may be readily distinguished by the stems, the stems of Asphodelus being leafless, while in Asphodeline the leaves are always produced on the erect stems. About six species are in cultivation, the best known being A. lutea (tall yellow Asphodel), growing about 3 ft. high, with the yellow flowers produced in dense clustered spikes. A. taurica has white flowers, produced in a similar manner on stems 1 ft. to 2 ft. high. A. liburnica (A. cretica) and A. tenuior have both yellow flowers borne in loose racemes. A. damascena has white blossoms in dense racemes, and A. brevicaulis has yellow flowers in loose racemes. These are all of vigorous growth and easy culture, thriving in any common garden soil, and may be used with good effect in bold masses among other tall-growing plants.

Aspidistra lurida *(Parlour Palm).*—A native of Japan, with long and graceful evergreen leaves, and bearing dull purple flowers on the surface of the soil. One of the most lasting of fine leaved plants in rooms. There is a variegated form also. Both are used out of doors with good effect in some gardens, and the plant is hardy enough in certain districts. Of the two kinds we much prefer the evergreen one. Division. 1½ ft. to 2 ft. high.

Aspidium *(Shield or Wood Fern).*—As now arranged, this genus embraces the Polystichum and some of the species of Lastrea. Tnere are numerous hardy kinds, including some of the noblest of hardy Ferns, of which the common Male Fern (A. Filix-mas) and Prickly Shield Fern may be taken as the types. These are a most useful class of plants, thriving everywhere, even in small town gardens and similarly confined places. All they require is a plentiful supply of water during hot, dry weather. Either alone or in groups these Ferns have a fine effect, particularly as an undergrowth to trees in the pleasure ground or in the shadier parts of the garden. These are both evergreen, therefore are invaluable for the garden. Their varieties are endless, no fewer than a hundied named sorts being enumerated in trade lists of A. aculeatum and fifty of A. Filix-mas. The larger growing varieties of these Ferns are noble subjects when fully developed, and they have a fine effect planted under trees or in similar situations, but the smaller and more delicate kinds require more careful treatment. A. aculeatum succeeds best in rich loam mixed with sand and leaf mould, and the situation should be thoroughly drained, and the same treatment suits the Male Fern. Both are easily grown in pots, and in such a state are very useful.

A. dilatatum (Broad Buckler Fern) is a pretty Fern, and of which there are some handsome varieties, especially the crested fronded sorts. It requires a fibrous peat loam and sand, and moist situation. A. cristatum (Crested Shield Fern), a British species, is a handsome and easily grown kind. A. Lonchitis (the Holly Fern) is one of the finest evergreen hardy Ferns. It grows naturally in clefts of rocks, and in order to cultivate it successfully it should be planted between pieces of grit rock in a mixture of loamy turfy peat and sand. A. munitum, a North American species, is a beautiful kind, requiring but little care to grow it well if planted in a peaty soil in a shady situation. Other valuable kinds are A. rigidum, A. Oreopteris, A. Thelypteris, A. spinulosum, A. cristatum, A. acrostichoides, A. montanum, which all succeed in a mixture of loam, turfy peat, and sand, in rather a moist and shady situation.

Asplenium *(Spleenwort).* — Amongst hardy evergreen Ferns few are more useful or diversified than these; the fine dark green colour of most of the varieties and their free-growing character render them worthy of culture. The soil best suited for them in a general way is a well drained mixture of peat, sand, and loam, just the sort of material in which the finer kinds of flowering shrubs, such as Kalmias, Andromedas, Rhododendrons, Azaleas, &c., would thrive to perfection, and with which Aspleniums might be advantageously associated. A. Adiantum-nigrum (the Black Spleenwort) would be especially interesting amongst hardy Azaleas, because these lose their foliage in winter, when the value of the Spleenwort would become apparent, carpeting, as it would, the surface of the soil with verdure. The shade, too, which the Azaleas would afford in the summer, if not planted too thickly, would just suit this Spleenwort, as it is generally found in a wild state fringing copses or on hedge banks, where it gets just a little protection from the scorching rays of a summer's sun. There are several distinct forms of this Asplenium, the most remarkable being perhaps grandiceps and microdon, both valuable and useful kinds. There is also a variegated form. Asplenium fontanum is a lovely alpine Fern. In cultivating it the conditions under which it was found growing in its native haunts should be as far as possible imitated. It loves to hide beneath overhanging rocks. It also does well in pots, planted in fibrous loam, with a good mixture of calcareous chippings about the size of Walnuts. Asplenium marinum is one of the most beautiful of evergreen Ferns, but it is far from being generally hardy; still it will succeed under conditions similar to those recommended in the case of the British Maiden-hair (A. Capillus-veneris). A. marinum imbricatum is a particulary fine variety having beautifully fringed and crisped fronds.

A. germanicum (alternifolium) is a rare dwarf-growing kind, admirably adapted for a shady spot in the alpine garden. It is perfectly hardy, and, although somewhat difficult to grow, is worthy of any care that may be bestowed upon it. A fine variety of it is A. g. acutidentatum, producing elegant little fronds about 2 in. or 3 in. in length. A. lanceolatum produces abundance of dark shining green lanceolate fronds, which frequently grow 15 in. high. The variety microdon is very distinct and ornamental. A. Ruta-muraria, the Wall Rue, is a pretty little plant for walls, and for placing in the chinks and fissures of rockwork. Its fronds are deep green, and remarkable for the profusion of brown sori. The varieties cristatum and crispum are pretty varieties of this little Fern. A. Trichomanes, the common Maiden-hair Spleenwort, is a handsome hardy species, the fronds being of a dark green colour. The variety crispum is finely tasselled, and incisum is also a remarkably handsome variety, the fronds of which are much serrated and divided, and multifidum is very distinct, each branchlet ending in a little crest. Like all the Spleenworts, it requires good, free, well-drained soil in which to grow, and, given this essential and ordinary care, success is pretty certain. A few pieces of stone placed on the soil around the roots prevents to a certain extent the evil effects that excessive evaporation has on all Ferns. The stones might be partly buried in the soil, and the portions of them left above it would be covered by the Ferns themselves; therefore no unsightly appearance would be presented. It is to be desired that our smaller Ferns be used amongst hardy flowers as carpets among the taller plants in shady parts of the rock

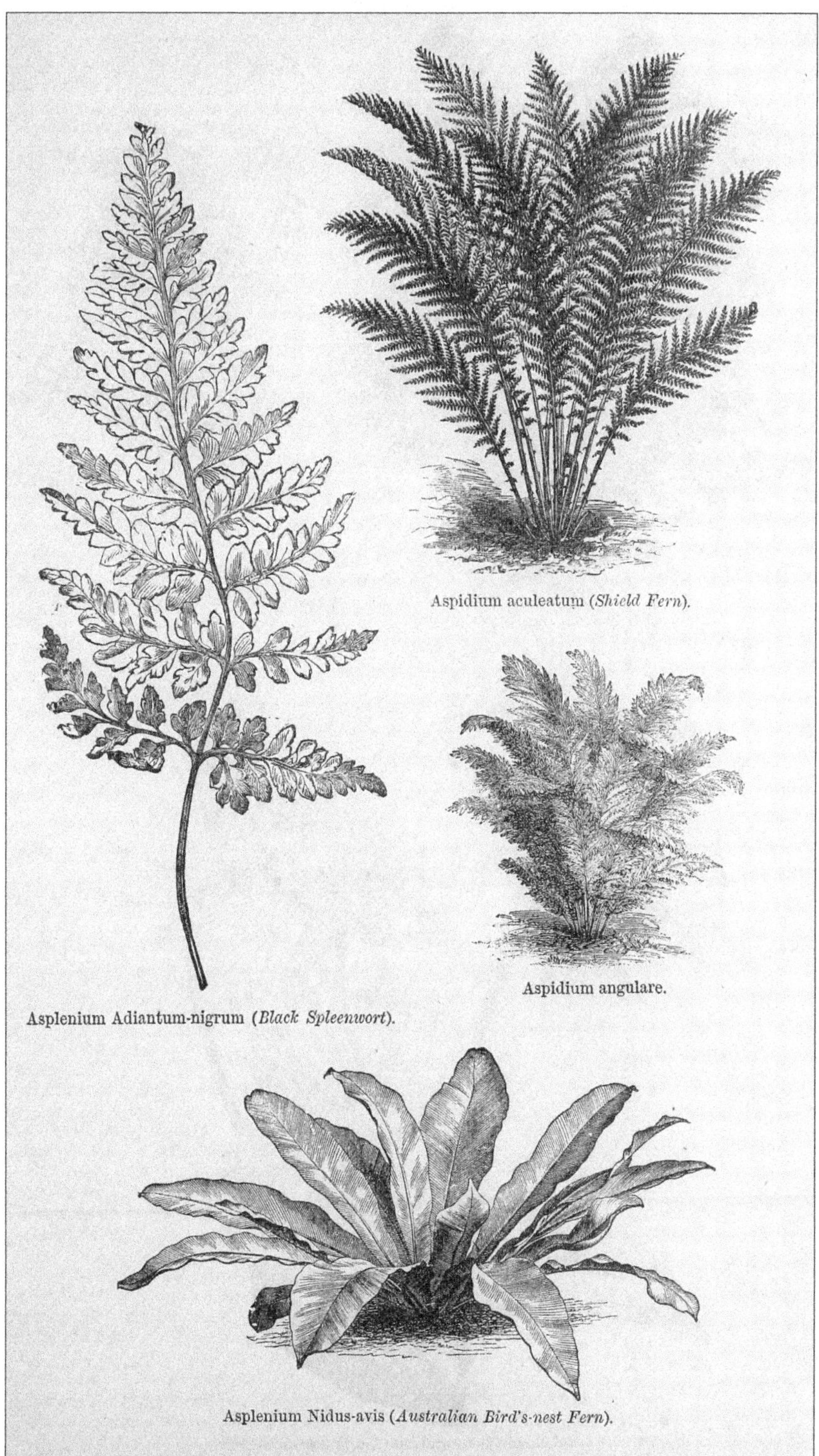

Aspidium aculeatum (*Shield Fern*).

Aspidium angulare.

Asplenium Adiantum-nigrum (*Black Spleenwort*).

Asplenium Nidus-avis (*Australian Bird's-nest Fern*).

[*To face page* 40.

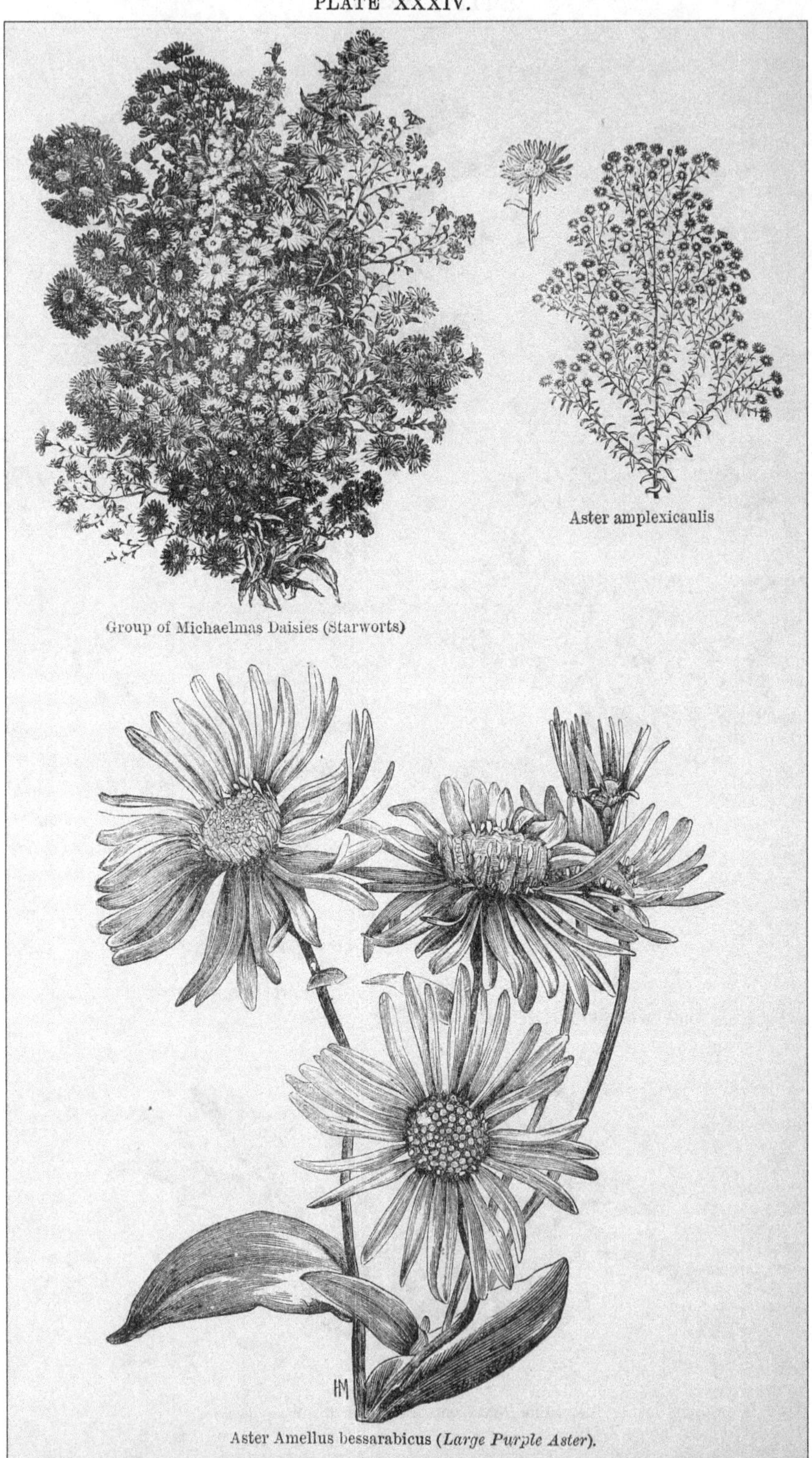

Group of Michaelmas Daisies (Starworts)

Aster amplexicaulis

Aster Amellus bessarabicus (*Large Purple Aster*).

[*To face page* **41.**

garden. In the wild garden there are various opportunities of associating them with hardy spring or other flowers. So placed they would seem more graceful than ever.

A. Nidus-avis.—This is a very large Fern, which has been placed out-of-doors in the garden in summer, from early in June to October, but it is not vigorous or hardy enough to be generally recommended for this purpose. It is a favourite subject in places where large collections of tropical Ferns are grown, and in such places plants may be tried in the open air in a warm, shady, and perfectly sheltered position. E. Indies. At Battersea Park the young plants as well as the large ones are grown out-of-doors during the summer in shady places and do very well. A. australasicum, a similar kind, is used in the same way.

Aster *(Michaelmas Daisy, Starwort).*—When we speak of Asters, we are apt to associate the name with the annual kinds popularly known as China Asters, and are thus led to overlook the rightful owners of the name—those hardy flowers, literally stars of the earth, which shine out all the brighter, owing to the time of the year during which they are in perfection. When our gardens are nearly devoid of bright colour, and when hardy flowers, of nearly all kinds, are at their lowest ebb, Michaelmas Daisies will bloom bravely during the last days of autumn. They are not quite so showy as Chrysanthemums at a distance, but when closely examined, they are more beautiful, their slender-rayed flowers possessing all the soft and delicate tints between white, rose, and purple, while the bright yellow disc gives them an additional charm. Nearly all the species, of which there are at least two or three hundred in books, are natives of North America, and are perfectly hardy in this country, thriving in any soil or climate, and requiring little attention. Their freedom of growth indeed is almost a fault, and they destroy other plants near them if care be not taken as to placing them properly. The worst kinds are generally the best growers. Scores of species have no merit whatever for garden borders and would speedily overrun them.

SELECTION.—Three distinct types of Aster may be readily distinguished by their habit of growth; first, there is the alpine or dwarf-growing kinds, from 6 in. to 3 ft. in height, such as A. alpinus; then the tall-growing kinds, varying in height from 3 ft. to 7 ft. in height; and lastly, those with spreading slender stems, such as turbinellus and patens. A representative selection should include some of each of these sections; therefore, of the dwarf kinds, we would recommend A. alpinus, a handsome sort, about 9 in. high, which produces large purple-blue flowers during summer. A. Amellus is also one of the most beautiful of hardy perennials; it grows about 2 ft. high, and produces a profusion of bright purple blossoms. There is also a white-flowered variety of it, named bessarabicus, which is even much finer than the type, but a little taller in growth. A. longifolius var. formosus is an extremely fine plant, remarkable for the profusion with which its heads of deep rosy pink flowers are produced on stems about 2 ft. high. The Pyrenean Starwort (A. pyrenæus), though not so showy as the last, is a desirable early autumn-flowering plant; it grows about 2 ft. high, and bears large heads of lilac-blue flowers. A. Reevesi is a pretty little variety, with slender stems laden in early autumn with tiny white flowers. A. cordifolius, A. Lindleyanus, and the Galatella section, A. dracunculoides, linifolius, may be added to the dwarfer-growing kinds, as well as A. sericeus, a remarkably distinct and pretty kind, and versicolor, a dwarf sort, with flowers an inch across, changing as they grow older from white to mauve.

The kinds with a spreading habit of growth are extremely graceful, and ought to be in every garden, as the slender sprays are specially useful as cut flowers, in which state they last a long time in perfection. There are not many, the best being A. turbinellus, with large mauve flowers; A. patens, with a similar spreading growth, but smaller flowers; A. laxus a fine species, with pale purplish flowers produced freely early in autumn. The tallest growing kinds are very numerous, and most of them possess a striking similarity in habit of growth and flowers. The most distinct are the largest kinds—A. Novæ-Angliæ and its varieties roseus, pulchellus, and A. Novi-Belgii; somewhat smaller are A. Chapmani, A. Drummondi, A. puniceus, and A. Shorti, all about 5 ft. high; they should be included in a selection as well as some which bear doubtful names, such as A. purpuratus, A. amethystinus, A. elegantissimus, A. multiflorus, which may be found in trade lists, and not omitting the elegant A. ericoi des.

CULTURE.—As regards their culture it is simple enough; they like a good soil, but do not refuse to grow in any kind. The question to be considered is rather how the plants should be placed. The dwarf kinds are excellent for the rougher parts of the rock garden or the front rows of a mixed border, and the taller and more vigorous sorts do well for naturalising, and many are valuable as cut flowers. In fact, so beautiful are these Starworts in the autumn, that wherever in gardens there may be no place for them in the general scheme of garden decoration, many will find it worth their while to grow the best kinds for the sake of their cut flowers alone. They add a new grace to the autumn days, but the better kinds are well worthy of a place in the finest gardens, and will do admirably with groups of the later flowering perennials such as the Flame flowers (Tritoma), the tall late Anemones, and the Cardinal flowers (Lobelia). But selection must be rigidly kept in view, and any stiff ugly mode of staking the plants avoided. Good culture and early thinning of the shoots will make this easier. The plants should not be left too long in the same position lest the tufts get weak and too thick. The following are the best kinds so far as there is any knowledge of the names of a genus so large and so confusing. All are easily increased by division of the root or by seed.

A. alpinus (*Blue Mountain Daisy*).—A pubescent dwarf kind, the single blue heads of which scattered over the Grass in high alpine meadows look like blue Daisies. In gardens it grows larger, and forms vigorous leafy tufts from 6 in. to 10 in. high. Flowers early; nearly 2 in. across. There is a white variety. Mixed borders and rough rock gardens; it is occasionally used as a bedding plant on the Continent. An interesting subject for naturalisation in an upland meadow. Division and seed like all the kinds.

A. Amellus (*Italian Starwort*).—A very handsome kind, 1½ ft. to 2½ ft. high. Flowers in late summer and autumn, being many, showy, blue. The variety bessarabicus very much resembles the species. A good border plant, and also fitted for the wild garden; in copses and by hedgerows. An excellent plant for cutting from.

A. cordifolius (*Drooping Starwort*).—A tall and graceful kind, with stems often zig-zag below, and inflorescence gracefully drooping; 3 ft. to 4 ft. high. There is some doubt as to the name of this kind, but under it at Kew some years ago there was a very graceful plant grown. In the Botanic Gardens in the Regent's Park the same plant bore the name elegans. It is valuable for its soft racemes of nearly white or pale lilac flowers, that hang soft as snow-laden branchlets. Beds of the finer autumn plants and borders.

A. discolor.—This is a little gem, rarely exceeding 1 foot in height, its numerous stems being often prostrate from the crowd of flowers they bear. The older flowers change to a rosy purple with a dark brown disc. A bed of this plant would be quite a feature in any flower garden during October. It is useful for rockwork, the margins of beds of hardy flowers, and for borders.

A. ericoides.—One of the most distinct of all the species. Its flowers are barely ½ in. in width, and are clear white in colour, with a yellow eye. The plant is dense in habit, attaing a height of from 2 ft. to 3 ft., and is a profuse flowerer.

A. grandiflorus (*Christmas Starwort*).—A peculiarly handsome species, with somewhat stiff and wiry stems, and with little of the vigour or coarseness of the other kinds; 2½ ft. to 3 ft. high; flowering in November and December; purplish, large. Except on warm soils and in favourable situations, this plant should be trained against a low south wall. One of the latest-blooming kinds, this Daisy is well adapted for lifting and flowering in pots in the conservatory. Its large flowers, showy, supply a colour wanting in Chrysanthemums, and well-grown plants of it would group well with a collection of Chrysanthemums, scarlet Salvias, and the like.

A. lævis (*Smooth Starwort*).—A variable and elegant kind, smooth throughout, growing about 3 ft. high. Flowers in autumn with yellow centre, in a large, close panicle; bluish lilac. Borders, naturalisation, and margins of shrubberies, in any soil.

A. laxus (*Loose Starwort*).—A very fine and pleasing species, with rather slender stems upwards of 3 ft. to 4 ft. high. Flowers early in autumn; pale purplish blue, about 1 in. across, in loose irregular clusters. Borders and groups, or for naturalisation in copses and shrubberies.

A. longifolius (*Rosy Starwort*).—A very handsome species, somewhat dwarf, producing masses of good rosy lilac flowers. It is said to vary much in a wild state, but the kind usually grown in gardens is compact in habit and not over 2 ft. high. It is an admirable border plant also fitted for beds or groups of the finer autumn hardy flowers.

A. Novæ-Angliæ (*New England Starwort*).—A very tall and robust perennial species, 5 ft. to 8 ft. high, with violet-purple flowers, appearing late in autumn. Among the tallest plants at the back parts of the mixed border, in warm soils and positions. It grows anywhere, but in cold soils and positions often blooms only in time to be destroyed by frosts. Naturalised in sunny sheltered positions in woods and copses, it would probably be seen to greater perfection than in a garden, and it would form a good covert. It is seldom satisfactory in a garden, and takes up much room for a late and uncertain bloom. The var. pulchellus is said to be the best of all the late-blooming bluish-purple kinds. It forms strong tufts from 4 ft. to 5 ft. in height, with deep blue, orange-eyed flowers, which form a dense head. There is a deep rose-coloured variety, which is of more value as a garden plant. Division.

A. Novi-Belgii (*New York Starwort*).—Another tall and vigorous perennial of the same character and value as the preceding, though the colour of the flower is blue. Position, soil, &c., the same as recommended for A. Novæ-Angliæ.

A. ptarmicoides.—A very distinct species, growing about 15 in. high, and bearing a profusion of rather small pure white flowers, similar to those of Achillea Ptarmica. It flowers a fortnight or so before the majority of the other kinds.

A. pyrenæus (*Pyrenean Starwort*).—A large, stout, handsome, and early autumn-flowering kind, 2 ft. to 3 ft. high, with large, lilac blue flowers, with yellow disc. A Pyrenean plant, remarkable for its blooming earlier than the finer American kinds; it may therefore have a peculiar value in some cases. Borders, fringes of shrubberies, and naturalisation on banks, copses, &c., in any soil.

A. Shorti (*Short's Starwort*).—A very pretty species, with a slender, spreading, nearly smooth stem, 2 ft. to 4 ft. high. Flowers in autumn; purplish-blue, about 1 in. across, numerous, in long panicles.

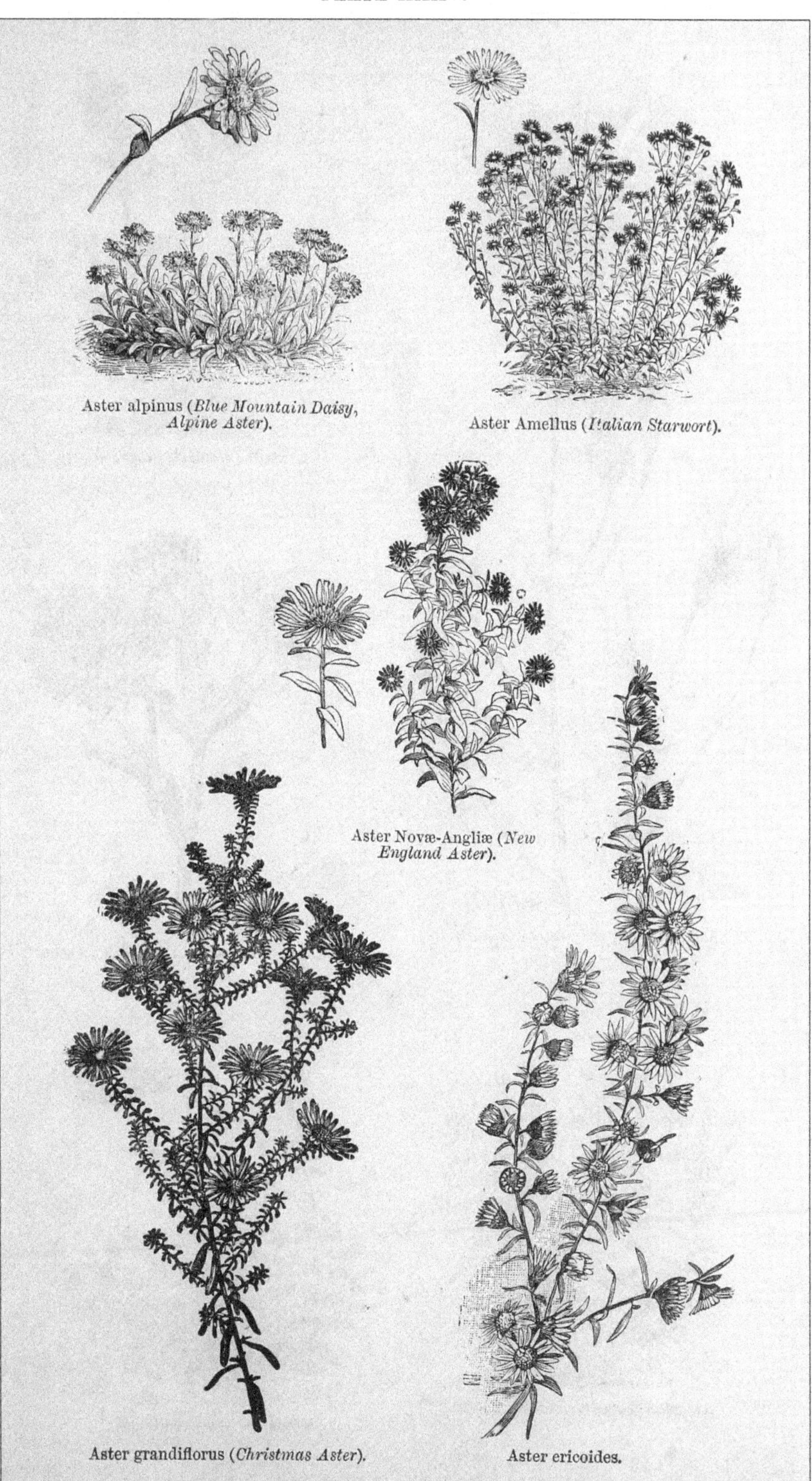

Aster alpinus (*Blue Mountain Daisy, Alpine Aster*).

Aster Amellus (*Italian Starwort*).

Aster Novæ-Angliæ (*New England Aster*).

Aster grandiflorus (*Christmas Aster*).

Aster ericoides.

[*To face page* **42.**

Aster turbinellus (*Mauve Starwort*).

Astilbe rivularis (*Goat's-beard*).

Flowers of Astrantia major (*Masterwort*).

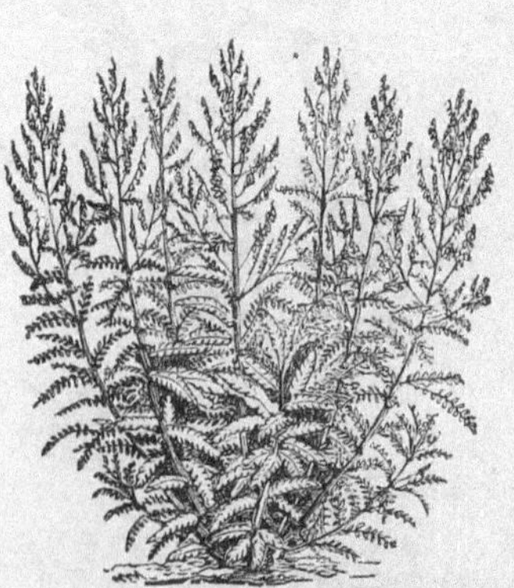

Astragalus galegiformis (*Milk Vetch*).

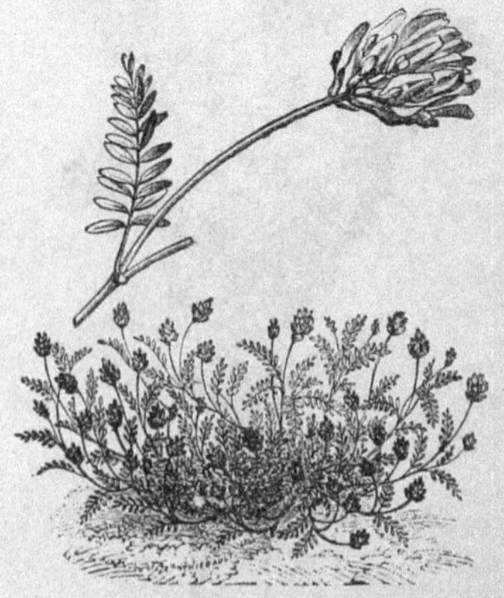

Astragalus monspessulanus.

[*To face page* 43.

Borders and groups of late flowering perennials.

A. turbinellus *(Mauve Starwort).*—A free showy kind, 2 ft. to 3 ft. high. Flowers late in summer and in autumn; delicate mauve, produced in panicles. Borders, with groups of handsomer autumn-flowering perennials, and naturalisation. Grows freely anywhere, but is worthy of good soil.

A. versicolor *(Changing Starwort).*—A compact growing species, sometimes confused with A. discolor, though as garden plants at least they are different Flowers in summer; showy, white at first, changing into pink or purple. Said to have been found in North America, but not recorded by recent American botanists. Borders in sandy loam, or beds of the finer perennials.

It were easy to add many species to the above specially selected kinds, but the reader may be assured that among them he has the most satisfactory, so far as they have been proved in cultivation. In the future the nomenclature of this family will probably be much altered; the above are the names by which they are best known in gardens.

Aster, China (See *Callistephus chinensis*).

Asteriscus mauritanicus.—A dwarf, half-hardy, perennial composite, allied to Buphthalmum, and producing an abundance of deep yellow flower-heads, about 1½ in. in diameter, throughout the summer. It is closely allied to A. maritimus, and may possibly be only a form of that plant; both plants are of doubtful value.

Astilbe rivularis *(Goat's-beard).*—A large-leaved and striking plant from Nepaul, with the habit and general appearance of a Spiræa, growing to a height of more than 3 ft., and of a free and graceful habit, which makes it useful for association with the finer-foliaged herbaceous plants, and for dotting here and there in the wild or picturesque garden. It keeps its foliage well through the season, but the flowers are not showy. Being hardy, the Astilbe usually succeeds well in any cool rich soil, and best in half-shaded positions. Easily multiplied by division. It is suited for isolation, borders, fringes of shrubberies, or for groups of hardy plants. A. rubra also grows from 4 ft. to 6 ft. high. The flowers are numerous, in dense panicles, and of a rose colour, appearing late in summer and in autumn. America and Japan. A. decandra is an American kind. These plants are not very important for gardens, inasmuch as Spiræa Aruncus has as good a habit, while a fine plant as regards its flowers.

Astragalus *(Milk Vetch).*—An enormous and numerous genus, but of which comparatively few attract cultivators. The best of those in cultivation are rock plants, but grow freely on the level ground in borders.

A. monspessulanus *(Montpellier Aster).*—A very valuable plant for the fronts of borders or the rougher portions of rockwork. The shoots, though vigorous, are prostrate, which causes it to be seen to greater advantage when drooping down the edge of rocks, and bearing its long heads of crimson and rosy flowers. It seems to grow well in any soil, and though it does not flourish so vigorously when in very gravelly or poor soil, yet in the very poor soil it will appear more floriferous in consequence of the small development of leaves. A native of the south of France, easily raised from seed. There are several varieties.

A. onobrychis *(Saintfoin-like Milk Vetch).*—A handsome species, in some varieties spreading, and in others growing about 18 in. high, having racemes of purplish crimson flowers. It is a hardy, good perennial, will thrive well on any good loam, and is a capital subject for the mixed border. Europe and Siberia, flowering in June. There are several varieties enumerated, three of which, alpinus, moldavicus, and microphyllus, are prostrate in habit, and, if introduced, would probably prove valuable for rockwork, and one, major, which grows erect. The plant is particularly suited for the rougher parts of rockwork, and for positions where a rich effect rather than rare and minute beauty is sought. There are white forms of all the varieties.

Other good kinds are A. vimineus, A. diphysus, and our own Purple Milk Vetch (A. hypoglottis), and its white form are also good plants for the rock garden. A. pannosus (Shaggy Milk Vetch) is a singular kind, from its very silvery and woolly pinnate leaves, which, growing in compact and luxuriant tufts about a span high, give the plant somewhat the appearance of a silvery Fern. Astragalus dasyglottis is a very fine and desirable species, well adapted for the rock garden. Its showy flower-heads, of a clear, bright purple colour, are very numerous, and are well set off by the fresh green foliage. Astragalus adsurgens is a dwarf species, producing numbers of violet-carmine flowers, and is well worthy of a place in the rock garden. A. vaginatus is a free grower, an abundant bloomer, and succeeds apparently in any ordinary border in an exposed position. The flowers are borne in dense, erect clusters, and are of a showy, deep violet-purple colour. It flowers for a long time.

Astrantia *(Masterwort).*—An interesting family of the Parsley Order, but best in botanic gardens. In choice borders some of the kinds are apt to over-run and exhaust the soil. If grown at all it should be in rough or wild places, or in the back part of a shrubbery. The curious flower-heads are pretty in the cut state. Grows in any soil.

Atamasco Lily *(Zephyranthes).*

Athamanta *(Spignel).*—Graceful perennials of the Parsley Order with finely cut Fern-like leaves. They are worth a place among fine-leaved plants in borders or groups, or naturalisation on banks, in ordinary soil. Division or seed.

Athanasia.—A genus of the Compositæ, of which the annual kind has been frequently grown out-of-doors, but it is not a plant of much importance. It is of easy culture.

Athyrium *(Lady Fern).*—A very beautiful genus of Ferns, of which A. Filix-fœmina may be taken to fairly represent the numerous forms and variations. They all like a compost consisting of loam, leaf mould, and peat, mixed in about equal proportions with the addition of some sharp sand; they require abundance of water during their growing period, but of course artificial moisture should be discontinued in winter, because all the varieties of this species are deciduous, and the ground at that period of the year is sure to be wet enough naturally to suit all their requirements. There are many fine hardy evergreen herbaceous plants amongst which Lady Ferns might be planted with advantage; they will thrive admirably anywhere provided they receive a little shade and protection from drying winds, being very impatient of drought; therefore plants that afford shade and shelter and that like moisture should always be chosen for intermixture with them. In variety of form this genus exhibits a greater variety than any other of our native Ferns, except the Scolopendriums. The principal varieties are A. Filix-fœmina plumosum, one of the most lovely hardy Ferns in existence, its fronds reminding one of a plume of feathers possessing a lovely shade of green. A. Filix-fœmina (Vernoniæ) corymbiferum is a wonderfully fine crested variety, and A. Filix-fœmina Victoriæ is one that should be in the possession of all hardy plant lovers, as it is considered the finest of all the varied forms of Athyrium. There is a Japanese kind of Athyrium called A. Goringianum pictum, a remarkably pretty variety and quite hardy.

Atragene.—The Atragenes very much resemble Clematises, but differ from them in having numerous petals, which are shorter than the four sepals, in being destitute of an involucre, in the carpels being terminated by a bearded tail, and in their producing leaves and one flower from the same bud contemporaneously in the spring.

A. alpina *(the Alpine Atragene).*—This plant forms a slender climber, about 8 ft. or 10 ft. high, found on the mountains in various parts of Europe, especially on calcareous soil in Austria, Carniola, Piedmont, Genoa, Dauphiny, the Eastern Pyrenees, and South Switzerland. The Alpine Atragene is a very ornamental, slender-climbing plant, and is quite hardy. A graceful plant for trailing over a low bush or stump. *Syns.*—Clematis cœrulea, C. alpina, and Atragene austriaca.

Atriplex *(Purple Orach).*—A. hortensis atro-sanguinea is a dark form of the common Orach. It is a hardy annual often used for its ornamental coloured foliage. Sown in open air in April or May, does well in almost any soil, and has been used among bedding plants, but has for the most part been given up in favour of other plants, but growing to a height of 4 ft. or even considerably more when well treated it has peculiar merit for those who know how to use it. It associates well with the Amarantus, annual Wormwood, Persicaria, and other bold and graceful annual plants grown for the flower garden.

Aubrietia *(Purple Rock Cress).*—A valuable little group of alpine plants from the mountains of South Europe. There are many varieties of Aubrietias in gardens, but probably all of them may be reduced to some half-dozen species; all of them are, however, beautiful, some of them eminently so. The oldest variety is the one called by some people Aubrietia purpurea, a pretty flower enough, but thrown into the shade by some of the other kinds. There is a well-known variegated variety of this, not of much value, except as a neat little rock plant. Then there is what is called deltoidea, which is very near, if not identical with purpurea. We have also grandiflora, a kind similar in colour to purpurea, but twice or thrice its size. This has a lax diffuse habit, which makes it a charming rock plant. There is a variety of this called græca, a fine plant, opening out at first a full purple, and dying off a lavender colour. Masses of this, with its various shades of colour, are very pleasing. There is also a fine variegated, large-flowered form of it. Then we have A. spathulata, erubescens, and hesperidifolia. The rest are Mooreana, Columnæ, and Campbelli. Mooreana is a compact little cushion-like plant, which, in its flowering season, is literally smothered with bloom. This has a shade of blue in it. Campbelli and Columnæ appear to me identical; but, like Mooreana, they are among the loveliest of Aubrietias. The last three are well adapted for spring gardening. They are perfectly hardy, and will flower from March until June.

Some find Aubrietias difficult to propagate; my practice is to pull off all the straggling side shoots in June or July from the old plants, securing as much stem as possible, and breaking it off close to the main root; then a piece of ground is dug in a cool, shady border, into which is worked plenty of rough sand and leaf-mould; the shoots are then planted in lines, a little sandy soil being placed about the portion put into the ground, and all this is trodden down firmly. The cuttings are then occasionally sprinkled and kept shaded from the sun, and, thus managed, but few failures occur. One great advantage in getting the cuttings started in June or July is that the plants become strong and dense by the end of the summer, and are well fitted for planting out.

The Aubrietia is excellent as a wall plant; I have seen a wall almost covered with long wiry-like tufts of this plant rooted firmly in its surface. The effect in spring is very beautiful. Imagine some of our old Fern-covered walls or sunk fences draped with the tufts of lovely blue of the "purple Arabis," as they call it in London! To establish it we need only sow the seed in any mossy or earthy chinks in autumn or spring. Rock-works, ruins, stony places, sloping banks, and rootwork will suit Aubrietias perfectly; and no

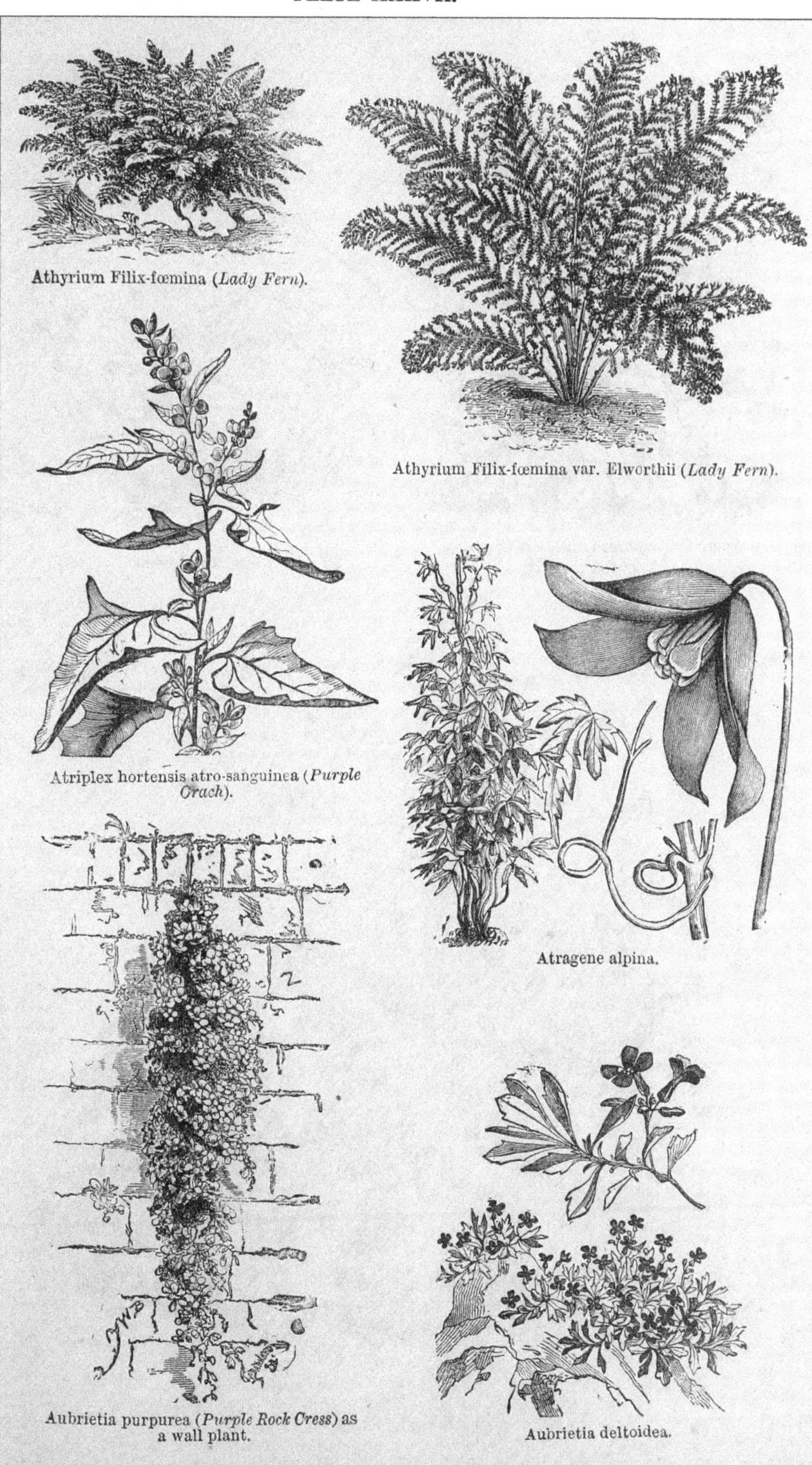

Athyrium Filix-fœmina (*Lady Fern*).

Athyrium Filix-fœmina var. Elworthii (*Lady Fern*).

Atriplex hortensis atro-sanguinea (*Purple Orach*).

Atragene alpina.

Aubrietia purpurea (*Purple Rock Cress*) as a wall plant.

Aubrietia deltoidea.

[*To face page* 44.

PLATE XXXVIII.

Aubrietia purpurea (*Purple Rock Cress*).

Bæria chrysostoma.

Bambusa falcata (*Bamboo*). See page 46.

[*To face page* 45.

plants are so easily established in such places, nor will any other alpine plant so quickly clothe them with the desired kind of vegetation. They make neat edgings, and may be used as such with good taste in any garden. Growing in common soil in the open border, or on any exposed spot, it thrives as luxuriantly as on the best made rockwork, forming round spreading tufts; and on fine days in spring the blue flowers come out on these in such crowds as to completely hide the leaves, making, in fact, hillocks of colour. For covering bare ground beneath Roses and shrubs it might also be tastefully employed. It is quite easy to naturalise it in bare rocky places. It is easily propagated by seeds, cuttings, or division—the last mode the most facile. Aubrietias, as we know them in gardens, are so near each other in character, that it is hopeless to try to identify them specifically, the more so since so many good seedling forms have been raised. To A. purpurea (the A. macrostyla of Boissier) belong A. grandiflora, A. græca, and the fine variety A. Campbelli. Aubrietias vary a good deal from seed, but their differences make them all the more valuable as garden plants, and they all agree in carpeting the earth or rocks with dense dwarf cushions of compact rosettes of leaves, profusely clothed with beautiful purplish-blue flowers in spring, and, in the case of young plants in moist and rich soils, almost throughout the year. There are one or two pretty variegated varieties.

Auricula *(Primula Auricula).*

Azalea.—A name for the interesting little alpine shrublet, now usually called Loiseleuria procumbens.

Azolla caroliniana.—A very small and curious water plant, which floats on water quite free of soil. The tufts of delicate green leaves appear like tiny emeralds. During summer it will grow out-of-doors, but then it becomes bronzed, and perhaps it is prettier when light green, as it is in the greenhouse or window. *Syn.*—A. pinnata.

Babiana.—A charming genus of bulbous Iridaceous plants from South Africa, allied to Sparaxis and Tritonia, but differing in having much broader foliage, which is often hairy and plaited. They grow from 6 in. to 12 in. high, and produce spikes of brilliant flowers ranging in colour from blue suffused with white to the richest crimson-magenta; many of them are sweet scented. They should be planted from September to January in light loamy soil thoroughly drained, in a due south aspect, about 4 in. deep and 2 in. to 4 in. apart. The early plantings make foliage in autumn, and require protection of hoops and mats from frost. The bulbs planted in December and January will only require the protection of a covering of Fern, which should be gradually removed as the foliage appears in spring. In stiff or wet soils the bulbs should be surrounded with sand, and the beds raised slightly above the surrounding level. These are mostly grown in mixed varieties, but can be obtained in separate named colours. One of the most distinct of the named varieties is B. rubro-cyanea, which has more fully expanded, larger blooms and a conspicuous ring of intense crimson dividing the centre from the outer blue. Among the varieties in cultivation are—Atro-cyanea, purple blue, marked white; Attraction, dark blue, vigorous habit; Celia, rose, marked white; General Froome, violet, spotted white; General Scott, lavender, suffused white; General Slade, magenta; Hellas, pale yellow, outside suffused purple; Julia, petals alternately white and blue; Kermesina, rich crimson-magenta; Lady Carey, rose, marked white; Rosea grandis, rose-purple, marked white; Rubro-cyanea, blue, crimson centre; Speciosa, mauve, suffused blue; Villosa, crimson. The Babianas being tender and difficult unless in careful hands are best fitted for the choice bulb garden or border in a sheltered spot. In the south of England a better climate may make their culture more easy, but, as a rule, they are unsatisfactory, even in the best part of the country.

Bachelor's Buttons.—A term applied to some double hardy flowers, chiefly to double Ranunculus and to the common Daisy.

Bæria chrysostoma, a Californian annual of the Composite family. It grows about 1 ft. high, forming a dense tufted head, which in early summer is covered with a profusion of bright yellow attractive flowers. It should be treated as a half-hardy annual.

Bahia lanata *(Woolly Bahia).*—A greyish herb, mostly much branched from the base of the stem, 6 in. to 15 in. high. It flowers in summer; one on each stalk, yellow, in great abundance. Suited for borders or banks, in light, sandy, well-drained loam, on which, flowering much more abundantly, it is a much more ornamental plant than when on cold clay soils. Fitted for groups or beds of silvery or variegated plants. Seed, or more readily by division. America. Compositæ.

Bald-money *(Meum athamanticum).*

Baldwinia uniflora.—A North American swamp plant of the Composite family. Of little garden interest.

Balm *(Melissa officinalis).*

Balsam *(Impatiens).*

Balsamita *(Costmary).*—A genus of Compositæ, to which belongs the Costmary, an old herb-garden plant. There is also a rather showy and singular free-growing perennial belonging to the same genus, B. grandiflora. It is of easy culture, and may be found worth a place among the stouter herbs in borders.

Balsamorhiza deltoidea *(Balsam Root).*—A North American plant of the Composite family. Of no garden value.

Bambusa *(Bamboo).* — The Bamboos have a claim to be included among trees and shrubs, being, in fact, tree Grasses, but they are so often used with good effect in the flower garden, that they are treated of here. These giants amongst Grasses are widely distributed in the warmer regions of the globe, the larger species reigning in the torrid zone. Humbler, but still remarkable representatives of the tribe are, however, pushed like advanced sentries to the northern parts of Japan and China, and to the south as far as the more southern parts of Chili, many plants from which regions are known to be quite hardy in England, as well as in continental Europe. The hardy kinds introduced into modern gardens are natives (with a single exception, perhaps) of Japan and Northern China, and for their introduction and distribution we are chiefly indebted to French horticulturists. The so-called subtropical gardening, which originated in Paris, and which has subsequently extended all over Europe, has indeed been the principal cause of researches having been made for plants of this tribe, many of which were first introduced at Paris. Such plants as Bamboos are specially to be prized, as they retain in an eminent degree the luxuriant gracefulness of tropical vegetation, even in the middle of winter, when their foliage shines as bright as ever. Nothing can exceed the grace of a Bamboo of any kind if freely grown. In cold bad soils and exposed dry places in the British Isles Bamboos have little chance; but, on the other hand, they will be found to make most graceful objects in many a sheltered nook in the south, south-western, and milder parts of England and Ireland. Those who wish to begin cautiously had better take B. Simoni, viridi-glaucescens, Metake, and falcata to commence with, as they are the hardiest. The best way to treat any of these plants, obtained in summer or autumn, would be to grow them in a cool frame or pit till the end of April, then harden them off for a fortnight or so, and plant out in a nice warm spot, sheltered also, with good free soil, taking care that the roots are carefully spread out, and giving a good free watering to settle the soil. There are no plants more worthy of attention than these where the climate is at all favourable, and there are numerous moist nooks near the sea-side where they will be found to grow most satisfactorily, as well as in the south. It is well to bear in mind that the first year or two of a young Bamboo after planting only thin poor shoots are produced. The following notes on the cultivated kinds, save B. Metake and B. falcata, are from an article in the *Gardeners' Chronicle* by M. Fenzi, of Florence, who has paid much attention to the genus.

B. aurea.—This species is easily distinguished by the stems assuming with age a clear golden colour, and its foliage also possesses a general golden hue, from which it has derived its appropriate name of Golden Bamboo. In general appearance it looks like a diminutive form of B. mitis in all its parts, the leaves being smaller and the stems thinner, differing, however, from it in having a rather straggling habit of growth, new shoots appearing often several yards from the parent plant. It appears to be wild both in Japan and in China.

B. falcata (*Arundinaria falcata, Thamnocalamus*).—A very ornamental species from Nepaul and the Himalayas, and at present the only kind of Bamboo much planted with us. It is hardy over the greater part of England and Ireland, but only attains full development in the south and west. It attains great luxuriance in Devon, and grows 20 ft. high near Cork, though in many districts it is stunted. It loves a deep, sandy, and rich soil, and plenty of moisture when growing fast. This is the most generally distributed Bamboo in British gardens, and is quite at home in them in many districts, though perhaps scarcely so hardy as B. Metake.

B. Fortunei fol. var.—This is a dwarf kind of Bamboo, so small, indeed, that at first appearance it would be taken for some humbler Grass. It is really a good plant, never attaining much more than 1 ft. in height. The comparatively large leaves are richly striped with white, and remain so all the year round, whilst other variegated grasses generally lose their variegation in summer. It is of an exceedingly spreading nature, and once established very difficult to eradicate, its numerous rhizomes running freely through the soil at the depth of 2 ft. and more. Thus it can defy the strongest cold, which will, at most, only slightly injure its elegant foliage.

B. gracilis.—Unluckily this species is hardy only in favourable situations, owing mostly to the peculiarity that it puts out new shoots in autumn, which are of course too tender when the first cold appears. If treated like a perennial plant it will put out early in spring new shoots full of vigour, which become quickly covered with their rich foliage, and form such a beautiful ornament as to well repay the slight additional trouble of covering it in winter. Magnificent specimens of this Bamboo are to be seen on the shores of the Lake Maggiore and that of Como. The B. graminea of many catalogues must probably be referred to this species.

B. Metake (*Metake*).—A large-leaved and rather dwarf species from Japan, growing from 4 ft. to 9 ft. high, with erect thickly-tufted stems, which are entirely covered by the sheaths of the leaves; the branches are also erect. It is a fine species as tested in London and Paris gardens, and seems with us to be the hardiest. It even thrives in London. The plants are of very free growth and well furnished, and the leaves, which are considerably broader and larger than those of the other hardy Bamboos, are of a fine green. It would be difficult to find more graceful ornaments for the flower garden or pleasure ground. It loves a peat soil, or a very free, moist, and deep loam, and runs a good deal at the root.

B. mitis, or edulis, is undoubtedly worthy the honour of being placed first, as it is the largest among the hardy kinds, the stems even of young plants soon attaining a thickness of more than 1 in., and a height of about 16 ft. It has a more compact habit of growth than many other kinds, its sub-

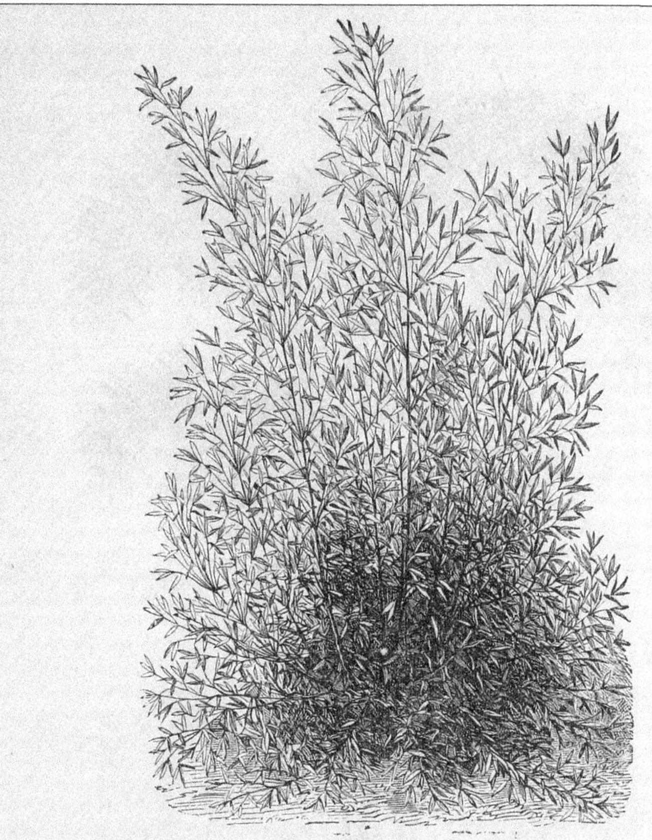

Bambusa aurea (*Golden Bamboo*).

Bambusa Metake (*Metake*).

[*To face page* 46.

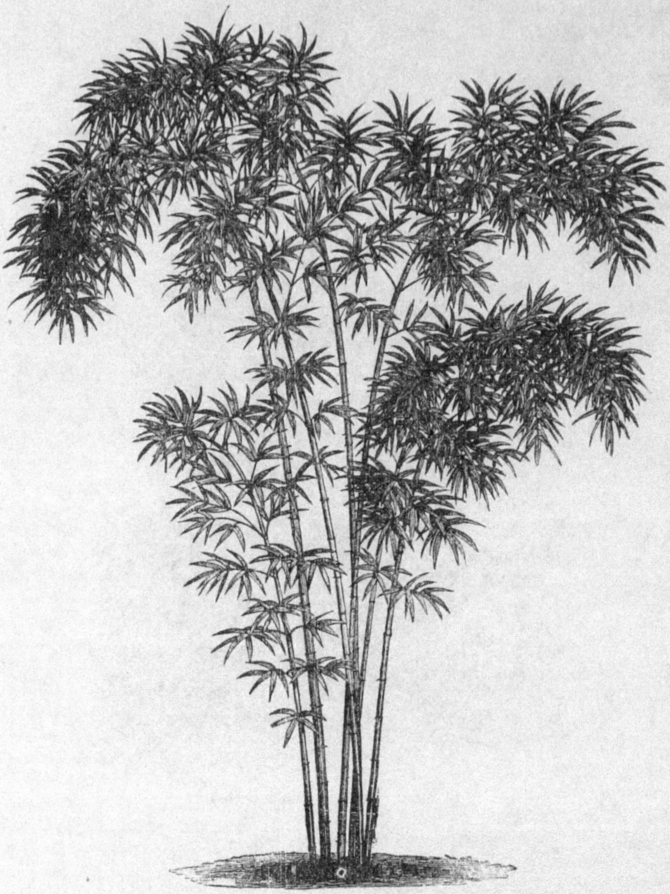

PLATE XL.

Bambusa viridi-glaucescens.

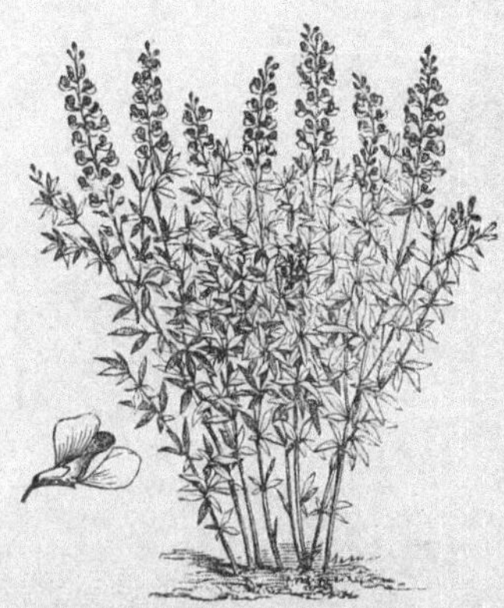

Baptisia australis (*False Indigo*).

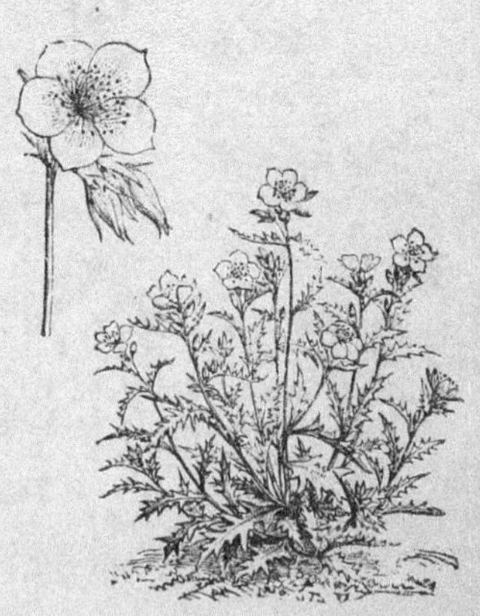

Bartonia aurea.

[*To face page* 47.

terranean rhizomes not expanding so much, and the shoots being inserted at a very small angle on the principal stem. According to Mr. Fortune, this kind attains in the northern districts of the Chinese empire the wonderful height of 60 ft. and a proportionate thickness. Is this the species alluded to by Mr. Fortune? Has the true species ever been tried in England? B. mitis is a late grower, new shoots appearing only in the month of June, and this peculiarity alone would confirm its comparatively northern origin.

B. nigra.—Among the Japanese species this well deserved the attention of early botanical travellers and collectors for its elegant stems and fine shoots, all of a shining jet black colour. The young stems when first developing from their spathes are greenish, changing gradually to a brownish yellow, which towards the end of the second year changes again into the beautiful dark colour characteristic of this species, and which makes it so well adapted for umbrella handles and riding whips. The foliage is very dense and of a deep green colour, the leaves being also more narrow than those of the above-described species.

B. Simoni, or Maximowiczi.—This kind, introduced recently, and, I think, at the same time into France, by M. Eugène Simon, and into St. Petersburg by Professor Maximowicz, appears to be a very hardy one, having been found in Manchuria.

B. violascens.—Under this name a new species has been recently introduced, which in its mode of growth and general appearance looks like an intermediate form between B. aurea and B. viridi-glaucescens. The young stems and shoots are beautifully tinged with a bright violaceous hue, and it seems to attain at least the size of the two species mentioned, and even to be not less hardy than them. It was first introduced from China into the Paris Garden.

B. viridi-glaucescens.—In its habit it is widely different from the other kinds in cultivation. Its slender culms, with rather long internodes, have their shoots inserted nearly at right angles, so that every one assumes a graceful drooping appearance, which is still more embellished by the hue of the rather large leaves, bright green above and glaucous below. The young stems are of a darkish green, changing gradually to a beautiful golden colour. Among the kinds as yet introduced this is certainly the most vigorous in growth. Clumps of this Bamboo (say eight or ten stems together), carefully taken up and potted, are of immense decorative value, the more so if seen by artificial light. They will look then like a group of finely-cut Chamædoreas, their foliage being more elegant and feathery than that of any Palm.

Baneberry *(Actea spicata, L.).*

Baptisia *(False Indigo).*—A free and hardy Lupine-like group of plants seldom satisfactory in gardens, though often grown. They are hardy perennials from N. America, forming strong bushy tufts from 3 ft. to 5 ft. high, with sea-green trifoliate leaves. The flowers are mostly of a delicate blue, and are borne in long erect spikes. Grows well in ordinary, deep, well-drained soil, preferring a sandy loam. B. australis, exaltata, and alba are the best known kinds.

Barbarea.—In the natural state none of the species of this genus are ornamental, but there are two varieties of B. vulgaris, an indigenous species, which are worthy of cultivation in any garden.

B. vulgaris fl.-pl. *(Yellow Rocket).*—This is a beautiful and curious plant. The process of doubling is, according to Mr. Sutherland, in his book on "Herbaceous Plants," "very peculiar in the flowers; I am not aware of any parallel to it in other double flowers. There appears to be no attempt made at any time to form either stamens or pistils. But the axis of the flower has the power of extending itself and producing numerous whorls of petals as it grows in length. A lengthened succession of flower is kept up both by this peculiar extension of the axis, and by the natural process of development of the inflorescence, which is open, but rather rigidly panicled. Height about 18 in. Flowers bright yellow, appearing from June till late summer, and often into autumn. B. vulgaris is a native of many parts of Britain, and this peculiar variety is probably an accidental variation." It is an excellent ornament of the mixed border, succeeds in almost any kind of soil, but prefers a rich moderately light loam. Propagated by division of the root-stock. There is a variegated form of the single plant called the Blotch-leaved Winter Cress, but it is not a very important plant. It is said to come true from seed.

Barnardia scilloides.—A pretty Scilla-like bulbous plant, growing about 6 in. high, and producing in summer dense clusters of small flowers, which are flesh tinted within and greenish on the exterior. It succeeds in any light sandy loam in warm borders or in the rock garden. A native of Macao.

Barren-wort *(Epimedium alpinum).*

Bartonia aurea.—A very showy golden-flowered hardy annual, 1 ft. to 2 ft. high. Should be sown in groups or patches where it is to remain in light soil in warm situations in April, thinning out the plants to about 1 ft. apart. As the seed is very small, care should be taken not to bury it too deep. The Bartonia is seldom used in any way but as a patch in a border, but well grown it is one of the annual plants which might be used as a bold mass or bed, relieved by tall slender plants through it here and there. Chili. Loasaceæ.

Bastard Balm *(Melittis Melissophyllum).*

Bear's-breech *(Acanthus).*

Beaucarnea. — A genus of graceful Dracæna-like plants, with swollen stem-bases, which grow freely in a cool house, but which have been occasionally used in the flower garden for a few months in the summer.

Beech Fern (*Phegopteris*).

Bee Balm (*Monarda didyma*).

Bee Larkspur (*Delphinium grandiflorum*).

Bee Orchis (*Ophrys apifera*).

Begonias.—A very extensive and beautiful genus till lately almost confined to hothouses, but the introduction of some hardy kinds and the raising of many hybrids and varieties by their aid has added a new feature to the flower garden. Beautiful, free blooming, and easily cultivated flowers, they are rapidly becoming great favourites, as they are the best plants that can be obtained—in the first instance during the months of June, July, August, and September as ornaments of the summer bedding garden; and secondly, as soon as there is any danger of frost at night, as ornaments of the conservatory. By lifting them into pots from the beds, which can be done without the slightest check, as they lift with capital balls and masses of fibrous roots closely and compactly surrounding the tuber, they afford five months of continuous bloom, which is more than any other plant with which I am acquainted will do. All they require to make them thrive and grow luxuriantly and quickly into large, free-branching tufts is a light, rich soil, and copious waterings while making their growth if the summer be at all hot and dry. The beds should not be at all hot or dry, and the plants should not be placed too thickly, as if the tubers be at all strong and of good size they make such vigorous and abundant growth, that if put too close to one another the plants will soon get crowded; when this inadvertently happens, every second one may with perfect safety be taken up even in the middle of summer and removed to another bed, where after a good watering they will next day appear as if they had never been transplanted at all, and the room afforded by their removal will be most welcome and serviceable to those left in the bed, allowing them fully to develop themselves. They soon fill up the gaps made by the removals in question. Begonias require a sheltered situation, fully exposed to the sun, as they dislike exposure to high winds, which are apt to break off the soft succulent stems at their junction with the tuber. No amount of heavy rains does them the slightest injury; the blooms are not knocked off by it, and though the heads droop somewhat and bend under it, they rise again absolutely uninjured as soon as the sun shines out after the shower. Those who wish to increase their stock of these lovely plants should begin in good time to take off their cuttings, just as they would those of a zonal Pelargonium, about the middle of July. These cuttings will root freely in silver sand and water in from a fortnight to three weeks, and before the end of the season make good-sized tubers, which often give a little bloom towards the end of autumn, and flower abundantly in the open ground the following year. Indeed, tubers in their second year often yield finer blooms than older and larger ones. The following are selected distinct varieties, but innumerable new seedlings are constantly being sent out.

SIX SELECTED DOUBLE BEGONIAS.—William Robinson, white; Marie Lemoine, rosy blush; Edouard Morren, light scarlet; Gloire de Nancy, deep scarlet; President Burelle, bright scarlet; Dinah Felix.

SIX SELECTED SINGLE BEGONIAS.—Countess of Kingston, Exposition de Sceaux, Jeanne d'Arc, Albert Crousse, Abbe Froment, Le Phoceen.—W. E. G.

Seed should be sown as soon as perfectly ripe, not later than middle of August, or then deferred to following February. Sow in small pots, carefully prepared on fine sifted sandy loam and leaf soil, on a perfectly flat surface. As the seed is very small no covering is required, excepting a slight sprinkling of silver sand. Plunge the pots in slight bottom heat, and as soon as the seedlings are large enough to be handled prick out into pans and keep in warm house near light until spring, transplant again in March into single pots, and many will show flower in May.

Tubers wintered in a cold house potted in April should be grown on without forcing for culture in outdoor beds in a free, light, fibrous mixture of loam and leaf-mould with a little coarse sand. They are very impatient of damp. Prepare the beds with plenty of well-rotted hot-bed manure, leaf-mould, and coarse sand, as Begonias require open compost to encourage free root action, and after planting they would much benefit by having a mulching of rotted manure or Cocoa fibre to keep the soil moist and assist surface rooting. Thus treated they will bloom from June to September, and if lifted before they are affected by frost and put in a warm greenhouse they will continue to bloom until December.—J. L.

Some of other species of Begonia are suitable for planting in the open during summer. The best of these are B. ricinifolia and heracleifolia, both with large palmately divided leaves; B. ascotensis and castanæfolia, erect growing kinds with narrow leaves and bright red flowers. The first named are used effectively in bedding out arrangements as an undergrowth to taller growing subjects, provided the latter are not too dense.

Bellevallia (*Hyacinthus*).

Bellflower.—A name for the various larger kinds of Campanula.

Bellis perennis (*The Daisy*).—Daisies deservedly rank amongst the most popular of garden flowers. They need only simple culture, increase rapidly, and freely accommodate themselves to soils and situations. In the decoration of the spring garden they are of great service, the several varieties and hues of colour in the flowers, when grown in large clumps or masses, producing striking effects. The common garden double Daisy

Tuberous Begonia.

Group of double and single tuberous Begonias.

[*To face page* **48.**

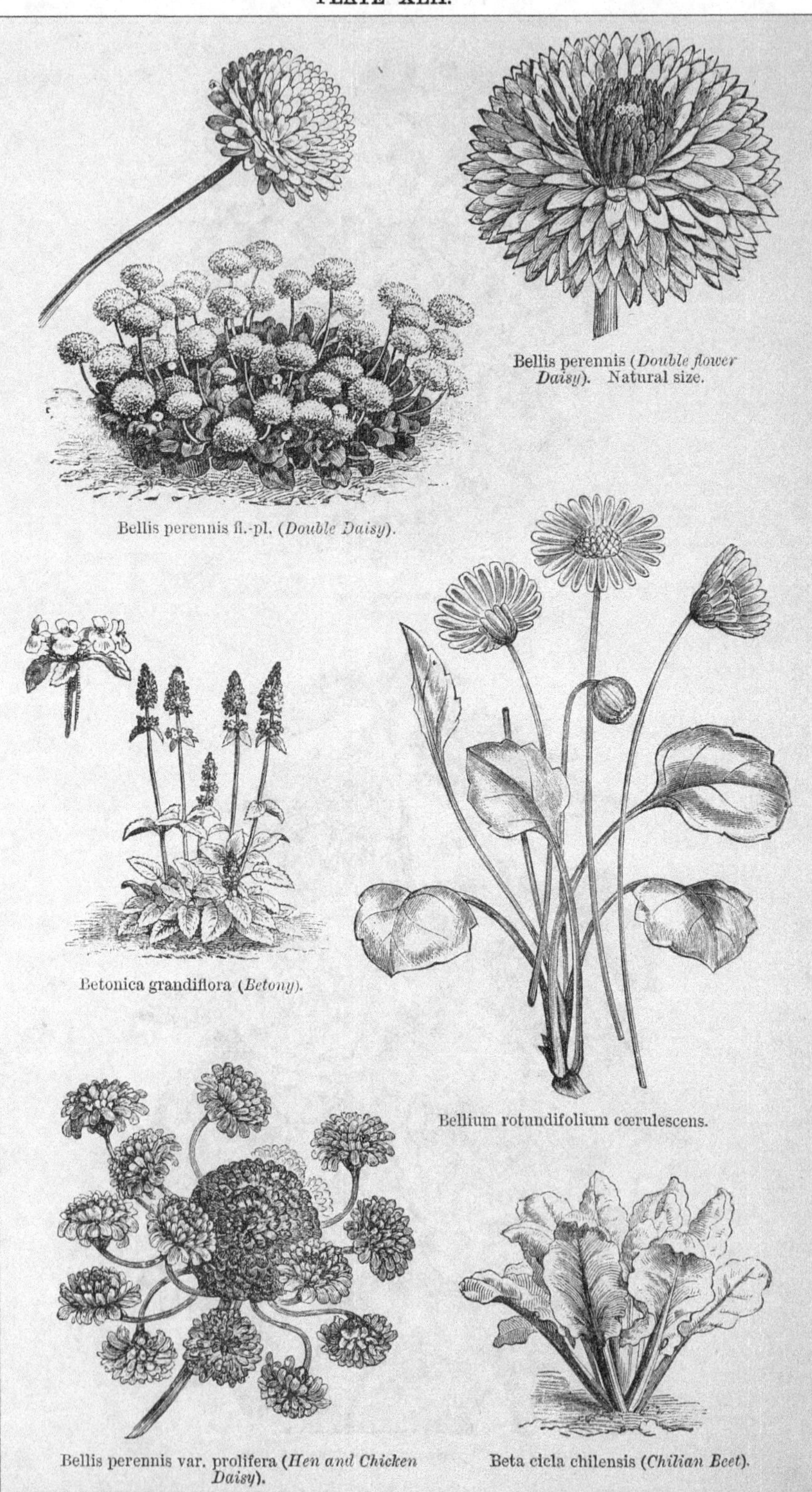

Bellis perennis fl.-pl. (*Double Daisy*).

Bellis perennis (*Double flower Daisy*). Natural size.

Betonica grandiflora (*Betony*).

Bellium rotundifolium cœrulescens.

Bellis perennis var. prolifera (*Hen and Chicken Daisy*).

Beta cicla chilensis (*Chilian Beet*).

[*To face page* 49.

has sprung from the wild Daisy of the field, just as the double Primrose has come from its wild single parent; and both owe their origin to some of the Continental florists of previous generations. Though we have numerous kinds, popular favour has adhered most closely to the old flat-petalled white and quilled red, both of which are grown by millions as market plants, and may be found in the spring in every costermonger's barrow. Added to these are the flat-petalled pink known as Pink Beauty, a charming pink of the quilled class; a deep rich red or crimson quilled kind, called Rob Roy; White Globe, large white quilled petals; an almost perpetual-flowered kind of medium size, having flat white petals, named Virginia, most useful for bouquets; and several others. As the double Daisy both sports and on the Continent seeds freely, new kinds of good quality are often cropping up; and it is by sporting that the yellow-blotched or Aucuba-leaved kinds have originated. Of these there are red, pink, and white-flowered sorts, the most effective being those having dark red flowers, as the comparison with the yellow leafage is very striking. These variegated forms, curiously enough, regain the usual green leafage in the summer, and exhibit their variableness only in the winter and spring months. Though somewhat more tender than the green-leaved or normal-leaved section, they will do well on a free porous soil in the winter, and in a cool shady border in the summer if transplanted there. Intense heat and drought or excessive moisture standing about the plant and foliage are the chief enemies ofthe double Daisy.

Propagating is simple, and may be done both in the spring and the autumn. Well dug soil, into which the plants, pulled to pieces, are dibbled at 6 in. apart each way, suits well, and if a good mass of bloom is desired, either in beds or as edgings, the plants may be put a little closer. Where the soil is good, the Daisy increases so rapidly that it may be transplanted twice in the year. One of the prettiest effects we have seen was a bed of white and red Daisies, planted in diamonds, with a strong plant of Blue King Pansy dotted into the centre of each. The giant or crown-flowered Daisies form almost a distinct section, and, though very vigorous, are not nearly so free of bloom as the better known kinds. These produce large and usually mottled red flowers upon long stalks, and are best suited for mixed borders. A very old favourite is the proliferous or Hen-and-Chickens Daisy. It differs in no respect of habit or foliage from its double congeners, except that when the flowers are at their best they send out numerous lesser ones from the axils of the involucral scales, hence the designation Hen-and-Chickens. It is worth a place as a floral curiosity; its flowers are usually rosy pink, and before starting its progeny very pretty. It may be propagated and cultivated as easily as the common garden kinds.

Bellium.—A genus closely allied to the Daisy, of which some three or four species are in cultivation. Although they come from the south of Europe, they are hardy on the rock garden. Like the beautiful Houstonia cœrulea, they are apt to exhaust themselves in flowering. B. bellidifolium, B. crassifolium, and B. minutum are much alike; B. minutum the best. Its numerous flowers are nearly as large as those of the Daisy. It is the best perennial of the three, and is easily grown in any light soil.

B. rotundifolium cœrulescens is a native of Morocco, and found in abundance on rich soils on the hills about Tangier, and occurring in great profusion by the water-courses of the Greater Atlas. This variety was first found by Balansa during his journey in Morocco in 1867, and has become well established in many gardens—not as a decorative plant in the same way that our favourite double Daisies are, but as a pretty rock plant, or filling a place in the alpine garden, where its profusely-produced cœulean blossoms invariably attract attention. It may be freely propagated at almost any time, either by division or by seed.

Bellwort (*Uvularia*).

Belvedere (*Chenopodium Scoparium*).

Bergamot (*Monarda fistulosa*).

Berkheya (*Stobæa*).

Beta cicla variegata (*Chilian Beet*).—A very showy plant and capable of being used with good effect. When well grown the leaves are often more than 3 ft. long, and present a vivid and most striking colouration. Their midribs are 4 in. or more across, and vary from a dark deep waxy orange to vivid polished crimson. The splendid hue of the lower part of the leaf-stalks flows on towards the point, and spreads in smaller streams through the main veins and ramifications of the great soft blade of the leaf, which is often 1 ft. and even 15 in. in diameter, if the plant be in rich ground. The under sides of the leaves are most richly coloured, and the habit such that these sides are well seen. It requires the treatment of an annual—to be raised in a gently heated frame, and afterwards planted out in very rich ground, though it may also be kept over the winter in pots. It varies a good deal from seed, and the most striking individuals should be selected before the plants are put out. Used sparingly, its effect would perhaps be more telling than if in quantity.

Betonica (*Betony*).—A large family of the Sage Order; not of much garden value. Betonica grandiflora is the only one amongst the numerous kinds of Betony in gardens that is really worth a place in a border. It is not at all fastidious as regards soil, though it attains the greatest vigour and produces the finest flower-spikes in rich loam which is not too damp. 12 in. to 18 in. high. Division.

Bidens (*Bur Marigold*).—A genus of North American annual plants of the Composite family. Only a few of these are worth cultivating, as there are several others of the same family superior in point of attractiveness. The species that are desirable are B. ferulæfolia, B. frondosa, and B. tripartita, all of which grow from 1 ft. to 1½ ft. high, and bear a profusion of yellow flowers and deeply cut foliage. They require to be treated similarly to other half hardy annuals.

Biebersteinia Orphanides.—A rare plant from South-eastern Europe, belonging to the Geranium family. It grows about 1 ft. high, has long finely divided leaves and erect spikes of small rose coloured flowers. It is sometimes met with in botanical collections.

Bindweed.-The English name for various kinds of Convolvulus.

Bird-foot Violet (*Viola pedata*).

Birthwort (*Aristolochia*).

Bitter Root (*Lewisia rediviva*).

Black Lily (*Lilium kamtschathense*).

Bladder Fern (*Cystopteris*).

Blechnum (*Hard Fern*).—There are several species in this genus of evergreen Ferns that are hardy and worthy of culture. The common British kind (B. Spicant) is pretty in itself, and yields several handsome varieties. Of these crispum has the lobes of the fronds undulated and curled and their points finely crested; cristatum is dwarfer in growth and the crested fronds are more forked and branched; imbricatum has lance-shaped fronds of thick texture, and the top of pinnæ distinctly imbricated; serratum rigidum, an erect growing variety with crested fronds, is distinct and handsome; multifurcatum has the fronds much divided and forked; and trinervium has the lowermost pinnæ so arranged as to give the frond a tripinnate appearance These are a few of the numerous varieties of B. Spicant, which now number nearly half a hundred. Among the hardy exotic kinds L. alpina is most desirable, as it is a rapid grower and soon covers broad spaces on rockwork by its dense growth. The hardy Blechnums may be grown in either loam, or loam and peat, or in a stiff clayey soil, but they dislike chalky soils and dry situations. They love shady and moist spots and abundance of water in the growing season. Under these conditions they attain a far greater size and vigour than when growing wild. They are alike suitable for the hardy fernery and shady spots or rockery and in borders.

Blennosperma californicum. — A dwarf growing annual from California, belonging to the Composite family. It is not much known in cultivation.

Blessed Thistle (*Carduus benedictus*).

Bletia hyacinthina. — A beautiful Chinese Orchid, having ribbed leaves and slender flower-stems 1 ft. or more high, bearing about half-a-dozen showy flowers of a deep rosy pink colour. It is only recently that it was found to be hardy. It thrives well in any sheltered and partially shaded situation in a peaty border well enriched by decayed leaf-mould. In some localities it would be advisable to cover the roots with a handful of protective material during severe cold. It is also known as B. japonica. A very interesting plant for the bog garden or bed of hardy Orchids.

Blitum capitatum (*Strawberry-blite*).—A hardy annual of the Spinach family, 1½ ft. to 2 ft. high. The flowers are small but followed by highly-coloured fruit calyxes which resemble small Strawberries. Treat as a hardy annual, sowing in April in the open air.

Bloodroot (*Sanguinaria canadensis*).

Bloomeria aurea.—An attractive little Californian plant, having umbels of small orange-coloured flowers striped with a deeper hue. It grows from 6 in. to 18 in. high, and is quite hardy in light garden soil in sunny positions, but for safety it should be protected during severe cold. *Syns.*—Nothoscordum aureum, Allium croceum.

Bluebell. — Applied to various plants, principally Campanula rotundifolia and the Wood Hyacinth (Scilla nutans).

Blue Cupidone (*Catananche cœrulea*).

Blue Spider-wort (*Commelina cœlestis*).

Bluets (*Houstonia cœrulea*).

Blumenbachia coronata.—This has much to recommend it, the flowers being showy, the foliage elegant, and the habit of growth dwarf and compact. Like all the Loasa family, the structure of the flowers is somewhat singular. The boat-shaped petals and the peculiar small scales between them, together with the brush-like bundles of stamens, render the blossoms very attractive. Its culture is simple enough; it merely requires to be treated as a hardy annual, but, like others from Chili, it is better to sow the seeds of it in spring than in autumn. It continues to flower from July to September if grown in warm light soils. The other species of Blumenbachia in cultivation are B. insignis and B. multifida. The blossoms of B. insignis are pure white, 1 in. across, with compressed keeled petals, furnished with a large serrated tooth on each side. B. multifida is a plant of much stronger growth, more hispid with stings, and with much larger five-parted leaves, longer two-bracted flower-stalks, and broader obtuse petals. Both species are natives of the southern parts of South America.

Bobartia aurantiaca.—A pretty Iridaceous bulbous plant from the Cape of Good Hope. It grows about 1 ft. high, and produces numerous rich yellow blossoms. As it is somewhat tender, it should be planted in light warm soil in a border, or on rockwork, and protected in winter if necessary. The bulbs should be lifted and separated after flowering, and replanted in autumn.

Bocconia cordata (*Plume Poppy*).—This forms handsome erect tufts from 3 ft. to over 8 ft. high, and is an admirable plant properly placed. The stems grow rather closely together, and are thickly set with large, reflexed, deeply-veined, oval-cordate leaves, the margins of which are somewhat lobed or sinuated. The flowers, which are very numerous, are borne in very large ter-

PLATE XLIII.

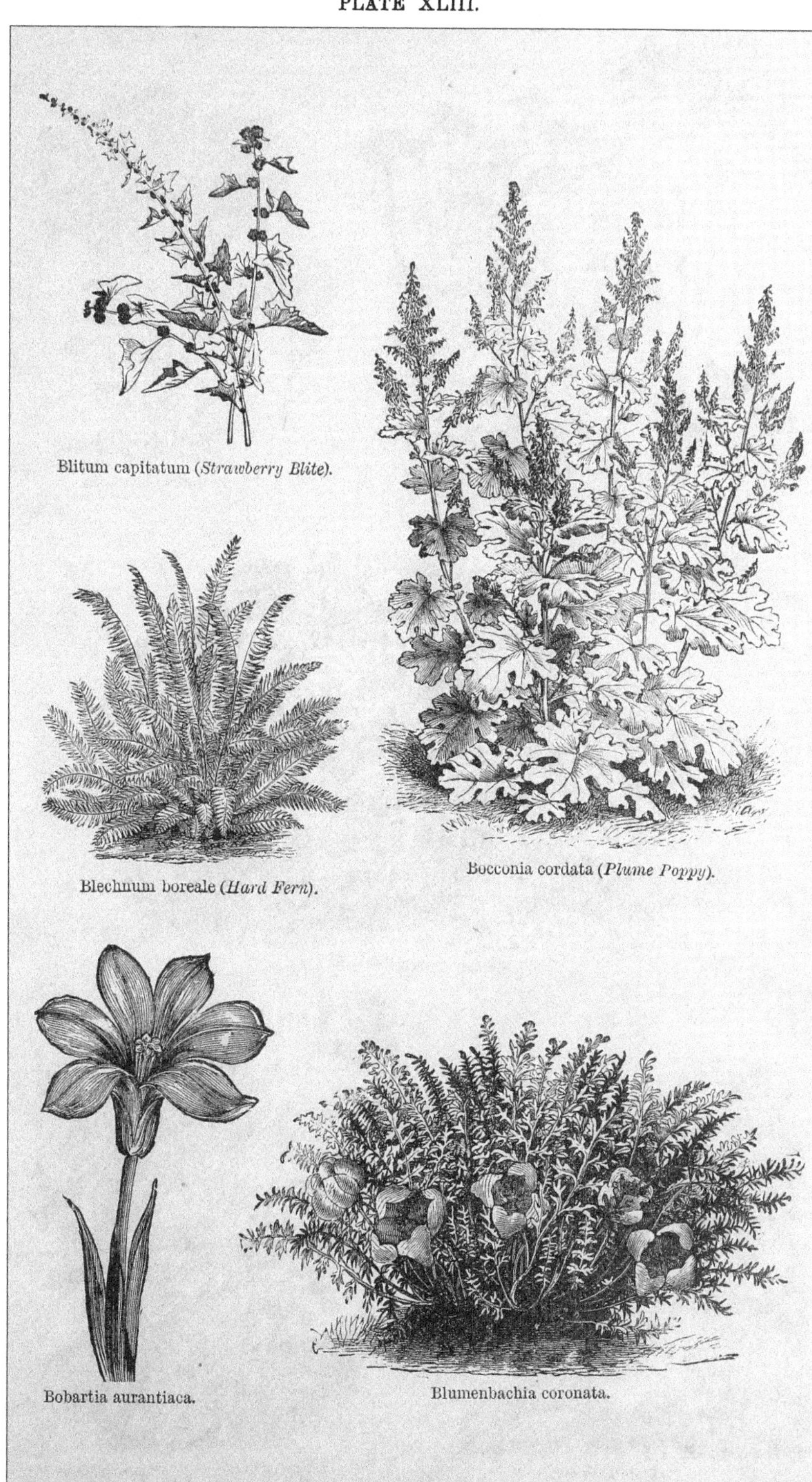

Blitum capitatum (*Strawberry Blite*).

Bocconia cordata (*Plume Poppy*).

Blechnum boreale (*Hard Fern*).

Bobartia aurantiaca.

Blumenbachia coronata.

[*To face page* **50.**

PLATE XLIV.

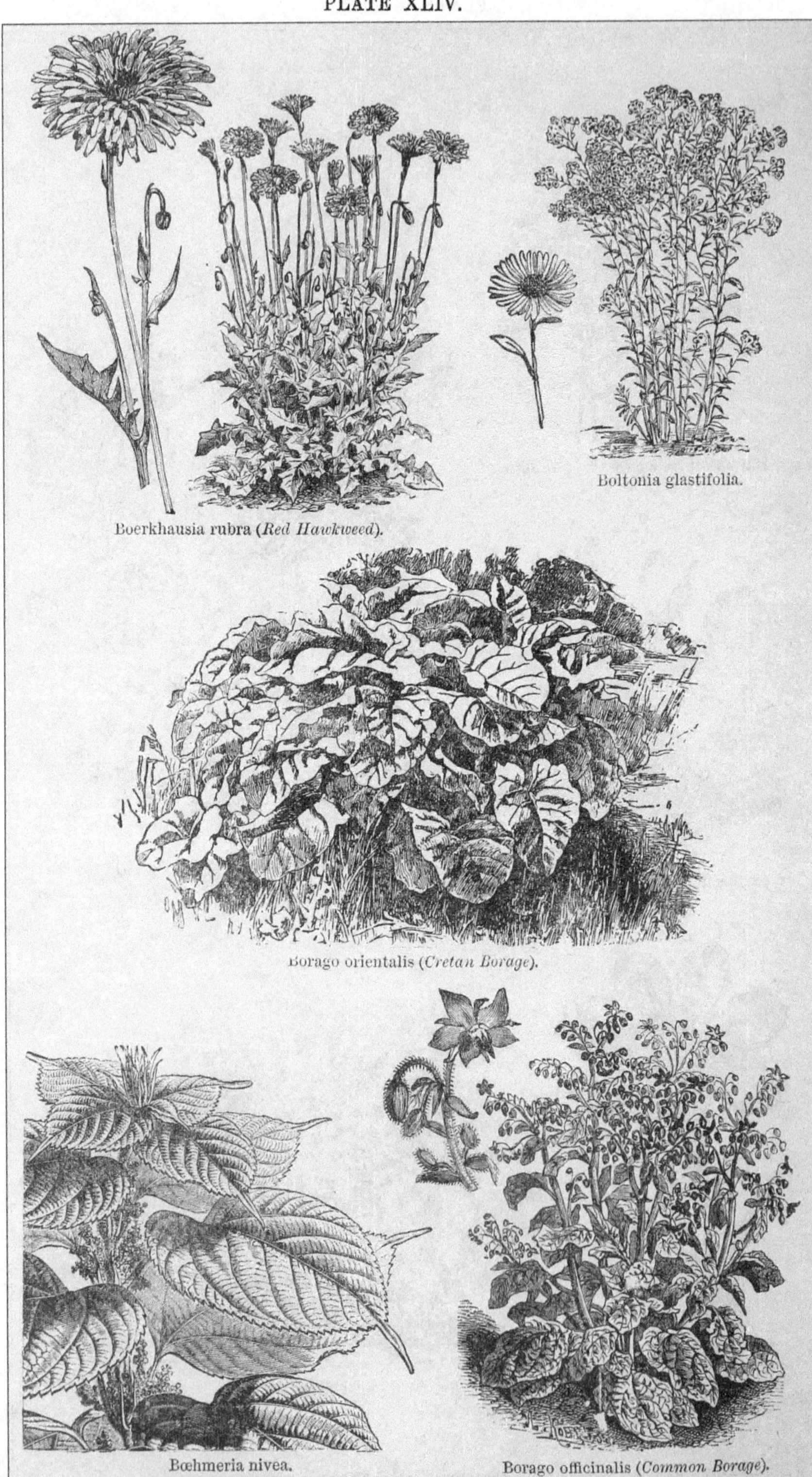

Boerkhausia rubra (*Red Hawkweed*).

Boltonia glastifolia.

Borago orientalis (*Cretan Borage*).

Bœhmeria nivea.

Borago officinalis (*Common Borage*).

[*To face page* 51.

minal panicles. The flowers are not showy, but the inflorescence, when the plant is well grown, has a distinct and pleasing appearance. The plant is seen to best effect when isolated, and does well in ordinary garden soil or free sandy loam. It is a good plant for the wild garden, the leaves being good in form, and the plant prettier when growing in a scattered way than as a stiff specimen. Division. China.

B. frutescens.—A vigorous-growing Mexican shrub, 3½ ft. to nearly 6 ft. high, with few and very brittle branches, large, sea-green, handsome leaves, and greenish flowers. Very effective when placed on Grass plats, either in groups or as isolated specimens. It requires a somewhat warmer climate than ours to thrive well, though it is sometimes seen in fair condition in the London parks. A mixture of free sandy loam and peat, well drained ground, and an airy position are necessary. Multiplied by sowing in a hot-bed in spring, and may be placed out from June to the end of September. It is difficult to propagate by cuttings.

Bœhmeria nivea.—A stout shrubby looking perennial of the Nettle family, occasionally grown in botanic gardens, but scarcely suitable for the private garden, though very distinct in habit, and with the under side of the leaves silvery. On warm soils about London it grows 3 ft. to 4 ft., and is increased by division.

Bœnninghausenia (*Ruta*).

Bog Arum (*Calla palustris*).

Bog Asphodel (*Narthecium ossifragum*).

Bog Bean (*Menyanthes trifoliata*).—Known also by the English names of Brookbean and Buckbean.

Bog Pimpernel (*Anagallis tenella*).

Bog Rhubarb.—A name for Petasites vulgaris.

Bog Violet (*Pinguicula vulgaris*).

Boltonia.—Plants very like the perennial Asters, and scarcely worth considering as distinct in a garden point of view. A recently introduced species, B. latisquama, is said to be a great improvement on the others in size of flower and brilliancy of colour, producing blossoms of a delicate rose colour.

Bomarea.—A most valuable genus allied to Alstrœmeria. South American, and requiring greenhouse temperature so far as now known, but Mr. Archer Hind, of Newton Abbot, states that he has B. edulis (syn., oculata) out of doors, and that this year it has flowered well after surviving a temperature last winter of 25° below freezing. If any of the other species should prove hardy in the southern counties it would be a gain, as their bold twining habit is so distinct, and their flowers are handsome.

Bongardia Rauwolfi.—This is a member of the Barberry family, though remarkably unlike one, as it has a depressed, Cyclamen-like stem, from the apex of which spring the leaves, which are divided into from three to eight pairs of leaflets, each of which is again divided, presenting the appearance of being arranged in whorls. These are wedge shaped, of a pale glaucous green hue, and each has a conspicuous reddish-purple blotch at the base, thus rendering them very attractive. The flower-stem is much branched, rising 6 in. high, and bearing roundish, golden-yellow blossoms from ¾ in. to 1 in. across, which droop gracefully from slender stalks. Though now rare, this beautiful plant is amongst the earliest recorded occupants of the garden, and is mentioned by all the early writers on garden plants. It is found from the Greek Archipelago to Afghanistan. Hardy on dry soils, and raised from seed.

Borago orientalis (*the Cretan Borage*) is a vigorous perennial often cultivated, but of slight value, bearing pale blue flowers early in spring, and in good soil having large and imposing foliage throughout the summer. Easily naturalised in any rough place, it is not worth a place in the garden proper, being very coarse in habit and occupying too much space; it should be associated with the early spring flowers. The common Borage is a very pretty plant, naturalised in dry places or banks, where it might be often welcome to those who sought it for other purposes than its beauty. There is a white variety. Borago laxiflora is pretty with its numerous suspended blue flowers; it grows very freely on sandy soils, but is not a good garden plant. On the whole, the genus as now known in gardens is of little interest, the common Borage in flower being really the most beautiful plant belonging to it.

Borkhausia (*Hawkweed*).—Of this genus of Compositæ, only one, B. rubra, the Red Hawkweed, is worthy of culture. It is a hardy annual from Italy, bearing pretty pink flowers about the size and form of the Dandelion. It should be sown in spring or autumn like other hardy annuals in any ordinary garden soil. It flowers from June to September, and is suitable for borders or beds of annual flowers. There is also a variety with white flowers.

Botryanthus (*Muscari*).

Botrychium (*Moonwort*).—A somewhat inconspicuous genus of Ferns, of which there are but a few species cultivated. The common Moonwort (B. Lunaria) is a widely distributed native species, generally found in moist, sheltered meadows. B. lunarioides, B. virginicum, B. lanceolatum, B. simplex, and B. ternatum are North American species perfectly hardy in this country. All the Botrychiums are deciduous, putting forth their young fronds about the end of April, and dying down somewhat early in the autumn. The fronds are of a beautiful dark green colour, and the plants produce small panicles of inflorescence. The soil best adapted for these Moonworts is a moist well-drained sandy mixture of loam and peat, just the kind of compost in which most of the American plants delight. With Spiræas, Lilies, &c., where the soil is little disturbed, and abundance of moisture is obtainable during their growing season, they would afford a very pleasing "groundwork."

Boussingaultia basselloides.—A very luxuriant trailing plant, 16 ft. to nearly 20 ft. long, sometimes more. Flowers, late in autumn, small, white, fragrant, becoming black as they fade, disposed in clusters 2 in. to 4 in. long, which spring from the axils of the leaves at the ends of the branches. The leaves are of a fine green, smooth, shining, fleshy, slightly wavy; stems very twining, tinged with red, growing with extraordinary rapidity, and producing numerous tubercles. Native of Quito. It is only suited for dry banks and chalk-pits, associated with climbing and trailing plants. Propagated by means of the tubercles of the stem; these are extremely brittle, and break with the least shock, but the smallest fragment will vegetate as well as the entire tubercle.

Bowiea volubilis.—A Cape bulbous plant, interesting only from a botanical point of view, as it is one of the few of the Liliaceæ having a climbing habit. The plant is devoid of foliage, but has numerous fleshy terete branches, rising several feet in height, proceeding from a large Turnip-like green bulb, half raised above the surface. The flowers are small and inconspicuous. It is a hardy perennial when planted in a sheltered situation against a wall, preferably with a southerly aspect. It should be allowed to ramble over dead branches placed near it. Propagated by seeds in this country.

Boykinia aconitifolia.—This is a North American hardy perennial plant nearly allied to the Saxifrages. It has a dense tuft of root leaves somewhat resembling those of an Aconitum, and tall stems with blossoms about the size of a threepenny-piece. Pure white and produced rather plentifully in loose clusters more curious than beautiful.

Brachycome iberidifolia *(Swan River Daisy)*.—A pretty annual of simple culture. It grows from about 8 in. to 12 in. high, and as its habit of growth is somewhat straggling, it is particularly suitable for grouping in masses. The flowers, about 1 in. across, in loose terminal clusters, are bright blue, with a paler centre, and resemble those of the common greenhouse Cineraria. There are other sorts with flowers of various shades of blue and purple, and one is pure white. Sow in cool house in September, prick off soon as ready 4 or 5 in a 4-inch pot, keep in cold pits during winter, guarding against damping. Pot on again in March singly into 4-in. pots, and finally plant out end of April into open borders; or sow on slight hotbed in March, prick out into pits for transplanting into open in May; or sow in the open in April and May.

Bracken *(or Brake)*.—Applied to large Ferns generally, and more particularly to Pteris aquilina.

Brasenia *(Water Shield)*.—A North American water plant. The flowers are of a dull purple, and come to the surface to perfect themselves, but make little show. This aquatic plant has a remarkably wide distribution, it being found not only in North America, including the north-west coast, but in Japan, Australia, and India.

Brassica.—Some forms of the Cabbage are used in the flower garden for winter effect, particularly the variegated Kales. The great secret in cultivation of these is poor soil. This brings out their colours and keeps the plant dwarf and compact in habit. Seed should be sown in May, and when fit to handle the plants should be pricked out into a very sunny spot, wide apart, and finally bedded out in October after the summer stuff is cut down by frost, a clump of them on a lawn well planted and varied in colour having a charming effect on clear days. In planting the stem should be completely hidden by sinking in the ground up to the lower leaves, and care should be taken that they are of equal height. Dwarf curled variegated makes the finest effect in beds. The colours are very varied, being from pure white to deep purple, passing through cream pink, pale and dark green, with variously coloured veins and laced edges. The Ragged Jack stands next in order of beauty; these are better seen on a close inspection, the beautifully cut leaves adding an additional charm. The new proliferous variegated is a curious variety raised by Messrs. Stuart, Mein, and Allan, of Kelso, who have long made a speciality of Kales; it is exactly like the dwarf curled variegated, but the midribs and upper surface of the leaves are completely studded with secondary leaflets which have a most beautiful effect when closely inspected. To have a good display of these Kales great attention must be paid to the fact that the seed must be of a good strain or selection, as the want of the finer laced and fringed light coloured varieties is totally against a good bed, the commoner colours being sombre, dark greens with purple veins. The Carnation striped Red Cabbage makes a splendid bed, but is not so hardy as the Kales.

B. oleracea crispa is a handsome kind of Cabbage nearly 4 ft. high, with elegantly cut arching leaves, the divisions of which are finely curled or frizzled. In autumn and winter it may be advantageously employed in the embellishment of winter gardens, the leaves being at their best during that part of the year. A still more striking subject is B. o. palmifolia, which attains a height of 6½ ft., and bears its leaves near the summit of the stem, having quite a Palm-like appearance in the end of the summer and in autumn. This kind might be used with good effect in various positions, as its "Cabbage" character is not so evident. The fact of their being Cabbages prevents many people from using these plants.

Bravoa geminiflora *(Scarlet Twin Flower)*.—A pretty Mexican bulbous plant of the Amaryllis Order. It grows from 1 ft. to 2 ft. high, the flower-stems being stout and erect, bearing on the upper part numerous

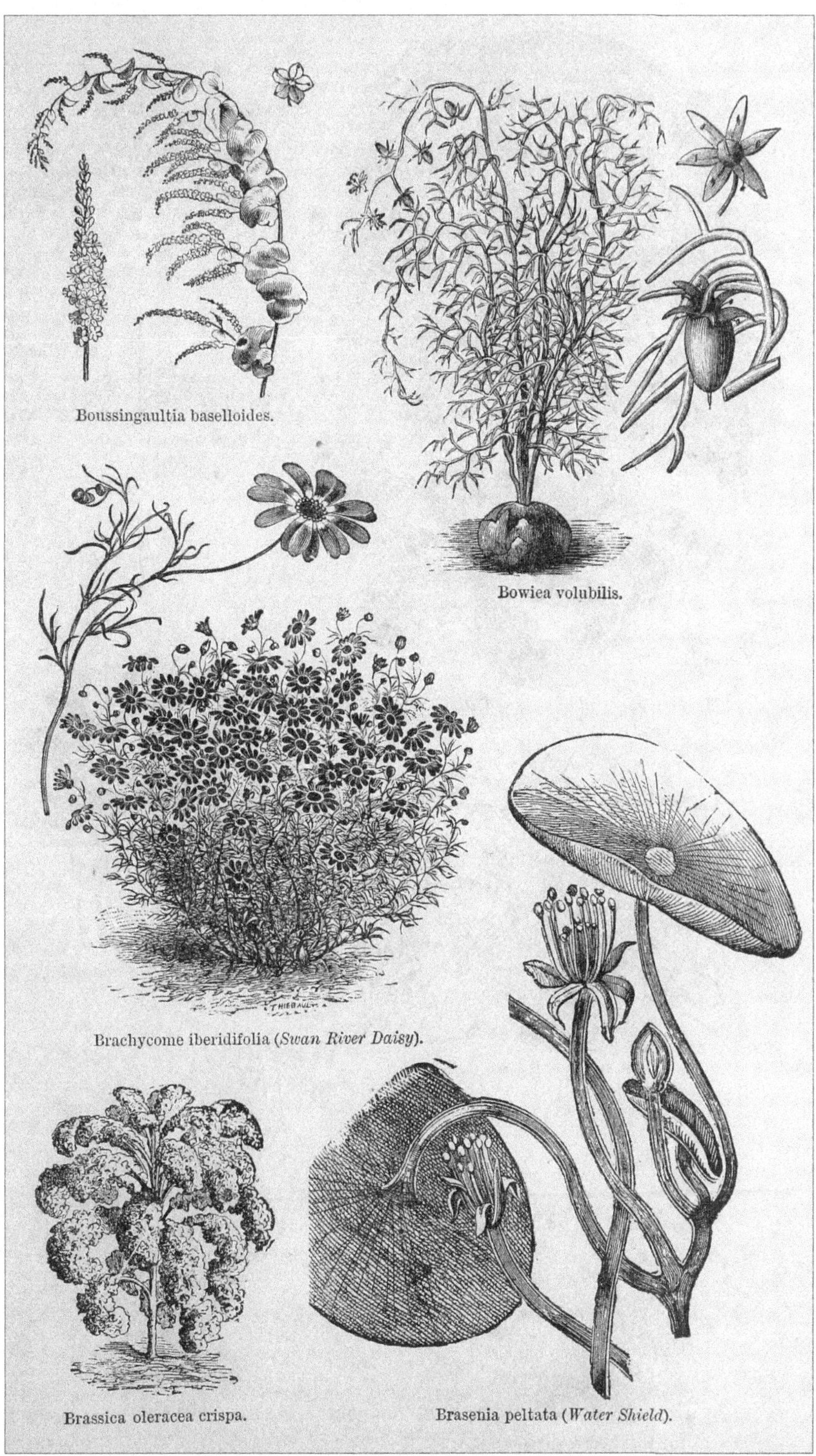

Boussingaultia baselloides.

Bowiea volubilis.

Brachycome iberidifolia (*Swan River Daisy*).

Brassica oleracea crispa.

Brasenia peltata (*Water Shield*).

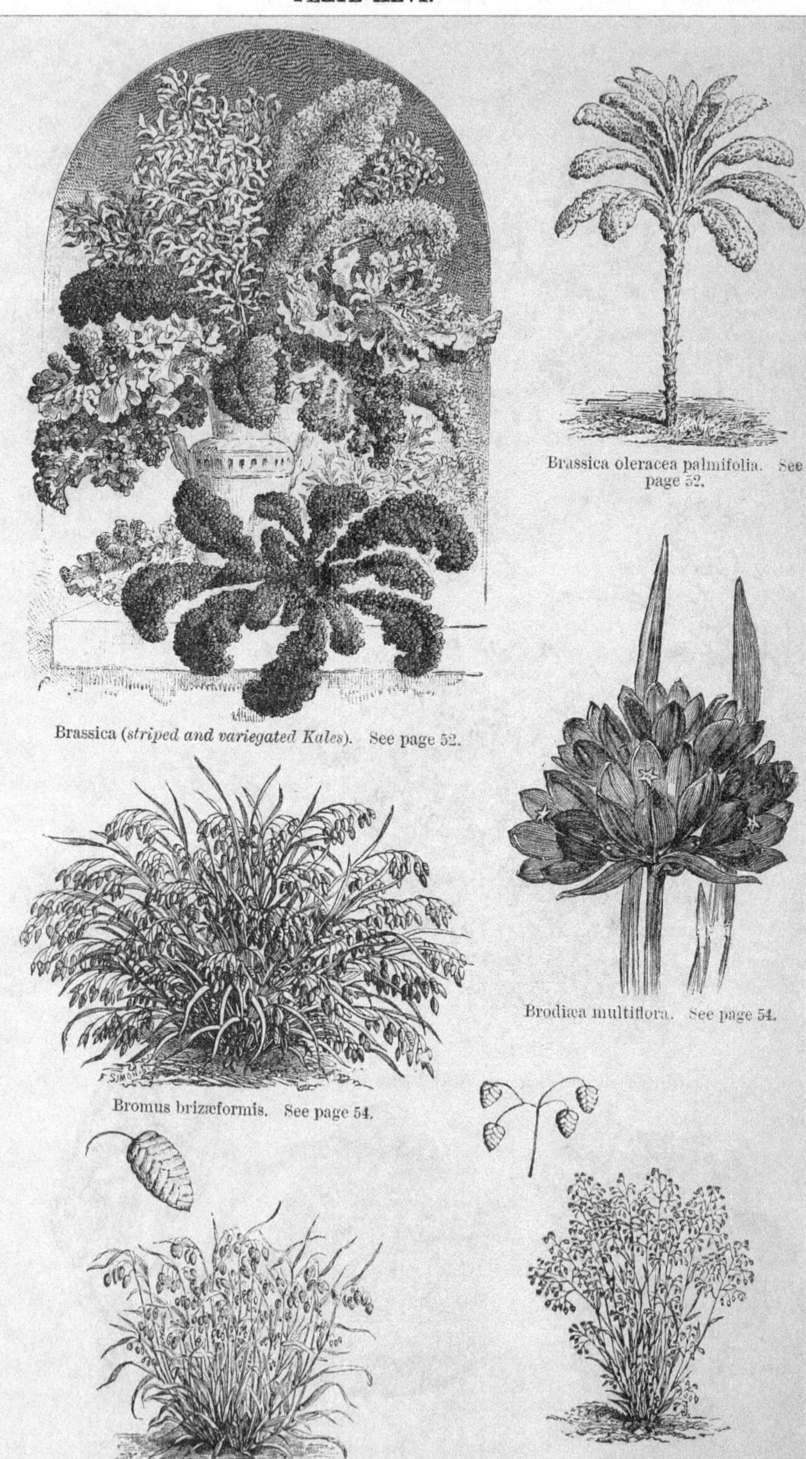

Brassica (*striped and variegated Kales*). See page 52.

Brassica oleracea palmifolia. See page 52.

Brodiæa multiflora. See page 54.

Bromus brizæformis. See page 54.

Briza maxima (*Quaking Grass*).

Briza minor.

[*To face page* **53.**

nodding tubular flowers in pairs. The colour is a rich scarlet on the exterior, inclined to yellow within. It succeeds well in warm sheltered situations in borders of light and well drained soil, but it generally requires some little protection over the bulbs in winter, though in mild seasons that is not required. It flowers in autumn and remains a long time in bloom.

Brevoortia coccinea *(Crimson Satin Flower).*—The name now applied to the plant generally grown as Brodiæa coccinea, one of the prettiest Californian plants in cultivation. It is of slender growth, the flowers being produced in drooping umbels on terminating stems, 1½ ft. to 2 ft. high. The flowers are tubular and of a deep crimson red, the lips being a vivid green colour. This plant is not only uncommonly showy—a feature which it shares with some other beautiful Liliaceæ from the same State—but, unlike some of them—for instance, the Calochorti—it is a most manageable bulb, and thrives in our fickle climate. It appears to succeed in soil of almost any description between the extremes of very light and very heavy, but makes the most vigorous growth in substantial, but friable loam; the bulbs may be kept out of the soil for at least six months, that is from August to February, without injury, though it is not desirable to keep them so long. It is not desirable to delay planting beyond October, as the flowers of the late-planted bulbs are scarcely so fine as they otherwise would be; and once planted they may remain undisturbed for several years, unless raised for the purpose of division. Offsets are produced in some quantity, and the plant may also be increased by seed, which vegetates pretty freely after an interval of some weeks, the resulting bulbils attaining a flowering size in from three to four years. As the plant is of slender habit, the flowers being usually produced on a naked space 1½ ft. high or more, not less than three bulbs should be grouped together, and five or six will produce a still better effect; a single Osier rod placed in their midst will suffice to support the somewhat fragile stems. In soils, notably those of a sandy character, the foliage is liable to wither partially by the time the flowers are in perfection. This may partly be remedied by sowing a few seeds of some neat annual of dwarf habit, which would carpet the soil around the stem, and bloom subsequently to its removal, or in the case of the early-flowering annuals, such as Limnanthes and Leptosiphon simultaneously with the bulbs.

Brexia madagascariensis.—A handsome stove shrub with a slender erect stem, clothed with leathery, long leaves. It is one of the tropical stove plants that have stood well in the open air from June to early in October, but very few places can spare it for this purpose. It requires ordinary stove culture during winter and spring, and should only be placed out after having made a strong growth, and having that growth hardened off. Madagascar.

Briza *(Quaking Grass).*—A graceful family of Grasses, mostly American, but some European. B. maxima is one of the handsomest of the smaller ornamental Grasses, growing 12 in. to 18 in. high; may be sown in the open ground in March, being quite hardy and exceedingly graceful while growing, and most useful for decoration either green or dried. B. media (Common Quaking Grass) is a smaller kind, 9 in. to 15 in. high. Borders, and naturalised on bare banks and slopes. Seed.

Brizopyrum.—A genus of Grasses, none of the species of which are ornamental enough to cultivate.

Brodiæa.—A genus of North American bulbous plants belonging to the Lily family. This genus has lately been thoroughly revised by the American botanists, whose nomenclature is followed in the subjoined account. It will be observed that the genera Calliprora, Triteleia, Milla, Seubertia, Hesperocordum are partly or entirely included among the Brodiæas, whilst B. coccinea (Gray) is called Brevoortia coccinea (Watson), and Brodiæa volubilis (Baker) is Stropholirion californicum. Brodiæa as now arranged contains fourteen species, of which about ten are in cultivation in this country. These are

B. capitata.—A pretty species, with flower-stems 1 ft. to 2 ft. high, supporting a dense many-flowered umbel of violet blue flowers. Flowers, like the preceding, in early summer.

B. congesta has stems from 1½ ft. to 4 ft. high, having a dense cluster of bluish purple blossoms, produced plentifully in early summer. The white variety (alba) is a pretty plant, and, like the type, is excellent for cutting purposes, as they last such a long time in good condition. Common in California.

B. gracilis.—A slender growing plant, growing from 2 in. to 4 in. high, bearing deep yellow flowers with brown nerves. Not very showy.

B. grandiflora.—A handsome plant, with the flower-stem about 1 ft. high, terminated by an umbel of from 3 to 10 flowers, varying from purple to a light rose colour. There is a variety, taller and stouter in growth and with larger blossoms, called major flowers, in early summer lasting in bloom several weeks. The variety minor of Baker is ranked as a species by Watson. Various parts of California.

B. ixioides.—A pretty plant, generally cultivated as Calliprora lutea. It has flower-stems 1 ft. to 2 ft. high, bearing a large loose umbel of yellow flowers tinged with brown. Known also as Milla ixioides.

B. lactea.—An elegant plant, producing large umbels of white star-shaped blossoms on stout stems 1 ft. or more high. The variety lilacina has lilac-tinted flowers. It is synonymous with Hesperocordum hyacinthinum and H. lacteum.

B. laxa.—A beautiful plant, having a large umbel of flowers produced on slender stems from 1 ft. to 2 ft. high. The colour of the blossoms is generally a bright violet-purple, with a deeper stripe running through each division. There is also a pure white variety, and another with delicate pink-coloured flowers. The variety Murrayana is much finer than the type as regards size, and the plant is more floriferous. B. laxa is also called Milla, Triteleia, and Seubertia laxa.

B. multiflora resembles the preceding, but flowers several weeks in advance. The flower-stem rises from 1 ft. to 2 ft. high, terminated by lilac-tinted blossoms; called also B. grandiflora var. brachypoda.

The kinds that are either not in cultivation or rare are, B. peduncularis, Douglasi, Bridgesi, crocea, and terrestris.

CULTURE AND POSITION.—The place to grow these plants in should be a sunny open spot, never shaded. The border should be thoroughly well drained, and if raised above the surface so much the better. The soil should consist of a rich garden mould, light and friable, and the bulbs planted at a good depth, so that they would be out of harm's way in the winter, for, though most of them are perfectly hardy, our recent severe winters have injured them in many places. Sunny borders or beds of choice shrubs offer a suitable position for this plant, apart from the ordinary borders, and devoted to the newer and choicer hardy bulbs. When practicable they should be planted in bold masses, as they have a much finer effect when in flower than just one or two together, and another advantage of massing is that the bulbs can be readily protected in winter if necessary. The propagation and other details of culture are given more fully under Brevoortia above.

Bromelia sphacelata *(Hardy B.)*.—A rough curious plant, forming tufts of harsh, rigid, spiny leaves 1 ft. to 2 ft. high. Flowers in summer; purple, sessile, crowded, and overlapping each other, in axillary spikes. Leaves numerous, erect, sword-shaped, long-pointed, fringed with stiff spines pointing upwards. Chili. The rock garden and warm borders in light perfectly-drained soil. This plant has stood out of doors for several winters, and will probably prove quite hardy on rockwork in the southern counties.

Bromus.—A large genus of Grasses, at least one of which is a very graceful plant, well worthy of culture—that is B. brizæformis, a very hardy biennial about 2 ft. high, with graceful and large drooping heads. This we have found to be more valuable for cutting and drying than any of the Quaking Grasses (Briza). We have grown it as an annual sown out of doors in spring and in pots, but probably autumnal-sown plants would give the best results. It is well worth a place for the sake of cutting for preservation if for no other reason.

Browallia elata.—This has generally been considered only as a beautiful pot plant in summer and autumn, but it does well in the open air, either in a bed by itself or in large patches mixed with other things. It supplies a shade of colour that is difficult to obtain, and it may be safely added to the list of bedding annuals that are continuous bloomers. Sow the seeds in March, prick off the young plants when large enough to handle, and grow them on till they are strong, and plant out in May. In addition to its other good qualities and its adaptability for massing, it is very useful to cut from, and the flowers are light and elegant. There is a white variety of it equally useful. B. grandiflora, a fine form of the old B. elata, is most useful for conservatory and indoor decoration. It comes true from seed. The flowers, being of a beautiful shade of blue, are very valuable for associating with those having bright colours that are usually employed in such structures. It also succeeds well out-of-doors. The new B. Roezli forms a dense compact bush, 16 in. to 20 in. in height, clothed with shining green leaves. The flowers are either of a delicate azure blue, or white with yellow tube, and are unusually large for the genus. They are produced in uninterrupted succession from spring until autumn. A native of the Rocky Mountains.

Bryanthus erectus *(Hybrid B.)*.—A dwarf evergreen bush, from 8 in. to 1 ft. high, bearing pretty pinkish flowers. It is said to be a hybrid, and is in appearance somewhat intermediate between Rhododendron Chamæcistus and Kalmia glauca. In very fine sandy soil or in that usually prepared for American plants, it grows well, and is worthy of a place on rockwork or in collections of very dwarf alpine shrubs, whether planted in the rock garden or in neat beds. There is at least one pretty American alpine shrub of the same genus which was once introduced, but now lost to our gardens.

Bryonia dioica *(Bryony)*.—Modest merit is acknowledged in the case of this wild British climber in various German gardens. It is chosen for its enduring verdure as well as for its graceful growth and foliage. It forms many wreaths in the town gardens of Vienna, and it is also trained up at the angles of large statue pedestals in the garden of Count Schwarzenberg. So trained, it looks well, and pleasantly relieves the mass of stone, but it need hardly be said that so common a plant will scarcely find a place in many gardens.

Buphthalmum cordifolium *(Oxeye)*.—A stout Composite plant with large and handsome foliage and large Sunflower-like blooms of a good orange colour. It is hardy on any soil, and its true place is the wild garden among the larger perennials, or, still better, in a small spreading colony by itself. The species belong to a very small family, some of which are tropical and others European, but none is so suited for the wild garden as B. cordifolium, which attains a height of from 3½ ft. to 5 ft. Hungary. There is no trouble in either cultivating or increas-

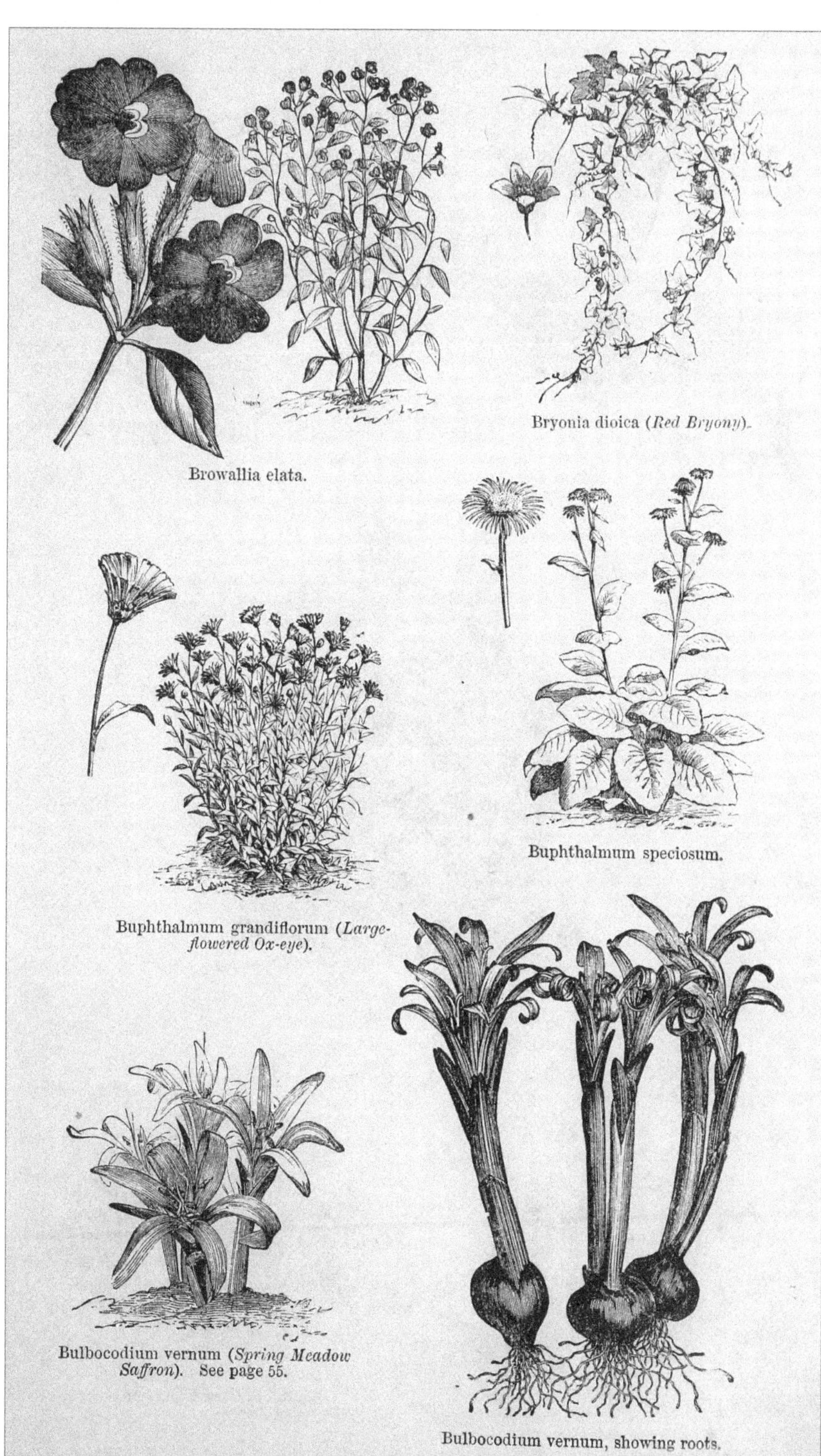

Browallia elata.

Bryonia dioica (*Red Bryony*).

Buphthalmum grandiflorum (*Large-flowered Ox-eye*).

Buphthalmum speciosum.

Bulbocodium vernum (*Spring Meadow Saffron*). See page 55.

Bulbocodium vernum, showing roots.

[*To face page* 54.

PLATE XLVIII.

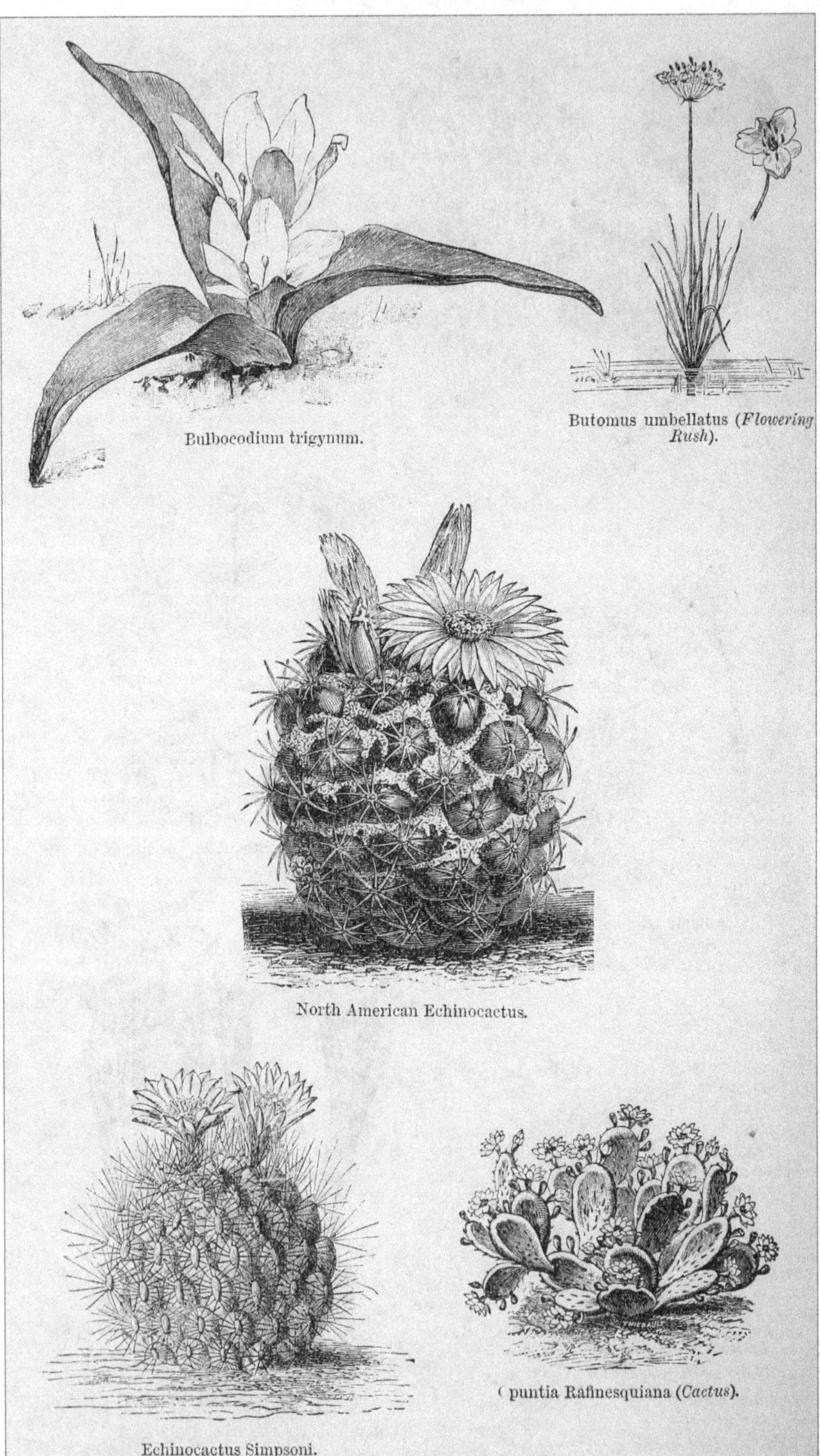

Bulbocodium trigynum.

Butomus umbellatus (*Flowering Rush*).

North American Echinocactus.

Echinocactus Simpsoni.

(puntia Rafinesquiana (*Cactus*).

[*To face page* **55.**

ing it. It is the best plant of its family. *Syn.*—Telekia speciosa.

Bugle *(Ajuga).*

Bugloss *(Anchusa officinalis).*

Bulbine.—A genus of Liliaceæ of which only one species, B. annua, from South Africa, can be grown successfully in the open air—an annual, growing about 1 ft. high, with narrow erect leaves and spikes of small yellow flowers that are not very showy. Only suitable for botanical collections.

Bulbocodium vernum *(Spring Meadow Saffron).*—A very pretty bulbous plant, growing from 4 in. to 6 in. high, flowering in early spring. If several tufts of it are put in good sandy soil on the rockwork or choice spring garden, it will prove one of the best as well as earliest of spring bulbs, sending up its fine large rosy purple flower-buds, distinct in colour from any other spring flower, earlier than Crocus susianus; in fact, they often show for several weeks ere the snow takes leave of us. The flowers are tubular, nearly 4 in. long, and usually most ornamental when in the bud state, the colour being a sweet violet-purple, the large buds appearing before the concave leaves, which attain vigorous proportions after the flowers are past. Associated with very early flowering plants like the Snowflake, Snowdrop, and Anemone blanda, it is very welcome indeed in the rock garden or in warm sunny borders. A native of the Alps of Europe, easily increased by dividing the bulbs in July or August, replanting them at distances of from 4 in. to 6 in. There is a variegated leaved variety of B. vernum. One other species, B. trigynum, is sometimes met with in cultivation, but it is rare, and little known.

Bullrush *(Scirpus lacustris).*—Erroneously applied to Typha latifolia, the Reed Mace.

Burning Bush *(Dictamnus Fraxinella).*

Butomus umbellatus *(Flowering Rush).*—A handsome native water plant, often very fine in a rich muddy soil. It should form a feature in all pieces of natural or artificial water, into which it is easily introduced. It flourishes best in the deep soil that occurs near many rivers and in the full sun.

Butter-and-Eggs.—An English name for a double form of Narcissus incomparabilis; also applied to the common Toad-flax.

Buttercup.—The English name for various native yellow Ranunculi, some of which in their double form are cultivated in gardens.

Butterfly Orchis *(Habenaria bifolia).*

Butterwort *(Pinguicula).*

Butterbur *(Petasites).*

Cacalia.—Composite plants of little garden value. The pretty Emilia sagittata is also called C. coccinea.

Caccinia glauca.—A dwarf hardy perennial belonging to the Borage family, from the highlands of Persia. It grows about 9 in. high, and bears short racemes of flowers, each about ½ in. in diameter, resembling in form that of Borage, pinkish when first expanding, but changing to blue. It is a neat border plant, which will interest the botanist and general lover of plants, but can hardly be termed a showy plant.—W. T.

Cactus.—Under this name various plants belonging to the Cactus family which have proved hardy in Europe and in England, have been much talked of. Opuntia, Echinocereus, Mammillaria, and Echinopsis are the genera already proved hardiest, and the following general remarks on the subject may be useful and suggestive. We have seen some charming effects afforded by Cacti in the open air in Southern England, the plants proving hardy and blooming freely fully exposed to the sun on a low rock garden. Miss Rose G. Kingsley wrote to *The Garden*: "In Dr. W. A. Bell's garden at Manitou, Colorado, I saw a collection of eight or ten species of hardy Cacti, which had been found in the neighbourhood, and which were grouped together on a bit of rough rockwork. A finer display I have seldom seen of all shades of yellow, crimson, and scarlet than these presented, and I cannot but think that the Colorado Cacti would succeed perfectly in England. They are undoubtedly hardy, as I have seen the thermometer at Manitou, seventy miles south of Denver, go down to 22° below zero (Fahrenheit), and in parts of the State where Opuntia Rafinesqui grows freely, the snow sometimes lies for two or three months. The moisture and want of actual sunlight of our English climate would of course be somewhat against them, as in Colorado the sun is powerful even in winter, and is seldom obscured by clouds or mist for more than five or six days from November to April; the atmosphere, too, is extraordinarily dry. As, however, O. Rafinesqui succeeds in England, I can see no reason why other kinds of Cacti growing in their native country under exactly similar circumstances should not thrive equally well with us. Professor Porter and Mr. Coulter, in their 'Synopsis of the Flora of Colorado,' mention the following kinds of Cactaceæ, viz., Echinocactus Simpsoni, Mammillaria Nuttalli var. cæspitosa, M. vivipara (this latter species grows freely over the plains and foothills east of the Rocky Mountains, round Colorado Springs, and has been, I believe, successfully cultivated in England); Cereus viridiflorus, yellow; C. Fendleri, flowers deep purple and fruit edible; C. gonacanthus, scarlet, open day and night; C. phœniceus; C. conoideus; C. paucispinus; Opuntia Camanchica; O. arborescens; O. Rafinesqui; and O. missouriensis. All the above have been described or verified by Dr. Engelmann, of St. Louis, but the Cactaceæ of Colorado had been by no means thoroughly worked out when Professor Porter's 'Synopsis' was published three years ago, and I do not doubt that by this time many additional species have been added to the list which I have just given."

When the foliage of a plant, or what answers to it, is perennial, as in Cacti, it is most important to so place it that it may be safe from various injuries, quite apart from those of climate. The best place for Cacti is, as a rule, on well-drained ledges in the rock garden. The Chilian species have, so far as we know, not been tried, consequently at present we have only a few of the northern outliers which may be termed hardy representatives of the family, and these should be planted on rockwork in open, airy situations, where they are free from dripping water, and where the drainage is perfect. It is probable alpine species will be found further south perfectly hardy, and that we may yet see a good collection of bright flowered Cactaceous plants on warm rocky borders or banks in our warmer counties.

Caladium esculentum.—This species for outdoor decoration is the best of a large genus with very fine foliage. It is only in the midland and southern counties of Great Britain that it can be advantageously grown, but its grand outlines and aspect when well developed make it worthy of all attention and of a prominent position wherever the climate is warm enough for its growth. It may be used with great effect in association with many fine foliage plants; but Ferdinanda, Ricinus, and Wigandia usually grow too strong for it, and, if planted too close, injure it. This plant requires, above all others, a thoroughly drained, light, rich, warm soil. In times of great heat it should be plentifully watered, and occasionally with liquid manure. The month of May is the best time for planting it out, and if groups are formed the plants should have an interval of 2 ft. or 2½ ft. between them. The foliage generally arrives at its full beauty and development in August and September. At the approach of cold frosty weather all the leaves, or all but the central one, should be cut down to within 1 in. or 2 in. from the crown, and a few days afterwards the tubers should be taken up and left on the ground for a few hours to dry; they should then be stored on the shelves of a greenhouse, or in a cellar or other place where they will be sheltered from frost and moisture. By placing the tubers in a hotbed in March plants may be obtained with well-grown leaves for planting out in the open air about the end of May or the beginning of June. New Zealand. It has been used with good effect in London gardens, but its culture is given up to a great extent owing to cold seasons.

C. odorum (*Colocasia odora*).—A very fine plant, with stout stems usually from 3 ft. to 8 ft. in height, but growing much taller in a warm stove. The leaves are erect, very broad, and heart-shaped, marked with strong veinings, and frequently measure more than 3¼ ft. in length. The flowers are exceedingly fragrant. It is a fine subject for isolation on Grass plats, its tall arborescent habit distinguishing it from all the other species; but it is unfortunately too tender to thrive in our climate except in sunny, sheltered dells in the southern parts, and should not be planted out until June. This, as grown in the Paris gardens, is a fine plant. E. Indies.

C. virginicum (*see Peltandra*).

Calamintha.—A genus of Labiatæ, numbering some three or four cultivated species, none of which can be recommended for the garden except C. glabella, which is a charming, minute, herbaceous plant, forming compact and neat little tufts about 3 in. high. Flowers in summer, bearing lilac-purple, tubular, scented blossoms, very numerous and large for the size of the plant. May be grown on the rock garden in sandy loam and among the very dwarfest plants. Division.

Calampelis (*Eccremocarpus*).

Calandrinia—The genus Calandrinia is a large one, and many of its species have been introduced to cultivation, though comparatively few of them have been found to be sufficiently effective for general culture, though where well grown and placed they are pretty and sometimes brilliant as flower garden or rock plants.

C. discolor.—A beautiful South American plant, growing from 1 ft. to 1½ ft. in height. Its leaves, which are fleshy, are pale green above and purple beneath. The flowers are produced in a long raceme, and measure 1½ in. across and bright rose in colour, forming a beautiful contrast to the tuft of golden yellow stamens in their centre. They only expand in sunshine, but open several weeks in succession in July and August. This species requires a dry soil and a warm exposure to succeed well. It may either be sown in the open border or in pots in April, but care must be taken in transplanting it.

C. grandiflora.—This somewhat resembles C. discolor, but is larger in all its parts, and differs in having more showy blossoms and in the shape of the leaves. It requires a situation similar to C. discolor, and attains the greatest perfection in a warm and moderately stiff soil. It flowers a little earlier than C. discolor, and continues to produce its flowers throughout the autumn. It is a handsome annual.

C. nitida.—Closely allied to C. discolor, but differs from it in dwarfer habit. It is a succulent annual, forming a tuft from 4 in. to 6 in. across in strong specimens, bearing numerous large rose flowers in a leafy raceme, borne on stems 6 in. to 9 in. high. The flowers are fully 2 in. across, expanding best in bright sunshine, as do all the species of this genus, when thus presenting a striking effect. It is best treated as a half-hardy annual, as it then blooms much earlier, but may be sown in the open air in May. Chili.

C. speciosa.—This has slender stems much branched and prostrate; the flowers, which are very showy, measure from ½ in. to 1 in. across, and are of a deep purple-crimson. On sunny mornings they expand widely, and close early in the afternoon. It flowers continuously from June to September. It succeeds well on dry soil, generally sows itself

PLATE XLIX.

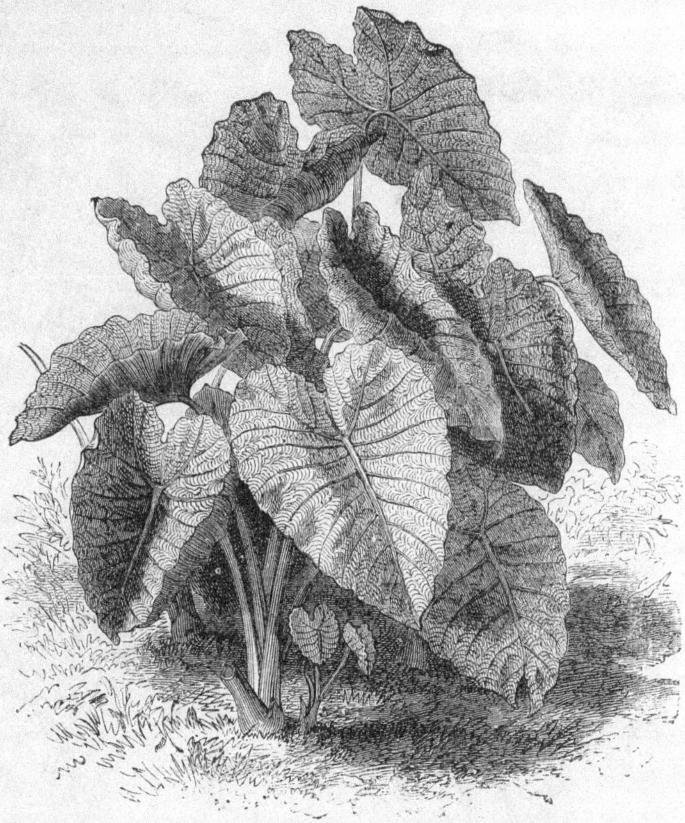

Caladium esculentum.

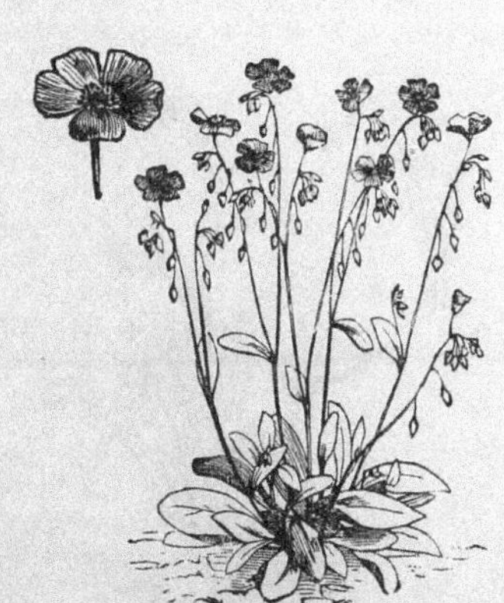

Calandrinia discolor.

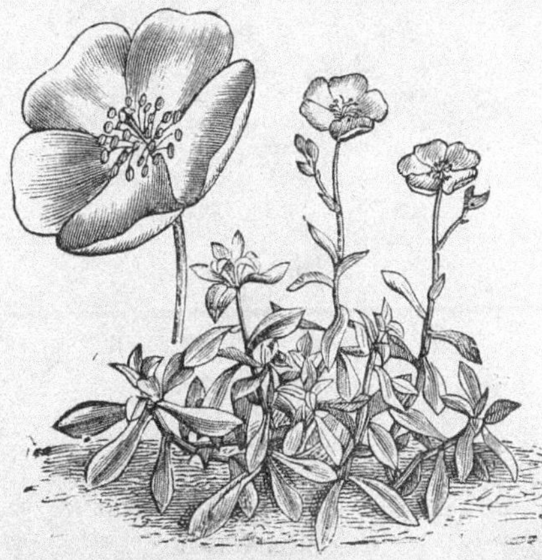

Calandrinia speciosa.

[To face page 56.

Calandrinia umbellata.

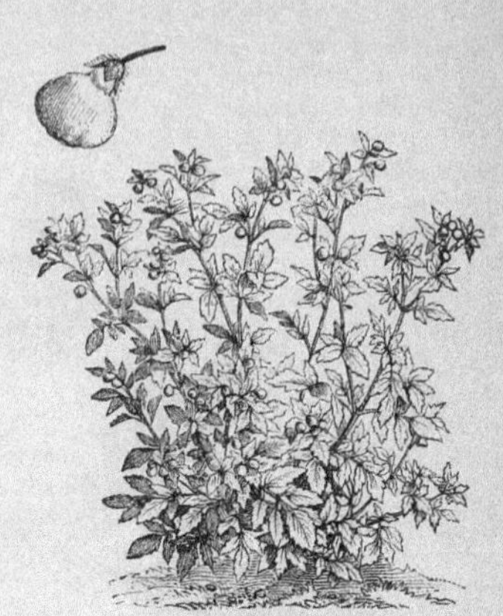

Calceolaria scabiosæfolia.

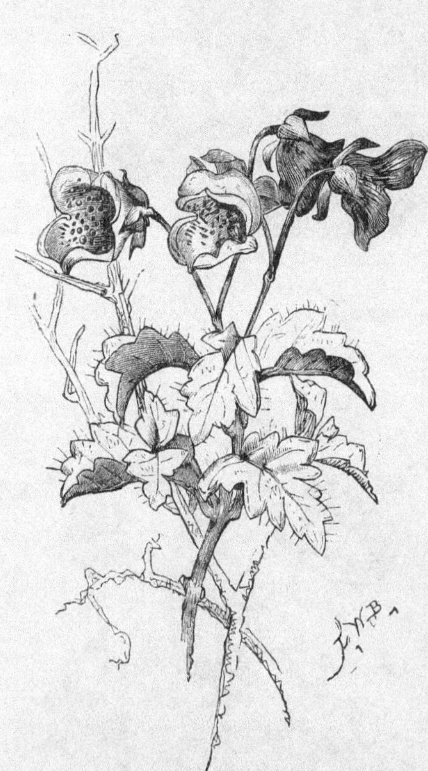

Calceolaria violacea.

Calimeris incisa. See page 59.

Calla æthiopica (*Arum Lily*)
See page 59.

[*To face page* 57.

plentifully, and is suited on account of its dwarf habit for the rock garden. California.

C. umbellata, a native of Chili, has reddish, much branched, little stems half shrubby at the base, and rarely grows more than 3 in. or 4 in. high. For vivid beauty and brilliancy of colour there is nothing to surpass it, the flowers being of a dazzling magenta-crimson glow, yet soft and refined. In the evenings and in cloudy weather it closes, and nothing is then seen but the tips of the flowers. It does very well in any fine, sandy peat or other open earth, is a hardy perennial on dry soils and well-drained rockwork, and looks best in small beds, but may be used with advantage as a broad edging to large ones. It seems to live longest in chinks in well-made rockwork. It is as readily raised from seed as the common Wallflower either in the open air in fine, sandy soil, or in pots. As it does not like transplantation, except when done very carefully, the best way for those who wish to use it for neat and bright beds in the summer flower garden is to sow a few grains in each small pot in autumn, keep them in dry sunny pits or frames during the winter, and then turn the plants out without much disturbance into the beds in the end of April or beginning of May. As its beauty is concealed during dull or rainy weather, this may prove a drawback to its use in the flower garden, but by employing it as a groundwork for some taller plants this defect may not be so noticeable. Young plants flower more continuously than old-established specimens. It may also be treated as an annual, sown in frames very early in spring.

There are other kinds, such as C. Lindleyana, C. Burridgi, C. procumbens, C. compressa, and C. micrantha, but their flowers are not so pretty as those of the kinds just mentioned.

Calceolaria.—A valuable genus of plants and popular, though not so much so as they have been. In the London district they are very much less employed than was the case some years ago, many of the varieties perishing from disease, or being short-lived as regards bloom. We used to see masses of the fine old Sultan (crimson) and Kentish Hero (rich brown) in almost every garden, and the many varieties sent out by Mr. Burley were at one time very popular, but they got a bad name, being liable to decay from a peculiar disease. We seldom now see the crimsons and maroons and crimsons with gold caps. The golden yellows, however, still hold their own, and the old variety Gaines' yellow has still perhaps the best constitution of any. C. aureo-floribunda is everywhere known, and the fine amplexicaulis with its good habit and pure lemon flowers is always a favourite. We think the main cause of their deterioration has been neglect in carefully cultivating them, and that the plants have not been generously treated. Both before and immediately after transplanting they should be kept always in a vigorous state in good rich soil through the winter, and near the glass, but as far from fire heat as possible—indeed, in cool pits where air is given at all times when possible, by entirely uncovering or tilting the lights when the weather will allow, the great object being to keep the foliage free from curl and free from insects to which, when fire heat is applied, they are very subject. Many of the newer highly-coloured varieties are most desirable additions to the flower garden, and will amply repay any extra trouble necessary to succeed with them.

PROPAGATION.—The shrubby varieties of Calceolaria bloom mostly in summer, and scarcely can a shoot be then found that does not contain an embryo flower-spike. Such shoots make bad cuttings, and are then so soft and watery that they do not strike well. Although we might like to propagate our next year's stock at that season, they would grow too much before winter, and would not only require transplanting, two months' extra attention, and pinching, but they would also occupy more frame room during winter than later struck cuttings would do. The proper time for propagating such Calceolarias is the end of September and October. For this purpose prepare a cold wooden or turf frame on a dry basis, fill it to within 6 in. of the top with sandy loam, and over that spread some clean silver sand. Then select nice stubby side-shoots, as firm as possible, pick out any flower-spikes that are visible, remove one or two of the base leaves, cut horizontally below a joint with a sharp knife, and dibble them thickly, regularly, and firmly into the frames, giving them a sprinkling of water through a fine rose to settle the soil and to prevent the cuttings from flagging. Keep the frames close and shaded for a day or two, but afterwards remove the shading, and only use it again during the succeeding month to counteract the effects of bright sunshine. Such cuttings take a long time to root, but if the atmosphere of the frame is kept dry, and the plants free from damp by dredgings of wood-ashes, dry dust, and old lime rubbish, they will root satisfactorily during winter, and in spring will yield tops for additional cuttings; whether these are employed for cuttings or not, they are best pinched off.

WINTERING.—These frames require no further care beyond protection, by means of covering the sashes, in the event of frost, and banking up the sides, if of wood, with soil. Wooden boxes, seed pans, or pots, might also be used for striking Calceolarias in; in which case they might be wintered in any pit, greenhouse, or conservatory. Whether they are propagated in frames or boxes, they should be transplanted, further apart than they previously were, into other frames, filled with rich open soil, in which they will become fit for planting out by the middle of May, if the shoots have been attentively pinched when they required it. In autumn, too, it is a good plan to draw some earth around the necks of the old plants, so as to induce the emission of roots from that portion of their stems; and at lifting time, in November, to separate every rooted branch and plant them as independent plants in frames. They will yield abundance of cuttings in spring, but it is from a few old plants, lifted and saved in frames for the purpose, we get our chief supply of such young-rooted plants.

SPRING PROPAGATION is often a matter of uncertainty, but I once saw a very fine lot which had been struck in February. These cuttings were selected from old plants wintered in frames, and fully exposed every fine day by drawing the sashes completely off them, consequently they were pretty hard and well seasoned. They had been inserted in cold frames, precisely as advised for autumn cuttings, and a failure could scarcely be found even in four lights of them, and by bedding-out time they had formed very serviceable, well-rooted stocky plants. Spring cuttings, however, are mostly rooted in hot-beds, in boxes, or pans, and often as many of them damp off as survive to become plants; nevertheless, where the stock is deficient, this mode must be resorted to. It is best to strike them after the middle of March in pure sand in a hot-bed or propagating pit where there is no stagnant atmospheric moisture, and when well rooted to pot them, or put them in boxes in light sandy soil, still keeping them in warm quarters for a few days. After that gradually shift them into places in which there is less heat. Powdered charcoal or wood-ashes strewed on the surface of the soil amongst the cuttings are great preventives of damp, and a judicious use of the watering-pot should be exercised. —W.

SPRING MANAGEMENT.—Mr. David Thomson in his "Handy Book of the Flower Garden" has some useful remarks on them. "When grown in beds or boxes, they generally thrive best after they are planted out, without ever having been put into small pots at all. What I would recommend, and have adopted sometimes when suitable, in the absence of cold pits and frames, is to throw out trenches like those generally used for Celery beds, put 6 inches of rotten leaves in the bottom, and then 6 inches of light rich soil. Here the young plants, lifted with as little injury to their roots as possible, should be pricked out about the end of March, 6 inches apart each way. When all are planted, water well, and lay some trellis-work or common stakes across the trench, and cover with mats or canvas when cold weather renders it necessary. For the first fortnight after being transplanted they should be shaded through the day when the sun shines. Thus managed, they make fine strong plants with very little attention. As they grow they should be looked over at intervals and topped, so as to keep them dwarf and well furnished. About three weeks before they are to be planted out, a spade, or any other sharp-edged tool, should be run along between the lines each way, cutting to the depth of 6 in. This cuts off the roots of each plant from its fellow, checks them for the time, but causes them to make fresh roots 'nearer home'; and the result is, that they lift with good balls, and scarcely receive any check when planted out. Should they droop when thus operated upon, give them a good soaking of water.

"Amateurs and others requiring small quantities of plants may adopt a similar plan to this, by transplanting them into boxes 6 in. deep, prepared much the same as directed for the trench, and otherwise managing them in the same way. In this case they can be lifted into any outhouse, or even covered over outside in case of spring frosts. When from any cause the stock is not equal to the demand, the points of the young plants strike freely in March and April in a gentle heat; but autumn-struck plants are in all respects to be preferred.

"SOIL.—Like the Verbena, the Calceolaria requires a deep, rich, loamy soil to grow and flower it well throughout the season. They are very subject to die off in hot sandy soils, and at best do not bloom for any length of time. Tagetes signata pumila is the best substitute for Calceolarias on light sandy soils, that are also too poor for Violas.

"THE VARIETIES MOST SUITABLE FOR BEDS.—Those marked * are best. *Ambassador, bronze crimson, the best of the crimsons, 1 ft.; *aurantia multiflora, orange yellow, extra fine, 1½ ft.; *aureo-floribunda orange yellow, suitable for damp localities, 1 ft.; *Golden Gem, deep yellow, fine constitution and free bloomer, 15 in.; Invincible, lemon, very dwarf, compact habit, free bloomer; *Prince of Orange, orange brown, compact habit, 1 ft.; Princess Alexandra, buff, dwarf compact habit, 9 in.; Princess Louisa, sulphur yellow, dwarf, good habit; Victor Emanuel, scarlet spotted with crimson, 15 in."

SPECIES.—Apart from the varieties, there are a number of species of Calceolaria of some merit for the flower garden, and some neglected and unknown that would be excellent. The genus Calceolaria is rich in species, and presents a wealth of variety in form and colour. It comprises upwards of 100 distinct species, including annual and perennial herbs and dwarf shrubs. Its geographical area almost exactly coincides with that of the genus Fuchsia, ranging, as it does, from Mexico to the southernmost point of South America, and re-appearing in New Zealand; but, unlike the Fuchsia, it is not represented on the eastern side of South America, at least only in the extreme south. The greater number of the species inhabit mountain valleys, ascending to an elevation of 13,000 ft. to 14,000 ft. within the Tropics, where they enjoy a most equable climate, which is never very hot and never very cold. Some few occur in the arid districts of Chili; they find their greatest concentration in Chili, Peru, and Ecuador. Only about four species are known to grow in Central America and

Mexico, and two have been discovered in New Zealand. A full account of the family is published by Mr. Hemsley in THE GARDEN for March 29, 1879, accompanying a plate of C. fuchsiæfolia.

C. amplexicaulis.—A few years ago this was almost as familiar in the flower garden as any bedding Calceolaria. It is the one with soft, dark green leaves clasping the stem, and a profusion of lemon-yellow flowers. C. amplexicaulis inhabits Peru and Ecuador, was introduced by Mr. W. Lobb about thirty years ago, and is one of the handsomest plants we have. Its colour is a very soft lemon; it is very free in habit, and, where the soil suits it, sends up a quantity of suckers which keep up a constant succession of flowers till quite late in the autumn. The long period of time during which I have known this variety I have never once seen it diseased, a failing to which all others are exceedingly subject, and which makes them so uncertain. Owing to its tall habit it groups well with the numerous bold-habited plants that are now invading our flower gardens, and it is usually handsomer in autumn than any of the newer kinds. It used to be in great esteem in Hertfordshire for bedding, a purpose for which it is one of the best and most continuous blooming kinds that can be grown. The way in which it is managed in gardens in that county is to plant it in small beds on the Grass, where it is tied and trained to sticks placed and brought together at the top, something after the manner in which they are placed over a basket of plants when sent out from a nursery. Thus managed, they form fine pyramids, and soon hide the supports; but it is only in moist, cool soils that they do this, although this plant will stand heat and drought better than any other.—S. D.

C. hyssopifolia is one of the finest of the small-growing kinds. It bears large, loose clusters of lemon-yellow blossoms in profusion, and continuously from early summer till autumn. The foliage is distinct, being narrow, resembling that of the Hyssop.

C. Kellyana *(Kelly's Calceolaria)*.—A curious hardy hybrid, with short downy stems, 6 in. to 9 in. high. Flowers in summer, nearly ¾ in. across, deep yellow with numerous small brown dots, two or three together on the top of the stems. Leaves, in a rosette, with a toothed margin, and more or less covered on both sides with soft white hairs. It was raised by a Mr. Kelly, propagator in the nurseries of Messrs. Dicksons & Son, of Edinburgh, about the year 1840. Its foliage resembles that of one of the Mimuli, creeping along the ground. It is a very interesting dwarf rock garden plant.

C. Pavonia is a noble species, the largest of any in cultivation. It grows from 2 ft. to 4 ft. high, has large light green, much wrinkled foliage, and bears from June to September a profusion of large, pale yellow, slipper-shaped blossoms. If planted against a warm south wall it is a fine object, but it should either be lifted at the approach of winter, or the stool protected with a thick covering.

C. violacea is an extremely pretty species with small, helmet-shaped flowers, rich purple and copiously spotted; succeeds well on warm borders or rockwork, and withstands mild winters in the south if slightly protected.

Calendula officinalis *(Pot Marigold)*.—An interesting old hardy biennial; one of the best for autumn and winter flowering. It is found in almost every garden; the petals of the flowers were used formerly to flavour various dishes in old English cookery, hence its name. There has recently appeared simultaneously in England and France a very distinct double variety called Meteor, the flowers of which are much more double than the old variety, and the petals striped with lemon on an orange ground, the two colours varying in body and producing a very pretty effect. It comes true from seed. Some variety of kinds is offered by the German seed houses. The plants are among the best biennials for autumn and winter flowering. Nothing short of severe frost will prevent them yielding abundance of blossoms. For late blooming seed should be sown in July. They usually sow themselves freely, and may be sown in the open ground either in spring or autumn.

Calimeris.—A genus of Compositæ, of no importance in the garden.

Calla æthiopica *(Ethiopian Lily, Arum Lily)*.—This well-known greenhouse plant may be grown either as an aquatic in pieces of ornamental water and fountain-basins, or in the open ground in cool, moist soil, and equally well in positions exposed to the full sun and in those which are shaded. Being so very distinct in leaf and beautiful in bloom, this old favourite will be seen to as much advantage grouped with the smaller fine-leaved plants in beds as ever it has been in our stoves or windows. It will do best in rich and well moistened borders or beds. It is hardy when treated as an aquatic in some of the milder districts, but as an outdoor plant it is never likely to have the importance it has when flowered in the greenhouse. Some, however, succeed out of doors. Mr. Burwood Bodlee writes to us from Lewes, Sussex, concerning it: "I have had in my garden pond for more than twenty years past a large number of Arum Lilies (Callas) all the winter long, and I have never lost a plant. A moderate spring flows through my pond, which is not frozen over, unless with a sharp frost. They increase from year to year spontaneously from seedlings, as well as laterally from the old roots, which are planted in the natural soil and kept down by a few large stones. We have had from fifty to one hundred blossoms out at one time, and the fine tropical-like leaves are highly ornamental."

Another correspondent writes: "A few years ago I planted some pieces of this amphibious plant in water by the side of an embankment, and they have so increased that at the present time they form an immense mass, 120 ft. long and several feet broad, and when in flower, they are a grand sight, hundreds of magnificent blooms being open at one time, and much finer than when carefully grown under the directions given in most gardening books. In severe winters they are cut down to the water's edge, but they always came up again as strong as ever when mild weather returned, and flowered well, though much later than when not cut down by severe frosts. In very mild seasons they commence flowering before Easter. A good plan as regards planting them is to fill some worn-out bushel baskets with any refuse soil that may be at hand; into this plant the Arums, and plunge them into about 2 ft. of water, then throw a few barrowfuls of soil and stones around to keep them steady and to form mounds. The baskets soon rot, and the plants become masses of foliage and flowers, as has just been stated."—S.

Many who grow this plant for the greenhouse or market plant it out in trenches in early summer, taking the plants up in autumn—a simple and good way of growing them. S. Africa. (*Syn.*, Richardia.)

C. albo-maculata is a similar kind, having the leaves handsomely spotted with white on a deep green ground. It is almost as hardy as C. æthiopica.

C. palustris (*Bog Arum*).—More beauty than any native bog plant affords results from planting in boggy places this small trailing Arad, which has pretty little spathes of the colour of those of its relative, the Ethiopian Lily. It is thoroughly hardy, and though often grown in water, likes a moist bog much better. In a bog or muddy place, shaded by trees to some extent, it will grow larger in flower and leaf, though it is quite at home even when fully exposed. In a bog carpeted with the dwarf dark green leaves of this plant, the effect is very pleasing, as its white flowers crop up here and there along each rnizome, just raised above the leaves. Those having natural bogs, &c., would find it a very interesting plant to introduce to them, while for moist spongy spots near the rock garden, or by the side of a rill, it is one of the best things that can be used. This, like many other bog plants, is often starved in cultivation. Its beauty is only felt when allowed to ramble over rich muddy soil. A native of the north of Europe, and also abundant in cold bogs in North America, flowering in summer, and increasing rapidly by its running stems; 3 in. to 8 in. high.

Callichroa platyglossa.—A showy Californian plant of the Compositæ. It is a half hardy annual; the seeds may either be sown in slightly heated frames in March, or in the open border, in the middle or latter part of April, in ordinary light soil not too rich, and the situation should be open. The seedlings should be well thinned, or if transplanted should not be less than 6 in. apart in order for each plant to fully develop itself. Treated thus, it flowers from July to September. It may be also sown in autumn on light warm soils to remain the winter, and so flower in spring. In this case the plants need protection during severe weather.

Calliopsis (see *Coreopsis*).

Calliprora lutea (*Brodiæa*).

Callirhoe.—A small, but handsome genus of North American Malvads, some half-dozen kinds of which are known in our gardens. They are hardy herbaceous perennials, and succeed well in the open border in rich light soil.

C. digitata.—A distinct-looking perennial and glaucous herb, growing 2 ft. or 3 ft. high, and producing reddish purple flowers in summer; it is not so showy as the others, but it succeeds under similar conditions of culture.

C. involucrata is a charming dwarf prostrate perennial, producing large violet crimson flowers 2 in. in diameter; indigenous to California. It is excellent for the rock garden, as it produces a continuous crop of its showy blossoms from early summer till late in autumn. It has the best effect when allowed to fall over the ledge of a rock.

C. macrorhiza alba is a pure white form of a species with purplish carmine flowers. It is of a neat habit of growth, producing from a tap-root, which ultimately attains some size, an erect stem from 1½ ft. to 2½ ft. high, which bears a corymbose raceme of pure white flowers. The plant appears to occur in several shades of colour, varying from rosy purple to pale rose and white. Sown early, it will bloom the first year. It is a native of the South-western States of North America.

C. pedata is one of the prettiest. It has a perennial root, from which arise several stems of a trailing habit when unsupported, bearing lobed foliage, variously incised, and very handsome crimson flowers, fully 2 in. in diameter, on long foot-stalks, from the upper axils. When sown early the plants bloom easily the first season, and continue in flower until a late period of the autumn. The roots will survive our average winters in dry soils, and increase in strength each season, though were it otherwise, the plant might be cultivated as a half-hardy annual. It is important to transplant the seedlings into the open ground as early as the season permits, *e.g.*, by the middle of May, as, having a tap-root, they do not appear to succeed well in pots. The shoots may either be allowed to trail, or may be supported by sticks. It succeeds better in dry soils than in those of a wet or highly enriched character, in which it grows too luxuriantly, producing foliage at the expense of flowers. The varieties nana and compacta are desirable varieties of dwarfer habit of growth than pedata.

Callisace dahurica.—A large and ornamental umbelliferous plant growing 8 ft. to 10 ft. high. The flowers, produced late in summer, are white and in umbels frequently 2 ft. across. The lower leaves are about 6 ft. long, and 4 ft. wide. It is an imposing plant in

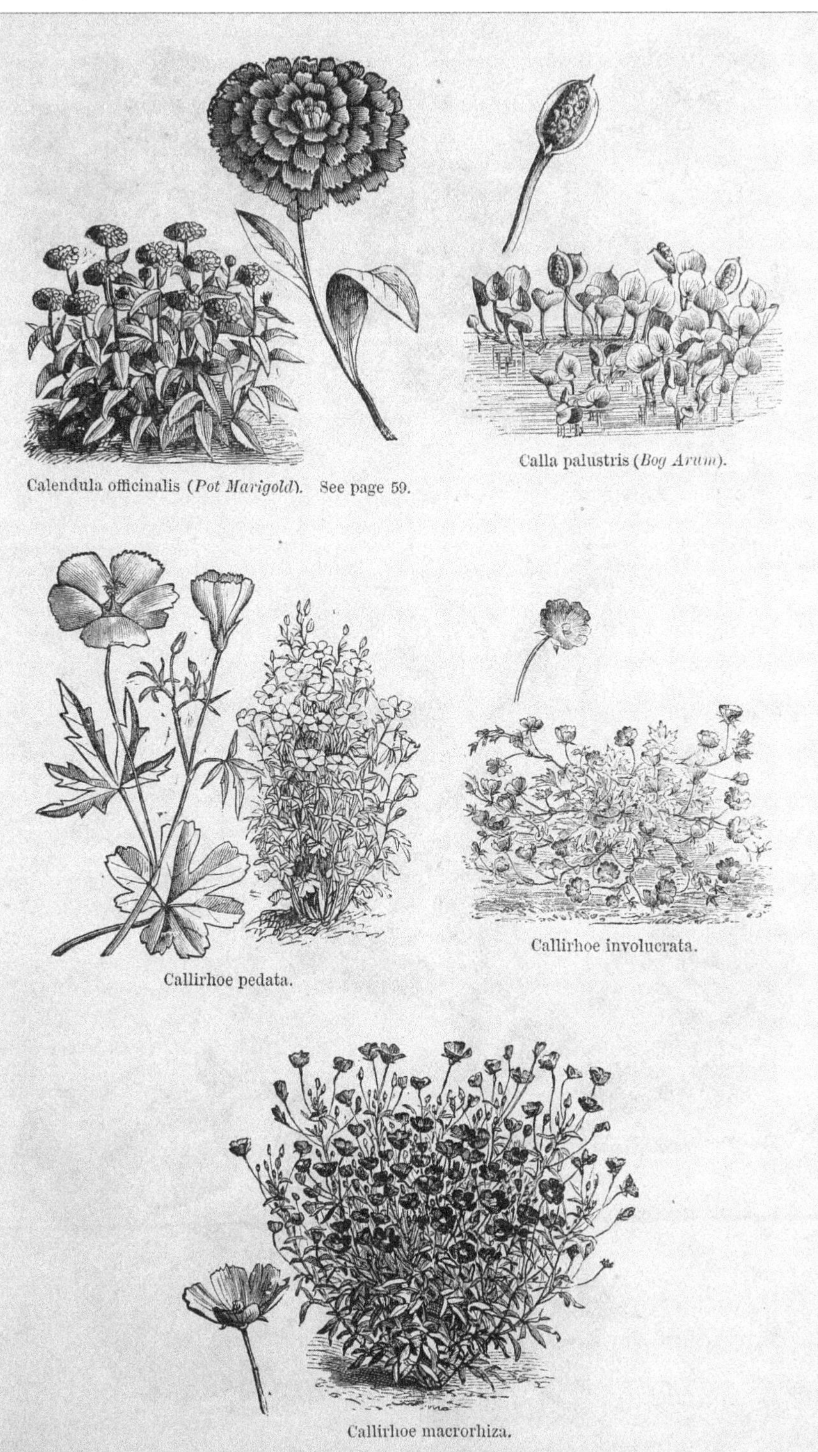

Calendula officinalis (*Pot Marigold*). See page 59.

Calla palustris (*Bog Arum*).

Callirhoe pedata.

Callirhoe involucrata.

Callirhoe macrorhiza.

[*To face page* 60.

PLATE LII.

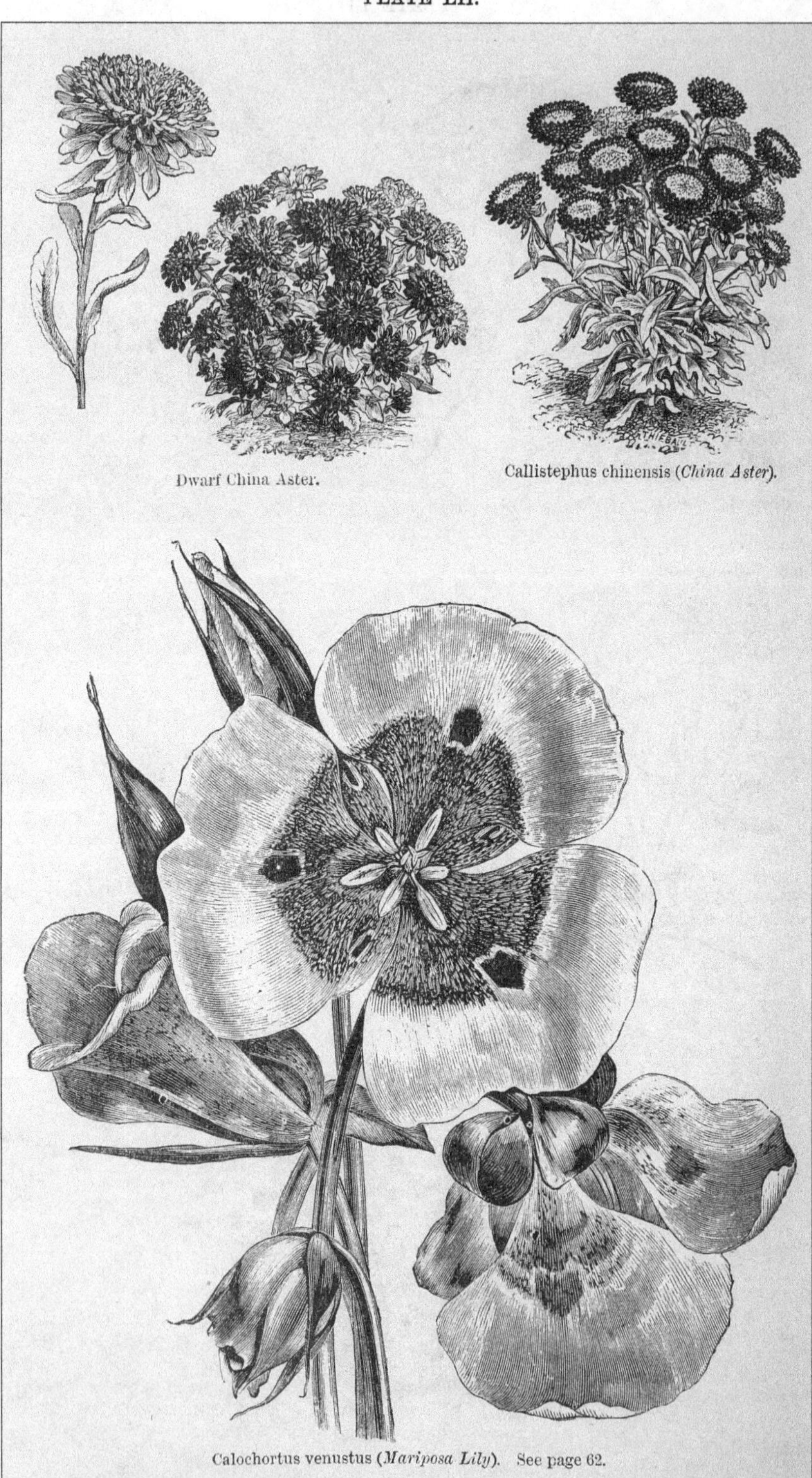

Dwarf China Aster.

Callistephus chinensis (*China Aster*).

Calochortus venustus (*Mariposa Lily*). See page 62.

[*To face page* **61.**

isolated positions on turf or in groups in deep rich loam. Propagated by division or seed. Native of Dahuria.

Callistephus chinensis (*China Aster*).—Among the many annuals now in cultivation Asters are among the best, and when well grown and cared for, do as much at a small cost in rendering a garden gay during summer and autumn as any other plant. To see Asters in their beauty, however, they must be grown in masses, and specially and well cultivated—not in any way left to hap-hazard or poor culture at any stage. The China Aster was introduced towards the end of the last century, and was raised in the Jardin des Plantes of Paris from seeds sent from China. Of annual duration numerous varieties have resulted from its extensive cultivation. In the wild state the flowers are single—that is to say, only the outer florets are strap-shaped, and usually of a rosy-lilac tint, with yellowish disc florets; but under cultivation all the florets have become ligulate or quilled, and a richness and variety of colouring has been developed scarcely surpassed in any one species, ranging from pure white to deep carmine, and violet and blue, though the yellow of the disc in the single form has not been reproduced in the double. We are mainly indebted to the French horticulturists, notably Truffaut, Fontaine, and Vilmorin, for the great perfection to which the different races have been brought, and in this country to Mr. Betteridge. It is worthy of remark that the different varieties are so far fixed that they will come true from carefully selected and well-ripened seed.

VARIETIES.—There are several distinctive methods by which the Aster may be classed, such as comparative height, habit, character of flower, suitability for exhibition, for pot culture, or for bedding. Tall Asters comprise the fine Pæony-flowered, the tall Chrysanthemum, the Emperor, the tall Victoria, the Quilled, and a few others. Kinds of medium height are found in the dwarfer forms of the Victoria, the fine Cocardeau, the Rose, and the Porcupine; the dwarf forms comprise the short Chrysanthemum, the dwarf pyramidal, and specially the dwarf bouquet, which is one of the most beautiful of all for pot culture. The best bedding kinds are found in the medium growing Victoria, the Rose, and the dwarf Chrysanthemum, as these vary from 9 in. to 12 in. in height, and form good bunches of bloom on each plant, and collectively fine masses of colour. The dwarf bouquet kinds, whilst specially good for pot culture, are also valuable as edgings to beds of taller kinds. For pot culture for exhibition the best are the medium growing Victorias, as these, if of a good strain, possess quality, and handsome and even heads of bloom. The distinctive forms, as found in these flowers, are those in request for exhibition, as it is the custom in schedules to invite exhibits under the heads of incurved, reflexed, and quilled flowers, or in some cases it is expressed as flat petalled and quilled, and in others as French and German, the French including the flat-petalled kinds and the German the quilled. These latter distinctions are, however, too arbitrary, and are giving place to those previously named, being more easily comprehended. The best examples of the incurved flowers are found in what is popularly known as Truffaut's Pæony-flowered Aster, formerly esteemed a grand exhibition strain, but which has given way to some extent to the more even and massive Victoria. An incurved flower shows all the tips of the petals converging inwards, as is seen in a Chrysanthemum, and the more solid and rotund the form of the flower, the better its prospect on an exhibition stand. All of these have flat petals, and, as a rule, have broader petals than any other kinds have. The most complete examples of solidity and perfection of form are found in the quilled varieties. Though not more than half the size of the flat-petalled kinds, they present a mass of quilled petals most evenly set, and in outline almost semi-circular. One is almost inclined to believe that such perfection of evenness could only be obtained by artificial means. It is, however, but the result of that constant care in seed selection which is ever the aim of the grower who prides himself in putting before the public things that are of the finest kind. The Cocardeau, or crown-flowered, Asters are partly flat-petalled and partly quilled, the outer margin of the petals being of a dark red or blue, and the centre disc of pure white. These are grown more for their novelty than merit. This bi-coloured form is, however, also found in some of the finest show quilled kinds, rendering them both valuable and attractive. The charming bouquet Asters are also of a semi-quilled character, the petals being stiff and closely set. The blooms are of medium size and very compact, and most acceptable for nosegays. For this latter purpose, however, those who grow the Victoria forms will find that the later blooms that start out from the main stems are admirable, the clear bright hues and excellent forms being exceedingly effective however employed.

Mr. James Betteridge, of Chipping Norton, an excellent cultivator of the Aster, speaks as follows of his mode of cultivation:

SOWING THE SEED.—For several years after I commenced the culture of quilled Asters I always sowed the seed in bottom-heat; but during the last decade the plan adopted has been to sow in a cold frame, under glass, some time between March 26 and April 26, in drills 6 in. apart, and not too thick in the drills. A few days suffice to bring them above the soil, when a liberal supply of air must be given, or the plants will be weak. When large enough they should be pricked out into another cold frame, slightly shading, where they will soon be established, and after they have attained strength enough to handle well plant them out into the beds or quarters where they are to bloom, in well-manured soil, being careful not to break the tender fibres of the roots. Let the rows be 1 ft. apart, and plant the strongest plants 12 in. from each other; this should be done in showery weather, when the plants soon get established. If the weather be hot and drying, a little watering will be necessary till they are rooted; afterwards keep them clear of weeds by hoeing among the plants. About the first week in August top-dress with rotten manure from an old hot-bed, giving a good soaking all over if the weather continues dry. After this, if the blooms are required for exhibition, the plants must be tied out to small stakes. As soon as it can be determined which buds will produce the best blooms, thin out or disbud, leaving about five or six blooms on each plant. Exhibition blooms should be of large size, full, high centre, deep, distinct colour, with solid petals. To secure these qualifications in this England of ours, shading of some kind is necessary. For this purpose we have tried many kinds; one of these, and the most useful during the past stormy season, has been a "tin shade," about 12 in. in diameter, with a spring socket to slide up a square stake, one which we formerly used for shading Dahlias, and was, we believe, sent out by Mr. C. Turner, Slough, many years since. Wire frames covered with linen or other light material will do as well; all that is required is perfect security against rain and hail-storms. In arranging blooms for exhibition boxes, or stands should be 6 in. or 7 in. high at the back, and 3 in. or 4 in. in front, painted green, and if the blooms are set in a frill of embossed or ornamental edged white paper, the effect is improved, lending an air of elegance and refinement.

SOIL.—Asters like a deep rich soil, and it is only under such conditions that really fine flowers can be obtained and the plants induced to hold out should dry weather set in. Planted in the ordinary way, they are mere weeds compared with such as are well fed and can get their roots down deep in search of moisture; and when they can do this, the hotter the weather the better it suits them. Confined to the top shallow crust of earth, they are soon dried up and the blooms starved; and this is why we so frequently see the poor puny plants that are to be found in borders, where, instead of being able to grow and develop themselves, all they can do is to struggle on for existence. The best way to manage them is to dig and cast off the top spit to one side, handy to be returned to its place again, and then trench and break up the soil below, working in with it at the same time plenty of short manure, thoroughly decomposed, which will have the effect of attracting the roots and affording them ample assistance just as they most require it, when expanding and perfecting the bloom. Trenching, as usually done, brings the crude earth to the surface, and buries that which has been exposed to the ameliorating influences of the atmosphere—a fact that should be borne in mind, as it takes years to get it in the condition in which plants will lay hold of it and start away freely. In very light soils a few barrow-loads of clay, chopped up finely and mixed well in, has a capital and lasting effect, and will do more in producing fine Asters than any other help that can be afforded. The thing to aim at is to keep the bottom cool and moist; then, if the weather be favourable, the plants will take care of themselves. When grown in groups of three in a border, similar preparation must be made, or neighbouring plants already in possession are sure to rob them and cut short their beauty long before the autumn sets in.

SAVING SEED.—If the autumn be fairly genial, there is no difficulty in saving Aster seed. Do not allow any one plant to carry more than three or four blooms, and these the finest. If needful to protect from heavy rain, the covering should admit of free circulation of air among the flower-heads. Gather when ripe, and clean through a wire sieve. Carefully selected seed usually produces as good blooms as those from which the seed was saved. Our best kinds, and especially all the new forms and colours, are grown in beds, over which temporary lights are fixed, by which means we are enabled to cut blooms of the purest shade or colour.

POT CULTURE.—Many have written about the kinds best suited for pots; but our practice has been to sow in the open in May, keeping the varieties separate, and then lift about the middle of September (when the buds have partially expanded); we put three plants into an 8-in. pot, pressing the soil firmly, shading for a few days, and then placing them in a cold greenhouse, where they will bloom late on through the autumn. Such as these have a splendid effect in a conservatory among the small kinds of foliaged plants, the only other plants equal in point of colour being the Chrysanthemums. If for exhibition purposes, the incurved and reflexed varieties, commonly described as French, are in their way most effective; but among the other sections there are few better than, if equal to, the best quilled Asters.

Calochortus (*Mariposa Lily*).—A lovely and valuable genus of bulbous plants of the Lily family. All natives of Western North America, but are not thoroughly hardy in the sense that plants from the Eastern States are. Many species are in cultivation, but only those that are known to be good garden plants are enumerated.

C. albus (*Cyclobothra alba*).—Branches deeply forked and spreading, 1 ft. to 2½ ft. high, with three to twelve heads of flowers; flowers, globose, white, nodding. The variety paniculatus differs from the type in being dwarfer, and in having smaller flowers. Of easy culture.

C. elegans (*Cyclobothra cærulea*).—Stem, slender, 3 in. to 6 in. high, with three to six heads of flowers, arranged in an umbellate manner; flowers, erect, white or pale lilac, bearded with pale hairs on the margin and face. The variety nanus has dwarfer and more slender stems, and petals more hairy, and C. Tolmiei is stouter and taller (about 1 ft. high), the flowers larger, and tinged or marked with lilac. One of the most satisfactory kinds to grow, as it rarely fails, and it bears a profusion of blossoms that remain in perfection for a week or two.

C. pulchellus (*Cyclobothra pulchella*).—1 ft. to $1\frac{1}{2}$ ft. high, with six to twelve nodding flowers of a deep yellow. A most beautiful plant when well grown. It succeeds well in the open border, and is one of the best for pot culture.

C. splendens.—Stem, flexuose, $1\frac{1}{2}$ ft. to 2 ft. high, with three to five flowers of a deep lilac, $1\frac{1}{2}$ in. to 2 in. long and broad, of the same colour in the middle, or occasionally with reddish spots, the third lower one furnished with a purple beard around the glandular pit. Similar to venustus, but well worth growing, as it has a good constitution and is a profuse bloomer.

C. venustus.—A lovely species, growing 1 ft. to 2 ft. high, with two to six flower-heads; petals broad, white or pale lilac, of the same colour in the centre, or spotted with orange-purple; claw, with an obscure, bearded, glandular pit, and a few hairs and red-brown spots above the pit. There are several varieties of this beautiful species—purpureus, purpurascens, lilacinus, each name indicating the shade of colour. This is the finest of all the species, and is one of the most beautiful open air flowers in cultivation.

The above are the best for ordinary culture out of about a dozen and a half kinds in cultivation; they are also the commonest, and may be readily procured. They represent all the types of flowers and growth to be found in the genus, and all are beautiful and well worth cultivating. The other kinds in cultivation are the following, though the majority of them are extremely rare and somewhat difficult to cultivate. These are—C. Benthami, Maweanus, Leichtlini, Gunnisoni, lilacinus, uniflorus, luteus, citrinus, macrocarpus, and flavus.

CULTURE AND POSITION.—The nature of these plants seems to indicate that they love a sunny and well-drained situation, such as rockwork, or a well-drained warm border. Planted well up on rockwork, where they are dry in winter and sufficiently moist in summer to do without watering, they make vigorous growth and flower freely. The principal requirement is, doubtless, protection from wet. For ordinary beds and borders it is expedient to lift the bulbs and store them until they begin to grow. For pot culture, a fibrous loam with a little leaf mould and plenty of silver sand has been found to answer admirably. In short, although rather delicate, these attractive plants will succeed very well if the precautions indicated be taken; and they will certainly amply repay for the little extra trouble they may exact. The best kinds for pot culture are C. pulchellus, venustus, elegans, splendens, luteus, and Tolmiei.

The New Plant Company, of Colchester, who have grown these plants very well, speak as follows of their mode of cultivating them:

With the exception of Calochortus flavus, we grow all the species of this family in a cold frame with a south aspect; the frame has a depth of soil of some 12 in. It consists of ordinary peat (taken from a neighbouring common, from which the stones are sifted), fibrous loam, and leaf soil, with a good proportion of red sand. We plant the bulbs in November, and then to protect from the heavy rains chiefly place the lights on, but when the weather is suitable they are removed. During severe frosts the soil in this frame is frozen as hard as a brick. In the spring all the lights are removed entirely; the plants are then subject to whatever weather we may get. If too dry we water them. Grown in this way we get fine displays of flower, while for vigorous, healthy growth the plants are not to be surpassed. The bulbs after maturing are removed from the soil, to be sold or replanted, in November. The exception to this treatment is Calochortus flavus. We plant this out-of-doors in a bed of ordinary soil, with a little peat and sand around the bulbs, in April. It seldom commences growth until June, but in August and September the bed is literally covered with bright yellow flowers, although great demands are made on them for cut flowers, for which they are invaluable. The bulbs we remove in October, and keep in a dry room until planting time comes round.

Although we give a slight protection to most of these plants, it is not absolutely necessary. The great obstacle to their successful cultivation out-of-doors is the fact that suitable spots are not always selected for them, and the nature of the subsoil not taken into consideration; stagnant moisture in the ground, and soil of a wet, retentive nature we have found to be most injurious; but in a well-drained sunny border we have grown most of the family well. A good plan to adopt in a border not well drained is to take out the soil to the depth of 1 ft., place 3 in. or 4 in. of broken brickbats in the bottom, and fill up with soil composed of half peat, one-fourth fibrous loam, and one-fourth common red sand, well mixed together. Here not only Calochortus, but many difficult to manage bulbs would present a very different appearance from what we frequently see, and well repay for the extra trouble. Try a space in a bed of 2 ft. by 1 ft. as an experiment—in such a space several dozens of Calochorti would thrive.—F. H.

Calopogon pulchellus.—A beautiful hardy Orchid suitable for moist or boggy places. The flowers are pink, 1 in. in diameter, and are borne in clusters of two to six upon each stem. They are beautifully bearded with white, yellow, and purple hairs, rivalling in beauty some of the tropical Orchids. It must have a moist spot, as it will fail in dry soil. Well suited for a good position on the rockwork, or for an open spot in the hardy fernery in peaty soil. Native of many parts of the United States, in wet prairies, or the edges of Pine woods. It is most abundant and thrifty in Cranberry swamps in the moss in its native country, but is also found in some grassy marshes, and is occasionally seen on solid ground in low, wet, woody ground.

Caltha *(Marsh Marigold)*.—The Marsh Marigold (Caltha palustris) that in early spring "shines like fire in swamps and hollows grey," is one of the finest hardy plants we possess, though it is so frequently met with in wet meadows and by the stream side that there is little need to give it a place in the garden, except on the margin of water, where it is always welcome. Its double varieties, however, are garden plants of much value, and well worth a place in a moist rich border, or, like the single form, by the water side, or in the artificial bog. There is a double variety of the smaller Caltha radicans, which differs from the common plant in being about half the size, and by having a creeping habit. In addition to the common species (C. palustris) and the rarer variety C. radicans, there are the double-flowered forms C. monstrosa, bearing beautiful golden rosettes, and C. minor fl.-pl., a small-growing kind, a free flowerer, and a highly desirable plant. There are also C. leptosepala, the new Californian kind, and another new kind named C. purpurascens, a very distinct and handsome sort, which grows about 1 ft. high, has purplish stems, and bears a profusion of bright orange flowers, the outside of the petals being flushed with a purplish tinge.

The various forms of the Marsh Marigold are splendid plants in their glowing and massive golden blossoms, and in groups or bold masses they will be found valuable by the tasteful gardener, and they are easily grown and increase freely.

Calypso borealis.—A pretty and interesting little hardy Orchid from the cold regions of North America, with rosy-purple sepals and petals, and a white lip, heavily blotched with cinnamon-brown. It succeeds in half-shady spots on the margins of the rock garden or artificial bog, or in a select spot among choice shrubs in light, moist, vegetable soil mulched with Cocoa fibre or other material to keep the surface open.

Calystegia *(see Convolvulus)*.

Camassia *(Quamash)*.—A North American genus of Liliaceæ, with at least one first-class plant, hardy and extremely valuable for cutting. There are three species known, and all of these are in cultivation, but with regard to two of them, C. Fraseri and C. Leichtlini, botanists are somewhat at variance as to the genus to which they belong, some placing the former with the Scillas and the latter with the Chlorogalum.

C. esculenta *(Quamash)* is a native of meadows and marshes from Middle California to Washington Territory and northward, where it grows from 1 ft. to 3 ft. high. Its stalks, which are stout, rising from bulbs 1 in. or more in diameter, bear a loose raceme of from ten to twenty flowers about 2 in. across. The colour varies from a deep to a pale blue, the deepest being Browni and atro-cœrulea, the latter the finest of all the forms. It thrives best in a deep, rich soil of a light sandy character in a moist situation. A bold group of it in flower has a fine effect in July, and it is excellent in the cut state, the buds of the spike opening in the house. A valuable summer flower for borders or groups and the bulb garden. The bulbs are used by the Indians as food.

C. Fraseri *(Eastern Quamash)*.—This is a native of the States east of the Mississippi. Its flowers are rather smaller than those of the western species. It grows about 1½ ft. high, with from six to eight leaves rather shorter than the scape, the latter bearing a raceme of from ten to thirty pale blue flowers, each about 1 in. across when expanded. All the parts of the flower are smaller than in C. esculenta. Though of comparatively recent introduction to our collections, this plant will not compare with the older kinds in point of beauty, and is much inferior to the old Camassia esculenta, the flowers being considerably smaller and of a pale purplish shade, and produced in short, dense racemes. It is, however, later in flowering than other Scillas and Camassias, which, perhaps, is a desirable quality. It grows well in a light rich soil if sufficiently moist during the growing season. All of the Camassias may be propagated either by dividing the bulbs or seeds, the latter a slow process.—W. G.

C. Leichtlini *(White Camassia)* is nearly related to the preceding; it grows often on open sandy ridge tops, but is also found in dry spots in ravines; its bulbs are generally deep in the ground, the base in some stiff, moisture-retaining soil. In lighter soils the flower-spike is large, being 9 in. long by 4 in. in diameter, while in heavier soils the spike is sometimes compound, and contains several hundred florets, which are creamy white and about 1 in. in diameter. The spike often reaches 3 ft. or 4 ft. in height. This is vigorous in growth, but not so handsome as the common Quamash. It is a native of British Columbia, where it was first discovered by Mr. John Jeffrey in 1853.

Campanula *(Hairbell, Bellflower)*.—A large, very beautiful, and in all ways most important family of plants. In Nature they have a very wide distribution, and their uses in gardens are wide too. The alpine species have obvious advantages for all kinds of rock gardens, being as a rule not difficult to cultivate. Some of them are very easy and free indeed; a group of species somewhat larger than the true alpines may be seen adorning rocks and old bridge walls in various countries, and these may well be used for similar positions in this country. Some make charming window plants,

Caltha palustris (*Marsh Marigold*).

Double-flowered Caltha palustris.

Camassia esculenta (*Quamash*)

Campanula Rapunculus (*Rampion*).

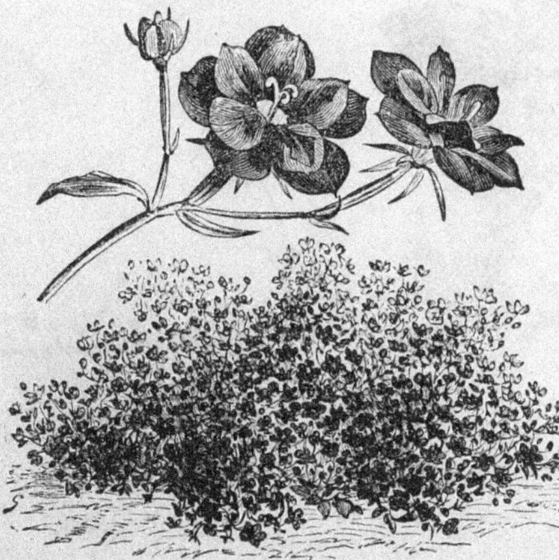

Campanula (Specularia) Speculum fl.-pl. (*Double-flowered Venus Looking-glass*).

Campanula Trachelium. See page 70.

[*To face page* **64.**

Campanula alliariæfolia.

Campanula cæspitosa.

Campanula celtidifolia.

Flowers of Campanula celtidifolia. Natural size. See page 66.

Campanula carpatica.

thriving freely in dry rooms and flowering gracefully. Numbers are good border and edging plants of easy culture; the tall and straggling kinds are admirable for growing in the wild garden, or in rough woody places or hedgerows. Some of the annual kinds well grown are showy and pretty treated as annuals, and certain groups, such as the finer forms of the Peach-leaved Bellflower, deserve good culture in the flower garden proper, flowering freely in early summer. The Canterbury Bell is one of the most useful and conspicuous of biennials, while the tall chimney Campanula is one of the most striking and valuable of our hardy flowers. But the aspects and values of the plants are so different, that it is only by looking at the species individually that one can estimate them fairly.

All are natives of temperate and most tropical countries. There are about 280 species, mostly European and Asiatic; 110 species are found in various parts of Europe, principally the Mediterranean zone and the isles of that region. Greece and its Archipelago, Italy, the isle of Crete, and Dalmatia offer a number of good kinds; Spain and Portugal also possess species peculiar to them. Fewer are found in Central Europe; still, Saxony, Bohemia, Hungary, and several provinces of Austria have their types. They shun cold countries, but at the same time occupy elevated mountain zones, and several charming Campanulas have selected the Alps, Piedmont, the Pyrenees, the Carpathians, and the mountains of Central France for their homes. Asia has at least 80 species; America is poor, with about a dozen; Africa has fewer; one only is found in Australia.

C. alliariæfolia is a strong plant with large heart-shaped leaves, covered with short hairs. The stems are somewhat erect, branching and leafy; the flowers white, drooping, and generally disposed on one side of the stem. It is a handsome plant, growing to a height of about 4 ft., and flowering in July and August. A native of the Caucasus, and sometimes cultivated under the name of C. lamiifolia. It ripens its seeds freely, and by this means may be readily increased.

C. Allioni forms an underground network of succulent roots, or stems, surmounted by stemless rosettes of leaves, about an inch long, from which arise stalkless, erect flowers, much resembling in shape and colour those of the common Canterbury Bell. Succeeds well in a moist, but free, sandy loam in exposed positions in the rock garden, but decidedly objects to limestone. It flowers in early spring; is increased by division. Alps of Europe.

C. alpina *(Alpine Hairbell)*.—Covered with stiff down, giving it a slightly grey appearance; habit, erect, not spreading; flowers of a fine dark blue, scattered in a pyramidal manner along the stems. Carpathians. Valuable for the front margins of the mixed border, as well as for the rock-work. In cultivation it grows from 5 in. to 10 in. high, and may be readily increased by division or seeds.

C. barbata *(Bearded Hairbell)*.—One of the sweet blue flowers that abound in the rich green meadows of Alpine France, Switzerland, Italy, and Austria, and readily known by the long beard at the mouth of its pretty pale sky-blue flowers, nearly 1¼ inch long, nodding gracefully from the stems, which usually bear two to five or more flowers. In elevated places in its native habitats it sometimes grows no more than from 4 in. to 10 in. high, but it grows nearly twice as high in the lower parts of the valleys in Piedmont. There is a white-flowered form which, like the type, is well worthy of culture, thriving freely in rather moist, well-drained loam on the rock garden or front margin of a mixed border. Easily increased by seed or division.

C. Barrelieri, allied to C. fragilis, has procumbent, one-flowered stems and roundish, heart-shaped leaves. Blossoms blue. On rocks by the seaside about Naples; a good trailing rock plant, thriving also in baskets or pots in windows.

C. cæspitosa *(Tufted Hairbell)*.—This we take to be the true specific type, of which C. pumila is a dwarf variety. Its roots ramble very much, and it soon forms large patches in a border consisting of moderately light garden soil. Its flowering stems, bearing, when growing vigorously, five or six gracefully-pendent light blue flowers, rise from a mass of leaves to a height of 6 in. or 8 in. It flowers in July and August, both abundantly and continuously, and may be increased readily by division of the root. European Alps. C. cæspitosa var. pumila, of dwarfer stature, scarcely exceeding 3 in. or 4 in. in height, flowers usually in pairs on a stem. There is also a white form of this variety, equally pretty as the blue. It is most easily increased by division and also by seed, but as a few tufts may be divided into small pieces, and quickly form a stock large enough for any garden, it is scarcely worth while raising it from seed, except where plants cannot be got. This variety is also known as C. pusilla, which, however, is a distinct species.

C. carpatica *(Carpathian Hairbell)*.—An old and deservedly popular garden plant of free-flowering habit. The flowers are large, cup shaped, borne on foot-stalks 12 in. to 15 in. in height, erect, and light blue in colour. Through July and August there is a succession of blooms, produced so constantly that at one time it was considered to have established a rightful claim to the flower garden. It is, however, a border plant of such value that no garden should be without it. It is readily increased by seeds, which are produced in great abundance. Pallida, a very pale blue, and alba, a pure white, are both desirable varieties; these, however, must be propagated by spring cuttings or root division, as the variation does not perpetuate itself from seed. There is also a blue-and-white variety, an excellent and showy plant, but barely so free blooming as the species; and a rare dark-blue sort named Bowoodiana.

C. celtidifolia. — A first-rate perennial species. The branches rise from a somewhat fleshy underground stem, of by no means a rambling disposition. The flowers are well expanded, of a light blue colour. When well established, it attains a height of from 3 ft. to 4 ft., but it requires three years at least before it assumes its true character. Native of Siberia, and flowers through June and July.

C. cenisia *(Mont Cenis Hairbell).*—A high alpine plant that grows among the fine Saxifraga biflora at the sides of glaciers on the high Alps, scarcely ever making much show above the ground, but very vigorous below, sending a great number of runners under the soil. Here and there it sends up a compact rosette of light green leaves. The flowers are solitary, blue, somewhat funnel-shaped, but open, and cut nearly to the base into five lobes. It is sufficiently interesting to merit a place on the rockwork. It should have a sandy or gritty and moist soil, and be somewhere near the eye if the rockwork be on a large scale. European Alps.

C. collina *(Sage-leaved Hairbell)* is a distinct species, with a creeping root, sending up stems to a height of 12 in. or 15 in., each stem bearing a few narrow leaves, from whose axils the pendent flowers are produced; the flowers are blue, appearing in May and June. It blooms somewhat sparsely, but is a desirable border plant. Caucasus.

C. fragilis *(Brittle Hairbell)* is a glabrous plant, excepting the young branches, which are coated with soft down; the flowering branches are either prostrate or hanging downward, where the position of the plant will admit, and where growing freely they attain a length of 12 in. or 15 in.; the flowers are 1 in. or more in diameter, of a delicate blue tint of colour. Of all the species this is the best adapted for a suspended basket, and it does admirably in a cottage window. It is rather tender, coming, as it does, from the seaside rocks in the south of Italy and Sicily. It should be increased by cuttings of the young shoots in spring. To divide the plants is a rash experiment, and one which will at once suggest how appropriate is the name. If planted on the rockery, a watchful eye must be kept against snails, which, while fond of all the family, have a special liking for this. C. fragilis hirsuta is a variety quite coverèd with hair or stiff down in all its parts, so much so as to look almost woolly. It is of about equal value with the normal smooth form.

C. garganica *(Gargano Hairbell).*—In compact, tufty habit of growth this plant bears a considerable resemblance to C. muralis, but it is larger in all its parts. The flowers are produced in irregular branching racemes, pale blue, shading off to white towards the centre. It blooms abundantly, and thrives well either on a rockery or in a border; owing to the pendent character of its flowering branches, its proper place is, however, against a rocky ledge, over which its masses of flowers may hang. It blooms in June and forms a valuable plant either for suspension in a window or conservatory. It may be readily increased by cuttings in early spring or by division. C. garganica var. hirsuta is a hairy plant that differs from the true Garganian type by the root leaves being more pointed and by the whole plant being downy. C. garganica var. floribunda receives its name from its free-flowering propensities. It is smaller in growth, and the corolla is more campanulate, and the racemes more erect. There is also a pure white form of this variety. The Gargano Hairbell is frequently met with under the name of fragilis, from which, however, it is perfectly distinct.

C. glomerata *(Clustered Bellflower)* grows about 2 ft. high, the erect stems being terminated by a dense cluster of flowers. They are pretty, varying in shades of colour between a deep intense purple and pure white; the latter form is somewhat rare. It is generally found throughout Europe growing in scattered isolated tufts in nearly all calcareous soils, flowering in June and July. In the border it forms a very compact and useful plant. C. glomerata var. dahurica is larger in all its parts, and is sometimes cultivated under the specific titles of C. speciosa and C. Cephalotes. This variety was introduced from Siberia by Fischer under the impression that it was a distinct species, but in cultivation it proves to be nothing more than a very showy variety. One of the finest of all the Bellflowers. The variety cervicarioides differs by its taller and more straggling growth.

C. grandiflora *(Platycodon).*

C. grandis.—A distinct and beautiful species, producing large masses of barren shoots, each forming a rosette of numerous laxly-arranged leaves, growing close to the ground, and so rapidly extending itself that by the third year a small plant will have covered with a dense leafy carpet a circle fully 3 ft. in diameter, from which a dozen or more flower-stems will have been developed to a height of 2 ft. or even more. The flowers are wide, somewhat shallow, densely arranged along the flowering branch. It flowers in June and July, and is exceedingly partial to a slight shade; the flowers are blue, fully 2 in. in diameter. It is a border plant of somewhat short duration in flower.

C. hederacea *(see Wahlenbergia).*

C. Hosti.—A species much resembling C. rotundifolia, producing copiously small drooping flowers of a deep violet-blue. Central Europe and of easy culture.

C. isophylla *(Ligurian Hairbell).*—A very ornamental and profusely flowering Italian species. The leaves are roundish or heart-shaped, and the flowers a pale, but very bright blue with whitish centre. It is a charming ornament for the rock garden, and

Campanula collina.

Campanula fragilis.

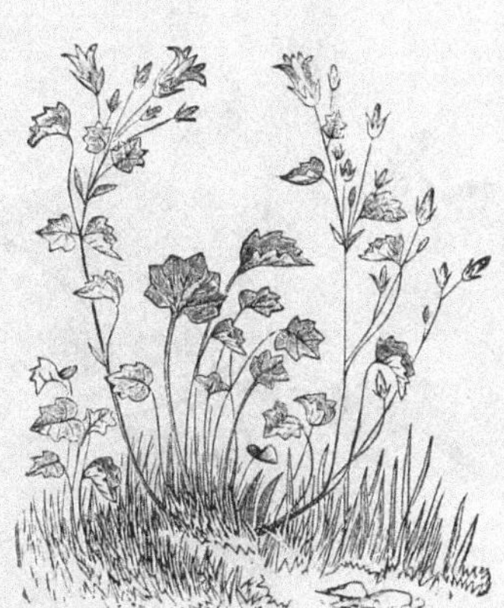

Campanula garganica.

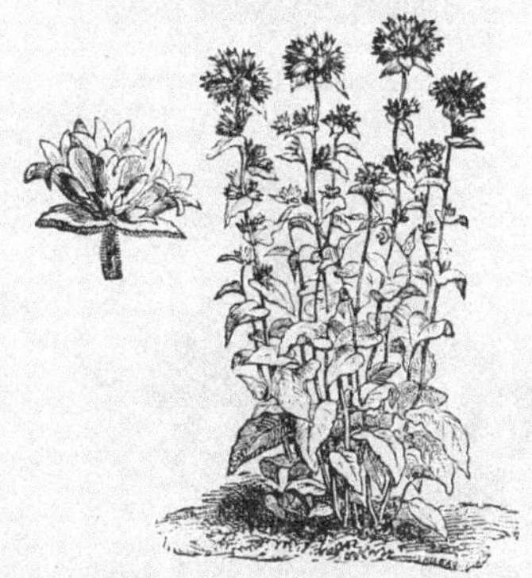

Campanula glomerata.

Campanula grandis.

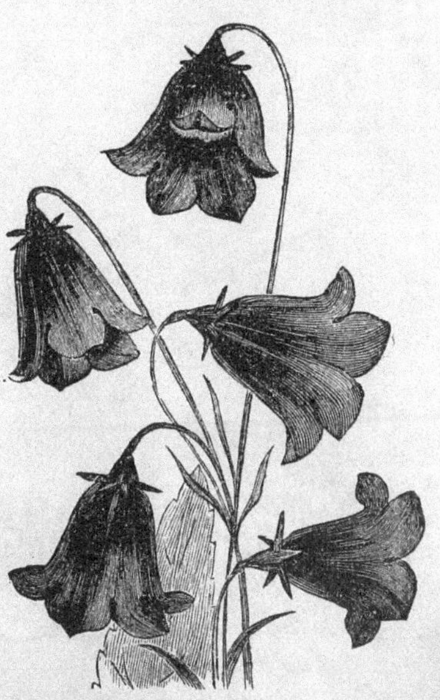

Flowers of Campanula Hosti. Natural size.

[*To face page* **66.**

PLATE LVI.

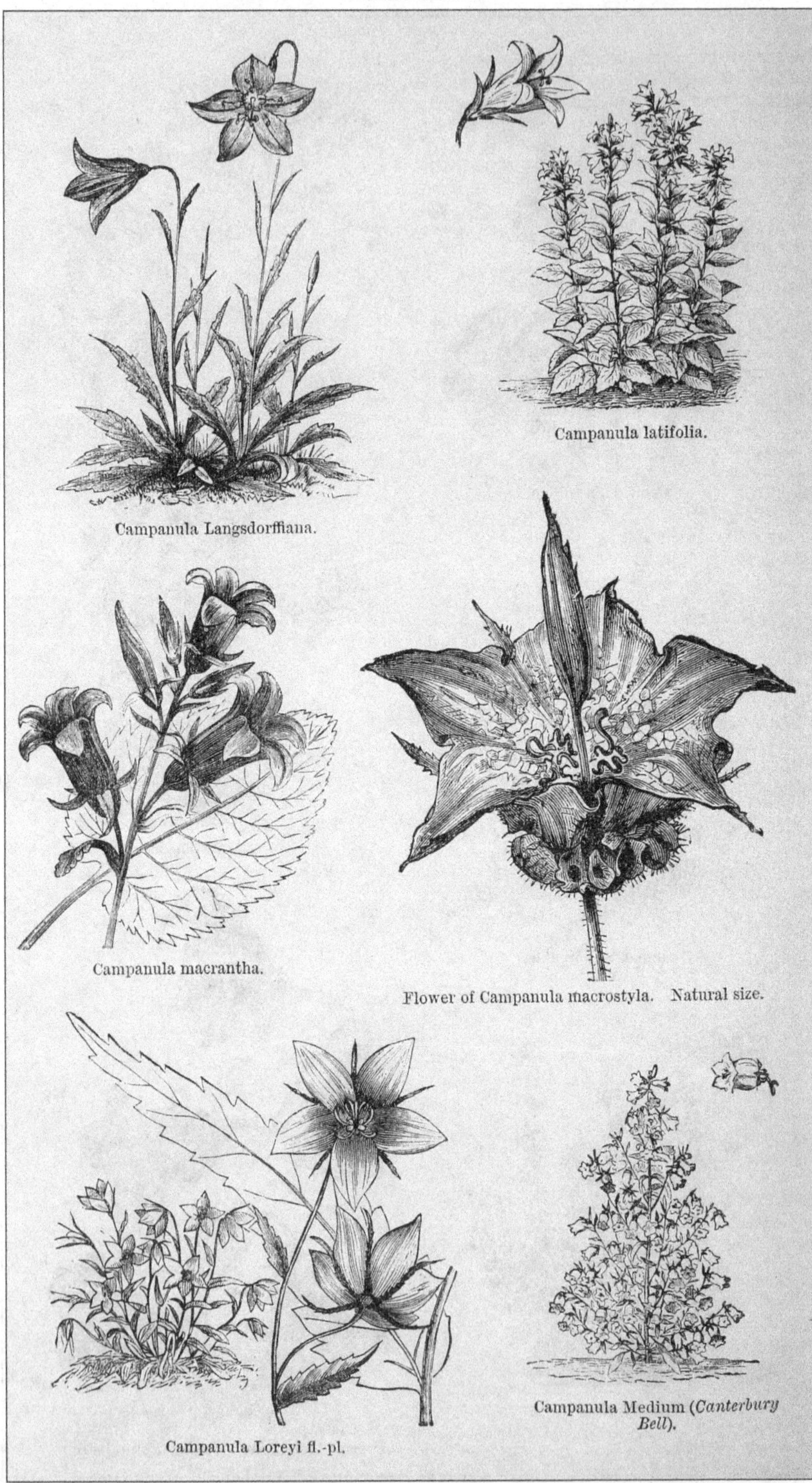

[*To face page* 67.

should be placed in sunny positions in well drained, rather dry fissures in sandy loam, and then it will repay the cultivator by a brilliant bloom. It is one of many kinds of Campanula that might with great advantage be naturalised in rocky spots, the sunny walls of old quarries, chalk pits, and like places.

C. lactiflora *(Milky Bellflower)*.—This plant, which is of Caucasian origin, is one of the bolder and larger Bellflowers, better suited for naturalisation than for the flower garden or border where its time of flower is short and habit stiff. The flowers are large and of a milky-white colour, produced in loose pyramidal manner,

C. Langsdorffiana *(Langsdorff's Hairbell)*.—A plant scarcely exceeding 4 in. in height. The flowers are blue, borne singly, nearly pendent. It is rather a shy grower, and does not increase rapidly at the root; but, no doubt, cuttings made from the early spring growth would strike freely. Russia.

C. latifolia *(Broad-leaved Bellflower)*.—The stems of this plant rise to a height of 3 ft. or 4 ft. The flowers are large, pendent, with slightly reflexed segments; they vary between white and blue in colour, and it is certainly one of the stateliest of our wild Bellflowers. It is a native of Britain, and generally found in woods; it is also abundant through the whole of Northern and Central Europe. It flowers in June, and is a very good plant for woods, copses, and rough places or hedgerows, and above all the water side or ditch.

C. Lœflingi *(Lœfling's Hairbell)*.—This is a showy species, dwarf in habit, rarely exceeding 4 inches high, producing a profusion of violet-blue blossoms ¾ in. across. It is a capital plant for the rock garden, or for front spaces in ordinary borders. Portugal and Spain in cornfields.

C. Loreyi *(Lorey's Hairbell)*.—An annual species from Mount Baldi, in North Italy. It grows 9 in. to 1 ft. high, has numerous branches and small, stalkless, shining leaves. Its blossoms are of a blue-violet colour, and are produced in sufficient quantity to render it a pretty border annual; it is seen to the best advantage when planted in masses. Its variety called alba, with blossoms of silvery grey, is also very attractive. It is an annual of easy culture sown out-of-doors in spring.

C. macrantha *(Large Bellflower)*.—The stems of this handsome plant rise to a height of 5 feet, and are well furnished at the base with ovate leaves. The vigorous stems are terminated by clusters of large deep blue flowers of a size almost equalling that of the Canterbury Bell, but less contracted at the mouth of the tube. It is a free, vigorous-growing perennial plant, a native of Russia, and perpetuates its giant character when grown from seeds. C. eriocarpa is probably synonymous with this species.

C. macrostyla *(Candelabra Bellflower)*.—A singular plant with large flowers, with blue netted veins on a white ground getting purple at the edges, and having a huge stigma. It is wholly distinct from any of the Campanulas hitherto in our gardens, and well deserves culture. It is readily recognised by its candelabra habit of growth. A native of Asia Minor, and an annual plant of easy culture.

C. Medium *(Canterbury Bell)*.—A familiar old biennial that should find a place in every garden. It is too well known to need description, but there are now a host of beautiful varieties bearing flowers with a great diversity of colours. These may be classed into three sections—single flowers, as seen in the old-fashioned single Bells; doubles, as found in the stout massive flowers in which two, three, and even four bells seem to be compressed into the outer one; and duplex flowers, as seen in the calycanthema forms, in which one bell grows in the other, and the combined two resemble a cup standing in a saucer. The single varieties are still to be had, but no one would care to be dependent upon these only who have once seen the beautiful products of a fine modern double and semi-double strain, and which will also produce some single flowers. Whilst the old single strains comprised only white and blue colours, the modern strains include nearly a score of hues, inclusive of white, lavender, mauve, several shades of purple, pink, rose, salmon, and also blue. The duplex strains have hitherto been chiefly confined to white and blue, but other colours are now being introduced. It is a curious fact in relation to this section that whilst the seed of the white form is of a pale hue and the seed of the blue form is of a dark tint, the seed of the double and semi-double forms of all colours is the same, viz., a glossy brown. The habit of the latter section is compact and good, the plants when in bloom ranging from 18 in. to 24 in. in height, forming perfect pyramids of flowers. They may be lifted and placed in pots even when in full bloom without injury, and as they invariably flower from the middle of June to the middle of July, according to place and season, it would be possible, in the shape of a large group of these in pots, to introduce to some of the midsummer shows a very novel and interesting feature. If so required, or if to be used for corridor, house, or conservatory decoration, it would be best to lift the plants and pot them in 6-in. pots early in May; this would enable them to get well established before the blooming period. The Calycanthema section usually exhibits both a taller and a looser growth, and should be planted in borders behind the double and single kinds.

March or April is the best time to sow seed of Campanula Medium and its varieties if it be desired to secure large massive plants that will produce several spikes of flowers, and display the beauties of one of our finest hardy border plants to the best advantage. The seed is small, and may, if plentiful, be sown in a warm spot in the open ground, but it is much safer to sow it in shallow pans or boxes placed in a frame or on a shelf in the greenhouse, as in the latter case all the seed will probably germinate and an abundance of plants be secured. When these are large enough to handle conveniently, prick them out into some shady spot, and keep them watered until the plants are well rooted. From that time they may safely be left to take care of themselves until September, when they should be transplanted into their permanent places in the flower borders, and so treated will not only get well established before the winter arrives, but will also develop their blooming crowns for the next year. The following are the principal varieties: Rosea, r. plena, cœrulea, c. plena, alba, a. plena, cœrulea striata, albo lilacea, a. l. plena, rosea lilacina plena, azurea, a. plena, calycanthema, c. alba, c. lilacina, c. l. plena.

C. muralis *(Wall Hairbell).*—A dense tufty-growing evergreen species, with small broadly cordate, irregularly-notched, bright green leaves, cucullate in shape, supported on foot-stalks 1 in. or more in length, but so numerous are the leaves, and so dense their arrangement, that the foot-stalks are scarcely visible. The flowers are pale blue. The plant has been long in cultivation, and is known also by the name of C. Portenschlagiana. It is a plant of free growth, equally adapted for pot culture or for rockwork. It spreads but slowly by the development of underground stems, and hence succeeds well when planted in fissures or crevices of the rockery. Dalmatia. Flowers in August or September.

C. nitida *(Shining Bellflower).*—A rigid-growing plant, both as regards stem and leaves; the latter are produced in dense rosettes, from the centre of which rises the stem to a height of about 9 in. The flowers are of a deep slaty-blue colour. This plant is one of the few species that are natives of North America; it sometimes goes by the name of C. americana, a species which is only of annual duration; whereas C. nitida is a long-lived perennial. There is a pure white and a double form, both of which are desirable varieties. It likes a well-drained sunny corner of the rockery, being rather liable to damp off in the winter. Increased by division of the rosettes that form at the base of the flowering-stem.

C. nobilis *(Noble Bellflower).*—A very large-flowered kind, in general habit of growth allied to C. punctata, having creeping underground stems. The flower-stems rise to a height of 18 in. to 20 in., with a few narrow leaves, and from the axils of these the large flowers, almost 3 in. long, hang pendent. The colour is a light chocolate with a shade of blue in it; there is also a creamy white variety. It blossoms in May and rarely seeds, but it may be readily increased by the underground stems. Native of China. Of easy culture in any soil.

C. persicifolia *(Peach-leaved Bellflower).*—A beautiful species, the flowers of which are large, fully 2 in. across, cup-shaped. The stems, though wiry in texture, are often weak at the base, owing to the rosettes of the previous year, whence they spring, losing their vigour during the process of growth, and lacking root hold—hence, when in full bloom, if not provided with some artificial support, the stems lie about in an untidy manner. It grows from 1½ ft. to 2 ft. high, and flowers in July and August—the early removal of the flower-stems causing not unfrequently the production of a few scattered autumnal blooms. It is found abundantly through Northern and Central Europe, and appears to thrive best and bloom most freely on calcareous soils. C. persicifolia maxima is a giant form, also called C. suaveolens. Besides both the double blue and double white forms, there is a very interesting variety named coronata, in which the corolla appears to be doubled. These three varieties are admirable for borders, and should find a place in every selection of herbaceous plants, the double white variety being particularly desirable not only for the border, but for growing in pots. All the varieties well repay for good culture. Plants occasionally divided and kept in rich beds give very fine crops of flowers.

C. primulæfolia *(Primrose-leaved Bellflower).*—Like the Canterbury Bell, this species is little better than a biennial, the flowering process almost invariably exhausting the entire energies of the plant. In general contour it is, however, quite distinct, as the stem of this species rises erect, and unbranched to a height of 2 ft., from a mass of broadly ovate, roughly-corrugated and serrated leaves, which resemble those of a coarse growing Polyanthus; the flowers are wide, of a slaty-blue colour, gradually shading to white at the base of the corolla; a native of Portugal. Flowers in July and August, and may be increased by seeds, which they produce freely. C. peregrina is a synonym.

C. pulla *(Austrian Hairbell).*—This, when grown freely, is one of the most charming of the whole group; its neat habit of growth, its deep purple drooping bells, and its simple unifloral character are all special elements of beauty. It is a native of the Austrian Alps, and is generally found growing among the Grass in the high mountain pastures, where its tendency to produce rambling ground stems is not checked or limited; therefore, if planted on the rockery, it should have a good wide shelf of moderately flat soil to itself in which a little peat and sand has been previously mixed. This plant is usually looked upon as a shy grower, and, consequently, is generally grown in a pot, which for various reasons is not at all adapted to its peculiar mode of growth; the first season it will do very well, but the second will bring disappointment. It may be increased by hundreds in the spring by taking off the young shoots and placing them in a gentle

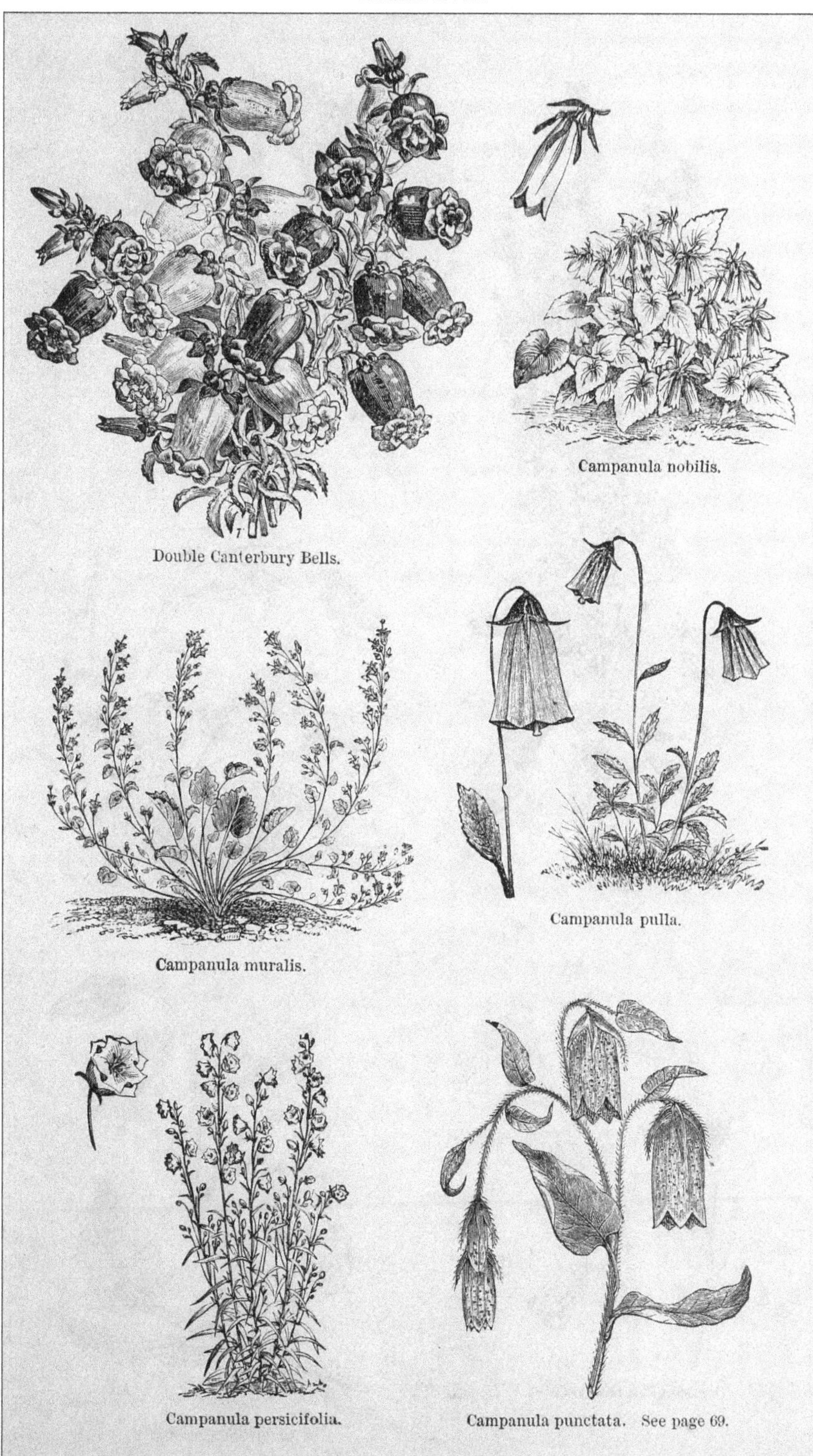

Double Canterbury Bells.

Campanula nobilis.

Campanula muralis.

Campanula pulla.

Campanula persicifolia.

Campanula punctata. See page 69.

[*To face page* 68.

PLATE LVIII.

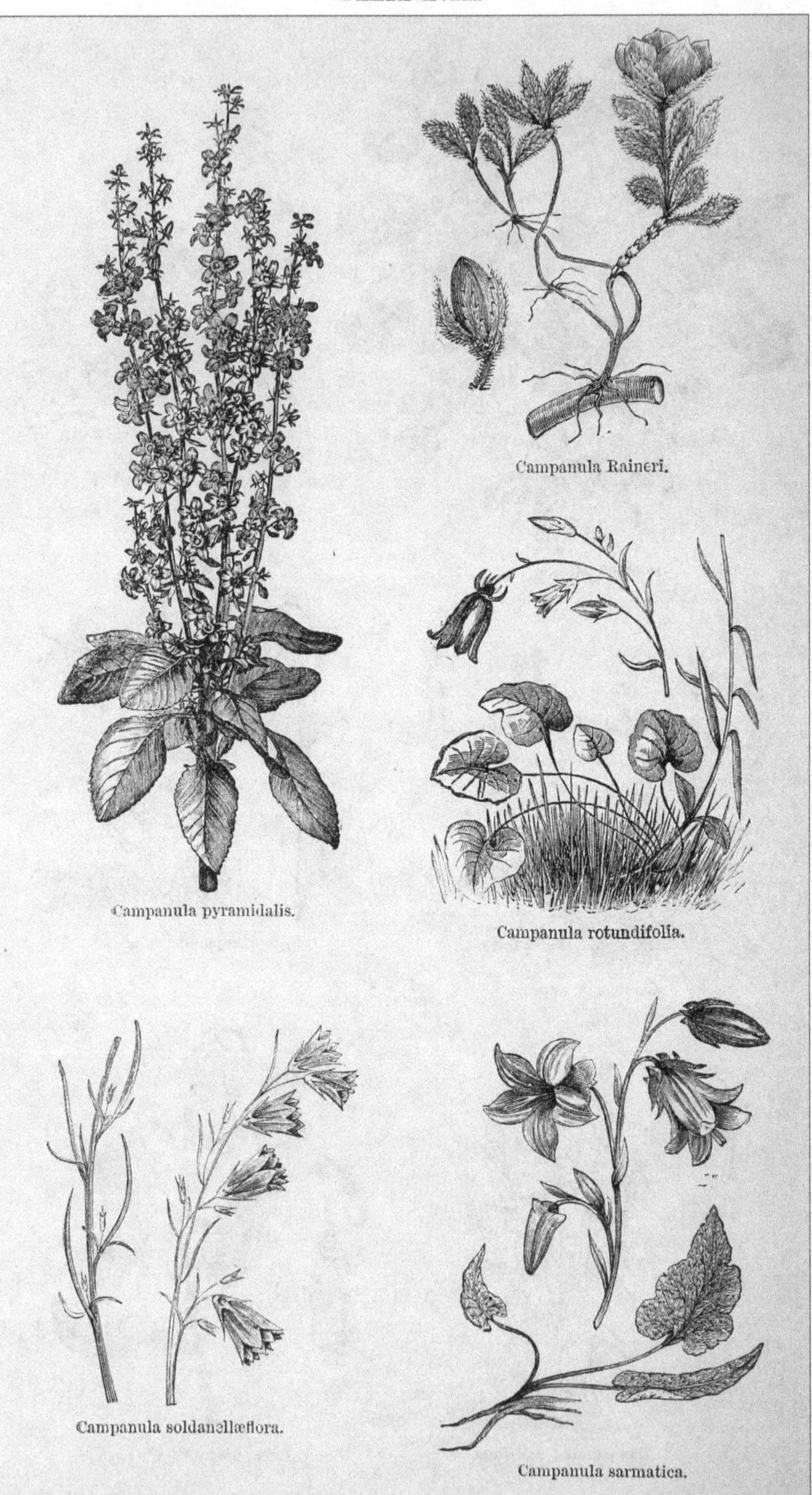

Campanula pyramidalis.

Campanula Raineri.

Campanula rotundifolia.

Campanula soldanellæflora.

Campanula sarmatica.

[To face page 69.

bottom-heat. They will strike as freely as those of a Lobelia, and make fine compact autumnal flowering plants, but observe that success depends greatly on the vigorous character of the plant from which the cuttings are taken. After blooming the foliage disappears and the plant goes to rest. An excellent rock plant.

C. pumila *(= C. cæspitosa).*

C. punctata *(Spotted Bellflower)* is of creeping habit, and, like many similar creeping plants, difficult to get established, but when once so, it is equally difficult to keep within bounds. Its leaves are rough; from amid the mass of foliage rise the flower-stems to a height of 8 in. or 9 in., each stem carrying two or three pendent milk-white flowers, fully 2 in. long; the spotted character, though slightly shown externally, is well seen on the inside. Siberia. Blooms in July. It does not produce seeds, but its creeping roots afford a ready means of propagation.

C. pusilla.—Smaller than C. cæspitosa, rarely exceeding 4 in. The leaves are roundish, heart-shaped, smooth, shining green, and toothed. The flowers are pale blue, in few-flowered racemes, and appear in June and July. Suitable only for rockwork. Native of Switzerland. Moist, very gritty loam is best for it. *Syn.*, C. modesta.

C. pyramidalis *(Steeple Bellflower).*—A strong, vigorous growing plant; thick and fleshy flower-stems, rising to a height of 4 ft. or 5 ft., furnished with numerous broadly ovate leaves; the flowers are produced on peduncles varying in length, but all closely adpressed to the stem, thus giving the inflorescence much more the character of a steeple than a pyramid. The flowers are either blue or white; the corolla widely expanded, its segments broad; they are rather erect than pendent, and, owing to the secondary floral branches developing their flowers gradually in succession, extend the blooming period over a considerable length of time, in the months of July, August, and September. This is a plant that dearly loves the shade, and is well adapted, not only as an ornament for the flower border, but also for the conservatory, where, during the latter summer months, a blue colour, and the bold character of this plant, is desirable, where its columnar spikes contrast with the lines of the Ferns or Palms. In general cultivation, though not absolutely a biennial, it is better to consider it as such, as from seedling plants, well grown on during the first year, the finest stems are developed. Carolina. It matures its seed freely, but care should be taken that it be not sown too thickly, as it is very liable to damp off; and, owing to this cause, one may lose the crop altogether. A border plant of the highest merit in favourable soils; perennial, but so important that occasional batches of seed should be sown to keep up a vigorous supply of young plants.

C. Raineri *(Rainer's Bellflower).*—A dwarf, compact, sturdy plant, the height varying from 3 in. to 6 in., each branch bearing a large solitary dark blue flower. It is a native of the Alps of Northern Italy. It thrives best planted in sunny positions in loam freely intermingled with pieces of stone, and well watered in dry weather. A gem for the rock garden.

C. Rapunculus *(Rampion Bellflower).*—A biennial plant, with foliaceous stems, rising to a height of 2 ft., and thick fleshy roots, which were once much cultivated as a culinary vegetable. The flowers are arranged loosely on the stem, and are widely expanded, of a purplish blue. It is a native of Britain, and extends across the whole of Central and Southern Europe, where it is generally found in hedgerows and open country districts, flowering in May or June, though not as showy as some of its congeners. It is a desirable plant for the woods or rough places.

C. rotundifolia *(English Hairbell).*—Although a wild plant, this loses none of its beauty by reason of its familiarity: in it we have the true type of the Hairbell. In early spring its round-leaved character is fully developed, as then the little tufts of radical leaves indicate at a glance the origin of its specific title, but as the flower-stems become developed, these gradually die away, and the leaves which accompany the flowers are long and narrow; hence it is a subject of surprise to those who only know it when in bloom that such an apparently inappropriate name should have been given to it. So well known is the plant that it is almost needless to give a description. There is a white variety, but it is generally of a dwarfer and more slender habit. There are several forms all of easy culture in any soil, and all beautiful.

C. rotundifolia soldanellæflora.—*(Double fringed Hairbell).*—A double flowered variety reminding one of the beautifully-fringed flowers of the Soldanella. When planted out in the border or rockery, and growing freely, it attains a height of fully 15 in. or even more. The upper portion of each of the stems is well furnished with pendent dark blue flowers, arranged on a somewhat irregularly branching raceme. All the leaves are linear, even those which spring from the root, which scarcely at all exceed in breadth those of the flowering stems.

C. sarmatica *(Sarmatian Bellflower).*—A free-growing herbaceous plant, with flower-stems rising to a height of about 2 ft., furnished with flowers through almost their entire length; the lower blooms are supported singly on longish foot-stalks, the upper ones are almost stalkless; the colour is a light blue. It is a native of the rocky sub-alpine regions of the Caucasus, where, owing to the weight of the masses of flowers it produces, it not unfrequently assumes a decumbent growth, a habit which we find holds good in cultivation; hence it requires the help of a stake to guide its early growth It flowers in July and August, and increases readily from seed, which come perfectly true. Synonymous with this species is C. gummifera, whose only distinction is apparently slightly narrower leaves.

C. Trachelium *(Rough Bellflower).*—A sturdy herbaceous plant of compact growth, producing a number of erect flower-stems attaining a height of 2 ft. or more; or the flowers are in twos or threes; they are of large size and blue, but there are white and double varieties; it blooms in July. A native of Britain.

C. turbinata *(Turban Bellflower)* is a dwarf compact-growing plant with deltoid leaves, of a greyish-green. In the type the flowers are borne singly on naked stems, about 6 in. long, of a deep shade of blue, and fully 1½ in. across. The flowering extends over the months of June and July, in a continuous succession; and if the plant be cut back, a second autumnal blooming not unfrequently follows. It is a native of the mountains in Transylvania; a charming plant, well adapted for either border or rock culture. Though distinct from the Carpathian Bellflower in its normal state, if propagated by seed, which it produces pretty freely, a vigour of growth is developed that, were it not for its larger leaves and dwarfer habit, would render its true specific identity doubtful; this, too, occurs where there is no possibility of hybridisation taking place. C. turbinata elegans is a hybrid between C. turbinata and C. carpatica, and it is superior to both its parents as a summer flowerer The variety pelviformis has flowers saucer-shaped, produced very freely.

C. urticæfolia *(Nettle-leaved Bellflower)* has much the aspect of C. Trachelium, but the whole plant is of smaller size; the flowers are either blue or white, and there are double forms of each. It is a native of Germany, loving the partial shade of woods. The double white variety is particularly desirable to cultivate on account of its distinctness and chaste beauty.

C. Van Houttei is a handsome and desirable variety. It is a strong grower, attaining a height in its flower-stems of 1 ft. to 2 ft.; its large pendent leaden-blue flowers are in the way of those of C. nobilis.

C. Waldsteiniana *(Waldstein's Hair-bell).*—A charming little free-flowering species of compact habit, and producing wiry stems from 3 in. to 6 in. high, bearing pale blue flowers. A native of Hungary, originally introduced in 1824, but for years lost to the country, till re-introduced recently. It appears to be a free-growing cultivable species, and may be readily increased by cuttings taken from the early spring growth and placed in a gentle bottom-heat. It thrives luxuriantly in limestone soil in a sunny position.

C. Zoysi.—A dense tufty plant allied to C. cenisia, but more compact in habit. Flower-stems about 2 in. or 3 in. high, terminated by one perfect bloom. Flowers large in proportion to the size of the plant, azure blue in colour. A native of Carinthia and Styria. Luxuriates in sandy loam in exposed parts of the rock garden, for which it is best suited.

Besides the above the following may also be found in catalogues, but which are either not much known yet or are too much like others to be of value: C. aggregata, azurea, altaica, americana, attica, carnica, Cervicaria, divergens, Elatines, laciniata, linifolia, macrorhiza, peregrina, retrorsa, speciosa, strigosa, Scheuchzeri, tenella, Tenori, Tommasini, thyrsoidea, Vidali.

Camptosorus rhizophyllus *(Walking Leaf).*—A small North American Fern, remarkable for its narrow fronds, which taper into slender prolongations, and take root at the tips like runners, giving rise to new plants. It is of easy growth in a somewhat shaded position in gritty loamy soil. Suitable for a shady nook in rock garden or for the select hardy fernery.

Candytuft *(Iberis).*

Canna *(Indian Shot).*—A very important genus of plants with fine foliage, which, though tropical, made good growth out of doors in our gardens in the southern counties and in warm positions. The larger kinds make rich masses of foliage, yet all may be associated intimately with flowering plants —an advantage that does not belong to some free-growing things like the Castor-oil plant. The general tendency of most of our flower garden plants is to flatness, and it is the special quality possessed by the Cannas for counteracting this that makes them so valuable. Another good quality of these most useful subjects is their power of withstanding the cold and storms of autumn. They do so better than many of our hardy shrubs, so that when the last leaves have been blown from the Lime, and the Dahlia and Heliotrope have been hurt by frost, we may see them waving graceful and verdant. Sheltered situations, places near warm walls, and nice snugly-warmed dells are suitable positions for them. They are generally used in huge ugly masses both about Paris and London, but their true beauty will never be seen till we learn to place them tastefully here and there among the flowering plants. A bed or two solely devoted to them will occasionally prove very effective; but enormous flat masses of them, containing perhaps several hundred plants of one variety, are bad. As to culture and propagation, nothing can be more simple; they may be stored in winter under shelves in the houses, in the root-room, or, in fact, anywhere if covered up to protect them from frost. In spring nothing is easier than pulling the roots in pieces and potting them separately. Afterwards it is usual to bring them on in heat, and finally harden them off previous to planting out in the middle of May. A modification of this practice is desirable, as some kinds are of a remarkably hardy constitution, and make a beautiful growth if put out without so much as a leaf on them. The soil for all Cannas should be deep, rich, and light. Cannas, protected by a coating of litter, have been

PLATE LIX.

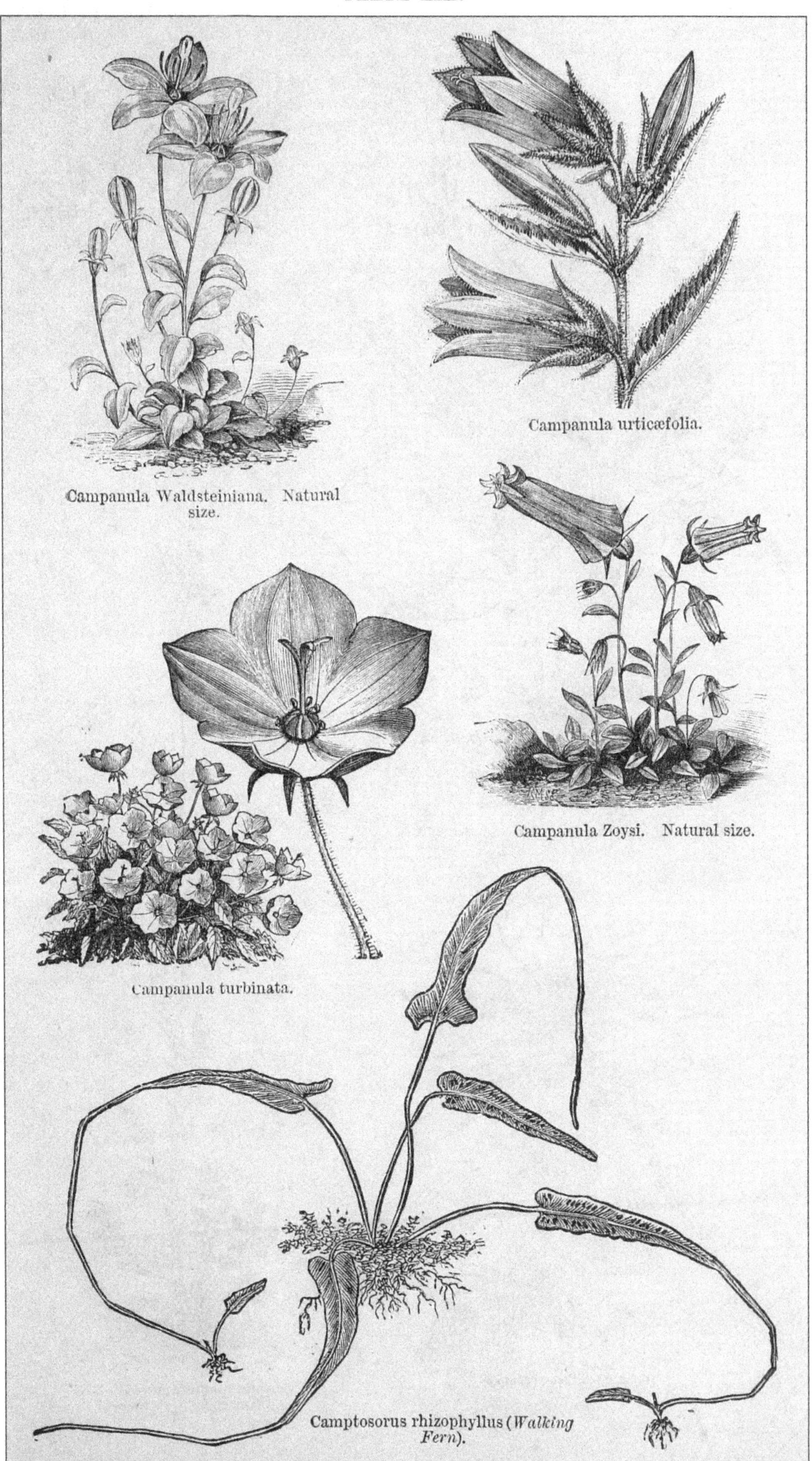

Campanula Waldsteiniana. Natural size.

Campanula urticæfolia.

Campanula Zoysi. Natural size.

Campanula turbinata.

Camptosorus rhizophyllus (*Walking Fern*).

[*To face page* 70.

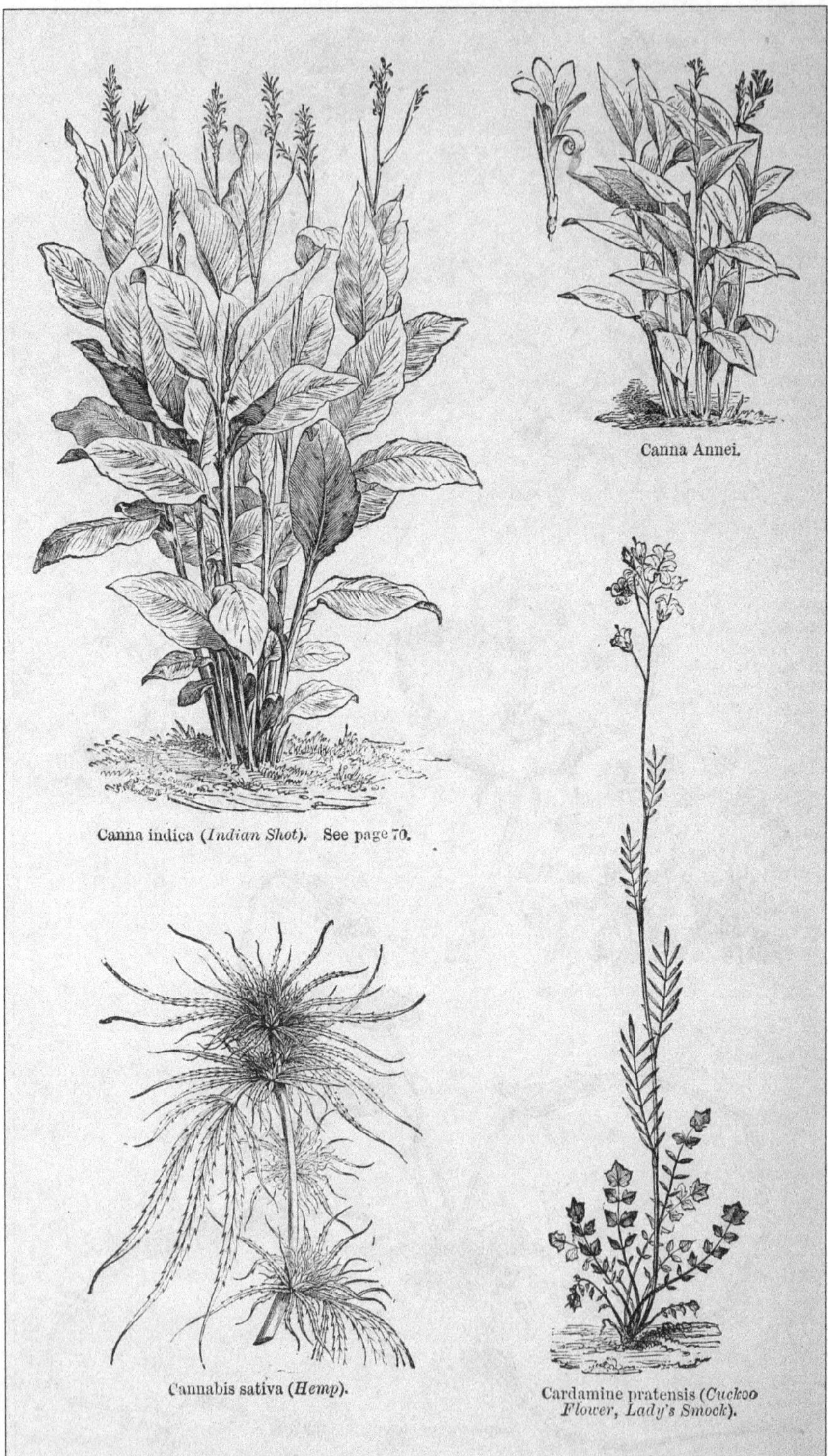

Canna indica (*Indian Shot*). See page 70.

Canna Annei.

Cannabis sativa (*Hemp*).

Cardamine pratensis (*Cuckoo Flower, Lady's Smock*).

[*To face page* **71.**

left out in Battersea Park through severe winters, and during the unfavourable summer of 1867 attained a height of nearly 12 feet. Where it is desired to change the arrangements as much as possible every year, it may not be any advantage to leave them in the ground, and in that case they may be taken up with the bedding plants and stored as easily as Carrots. Wherever they are grown as isolated tufts, in small groups, or in small beds, it will be best not to take them up oftener than every second or third year if the ground be warm and well drained. These noble plants would also adorn the conservatory, which is often as devoid of any graceful vegetation as the unhappy flower gardens which are seen all over the country. Few subjects would be more effective, none more easily obtained.

The only really hardy species is C. Achiras, a rather handsome plant, but not so desirable as the more tender species and varieties. The species and varieties available for massing or mixed planting are exceedingly numerous. In the first place, the wild forms described amount to nearly 100, whilst the garden hybrids and seminal varieties are practically innumerable. Foremost among the raisers of new varieties of Canna, M. Année, a French amateur, may be named. He commenced operations many years ago, and there are several varieties bearing his name, as C. Annei rosea, C. Annei floribunda. The species are fully described in THE GARDEN for October 21, 1878, and the varieties in the "Sub-tropical Garden." The Iris-flowered Canna is a lovely kind which we have not seen out of doors in England; it has flowered in Paris gardens out of doors, but requires a warm moist house to develop its full size and beauty.

Cannabis sativa *(Hemp Plant).* — A well-known annual, native of India and Persia, and largely cultivated in Europe for the sake of its fibre. In ordinary situations it grows from 4 ft. to 10 ft. high, but in Italy, under favourable circumstances, it is sometimes as high as 20 ft. In plants growing singly the stem is frequently much branched, but when in masses it is generally straight and unbranched. The Hemp is useful where the tenderer sub-tropical plants cannot be enjoyed. Single well-grown plants of it look very imposing and distinct, and are good for the backs of borders, mixed groups, and a few well-grown plants look well as a separate group. For these purposes it should be sown early in April in the open ground. To get large plants it would be worth while raising it in frames. It loves a warm, sandy loam, and is one of the few plants that thrive in small London gardens, where their effect is both distinct and graceful. There are several varieties offered in the seed catalogues, but, as regards habit, not differing much—at least in our climate.

Cardamine *(Cuckoo Flower or Lady's Smock).* — A large genus of the Crucifer family, few of which are cultivated, the best being the native Cuckoo Flower and its double form. Amongst hardy flowers few will be found to give more satisfaction than this, and although, like the type, which in many districts colours the meadows with its soft-hued flowers, it delights in damp, almost swampy situations, it will grow well almost anywhere. The single kind is too common to need cultivation; the double kind is a pretty subject for the spring garden and for borders. Division. C. trifolia is a pretty species, with white flowers, from Switzerland; about 9 in. to 12 in. high; a border or rough rock plant. C. latifolia, C. asarifolia, and C. rotundifolia are pretty dwarf plants when in flower, but not likely to be popular in gardens.

Cardinal Flower *(Lobelia cardinalis).*

Carduus.—A genus of Composites of Thistle-like appearance. The finest is C. eriophorus, the Woolly-headed Thistle, a remarkably conspicuous native plant, with a much-branched, hairy stem 3 ft. to 5 ft. high, and very deeply cut, undulated; spiny leaves, the lower ones often 2 ft. long. The flower-heads are large, of a purplish red colour, and surrounded on the under side with a dense white cottony web. There are few plants more handsome or novel in appearance than this. It is suitable for borders, or groups of hardy fine-foliaged plants, and grows well in any common soil. Being a native of the limestone districts of the south of England, it will be an interesting plant to naturalise in others. C. altissimus and acanthoides are also met with in botanical collections. Various plants synonymous with this genus are described under other names in this book.

Carex *(Sedge).*—An enormous genus of plants well known in all northern and temperate countries, few of which have a place in the garden, and yet some have much merit, quite apart from their botanical interest.

C. paniculata is a very large Sedge, growing somewhat like a dwarf Tree Fern, with strong, thick stems, and with luxuriant masses of drooping leaves. The roots form dense tufts, frequently elevated from 1 ft. to 3 ft. above the surface of the ground; and when the plant is in flower, it generally has a large and spreading panicle. A few tufts of this are very effective on the margins of water near groups of picturesque plants. The finer specimens are of great age, and must be procured from the bogs where the plant occurs wild.

C. pendula.—A very graceful plant, unlike any of the other British Carices, growing in large round tufts, with numerous flowering-stems and barren shoots, which attain a height of from 3 ft. to 6 ft. The leaves are often 2 ft. or more in length, and are chiefly at the base of the plant. It is most attractive when in flower, from the graceful disposition of its pendent spikes, usually about half a dozen in number, and each from 4 in. to 7 in. in length. Very suitable for the margin of water or for boggy or moist spots. Some of the kinds are variegated, as C. riparia, and there are some striking forms, such as C. Fraseriana, but compared with the above mainly of botanical interest.

Carlina acaulis.—A dwarf perennial, Thistle-like, and a rather interesting plant from its foliage, which has some resemblance to the leaves of a miniature Acanthus, and is disposed in a broad, handsome, regular rosette very close to the ground. Its single yellowish flower, 3 in. or more across, is borne on a very short, erect stalk in the centre of the rosette. Although too dwarf for association with plants of more imposing stature, it is worthy of a place on a bank or slope, or on the margins of low beds or groups, where its distinct habit will be seen to great advantage. It thrives best in dry, stony, calcareous soil, and is easily multiplied by seeds. Central Europe.

Carnation *(Dianthus Caryophyllus).*

Carthamus tinctorius *(Safflower).*—An annual of the Composite family of little ornamental value. The yellow flowers yield Safflower, largely used as a dye. Easily grown as an annual plant, but mainly of interest for the botanical or economical collection.

Cassia marilandica *(American Senna).*—A hardy, graceful perennial, 3½ ft. to 5 ft. high, with pinnate leaves, resembling those of the Acacia, and slender stems, bearing yellow flowers in numerous small clusters in autumn. It is somewhat late in growth, but once commenced grows with great rapidity. It thrives best in a position with a south aspect, and may be multiplied either by division in spring or by seed. It should always be planted in a warm, deep, sandy loam, and is very suitable for borders or association in groups with the finer hardy subjects, its graceful leaves qualifying it for a place in a group of hardy foliage plants. Recent experience seems to show that our climate is not warm enough for it. N. America.

C. corymbosa is a pretty free-flowering greenhouse variety often used with good effect in the flower garden in summer, and for that purpose requiring to be stored in the winter in the greenhouse.

Castilleja.—A curious and showy genus of Californian herbs which will not, we think, bear cultivation in the open air of this country for any length of time. We introduced seeds which grew, and the plants flowered, but these were not very ornamental, and they soon disappeared. Seed imported yearly seems the only way of increasing them.

Catananche cœrulea *(Blue Cupidone).*—An old border plant, about 2 ft. high, flowering in summer; fine blue, each stalk being terminated by a single head. South of France and Italy Suitable for borders, margins of shrubberies, or naturalisation in a well-drained warm soil. Seed. Generally a "middling" plant only, but good when well grown. There is a white variety common as the blue, and a "bicolor" one. It is easily grown in any soil, and easily raised from seed. Compositæ.

Catchfly *(Silene).*

Cathcartia villosa.—A beautiful Poppy from the eastern part of the Himalayas, somewhat resembling the Welsh Poppy (Meconopsis cambrica). It is a perennial herb having densely hairy and lobed leaves produced in flat, dwarf tufts from which arise the slender flower-stalks from 6 in. to 12 in. high, bearing drooping cup-shaped blossoms of a golden yellow. It is quite hardy on well-drained rockwork, and succeeds best in exposed and dry spots, damp being prejudicial to it.

Caulophyllum thalictroides.—An interesting perennial herbaceous plant of the Barberry family, of dwarf growth, having finely-cut foliage and small white flowers, succeeded by deep blue berries. Thrives in partially shaded borders in peaty soil. North America.

Celandine *(Chelidonium).*

Celandine, Lesser *(Ficaria).*

Celosia *(Cockscomb).*—Indian annuals of the Amaranth family. They are generally too tender for the open air, though we have seen the Cockscomb of the dwarf and tall varieties occasionally used with fine effect in bold groups. For this purpose they should be sown in pans in March and kept near the glass to prevent the seedlings being drawn. As soon as large enough to handle they should be pricked off into small pots, grown on fast in gentle heat until the crowns are formed when they may be planted out in June in rich soil, and liberally supplied with water; thus treated, they will continue in good condition for a long time.

Celsia cretica.—A plant of the Figwort family. It is a fine plant when well grown. Once it was seldom seen away from the houses or pits of botanic gardens; now, grown freely, it is really a distinct and handsome plant with its rich yellow flowers and massive polished yellow buds. Treated as an annual, and well grown in good soil, it is a stout and effective plant. Candia, N. Africa.

Cenia turbinata.—Low growing, half-hardy annuals of the Composite Order, from the Cape of Good Hope. The white and yellow varieties are sometimes grown, but are not ornamental.

Centaurea *(Knapweed).*—A very large genus of plants, a great number of which inhabit Southern and Middle Europe, some being good garden plants. Most are hardy; but some of the southern species require the greenhouse in winter, but make free growth out-of-doors in summer, and are freely used for their distinct form and the silvery colour of the foliage. As to the flowers, the most valuable plant of the family is the Cornflower (C. Cyanus), which as a market flower has merits possessed by few plants, and by no other hardy annual.

C. argentea has very elegant silvery-surfaced Fern-like leaves, and a compact, good habit, and when planted out or plunged in pots it has a good effect either as a bedding plant or for conservatory decoration; for bedding it must be plunged

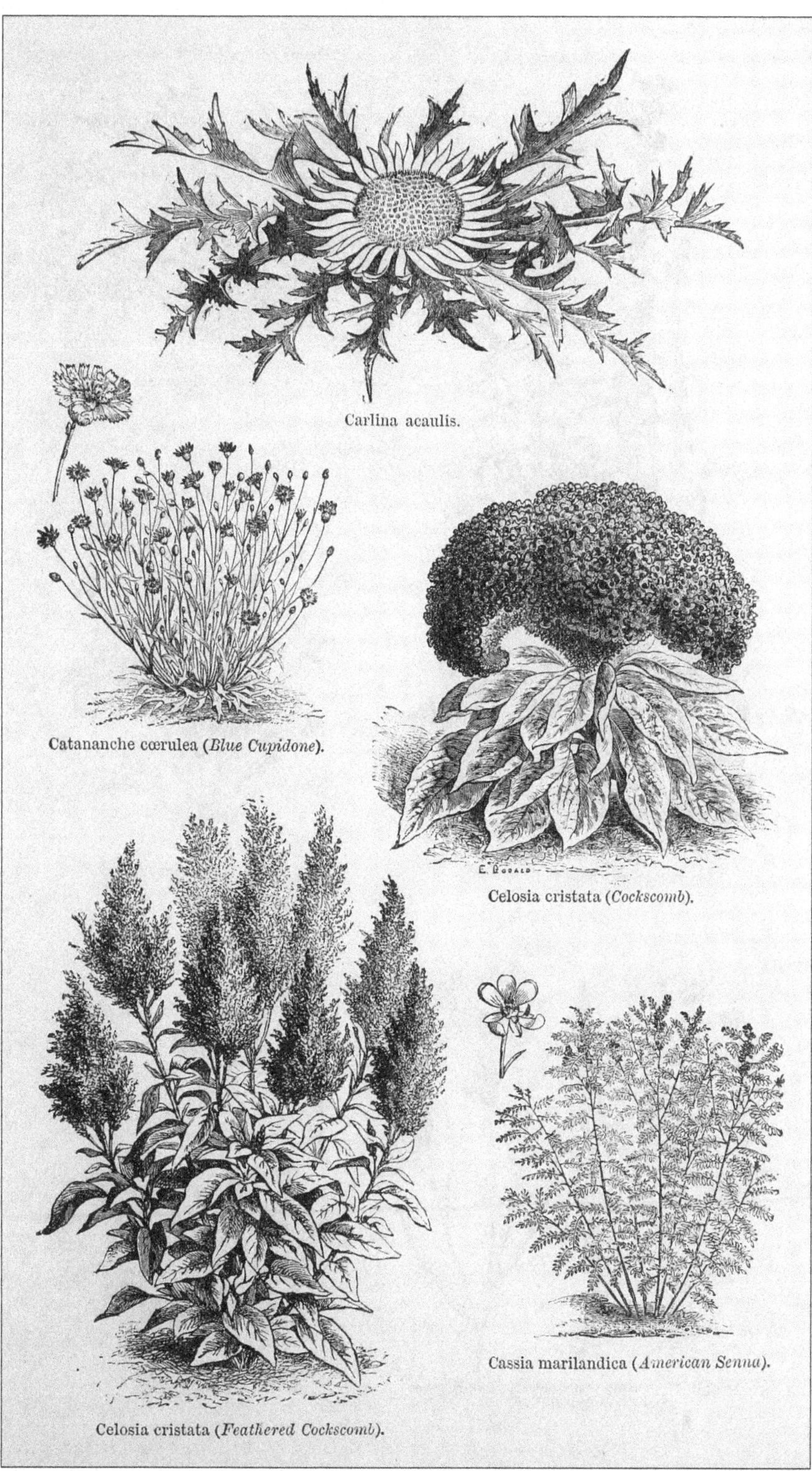

Carlina acaulis.

Catananche cœrulea (*Blue Cupidone*).

Celosia cristata (*Cockscomb*).

Cassia marilandica (*American Senna*).

Celosia cristata (*Feathered Cockscomb*).

[To face page 72.

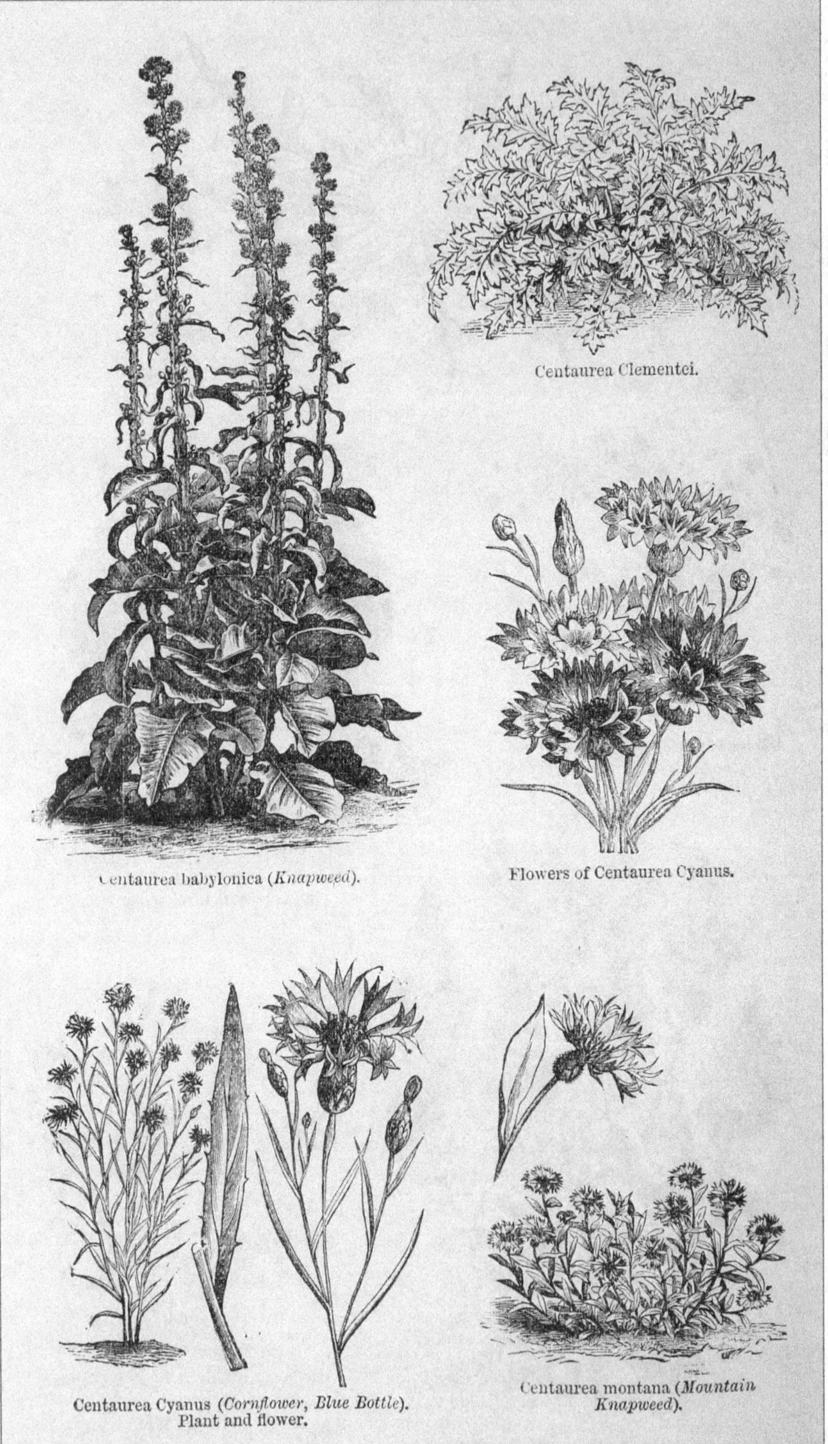

Centaurea Clementei.

Centaurea babylonica (*Knapweed*).

Flowers of Centaurea Cyanus.

Centaurea Cyanus (*Cornflower*, *Blue Bottle*).
Plant and flower.

Centaurea montana (*Mountain Knapweed*).

[*To face page* 73.

and partly starved to bring out its whiteness. Plumosa is an excellent variety.

C. babylonica.—Among the Centaureas there are a few subjects which might be used among hardy fine-leaved plants, the most distinct and remarkable being the very silvery-leaved C. babylonica. This is quite hardy, and when planted in good ground sends up strong shoots clad with yellow flowers to a height of 10 ft. or 12 ft. The bloom, which continues from July to September, is not by any means so attractive as the leaves, but the plant is at all times picturesque. In groups or in rough or undulating parts of pleasure grounds it has a very fine effect. Suited for association with the tallest plants. A free sandy loam suits it best. Seed. Levant.

C. Clementei.—This fine silver-grey-leaved plant forms a companion to the Acanthuses, the deep green of which helps to enhance the peculiar beauty of the Centaurea. It is preferable to the Artichoke, which is sometimes recommended as an early, grey-leaved plant, as it is of better habit and retains its foliage throughout the season. Small plants from seed are most serviceable for edging sub-tropical or other beds, and when too large for that purpose, they may be transferred to shrubbery or mixed borders, or planted out singly on Grass. The blossoms are best picked off, as they rather detract from than add to the beauty of the plant.

C. Cyanus *(Cornflower)*.—A beautiful native flower, which deserves a place in every garden. It is an annual plant and of very easy culture; in fact, it sows itself, and the young plants standing our hardest winters flower better grown in this way than when sown with the annuals in spring. The spring plants are weaker in growth and shorter than the autumn or summer raised ones, attaining a height of 2 ft. The typical colour is a beautiful blue tint; there are white and purplish forms. The many garden varieties now in cultivation range through white, rose, sky-blue, striped, to dark purple, the delicate tints of which are most attractive. In Prussia this plant is called Kaiser Blume, and is a great favourite. They are all much esteemed in the flower market, the long wiry stems being so manageable. The plants will flourish in almost any soil or position, doing best, perhaps, in strong soil.

C. dealbata.—A very hardy perennial with graceful and somewhat silvery leaves, 15 in. to 18 in. high. Flowers in summer; rose coloured. Caucasus. Borders. Division.

C. gymnocarpa.—A half-shrubby plant from the south of Europe, nearly 2 ft. high, with hard, branching, bushy stems, and elegantly cut, arching leaves, which are covered with a short, whitish-satiny down. A variety (C. plumosa) has the leaves much more divided and not so white. This plant is somewhat hardier than C. ragusina, but both require greenhouse treatment in winter. Same soil, positions, and treatment as for C. ragusina. Useful as this is as an edging or bedding plant, it is when grown as fine single specimens that its beauty is most seen.

C. macrocephala *(Great Golden Knapweed)*.—A strong habited plant growing from 4 ft. to 5 ft. in height, and bearing a great golden head of bloom. For the back part of a herbaceous border or for naturalisation in semi-wild nooks and corners, or in positions where herbaceous plants must perforce compete with the roots of trees and shrubs, this robust and vigorous species well deserves a place as the largest and most effective of all the yellow flowered Centaureas. In deep rich soils it forms an effective mass, and the flower-heads when cut are very effective arranged with other flowers, or even alone in a vessel with their own leaves.

C. montana *(Mountain Knapweed)*.—A handsome border plant, 1 ft. to 2½ ft. high, with slightly cottony leaves, and flowers resembling those of the Cornflower (C. Cyanus), but larger. There is a white and a red variety in cultivation. Europe. Borders, margins of shrubberies, or naturalisation in any soil. Division and seed. These plants are somewhat coarse in borders and scarcely worth a place therein, but they will grow anywhere, and when cut their flowers are pretty. They are grown for their flowers by the market gardeners, being larger than those of the Cornflower (C. Cyanus).

C. moschata *(Sweet Sultan)*.—A valuable old annual flower, well deserving of a place if only for affording cut flowers. There are two shades of colour—delicate purple and creamy white. The first produces the finest flowers; both are valuable annuals during summer and winter. They are somewhat "miffy" in growth until well established in the soil, and they are also fastidious to please in the matter of soil. Aphides are very partial to them when the seedlings are young, and unless they are quickly cleared off them the plants soon dwindle away. The first essential to healthy growth is a calcareous soil, and any soil deficient in lime should have some lime rubble worked into it to make it acceptable to these Centaureas. The best time to sow them is about the middle of April, and the position should be open and sunny to insure them doing well. Sow the seed where the plants are intended to remain, as they are, like many annuals, very impatient of being transplanted. After the seedlings are up they should be thinned out early, leaving three plants in each patch, giving them a distance of 1 ft. each. Water them when necessary. They are now grown for market to a large extent. The plants attain a height of somewhat over 1 ft., and are natives of Persia. (*Syn.*, Amberboa.)—J. R.

C. ragusina.—A favourite silvery-leaved plant, tender, but making strong growth out-of-doors in summer and autumn. For toning down glaring colours nothing could possibly be finer than this. Solitary plants of it placed in centres of small circles and surrounded with any bright colour look well. Where large groundworks of scarlet, purple, or blue are made an elevated plant of this, placed here and there, break up the monotony which would otherwise exist, and enhances the appearance of the outlying colours in a very marked degree. Wherever any vivid or intense colour appears this plant should never be far from it. The pale silvery foliage of isolated plants judiciously dotted on lawns on the bright green Grass, and amongst dwarf, dusky-leaved shrubs has also a good effect. It is never injured by wet, and its robust compact-growing habit prevents the wind from having any tarnishing influence on it. It thrives in the coldest situation throughout the summer. The propagating of Centaurea ragusina is a matter often attended with very indifferent results. It is, however, as easily, and may be as successfully rooted as Zonal Pelargoniums. When taking cuttings from the old plant they should not be cut away, but pulled off with a heel so that they may have a hard base. The small firm shoots should be chosen in preference to large ones. In making them the knife should be used as little as possible. Each cutting should be put singly into a small 2½-in. pot filled with a mixture of loam, leaf-mould, and sand, and put in a cold frame. One good watering is sufficient until they are rooted; and if the weather is excessively damp, the lights may be drawn over them, and tilted up back and front, otherwise they may remain fully exposed. Treated in this manner I have rooted a batch of 2000 without losing twenty. Autumn is the best time to propagate them. They will fill their pots with roots in three weeks, and if there is plenty of house room they will make all the better plants by being shifted into 4-in. pots, where they will grow a little in autumn, and be nice and strong for bedding out next year. A cold frame from which frost can be excluded is a suitable place for their winter quarters. The leaves should be kept dry, as they are rather liable to damp during the short days. Every favourable opportunity should be taken for air giving; they also winter well in an airy vinery or greenhouse. Old plants are sometimes lifted and kept over the winter. Where large plants are required this is a sure means of obtaining them; but for ordinary bedding purposes autumn-struck cuttings are the best.—J. M. (*Syn.*, C. candidissima.)

C. ragusina compacta.—This is more compact in habit and shorter in the leaf than C. ragusina. It is, however, not so easily increased from cuttings, but it produces seeds much more freely, and although the seedling plants from them are not all of the same habit, they are easily sorted out and classified as regards size. Cuttings put in in March root freely, and make good plants by May. Those kept for stock should be placed in a greenhouse, the temperature of which is about 55° in February, and as soon as they make a few leaves nip out the points. This will induce them to throw out a number of young shoots, which if put in a sandy soil on a brisk bottom heat will emit roots in about ten days. Care must be taken not to water over the leaves, and the soil must not be allowed to get dry at the bottom.

C. rutæfolia.—A hardy perennial, growing to the height of 2 ft., with branching stems amply clothed with deeply pinnatifid foliage, the entire plant being covered with a dense white tomentum, as in C. ragusina. It is, however, of freer growth than that species, increasing in size throughout the summer till checked by the cold of autumn. The flower-heads are pure white, about one-half the size of those of C. ragusina, but much more abundant. The plant derives its ornamental value from the persistent whiteness of its foliage, which suffers less from rain than that of the species previously named. It succeeds in any soil, but its growth is more compact in rather dry soils or on rockwork.—W. T.

C. suaveolens (*Yellow Sweet Sultan*).—*Syn.*, Amberboa odorata.—A desirable citron yellow hardy annual and favourite border flower, thriving best in light dry soil; should be sown and treated same as C. moschata. Sow in beds with flowers grown for cutting in April, raising a batch in frames, and sowing one in light, rich earth in the open air where it is to remain.

C. uniflora.—The flower-heads of this, previous to opening, look like withered balls, in consequence of each of the scales being terminated by a dark brown feather-like point; and as these become developed, they lie down close upon the head, appearing to enclose it in a net. The stems rise 6 in. to 15 in. high, each bearing a solitary flower 2 in. or more across, of a lilac-rose. It is distinct and curious, grows freely in well-drained and sandy soil, and merits a place on the rockwork or borders.

Centauridium Drummondi.—A showy half-hardy annual Composite from Texas, growing from 2½ ft. to 3 ft. high, and flowering from July to September. It should be sown on slight heat in a frame in April, and planted out in May. It has large citron-yellow flowers, much resembling those of Centaurea.

Centranthus ruber (*Red Valerian*).—A handsome hardy border plant, and an old inhabitant of British gardens. It belongs to the Valerian family and comes from the Mediterranean. There are two or three varieties—a white, a purple, and a red or crimson. The plant produces stout erect stems, rather woody at the base, terminated in bold clusters of numerous small flowers. It begins to bloom in June, and continues through the summer. It requires only ordinary garden soil, and occasional lifting and cutting in, without which it is apt to become weak and die out in the lapse of a few years. It grows well on the crumbling walls of an old ruin, and is useful for planting on dry rockwork and in rocky or stony banks. Propagated by division and cuttings of side shoots in autumn under a hand-glass, and from seeds.

Centaurea moschata (*Sweet Sultan*).

Centaurea suaveolens (*Yellow Sweet Sultan*).

Centaurea macrocephala (*Golden Knapweed*). See page 73.

Centranthus ruber (*Red Valerian*).

Centauridium Drummondi.

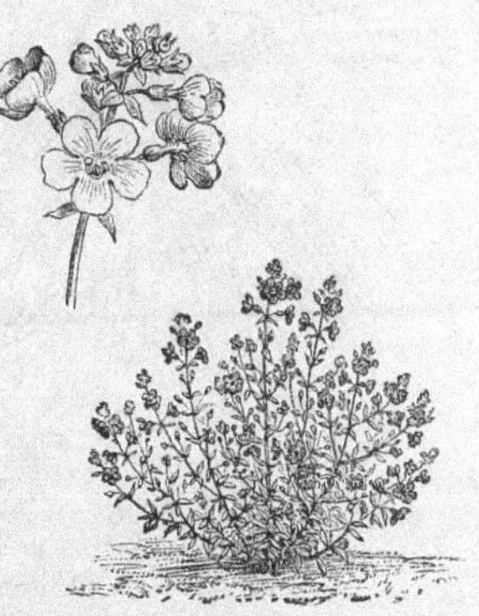

Chænostoma fastigiatum. See page 75.

PLATE LXIV.

Centranthus macrosiphon nanus. See page 75.

Cerastium grandiflorum (*Mouse-ear Chickweed*).

Chamærops humilis. See page 76.

[*To face page 75.*

C. macrosiphon is a hardy annual with pretty rose coloured flowers. It is useful for the rock garden or flower border. It may be sown in September and pricked off into pots for winter for transplanting in spring. It may be again sown in the open ground in March and April, thinning the seedlings out about 1 ft. apart. There are several varieties—white, red, and two-coloured, besides a dwarf variety (pygmæus).

Cephalanthera.—A small genus of Orchids, of which two or three are natives, but none are worth the trouble of cultivating.

Cephalaria.—Plants of the Valerian family, all of large coarse growth, and only suitable for the wild garden. The handsome large flowers are, however, desirable for cutting from, and for this purpose a plant might profitably occupy a place in the rougher parts of the garden. The best species are C. tatarica and C. procera, with white or yellowish flowers like a Scabious; of easy culture.

Cerastium *(Mouse-ear Chickweed).*—A numerous genus of the Chickweed family, containing but a few garden plants of value in proportion to its number.

C. alpinum.—An interesting British plant, found on Scotch mountains, and also more sparsely on those of England and Wales. Dwarf, seldom more than a couple of inches high, with leaves clothed with a dewy-looking silky down, which gives the plant a singularly shaggy appearance. It bears large white flowers in early summer. It is at all times a pretty and distinct-looking object on those parts of rockwork that come near the eye. It is not, like the common garden species, a plant fitted for forming edgings. Readily increased by division, by cuttings, or seeds.

C. Biebersteini.—A very silvery species, closely allied to the common C. tomentosum. Useful for the same purposes, and propagated and cultivated with the same facility as C. tomentosum. It was once expected that it would surpass in utility the common kind, but this it has failed to do. A very good plant for borders or rough rock or root work, and, being seldomer seen than the common one, deserves a little more attention, and a better position in the mixed border or the lower and rougher parts of rockwork. A native of the higher mountains of Tauria, flowering with us in early summer.

C. grandiflorum.—Less downy and silvery than the following, producing pure white flowers in great abundance in early summer. A fine plant for the front margin of a mixed border, or for the rougher parts of rockwork, but only in association with other strong and fast growing things, as it spreads about so quickly that it would overrun and injure delicate and tiny plants if placed near them. Native of Hungary and neighbouring countries, on dry hills and mountains.

C. tomentosum *(Snow in Summer).*—Now used in almost every garden for forming compact silvery edgings to flower beds and borders. Its hardiness, power of bearing clipping and mutilation, and great facility of propagation make it worthy of all the attention it receives. It is also very useful as a border plant, and for rootwork or rough rockwork, but it is too common to be permitted a place on small or choice rockwork that might be devoted to some of the many rarely seen and beautiful alpine plants. A native of mountains in the south of Europe, flowering freely with us in early summer. The preceding include all the kinds that are worth growing, except in botanical and unusually extensive collections.

Cerinthe *(Honeywort).*—A small genus of the Borage family, of which there are two or three interesting plants. C. aspera is probably the best species, producing an abundance of yellow flowers, the tube of which is black at the base. In general habit it closely agrees with the other species of the genus, but the seed is somewhat larger. Cerinthe minor has its growth curved and branching, the flower-stems arching over considerably, so much so that the delicate yellow tube-shaped bloom is entirely hidden at the apex of the stem by the long and closely-imbricated pale green leaves with which the stem is furnished. Both are half-hardy annuals, and require to be sown in early spring and frames, and afterwards planted out in good soil. They are, however, not likely to be much in favour generally, owing to their quiet colours. Natives of Greece, but occur pretty generally in Italy and other countries of Southern Europe.

Ceterach officinarum.—A distinct and beautiful little native Fern, admirably adapted for rock or alpine gardens, as it thrives best when planted between the chinks of rocks or stone walls. It is a Fern that, unlike most other varieties, dislikes a confined, damp position; hence it can be planted in the most exposed places with good effect, and, with a little careful attention to its simple requirements at the outset, with almost certain success. The chinks and crevices wherein it is proposed to plant this Ceterach should be filled with a mixture of sandy peat and pounded limestone. It might be associated in such positions with some of the little flowering Sedums, and various other plants for walls and stony places.

Chænostoma.—A small genus of the Figwort family, natives of the Cape of Good Hope. They are naturally perennial, but in the open air in this country they must be treated as half-hardy annuals. C. fastigiatum is the prettiest. It grows from 6 in. to 9 in. high, forming a dense compact tuft, and produces an abundance of small pinkish, sometimes white blossoms. It is a neat annual plant for the front margins of borders. The seeds should be sown either in spring in warm frames or in August, when the seedlings require to be wintered in a pit. Flowers from June to November.

Chamæbatia foliolosa *(Tarweed).*—A little shrubby plant of the Rose family remarkable for the Fern-like beauty of its leaves. The flowers are white and somewhat like those of a Bramble. It grows about 1 ft. high, forming a dense spreading tuft, and covering the ground in its native country, California. It has not proved hardy in our climate, but we have seen it growing freely in mountain districts covered with snow through the winter, and believe that it is worth trial on the rock garden in the milder part of the country.

Chamælirium luteum *(Blazing Star).*—A North American plant of the Lily family, but not important for the garden. It has unattractive yellowish flowers.

Chamæpeuce *(Fish-bone Thistle).*—Spiny leaved plants allied to the Thistle, valuable for the flower garden, as their foliage is distinct and handsome. There are two kinds in cultivation—C. diacantha and Cassabonæ. The former has foliage of a shining green, marking with silvery lines, and the spines are ivory white. C. Cassabonæ has deep green leaves veined with white and with brown spines. Both grow in compact rosette-like masses about 9 in. high before they produce their Thistle-like flowers the second year, when the flower-stems attain 2 ft. to 3 ft. high. They require light, well-drained soil and a warm position, and should seldom be watered. As the stems are not produced until the second year, the radical rosettes of the first year may be advantageously used in forming edgings, or on the margins of groups, for which their light green, silver-veined leaves are very suitable, or they may serve to fill a vacant space in the mixed border. Seed sown in February and carefully attended to will furnish good plants by May, but the best for immediate effect are those sown in September in a border in the open ground, potted up carefully, and given greenhouse treatment during the winter.

Chamærops excelsa.—A genus of Palms remarkable for their hardiness in our country, and, judiciously used, capable of good and distinct effects in the garden.

C. Fortunei *(the Chusan Palm).*—A most valuable Palm often confounded with C. excelsa, from which, however, it differs in being of a stouter habit, having a more profuse matted network of fibres around the bases of the leaves and crown, the segments of the leaves much broader, and the leaf-stalks shorter and stouter, from 1 ft. to 2 ft. long, and quite unarmed. It grows 12 ft. or more high, and has a handsome, spreading head of fan-like leaves, which are slit into segments about half-way down. This Palm is perfectly hardy in this country. A plant of it in the garden at Osborne has stood out for many winters and attained a considerable height. It is also placed out at Kew, though protected in winter. On the water-side of the high mound in the Royal Botanic Gardens, Regent's Park, it is in even better health than at Kew, though it has not had any protection: the severe frosts of recent winters have not hurt it. If small plants of this are procured, it is better to grow them on freely for a year or two in the greenhouse, and then turn them out in April, spreading the roots a little and giving them a deep loamy soil. Plant in a sheltered place, so that the leaves may not be injured by winds when they grow up and get large. A gentle hollow, or among shrubs on the sides of some sheltered glade, will prove the best place for it. The establishment of the Palm among our somewhat monotonous shrubbery and garden vegetation is surely worthy of a little trouble, and the precautions indicated will prove quite sufficient. C. humilis is also hardy—at least on sandy soil in this country.

Chamomile *(Anthemis nobilis).*

Checkered Lily *(Fritillaria Meleagris)*

Cheddar Pink *(Dianthus cæsius).*

Cheiranthus *(Wallflower).*—A beautiful genus of cruciferous plants made familiar to us by the favourite Wallflower (C. Cheiri), which is too well known to need description. It is almost the only species of the genus grown in gardens, as other well known plants that bear the name of Cheiranthus belong to the genus Erysimum. The Wallflower is a native of Southern Europe, growing on old walls, in quarries, and on sea cliffs. It loves a wall better than any garden; while it grows coarsely in garden soil, it forms a dwarf enduring bush on an old wall, and grows even on walls that are quite new, planted in mortar. There is no variety of the Wallflower yet seen that is not worthy of cultivation; but the choice old double kinds—the double yellow, double purple, double orange, dark, &c.—are plants worthy of a place among the finest border plants, being ornamental in a high degree, and endeared to us by many associations. These are the varieties most worthy of a place on dry stony banks near the rockwork, and also on old ruins, on which the common kind is likely to find a home for itself. The fine mixed "German" kinds, that are so easily raised from seed, would also be worthy of introduction to ruins and stony places.

WALLFLOWERS are classed thus: Single biennials, double biennials, and perennial double kinds. Of these the first and last are well known; the second, which are of Continental origin, are grown here almost, if not entirely, from imported seed. The single biennial section is the most popular, and our best strains of it are unequalled for richness of colouring and sweetness of perfume. The superb, dark crimson-marked kinds grown around London are too well known to need description, and can be bought in the seed trade under the various designations of Young's Blood-red, Harbinger, and Saunders' Dark, all good strains of this popular Wallflower. The Belvoir Castle Yellow is a close, compact growing kind, which produces bright yellow flowers, and is specially suitable for spring bedding. It is easily recog-

PLATE LXV.

Hardy Palm, Chamærops excelsa.

Hardy Palm, Chamærops Fortunei.

[*To face page* 76.

Chamæpeuce diacantha (*Fish-bone Thistle*).

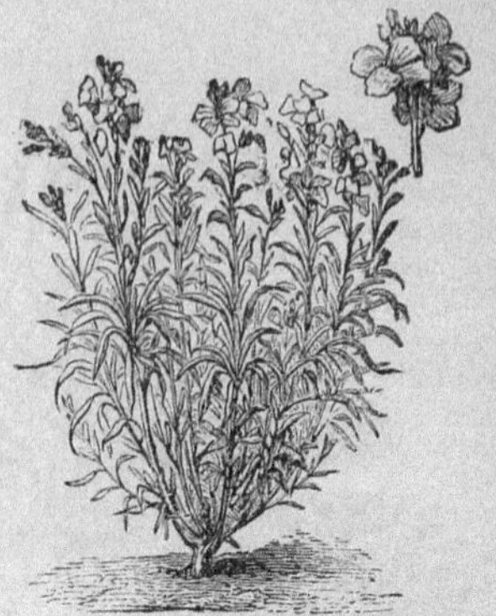

Cheiranthus Cheiri (*Wallflower*).

Cheiranthus Cheiri fl.-pl. (*Double Wallflower*).

Cheiranthus maritimus.

Cheiranthus annuus grandiflorus fl.-pl. (*Ten-week Stock*).

[*To face page 77.*

nised by its dwarf growth, pale green foliage, and pale flower buds. The Golden Yellow—or, as it has been misnamed, Tom Thumb Yellow—is as robust and tall as the crimson kinds, and produces in great abundance masses of rich, orange-yellow bloom. This is a grand kind well worthy the attention of market growers. In country districts the common single Wallflower is usually a compound of all these, and, therefore, deficient in that rich and effective colouring so peculiar to good, pure strains of the single Wallflower. The biennial double kinds are remarkable for the variety of colours which they furnish, and the stout woodiness of habit which they present; indeed, in these respects they resemble a distinct species, but probably this is due to their hybrid character.

The double perennial kinds familiar to us are the yellow, dark crimson, red, and dwarf yellow. The yellow is most common, and a beautiful clear-coloured kind it is, and a great favourite it is with cottagers, who propagate it by putting in slips of it in their peculiar and usually successful fashion about the time the plants are in flower. It can be propagated freely by means of slips put in under hand-lights in sharp, sandy soil, and will produce plants capable of flowering the next spring. The other kinds are amenable to the same mode of propagation, the most common of these being the red, an intermediate form that resembles the tall yellow, but which has a shade of colour that is neither striking nor pleasing. The old dark crimson is now apparently almost extinct; in colour the flowers are almost black and very striking. The dwarf yellow has flowers of a dull, almost buff tint, but it is fairly well adapted for pot culture. The Raby Castle is a valuable and sturdy kind.

Increase and culture.—Many persons sow their Wallflower seed too late in the year, in June and July instead of April and May. If dry weather follows close on the act of sowing the seed, or after the plants have grown 2 in. or 3 in., they receive a check in their development, and the plants, instead of being dwarf, vigorous, and bushy, are thin and poor. The winter will sometimes injure the Wallflower severely, especially when very severe frost follows close on heavy rains, and the stronger and better rooted the plants are, the better are they likely to stand the effects of the weather. The plants used for filling beds should be those that had been once transplanted at least, because the moving induces them to throw out fibry roots near the surface, and they can be lifted with soil adhering to them. When the Wallflower is allowed to grow where it is sown, a strong tap-root is formed, which strikes deeply into the soil, and but few surface roots are put forth. In transplanting from the seed beds it is well to pinch off the tap-root, thereby inducing fibry roots. The great advantage of plants with fibry roots is that they can be transplanted at any time during the winter when the weather is open without harm. Mr. Ingram, of Belvoir Castle gardens, used to take precautions to induce his Wallflowers to form fibry roots previous to transplantation, and would transplant them into trenches and lines in soil, with slates or bricks buried a little in the soil to prevent the formation of the tap-roots, as he found that plants with tap-roots and little fibre stood badly.

In London market gardens, where the Wallflower is invariably well cultivated, seed is sown in the open ground early in February; the young plants are put out into their permanent quarters in May, and by Christmas, if the winter be mild, they produce bloom, and are so large that they would not be covered by a bushel basket. If required for spring bedding, they should be sown early and planted out in some spare ground about 12 in. apart. These plants will lift with good balls of earth early in winter, and in spring will produce superb masses of bloom. Some market growers sow seed late in the summer, allow the young plants to remain in the seed-bed all the winter, plant out in March, and, if the season be favourable, reap a good crop of flowers all through the next winter.

Seeds should be saved from plants possessing the best branching habit, and which bear the darkest coloured blossoms. When in flower go over the plants and mark those possessing these qualities by placing a stake by the side of each. Allow these to remain undisturbed until the seed is ripe, when they may be pulled up, roots and all, and housed in a dry place until a convenient season arrives for threshing out the seed. Cuttings of the double kinds may be put in as soon as they can be got after the plants go out of bloom. Put them in firm sandy soil under a handlight, and when struck plant them out. Cuttings put in in August, September, or October will strike freely without any protection in a shady border, or in pots or boxes filled with sandy soil.

Besides the Wallflower there are several perennial species of doubtful hardiness, such as C. arbuscula and mutabilis, natives of a warmer climate than ours, therefore cannot be recommended for open air culture, though they make pretty pot plants. Besides these there are various hybrids, such as C. Sermoneri, Delahaynus, Bocconi, and Marshalli, which is the finest of the hybrid kinds said to have been raised between C. Cheiri and Erysimum ochroleucum. It is a compact plant, growing from 9 in. to 1 ft. high, and producing a profusion of bright orange-coloured scented blossoms. It is a brilliant ornament of the border, and a good plant for massing in spring bedding; few plants indeed, so hardy and easy to cultivate, equal it in showy beauty and display from April till July.

All these perennial varieties prefer dry and slightly elevated positions during winter, but like a cool, moist soil in summer. Propagation necessarily is by means of cuttings, but it is often found that by top dressing with fine soil the summer wood is induced to root freely, and then in autumn the production of a good stock is easily accomplished. Cuttings should be taken off just as the plants are passing out of flower, put in under a hand-light, and treated as Pink pipings usually are until rooted.

C. alpinus, ochroleucus, and others belong to the genus Erysimum, which see.

Chelidonium *(Celandine).*—The best known species of these Poppyworts is the native C. majus, which is a showy plant with yellow flowers, and a desirable plant for growing in rough places, where little else will grow. The variety laciniatum, with deeply cut Fern-like foliage, is an elegant plant, and there is also a variety with double flowers, and another with variegated foliage. The Japanese Celandine (Stylophorum diphyllum) is known also as C. japonicum. The native kinds are of easy culture, but not fitted for a place in the garden proper.

Chelone *(Turtle-head).*—A small genus of North American plants nearly allied to the Pentstemon. The two species in cultivation are handsome border plants, flowering in late summer and autumn. C. Lyoni grows from 2 ft. to 3 ft. high, forming a dense mass of stems, with deep green foliage, and bearing dense clusters of showy pink blossoms plentifully from July to September. C. obliqua is a taller plant of more slender growth. It is similar to the preceding, but the colour of the typical form is a richer pink, and there is also a white-flowered variety. It generally flowers earlier than C. Lyoni, and continues till the autumn. Both are of easy culture, thriving in open borders of good deep soil, and may be readily propagated either by seeds, cuttings, or division of the roots. These plants, though bearing pretty flowers, and though free in growth, do not possess any particular value for the garden. The more striking and graceful Pentstemon barbatus is also known as C. barbata.

Chenopodium *(Goosefoot).*—A large genus, consisting chiefly of weeds, and of no value in the garden, except it be C. Atriplicis, which is a vigorous-growing Chinese annual, with erect, slightly branched, reddish stem, over 3 ft. in height, the young shoots and leaves covered with a rosy violet powder. The leaves are very numerous, nearly spoon-shaped, and long-stalked. This plant is ornamental in foliage, and well adapted for planting on grass-plats or grouping with plants with foliage of an ornamental character.

Chionodoxa *(Glory of the Snow).*—This is a small genus allied to Puschkinia and Scilla, including perhaps two or three species found in the high mountains of Western Anatolia and Crete, not easily distinguishable by any well-marked characters.

C. Luciliæ.—This and C. Forbesi are considered to be forms of one plant, which, without doubt, is among the loveliest of hardy flowers. Its habit of growth is similar to the two-leaved Squill (Scilla bifolia), as it rarely develops more than a pair of leaves. The blossoms, from five to ten in number, are produced on gracefully arching stems, from 4 in. to 8 in. high, and are nearly 1 in. across, star-like in form, and of a beautiful clear blue tint on the outside, gradually merging into pure white in the centre, as in the Nemophila insignis, but is even brighter. There is a pure white variety, which, however, is very rare even in its native habitat. C. Luciliæ has only been introduced a few years, but it has proved perfectly hardy during the late severe seasons, and as it readily produces seed in cultivation we may hope soon to see it widely distributed, and taking a place amongst the early spring flowering bulbs of English gardens. It flowers in this country in the beginning of May. It succeeds well in any ordinary garden soil provided it be not too heavy and damp. It may be propagated by seed or by separating the bulblets from the parent bulbs in autumn, and replanting them in fresh soil. A native of the Nymph Dagh and other mountainous districts near Smyrna, at elevations of from 3000 ft. to 4300 ft., whence it was introduced in 1877 by Mr. Maw, of Broseley.

C. nana.—This pretty little species has been a long time in cultivation, but under the name of Puschkinia scilloides. It has slender flower-stems bearing from four to ten delicate azure-blue blossoms. It flowers in March and April, according to the season. It is a native of the hilly woods of the Caucasus (Armenia), and is perfectly hardy, thriving in light, rich soil in an open sunny situation. C. cretica is not in cultivation. It is similar to the last, and probably would be quite as hardy.

Chimaphila *(Pipsissewa).*—Small shrubby plants, natives of the dry woods of North America. There are two species in cultivation somewhat difficult to cultivate successfully. C. maculata (Spotted Wintergreen) has small leathery leaves variegated with white. It attains a height of from 3 in. to 6 in., and is a very suitable subject for a half shady and mossy, but not wet, place in the rock garden, associating well with such plants as the dwarf Andromedas and the Pyrolas, and succeeding best in very sandy leaf-soil. C. umbellata, with glossy unspotted leaves and somewhat larger reddish flowers, is suited for like positions. Both are rare in cultivation and very seldom seen well grown. They flower in summer, and are increased by careful division. We believe the difficulty lies chiefly in the fact that rarely if ever is a good patch imported, or if imported as such it becomes infinitessimally divided to an extent, that even life itself is endangered, and it has no strength for vigorous growth.—N.

Chlora perfoliata *(Yellow-wort).*—A pretty native plant found abundantly in some places by the seaside. It is a slender plant about 1 ft. high, having greyish green leaves and bright yellow flowers as large as a shilling. There is a large flowered variety called grandiflora which is not so common as the type.

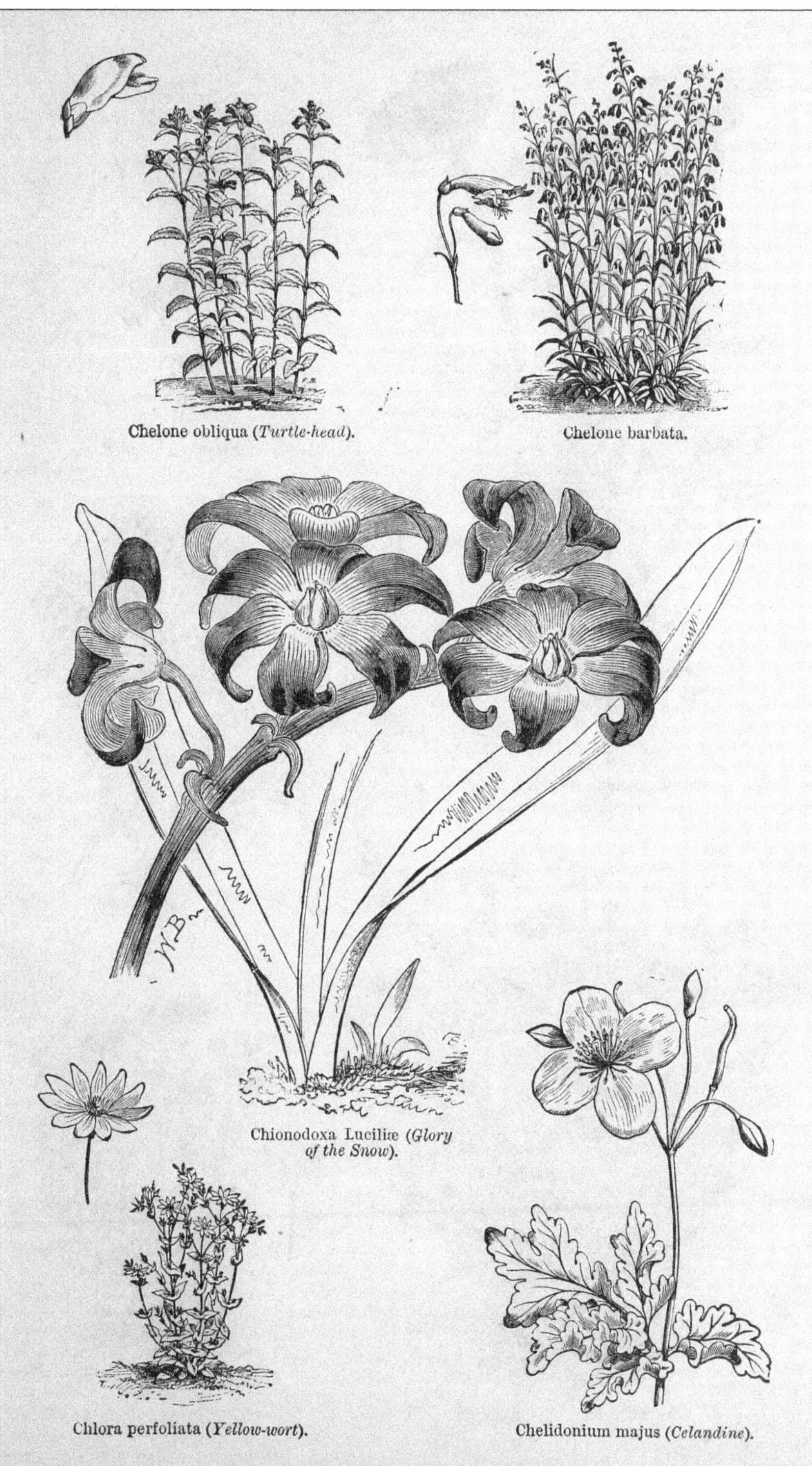

Chelone obliqua (*Turtle-head*).

Chelone barbata.

Chionodoxa Luciliæ (*Glory of the Snow*).

Chlora perfoliata (*Yellow-wort*).

Chelidonium majus (*Celandine*).

[*To face page* **78.**

PLATE LXVIII.

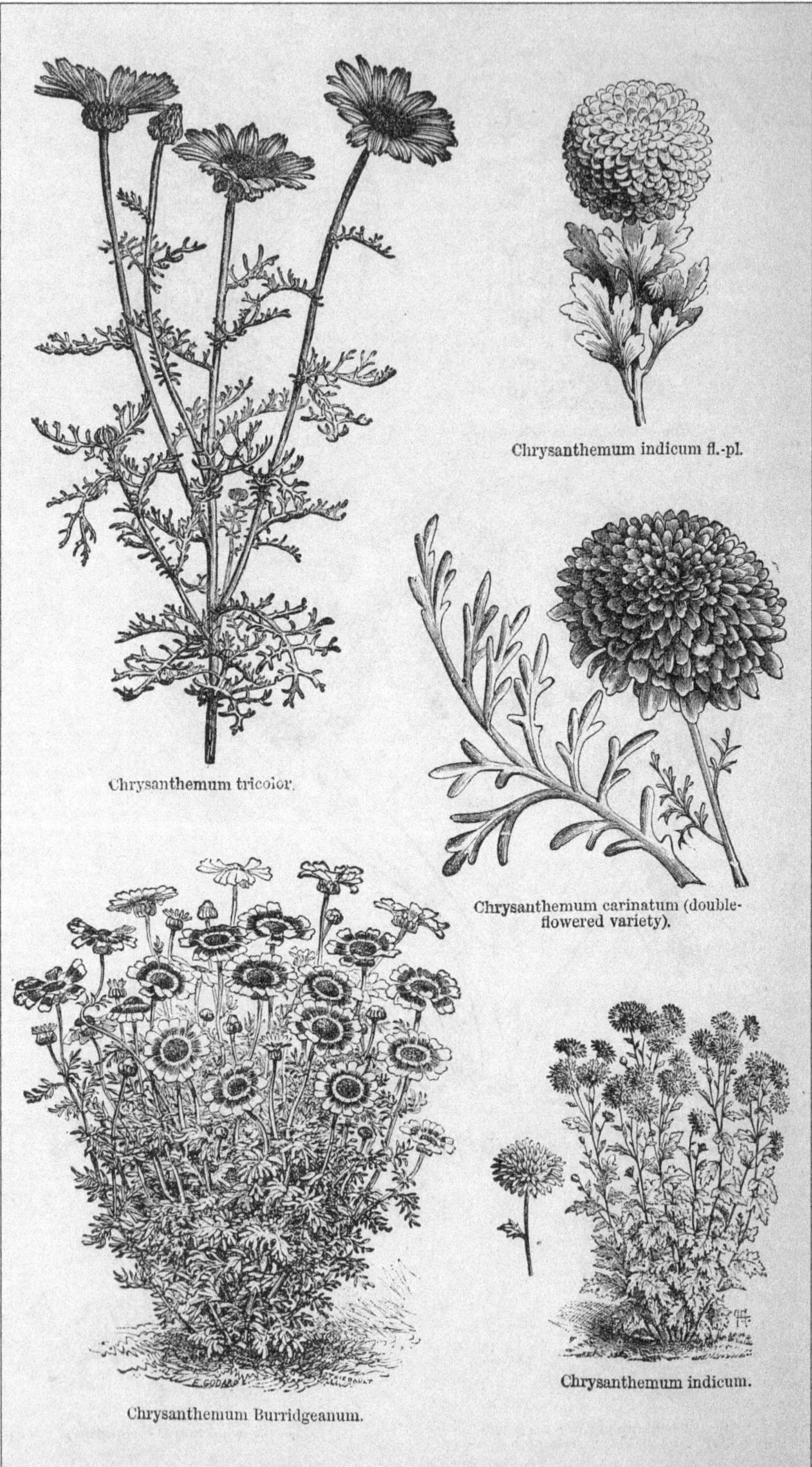

Chrysanthemum tricolor.

Chrysanthemum indicum fl.-pl.

Chrysanthemum carinatum (double-flowered variety).

Chrysanthemum Burridgeanum.

Chrysanthemum indicum.

[To face page 79.

It delights in a stiff, rich soil with plenty of moisture. It is only of biennial duration, but it yields seeds freely, which either sow themselves, or may be sown as soon as ripe or in early spring. Not a garden plant of high value, though beautiful in a wild state.

Chlorogalum Pomeridianum *(Soap Plant).*—A Liliaceous bulbous plant, native of California. It is not a showy plant by any means, the flowers being small and white on slender branching stems a yard or so in height. It requires a warm dry soil to thrive well in the open border. Propagation by bulblets or seeds. A botanic garden plant if grown at all.

Christmas Rose *(Helleborus niger).*

Chrysanthemum. — An extensive genus of perennials and annuals, a few of which are valuable garden plants. Besides those mentioned below there are several other kinds in cultivation, such as C. arcticum absinthifolium, which have little claim for general culture, and are therefore omitted.

C. carinatum *(Tricolor Chrysanthemum).*—A showy annual from N. Africa, which varies much in cultivation and is a valuable plant if only for its yield of flowers for cutting. There are double white and yellow forms, and the showy ones known as C. Burridgeanum, raised from seeds by Mr. Burridge, a seed grower in Suffolk. This variety is a compact plant, and bears large golden-yellow flowers, each having a dark purplish brown zone and a purple eye or disc, the colours being bright and well defined. Dunnett's varieties of the same plant are also good. It is propagated from seeds sown in April or early in May in open beds or borders where the plants are to flower, or they may be sown earlier in pans or boxes of light, rich earth in a pit or frame, from which they can be transplanted after all danger from frost is over. Planted singly in rich soil in an open and sunny position, it forms one of the most beautiful of all composite-flowered annuals, and well deserves culture. It generally blooms in August, and lasts several months in beauty, or until cut down by frost.

C. coronarium *(Crown Daisy).*—A handsome and bold annual 2 ft. to 3 ft. high in its wild form in Algeria, and in cultivation breaking into a number of forms, few of them, to our thinking, so fine as the single wild flower, pale yellow or buff. There are white and yellow double and dwarf forms. Treat as a half-hardy annual, sowing in good ground in April or early in May. In warm soil and mild winters one might hope to have autumn sown plants survive, in which case there would be a stronger and better bloom.

C. frutescens *(Paris Daisy, Marguerite).*—A vigorous half-hardy plant; in one season forming, when planted out, bushes 3 ft. in height and as much in diameter. It is much branched, and very symmetrical in habit; the foliage is pinnate, fleshy, and glaucous; the flowers are large, pure white, with a yellow centre, and produced in the greatest profusion from June until cut down by frost. The plant is extensively grown on the Continent for summer bedding as well as for pots. It is a fine Daisy-like plant, and several useful forms or allies are also valuable, such for example as the yellow Etoile d'Or, Comtesse de Chambord, and C. frutescens aureum. These plants are of very easy culture and propagation, being for the outdoor garden best treated as bedding plants and put out in May though in certain sea-shore and favoured districts they survive the winter.

C. lacustre *(Marsh Ox-eye Daisy)* resembles the large Ox-eye Daisy of our pastures, but much larger in every way. It grows about 2 ft. high, and thus may be distinguished at a glance from another plant with flowers strikingly similar, but which is much taller in stature, growing in a good soil from 6 ft. to 8 ft. high (Pyrethrum serotinum). The Marsh Ox-eye Daisy is a stout perennial and fit for a large collection, but somewhat coarse for choice positions. Some may give it a place for its yield of cut flowers.

C. segetum *(Corn Marigold).*—A showy yellow native plant, as well worthy of cultivation as many an exotic, and in certain cases worth growing for cutting. The flowers are yellow. Give it the treatment of a hardy annual, preferring autumn-sown plants.

C. indicum *(Common Chrysanthemum).* —The numerous beautiful varieties of this species, so popular as pot plants in autumn, deserve to be more extensively grown in the open air, as many varieties are quite suited to the purpose, and they add such a wealth of beauty to the open-air garden in October and November, when little else is in bloom. The varieties for open-air culture require to be well selected, and then their culture is simple.

CULTURE.—The Chrysanthemum is quite hardy and will grow in almost any soil or situation, and therefore it is impossible to lay down any general rule for its cultivation. The varieties, however, vary a good deal, and the treatment necessary for one is often unsuitable for another; for instance, the summer-flowering varieties are spoiled by being exposed to hot sunshine; they succeed best in the north of England and Scotland, while the large flowering kinds and Pompones require all the warmth that the south can furnish. As the climate of some parts of China is much warmer than that of this country, particularly the autumn months, the thermometer often standing at 100° Fahr. in the shade, a south or south-westerly aspect should be selected for the site of a Chrysanthemum plantation, and the shade of trees or buildings should be avoided. In town gardens this is not always possible, but the nearer we can approach the conditions just alluded to, the greater will be our success. None need, however, despair, for in spite of such obstacles as smoke and fog, some of the finest flowers exhibited at the autumn shows are grown in the small gardens and yards of densely populated cities. Much, however, depends upon soil; the Chrysanthemum is such a voracious feeder that where vigorous foliage and fine flowers are desired it cannot well be made too good. If practicable, it should consist of equal parts of fresh loam, rotten manure, leaf-mould, and sand. These, well mixed in autumn, and allowed to remain in ridges during the winter, will be in good condition to receive the plants in March or April. Many prefer autumn-made cuttings, and if intended for pot culture they are best, but for open-air culture strong suckers, if they can be procured, should be selected. They receive no check when planted, and in order to allow a free circulation of air the plants should stand at least 3 ft. apart. The same roots left undisturbed for two seasons in succession never produce fine flowers. They should, therefore, be replanted every year towards the end of May. Strong stakes should be placed to each plant, and the main stem should be firmly secured thereto during the growing season, the laterals being neatly tied into their proper places—but not too stiffly, and all unsightly branches should be removed. The beauty of the foliage and size of the flowers may be increased by the application of liquid manure during the summer months. Early in October disbudding should be commenced. The centre or crown bud, if perfect, should be left, and all the others carefully removed. One fine flower on each branch is preferable to several small ones.

VARIETIES.—The following varieties are good in habit, free flowering, and well adapted for open-air culture :—

White.—Vesta, Mrs. Rundle, Virgin Queen, Eve, White Venus, Mrs. Haliburton, Beverley.

Red.—Cardinal Wiseman, Julia Lagravère, Prince Albert, Triomphe du Nord, Pio Nono, Jewess, Duc de Corregliano.

Yellow and Orange. — Aurea multiflora, Chevalier Domage, Golden Beverley, Jardin des Plantes, Sulphurea superba, Guernsey Nugget, Abbé Passaglia, John Salter.

Blush and Rose.—Belladonna, Ariadne, Hermine, Christine, Princess of Teck, Venus, Lady Talfourd, Lady Harding, Lady Slade.

Crimson - Purple. — Alma, Mr. Murray, Prince of Wales, Progne, Prince Alfred, Dr. Rozas, Lord Derby, Mulberry.

WALL CHRYSANTHEMUMS. — In many well-kept town gardens the eye is offended by unsightly brick walls, and the question is often asked, What can be done to hide them? The answer is, Train Chrysanthemums upon them; if well nailed in they take up but little room and afford a pleasing background to the other occupants of the borders. Strong cuttings or suckers planted at the base of the wall 1 ft. apart early in March, in soil similar to that just recommended, will make rapid growth, and if kept neatly nailed in and all the side shoots removed as they appear, will soon cover a wall of ordinary height. Liquid manure from time to time will strengthen the plants and improve the foliage, and if carefully disbudded in September a grand show of large flowers will be the result. Should it be desired to protect the blossoms from wind and weather it can be effectually done by nailing a 12-in. board on the top of the wall, so as to form a coping. This supported by a few poles in front is all that is required, and if practicable a canvas covering fastened in front when the nights are cold will generally prove sufficient protection; with these simple precautions the duration of the flowers will be greatly prolonged. The varieties named below are best for wall culture :—

White. — Beverley, White - Globe, White Queen of England, Vesta, Mrs. Rundle, Virgin Queen, Princess of Wales, White Venus.

Yellow and Orange. — Mr. G. Glenny, Guernsey Nugget, Jardin des Plantes, Golden Beverley, John Salter, General Slade, Nil Desperandum, Dr. Brock.

Blush and Rose.—Aimée Ferrière, Belladonna, Princess of Teck, Venus, Pink Perfection, Duchess of Manchester, Lady Harding, Lady Slade.

Crimson.—Prince of Wales, Prince Alfred, Lord Derby, Lady Talfourd, Alma, Progne, Aregena.

Red and Brown. — Triomphe du Nord, Prince Albert, Bernard Palissy, Mr. Brunlees, Julie Lagravère, Hercules, Sanguinea, Pio Nono.

POMPONES. — These require the same treatment as the large flowering varieties. They may be advantageously planted, either in front of tall growing kinds or in beds by themselves. If for the latter purpose, they should be planted in April. Each root should be set about 12 in. apart, and the head or leader should be taken off when about 8 in. high, and all lateral branches encouraged in growth. These when sufficiently developed should be pegged down in the same way as ordinary bedding plants, and this must be continued during the summer, all strong shoots being shortened or regulated in order to ensure an equal surface of bloom. Liquid manure may be given them occasionally, as to effect that greatly depends upon a judicious arrangement of the colours. The following are all good free flowering kinds :—

White. — Mdlle. Marthe, Miss Talfourd, Argentine, White Trevenna, Cedo Nulli, Mme. Domage, Modèle.

Blush and Rose. — Adonis, Andromeda, Mme. Roussilon, Mrs. Dix, Hélène, Rose Trevenna, Trophée.

Red and Brown.—Bob, Brilliant, Miss Julia, Mustapha, Aureole, Firefly.

Yellow and Orange. — Aigle d'Or, Drin-Drin, Mr. Astie, General Canrobert, Golden Cedo Nulli, Aurore Boreale, La Vogue.

Crimson. — Salomon, Duruflet, President Decaisne, Miranda, Crimson Perfection, James Forsyth.

SUMMER FLOWERING KINDS.—This is a most important class, furnishing a number of

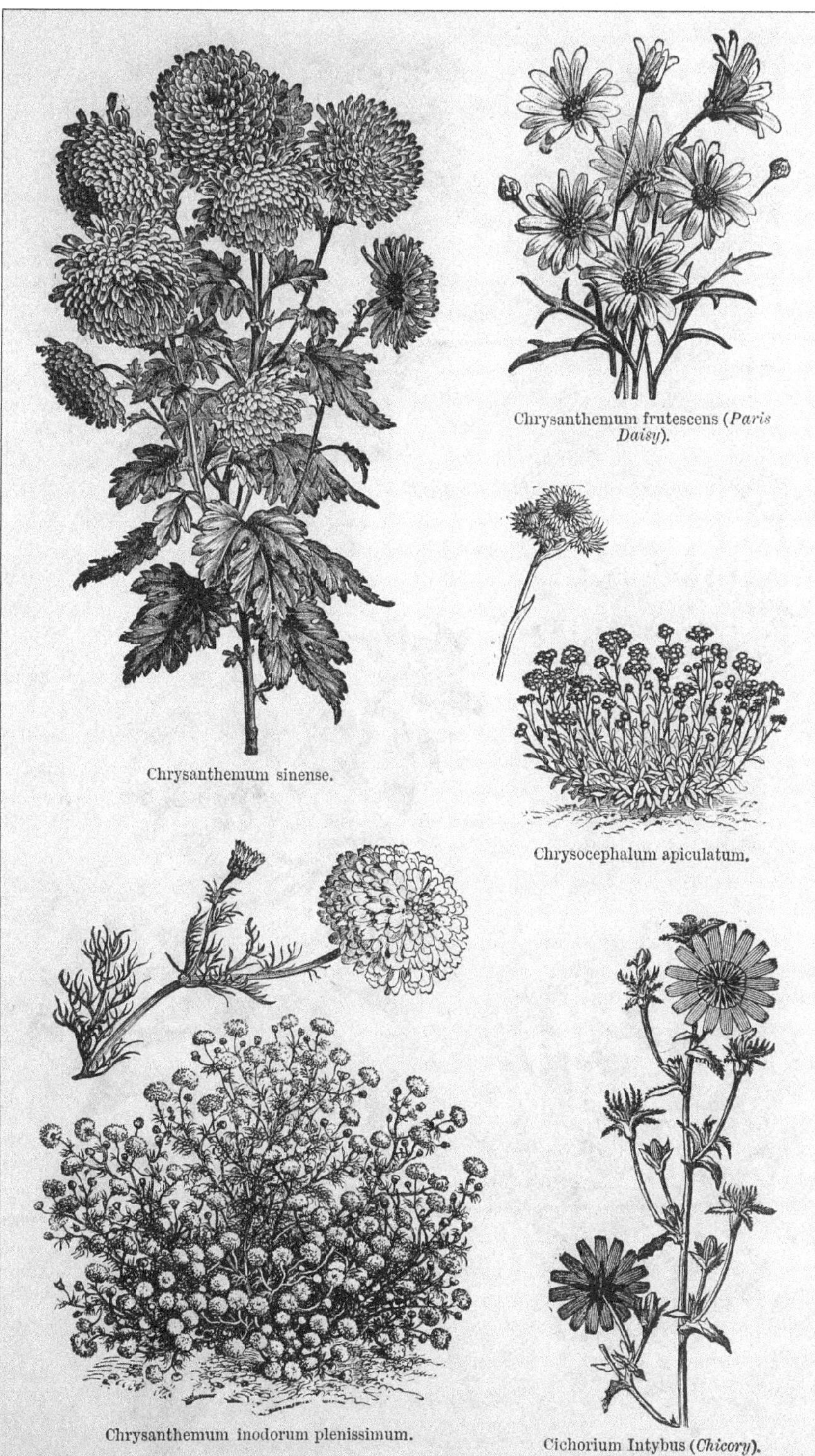

Chrysanthemum frutescens (*Paris Daisy*).

Chrysanthemum sinense.

Chrysocephalum apiculatum.

Chrysanthemum inodorum plenissimum.

Cichorium Intybus (*Chicory*).

[*To face page* **80.**

PLATE LXX.

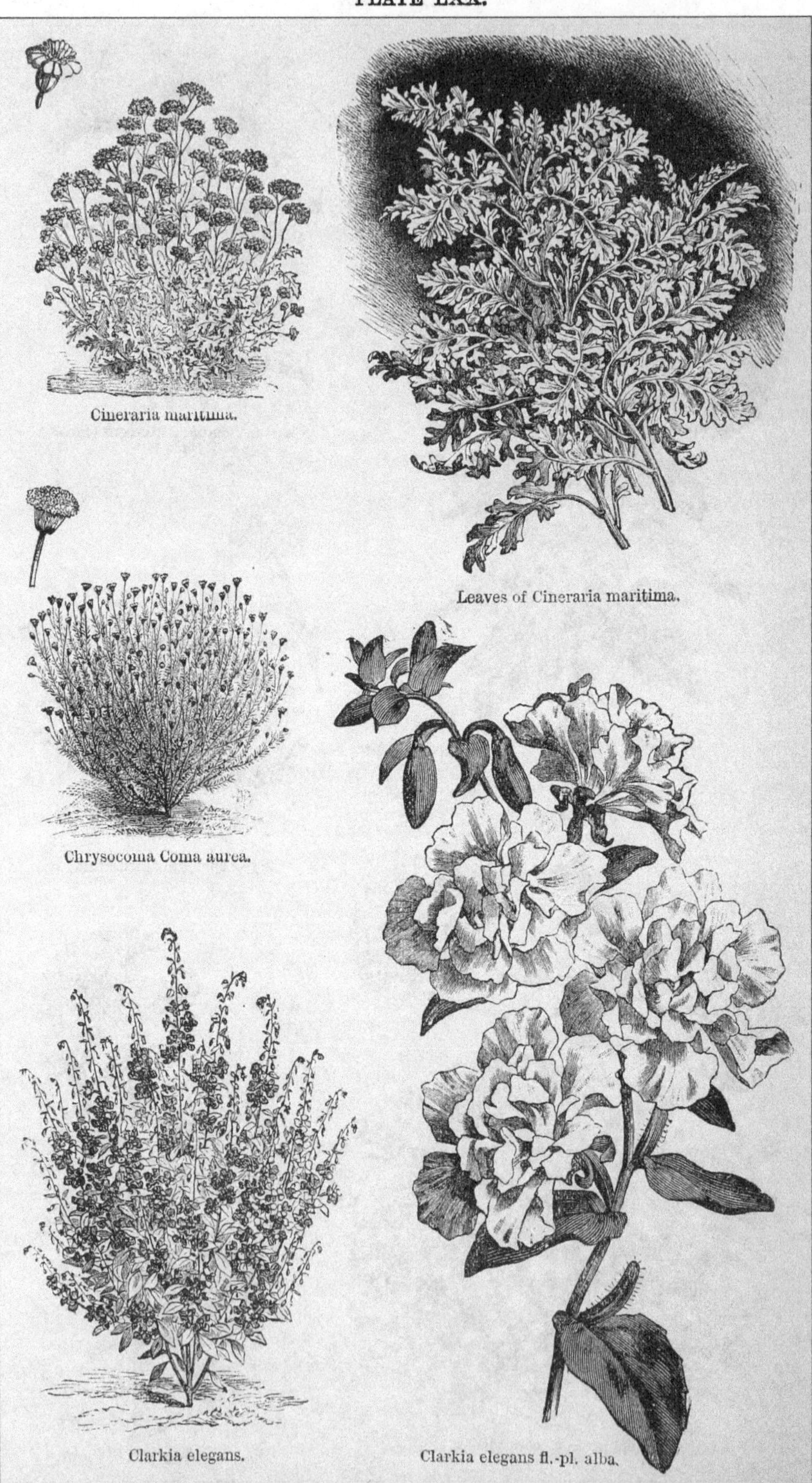

[To face page 81.

beautiful varieties that are valuable for cutting from, and they enliven the borders when other hardy flowers are on the wane. They do not appear to advantage in open borders unless shaded from hot sunshine in August and September. They are very dwarf and of various shades, but for outdoor decoration in the south of England they are of but little value. The climate of the north and Scotland suits them admirably, and there they are invaluable mixed with herbaceous plants or in ribbon borders. Their free habit and profusion of brightly-coloured flowers render them very attractive. The following are worth a place in every garden :—

Early Pompones.—Nanum, white; Mme. Dufoy, white; Souvenir d'un Ami, white; St. Mary, white; Chromatella, yellow; Hendersoni, yellow; Le Luxembourg, yellow; Mdme. Pecoul, rose; Adrastus, rose; Delphine Cabouche, rose; Frederick Pele, red; Scarlet Gem, red.

The Japanese varieties are too late for open-air culture.—A. S.

Chrysobactron Hookeri *(Anthericum).*

Chrysocephalum.—Composite plants; of slight value for gardens generally.

Chrysocoma.—A small genus belonging to the Composite family; of little garden value.

Chrysopsis *(Golden Aster).*—North American plants of the Composite family, suitable mainly for botanical collections.

Chrysurus *(Lamarckia).*

Cichorium Intybus *(Chicory).*—A well-known native plant, growing from 2 ft. to 5 ft. high, producing in summer and autumn handsome bright blue flowers. It is worth introducing as a wild plant in localities where it is not found native. It is a rampant grower, and will take care of itself under almost any condition.

Cimicifuga *(Bugbane).*—A genus of the Crowfoot Order, nearly allied to the Baneberry (Actæa). They are all tall growing herbaceous plants; one at least is very handsome and well worth cultivating. This is C. racemosa (Black Snakeroot), which grows from 3 ft. to 8 ft high, with feathery racemes of white blossoms from 1 ft. to 3 ft. long, which, being slender, gracefully droop. C. Serpentaria, a variety of the preceding, is likewise a handsome plant. Both are of easy culture in ordinary good garden soil, and make striking objects for the mixed shrubbery border or better as groups in the wild garden. Being strong, vigorous growers, they are well fitted for naturalising on the outskirts of woods, by woodland walks, and like places. Some of the plants have, it is fair to state, an odour which does not recommend them for the garden. North America.

Cineraria maritima.—A very handsome bushy perennial with finely-cut leaves, covered on the under side with a silvery down. It bears numerous heads of bright yellow flowers in summer. When the effect of its foliage only is desired the flowering-stems should be pinched off on their first appearance. The plant then becomes more leafy and more branching. Useful on the margins of shrubberies or isolated on banks, or on the turf of the pleasure ground, where it would form an agreeable variety among the Acanthuses and various other dark green subjects recommended for groups. It is also used with good effect for edgings to flower beds, but the superiority of the silvery Centaureas has lessened its value in this respect. The best way to increase the plant is from cuttings in spring, choosing the twiggy side shoots; but the quickest way to obtain plants is to sow seed in heat in February and transplant in the open in May. C. acanthifolia is a variety of C. maritima.

Cinna.—A small genus of Grasses of no ornamental value.

Cinquefoil *(Potentilla).*

Cirsium.—Thistle-like plants, none of which can be recommended for the garden proper.

Cladium Mariscus.—A vigorous native fen plant; grows from 2 ft. to 6 ft. high, and when in flower is crowned with dense, close, chestnut-coloured panicles, which are sometimes 3 ft. in length. The radical leaves are glaucous, rigid, and often 4 ft. long. Worthy of a place near such subjects as Carex pendula or the Typhas on the margin of water.

Clarkia.—These Californian plants are among the prettiest of hardy annuals. They are robust in growth, of easy culture, and flower profusely for a long time. There are two species from which the numerous varieties now in cultivation have been obtained. C. elegans grows about 2 ft. high, erect, and much branched, bearing long, leafy racemes of flowers with undivided petals, varying from purple to pale red or a salmon colour. The principal varieties of this species have double flowers, and two—Purple King (deep purple) and Salmon Queen (salmon pink)—both having flowers produced freely on strong branching plants, are very effective border flowers. The other species is C. pulchella, which may be readily distinguished from the preceding by the deeply cut petals. It varies in height from about 1 ft. in the Tom Thumb sorts to 2 ft. It has normally magenta flowers, but there is every variation between deep purple and pure white. There are also several double-flowered forms of C. pulchella that are very pretty. There is also a very distinct variety called integripetala, from which some beautiful varieties have been obtained, notably limbata and marginata, the former dark rose and white, the other pale rose and white; both extremely pretty. Altogether there are about a score of varieties mentioned in seed lists, most of which are distinct from each other, and are well worthy of culture.

CULTURE.—All the varieties are suitable for borders, the dwarfer kinds, by their habit of growth, being well adapted for sunny spots in the rock garden or on the ridge of sloping banks. Their growth is much affected by the nature of the soil they are grown in. Like all other hardy annuals, they may be either sown in autumn or spring. By sowing in the beginning of September the seedlings gain strength before the winter, and flower well in early spring. The first spring sowing should take place in the middle of March, when the plants would flower in July. Other sowings may be made until about the middle of June for flowering in September and October. The best soil is ordinary garden mould, not too rich or too dry.—G.

Clary (*Salvia Sclarea*).

Claytonia.—A small genus of the Purslane family, of which three species are pretty garden plants. C. caroliniana, is a spreading dwarf species producing in spring loose racemes of pretty rose-coloured flowers, and C. virginica (Spring Beauty), a slender, erect growing plant, with pink blossoms, are suitable for growing in warm spots in the rock garden in loamy soil, but C. sibirica, also a dwarf species with pink flowers, requires to be grown in a damp peaty soil like that of an artificial bog. C. perfoliata and C. alsinoides are weeds in many localities.

Clematis (*Virgin's Bower*).—Although these are with few exceptions shrubby climbing plants, yet their use in the flower garden, often in a dwarf or frequently cut down condition, is so frequent and so important, that they cannot well be omitted from the "English Flower Garden." There are considerably over a hundred species, mostly from cold or temperate climates. They are widespread, in Europe from Russia to Portugal and the Balearic Isles; in Asia from the Ural range, in Siberia to India and Java, and even to China and Japan. They show themselves in both Americas, in several Polynesian islands, and even in New Zealand. Thus they extend from pole to pole, from the seashores to the slopes of the highest mountains. Scarcely a species can be said to be without beauty, so graceful is their habit, or so bold and showy their flowers. The earliest-flowering hardy species commence unfolding their blossoms in April, and these are succeeded by other species and varieties throughout the summer and autumn, some of them continuing in bloom up to Christmas in mild localities. In colour they present almost every shade and combination of red and blue, and pure scarlets and crimsons are not wanting. The lilac, pale blue, purple, mauve, claret, violet-purple varieties are connected by every intermediate shade. There are also yellow and many pure white-flowered species and varieties, the flowers varying from less than 1 in. to 8 in. or 9 in. in diameter. It is not only as climbers on trees or for covering walls, trellises, &c., that Clematises are useful. They trail or creep equally as well as they climb, and are also admirably suited for bedding, festooning, and other garden purposes. With all the wealth of variety in wild species, and the Japanese varieties in our gardens in 1863, Clematises were not extensively planted; but about that time Mr. Jackman, of Woking, commenced hybridising, employing C. Viticella and the large-flowered Japanese sorts. Among the earliest acquisitions were the beautiful C. Jackmani and rubro-violacea, and since then other raisers have reared an almost endless variety. Special descriptive catalogues of varieties are now issued, so that it is unnecessary to say more on this point than that every year new varieties are sent out, and intending purchasers should contrive, if possible, to see and make their own choice. The original large-flowered Japanese kinds are more or less tender, and in severe winters are often cut to the ground; but many of the hybrids—Jackmani, for example—will bear without injury the severe frosts. However, whether cut down by frost or knife, they will throw up again in the spring and flower perhaps even more profusely. They never form very thick stems, and are not so suitable for permanently covering large spaces, particularly where the height is considerable; to obtain their flowers in perfection the plants must be freely pruned. On the other hand, such species as C. Flammula, montana, and Vitalba grow quickly to a great height or length, especially the two last. A few species, such as C. erecta, tubulosa, and integrifolia are herbaceous.

The sections into which this family may be grouped for garden use are well treated of in Moore and Jackman's "Clematis as a Garden Flower," which we quote.

THE MONTANA, PATENS, AND FLORIDA TYPES.—These sections include the earliest or spring-flowering divisions of the family. The majority of the species and varieties of which they consist come into blossom naturally about May; but some few of them, such as C. calycina and its allies, are much earlier than this, and blossom from the commencement of the year onwards. These latter are best suited for planting against walls, in warm sheltered situations, where their opening flowers may be in some degree protected against inclement weather. The varieties belonging to C. patens and C. florida (represented by C. Sieboldi, one of the same type) are perfectly adapted for planting against walls or in corridors, and some of them make elegant early-blooming beds, especially in positions where their blossoms are thoroughly sheltered either naturally or artificially, from severe spring frosts, which occasionally, though rarely, may somewhat injure them. C. montana, also a spring or May bloomer, is of vigorous growth and perfectly hardy, and is specially adapted for covering walls, or trellises, or arbours, or in fact for planting in any position where rapidity of growth is desired. None of these plants, especially those of the montana type, are very particular as to soil, but will grow in any good garden earth which is fairly enriched, efficiently drained,

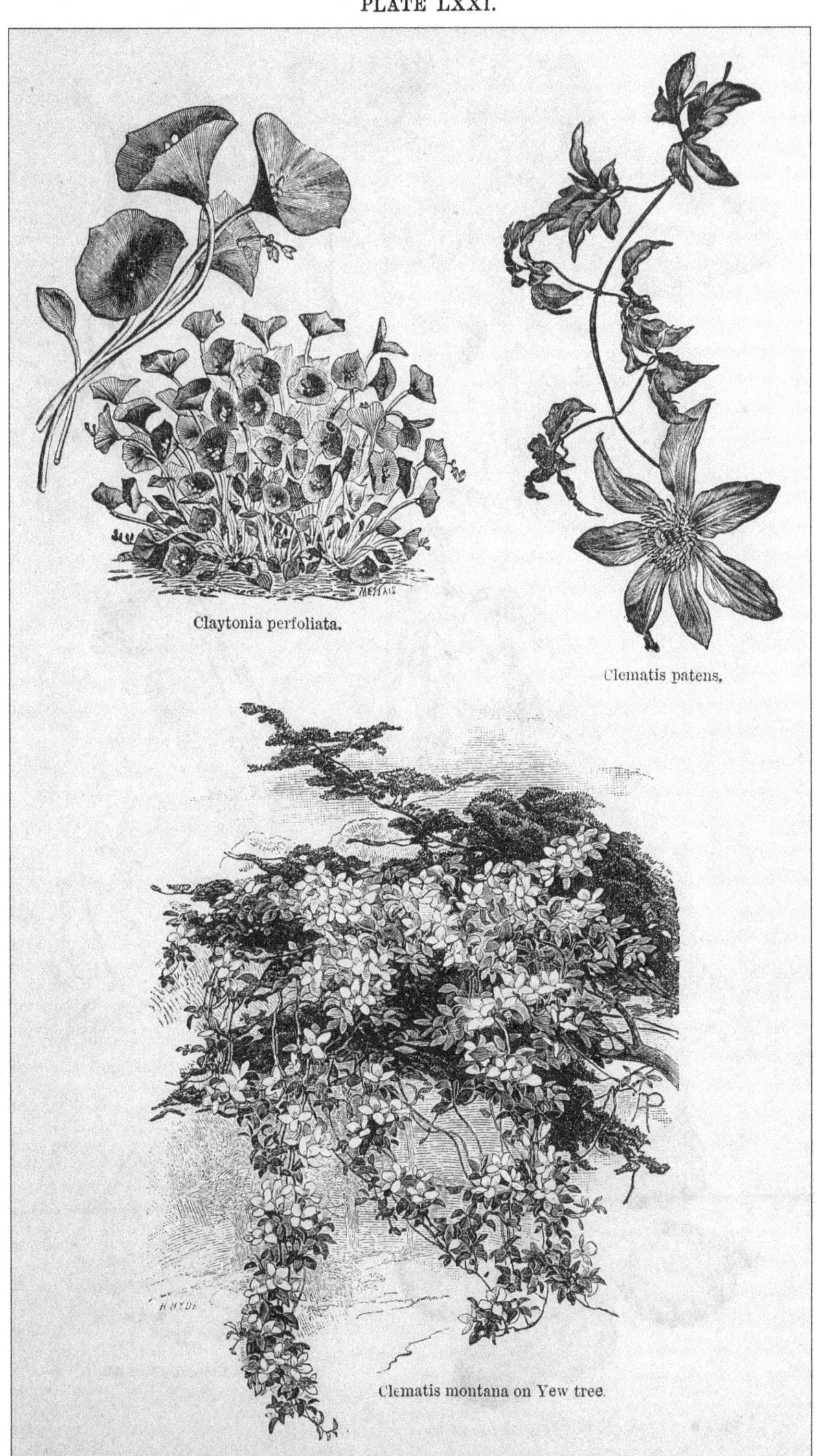

Claytonia perfoliata.

Clematis patens.

Clematis montana on Yew tree.

[*To face page* **82.**

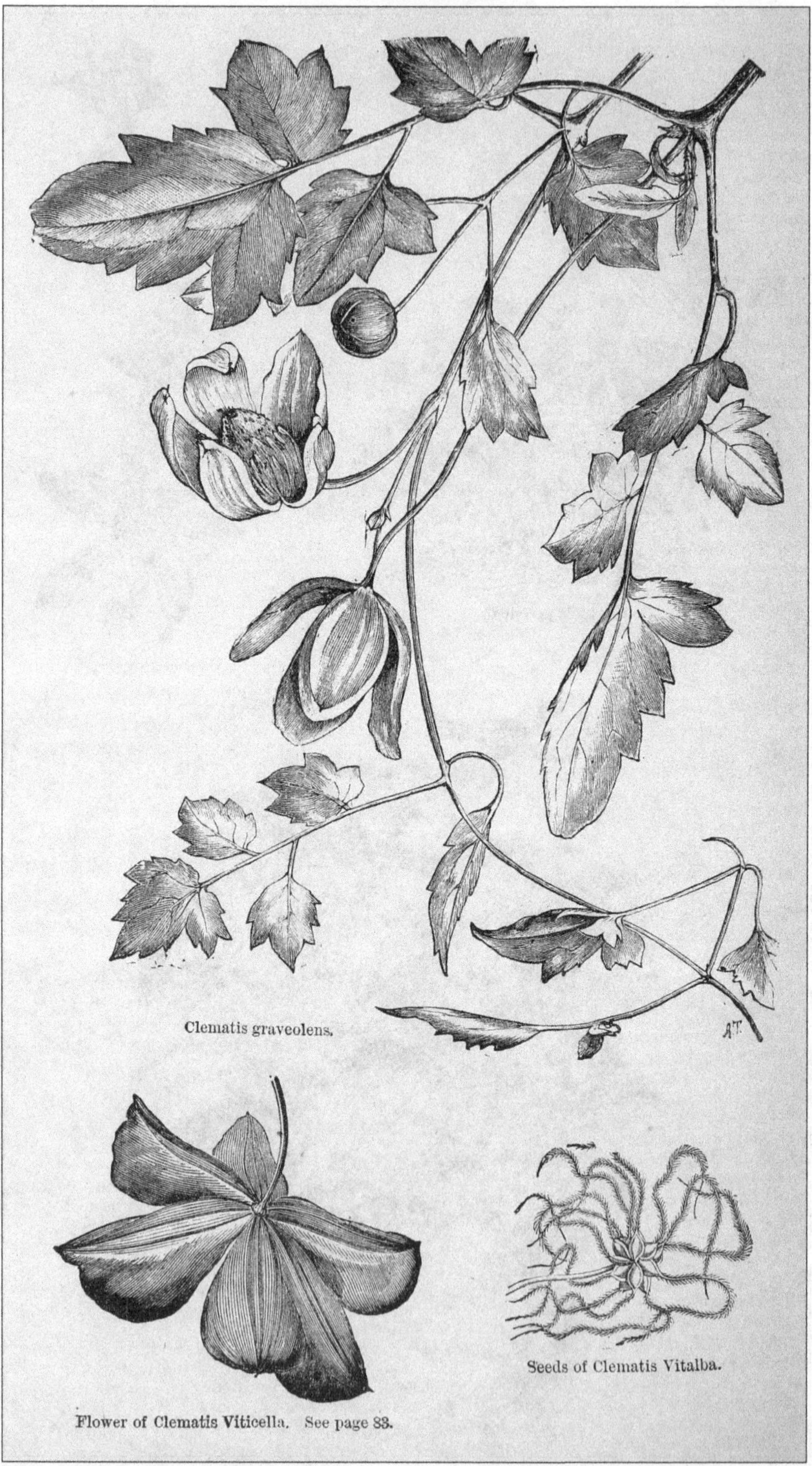

Clematis graveolens.

Flower of Clematis Viticella. See page 83.

Seeds of Clematis Vitalba.

[*To face page* **83.**

and maintained in a healthy state. Where it can be provided, a rich soil of a light loamy texture is the best for all these plants, and if this be mixed, either naturally or artificially, with chalk or lime, so much the better for the Clematises. Mulching, which consists in covering the surface of the soils for some little distance round about the plants with half-rotten manure, is a way of manuring from which the plants derive much benefit. When growth commences in the spring-time the young shoots must be attended to, and trained around or against the supports provided for them, whatever these may be. The weaker shoots may, if necessary to prevent entanglement, be cut away during the summer; but all the vigorous shoots for which there is space should be trained in, since it is these, when thoroughly developed and matured, which furnish the flowers for the ensuing year. The strong growing montana is a good verandah plant, and suited for rapidly covering any bare spaces which requiring clothing; but the less robust plants of the florida and patens types are better suited for training on walls or for furnishing corridors, or any similar positions where their blossoms may be brought more closely into view. The pruning should take place in the month of February or March, after the severe frosts of winter have passed away.

THE GRAVEOLENS TYPE.—This small group comprises a series of hardy fast-growing species, which may be said to require scarcely any cultivation. They grow freely enough in any ordinary garden earth which is of sound texture and fairly drained, though many, probably most, of them have a preference for soil of a calcareous character. They will scramble over trellises or thickets, or clamber up snaggy poles, or amongst the boughs of trees, and therefore may be employed in any position where a summer screen is wanted; but being deciduous they are not adapted to form winter screens. As to training, they make holdfasts of their leaf-stalks, and thus take care of themselves; and the natural growth thus made and thus disposed would in most cases be of a more picturesque character than would result from artificial training. When, however, they are planted near to or beneath large trees, or at all within their influence, it is desirable, in order to promote vigorous and rapid growth, to mulch the surface of the soil during winter, forking in the manure about the month of March. In such situations, too, it may chance that drought may overtake the roots, since not only will the trees themselves suck up much of the available moisture, but in a greater or less degree they will keep off the natural supply which comes in the shape of rain. In dry weather, therefore, and especially until the plants become tolerably well established, the artificial application of water, in such quantities as circumstances may render necessary or desirable, should not be forgotten.

THE LANUGINOSA TYPE.—These plants are hardy and of a tolerably vigorous habit of growth, and they produce blossoms of enormous size, so that liberal cultivation is for them an absolute necessity. They will, indeed, succeed in any good, sound, well-drained garden soil which is freely and annually manured, but they would no doubt prefer a light mellow loam to any other basis, and therefore in the case of very light soils it would be a material benefit to them to resort to the admixture of the best loam that may be available in trenching up and preparing the ground before setting out the plants. The more fertile the natural soil the less manuring will be necessary, and *vice versâ*; but it should be understood that in any case a really well-enriched soil, either natural or artificial, should be secured if the full beauty of this race of the Clematis is sought to be developed.

C. lanuginosa and the varieties of this type of growth are exceedingly well adapted for planting against conservatory walls or trellis work, whether the latter be put up in the form of a screen or a verandah, and they are also suitable for poles or pyramids. In the latter cases, especially, they should be annually pruned down to about 3 ft. from the ground surface, to prevent them becoming lanky and bare of new shoots near the base, the tendency of the new growth being to develop itself with excessive vigour at the extremities. When thus cut rather low, so as to secure a supply of foliage at or near to the base, the beauty of the plants is much enhanced. The same remarks apply to those on walls and trellises, if they are required to cover an allotted space; but in this case it frequently happens that the lower part can be filled out by less aspiring subjects, and then it is as well to secure and utilise the more vigorous growth of the plants towards the top. In any case the successional summer growths should be trained in so as to secure the later crops of blossoms, the habit in this race being to throw out a sprinkling of flowers at intervals till the frost comes to arrest further growth.

PRUNING.—It will be evident from what has already been said that comparatively slight pruning is here required. The type itself and those varieties which come nearest to it in habit, indeed, die back almost sufficiently to render pruning unnecessary; but in those instances where a mixture of blood has led to a more extended growth it will be necessary to cut so as to remove the weakly and ill-ripened portions of the year-old wood. Under favourable conditions the plants will make an annual growth of from 8 ft. to 10 ft. in length, and of this the unripened extremities, together with the weak or superfluous shoots and the dead wood, are the only parts which ought to be removed. This pruning is best done in February, after the severe winter frosts are past, and before the plants burst out into new growth. We have said that these plants are hardy, and for all practical purposes they may be so regarded. They are, however, less robust in constitution than some of the allied groups, and hence in their case the mulching which has been recommended as an advantage to all may be looked upon as being rather more of necessity, provided the plants occupy positions where such an application would be at all admissible. The annual feeding, by working in some half-decayed manure during the early spring, should on no account be admitted, as the size and succession of the blossoms depend entirely upon the vigour which is kept up in the plants; but where the mulching of manure would be objectionable because unsightly, a surface covering of some other protective material, such as the refuse of the fibre of the Cocoa-nut, would be a desirable substitute for it.

THE VITICELLA AND JACKMANI TYPES.—These groups represent some of the hardiest as well as some of the noblest of the whole family. The severest winters do not injure them in any material degree, and from their wonderful fertility of flowers the plants in the late summer and autumn months literally become masses of blossoms, successively and continuously renewed. In regard to soil, the same free, well-drained, deep, and well-enriched staple, which has been noted as suitable for the preceding groups, will be found equally adapted for these. A friable loam is the best soil they can have; if it is not so suitable as this in quality, it should at least be deep that the roots may penetrate freely. A loamy soil is the best, because the plants must have manure liberally supplied to them, in order to keep up their strength ; and in a loamy staple the fertilising properties of the manure are not liable to be dissipated, as they are in one which is poor and porous. When, however, the soil approaches this latter description, it is all the more necessary for the plants that manure should be abundantly applied, to make good the natural deficiency in fertility. In the case of light soils, a good proportion of loam—made friable by frosts, if at all of a heavy or clayey character—should be incorporated, since this will render it the more holding; deep trenching should also be resorted to for the same purpose. In the case of heavy soils, they should be ameliorated by the free intermixture of friable soil or of any sharp gritty material which may be available, the drainage being made efficient, and the soil well aërated before planting.

PLANTING.—When the ground has been prepared, the plants may be put out during any open weather which may occur between the middle of September and the end of April. After planting it is beneficial, though not absolutely necessary, to apply a mulching of a few inches of partially rotten manure on the surface; this will both serve to protect the newly-disturbed roots and also tend to fertilise the ground. The varieties of these types of Clematis are essentially outdoor or border plants, since they require abundance of root space; they may, indeed, be grown into exhibition specimens, as will hereafter be explained, but even then they require a liberal supply of root accommodation and high feeling. To sum up this branch of our subiect, the Viticella and Jackmani types of Clematis require to be grown in rich deep soil, to be manured freely every season, and to bc planted out in the open ground, that their roots may have free pasturage.

These forms of Clematis flower on the vigorous summer shoots, forming dense masses of blossoms. The object, therefore, in pruning should be that of favouring to the utmost the development of these vigorous young shoots, and this is done by cutting the summer growth back early in the season as soon as the frosts have disfigured the plants, say about November, to within about 6 in. of the soil. The mulching, which is then to be applied with a liberal hand, serves to prevent the soil becoming severely frost-bound, and should, about the middle or end of February, be neatly forked in along with, in some cases, an additional supply of rotten manure, the latter being regulated by the manurial qualities of the original mulching and by the natural strength of the land. Thus treated, the plants will commence flowering about the first week in July, and they will go on yielding flowers as long as their strength will enable them to throw out lateral growths. To obtain a later bloom a portion of the plants should be left over at the November pruning, and not be cut back till the end of April, after they have commenced to grow.

POSITION.—The use to which these types of Clematis may be applied are exceedingly various. They may be trained up snaggy poles, either singly or several together, to form pillar plants, or they may be allowed to scramble over masses of rockwork or root-work. They may be festooned, or they may be trained over verandahs, or fastened to walls or trellis-work, or led over ornamental iron supports as single standard specimens for lawns. In either way and in every way they are found to be thoroughly effective as flowering plants, many of them indeed, and especially those of the true Jackmani type, being capable of producing a startling impression in consequence of the gorgeous masses in which their rich Tyrian hues are displayed. One of the most useful purposes to which these varieties of Clematis could be put would be to drape a mural ruin or to cover an unsightly bank or slope. They will grow in almost any situation if the soil is not absolutely deficient of food, or if the roots of other plants do not rob them of a fair supply of nutriment; and in such situations nothing would be required but to throw down a few tree-roots or rough branches for them to scramble over. Thus planted, a layer of manure worked in annually with the fork, and a supply of water in very dry weather, would secure a good result. Again, they rank amongst the noblest of ornaments for low walls, trellises, &c., to which they

PLATE LXXIII.

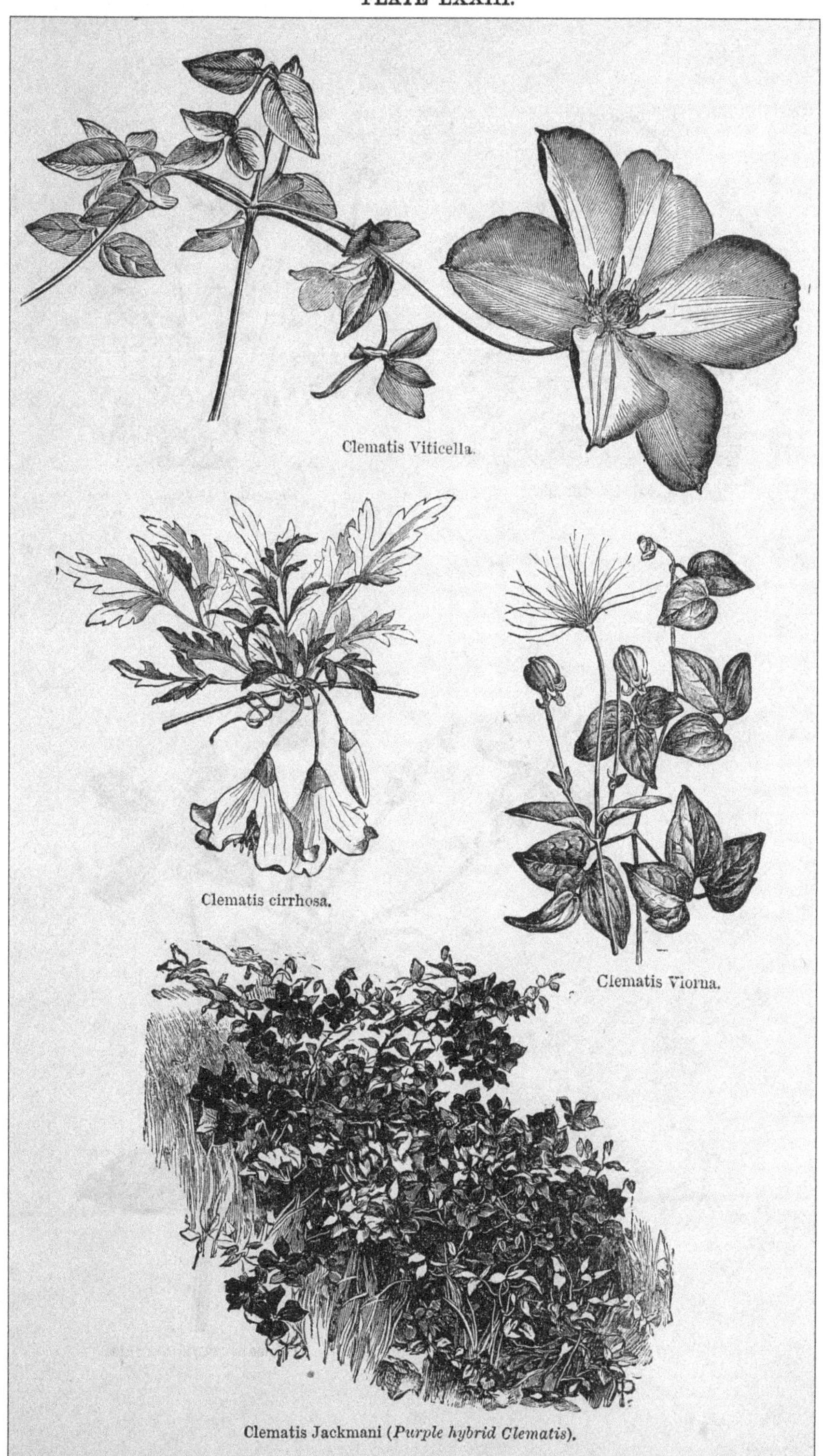

Clematis Viticella.

Clematis cirrhosa.

Clematis Viorna.

Clematis Jackmani (*Purple hybrid Clematis*).

[*To face page* 84.

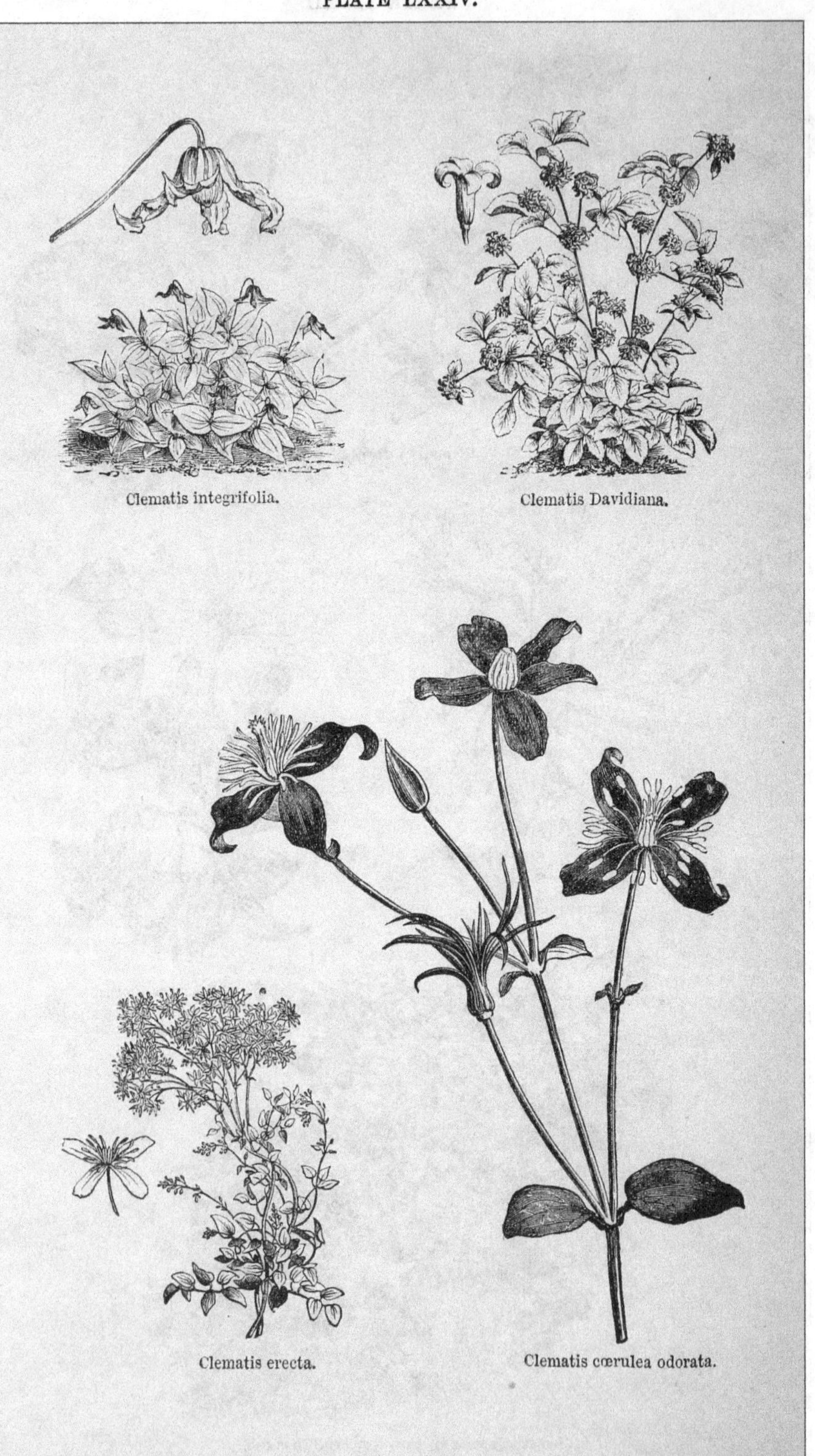

Clematis integrifolia.

Clematis Davidiana.

Clematis erecta.

Clematis cœrulea odorata.

[*To face page* 85.

must necessarily in the first instance be nailed or tied; but once firmly fixed, they should be allowed to fall down in rich picturesque masses. Probably, however, the simplest and grandest use that could be made of them would be to plant them on large masses of rockwork, giving them a good depth of rich, light, and sandy earth, and allowing their shoots to fall over the face of the blocks without any training or pruning. We shall here find among the varieties of the Viticella and Jackmani types those kinds of Clematis which are especially adapted for bedding out for summer and autumn flowering.

NON-CLIMBING TYPES.—The non-climbing species and varieties of Clematis consist of two small but distinct groups, the one herbaceous, the other sub-shrubby. The herbaceous species thrive best in good rich deep loamy soil, and when they become well established they form somewhat striking plants for the mixed border, though scarcely any of them fall into the very front rank of herbaceous perennials.

The single and double-flowered varieties of C. erecta and C. maritima are most ornamental, and are well worth introducing even into a select collection. These herbaceous species grow freely enough in any tolerably fertile garden soil, but it should be of good depth, as the roots are strong and strike downwards. A free application of manure is beneficial to them, especially if the soil is not naturally of a fertile character.

The sub-shrubby varieties of the non-climbing group include some exceedingly ornamental plants. They are especially adapted for the back rows in mixed flower borders where plants trained to a height of 5 ft. to 6 ft. would not be obtrusive; for prominent positions in the front parts of shrubbery borders, or for dwarf standards or iron trainers in beds, whether of Clematis or of other plants. They require a good preparation of the soil, which should be deep and rich, exactly as recommended for the varieties of the Jackmani and lanuginosa groups. In spring, before growth recommences, the plants should be pruned back to the well ripened wood at from 1½ ft. to 2 ft. from the ground, and a firm stake or support provided for each. To this support, as they grow, the young shoots require to be tied. When the branches begin to ramify, which they do at a height of 3 ft. or 4 ft., they may be allowed to fall down on all sides, and in this way they ultimately form a mass of flowers like a huge bouquet. C. cœrulea odorata is a most desirable plant of this group, not only for its abundant well-contrasted blossoms, but also for their fine scent.

The various forms and allies of the erect Clematis, C. erecta, are very fragrant and hardy; grow from 3 ft. to 4 ft. high, and bear large umbel-like clusters of white blossoms, which emit an agreeable perfume. They are most admirably adapted for planting by the sides of frequented walks, &c., but delight in full exposure. The principal varieties are the double-flowered kind, which bears a profusion of small button-like blossoms, lasting longer in perfection than the single blossoms; pumila, which differs from the type in being much dwarfer in growth; pauciflora, producing smaller flowers in less dense clusters. It inhabits various parts in South Central Europe.

THE CLEMATIS AS A BEDDING PLANT.—When employed as a bedding plant the Clematis should be permanently planted out, so that the roots may not again be disturbed. The soil should be rich, open, and deep, and of a calcareous loamy character if possible. If it be of a heavy texture it must be well drained and ameliorated by admixture of gritty matter, such as road-scrapings, the sand washed up by the road-side, burnt clay, or even ashes, in moderate quantities; an admixture of half-rotten leaf-mould would also be beneficial. If, on the contrary, it be light and dry, the soil should be strengthened by the admixture of good and rather heavy loam. Deep trenching and a liberal manuring should be resorted to before planting; and a thorough dressing of good sound manure should be forked in annually in November, when, in ordinary cases, the summer growth may be cut back.

The young plants of Clematis, when planted out, should be set at about 2 ft. apart, so that they may cover the surface quickly. When they become strong and well established a portion of them may be removed if desired, as from the more vigorous growth of established plants they will branch more freely and cover more quickly. The growing shoots should be looked to at least once a week, and pegged down or trained where most required to cover the surface. They cling together so firmly by their clasping leaf-stalks that this should always be done before they get at all entangled, for the young shoots would be certain to suffer injury in the process of disentanglement. The plants should be raised by some means so as to give a convex surface to the bed, and thus the better to display their flowers. This may be done in a variety of ways, either by raising the surface of the bed itself to the desired shape, by pegging down a layer of twiggy branches, such as pea-sticks, for the plants to grow over, by fixing a common hooped trellis of rods to which the shoots should at the first be tied, or, what in many situations would be the best plan of all, but which would be scarcely admissible in a dressed parterre, by arranging root-masses of suitable bulk on the surface of the beds for the plants to scramble over and amongst. Whatever plan may be adopted, the plants must be trained as already recommended till they have furnished the space to be covered, when they may be allowed to grow more at random. As the plants do not throw up flowers from the lower portion of their stems, it is desirable in training them to cover the beds, that the points of one series of plants should be so arranged that that they may overlap those portions of the adjoining ones which remain bare. This point should be borne in mind from the first and until the whole surface is evenly covered with flowering wood.

Continuity of flowering is dependent upon continuity of growth. Now this at once suggests summer feeding. Thus in dry weather manure water should be given alternately with pure water, the water not being applied over the leaves and flowers, but beneath them. It is to be recommended, if the summer is at all a dry one, to have the beds thoroughly saturated with pure water just as the buds are being developed and begin to acquire size; if this is done thoroughly it will increase the size of the flowers, and will carry the plants on for a considerable period. One or two such thorough waterings may be given subsequently, if the season is such as to require it, ayplying at least one dose of liquid manure when the plants have been flowering for a considerable period. No other attention is required till the frosts of November come, after which the plants may be pruned hard back.

THE CLEMATIS AS A WALL PLANT.—When grown on walls the plants of the early-flowering section are not liable to suffer injury from frost, being hardy, and flourishing even in exposed situations. The chief risk they incur is that arising from the incidence of the morning sun upon them when a sharp late spring frost may have caught the expanded flowers. In any moderately sheltered position, however, this risk is but slight indeed, and the plants will grow freely and flower satisfactorily, coming into blossom about the middle or end of May, and continuing to flower more or less abundantly, according to the situation they occupy, up to the end of June or beginning of July. In very sheltered situations, some of the varieties of the montana group, notably C. calycina, may be had still earlier than the foregoing; while as a May bloomer, vigorous in growth, hardy in constitution, and most prolific of flowers, C. montana—with its variety—is strongly to be recommended as a distinct type of the genus.

The summer and autumnal-flowering groups are gorgeous wall plants, and include not only the nobler forms bred from C. lanuginosa, but the descendants of C. Viticella. They bloom in July, and continue on till October or November. The lanuginosa breed is specially effective when thus grown, on account of the immense size to which the blossoms of many of the varieties attain. They require in this position the most liberal feeding and moderate pruning, the summer growths being carefully trained in to secure the successional flowers they produce.

The Viticella and Jackmani varieties attract rather by the profusion than by the individual size of their flowers. These plants, while fed to the utmost in order to meet the demand upon their powers, should be pruned hard back every autumn, unless a considerable space is intended to be covered, and the young shoots should be trained up to their full extent in the early part of the summer, until flowers begin to appear, when, as all the lateral growths develop flowers, it may be better to let them fall in graceful wreaths of pendent spray. Very good results have been obtained for a time by not pruning back.

PILLARS, ROCKERY, AND ROOTWORK.—For pillar plants we recommend the varieties of the true C. Jackmani type—hardy, free flowering, and continuous blooming. Such pillars of purple are amongst the most beautiful objects which can be dotted about the garden. Deep soil and thorough drainage are greatly conducive to a successful issue. The plants, being perfectly hardy, may be planted out during any open weather between September and April, and will be all the better if mulched, not merely to afford shelter to the newly embedded roots, but also to furnish, as the rains descend, by continuous infiltration, a supply of congenial food, which may result in free rapid growth. Before starting, the plants should be cut back to within a few buds of their base—say within from 4 in. to 6 in. of the ground—and then all is ready for the spring growth. At some convenient time after planting, and before the necessity for training has arisen, the poles should be placed about the plants. A single snaggy pole, well provided with short lateral branches, may be inserted, and the plant trained about this, or three or four poles may be set in so as to form a pyramid, should this style be preferred; but that which is best suited to the habit and character of the plants is to set three or four poles erect at a little distance apart, much as Hop poles are set up, and to train the plants over and about them. At home in dressed ground, as bedding plants, as pillar plants, as umbrella plants, as single specimens or in masses, they are no less at home in wilderness scenery, about ruins or rockwork, or amongst those grotesque arrangements of old tree stumps to which the term rootery is commonly applied. In fact, the bed, the pillar, the wall, the rock, or whatever it may be, is merely the skeleton or foundation on which the glorious blossoms of the Clematis are to be displayed. A deep rich soil must be provided, and this will, in the present case, be facilitated by the inducement there may be to throw up irregular mounds, on which to arrange, as taste may direct, the stumps or stony masses which are to give name to the spot. The same necessity will exist, in both cases, for ample feeding. The same general rule as to close pruning must also be followed, unless indeed the rootery requires more filling up, in which case it may be desirable to leave the whole of the matured bine of the previous season until

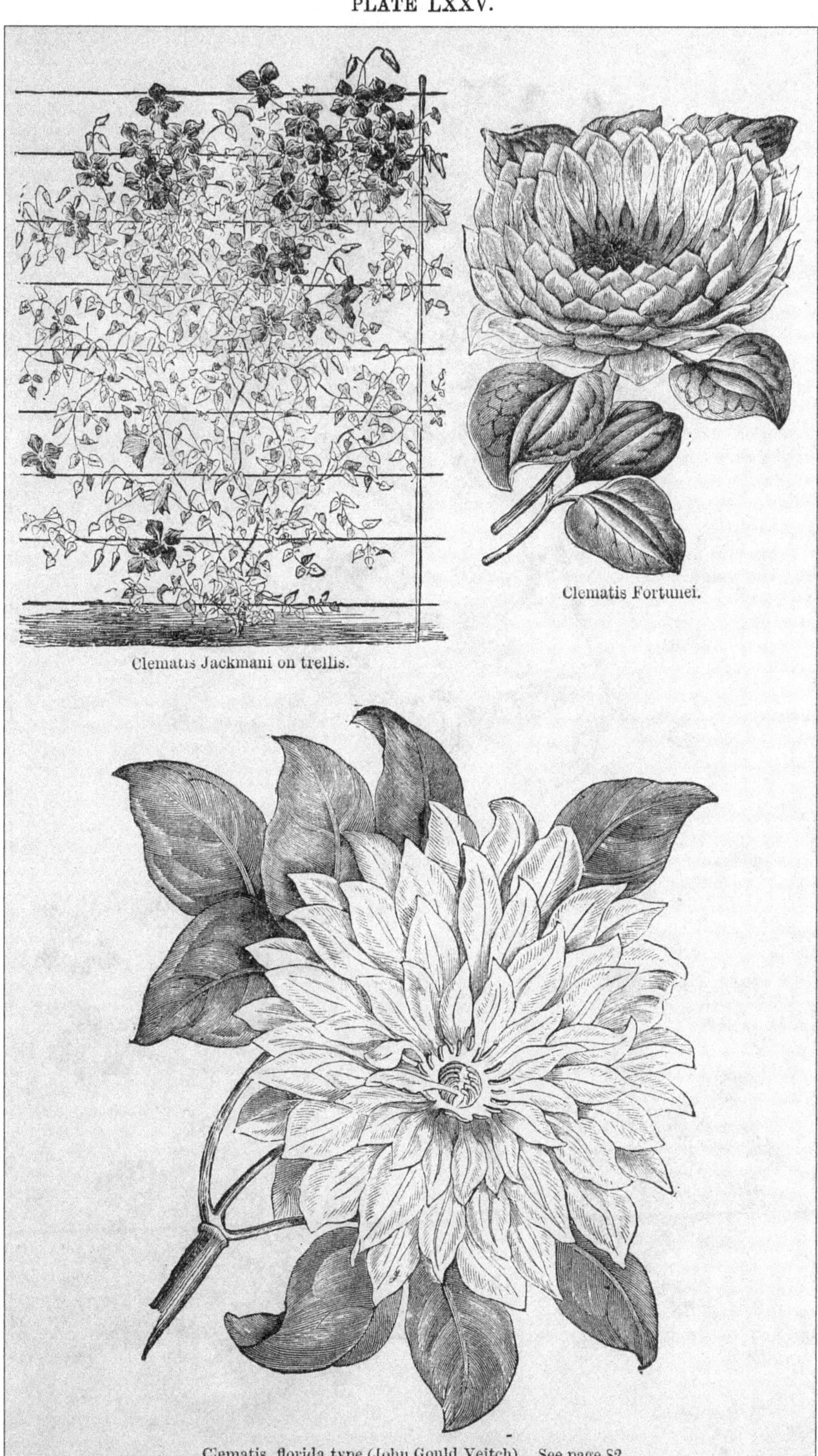

Clematis Jackmani on trellis.

Clematis Fortunei.

Clematis, florida type (John Gould Veitch). See page 82.

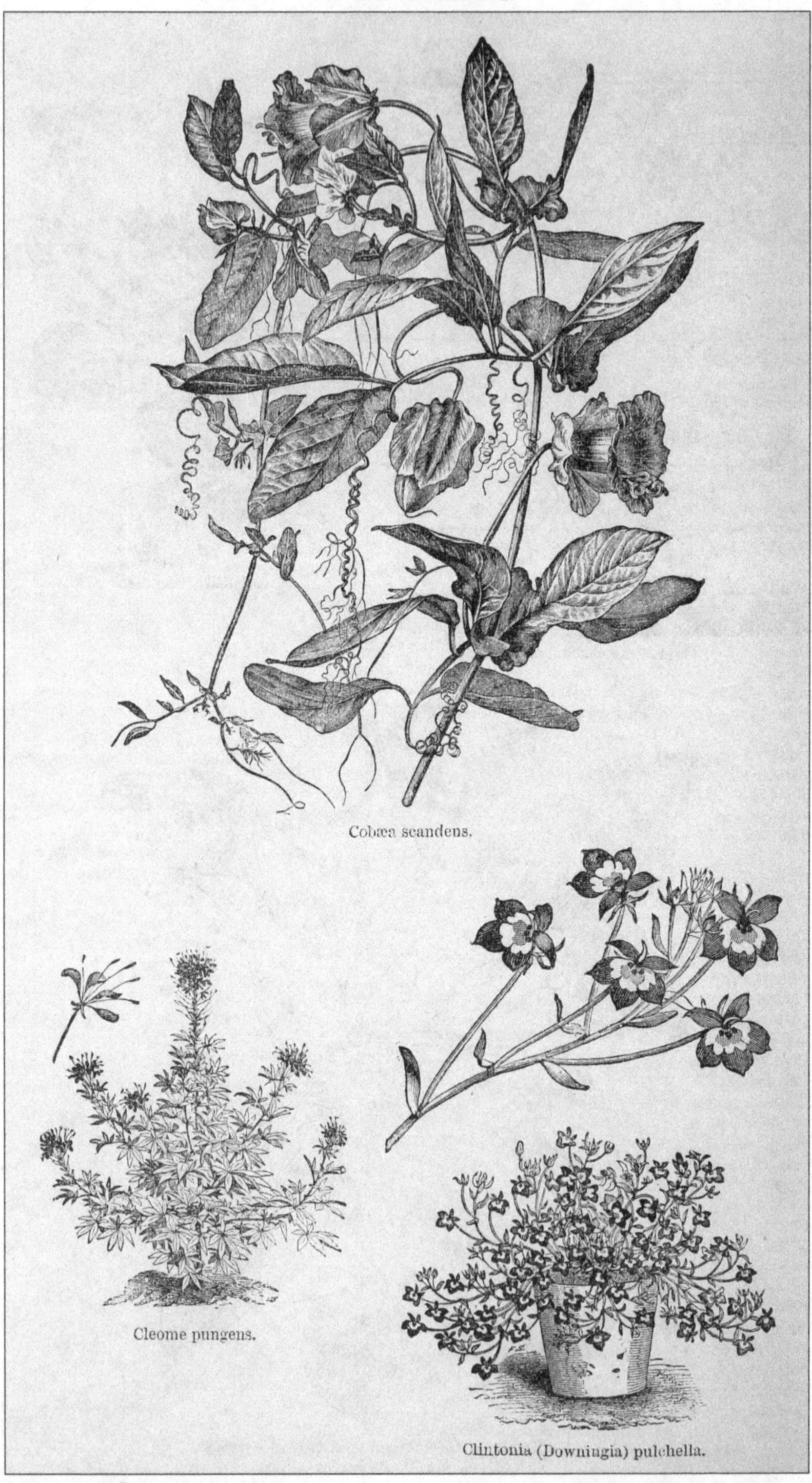

Cobæa scandens.

Cleome pungens.

Clintonia (Downingia) pulchella.

[To face page 87.

the plants are sufficiently extended to entirely cover the prescribed space with their annual growth. In regard to training, all that will usually be required will be to lead the young shoots, during their spring and early summer growth, as evenly as possible, over the masses of roots—or rocks if planted on a rockery—leaving them afterwards to fill out the picture in their own natural way. The result, unshackled by formality, will certainly not be the least pleasing of those realised in the several departments of the garden to which the Clematis may be introduced. Catalogues abound with the names of the newer and finer hybrid forms. The following selections for outdoor work are given by Messrs. Moore and Jackman:—

FOR PLANTING AGAINST CONSERVATORY WALLS AND IN CORRIDORS.—*Spring-flowering sorts* (for S., W., or N.-W. aspects only).—C. Albert Victor, barbellata, calycina, Countess of Lovelace, Edith Jackman, Fair Rosamond, George Cubitt, Lady Londesborough, Lord Derby, Lord Londesborough, Lord Mayo, Maiden's Blush, Miss Bateman, monstrosa, montana (any aspect), montana grandiflora, patens floribunda, Sophia plena, Standishi, Stella, The Queen, Vesta.

Late summer-flowering and autumn-flowering sorts.—C. Alexandra, cœrulea odorata, Flammula, Gem, Gloire de St. Julien, Henryi, Jackmani, Jeanne d'Arc, Lady Bovill, Lady Caroline Nevill, Lady Maria Meade, lanuginosa, lanuginosa candida, lanuginosa nivea, Lawsoniana, Madame Van Houtte, magnifica, marmorata, Mrs. James Bateman, Otto Frœbel, perfecta, Prince of Wales, purpurea hybrida, reginæ, Renaulti cœrulea grandiflora, rubella, rubro-violacea, Sensation, splendida, Star of India, Thomas Moore, tunbridgensis, velutina purpurea, Victoria, Viticella rubra grandiflora, Viticella venosa.

FOR PERMANENT BEDDING-OUT. — C. Helena patens, Sophia, Standishi, Alexandra, Jackmani, magnifica, Prince of Wales, rubella, rubro-violacea, Star of India, tunbridgensis, velutina purpurea.

FOR GROWING ON PILLARS AND POLES. —C. Alexandra, Beauty of Surrey, Excelsior, Flammula, Gem, Gloire de St. Julien, Henryi, Jackmani, Jeanne d'Arc, Lady Bovill, Lady Caroline Nevill, Lady Maria Meade, lanuginosa, lanuginosa candida, lanuginosa nivea, Lawsoniana, Madame Van Houtte, magnifica, Marie Lefebvre, marmorata, Mrs. James Bateman, Otto Frœbel, Prince of Wales, reginæ, rubella, Sensation, Sir Robert Napier, Star of India, Thomas Moore, tunbridgensis, velutina purpurea, Victoria, Viticella rubra grandiflora, Viticella venosa.

FOR PLANTING ON ROOTERIES.—C. Alexandra, Flammula, graveolens, Hendersoni, Jackmani, Lady Bovill, magnifica, marmorata, modesta, Mrs. James Bateman, orientalis, Prince of Wales, purpurea hybrida, Renaulti cœrulea grandiflora, rubella, rubro-violacea, splendida, Star of India, Thomas Moore, tunbridgensis, velutina purpurea, Viticella atrorubens, Viticella major, Viticella purpurea plena, Viticella rubra grandiflora, Viticella venosa.

The only remarkable novelty recently added to the genus is the Clematis coccinea, a red flowered kind, very distinct, which promises to be a beautiful species.

Cleome *(Spider-flower)*.—A genus of the Caper family, of which there are several species often enumerated in seed catalogues; though vigorous and distinct they are not likely to take any useful place in the flower garden. They are mostly annuals, getting on best with the treatment usually given to half-hardy annuals.

Climbing Fern *(Lygodium)*.

Clintonia.—The pretty annual species of this duplicate genus are now referred to Downingia, which see.

Clintonia.—A small genus of the Lily family, natives of North America. C. borealis is an unattractive plant with greenish yellow flowers, but the other, C. Andrewsiana, is a pretty plant, having a tuft of ample bright green leaves about 1 foot long and about 4 in. wide, from which arises the stout flower stem some 6 in. high, terminated by a dense umbel of deep rose-coloured blossoms, produced from April to June. It thrives well in a peaty soil in a partially shady position, just such a place as suits the Mocasson flower. It is yet scarce in cultivation, not having been introduced long.

Cnicus benedictus *(Blessed Thistle)*. —A handsome biennial plant, having bold deep green foliage, blotched and marbled with silvery white. It is useful for associating with plants of sub-tropical growth, but it must always have good deep soil and plenty of space to develop properly. It grows freely in a thin shrubbery and luxuriantly, or on any rich bank of rubbish or manure.

Cobæa scandens.—This well-known greenhouse plant thrives against an outside wall in favourable localities in the southern and western counties, and it will cover a considerable space of trellis-work during summer. It should be planted in light rich soil, and if watered liberally during the growing season it will soon cover a large space and flower freely. If afforded some protection it will survive an ordinary winter. Plants of it may be easily raised from seeds, which should be sown during spring in a frame or hand-light. Cuttings also strike readily in a brisk heat in spring. The variegated form must be raised from cuttings.

Colchicum *(Meadow Saffron)*.—A genus of hardy bulbous plants that flower in the autumn just when summer-blooming plants have lost their freshness, and when the days begin to shorten, and often but erroneously termed "autumn Crocuses." Unlike many bulbous plants, their presence in the ground early in spring, when the borders are being prepared for summer-flowering plants, is not likely to be overlooked, for their leaves are amongst the earliest harbingers of spring, bearing with them the seed-pod. The individual flowers of Colchicums do not, as a rule, last long, but as they are produced in succession, a long season of bloom is the result. The flowers are often destroyed on account of the plants being grown in unsuitable positions, viz., in bare beds of soil, which, splashing the blooms during heavy rainfalls, impairs their beauty. The best places for Colchicums are grassy places near shrubberies and trees, where the soil is well drained and rich. On rockwork, too, planted among dwarf Sedums and similar subjects, Colchicums would thrive well and make a show in autumn, when rock gardens are comparatively flowerless. They are typical of the plants that will look better in grassy places or in what we call the wild garden than in any formal bed or border. Their naked flowers want the relief and grace of the grass and foliage around them. Choice sorts and rare, or those we want to increase, should be grown in nursery beds, but the true use for the plants when they can be spared is in pretty groups and colonies in turf not often mown. There are about 30 different kinds of Colchicum, though only about half of this number are in cultivation. Among these there are some whose differences are so slight as to be scarcely worth growing, except in cases where a very full collection is desired. The genus has rather a wide geographical range, some species extending to the Himalayas; others are found in North Africa; but the majority are natives of Central and Southern Europe. Though there are so many names to be found in catalogues, the really distinct kinds are few, and there is such a striking similarity among these few, that for cultural purposes they might be conveniently classed into groups. The best known is

C. autumnale, commonly known in gardens as the Autumn Crocus. The flowers appear before the leaves, are of a rosy purple colour, and rise 2 in. or 3 in. above the surface in clusters of about six. It flowers from September till November. There are several varieties of this plant, the chief being the double purple, white and striped; roseum, rose lilac; striatum, rose lilac, striped with white; pallidum, pale rose; album, pure white; and atropurpureum, deep purple. Similar to C. autumnale is C. arenarium, byzantinum, montanum, crociflorum, lætum, lusitanicum, neapolitanum, alpinum, hymetticum; all, like autumnale, are natives of Europe, and from a garden standpoint identical with each other.

C. Parkinsoni.—This is a most distinct and beautiful plant, distinguished readily from any of the foregoing by the peculiar chequered markings of the violet-purple flowers. It produces its flowers in autumn and its foliage in spring. Other kinds similar to this are Bivonæ, variegatum, Agrippinum, chionense, tessellatum, all of which have the flowers chequered with white on a dark purple ground.

C. speciosum, from the Caucasus, is the noblest of the genus, and a beautiful and valuable plant for the garden in autumn, when it produces its large rosy purple flowers, nearly 1 foot above the surface. The foliage, too, is broad and handsome, but as it does not appear till spring it is not so noticeable. Like the rest of the Colchicums, it is as well suited for appropriate places in the rock garden as the border, and it thrives well in any soil, provided it be not too poor or too heavy; but to have it in perfection, choose a situation fully exposed to the sun, with the soil of a sandy character—in fact, such a spot as is likely to dry up during summer; here they will luxuriate, and enjoy the autumn, winter, and early spring rains.

Coleus.—These handsome-leaved plants afford a few kinds that succeed well in the open air in summer, and when used judiciously produce a fine effect. In some of the London parks they are arranged in large masses by themselves and generally of one kind only. Though there is such a host of varieties, there are but few that succeed satisfactorily in the open air. Mr. Wildsmith, of Heckfield, writes: "We have tried at least a score of varieties in bedding out, with the result that the first kind that was recommended for the purpose (Verschaffelti), is still the only one that succeeds well. The culture of all the varieties is of the simplest nature; cuttings strike freely in any sandy soil, in a moist heat of 70°. As soon as struck, pot them in light loam, containing a tenth part of well rotted manure; they should be grown on in a moist heat of from 65° to 70°, and not be allowed to get pot-bound till the plants have attained the size desired, and should at all times be well exposed to the light to bring out their colour. For bedding out, the plants should be struck in March, be grown on in warmth till about the middle of May, and then gradually inured to bear full exposure in readiness for planting out the first week in June."

Collinsia.—A very pretty family of N. American annuals, belonging to the Pentstemon Order, and very dressy and pretty in spring where well grown and tastefully used. If sown in autumn they will on many soils survive the winter and flower much better than spring-sown plants. They are of the easiest culture sown in the open air; if in spring, the plants flower some twelve weeks after they are sown. If sown in autumn they flower very early, if the spring is at all favourable. There are about nine to a dozen species or varieties in cultivation and enumerated in the catalogues, the only one requiring special treatment being C. verna, which *must* be sown in autumn. "The seeds (after Mr. W. Thompson) must be sown about the end of August, or from that time to the middle of September, in pans of light vegetable soil, which should be kept thoroughly damp. In a fortnight the seedlings will show themselves, and when these have made their first pair of leaves (besides the seed-lobes) are to be pricked out singly in pans, boxes, or pots, the latter if intended for blooming under glass. From this moment it is impor-

PLATE LXXVII.

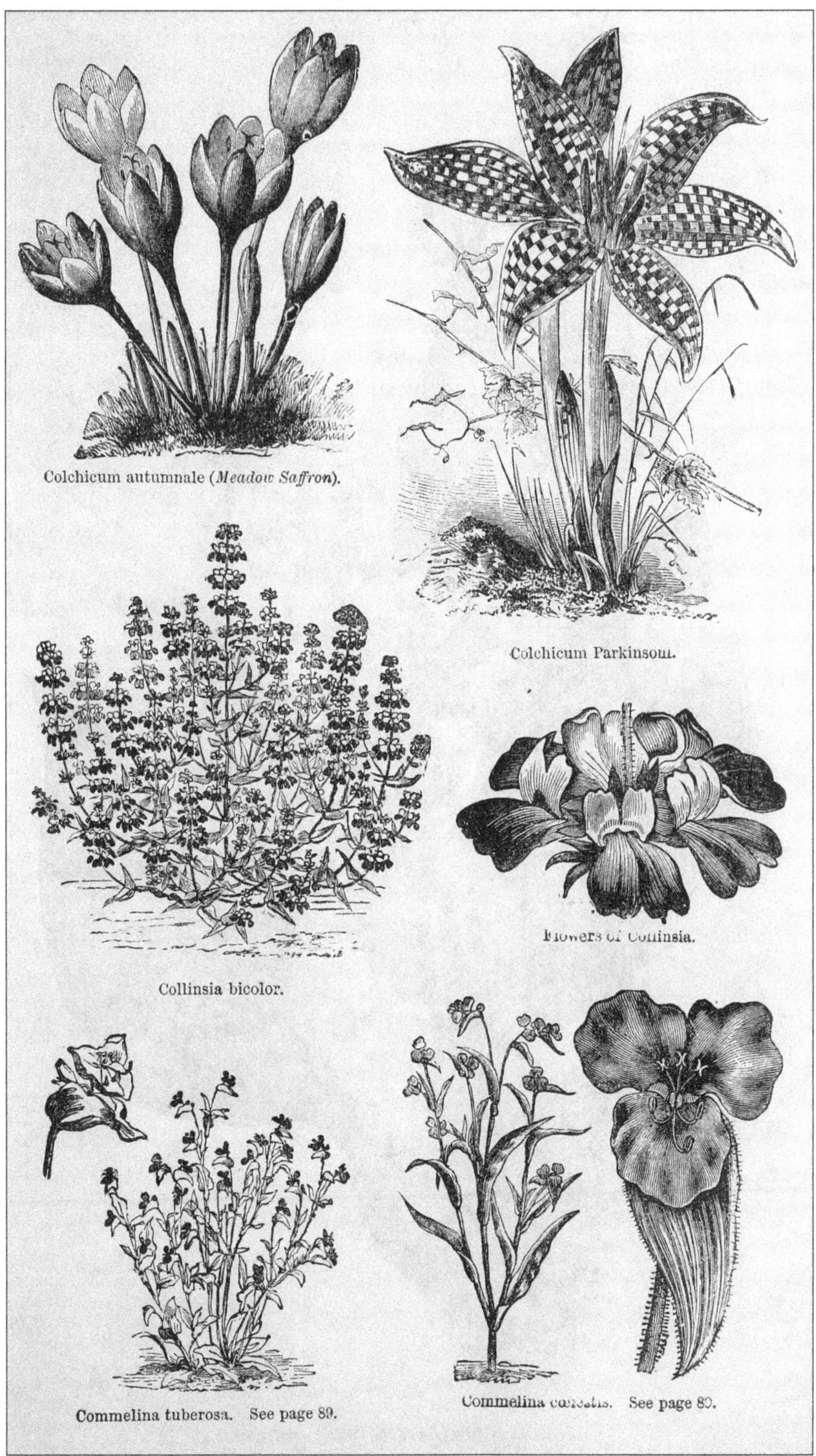

Colchicum autumnale (*Meadow Saffron*).

Colchicum Parkinsonii.

Collinsia bicolor.

Flowers of Collinsia.

Commelina tuberosa. See page 89.

Commelina cœlestis. See page 89.

Convallaria majalis (*Lily of the Valley*), well grown.

[*To face page* **89.**

tant that the seedlings should be kept cool and damp, and as near the glass as piacticable, that their growth may not be forced. If intended to bloom in the open ground, they may be planted out in light, rich soil, in partially shaded situations, or in a north or north-west aspect, and will require no other attention but slight protection during severe weather. Thus treated, they will, in ordinary seasons, commence blooming early in April, and continue in flower six or eight weeks." Notwithstanding its charm of colouring, this has never become so well known as C. bicolor, grandiflora, violacea, heterophylla, and the various other forms. The prettiest use for these plants is in the spring garden in beds, or occasionally as a broad edging.

Collinsonia canadensis *(Horse Balm)*.—An unattractive plant of the Sage family; native of North America.

Collomia coccinea.—A bright coloured annual, but not one of great value or beauty in the presence of the many good things we have. Better culture than it usually gets may improve it. Sow in April in open ground, or sow in a dry pit in autumn and protect during winter if good plants, either for pots or planting out, are desired. Polemonium Order, growing from 1 ft. to 18 in. high, and flowering in summer and autumn. On warm soils it grows higher and sows itself every year, surviving the winter, and growing much stronger and taller in consequence.

Comfrey *(Symphytum)*.

Commelina cœlestis *(Blue C.)*—A pretty and continuous-blooming Gentian blue flowered plant. There is a white variety, C. c. alba, about 2 ft. high, more or less profusely covered with blossoms from early summer till autumn. It delights in light, sandy, well-drained soils. The roots are fleshy, and liable to suffer from severe frost; it is advisable in some districts to cover them on the approach of winter with coal-ashes. In cold wet districts the roots may be lifted and stored in dry leaf-mould, or the plant may be treated as a half-hardy annual, sowing the seed in heat, and pricking off into small pots as soon as the seedlings are fit to handle, and planting them out about the end of May. The roots, if lifted and stored in winter, will be the better for a start in slight heat in spring, and may be increased by division. On some warm or stony soils and in districts near the sea where soil of a light nature prevails, it grows like a weed. It is so fine in colour that a group or small bed of it is always welcome. Mexico. Tradescantia Order.

Compass plant *(Silphium laciniatum)*.

Conandron ramondioides.—A small Japanese plant allied to Ramondia, having thick wrinkled leaves, produced in flat tufts, from which arise erect flower-stems some 6 in. high, bearing numerous lilac-purple blossoms. Though said to be quite hardy, this requires a sheltered position, such as is afforded by snug nooks in the rock garden. Plants placed between the blocks of stone thrive capitally, provided there is a good depth of soil in the chink and the position is a tolerably moist one.

Convallaria majalis *(Lily of the Valley)*.—This universal favourite delights in partial shade and moisture, and in a rich light soil that can be readily permeated by its fibrous roots. Naturally it is found growing in the valley sheltered by shrubs, and in the forest luxuriating under trees. The best situation to plant it is where it will receive partial shelter and shade, either from wall, fence, or trees. It is advantageous to have a plantation of Lilies of the Valley upon a south aspect for the sake of earliness and producing them in succession, for by this means blooms may be gathered a fortnight or three weeks earlier, particularly under shady walls, with liberal surface-dressings of rotten manure, and an abundant supply of moisture throughout active growth. The chief point to guard against in such an aspect is frost, which is so destructive to the blooms, as they appear simultaneously with the leaves. Placing a few Spruce or other evergreen branches sparsely over the beds is a very efficient protection against frost, and such a course of treatment will also be found beneficial in affording shelter and encouraging growth. Preference should be given to a soft, loamy soil well enriched with rotten manure, though fine Lilies may be grown upon rather heavy loam. In preparing loam of this texture for the reception of the plants, it will be greatly benefited by a liberal admixture of leaf-soil and sharp sand; whatever the nature of the soil, it should be moderately firm before planting.

The best time to plant is early in autumn, immediately after the foliage decays, dividing and selecting the crowns singly. For permanent beds that are likely to remain for several years without being disturbed, they may be planted 2 in. or even 3 in. apart, as they do not become so soon crowded as to require thinning out. Cover the surface after planting with 1 in. or 2 in. of rotten manure, bearing in mind that thorough maturity can only be insured by repeated applications of water—manure water being the most exhilarating. Treated thus, with annual surface-dressings of manure, the beds will keep in good condition for years, and produce fine blossoms in abundance. When the plants become crowded with shoots they should be thinned out, or, better still, lifted and replanted, for weakly, abortive crowns tend only to retard vigorous development. It is now largely forced into flower early, for which purpose the roots are usually imported from the Continent where they are grown and prepared expressly for the purpose. It may be naturalised on any place if sufficiently moist and shaded, and soon spreads its broad masses. There is a variety called rosea, the blossoms of which, however, are not rosy, but a purplish tinge, not nearly so pleasing as the white kind, and there is also a variety with gold-striped foliage, and another with double flowers.

Convolvulus *(Bindweed)*.—A handsome family of climbing herbs; very hardy and gracefully effective where properly used.

C. althæoides *(Riviera Bindweed)*.—This pretty Bindweed grows in many places around the basin of the Mediterranean, but is no less happy in an English garden. It is cheery in the colour of its rosy pink cups, and graceful and distinct in leaf and growth. It is hardy with us, and a good border or bank plant, or for the rough rock garden.

C. dahuricus *(Dahurian C.)*.—A showy twining perennial, producing in summer rosy purple flowers. Excellent for covering bowers, railings, stumps, cottages, &c., and also for naturalisation in hedgerows and copses. It grows in almost any kind of soil, and, like its relation the Bindweed, is readily increased by division of the roots, which creep. Native of the Caucasus.

C. mauritanicus *(Blue Rock Bindweed)*.—A prostrate, twining plant, with very slender stems. Flowers, a good blue, with a white throat and yellow anthers 1 in. across. The rock garden, and raised borders—always in sunny, somewhat raised positions, and in sandy, well-drained soil. A lovely N. Africa plant, and well repays attention. Division or cuttings.

C. pubescens fl.-pl. *(Double Bindweed)*.—A very handsome and useful plant for clothing trellises, stumps, porches, and rusticwork. It grows rapidly to the height of 6 ft. The flowers are large, double, and pale rose, and appear in June and onward, continuing for some months. It likes a light, rich soil and warm aspect. It may be grown in large pots, tubs, or boxes, for forming small bowers on balconies, or it may be grown to hide small fences, or to climb round posts. Division. China.

C. Soldanella *(Sea Bindweed.)* — A distinct trailing species with fleshy leaves, flowering in summer; pale red, large, handsome. The rock garden, planted so that its shoots may droop over the brows of rocks. Also in borders, in ordinary soil. Division. Europe and Britain.

C. sepium *(The Greater Bindweed)*. — This, although generally a great pest in gardens, may nevertheless be rendered a most useful and beautiful auxiliary in certain spots where it would not become a nuisance; but it is surpassed by the larger and stouter white.

C. sylvaticus, than which no plants form more beautiful and delicate curtains of foliage and flowers, and none grow more vigorously in any soil. The wild garden is the place where these are most at home, and where their vigorous roots may ramble without injury to other things. Among bushes or hedges, or over railings, or on rough banks, they are charming, and take care of themselves.

C. Scammonia *(Scammony)* is a beautitul plant of slender growth, producing throughout the summer a profusion of large white blossoms. It grows well in any position or any soil, and is a capital plant for rambling over roots, low shrubs, and such like objects. There are other species in gardens such as C. Cneorum, lineatus, Cantabrica, dorycnoides, and erubescens, but they are either too tender or unconspicuous for general cultivation.

C. tricolor.—One of the most beautiful of hardy annuals, too well known to need description. There are now numerous varieties in cultivation, varying more or less in the colour of the flowers or habit of growth. The colours of the flowers of the type are blue, yellow, and white, but there are varieties entirely white (albus), another striped violet and white (striatus) and another (splendens) with blossoms of a rich purple. Others again differ in habit of growth, such as unicaulis, which has the flowers crowded at the tips of the branches. The forms monstrosus, undulatus, quadricolor, kermesinus are desirable; in fact, every variety is worth growing. Being perfectly hardy, seeds may be sown in the open ground in September for flowering in spring, or sown in February in a heated frame for transplanting in May for midsummer flowering, and in the open ground from April to the end of May for flowering in late summer and autumn. It likes good warm soil not too dry or too wet, especially for the autumn-sown plants which have to stand the winter. (= C. minor.)

C. major *(= Ipomæa)*.

Cone Flower *(Rudbeckia)*.

Cooperia.—A small genus of bulbous plants of the Amaryllis family. There are two kinds in cultivation—C. Drummondi and C. pedunculata. The former has small bulbs, narrow leaves, and flower-stalks produced in late summer, terminated by a single flower having long tubular blossoms, pure white, and fragrant like a Primrose. The other (C. pedunculata) is somewhat similar, and is also a night-flowering plant. These are both natives of Texas, and are not perfectly hardy. They, however, thrive in the warmer parts of England, planted in sunny sheltered borders of light soil, but it is advisable to give these bulbs frame protection in winter. They are lovely plants and certainly deserve any care bestowed upon their culture. They thrive well in pots and in frames, but are as yet little known in general cultivation in this country.

Coprosma Baueriana variegata is a shrubby plant (tender), with shining green leaves broadly banded with creamy yellow, and is useful for margins of large beds or borders. It requires to be pegged down, when it soon grows into shape, and forms a very pretty margin. It is easily increased by means of cuttings put in in spring and early in summer, on a gentle bottom-heat under bell-glasses, and not allowed to get too dry. Often used in the flower garden with good effect. It should be taken up and housed in autumn to keep from severe frosts. New Zealand. Cinchona Order.

PLATE LXXIX.

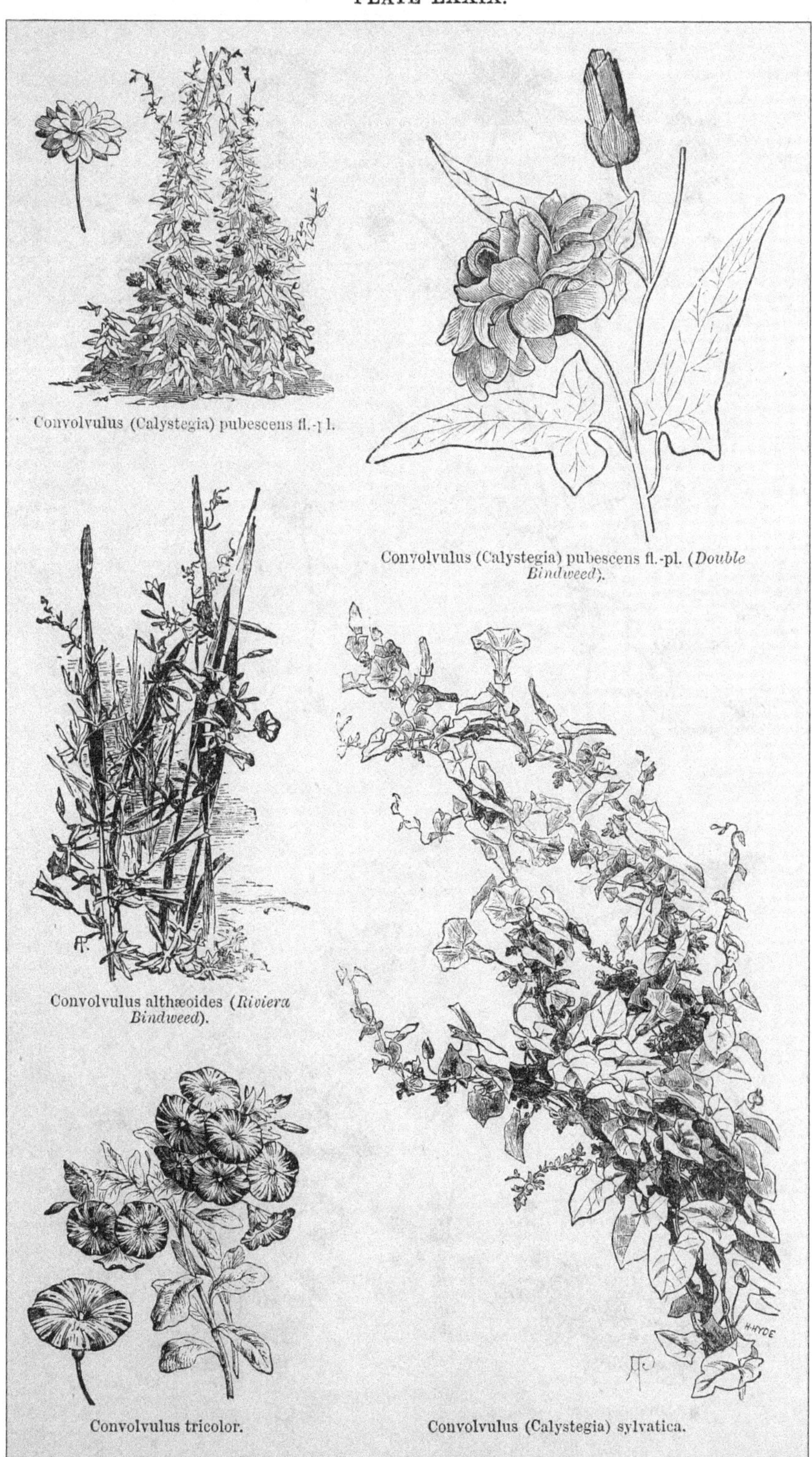

Convolvulus (Calystegia) pubescens fl.-pl.

Convolvulus (Calystegia) pubescens fl.-pl. (*Double Bindweed*).

Convolvulus althæoides (*Riviera Bindweed*).

Convolvulus tricolor.

Convolvulus (Calystegia) sylvatica.

[*To face page* **90.**

PLATE LXXX.

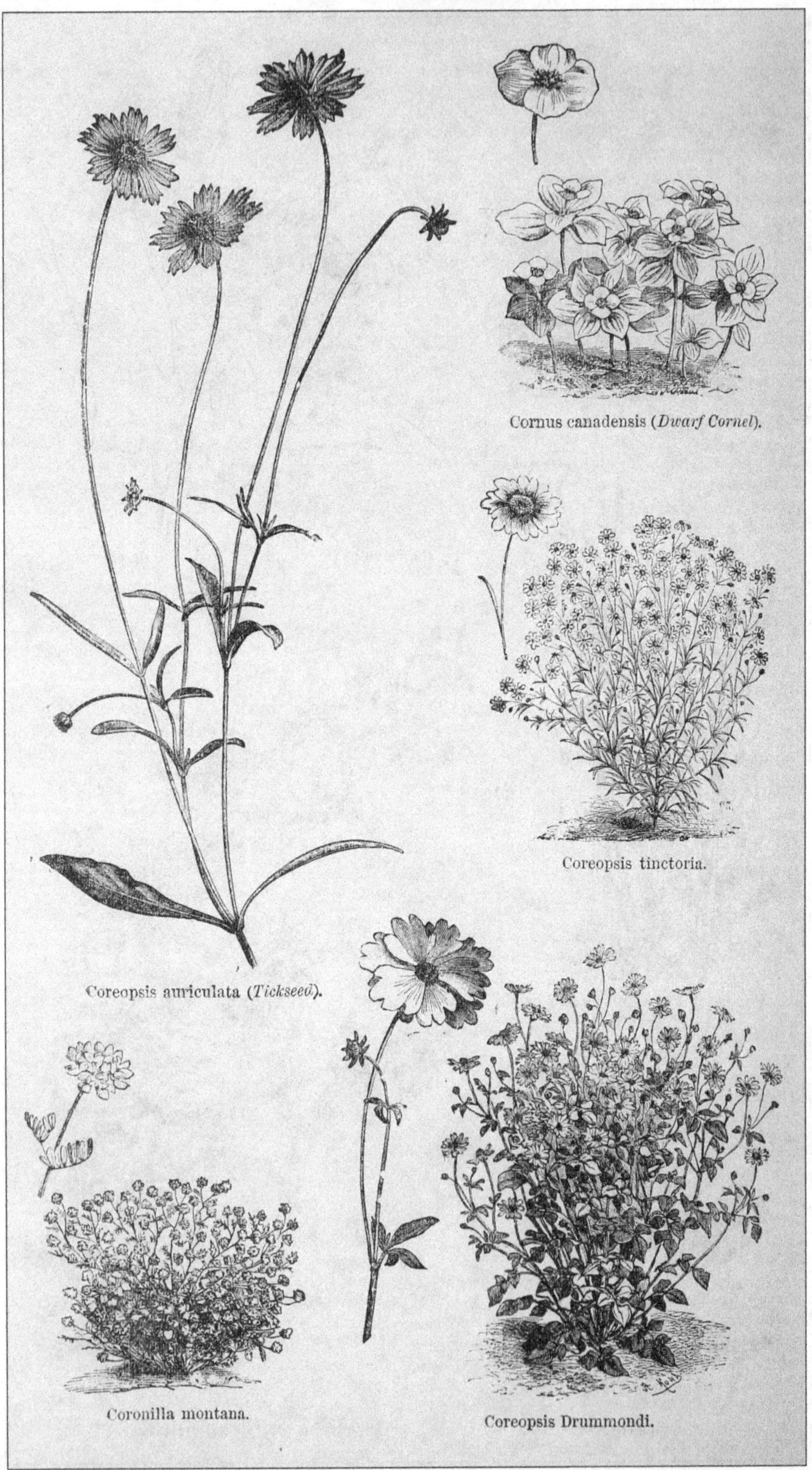

[*To face page* 91.

Coptis trifoliata *(Gold Thread)*.—A little evergreen bog plant 3 in. or 4 in. high, with three-leafleted or trifoliate shining leaves. It derives its common name from its long bright yellow roots. It is occasionally grown in botanic gardens. A native of the northern parts of America, Asia, and Europe, flowering in summer; white, and easily grown in moist peat or very moist sandy soil. Division.

Coral Root *(Dentaria bulbifera)*.

Coreopsis *(Tickseed)*.—This genus of North American Composites, which now includes Calliopsis, contains several important garden plants—the annuals showy summer flowers, the perennials valuable late-blooming plants. The choicest among the perennials is C. auriculata, a plant growing about 2 ft. high, with a spreading habit of growth, producing in autumn an abundance of rich yellow blossoms on slender stalks. A useful plant for cutting from; grows well in any ordinary soil, and freely propagated by seed or division. Nearly allied to this species and very similar to it is C. lanceolata, an equally showy plant, also delighting in a rich, damp soil. C. tenuifolia is a pretty plant, having elegant feathery foliage and rich golden yellow blossoms, produced from summer till autumn. C. verticillata is similar to it, and is likewise a showy border plant. Neither of these are so robust as the preceding, and therefore require more select spots, such as the front rows of a mixed border in the rougher parts of the rock garden. Other perennial kinds in cultivation are C. palmata, senifolia, nudata, maritima, tripteris, gladiata, and delphinifolia, but none can be recommended for general cultivation. The annual species are among the showiest of summer flowers, and most valuable for garden adornment. Being quite hardy, they are capable of making a fine display in spring from seeds sown in September, while an almost continuous display of bloom may be had from July to October by sowing successively from early March till the middle of June, in ordinary garden soil, that of a moist description being most preferable for the spring sowings. The following are the principal annual kinds; C. aristosa grows from 2 ft. to 3 ft. high, and bears large golden-yellow blossoms; C. Atkinsoniana, 1 ft. to 3 ft., flowers orange-yellow spotted with brown in centre; C. coronata, orange-yellow, with a circle of brownish crimson in centre; C. Drummondi, 1 ft. to 1½ ft. high, golden yellow; C. tinctoria, 1 ft. to 3 ft. high, flowers crimson-brown tipped with orange-yellow.

Coris monspeliensis *(Montpelier C.)*—A rather pretty dwarf branching plant, about 6 in. high, usually biennial in our gardens. South of France. Does on dry and sunny parts of rockwork, in sandy soil, and among dwarf plants. Seed.

Cornish Moneywort *(Sibthorpia europea)*.

Cornish Heath *(Erica vagans)*.

Corn Flower *(Centaurea Cyanus)*.

Corn Marigold *(Chrysanthemum segetum)*.

Cornus canadensis *(Dwarf Cornel)*.—A pretty miniature shrub, of which each little shoot is tipped with white bracts, pointed with a tint of rose. It is lost among coarse herbaceous plants, and totally obscured by ordinary shrubs, and should therefore be placed among alpine plants on a rockwork near the edge of a bed of very dwarf Heaths or American plants, or in the bog garden. It grows about the size of the Partridge Berry, or somewhat larger. North America, in damp cold woods. Growers of British plants may like to possess Cornus suecica, a tiny native species, which grows well in a bog or moist peat bed among very dwarf plants.

Coronilla, a small genus of the Pea family, consisting chiefly of shrubs, but containing at least two really good herbaceous plants, both of which are valuable for the rock garden, as well as for the mixed border. They are

C. iberica, growing about 1 ft. high, having a dense tuft of slender stems that trail on the ground or fall over the ledge of a rock in a graceful way. It makes a glorious show in early summer by its profusion of bright yellow blossoms, that rest on a cushion of deep green foliage. Its place is in the rock garden, where it delights to send its roots down the side of a big stone, in plenty of good soil, not less than 18 in. deep. A dozen or so of small plants, placed in groups of three or four in this position, will, in a year or two, result in glowing masses of colour that will compare with any alpine flower. It does well also on the margins of borders, but does not show itself so well as on a bank or in the rock garden. Propagated by cuttings inserted in early spring.

C. varia.—A very handsome, free, and graceful plant, with a profusion of pretty rose-coloured flowers. It is widely distributed on the Continent, and is found on many of the railway banks in France and Northern Italy. It forms low, dense tufts, sheeted with rosy pink, which attract the traveller's eye, their beauty and dressy appearance marking them among the weeds that inhabit such places. It ought to be grown in every garden as a border flower, or for naturalising in semi-wild spots. Perhaps, however, the most graceful use that could be made of it would be to plant it on some tall, bare rock, and allow its vigorous shoots and bright little coronets to teem over and form a lovely curtain down the face of the stone. It is also admirable for chalky banks, or for running about among low trailing shrubs like the common Cotoneaster. When in good soil the shoots grow as much as 5 ft. long, and it thrives on almost any sort of soil. It is readily increased by seeds, which are frequently offered in our seed catalogues. There is a fine deep rose-coloured variety named Hausknechti well worthy of culture, and another of compact growth called compacta. Other species are in cultivation, such as montana, libanotica, minima, valentina, ramosissima, and vaginalis, but they are only suited for botanical collections.

Cortusa *(Alpine Sanicle).*—Plants resembling Primroses of the cortusoides type, but, being inferior, are not so important since the introduction of the finer forms of this race. C. Matthioli is a Piedmontese plant, growing about 1 ft. high, and thriving in peat, in shady or half-shady spots. C. pubens is a more recently introduced kind, requiring the same treatment.

Corydalis *(Corydalis, Fumitory).*—A numerous family, but not many species worth cultivation, though some of these are important.

C. Ledebouriana *(Ledebour's Fumitory).*—The newest species of Corydalis in cultivation; it is unlike any other kind, on account of the peculiar glaucous leaves. These are arranged in a whorl about half-way up the stem, which grows from 9 in. to 12 in. high. The flowers are a deep vinous-purple, with pinkish spurs. It is an early and a hardy plant.

C. lutea *(Yellow Fumitory).*—This well-known plant is not so much esteemed as it deserves, for not only are its graceful masses of delicate pale green leaves profusely dotted over with spurred yellow flowers, pretty in borders, but it grows to perfection on walls, the tufts often being as full of flower when emerging from some chink in a fortress wall where a drop of rain never falls upon them, as when planted in fertile soil. It is well suited for the rougher kind of rock and root work. A naturalised plant in England, and widely spread over Continental Europe. Division or seeds. In any stony position it spreads about with weed-like rapidity.

C. nobilis *(Noble Fumitory).*—A distinct and handsome plant, 10 in. or 1 ft. high; the flower-stems are stout and leafy to the top, and bear a massive head of flowers of a rich golden yellow, with a small protuberance in the centre of each of a reddish chocolate colour, and the effect of this, with the yellow and the green rosette when the bloom is young, makes the plant very ornamental. It is easy of culture in light borders, but is rather slow of increase; where it does not thrive as a border plant, it should be placed in deep, light, and rich soil on the lower flanks of rockwork, associated with plants of the vigour and stature of the Vernal Adonis, the American Cowslip, and the Rocky Mountain Columbine. Siberia. Division; flowering in early summer.

C. solida *(Bulbous Fumitory).*—A compact tuberous-rooted species, from 4 in. to 6 in. or 7 in. in height, and freely producing dull purplish flowers. It has a solid bulbous root, is quite hardy, and of easy culture in almost any soil. A pretty little plant for borders, for naturalising in open spots in woods, and also for use in the spring garden. It is naturalised in several parts of England, but is not a true native, its home being the warmer parts of Europe. Division, flowers in April. (*Syn.*, Fumaria solida). Very like this species in appearance is C. tuberosa (= C. cava), but having the small floral leaves entire. It has a good white variety, C. tuberosa albiflora.

The preceding are the most distinct and useful species so far as yet known. C. aurea, C. bracteata, C. pallida, C. Marschalli, and C. Semenowi are among the other cultivated kinds, but less desirable for general cultivation.

Cotton Thistle *(Carduus eriophorus).*

Cotyledon Umbilicus *(Wall Navelwort).*—A native of Britain and Ireland and many parts of Western Europe, and very common on walls. Of little importance for cultivation, except perhaps now and then as a hardy fernery or bog plant.

Cousinia Hystrix.—A bold and singular plant, of botanical interest mainly. Compositæ.

Cowslip *(Primula veris).*

Cosmidium filiforme. — Similar to Coreopsis, having erect stems from 1 ft. to 2 ft. high, and producing in autumn numerous flowers 2 in. across, of a light orange-yellow. It thrives best in a light, sandy loam, and seeds should be sown in heat in February or March. It is a pretty border flower, but as there are so many perennial plants of the same stamp it may be dispensed with. A native of Texas. C. Burridgeanum is another species with yellow flowers, and a variety of it (atropurpureum) with dark purple flowers is a showy half-hardy annual of easy culture, and well worth growing.

Cosmos.—Mexican plants of the Composite family. One species, C. bipinnatus, is a handsome annual, growing from 3 ft. to 5 ft. in height, having finely-divided feathery foliage, and large, Dahlia-like blossoms of a bright red-purple with yellow centres. It requires to be treated as a tender annual. The seeds should be sown in February or March in a heated frame, and the seedlings transplanted in May in good, rich, moist soil, with a warm exposure. Flowers from August to October. Good for grouping with bold and graceful annuals, and better than many more popular ones.

Costmary *(Balsamita vulgaris).*

Cranesbill *(Geranium).*

Cranberry, American *(Vaccinium macrocarpum).*

Crambe cordifolia.—One of the finest of hardy and large-leaved herbaceous plants. It is as easily grown as the common Seakale, and in heavy rich ground makes a fine rich growth of leaves, surmounted in summer by a dense spray of small white flowers. In planting it, the deeper and richer the soil the finer the result. It may be planted wherever a bold though low type of vegetation is desired. There is another species, C. juncea, a dwarf kind, with white flowers and much-branched stems, the ramifications of which are very slender and elegant. This is also effective, but not so valuable as C. cordifolia.

Cortusa Matthioli (*Alpine Sanicle*).

Corydalis nobilis.

Corydalis bulbosa (*Bulbous Fumitory*).

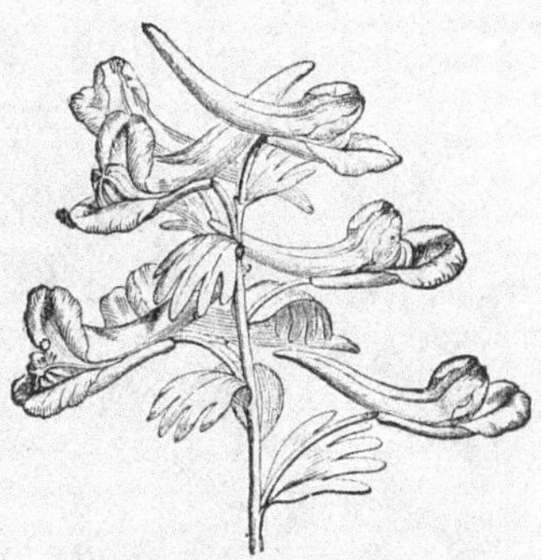

Flowers of Corydalis solida.

Corydalis lutea (*Yellow Fumitory*).

Cosmidium Burridgeanum.

[*To face page* **92.**

PLATE LXXXII.

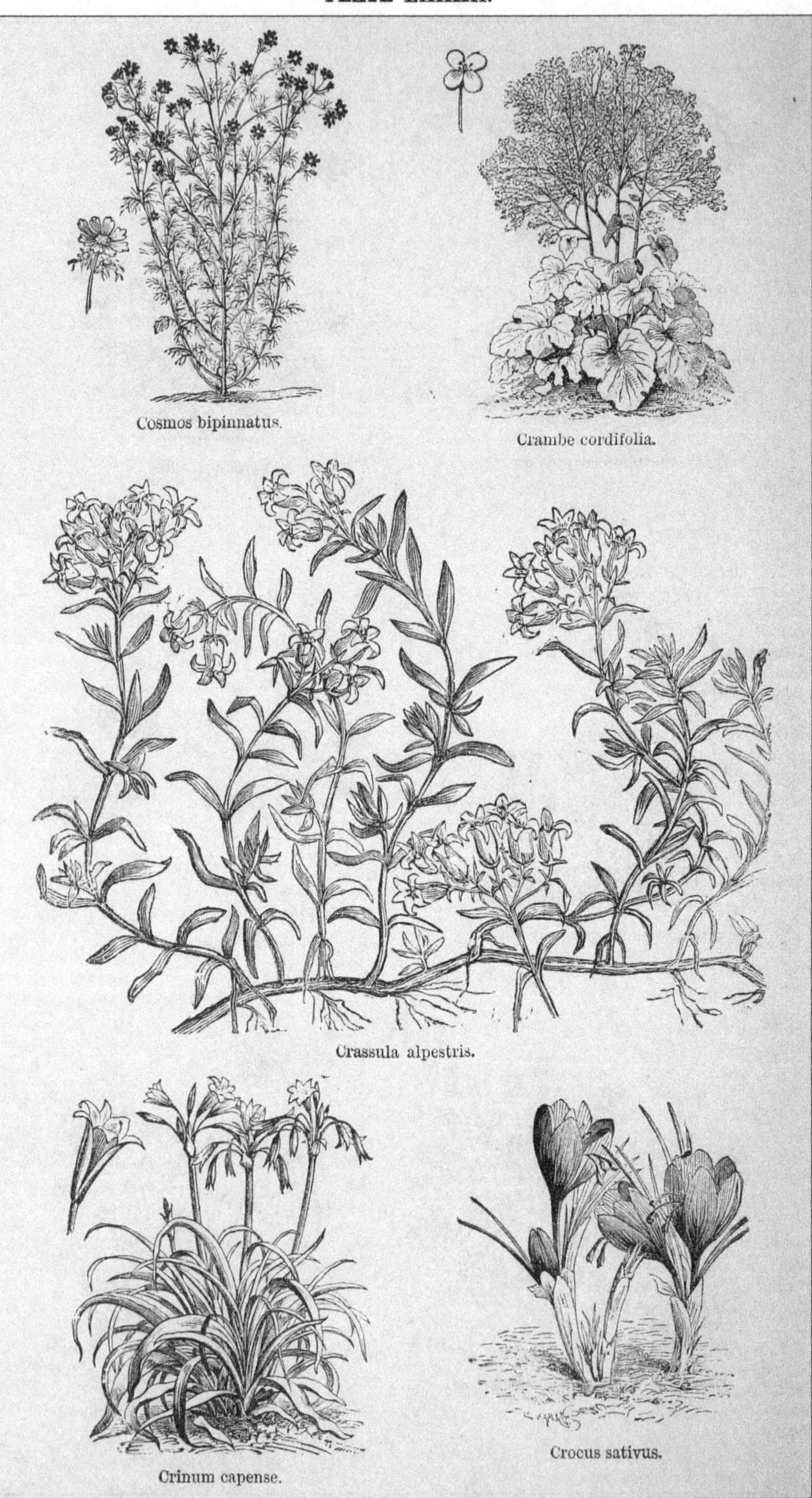

Cosmos bipinnatus.

Crambe cordifolia.

Crassula alpestris.

Crinum capense.

Crocus sativus.

[*To face page* 93.

Crassula alpestris.—A pretty rock plant which has lately been brought into notice in this country by Herr Max Leichtlin, of Baden-Baden, who sent it to Kew a year or two ago under the name of Sedum alpestre. It is a neat-habited perennial, with creeping and slightly branched stems, which are of a bright reddish tint. The flowers are borne in terminal clusters; they are white, of wax-like substance, and have orange-red tipped stamens, producing a pretty effect. Though it is a native of the Cape of Good Hope, from whence we have so many tender species of Crassula, it is quite hardy enough to live unprotected in this climate, as was proved in the Kew collection. It thrives well in a thoroughly drained and open position on rockwork, and continues to flower throughout the summer.

Crepis.—A small genus of Composites, of which there is but one that is worth growing. This is C. aurea, a perennial, growing from 6 in. to 12 in. high, and bearing small orange-coloured blossoms, but seldom more than one on each slender stem. Borders or banks. Division and seed.

Creeping Jenny (*Lysimachia Nummularia*).

Creeping Vervain (*Zapania nodiflora*).

Creeping Wintergreen (*Gaultheria procumbens*).

Crinum.—A few of the South African species of this genus are quite hardy in this country, and very beautiful plants they are. One of the best known and the hardiest is C. capense, a noble bulbous plant, growing from 2 ft. to 3 ft. high. It flowers late in summer, the blossoms being large, funnel-shaped, pink, produced in umbels of ten or fifteen blooms terminating a stout erect stem. There are several varieties: album, pure white; riparium, deep purple; fortuitum, white; and striatum, striped pink and white. Several fine hybrids have also been raised. They are specially adapted for growing in isolated tufts or small beds in the pleasure ground, arranged with groups of hardy fragrant plants or with the nobler herbaceous subjects, particularly those that flower in late summer and early autumn; also for grouping and massing on small islands or parts of islands on which a distinct and choice type of vegetation is sought, and near the margin of water. Few plants repay better for a sheltered and warm position, and deep, very rich soil, with abundance of water in summer. In very cold situations a little pile of leaves may be desirable over the roots in winter, and by planting the top of the bulbs ½ ft. beneath the surface there need not be any fear of injury from the weather. Division and seed. C. campanulatum is also a hardy species, but as it is rare, and scarcely ever flowers, it is not worth growing in the open. C. Moorei and ornatum are said to be hardy in Ireland, and there are certain hybrids of recent origin which may prove hardy.

Crocosma aurea (*Tritonia*).

Crocus.—Of a genus consisting of nearly seventy species, it is a matter of surprise that but three or four only are generally used for garden decoration; and these—Crocus aureus and Crocus vernus and their varieties, and perhaps one or two other species—appear to have been in cultivation at least two or three hundred years. Crocuses, both vernal and autumnal, flower at a time when every flower is of value; and we do not doubt that ere long the little known species of the genus that have been recently introduced will add largely to our means of garden decoration during the dull months from late autumn to early spring.

CULTURAL DIRECTIONS for a genus so well known and so easily grown seem almost superfluous; but there are a few points to which it may be convenient to refer in dealing with the Croci as decorative plants. Taking the whole genus of about seventy species, they must be viewed as in continuous succession, from the beginning of August till April; but of these it is only the earlier autumnal, or the distinctly vernal, species that can be relied upon in our climate for open air garden decoration. Although all are hardy, and most of the winter flowering species will flower in the open ground, those that flower in November, December, and January are so liable to injury by frost and rain, that they are practically worthless as decorative plants for the open garden.

Crocuses are easily multiplied by seed, which should be sown as soon as ripe in July, though germination will not take place till the natural growing period of the species. Seedlings take from two to three years to arrive at maturity, and should be left for the first two years undisturbed in the seed bed, and then taken up and replanted. Holland, with its rich, light, alluvial soil and Lincolnshire, with its "Trent warp," have been for many generations the sources from which the English market has been supplied with the varieties of the three or four species grown in English gardens. The last five or six years have put us in possession of nearly the whole of the known species of the genus, and we must commend them to the care of the Dutch and Lincolnshire bulb growers wherewith to further enrich our collections.

For the less robust and less floriferous species, the protection of a brick pit is necessary. The bottom of this should be well below the level of the ground, and it should be filled up with about 1 ft. in depth of fine river silt or sandy loam, the surface of which should be a little below the level of the surface of the ground adjacent to the pit. Proper drainage is essential, but this being attained, Crocuses during their period of growth delight in a uniformly moist subsoil. It is convenient to separate each species by strips of slate or tiles, which may be buried below the surface, and the corms planted about 3 in. deep. A mulching of rotted Cocoa-nut fibre or finely-sifted peat keeps the surface uniformly moist, and prevents the substratum of loam from clogging or caking on the surface. At the time of the maturity of the foliage, which generally takes place about the end of May, water should be withheld and the Crocus bed covered up and allowed to get quite dry till the end of July, when a copious watering may be given, or the pit may be exposed to natural rainfall.

Of the earlier autumnal species suitable for the open border the following may be enumerated for successional flowering:—

C. Scharojani, orange; early in August.
„ vallicola, straw coloured; late in August and early in September.
„ nudiflorus, blue; September.
„ pulchellus, lilac; Sept. and Oct.
„ speciosus, blue; Sept. and Oct.
„ iridiflorus, blue; Sept. and Oct.
„ Salzmanni, „ Clusi } lilac or blue; October and November.
„ cancellatus, „ Cambessidesii, „ hadriaticus } in the early autumn.

These are succeeded by a long series of late autumnal, winter, and early vernal species, which are best grown to advantage under the protection of a brick pit.

Of the vernal species suitable for the border, the earliest is C. Imperati, flowering in February, followed by

C. susianus, or Cloth of Gold, in February.
„ biflorus
„ etruscus
„ suaveolens
„ versicolor
„ vernus
„ Tommasinianus
„ dalmaticus
„ banaticus
„ Sieberi and var. versicolor
„ chrysanthus
„ aureus
„ sulphureus
„ vars. pallidus and striatus
„ stellaris
„ Olivieri
„ minimus
} Flowering from the end of February to the first week in April.

Of the Croci but recently introduced, many of the vernal species will probably be found suitable for spring garden decoration, but as they are rare and scarcely procurable we give those which are more generally known and easily obtainable.

C. alatavicus.—The flowers of this new Asiatic species are white, yellow towards the throat, the outer surface of the outer segments being freckled with rich purple. It is a free flowering species, but from its early flowering time, January and February, it can only be grown to advantage under the protection of a cold frame. A white variety without the external purple freckling is not uncommon. The leaves are produced at the flowering time in the early spring.

C. aureus.—A handsome plant; native of the Banat, Transylvania, European Turkey, Greece, and Western Bithynia, generally appearing at low elevations, flowering in February. It was one of the first species introduced to cultivation, and the parent of our yellow garden or Dutch yellow Crocus, and of a number of old horticultural varieties—lacteus, sulphureus, sulphureus pallidus, sulphureus striatus, &c., the history of which is unknown; they are not known in a wild state, and are all sterile. The wild plant varies considerably, from unstriped orange to varieties striped with grey lines, like those occurring in the Dutch yellow Crocus. The stigmata are short, unbranched, pale yellow, and much shorter than the anthers; in the Transylvanian plant the stigmata are occasionally orange. The anthers are wedge-shaped, tapering towards the point, and notably divergent. C. aureus, the unstriped form, produces seed readily in cultivation, but the striped Dutch yellow is always sterile, though effete capsules are occasionally formed. C. Olivieri resembles C. aureus, but is smaller.

C. banaticus. — This is a common species in the Banat, Hungary, and Transylvania, where it takes the place of C. vernus, to which it is allied. It is a highly ornamental plant; the flowers are of a deep rich purple, occasionally varied with white, with a darker purple blotch near the end of the segments. The throat is glabrous, by which it is easily distinguished from C. vernus. It is cultivated in several Continental and English gardens under the name of C. veluchensis, which is, however, a distinct species. The flowers are produced in February and March.

C. biflorus.—The Scotch, or Cloth of Silver, Crocus, is a large variety of the typical form of the species, which is abundant throughout a large portion of Italy. The segments vary from white to a pale lavender, the outer surface of the outer segments being distinctly feathered with purple markings. In the var. estriatus, from Florence, the flowers are of a uniform pale lavender tint, orange towards the base. In var. Weldeni, from Trieste and Dalmatia, the outer segments are externally freckled with bright purple. C. nubigenus is a very small variety from Asia Minor, in which the outer segments are suffused and freckled with brown; C. Pestalozzæ is an albino of this variety. In C. Adami, from the Caucasus, the segments are pale purple, either self-coloured or externally feathered with dark purple. C. biflorus is an early flowering spring species, and a highly ornamental plant for border decoration.

C. Boryi.—In this species the flowers are white, bright orange at the throat. It is an abundant species at Corfu and in the neighbourhood of Patras, flowering in October, but it does not bloom freely in cultivation, requiring the protection of glass for the proper development of its flowers.

C. cancellatus.—A beautiful autumnal species, varying in the colouring of the flowers, from white to pale bluish purple. The flowers are generally veined or feathered towards the base of the segments. The flowers appear without the leaves, which succeed in the spring. The flowering time of C. cancellatus is from the end of October to December. It is a robust species, easy of culture, but like many other late autumnal species it is best seen to advantage under the protection of a cold frame. C. cancellatus is known under a variety of names, viz., C. Schimperi, C. Spruneri, C. cilicicus, and C. damascenus. The western forms are nearly white, and those occurring in the east of the area of distribution either blue or purple; but the differences of colour are not sufficient to distinguish them as species.

C. chrysanthus.—A vernal Crocus, flowering from January to March according to its range of elevation, which is considerable, varying from a little above the sea-level to an altitude of 3000 ft. or 4000 ft. The flowers are smaller than those of C. aureus, usually bright orange, but occasionally bronzed and feathered externally. A white variety is also found in Bithynia and on Mount Olympus above Broussa; this species also varies with pale sulphur-coloured flowers, occasionally suffused with blue towards the ends of the segments, dying out towards the orange throat. There are four varieties of this Crocus distinct in colouring; they are fusco-tinctus, fusco-lineatus, albidus, and cœrulescens.

C. Imperati.—This is one of the earliest vernal species, abundant in the district south of Naples, and said to extend into Calabria. It is very variable in its colour and markings. A self-coloured white variety occurs near Ravello, and one in which the normal lilac tint of the flower is exchanged for a clear rose colour. The outer surface of the outer segments is coated with rich buff, suffused with purple featherings. C. Imperati, from its robust habit and early flowering time, is one of the most valuable species for spring decoration. It flowers from a fortnight to three weeks before C. vernus. Similar to C. Imperati is C. minimus, an abundant species on the sea-board of the western coast of Corsica, the neighbouring islets, and in parts of Sardinia; it flowers from the end of January to March. The flowers resemble those of C. Imperati in miniature, but are smaller, of a darker purple colour, and more heavily suffused with external brown featherings. Although perfectly hardy, it is not a robust plant capable of use for garden decoration. C. suaveolens is also closely allied to C. Imperati, flowering in February. The flowers are somewhat smaller and the segments more acute than in C. Imperati. It is a species of hardy and free-flowering habit, and under bright sunshine forms a highly ornamental object in the early spring garden.

C. iridiflorus.—A highly ornamental species from the Banat and Transylvania, producing its bright purple flowers before the leaves in September and October; remarkable for its purple stigma and the marked difference in size of the inner and outer segments of the perianth. This is a beautiful plant, and should always be secured if possible. It is often sold under the name of C. byzantinus.

C. lævigatus.—A pretty species from the mountains of Greece and the Cyclades. The flowers vary from white to lilac, being distinctly feathered with purple markings. Its usual flowering time is from the end of October to Christmas, but under cultivation through the winter up to March. It does not flower freely in cultivation, and, like the other allied species, it is best seen to advantage under the protection of a cold frame.

C. longiflorus is abundant in the south of Italy, Sicily, and Malta; it flowers in October. The flowers are light purple, yellow at the throat. In general aspect this species somewhat approaches the character of C. sativus, especially in the stigmata, which are usually bright scarlet and entire, but occasionally broken up into fine capillary divisions. The stigmata are collected in Sicily from the wild plant for saffron. It is a free flowering and very ornamental species.

C. medius.—A beautiful purple autumn flowering species, limited to the Riviera and the adjacent spurs of the Maritime Alps. The flowers, produced in October before the leaves, which appear in the following spring, are rarely more than two to three to a corm; the blossoms are bright purple, veined at the base; the stigmata bright scarlet and much branched.

C. nudiflorus.—A pretty and well known species, a native of the Pyrenees and north of Spain, which has become naturalised at Nottingham and in other localities in the midland counties. Its large bluish-purple flowers are produced in September and October before the leaves. Where once established it is difficult to eradicate; the corms produce long stolon-like shoots, which form independent corms on the death of the parent, and by this means the plant soon spreads to considerable distances from where originally planted.

C. ochroleucus has creamy white flowers with orange throat, abundantly produced from the end of October to the end of December. It well repays the protection of a cold frame, which preserves its showy flowers from injury by frost and rain.

C. pulchellus.—An autumnal species, invaluable for decorative purposes. The flowers, of a pale lavender colour, with a bright yellow throat, are freely produced from the middle of September to early in December. It is readily multiplied from seed.

C. serotinus.—A native of the south of Spain, flowering in November. The blossoms are more or less distinctly feathered with darker purple. C. Salzmanni is closely allied to C. serotinus, but of larger stature, flowering with the leaves in October and November. It is a plant of robust habit and ready of multiplication. As the flowers are liable to injury by frost and snow in our climate, it is best seen to advantage under the protection of a cold frame. C. Clusi closely resembles in aspect C. serotinus, and flowers with the leaves in October.

C. Sieberi.—A common vernal species in the mountains of Greece and the Greek Archipelago. The flower is usually bright lilac, orange at the base, but the form found in Crete and the Cyclades presents a great variety of colouring, from white to purple, and these colours are mottled, intermixed, and striped in endless variety, contrasting with the bright orange throat. The Cretan variety is a plant of exceptional beauty. It flowers in cultivation from the end of February to the middle of March.

C. speciosus.—This is among the handsomest of the autumn Crocuses, flowering at the end of September and early in October. It has a wide geographical range, extending from North Persia, through Georgia, the Caucasus, and the Crimea into Hungary. The perianth segments, 2 inches high, are of a rich bluish purple, suffused with darker purple veins, with which the bright orange, much divided stigmata form a beautiful contrast. It has been long in cultivation, and readily multiplies by means of small bulbels produced at the base of the corm.

C. susianus.—This, the well-known Cloth of Gold Crocus, was an early importation from the Crimea. Both the orange and bronzed susianus are among the earliest vernal Crocuses, flowering in the open border in February. C. stellaris is an old garden plant, somewhat resembling C. susianus. The flower is orange, distinctly feathered with bronze on the outer coat of the outer segments. It is sterile, never producing seed. It flowers early in March.

C. vernus was one of the earliest cultivated species. It is a native of the Alps, Pyrenees, Tyrol, Carpathians, Italy, and Dalmatia, and has been naturalised in several English localities, but is not truly indigenous. It is remarkable for the great range of the colouring of its flowers, the endless varieties from pure white to deep purple being generally intermixed in its native habitats, and correspond with the horticultural varieties which decorate our gardens. The flowers are produced from early in March at low elevations to as late as June and July in the higher Alps. C. vernus is the parent of nearly all the purple, white, and striped Crocuses grown in Holland.

C. versicolor.—This well-known species has been for many years in cultivation. The flowers present a great variety of colouring, from purple to white, and various kinds of striping and feathering. It differs from the two preceding species in having the whole of the perianth segments similarly coloured, in which the external buff coating found in C. Imperati and C. suaveolens is absent. Its flowering time is March.

C. zonatus.—A species from the mountains of Cilicia, with bright vinous lilac flowers, golden at the base, abundantly produced about the middle of Sepember. It is a highly ornamental and free flowering species easy of culture. The flowers are produced before the leaves, which do not appear till the spring. It has been in cultivation about seven or eight years.

The above account of the genus is condensed to suit the plan of this work, from an article on the family, by Mr. Geo. Maw, of Benthall Hall, near Broseley, in *The Garden* of January 28, 1882. The paper contains a full account of the family with descriptions of other species not at present in cultivation, botanical authorities, and fuller technical descriptions.

Croomia pauciflora.—A North American plant of little garden value.

Crucianella stylosa *(Fœtid C.)*.—A A pretty dwarf herb, with leaves arranged in whorls 9 in. to 12 in. high. Flowering in summer; pale rose, small, but freely produced. Persia and the Caucasus. Thrives on borders, rockwork, or bare banks, in sandy or calcareous soil, but is offensive from its fox-like odour in certain states of the atmosphere, and for that reason does not deserve a place in the choice garden. Division.

Cuckoo Flower *(Cardamine pratensis)*.

Cucumis.—There are certain of the hardier species of this genus worthy of a place in the garden, as, for example, the Gooseberry Gourd (C. grossularoides), but they have no essential place in gardens.

Cucubalus.—A small genus of the Catchfly family, only suitable for botanic gardens.

Cucurbita *(Gourd)*.—The Gourd tribe is capable, if properly used, of adding much remarkable beauty and character to the garden; yet, as a rule, it is seldom used. There is no Natural Order more wonderful in the variety and singular shapes of its fruit than that to which the Melon, Cucumber, and Vegetable Marrow belong. From the writhing Snake Cucumber, which hangs down 4 ft. or 5 ft. long from its stem, to the round enormous Giant Pumpkin or Gourd, the grotesque variation, both in colour and shape and size, is marvellous. There are some pretty little Gourds which do not weigh more than $\frac{1}{4}$ oz. when ripe; while, on the other hand, there are kinds with fruit almost large enough to make a sponge bath. Eggs, bottles, gooseberries, clubs, caskets, folded umbrellas, balls, vases, urns, balloons, all have their likenesses in the Gourd family. Those who have seen a good collection of them will be able to understand Nathaniel Hawthorne's

PLATE LXXXIII.

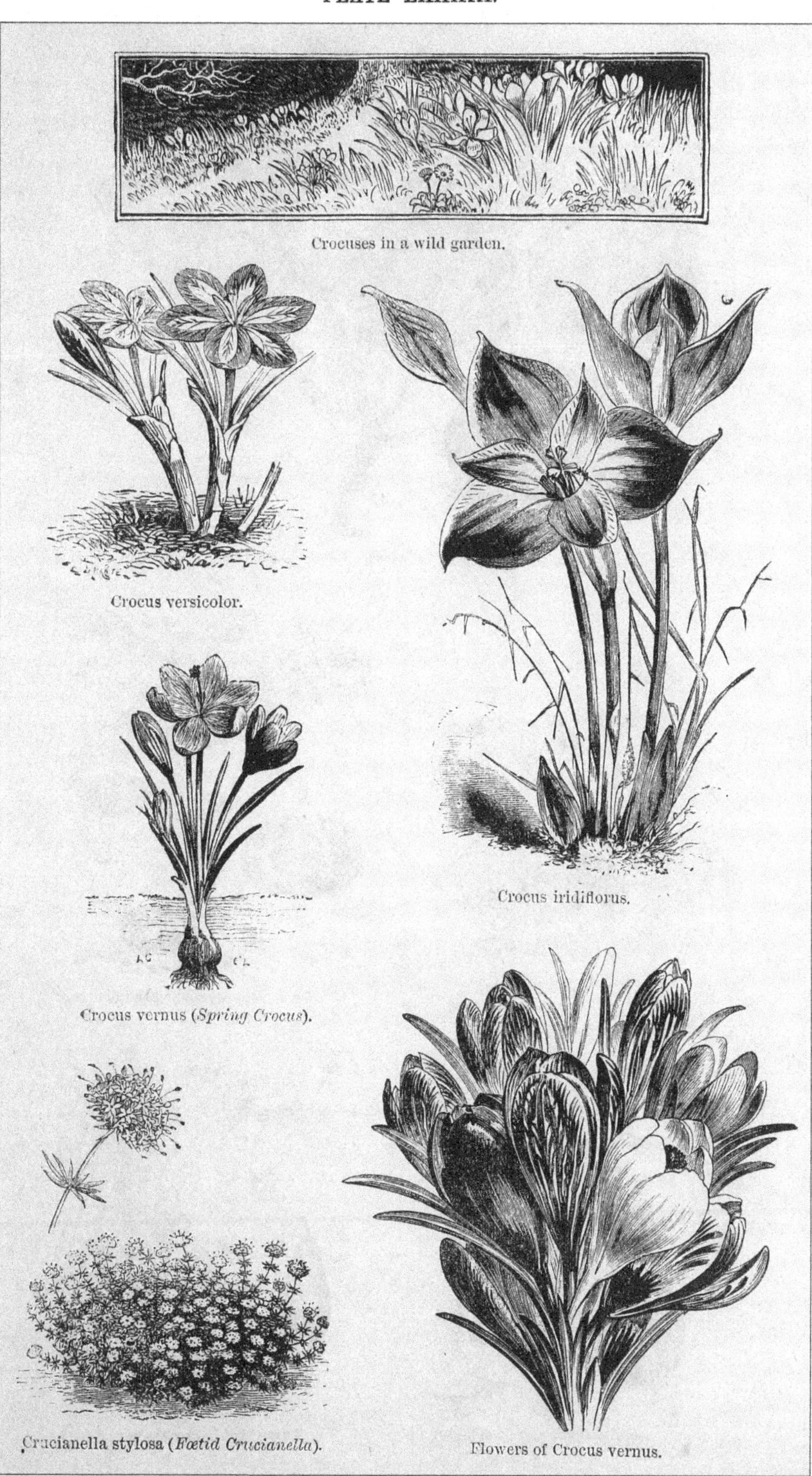

Crocuses in a wild garden.

Crocus versicolor.

Crocus iridiflorus.

Crocus vernus (*Spring Crocus*).

Crucianella stylosa (*Fœtid Crucianella*).

Flowers of Crocus vernus.

[*To face page* **96.**

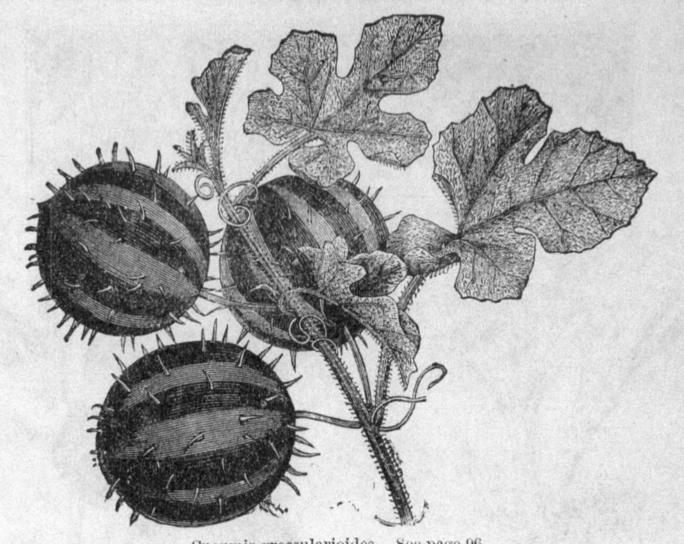

Cucumis grossularioides. See page 96.

Cucurbita versicolor.

Cucurbita depressa striata.

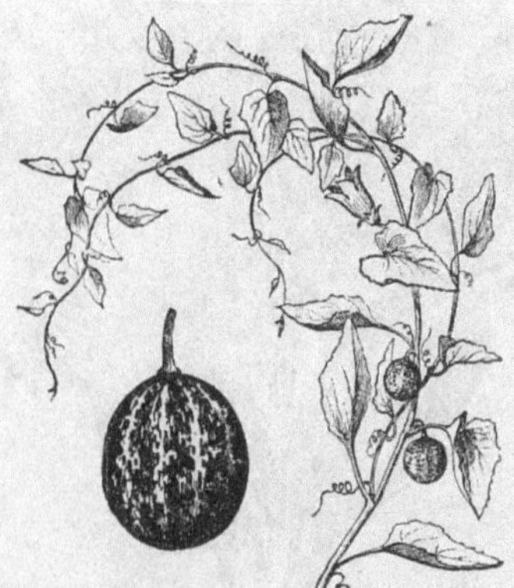

Cucurbita perennis (*Perennial Cucumber*).

Cucurbita Lagenaria var.

[*To face page* 97.

enthusiasm about these quaint and graceful vegetable forms when he says: "A hundred Gourds in my garden were worthy, in my eyes at least, of being rendered indestructible in marble. If ever Providence (but I know it never will) should assign me a superfluity of gold, part of it should be expended for a service of plate, or most delicate porcelain, to be wrought into the shape of Gourds gathered from Vines which I will plant with my own hands. As dishes for containing vegetables they would be peculiarly appropriate. Gazing at them, I felt that by my agency something worth living for had been done. A new substance was born into the world. They were real and tangible existences which the mind could seize hold of and rejoice in." It is satisfactory to know that they may be readily and beautifully grown in this country. There are many positions in gardens in which they might be grown with great advantage—on low trellises, depending from the edges of raised beds, the smaller and medium-sized kinds trained over arches or arched trellis-work, covering banks, or on the ordinary level earth of the garden. Isolated, too, some kinds would look very effective, and in fact there is hardly any limit to the uses to which they might be applied. In the Royal Botanic Gardens at Dublin there is a singular wigwam made by placing a number of dead branches so as to form the framework, and then planting Aristolochia Sipho all round these. It runs over them, and the large leaves make a perfect summer roof. A similar tent might be made with the free-growing Gourds, and it would have the additional merit of suspending some of the most singular, graceful, and gigantic of all known fruits from the roof. A very good way to grow them is trained over sheds or outhouses. These "carry" the foliage and showy fruit so well that the eye enjoys the picture. A bold and effective use may now and then be made of them on walls.

A SELECTION OF GOURDS.—Amongst the most beautiful are the Turk's-cap varieties, such as Grand Mogul, Pasha of Egypt, Viceroy, Empress, Bishop's Hat, &c.; the Serpent Gourd, Gooseberry Gourd, Hercules' Club, Gorilla, St. Aignan, Mons. Fould, Siphon, Half-moon, Giant's Punchbowl, and the Mammoth, weighing from 170 lb. to upwards of 200 lb.; while amongst the miniature varieties the Fig, Cricket-ball, Thumb, Cherry, Striped Custard, Hen's-egg, Pear, Bottle, Orange, Plover's-egg, &c., are very pretty examples, and very serviceable for ornament. All these are well adapted to the climate of England, and there are many others equally suitable—a fact sufficiently indicated in one collection shown by Mr. W. Young, which consisted of 500 varieties, all English grown, the greater number of which were sown where grown, and came to maturity without protection. The ground being manured and dug one spit deep, the seed was sown the second week in May, and from first to last many of the plants had no water supplied to them through the season. Others, by way of experiment, had it in various quantities—the more water was given, the larger, the freer, and the better the produce. Sowing in a frame at the end of April, and exposing them to the free air during the day so as to prevent them being drawn, and then removing the frame altogether to harden them off before planting out, would be the best way to secure an early growth of Gourds. Sowing in the open ground under hand-lights would also do, but not so well. Where there are waste heaps of rubbish or manure it is a good plan to cover them with a free growth of Gourds. Although they will grow under the conditions described above, they do best with plenty of manure, and should be mulched or well watered if the soil be not deep and rich.

C. perennis *(Perennial Cucumber)*.—A remarkable-looking and vigorous hardy trailing plant, with large hoary leaves, chiefly valuable for the botanical collection. Seed and careful division.

Cuphea.—An interesting genus of plants of the family Lythraceæ, of which C. platycentra (Cigar Plant) is perhaps the most useful for the summer flower garden. It is a dwarf compact bush, about 12 inches high, and bears a profusion of vermilion tube-shaped blossoms, with black lower lip and edged with white on the tip. Easily propagated by cuttings taken in September or April, and put in slight heat like Verbenas. It may also be raised from seed sown in heat in spring. C. strigulosa is a pretty variety, useful for planting out in single plants in the mixed border for cutting from, but chiefly used as a pot plant for autumn greenhouse flowering. C. Zimapani is a most useful annual, nearly allied to the old-fashioned C. silenoides, but better. It grows about 1½ ft. high, has numerous branches arising from the base, and when but a few inches high in early summer it begins to produce its flowers, which are of a rich deep purple colour bordered with a lighter hue, resembling those of an ordinary Sweet Pea, and are about the same size, and appear till the frosts cut the plants down. They are well fitted for cutting, as the branches continue to lengthen and the flowers to expand for a long time when placed in water. Altogether, it is a very desirable plant, especially in autumn, when most other annuals are gone, and flowers even of perennials are scarce. Other kinds are C. eminens, Galeottiana, miniata, ocymoides, purpurea, Roezli, and silenoides, all plants of not so great importance for the flower garden as those before named.

Cushion Pink *(Silene acaulis)*.

Cyananthus lobatus *(Lobed C.)*.—A brilliant and remarkable Himalayan rock plant, about 4 in. high. Flowering in August and September; purplish blue with a whitish centre. Thrives in the rock garden in sunny chinks. It grows best in a mixture of sandy peat and leaf-mould, with plenty of moisture during the growing season. Increased freely by cuttings. The seed requires a dry, favourable season to ripen it; in wet weather the large, erect, persistent calyx becomes filled with water, which remains and rots the included seed vessel. Polemoniaceæ.

C. incanus.—This differs from C. lobatus, being much more floriferous; like that species, it should be planted in a dry, sunny, well-drained position, as, if the situation be too damp, the fleshy rootstock is liable to rot. It is a good plan even to place something over the plant during the resting season. The flowers are not so large as those of the other species, but they are more charming in colour, which is enhanced by the white tuft of sericeous hairs in the throat of the corolla. Polemoniaceæ.

Cyclamen (*Sowbread*).—We are so much accustomed to see the bright flowers of the Persian Cyclamens in our greenhouses, that many never think of Cyclamens as hardy flowers. They are, however, excepting the Persian one, as hardy as Primroses; but they love the shelter and shade of low bushes or hill copses, where they may nestle and bloom in security. That they are hardy is not a reason why they should thrive in a bare exposed border. In such places as they naturally inhabit there is usually the friendly shelter of Grasses or branchlets about them, so that the large leaves are not torn to pieces by wind or hail. Thus, for example, the Ivy-leaved Cyclamen is in full leaf throughout the winter and early spring, and for the sake of the beauty of the leaves alone it is desirable so to place the plants that they may be saved from injury. By acting on these considerations it is easy enough to naturalise the hardier kinds of Cyclamen in many parts of the country. Good drainage is necessary for the successful culture of Cyclamens in the open air. The species grow naturally among broken rocks and stones mixed with vegetable soil, grit, &c., and are therefore not liable to be surrounded by stagnant water. Mr. Atkins, of Painswick, who has paid much attention to the culture of Cyclamens, and has succeeded with them in a very remarkable degree, thinks that the tuber should in all cases be buried beneath the surface of the earth, and not exposed as in the case of the Persian Cyclamen grown in pots. His chief reason for this opinion is that in some species the roots issue from the upper surface of the tuber only. They enjoy plenty of moisture at the root at all seasons, and thrive best in a rich, friable, open soil, with plenty of well-decayed vegetable matter in it. They are all admirably adapted for the rock garden, enjoying warm, sheltered nooks, partial shade and shelter from dry, cutting winds. They may be grown on any aspect if the essential conditions above mentioned be secured for them, but an eastern or southeastern aspect is best, always provided there is partial shade. We have seen them growing under trees among Grass, where they flowered profusely every year without attention.

Cyclamens are best propagated by seed sown, as soon as it is ripe, in well-drained pots of light soil. Cover the surface of the soil after sowing with a little moss, to ensure uniform dampness, and place them in a sheltered spot out-of-doors. As soon as the plants begin to appear, which may be in a month or six weeks, the Moss should be gradually removed. When the first leaf is tolerably developed, they should be transplanted about 1 in. apart in seed pans of rich light earth, and encouraged to grow as long as possible, being sheltered in a cold frame, with abundance of air at all times. When the leaves have perished the following summer, the tubers may be planted out or potted, according to their strength.

From the earliest times there appears to have been great difficulty felt by our best botanists in clearly defining the species of Cyclamen, from the great variation in shape and colouring of the leaves both above and below. Too much dependence on these characters has been the cause of much confusion and an undue multiplication of species. Some of the varieties of this genus become so fixed, and reproduce themselves so truly from seed, as to be regarded as species by some cultivators. The following are the more important species and varieties.

C. Atkinsi.—A hybrid variety of the Coum section. The flowers are larger than in the type, varying in colour from deep red to pure white; produced plentifully in winter.

C. cyprium.—This well-defined species has rather small heart-shaped leaves of a dark green colour, marbled on the upper surface with bluish-grey, and of a deep purple colour beneath. The flowers, which are pure white, tinted with soft lilac (the restricted mouth being spotted with carmine-purple), are well elevated above the foliage, a character which distinguishes it from most of its allies, except C. persicum, from which its foliage serves to distinguish it at a glance. It is one of the most chaste and beautiful of all the hardy kinds. A native of the isle of Cyprus and other places in South Europe. It is found in mountainous districts growing on shaded rocks. (=C. neapolitanum.)

C. Coum (*Round-leaved Cyclamen*).—This, with the others of the same section, though perfectly hardy and frequently in bloom in the open ground before the Snowdrop, yet, to preserve the flowers from the effects of unfavourable weather, will be the better for slight protection, or a pit or frame devoted to them in which to plant them out. Grown in this way during the early spring, from January to the middle of March, they are one sheet of bloom. When so cultivated it is best to take out the soil, say 1½ ft. to 2 ft. deep, place a layer of rough stones 9 in. to 12 in. deep at the bottom, and cover them with inverted turf to keep the soil from washing down and injuring the drainage; then fill up with soil composed of about one-third of good free loam, one-third of well-decayed leaf-mould, and one-third of thoroughly decomposed cow manure. Plant 1½ in. to 2 in. deep, and every year, soon after the leaves

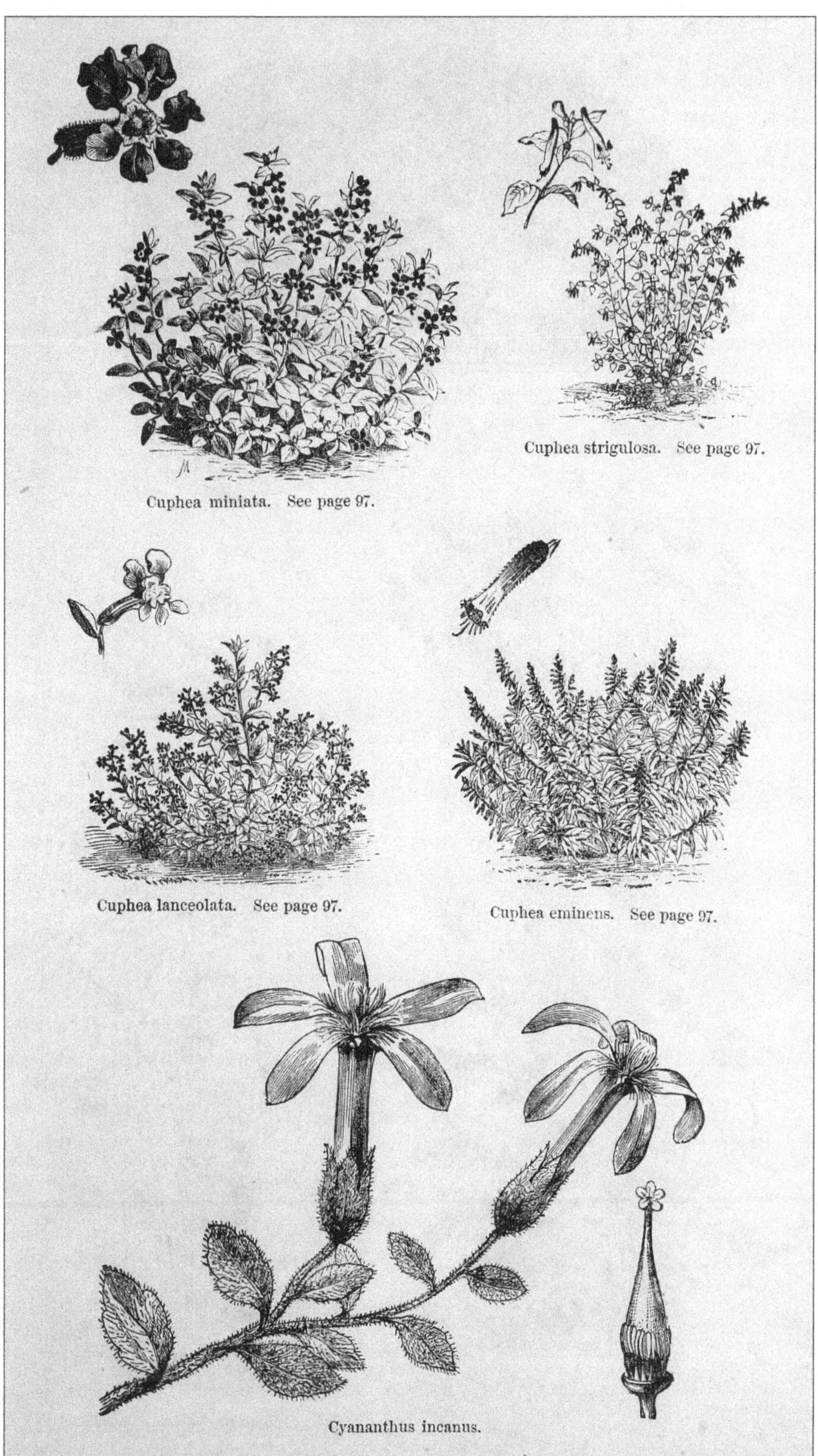

Cuphea miniata. See page 97.

Cuphea strigulosa. See page 97.

Cuphea lanceolata. See page 97.

Cuphea eminens. See page 97.

Cyananthus incanus.

Cyclamen europæum (*Sowbread*) in Grass.

Cyclamen hederæfolium (*Ivy-leaved Cyclamen*).

Cyclamen europæum.

Cynara Scolymus (*Globe Artichoke*).
See page 100.

Cyperus longus (*Galingale*). See page 100.

[*To face page* 99.

die down, take off the surface as far as the tops of the tubers, and fresh surface them with the same compost, or in alternate years they may only have a dressing on the surface of well-decayed leaves or cow manure. During summer, or indeed after April, the glass should be removed, and they ought to be slightly shaded with Larch Fir boughs (cut before the leaves expand) laid over them, to shelter from the extreme heat of the sun. As soon as they begin to appear in the autumn gradually take these off, and do not use the glass until severe weather sets in—at all times, both day and night, admitting air at back and front—and in fine weather draw the lights off, remembering that the plants are perfectly hardy, and soon injured if kept too close. They do not like frequent removal. There is a pretty white variety of C. Coum, which is extremely desirable. (= C. hyemale.)

C. europæum *(European Cyclamen).*—The leaves of this species appear before and with the flowers, and remain during the greater part of the year. Flowers from June to November, or, with slight protection, until the end of the year. The flowers are of a reddish-purple colour. Some of the southern varieties, by attention to cultivation under glass, may even assume a perpetual flowering character. The varieties C. Clusi, littorale, and Peakeanum are of this section. In these varieties the flowers become much longer, of a more delicate colour, often approaching peach colour, and are almost the size of those of C. persicum; pure white are rare, but pale ones are not uncommon: they are very fragrant. C. europæum thrives freely in various parts of the country in light, loamy, well-drained soil, as a choice border and rockwork plant. Where it does not do well in ordinary soil it should be tried in a deep bed of light loam, mingled with pieces of broken stone. In all cases it is best to cover the ground with Cocoa fibre. It is a very desirable species on account of its delightful fragrance and long succession of flowers. They luxuriate in the *débris* of old walls and on the mountain side, with a very sparing quantity of vegetable earth to grow in. The bulb of this species varies considerably in size and shape; sometimes it is much elongated and irregular, and is then the C. anemonoides of old authors. (= C. odoratum, C. æstivum.)

C. hederæfolium *(Ivy-leaved Cyclamen).*—A native of Switzerland, South Europe, Italy, Greece and its isles, and the north coast of Africa. Tuber not infrequently 1 ft. in diameter, and covered with a brownish, rough rind, which cracks irregularly so as to form little scales. The root fibres emerge from the whole of the upper surface of the tuber, but principally from the rim; few or none issue from the lower surface. The leaves and flowers generally spring direct from the tuber without the intervention of any stem (a small stem, however, is sometimes produced, especially if the tuber be planted deep); at first they spread horizontally, but ultimately become erect. The leaves are variously marked, and the greater portion of them appear after the flowers, continuing in great beauty the whole winter and early spring, when, if well grown, they are one of the greatest ornaments of our borders and rockeries. Often they are as much as 6 in. long, 5½ in. in diameter, and 100 to 150 leaves springing from one tuber. They are admirably adapted for table decoration during winter. The flowers begin to appear at the end of August, continuing until October, and are purplish-red, frequently with a stripe of lighter colour. There is a pure white variety, and also a white one with pink base or mouth of corolla; these reproduce themselves tolerably true from seeds. Strong tubers will produce from 200 to 300 flowers each. Some of them are delightfully fragrant. They are quite hardy, but are worthy of a little protection to preserve the late blooms, which often continue to spring up till the end of the year. This species is so perfectly hardy as to make it very desirable not only for the rockery, but also for the open borders. It will grow in almost any soil and situation, though best (and it well deserves it) in a well-drained rich border or rockery. It does not like frequent removal. It has been naturalised successfully on the mossy floor of a thin wood, on very sandy, poor soil, and it may be naturalised with perfect success almost everywhere in these islands. It would be peculiarly attractive when seen in a semi-wild state in pleasure grounds and by wood walks. C. græcum is a very near ally, if more than a variety, of C. hederæfolium; it requires the same treatment. The foliage is more after the C. persicum, or the southern form of C. europæum. C. africanum (algeriense macrophyllum), much larger in all parts than C. hederæfolium, otherwise very nearly allied, is hardy in warm sheltered situations.

C. ibericum *(Iberian Cyclamen).*—This belongs to the Coum section. There is some obscurity respecting the authority for this species and its native country. The leaves are very various. It flowers in spring, the colour varying from deep red-purple to rose, lilac, and white, with intensely dark mouth; produced more abundantly than by C. Coum.

C. vernum *(Spring Cyclamen).*—The leaves of this species rise before the flowers in spring; they are generally marked more or less with white on the upper surface, and often of a purplish cast beneath. This, though one of the most interesting species, and perfectly hardy, is seldom met with cultivated successfully in the open border or rockery; it is very impatient of excessive wet about the tubers, and likes a light soil, in a nook rather shady and well sheltered from winds, its tender fleshy leaves being soon injured. The tubers should also be planted deep, say not less than 2 in. to 2½ in. beneath the surface. C. vernum of Sweet is considered by many as only a variety of Coum, and it is known as C. Coum var. zonale. It is also known as C. repandum. There is a white-flowered variety.

Cyclobothra *(Calochortus).*

Cynara Scolymus *(French Artichoke).*—This plant, although chiefly grown for culinary purposes, possesses sufficient merit as a foliage plant to entitle it to a place. Its long, deeply-divided leaves, white and downy beneath, its height (4 ft. to 5 ft.), its purplish flower-heads, and distinct habit render it very suitable for planting on the irregular and rougher parts of pleasure grounds, grass plats, &c., which are often occupied by objects far less striking.

Cypress Spurge *(Euphorbia Cyparissias).*

Cynoglossum *(Hound's-tongue).*—A genus of the Forget-me-not Order, some of which are in cultivation, but we have never noticed them of any essential use in the flower garden; therefore we say little of them.

Cyperus longus *(Galingale).*—The stiff, erect, tapering, triangular stem of this plant, which is from 2 ft. to 3 ft. high, is crowned by a handsome, loose, umbellate panicle of chestnut-coloured flower-spikes, at the base of which there are three or more leaves. These are often 1 ft. or 2 ft. long, the lower ones arching gracefully, and of a bright shining green, giving the plant a very pleasing aspect. The rootstock is thick and aromatic, and was formerly much used in medicine as a tonic. A rare native plant, suitable for the bog bed or the margin of water.

Cypripedium *(Lady's Slipper).*—A genus of hardy Orchids, containing several beautiful species that are perfectly hardy in our climate, and of which the Mocasson Flower (C. spectabile) is by far the most important; indeed, it is one of the handsomest of all hardy flowers, and, fortunately, of easy culture. The following are a few of the cultivated kinds, of which most are worthy of general culture.

C. acaule *(Stemless Lady's Slipper).*—A dwarf species with a naked, downy flower-stalk from 8 in. to 12 in. high, and bearing a green bract at the top. Flowers early in summer, large, solitary, of a purplish colour, with a rosy purple (rarely white) lip, which is nearly 2 in. long, and has a singular closed fissure down its whole length in front. Leaves two, at the base of the flower-stem, oblong, obtuse, downy. Northern States of North America, in woods and bogs. Requires a shady position in moist peaty soil or leaf-mould.

C. Calceolus *(English Lady's Slipper).*—The only British species of Cypripedium, and the largest flowered of our native Orchids. Grows from 1 ft. to 1½ ft. high. Flowers in summer, solitary (sometimes two), large, of a dark brown colour, with an inflated, clear yellow lip netted with darker veins, and about 1 in. in length. Leaves, generally three or four in number, large, ovate, pointed, veined. North Europe, and occasionally found in the northern counties of England, where, however, it is now almost exterminated by unscrupulous plant gatherers. A very ornamental plant for the rock garden, where it should be planted in sunny sheltered nooks in calcareous soil, or in narrow fissures of limestone rock, well drained, in rich fibrous loam. It prefers an east aspect.

C. guttatum *(Spotted Lady's Slipper).*—A handsome, rare kind, seldom seen in gardens. Grows from 6 in. to 9 in. high. Flowers in summer, solitary, rather small but beautiful, white, heavily blotched or spotted with deep rosy purple. Leaves two, alternate, oval-elliptical, pointed, downy. Canada, N. Europe (near Moscow), and N. Asia, in dense forests amongst the roots of trees in moist, black, vegetable mould. Requires a shady position in leaf-mould, moss, and sand, and should be kept rather dry in winter.

C. japonicum *(Japanese Lady's Slipper).*—About 1 ft. in height, and its hairy stems, which are as thick as one's little finger, bear two plicate fan-shaped leaves of a bright green colour, rather jagged or erosely cut around the margins. The flowers are solitary, on hairy scapes, the sepals being of an apple-green tint; the petals, too, are of the same colour, but are dotted with purplish crimson at the base; the lip is large, and curiously folded in front, as in the better-known C. (humile) acaule, to which it seems most nearly allied; the colour of the lip is a soft creamy-yellow, with bold purple dots and lines.

C. macranthum *(Large Lady's Slipper).*—This species bears a considerable resemblance to C. ventricosum, but has lighter-coloured flowers. It grows about 1 ft. high. Flowers early in June, large, of a uniform purplish rose colour with deeper coloured veins. Lip globose, inflated, finely marked with deep purple reticulations. Siberia. This handsome, and at present rare, species may be treated in the same manner as C. guttatum.

C. pubescens.—A dwarf species with a pubescent stem, seldom more than 2 ft. high. Flowers early in summer, one to three on each stem, scentless, greenish yellow, more or less spotted with brown, with a pale yellow lip from 1½ in. to 2 in. long, and flattened at the sides. Leaves broadly oval, pointed, pubescent. America, in bogs and low woods, from Pennsylvania to Carolina. Does well on dry sunny banks, among loam, stones, and grit.

C. spectabile *(Mocasson Flower).*—The most beautiful of the hardy plants of this genus; 15 in. to 2½ ft. high. Flowers in summer, one or two (rarely three) on each stem; large, handsome, white, and much inflated, rounded lip, about 1½ in. long, white, and marked with a large blotch of bright rosy carmine in front. A variety (C. s. album) has the lip entirely white. In America it naturally luxuriates in woods, moist meadows, and also peaty bogs in the Northern States. Good native specimens produce from fifty to seventy

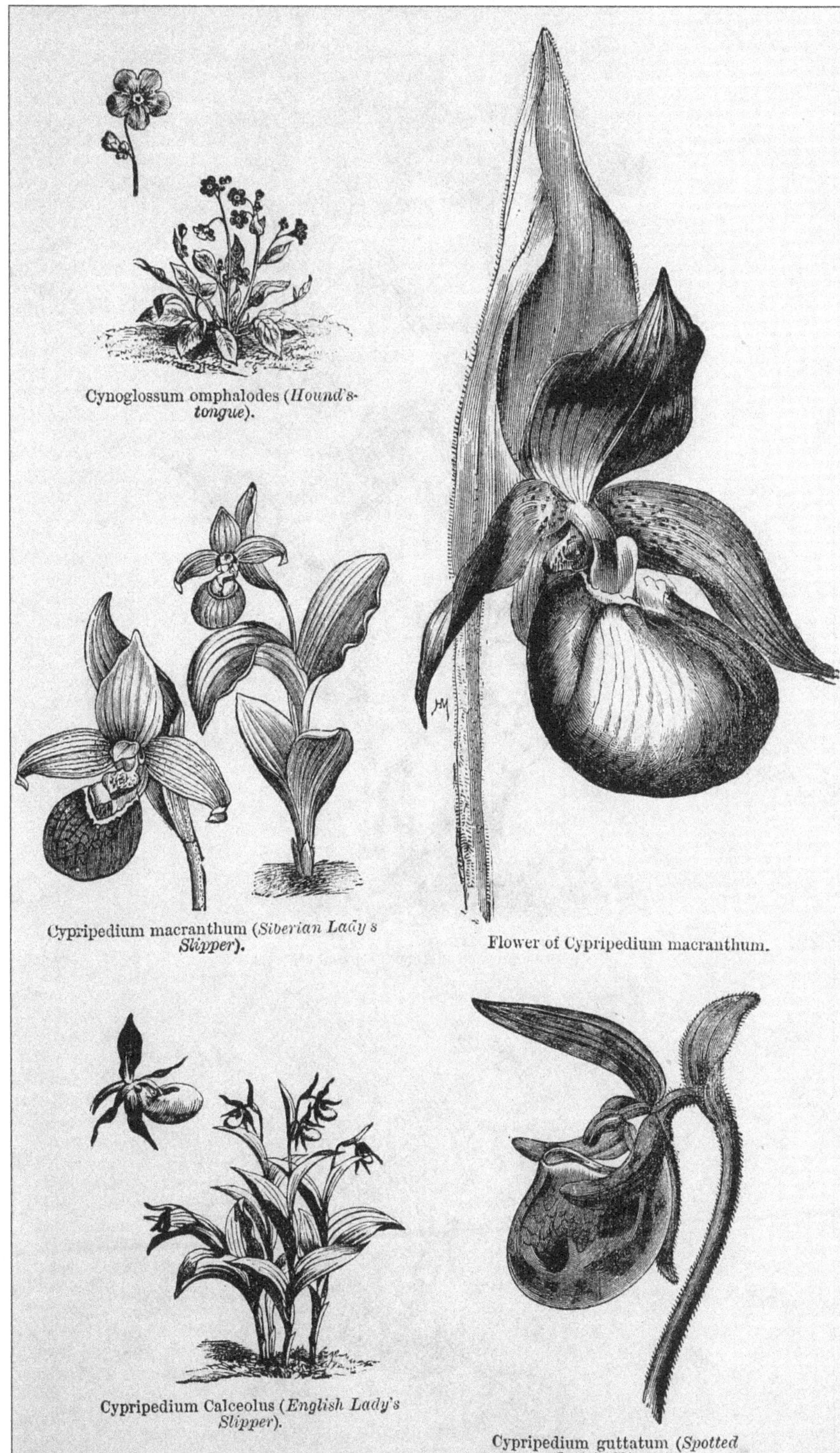

Cynoglossum omphalodes (*Hound's-tongue*).

Cypripedium macranthum (*Siberian Lady's Slipper*).

Flower of Cypripedium macranthum.

Cypripedium Calceolus (*English Lady's Slipper*).

Cypripedium guttatum (*Spotted Lady's Slipper*).

PLATE LXXXVIII.

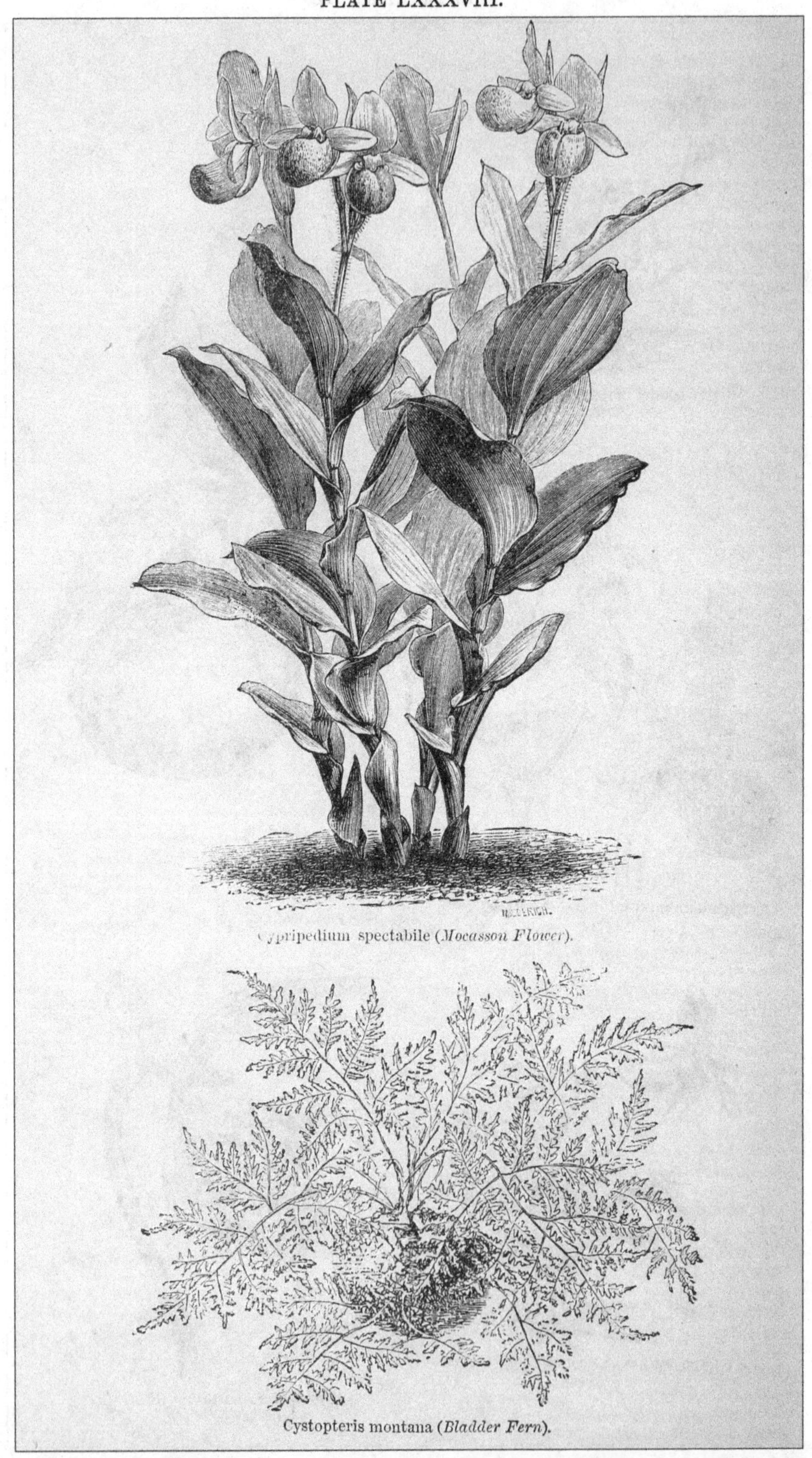

Cypripedium spectabile (*Mocasson Flower*).

Cystopteris montana (*Bladder Fern*).

[*To face page* 101.

flowers on a single tuft, 3 ft. across, formed on a thick mat-like mass of fleshy roots. The plant is perfectly hardy in this country. It succeeds perfectly if planted out in a deep, rich, peaty soil, and if a few nodules of sandstone or rough sandstone grit be mixed with the soil, so much the better. We have also seen this Lady's Slipper thrive well in turfy loam on a moist peaty bottom; in either case, however, deep planting is necessary, as the roots are then cool and moist during the hot summer weather, and they do not suffer from frost in the winter. It also succeeds in a deep bed of peaty compost, and in sunny districts would be the better of a shaded position where, however, it must not be robbed at the root by trees or shrubs.

Cynanchum.—A genus of very little value for the flower garden, though sometimes grown.

Cystopteris (*Bladder Fern*).—The cultivated species of this genus are small, elegant Ferns of a very delicate fragile texture, and all natives of Britain. They grow naturally on rocks and walls, chiefly in mountainous districts. The best known are C. fragilis, which has finely-cut fronds about 6 in. high. It is of easy culture, succeeding in an ordinary border of garden soil, though it is seen to the best advantage on shady parts of the rock garden in a well-drained soil. There are two or three varieties, Dickieana being the best. C. alpina is a much smaller plant, and not at all difficult to cultivate or increase when once established, but is more susceptible to excessive moisture than C. fragilis. A sheltered situation in a well-drained part of the rock garden suits it well. C. montana is another elegant plant amenable to the same treatment as C. fragilis.

Dactylis (*Cocksfoot*).—The variegated forms of this native grass are excellent. D. glomerata variegata is one of the most useful of edging plants. It is easy of increase by division, either in autumn or spring. It likes a heavy soil, but thrives in almost any kind. If the soil be too poor it is apt to get rusty-looking in dry autumns. It bears clipping to keep it dwarfer, but is prettier let alone, except cutting of flower-stems. D. g. v. elegantissima is said to be superior to this variety. The yellow variegated Cocksfoot is a new form which is likely to be as popular or more so than the old well-known variegated form of this plant. The variegation is soft yellow, regular, and the growth free. These grasses are graceful as edgings to beds, and as carpets or mixtures, or as tufts in borders. The graceful pointed leaves should not be clipped.

Daisy (*Bellis perennis*).

Dahlia.—A noble race of plants not at all so much or so well used as they deserve to be. The "bedding-out fever" threw them into the background, and led also to the origin of the dwarfest of the varieties—the "bedders." The old Dahlia bank or large bed—one of the best features in gardens of twenty-five years ago—disappeared in the rage for lines and cakes of colour! It deserves to be seen again. The compact form sought by the raiser of "bedding" Dahlias is a poor quality. The bolder the form the better. There are now various classes of Dahlias, and the more variety the better. It is not well to heap up the various artificial distinctions into classes. One evil of it is that some in saving seedlings are led to destroy the forms that go away from the various "classes." The aim should be to keep these deviations rather, and break down the hard and fast lines of "bedders," "pompones," "show," "fancy," and so on. A plant so vigorous in growth and profuse in bold masses of flower cannot fail to be well grouped by those who care for it, if they will only remember that the highest beauty in gardens comes to us in broken or tossing, not in flat or shorn, surfaces! The new and rightly popular race of single Dahlias helps in securing variety of bloom, and in affording flowers for cutting. To get a good result with the Dahlia, rich, deep, and moist soil is essential, and care in putting out strong plants as early as may be safe, so as to get a good growth in time for a rich and early autumn bloom. Where the plants are put out a little too late and are weak, they may only begin to give few and poor blooms when the frost comes to kill. If planted in May, when or where frost is feared, it is easy to protect the young plants at night by turning a garden pot over them. If the soil is not deep, and rich, and moist, manure water should be used. Watering in early growth is usually necessary. Afterwards it is not so in moist districts, where the plant is well treated as regards depth and quality of soil. In dry places water is essential most seasons. Staking and tying out the shoots must be particularly attended to, as the stems are brittle, and break under little wind-pressure.

Earwigs are great enemies to Dahlias, but they may be trapped by using small round troughs, which may be procured at any pottery. They may also be caught in pieces of Hemlock stems, 6 in. long, leaving a joint on one end, and sticking the pieces here and there through the Dahlias. Small pots, with a little bit of dry Sphagnum Moss inside, inverted on the tops of stakes, likewise form good traps. Manure-water may be safely used after the plants show flower, provided it be given in the evening. Some Dahlias will produce all their flowers, or nearly so, perfect; others require their shoots to be well thinned, and all blooms except three or four to be cut off, in order to be eligible for show purposes; but, as a general rule, all may be more or less thinned.

INCREASE.—Dahlia roots have kept well and flowered for years in succession, in dry soils especially, when the plants have been cut down to the ground and covered with some material to protect them from frost, but the usual practice is to take up the roots and store them in winter. Dahlias may be propagated by means of cuttings, layers, root-division, or seeds, the last being only used where new varieties are sought. Cuttings are the best means by which Dahlias may be multiplied; division of the roots is commonly practised, but cuttings are best. When cuttings are used, each has its own self-formed tubers, and they are on that account less liable to rot during the winter. If started in February or March, in a temperature of 65° to 70°, each crown will produce three or four cuttings every two or three days. These may be taken off even as early as March, removing them close to the crown, but taking care not to injure it, as others will come up at the base of those removed. It is also necessary for the growth of good plants not to let the cuttings become too long before taking them off the tubers, as they are more apt to flag, need more room, and are not so convenient to plant out as short cuttings. When the crowns have supplied all the cuttings that can be got from them, they may then be divided, if required; consequently nothing is lost.

To propagate from layers the lowest branches of the plant should be pegged down in the soil, in which, if of a sandy nature, they will root freely; in the absence of sandy soil, a quantity of leaf-mould, with a mixture of sand, may be laid down for them to root in. Pure white sand alone is best suited for striking them in, and a mixture of leaf-mould and sand is very good for starting the crowns in. Cuttings may be successfully struck during the summer months; but this is seldom done except in the case of choice varieties. Three-inch pots are best for putting the cuttings into, six cuttings being put in each pot, which should be plunged in a brisk bottom-heat, covered with hand-glasses, and shaded from bright sunshine; in less than a fortnight they will all be rooted, and may be potted off singly into large 3-in. pots, which may be put into lower temperatures gradually until, say, May 1, when they may be inured to the open air for planting. To raise seedlings the seed should be sown in February in heat, and the young plants should be treated in the same way as cuttings. The sowing of one seed the right way upwards in a 2½-in. pot saves the after trouble of pricking out, and secures the plant in its entirety, not a fibre of a root being lost.

WINTERING.—As long as the weather keeps mild, Dahlia roots are best in the soil, and need not be taken up till the end of November; but should sharp frosts be followed by heavy rain, their removal from the ground should be prompt. A dry day should be chosen for lifting the roots, the stem of the plant cut off to within 2 in. or 3 in. of the crown of the roots, and the roots placed on some sticks or boughs to dry, with the neck downwards, and so arranged that the air can pass underneath them. Soil may be allowed to adhere to the tubers, but the greater portion is best removed. If the weather be fine and dry the roots may remain in this position for about three days, but covered with a mat at night to screen them from frost. The floor of a greenhouse from which frost can be excluded, or a dry cellar is a capital place wherein to store the roots. A little ventilation is necessary to keep the roots from getting mouldy; and on the other hand a hot dry atmosphere must be avoided, as in it the tubers might shrivel. By lifting the roots with some soil adhering to them, they are kept in plump condition during the winter, and this is desirable in cases where the roots are required for early forcing. On the floor of a greenhouse they will generally keep remarkably well, it being light and airy, and during the depth of winter much water should not, as a rule, be given to the plants. In the case of nurserymen who cultivate the Dahlia largely for sale, it is customary to winter the roots in a close shed that is airy without being very light, and from which frost can be excluded at will. Broad shelves form receptacles for stowing away the roots, and these are carefully looked over at intervals. The tubers of some sorts are more difficult of preservation than others, and it frequently happens that choice varieties are bad keepers.

CLASSES AND VARIETIES.—These are convenient, but very artificial, and the sooner their arbitrary lines are broken down the better. The two leading classes a few years ago were the show and fancy Dahlias—distinctions confusing to some, as a white or yellow Dahlia, edged or tipped with a dark colour, is classed as an edged, tipped, or laced Dahlia, and is included among the show flowers; but when the disposition of colours is reversed, and the flowerets, being of a dark colour, are tipped with a light colour, it is then denominated a fancy Dahlia, as are all the kinds with Carnation-like stripe. The catalogues abound with names of varieties, and where all are good it is best for the grower to make his own selection, the more so as new forms are often raised. In the case of plants where weedy or useless things are offered we note them as such, but here all are good. Now and then varieties that do not meet the ideal of the hard-shell florist please best the artist or the gardener. All the show and fancy Dahlias are splendid flowerers well grown, and a wonderful illustration of the range and change beginning from one or two single species.

BOUQUET OR POMPONE DAHLIAS during these last few years have become very popular; the blooms being small and compact, resembling a Persian Ranunculus more than those of a Dahlia, make them more useful in some cut-flower decoration than large Dahlias. They are effective as a background in mixed borders or for large beds. The roots, left in the open ground all winter, are quite safe if a good coating of coal-ashes be put over them when the tops are cut down. If lifted for purposes of propagation they may be safely stored in any shed secure from frost, and if covered with any partially dry material, such as old tan, Cocoa-nut fibre, or leaf-

Dactylis glomerata marginata (*Variegated Cocksfoot Grass*). See page 101.

Type of Florist's Dahlia.

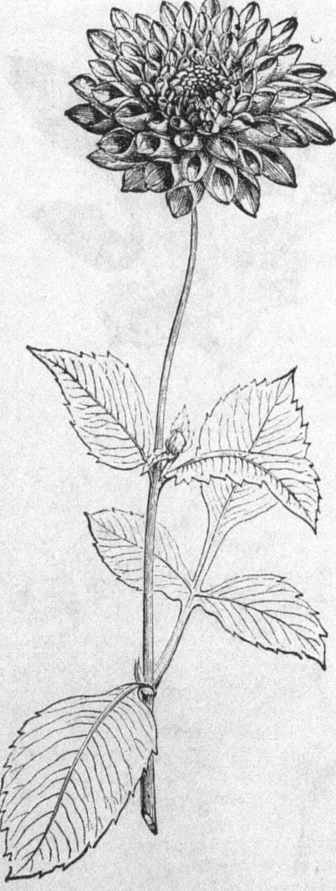

Dahlia Juarezi.

Dahlia imperialis, showing habit of growth.

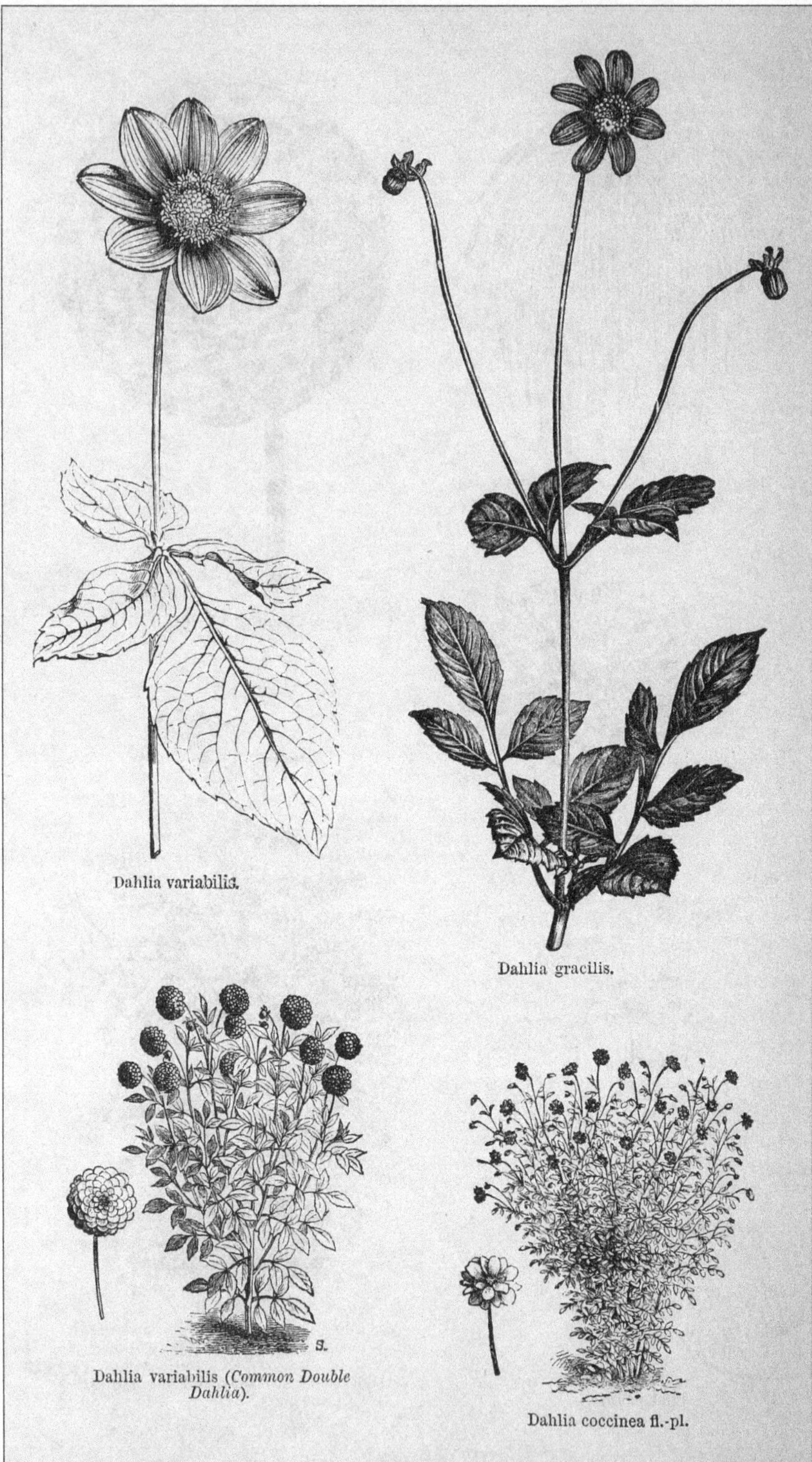

Dahlia variabilis.

Dahlia gracilis.

Dahlia variabilis (*Common Double Dahlia*).

Dahlia coccinea fl.-pl.

[*To face page* 103.

mould, they will start more strongly into growth than if over-dried. In autumn the flowers are especially useful in floral decorations of a large character, where delicate or fragile blossoms are not nearly so effective. We do not name varieties, as they change often and new ones are frequently raised. The planter will consult his own taste as to varieties, and the catalogues that name them are numerous. He may also raise seedlings with the hope of having new forms as well as good flowers. The same may be said of the bedding Dahlias, some of which have valuable and distinct qualities.

SINGLE DAHLIAS.—The single varieties are the great gain of the day, and show a good variety of colour and a better form than the double kinds. We do not mean by this that one race should be preferred to the other, except for certain uses; all are worth growing. Half-a-dozen good single kinds are alba, Paragon, Yellow Gem, Pink Beauty, Morning Star, Glory, and White Queen. As single Dahlias seed freely, and seedlings are easily raised, the named kinds cannot expect to have any lengthened popularity. We hope many more good kinds may be raised.

D. Juarezi is one of the most important of the flowers that have been introduced of late years, and it is the more desirable because of its easy culture, requiring, as it does, no different treatment from that given to ordinary Dahlias. It is not quite double, but is very fine in form and brilliant in colour. It flowers somewhat sparsely. Probably if pot roots of the preceding year only were planted, the plants would bloom earlier and in greater profusion. There are about six or seven species of Dahlia, all natives of Mexico, and from which the different races of garden varieties have been obtained. Those in cultivation are

D. coccinea, a tall growing plant with bright scarlet flowers that rarely vary. Nearly related to this and only differing in some slight points is D. Cervantesi, also with showy scarlet flowers.

D. glabrata is a beautiful plant of dwarf spreading growth, more slender than any of the other species. The flowers are smaller than those of other kinds, and vary from pure white to deep purple. It is hardier than any, and plants that are left in the ground are generally uninjured throughout average winters. Its dwarf growth adapts it for positions not suitable for the taller kinds, and it has a good effect in masses, the colour being quite unlike that of any other Dahlia. It is known also as D. Merki, repens, and Decaisneana.

D. gracilis is a distinct and graceful plant having slender stems and finely divided foliage, which gives the plant a more free habit than any other. The flowers are of the ordinary size and of a bright scarlet.

D. imperialis has large and graceful, much-divided leaves, and flowers, of a beautiful French white, thrown up in a great cone-like mass. It rarely flowers in the open air in this country, but it is of service both in the flower garden and conservatory. Planted in rich soil, and placed in a warm, sheltered position in the open air at the end of May, it grows well with us in summer, and, in consequence of its large and graceful leaves, is an ornament worthy of being used as a "fine-foliaged" plant. Similar to this, but not so fine, is D. Maximiliana.

D. variabilis is the supposed parent of all the garden varieties. The wild plant has scarlet flowers like coccinea, and is of similar growth. A packet of seed, however, will yield plants with flowers of all shades, from crimson to white and yellow.

Dalibardia fragarioides.—A dwarf creeping plant of the Rose family, bearing clusters of bright yellow flowers in summer. Grows freely anywhere, but is not of any distinct value as a garden plant. N. America. *Syn.*, Waldsteinia.

Daphne.—Though this is a genus of shrubs, there is a group of small-growing species among them that claims a place in the rock garden. The best of the group are the following: D. Cneorum (Garland Flower), a very neat evergreen shrub, 6 in. to 12 in. high, that bears a profusion of rosy lilac flowers, the unopened buds being crimson. It flowers in April and September, in dense terminal umbels, deliciously fragrant, and often twice a year. A native of most of the great mountain chains of Europe and suitable for the rock garden, front margin of the mixed border, or as an edging to beds of choice low shrubs. Thrives best in sandy peaty soil, kept rather moist in summer. Increased by layers. D. rupestris (Rock Daphne) is a neat and diminutive shrub, having erect shoots forming dense, compact tufts, 2 in. high and 1 ft. or more across, covered with a mass of bloom which sometimes almost eclipses the plant. Its colour is a soft shaded pink or rose, and its flowers are individually larger and more waxy than those of D. Cneorum, but form clustered heads in the same way. It is essentially a rock plant, growing wild in fissures of limestone in peaty loam. In cultivation it is of slow growth, taking some years to form a moderate-sized tuft, but it is a gem worth waiting for. It seems to thrive in very stony and peaty earth with abundance of white sand, and should be planted in a well-drained, but not a dry, position. D. Blagayana is a beautiful new alpine shrub, growing from 3 in. to 6 in. high, and of loose growth. The leaves form rosette-like tufts at the tips of the branches, and encircle the dense clusters of deliciously-scented white flowers. It blooms in spring continuously for several weeks. It is of easy culture, thriving well on the rock garden in spots well drained and surrounded by stones, so that its wiry roots may ramble among them. It may be increased by layers pegged down in spring and separated from the plants as soon as roots are emitted. Carniola.

Darlingtonia californica *(Californian Pitcher Plant)*.—A most singular plant, resembling the Sarracenias, but very distinct. The leaves, which rise 2 ft. or more high, are hollow, and terminate in a curiously-shaped hood, from which hang two ribbon-like appendages. The greater part of the leaf is green, but the hood is in a mature state a deep crimson-red. The flowers are almost as curious. This remarkable plant is now better understood, and found to grow in our climate if care be taken with it, and it would be difficult to name a more interesting plant than this for a sheltered bog garden. It is not so difficult to manage out of doors as under glass; indeed it only requires planting in a moderately wet bog in a light spongy soil consisting of fibrous peat and chopped Sphagnum moss. A place should be selected for the plants by the side of a stream, in an artificial bog or in any moist place, and they should be fully exposed to direct sunlight, but sheltered from the cold winds of early spring when they are throwing up their young leaves. The plants will require frequent watering in dry seasons, unless in a naturally wet spot. When they have attained a large size they develop side shoots, which, if taken off and potted, make good plants in a short time. It may also be raised from seed, but to do this several years are necessary.

Datisca cannabina. — A tall and graceful herbaceous perennial from 4 ft. to 7 ft. high. The long stems are clothed with large pinnate leaves; the yellowish-green flowers appear towards the end of the summer. The male plant is very strong and graceful in habit. The female plant remains green much longer than the male, and, when profusely laden with fruit, each shoot droops and the plant is very effective. It should not be forgotten in any selection of hardy plants of fine habit. Seed will probably be found the best way to increase it, and then one would be pretty sure of securing plants of both sexes. The border is not its proper location; it is, above all others, a plant adapted for the grassy margin of an irregular shrubbery, and will be rendered all the more effective if planted on the face of a grassy slope, where its deep-seeking roots will soon defy even the most protracted drought; nor is any margin of open soil between it and the adjacent turf requisite. It grows to a height of 5 feet in clay soil, but attains a greater size in light and warm soil.

Datura *(Thorn Apple)*.—A genus of the Nightshade family, containing several handsome garden plants that well merit cultivation. As they are natives of Mexico and similar climates, none are hardy with us, but owing to their rapid growth they succeed well and make bold, effective plants in a short season, if treated as half hardy annuals. The best kinds to cultivate are the following: D. ceratocaula, which grows from 2 ft. to 3 ft. high, producing large, sweet-scented, trumpet-like flowers, often 6 in. in length, and 4 in. or 5 in. across. They are white, tinged with violet-purple, expanding in the afternoon and closing the following morning. D. fastuosa is a handsome species, having white blossoms not so large as the preceding. There is a fine variety of it, with the tube of the flower violet and the inside white. The most striking forms of this species are those bearing double flowers, the primary corolla having a second and sometimes a third arising from its tube, all perfectly regular in form, and often parti-coloured, as in the single variety with violet flowers. D. fastuosa Huberiana, of the seed catalogues, and several varieties of it that are offered, are reputed to be hybrids of this species with the dwarf D. chlorantha flore-pleno or D. humilis flava of the gardens; but although they offer a greater variety of colour, they are less hardy than the older forms just described, and appear to require a warmer climate for their complete development. D. meteloides is a handsome Mexican plant, called in gardens Wright's Datura. It has a fine aspect as isolated specimens in sunny, but sheltered nooks. It grows from 3 ft. to 4 ft. high, and has wide-spreading branches that bloom profusely from the middle of July till frost sets in. The flowers are white, tinged with mauve; they measure from 4 in. to 6 in. across, and are showy and sweet, but the leaves emit a disagreeable odour. Besides these there are several others in cultivation, such as D. ferox and quercifolia, but those we describe are the finest. The culture of these Daturas offers no especial difficulty. Fresh seeds are readily raised in an ordinary hotbed, and the young plants should be pricked out singly in pots while small, and finally planted out where they are to stand. They need ample space for their full development, and should be grown in light warm soils.

Day Lily *(Hemerocallis)*[1]

Delphinium *(Larkspur)*.—Few plants contribute so much to the beauty of the flower garden as the Larkspur. There are many species in cultivation, both annual and perennial, but the most important are the tall hybrid perennials, of which there is now such a wealth of varieties embodying a wonderful range of lovely colour. Their stateliness of habit and fine blue and purple shades are unequalled. The great variety of their heights, varying from 1 ft. to 9 ft. high; the greater variety of their shades of colour, from almost scarlet to pure white, from the palest and most chaste lavender up through every conceivable shade of blue to deep indigo; and the very considerable variety of size and form of their individual blooms, some of which are single, semi-double, and perfectly double, and set on spikes ranging from 1 ft. to 6 ft. in length, render them objects of great value as garden plants. About a dozen species have given rise to the cultivated varieties, the chief of which have been D. grandiflorum, formosum, lasiostachyum, cheilanthum, elatum, peregrinum, and others.

CULTURE AND POSITION.—The combinations in which they can be placed in gardens are numerous. They are splendid objects in

PLATE XCI.

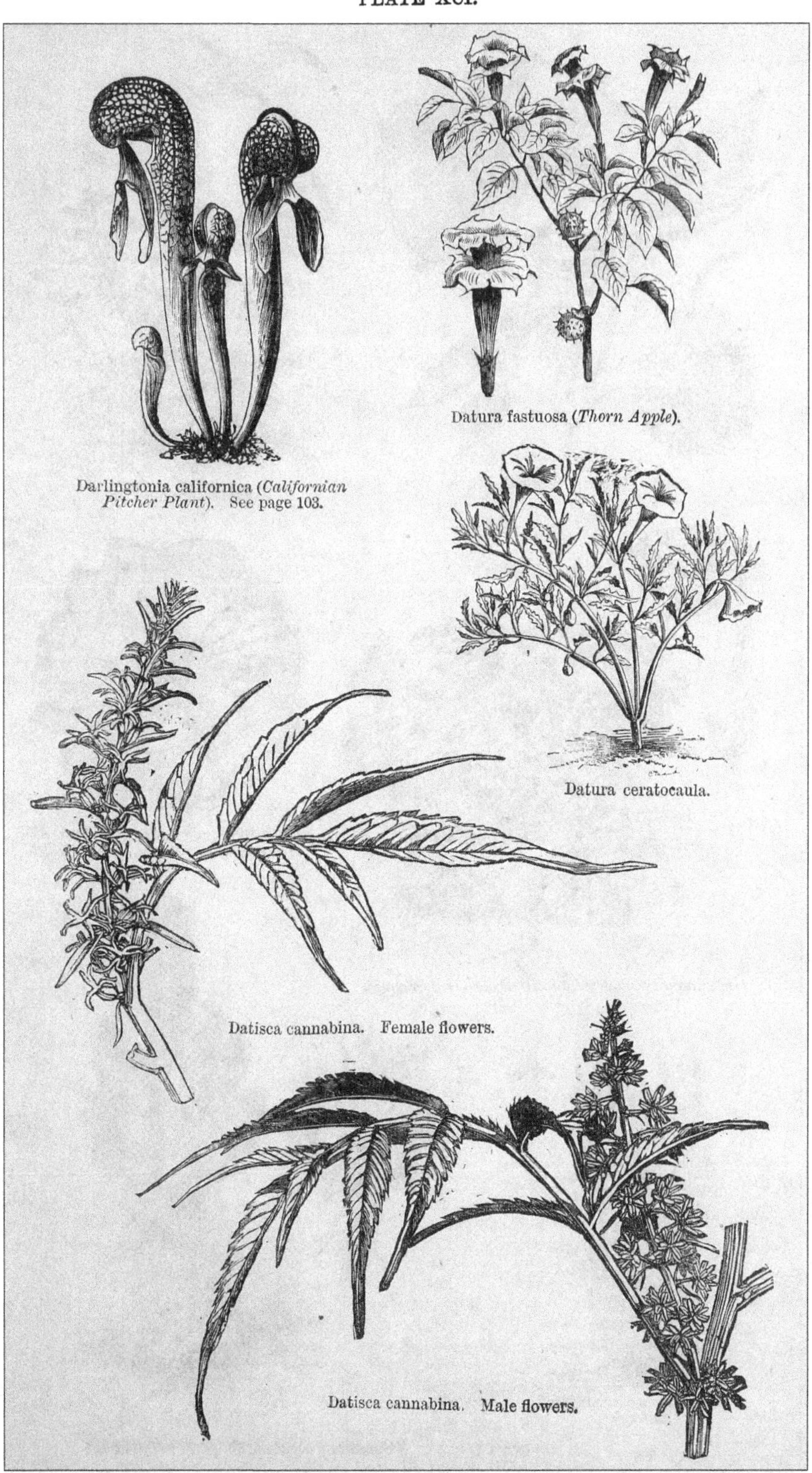

Darlingtonia californica (*Californian Pitcher Plant*). See page 103.

Datura fastuosa (*Thorn Apple*).

Datura ceratocaula.

Datisca cannabina. Female flowers.

Datisca cannabina. Male flowers.

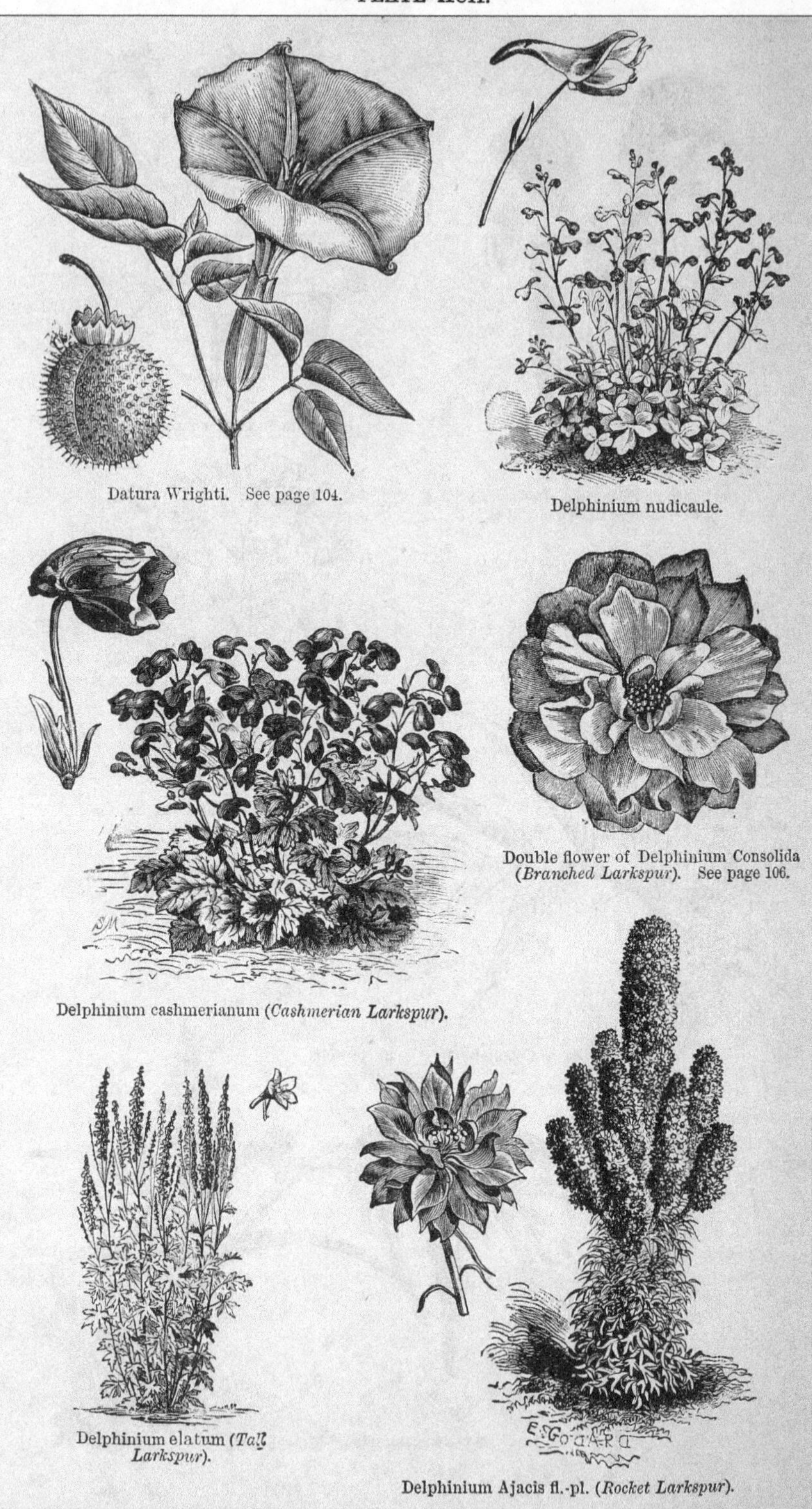

Datura Wrighti. See page 104.

Delphinium nudicaule.

Double flower of Delphinium Consolida (*Branched Larkspur*). See page 106.

Delphinium cashmerianum (*Cashmerian Larkspur*).

Delphinium elatum (*Tall Larkspur*).

Delphinium Ajacis fl.-pl. (*Rocket Larkspur*).

various positions, and may be used in various ways—in the mixed border, in masses or groups either in one or several colours, and associated with other flowering plants or with shrubs. Perennial Larkspurs thrive in almost any situation or soil; they are easily increased, and are quite hardy. A deep, friable loam, enriched with rotten manure, is their favourite compost, but the character of the soil apparently is of but little importance, as they will grow luxuriantly in hot, sandy soil if it be heavily manured and watered. Good loam requires less attention in that way. Every three or four years they should be replanted and divided at the same time. This is best done in spring, just as they are started into growth, or in summer; if done in summer, cut the plants down that are intended for division, and let them remain for a week or ten days until they start afresh; then carefully divide and replant, shading and watering until they have become established. Late autumn division in the case of Larkspurs is not advisable. Delphiniums can be made to bloom for several months by continually cutting off the spikes immediately they have done flowering. If the central spike be removed the side shoots will flower, and by thus cutting off the old flowers before they form seeds fresh shoots will issue from the base, and keep up a succession of bloom. Another plan is to let the shoots remain intact until all have nearly done flowering, and then cut the entire plant to the ground, when in about three weeks or so there will be a fresh bloom. In order, however, to keep the plants from becoming exhausted they must in this case have a heavy dressing of manure or a liberal supply of manure water. Top-dressings keep the soil cool and moist, give the plants a healthier growth, and greatly increase the quantity and improve the quality of the flowers.

The following is a selection of the finest kinds: *Single Varieties.*—Belladonna, Hendersoni, Cambridge, Granville, Gloire de St. Mande, Barlowi, versicolor, Coronet, magnificum, Lavender, pulchrum, formosum, lilacinum, Celestial, Madame Hock, mesoleucum superbum, Montmorence, Defiance, and Attraction. Here, as usual, we counsel the grower to consult his own taste, and to raise seedlings of his own, while taking care to have a good stock of the standard varieties he likes best. *Double Varieties.*—Madame E. Geny, Madame Henri Jacotot, Madame Richalet, Pompon Brilliant, Roi Leopold, Hermann Stenger, Claire Courant, George Taylor, Roncevaux, Le XIXe. Siècle, Keteleeri, Prince of Wales, General Ulrich, Arc en Ciel, Sphere, Michael Angelo, Delight, Glynn, Barlowi vittatum, Star, Perfectum novum, Triomphe de Pontoise, Pompon de Tirlemont, Victor Lemoine, Trophée, Madame Henri Galotat, Louis Figuer, Azureum plenum, and Madame Ravillana.

The best of the numerous perennial species and distinct from the hybrids are—D. cashmerianum, with flowers nearly equal in size to those of D. formosum, and with stems about 15 inches in height. The flowers are 1 inch in diameter, and are usually of a light blue-purple, but vary in shade to mauve and dark-blue, and are produced in terminal corymbs of six or more. It is well suited for the border or large rockery; in either case, perfect drainage is essential to its successful cultivation, and this is best attained in rock garden culture; its branches have a prostrate habit, which appears to adapt it for such conditions. It is best increased from seed. D. cardinale, a beautiful species of tall, elegantly-spreading growth, having large flowers of a bright scarlet, like those of D. nudicaule. It blossoms at a later period of the summer, and continues longer in flower than D. nudicaule, owing in part to its slower development. It is a most desirable plant, as hardy as D. nudicaule. Seedlings will probably not flower till the second season. In very damp soil it would be prudent to protect the root with a hand-light or inverted pot in winter. D. chinense is distinct from other Larkspurs, and neat and rather dwarf in growth, having finely cut, feathery foliage, and producing, freely, spikes of large blossoms, usually rich blue-purple, but sometimes white. It is a good perennial, easily raised from seed, and continues to flower throughout the summer till late in autumn. Borders and beds. D. nudicaule is another species, with scarlet blossoms, having a dwarf, compact, branching growth, a hardy constitution, and free blooming habit. Its usual height is about 12 in. to 15 in., but it is occasionally dwarfer, and sometimes reaches $2\frac{1}{2}$ feet or 3 ft. The flowers are produced in loose spikes, each blossom being about 1 in. in length; the colour varies from light scarlet to a shade verging closely on crimson, and when seen in the open air, especially in sunshine, dazzles the eye by its brilliancy. It is perfectly hardy, and commences growth so early in the season that it may almost be termed a spring flower, but it may be had in bloom during several of the summer months. It is a handsome plant for warm borders. Although somewhat apt to damp off on level ground, it is a perennial on raised ground, and keeps up a succession of bloom. It is as easy to raise from seed as other Larkspurs. A tall variety of nudicaule is called elatius. D. tricorne is a new dwarf perennial species, 6 in. to 12 in. high; apparently of little value for general cultivation.

THE ANNUAL LARKSPURS.—In these well known hardy annuals there is, as in the perennial kinds, a wealth of beauty for the summer flower garden. We have a host of beautiful sorts having a wide range of colour. There is a great diversity, too, in the habit of growth, some being as dwarf as a Hyacinth, others with a branching habit resembling a candelabrum, while others again grow 3 ft. or $4\frac{1}{2}$ ft. high. The species which have given rise to these varieties have been D. Ajacis (Rocket Larkspur) and D.

Consolida. D. Ajacis has the flowers arranged in long loose spikes forming an erect and spreading panicle. The stem, which rises to a considerable height, is vigorous and has open spreading branches. All the varieties of the Rocket Larkspur may be ranged under three great groups: 1. D. Ajacis majus (large Larkspur).—The stem of this is single, and varies from 3 ft. to 4 ft. 6 in. in height; the flowers are double, and form a long, single, and compact spike, generally rounded off at the extremity. This kind has produced the following varieties—white, flesh-coloured, rose, mauve or puce-coloured, pale violet, violet, ash-coloured, claret, and brown. 2. D. Ajacis minus (dwarf Larkspur).—The stem of this is from 20 in. to 24 in. in height, and even shorter when sown thickly, or in dry or poor soils. The flowers are very double, and are produced in a single well-furnished spike, which is usually cylindrical, rounded off at the extremity, and rarely tapering. The principal varieties are the following—white, mother-of-pearl, flesh colour, rose, mauve, pale mauve, peach-blossom, light violet, violet, blue-violet, pale blue, ash-grey, brown, light brown, white striped with rose, white striped with grey, rose and white, and flax-coloured and white. 3. D. Ajacis hyacinthiflorum (dwarf Hyacinth-flowered Larkspur).—The varieties of which this group consists have been for the most part produced in Belgium and Germany. They do not differ at all from other kinds in the form of their flowers, but only in the disposition of the inflorescence, the spike on which the flowers are set being more tapering, and the flowers themselves farther apart, than those in the two previously-mentioned groups. There is a strain called the Tall Hyacinth, growing about 2 ft. high. Other strains to be found in catalogues are the Ranunculus-flowered (ranunculiflorum) and one called the Stock-flowered, all of which are worth cultivating.

D. Consolida is the Branched Larkspur. This species has branching stems and flowers, of a beautiful violet-blue, hung on slender elongated peduncles, produced later than those of D. Ajacis. It embraces several varieties, both single and double, all of which may be reproduced from seed. The principal sorts are white, flesh-colour, red, lilac, violet, flaxen, and variegated. The varieties of it especially worthy of being cultivated are candelabrum, bearing pyramidal spikes of flowers of various colours and the Emperor varieties, of symmetrical bushy habit, forming fine, compact, well-proportioned specimens, 1½ ft. high by 3½ ft. in circumference. In habit and doubleness of flowers this strain possesses great constancy. There are three colours, viz., dark blue, tricolored, and red-striped. In D. tricolor elegans the flowers are rose coloured, streaked with blue or purple, and growing about 3 ft. high.

CULTURE.—These Larkspurs should be sown where they are to remain any time after February when the weather will permit, but usually from March to April. They may also be sown in September and October, and even later when the ground is not frozen. The produce of these winter sowings is, however, always liable to be devoured by slugs and grubs. The sowing may be made either broadcast or in rows from 4 in. to 8 in. apart, and the plants should stand 4 in. or 5 in. asunder. The branching varieties may be sown in reserve beds, and should be transferred to the beds in March when about 12 in. or 16 in. high, the plants being lifted carefully with balls of earth round the roots, so that they may not suffer. These branching varieties are well adapted for garden decoration either in masses of one or various colours. It may be planted in borders or among clumps of young thinly-planted trees. One great advantage belonging to this class is that it flowers for a longer time and is also earlier than Ajacis—that is to say, throughout the summer, and, according to the period of sowing, from the end of June or July to September, and even to October if care be taken to cut off the flower-stems that have shed their blossoms. They succeed, moreover, in the dryest calcareous soils, and even upon declivities of hills. By pinching, dwarf plants may be obtained that are useful under certain circumstances. In gathering the seed care should be taken only to save it from flowers that are perfectly double; single-flowered plants should, therefore, be carefully weeded out if good seed is to be obtained. Larkspurs are at their best in June and July; they will bloom almost anywhere, especially in dry localities, and do not require much attention. They look well either of one colour or of all the colours mixed, and, by using varieties possessing different colours separately, striking contrasts may be produced. Other annual Larkspurs, such as D. orientale and D. cardiopetalum, are scarcely worth cultivating now that we have so many beautiful varieties.

Dentaria *(Toothwort)*.—Pretty and interesting spring-flowering plants of the Crucifer family, not so much grown as they deserve. There are some half-a-dozen species in cultivation, all worth growing in half-shaded positions in peat beds among shrubs, margins of borders, or in the cool shrubbery. They grow best in a light sandy soil well enriched by decayed leaf-mould, or in soil of a peaty nature. Their flowers are welcome in the early days in spring, and remain in beauty for some time. They are of easy propagation by the small tuber-like roots. Some, such as D. bulbifera, bear bulblets on the stem, and from these the plant may be increased. None of them ripen seed freely with us. The species are—D. bulbifera, 1 ft. to 2 ft. high. Flowers in spring; purple, sometimes nearly white, rather large, produced in a raceme at the top of the stem. D. digitata, a handsome dwarf kind, about 12 in. high, flowering in April; rich purple, in flat racemes at the top of the stem. A native of Europe.

Delphinium grandiflorum.

Dentaria digitata (*Toothwort*).

Delphinium Consolida (*Branched Larkspur*).

[*To face page* 106.

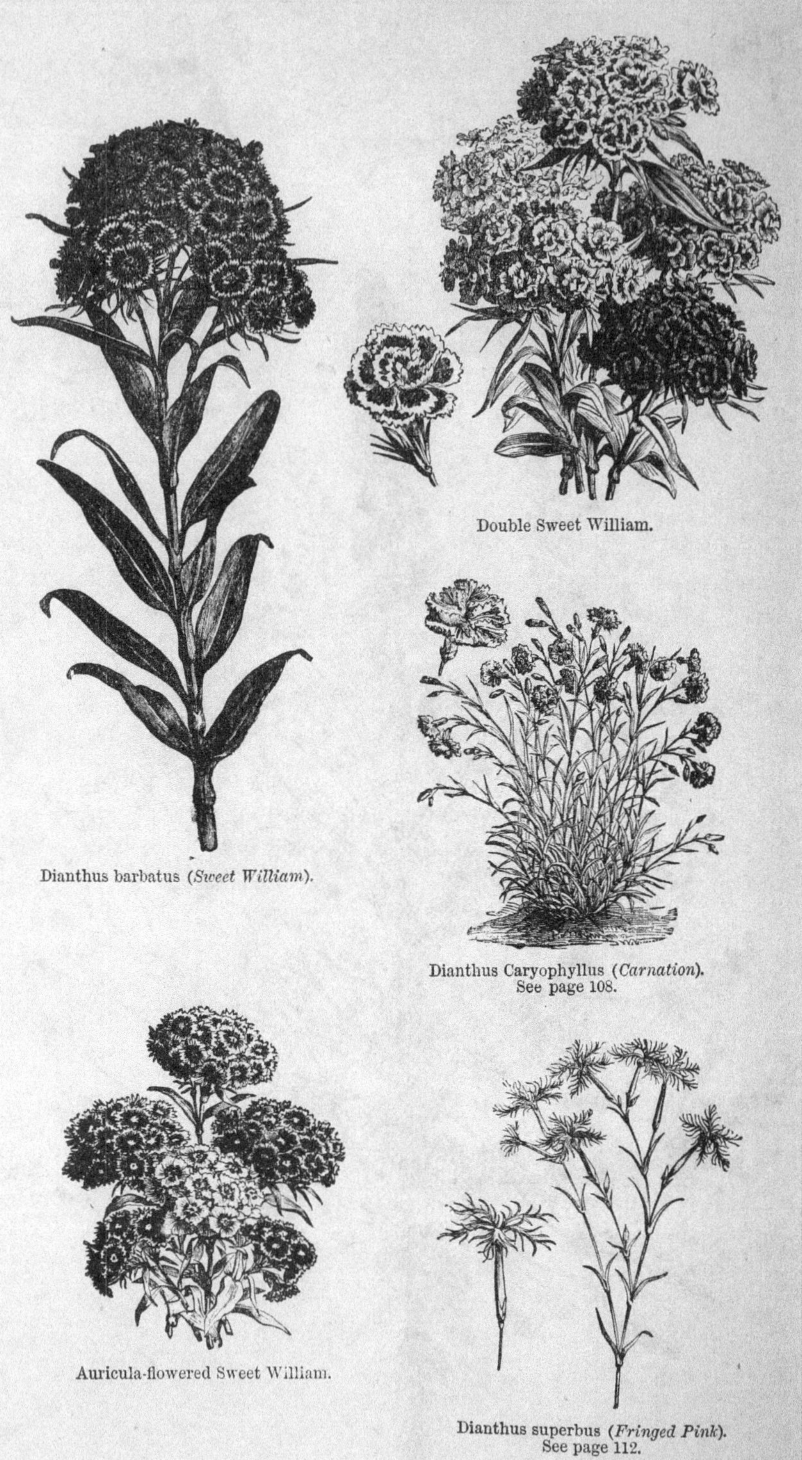

Dianthus barbatus (*Sweet William*).

Double Sweet William.

Dianthus Caryophyllus (*Carnation*).
See page 108.

Auricula-flowered Sweet William.

Dianthus superbus (*Fringed Pink*).
See page 112.

D. diphylla is a pretty plant, growing from 6 inches to 12 inches high, and bearing but two leaves. The flowers are purple, sometimes white, and occasionally yellowish. Woods in North America. D. enneaphylla is about 1 ft. high; flowers creamy white, produced in clusters in April and June. A pretty plant for a shady border. Mountain woods in Central Europe. D. maxima is the largest of all the species, growing 2 ft. high. Flowers pale purple in many flower-heads. N. America. D. pinnata is a stout species at once distinguished by its pinnate leaves; 14 in. to 20 in. high; flowering from April to June; large, pale purple, lilac, or white, in a terminal cluster. Switzerland, in mountain and sub-alpine woods. D. polyphylla, similar to D. enneaphylla, is about 1 ft. high, with flowers in clusters, cream coloured. A handsome plant; from woods in Hungary.

Deptford Pink *(Dianthus Armeria).*

Desmodium *(Tick Trefoil).*—A few of the North American species are cultivated, but they have a weedy appearance that precludes them from general culture. These are D. canadense, marilandica, and Dilleni, all growing from 2 ft. to 4 ft. high, with slender stems, terminated by dense racemes of small purplish flowers. A really pretty plant is D. penduliflorum, a shrub that is quite hardy if the stems are cut down annually. It has long, graceful shoots, bearing along their upper portions numerous rich violet-purple blossoms. It blooms in September, and has a graceful effect at that season, though it is not a showy plant.

Dianthus *(Pink).*—A genus of the highest garden value, containing as it does several of our finest families of hardy flowers—the Carnation, Pink, and Sweet William, besides numerous alpine and rock species that are among the most charming of mountain plants. The genus is a very large one. A good many of the species are heath, dry meadow, or maritime Alps plants, or shore plants, such, for example, as the Fringed Pink (D. superbus); and, in consequence, so far as our climate is concerned, they are almost at home in lowland gardens. But some, on the other hand, are among the very highest alpine plants, as, for example, the Glacier Pink and the Alpine Pink.

The following is a selection of the finest species—

D. alpinus *(Alpine Pink).*—A beautiful and distinct plant, recognised at a glance from any other cultivated Pink by its blunt-pointed, shining green leaves. The stems bear solitary flowers, of a circular form, deep rose spotted with crimson, and when the plant is in good health are so freely produced as to hide the leaves. In poor, moist, and very sandy loam on rockwork it thrives and forms a dwarf carpet, though the flower-stems may rise little more than 1 in. in height; both leaves and stems are much taller and more vigorous in deep, moist, peaty soil. Wireworms, rather than unsuitable soil, often cause its death. It should be placed in an exposed position, and carefully guarded against drought, especially when recently planted; comes true from seed, and is not difficult to increase in that way, or by division where it grows freely. Alps of Austria, flowering in summer. A choice rock garden plant.

D. barbatus *(Sweet William).*—One of the most admired of our garden flowers. Few flowers better repay for care, and it should be seen in every garden. It is hardy, of vigorous growth, and has a rich profusion of bright coloured flowers, forming sheets of bloom, the colours vivid and chaste, the flowers finely and distinctly marked. What renders the Sweet William of such high value to the owners of small gardens is that it is of extremely easy culture, and may be raised from seed without the aid of glass.

The Sweet William has been greatly improved of late years, and the old varieties are far surpassed. The points the "florist" improver aims at are a circular flower, with no indentation where the petals meet, thick in petal, and all the petals marked alike, the colours meeting each other in clearly-defined lines without any feathering or flushing into each other; but in this, as in other flowers, the more variety of form the better. There is a great variety of colour in the Sweet William, and they may be classed under two heads—dark and light kinds. Of the latter there is a strain known as the Auricula-eyed, the blooms of which have a clear white eye in a setting of red or purple or some other rich dark colour. Smooth-edged flowers, such as Hunt's strain, have their admirers. Fine, even-rounded trusses are always found in a good strain, but size is generally allied to high culture. Except for show purposes, however, very large trusses are not to be desired, as these usually need some support. The finest strain is usually to be found where there has been year after year the greatest care exercised as regards the selection of only the finest flowers, largest trusses, and most varied markings. The only self-coloured flowers are those that are pure white, pink, or crimson; all others are parti-coloured or variously marked, some very prettily mottled, others more or less edged with white or pale pink.

CULTURE.—This is very simple in an ordinary way, for all that one needs to do is to sow the seed in April, in a well-prepared bed of soil, in a sunny situation, thinning out the young plants when large enough, or, if a large stock is required, planting them out some 6 in. apart in good soil. About the latter end of September transplant them to their permanent quarters, and they will the following summer yield an abundance of bloom. When, however, any particular strain is required to be rapidly increased the following plan is a good one: Sow in pots, and allow the seedlings to become a little drawn and lanky before planting out. Plant out in light loam, dressed only with a little leaf-mould or loam from rotted turves, placing the seedlings so that a few of the lower joints are under the surface of the soil. When the blooming stems are well above the foliage prick in a dressing of guano all round the plants, give plenty of water in dry weather, and a further slight dressing of guano just before the flowers begin to open. The effect of this treatment will be vigorous stocky shoots from the buried joints of the plant, all rooted and ready to plant out as soon as the bloom is over. Sweet Williams may also be propagated by means of cuttings taken off in early summer, for the main stems, which should rise for bloom, creep along the ground, and throw up shoots suitable for cuttings from every joint, and a little sheaf of them may be taken from the tips of the main stems, so that each plant would furnish over a hundred cuttings.

Double-flowered kinds, as a rule, are not desirable, except the double dwarf magnificus, the deep velvety crimson flowers of which are by far the finest we have met with among the double kinds. The heads of flower, which are large, are produced in profusion, and the colour is rich and effective. Its dwarfness, too, is a recommendation, and it is, moreover, a vigorous grower, and soon forms a strong tuft.

D. Caryophyllus *(Carnation)* is so well known, and is such a universal favourite, that it seems superfluous to write a description of it. So early as 1769 we find that the Carnation was divided into four classes, viz., Flakes, Bizarres, Picotees, and Painted Ladies, The Flakes were those having two colours only, the stripes going the whole length of the petals. Bizarres (from the French, meaning odd or irregular) were spotted or striped with three distinct colours. Picotees (from the French, *piquotée*) were freckled or spotted, and had a white ground with the additional colours in spots, giving the flowers the appearance of being dusted with the colours. Painted Ladies were those having the under side of the petals white and the upper surface red or purple, so laid on as to appear as if really painted. Unfortunately this last class has so entirely disappeared from cultivation that many of the growers of the present day are not aware that it ever existed. The first two classes still remain unchanged; the Picotees, however, instead of being spotted, have the colours now confined to the outer edge of the petals, and any spot on the ground colour (which may be either white of yellow) would detract from its merits as an exhibition flower.

Another class is the Clove, which, although not recognised by florists of the "hard shell" school, is, without doubt, the greatest favourite with almost everyone. The Clove is self-coloured, but may be of any shade, though the old crimson and the white are still the best known. It seems strange that at no flower show in the country does one ever see prizes offered for a collection of self-coloured or Clove Carnations. The Tree Carnation is another class of recent introduction, and is very valuable as a pot plant; or, if planted out in a greenhouse border, it produces its flowers in winter and spring, when none can be had out-of-doors. The most popular of this class is Souvenir de la Malmaison, with its large cream-coloured blossoms and delightful fragrance, and from this have been obtained sports of different colours; so that, with these and other varieties, there is now no difficulty in having them of all colours, from pure white to bright scarlet.

As a rule, the choice-named varieties of Carnations and Picotees for show purposes are grown in pots, but as our subject in this book is the open-air garden only, we confine our remarks to their culture in the garden, but also treat of it, shortly, from the exhibiting florist's point of view.

CULTURE FOR BORDERS AND BEDS.—First, then, of the wants of the general public, who rightly esteem a good crimson or white Clove as it grows in the open garden as much as the most exact staged flower. We say rightly, because the superiority in form is, from the point of view of those who have thought and studied most about form, wholly with the bold, free, undressed flower. The Carnation and Picotee are best propagated by layers very late in July, or very early in August. The florists do not consider two-year-old plants good enough to furnish what are called exhibition blooms, but for the flower garden they are better than young plants; and even old tufts in a suitable spot on rockwork, &c., will furnish good flowers for years. It is when the plants are about two years of age that they are most valuable for general garden use. Wherever many flowers are required for the house, the various kinds of border Carnations are worth growing to a considerable extent, merely for the sake of cutting their flowers, even if they are not desired as flower-garden ornaments. The special beds where the florists' kinds may receive attention should be in the kitchen garden or some by-spot, but some of all kinds may be judiciously planted in the mixed border and in the flower garden, and will there prove highly ornamental. From the borders, when they get a little scraggy, as they are wont to do when a few years old, they must be removed, and the stock kept up with young plants. Therefore it is desirable to propagate a stock of Carnations every year. They are too often left in the hands of a few enthusiastic florists, whereas they ought to be in every garden. Some of our nurserymen are now beginning to see the mistake of neglecting a noble flower like this, and are trying to raise a race of bold, free, and varied border flowers which may be easily grown in every garden in the open air. They will succeed, and our gardens be very much the better of it. In specially cultivating the better kinds in beds, it is usual to cover the surface with 1 in or more of fine rotten manure that has been passed through a sieve, and plenty of water given in dry weather; but as many will not care about paying more attention than is necessary, it may be stated that neither water nor top-dressing is usually required in ordinarily good garden soil, and the result will be quite as valuable from an ornamental point of view. But when a good collection is grown in special little beds in some warm border in the kitchen garden, a top-dressing, composed of one barrow of mould to three of decayed manure, could be given in a very short time, and if the weather or soil were very dry, an occasional heavy watering would of course improve matters. As to varieties, they are endless; and as home, Continental, and American florists are busy raising seedlings, they are likely to be much added to, though enough attention has not as yet been paid to the raising of vigorous border and flower-garden kinds offering as great a range in colour, form, and fragrance as possible. In ordering from nurserymen, the public should distinctly make known their wishes in this respect. The Carnation is not a flower that depends for its beauty on the carrying out of an elaborate code of instructions, which only the special grower for exhibition may care to master.

Border Carnations and Picotees.—Although we prefer in this book to leave selections of varieties to the reader and the catalogues of the day rather than mention those of any one year or grower in the case of these border kinds, we mention some in the hope that the number will be largely added to by the efforts of raisers in all gardening countries: Captain Webb, Colonel Arnet, Countess of Ellesmere, Delicata, Dora, Duke of Wellington, Emperor, Erebus, Fanny Archer, Flamer, Garibaldi, Knight, Lady Armstrong, Maximilian, Miss Wheeler, Mrs. Mathews, Mrs. Raynor, Nigger, Novgorod, Old Clove, Pilrig Park, Prince of Orange, Prince of Wales Clove, Quadricolor, Redbrae's Picotee, Red Rover, Red King, Rosy Gem, Scarlet Gem, Stranger, Taglioni, The Bride, Turk, Utility, Violette. Although the above are described as "border" kinds, they are also well worthy of special culture in good beds, either for cutting their flowers, or for their beauty where they grow.

Perpetual Carnations in the open air.—These, if from a good strain of French seed, are very satisfactory plants and useful for cutting. The drawback they have is the habit of flowering in the winter, but this can be obviated by sowing early, so as to get the plants to a good size by autumn, when they will begin flowering in the spring and continue in bloom all the summer. Pipings struck in the spring and planted out in the autumn will behave in the same way. Old plants are difficult to manage in the open air, but will survive the winter if well thinned out; the only danger is damp cold, which rots them at the surface of the ground. They grow very well in a light, rich soil on chalk. On tne chalk hills, far from any damp soil and where no damp rises in winter, they should survive most winters. "My plants were in a light soil on chalk, but so near clay soil that the wind brought the cold air from the clay fields. It is impossible to say what colours will be produced from Carnation seed. My seed was of self kinds, but quite a third of the plants turned out flakes and Picotees; only four were single, and three of these were worth keeping for their brilliancy and free flowering as well as their smooth-edged petals. One plant was merely a perpetual flowering form of the common white Pink of the cottage garden; another assumed a dwarf, almost creeping, habit, with very large flowers on short stalks; some developed a climbing habit, but the greater number spread into large bushes 3 ft. across, sending up 60 or 70 flowering shoots at one time; nearly all were Clove scented. The best flowers were produced from plants of moderate and sturdy habit. Most of these had very large petals, and resembled a Gardenia in general appearance. Their free-rooting habit makes them unsuitable for pots. Many of my plants filled almost 3 ft. of soil with their roots; it is manifestly waste to cramp such free-growing plants in pots."—J. D.

Garden culture for exhibition.—About the end of July cover the bed intended to be devoted to Carnations, &c., about 2 in. with good rotten manure, and add to this, if the soil is sandy, 2 in. of good mellow loam, or, if it is stiff, the same quantity of sand. Then whenever time can be spared fork it in well and dig it over. Then put the plants in firmly, all of one sort in a row with a good legible label at the end. Thus they will need no attention till next spring, being perfectly hardy. At the same time take up and put in in the same way any seedlings sown in the spring, which will now be fine strong plants. The next spring, when the severe cold has ceased (about March or April), fork over carefully between the rows, and water them if dry in fine weather. When the flower-stems begin to rise place a stick about 30 in. long to each plant. They should be painted a light whitish green. The flower-stems must be kept well tied up as they grow, tying quite loosely, for if the stems are tied tightly they will knee and bend, and finally break. About June 20 (or later), when the buds have appeared, take off all the buds but three on each shoot, so as to leave each bud a little footstalk to itself when it grows (what is lost by this in quantity will be repaid twenty-fold in quality). From this time until the buds are nearly showing colour give them occasionally a little weak manure water, *i.e.*, a handful of well-rotted stable manure to a large pot of water. As soon as the buds show colour at the top tie them round, about half way down, with a little strip of bass. This should be done every morning in July, as it will save much trouble and the unsightly peculiarity termed a "split pod." If in spite of this the pods split on one side, carefully open the bud all round at the other segments. For this purpose use the flat wedge handle of the knife that is used for layering.

Unless it is intended to save seed, cut off dead blooms as soon as they wither and the flower-stems as soon as no more buds are left to come out, which will be about the end of August or beginning of September. The last week in July (not later) see about layering. As soon as the layers are rooted, which will be early in September, take them off and lay them in by the heels for a time whilst taking up and throwing away the old stools; top-dress and fork over the bed with 2 in. of well rotted stable litter or cow-house sweepings, replace the layers, and then they are in the same condition as at the beginning.

PROPAGATION BY SEED.—The proper time to sow is about April or May. Prepare a compost of equal parts loam, leaf-mould, and silver sand, sift it fine, and fill a quantity of 3-in. pots (as many as you have sorts of seed) to within 1 in. of the rim. Sprinkle each with a fine rose, flatten the surface, and put down each seed separately, with the point of a knife, about ½ in. apart. Cover them very lightly with finely-sifted compost, and put them in a cold frame or house out of danger of frost. When they have made three pairs of leaves, prick them out round the edges of 5-in. pots filled with the same compost, about 2 in. apart, and keep still in the cool house till there is no longer fear of frost. When about 3 in. high, prick them out into beds, about 4 in. apart. The beds might be enriched with a little sand and manure if convenient. In the autumn they will be nice little plants, and may be put into beds when they flower, which will be the next year. Keep and name any kind really good, discarding all singles, and using the rest for borders or beds for cutting from.

BY PIPINGS.—When the plants throw up too many shoots to layer, or the root is attacked by disease, the shoots may be taken off as follows: Take the shoot just above the fourth or fifth joint from the top, and with a sharp pull draw it out from the socket formed by the next joint, which it will pull away with it. Make a little upward slit in the cutting just through the joint, and thrust it firmly into a pot filled with the compost described above for seeds to within 1 in. of the top of the pot, and the rest with silver sand. Water the pot and plunge it in fibre under a hand-light for three or four weeks, when the pipings will be rooted. They may then be potted off singly or bedded like layers, and will flower the next year. Plants struck in this way are never so good as those propagated by layers, but to save a good sort or get up a good stock it is a useful expedient.

BY LAYERS.—This is the best and most generally accepted method of propagating Carnations and Picotees. It should be commenced at latest the last week in July, and should be all finished by the second week in August. It is performed as follows: Having prepared the compost prescribed, first scrape away the earth all round the plant to the depth of 2 in., and replace it with the fresh compost. Then strip each shoot up to the top three or four joints, going all round the plant before proceeding further. Then with a fine sharp knife cut half through a shoot just above a joint, make a slanting cut down through the joint, bringing the knife out just below it; take a peg with a hook in it, and thrust it into the fresh compost just above the tongue, so that it catches it as it comes down and pegs it into the earth. Then cover it with a little more compost placed firmly. Proceed thus all round the plant, finally carefully watering with a fine rose water-pot to settle the soil around the layers. In about a month the layers will be rooted, and by the second week in October all the young plants ought to be in their winter quarters.

DISEASES.—Carnations are subject to three fatal diseases—canker, mildew, and gout. Canker is produced by damp, and the first symptom is either the grass turning colour, or the non-formation or stoppage in the growth of shoots. The roots will be found rotted through, and decay creeping up the centre of the shoots, which should be taken off at once, well above the affected part, and piped. The only safeguard against it is to have the compost sweet and well mixed, and not to give too much water early in the year. Mildew is of two sorts, the white and the black. White mildew, which appears like flour on the leaves, should be sprinkled with water, and then dusted with sulphur one morning, and the sulphur carefully washed off the next morning. Black mildew appears as little black specks on the grass, and means death. It is the Carnation's cholera. The instant it appears every shoot bearing the slightest spot should be cut off and burnt with the root and earth, only the shoot quite untainted being saved for piping. If not, it will spread like wildfire, and ruin the whole stock. Gout is generally the result of twisted stems and damp atmosphere. The soft wood just above the ground swells into a cavernous goître. It does not seem to spoil the flowers. Though a gouty plant can never be layered, all the shoots should be cut off and piped. Damp is the greatest enemy of the Carnation, producing in the first instance all these ills, and so must be particularly guarded against, especially when the young plants are in the frame in winter. Hundreds have been lost by a single injudicious watering.—A.

D. cæsius *(Cheddar Pink)*.—One of the neatest and prettiest of the dwarf Pinks. The large, fragrant, rosy flowers, showing in spring, are supported on stems 6 in. in height, and sometimes taller if in rich soil. It requires peculiar treatment, as in winter it perishes in the ordinary border, but flourishes freely and flowers abundantly on an old wall; a native of Europe, and on the rocks at Cheddar, in Somersetshire. To establish it on the top or any part of an old wall the best way would be to sow the seeds on the wall in a little cushion of moss, if such existed, or, if not, to place a little earth with the seed in a chink. It may also be grown upon a rockwork in firm, calcareous, sandy, or

PLATE XCV.

Flowers of Dianthus Caryophyllus (*Carnation*). See page 108.

Dianthus superbus grandiflorus (*Double-fringed Pink*). See page 112.

Dianthus chinensis fl.-pl. (*Double Chinese Pink*). See page 112

PLATE XCVI.

Dianthus cæsius (*Cheddar Pink*).

Single Pink (natural size). See page 112.

Dianthus deltoides (*Maiden Pink*).

Dianthus dentosus hybridus (*Amoor Pink*).

Dianthus plumarius fl.-pl. (*Double Pink*).

Dianthus plumarius (*The Pink*).

[*To face page* 111.

gritty earth, placed in a chink between two small rocks.

D. deltoides *(Maiden Pink)*.—A pretty native plant, with bright pink-spotted or white flowers, rather freely produced on stems from 6 in. to 12 in. long; it will grow almost anywhere, on border or on rockwork, and does not appear to suffer from wireworm, as most other Pinks do. It frequently flowers several times during the summer; may be readily raised from seed, or easily increased by division. The variety glauca has white flowers, with a pink eye. It is abundant on Arthur's Seat, near Edinburgh, and forms a charming contrast to the crimson kind.

D. dentosus *(Amoor Pink)*.—A distinct and pretty species; dwarf, with violet-lilac flowers more than 1 in. across, the margins toothed at the edge, the base of each petal having a regular dark violet spot, which produces a dark eye nearly ½ in. across in the centre of the flower. Flowers in May and June, continuing till autumn, and thrives well in sandy soil, in borders, or on rockwork; seed; South Russia.

D. neglectus *(Glacier Pink)*.—It is impossible to exaggerate the beauty of this plant. It forms, very close to the ground, tufts resembling short wiry grass, from which spring many brilliant flowers, 1 inch across, and of a deep and bright rose. It grows with freedom in very sandy loam, either in pots or on rockwork, rooting through the bottoms of the pots into the sand as freely as any weed, and is perfectly hardy and easily grown. Alps and Pyrenees. Easily increased by division and seed. *Syn.*, D. glacialis.

D. petræus *(Rock Pink)*.—A charming species, forming hard tufts, 1 in. or 2 in. high, from which spring numerous flower-stems, each bearing a fine rose-coloured flower. It seems to escape the attacks of wireworm. It flowers in summer, and is well worthy of a position on the rock garden, where it ought to be planted in sandy and rather poor moist loam. Hungary; seed or division.

D. plumarius *(The Pink)*.—This plant is considered the parent from which our numerous varieties of Pinks have sprung. It has single purple flowers, rather deeply cut at the margin, and is naturalised on old walls in various parts of England, though it is not a true native. The wild plant is rather handsome when grown into healthy tufts, but on the level ground it is apt to perish or get shabby. But the many fragrant double varieties are welcome everywhere, and there is no reason why these should not be cultivated as rock or bank plants, particularly as they live much longer and thrive better on such elevations, though they can be grown well in any ordinary soil or position. They are such well known flowers, so hardy and fragrant, that one would think a few words in praise of them quite unnecessary; but, with many other charming plants, they have been almost driven away by the taste for common bedding plants. They have for many years been amongst the most favourite "florists'" flowers in European countries. They are hardier and dwarfer than the Carnation. In order to grow the finest kinds well, the ground must be rich and well prepared beforehand, and they should be planted in August. Planted 9 in. by 9 in. plant from plant every way. If the winter is very severe, they are the better for a little litter being put over them. In spring the beds in which they are grown will be improved by stirring their surface a little, and giving them a top-dressing of fine old manure and a slight dusting of guano. As they push up their flower-spikes they should be staked, and if for competition the buds should be thinned, as a number of varieties produce buds too freely. The culture of Pinks, however, either for exhibition or the garden, is very simple, and the outlay small. We can get newly-struck pipings in August and September—the best months to plant. Place them at once in a sunny situation; if situated in a smoky town a cold frame will be needed; if the air is clear an open bed will do; and when once planted in the open garden, scarcely any care is needed till they begin to push up their flower-stems. In planting in the spring, do so as early as the weather will permit, and, as soon as they begin to grow, mulch the top of the bed with equal quantities of well-rotted horse manure and leaf mould, the whole about 1 in. deep. They will then push on their new growth fast.

Increasing stock.—If the plants have made good growth in July, cut with a sharp knife the shoots that are the strongest, cut the ends of the grass off, and cut the shoot two or three joints below the grass or leaves. Prepare a bit of ground as follows: Scatter on the surface a little salt, then riddle on 2 in. deep of fresh soil, prick the pipings in, and put a light or hand-glass over them; they will then be rooted in a few weeks. Where seed is wanted the flowers should be protected from wet, and as the flowers decay the withered petals should be removed, as they encourage damp and also form a harbour for insects. Seed may be sown early in June in pots, or in the open ground, and saved only from some of the finest and most constant varieties, such as have a vigorous and hardy habit of growth.

Garden or border Pinks.—The show Pinks may be left to the taste and care of the exhibitor. There are certain old and new kinds which must be taken good care of by the "general lover" of flowers, and these are the vigorous and hardier border kinds, which we grow for their beauty in the garden and for their fragrance. As in the case of the hardier border and Clove Carnations, so we must encourage these. We give the names of some of the best of the hardier kinds—Anne Boleyn, Ascot (soft pink), Fimbriata major, Fragrans (pure), George White, Hercules, Lady Blanche, Mrs. Moore, Mrs. Pettifer, Mrs. Sinkins (Mule Pink), Marie Paré (Mule

Pink), Napoleon III., Multiflora, Newmarket, Pluto, Purity, Robusta, Rubens, Thalia, White Queen, Wm. Bruce, High Clére, Multiflora flore-pleno, Multiflora rosea, Striatiflorus, Speciosa fl.-pl., Coccinea, Early Blush, Fimbriata alba (old white), Lord Lyons, Miss Joliffe, Mrs. Moore, Nellie, White Perpetual, the Clove Pink.

DWARF SINGLE AND DOUBLE PINKS.—Messrs. Dicksons & Co., of Edinburgh, have lately been raising some dwarf profuse-blooming Pinks so compact in habit and stiff in the stem that they are said to do without stakes. We think most Pinks are better without stakes, especially when their foliage is healthy and in such wide tufts as will keep the flowers free of the splashed earth; but these new dwarf sorts may be compact enough for such effects as we seek on the rock garden. Mr. J. Grieve, who raised them, says: "Both the single and the dwarf double varieties will prove quite a boon to the flower gardener and for bouquets. To the ordinary eye all florists' Pinks consist of but one variety; whereas amongst the single and dwarf sorts there are endless colours, and the petals of many of the flowers are so finely cut and varied in tint as to render them easily mistaken by the uninitiated for other plants. Numbers of the single sorts look like miniature Petunias, and we have a few not unlike Calochortus venustus. What flower, too, can rival single Pinks of any colour for table decoration?" Carnea, Beauty, Delicata, Rosea, Spicata, and Odorata are among the best of these new dwarf Pinks, a class which will no doubt be added to.

D. sinensis *(Chinese Pink)*.—This species has given rise to a beautiful race of flowers. It is an annual, biennial, or perennial, according to the way it is sown and grown. If sown early, the plants will flower the first year; if late, the second. On dry soils, and if the winters be mild, it will live for two or three years. The varieties, both single and double, are now very numerous and beautiful; they may be classed under D. Heddewigi and D. laciniatus. The forms of Heddewigi, the Japanese variety, are dwarf and very handsome, while there are also the double-flowered forms of it, particularly Diadematus, the flowers of which are large and very double. The laciniated section have the petals very deeply cut into a fine fringe, and of this class there are also double-flowered forms. The colours of both are much varied, including striped crimson and white sorts. There is a dwarf class (nanus), growing about 6 in. high, that is very desirable, but not so useful for cutting from as the taller varieties. Two beautiful and distinct selected sorts, Crimson Belle and Eastern Queen, are among the best varieties. Sow under glass in February, with very little or no bottom-heat; give air freely during open weather, and plant out in April in well-cultivated soil, which need not be rich. Place the plants from 9 in. to 12 in. apart each way, and they will form compact tufts, which will be covered with blossom. Encourage the formation and growth of laterals by pinching off decayed flowers, and the result will be a mass of blossom through the whole summer, probably till November. Some sow the seeds in autumn, and winter the young growing plants in frames or under hand-glasses, hardening them off by degrees in spring, until they have become fully established. These Pinks are admirable for the flower garden, either in beds by themselves, mixed, or in decided colours; they may be well used with taller plants of a different character dotted sparsely among them.

D. superbus *(Fringed Pink)*.—A handsome and fragrant species, easily known by its petals being cut into lines or strips for more than half their length. It inhabits many parts of Europe from the shores of Norway to the Pyrenees, and is a true perennial, though it perishes so often in our gardens, when very young, that many regard it as a biennial. It is more apt to perish in winter on rich and moist soil than on that which is poor, light, and well drained, and it should be planted in fibry loam, well mixed with sand or grit, where it is desired to establish it as a perennial. It, however, grows on nearly every description of soil; and by raising it every year from seed an abundant stock may be kept up even where it perishes in winter. It comes true from seed, generally grows more than 1 foot high, flowers in summer or early autumn, and is better suited for mixed beds and borders than for the rock garden.

Dianthera americana *(Water Willow)*. — A perennial herb, growing from 2 ft. to 3 ft. high, with slender erect stems and small purplish flowers. It grows in streams and ponds in North America, and being quite hardy it is suitable for cultivating in like positions here. Acanthaceæ.

Diapensia lapponica.—A sturdy and very dwarf evergreen alpine shrub, growing in dense rounded tufts, with narrow, closely-packed leaves and solitary white flowers, about half an inch across, the whole plant being often under 2 in. high. It may be grown well on fully exposed spots on the rock garden, in deep sandy and stony peat, kept well moistened during the warm season. A native of N. Europe and N. America, on high mountains or in arctic latitudes, flowering in summer.

Dicentra *(Bleeding Heart)*.—A genus of the Fumitory Order, including several important garden plants. There are about half-a-dozen cultivated species, of which the following are the finest:—

D. chrysantha.—A fine plant, forming a spreading tuft of glaucous rigid foliage, from which arises a stiff leafy stem, 3 ft. to 4 ft. high, bearing long branching panicles of bright golden-yellow blossoms, each about 1 inch long. It flowers in August and September; the seedlings do not bloom till

Group of Border Pinks. See page 111.

PLATE XCVIII.

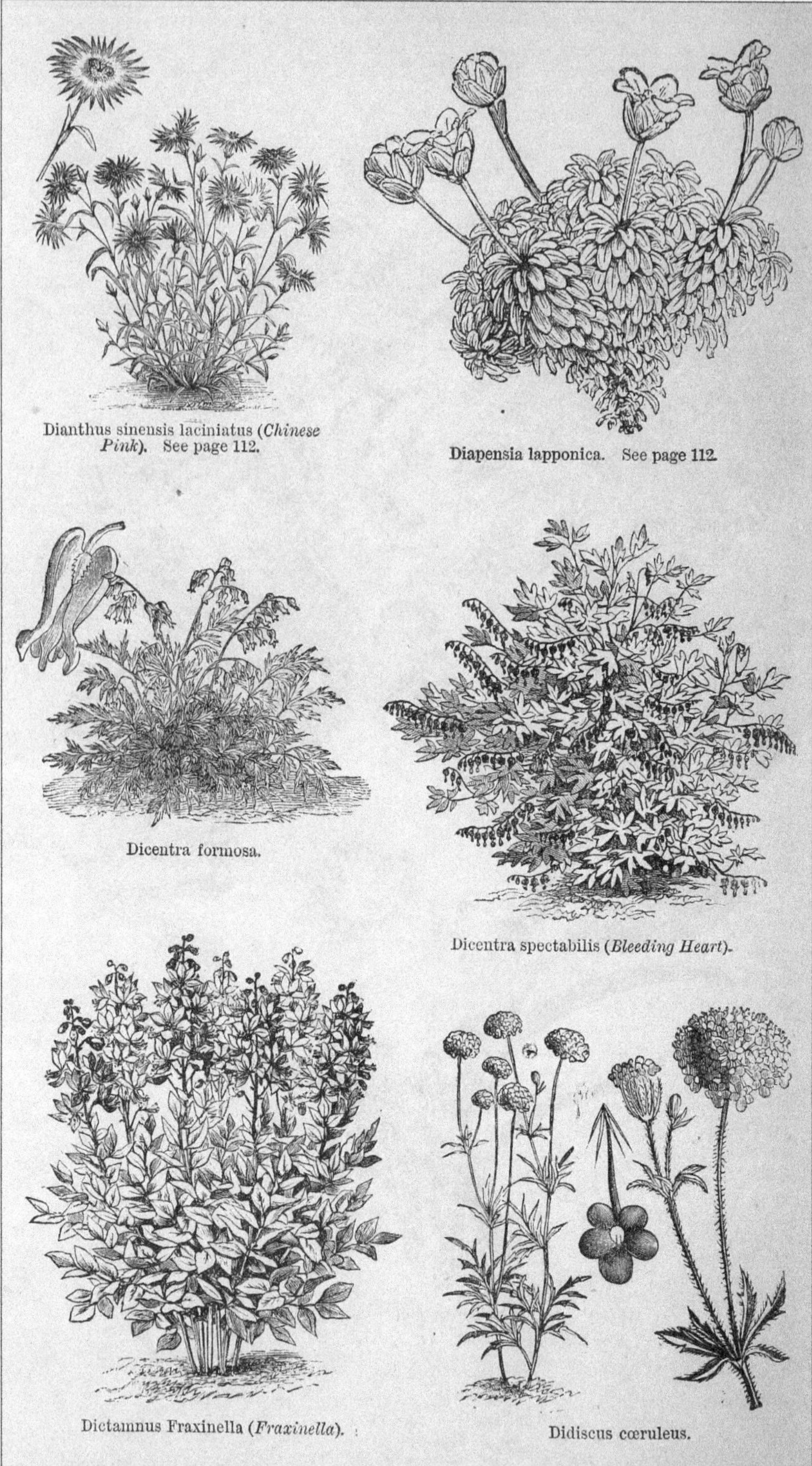

Dianthus sinensis laciniatus (*Chinese Pink*). See page 112.

Diapensia lapponica. See page 112.

Dicentra formosa.

Dicentra spectabilis (*Bleeding Heart*).

Dictamnus Fraxinella (*Fraxinella*).

Didiscus cœruleus.

[*To face page* 113.

the second year. The hardiness of this handsome plant has not yet been fully tested, but it seems to be quite hardy in light rich soil in warm, sheltered positions. It is worthy of a trial. California.

D. eximia combines the grace of a Fern with the flowering qualities of a good hardy perennial. It grows from 1 ft. to 1½ ft. high, and bears its numerous reddish-purple blossoms in long drooping racemes. It is useful for the rock garden, and mixed border, or for naturalising by woodland walks. It enjoys most a rich sandy soil, but it will grow anywhere. N. America. Division.

D. formosa is similar to the preceding, having also Fern-like foliage, but is dwarfer in growth, the racemes are shorter and more crowded, and the colour of the flowers is lighter. Suitable for same positions as D. eximia. Native of California.

D. spectabilis is a beautiful and most important plant, now too well known to need description, nearly every garden in the country being embellished with its singularly beautiful flowers, which open in early summer, gracefully suspended in strings of a dozen or more on slender stalks, and resemble rosy hearts. It succeeds best in warm, light, rich soils, in sheltered positions, as it is liable to be cut down by late spring frosts. Besides a position in the mixed border, it is of such remarkable beauty and grace that it may be used with the best effect near the lower flanks of, or in bushy places near, rockwork, or on low parts where the stone or "rock" is suggested rather than exposed. It is worthy of naturalisation by wood walks, on light rich soils. It is also an excellent mixed border plant, and better still for snug corners on the fringes of choice shrubs in peat, which soil suits this well. There is a "white" variety, by no means so ornamental as the common one, though worth growing for variety's sake. Propagated by division in autumn.

D. cucullaria *(Dutchman's Breeches)* and D. thalictrifolia are less important, belonging more to the curious garden.

Dichondra repanda.—A small trailing plant resembling a Convolvulus, of no general value.

Dicksonia antarctica. — A noble evergreen Tree Fern, with a stout trunk, attaining a height of 30 ft. or more. The fronds form a magnificent crown, often 20 ft. to 30 ft. across, are from 6 ft. to 20 ft. long, beautifully arched, and become pendulous with age. The hardiest of Tree Ferns, and therefore most suitable for placing in the open air, in sheltered shady dells, from the middle of May to the beginning of October. In favourable localities it may even be left out all the winter.

Dictamnus fraxinella *(Fraxinella).* —A favourite old plant, worthy of a place in every garden. It grows about 2 feet high, the erect stems forming dense tufts. The flowers are large, pale purple, pencilled with darker lines, and are borne in racemes in June and July. There is a white flowered variety. It succeeds best in a light dry soil, in partially shaded borders. Propagated by seeds sown as soon as ripe, or by means of its fleshy roots, which, if cut into pieces in the spring, will strike freely and form good plants much quicker than seedlings. It is a proverbially slow-growing plant in most gardens, though more free in some warm, chalky soils. Native of the Taurian and Caucasian Mountains.

Didiscus cœruleus.—A native of New Holland, growing from 1 ft. to 2 ft. high. Its stems are erect and much branched, each branch being terminated by a flat umbel of small flowers, of a pleasing clear blue colour, borne freely from August to October. It is a half-hardy annual, and requires rather careful treatment, inasmuch as it is impatient of excessive moisture, particularly in the early stages of its growth. It requires to be raised in a gentle hotbed, and the seedlings should be transplanted in May into a warm, friable soil, in which they will flower freely. Those who seek distinct and novel effects might use this, its peculiarity being that it is a pretty blue flower of the Parsley order —one usually having pale or poor flowers. A little bed or ground work would be charming if only as a change. *Syn.*—Trachymene cœrulea.

Dielytra *(Dicentra).* — An erroneous name given to well-known plants which we treat of under their proper name, Dicentra.

Digitalis *(Foxglove).*—The most important plant in this genus is our native Foxglove, for though there are several species in cultivation none are so handsome. The best of the exotics is D. grandiflora, a tall slender plant, bearing large, bell-shaped, yellow blossoms in long racemes. The other kinds in gardens are D. ferruginea, aurea, eriostachys, fulva, lævigata, lanata, lutea, ochroleuca, parviflora, Thapsus, tomentosa, but these are mainly fit for botanical collections.

D. purpurea *(Foxglove).*—Wild Foxgloves seldom differ in colour, but when cultivated they assume a variety of colours, and include white, cream, rose, red, deep red, and other shades. The charm, however, of these varieties lies in pretty throat markings, spots and blotchings of deep purple and maroon, and these, when seen in large flowers, make them resemble those of a Gloxinia; hence the name gloxiniflora, applied to some finely-spotted kinds. The garden plants are more robust, the stems stouter, and the flowers much larger than those of the wild plant, and they make grand border flowers. They look well as a background to mixed borders, associated with other tall growing subjects; and the improved varieties are desirable additions to the wild garden, where, if planted or sown in bold masses, they have a fine effect. They are good, too, among Rhododendrons, where these bushes are not too thickly planted, and they break the masses of foliage charmingly. The seed being small, it is best sown in pans or boxes, under glass, early in May, and when the young plants are well up they should be placed out of doors to get thoroughly hardened before finally planting out. Where planted in shrubbery borders it is well to make varied clumps of several plants, as they produce a finer effect than when set singly. Not unfrequently the Foxglove blooms two years in succession; but in all cases it is well to sow a little seed annually; and if any be to spare, it may be scattered in woods or copses where we may desire to establish the plants. Those who do not require to save seed should cut out the centre spike as soon as it gets shabby, and the side shoots will be considerably benefited thereby, especially if a good supply of water be given in dry weather. In the case of a good variety, a side shoot will supply an abundance of seed.—D.

Digraphis *(Ribbon Grass)*.—A genus of Grasses, of which the Ribbon Grass (D. arundinacea variegata) is the most familiar example. Being perfectly hardy and perennial, it is valuable for producing good effects in the flower garden in a variety of ways. To grow it well it should be treated liberally and renewed by young plants every other year. If it be not desired to use it in the flower garden proper, a few tufts by a back shrubbery will suffice. It grows anywhere.

Dimorphanthus mandschuricus.—A handsome hardy shrub, with very large, much-divided, spiny leaves, which very much resemble those of the Angelica tree of North America. In this country it attains a height of 6 ft. to 10 ft., which it will probably much exceed when well established in favourable positions. It is of the highest importance for the sub-tropical garden. As to its treatment, it seems to thrive with the greatest vigour in a well-drained deep loam, and would grow well in ordinary garden soil in some sheltered, but sunny, spot; it may also be grouped with like subjects, always allowing space for the spread of its great leaves. We group it here because, although a shrub, some may wish to use it, as the Ailantus is used, for the sake of its fine foliage, in the flower garden in districts where tenderer plants may not thrive in the open air.

Dimorphotheca pluvialis *(Cape Marigold)*.—A handsome hardy annual from the Cape of Good Hope; the flowers are white and purplish-violet beneath, expanding during fine weather. It thrives in any good soil. Plants from spring-sown seed flower from July to September. It is a bold and free annual, and might be made an effective ground or carpet plant for the larger class of flower garden subjects; alone, however, and in a good position it is well worth growing. Compositæ. It grows from 18 in. to 24 in. high.

Dioscorea *(Yam)*.—Climbing plants with inconspicuous flowers, but their elegant foliage and graceful growth make them suitable for covering trellises and the like. D. Batatas, D. villosa, and D. japonica are the species in cultivation; all of these grow in good garden soil. They may occasionally find a place in the curious garden, but we have so many climbers with fair flowers and good foliage, too, that there is scarcely room for these in the select garden.

Diotis maritima.—A dwarf cottony herb suitable for the rock garden, and sometimes employed as an edging plant in the flower garden, though it is apt to grow rather straggling, and should be kept neatly pegged down and cut in well to prevent this tendency. It should have a sandy and deep soil. Increased by cuttings, as it seldom seeds in gardens. Native of seashore in south of Britain.

Dipcadi serotinum.—A Liliaceous plant with unattractive flowers, in habit like the Wood Hyacinth.

Diphylleia cymosa.—An interesting perennial of the Barberry family, about 1 ft. high, having large umbrella-like leaves, produced in pairs. Flowers white, in loose clusters in summer, and succeeded by bluish-black berries. N. America, on the borders of rivulets and on mountains. Thrives in peat borders and fringes of beds of American plants, in the most moist spots. Division in spring. Hitherto only seen as single weak specimens, this plant, if more plentiful, might be made good use of in a choice garden of American plants and flowers.

Diplopappus.—Perennial plants, much resembling the Michaelmas Daisy. Natives of N. America; but as there are so many fine Asters that flower at the same time, the species of Diplopappus are scarcely worth growing, except, perhaps, D. linearifolius, an erect-growing plant, from 1 ft. to 2 ft. high, having stiff narrow foliage and dense heads of violet blossoms. Other cultivated kinds are D. umbellatus, amygdalinus, and cornifolius. They grow freely in any soil.

Diplostephium umbellatum *(= Diplopappus)*.

Diplotaxis tenuifolia.—Plants sometimes mentioned in seed catalogues, but all are weeds, except, perhaps, a variegated-leaved form of D. tenuifolia. This is superseded by other variegated plants.

Dipsacus *(Teasel)*.—Coarse growing plants, but fine in form; worthy of a place in the wild garden, where their fine foliage and habit have a good effect. There are three native species, D. Fullonum, pilosus, and sylvestris; but the boldest is D. laciniatus, a European species growing 5 ft. to 8 ft. high, and with large deeply-cut foliage; seeds.

Disporum.—A genus allied to the Solomon's Seal, of no merit as garden plants.

Diuris.—Australian terrestrial Orchids, about which little is known at present in gardens. D. lilacina has recently been introduced, and is an attractive plant.

Dodecatheon *(American Cowslip)*.—A beautiful genus of the Primrose family, all

Dicksoni antarctica (*Tree Fern*). See page 113.

Dimorphanthus mandschuricus.

[*To face page* **114.**

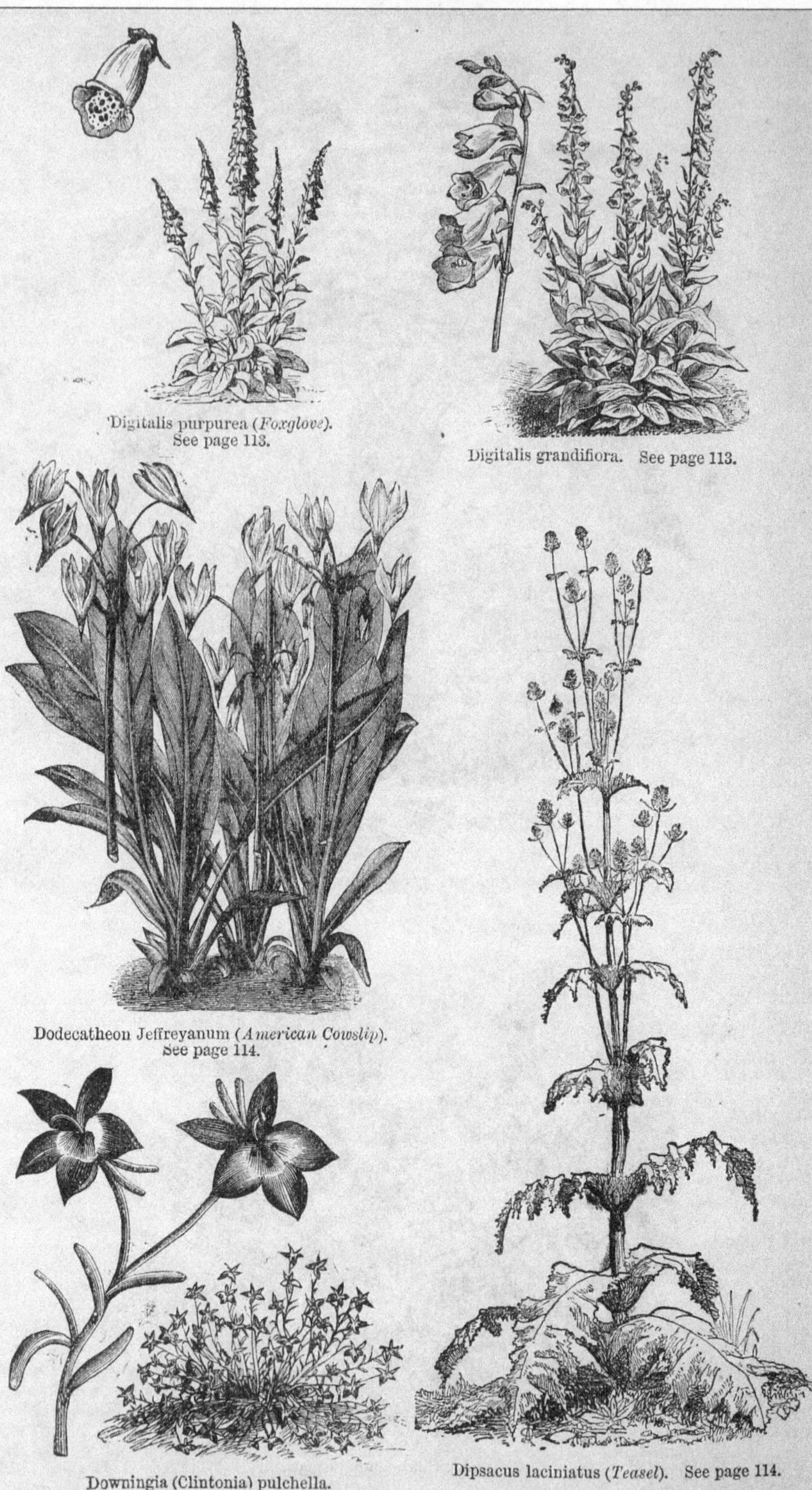

Digitalis purpurea (*Foxglove*). See page 113.

Digitalis grandiflora. See page 113.

Dodecatheon Jeffreyanum (*American Cowslip*). See page 114.

Downingia (Clintonia) pulchella.

Dipsacus laciniatus (*Teasel*). See page 114.

perennial and natives of N. America, where they were called Shooting Stars. They are all perfectly hardy in this country, requiring a cool situation and light, loamy soil; the nature of the latter is, however, of but small importance, as they grow almost as freely in peat or leaf-mould as they do in loam; situation is the principal point in their cultivation. In borders where Primulas and Soldanellas thrive, Dodecatheons will be found to soon establish themselves, and in spring to form lovely and attractive objects. All the species and varieties grow freely in sandy loam, and soon form large tufts, which require dividing every third or fourth year. The best time for transplanting them is the latter end of January or the beginning of February, when the roots are becoming active; care must be taken not to divide them into too small pieces, as in that case there is danger of losing the plants while in a weakly condition. All may be easily raised from seed.

D. integrifolium.—A lovely and gaily-coloured flower, deep rosy crimson, the base of each petal white, springing from a yellow and dark orange cup, and appearing in May on stems from 4 in. to 6 in. high. The leaves are much smaller than those of D. Meadia, oval, and quite entire. A native of the Rocky Mountains, a gem for the rock garden, planted in sandy peat or sandy loam with leaf-mould, and increased by careful division of the root and by seed. It is easily grown in pots, placed in the open air in some sheltered and half shady spot during summer, and kept in shallow cold frames during winter. Strong, well-established plants of this produce abundance of seed, which should be sown immediately it is gathered.

D. Jeffreyanum.—A stout kind, which grows more than 2 ft. in good soil. It has larger and thicker leaves than D. Meadia, and very strong and conspicuous reddish midribs, the flower being like that of the old kind except that it is somewhat larger and darker in colour. It is a hardy and first-class plant, flourishing freely in light rich and deep loam, and thriving best in a warm and sheltered spot, where its great leaves may not be broken by high winds. *Syn.*, D. lancifolium.

D. Meadia *(American Cowslip)*, bright, graceful, and hardy, is second to none of our old border flowers. It is supported in umbels on straight slender stems from 10 in. to 16 in. high, each flower drooping elegantly, and the purplish petals springing up vertically from the pointed centre of the flower, much as those of the common greenhouse Cyclamen do. It loves a rich light loam, and is one of the most suitable plants for the rock garden, well-arranged mixed borders, or the fringes of beds of American plants. In many deep light loams the plant flourishes without any preparation, but where a place is prepared for it, as is often necessary, it is very desirable to add plenty of leaf-mould. In a somewhat shaded and sheltered position it attains its greatest size and beauty, though it often thrives in exposed borders, and is best increased by division when the plants die down in autumn; when seed is sown, it should be soon after it is gathered. Of this there are numbers of pretty and distinct varieties, differing more or less in size of flower, colour, and height of plant. Among the best of these may be mentioned D. giganteum, elegans, albiflorum, splendidum, and violaceum. D. californicum, thought by some to be a species, is probably only a variety of D. Meadia. It is, however, a distinct and pretty plant, and worth growing.

Dog's-tooth Violet *(Erythronium Dens-canis)*.

Dog Violet *(Viola canina)*.

Dolomiæa macrocephala.—A bold and curious stemless plant somewhat resembling a Thistle, having a tuft of deeply-cut leaves, from the centre of which springs a cluster of short-stalked flowers, an inch or more across and purple in colour. It is a rare plant in cultivation, and where grown is difficult to grow well, as it appears to be somewhat tender. N.W. India. This and another species, D. lucida, are best in botanical collections.

Dondia Epipactis.—A singular and pleasing little herb, growing from 3 in. to 6 in. high, and producing in early spring small heads of greenish-yellow flowers. Suitable for the rock garden, margins of borders, banks, or any place where such a small and modestly-pretty plant may be seen. Increased by division after flowering. Carinthia and Carniola. *Syn.*, Hacquetia Epipactis.

Dorema.—Umbelliferous plants of little use in gardens. D. ammoniacum is a rare cultivated plant, yielding the drug ammoniacum of commerce.

Doronicum *(Leopard's Bane)*.—A genus of Composites, of which half a dozen species are in gardens. They are all of somewhat coarse growth, their chief merit being that they flower in early spring. All are of vigorous growth, thriving in any soil or position, and are therefore excellent for dry banks, where little else will thrive, for rough places or for naturalising. All readily increased by division of the roots. They range in height from 9 in. to 12 in., and all have large, bright, yellow, Daisy-like flowers. The best species are D. austriacum and caucasicum, both of which are neater in growth than the rest and produce in early spring a profusion of blossoms that enliven the borders and are useful for cutting. The other kinds are D. Clusi, carpatanum, Columnæ Pardalianches, and plantagineum, all of them natives of Europe.

Downingia.—Charming little Californian half-hardy annuals, generally known as Clintonia. There are two species, D. pulchella and elegans; both are similar to each other, and resemble in habit the common bedding Lobelias, but are more brilliant in colour. D. pulchella is the most desirable species. It is of dwarf habit, rarely exceeding 6 in. in height, and is suitable for edging small beds or borders. When covered with its bright blue flowers it is very pretty. It likes a free, well-enriched soil and an open situation. The seed should be sown in March and April in the open ground, and some two months earlier for pot culture. Each plant should be allowed quite 8 in. for development, and those from the last sowing should be well mulched and watered in hot weather. There are several varieties of D. pulchella differing in the colour of the flowers, the best being alba (white), rubra (red), and atropurpurea (dark purple).

Draba (*Whitlow Grass*).—Minute alpine plants, mostly having bright yellow or white flowers, the leaves being often in singularly neat rosettes. They are too dwarf to take care of themselves among any plants much bigger than mosses, and, therefore, there are few positions suitable for them; but wherever there are mossy walls, ruins, or bits of mountain ground with sparse vegetation, it would be very interesting to try them. The best known and showiest is D. aizoides, found on old walls and rocks in the west of England. It forms a dwarf, spreading, cushion-like tuft, which in spring is covered with bright yellow blossoms. D. aizoon, alpina, ciliaris, cuspidata, lapponica, rupestris, frigida, and helvetica are very dwarf, compact-growing plants. The flowers in each case are small, white or yellow, produced abundantly. Rarer kinds are D. Mawi, glacialis, and bruniifolia, all worth growing in a full collection of alpine flowers for a choice rock garden.

The Dracæna (*Dragon's-blood*).—These are among the best of those plants which may be brought from the conservatory or greenhouse in early summer, and placed in the flower garden till it is time to take them in again to the houses in which they are to pass the winter months. And even if it were not necessary to protect them through the winter, it would be almost worth our while to bring them indoors at that season, so graceful are they, and so useful for adding to the charms of our conservatories. The hardier kinds, like indivisa and Draco, may be placed out in the summer very far north. D. indivisa grows well in the open air in the south of England and Ireland. It is a very graceful plant, with leaves from 2 ft. to 4 ft. long, and 1 in. to 2 in. in breadth, tapering to a point, pendent, and dark green. It should not be confounded with the conservatory plant known as Cordyline indivisa, which is too tender to succeed well in the open air, and somewhat difficult to grow. It is, on the contrary, perfectly hardy in the south and west of England and Ireland. D. indivisa lineata is a fine variety, the leaves of which are much broader than those of the type, measuring sometimes 4 in. across, and coloured with reddish pink at the sheathing base. Other good varieties are—D. indivisa atro-purpurea, which has the base of the leaf and the midrib on the underside of a dark purple; and D. indivisa Veitchi, in which the habit and size of the leaf are the same as in the species, but in addition it has a sheathing base, and the midrib on the under side is of a beautiful deep red. It would be difficult to find a plant more worthy of cultivation than this. Where it does well in the garden or pleasure-ground in the southern or western counties it surpasses any Yucca in distinctness and grace. At Knockmaroon Lodge, near Dublin, a plant, 16 feet high, with a stem some 6 in. in diameter, annually flowered and bore abundance of seeds, from which Mr. Pressly, the gardener, raised seedlings easily in a cold frame. In the Scilly Islands the plant becomes a great tree, luxuriating in the mildness and moisture from the Gulf Stream. The fact that it annually flowers on young plants in the climate of Dublin, and ripens its seeds, is sufficient proof of its hardiness and that it will succeed in many districts. It is readily increased from seed, from pieces of the stem, and offsets. If a plant is cut over close to the ground, a number of young shoots soon spring up, which can be taken off as cuttings, and which strike with freedom. Recent severe winters may have hurt it in many places; but after so many years' success no one in a likely district will give up its culture.

Dracocephalum (*Dragon's-head*).—A genus of the Sage family, among which are a few choice perennials, suitable either for the rock garden or mixed border, as they succeed well in light garden soil and are easily increased by division or seed. D. altaiense has leaves of a bright green colour, and axiliary clusters of large tubular flowers of a dense Gentian-like blue colour, spotted or dotted with red in the throat. D. austriacum has flower-stems nearly 1 ft. in height, densely covered with rich purple blossoms; D. Ruyschiana, a handsome species, has narrow Hyssop-like leaves and purplish-blue flowers, but its variety japonicum, a new introduction from Japan, is even more showy. D. peregrinum, with pretty blue flowers always produced in pairs, is a desirable kind; and so is D. argunense, which is in the way of D. Ruyschiana. The most beautiful of all is D. grandiflorum, which is the earliest to flower. It is very dwarf, and has large clusters of intensely blue flowers, which scarcely overtop the foliage—a beautiful rock garden plant. In D. speciosum, a Himalayan species, the flowers are small, and, though of a deep purple, are nearly smothered by the large green bracts. The hardy annual kinds, such as D. moldavicum and D. canescens, are ornamental, and worth a place where a full collection is desired.

Dragon's-mouth (*Arum crinitum*).

Dragon's-head (*Dracocephalum*).

Drop-wort (*Spiræa Filipendula*).

Drosera (*Sundew*).—Most interesting little bog plants, of which all the hardy species, except one, are natives of Britain. All are charactèrised by tufts of leaves and have their

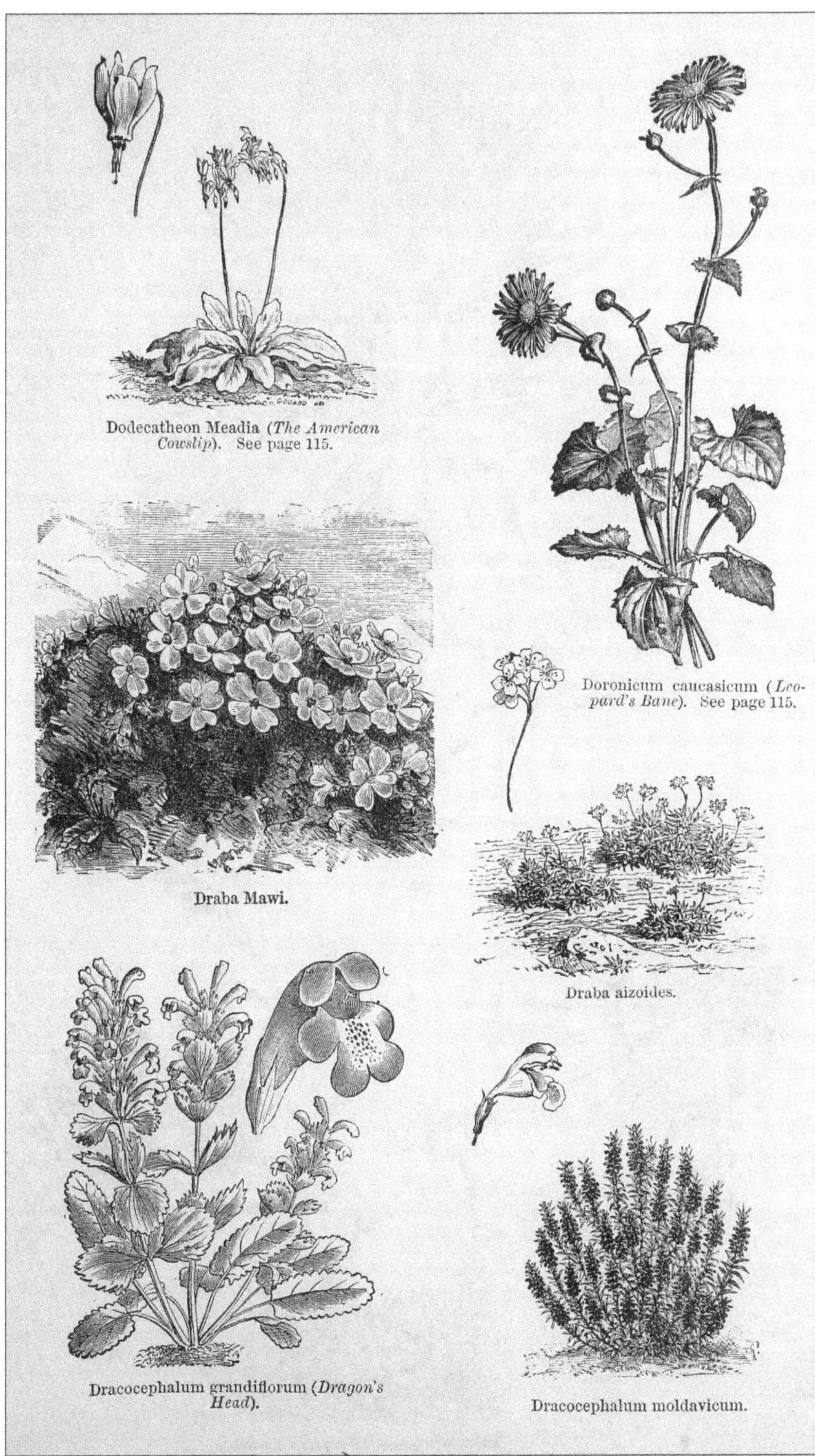

Dodecatheon Meadia (*The American Cowslip*). See page 115.

Doronicum caucasicum (*Leopard's Bane*). See page 115.

Draba Mawi.

Draba aizoides.

Dracocephalum grandiflorum (*Dragon's Head*).

Dracocephalum moldavicum.

[*To face page* 116.

PLATE CII.

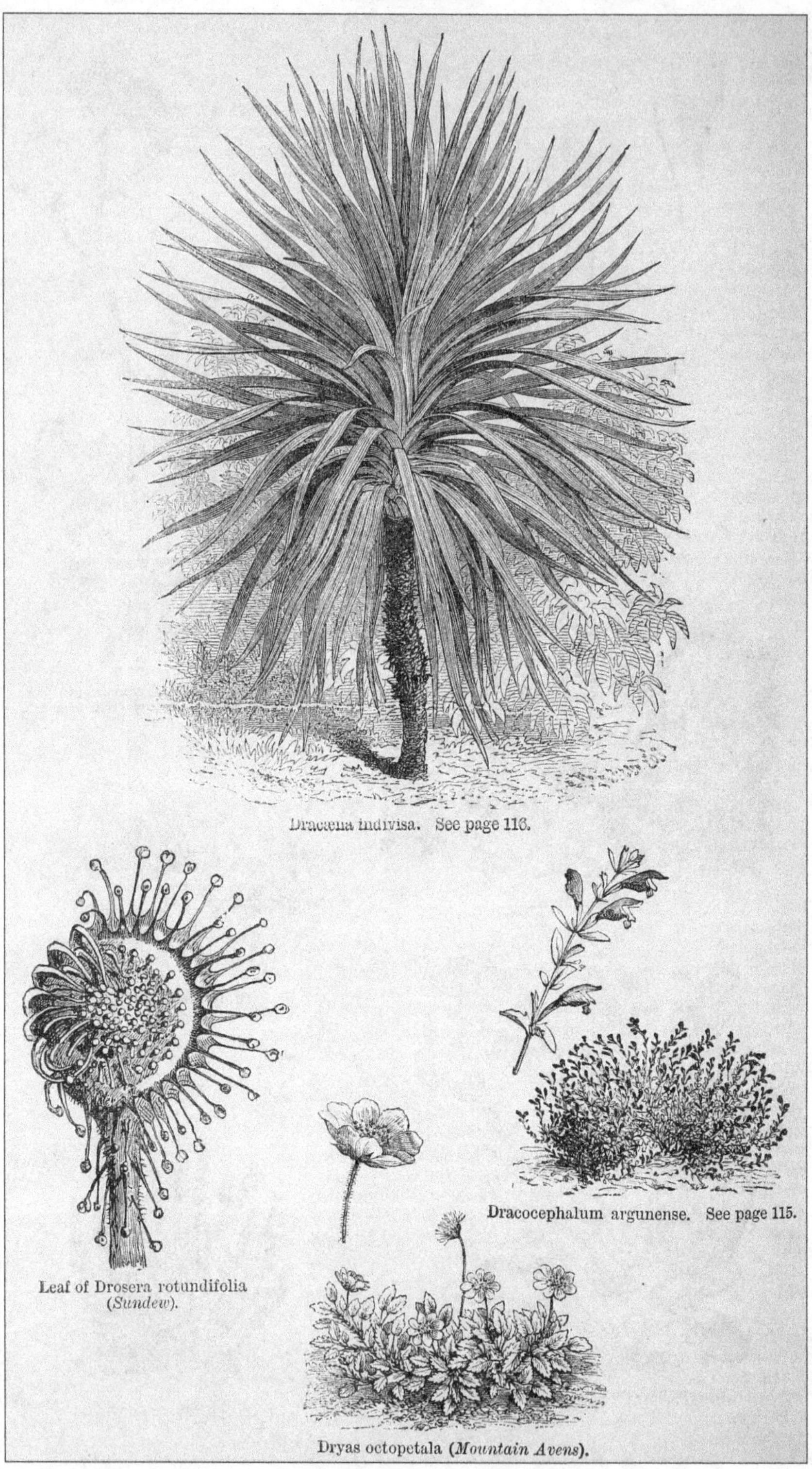

Dracæna indivisa. See page 116.

Leaf of Drosera rotundifolia (*Sundew*).

Dracocephalum argunense. See page 115.

Dryas octopetala (*Mountain Avens*).

[*To face page* 117.

surfaces covered with dense glandular hairs. When the native kinds are grown artificially the condition of their natural home should be imitated as far as possible. In a bog on a very small scale it is not easy to secure that humidity of atmosphere the plants have at home, but where Sphagnum grows they will grow. The native kinds are intermedia, longifolia, obovata, and rotundifolia. The North American Thread-leaved Sundew (D. filiformis) is a beautiful bog plant, with very long slender leaves covered with glandular hairs. Flowers purple-rose colour, half an inch wide, and opening only in the sunshine. It is quite hardy, but appears to be difficult to cultivate.

Dryas *(Mountain Avens)*.—A genus of the Rose family, containing two or three dwarf alpine plants of spreading growth and with neat evergreen foliage. They succeed well in cultivation, thriving on level borders in light soil, though they are seen to best advantage in the rock garden, where they can spread over the brows and surfaces of rocks. They like an exposed, but not too dry, spot, though when well established they will flourish under almost any conditions. Propagated by seed or division in spring. The kinds are D. Drummondi, a dwarf, hardy, evergreen trailer, with flower-stems 3 in. to 8 in. high. Flowers 1 in. across, golden yellow, appearing in summer. N. America. D. octopetala, a creeping evergreen, forming dense tufts. Flowers white, yellow centre, 1 in. or more across, borne on stalks 3 in. to 8 in. high. A British mountain plant.

Dutchman's-breeches *(Dicentra)*.

Dwarf Cornel *(Cornus suecica)*.

Dyckia rariflora. — An interesting plant of the Bromeliad family, which is for the most part tropical. Though this is a native of Brazil, it is hardy in sheltered localities if planted in dry, snug nooks of a well-drained rock garden, in sandy loam. It grows about 2 ft. high, with a rosette of spiny leaves and spikes of orange-coloured flowers, which, however, are but seldom produced in the open air here.

Eccremocarpus scaber.—A delightful old climbing plant for adorning walls, trellises, and pillars, and for association with the Canary Creeper, Clematis, and the best climbing plants. The orange-red flowers are beautiful and borne freely, and the foliage is good. If the roots are protected during winter it is uninjured, and the plant annually increases in size. Increased freely by seed, which are plentifully produced.

Echeveria.—A genus of dwarf succulent plants, much used in the flower garden, particularly the semi-hardy species like secunda; others are more tender, and need a greenhouse to keep them through the winter, and a warm house or frame to propagate them in the spring. Echeveria secunda is well known by its pale-green rosette, the tips of the leaves being marked with red. E. secunda major is but a mealy form of the same. E. secunda glauca differs only in having leaves rather more pointed and still more glaucous. E. secunda pumila is a smaller form, with narrow leaves that are of the colour of secunda major. Echeveria glauca metallica is intermediate between the well-known metallica and secunda glauca. It has a dwarf, massive habit of growth, the leaves very solid and fleshy, and is a very distinct and useful kind. Echeveria metallica is a noble species, and distinct in the size of its leaves and their rich metallic hue.

Increase.—As soon as the seed is ripe prepare to sow it. Fill some 4-in. pots to within ½ in. of the rim with well-sanded loam and leaf-mould in equal proportions. Make the surface very firm, and water the soil so that the whole body of it becomes thoroughly moistened. Having allowed the moisture to drain away, scatter the seed thinly on the surface and cover thinly with silver sand. Place the pot in a hand-light or in a close frame; cover with a pane of glass and shade. The seed will germinate before the soil can dry, and by sowing as soon as ripe every seed will come up. As soon as the seedlings are large enough to handle prick them out thinly into pans or 6-in. pots; keep them close until fairly established, and then allow them to receive the full benefit of sun and air. After the middle of September no water must be given, and care must be taken to remove all decay as soon as perceived. These little plants will, if planted in well-worked and fairly-enriched soil early in April, make good strong plants by the autumn. There is another method by which they may be increased. With a sharp knife cut out the heart of the plant, and this will have the effect of inducing the formation of offshoots. These taken off will speedily make good specimens. E. metallica may be increased in the following manner: Take off the flower-stems which come early in the season; cut off the embryo flowers and place the stems in pots of sandy soil. These will strike and will produce numerous little offsets from the axils of the flower-stem leaves. Taken off they will readily strike. This kind may also be raised from seed in the manner above described. The dwarfer kinds are mostly used as edgings or panels. The fine Echeveria metallica is very effective on the margins of beds and groups of the dwarfer foliage plants, or here and there among hardy succulents, and should be planted out about the middle of May.

Echinacea.—Stout and singular perennials of the Composite Order, requiring some care as to position to secure a good effect. E. purpurea has stems that rise from 3 ft. to 4 ft. high, and are terminated by great flower-heads, the central florets of which are of a dark brown colour, and form a close, conical boss on which the golden anthers are shown off to advantage, while the long pendulous straps belonging to the ray florets hang down in the form of a long purplish-pink fringe, thus giving the plant a singular aspect. In order to grow it well, a deep, warm soil is required. Southern and Western States of N. America. Associated with the Tritomas, Eryngiums, or plants of like stature and merit, in mixed borders, or as groups in warm spots, always in good soil. Propagated by seed or division of the root in autumn or very early spring; the plants must not be often disturbed for this purpose. With the general habit of the well-known E. purpurea, E. angustifolia has narrower foliage, more numerous ray florets of a lighter shade of red-purple, and the flower-heads are more freely produced. It also blooms about a month earlier. The stems vary from 3 ft. to 5 ft. in height, each bearing a single conspicuous head with drooping riband-like rays. From N.-W. States of America.

Echinochloa *(Panicum)*.

Echinops ruthenicus *(Globe Thistle)*.—A fine hardy plant from S. Russia, with stems 3 ft. or 4 ft. high, much branched in the upper part, and covered with a silvery down. The flowers are blue, and borne in almost spherical heads on the tops of the erect branches. The plant thrives in ordinary soil. Easily multiplied by seed, division of the tufts, or by cuttings of the roots in spring. This is the most ornamental of its distinct family, and is highly suitable for grouping with the bolder herbaceous plants. It would also look well isolated on the turf. There are a number of other species, mostly from S. Europe and the Levant, among which may be named E. Ritro and E. banaticus; but we have never seen any kind so good as E. ruthenicus, and, as the species are very much alike, it is enough to grow the best.

Echinocactus Simpsoni.—A beautiful little Cactaceous plant, a native of Colorado, at great elevations on the mountains, and believed to be hardy enough for our climate. It grows in a globular mass, 3 in. or 4 in. across, covered with white spines. It flowers early in March in this country, the blossoms being large, of a pale purple tint, and very beautiful. No one appears to have had any lengthened experience in cultivating it, but, so far, it seems to thrive. The conditions under which the plant grows naturally should be imitated as far as may be. In its native habitat it enjoys a dry climate, and it is, in some seasons at least, more or less protected from the effects of frost by a covering of snow. But in this country it has withstood 32° of frost without injury, and therefore, if in a dry spot, it may escape and flourish.

Echinospermum.—Sometimes mentioned in seed catalogues, but the species introduced at present are worthless.

Echium *(Viper's Bugloss)*.—A noble genus of the Forget-me-not Order, the finer kinds of which, superb in the open gardens of S. Europe, are too tender for ours. E. plantagineum is one of the handsomest of the hardy annual or biennial species. It has showy flowers, of a rich purplish-violet, in long slender wreaths that rise erect from the tuft of broad leaves. It is handsomer than our indigenous species, E. violaceum and E. vulgare, which, were they not so common in a wild state, would be included among our hardy flowers. E. rubrum is a scarce species, and handsome. The habit of growth is similar to those above mentioned, but the colour is a reddish violet, similar to E. creticum—also an attractive plant. The Salamanca Viper's Bugloss (E. salmanticum) is another fine species, and difficult to obtain, except from its native locality. These five species, which are now in cultivation, are representative of the annual and biennial Echiums, and all are showy and worthy of a place in a full collection. They are of the simplest culture. The seeds should be sown in ordinary garden soil, either in spring for the current year's flowering, or late in autumn for flowering in early summer. Our native E. vulgare is a good plant in certain positions; its long racemes of blue flowers are handsomer than those of the Italian Anchusa. Against a hot wall, where nothing else would grow, at the Grammar School, Colchester, Dr. Acland planted some, which furnished a beautiful bloom. It is valuable for similar positions, particularly on hot gravelly or chalky soils.

Echinocystis lobata *(Wild Balsam Apple)*.—A tall, climbing annual of the Gourd family, with greenish-yellow flowers, succeeded by large oval fruits. Useful for covering arbours, but scarcely worth growing in our own country. N. America.

Echinocereus.—Plants of the Cactus family (from arid regions in N. America), some of which have lately been said to be hardy in this climate. Mr. E. G. Loder, of Weedon, Northamptonshire, grows and flowers them successfully. He thus writes to *The Garden:* "I have a wall here where the Ivy hangs over in such a way that it keeps off a large portion of the winter's snow and rain from the plants growing underneath. In this position I have grown several species of Echinocereus and Opuntia, an Echinocactus, and a Mammillaria. Only small plants were tried, yet several of them have flowered in spite of our very severe winters and not favourable summers. We had 41° of frost here last winter, but none of these Cacti were at all injured by it. No species of Cactus which I have tried does well when planted in a level border. A narrow rock border, raised about 1 ft. high, against a south wall, would be a capital position, but it is much improved if the wall has a good wide coping. The most sightly is a natural one of Ivy. What success I have in the culture of these plants has amply repaid me all the trouble and care that I have spent upon them; but much greater success may reasonably be expected by any one who will undertake their cultivation in a more sunny part of England. All of them are beautiful, and some quite splendid when in flower. Echinocereus Fendleri bears some of the brightest coloured flowers that I have ever seen—a rich purple." The species of Echino-

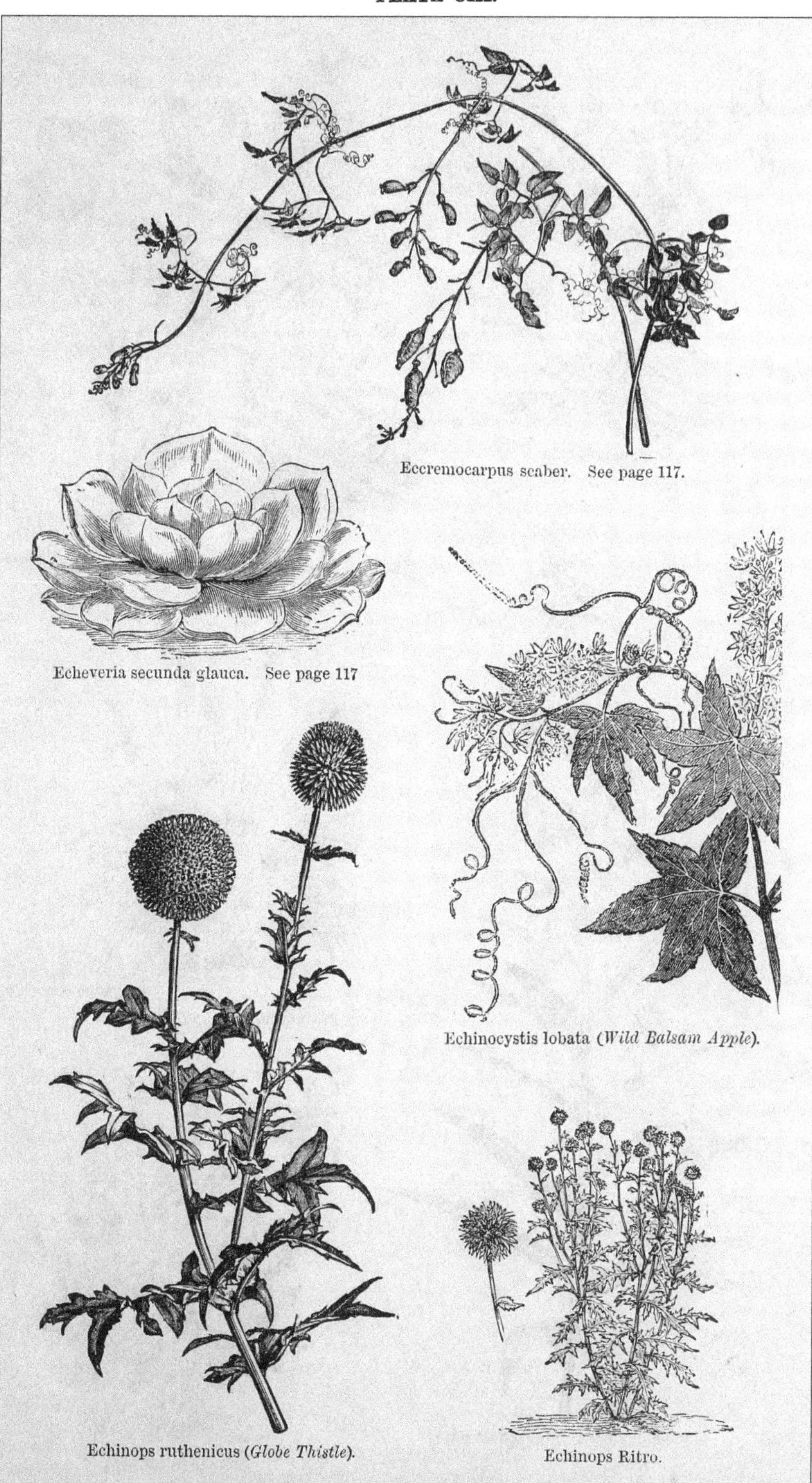

Eccremocarpus scaber. See page 117.

Echeveria secunda glauca. See page 117

Echinocystis lobata (*Wild Balsam Apple*).

Echinops ruthenicus (*Globe Thistle*).

Echinops Ritro.

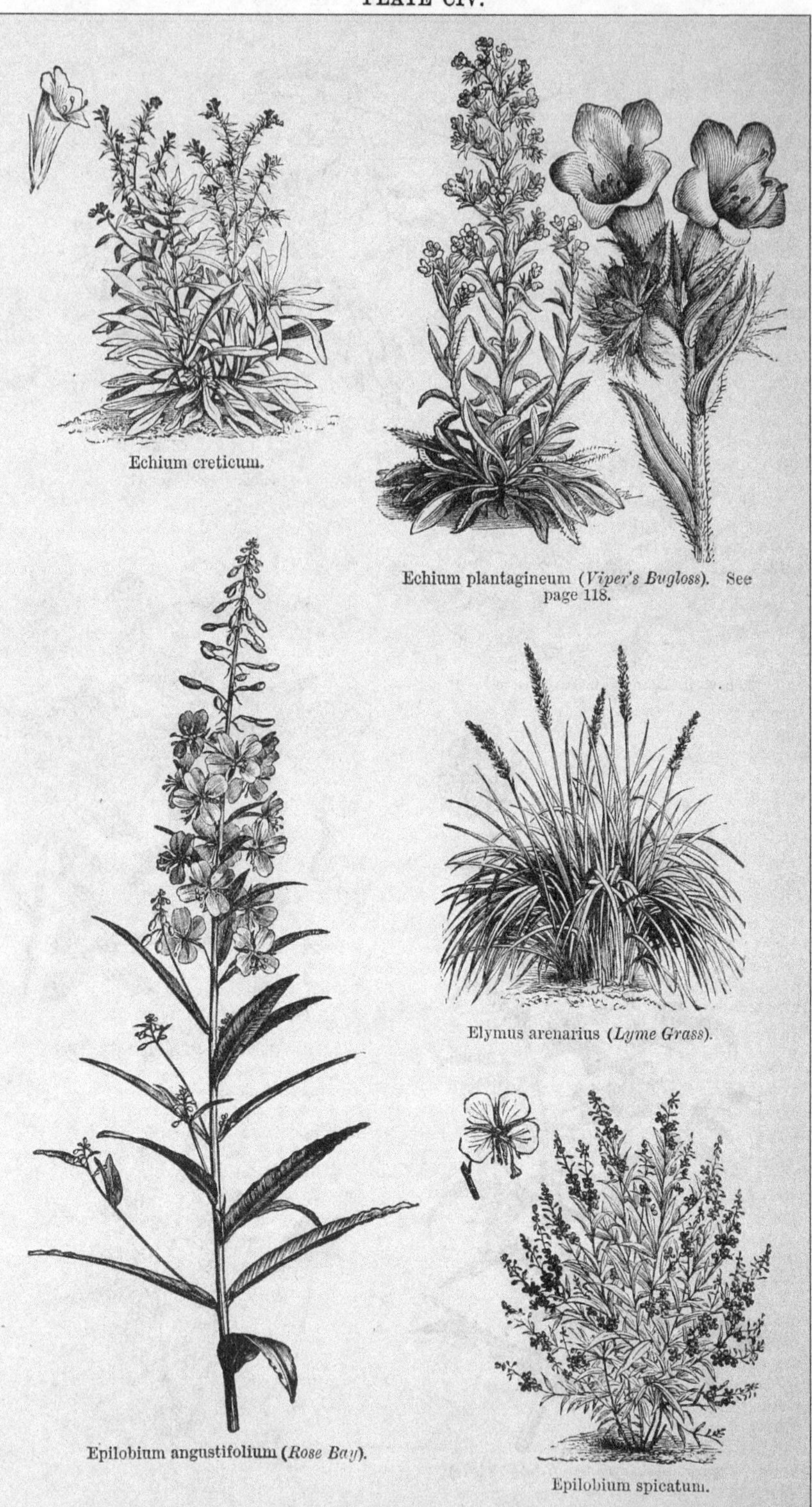

Echium creticum.

Echium plantagineum (*Viper's Bugloss*). See page 118.

Elymus arenarius (*Lyme Grass*).

Epilobium angustifolium (*Rose Bay*).

Epilobium spicatum.

[*To face page* 119.

cereus that Mr. Loder grows are E. nœphiceus, gonacanthus, Fendleri, viridiflorus, and paucispinus. We have no doubt that on raised stony borders and sunny banks in rock gardens various forms of the hardy Cacti of N. America would not only be found hardy, but would flower well. Give them well-drained soil and sunny, but exposed, spots, trying them away from all coping or artificial protection, but taking great care so to place them in relation to surrounding objects that their stems cannot easily be hurt in clearing or passing. A few friendly stones and low evergreens can be so grouped around them that the digger and other dangerous animals may be kept off. A close turf of some dwarf, clean-growing, alpine plant will prevent earth-splashings and improve the effect.

Edelweiss *(Leontopodium alpinum).*

Edraianthus comprises about half-a-dozen species, inhabiting the mountains of South-eastern Europe and Asia Minor, and resembling alpine Hairbells. They all have the tufted habit and narrow leaves of E. Pumilio, but this is the smallest of the series. In habit it closely resembles Silene acaulis, the wild specimens being only 2 in. to 3 in. high, including the flowers. The shining foliage is hairy and of a silvery grey hue. It is in the true sense of the word a rock plant, and requires to be grown with the choicer alpine plants in a well-formed rock garden or border devoted to dwarf plants. E. tenuifolius and graminifolius, natives of Greece and Bosnia, grow from 3 in. to 6 in. or more high, and have narrow, grass-like leaves, chiefly radical, and flowers nearly double the size of those of E. Pumilio. E. serpyllifolius has, as the name indicates, Thyme-like foliage; and E. Kitaibeli, a native of Bosnia, is about 6 in. high, and has proportionately large flowers. E. dalmaticus is one of the best of the cultivated kinds.

Elecampane *(Inula).*

Eleusine *(= Poa).*

Elsholtzia cristata.—A coarse, weedy plant of the Sage family, of no garden value.

Elymus arenarius *(Lyme Grass).*—This wild British Grass, a strong-rooting and distinct plant, adds a feature to the garden here and there. Planted a short distance away from the margin of a shrubbery, or on a bank on the Grass, and allowed to have its own way in deep soil, it makes an effective plant. It is hardy in all parts of these islands. In very good soil it will grow 4 ft. high; and as it is for the leaves we should cultivate it, there would be no loss if the flowers were removed. It is found frequently on our shores, but more abundantly in the north than in the south. E. condensatus (Bunch Grass) is a vigorous perennial Grass from British Columbia, forming a dense, compact, column-like growth, more than 8 ft. in height, covered from the base almost to the top with long arching leaves, and crowned in the flowering season with numerous erect, rigid spikes, each 6½ in. long, and resembling an elongated ear of wheat in form. It is a very ornamental plant, and may be grown in the same way as the Lyme Grass. Other varieties and species might be mentioned, but one or two kinds give us the best effect of this race of Grasses.

Emilia sagittata.—A pretty half-hardy annual, belonging to the Compositæ, growing from 9 in. to 18 in. high, with slender flower-stems, terminated by bright orange-scarlet heads nearly 1 in. across. It likes a light, rich soil, and flowers from July to September if the seed be sown in heat in February and March, the seedlings being put out in May. East Indies. *Syn.*, Cacalia.

Empetrum nigrum *(Crowberry).*—A small evergreen Heath-like bush, of the easiest culture. May be planted with the dwarfer and least select rock shrubs. It is a native plant, and the badge of the Scotch clan McLean.

Emmenanthe penduliflora.—A Californian plant, an annual of low growth, about 1 ft. in height, and of neat tufted habit, like that of the Nemophila. The flowers are produced in loose terminal racemes, of pale primrose yellow. Succeeds best as a hardy annual on any ordinary soil. Not a showy plant.

Epigæa repens *(Ground Laurel, Mayflower).*—A small trailing evergreen found in sandy soil in the shade of Pines, common in many parts of North America, and remarkable for its delicate rose-tinted flowers in small clusters, exhaling a rich odour, and appearing in early spring. In planting it, it would be well to bear in mind that its natural habitat is under trees, and plant a few specimens in the shade of Pines or shrubs. It was at one time lost in our nurseries and gardens, owing to the habit of planting all things in the same kind of situation—an exposed one. It is a charming shrublet for the rock garden, under the shade of larger shrubs, and also for the wild garden, rambling about in sandy or peaty soil under Fir trees. It only grows a few inches high. Caprifoliaceæ.

Epilobium *(French Willow).*—A large genus of plants, of which few are worthy of cultivation, but these are important, the best perhaps being the showy crimson native E. angustifolium, of which there is a pure white variety. It is a native of Europe and Siberia, and is found in many parts of Britain. This fine plant, which runs so quickly in a border as to soon become a most troublesome weed, is magnificent when allowed to run wild in a rough shrubbery or copse, where it may bloom along with the Foxglove. Division. Other kinds somewhat less vigorous in habit are E. angustissimum, E. Dodonæi, and E. rosmarinifolium. The common native E. hirsutum is stouter than the French Willow. It is only useful by the margins of streams and ponds, associated with the Loosestrife and like plants. There is a variegated form. The Rocky Mountain Willow Herb (E. obcordatum) is a beautiful rock plant. The Willow

Herbs of our own latitudes are very tall and vigorous, but on the dreary summits of the Rocky Mountains and Californian Sierras, one species has succeeded in contending against the elements by reason of its very dwarf stature; it has, in fact, imitated the Phloxes and Pentstemons of the same region; but though not more than 3 in. high, it has retained the size and beauty of the flower of the finest species, the colour being rosy-crimson. It is hardy, and thrives in ordinary sandy soil in the rock garden.

Epimedium (*Barren-wort*). — An interesting and, when well grown, elegant genus of plants of the Barberry Order, although not shrubby. E. pinnatum is a hardy dwarf perennial from Asia Minor, from 8 in. to 2½ ft. high, forming handsome tufts of pinnate leaves, and bearing long clusters of yellow flowers. The handsome leaves remain on the plant until the new ones appear in the ensuing spring. It is not a good plan to remove them, as they serve to shelter the buds of the new leaves during the winter, and the plants flower much better when they are allowed to remain. Cool, peaty soil and a slightly shaded position will be found most suitable. Other species are alpinum, macranthum, Musschianum, purpureum, rubrum, and violaceum, all loving half-shady spots in peat or moist, sandy soil. None of them are so valuable for general culture as the first mentioned.

Epipactis palustris (*Marsh E.*).—A somewhat showy hardy Orchid, 1 ft. to 1½ ft. high, flowering late in summer, bearing rather handsome purplish flowers. A native of all parts of temperate and Southern Europe, in moist grassy places. A good plant for the artificial bog, or moist spots near a rivulet, in soft peat. In moist districts it thrives very well in ordinary moist soil.

Equisetum Telmateia (*Giant Horse-tail*).—A British plant, of noble port and much grace of character when well developed, growing from 3 ft. to 6 ft. high in favourable spots. The stem is furnished from top to bottom with spreading whorls of slender, slightly-drooping branches, the whole forming a pyramid of very distinct effect. It is a highly ornamental subject for planting in the hardy fernery, the artificial bog, shady peat borders, near cascades, or among shrubs, and grows best in moist vegetable soil. Multiplied by division. E. sylvaticum is another native Horse-tail, much dwarfer, but of exquisite grace when well grown, the stem standing from 8 in. to 15 in. high, and being well covered with numerous slender branches. Suitable for rockwork, margins of water, and the hardy fernery.

Eragrostis (*Love Grass*).—A genus of Grasses, some of which are very ornamental and worth cultivating for their elegant feathery panicles. E. ægyptiaca, with silvery white plumes, maxima, elegans, pilosa, amabilis, pellucida, capillaris, plumosa, are all elegant annual kinds. They are useful for cutting from during summer for arranging with flowers. Seed may be sown in autumn or spring in the open air, on or in a slightly heated frame. For preserving, the stems should be gathered before they are too ripe.

Eranthis hyemalis (*Winter Aconite*).—A valuable small plant, with yellow flowers surrounded by a whorl of shining-green, divided leaves, and a short, blackish underground stem, resembling a tuber; 3 in. to 8 in. high; flowering from January to March. It is seen to best advantage in a half-wild state, under trees on slopes in woody places, though it is occasionally worthy of a place among the earliest border flowers. It grows in any soil. It often naturalises itself freely in Grass, and is very beautiful when the flowers peep in early spring, looking like gold buttons. When the branches of large trees are allowed to rest on the turf of lawn or pleasure-ground, a few roots of these scattered beneath will soon form a carpet, glowing into sheets of yellow in winter or spring. We may so enjoy it without giving it positions suited for rarer and more fastidious plants, or taking any trouble about it.

Eremostachys.—Distinct looking perennial plants of the Sage family, but suitable only for botanical collections. E. laciniata has deeply cut leaves, and flower-stems about 4 ft. high, with numerous leafy bracts that enclose large purplish flowers arranged in whorls. E. iberica differs from E. laciniata in having the flowers yellow. Both seem difficult to cultivate, and we have never seen them well grown.

Eremurus.—Plants of the Lily family, allied to Asphodelus, and native of Central Asia, Siberia, and the Caucasus. There are several species in cultivation, presumably quite hardy, but only two or three have proved satisfactory. The finest of all is E. robustus; it has a gigantic flower-spike, 7 ft. to 10 ft. high, the upper part of which is covered with large pink flowers from 2 ft. to 3 ft. of its length. It has a large tuft of stout leaves about 1 yd. high, and is so hardy that it can force its shoot through the frozen ground, and will endure heat, cold, and wind with equal indifference. It flowers in June, and lasts in beauty about a month, presenting a most striking appearance. Any warm corner open to the sun suits it, and a deep loamy soil, somewhat moist, as it suffers from drought. It ripens seed plentifully, from which any number of seedlings may be raised. Other cultivated kinds that have flowered in this country are—E. himalaicus, with tall spikes of white blossoms, and E. turkestanicus—from the same habitat as E. robustus, though by no means such a strong growing ornamental plant. It starts earlier into growth, and is done blooming by the time E. robustus begins. Its blossoms are dark red and brown, on a 16-in. raceme, terminating a scape 4 ft. high. E. inderiensis, Kaufmanni, Olgæ, spectabilis, Korolkowi

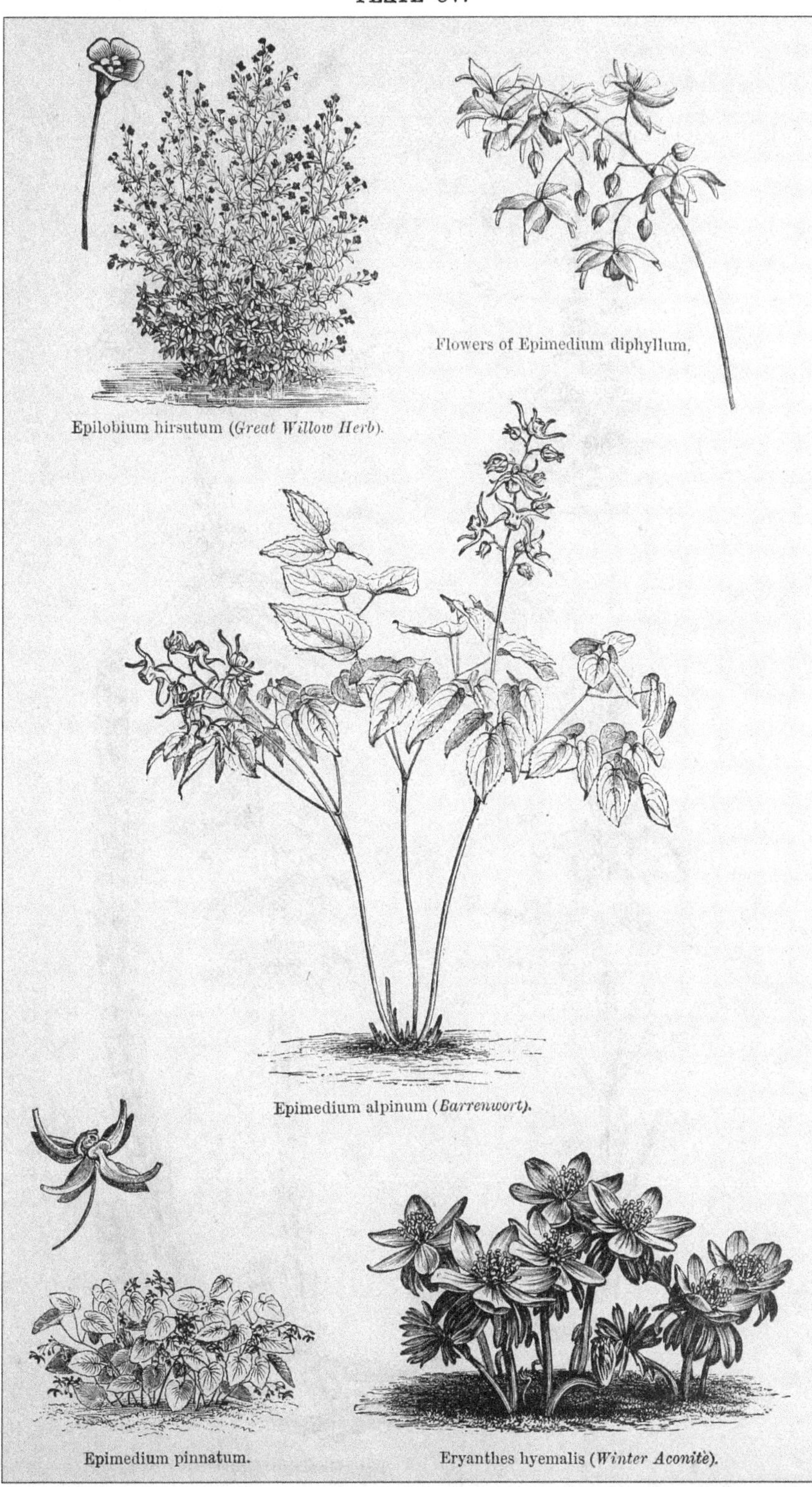

Epilobium hirsutum (*Great Willow Herb*).

Flowers of Epimedium diphyllum.

Epimedium alpinum (*Barrenwort*).

Epimedium pinnatum.

Eryanthes hyemalis (*Winter Aconite*).

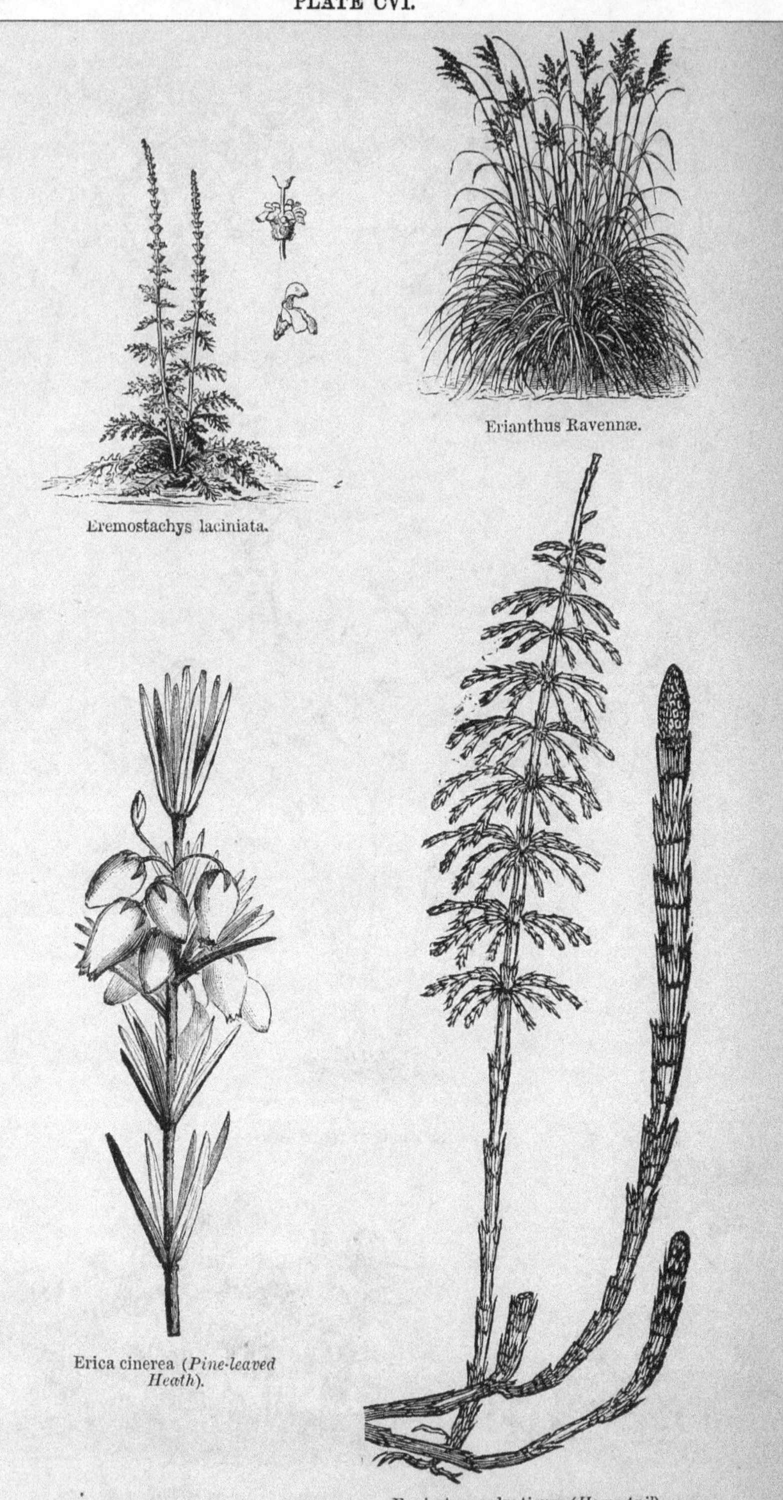

Eremostachys laciniata.

Erianthus Ravennæ.

Erica cinerea (*Pine-leaved Heath*).

Equisetum sylvaticum (*Horse-tail*).

tauricus, kurdistanicus, are likewise in cultivation, but at present very little is known of their merits or hardiness.

Erianthus Ravennæ.—An ornamental Grass from S. Europe, somewhat like the Pampas Grass in habit, but smaller in size, having violet-tinged leaves. The flowering-stems grow from 5 ft. to 6½ ft. high, but as it only flowers with us in a very warm season, it must be estimated for its foliage alone. Its dense and handsome tufts thrive well in light or calcareous soil, in positions with a south aspect. It thrives but poorly on cold soils, and will probably not grow well north of London except in peculiarly favourable positions, and in well-drained free loams. It is fitted for association with such Grasses as Arundo conspicua. Multiplied by division of the tufts in spring or autumn. E. strictus is another species, but is not so ornamental as E. Ravennæ.

Erica *(Heath)*.—Though shrubby, these are so neat and dwarf in habit, so essential in any scheme of bold rock or hardy gardening, and so valuable for the flowers they yield, that we cannot separate them in this book from the flowers, with which, indeed, they group well in many positions. The late Mr. James McNab, who was fond of these plants, wrote to us concerning hardy Heaths in gardens: "The beauty belonging to clumps of hardy Heaths, some of them of great age, as seen on our highland moors and pastures, is owing to the plants being annually eaten over by sheep, particularly during winter and spring—a circumstance that makes them branch and flower in dense hemispherical masses. When under cultivation clipping has exactly the same tendency. If hardy Heaths are regularly cut they will keep in good flowering condition for many years; but if neglected or left to themselves, a very few years will see them long and lanky. If the autumn-flowering hardy Heaths are regularly clipped over in spring they will form neat patches and flower abundantly, while the spring or summer-flowering varieties should always be cut over as soon as possible after the flowering is past; if this is done they will branch out freely and regularly and flower abundantly the following season. When the plants become old they are apt to assume a tall and wiry condition and to flower imperfectly. It is therefore necessary that young plants should always be coming on to take their place. It is not absolutely necessary that hardy Heaths should be grown in peat soil. A mixture of peat and loam, or leaf-mould and sand, will answer the purpose. Where peat is scarce sand should predominate. In nurseries hardy Heaths are generally propagated by means of layers, an operation easily accomplished by placing some peat, freely mixed with sand, in the middle of the plants, which, when rooted, are torn in pieces and planted out in nursery rows. Many kinds do well in this way, but most of them flower better when raised from cuttings, besides being better shaped. The single stem of the cutting-made plants seems to favour the shape and the flowering, while layers in many cases have several stems, and scarcely two plants are of exactly the same shape.

"All the hardy Heaths are readily propagated by means of cuttings or layers. Cuttings make by far the best-shaped plants, and are readily struck if placed any time during the autumn months in a sand bath in a cold pit, provided they are not exposed to the sun. In such a situation they soon root and start into growth. When rooted and hardened they could be planted out in lines, but I prefer putting them round the edge of pots, and placing them in a cold frame partially shaded for a time from the sun. As soon as established, all the points of the top shoots should be cut off, when side branches will be freely produced, and thus a compact habit of growth will be induced. After being some months established in pots and sufficiently hardened, they can then be planted out in rows in open air beds, leaving about 6 in. or 7 in. between plant and plant. When the young shoots have pushed about a couple of inches, cut the tops off again. This is generally done during the following spring with a pair of spring or sheep shears."

Growing with the greatest luxuriance in sandy peat, which for the most part forms their natural soil, there are, at the same time (says Mr. Hugh Fraser), few loams in which, if rich in vegetable matter and free from chalk or lime, they will not succeed; while the worst for the purpose may be adapted for their wants by the application of peat or old leaf-soil, and even a liberal allowance of well rotted manure. Several of the showiest sorts—such as the varieties of herbacea, mediterranea, and australis—flower in the order indicated from February till April, giving a gaiety and effect which no other plants could at that season, and contrasting admirably with the early bulbs with which they are associated. The other sorts—varieties of tetralix, cinerea, and vulgaris—are in perfection from May to September, the one succeeding the other, when vagans begins to develop itself and continues till late in autumn. Transplanting may be safely carried out between September and April, preferably in February and March if the weather is open and the soil in good working condition, as then the work is done before growth commences. For a list of species and varieties we adopt that of Mr. Fraser in his book on trees and shrubs.

E. ciliaris, indigenous to Portugal, the south of England, and some parts of Ireland, is a neat dwarf species from 9 in. to 12 in. in height. It produces its pale red flowers in terminal racemes from June to July. It is one of the finest of the hardy Heaths.

E. cinerea.—This species is found in great abundance in many of the northern countries of Europe and all over Britain, rarely rising above 1 foot from the ground.

The flowers are reddish-purple, changing to blue, and begin to expand early in June. Among its varieties are—alba, atropurpurea, bicolor, coccinea, monstrosa, pallida, purpurea, rosea, and spicata.

E. herbacea, indigenous to a wide area in Central Europe and to some localities in North Wales, is one of the finest of our hardy Heaths. It produces its lovely pale red blossoms from the beginning of March (and in some seasons much earlier) till the beginning of April; it is a spring-flowering plant, and as it may be clipped freely without damage, it is valuable as an edging in flower gardens. It grows about 1 ft. high. The var. carnea, according to some botanists, is the type of the species, from which it only differs in having brighc red or flesh-coloured flowers.

E. mediterranea is so named from being found abundantly in the countries bordering the Mediterranean Sea. It is also found in several districts in Ireland. In habit of growth and general appearance it resembles the preceding, which, as has been suggested, is probably only a form of this species. Its flowers are pale red, the anthers of a darker colour and very prominent, and usually in perfection in April. The vars. are —alba, carnea, glauca, nana, stricta, and rubra.

E. Mackiana.—This kind is indigenous to the Continent, and is also found in Connemara, Ireland. It has broad, ovate leaves, silvery on the under surface, possibly a variety of tetralix; it grows about 1 ft. high. The flowers are pale red, expanding in July and August. It is a remarkably showy plant.

E. tetralix.—This beautiful species is found wild in all the northern countries of Europe, and very abundantly on the moors and heaths of Britain, growing to heights of from 1 ft. to 2 ft. It is readily distinguished by its ciliated leaves, arranged in four whorls round the stems. The flowers are in terminal racemes, of a delicate pink colour, and generally in perfection from July to August. Among its varieties are—rubra and alba.

E. vagans.—This species is found wild in the south of France, in some parts of Ireland, and very abundantly on the moorlands of Cornwall. It grows from 6 in. to 1 ft. high, forming a neat bush. The flowers are pale purplish-red, and produced in great abundance along the branches. They are generally in perfection in August and September. It is an exceedingly showy plant, invaluable for margins to clumps of the larger peat-soil shrubs, and forms a neat close edging for flower gardens. The varieties are—alba, alba nana, carnea, and rubra.

E. vulgaris *(Calluna)*.—This is the common Heather or Ling of our moors. Its numerous varieties are so very beautiful that they should never be overlooked in forming a collection of hardy Heaths. They are all sports from the species, and have been found from time to time either associated with it in a wild state or in cultivation in gardens. The varieties are—alba, Alporti, aurea, argentea, coccinea, decumbens, dumosa flore-pleno, pygmæa, pumila, Hammondi, rigida, Searlei, and tomentosa. The tall Heaths, more strictly shrubs, are omitted from these selections.

Erigeron *(Fleabane)*. — Michaelmas Daisy-like plants of dwarf growth, all of which are somewhat alike in general appearance, all having pink or purple flowers with yellow centres. They flourish in any kind of garden soil, but one or two are best adapted for the rock garden. Of these, E. alpinum grandiflorum is the finest. It is similar to the alpine Aster, having large heads of purplish flowers in late summer, and remaining in beauty for a long time. Suitable for the rock garden, and also for well-drained borders. Division and seed. E. Roylei, a Himalayan plant, is another good alpine, of very dwarf, tufted growth, having large blossoms of a bluish-purple, with yellow eye. By far the best of the taller kinds is E. (Stenactis) speciosum, which is a vigorous species, with erect stems, that grows about 2½ ft. high, and bears in profusion, during June and July, large purplish lilac Aster-like flowers, with conspicuous orange centres. E. macranthum is another showy species, of a neat habit of growth, about 1 ft. high. It bears an abundance of large, purple, yellow-eyed blossoms in summer, and, like E. speciosum, will grow in any soil. E. mucronatum, known also as Vittadenia triloba, is a valuable border flower, with neat and compact growth, for several weeks in summer being literally dense rounded masses of blooms about 9 in. high. The flowers are pink when first expanded, and afterwards change to white, so that a plant presents every intermediate shade. Other kinds to be found in gardens are E. multiradiatum, glabellum, glaucum, bellidifolium, strigosum, and philadelphicum—the two last being the prettiest. All are easily increased by division in autumn or spring. The most effective and useful plant of the genus, so far as known, is E. speciosum, which is excellent for groups or borders.

Erinus alpinus *(Wall E.)*.—A distinct little plant, with violet-purple flowers, in racemes, abundantly produced over very dwarf tufts of leaves. A native of the Alps and the Pyrenees, perishing in winter on the level ground in most gardens, but quite permanent and producing masses of flowers when allowed to run wild on old walls or ruins. It is easily established on old ruins or walls by sowing the seeds in mossy or earthy chinks, and is of course well suited for rockwork, growing thereon in any position, often flowering bravely on earthless mossy rocks and stones. E. hirsutus is a variety covered with long and whitish pubescence. There is a white variety.

Eriogonum.—North American plants which, as seen on the Rocky Mountains and alpine regions in California, are of much beauty, but which we have never seen good

PLATE CVII.

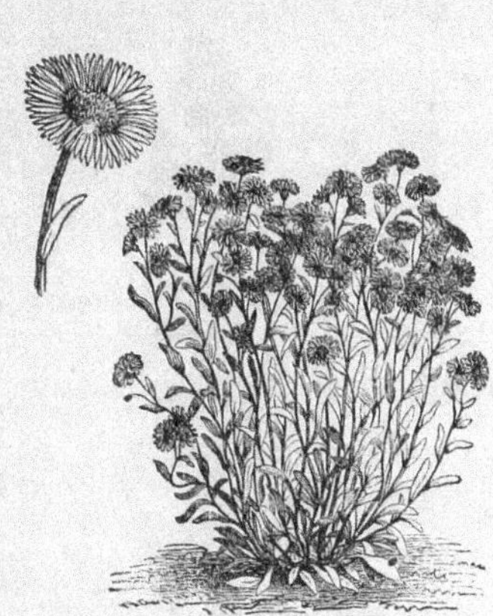

Erigeron speciosum (*Fleabane*).

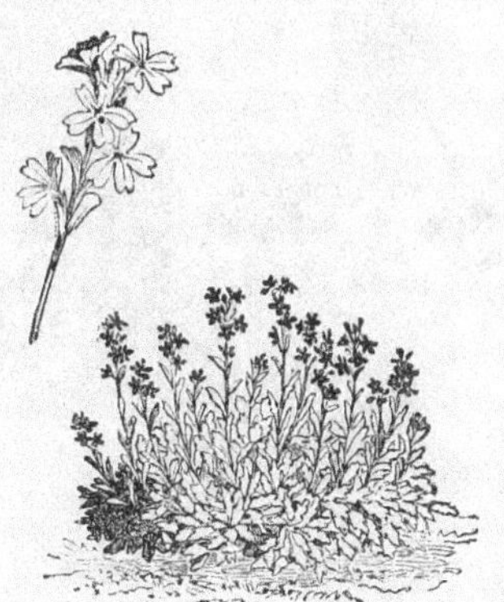

Erinus alpinus (*Wall Erinus*).

Erodium petræum (*Stork's-bill*). See page 123.

[*To face page* **122.**

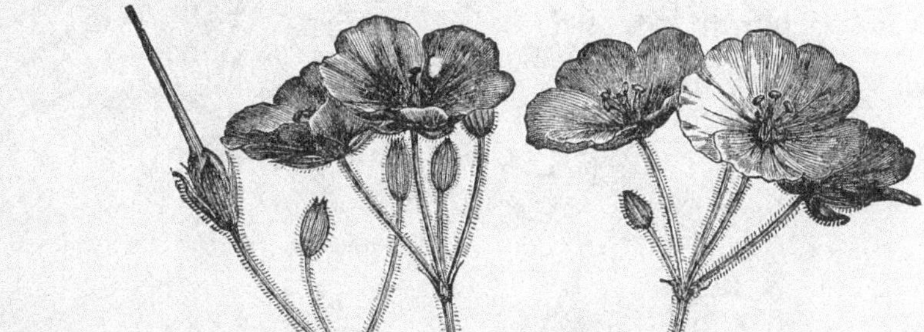

PLATE CVIII.

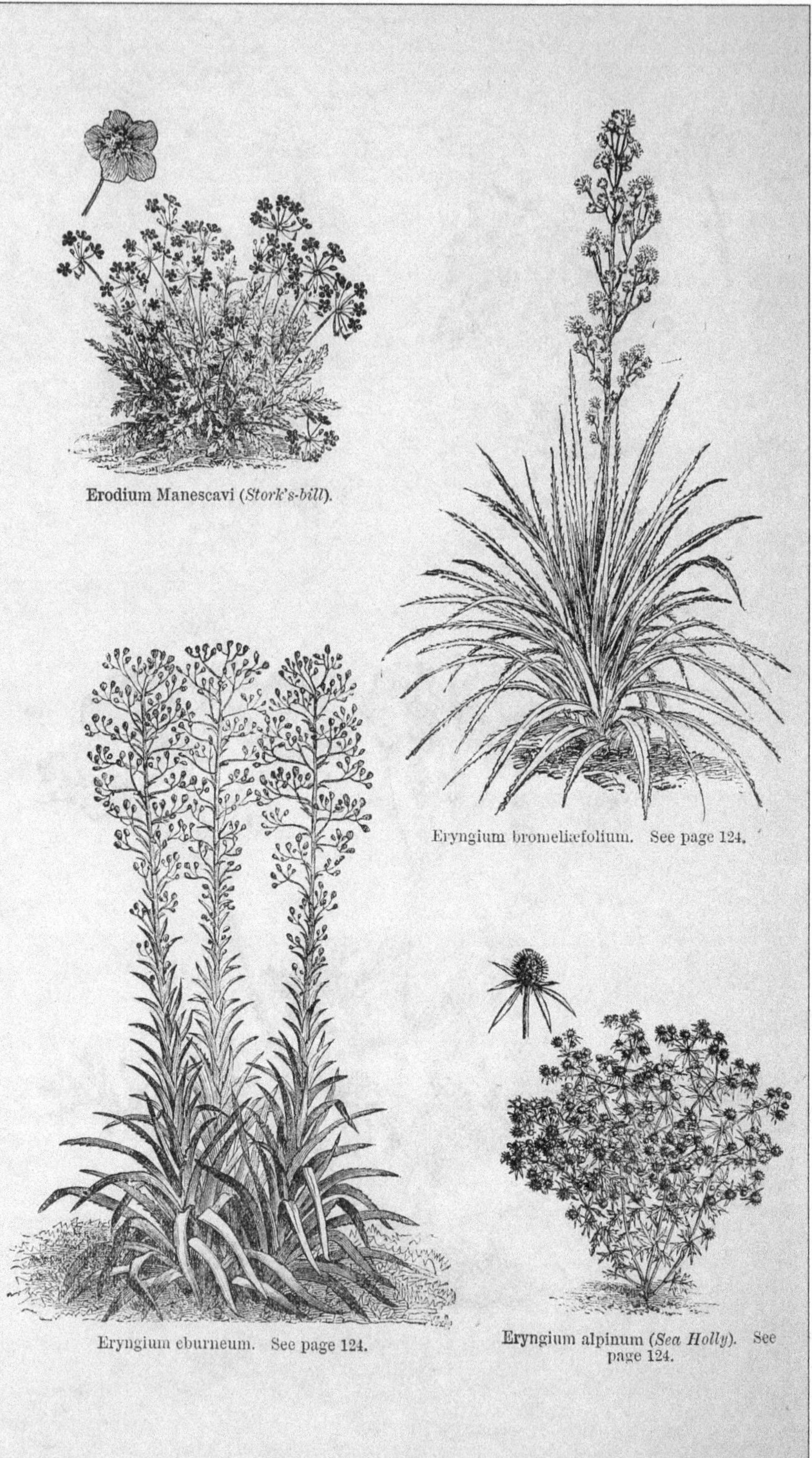

Erodium Manescavi (*Stork's-bill*).

Eryngium bromeliæfolium. See page 124.

Eryngium eburneum. See page 124.

Eryngium alpinum (*Sea Holly*). See page 124.

[*To face page* 123.

in cultivation, except, perhaps, E. umbellatum, which, from a dense spreading tuft of leaves, throws up numerous flower-stems, 6 in. to 8 in. high, on which are produced golden yellow blooms in umbels 4 in. or more across, forming a neat and conspicuous tuft. It is worthy of a place on any rockwork or border, in light, sandy soil, in which it has never failed to bloom profusely. The variety Sileri is much better than the type. Other species are E. compositum, flavum, racemosum, ursinum, which, however, are suitable only for botanical collections.

Eriophorum *(Cotton Grass)*.—Sedge-like plants with heads of white cottony seeds, on which account they are interesting in the artificial bog or by lake margins. E. polystachyon is the best kind for a garden; it may be found plentifully in some marshy districts.

Eritrichium *(Fairy Forget-me-not)*.—A minute alpine gem, closely allied to the Forget-me-nots, which, however, it far excels in the intensity of the azure-blue of its small blossoms. Though it has the reputation of being somewhat difficult to cultivate, a fair amount of success may be ensured by planting it in a compost consisting of finely-broken limestone or sandstone, with the addition of a small quantity of rich fibry loam and peat, and selecting a spot for it in the rock garden, where it will be fully exposed and where the roots will be in proximity to masses of half-buried rock, to the sides of which they delight to cling. The chief enemy with which one has to contend in growing this little plant, and indeed all other alpine plants which have silky or cottony foliage, is moisture during winter, which causes them to damp off in a short time; in their native habitats they are covered with dry snow during that period. To obviate this some recommend planting it where an overhanging ledge protects it from rain, &c., but if such protection be not removed during summer, it has an injurious effect, as it causes too much shade and dryness. A better plan is to place two pieces of glass so as to form a ridge over the plant, thus keeping it dry and at the same time allowing a free access of air at the ends, but these should be removed early in spring. Alps, at high elevations.—G.

Erodium *(Stork's-bill)*.—Like the Geranium, but usually smaller and more southern in origin than the hardy Geranium. They are suited for chalky banks or the rock garden, and some for borders, while some may be naturalised in the Grass in warm soil. Among the better species are—

E. macradenum.—A pretty Pyrenean species, 6 in. to 10 in. high, with the blooms French white, delicately tinged with purple, and veined with purplish rose; the lower petals are larger than the others; the two upper ones have each a dark-coloured spot, which at once distinguishes this species from the other Erodiums. The best position for this plant is in a crevice, tightly placed between two rocks, exposed to the hottest sun, where the roots can penetrate into dry sandy or stony soil to the depth of at least 3 ft. When grown in this way, this Erodium is an extremely pretty object; the leaves, by the dryness of the situation, are kept in a dwarf state, nestling to the rock, and the flowers are produced in great abundance during the summer months. The plant has an aromatic fragrance.

E. Manescavi is a vigorous herbaceous plant, and the most showy species of Erodium. It grows from 1 ft. to 1½ ft. high, and throws up strong flower-stalks, which rise above the foliage, each bearing from seven to fifteen showy purplish flowers from 1 in. to 1½ in. across. The plant is not at all fastidious as to soil or situation, but the best position for its successful cultivation is in dry, hard soil, fully exposed to the sun. If the soil be too rich, the plant produces its leaves in such abundance that the flowers are hidden by them. Seed, or by careful division.

E. petræum.—This has purplish-rose flowers, from three to five on each of the stalks, which are 4 in. to 6 in. high. The leaves and flower-stalks are densely clothed with minute hairs. It thrives best when planted in deep sandy or gravelly soil, in warm positions, amongst the dwarfer kinds of alpine plants.

E. Reichardi.—A miniature species, growing about 2 in. to 3 in. high when in flower; the leaves are small, heart-shaped, and lie close upon the ground, forming little tufts from which arise slender stalks, each bearing a solitary white flower, marked with delicate pink veins. It often continues in flower for many weeks. This Erodium should be grown amongst the dwarfest and choicest alpines, such as the Androsaces and Gentians, in gritty peat mixed with a small portion of loam.

To the foregoing may be added: E. caruifolium, a plant from 6 in. to 10 in. high; flowers, red, about ½ in. in diameter, and borne in umbels of nine or ten blossoms. E. alpinum, which resembles E. Manescavi, but is much dwarfer, growing 6 in. to 8 in. high, and flowering continuously from spring to autumn. E. romanum, allied to the British E. cicutarium, but with larger flowers, grows 6 in. to 9 in. high; flowers, purplish, appearing in spring and early summer. E. trichomanefolium, a very pretty dwarf kind, 4 in. to 6 in. high, with leaves so deeply cut as to resemble a Fern; flowers, flesh-coloured, marked with darker veins. All the preceding, with the exception of E. Manescavi and E. hymenodes, are suited for rockwork or borders, in light, sandy, or calcareous loam. E. Manescavi should, perhaps, be confined to the border, as it is somewhat too tall and spreading for the rockwork.

Erpetion reniforme *(New Holland Violet)*.—This mantles the ground with a mass of small leaves, has numerous slender, creeping stems, and bears blue and white flowers of exquisite beauty, rising not more than a couple of inches from the ground, and borne throughout the summer. It is peculiarly fitted for planting out over the surface of a bed of peat or very light earth, in which some handsome plants would be put out during the summer in a scattered manner, and the little herb allowed to crawl rapidly over the surface. Being very small and delicate, as well as pretty, it should not be used under or around coarse subjects. It must be treated like an ordinary tender bedding plant—taken up or propagated in autumn, and put out in May or June. Australia. Division.

Eryngium *(Sea Holly)*.—A remarkably distinct family of the Umbelliferæ, some of the varieties being beautiful and hardy. Some of the S. American kinds are very striking in port, but the palm of beauty, so far as our gardens are concerned, goes with the northern and mountain kinds.

E. alpinum.—A singularly fine plant, from 2 ft. to nearly 3 ft. high, forming a rather stiff bush, with leathery and very spiny leaves of a sea-green colour, and bearing numerous roundish heads of bluish flowers, the stems beneath them being also, for some inches down, of a very handsome blue. Suitable for planting in groups in pleasure-grounds, for borders, or grouping with the finest and most distinct hardy plants.

E. amethystinum is not so tall as the preceding, seldom growing more than 2½ ft. high. It is remarkable for the beautiful amethystine bloom which the leaves assume in July, and which they preserve until the approach of frost. It is suitable for the positions recommended for the preceding kind. Various other members of this family are useful in like manner; indeed, there is scarcely one of them that is not so, including our own common Sea Holly (E. maritimum). Other kinds are E. spina-alba (dwarf and excellent), E. Bourgati, E. giganteum, and E. planum. The plants are frequently misnamed, and it is probable other species of the genus are worthy of cultivation.

E. Leavenworthi.—The species of annual duration are few in number, and include probably but one possessing any value for the garden—E. Leavenworthi. It grows from 2 ft. to 2½ ft. high, with an erect, grooved stem, branched in its upper half, and bearing flower-heads that in their early stages are green, but, after attaining their full growth, assume a rich deep violet tint, as well as the involucrum, coronal tuft, and foot-stalk, which have, in addition, a fine metallic lustre. As in the case of E. amethystinum and some other species, the heads preserve their colour for some time after being cut. This species requires the treatment of half-hardy annuals, and, being a somewhat late bloomer, should be sown early in heat, or in a greenhouse, in which case it will produce its flower-heads in September and October. Texas.

THE LARGE S. AMERICAN SPECIES.—The most extraordinary forms among the Sea Hollies are a group of species belonging exclusively to America, and particularly to Brazil, the Argentine Republic, and Mexico. Of this class are the three following, viz., E. aquaticum (E. yuccæfolium), from Pennsylvania and Virginia; E. paniculatum, originally from Brazil, and frequently cultivated under the name of E. bromeliæfolium; and, lastly, the true E. bromeliæfolium, from the humid forests of Mexico. But of all the long narrow-leaved kinds none equal those raised from seeds sent to Paris some years ago by M. Lasseaux. These seeds, which were sown immediately, soon germinated, and the young plants obtained from them soon developed their true character. Three years afterwards many of them flowered, and were successively described by M. Decaisne, who recognised, besides E. pandanifolium and E. bracteatum, three new species, viz., E. Lasseauxi, eburneum, and platyphyllum. E. Lasseauxi differs from E. pandanifolium, which it much resembles in foliage, inasmuch as it is more upright and glaucous underneath, but, above all, inasmuch as its flowers are whitish, instead of reddish, violet. The most remarkable kinds are E. Lasseauxi, pandanifolium, and eburneum; E. platyphyllum, characterised by its large leaves, spread in a rosette upon the soil, might yet, if its foliage were more abundant, be classed among the preceding. So far as these plants have been tried in our gardens they afford striking effects from seedling plants, but more by their vigorous habit than their flowers, which have not the charm of the hardy European kinds. Moreover, they are not really hardy in our climate, though, by means of protection and planting in warm soils, some success may be attained.

Erysimum. — Wallflower-like plants, mostly of dwarf growth, of perennial, biennial, and annual duration. Of the perennials the following are the finest:—

E. ochroleucum *(Alpine Wallflower)*.—This handsome plant forms, when under cultivation, neat, rich green tufts, 6 in. to 12 in. high, and is in spring covered with a profusion of beautiful sulphur-coloured flowers. Rockwork will be found to offer the most congenial home for it; it does very well on good level ground, but is apt to get somewhat naked about the base, and will perhaps perish on heavy soils during an unusually severe winter. It thrives best when rather frequently divided. It is propagated by division and by cuttings. It is capital as a dwarf border plant on light soils. Alps and Pyrenees, flowering in spring. There are several varieties. (= Cheiranthus alpinus.)

E. pumilum *(Liliputian Wallflower)*.—A very small plant, rare in cultivation, resembling in the size and colour of its flowers the alpine Wallflower, but without the vigorous and rich green foliage of that species; producing flowers very large for the

PLATE CIX.

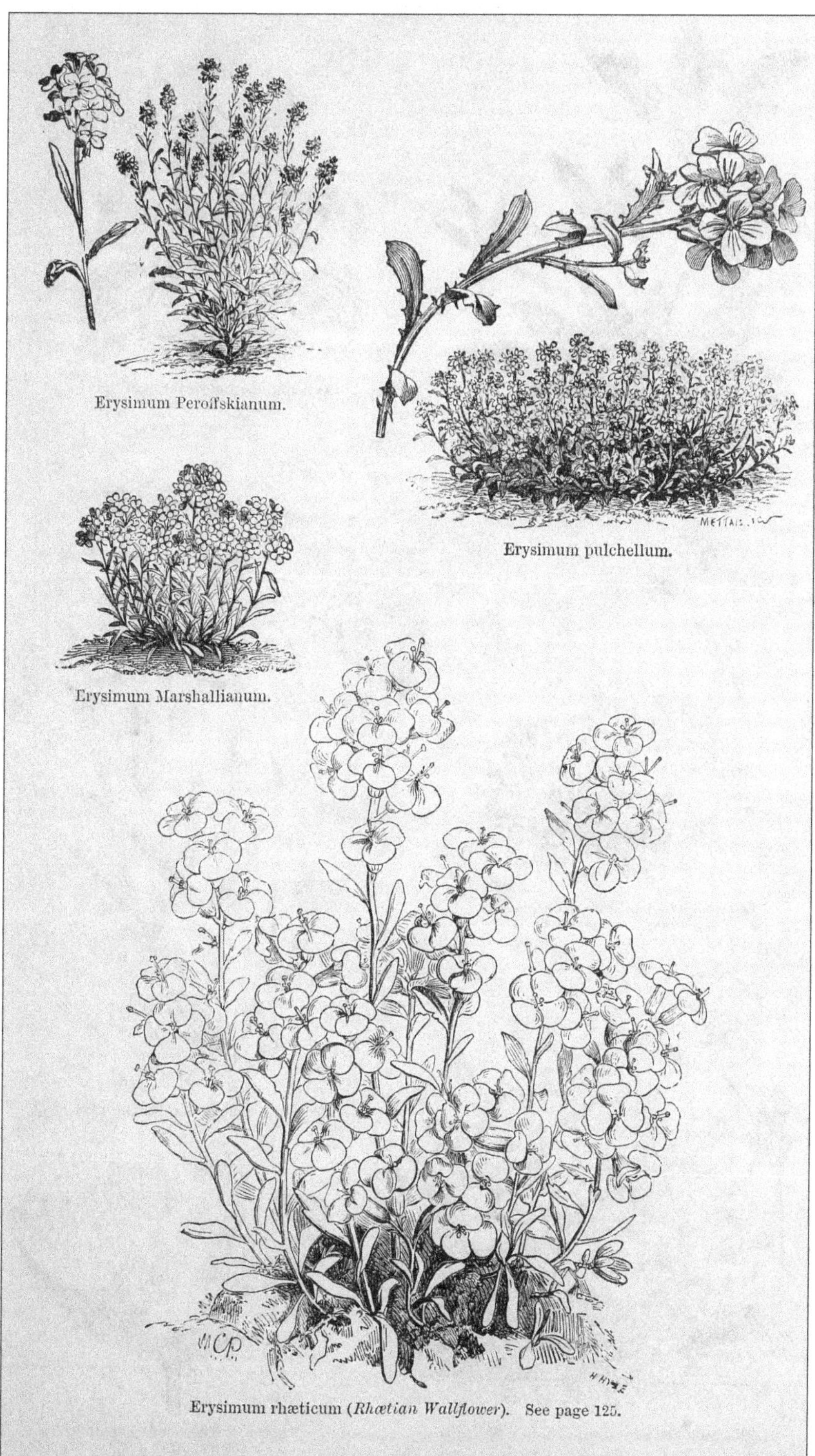

Erysimum Peroffskianum.

Erysimum pulchellum.

Erysimum Marshallianum.

Erysimum rhæticum (*Rhætian Wallflower*). See page 125.

[*To face page* **124.**

PLATE CX.

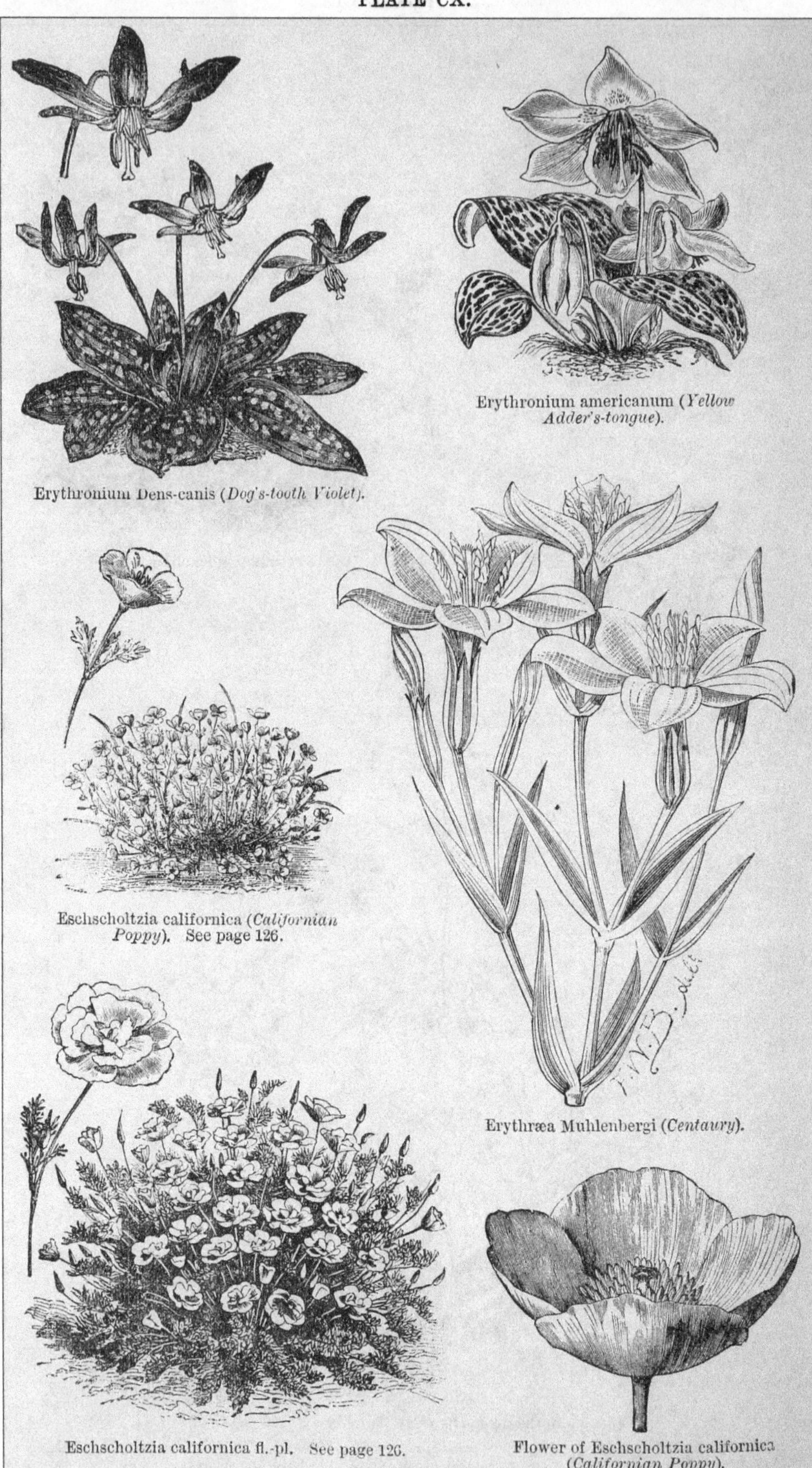

Erythronium Dens-canis (*Dog's-tooth Violet*).

Erythronium americanum (*Yellow Adder's-tongue*).

Eschscholtzia californica (*Californian Poppy*). See page 126.

Erythræa Muhlenbergi (*Centaury*).

Eschscholtzia californica fl.-pl. See page 126.

Flower of Eschscholtzia californica (*Californian Poppy*).

[*To face page* 125.

plant, which is often only 1 in. high, above a few narrow, sparsely-toothed leaves, barely rising from the ground. A native of high and bare places in the Alps and Pyrenees, requiring to be grown on the rock garden in an exposed spot, in very sandy or gritty loam, surrounded by a few small stones to guard it from excessive drought and accident, and associated with the most minute alpine plants. It is very nearly related to the alpine Wallflower, E. ochroleucum, but is at once separated from that plant by its minuteness and the greyish-green colour of its leaves.

E. rhæticum *(Rhætian Wallflower).*—A pretty mountain flower which, though rarely to be met with in cultivation, is a common alpine in Rhætia and the neighbouring districts, where in early summer broad dense-tufted masses of it are aglow with pretty, clear yellow blossoms. It is somewhat similar to the Dwarf Alpine Wallflower (E. pumilum) of the Swiss mountains, and the dense tufts formed by it are studded with bright lemon-coloured blossoms. The Rhætian kind bears a resemblance also to the Lance-leaved Wallflower (E. lanceolatum), a common European plant which has major and minor varieties, neither of which, however, is so desirable as the kind under notice. E. canescens, a South European species with yellow, scentless flowers, is also a neat-growing alpine, and so is E. rupestre, than which few plants are more to be desired for adorning rockwork. All of them are easy plants to grow, and they all delight in gritty soil in a well-drained and sunny position on the rockery. Among the biennal and annual kinds the best is E. Perofskianum, which grows from 1 ft. to 1½ ft. high, producing dense racemes of orange-yellow flowers. For early flowering it should be sown in autumn, and again in March and April for later bloom. E. arkansanum and pachycarpum are similar to E. Perofskianum, but inferior and now rarely met with.

E. Barbarea *(Barbarea vulgaris).*

Erythræa *(Centaury).*—A small genus of the Gentian family, all rather pretty dwarf plants of biennial duration. The native species, E. littoralis, common in some seashore districts, is worth cultivating. It grows from 4 in. to 6 in. high, and bears in abundance rich pink flowers, which last a considerable time in beauty, and will withstand full exposure to the sun, although partial shade is beneficial. A similar and very beautiful species is E. diffusa, which is a rapid grower and bears in summer a profusion of pink blossoms.

E. Muhlenbergi is another beautiful plant. It is neat in habit and grows about 8 in. high, producing many slender branches. It is very floriferous, the blossoms measuring 3½ in. across, and being of a deep pink colour, with a greenish-white star in the centre. To grow these plants well, seeds should be sown in autumn, and grown under liberal treatment till the spring; they will then flower much earlier and finer than spring-sown plants. These are excellent for the rock garden or margins of a loamy border, but the soil must be moist. Altogether, a group of plants of more botanical than garden importance, from the fact that on account of their duration or other peculiarities they do not easily make good garden plants.

Erythrochæte palmatifida *(Senecio japonicus).*

Erythronium *(Dog's-tooth Violet).*—These Lilaceous bulbous plants are among the loveliest of our hardy garden flowers, though only one of them, the old favourite Dens-canis, is commonly cultivated. The genus is not a large one—there are about a dozen species and varieties. These belong to the North American continent, with the exception of

E. Dens-canis, a beautiful plant found in various parts of Europe. It has handsome oval leaves, marked with patches of reddish-brown. The flowers are borne singly on stems 4 in. to 6 in. high, drooping gracefully, and are rosy purple or lilac. There is a variety with white, one with rose-coloured, and one with flesh-coloured flowers. There is also a form called E. longifolium, which has longer and narrower leaves and larger flowers, and it is from this variety that the sorts enumerated in trade catalogues under the name of majus are apparently derived. E. Dens-canis thrives in moist sandy or peaty soil, in positions fully exposed to the sun. It is one of the most valuable subjects for the spring or rock garden, or border of choice hardy bulbs, and, where sufficiently plentiful, for edgings to American plants in peat soil. The bulbs are white and oblong, resembling a dog's tooth, hence its common name. It is increased by dividing the bulbs every two or three years, replanting rather deeply. A native of Central Europe. The varieties of sibiricum, a robust-growing plant from the Altaian Mountains in Siberia, and japonicum, with violet-purple flowers, are not, so far as we are aware, yet in cultivation.

E. americanum *(Yellow Adder's-tongue)* is common in low copses and the woods in the Eastern States of North America, where it flowers in May. The leaves are pale green, mottled, and commonly dotted with purple and white. Flowers 1 in. across, pale yellow, spotted near the base; produced on slender stalks 6 in. to 9 in. high. A variety (E. bracteatum) differs in having a bract developed, as E. grandiflorum sometimes has. This is a very pretty kind, but is seldom seen in cultivation in consequence of its being a somewhat shy flowerer. The late Mr. M'Nab was very successful with it in the Edinburgh Botanic Garden, and he thus writes of it in an early volume of *The Garden*: "This interesting plant formerly grew in the open border here, but its flowers were rarely seen. Some years ago I put a tuft of the bulbs in one of the stone compartments of the rock garden, having a southern aspect, the soil being a mixture of peat and loam. As soon as the space became filled with roots, flowers were freely produced, and on the 20th of April it was covered with yellow blooms. In these confined spaces the bulbs are better matured for flowering than they can possibly be in open borders, where the surface of the ground is generally seen covered with a mass of small green leaves proceeding from numerous unmatured bulbs, having but few of the larger spotted leaves which generally accompany the flowers." It is probable the rich soil of our gardens develops growth at the expense of flower. Tried in poor sandy soil in copses or in the wild garden, the little plant may bloom better.

E. giganteum.—This, the noblest of the genus, is considered a variety of E. grandiflorum. It is a showy flower, pure white, with a ring of bright orange-red, and measures 3 in. in diameter. It is found in California at an elevation of 6000 ft. to 10,000 ft., and also in Vancouver's Island. It was called E. maximum by Douglas, and E. speciosum by Nuttall. Rarer or little known kinds are E. revolutum, albidum, purpurascens, propullans, and Hartwegi; but none of these are so handsome as those described above.

E. grandiflorum.—The only cultivated kind that bears more than one flower on a stem. It is an extremely handsome plant when well grown. In a peat bed, with Lilies and other peat-loving plants, it attains fine proportions, as many as five flowers being produced on the stems. The late Mr. M'Nab used to grow the larger of the American kinds as well as the European Dens-canis very successfully in grass. Writing of them in spring, he says, "Many Dog's-tooth Violets are in bloom on the northern grassy slopes of the rock garden; they were thickly dibbled in, here and there, when the turf was first laid, and, being placed in all exposures, a longer flowering season of these interesting plants has been obtained. In such places they do not seem to multiply fast, as single flowers proceeding from the two or three spotted leaves are only produced. On grass banks with a southern aspect the leaves are all ripened off before the first grass cutting takes place, which is not the case on grass slopes having a northern aspect."

Erythrina.—These in a wild state are very beautiful trees or shrubs, pretty generally distributed through the tropics of both hemispheres. Some attain great dimensions, while others are dwarf bushes with woody rootstocks. Many of the species produce beautiful large Pea flowers, usually of a blood-red or scarlet colour, in terminal racemes. The varieties of these have proved very hardy and useful in the summer garden, flowering freely, and showing considerable beauty of foliage. E. ornata, Marie Belanger, laurifolia, Crista-galli, profusa, Madame Belanger, ruberrima, Hendersoni, have stood out with slight protection in England. The common old Erythrina Crista-galli will thrive for years against a warm south wall in a light soil if protected about the roots in winter, and, so grown, is often a very handsome object in the warmer countries. How far E. herbacea will prove an efficient substitute for the older and better known species remains to be seen, but, having been found to resist a New York winter, it may be fairly assumed to be sufficiently hardy for open-air cultivation in England, and it is deserving of a trial. It is of rather dwarfer habit than the old species, and has a woody root-stock which throws up in summer, under favourable conditions, stems from 2 ft. to 4 ft. high, these stems being of two kinds, one sort bearing leaves only, the other flowers with a very few leaves. The flowering stems bear a raceme, 1 ft. to 2 ft. long, of narrow flowers, each about 2 in. in length, the deep scarlet standard, which in so many genera is erect, being here horizontal and folded over the wings and keel. The seeds are of bright scarlet, and should be sown in heat as early as practicable, the seedlings being kept in a frame the first winter. This species is a native of Texas, and is found as far north as Carolina, and westward to Sonora.

Eschscholtzia (*Californian Poppy*).—Showy annuals long and favourably known. The beautiful new forms of this glowing annual, which we have recently seen, are acquisitions; the rich reddish-orange of Mandarin and the unique form of the double crocea are of real value, and, with crocea, alba, and the orange aurantiaca, give us a batch of most attractive flowers. What is known as E. rosea is pretty, but is apt to revert to the white form from which it sprang. To have these showy flowers in all their beauty, they should be sown in August and September to bloom in early summer. They may also be sown later—allowing the plants to bloom where they are sown. The plants get deeply and firmly rooted, and flower much longer than those sown in spring. They are very hardy, and snails and slugs do not molest them. There are some half-a-dozen kinds, all of which are well worth growing, viz., E. californica, orange, very strong; E. crocea, saffron colour; E. c. alba, white; E. c. Mandarin, orange and crimson, very fine; E. crocea fl.-pl., double; E. c. rosea, and E. tenuifolia; new forms are raised from time to time.

Eucalyptus.—Large handsome Australian trees and shrubs, of which there are a number of species, many growing to a great height. The leaves are of a thick leathery texture, and very variable in shape. They are most likely to be seen in the south of England and Ireland, where a few of the species will be found to live in the open air. Some grow them, for the aspect they present after a single year's growth, in the open air about London, in which case they should be put out about the middle of May. Owing to some letters that appeared in the *Times*, written by persons unaware of the results of

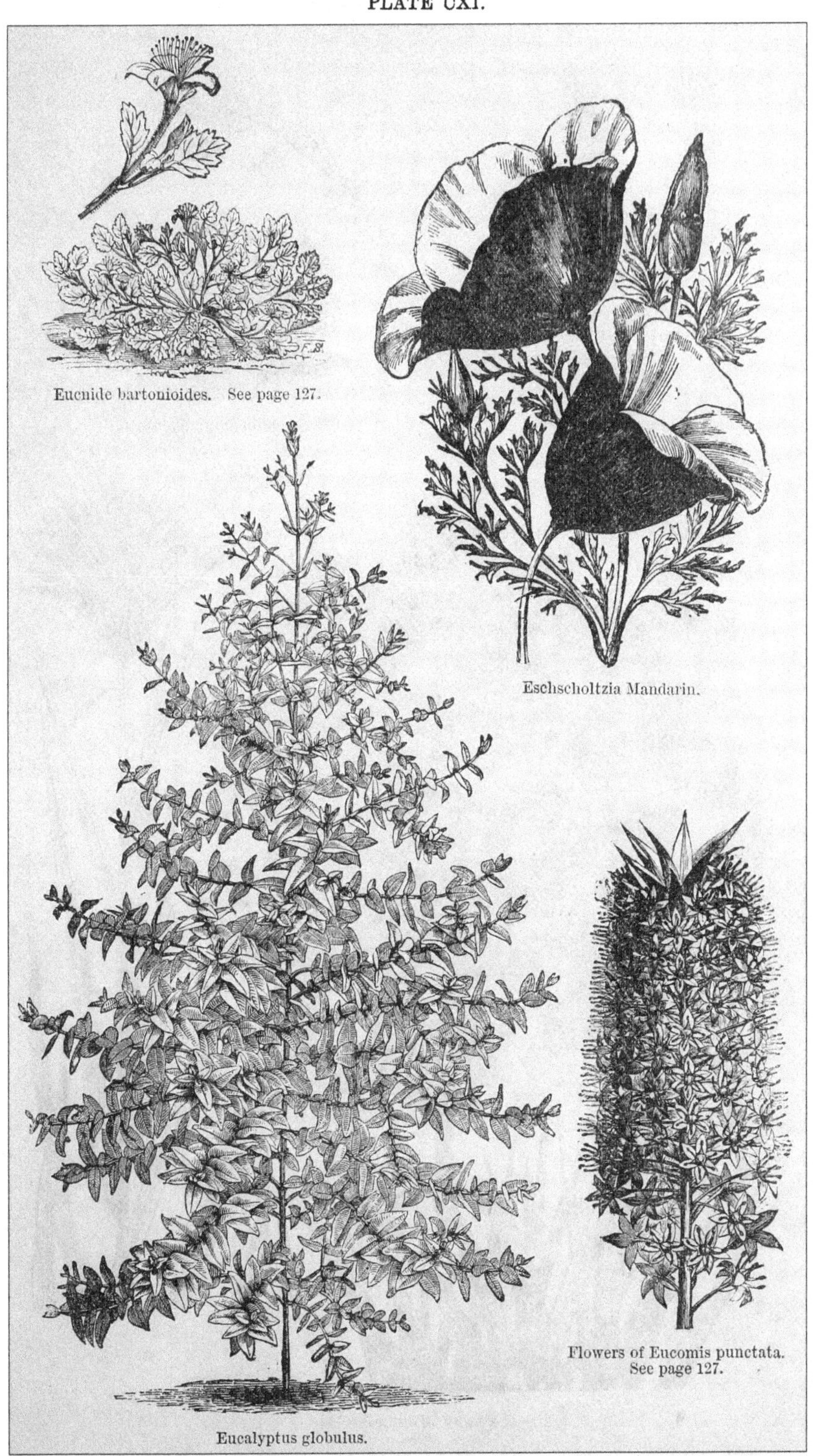

Eucnide bartonioides. See page 127.

Eschscholtzia Mandarin.

Flowers of Eucomis punctata. See page 127.

Eucalyptus globulus.

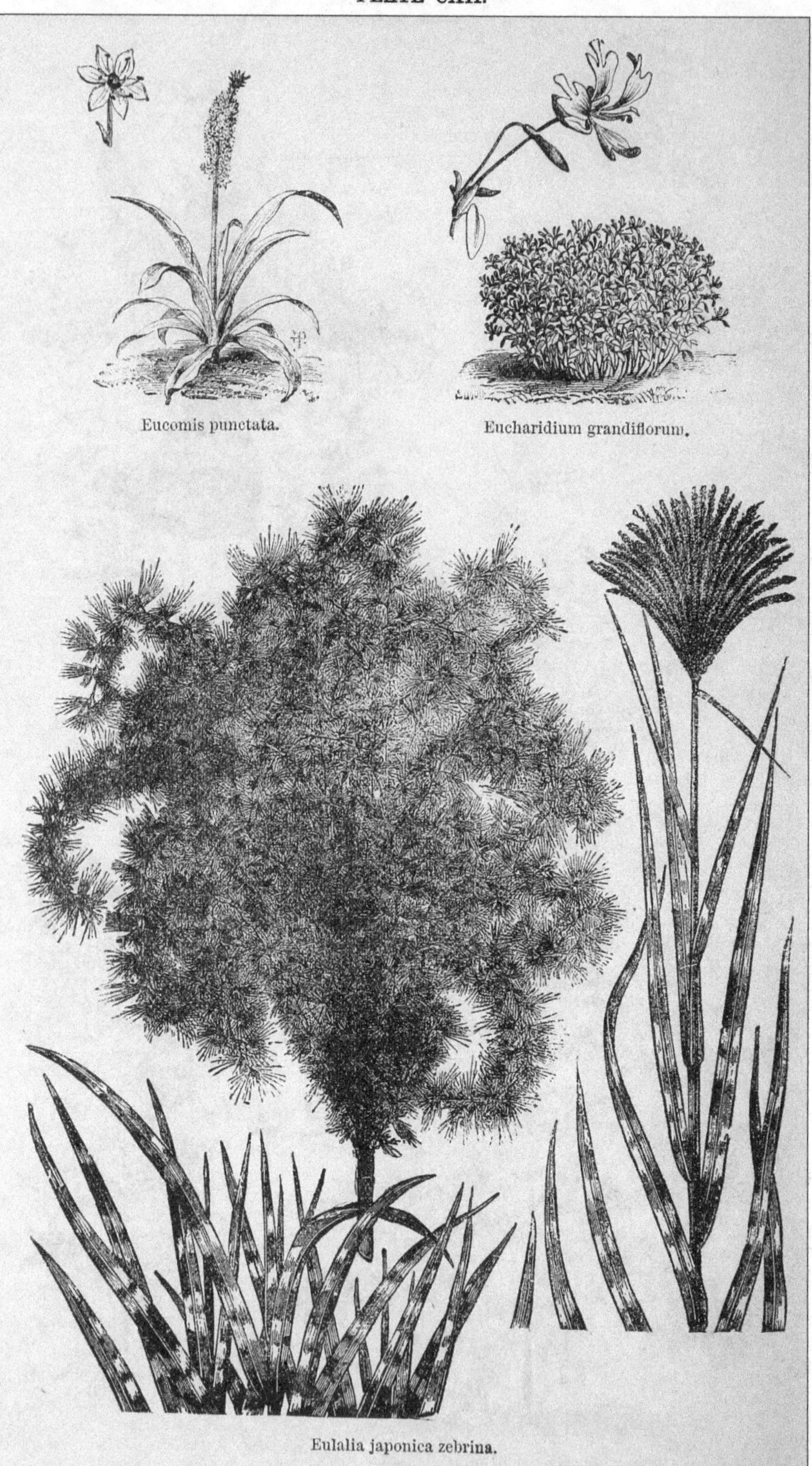

Eucomis punctata.

Eucharidium grandiflorum.

Eulalia japonica zebrina.

the planting of the Gum tree in this country, a great many were induced to plant the common Gum tree, which perished in the first severe frost. It is only in the more favoured districts these trees have the least chance, and there they never present the graceful and stately port they show in countries that really suit them, as parts of Italy and California. What the higher mountain species may do yet remains to be seen, and the common Gum tree is sometimes made fair use of among the larger plants put out for the summer in the London parks.

Eucharidium.—Pretty hardy annuals of the Evening Primrose family, from California. They require the same treatment as all annuals from that region. They may be sown in autumn for early summer flowering, or from March to June for late summer and autumn bloom. They flower in about eight weeks after sowing, and remain in flower a long time. Three species are in cultivation. E. concinnum grows about 9 inches high, bearing a profusion of rosy purple blooms; E. grandiflorum has larger flowers, rosy purple, streaked with white. There is a white variety (album), and one with pink flowers (roseum). E. Breweri is an elegant new annual, of a more robust habit, and having red flowers of a deeper, richer colour than those of E. grandiflorum. They are of secondary importance in the flower garden, but may be used occasionally as surface plants or in bold masses. Like many other annuals, they suffer in general estimation in being judged of from spring-sown plants, the bloom of which is poor and short-lived.

Eucnide bartonioides.—A half-hardy annual of the Loasa family, from Mexico. The stems grow about 1 ft. high, and bear sulphur-yellow flowers, 1½ in. across, which are showy when several are expanded in August and September. Seeds should be sown in heated frames in early spring, but the seedlings should be very carefully transplanted in May to the open borders, as they are very susceptible to injury at that period. (= Microsperma.)

Eucomis. — Cape bulbous plants, not very showy, though the broad, handsome foliage, more or less spotted at the base with purple; from which rise the tall cylindrical spikes of blossoms surmounted by a crown-like tuft of leaves, renders them interesting objects, deserving of cultivation in the outdoor garden; they have proved of sufficient hardiness to survive our ordinary winters on light and dry soils, as in the case of many other Cape plants. There are but four species, all of which are in cultivation. E. undulata has leaves 18 in. long, wavy at the margins, and profusely marked with dark purple blotches on the under surface, which, in the variety striata, assume the form of stripes. The flower-spike is 2 ft. to 4 ft. high. On the upper half are densely arranged, in a cylindrical manner, the numerous greenish-white blossoms, with purplish centre, crowned by a tuft of narrow green leaves. E. punctata is the largest growing kind, having leaves about 3 ft. long. E. regia is dwarfer than either of the preceding. The raceme of flowers is about 1 ft. high, and the tuft of leaves at the top is larger than in the other kinds. E. nana is the smallest kind. The leaves are spreading, and lie horizontally, whilst in the others they are more erect. They thrive best in light, sandy soil, with the roots protected by a covering of some kind during winter. The foot of a south wall, associated with the larger hardy bulbs, suits these plants, which are not among the most effective or graceful of the Lily family.

Eulalia japonica. — A valuable and hardy ornamental perennial Grass of robust growth, attaining 6 ft. to 7 ft. in height, established plants forming clumps 16 ft. to 18 ft. in circumference. The flower panicles are at first brownish-violet, with erect branches, but as the flowers open the branches of the panicle curve over gracefully, and bear a resemblance to a Prince of Wales' Feather. Each of the individual flowers, which are very numerous, has at its base a tuft of long, silky hairs, and these contribute greatly to the feathery lightness of the whole. For isolated positions on lawns it is an excellent plant, or it might be used in groups, or on the margin of the shrubbery. Even more valuable than the type are the two variegated forms of this Grass. One, called variegata, has the leaves longitudinally striped with white and green; the other, zebrina, has distinct cross bars of yellow on the green, rendering it singularly attractive. These variegated forms are not quite so hardy as the type, particularly zebrina. Division or seed. Japan.

Eupatorium (*Thorough-wort*).—Coarse-growing Composite plants, most of which are better suited for naturalising in the wild garden than for borders, though two or three are worth a place for supplying cut flowers in autumn. The most suitable for this purpose are E. ageratoides, altissimum, and aromaticum, all of which grow from 3 ft. to 5 ft. high, and bear a profusion of white blossoms in dense, flat heads. E. purpureum grows as much as 12 ft. high, the stems being terminated by huge clusters of purple flowers, thus forming a fine object in the rougher parts of a garden. All grow in any kind of soil.

Euphorbia (*Spurge*).—This is a very largely cultivated genus, but contains few hardy species of value for the flower garden. The foliage of some, such as E. Cyparissius, is very elegant. In spring, E. pilosa and amygdaloides are made somewhat attractive by the yellow flowers, when little else is in bloom, but they are scarcely worth growing in a general way. Some of the dwarf kinds, such as E. Myrsinites, portlandica, capitata, and triflora, are neat in habit. All grow in any soil or position. There are a few variegated forms. The well-known Cape Spurge (E. Lathyris) is often seen in cottage gardens, and in habit is a distinct plant, with a certain beauty of foliage and port. A few plants on a bank or rough place are not amiss.

Eustylis (= *Nemastylus*).

Eutoca.—These Californian plants, numbering about four species, are pretty annuals, amongst which E. viscida is the best. It grows about 1 ft. high, and has erect branches with coarsely toothed leaves. The flowers, which are borne on curling racemes, are a deep blue, flushed in the centre with violet, and appear from July to September. There is a variety with white (albida) and another with lilac (lilacina) flowers, both pretty. They are so modest in colour that they are not much used, though a surfacing of the best kinds would be pretty in certain circumstances. But the spring-sown plants are too short in duration to allow of much use being made of them in gardens generally. The plant thrives best in sandy loam. It is best sown early in the season under glass, or in the preceding autumn. The other kinds are E. divaricata, E. Menziesi, and E. Wrangeliana, but they are scarcely worth cultivating.

Evening Primrose (*Œnothera*).

Everlasting Pea (*Lathyrus latifolius*).

Exogonium Purga (*Jalap Plant*).—Of autumn-flowering hardy plants, there is, perhaps, none more beautiful than this. Of its hardiness there can be little doubt. It has lived at Bitton, Gloucestershire, without any protection for years, and each year it has flowered beautifully. It has also done well at Drayton-Beauchamp, Kew, Fulham, and in the Edinburgh Botanic Gardens. Mr. Ellacombe grows it in a sheltered corner, and gives it a tall wire trellis to grow up, with a spreading top. It does not flower in the lower parts, but the entire top and the pendent shoots become a mass of most lovely blossoms. At Bitton, if not checked by late spring frosts, it comes into blossom early in September, and continues to flower till cut down by frost. Mr. Ellacombe states that if he were to plant another, he should place it under a south wall, near a Peach or Apricot tree, and let it wind its way through the branches. With a very little training it would do no injury to the tree, and in such a situation it would probably flower earlier, and perfect its seeds. E. Purga has roundish tubers of variable size, those of mature growth being about as large as an Orange and of a dark colour. These, as we have said, are the true Jalap tubers. The plant gets its name from Xalapa, in Mexico, its native region. Division of tubers.

Farfugium grande.—A vigorous perennial, with fleshy stems from 1 ft. to 2 ft. high, and broad leaves of a light green colour, variously streaked, and spotted with yellow in one variety, and with white and rose in another. It does best in free, moist soil, with plenty of leaf-mould or peat, in a half-shady position. During the heats of summer it will require frequent watering. At the approach of winter it should be removed to the greenhouse, except in the milder districts, where it survives an ordinary winter. In colder parts it is scarcely worth planting out, as it grows so slowly; but where it thrives it is handsome in borders, or near the margin of beds. Multiplied by division in spring; the offsets to be potted and kept in a frame until they are well rooted.

Feather Grass (*Stipa pennata*).

Felicia tenella.—A neat little half-hardy plant of the Starwort family. It has dwarf slender stems, terminated by flower-heads, ½ in. across, of a pale violet-blue with yellow centres. It should be raised as a half-hardy annual, but it becomes shrubby, and may be kept over the winter like half-hardy bedding plants. Flowers in light soil from July to September. Cape of Good Hope. Would form a graceful carpet of blue Daisies beneath Gladioli, Tuberoses, standard Fuchsias, or other tall plants of slender habit or form. *Syn.*, Aster tenellus.

Fennel Flower (*Nigella*).

Fenzlia dianthiflora.—A charming Californian annual, forming compact tufts from 1 in. to 4 in. in height. Its flowers are large, and borne in abundance; they vary in colour from purple and lilac to almost white. The plant is perfectly hardy, and, like several other Californian annuals, does best when sown in autumn. It thrives in any ordinary soil, but the warmer and more sheltered the situation is the better. Would make a lovely ground or carpet plant, scattering some slender bulbs or other taller plants through it. *Syn.*, Gilia dianthoides.

Ferdinanda eminens.—One of the tallest and finest sub-tropical plants, growing well in the southern counties when supplied with rich soil and abundant moisture. It is the better for being sheltered. Where the soil is rich, deep, and humid, and the position warm, it attains large dimensions, sometimes growing over 12 ft. high, and suspending pairs of immense opposite leaves. It forms a good companion to the Castor-oil plant. It requires to be planted out, in a young state, about the middle of May, and grows freely from cuttings; greenhouse treatment will do in winter. It is better to keep a stock in pots through the summer to afford cuttings, though the old ones may be used for that purpose.

Ferraria.—Plants of the Iris order from the Cape of Good Hope. Only one species is hardy enough to grow in the open air. It is F. undulata, a curious plant with flowers in shape like the Tiger flower (Tigridia) but small, of a dull plum colour and wavy edged. It requires a light, sandy soil, on a warm, sunny border, and if close under a south wall all the better. The bulbs require either protection during winter or lifting in autumn, when they may be divided.

Ferula (*Giant Fennel*). — These are among the finest of the umbelliferous plants that have so long remained unnoticed in our botanic gardens, their great value for garden embellishment being unrecognised. Their chief charm consists in furnishing large tufts

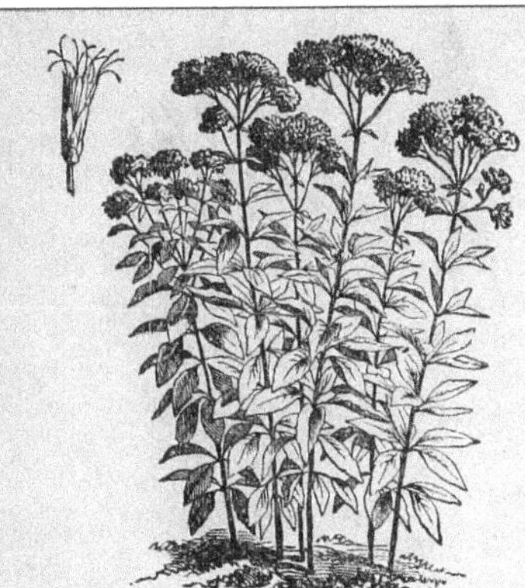

Eupatorium purpureum (*Thorough-wort*). See page 127.

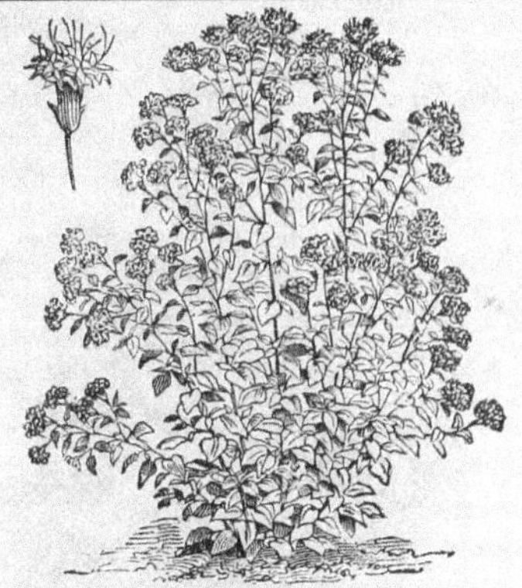

Eupatorium aromaticum. See page 127.

Eutoca viscida.

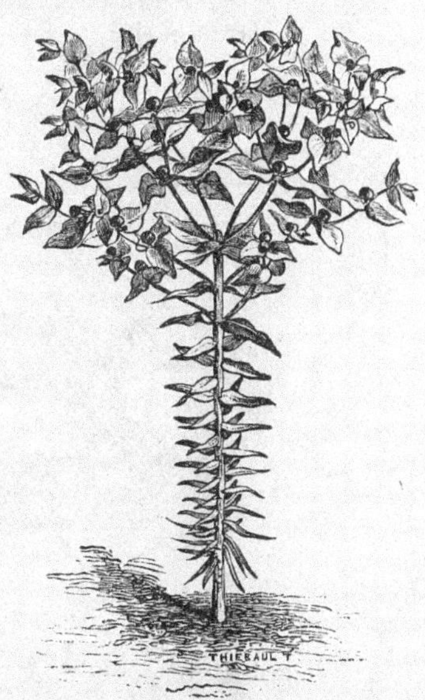

Euphorbia Lathyris (*Cape Spurge*). See page 127.

Fenzlia dianthiflora (syn., Gilia dianthoides).

Felicia tenella

PLATE CXIV.

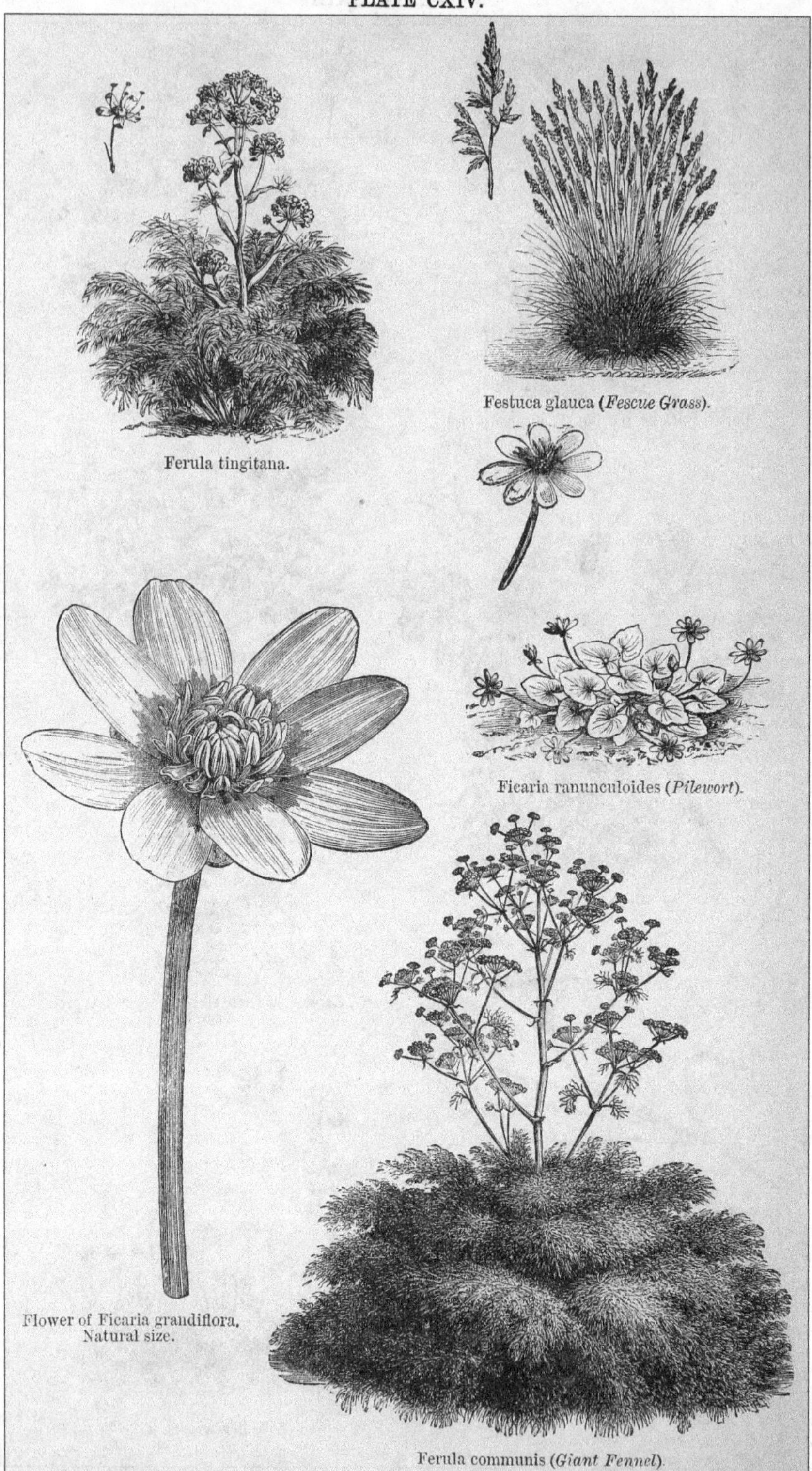

Ferula tingitana.

Festuca glauca (*Fescue Grass*).

Ficaria ranunculoides (*Pilewort*).

Flower of Ficaria grandiflora, Natural size.

Ferula communis (*Giant Fennel*).

[*To face page* **129.**

of the freshest green in early spring. The leaf is apt to lose some of its beauty and fade away early in autumn, but this may to some extent be retarded by cutting out the flower-bearing shoots the moment they appear. Not that these are ugly; on the contrary, the plants are fine and striking when in flower. The Ferulas should be planted well at first, and it is only when they are established that their good effect is seen. Where bold spring flowers are naturalised or planted in colonies, the presence of a group of these fine-leaved, hardy plants will be valuable, as their fine plumes rise in early spring. They are among the true hardy plants of the northern world, never, so far as we have noticed, suffering from any cold. Their fine forms in summer or autumn, when they throw up their flowering-shoots to a height of 10 ft. or so, are remarkable enough; but it is their appearance when breaking up in spring that charms us most. A good way is to place them singly or in small groups, a little without the margin of a shrubbery, or isolated on the Grass, so that their verdure may be seen in the garden landscape in early spring. Deep, free soil should be supplied before planting, if the soil be not naturally good and deep. The Ferulas are most readily raised from seed, which should be sown as soon as gathered in a nursery bed in the open air. The plants, even when well established, do not bear division well, though with care they may be transplanted. One of the best known and most valuable kinds is F. tingitana, which is elegant in habit and as vigorous as it is graceful. It takes several years to form strong plants, which have the appearance of massive plumes of the larger filmy Ferns. F. communis is a similar good species, and a few others, including F. glauca, neapolitana, Ferulago, and persica, may with advantage be added where much variety is sought, but the effect of the first two is not surpassed. The flower-stems developed the second or third year from seed are from 6 ft. to 10 ft. high, branched, and bear numbers of small inconspicuous flowers. S. Europe and N. Africa.

Festuca *(Fescue Grass)*.—A large genus of annual and perennial Grasses, containing but few for the garden. A variety of the Sheep's Fescue (F. ovina), named glauca, is a pretty dwarf, hardy Grass, forming dense tufts of leaves of a glaucous hue or soft blue, and on this account it is sometimes called "blue" Grass. It makes good edgings, and, when so used, the flower-spikes should be cut away. Another variety of F. ovina, named viridis, is also a pretty edging plant, and, being of slow growth, does not require renewal for years.

Feverfew *(Pyrethrum Parthenium)*.

Ficaria (*Pilewort*).—Plants of the Crowfoot family, nearly allied to and much resembling some kinds of Buttercup. F. ranunculoides (Lesser Celandine) is a very common British plant from 3 in. to 6 in. high, producing in early spring golden yellow flowers. It is so very common that it would not have been mentioned here but for its double and white varieties, which are pretty plants. Moist borders, in any soil. A good plant for growing under trees. Division.

F. grandiflora.—A large flowered kind, about twice the size of the British kind, the flowers being nearly 2 in. across. It is an easily-grown, showy plant, and could be naturalised. Southern Europe and Northern Africa.

Ficus elastica *(India-rubber Plant)*.—This not only exists in fair health in the open air in summer, but sometimes makes a good growth under the influence of our weak northern sun. It is best adapted for select mixed groups, and in small gardens as isolated specimens amongst low bedding plants. It will best enjoy stove treatment in winter, and is propagated from cuttings. It should be put out at the end of May. In all cases it is better to use plants with single stems. The trailing F. repens and F. stipulata also thrive in the open air in summer, and have a pretty effect, trailing up stems of trees in the subtropical garden. In mild districts these are hardy against walls or rocks.

Flame Flower *(Kniphofia)*.

Flax *(Linum)*.

Flowering Rush *(Butomus umbellatus)*.

Fluggea japonica *(Ophiopogon)*.

Fly Orchis *(Ophrys muscifera)*.

Fœniculum *(Fennel)*.—The common Fennel is a graceful plant, and were there not others of the family of much grace of foliage, it would be of value for the effect of its leaves. F. dulce is a nearly allied kind. Grows in any soil or on any waste bank.

Forget-me-not *(Myosotis palustris)*.

Foxglove *(Digitalis purpurea)*.

Fragaria indica *(Indian Strawberry)*.—An interesting little trailing herb, bearing an abundance of deep red berries, flowering late in summer. In borders and rough rockwork; in any (not over wet) soil. Nepaul.

Francoa.—Chilian plants of the Saxifrage family, alike in general appearance. They are rather tender plants, only suitable for dry, sheltered positions on warm sunny borders or banks. They prefer for soil a light loam. They are good for cutting, the long branching stems bearing numerous white or pink blossoms on stems 18 in. to 2 ft. high. The plants are raised from seed, and in spring they furnish a supply of flowers for a long time. Francoa ramosa has white or pink flowers; it has a short stem, and in this respect differs from F. appendiculata, which is stemless, and has flowers deeper in colour than the others. F. sonchifolia has also, like F. ramosa, a short stem, but its leaves are sessile and not stalked as in that species; its flowers are rose coloured. They are often grown as window plants, and may be seen as such in parts where they do not thrive in the open garden.

Frankenia lævis *(Sea Heath)*.—A very small evergreen herb with crowded leaves like a Heath. Common in marshes by the seaside in many parts of Europe and on the eastern coast of England. Best for the rock garden, but of botanical interest mainly.

Frasera *(American Columbo)*.—North American plants of the Gentian family, biennial, rare, and little known in cultivation.

Fraxinella *(Dictamnus)*.

French Honeysuckle *(Hedysarum coronarium)*.

Fritillaria *(Fritillary)*.—A large genus of bulbous plants of the Lily family, several of which are valuable garden plants, some being of stately growth, such as the Crown Imperial, others delicate and pretty, as F. recurva; but the larger portion have dull-tinted, curiously interesting flowers. They may be put to a variety of uses; the Crown Imperial is a fine plant for the mixed border or the shrubbery, and, being a vigorous grower, it is well able to take care of itself if naturalised in the wild garden. The early spring growth of this plant makes it a valuable one. The Snake's-head (F. Meleagris) and others, such as F. latifolia, pyrenaica, as well as the choicer kinds, are fitted for the bulb border and for planting in grassy places. There are only one or two that require special treatment; all the others thrive in any ordinary garden soil. They may all be readily increased by offsets from the old bulbs, which should be lifted every three or four years and planted in fresh soil; the plants will be greatly improved thereby. The lifting should be done in autumn, and the bulbs replanted without delay. The following are among the most desirable for general cultivation:—

F. aurea, one of the prettiest of the genus, quite hardy, about 5 inches in height, bearing a stem of four to six thick, fleshy, deep green leaves, and a solitary nodding flower; the latter is pale yellow in colour, spotted, or chequered with brown. Silesia.

F. Burnati, a handsome plant, growing about 9 in. high, having solitary, drooping blossoms, 2 in. long, of a plum colour chequered with yellowish-green markings. A native of European Alps, and quite hardy. It flowers with the Snowdrop, and is as easy to grow.

F. imperialis *(Crown Imperial)*.—A showy and stately plant, growing from 3 ft. to 4 ft. high, the bright green and stout shoots being crested by large, dense whorls of drooping, bell-like flowers and a crown-like tuft of foliage. There are several varieties, differing chiefly in the colour of the flowers. The principal are, lutea (yellow), rubra (red), double red and double yellow, rubra maxima (very large red flowers), Aurora (bronzy orange), sulphurine (large sulphur-yellow), Orange Crown, orange-red, Stagzwaard (a fasciated stem form, with very large deep red blossoms), and aurea marginata (gold-striped foliage); every leaf being margined with a broad golden yellow band, it blends with the remaining portion of the foliage. This plant thrives best in a rich, deep loam, and will be better if the bulbs remain undisturbed for years. It is best, perhaps, in a group on the fringe of the shrubbery or group of American plants. Its strong odour is somewhat against it when gathered: and for artistic effects in the garden it is not so valuable as the common Snake's-head.

F. latifolia.—A most variable species as regards the colour of the flowers, which are in every case larger than those of our native F. Meleagris. The flowers are borne on stems about 1 foot high, are pendulous, and vary in colour through various shades of purple, black, lilac, and yellow. The principal named varieties are—Black Knight, Captain Marryat, Caroline Chisholm, Cooper, Dandy, Jerome, Maria Goldsmith, Marianne, Mellina, Pharaoh, Rembrandt, Shakespeare, Van Speyk, each of which represents a different shade of colour. They grow freely in any soil in an open situation, and are excellent for naturalising. Caucasus.

F. Meleagris *(Snake's-head)* is an elegant native species, of which there are numerous varieties. It grows from 9 in. to 18 in. high, and bears in early summer a solitary, drooping flower, beautifully tesselated with purple or purplish-maroon on a pale ground. The chief varieties are the white (alba), in which there are scarcely any dark markings; nigra, a deep purplish-black; pallida, light purple; and angustifolia, with long, narrow leaves; major, with larger flowers than the type; præcox, flowers a week or so earlier than the other forms; flavida, flowers yellowish; and the rare double variety. This beautiful plant, in all its forms, may be used with excellent effect by the tasteful gardener. It grows freely in grass not mown early, and is therefore admirable as a wild garden plant; its various forms are among the most beautiful inhabitants of the hardy bulb garden, and tufts of the chequered or white-flowered are among the most graceful plants seen in English cottage gardens.

F. Moggridgei *(Golden Snake's-head)*.—A beautiful plant from the European Alps, having pendulous blossoms, 2 in. long, of a fine golden-yellow, chequered with brownish-crimson on the inner surface of the bell. It may be seen on its native mountains growing among the short-stunted Grass, accompanied by alpine plants, at an elevation of from 5000 ft. to 7000 ft., giving the slopes the appearance of a golden sheet of bloom. It flowers in early spring, and is hardy. It is a lovely plant for the choice bulb portions of the rock garden, and, when plentiful, for dotting in groups in Grass where it may escape the mower.

F. pudica is one of the most charming of hardy bulbs, taking a place among yellow flowers similar to that of the Snowdrop among white ones. It is a native of the Rocky Mountains and the Sierra Nevada of California, where it grows in a dry, barren soil, and is one of the principal spring ornaments of the flora. It grows nearly 6 in. high, and

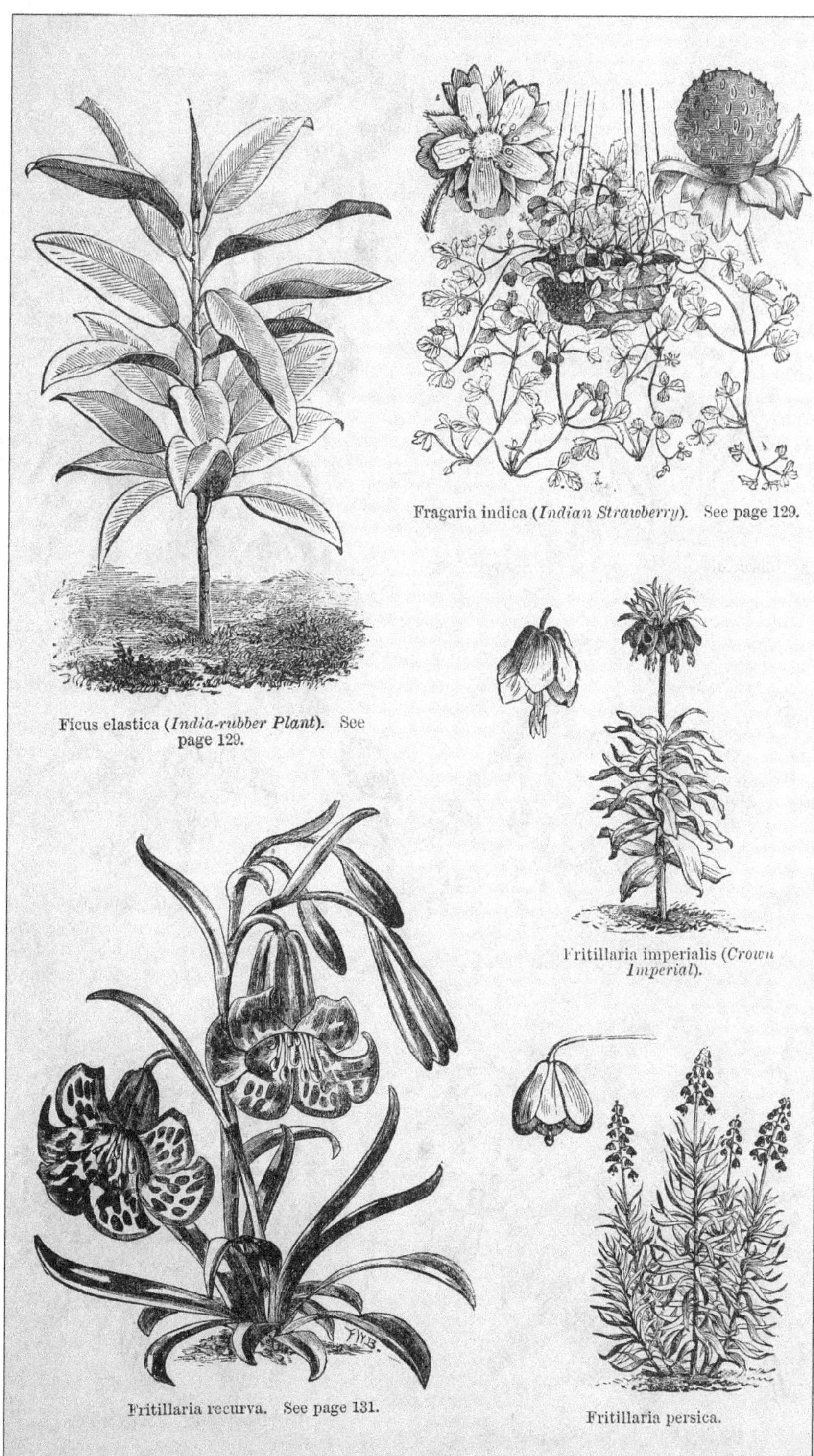

Ficus elastica (*India-rubber Plant*). See page 129.

Fragaria indica (*Indian Strawberry*). See page 129.

Fritillaria imperialis (*Crown Imperial*).

Fritillaria recurva. See page 131.

Fritillaria persica.

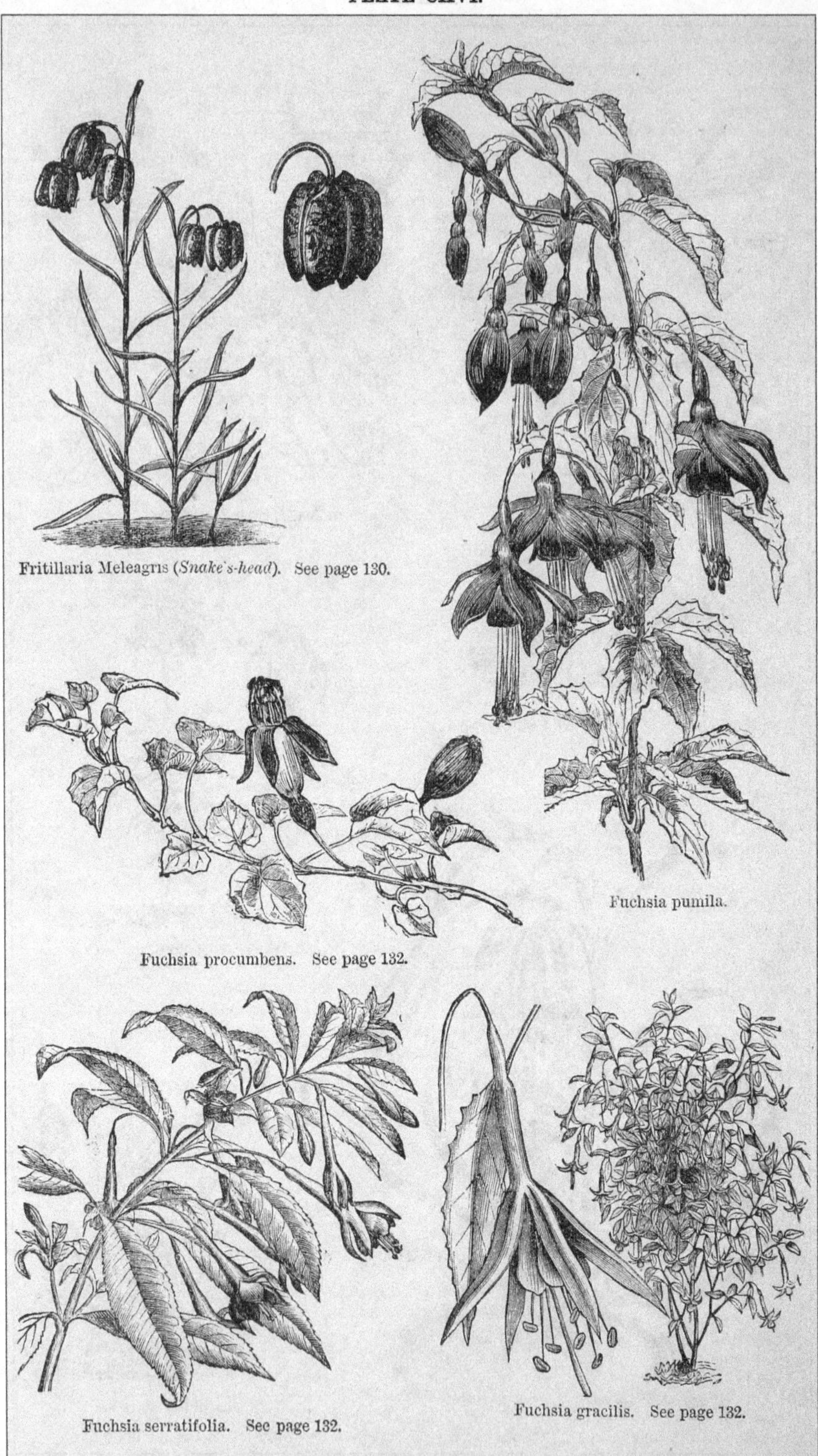

Fritillaria Meleagris (*Snake's-head*). See page 130.

Fuchsia pumila.

Fuchsia procumbens. See page 132.

Fuchsia serratifolia. See page 132.

Fuchsia gracilis. See page 132.

has bright golden-yellow flowers, peculiarly graceful in form and drooping like a Snowflake. It is perfectly hardy, thriving in warm, sunny borders of loamy soil.

F. recurva. — In colour this is the showiest of the Fritillaries, the red being as bright as that of some Lilies, and intermixed, especially in the inside of the flower, with bright yellow. In England it flowers early in May or in the latter part of April. The bulbs of this pretty plant consist of a slightly flattened tuberous stock, covered by somewhat widely-placed, articulated scales, which at first sight closely resemble those of Lilium philadelphicum. A tuft of linear bright green leaves is produced above the soil, and from this rises a slender, purplish stem, from 6 in. to 2½ ft. high, bearing several pendent Lily-like flowers. It is not a robust plant, and it is under careful cultivation only that it has succeeded in this country, growing best in fibry loam, on a warm, sunny border, near a wall. In winter it is advisable to cover the bulbs with some protective material or a hand-light. California.

Besides these there are many others in cultivation, but the majority are unattractive, though some are useful for naturalising in the wild garden among Grass; for this purpose the most suitable are F. delphinensis, a robust plant with stems, 1 ft. or more high, bearing brownish-purple flowers, more or less chequered with greenish-yellow; F. pyrenaica, a similar species, but more robust; F. liliacea, lusitanica, pallidiflora, tulipifolia, lanceolata, ruthenica, and tristis, all of which have dull brownish-purple or greenish flowers.

Frog-bit *(Hydrocharis Morsus-ranæ).*

Fruiting Duckweed *(Nertera depressa).*

Fuchsia.—This, one of the most beautiful ornaments of the garden when well grown, is too little seen in our flower gardens; but near the sea all round our coasts, and especially in the southern and western parts of England and Ireland, several species of Fuchsia are hardy, and are perhaps the most beautiful objects in the gardens. In other districts the species and varieties are cut down by frost, but spring up again vigorously and, in fact, live the life of herbaceous plants. Hence they find a place in this book. But in mild districts and near the sea coast they frequently escape being cut down for years, and become large and handsome bushes. There are no plants more calculated to improve the flower garden than Fuchsias. Not showy in mass of flower, they are of the highest order of beauty; the drooping habit of the shoots of most kinds gives the plant a grace that is valuable indeed, and that no flower garden should be without. Even in dwarf lines, where this drooping tendency is not seen to such advantage, or, it may be, seen at a disadvantage, the Fuchsia is very valuable; but it is when we use plants with rather tall stems or pyramids that the full beauty of the Fuchsia as a flower-garden plant is seen. In the milder districts, where the Fuchsia is a shrub, we see the habit to perfection; in others the tall-stemmed or pyramidal plants will have to be placed out in summer. The right way to manage Fuchsias put out for the summer only is to induce them, as far as possible, to produce all their growth in the open air. That is the secret; start them, nurture them, and make them full of leaves and strong young growth in the spring, so as to go out strong, and they will be disappointing indeed; but keep them back and do not let them burst forth into leaf until put in the open air in May, and they will then go on and retain all the strength they gather, suspending quantities of graceful blossoms until the leaves have deserted the trees, when they should be taken up and put in a dry cave, cellar, or shed for the winter. In a cool position of that kind it would not be difficult to "keep them back" in spring. And supposing they seemed inclined to push forth too much before the time had quite arrived when it would be convenient or desirable to put them in the flower garden, there should be no difficulty in placing them in some quiet, sheltered nook, where they might receive more protection than in the flower garden proper, and yet have full opportunity to make growth in the open air—the great point to be attained. In many places refuse plants may be turned to good account in this way. Nothing is simpler than to make of these standards for the flower garden by cutting away the lower and middle side shoots and leaving the head to form a standard. All may be freely propagated from cuttings in spring or autumn. There are about a dozen kinds more or less hardy that succeed well in the open air in the south and midland counties, and many more may be added to the list in warm, seaside localities; in fact, there is not a Fuchsia in cultivation that will not thrive in the open air in summer, and if used judiciously they give an air of grace to the flower garden afforded by no other plants. The following are among the hardiest kinds :—

F. coccinea.—A well known bushy plant, graceful and beautiful both in growth and bloom. It readily adapts itself to any locality, provided the soil be not of the wettest and coldest description, and even then a slight covering of coal ashes after the stems have been cut down in autumn will suffice to protect the roots in winter. In favourable situations it often attains a height of 6 ft. From the axils of the leaves, which are of a fine green colour, beautifully tinged or veined with red, the flowers, which before they fully open are not unlike crimson drops, are produced in profusion during the greater part of the summer. Chili.

F. conica.—A vigorous species, growing from 3 ft. to 6 ft. high, of compact habit, but not such a free flowerer as some of the others. The flowers have scarlet sepals, and dark purple petals. Chili.

F. corallina.—A beautiful plant of taller and more slender growth than the others, and therefore specially adapted for walls and houses. The flowers are large and of a showy red colour, and the plant is a vigorous grower and free bloomer.

F. discolor is a dwarf variety, producing a profusion of small scarlet flowers. It is the hardiest of all, not being injured by the winters of Scotland, in the milder parts, if treated as a herbaceous plant. F. pumila is a similar, but more slender variety, equally desirable.

F. globosa.—One of the best of the hardy Fuchsias. The flowers are globose in bud, and retain that shape for some time after they begin to expand on account of the petals continuing to adhere to the tips. This is an extremely profuse bloomer, and the flowers are richly coloured. It forms a sturdy and often a large shrub in seashore districts. There is no reason why it should not be grown in drier districts, even if cut down by frost every year, as it is always a handsome plant.

F. gracilis.—A very distinct plant of slender habit, having its flowers borne on remarkably long, slender stalks. The young shoots are of a purplish red, the calyx is of a brighter scarlet, and the corolla has a greater infusion of red than other hardy kinds. In mild and moist districts it grows nearly 7 ft. high, from 12 ft. to 15 ft. in circumference, and is of rapid growth. Some winters it is not cut down by frost. There is a variety called multiflora, which is a very free-flowering kind, with flowers of shorter, darker crimson than gracilis. F. tenella is a seedling variety of F. gracilis. Chili.

F. Riccartoni.—One of the prettiest and hardiest sorts, growing well even in parts of Scotland without protection. The plant is of a compact twiggy growth, and in summer bears many bright red blossoms. A garden hybrid.

Besides these, various other kinds are in cultivation, such as F. procumbens—a curious little New Zealand species—serratifolia, magellanica, thymifolia, and microphylla, and nearly all the hybrid kinds will do out-of-doors in summer, and bloom well, though they may be cut down in the winter. Among the most distinct and pretty are the several dwarf and fragile kinds, such as F. microphylla, F. pumila, and several hardy hybrids of the globosa section, all of which seem to flourish unusually well near the sea, but may be grown almost anywhere.

Fumaria *(Fumitory).*—Mostly annual plants of a weedy nature. One perennial. The species of Corydalis are sometimes placed under this genus, especially in Continental gardens and nurseries.

Funkia *(Plantain Lily).* — A most valuable genus of Japanese plants of the Lily family, containing about half-a-dozen species and numerous varieties. It is to the Funkia that we owe some of the boldest effects seen in our gardens and public parks. Their chief value lies in their handsome foliage, but they are also free-flowering herbaceous plants, producing spikes of bell-shaped flowers. They are noble and most useful plants for many positions in the garden. Few plants lend such a tropical effect in summer as F. Sieboldi when finely developed. For grouping they are highly suitable, and few plants thrive better in open places in shrubberies. The bold and striking foliage of some of the strongest plain-leaved section of Funkias renders them very effective as edging plants for large beds, while the several kinds that have variegated foliage, such as F. undulata variegata make good groups, or are suitable for edgings. They are seen to best advantage when grown in well-drained deep soil. All the species are easily multiplied by means of division in spring or in the autumn. The best species are the following:—

F. Fortunei. — This strong-growing species has smaller and more leathery leaves than F. Sieboldi, and they are of a much more bluish or glaucous tint. The flowers are pure white or pale mauve.

F. grandiflora grows from 12 in. to 18 in. high, producing numerous large, handsome, pure white, sweet-scented flowers in August and September. In some places this species is used as an edging plant, but it is seen to greatest advantage when planted in tufts, in beds or borders, in a well-drained sandy loam. About Paris this plant is grown in quantities as a flower garden plant, but with us it does not appear to flower regularly unless in sunny spots and in warm, well drained, and very sandy loam. The young leaves are a favourite prey of slugs and snails. It is also known as F. japonica.

F. lancifolia is a small-growing species, producing tufts of lance-shaped leaves, narrowing towards both ends from the middle. There are some interesting varieties of this species, chief amongst which are the white-flowered variety alba (or speciosa as it is more commonly called—a beautiful plant), spathulata, and plantaginifolia, with long narrow leaves. There are some very pretty varieties with leaves of different forms of variegation, all of which are well worth growing, notably albo-marginata, having a narrow line of white along the margin of the leaf; undulata variegata, in which the leaves are undulated on the margin and variegated on the greater part of the surface; and umvittata, with a broad white midrib to the leaf.

F. ovata produces large tufts of broad, deep, shining, green leaves. Flower-stems 1 ft. or 18 in. high, terminating in a short raceme of lilac-blue flowers, which appear in late summer and autumn. This is one of the strongest growing species, and when in flower is very handsome. There is a variegated leaved form of this species.

F. Sieboldi is the most ornamental of all the species. It grows from 18 in. to 3 ft. in height, and has large, somewhat heart-shaped, glaucous leaves that often measure over 1 ft. across. The flowers are borne in tall, one-

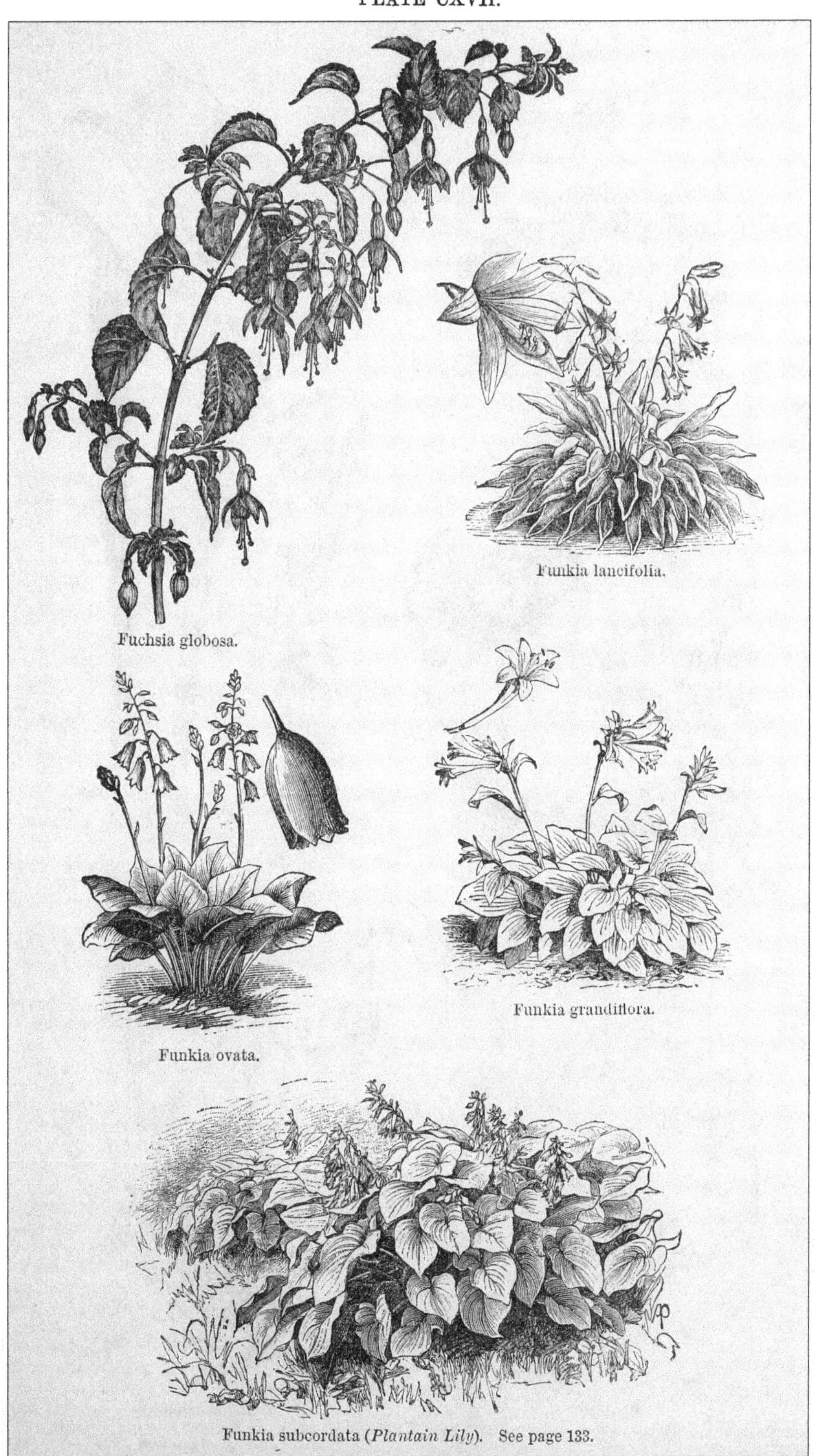

Fuchsia globosa.

Funkia lancifolia.

Funkia ovata.

Funkia grandiflora.

Funkia subcordata (*Plantain Lily*). See page 133.

[*To face page* **132.**

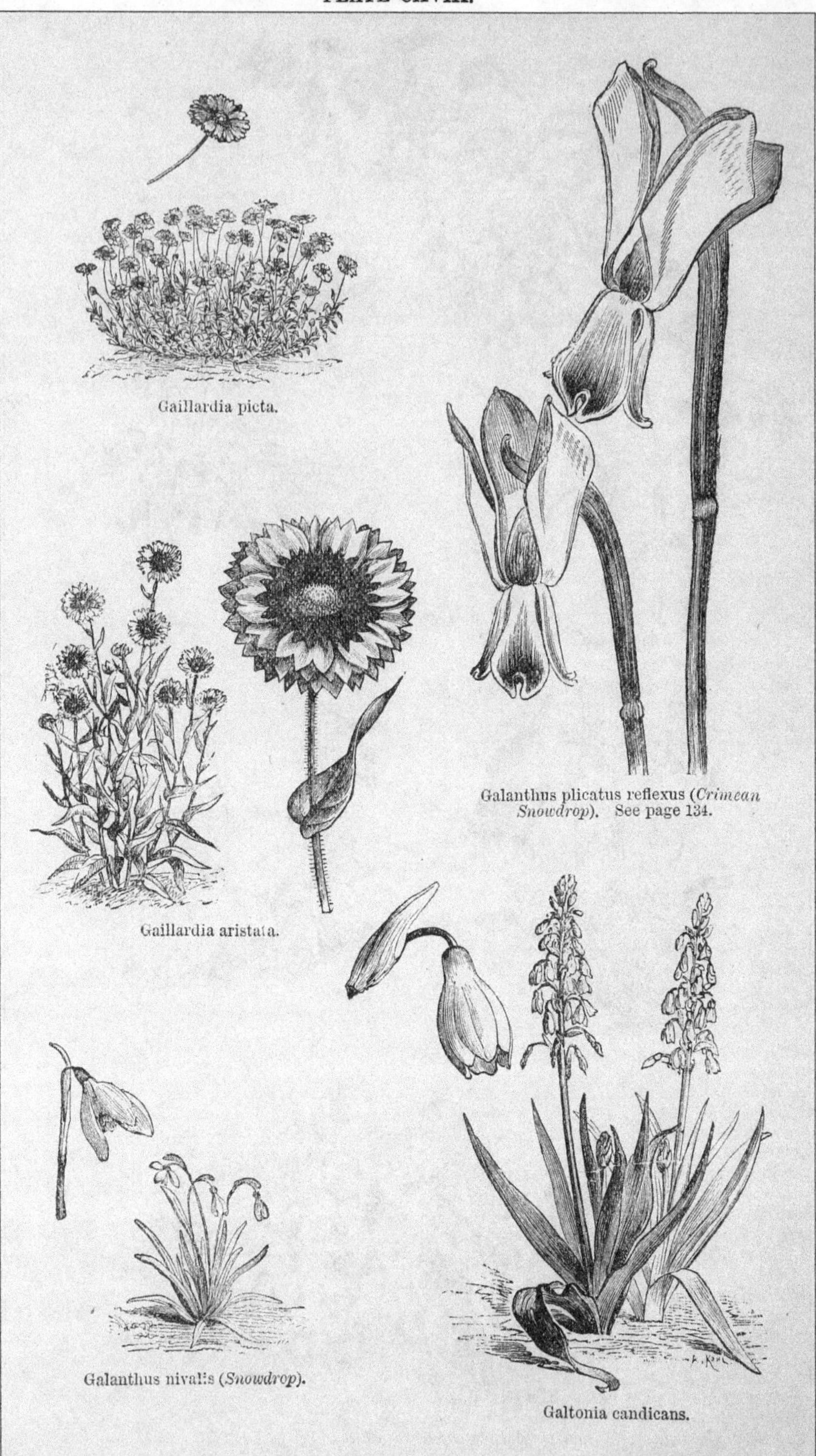

Gaillardia picta.

Galanthus plicatus reflexus (*Crimean Snowdrop*). See page 134.

Gaillardia aristata.

Galanthus nivalis (*Snowdrop*).

Galtonia candicans.

sided racemes well above the foliage, and are of a creamy-lilac colour. There is a variety with yellow-margined foliage, which is interesting. An admirable plant in tasteful hands for picturesque groups or massive edgings.

F. subcordata.—A handsome species having heart-shaped leaves, slightly glaucous on the upper surface, but not so deeply ribbed as F. Sieboldi. It is a rapid grower, forming dense spreading tufts in a few years. This species yields many interesting varieties, differing principally in the markings of the foliage, the most notable being marmorata and argentea, two variegated kinds. The flowers are white, borne in slender racemes 18 in. or more high.

Gagea.—Small plants of the Lily order, with flowers somewhat like a Star of Bethlehem, but yellow. They are not likely to be generally admired, but would be pretty in grassy places in sandy soil. Europe, G. lutea being British.

Gaillardia.—A genus of great importance in the flower garden, including, as it does, some of the showiest of flowers, valuable for their long duration on the plants and in a cut state. It is wholly confined to North America, and numbers some half-a-dozen species and numerous garden varieties. The numerous kinds now in English gardens appear to fall under three species, but there is a strong family likeness running through the series. The kinds are

G. aristata. a perennial, growing from 1 ft. to 1½ ft. high, with narrow leaves, sometimes deeply cut. The flowers are from 1½ in. to 4 in. across, the ray florets having an outer zone of orange yellow and an inner one of brownish-red, while the centre is a deep bluish purple This is is the commonest kind, and as it has been raised largely from seed, there are many varieties, differing more or less widely from the type, and variously named. G. picta somewhat resembles this, but may be known by its smaller flowers, and by its being of biennial duration. It is dwarfer than G. aristata, and the colours of the flowers are brighter. G. amblyodon is a beautiful annual from Texas, introduced to our gardens a few years ago. Its flowers are smaller than those of the preceding, and the colour is a deep cinnabar-red. They are produced plentifully on strong plants for several weeks towards the close of summer. G. pulchella is the oldest cultivated species, having been introduced about a century ago. It grows from 1 ft. to 1½ ft. high, and bears flowers 2 in. across of a bright yellow and purplish red colour. It also is only of annual duration, and now seems to have become very scarce. G. bicolor and pinnatifida are seldom met with in gardens, owing probably to their being somewhat tender. The garden varieties, as has been stated, are numerous, but the most distinct of those that have been named are

G. grandiflora, said to be a hybrid, presumably between G. picta and G. aristata. It is a beautiful plant, of vigorous growth, with large and brightly-coloured flowers, which are only surpassed by its variety maxima. It is by far the finest of all the Gaillardias.

G. hybrida is another garden cross, much resembling the preceding; the variety splendens differs in having brighter tinted flowers. G. Telemachi, Drummondi, Loiselli, and Bosselari appear to be synonymous with some of the preceding, and G. Richardsoni, another so-called species, scarcely differs from them.

All thrive in good garden soil of a friable character, but they will not flourish on a cold stiff soil or one that is too light or dry. Where possible they should be grown in bold groups, for they thrive better so placed than as solitary plants in a parched border, and there are no plants that have a finer effect in a bed by themselves than the various kinds of these showy Composites. Where they are apt to die in winter they may yet be used in mixed borders, treated as half-hardy annuals, for if sown in a mild hotbed at the end of February or the beginning of March they may be grown into good plants, and a full display of their fine flower-heads obtained as early as upon those that may have withstood the winter in the borders. They may be propagated by cuttings in autumn or spring, and by division in spring, assisted afterwards by slight heat if the locality is a cold one.

Galtonia *(Hyacinthus)* **candicans.**—A noble bulbous plant from the Cape of Good Hope, having in autumn flower-spikes from 4 ft. to 6 ft. high, furnished with spires of waxy white bell-like blossoms, 1½ in. long. It is of easy culture, and quite hardy in light soils, and is valuable for bold groups in the mixed border or the flower garden. Increased by offsets from the bulbs, or from seeds, which flower about the fourth year. The distinct habit of this plant makes it valuable for the flower garden. When a group is well grown in good deep soil it is striking and novel in the late summer garden. A group rising from a bed of choice dwarf shrubs tells well, as indeed the plant does in almost any position when well grown.

Galactites tomentosa.—A Composite, indigenous to the shores of the Mediterranean, growing from 2 ft. to 3 ft. high, with spiny foliage, blotched with white, and cottony-white on the under surface. The flower-heads are lilac-purple. If sown as early as February, the plant blooms the first season, but stronger plants are obtained by sowing in autumn. It succeeds best in good loamy soil, and is effective in small masses in the sub-tropical garden.

Galangale *(Cyperus longus).*

Galanthus *(Snowdrop).*—Although this is one of the most abundant of hardy bulbs, one never tires of its modest beauty whatever may be its surroundings. The Snowdrop never looks better than when naturalised amid tender herbage in old-fashioned orchards and paddocks, on the margins of lawns, or beside woodland walks, and it is one of the hardy bulbs sufficiently vigorous to propagate itself in a semi-wild state or on Grass. The leaves complete their functions so early in the year that it may be planted in Grass that is repeatedly mown, as well as on banks in pleasure-grounds or half wild places. The bulbs may be inserted a couple of inches into the turf, and the spot afterwards made firm and level, especially if it be on a trimly-kept lawn. Almost any soil suits the Snowdrop, but deep, rich, well-drained gravelly-bottomed soils are most suitable, although in some parts of Nottinghamshire and Leicestershire it grows luxuriantly on hedge-banks and in old orchards on the coldest and wettest of clay soils. They are all quite hardy in our gardens, where they may be used in an infinite variety of ways, not only in isolated masses on the Grass, but grouped on well-made rockeries or rootwork, or in the wild garden, where they may be tastefully associated with Anemone fulgens, A. coronaria, A. hortensis, A. blanda, early purple and yellow kinds of Crocus, Winter Aconites, and Irises of various kinds, such as I. reticulata, I. Histrio, and I. stylosa, all of which bloom in January and February in mild winters. As cut flowers all the Snowdrops are most attractive, since they may either be grouped with sprays of Box or Fern, or associated in bouquets and wreaths with Orchids, Rose buds, Forget-me-nots, and other choice cut flowers. To cull the flowers in the bud state is, however, essential in order to ensure their most perfect beauty, as they then bear carriage better and open fresher in water than when cut from the plants after they are fully expanded. Snowdrop buds so gathered will remain beautiful for ten days or even longer, while those cut after expansion on the plant will fade in about a week at the latest.

There are about half-a-dozen species in cultivation, all of which bear a strong resemblance to each other. The common native species, G. nivalis, is recognised by its dwarf, narrow leaves and small flowers. There are two interesting varieties of this species—Shaylocki and virescens. The first has the outside tips of each outer petal green, the upper portions white. It has, moreover, a two-leaved spathe. G. virescens has the tip of each outer petal white, but the whole of the upper portion suffused with green. The Crimean Snowdrop, G. plicatus, has very broad leaves, the margins of which are curiously turned down or deflected, and the flowers are larger than those of nivalis. G. Imperati is the most stately of the whole group, varying from 6 in. to 12 in. in height, the sepals being about 1 in. in length. G. Elwesi is the finest, perhaps. It is distinguished by its very globular flower, which is twice as broad as that of G. Imperati, and by the green base of the inner segments, which show between the outer ones. The leaves are also very glaucous and can be distinguished at a glance from those of any other Snowdrop. G. Redoutei is a most distinct, robust-growing kind, yet rare in gardens, as is also G. latifolius and reflexus. Mr. Melville, of Dunrobin Castle, has recently raised some interesting varieties of Snowdrops, which differ chiefly in the time of flowering. They are named G. Melvillei, proculiformis, serotinus (a late-flowering kind), and præcox, which flowers about Christmas time. There is also a variety of G. nivalis, with yellowish flowers, called lutescens.

Galatella hyssopifolia *(Hyssop-leaved G.)*—A pretty, Aster-like North American plant, suited for borders or naturalisation in any soil. Division. From a garden point of view, it and its fellows are not distinct from the Starworts (Aster), and are inferior to them.

Galax aphylla *(Wand Plant)*.—One of the neatest little plants for rockwork; its stout, round leaves, which are evergreen, beautifully toothed and tinted, are borne on slender stems 6 in. or 8 in. high, though from a specimen sent from its native habitat in North America it would seem to grow to more than 1 ft. high; its white wand-like flowers must have suggested its common name. Of easy culture in peat or leaf-soil, somewhat moist, in the bog garden, or on the margins of beds of dwarf shrubs growing in peat.

Galega *(Goat's Rue)*.—Graceful and hardy plants of the Pea flower order, flourishing in any soil. On account of their free growth they are good subjects for the wild garden, and in colonies or groups they are very effective. They are herbaceous perennials, and grow from 2 ft. to 5 ft. in height, according to position and soil. All are excellent for affording an abundance of cut flowers of varied hues. Seeds or division. The species and varieties are—G. officinalis, or Common Goat's Rue, a native of Southern Europe. It averages from 3 ft. to 5 ft. high, and bears abundantly in summer dense clusters of Pea-shaped blossoms of a pretty pink colour. There is also a white-flowered variety named alba, which is desirable for cut flowers. A variety called africana has longer racemes of blossoms of a more purple tinge. G. orientalis is a handsome plant from the Caucasus. It varies in height from 2 ft. to 4 ft., and bears flowers of a bluish-purple colour. G. persica is a rather later flowering kind, and grows from 2 ft. to 4 ft. in height. The flowers, which are white, are produced in dense racemes on slender axillary stalks. G. biloba is a species with pretty bluish-lilac flowers, but though it is an old inhabitant of gardens, it is now scarce.

Galium *(Lady's Bedstraw)*.—These are numerous in botanical collections; but they are of a weedy nature, though the whites and pale yellows intermingled are beautiful in point of colour, and their odour is pleasant. They seem to delight in the sun, and all flower very freely. They do well on banks where many other things would not grow.

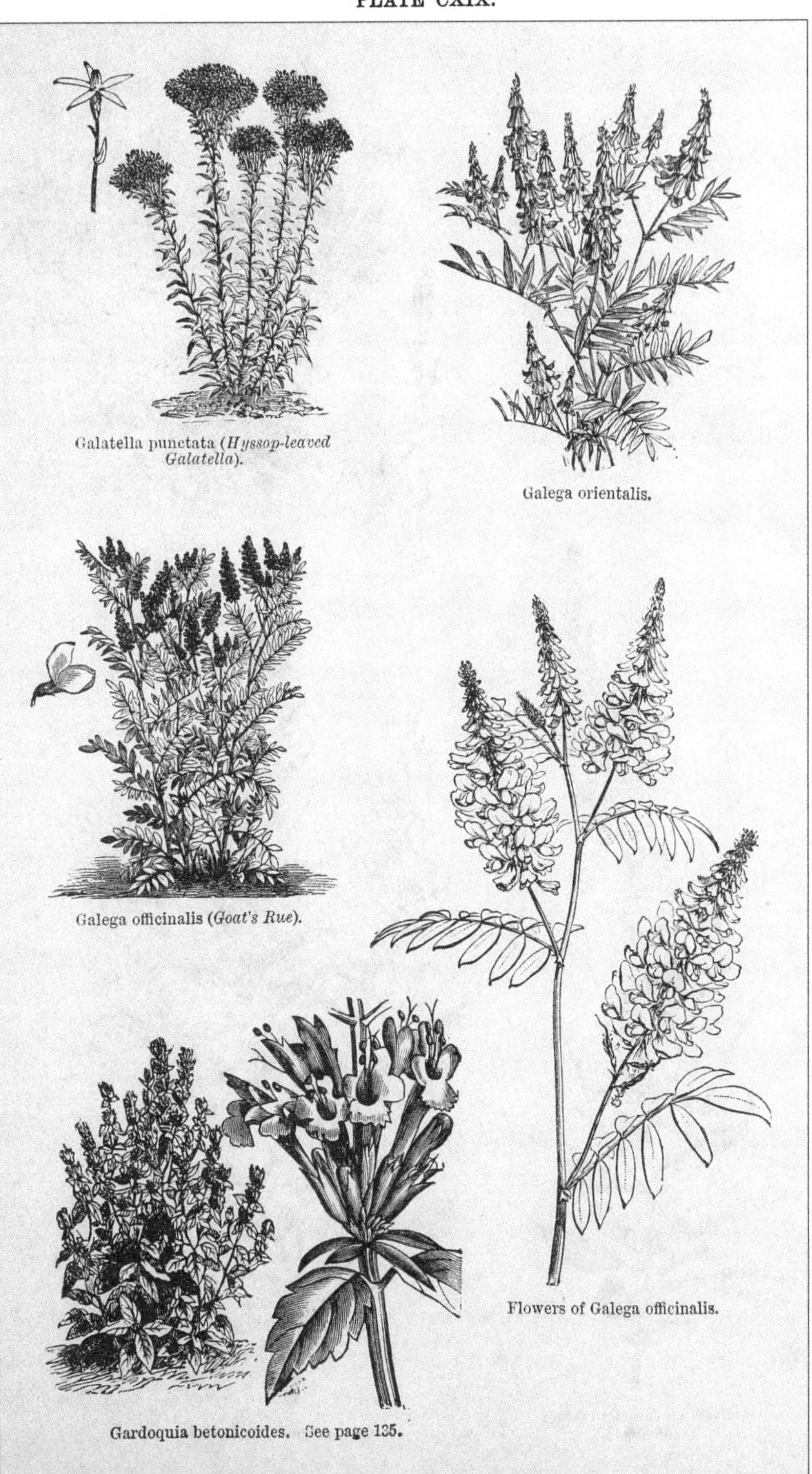

Galatella punctata (*Hyssop-leaved Galatella*).

Galega orientalis.

Galega officinalis (*Goat's Rue*).

Flowers of Galega officinalis.

Gardoquia betonicoides. See page 135.

PLATE CXX.

Gaultheria procumbens (*Creeping Wintergreen*).

Gaura Lindheimeri.

A flowering branch of Gaura Lindheimeri.

Genista tinctoria (*Rock Broom*).

[*To face page* **135.**

Gardoquia.—A rarely-grown genus of Labiates, mostly requiring greenhouse culture, G. betonicoides being the only kind fit for the open air. It grows from 2 ft. to 3 ft. high, has tubular flowers of a bright rosy magenta, and should be treated as a half-hardy annual.

Garland-flower *(Daphne Cneorum).*

Gaultheria procumbens *(Creeping Wintergreen).*—This neat little shrub is of itself pretty, but the berries give it quite a charm through the autumn and winter months, when it is, or rather ought to be, one of the most attractive objects on every well-made rockwork. Its drooping, white flowers are pretty, too. A native of North America, in sandy places and cool damp woods, often in the shade of evergreens, from Canada to Virginia. It does well in moist peat, and forms edgings to beds in some places where the natural soil is of that quality, or of a loamy nature. Easily increased by division or by seeds, and suitable for the rock garden, the front margins of borders, and occasionally as an edging to beds of choice and dwarf American plants, but it is best where well exposed. G. Shallon is too large a plant for all but the rougher flanks of the rock garden, being really a vigorous shrub where it has root room.

Gaura Lindheimeri.—A graceful perennial, 3 ft. to 4½ ft. high, flowering in summer and autumn; white and rose, slightly drooping, numerous, and in long, slender spikes. Borders, in sandy loam. Division and seed. May be also used with the larger forms of bedding plants.

Gazania.—Very handsome and distinct dwarf plants of the Composite order; of much value for the flower garden, though only hardy enough for summer growth therein. They strike freely in a cold frame in August, but later on require bottom-heat. Spring-struck plants are almost worthless unless struck very early; hence it is advisable to put in the stock in August and let them stand in their cutting-pots till potting-off time in spring. Such plants come well into flower when put out in May; whereas, when they are topped for a spring batch, both lots are small and late. Short young tops should be used for cuttings, and they may be inserted pretty thickly in the cutting pots. When fairly established, they must be just protected from frost, and kept in dry, airy quarters. If kept warm they grow too much, and are in spring poor, lanky specimens that will hardly bear handling; cool treatment keeps them short and sturdy.

G. rigens is the best known kind. It has long, deep green foliage, silvery beneath, and flowers, 2 in. across, of a bright orange-yellow, with dark centre. G. splendens is a fine variety of it, and there is also an effective variegated-leaved variety.

Geissorrhiza.—Handsome, Ixia-like, bulbous plants; natives of the Cape of Good Hope. G. violacea has flowers of a rich scarlet, borne on slender stems about 1 ft. high. G. Hookeri has large white flowers, with purple centre. They require the same treatment as Ixias.

Genista *(Rock Broom).*—A very large family of shrubs of the Pea-flower order, among which some are so dwarf as to be useful on the rock garden, the best for this purpose being G. sagittalis, G. pilosa, G. prostrata, and G. tinctoria—all small hardy shrubs of the easiest culture. Where not to be had in nurseries they are easily raised from seed sown in the open air. The above are from 6 in. to nearly 18 in. high, and in groups or masses are very effective. G. sagittalis is a good marginal plant in borders.

Gentiana *(Gentian).*—Of these beautiful hardy plants there are two sections—the first, of strong, easily-grown kinds, suitable for borders; and the second, of the dwarfer kinds, which should be grown on the rock garden, or in borders or beds devoted to choice dwarf plants. The Willow Gentian and some of the American perennial kinds, and those with herbaceous shoots, generally grow freely in good moist soil in borders. So does the well-known Gentianella (Gentiana acaulis), which should be grown in every garden; being, however, dwarf in habit and large and splendid in bloom, it is used as an edging plant. It is well to form carpets of this in parts of the rock garden or on raised borders, in which position it will be seen to the best advantage when planted in moist loam. The other type of Gentian, represented most familiarly by the Vernal Gentian (Gentiana verna), is by no means so difficult of cultivation as is commonly supposed. Want of moisture in the soil and of free exposure to sun and air are the main causes of failure. In the cool pastures and uplands, where the plant thrives in a wild state, it is rarely subjected to such drought as in a parched border. Deep, moist, sandy loam will suit it perfectly; if the surface be strewn with bits of broken stone it is prevented from cracking and parching, as it often does when bare. Well-rooted plants should be secured to begin with. It is important that the plant be not overshadowed or overrun by tall or straggling border flowers. This is easily guarded against by associating with it plants somewhat resembling it in stature. These various conditions being observed, the Vernal Gentian will soon spread into healthy little tufts and take care of itself from year to year, forming carpets on the rock garden and small beds or edgings on the level ground.

G. acaulis *(Gentianella).*—A well-known old inhabitant of our gardens, and, while among the most beautiful of the Gentians, it is easily cultivated, except on very dry soils. In some places edgings are made of it, and where the plant does well it should be used in every garden to some extent, as, when in flower, edgings of it are of great beauty, and when out of bloom the masses of little leaves, gathered into compact rosettes, form a dwarf and firm edging. It is at home on the rock garden, where there are good masses of moist loam into which it can root, and it is particularly well suited for rockwork, where the stone is suggested here and there rather than exposed. It is abundant in many parts of the Alps and Pyrenees. With us the flowers open in spring and early summer, but on its native hills their opening is regulated by position, like those of the Vernal Gentian. No garden should be without such an easily grown plant, so attractive from its associations as well as its great beauty. G. alpina is a marked variety, with very small broad leaves, and there are several other varieties. Mr. H. Gusmus, of Laibach, Austria, has numerous varieties of Gentiana acaulis. Their colours are remarkably beautiful, varying from the deepest azure blue to pure white, and in one flower of the latter colour the tips of the corolla are of a rich blue. In all the forms except the white the throat of the corolla is copiously spotted with blue on a greenish ground, and all have greenish marks on the outside. We hope soon to see them adorning our rock gardens and flower borders.

G. asclepiadea *(Willow Gentian)* prefers a sheltered position, and slight shade is beneficial to it. Being herbaceous, it gives no trouble, dying down out of harm's way during the winter. Properly grown, it will spring up to 2 ft. and freely produce flowers nearly the whole length of the stem, of a good size, and of a deep purple-blue. In Mr. Rawson's garden at Bromley, in Kent, it has grown 3 ft. to 4 ft. in diameter. The white variety is, with the exception of the colour of the flowers, almost the same, and requires similar treatment. This is the only Gentian we know that will grow in woods, and therefore it may be naturalised. The effect in a wood among the Grass is charming. Europe. Division. Flowering in autumn.

G. bavarica *(Bavarian Gentian)*.—In size and flower this species resembles the Vernal Gentian, but is very readily known from that by its smaller Box-like leaves of a yellowish-green tone, and by all its tiny stems being thickly clothed with foliage, forming close, dense little tufts, from which spring flowers of the most brilliant blue. While G. verna is found on dry ground, or ground not overflowed by water, G. bavarica in seen in perfection in spongy, boggy spots, where some diminutive rill passes by. We must imitate these conditions as far as possible if we desire to succeed with the plant in gardens. A moist peat bed, well watered occasionally, and with no coarse plants in it to give injurious shade, will enable us to grow this lovely alpine plant.

G. verna *(Vernal Gentian)*, the type of much that is beautiful in alpine vegetation. A few things are essential to success in its cultivation, and these are far from difficult to secure. They are good, deep, sandy loam on a level spot, abundance of water during the warm and dry months, if the soil be not deep and moist, and perfect exposure to the sun. Grit or broken limestone may be advantageously mingled with the soil; but if there be plenty of sand they are not essential; a few pieces half buried on the surface of the ground will tend to prevent evaporation and guard the plant till it has taken root. It is so dwarf that if weeds be allowed to grow round it, they soon injure it. In moist districts, where there is a good, deep, sandy loam, it may be grown on the front edge of a border carefully surrounded by half plunged stones. In all cases well-rooted specimens should be secured to begin with, as failure often occurs from imperfectly rooted, half-dead plants that would have little chance of surviving even if favoured with the air of their native wilds. In a wild state this plant is abundant in mountain pastures on the Alps of Central Europe, in Asia, and in Britain also.

G. septemfida *(Crested Gentian)*.—A lovely plant, bearing, on stems 6 inches to 12 inches high, flowers in clusters, cylindrical, widening towards the mouth, of a beautiful blue and white inside, greenish-brown outside, having between each of the larger segments of the flowers one smaller and finely cut. A native of the Caucasus, and one of the most desirable species for cultivation on the rock garden, thriving best in moist, sandy peat, and increased by division. In addition to the preceding kinds, there are various other Gentians in cultivation: G. caucasica, adscendens, gelida, cruciata, affinis, algida, arvernensis, crinita, and Andrewsi; but the species alluded to above represent the beauty of the family as at the present time known in cultivation. As new and beautiful kinds are introduced they may be added to the choice collection. Most of the Gentians may be raised from seed, but it is a slow process.

Gentianella *(Gentiana acaulis)*.

Geranium *(Crane's-bill)*.—The hardy Geraniums are usually stout perennials, natives of the fields and woods of Europe and Britain, though some are dainty alpine flowers. The handsomest of all hardy Geraniums is probably G. armenum. It is a robust perennial, sometimes 3 ft. in height, flowering abundantly in midsummer, and sometimes in a less degree till late on in the autumn. One of its merits is, indeed, the abundance of its large and handsome flowers; as to culture, it only requires planting in any ordinary garden soil; as to position, it will suit well for the mixed border, also for grouping with the finer perennials in beds and on the margins of shrubberies. Amongst the other kinds are some that are very ornamental, the best being the dwarf magenta-coloured G. sanguineum, and its beautiful Lancashire variety with rose-coloured blossoms finely pencilled with dark lines; G. pratense, a tall-growing kind, with large purple flowers, and its pure white variety. There is also an intermediate form with white

Gentiana verna (*Vernal Gentian*). See page 135.

Gentiana asclepiadea (*Willow Gentian*).

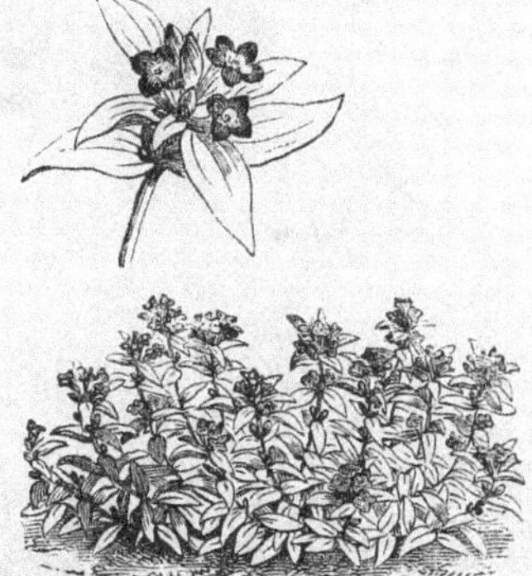

Gentiana cruciata.

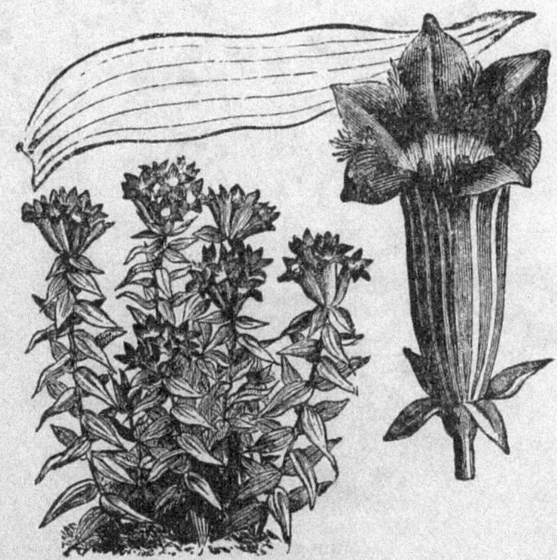

Gentiana septemfida (*Crested Gentian*).

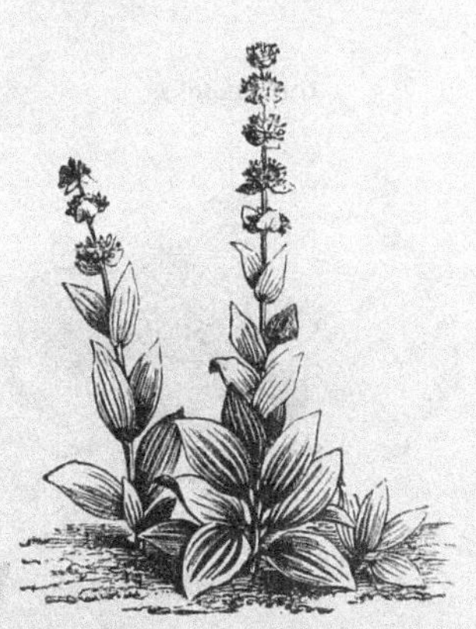

Gentiana lutea (*Yellow Gentian*).

[*To face page* 136.

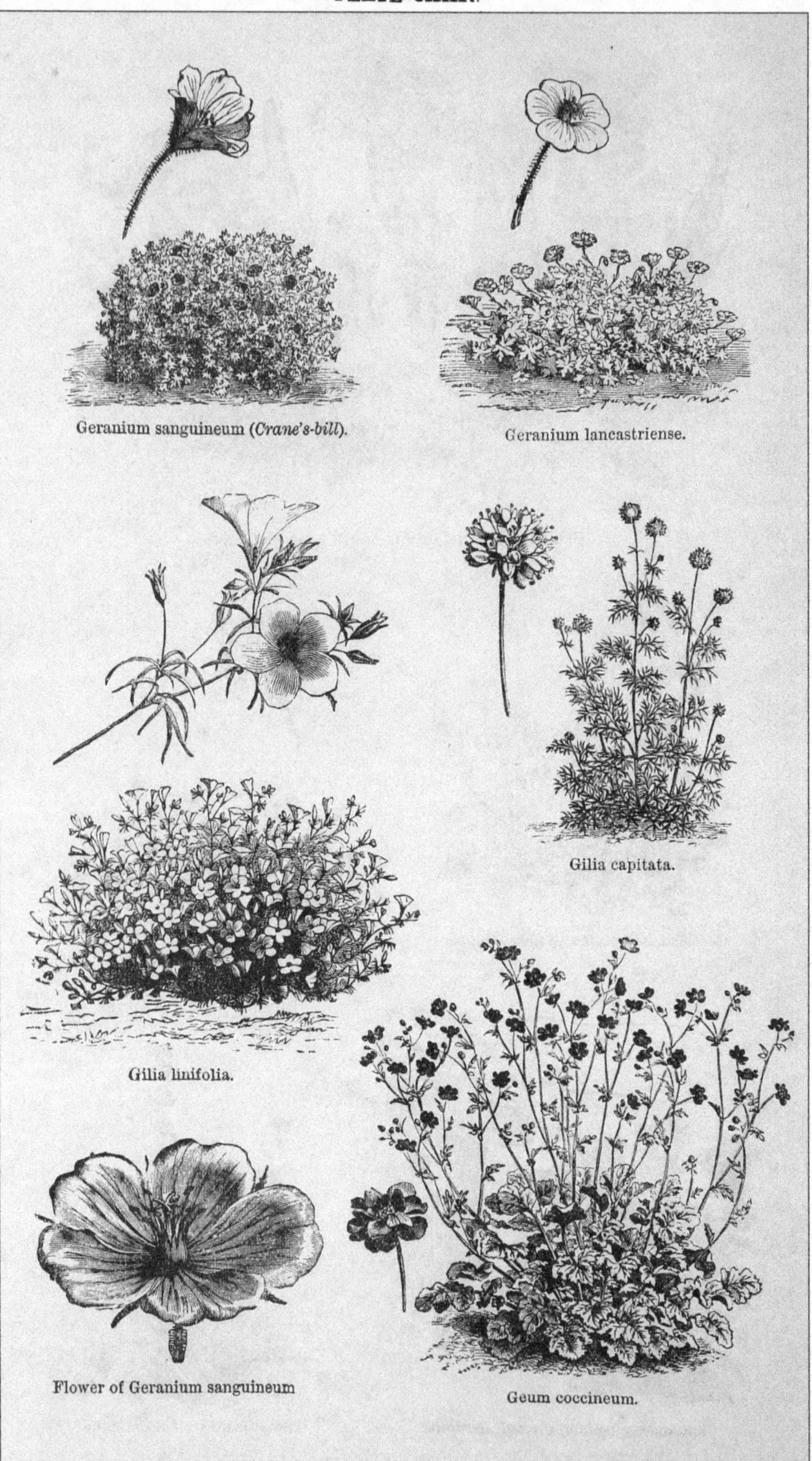

Geranium sanguineum (*Crane's-bill*).

Geranium lancastriense.

Gilia capitata.

Gilia linifolia.

Flower of Geranium sanguineum

Geum coccineum.

and purple flowers. The Caucasian species, G. gymnocaulon and ibericum, with rich purple blossoms, 2 in. across, delicately pencilled with black, are beautiful. G. platypetalum, striatum, ibericum, and Lamberti are good kinds, suited for shrubbery borders, being mostly free and vigorous enough for naturalisation. G. Endressi, a tall-growing kind with rose-coloured blossoms, is also very attractive. All the above kinds of Geraniums are perfectly hardy, easily cultivated, will grow in any ordinary soil, and should be included in every collection of hardy flowers. The pretty rock garden kinds are G. cinereum and G. argenteum, both dwarf and charming alpine plants, which, unlike the above-mentioned stout perennials, must be associated with very dwarf rock and border plants. Except perhaps these, all the Geraniums are very freely increased by division, and all by seeds.

German Ivy *(Mikania scandens)*.

Geum *(Avens)*.—This genus contains but very few ornamental plants. By far the finest species is G. coccineum, which inhabits the Bithynian Mount Olympus and has long been a popular garden favourite. Probably the Chilian Avens (G. chilense) is but a mere geographical form of it, only differing in the darker colour of the blossoms, which sometimes assume a coppery hue. The double variety of G. coccineum is a more valuable kind than the type, as its blossoms last much longer in perfection. It commences to flower in May, and produces a continuous succession of blossoms till October. There are several forms of the double kind. It is one of the most easily-managed of hardy plants ; it grows vigorously in ordinary garden soil, and it may be readily multiplied by seeds, which are freely borne, or by division in spring. The most desirable of the other kinds in cultivation are G. montanum and G. reptans. G. montanum is of dwarf, tufted habit, with deeply-lobed leaves. The flowers are large, of a yellow colour, and are borne singly on erect stalks about 1 ft. high. Central Europe. A good rock or choice border flower. G. reptans is a trailing kind with deeply-cut foliage and erect flower-stalks about 6 in. high, bearing a single blossom of the same colour, but larger than the preceding.

Gilia.—Hardy annuals growing from 1 ft. to 2 ft. high, and bearing for a long time in succession abundance of blue, white, lavender, or rose-coloured blossoms. The seed may be sown in autumn for spring blooming, and in April for summer and autumn display. They should be grown in masses to be effective, and the soil should be light and well enriched with decomposed manure. The flowers are useful for small bouquets or vases, and last for a long time in water. The best kinds are G. achilleæfolia major (blue), G. a. alba (white), G. capitata (lavender), G. tricolor (white and purple), G. rosea splendens (rose), G. nivalis (white), G. liniflora, and G. laciniata. A mixed packet of seed will also give a fine variety of colours. The species may be made occasional graceful use of as carpet plants, and also among the general class of annuals.

Gillenia trifoliata. — A Spiræa-like plant with numerous erect, slender stems, some 2 ft. in height, branching in their upper portion into a loose panicle of white flowers. G. stipulacea is another species ; though distinct and graceful, neither of them is of special value for the flower garden proper. They grow in peat or free loamy soil, and are hardy. As shrubbery plants or in the wild garden they may be given a place by some. North America. Division.

Gilliflower.—A term applied to various flowers—to the Carnation in Shakespeare's time and to Stocks in later days, but originally it belonged to the Carnation, or one of the groups of races which that flower has given rise to. In the west of England it is applied to the Wallflower by the people.

Gladiolus.—A beautiful genus of bulbous plants, for the most part natives of the Cape of Good Hope, every introduced species of which is of ornamental value ; all are easily grown, and suitable for a variety of uses in the open air garden. The chief beauty of the Gladiolus is derived from the numerous hybrid varieties now in cultivation, the species being few. G. gandavensis and brenchleyensis are the principal kinds from which the beautiful race of hybrids has come, affording such a wealth of beauty. They are by far the most important class, though the earlier flowering kinds (descendants of G. ramosus, Colvillei, trimaculatus, and others) are becoming very valuable for early summer flowering. The gandavensis section suffer from our cold autumn rains ; the bulbs, therefore, must be lifted in autumn.

In growing Gladioli for garden decoration it is necessary to select positions where they will present the most effective appearance, and there prepare the soil for them. They are happy planted in clumps between Dahlias, Phloxes, Roses, and other subjects of a somewhat similar character. They are very effective when planted in clumps alternately with Tritomas, and also when associated with large masses of Cannas. They are likewise in every way suitable for intermixing with American plants, the dark foliage of which shows off the richly-coloured flowers to good advantage. The positions for them should be marked out in the course of the autumn or winter, and two or three spadefuls of manure should be dug into them. As a rule, the space for each clump of bulbs should be 18 in. in diameter, and the soil should be turned up to a depth of 18 in. to 24 in. March and April are the best months in which to plant for garden decoration, as they are then at their best during August and the early part of September, but a succession of planting is desirable to secure a late bloom. Those who desire their gardens to be beautiful late in the autumn should not fail to largely employ the Gladiolus, as it is the handsomest of all late blooming garden plants, and the spikes are seen to great advantage about the time the heavy rains and gales usually come in autumn. When extra fine spikes of bloom are required for exhibition or other purposes, it is necessary to give special treatment, and for this it is of the utmost importance that an open situation be selected. There can be no doubt that a deep loamy soil, not too heavy in texture, is the most suitable for the production of spikes for exhibition, but by deep digging and liberal manuring very satisfactory results may be obtained in soils of even an uncongenial character. Early in autumn the soil should be dressed liberally with manure from an old hotbed. After it is spread regularly over the surface, trench the soil up to a depth of 2 ft., and leave the surface as rough as possible, so as to expose a large body of it to the direct action of the frost and rains during the winter; this is of importance in the case of heavy soils, for it is very desirable that they should be thoroughly pulverised by the action of the weather. If this is done it will be in good condition for working in spring, and a pricking over with the fork will suffice to reduce it to a fine tilth, and even in wet seasons admit of the bulbs being planted without unnecessary delay. The planting of the bulbs should commence in March, and be continued at intervals of a fortnight until June. By this means a succession of bloom will be obtained from the earliest moment at which the show varieties may be had in flower until quite the end of the season. If planted in beds, the rows should be 18 in. apart, and the beds must be 4 ft. in width. Beds of this size will admit of one row being planted down the centre, and a row on each side at a distance of 6 in. from the edge of the bed. As soon as the plants have made sufficient progress to require support, stout stakes should be put to them. The top of the stake must not be higher than the first bloom, and the stem should have one tie only, and that a strong one of bast. After they are staked the surface of the bed should be covered to a uniform depth of 4 in. or 6 in. with partly decayed manure. This dressing is of the greatest value, for it materially assists in keeping the soil cool and moist about the roots during hot weather. As soon as the plants show bloom, liquid manure in some form is most beneficial in promoting a full development of the flowers. If for exhibition the spikes should be cut when about two-thirds of the bloom are expanded, as the lower flowers are generally of a finer quality than those towards the top.

In order to insure a supply of a given number of spikes at any particular date. a large number of sorts should be planted. That is to say, instead of planting from six to twelve bulbs of a sort, it is preferable to plant from one to three of each variety, and increase the number of sorts accordingly. For example, in purchasing a hundred bulbs from fifty to seventy varieties should be selected. For decorative purposes it is also preferable to have a large number of sorts, because of the greater variety of colour which they afford. The improvement that has been made in this beautiful flower within the last ten years has proceeded at such a rapid rate that many of the sorts which, a few years ago, occupied a foremost position are now quite surpassed, and for exhibition purposes are comparatively worthless. Most of the large nursery and seed houses supply the finest exhibition as well as bulbs for ordinary planting.

Early-flowering kinds. — During the past few years these beautiful flowers have rapidly become popular on account of their great value for cutting purposes. They have been obtained by hybridising several of the South African species, particularly G. ramosus, the branching kinds which are a distinct group, and G. trimaculatus, blandus venustus, and Colvillei forming what is known as the nanus section. Of Gladiolus ramosus, there are a great number of varieties that are dwarfer in habit, more graceful in appearance, earlier in flower, and almost as variable in colour as those of G. gandavensis; they are, moreover, much hardier, and beds of them may be left without any protection during the winter, so as to afford an early supply of flowers for cutting, and, unless the weather is very severe, never require covering. This remark applies only to bulbs that have become established in the ground, for fresh planted bulbs are quite as tender as any other Gladioli, and must be kept entirely from frost. Amateurs often make a mistake in this matter. Many bulbous and other plants are hardy only after they are well established. There is a great number of varieties of the nanus section of almost every shade of colour, growing from 1 ft. to 2 ft. high, which invariably have the three characteristic blotches of G. trimaculatus on the lower segments of the flower. G. Colvillei is one of the prettiest and hardiest of all, and is most valuable for cutting purposes, particularly the white variety, which affords an abundance of beautiful white flowers in early summer. The time of flowering depends upon the time they are planted, but the dwarf sections are earliest. If the varieties of G. ramosus and these are planted at the same time, the dwarfs are in flower a fortnight before them.

These early-flowering kinds are of simple culture; they succeed best in well-drained, raised beds of good loamy soil, in a sunny position. Some varieties, such as Colvillei, are safe if allowed to remain undisturbed, but some persons prefer taking the bulbs up and thoroughly drying them, and then planting again at about November; they will then flower early in June. If the bulbs remain in the ground throughout the winter, care must be taken to apply some protective material in severe cold. Propagation may be effected

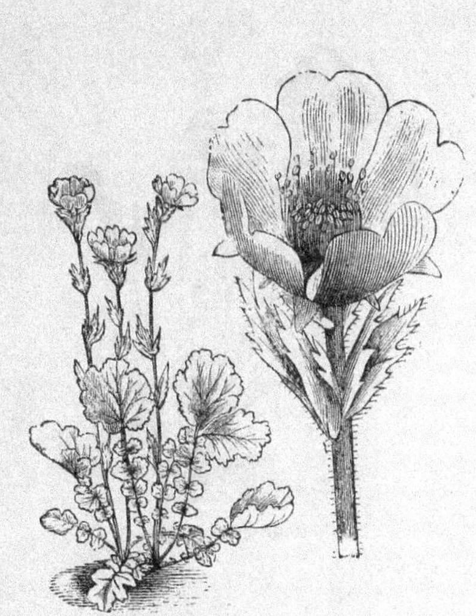

Geum montanum (*Avens*). See page 137.

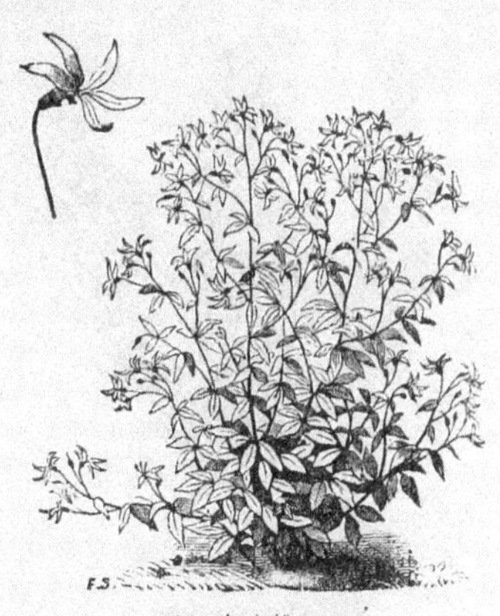

Gillenia trifoliata.

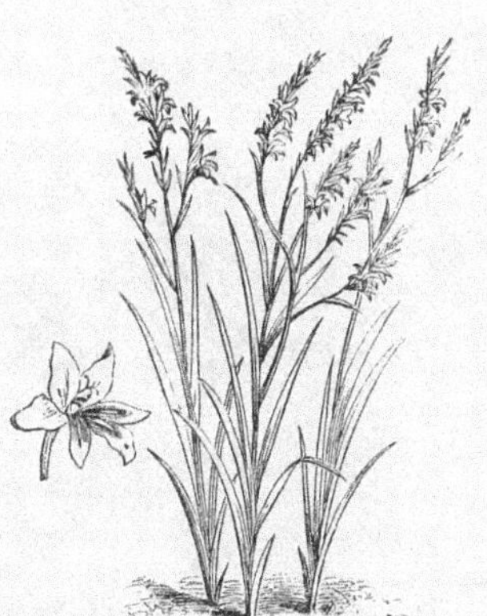

Gladiolus ramosus.

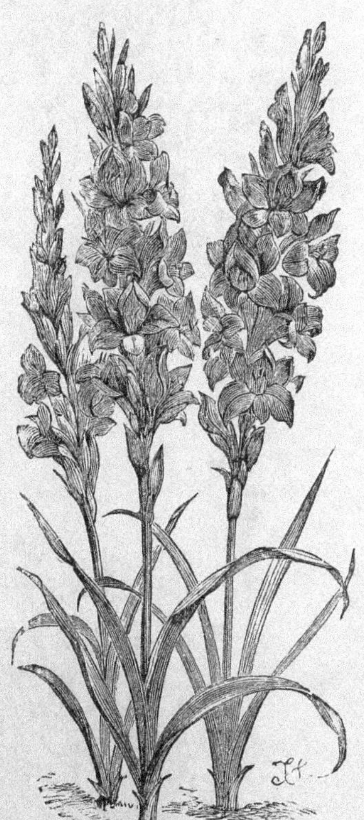

Gladiolus brenchleyensis.

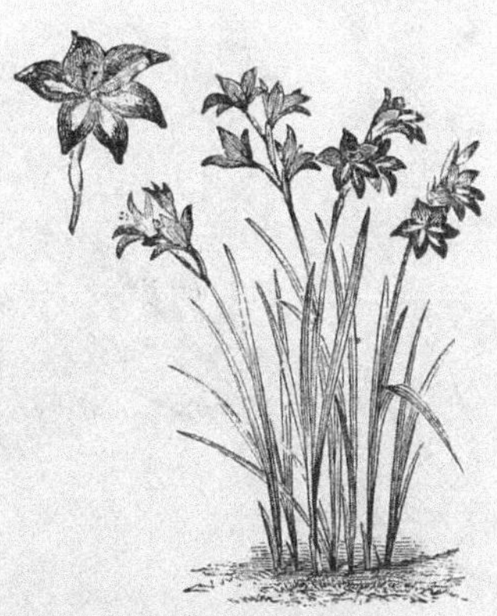

Gladiolus Colvillei.

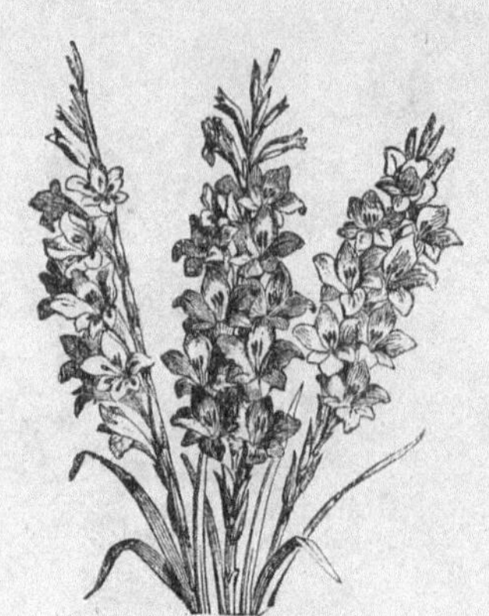

Gladiolus gandavensis.

Gladiolus communis.

Flower of Glaucium (*Horned Poppy*).

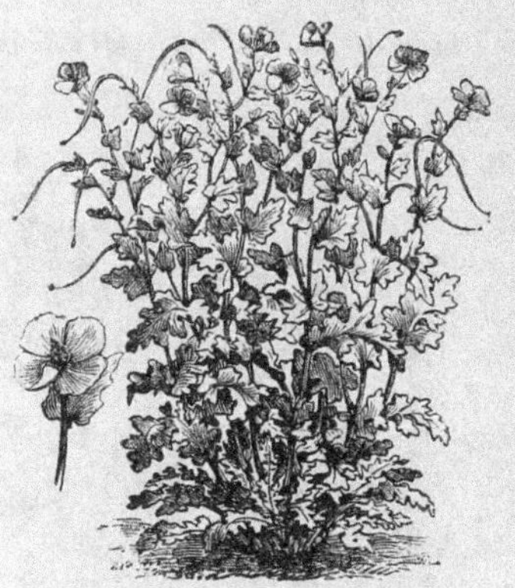

Glaucium luteum.

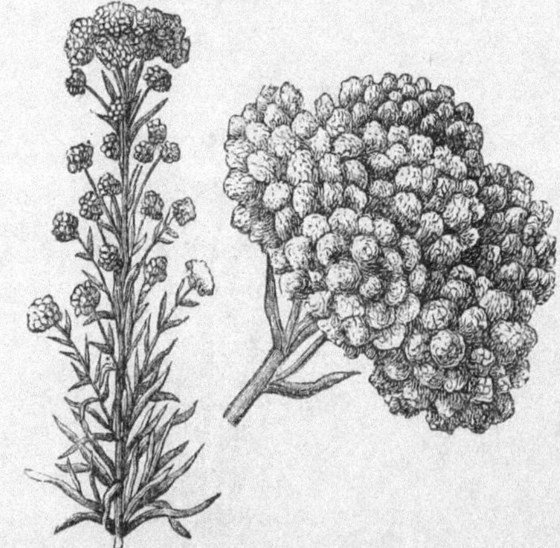

Gnaphalium lanatum. See page 140.

Godetia rubicunda. See page 140.

rapidly both by seeds and offsets. By the former method flowering bulbs are produced the second season, and can be left in the ground during the winter, provided the soil is light and dry and the bulbs protected from frosts. These Gladioli are extremely useful for pot culture, and can be had in flower at mid-winter by gently forcing; and for a succession of bloom between the forced plants and those in the open beds they may be grown in cold frames. For this purpose a bed of loam, leaf-mould, and sand in nearly equal proportions, in which to plant the bulbs, should be made up in October. The bed should be about 1 ft. in depth and well drained, and in this the bulbs may be planted thickly and 4 in. in depth. The lights should then be replaced, and air left on always except during severe frosts. No water should be given until the leaves appear about February, or earlier if the season be mild, and then only enough to keep the soil moist. The lights should be removed during mild weather, and in April altogether. During the latter part of May and in June abundance of bloom may be cut for decorative purposes. Besides those named, the following are a dozen of the best kinds: The Bride, Grootvoorst, Rubens, Maori Chief, The Fairy, Elvira, Rembrandt, Philip Miller, Beatrice, Baron von Humboldt, Sir Walter Raleigh, and Rose Distinctive.

Another interesting race of hybrids has lately been obtained between G. gandavensis and G. purpureo-auratus, a Cape species, with yellow and purple flowers. The hybrids have large flowers of a creamy-white and deep purplish-crimson. The named kinds are G. hybridus Frœbeli, G. h. Lemoinei, and Marie Lemoine. Although by no means so showy as many other kinds, they are yet most graceful and distinct in port, and also in the shape and colouration of their flowers. In deep sandy soil they attain a height of nearly 5 ft., and the flowers gradually developing themselves render the plant effective for at least five weeks from the time the first and lowermost blossom expands. As graceful plants, they well deserve culture, being perfectly hardy—hardier indeed than many home-raised hybrids; but a warm, deep soil and a sheltered position near the foot of a south or west wall are most congenial to their strong growth.

Among the true species of Gladiolus there are a few almost equal in beauty to the hybrid sorts, one of the finest being G. Saundersi, which grows about 2 ft. high, and has large showy flowers of a brilliant scarlet with a conspicuous pure white centre. It is not often grown, though it is hardy and of very easy culture, only requiring to be planted in a sunny position in a light rich soil. G. purpureo-auratus is another good species, though not so showy, the flowers being of a creamy-white, heavily blotched with vinous purple on the lower divisions. Also a native of the Cape.

There is another type of Gladiolus quite hardy, and pretty plants for the mixed border—the European sorts. There is a strong similarity among them, all growing from 1 ft. to 1½ ft. high, and bearing rather small rosy-purple flowers. The best known kinds are G. byzantinus, communis, segetus, illyricus, neglectus, serotinus—all natives of Europe. They like a warm, dry soil and a sunny situation. These European kinds are of particular interest from their free and hardy habit, which makes them as easy to grow as native plants. They come in admirably for the wild garden, thriving in copses, open warm woods, in snug spots in broken hedgerow banks, and on fringes of shrubbery in the garden.

DISEASE. — This is frequently, if not always, accompanied by some condition of the fungus known as Copper-web, the Rhizoctonia crocorum of De Candolle, known in France under the name of Tacon. The fungus attacks also the Narcissus, the Crocus, Asparagus, Potatoes, and other bulbs, roots, &c. A good deal of attention was paid to this disease in 1876, when Mr. G. W. Smith for the first time detected the curious fungus in abundance, named by him Urocystis Gladioli. The Urocystis and Rhizoctonia are probably two conditions of the same thing, the Rhizoctonia being possibly the spawn and the Urocystis the fruit; the latter is capable of remaining in a resting state for a year or more. The Urocystis can frequently be found in the decayed red-brown portions of the diseased corm. No attempts have been made in the direction of a cure, as far as we know. The disease is confined to certain localities and to certain gardens; it is quite unknown in some districts.

Gladwin *(Iris fœtidissima).*

Glaucium *(Horned Poppy).*—Plants of the Poppy family, mostly of biennial duration. G. luteum may be found a plant useful for winter decorative bedding, as it is quite hardy and has handsome ornamental foliage of a silvery hue, almost as white as that of the Centaurea. The leaves are much more deeply cut than those of that plant, and, planted closely together, are effective, either in masses or lines. To ensure strong plants for the winter, seed should be sown about May, as the plant is a biennial and should be treated as such. When in bloom it makes a striking border plant, the flowers being large and orange-red. G. Fischeri is in every sense a superior plant; its snow-white, woolly, undulated foliage is very telling, added to which the blossom is of an unusual flame colour. G. corniculatum is similar, but not so handsome. Both require the same treatment as G. luteum.

Glaux maritima.—A small, native seashore plant of no garden value.

Glechoma hederacea *(Ground Ivy).* —A well-known, small, creeping, British plant, almost too common to deserve notice here; but it has one or two varieties finely variegated that are quite worthy of a place in beds of plants with variegated leaves or on the edges of raised borders.

Globe Flower (*Trollius europæus*).

Globe Thistle (*Echinops*).

Globularia. — Interesting and dwarf alpine plants, good on the rock garden, always in light and peaty soils. G. Alypum is said to be the best; it inhabits dry rocks naturally. Other kinds are G. cordifolia, G. nana, G. nudicaulis, and G. trichosantha.

Glossocomia ovata.—A rare little perennial plant of the Bellflower family. Native of North India. It has showy blue flowers, but, being extremely rare and difficult to cultivate, it is scarcely worth including here.

Glycirrhiza (*Liquorice*).—Coarse growing plants of the Pea family, with pinnate leaves and bluish flowers. G. glabra yields the liquorice of commerce.

Gnaphalium.—A genus of Composites, usually with downy foliage but of no ornamental value, except G. lanatum, a silvery-foliaged plant, growing about 1 ft. high, suitable for edgings. It is very easily propagated by division in spring, and is whitest and most compact on dry poor soils. It bears pegging down, and should never be allowed to bloom. It is somewhat tender and requires to be propagated annually. (See Antennaria.)

Goat's Rue (*Galega*).

Godetia.—Among the most beautiful of hardy annuals for spring and summer flowering. They require to be sown in September in order to secure strong plants before severe weather sets in, and if well thinned out and the soil be rich they will give a grand display of blossom during May and June; if seed be sown again in April a succession of bloom may be easily kept up. Among the best kinds are G. Whitneyi, G. Lady Albemarle, G. Lindleyana, G. The Bride, G. rubicunda, G. Princess of Wales, G. Tom Thumb, and good new varieties are being added to these. These showy and admirable annuals are from 10 in. to 20 in. high, and so free and vigorous, and so profuse in striking and delicate colours, that they may be made effective use of in the early summer garden. Evening Primrose family. Natives of North America.

Golden Club (*Orontium*).

Golden Drop (*Onosma tauricum*).

Golden Rod (*Solidago*).

Golden Saxifrage (*Chrysosplenium*).

Goldfussia isophylla. — Respecting this, a correspondent writes: "We always grow a few, not so much on account of its pale blue tubular flowers, which are produced in the dull winter months, as because when put outside in the summer its leaves become a shining black, and, being narrow and numerous, the plant has a prettier appearance than Perilla and the usual funereal plants with dark leaves that one sees."

Goldilocks (*Linosyris vulgaris*).

Gold Thread (*Coptis trifoliata*).

Goodyera pubescens (*Rattlesnake Plantain*) is a beautiful little Orchid with leaves, close to the ground, delicately veined with silver; it is, in fact, a hardy variegated Orchid, particularly distinct and pleasing, although its flowers are not showy. It has long been grown here and there in botanic and other choice collections. It thrives well in any shady position, such as may be found in any good rock garden, planted in moist peaty soil, with here and there a soft sandstone for its roots to cling to and run among. It is quite hardy. Native of Eastern United States. G. repens and Menziesi are less desirable and much rarer.

Gourd (*Cucurbita*).

Grammanthes gentianoides. — A pretty, half-hardy annual, and a capital plant for growing on the dry parts of a rockery, or, indeed, in any other exposed situation. It grows about 2 in. high, and forms a dense, much-branched tuft with fleshy leaves about ½ inch long. It produces, in abundance, flowers about two-thirds of an inch across. Their colour, when first expanded, is orange, with a distinct V-shaped mark at the base of each petal, but they finally assume a deep red hue. It is sometimes used with good effect in geometrical beds, and succeeds well in dry, warm soil. Seeds should be sown in heat in February and March, and the seedlings planted out in May. It belongs to the Stonecrop family, and is a native of the Cape of Good Hope.

Grape Hyacinth (*Muscari*).

Grass of Parnassus (*Parnassia palustris*).

Gratiola.—Dwarf perennial plants of the Figwort family. Of little use outside of botanic gardens.

Great Reed (*Arundo Donax*).

Grindelia. — Yellow-flowered, North American Composites of little garden value. G. grandiflora grows 3 ft. or 4 ft. high, and G. squarrosa is dwarfer; both of these are hardy, and succeed in any kind of soil. They flower from July till late in autumn.

Ground Laurel (*Epigæa repens*).

Ground Nut (*Apios tuberosa*).

Ground Pine (*Lycopodium dendroideum*).

Guernsey Lily (*Nerine sarniensis*)

Gunnera.—South American plants remarkable for their large and fine foliage, which somewhat resembles that of gigantic Rhubarb. There are two kinds in cultivation—G. scabra and G. manicata. Both are handsome plants, especially the former, the leaves of which sometimes measure 6 ft. across. Both species are deserving of a place in any garden, for no plants in cultivation are so stately as well-grown samples of these two Gunneras. They may be termed quite hardy if a slight protection is afforded during the severest cold, such as a layer of dry leaves placed amongst the stems, with their own leaves bent down upon them. In spring these should be removed, and the tender

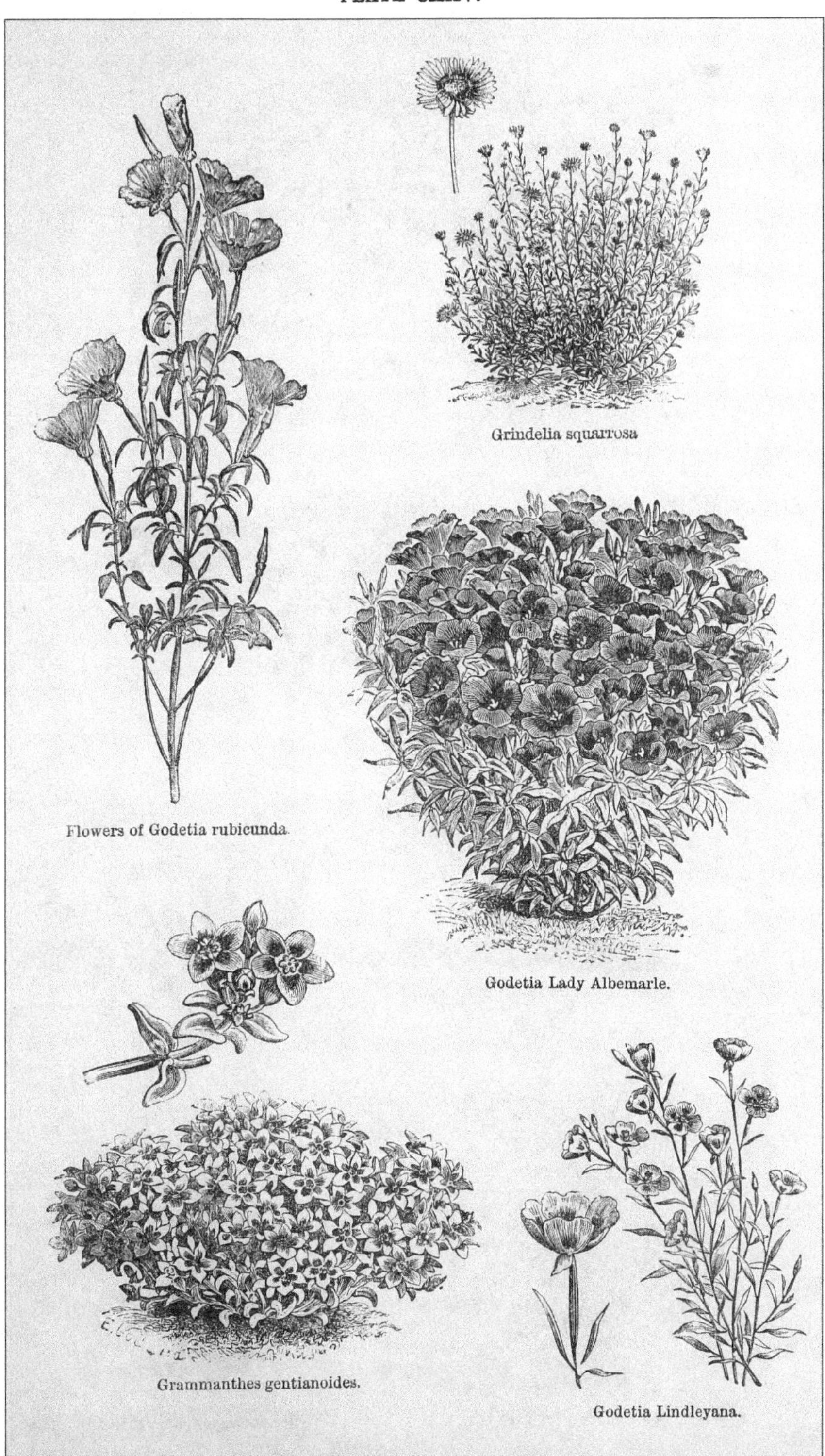

Grindelia squarrosa

Flowers of Godetia rubicunda.

Godetia Lady Albemarle.

Grammanthes gentianoides.

Godetia Lindleyana.

[*To face page* 140.

PLATE CXXVI.

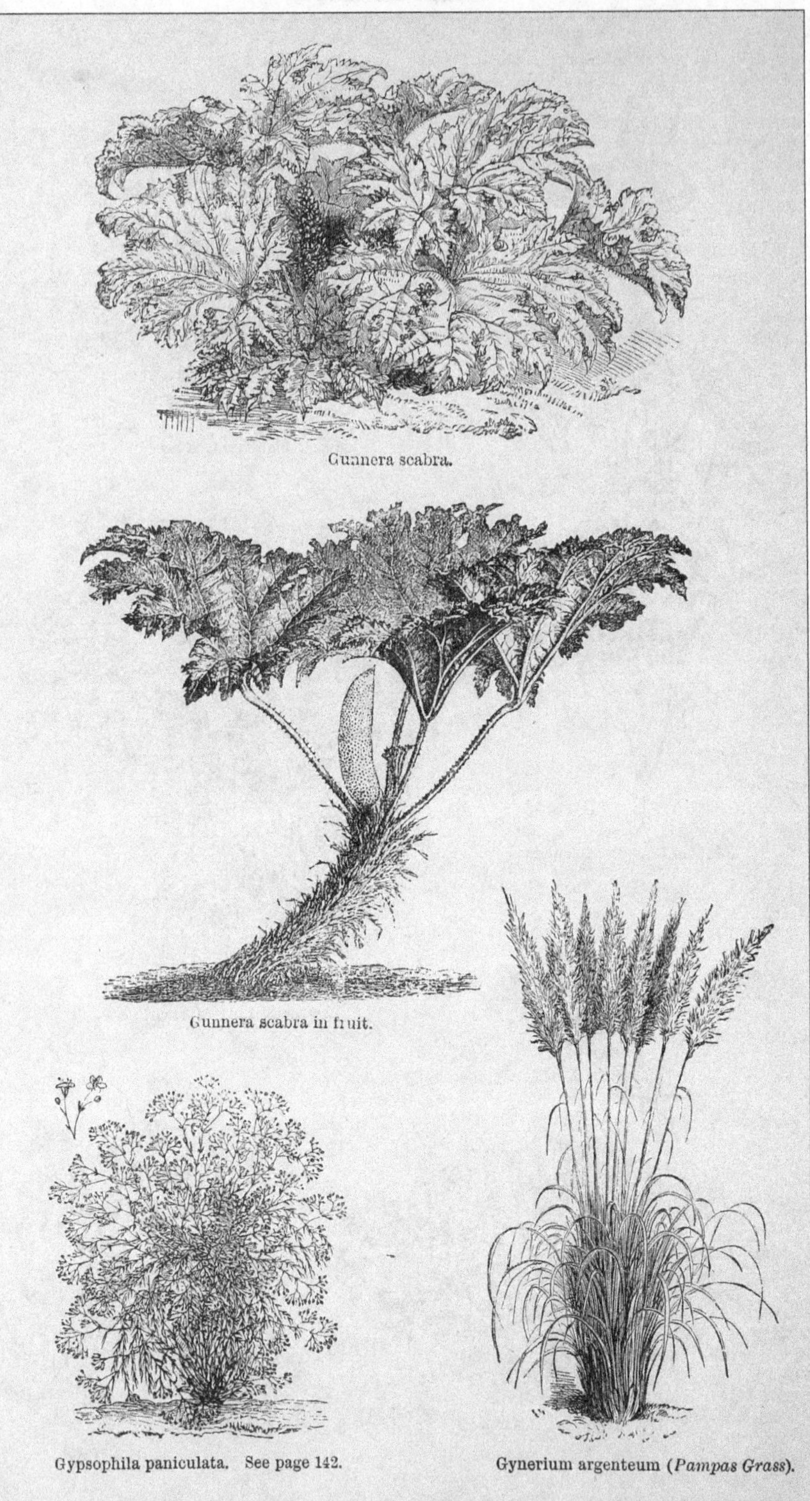

Gunnera scabra.

Gunnera scabra in fruit.

Gypsophila paniculata. See page 142.

Gynerium argenteum (*Pampas Grass*).

[*To face page* **141.**

growth afforded a slight protection by means of a piece of canvas shading or an ordinary mat. In mild winters this precaution is scarcely necessary, especially in the south and other more favoured localities. The Gunnera should be planted in some isolated spot, and not as a rule in the "flower garden proper," as it must not be disturbed after being well planted, and would associate badly with the ordinary occupants of the parterre. The plant cannot have too much sun or warmth, but makes little progress if its huge leaves are torn by storms. In places with any diversity of surface it will be easy to select a spot well open to the sun and yet sheltered by surrounding objects (shrubs and clumps). To grow these plants to perfection it is indispensable that they be planted well at the outset. A large hole, about 6 ft. by 4 ft. deep, should be dug out, a good layer of drainage material put at the bottom, and it should be filled with a rich compost of loam and manure. In summer the plants ought to be allowed a plentiful supply of water, and, in order that a large quantity may be given at each application, a circular ridge of turves should be placed round them, so that the water is compelled to sink down about the roots. They should have, in addition to this, an annual mulching of well-rotted manure in early spring. They thrive well on the margins of ponds or lakes where their roots can penetrate the moist soil, and in such a position, if judiciously placed, they have a fine effect. Perhaps the finest plant of G. scabra in the country is the noble specimen at the head of the herbaceous ground at Kew; it measures from 10 ft. to 15 ft. through and 8 ft. high, and often develops leaves from 3 ft. to 4 ft. across. This specimen has attained these dimensions in a comparatively short time, as it was only planted a few years ago. Though the two kinds bear a great resemblance to each other, they have well marked characteristics. The leaves of G. manicata are much more kidney-shaped than those of G. scabra, and they, moreover, attain a much larger size, often measuring as much as from 4 ft. to 6 ft. across. The spikes of fruits are also very different from those of G. scabra, being much longer, and the secondary spikes are long and flexuose, whereas in G. scabra they are short and stiff. Propagated by seed or division of established plants.

Gymnadenia conopsea.—A sweet-scented, native Orchid, 6 in. to nearly 2 ft. high. Its flowers are pale purple, in a dense, cylindrical, tapering spike, 1 in. to 4 in. long, and are produced in summer. It is suitable for growing in grassy places, drier parts of the bog bed, or borders; worth growing for its fragrance only. Increased by separation of the root knobs.

Gymnolomia.—A genus of Composites from N. America, of which one species, G. Porteri, a half-hardy annual, is introduced. Its value has not yet been tested.

Gymnothrix latifolia.—A fine perennial Grass introduced from Monte Video, but not hardy enough to show its true character in the open air in our country. Mr. Ellacombe, however, informs us that it is valuable on account of its rich green foliage and very excellent habit. It is particularly suitable for the centre of a bed. Mr. W. Gumbleton, writing from Cork, says it does not do well with him nor flower, only attaining a height of less than 3 ft.

Gynerium argenteum *(Pampas Grass)*.—This noble Grass, which grows 4 ft. to 14 ft. high, according to the strength of the plant and soil or district, is now well known. There is reason to believe that some varieties are better in habit than others and flower earlier. In such cases it would be better to patiently divide them than to trust to seedlings. There are a number of varieties, some of a delicate rosy colour, one variegated, and several dwarf and neat in habit. Some of these seem somewhat more tender than the original type. It is not enough to place it in out-of-the-way spots, but the landscape, so to say, of every garden and pleasure ground should be influenced by it. It should be planted even far more extensively than it is at present, and given very deep and good soil, either natural or made. The soils of very many gardens are insufficient to give it the highest degree of strength and vigour, and no plant better repays for a thorough preparation, which ought to be the more freely given when it is considered that one preparation suffices for many years. If convenient, give it a somewhat sheltered position in the flower garden, so as to prevent as much as possible that ceaseless searing away of the foliage which occurs wherever the plant is much exposed to the breeze. Also when backed with shrubs, its bright silvery plumes are less liable to be injured. We rarely see such fine specimens as in quiet nooks where it is pretty well sheltered by the surrounding vegetation. It is very striking to come upon noble specimens in such quiet green nooks; but, as before hinted, to leave such a magnificent plant out of the flower garden proper is a decided mistake. It should be planted about the beginning of April, and mulched with rotten manure, watered copiously in hot, dry weather. Gynerium jubatum is very well spoken of, but as yet has not been tried much except in certain favoured spots. The leaves resemble those of G. argenteum, but are of a deeper green, and droop elegantly at their extremities. From the centre of the tuft, and exceeding it by 2 ft. or 3 ft., arise numerous stems, each bearing an immense loose panicle of long filamentous silvery flowers, of a rosy tint with silvery sheen. It is a native of the republic of Ecuador, and is earlier in bloom than the Pampas. The sexes are borne on separate plants in all the species of Gynerium, and the plumes of male flowers are neither so handsome nor so durable as the plumes of female flowers.

Gypsophila.—Plants of the Stitchwort family, not without value. The larger kinds are usually very elegant, and bear tiny white blossoms in myriads on slender, spreading panicles. Of this type the best is G. paniculata, which forms a dense, compact bush, 3 ft. or more in height and as much across. The flowers are small, white, exceedingly numerous, and arranged on thread-like stalks in much-branched stems, with the light, airy, graceful effect of certain ornamental Grasses. Very useful for cutting. It thrives in any soil, and is suitable for borders and naturalisation in woods and copses, or by the margin of streams. Division or seed. G. fastigiata, perfoliata, altissima, Steveni are very similar. G. prostrata is a pretty species for the rock garden or the mixed border. It grows in spreading masses, and has white or pink small flowers, borne on slender stems in loose graceful panicles from midsummer to September. G. cerastioides grows about 2 in. high, and has a spreading habit. The leaves are about 1½ in. long, and from their axils are produced small clusters of blossoms, which are ½ in. across, white with violet streaks. It is from Northern India, and quite unlike any of the group now in our gardens, being dwarfer and having larger flowers. It is a rapid grower and soon spreads into a broad tuft if in good soil and in an open position on the rock-garden. Increased by seeds or cuttings in spring.

Habenaria *(Rein Orchis)*.—Terrestrial Orchids from N. America, some of which are inconspicuous, while others, including those named below, are pretty and interesting, and all grow from 1 ft. to 2 ft. high. To succeed in out-door culture, a spot should be prepared with about equal parts of leaf-mould, or peat, and sand, with partial shade; the soil should be well mulched with leaves, grass, or other material to protect the roots from the heat of the sun, and to keep it moist. H. blephariglottis flowers in July, in spikes, white and beautifully fringed. H. ciliaris is the handsomest species of the genus. The flowers are bright orange-yellow with a conspicuous fringe upon them, and are produced from July to September. H. fimbriata flowers in a long spike, lilac-purple, beautifully fringed. H. psycodes, flowers purple, in spikes 4 in. to 10 in. long, very handsome and fragrant. They are charming plants for the bog garden, or for a quiet nook with moist peaty soil.

Haberlea rhodopensis.—This is a pretty little rock plant resembling a Gloxinia in miniature. It forms dense tufts of numerous small rosettes of leaves which somewhat resemble those of the Pyrenean Ramondia (R. pyrenaica), each rosette bearing in spring from one to five slender flower-stalks with two to four blossoms each, nearly 1 in. long, of a bluish-lilac colour with a yellowish throat. Messrs. Frœbel, of Zurich, who grow it well, write to us: "We have treated this plant in the same manner as the Pyrenean Ramondia, *i.e.*, we have planted it on the north side of the rockwork; therefore the sun never directly reaches it. We grow it in fibrous peat, and fix the plants, if possible, into the fissures of the rockwork, so that the rosettes which it forms hang in an oblique position, just as they do in their native country. It succeeds well in this way; but if no rockwork be at hand it may be grown equally well on the north side of a Rhododendron bed. We have it thus situated quite close to a stone edging, a way in which we also grow the Ramondia, and the Haberlea flowers profusely every year in May and June. The plant is very hardy, having withstood our often very hard winters, without any protection, quite unharmed." It is a native of the Balkan Mountains, where it is found growing among moss and leaves on damp, shady, steep declivities at high elevations.

Hablitzia tamnoides. — A hardy, climbing, herbaceous plant, producing clusters of greenish-yellow flowers in the greatest profusion. When the plant is tied to a strong stake or trellis it reaches a height of 8 ft. or 10 ft., and has a graceful appearance. It continues in flower throughout the whole summer and the greater portion of the autumn; it requires a good soil, plenty of moisture in summer, and freedom from stagnant water in winter, and forms a good subject for planting in open situations in the wild garden, or on rock or root work, or where it could be used to clothe the stems of naked trunks of trees.

Habranthus pratensis.—A brilliant bulbous plant of the Amaryllis family, hardy, at least in the southern and eastern parts of the country. It has stout and erect flower-stems, about 1 ft. in height, and flowers of the brightest scarlet hue, feathered here and there at the base with yellow. The variety fulgens is the finest form of the plant. It blooms freely in the Rev. Mr. Nelson's garden at Aldborough, in Norfolk, in the open border, flowering at the end of May or beginning of June. It grows very freely in strong loam, improved in texture by the addition of a little leaf-mould and sand. Its propagation is too easy, for in many soils it is said to split up into offsets instead of growing to a flowering size. At Aldborough it makes numerous offsets. A choice plant for the select bulb garden or rock garden. Chili. H. Andersoni is a much inferior plant.

Hairbell.— The name applied to the slender growing Campanulas, principally our native C. rotundifolia, also spelled Harebell. There seems to be no authority for either name, but this, to us, appears to be the most appropriate.

Hand Orchis *(Orchis maculata)*.

Haplophyllum patavinum *(Ruta)*.

Hard Fern *(Blechnum boreale)*.

Harebell.—A name generally applied to the slender growing Campanulas, such as C. rotundifolia.

H rpalium rigidum *(Prairie Sunflower)*.—A very showy and beautiful Com-

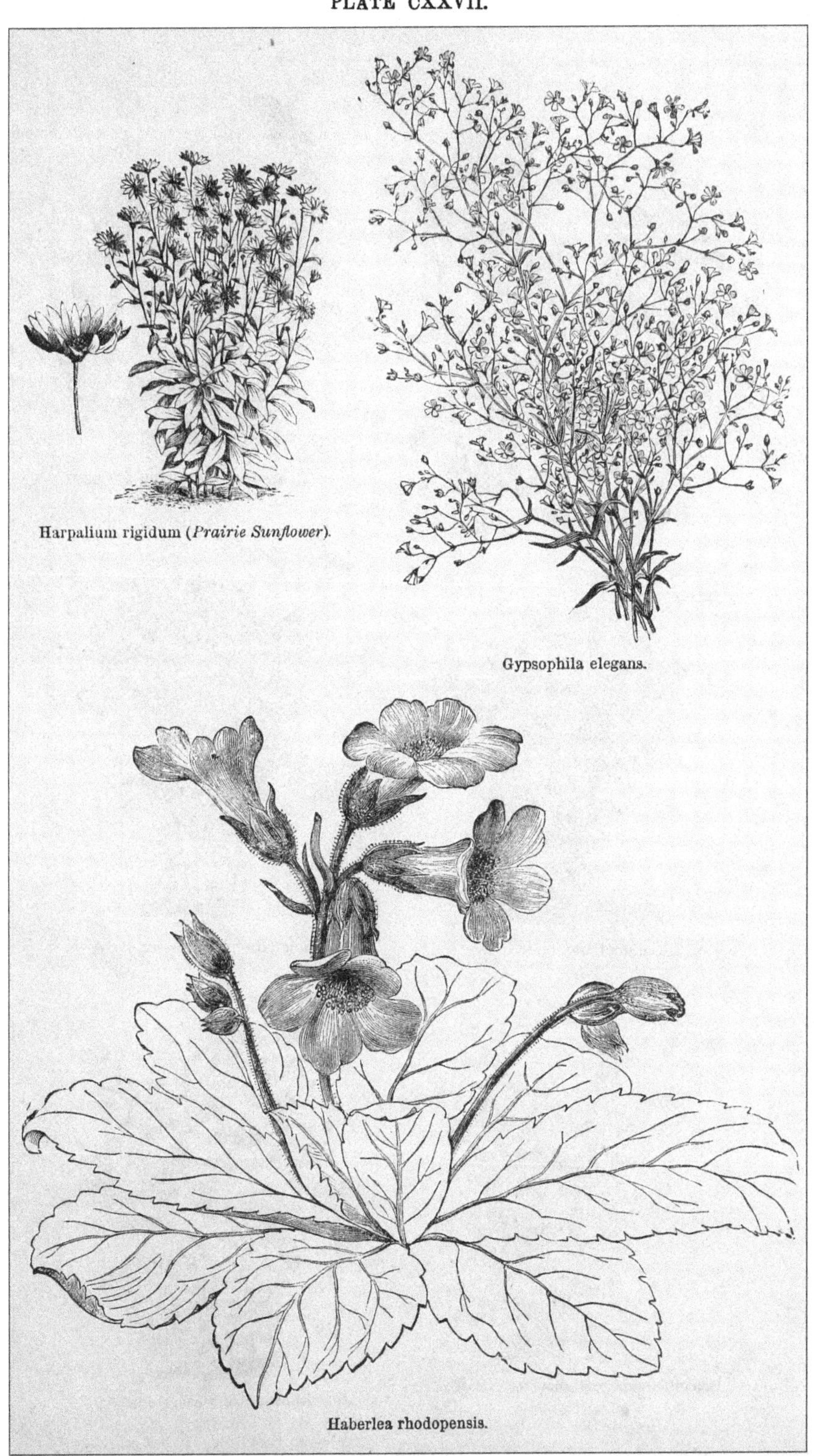

Harpalium rigidum (*Prairie Sunflower*).

Gypsophila elegans.

Haberlea rhodopensis.

[*To face page* 142.

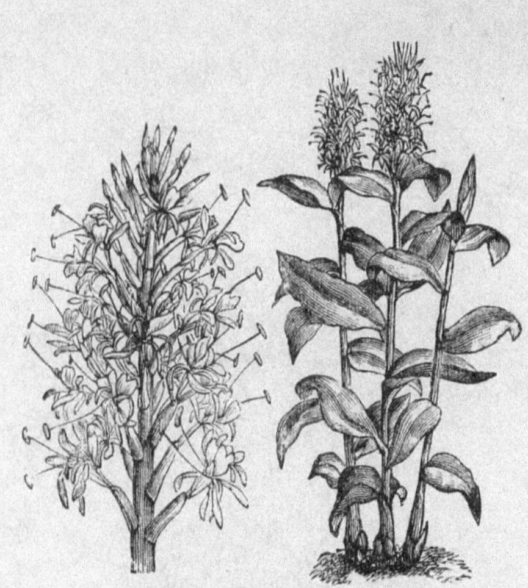

Hedychium Gardnerianum.

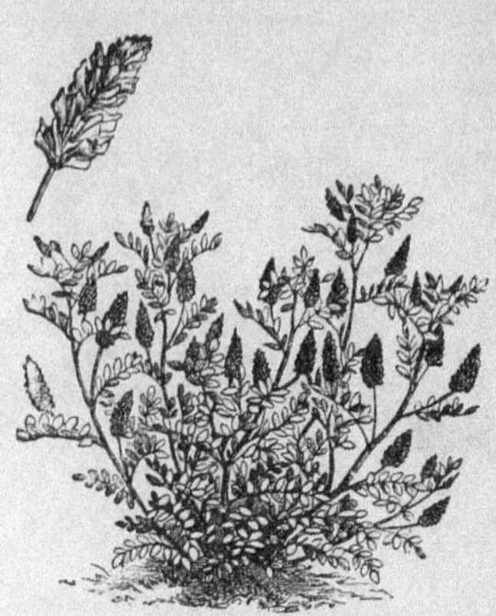

Hedysarum coronarium (*French Honeysuckle*).

Helenium tenuifolium.

Helenium autumnale (*Sneeze Weed*).

Helianthemum guttatum (*Sun Rose*).

Helianthus multiflorus (*Sunflower*)

[*To face page* **143.**

posite, free and hardy, growing from 3 ft. to over 4 ft. high, and running much at the root in good soils. Of all the Sunflower-like blossoms it is among the best, and in all ways a good perennial. Suitable for groups, beds, or the mixed border. N. America.

Heart's-ease (*Viola tricolor*).

Hebenstretia.—Interesting little plants allied to the Labiates, but only suitable for botanical collections.

Hedychium Gardnerianum.—This, though mostly grown as a greenhouse plant, will flower out-of-doors, and live through an ordinary winter without much protection. It should be planted out in May, so as to make its summer growth where it is to stand, as then its roots penetrate deeply and become established before the top dies off in autumn. It thrives best in a loose sandy loam, made rich with manure. A mulching should be given while the plant is making its growth, with an occasional watering of liquid manure in dry weather. It is excellent for choice groups in the sub-tropical garden, in warm, sheltered spots. A heap of cinder ashes or half-rotten leaves laid over the crowns in winter will ensure their safety; or the roots may be lifted in autumn and wintered in any dry place along with Dahlias and Cannas. Increased by dividing the roots in spring, but in doing so it is necessary that each piece should have a young crown attached to it.

Hedysarum (*French Honeysuckle*).—Plants of the Pea family, mostly of a weedy character, a few only of the perennial kinds being ornamental. H. coronarium, a popular and rather showy plant, grows 3 ft. or 4 ft. high, and produces in summer dense spikes of red flowers. There is a white-flowered variety; it grows in any ordinary soil, but is not a perennial, though it usually sows itself where established. Among the dwarfer kinds the two following are desirable plants: H. obscurum, a brilliant and compact perennial; 6 in. to 12 in. high, with showy, purple flowers in racemes. It is suitable for the rock garden, borders, and naturalisation amongst vegetation not more than 1 ft. high, chiefly on banks and slopes in sandy loam. Division or seed. H. Mackenzii—said to be the handsomest of the genus. It grows about 2 feet high, with long racemes of from seven to thirty rather large Pea-like flowers, of a rosy-purple colour. It is perfectly hardy in any situation, and flowers in June and July. It is rather too tall for the rock garden, and is better suited for the mixed border.

Helenium (*Sneeze-weed*).—Composites of a vigorous growth, flowering in autumn and thriving in any soil. There are two or three species, the most useful being H. autumnale, a plant about 6 ft. high, bearing yellow flower-heads. The varieties grandiceps and pumilum are very distinct, the former being of gigantic growth with a fasciated head of bloom, which makes the plant very showy; the latter is much dwarfer than the type and more desirable. H. atropurpureum grows 3 ft. or 4 ft. high, and has reddish-brown flower-heads. H. Hoopesi is desirable, as it flowers in early summer. It is rather a coarse grower, but has large yellow flowers. The first-mentioned species and its varieties are excellent border plants, and, though vigorous, remain long in bloom. They are very useful for cutting, as they remain a long time fresh in the cut state. N. America.

Helianthella tenuifolia.—A perennial Sunflower-like plant from Florida, not much known in gardens and of no great value.

Helianthemum (*Sun Rose*).—Though strictly shrubby plants for the most part, these dwarf evergreens possess so much the aspect of rock plants, that they cannot be well separated from them. They are extremely valuable ornamental plants, and there are few more brilliant sights than that afforded by masses of them when in full beauty. They are of the easiest possible culture, and are, with but few exceptions, dwarf and compact in growth, their flowers being borne in great profusion and presenting a remarkable diversity of colour. The common Sun Rose (H. vulgare) is extremely variable both in habit and colour, and from it has sprung the host of varieties enumerated in trade lists; indeed, we need but this species to represent, for garden purposes, the variation illustrated in all the dwarf shrubby species of the family. The colours range from white and yellow to the deepest crimson. There are also double flowered kinds and one with variegated foliage. Other pretty, dwarf, shrubby species, similar to H. vulgare, are H. rosmarinifolium, pilosum, and croceum. There is also a herbaceous perennial species, H. Tuberaria, which differs completely in aspect from the shrubby species, and is second to none in beauty. It grows 6 in. to 12 in. high, with flowers resembling a single yellow Rose, with dark centre, 2 in. across, drooping when in bud. It is suited for the rock garden on warm ledges, in well-drained sandy or calcareous soil. When sufficiently plentiful it should be used in the mixed border. Seed and careful division. When a full collection is required there are other introduced species, but the above fairly represent the beauty of the family.

Helianthus (*Sunflower*).—This genus is typical of the coarse-growing, yellow-flowered Composites that abound in North America, and of which not a few have found their way into English gardens. These coarse Helianthi have their place, but it is rather in the wild than in the flower garden. All the perennials are vigorous growers, and generally attain a great height. The most valuable border species is

H. multiflorus, which grows about 1 yard high, forming a dense thicket of stems which in autumn bear large, showy, bright yellow heads. There is a major variety of the single kind that is superior to the type. The double kind, H. m. flore-plenus, is a very popular plant, and deservedly so, as there are few more showy autumn perennials. It is, like the original, of easy culture in any soil or position, is one of the best plants for towns, and abounds in some London parks and squares. The best of the other perennial kinds suitable for the rougher parts of the garden or for naturalising are H. giganteus (with rather small flowers, but growing 6 feet to 8 feet or 10 feet high), divaricatus, occidentalis, decapetalus, lævis, and lætiflorus, all growing about the same height, much resembling each other, and flowering about the same period. H. angustifolius grows about 3 ft. high, and is a distinct and good plant. One of the most desirable of the larger species is H. orgyalis, a tall species 6 ft. to 8 ft. high, with slender stems clothed with most graceful, drooping, narrow foliage. It is suitable for growing among groups of fine-leaved hardy plants in the sub-tropical garden or pleasure-ground, margins of shrubberies, or wood-walks. The tips of the shoots for a length of 15 in. or so, cut off and placed in water indoors, are very ornamental. Division. The bold and handsome Harpalium rigidum is sometimes grouped under Helianthus.

H. annuus (*Common Sunflower*).—This is often regarded as suitable only for cottagers or for gardens to which little attention is given. As an ornamental plant it is, however, of much value, its robust growth and commanding aspect rendering it suitable for many situations where plants of smaller growth would be quite lost. If placed where protection is afforded from high winds, and allowed sufficient space, it assumes a somewhat dense, branching habit, and when covered with bloom presents a very distinct and by no means unattractive appearance. More than one plant should not be put in a place. Crowding Sunflowers destroys their distinctive characteristics. Seed may be raised easily if sown in pots or pans in March or April and placed in a window, frame, or greenhouse, but will germinate all the more rapidly if a little bottom-heat be given. When the plants are a few inches high, they should be pricked out into a frame and protected from snails, and be put out into the open ground when well hardened and strong. Sunflowers will thrive well in any good garden soil, the size of the flowers being large or small as the ground varies in quality. In exposed places the plants should have some support. There are now several varieties of the common one, the best being giganteus, 6 ft. to 12 ft.; californicus, 5 ft., cucumerifolius, 3 ft. to 5 ft.; grandiflorus, plenissimus, globosus, and fistulosus. The art in using the Sunflower consists in placing one or a few plants in spots where they can be well seen and will not interfere with other and more refined plants either above or below it. We have seen choice and carefully made borders ruined by a number of Sunflowers being placed along them. A few plants well placed near a shrubbery would have given much better effect.

Helichrysum (*Everlasting Flower*).—Composites, mostly native of the Cape of Good Hope, of which a few are cultivated. The most important garden plants are H. macranthum and H. bracteatum. They are generally treated as annuals, and, unless exceptionally well managed by sowing early under glass, the season is so far advanced before they commence flowering that the best period for laying on the brightest colours is lost, and early frosts find the plants just approaching their best. During mild winters they are uninjured, and the plants are the most continuous flowerers next year even in the driest and hottest of summers. They are particularly adapted for a background plant on dry borders. If sown in pans or boxes where they can be slightly protected during winter, and planted out early in April, they have a chance of producing a good crop of flowers for drying. The colours vary from deep crimson to yellow and white. The hardy perennial Helichrysums are not important plants, and seldom succeed well. H. orientale, which furnishes the Immortelle of the French, flowers poorly except in very hot seasons. None of the other hardy kinds are worth growing, except perhaps H. arenarium, which has bright, golden-yellow flowers.

Heliophila.—Small growing Cruciferous annuals. H. araboides is a pretty blue-flowered annual, of which occasional use might be made. It is dwarf, though free in habit of growth and flower.

Heliopsis.—Stout, yellow Composites not of essential value in the presence of the good plants of the same order now in cultivation. There are several kinds.

Heliotropium (*Cherry Pie*).—These are universal favourites for flower gardens on account of their delicate fragrance. They are easily increased by cuttings in August or September and again in spring. For the flower garden spring-struck plants are the best. It is a good plan to lift a few plants from the beds in September, winter them in a warm greenhouse, and in spring to put them into a warm place, where they will soon produce plenty of cuttings. These may be struck on slight heat like Verbenas, potted on, made to grow rapidly, and fit to plant out at the end of May when danger of frost is past. Heliotropes may be raised from seed and flowered the same year—in fact, treated as annuals. Sow early—February or beginning of March—and they will grow into sturdy little plants before planting time. When bedded out they should be placed in good dry soil. The following are good varieties of the different shades of colour, and new varieties are raised from time to time: Anna Turrell, Roi des Noirs, Triomphe de Liége, and the old species H. peruvianum, which many will like from its associations if for no other reason. Heliotropes, though quiet in colour, are charming

Helianthus annuus (*Common Sunflower*).

Helianthus globulosus fistulosus.

Helianthus annuus var. californicus.

Flower of Helianthus annuus (*Common Sunflower*).

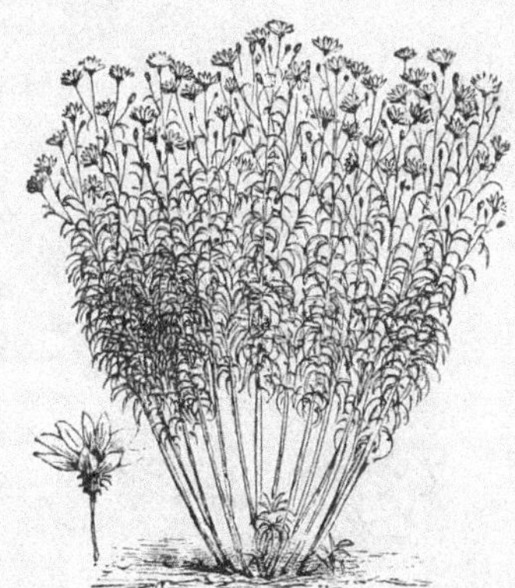

Helianthus orgyalis.

Helichrysum brachyrhynchum (*Everlasting Flower*).

Helichrysum macranthum (*Everlasting Flower*).

[*To face page* 144.

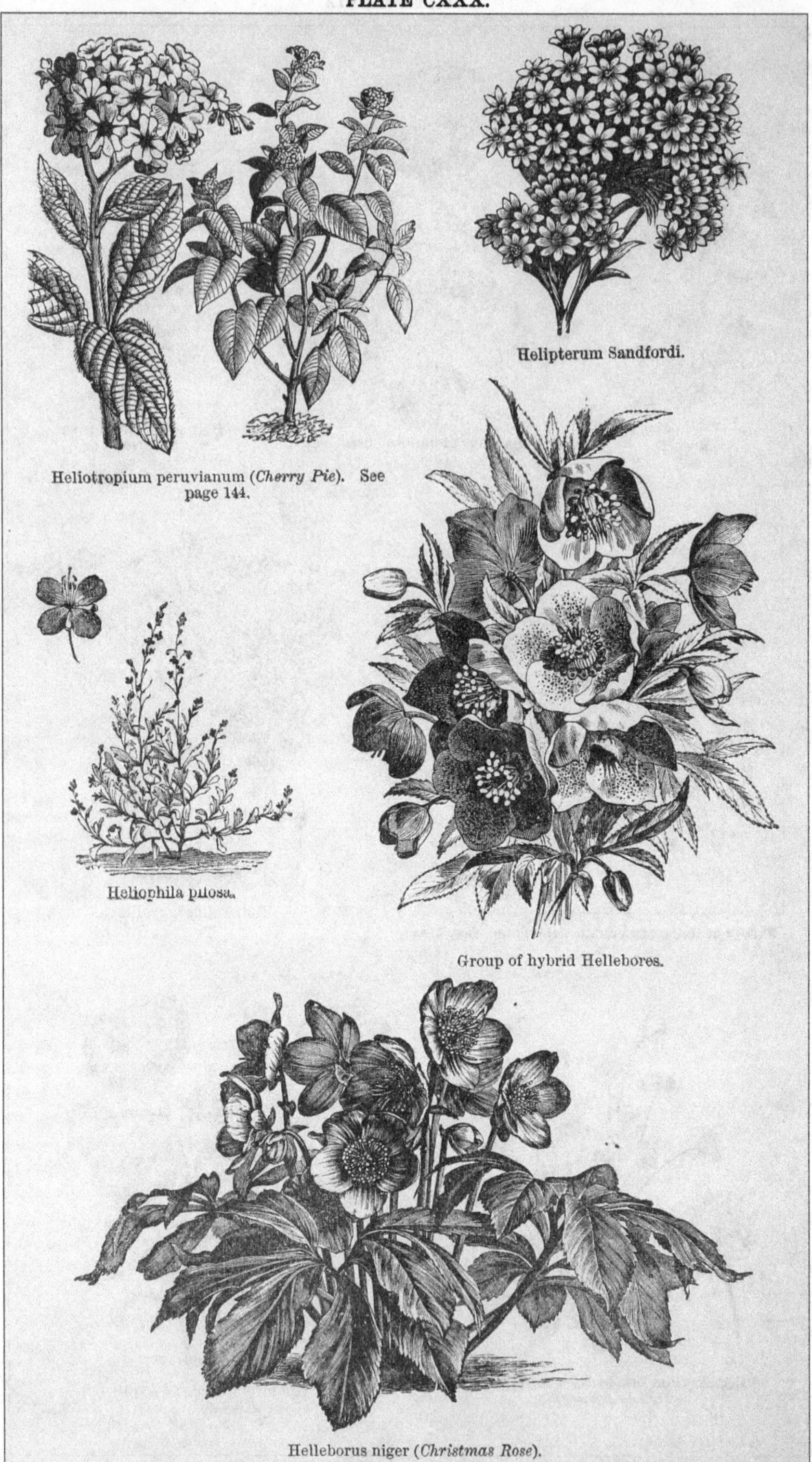

Heliotropium peruvianum (*Cherry Pie*). See page 144.

Helipterum Sandfordi.

Heliophila pilosa.

Group of hybrid Hellebores.

Helleborus niger (*Christmas Rose*).

flower garden plants, either grown for their own sakes as simple masses or associated with tall plants which grow above them.

Helipterum Sanfordi.—A pretty, bright yellow, half-hardy annual Composite of dwarf, branched habit of growth. Its merits have not been considered remarkable, though it has been well introduced. The short life of the spring-raised everlasting annuals leads to a poor result.

Helleborus (*Christmas Rose*).—One of the most valuable classes of hardy perennials we have, as they flower in the open air at a season when there is but little else in bloom. They appear in continuous succession from October till April, beginning with the Christmas Rose (H. niger) and ending with the handsome crimson-flowered kinds. The old white Christmas Rose is well known and admired, but the handsome kinds with coloured flowers have, hitherto, not been much known, and recently there have appeared some really beautiful hybrid varieties that add a deal of beauty to our winter and spring garden, for not only do their flowers withstand the rigours of winter, but the verdure and vigorous growth of their leaves distinguish these plants throughout the whole year.

The Hellebores, besides being excellent border flowers, are specially well adapted for naturalising. There are a few kinds—those with inconspicuous flowers, but handsome foliage—whose only place is in the wild garden. Of such is the native H. fœtidus, H. lividus, viridus, and H. Bocconi, which have elegant foliage when well developed in a shady place in rich soil, such as is usually found in woods. The Hellebores may be conveniently classed into three groups, according to the colour of the flowers—those with white flowers, those with red, and those with green, which latter will scarcely get much place in the garden. The white-flowered group is the most important, as it contains the beautiful old Christmas Rose.

H. niger is a well-known kind, scarcely needing description, as it may be at once recognised by its pale green, leathery, and quite smooth leaves, which are divided into seven or nine segments, 3 in. to 6 in. long and 1 in. to 2 in. broad. The blossoms, which are usually borne singly on stems 6 in. long, are about 3 in. across, and vary from a waxy-white to a delicate blush tint. The variety minor is smaller in every part than the type, and is also known under the name of H. angustifolius. H. altifolius, though considered by some to be a variety of H. niger, is a distinct kind, and much larger in all its parts than H. niger. It has leaf-stalks over 1 ft. in length, and blossoms from 3 in. to 5 in. across; the latter are borne on branching stems bearing from two to seven flowers, which have a stronger tendency to assume a rosy hue than the ordinary kind. Another characteristic is that the leaf and flower-stems are beautifully mottled with purple and green, whilst in H. niger they are of a pale green colour. It also flowers much earlier—in some seasons in the beginning of October. It has been known a long time under the names of H. niger var. major, maximus, giganteus, and grandiflorus.

Other white-flowered kinds are H. olympicus—a tall, slender species with cup-shaped blossoms that vary from pure white to greenish-white—also known as H. abchasicus albus, and kamtschatkensis albus. Flowers from February to March. H. guttatus is like the preceding, but has the inside of the blossoms spotted with purple. There are several forms of it; in some the markings assume the form of small dots, in others of thin streaks. This is one of the parents of the beautiful hybrids that have lately been obtained, the best of which are Herr Leichtlin, F. C. Heinemann, and C. Benary.

The finest of the red or crimson kinds are H. colchicus, one of the finest of all the species. It attains a larger size than any other, and may be readily recognised by its thick, dark green leaves, with five to seven broad and coarsely-toothed divisions, the veins of which are raised on the under sides, and are of a dark purple colour when in a young state. The blossoms, borne on forked stems that rise considerably above the foliage, are of a very dark purple colour. Under good cultivation the leaves attain the length of 1½ ft. and 2 ft., forming fine specimens, and flowers are produced from the end of January to the end of March. A fine hybrid has been obtained from it by crossing with H. guttatus, the result being a form with large spreading flowers of a lighter hue than in H. colchicus, and profusely marked with dark carmine streaks. Another hybrid between this and H. altifolius resulted in a form with larger flowers of a lighter purple. H. atrorubens has leaves much thinner in texture and smaller flowers than H. colchicus. The colour is dull purple on the outside and greenish-purple within. It is a native of Hungary, and is a common object of culture in gardens, but is often confused with H. abchasicus, a more slender and taller plant, the flower-stems of which are longer, and the blossoms nodding and smaller. H. abchasicus is a much superior kind to atrorubens, the colour of the blossoms—a deep ruby-crimson—making them very attractive. Other fine varieties of the red-flowered group are Gretchen Heinemann, James Atkins, and Apotheker Bogren, all well worthy of culture. Other reddish-flowered kinds, such as H. purpurascens and H. cupreus, are not worth growing.

All the kinds will thrive in ordinary garden soil, but for the choicer kinds a prepared soil is preferable. This should consist of equal parts of good fibry loam and well-decomposed manure, half part fibry peat, and half part coarse sand. Thorough drainage should always be given, as stagnant moisture is very injurious to them. A moist and sheltered situation, where they will obtain partial shade, such as afforded at the margins of high shrubberies, &c., is best suited to them, but care should be taken to keep the roots of the shrubs from exhausting the border in which the Hellebores are planted. In the flowering season a thin mulching of moss or like material should be placed on the surface of the soil round the plants, as this prevents the blossoms from being bespattered by heavy rains, &c. Anyone beginning to grow these useful plants should give the soil a good preparation, and if well trenched and manured they would not require replanting for at least seven years; but a top-dressing of well-decayed manure and a little liquid manure might be given during the growing season when the plants are making their foliage, as upon the size and substance of the leaves will depend the size of the flowers. The common white Christmas Rose is a favourite pot plant, and if required for this purpose care should be taken to protect the foliage from injury; when the blooming season is over they should be protected by a frame until genial weather permits them to be plunged in the open air. Hardy subjects like the Christmas Rose frequently suffer when removed from under glass, for although hardy enough to withstand our severest winters when continuously exposed, their growth, made under more exciting circumstances, will not withstand any sudden variations of temperature. For this reason it is advisable to keep them in as cool a position as possible when in flower, so that the growth of young foliage may not be excited before its natural season.

Propagation may be effected by division or by seeds, which, in favourable seasons, are produced plentifully; as soon as thoroughly ripened they should be sown in pans under glass, for they soon lose their vitality. As soon as the seedlings are large enough they should be pricked off thickly into a shady border, in a light, rich soil; the second year they should be transplanted to their permanent place, and in the third season the majority will produce blossoms. In dividing the clumps operated on must be well-established plants, that have formed root-stocks large enough to be worth cutting up. The divided plants, if placed in a bed of good, light soil, will, if undisturbed, be good flowering plants in another couple of years, but four years are required to bring a Christmas Rose to perfection. By July the Hellebore is in its strongest vigour, and during that month the operation of lifting and dividing the plants should be carried out.

Helonias bullata.—A distinct and handsome bog perennial, growing 12 in. to 16 in. high and having handsome purplish-rose flowers arranged in an oval spike. It is suitable for the artificial bog, or for moist ground near a rivulet. In fine sandy and very moist soils it thrives well as a border plant. N. America. *Syn.*, H. latifolia.

Hemerocallis *(Day Lilies)*.—Valuable border plants, having elegant grassy foliage and handsome flowers varying in hue from orange to clear yellow They are all perfectly hardy and thrive in almost any kind of soil, but prefer one that is rich and moist and in an open situation. The more robust growers, such as H. fulva and disticha, are well able to take care of themselves, and flourish in the wild garden, where their distinct aspect has a good effect. By the margins of ponds, lakes, or rivers the stronger kinds are quite at home. The first to flower is H. Dumortieri, the flowers of which are somewhat smaller than those of most of the other kinds, and are orange coloured, streaked on the outside with red. Succeeding it is the clear yellow-flowered H. flava, a showy and attractive plant and among the finest of hardy flowers. Resembling the latter in colour, and expanding its flowers a little later, is H. graminea, which has the narrowest foliage of all. The pretty yellow flowers of this kind are excellent to cut, as they last longer in that state than the others. The last to flower is the coppery-red-flowered H. fulva, with its several varieties, the principal of which is disticha, which produces flowers of the same colour as the original, but has more on a stem. Then there is the double-flowered form, which is considered the best for border purposes, as the flowers continue in perfection a much longer time. The variegated-leaved variety, known as H. Kwanso variegata, is probably a form of H. fulva, and a highly ornamental plant it is when fully developed and the variegation well marked. Other names exist in collections, and among them are H. Sieboldi, rutilans, and Middendorfiana, but they are either identical with some of those mentioned above or insufficiently known.

Hemiphragma heterophylla.—A dwarf trailing plant of the Figwort family, bearing inconspicuous flowers, succeeded by bright red berries about the size of small Peas, on slender creeping stems. It is rather tender and requires a sheltered and well-drained spot in the rock garden. Himalayas.

Hemp Agrimony *(Eupatorium cannabinum)*.

Hepatica.—See Anemone.

Heracleum *(Giant Parsnip)*.—Umbelliferous perennial plants, mostly of gigantic growth, having huge spreading leaves and tall flower-stems, with umbelled clusters, a foot or more across, of small white flowers. Though well-developed plants of the large kinds have a fine effect when isolated in a not too obstructive position in the garden, they are generally suitable only for the rougher parts of pleasure grounds, by the banks of rivers or lakes, or, in fact, anywhere where they can grow freely and well and show their stately growth to advantage. The finest species are H. giganteum, lanatum, sibiricum, eminens, Wilhelmsi, and pubescens, all of which attain when in flower a height of from 5 ft. to 10 ft. All are increased by seed, plentifully produced. A plant, or a group or colony, on an island or rough spot, is effective, but this will be quite enough. If allowed to seed and increase, the Giant Cow Parsnip becomes a nuisance and an eyesore.

PLATE CXXXI.

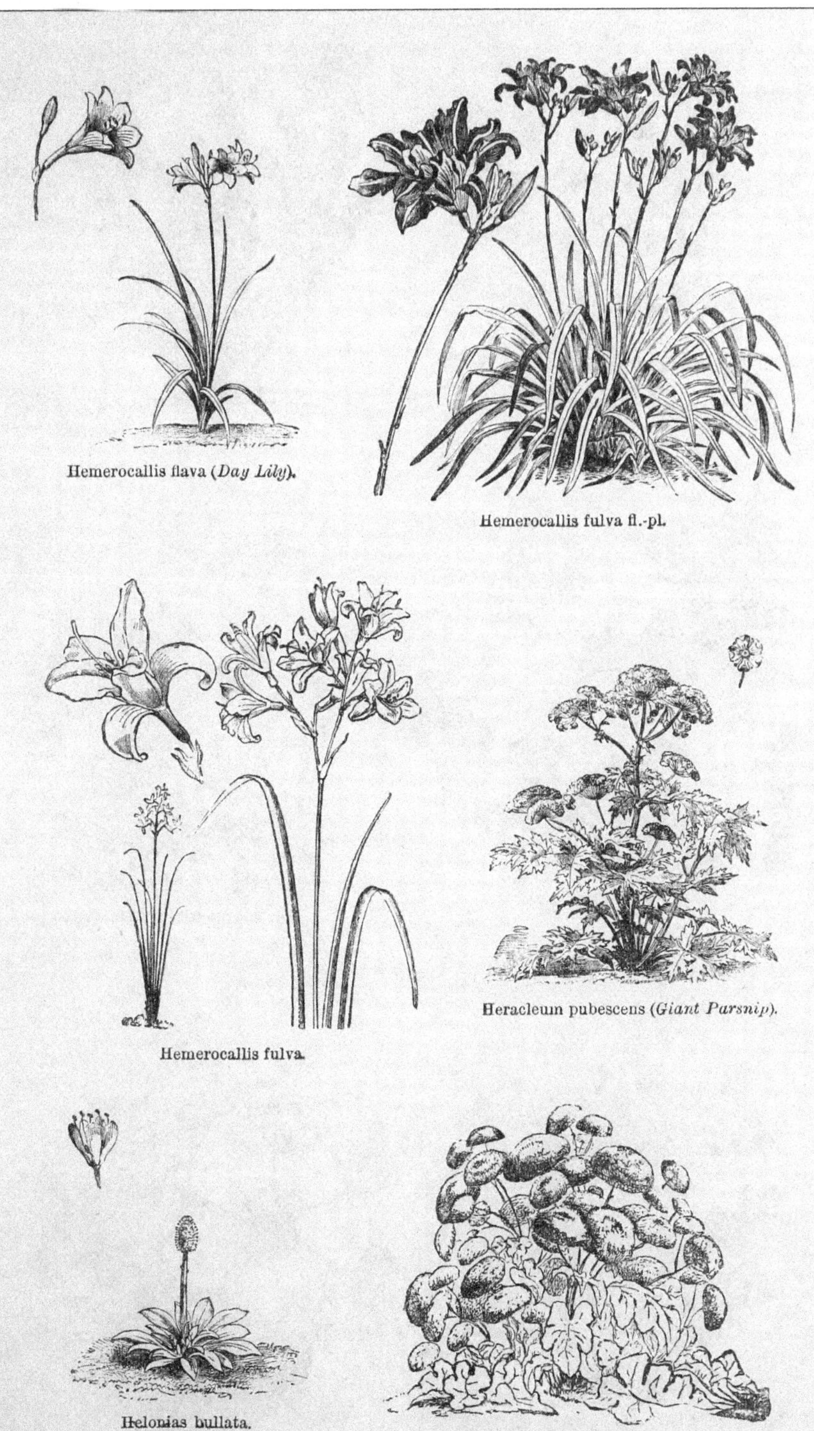

Hemerocallis flava (*Day Lily*).

Hemerocallis fulva fl.-pl.

Hemerocallis fulva.

Heracleum pubescens (*Giant Parsnip*).

Helonias bullata.

Heracleum absinthiifolium.

[*To face page* **146.**

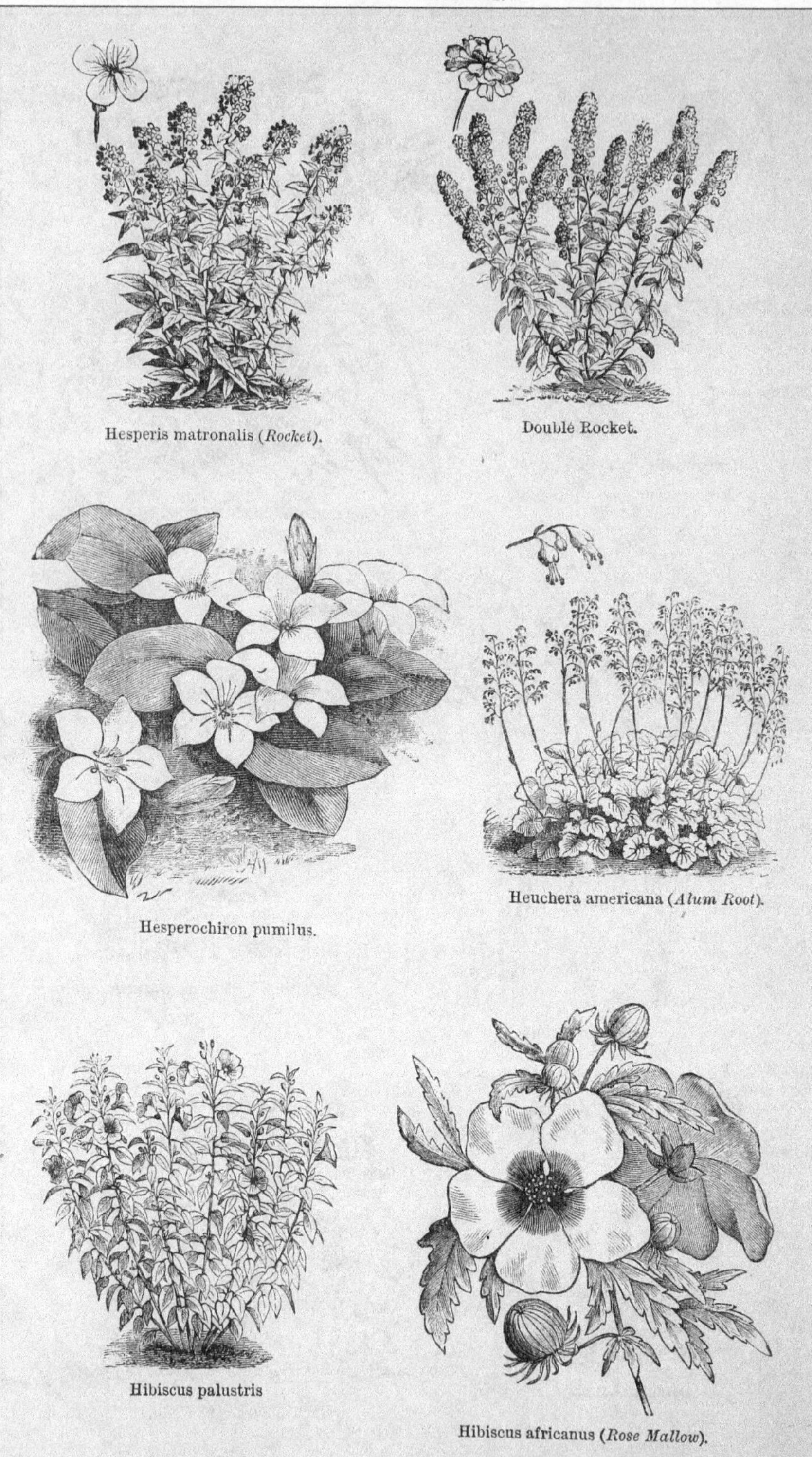

Hesperis matronalis (*Rocket*).

Double Rocket.

Hesperochiron pumilus.

Heuchera americana (*Alum Root*).

Hibiscus palustris

Hibiscus africanus (*Rose Mallow*).

[*To face page* 147.

Herb Christopher *(Actæa)*.

Herb Paris *(Paris quadrifolia)*.

Herminium Monorchis.—An inconspicuous terrestrial Orchid.

Herniaria.—Dwarf trailing perennial plants, forming a dense turfy mass that remains green throughout the year. There are two or three species, but the most important is H. glabra, which of late years has been largely used in geometrical gardening on account of its dwarf, compact growth. It rarely requires clipping and is always of a deep green, even in a hot and dry season. They grow in any soil.

Hesperaloe yuccæfolia.—A North American Liliaceous plant having Yucca-like leaves and a flower-spike, upwards of 4 ft. in height, furnished thickly on the upper half with orange-red flowers nearly 1 in. in length. It has not been tested as regards hardiness, and but very little is known of its requirements under cultivation.

Hesperis matronalis *(Rocket)*.—A popular old garden plant, and among the most desirable of hardy flowers. It bears showy, varied, and fragrant flower-spikes. The original single-flowered kind grows from 1 ft. to 3 ft. high, and has pinkish flowers, but the double kinds are much more valued. There are two distinct forms of the double white as well as of the double purple Rocket in cultivation. The former is a tall-growing white, which turns to a pale flesh colour with age; the other is the old white variety, of dwarfer growth, and with smaller and more compact flowers. It can sometimes be met with in the north, but it is little known in the south, where it does not flourish so well as does the common variety. There is the old purple double Rocket and a free-growing dwarf form known as Compactness, which has also larger and darker flowers. Rockets require care in cultivating and will soon be lost if left to themselves. They should be divided every second year at least and transplanted. They seem to get tired of the soil and to require change more than most perennials. They strike very freely in the open ground if the young shoots be formed into cuttings when they are about 3 in. long, and the spikes of bloom on the remaining stems are all the finer when some of the others have been removed. The cuttings do best when shaded from the sun for about three weeks with a few Laurel branches rather than when covered with a pot or box, as has been advised. They like a rich, rather moist soil and are the better for repeated applications of liquid manure in case the soil is not as deep and good as it should be. Double Rockets really belong to a most important set of garden plants, requiring annual and biennial attention, and therefore cannot well be used in groups of true perennials. It is always worth while having a bed of them in the reserve garden in case the plants should be lost or neglected in the borders. In the place we have seen them best grown there was a yearly transfer of plants from the reserve garden to the mixed border, where the groups looked very well. The single Rocket is easily naturalised, and is a showy plant in woods or shrubberies.

H. tristis *(Night-scented Stock)*.—A quaint plant with dull-coloured flowers, sweet-scented at night. It is rather tender, and requires a light, warm soil in a sheltered position.

Hesperochiron pumilus.—A pretty Californian rock plant. It is stemless, dwarf in growth, with leaves borne on slender stalks, forming a rosulate tuft. The flowers are bell-shaped, ½ in. across, and white, varying to a purplish tinge. It grows in marshy ground, and in damp places in the Rocky Mountains and Northern Utah, and is apparently quite hardy, as it thrives in ordinary soil in well-drained parts of the rock garden. H. californicus is a species of somewhat the same type.

Hesperocordon lacteum.—A Californian bulbous plant, now placed with the Brodiæas. H. lilacinum is a variety of H. lacteum

Heuchera *(Alum Root)*.—Perennial plants from N. America, all bearing a strong resemblance to each other. They form compact rounded tufts of foliage which in some cases are beautiful, being of a satiny deep green colour, but veined and washed with reddish brown. They all have white flowers on slender, erect spikes. The most desirable species are H. americana, Richardsoni, and glabra. They grow well in ordinary soil, best, perhaps, in peaty soil in an exposed place, and are well worthy of a good space for their foliage alone. May be used as tufts in borders, as "carpet" plants under bolder and larger hardy flowers, or, perhaps, best of all, as a spreading colony filling completely a space between shrubs. Saxifragaceæ.

Hibiscus *(Rose Mallow)*.—A genus of shrubby and herbaceous perennial and annual plants. They are numerous in hothouses, but few are suited for the flower garden. The splendid hardy, herbaceous Rose Mallows of the woods and swamps of N. America are, as a rule, large and handsome. They will grow with us, but probably our climate is not warm enough for them, though it would be very desirable to try tufts of them in warm sunny places in the southern parts of England, in deep, rich, moist soil. They have splendid crimson or rosy flowers, as large as saucers, and grow from 4 ft. to 7 ft. high. The finest are H. Moscheutos, H. palustris, H. grandiflorus, and H. coccineus. They seldom bloom in the open air in England, as they flower so late in the season. There are two or three annual kinds, the finest being H. Manihot, which forms handsome pyramidal specimens from 4 ft. to 6 ft. high, the flowers being 3 in. or 4 in. across, and in colour pale yellow with a dark centre. It should be treated as a half-hardy annual, should be sown in heat in February, and planted out in May in good, deep soil. H. africanus is a hardy annual with showy pale yellow flowers that only open in fine weather. In light soil it usually sows itself. H. Trionum is similar, but inferior to the preceding.

Hieracium (*Hawkweed*).—A very extensive genus of Composites, consisting chiefly of perennial herbs with yellow flowers. The kind best worth a place in the garden is H. aurantiacum, which is distinct among plants of a similar character on account of its colour—a deep orange-red. It is a vigorous grower, attains a height of 1 ft. or 1½ ft., and soon forms a spreading mass. W. Europe. Some of the yellow alpine and other kinds are valuable in botanical collections, and some of them are beautiful, but the prevalence of yellow flowers of the same type makes them less important.

Hierochloa.—A genus of Grasses of no ornamental value.

Hippocrepis comosa (*Horse-shoe Vetch*).—A small, prostrate, British evergreen herb about 6 in. high. Its flowers, 5 to 8 of which are borne together in a crown, are yellow, and resemble those of the Common Bird's-foot Trefoil, than which, however, they are paler and rather smaller. H. comosa grows freely in any exposed part of the rock garden and borders in any soil. Seed and division.

Hippuris (*Mare's-tail*).—A British water weed, worth growing with the Equisetums and the like by the sides of ponds where the growth near is not too rank.

Holcus.—British Grasses, the only value of which is that one, H. mollis, affords a pretty variegated-leaved form that is used in the flower garden for lines and edgings, as it retains its markings very well.

Holly Fern (*Aspidium Lonchitis*).

Hollyhock (*Althæa*).

Homogyne alpina.—A dwarf perennial Composite allied to the Coltsfoot. Only suitable for botanical collections.

Honesty (*Lunaria biennis*).

Hoop Petticoat (*Narcissus Bulbocodium*).

Hordeum.—Grasses, of which the Barley is the most familiar type. Of no ornamental value except H. jubatum (Squirrel-tail Grass), which has long feathery spikes, grows in any soil in open places, and is easily raised as an annual. It is, indeed, one of the most distinct of the dwarfer ornamental Grasses. Sow in autumn or in spring.

Horkelia.—N. American plants of the Rose family of a weedy nature.

Horminum pyrenaicum.—A Pyrenean plant, forming dense tufts of foliage and having purplish-blue flowers, in spikes about 9 in. high, which appear in July or August. It is quite hardy and of easy culture, but is not a plant of much character from a garden point of view, though botanically interesting. Labiatæ.

Horned Poppy (*Glaucium luteum*).

Hoteia japonica.—A fine herbaceous plant of tufted habit, growing from 1 ft. to 16 in. high. Its flowers, which appear early in summer, are silvery white and in a panicled cluster. It is an excellent plant for a shady border in a rich soil. Strong clumps planted in autumn will flower the following spring. Where forced plants are to spare they may be planted out when they have done blooming, but will not make much show the following season. Much and well used indoors, it is very seldom seen in a good state in the open garden. This is partly owing to the fact that it does badly in heavy and poor soils. Where it thrives and flowers well it would be a graceful aid in the varied flower garden. Increased by division in autumn. Japan. *Syns.*, Spiræa japonica, Astilbe barbata.

Hottonia palustris (*Water Violet*).—A pretty British water plant, which, however, thrives better on soft mud banks than submerged. The deeply-cut leaves form quite a deep green and dwarf tuft over the mud, and from this arise stems bearing at intervals whorls of handsome pale lilac or pink flowers. As water and bog may be associated with rockwork, this plant may with advantage be grown either in the water or on a bank of wet soil at its margin. It grows from 9 in. to 2 ft. high, flowers in early summer, and is found in abundance in many parts of England.

House Leek (*Sempervivum tectorum*).

Houstonia cœrulea (*Bluets*). — A diminutive and very pretty North American plant. It forms small, dense, cushion-like tufts, and bears, from late spring to autumn, crowds of tiny slender stems, about 3 in. high, the flowers being pale blue, changing to white. There is also a white variety. It succeeds best in peaty or sandy soil, in sheltered, shady nooks on well-drained parts of the rock garden. As it sometimes perishes in winter, it is advisable to keep reserve plants in pots. Propagated by careful division in spring, or by seed. H. serpyllifolia is not so good.

Houttynia cordata. — A curious Japanese plant, and usually grown as an aquatic under glass, but we have noticed it succeed as a border plant in the College Gardens at Dublin, though dwarf and altered in character. In that state it had a peculiarly offensive odour.

Humea elegans. — A very graceful half-hardy biennial, 3 ft. to 8 ft. high, having large leaves with a strong balsamic odour, and forming, when in flower, a most elegant pyramid of feathery blossoms of a reddish-brown colour. It is highly ornamental as a back line to a long border, as a single specimen to let into the lawn, for the centre of a bed or vase, or grouped in masses with other elegant foliage plants. Excellent effects may be obtained by combining it in masses or groups with other good plants. For cutting, the light, feathery sprays are useful. The proper time to sow seed is in July or August, as plants of it do not bloom the first year, and, if raised before they get too large to winter conveniently, often become leafless

PLATE CXXXIII.

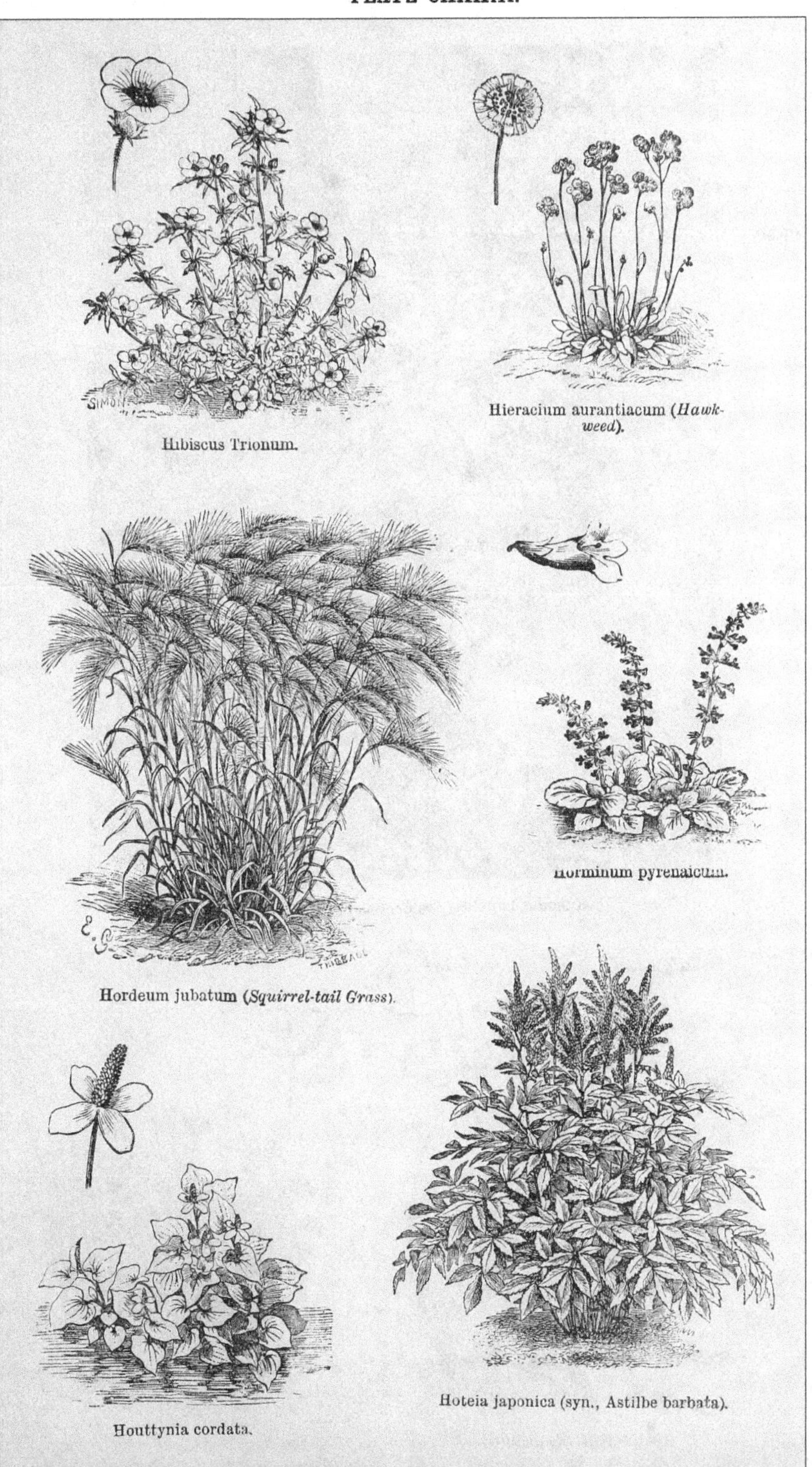

Hibiscus Trionum.

Hieracium aurantiacum (*Hawk-weed*).

Hordeum jubatum (*Squirrel-tail Grass*).

Horminum pyrenaicum.

Houttynia cordata.

Hoteia japonica (syn., Astilbe barbata).

[*To face page* **148.**

Humulus Lupulus (*The Common Hop*) on a tree.

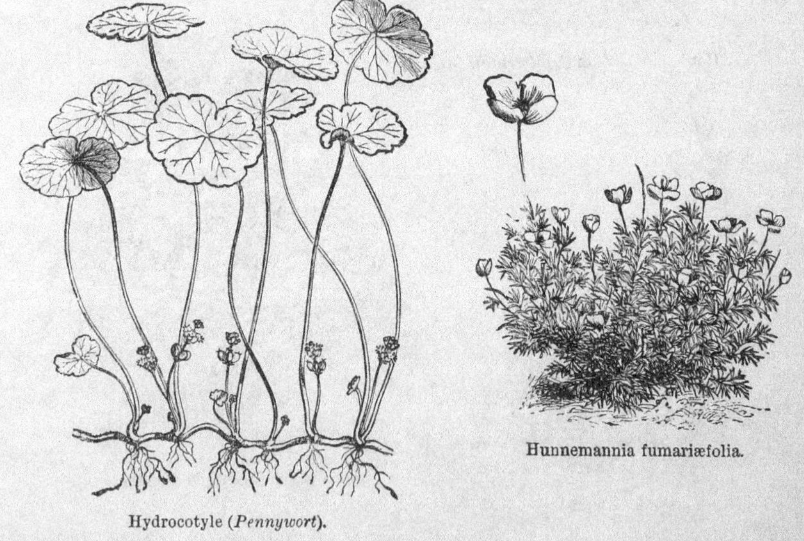

Hunnemannia fumariæfolia.

Hydrocotyle (*Pennywort*).

[*To face page* 149.

below, which nakedness of stem detracts from their beauty. To prevent this partial defoliation, they should be kept well fed during the winter with weak liquid manure, and receive a shift into larger pots early in spring; rich soil should be used for the purpose, as Humeas are gross feeders, and can only be kept in a healthy condition by good feeding. When planting them out in beds, which may be done by the first week in June, it is a good plan to put a spadeful of rotten manure under each, and mix it up with the soil. As the plants, when of large size, hold a good deal of wind, they must be securely staked, to prevent them from becoming damaged. Compositæ. Australia.

Humulus Lupulus *(Common Hop)*.—This well-known vigorous twining perennial is admirable for covering bowers, especially where vegetation that disappears in winter is desired; also when allowed to run wild among shrubs or hedgerows in almost any soil. A slender plant up an Apple or other fruit tree, near the mixed border, looks well. Division.

Hunnemannia fumariæfolia.—An erect-growing perennial, from 2 ft. to 3 ft. high, with glaucous foliage, resembling that of some of the Fumitories. Its flowers are large and showy, of a rich orange colour, in form like Eschscholtzia californica. They continue in perfection for a long time. Being a native of Mexico, it is rather tender, and not a satisfactory plant for open-air culture. Papaveraceæ.

Hutchinsia alpina.—A neat little plant, with shining leaves and pure white flowers, produced abundantly in clusters about 1 in. high. Quite free in sandy soil, and easily increased by division or by seeds. Planted in an open spot, either on rockwork or in good free border soil, it becomes a compact mass of white flowers. Its proper home is on the rock garden, though where borders of dwarf and choice hardy plants exist, it may be grown with success. Central and S. Europe. Cruciferæ. *Syn.*, Smelowskia alpina.

Hyacinthus *(Hyacinth)*.—The familiar garden Hyacinth is not generally included among hardy plants, though it is perfectly so, and, when treated in the manner it should be, is a most important plant. The parent of all the varieties is H. orientalis; this is as hardy as a Daffodil, and its varieties are scarcely less so. Hyacinths in the open air are generally the refuse, as it were, of preceding years' forced bulbs, and even these create a good display if planted in suitable positions. But to have a fine bloom of Hyacinths in the open air it is essential that some care should be taken in planting, that the bulbs should be good and sound, and due regard paid to a selection of colours, as tints massed by themselves are far more effective than a confusion of various colours. Now that bulbs in quantity may be obtained cheaply there is no difficulty in the matter. The Hyacinth will grow well in any good garden soil, but a light rich soil will suit it best, and the position of the bed should be effectually drained, for though the plant loves moisture, it cannot endure to stand in a bog during the winter. It is advisable to plant early and deep. If a rich effect is required, the bulbs should be 6 in. apart, but a good effect may be produced by planting 9 in. or even more apart. The time of blooming may be to some extent influenced by the time and manner of planting, but no strict rules can be given to suit particular cases. Late planting and deep planting both tend to defer the time of blooming, but there will be no great difference in any case, and as a rule the late bloom is to be preferred, because less liable to injury from frost. The shallowest planting should ensure a depth of 3 in. of earth above the crown of the bulb, but generally speaking they will flower better and a few days later, and come up better bulbs after flowering, if covered with fully 6 in. of earth over the crowns. The Hyacinth is so hardy that protection need not be thought of except in peculiar cases of unusual exposure, or on the occurrence of an excessively low temperature when they are growing freely. In any case there is no protection so effectual as dry litter, but a thin coat of half-rotten manure spread over the bed is to be preferred in the event of danger being apprehended at any time before the growth has fairly pushed through. The bulbs need no more attention until the flower-stems are much advanced, unless very severe weather intervenes, when they should have a mat or some oiled calico thrown over them. Waterproof calico is also useful in very wet weather, as excess of water, especially when iced by February frosts and March winds, is by no means good for Hyacinths, which will thrive all the better for being sheltered by a waterproof covering. Hyacinths in the open air hardly ever require artificial watering, the natural moisture of the soil and the strength of the manure mixed with it being sufficient to carry them through. When grown in beds, Hyacinths do not require sticks or ties; all they need is to be planted properly and they will take care of themselves until the cultivator lifts them to appropriate the ground to other plants. After blooming, the bulbs, if expected to flower again, must be left undisturbed until the leaves wither or die of their own accord. The bulbs should then be taken up, dried in a stack for a week or two, and finally placed in the sun for a few hours, the dry leaves being pulled off. Offsets should be removed likewise from the bulbs, and stored in dry sand or earth till planting time comes again. Some lay it down as a rule that the bulbs should be taken up every year, but we have seen handsome beds that were not disturbed for several years. Offsets, if carefully cultivated in a rich piece of light soil for two or three years, will produce a good many flowering bulbs, but, as a rule, imported bulbs are strongest. However carefully they may be cultivated in England, they seldom flower again so well as in the first season, but it is a great mistake to throw them away as worthless, as many people do. Selections for bedding in distinct colours of red, yellow, white, blue, or mixed are to be bought cheaply.

H. amethystinus is a beautiful plant with deep sky-blue bells, gracefully disposed on stems from 8 in. to 1 ft. high, taller when the plant is grown in very light rich soil, in a not too cold or exposed position. It is quite hardy, a native of the Pyrenees and Southern Europe, and flowers in early summer. In stature and general appearance it is somewhat like a graceful Scilla, but is at once distinguished, as its pretty bells are not divided into segments as they are in Scilla. It is worthy of association with the choicest hardy bulbs and inhabitants of the rock garden. The white variety (albus) is a lovely plant worth growing in every select collection.

H. candicans *(Galtonia)*.

Hydrocharis Morsus-ranæ *(Frogbit)*.—A pretty native water plant, having floating leaves and attractive yellow flowers, and well worth introducing in artificial water. It may often be gathered from neighbouring ponds or streams in spring, when the plants float again after being submerged in winter.

Hydrocotyle *(Pennywort)*. — Small creeping plants, usually with round leaves and inconspicuous flowers. There are several kinds grown, their only use being as a surface growth to the artificial bog. The most desirable are H. moschata and microphylla, two New Zealand species, and nitidula, though all of these are somewhat tender. The common H. vulgaris is rather too rank a grower. Umbelliferæ.

Hymenophyllum *(Filmy Fern)*.—Although these Filmy Ferns are hardy and very beautiful, yet the essential conditions for their successful culture are such that in a general sense they cannot be used with effect in the open air. Still, as two grow abundantly in certain hilly districts, in moist, shady, or rocky situations, there is no reason why they should not be inmates of the garden, at least in some places in the west or north, or in hilly districts.

Hymenoxys californica.—A hardy annual Composite, growing about 6 in. high, with a compact tufted habit of growth, and bearing in summer a profusion of bright yellow, Daisy-like blossoms. Sow in autumn in light dry soil for early flowering, or in spring for later bloom. (= Shortia.)

Hyoscyamus *(Henbane)*.—Plants of the Nightshade family, of no garden value. (H. physaloides = Nicandra.)

Hypericum *(St. John's Wort)*.—An extensive genus, consisting for the most part of shrubs and under-shrubs, but including a few herbaceous perennials and annuals. The Rose of Sharon (H. calycinum), so valuable as an undergrowth, is probably the most familiar of the genus, but as it and many other cultivated species are shrubby, they are not treated of here. Some of the herbaceous perennials are valuable border and rock garden plants, and the best of these is H. olympicum, one of the largest flowered kinds, though the plant does not grow more than 1 ft. high. It is known by its very glaucous foliage and erect single stems, with terminal flowers about 2 in. across, and of a bright yellow colour. It forms handsome specimens that flower early in the season, and its value as a choice border plant can scarcely be over-rated. It may be propagated easily by cuttings, which should be put in when the shoots are fully ripened, as the young plants will become well established before winter. H. Elodes is a pretty native for the banks of pools and lakes. H. nummularium and humifusum, both dwarf trailers, are also desirable for the rock garden. Among the shrubby species there are several that, owing to their dwarf, compact growth, are well suited for the rock garden. Of these, the best are H. ægyptiacum, balearicum, empetrifolium, Coris, patulum, uralum, and oblongifolium; the three last are of larger growth than the others, but as they are drooping in habit they have a good effect among the boulders of a large rock garden, or on banks near it.

Hypochæris pinnatifida.—A weedy Composite.

Hypolepis millefolium.—A very elegant New Zealand Fern, having a stout and wide-spreading rhizome, from which arise erect fronds, from 1 ft. to 1½ ft., very finely cut, and of a light green hue. As to its hardiness, there can be no doubt, as it has flourished for two or three years in a Surrey garden on an open rockery. It requires a sheltered nook, peaty soil, and partial shade.

Hypoxis.—Low-growing plants, mostly natives of South Africa. They have grassy foliage and yellow flowers, are tender, but are sometimes planted out in summer, in light, sandy soil, in warm borders.

Ianthe bugulifolia. —An interesting plant resembling a Mullein, having a rosette of root leaves, from which springs an erect flower-spike, 8 in. to 12 in. high, thickly set with curiously coloured blossoms of chocolate-brown and yellow. It is hardy, but not a true perennial, therefore it is advisable to raise seedlings yearly in frames, and afterwards transfer to the border or rock garden Turkey. Scrophulariaceæ. (= Celsia.)

Iberidella rotundifolia —A dwarf evergreen herb, with thick, leathery leaves of a glaucous olive-green. It grows 3 in. to 6 in. high, and flowers in early summer. The blossoms are rosy-lilac, sweet scented, and freely produced in terminal racemes. Suitable for the rock garden, succeeding best in rich, gritty loam in deep fissures; being taprooted, it will not readily increase by division, but it is easily raised by seed sown in small pots in a cold frame. This elegant little plant is a native of the European Alps. Cruciferæ. (= Thlaspi rotundifolium.)

PLATE CXXXV.

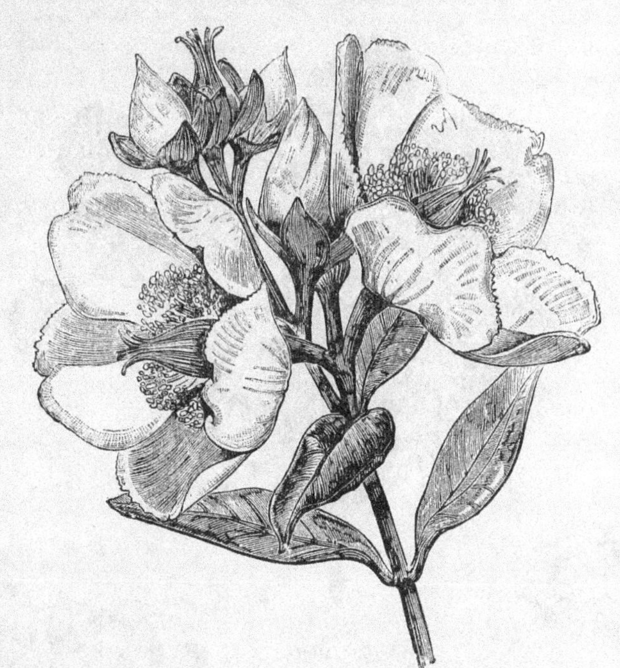
Hypericum Hookerianum (*St. John's Wort*). A low shrub.

Hypericum ægyptiacum.

Hypericum calycinum (*Rose of Sharon*).

Hypericum nepalense. A dwarf shrub.

[*To face page* **150.**

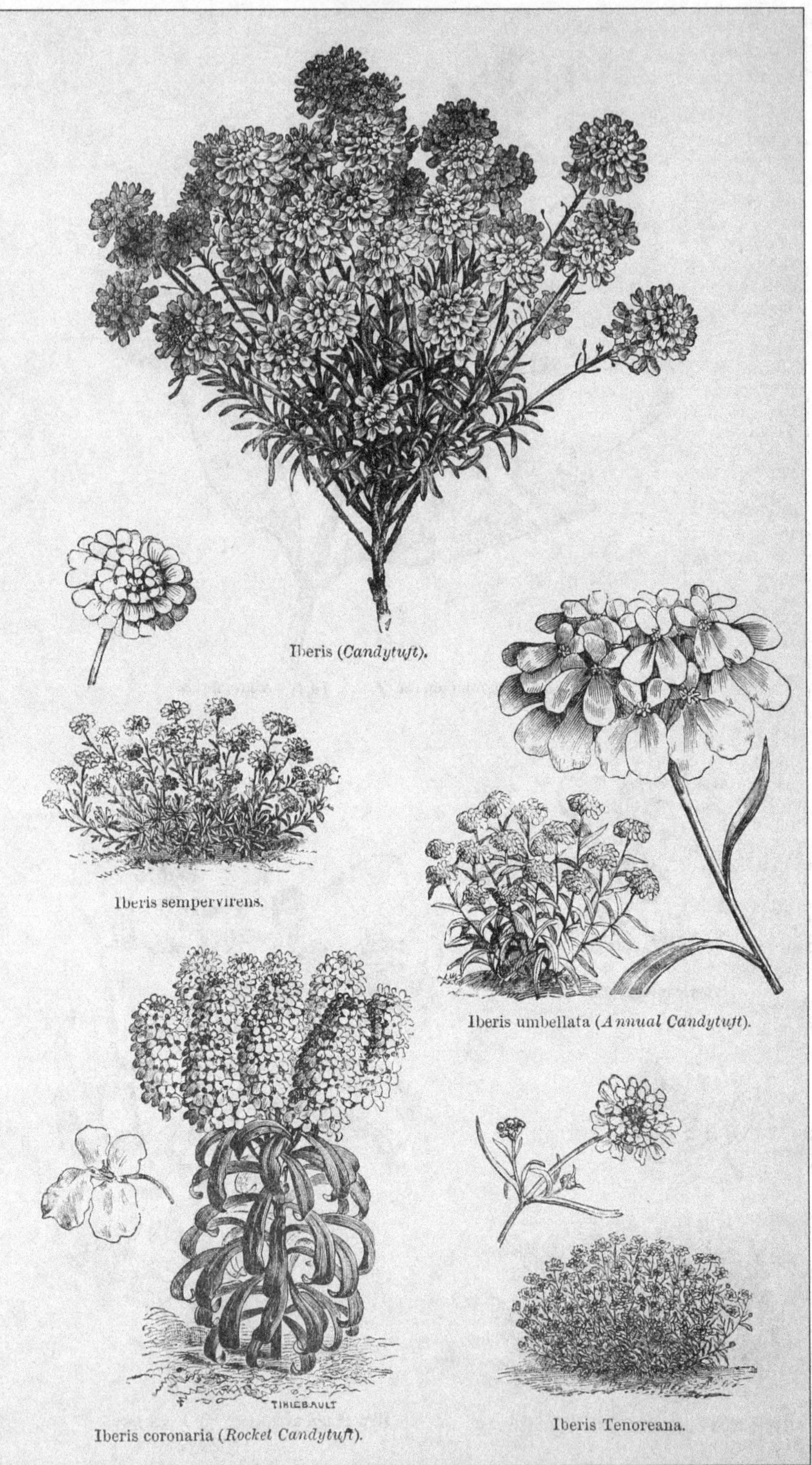

Iberis (*Candytuft*).

Iberis sempervirens.

Iberis umbellata (*Annual Candytuft*).

Iberis coronaria (*Rocket Candytuft*).

Iberis Tenoreana.

[*To face page* **151.**

Iberis *(Candytuft)*.—A well-known and valuable genus of hardy perennial and annual plants. All the perennials are somewhat shrubby in growth and have evergreen foliage, so that they are valuable winter plants. The principal kinds are those treated of.

I. correæfolia is readily known from any other cultivated species by its rather large leaves, its compact heads of large white flowers, and by flowering later than other common white kinds. Both the individual flowers and the corymb are larger and denser than in the other species. It is an invaluable hardy plant, and particularly useful in consequence of coming into full beauty about the end of May or beginning of June, when the other kinds are fading away. It is indispensable for rockwork, the mixed border, and the spring garden, and may also be naturalised with good effect in bare, rocky places. It is particularly well suited for planting on the margins of choice shrubberies, bringing them neatly down to the grass line, and it may also be used as an edging to beds. Supposed to be a hybrid. Readily increased by cuttings, but it appears to lose its true typical character when raised from seed.

I. corifolia.—A very dwarf kind, attaining a height of only 3 in. or 4 in. when in flower, and perfectly covered with small white blooms early in May. Few alpine plants are more worthy of general culture either on rockwork or in the mixed border—for the front rank of which it is admirably suited. It is probably a small variety of I. sempervirens, but is distinct and remarkably true and constant to its character. A native of Sicily and probably of other parts of Southern Europe, easily propagated by seeds or cuttings, and thriving in any soil.

I. gibraltarica is a beautiful plant, larger in all its parts than the other cultivated kinds. It has large flowers of a delicate lilac tint, arranged in low close heads, appearing in spring and early summer. It is an ornamental species, but will never rival the well-known white border kinds. Its hardiness is doubtful, and it should, therefore, be planted on sunny spots on rockwork or banks in light soil, and wintered in frames. A native of the south of Spain; increased by cuttings, as it rarely produces seeds in our climate.

I. jucunda is distinct from any other, as it only grows about 2½ in. high. The leaves are small, and the flowers are in small clusters, of a pleasing flesh colour, and prettily veined with rose in early summer. It does not possess the vigour of our common evergreen Iberises, but it is none the less valuable as a rock plant for being unlike them, and is fitted for association with a dwarfer and more select set of subjects. It should be planted on warm and sunny parts of the rock garden in well drained sandy loam. (=Æthionema.)

I. petræa, another pretty dwarf alpine species, bears a flat cluster of pure white flowers, relieved in the centre by a tinge of red. It only grows about 3 in. high; therefore it requires to be placed among the choicest subjects in the rock garden. Many cultivators cannot succeed with it, but in a well-drained position, with plenty of moisture at the roots, it thrives finely.

I. semperflorens is an upright-habited, shrubby plant, producing large dense corymbs of pure white flowers. It is hardy enough to stand our winters when grown at the foot of a south wall or in a very sunny corner of the rockery; it is not adapted for general border culture, but under those favourable conditions it forms a pretty evergreen bush rarely devoid of bloom all the year round. It is a native of Sicily and other Mediterranean islands. There is a pretty variegated form in cultivation, which appears to retain its variegation with much constancy.

I. sempervirens.—This is the common rock or perennial Candytuft of our gardens, as popular as the yellow Alyssum and white Arabis. Half-shrubby, dwarf, spreading, evergreen, and perfectly hardy, it escapes destruction where many herbaceous plants are destroyed; and in April and May its neat tufts of dark green are transformed into masses of snowy white. Where a very dwarf evergreen edging is required for a shrubbery, or for beds of shrubs, it is one of the most suitable plants known, as on any kind of soil it quickly forms a spreading band almost as low as the lawn-grass, finishing off the plantation very neatly at all times, and changing into dense wreaths of snowy-white flowers around the borders in spring and early summer. When in good soil and fully exposed, it forms spreading tufts nearly 1 ft. high, which last for many years. Like all its relatives, it should be exposed to the full sun rather than shaded, if the best result be sought. It is readily increased by seeds or cuttings. The more common garden name for this species is I. saxatilis. I. Garrexiana is a variety not sufficiently distinct to be worthy of cultivation; in fact, it and several other Iberises prove, when grown side by side, to be mere varieties, and very slight ones, of I. sempervirens; it, however, seeds more abundantly, and is less spreading in habit. I. superba, another variety, is the finest of all the perennial kinds, as it is of good bushy habit and bears a profusion of pure white flowers in large dense heads.

I. Tenoreana is a neat species, of dwarf growth, that produces in summer a profusion of white flowers, changing to purple. As the commonly cultivated kinds are pure white, this will be more valuable from its purplish tone, added to its neat habit. It, however, has not the perfect hardiness and fine constitution of the white kinds, and, so far as experience goes, is very apt to perish on heavy soils in winter; but on light sandy soils and also in well-drained positions on rockwork it will prove a gem. Where rockwork does not exist, it should be placed on raised beds or banks. A native of Naples, and easily raised from seed; it should be treated as a biennial.

I. umbellata *(Annual Candytuft)*.—This species and a near ally (I. coronaria) yield the popular hardy annual Candytufts, which are among the most beautiful of open-air flowers, being extremely varied in colour. They are all very hardy and seeds may be sown at all seasons, but, as in the case of most other hardy annuals, the finest displays are obtained from autumn-sown plants, and these flower from May to July. They require, to do well, a rich moist soil and plenty of room between the plants to develop themselves and flower freely. There is a great number of varieties, differing both in growth and colour. What are known as the dwarf or nana strain are very desirable, as they are neat and dwarf in growth, abundant bloomers, and very showy. I. umbellata nana rosea and alba are two of the most distinct, being only about 9 in. high; their names indicate their colours. The dark crimson, carmine, lilac, and purple sorts, which grow about 1 ft. high, are also very fine. The Rocket Candytuft is I. coronaria, which in good soil grows 12 in. to 16 in. high, and has pure white flowers in long dense heads. There is a dwarf variety of the Rocket Candytuft called pumila, which grows only from 4 in. to 6 in. high, and forms spreading tufts 1 ft. or more across. The Giant Snowflake is likewise an excellent variety. These Rocket Candytufts require the same treatment as the common varieties. I. pinnata and I. linifolia are not worth growing.

Impatiens *(Balsam)*.—The species that thrive in the open air are all annual and hardy, and sow themselves freely where they get a chance. The best kinds are—the common I. glandulifera, which attains a height of from 4 ft. to 6 ft., and bears numerous flowers, varying in colour from white to rose. This kind will soon take possession of shrubberies if not checked; it is seen to best advantage as an isolated specimen. I. longicornu is very beautiful. It has the habit of the last, but the lower part of its helmet-shaped flowers is of a bright yellow, marked by transverse lines of dark brown; the upper part is rose colour. I. Roylei is a much dwarfer kind than the preceding, and has blossoms of a deep rose colour. I. cristata has light rose coloured blossoms.

I. balsamina *(the common garden Balsam)* may be grown well enough in the open air, and makes a beautiful display in warm places. The plants should be raised in a frame and afterwards transplanted. Soil too rich should be avoided; such as was manured for a previous crop, and has been well pulverised and broken up by forking, suits best, as this produces the finest flowers and a less sappy growth. If it were proposed to plant a bed of Balsams in colours, say of half-a-dozen kinds, it is desirable to know what are the habits of the various colours, that the plants may be arranged to the best advantage. Some colours are produced on plants that are taller than others, and the habit is always the same from year to year, so that if the colours be obtained they may be relied upon to produce the desired effect if properly planted. Colours and markings in any good and valued strain of Balsams include the following, and probably a few others, as some sorts sport continually: Pure white, buff-white, rosy-white, lavender-white, pale mauve, peach, pink, carmine, scarlet, cerise, crimson, violet, purple, purple-white blotch, scarlet-white blotch, carmine-white blotch, crimson-white blotch, white-carmine flake, white-purple flake, carmine bizarre, and crimson bizarre. A bed of good Balsams is a novelty in the garden, and might be made tasteful use of, not so much for their effect of colour as for their beauty when seen near at hand.

Imperata sacchariflora.—A hardy Grass, from the region of the Amoor, with graceful, curved foliage, forming a tuft, about 3 ft. high, that throws up numerous flower-spikes, about 5 ft. in height, bearing silvery plumes of flowers. The leaves are of a lively green, with a broad white stripe down the mid-rib. It is scarcely ornamental enough for the garden proper.

Indian Pink *(Dianthus sinensis)*.

Inula.—Perennial plants of the Composite family, none of which are very important for the garden. I. Helenium (Elecampane) is a vigorous-growing British plant, about 3 ft. or 4 ft. high, with a stout stem, large leaves, and yellow flowers. It is well suited for planting along with other large-leaved plants, or as isolated specimens on rough slopes or wild places, in free, moist, good soil. I. Oculis Christi grows from 1½ ft. to 2 ft. high, and bears orange-coloured flowers in summer. I. salicina, montana, and glandulosa are similar, the last being the finest. All succeed in any soil, and are adapted for naturalising. Easily propagated by division or seed.

Ionopsidium acaule *(Violet Cress)*.—A charming little Portuguese annual, whose dense tufts, not 2 in. high, of violet flowers spring up in all directions where plants of it have existed the previous season. Like a weed it sows itself, thereby possessing the advantages of a perennial. Its peculiar beauty adapts it for various purposes and positions. On rockeries, associated with even the choicest of alpine plants, it holds its own as regards beauty, and it never overruns its neighbours. It is particularly suitable for sowing near pathways, rugged steps, or similar positions, places in which it grows freely; indeed, it would even flourish on a hard gravel walk. It flowers in a couple of months after sowing, and often produces a second crop of blossoms in the autumn. Portugal and Morocco. Cruciferæ.

Ipomæa *(Morning Glory)*.—Beautiful slender, twining plants of the Convolvulus family, for the most part tropical. A few succeed in the open air when treated as half-hardy annuals. The most popular of these is

PLATE CXXXVII.

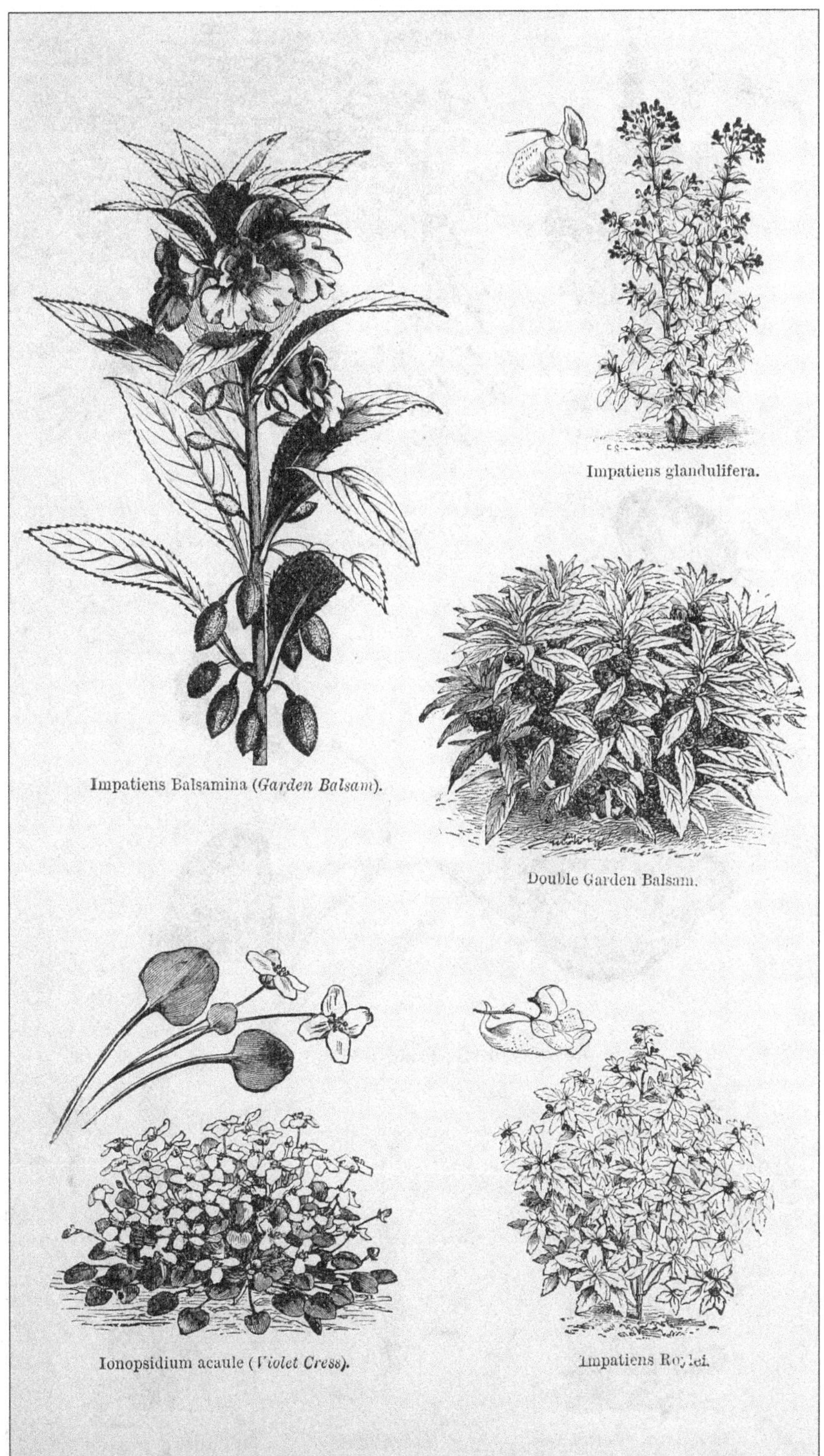

Impatiens Balsamina (*Garden Balsam*).

Impatiens glandulifera.

Double Garden Balsam.

Ionopsidium acaule (*Violet Cress*).

Impatiens Roylei.

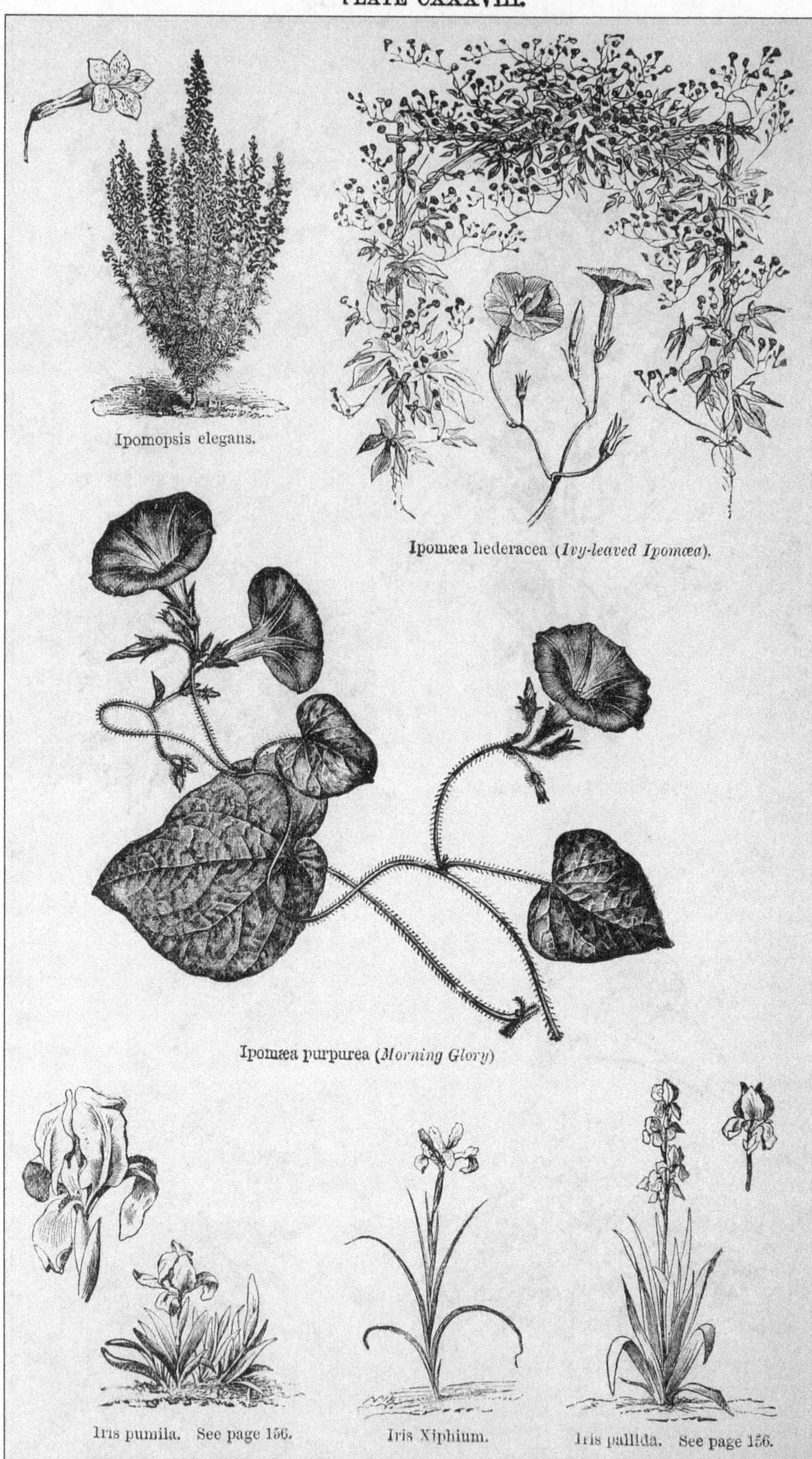

Ipomopsis elegans.

Ipomæa hederacea (*Ivy-leaved Ipomæa*).

Ipomæa purpurea (*Morning Glory*)

Iris pumila. See page 156.

Iris Xiphium.

Iris pallida. See page 156.

[*To face page* 153.

I. purpurea, or Convolvulus major, as it is called, which is too well known to need description, as it is one of the oldest of cultivated plants. The varieties of it are numerous; there are white, rose, and deep violet-coloured varieties, while Burridgei is crimson, Dicksoni deep blue, and tricolor striped with red, blue and white. A mixed packet of seed would contain most of these. This beautiful, though common, plant deserves much attention, as its uses are various. It may be used for the open border, festooning branches, for covering arbours, trellises, and the like, or for rambling over shrubs. It grows freely in any good ordinary garden soil. Seeds should be sown in heat in early spring, and the seedlings transplanted as soon as large enough in May. In some localities seed may be sown at once in the open border, but as a rule the plants raised under glass succeed best. It is known also as Pharbitis hispida. Tropical America and Asia.

I. hederacea *(Ivy-leaved Morning Glory)* is somewhat similar to the common Morning Glory (I. purpurea), but may be distinguished by the lobed leaves, which resemble those of the Ivy. The flowers, too, are somewhat smaller, of a deep blue and striped with red. The varieties grandiflora (light blue), superba (light blue, bordered with white), and atroviolacea (dark violet and white) are all worth cultivating, as are also the new Japanese variety, Huberi, and the variegated-leaved form of it. The Ivy-leaved Morning Glory is somewhat hardier than I. purpurea, and seeds may be sown in the open border in April in light, rich soil, where it flowers from July to September. It is also known as I. Nil. North America. Other kinds of Ipomæas for open-air culture are I. rubrocœrulea, a half-hardy annual, and I. leptophylla, a hardy perennial from North America, but neither are so desirable as those mentioned above.

Ipomopsis.—Graceful and beautiful biennials from California that are hardy in light, dry, and warm soils in the milder parts. There are three kinds, and each forms a tuft of finely-cut feathery foliage, and has slender flower-spikes from 2 ft. to 3 ft. high thickly set with flowers that open in succession. In I. elegans they are scarlet and thickly spotted, and in superba much the same, while in the rosea variety they are a deep pink. The seeds should be sown in spring in the open border in ordinary soil, or in pots. During the first year the plants make growth, and early the following summer they flower. If planted out to stand the winter it is advisable to give a little protection. Other kinds mentioned in catalogues belong to Gilia, of which Ipomopsis is really a synonym. On light soils early autumn sowing should be tried. These flowers are very seldom well grown.

Iresine.—Half hardy plants, of dwarf growth, remarkable for their handsome foliage; they are, therefore, much used with other tender plants in summer for the flower-garden. There are two types, from which several varieties have sprung. I. Herbsti grows from 1 ft. to 2 ft. high, and has crimson stems and rich carmine-veined foliage, the brilliancy of which continues until late in autumn, and is more effective in wet than in hot, dry seasons. It requires a moist, rich soil, and is readily increased by cuttings taken in September and wintered in pots in a warm greenhouse. In early spring the plants should be repotted and grown on in heat, and fresh cuttings taken in March and April will make them fit to put out in May. I. brilliantissima and Wallisi are two varieties possessing more brightness of colour in their foliage. Lindeni is quite distinct from the foregoing, having more pointed leaves, which are of a deep blood red. It is of a compact and graceful habit, and bears pinching back and pegging down to any required height. It makes a good edging plant, and requires the same treatment as I. Herbsti. Amarantaceæ.

Iris.—Everyone admires the common Iris, but it is not everyone who is aware of the beauty and the delight that may be found in the many members of the family now in cultivation. Taking them all through, no other class of hardy flowers possesses that union of grace of outline with delicacy of colouring which is the charm of the Irises. The majority are lovely, and they are all worthy the attention of those who love flowers as flowers, and who do not regard plants merely as material for constructing the gaudy or the grotesque patchworks sometimes spoken of as gardens. By some they have been compared to Orchids, and those who delight in singular and beautiful combinations of colour, and to whom the pleasures of greenhouses and hothouses are denied, may find a good substitute in the cultivation of a selection of the varieties of hardy Irises. The genus is well represented in our gardens, and by species for the most part hardy and possessing a considerable diversity of habit of growth as well as colour. They vary in height from a few inches to as much as 6 ft., and may be conveniently divided into two classes—those with bulbous roots, which are now called Xiphions, and those (the greatest number) that have creeping stems. In treating of the culture of Irises it is well to consider the bulbous and the non-bulbous kinds separately, as they require somewhat different treatment. The bulbous kinds, or Xiphions, should have a warm and sheltered situation, say under the protection of a south wall. They succeed in almost any kind of light garden soil, but prefer one that is loose, friable, and sandy This, however, should not be too poor, but well enriched with thoroughly rotten leaf-mould and manure. Sun they must have, and the protection or shelter must be without shade. They need an autumn drought to ripen, and a dry soil in winter to preserve the bulbs and keep them at rest, but in spring, when the leaves are pushing up, they love rain if it is not excessive. This treatment applies to the Spanish and English Irises as well as to the rarer species, but they will be found, as a rule, more robust than those of smaller growth. The great point to be observed is not to meddle with the bulbs as long as the plants are seen to be doing well, for they dislike disturbance, and when it is found necessary to transplant because of exhaustion of the soil care should be taken that the bulbs be not allowed to become dry or shrivelled. It is advisable to place a thin layer of Cocoa-nut fibre refuse or some similar material to serve the purpose of protection during severe weather, and to prevent the flowers from being bespattered by mud during heavy rain. Some kinds produce seeds very freely in some seasons; these should be carefully collected, and when well ripened should be sown at once. This will be found to be a wholesale mode of increasing the stock, as they will make strong flowering bulbs in about three years.

Most of the non-bulbous Irises like rich soil, full of decomposed vegetable matter. The coarser and stronger forms will feed on even rank manure, but to the more delicate ones this is almost poison; and all of them, indeed, thrive all the better if their food is given to them in a well-digested form. If it is thus well digested they can hardly have too much of it. As regards moisture, they vary a good deal. The common Pseudacorus and many of the spuria group thrive best in the damp. Others, again, hate the damp—at least in winter—and will stand very considerable drought in summer. The condition that suits the majority is comparative dryness in winter and an abundant supply of water in summer. Unfortunately, this is the very reverse of what they generally meet with. They also vary a good deal as to the nature of the soil they like best. Some like a deep, somewhat stiff, but rich loam, and their long, thong-like roots reach down for an amazing distance, while others are partial to a lighter, looser soil, which, however, must be proportionately richer in vegetable matter. Most of these plants have every good quality that one could desire as regards hardiness, freedom of bloom, and easy culture. Like a good many other plants, their season of bloom is not long enough to please all; but, by cultivating them in various places where they are not in the way when out of flower, this is not felt to be a drawback. The more vigorous growing kinds are well adapted for planting among large shrubs, which ought to be more widely apart than they are generally placed in thin copses; in tufts near water and in isolated groups on the Grass near they may be enjoyed, and also on mixed borders and beds, but they should not be placed in masses of shrubs. In such places they would mark the season agreeably, and not be in the way when their bloom was over, as they are sometimes found to be in borders. In the smallest gardens, where people have not the space to plant them in these various ways, one of the best modes would be to establish healthy tufts in the fringes of the shrubbery. Another good way is to place them here and there in carpets of low evergreens, above which their flowers would be seen in early summer. It is worth the while of all who care for hardy flowers to devote a little attention to establishing a good stock of these plants in their gardens. They have all the beauty of the finest tropical flowers without their cost, and will repay the trouble of first arranging and planting them so that their beauty may be seen to the best advantage, and so that they may not be in the way of the imperious needs of "bedding out." Tufts of the finest kinds look very beautiful here and there among dwarf Roses. The flowering season of the Iris extends over the greater part of the year—indeed, if the winter be mild, it is not an uncommon occurrence to have Iris flowers throughout the year. The following are among the finest kinds, and are more or less easily procured in nurseries:—

I. alata.—This is a beautiful bulbous species with fine large blossoms, the ground colour of which is a delicate lilac-blue, with showy blotches of bright yellow, copiously spotted with a darker hue. The foliage much resembles that of a Leek, and is produced at the same time as the flowers. It generally commences to bloom in October, and, if the weather is not too severe, it produces other flowers about Christmas time. It is very easy to grow, as it merely requires a warm, dry, sunny border; the bulbs should be planted in autumn in ordinary garden soil. The greatest drawback to its culture is that the flowers are spoilt by soil-splashings during heavy rains. If, however, the bulbs were planted among Stonecrop, or in Grass, on warm sheltered banks, their pale blue blossoms would remain uninjured, and would prove a source of attraction when our gardens are comparatively flowerless. The beauty of the flowers may also be much preserved by having a hand-light placed over them during severe cold, snow, or wet. Native of the Mediterranean region. *Syn.*, Xiphion planifolium.

I. biflora.—A handsome, rather dwarf plant, 9 in. to 15 in. high, bearing large violet flowers on stout stems. Similar to this are I. sub-biflora and I. nudicaulis, which latter is one of the finest of all the dwarf Irises. It grows from 4 in. to 10 in. high; its flowers are large, of a rich violet-blue, and are produced four to seven on a stem in early summer. It has the vigour of the German Iris and the dwarfness of the old Crimean Iris. It is, however, much sturdier than this, and is, in a word, second to no hardy plant introduced of late years. It is suited for the front ranks of the herbaceous border, and also well deserves a position among the more vigorous plants in the rock garden, being so dwarf. It should be in every garden where early summer flowers are valued. There are

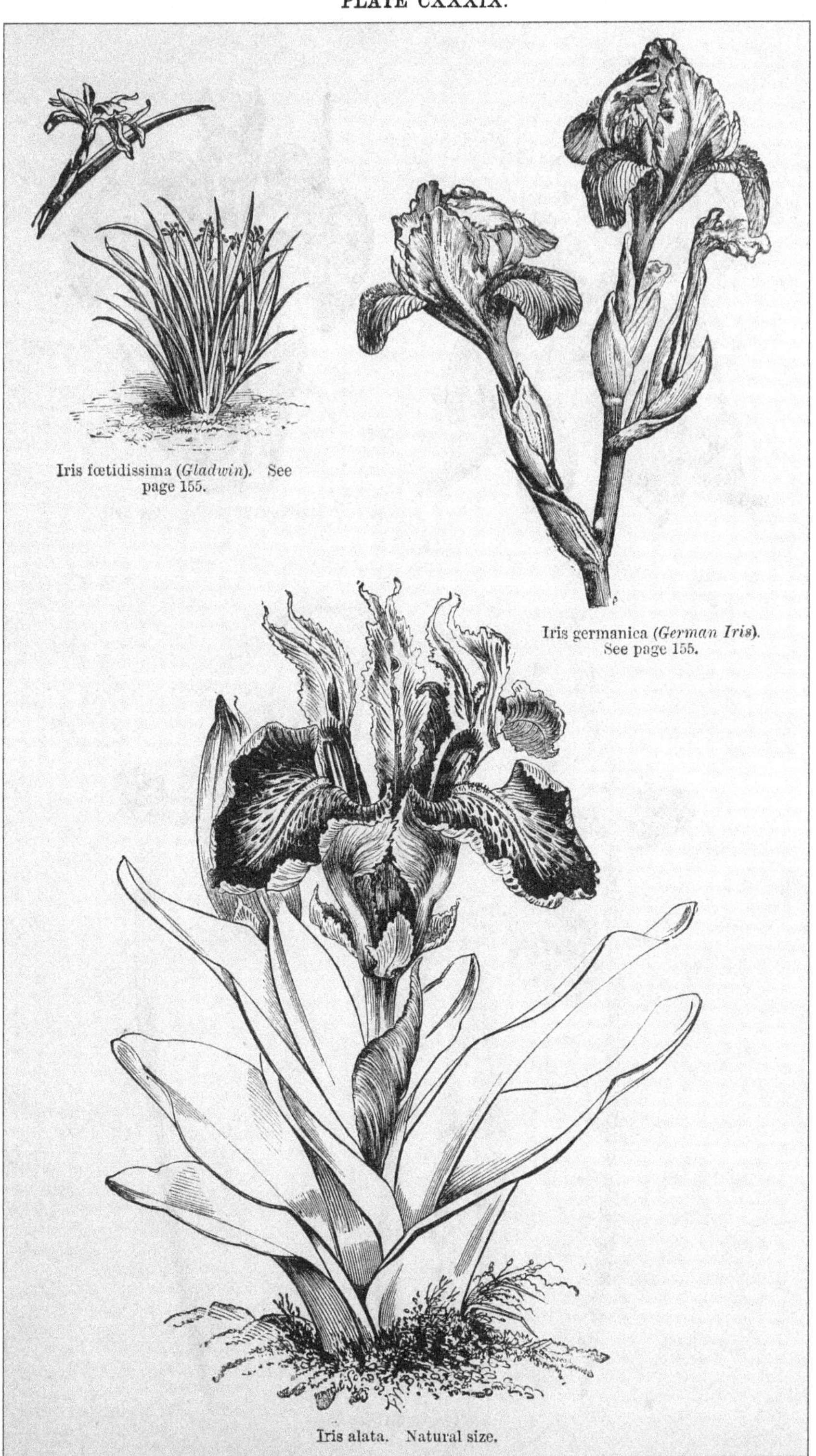

Iris fœtidissima (*Gladwin*). See page 155.

Iris germanica (*German Iris*). See page 155.

Iris alata. Natural size.

PLATE CXL.

Iris iberica.

Iris Kæmpferi (*Japanese Iris*).

Iris (Xiphion) Histrio.

[*To face page* **155.**

several other species similar to the three mentioned above, but they are not so beautiful, and are all rare in cultivation.

I. cristata is one of the best of the dwarf Irises, with flowers of a delicate blue and richly marked. It is a quaint, delicate plant, which grows from 4 in. to 6 in. high, with leaves that are very broad for the size of the plant. Its growth is peculiar; it throws out long, slender rhizomes, wholly above ground; hence, though a very humble plant, it soon covers a considerable space. The way to manage this Iris successfully is to plant it among stones, with little or no soil. Thus situated, it seems at home, and the blooms last longer and keep in better condition than when growing on the ground. It loves a warm sandy loam, such as that in the Edinburgh nurseries.

I. fœtidissima *(Gladwin)*.—A well-known, but undeservedly neglected British plant, 1½ ft. to 2 ft. high. Its flowers are lead-coloured or bluish, rarely yellow. A variety with variegated leaves forms a pleasing border plant, and is also an excellent house plant. The common green form is well worthy of being grown in semi-wild places for the sake of the effect of its brilliant coral-red seeds, which are plentifully produced in autumn in gaping capsules. Division and seeds, which, if thrown in semi-wild places, will soon spring into plants.

I. germanica *(German Iris)* is the commonest of all, and too familiar to need description. There are a few varieties of it, the principal being atro-violacea, of a deep, rich, violet-purple; Telfordi, with unusually large flowers of a velvety, dark purple; and major, which is similar to the last. I. squalens, flavescens, sambucina, neglecta, hybrida, and amœna come under the same group as the German Iris, all being about the same growth, but the colours of the numerous varieties are indescribable, so varied and rich are they. Like all the large robust kinds, these are all good border plants, and are also well adapted for naturalising in semi-wild places, as they take care of themselves in any soil or position, but thrive best in a good light soil in an open, warm situation. The majority of the named sorts of the so-called German Iris belong to this group.

I. Histrio.—This is a very beautiful bulbous species that, when seen peeping through the ground in winter or early spring, reminds one of I. reticulata, but its growth is rather taller, and the petals are broader and more conspicuously spotted or blotched. Its colour is rich bluish-purple, flushed towards the base of the petals with rose-pink, the markings being of the deepest purple, relieved by a crest of gold. The blossoms are delicately scented. It is one of the earliest of all Irises, being usually in bloom with the first Snowdrops. Its culture is simple, and it succeeds well under the same treatment as I. reticulata. Though a native of Mount Lebanon, it is perfectly hardy with us, but during severe frosts it is advisable to lay a thin covering of some protective material, such as Cocoa-nut fibre refuse, over the bulbs.

I. iberica.—One of the most singular of all the Irises, yet very handsome. The flowers are large, the upright petals white, pencilled and spotted with purple or violet, while the drooping petals or falls are veined with dark purple or purple-black on a yellowish ground, with a conspicuous dark blotch in the centre. This is the colour of the commonest form, but there are several, and one, ochracea, is very distinct. It is a handsome plant, and far easier to cultivate than most persons imagine, being perfectly hardy and not at all fastidious as to its requirements. It thrives best in a rich fibrous loam, in which it can send its long roots deep into the soil. The rhizome does not require to be planted deeply—just below the surface is sufficient. In most cases the roots perish when planted deeply. The rhizome, during the winter, is very impatient of moisture, and should be kept comparatively dry. Coarse river sand should be used, the rhizome being planted completely in it, in the same way as many Cape and other bulbs are planted. By this means it is kept rather dry during the winter and great assistance is also given to the plant in summer, as the young shoots can easily force their way through. Under this treatment the plant grows more freely, and can be easily multiplied by division of the rhizome. It is admirably adapted for rockwork, or for the select border, and, when better known, will find a place in every garden. It flowers in spring, and though the blossoms are of somewhat short duration, their extreme beauty amply compensates for this drawback.

I. Kæmpferi *(Japanese Iris)*.—The large number of varieties in cultivation under this name have sprung from I. lævigata and I. setosa. They form quite a new race of garden plants, and every year many beautiful sorts are added to the number, being obtained chiefly by direct importation from Japan, though many seedlings have been raised in this country. The flowers of these Irises are extremely variable both in size and colour, some measuring as much as 9 in. and 10 in. across. The large and double-flowered varieties, produced by the Japanese, and also, of late, in some European gardens, are truly magnificent; the flowers appear rather flat, presenting the outer segments in their entire dimension; the colours are very striking, and vary from white to lilac and violet, and from sky blue to indigo, brown, and blackish-maroon. The varieties of I. setosa differ from those of I. lævigata in having broader and less drooping petals, and often the three inner petals are of the same size as the outer, so that the flower is symmetrical in form. I. Kæmpferi will grow in almost any kind of soil, but it succeeds best in a good loam, provided a quantity of peat be added to it. The peat is not so much to afford nourishment as to retain moisture during the hot and dry summer months, for this Iris likes moisture, and its numerous roots will often go 2 ft. deep in search of it. It dislikes shade, and prefers a sunny, hot situation. Two-year-old seedling plants of it flower in June and July, and among them will be found endless variations of colour, from white to dark blue, and from pale rose and lilac to deep maroon and brown. Complaints are often heard that plants of these Irises are difficult to flower, and the fact is that, if not carefully transplanted, they do not flower the first year, but are, on that account, much finer the second season. In short, they must be well established before they can produce fine flowers. They may be propagated by division and by means of seeds, which should be sown as soon as gathered either in pots or in the open ground, when they will vegetate the following spring.

I. pallida.—A tall and noble Iris, from 2 ft. to 4 ft. high, having large, pale mauve flowers, with a whitish beard. The blossoms are sweetly scented, and appear in early summer. There are a few varieties of this Iris that vary slightly in the depth of colour. They are all very beautiful. Similar in growth and size is the common Florentine Iris (I. florentina), which has large white flowers, tinged with lavender. I. albicans is a pure white variety of it, and very handsome. I. plicata and Swerti come under the same group, being of tall growth, and having deeply-forked stems and large light-coloured flowers, mottled and veined with purple. These are fine plants for borders or groups on lawns, and they grow well in any description of soil.

I. Monnieri.—A bright and noble-flowered kind, quite distinct from any other in cultivation. The leaves are dark green, and the flower-stem stands nearly 4 ft. high. The outer divisions of the flowers, which are very fragrant, are recurved, and are of a rich golden yellow, margined with white. It is not by any means a common plant. I. aurea, an allied species, is a noble plant, 3 ft. to 5 ft. high, with large foliage and long, stout flower-stems, that bear large blossoms of a uniform bright golden-yellow. I. ochroleuca is of similar growth and habit, but has white and yellow falls and deep yellow standards. It is a most distinct plant, and one of the largest of all the Irises; on this account it passes under the name of I. gigantea. All of these kinds are suitable for borders among the taller herbaceous plants, and in tufts 2 ft. or 3 ft. within the margin of shrubberies. They thrive best in rather moist soil. Easily propagated by division or seed.

I. persica is one of the most attractive and distinct of all the early-blooming kinds, and one that well deserves a place in all collections where the soil is warm and dry. Its blossoms, which are produced from a tuft of bright green leaves that just peep over the soil, are white, suffused with a pale Prussian blue tint, and blotched with velvety-purple. It is a native of Persia, and therefore somewhat tender, but in warm, sheltered spots, in light sandy soil, it succeeds well enough, and flowers in winter and spring, according to the weather. (=Xiphion persicum.)

I. pumila.—The best of all the dwarf Irises, for to it we owe the many lovely varieties that create such a rich display of bloom in spring. It grows from 4 in. to 8 in. high, and has deep violet flowers, unusually large for the size of the plant. All the varieties flower profusely, but to be appreciated fully they should be planted in large masses (a good-sized bed of each sort) or in broad lines. There are several named varieties, the best of which are the white, straw-coloured, pale blue, and blush-coloured; all of these may be found described in trade catalogues under Latin names. The most attractive is the sky-blue (cœrulea), which in early spring forms sheets of bright colour. Similar in growth to I. pumila, but with larger flowers, is I. olbiensis, which normally has rich violet-purple flowers, but which also occurs with white, blue, yellow, and various other colours. I. Chamæris comes also in this dwarf group, as do I. verna (a handsome, but rare North American species), I. italica, and several others. These all flower about the end of April or the beginning of May, if in warm localities, in light soil.

I. Pseudacorus (*Common Water Flag*)—Common as is Pseudacorus, everyone who has grown it fairly will admit its beauty. Whoever has in his garden a pond or a ditch, or even a thoroughly damp spot, ought to plant this Iris largely. Few things, indeed, are more beautiful than a great clump of this yellow Flag, with the tall leaves starting up from the side of a pool, and the golden clusters of flowers gleaming bright in a mid-summer sun. Three things it loves—a rich soil, plenty of water, and abundance of sunlight. It is cruel to place it, as it is sometimes placed, in some dank, dark hole, where the sun's beams never reach it; it is disappointing to plant it in a dry and stony spot, where summer is to it one long continued thirst. But put it where its roots can run at will in rich black mud, and let its head raise itself to the full light of a summer sky, and it will be a golden glory throughout the long days of June.

I. reticulata.—One of the most beautiful of all hardy flowers. While the snow is still on the ground—in January, or even earlier—the leaves begin to shoot, and while these are only a few inches high, the bud opens out to the pale wintry sun its beauty of violet and gold. After the flower has faded the erect narrow leaves grow apace, attaining 1 ft. or more in height, and, as in the Crocus, the ripe ovary is in due time thrust upwards from the soil. This little treasure is indeed the Iris companion of the Crocuses, and those who have seen large clumps of it growing in some sheltered, but sunny, spot in the bright but gusty days of February or March, may well wish for the time when its netted bulbs shall be as plentiful as the Crocus corms. Two very distinct forms are known.

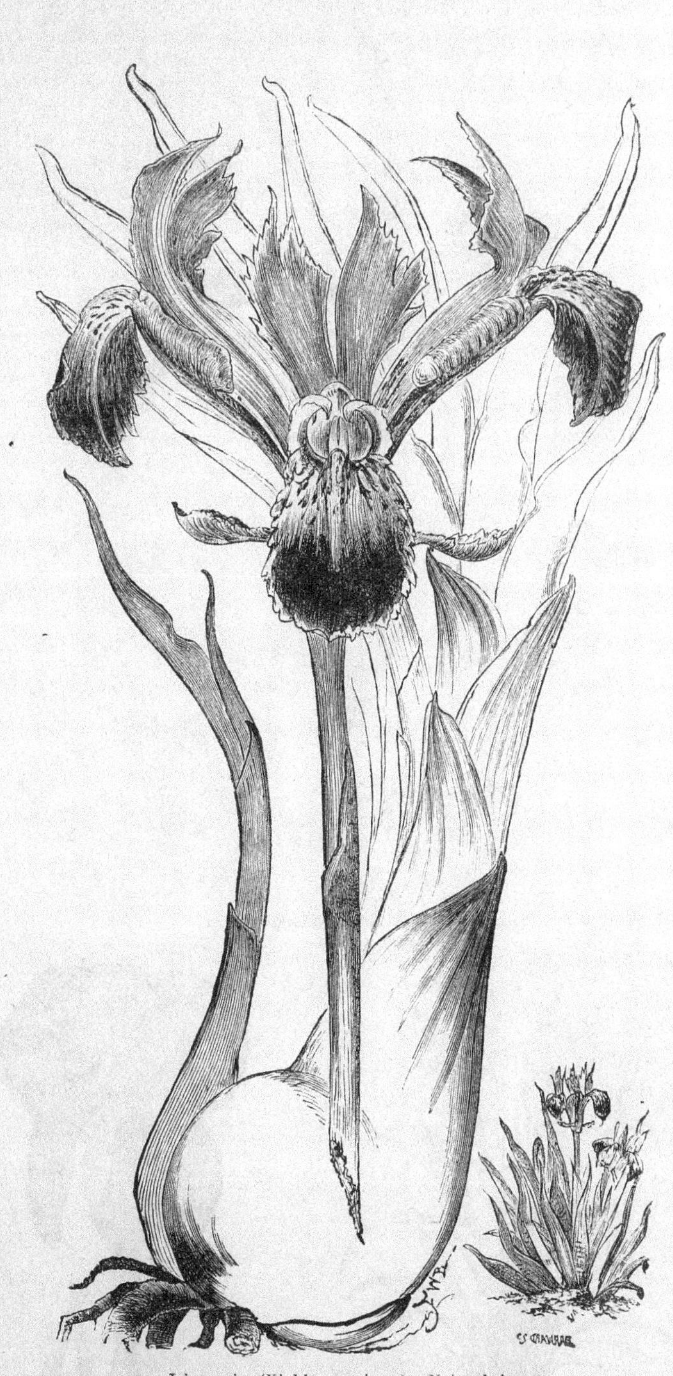

Iris persica (Xiphion persicum). Natural size.

[*To face page* **156.**

PLATE CXLII.

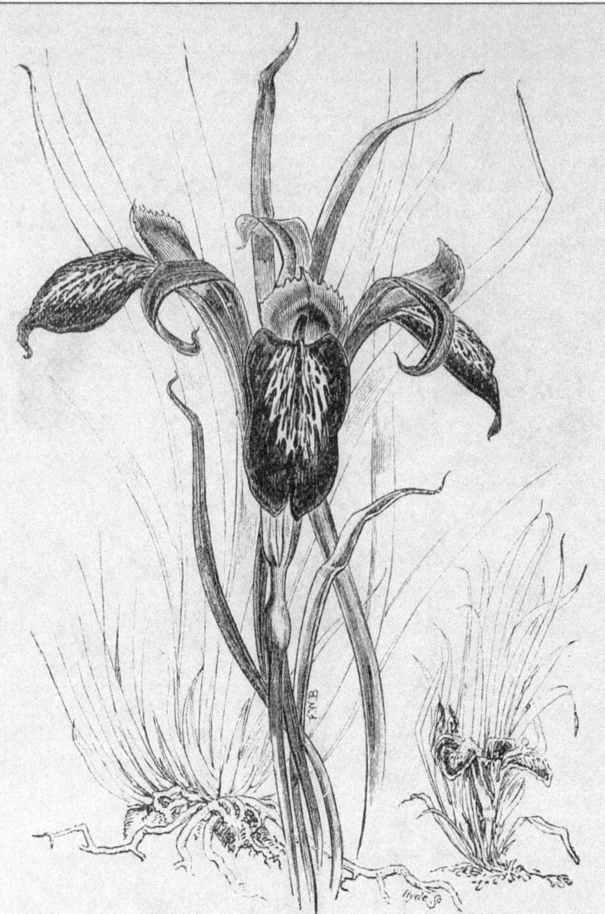

Iris ruthenica. Natural size.

Iris xiphioides (*English Iris*). See page 158.

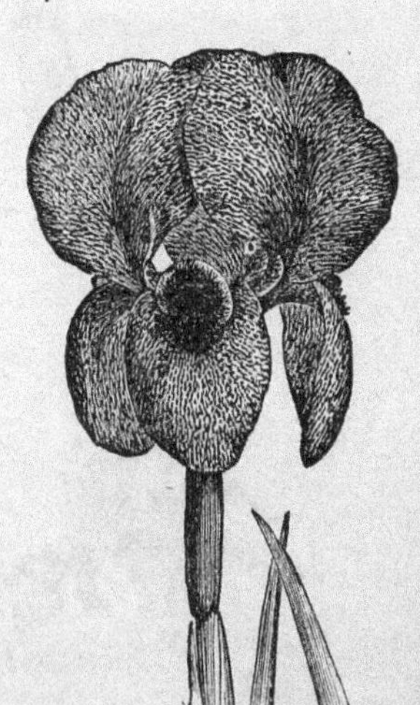

Iris susiana (*Mourning Iris*). See page 158.

[*To face page* **157.**

One, the typical form, may be recognised by its deep dark violet colour, and the brightness of the yellow. This kind comes from some parts of the Caucasus and from Palestine. The other form is the one known as Krelagei, attention having first been called to it by the well-known horticulturist, Krelage. It may be recognised by its flowers being of a purple or plum colour, with the yellow marking less vivid and the whole flower smaller. This kind seems to be more abundant than the typical one, and is represented in many of the illustrations as Iris reticulata. It is the form common in the Caucasus, the Transcaucasus, and Persia. It is a matter of some little importance to distinguish between the type and its variety, for the former is not only far more beautiful than the latter—its deep dark violet and lustrous yellow producing an exquisite effect, a long way surpassing the duller purple of Krelagei—but it possesses a delightful violet fragrance which the latter is absolutely without. One advantage, however, Krelagei possesses—it invariably flowers ten or fourteen days earlier than the type. Concerning its culture very little need be said. It is absolutely hardy, and may be planted wherever the Crocus is found. A sheltered spot is, however, advisable, in order that its tall narrow leaves may, after flowering is over, flourish protected from the wind—a spot sheltered, but sunny. It flourishes in sandy soil; it thrives in peat, which is dry and warm in summer; it does well in common garden ground, and also in a stiff heavy clay, so that it does not seem particular about soil. Sometimes, however, it refuses to grow, and in damp places the bulbs will rot in summer. Since the flowers come before the leaves grow tall it makes a good pot plant, and a well-grown clump of the fragrant sort is a charming addition to the Christmas table. Some care, however, is needed to make it bloom more than one year in pots, and it is advisable that those who cannot give it the requisite attention should harden off the pot plant when it has flowered and plant it in the open, seeking a fresh supply for the next winter from some old clump or from bought bulbs. I. Kolpakowskyana is also a lovely plant, smaller than, though similar to, I. reticulata. There is more white in the flowers, and this makes it distinct. It is a new introduction, but quite hardy, and thrives under the same conditions as I. reticulata.

I. ruthenica.—One of the smallest of Irises. Its flowers, which are about 2 in. across and of a bright lilac-purple colour with a white lip, are borne on stems about 1 in. above the surface, and nestle snugly among the bright green slender foliage. It blooms during nearly every month of the year, though mostly in the autumn, but as the blooms are hidden among the foliage, ordinary observers are unaware of its being in blossom at all. It is of easy culture in any open situation.

I. sibirica.—A tall, slender-growing plant, from 2 ft. to 3 ft. high, with narrow grassy leaves and somewhat small, but showy, blue flowers, beautifully veined with white and violet. There are several varieties of this Iris, some of which are very handsome. The white-flowered variety, also called I. flexuosa, is a pretty plant, and so is I. acuta, but the double-flowered form is not so. The finest variety is I. orientalis, the flowers of which are larger, of a deeper colour, with a different veining, and the falls especially broad and expanding. It derives its name from the bright scarlet or crimson tinge of the spathe sheaths, which gives the bud a remarkable beauty, even before the flower has opened. Melpomene and nigrescens are likewise very beautiful varieties in the way of orientalis, but finer than the type. The Siberian Iris thrives best in rich, damp soil, and this is, perhaps, especially the case with orientalis and some of the other large-flowered varieties. But if it is to flower well it must have abundance of sunshine; it will not show its real beauty in dampness begotten of shade; nor is it, like I. Pseudacorus and some other forms, absolutely devoted to it, for when it becomes thoroughly established it will bear without even flinching an amount of drought which would be fatal to it when newly planted. In planting it will be well to secure good ground to start with, for the plant has a very great objection to being moved. Transplantation will in most cases prevent the bloom of the succeeding summer. Moreover, its real beauty does not become apparent till it has grown into a good-sized tuft. Place your plant, then, in thoroughly good soil, with appropriate surroundings, so that its head of flowers may be seen above dwarfer plants, standing out against still taller foliage, and then leave it alone. The Siberian Iris is a capital subject for naturalising, for though it flourishes best in the soil indicated it will grow and thrive in almost any soil, even in the worst description of clayey soils. Flowers in May and June.

I. spuria.—Not a very attractive plant, though in some of the forms, as spuria major, and also in the Algerian variety known as Reichenbachi, the colouring is bright and handsome, especially when seen in masses; but the mixture which they offer of blue or purple and yellow is not pleasing, and, besides, there is a certain stiffness and want of elegance in their outlines. The smaller flowers of such varieties as that known as desertorum, with its paler flowers, narrow falls, and, in some cases, marked fragrance, or the white Güldenstädti, however, are pretty. I. graminea is of no great value as a border plant; the flowers are too much hidden by the overtopping leaves, and are themselves, singly, of no great beauty. Nevertheless, their mixed blue and purple tints will be found to render them of value as cut blooms; they can then be made to harmonise effectually with other flowers.

I. stylosa.—A beautiful plant, whose charm is enhanced by its flowering in midwinter, which no others, except one or two of the bulbous class, do. Though not a very attractive plant, on account of its flowers being hidden among its grassy foliage, this Iris should be grown in every garden, for when mixed with even the most delicate flowers of the stove or Orchid house, the silky, sky-blue flowers possess a charm and softness unequalled by perhaps any other flower of the same colour. They are fragrant, and are produced from Christmas onwards to the end of January. Although the plant is quite hardy, its flowers are so large and delicate in texture that some slight protection from rain and rough winds is desirable, unless, indeed, the position which it occupies is well sheltered and the weather mild. It succeeds in almost any soil. It is also called, but not often, I. unguicularis.

I. susiana *(Mourning I.)*.—One of the most singular of all the flowers of temperate and northern climes. It grows 1½ ft. to 2½ ft. high; the flowers, which are produced in early summer, are very large and densely spotted and striped with dark purple on a grey ground. It should be grown in sunny nooks in the rock garden, or on sheltered banks or borders, always in light, warm, and thoroughly drained soil. We have seen it thriving perfectly as a border plant, and flowering well in the Archbishop of Canterbury's garden near Broadstairs. Iris susiana may be treated as a perfectly hardy plant in some parts of the country; a dry bottom and free soil are, no doubt, essential to success. In cold districts or on heavy soil the protection of a handlight would be desirable in winter. Asia Minor and Persia. Increased by division.

I. tectorum is a charming species, second to none in point of beauty, and should find a place in the most select collections; its blossoms are large and of a bright purple, beautifully mottled with dark purple, and, unlike those of the majority of Irises, all the six divisions are on the same plane, this much enhancing their beauty. Being a native of Japan, it is somewhat tender, but if planted in a sunny, warm situation will flower very abundantly, and show off its strongly-crested blue and dark blue blotched flowers to great advantage. A rich soil, much sun, and some dampness are the conditions under which it succeeds best. In Japan it grows on the top of straw-covered house-roofs, feeding on the decomposing straw, and exposed to sun and weather all the year round.

I. tridentata.—This is a N. American form, the falls of which are largely developed, highly coloured, and manifest real beauty in their form and markings. It is an abundant bloomer, a strong grower, spreads rapidly, and is in every way a desirable plant. It occurs in the Northern States. The closely allied I. setosa from Asia is a far less beautiful plant. I. tenax, also a N. American form, is a close neighbour of sibirica, and is a very desirable plant. Allied to tenax is the Californian form I. longipetala. This, the various cultivated specimens of which appear to vary not a little, is a showy plant; but its rather long and straggling falls, in spite of their charming light violet or lavender colour and their graceful markings, give it a more or less unfinished look.

I. tuberosa *(Snake's-head)*.—This is an interesting, but dull-coloured kind, 12 in. or 13 in. high. Its flowers are small, of a brownish-green streaked with yellow, and with a purplish-brown tinge on the upper part. There are usually two tubers. It is not worthy of a place in the garden, but where admired it may be naturalised in light soil. S. Europe.

I. variegata is a handsome, richly-toned kind, growing 1 ft. to 2 ft. high, with large, slightly fragrant flowers, having bright yellow standards and claret-red falls beautifully veined. The varieties of this are very numerous, one of the most distinct and finest being De Berghi, which has deep brown-purple falls and bright yellow standards. Similar in aspect is I. aphylla, with deep lilac falls and white standards veined with purple. This, too, has numerous varieties, the colours of which are indescribable, but all beautiful. I. lurida and its varieties come also under this group. The same remarks as to culture and position apply to these as to I. germanica.

I. virginica.—This is a vigorous floriferous plant, spreading very rapidly when grown in a somewhat moist, rich soil. The flowers vary very considerably in tint, and some of the more deeply-coloured forms are not unhandsome. There is, however, a certain stiffness and formality about the blooms that prevent it from being considered a really attractive kind. More highly coloured, frequently very striking from the juxtaposition of a pure white and a deep rose tint, is the very closely allied I. versicolor; but this, too, lacks a certain elegance, so that one is, in looking at it, led to wonder why a flower so beautifully coloured gives one so little pleasure. Many seedlings, both of virginica and versicolor, are in cultivation; and though what may be perhaps considered as the typical forms of each are very distinct, almost every intermediate stage between the two may be seen. These thrive in any kind of moist soil, and are perfectly hardy. N. America.

I. xiphioides *(English Iris)*.—Like the Spanish Iris, this is one of the bulbous species. It is very handsome, possessing an infinite variety of tints in the blossoms, which are large and with broader outer petals than I. Xiphium. The colours vary from white, through pink and mauve, to the deepest purple, and some are beautifully flaked and spotted. All of these may be had in a mixed collection. It comes into flower a week or so later than the Spanish Iris, but requires exactly the same cultural treatment. Spain and the Pyrenees. (= Iris anglica, Xiphion latifolium.)

I. Xiphium *(the Spanish Flag or Iris).*—A very old inhabitant of gardens. It and the so-called English Iris are at present the bulbous Irises most commonly cultivated. The native habitat of each is extremely limited, the Spanish Iris being found in Spain and Portugal, and only here and there in the south of France, while the Pyrenean form, as its name denotes, comes from the Pyrenees. The misleading title of English Iris applied to the Pyrenean form appears to have arisen from the plant having been introduced to the Dutch gardeners in a roundabout way through some Bristol merchants, in whose gardens it was a favourite long ago. The Spanish Iris begins growth early, and having attained a few inches in height it will remain stationary during the winter, careless of even biting frosts. In the spring the tall, narrow leaves will grow apace, and a stalk, almost wholly ensheathed by clasping leaves and spathe valves, will, late in May or in June, bear one or two flowers whose great beauty consists in the vividness, and yet chasteness, of their colouring. The bulb, which produces offsets in great abundance, is much smaller than that of the Pyrenean Iris, and its coat, though it may be spoken of as fibro-membraneous, is not nearly so thick and rough as is the coat of that species. The plant having been for many generations in the hands of the Dutch florists, many seedlings have been raised, and much variety of colouring has been gained, though the number of named varieties is not so great as in the case of the Pyrenean Iris. The prevailing colours are blue, with various shades of purple or violet, yellow, and white. There is a form, known under different names in different florists' lists, the opaque snow-white hue of which renders it very charming. The blue tints of the cultivated seedlings seem to be derived from the typical native Spanish plant; the yellow hues may be traced to the Portugal variety known sometimes under the name of Iris lusitanica. The Portuguese plants are for the most part yellow, and generally have the more funnel-shaped flowers, while the Spanish plants are generally blue and have more spreading flowers; but the distinction is not absolute, and the cultivated varieties are mixed both in colour and form. Iris sordida is simply a form of the Portuguese variety, in which the yellow is blotched with purple or violet, a phase of colouring which is not at all uncommon in the Dutch seedlings.

SOIL AND SITUATION.—The Spanish Iris must not be waterlogged in autumn and winter. It prefers a loose, friable, sandy soil, which, however, should not be too poor, for it repays feeding with rich well-digested stuff—thoroughly rotten leaf mould or manure—by giving fuller and richer bloom. Sun it must have, but its slender stalks suffer from the winds; it should therefore have the protection of shelter without shade. It needs an autumn drought to ripen its bulbs, and a winter dryness to keep it at rest; but in the spring, when it is rapidly pushing its slender leaves and shoots, it loves any rain that is not excessive. On the whole it flourishes in dry places much better than the Pyrenean Iris, which will stand a much greater amount of wet. The Thunderbolt, or clouded variety, is much more robust than either the typical form or the Portuguese kind, and will flourish even grandly in stiff, damp soils, where the latter would speedily perish. The golden rule of not meddling over much applies most distinctly to the Spanish Iris. The new roots begin to shoot out almost before the old stalk has withered, and the bulb hates to be kept out of the ground. Plant, then, the Spanish Iris in clumps on some rich, loose, friable plot, where their bright colouring may be shown to advantage, and let them stay there year after year until the dwindling foliage tells you that they have exhausted their territory of soil; but it will be some time before that comes to pass.

Isopyrum thalictroides.—A graceful little plant allied to the Meadow Rues, but with white flowers prettier than those of the Thalictrums. It is, however, chiefly valuable for its Maiden-hair Fern-like foliage, and is worthy of culture in the flower garden for this alone. When it is grown for the sake of its dwarf elegant leaves only the flower-stems should be pinched out. It is well suited for rockwork, and particularly so for the front edge of the mixed border, is hardy, and easy to grow on any soil. It is easily propagated by division or by seed. Europe. Ranunculaceæ.

Isotoma axillaris.—A showy half-hardy plant, resembling some of the dwarfer Lobelias. Its habit of growth is dense and compact, and the flowers are borne so profusely as to form a mass. They are ½ in. across, star-shaped, and of a pale blue. It continues to flower for a long time, even till cut off by frosts. If preserved in a frame during winter, after the manner of bedding Lobelias, it is perennial, and may be propagated in spring by cuttings. New Holland. Lobeliaceæ.

Ithuriel's Spear *(Triteleia laxa).*

Iva.—Composite plants of no garden value.

Ivy Harebell *(Wahlenbergia).*

Ivy Toadflax *(Linaria Cymbalaria).*

Ixia.—Charming South African bulbs, very slender and elegant in growth, and brilliant in flower. They are not grown nearly so much as they deserve to be, probably because they are considered tender and require treatment under glass. They are most valuable open air plants, as they yield an abundance of bright bloom in summer for cutting. For culture outdoors, choose a light loamy soil, thoroughly drained, and with a due south aspect; if backed by a wall or greenhouse all the better. Plant the bulbs from September to January, at a depth of from 3 in. to 4 in., and 1 in. to 3 in. apart. As the early plantings make foliage during the autumn, it is necessary to give protection during severe frost, and this may be best accomplished by hooping the beds over and covering when necessary with mats; or if tiffany is used it may be allowed to remain till the danger of severe frosts has ceased. The plantings made in December and January require no protection in winter, but as they will flower later in the summer than the early plantings, an aspect should be selected where the sun's rays will be somewhat broken; attention to this will prolong the blooming period. On stiff soil, or soils that lay rather wet in winter, the beds should be raised and the bulbs surrounded with sand, care being taken that they are planted 1 in. or 2 in. above the level of the path; and where protection cannot conveniently be given, planting should not be made till December or January. There is a large number of varieties now in cultivation. The chief species from which they have been derived appear to be I. crateroides, patens, maculata, fusco-citrina, ochroleuca, columellaris, speciosa, and viridiflora, which last is of a beautiful sea-green, a colour that is quite unique among cultivated plants, and should in no case be omitted. A representative collection of varieties are the following: Achievement, Amanda, Aurantiaca, Cleopatra, Conqueror, Duchess of Edinburgh, Glory, Grachus, Hercules, Hypatia, Isabelle, Impératrice Eugénie, Lady Carey, Lady of the Lake, La Fiancée, Lesbia, Loela, Miralba, Nosegay, Pallas, Pearl, Prestios, Princess Alexandra, Sarnia's Glory, Sunbeam, Surprise, Theseus, Titian, and Vulcan.

Ixiolirion (*Ixia Lily*).—A small genus, numbering about four kinds. Natives of Western Asia. The varieties bear a strong resemblance to each other, and only appear to differ in the depth of colour of the flowers, which are of a bright violet-blue. They all grow from 1 ft. to 1½ ft. high, and have grassy foliage and large trumpet-shaped flowers that are borne in a loose, elegant manner. I. Pallasi has flowers of the deepest shade, and I. tataricum of the palest, the intermediate shades being those of I. montanum and I. Ledebouri. Such beautiful hardy plants as these are certainly deserving of a place in every collection, be it ever so select, for the flowers not only last long in good condition on the plants, but are also very enduring in a cut state. With regard to their culture, they should be treated in a similar manner to the rarer kinds of bulbs, such as Calochorti, Habranthi, Zephyranthes, and similar families, for though they may be quite hardy, it is not advisable to plant out such rare bulbs in ordinary borders. The position in which to grow these plants should be an open and dry one—a sunny border, for example, which is all the better if it has a wall at the back, so as to catch all the available sun heat possible in early spring, when the bulbs are pushing up their young leaves and flowers. The border should be well drained, and a bed of light, rich, loamy soil, say about 1 ft. in depth, should be placed upon the drainage. When the young growth appears, place a common handlight over the plants—even two panes of glass will be beneficial—and if similar protection is again afforded at the latter part of summer, it will tend considerably to keep the soil dry and warm, and so ripen the bulbs. A handful of dry, sharp sand placed in a layer beneath the bulbs and around them does good, as it is conducive to the formation of roots. It is probable that when more established with us they will be found as hardy and as free as some of our commonest hardy bulbs.

Jaborosa integrifolia.—An interesting dwarf perennial, allied to the Mandrake. It grows from 9 in. to 12 in. high, has broad, ample leaves, and large, pure white tubular flowers about 2 in. long, and fragrant and handsome. A native of Buenos Ayres, it is somewhat tender, and succeeds well only in light, warm soils in sheltered situations. It is best when planted close at the foot of a south wall in good loamy soil. Here it makes a good plant, but spreads so rapidly as to become troublesome. When flourishing it may be easily propagated by division of the long, creeping stems. Solanaceæ.

Jacobea (*Senecio elegans*).

Jacob's-ladder (*Polemonium cœruleum*).

Jasione (*Sheep's Scabious*).—A small genus, of perennial and annual duration, of the Bellflower family, all of which are of dwarf growth. J. humilis is a creeping, tufted plant about 6 in. high, bearing small heads of pretty blue flowers in July and August. Though a native of the high Pyrenees, it often succumbs to the damp and frosts of our climate, and therefore it requires a dry, well-drained part of the rock garden, and should be afforded a little protection during severe cold and wet in winter. In summer, however, it likes plenty of moisture. J. perennis is a taller kind, often growing above 1 ft. high. It has bright blue flowers, in dense heads, which appear from June to August; it is a rock garden plant of stronger constitution than the preceding, and flourishes best in good, light loam. Native of mountains of Central and South Europe. These perennial kinds may be propagated best from seed, as they do not divide well. J. montana is a neat, but not showy, hardy annual with small, pretty, bright blue flower-heads produced in summer in light, warm soils. Sow seed in autumn or spring. A native plant.

Jeffersonia diphylla (*Twin-leaf*).—A very interesting dwarf plant, allied to the Blood-root, and from 6 in. to 10 in. high. The flowers are white, about 1 in. across, and are freely produced in early spring when the plant is in vigorous health. It is a good plant for peaty and somewhat shady spots on rock-work, and in a minor degree for the margins of beds of dwarf American plants. Seed should be sown in sandy soil as soon as gathered, but careful division of the root in winter is the best way to increase the plant. A native of rich, shady woods in N. America.

Jerusalem Sage (*Phlomis fruticosa*).

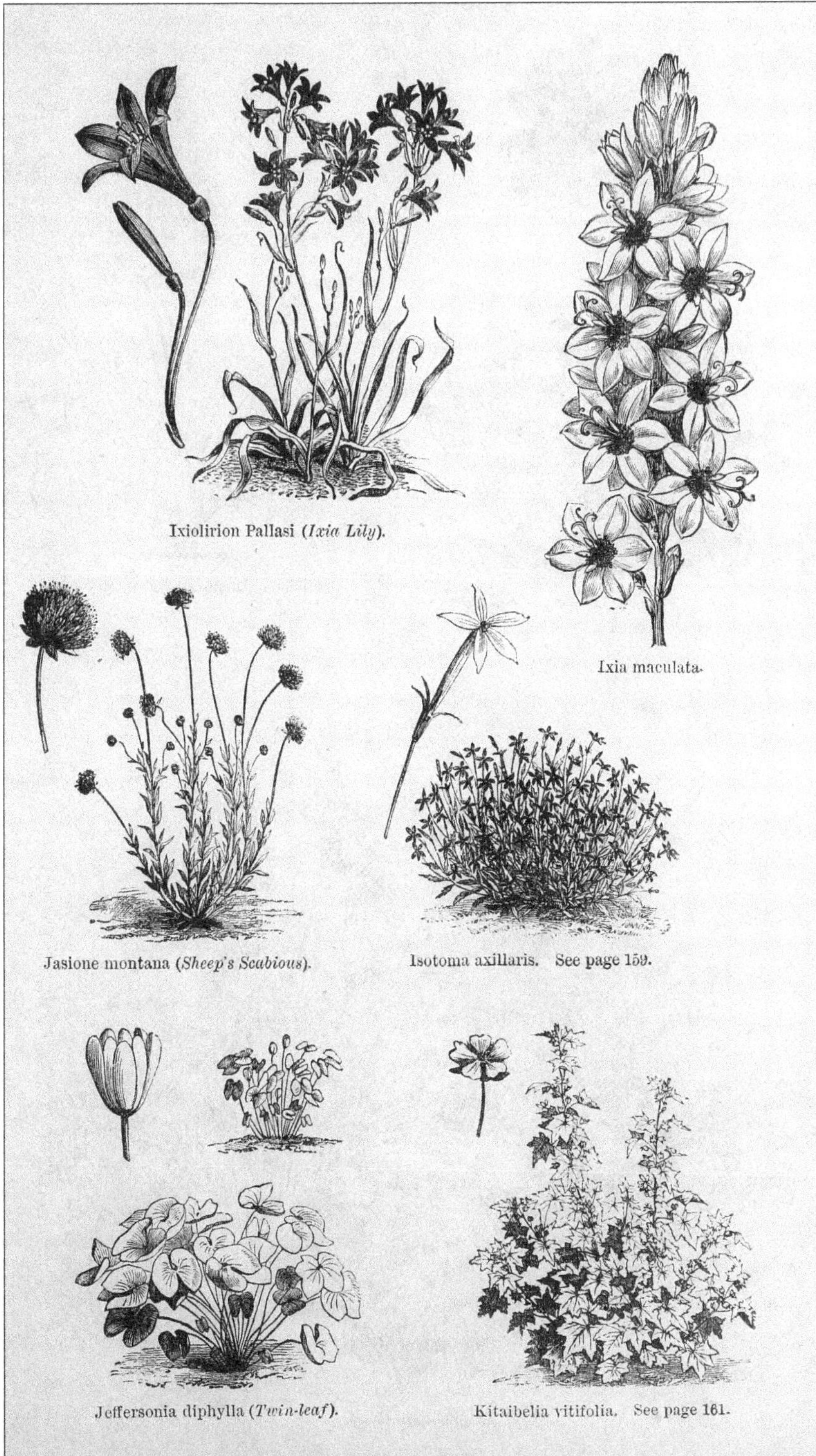

Ixiolirion Pallasi (*Ixia Lily*).

Ixia maculata.

Jasione montana (*Sheep's Scabious*).

Isotoma axillaris. See page 159.

Jeffersonia diphylla (*Twin-leaf*).

Kitaibelia vitifolia. See page 161.

[*To face page* 160.

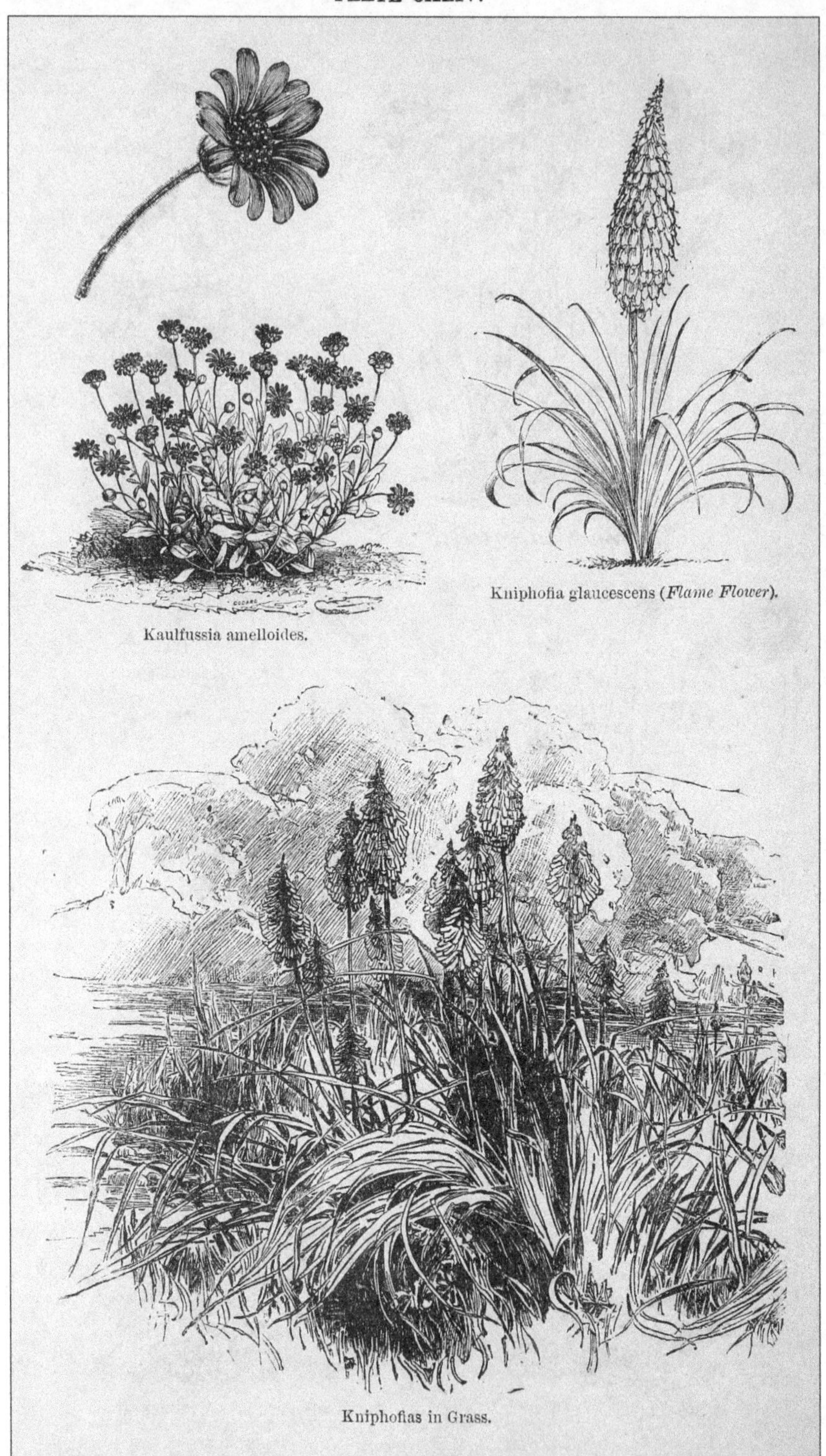

Kaulfussia amelloides.

Kniphofia glaucescens (*Flame Flower*).

Kniphofias in Grass.

[*To face page* **161.**

Jonquil *(Narcissus Jonquilla).*

Juncus *(Rush).*—Aquatic and marsh plants, generally with long, narrow, round leaves. J. effusus spiralis is a very singular-looking plant, forming spreading tufts of leaves, which, instead of growing straight like those of other kinds, are curiously twisted in a regular corkscrew form. From its very unusual appearance it is well worthy of cultivation, and may be planted with advantage on the margins of pieces of water, near cascades, &c., or in the artificial bog. It is easily multiplied by division of the tufts. J. zebrinus is apparently a form of the common Rush (J. communis). The long round leaves are transversely barred with alternate bands of yellow and green. It is a striking plant when associated with others, as its peculiarly rigid habit of growth and singular markings stand out in bold relief. Though it has not been grown entirely in the open air, it will most probably prove to be quite hardy and will retain its variegation; and if so, it will make a capital subject for planting amongst other moisture-loving plants. Native of Japan.

Jurinea.—A small genus of Compositæ, of no ornamental value. J. polyclonos and J. spectabilis, both Caucasian. plants, are mentioned in catalogues.

Jussiæa natans.—A curious aquatic plant that bears large yellow blossoms a few inches above the surface of the water. It is a valuable plant for still, artificial water, such as a pool or small lake, and, being hardy, it takes care of itself. Onagraceæ.

Justicia pedunculosa (= Dianthera americana).

Kaulfussia amelloides.—A charming hardy annual of dwarfish growth, with Daisy-like flowers, normally of a deep purple, but with white, rose, scarlet, and violet varieties, severally named in catalogues alba, rosea, kermesina, and atro-violacea. The plant forms a compact tuft, and is suitable for lines or masses, and a tuft here and there in the border would help to brighten it up. If sown in April, in the open, the plants will flower in June, and there is not much advantage in sowing earlier, as the seedlings soon develop themselves and flower. It would make a pretty ground or "carpet" plant with taller plants through it here and there. Cape of Good Hope. Compositæ. *Syn.*, Amellus annuus.

Kernera saxatilis.—A neat little plant very like the dwarf Scurvy Grass (Cochlearia). It forms a compact tuft of foliage, and in early summer is a dense mass of tiny white blooms. It grows in any soil in an open position in the rock garden, where it is an attractive plant in spring, and may be freely propagated by seeds. Europe. Cruciferæ.

Kitaibelia vitifolia.—A large, coarse-growing plant, from 4 ft. to 6 ft. or more high, with Grape-vine-like foliage. It produces in summer large white blooms from the upper parts of the stems. The plant is too coarse and not showy enough for border culture, but is well adapted for growing among shrubs, or for naturalising in the wild garden. Seed or division. Malvaceæ. Hungary.

Kleinia repens.—A small, dwarf, succulent plant, with cylindrical leaves of a bluish glaucous grey. It is used for geometrical beds in summer, but is not hardy. Propagated by division in early spring in heat, afterwards potted and planted out in May in light dry soil.

Knautia *(Scabiosa).*

Kniphofia *(Flame Flower).*—South African plants of the Lily family, of which there are about a dozen species and varieties in cultivation. All of these, however, bear a strong resemblance to each other. The most important kinds are those mentioned.

K. carnosa is a beautiful plant, different in shape and habit from the other species. It forms several low, spreading leaf rosettes, from the midst of which rise a number of flower-stalks to the height of 1 ft., producing a completely cylindrical flower-spike about 3 in. long and 1½ in. broad; but the comparative smallness of the flowers is compensated for by their glowing apricot colour, which is made still brighter by the bright yellow anthers. The individual flowers open first on the top side, and, taking it all in all, it is a lovely and striking plant. It flowers in September, and is a native of Abyssinia.

K. caulescens.—Though less brilliant in colour than K. Uvaria, its very distinct and robust habit, and the early period at which it blooms, will make this species welcome in any garden. The stem, at 5 in. to 6 in. from the ground, can just be spanned by two hands; the scape is about 4½ ft. in height, with a dense head of flower 6 in. in length. This is of a reddish-salmon colour in its earlier stages, but in the fully expanded flower the colour passes gradually to white, faintly tinged with greenish-yellow, producing an effective contrast. The foliage of this species affords a distinctive feature, being of a very glaucous blue-grey tint. It succeeds in light, warm soils, in open sunny positions, and, though hardy enough, it is advisable to give it a little protection in severe cold. K. Quartiniana is a similar, but less handsome species.

K. Macowani is a beautiful plant, a miniature of the preceding, being only from 1 ft. to 1½ ft. high, but its grassy foliage and oblong clusters of bright orange-red flowers make it very distinct. It is somewhat more tender than the common kind, though it succeeds well enough in light soil in a sheltered, sunny position.

K. triangularis, at first sight, reminds one of the charming K. Macowani, especially as regards the flower-spike, which is about the same size and of a similar tint. The foliage, however, is broader and longer, and in this respect it resembles K. Uvaria. It is a desirable kind because it is earlier in flower than the other varieties, and also because it is a free grower and flowerer.

K. Uvaria.—The common Flame Flower is a very important garden plant, being brilliant in colour, of noble habit, and very useful. It is a capital border plant in good light soil, and no better plant exists for groups in the picturesque garden, provided the situation be not too exposed or the soil too heavy. It begins to flower in late summer and lasts for many weeks in perfection. There are several varieties. K. pumila is a pretty dwarf plant, and K. præcox is distinct, as it flowers much earlier than the type. The variety nobilis is really a noble plant, the leaves being from 5 ft. to 7 ft. long, and the flower-stems over 5 ft. high, and surmounted by a flower-head 1 ft. or more long. The lower half is bright yellow, deepening through orange to bright red. It is a most conspicuous plant, and much earlier than the variety grandis, which is likewise very handsome. Saundersi is a variety that flowers intermediate between the preceding two, and is desirable on that account. K. Uvaria is synonymous with K. aloides and Tritoma Uvaria. K. Rooperi is much inferior as regards colour to the common Flame Flower, but for the sake of variety it is worth growing, as it flowers in summer. The same remark applies to K. sarmentosa and Burchelli.

PROPAGATION.—All the stemless kinds are easily propagated by division and seed when produced in favourable seasons; but the stemmed or caulescent kinds cannot be so treated. However, those who wish to increase their stock of these need not fear to behead them; in fact, it is the only way in which K. caulescens can be propagated, as it seldom develops offshoots unless so treated. When thus dealt with it will throw up a large number of shoots, which, if allowed to remain until a root or two are produced, may be taken off and kept in a close frame for a time and then potted in a sandy compost. K. sarmentosa is the easiest to increase, as it throws out underground shoots, which may be taken off at any time. K. Quartiniana develops small shoots at almost right angles with the base of the stem, and if these be taken off and treated as cuttings they may be relied upon to strike freely.

Knotweed *(Polygonum).*

Koniga *(Alyssum maritimum).*

Kœleria.—Grasses of no garden value.

Korolkowia Sewerzowi.—A singular-looking bulbous plant, allied to and much resembling a Fritillary. It grows from 1 ft. to 1½ ft. high, and has broad glaucous leaves and nodding flowers that are greenish outside and vinous purple within. A native of the mountains of Turcomania, quite hardy in our climate. Propagated by bulblets or seed.

Kuhnia eupatorioides. — A North American Composite of little value.

Lachenalia serotina (= Dipcadi and Uropetalum).

Lactuca *(Lettuce).*—Composite plants, more often seen in the kitchen than the flower garden. There are one or two blue or lilac-blue flowered kinds, grown in full collections, and not without beauty.

Lady Fern *(Asplenium Filix-fœmina).*

Lady's Fingers *(Anthyllis).*

Lady's Slipper *(Cypripedium Calceolus).*

Lagurus ovatus *(Hare's-tail Grass).* —An annual ornamental Grass, growing about 1 ft. high, and forming pretty hare's-tail-like plumes that are useful for bouquets. It should be sown in pots in August, wintered in frames, and divided and transplanted in spring or sown in open ground in April. It flowers from July to September.

Lallemantia.—A genus of the Labiatæ, nearly allied to Dracocephalus. L. canescens is the same as D. canescens, and is a plant of no ornamental value.

Lamarckia aurea. — A small hardy annual Grass, with silky plumes that become golden as they mature. It is suitable for bouquets, and may be dried for winter use. Seeds should be sown in spring or autumn, in the open border in light soil. (= Chrysurus cynosuroides.)

Lamium *(Dead Nettle).*—In this genus there are a few plants that are occasionally worth a place in dry poor soil where little else will grow—such as on dry banks or beneath trees. L. garganicum grows from 1 ft. to 1½ ft. high, and has whorls of purplish blossoms, produced in summer. L. Orvala is taller and has deep red flowers, which are produced in early summer. L. maculatum, a native plant, has leaves blotched with silvery-white. Of this species there is a variety called aureum which is one of the best of golden-leaved plants for edgings. It does not withstand such full exposure as that which suits the yellow Feverfew, but in sandy or moist soils its peculiar tint is unequalled by that of any other hardy plant, and its blooms are also pretty. It does not require to be constantly trimmed like the Feverfew. It is known also as L. striatum aureum.

Lancea tibetica.—A little alpine Labiate from the Himalayas. It is rare, and no one has proved its merits, though it is said to be a pretty plant.

Lantana.—Pretty S. American plants, usually grown in a greenhouse, but valuable summer border plants. The Verbena-like heads of bloom are rich and varied in colour, ranging from crimson through scarlet, orange, and yellow to white, the colours varying in the same head. They flower profusely for about three parts of the year, and are as simple to grow as a Pelargonium. They require the protection of the greenhouse during the winter after being lifted in autumn, and may be propagated in spring either by cuttings or seeds, the plants being grown on in rich light soil till planted out in a warm position. There are many sorts grown, a selection of which should include Phosphore, Don Calmet, Distinction, Eclat, Victoire, La Neige, Feu Follet, Pluie d'Or,

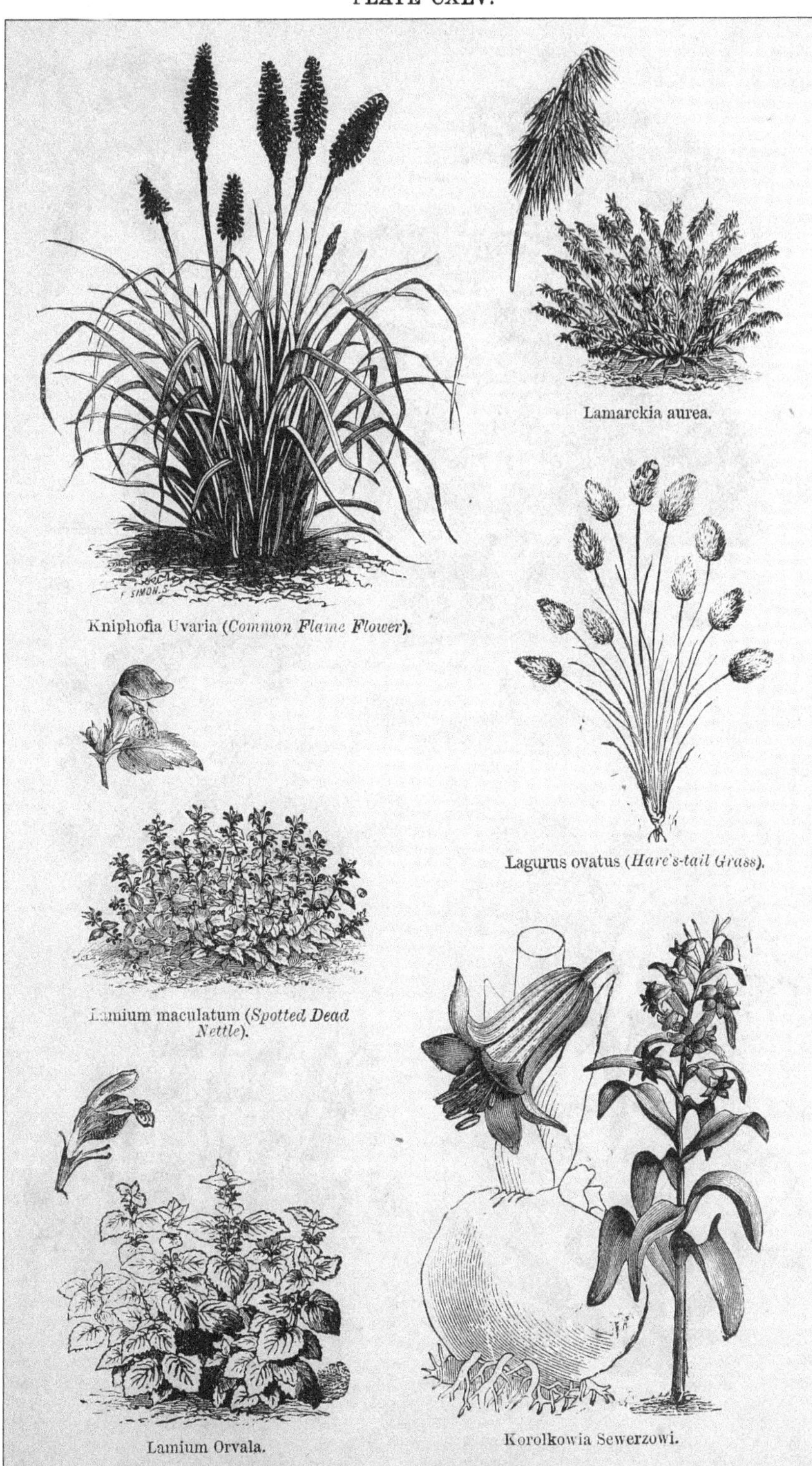

Kniphofia Uvaria (*Common Flame Flower*).

Lamarckia aurea.

Lagurus ovatus (*Hare's-tail Grass*).

Lamium maculatum (*Spotted Dead Nettle*).

Lamium Orvala.

Korolkowia Sewerzowi.

[*To face page* 162.

PLATE CXLVI.

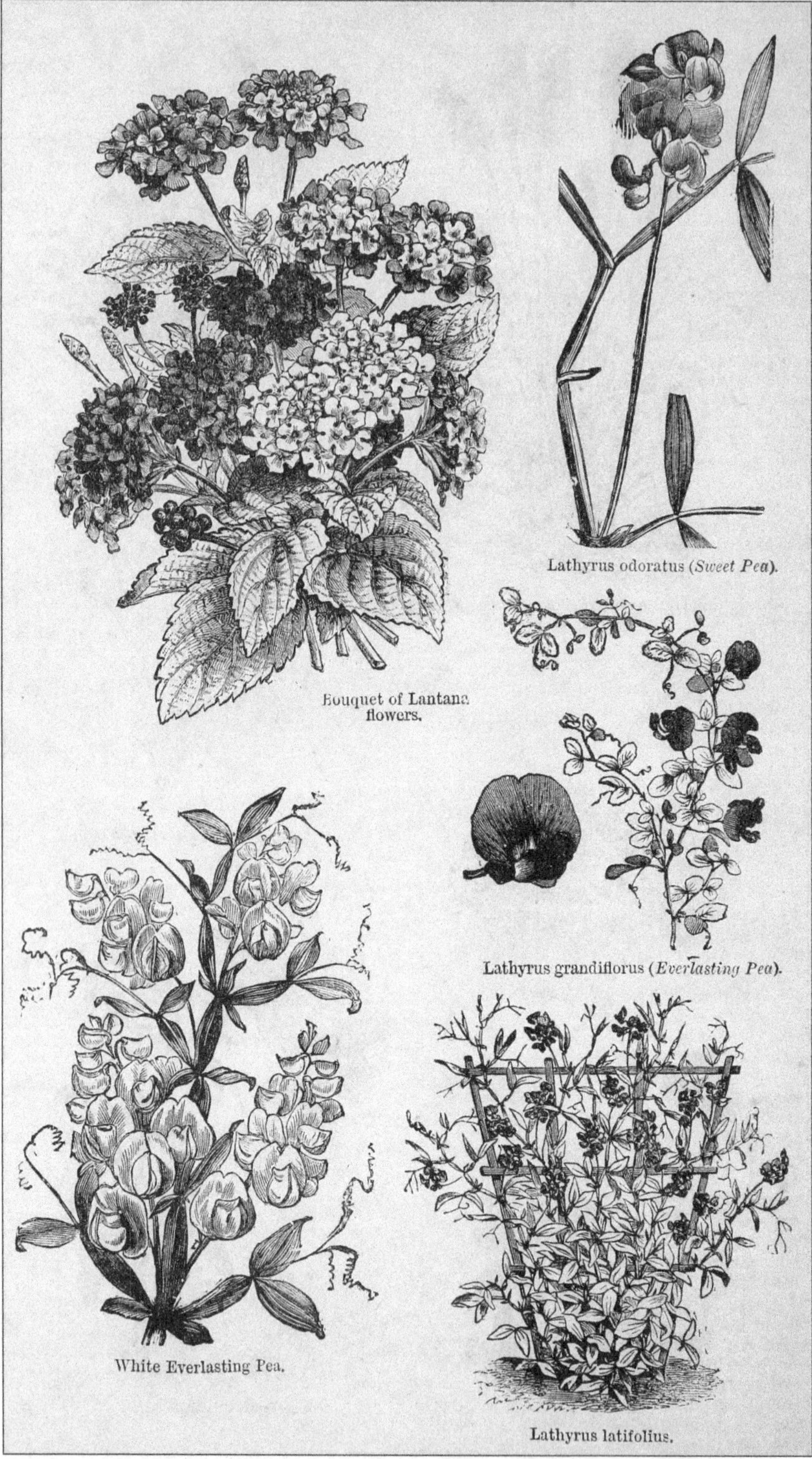

Bouquet of Lantana flowers.

Lathyrus odoratus (*Sweet Pea*).

Lathyrus grandiflorus (*Everlasting Pea*).

White Everlasting Pea.

Lathyrus latifolius.

[*To face page* **163.**

Ver Luisant, Ne Plus Ultra, Eldorado, and Heroine. Like many dwarf half-hardy plants, they have various uses in the flower garden, and may be trained as standards. The pretty L. Sellowi is a good dwarf plant. Their odour is somewhat objectionable, and they are not likely to be very much used. West Indies. Verbenaceæ.

Larkspur (*Delphinium*).

Lasthenia glabrata.—A pretty hardy annual, from 9 in. to 1½ ft. high, producing a profusion of rich orange-yellow blossoms. It should be sown either in autumn or early summer, or in spring for later bloom. Like other annuals, this looks best in broad tufts, but care must be taken that the plants are thinned properly. The autumn-sown plants come in with the Iberis, Wallflowers, early Phloxes, and the like. L. californica is a slight variety of the preceding. California. Compositæ.

Lastrea.—See Aspidium and Nephrodium.

Lathræa Squamaria.—A curious leafless parasite, often found in woods, growing from the roots of trees.

Lathyrus.—A genus of annual and perennial plants, several of which are very beautiful and valuable for the garden. The perennials—or Everlasting Peas, as they are called—are particularly valuable, for few plants are of more elegant growth. As a temporary screen in summer for shutting out unsightly objects, for rapidly covering arbours or trellis-work, or as floral fences for the protection of more tender plants, these Everlasting Peas are unsurpassed; and, lastly, they make charming plants for the wild garden associated with such companions as the Canary Creeper, the various kinds of Convolvulus and Clematis, and other climbing plants. The chief of the perennial kinds is

L. latifolius, one of the hardiest and most easily cultivated of plants, thriving almost anywhere, even in court-yards amongst flags and boulders. It may be made to ornament any dead, naked walls, for a few bits of it dibbled in among the stones under a wall will take care of themselves, as the long, leathery roots penetrate to a great depth. There is a pure white variety, also a striped one, and others with deeper coloured flowers than those of the type. All are peculiarly well adapted for wild, rough places, and ramble and scramble over bushes and stones. Staking, and tying, and training only spoil them. An old tree stump, or the side of a trellis or summer-house is where this Lathyrus delights to grow undisturbed.

Other Everlasting Peas are L. grandiflorus, which is not so rampant in growth, but handsomer in bloom than L. latifolius; the stems are weaker, and the flowers, which are produced in pairs, are twice the size. Another species, L. rotundifolius, is a lower-growing plant than any of the family, having some affinity to L. latifolius, but being neater in habit; it is well suited for stony banks. L. pyrenaicus is the most rampant of the family, growing 20 ft. in a single season, and bearing a prodigious quantity of blossoms, of a yellowish tint, veined with purple. This is a good kind for running over the trunks of dead trees and similar places. Lathyrus sylvestris belongs to our native flora. Its flowers are large and greenish, with purple veins, and it is a treat to see this plant festooning the bushes in open places in woods. L. Sibthorpi is remarkable for the early period at which it blooms, and no less so for the tender rosy-purple or deep mauve colour of its flowers. In habit it is dwarf, compared with L. latifolius and its varieties, rarely exceeding 3 ft. In very young plants the merits of this species are not so evident, but when thoroughly established in good, loamy soil it forms one of the greatest attractions of the spring border. L. Drummondi is a very robust grower, and perfectly hardy. The flowers are borne in clusters like those of L. latifolius, and are of a very bright carmine hue. The dwarf, non-climbing species are generally known under the name of Orobus.

L. odoratus (*Sweet Pea*).—This beautiful and popular annual needs no description. There are many ways in which the Sweet Pea can be turned to account in the adornment of a garden. A common method is to sow little patches in borders, the seed being generally that of mixed varieties, and then, by placing some stakes against them, pillars of flower are secured. Where it can be done, a hedge of Sweet Peas is an attractive sight, and sometimes they can be turned to account to hide an unsightly place during the summer. Many people grow a hedge in the same way in order to yield a supply of cut flowers, which are always produced abundantly. It is useless to grow the Sweet Pea unless planted in good soil. Some sow in late autumn, but this is not always a satisfactory plan, though when it succeeds the result is very fine. By sowing indoors in pots or boxes about the middle of February, and, when the young plants are 1 in. high, gradually hardening them off, Sweet Peas acquire a sturdiness and toughness which, when they are planted out in good, well-manured soil in April, conduces to rapid growth and immunity from the attacks of birds and slugs, which otherwise feast upon the tender shoots the moment they appear above ground. The soil should be well trenched and plenty of good stable manure worked in, and after the plants have been rather thickly dibbled in, they should have suitably-arranged supports of galvanised wire netting placed round them. Then, with a little attention during dry weather and regular removal of the incipient pods, they will produce an abundance of beautiful and deliciously fragrant flowers all through the summer and autumn. When getting rather past their best they should be cut down level with the tops of the sticks, and renewed growth will be the result. From the bottom to the top a new growth will spring up, and there will be an abundance of bloom until the end of October. There are many varieties of the Sweet Pea, varying chiefly in colour; the best of the named sorts are Butterfly, Black and Scarlet, Invincible, Painted Lady, Purple Striped, Princess of Prussia, and Fairy Queen.

Though none of the other annual kinds of Lathyrus can surpass the Sweet Pea, there are several others that make pretty border plants. Of these the best is the Tangier Pea (L. tingitanus), which grows about 3 ft. high, and produces small, dark red-purple flowers in abundance. The Chickling Vetch (L. sativus) is another pretty kind, with flowers varying from pure white to deep purple. The variety azureus is a remarkably elegant dwarf kind that bears a profusion of clear blue flowers; L. s. coloratus has flowers white, purple, and blue; L. Gorgoni, which grows about 2 ft. in height, is a very desirable plant, with pale salmon-coloured flowers; L. articulatus, Clymenum, and calcaratus are other pretty kinds suitable for borders.

Laurentia.—Small-growing plants allied to and resembling the Lobelia, but of doubtful value.

Lavatera *(Tree Mallow).*—These for the most part are vigorous and somewhat coarse plants, of biennial, annual, and perennial duration, few of which are desirable for the garden. The most useful is L. trimestris, a beautiful South European annual, growing from 2 ft. to 3 ft. high, and bearing in summer a profusion of large pale rose or white blossoms. It thrives in ordinary soil, preferring one that is rich and light. It may be sown in autumn or early spring in the open border. Among the taller kinds the best is L. arborea, which has the appearance of a small tree, and attains in the southern counties a height of nearly 10 ft. The stem branches into a broad, compact, roundish, and very leafy head. It may be used to adorn warm and sheltered parts of pleasure-grounds and rough places. In rich, well-drained beds it would prove a worthy companion for the Ricinus and the Cannas. It is most at home on dry soils, but during the summer months thrives on all. Being a biennial, it requires to be raised from seed annually. L. cashmeriana, unguiculata, thuringiaca, sylvestris, and others of a similar character are not worth growing except in the wild garden, or naturalised in other places.

Lavender Cotton *(Santolina).*

Lavendula.—The most familiar species of this genus, L. vera, the common Lavender, needs no description, but it is a shrub that should on no account be omitted. It succeeds best in an open sunny position, in light soil. The white-flowered variety is as sweet as the blue, and flowers at the same time. L. lanata, dentata, and Stœchas may also be seen in botanic gardens. Though a bush, the Lavender has been for centuries associated with our old garden flowers. For low hedges, as dividing lines in or around ground devoted to nursery beds of hardy flowers, it would be admirable, as in many other positions. It is excellent for the embellishment of dry banks and warm slopes. The Sage order. S. Europe.

Lawn Pearlwort *(Sagina glabra var. corsica).*

Leavenworthia aurea.—A pretty cruciferous annual of very dwarf habit, forming a neat rosette-like tuft, from which arise numerous stalks, 4 in. or 5 in. high, each bearing a single flower, about ½ in. across, with white petals stained at the base with deep yellow. Occasionally, when strongly grown, a short stem is thrown up, bearing the flowers on long pedicels. It blooms in May and June, and should be treated as a half-hardy annual. Arkansas.

Leersia oryzoides.—A native aquatic Grass—one of the few that thrive when half submerged. Of no garden value.

Leiophyllum buxifolium *(Sand Myrtle).*—A neat and tiny shrub, forming compact bushes, from 4 in. to 6 in. high, and densely covered with pinkish-white flowers in May, the unopened buds being of a delicate pink hue. It is particularly suited for grouping with diminutive shrubs, such as the Partridge Berry, the sweet Daphne Cneorum, the small Andromedas, and Willows like S. reticulata and serpyllifolia, that rise little above the ground. It likes peat. A native of sandy "pine barrens" in New Jersey, and easily to be had in our nurseries under the name of Ledum thymifolium. There is more than one variety in cultivation.

Leontice *(Lion's Leaf).*—A small genus of dwarf perennials of the Barbery family, native of Europe and Asia. There are some three or four species in cultivation, though they are only found in botanical collections. They are all pretty plants, about 1 ft. high, and bear in spring numerous bright yellow blossoms. L. Leontopetalum, Chrysogonum, vesicaria, and odessana are the kinds grown, and all of these are perfectly hardy; but, as they seem to be injured by the excessive moisture of our winters, they are generally cultivated in frames or under hand-lights, in order that they may be sheltered, and also that their flowers may be preserved from the effects of the weather in spring. A light friable soil is best to plant them in; the large depressed root-stocks, which resemble the corms of a Cyclamen, should not be placed beneath the surface of the soil, but fastened by some means, so that only the bases, from which the fibry roots are emitted, should be in contact.

Leontodon *(Dandelion).*—Besides the common Dandelion there are others of this genus enumerated in catalogues, but none of them are of value.

Leontopodium alpinum *(Edelweiss)* —This little hoary alpine herb, so well known to alpine tourists, is well worthy of culture, as the star-like heads of leaves surrounding the small, yellow, inconspicuous flowers are clothed with a dense white woolly substance that renders it distinct

PLATE CXLVII.

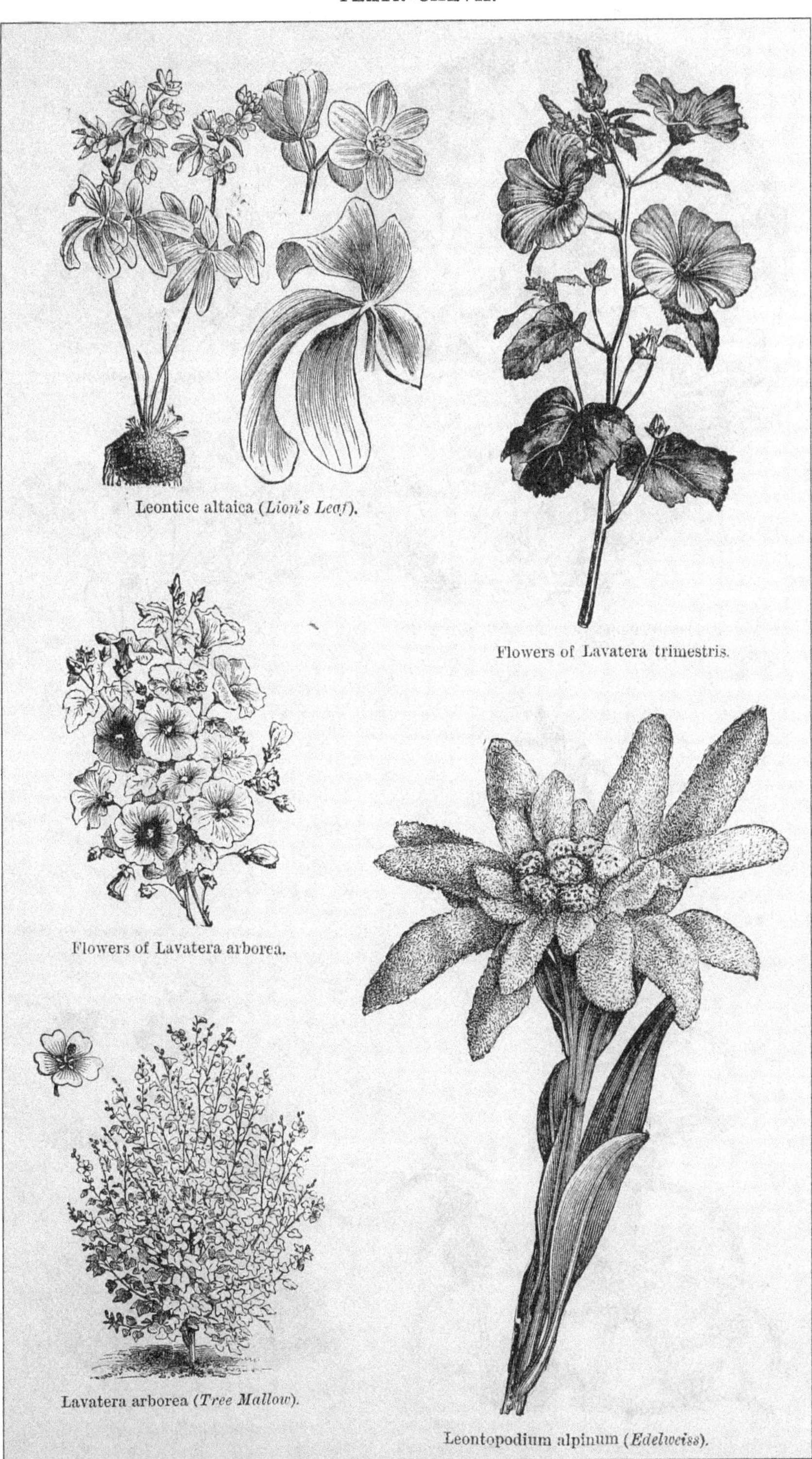

Leontice altaica (*Lion's Leaf*).

Flowers of Lavatera trimestris.

Flowers of Lavatera arborea.

Lavatera arborea (*Tree Mallow*).

Leontopodium alpinum (*Edelweiss*).

[*To face page* **164.**

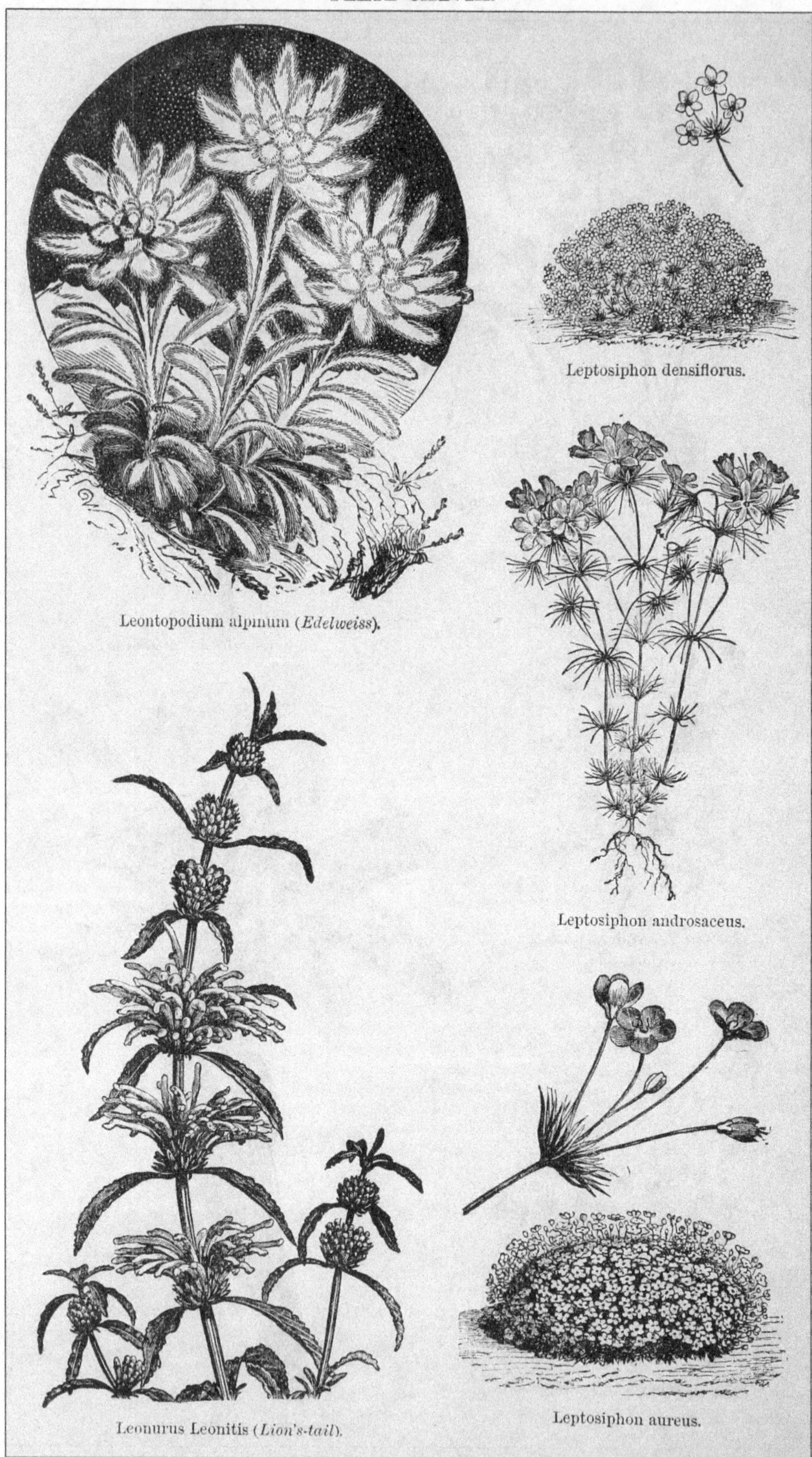

Leontopodium alpinum (*Edelweiss*).

Leptosiphon densiflorus.

Leptosiphon androsaceus.

Leonurus Leonitis (*Lion's-tail*).

Leptosiphon aureus.

[*To face page* **165.**

and pretty. Its culture is not difficult in any way, on sandy soils, even as a border plant, but it grows most luxuriantly in moist rich soils. In order to keep a good stock of flowering plants, the old ones should be divided annually or young ones raised from seeds, which in some seasons ripen plentifully. It succeeds well either in the rock garden on exposed spots or in an ordinary border, if not placed too near rank-growing things. (= Gnaphalium alpinum.) Compositæ.

Leonurus Leonitis *(Lion's-tail)*.—A plant of the Labiate family, allied to Phlomis, growing about 2 ft. high and bearing in summer whorls of bright scarlet flowers that are very showy. It is a Cape plant, and not sufficiently hardy to stand our climate during the winter, even when supplemented by the protection of a cold frame, though in warm light soils, in the southern parts of the country, we have observed it thrive well out of doors in summer. In districts where it will not bloom out of doors it is well worthy of a place as a cool greenhouse plant. About Paris, established plants placed out for the summer flower profusely. Wherever it can be grown well in the open air, it would form a valuable plant for association with the finer bedding and sub-tropical plants. Cuttings strike freely in spring—more freely than in autumn—in a slight bottom-heat, by merely removing the young shoots from a growing plant.

Leopard's-bane *(Doronicum)*

Leptinella scariosa.—A very dwarf Chilian plant, having small, deeply-cut foliage that forms a dense, compact carpet, on which account it is employed for flat geometrical bedding, for though a fast grower, it does not require clipping or other attention. It will grow in any soil or situation, and is useful for clothing places in the rock garden where choicer subjects would fail. Its flowers are inconspicuous. Propagated by division. Compositæ. L. filicaulis and L. dioica, if not synonymous, are very similar.

Leptosiphon.—All the species of this genus are deserving of cultivation. To produce the best results with these charming annuals, they must be strongly grown, and robust specimens can only be obtained by thin sowing. In light, dry soils early autumn sowing is strongly recommended. It is not enough to sow the seed at any period of autumn; it must be—and this applies to most annuals sown at this season—committed to the soil sufficiently early to permit the young plants to attain some size before the setting in of winter; in short, a good framework must be laid, which, on the return of spring, will develop into an effective tuft. When the sowing of the seed has been deferred until a late period of the summer, it is better to sow in pans or boxes, in a frame or orchard house, and plant out seedlings in April; or, in case this should be thought to involve too much trouble, the sowing may be deferred till early spring. A fair amount of success may be looked for, especially in good soils, in which spring sowing will often yield excellent results; the advantages of autumnal sowing are, in fact, best seen in light sandy soils. Of the numerous kinds in cultivation the best is L. roseus, which is one of the most charming of all hardy annuals. It forms dense, compact tufts, profusely studded with rosy carmine flowers. The very pretty L. luteus and its deeper-coloured variety aureus are scarcely inferior to L. roseus, which they closely resemble in their general habit, but have somewhat smaller flowers. The hybrid varieties of these, introduced by Messrs. Vilmorin, are no less interesting for the singular variety of shades occurring among them. The larger-flowered species, L. densiflorus and L. androsaceus should be too well known to need either description or eulogium. Both have lilac-purple flowers, and are most attractive annuals, never failing to delight their possessor. Of both species there are good white varieties which deserve especial recommendation; that of L. densiflorus is particularly desirable. All natives of California. Polemoniaceæ.

Leptosyne.—Californian plants of the Composite family, much resembling some of the Coreopsis. L. Douglasi is a pretty half-hardy annual, growing about 1 ft. high, and having large yellow flowers. L. Stillmanni much resembles it, but is smaller. L. maritima, though naturally a perennial, is somewhat tender, and requires to be treated as an annual. It is a showy plant, about 6 in. high, and produces large, bright yellow flowers. These all thrive best in a light warm soil in an open, sunny position. The seeds should be sown early in heat and the seedlings transplanted in May.

Leucanthemum alpinum *(Alpine Feverfew)*.—A very dwarf plant. The leaves are small, and the flowers, which are pure white with yellow centres and supported on hoary little stems from 1 in. to 3 in. long, are more than 1 in. across, and produced abundantly. It is a rather quaint and pretty plant, and well deserves cultivation on rockwork in bare, level places, on poor, sandy, or gravelly soil. It is sometimes known as Chrysanthemum arcticum and Pyrethrum alpinum. It is a native of the Alps of Europe, and is readily increased by division or by seed. For other species of Leucanthemum see Chrysanthemum.

Leucojum *(Snowflake)*.—Pretty bulbous plants allied to the Snowdrop. There are several species in cultivation, but only two and their varieties are common.

L. æstivum *(Summer Snowflake)* is a tall and vigorous plant, bearing flowers on stalks from 1 ft. to 1½ ft. high. The flowers are white, drooping bells, marked with green both inside and out, and are produced in clusters of from four to eight blooms on each stem. The leaves are very numerous, and in shape like those of Daffodils. It blooms early in summer (in many places before spring has ended), and forms a pleasing object either in the mixed border or on the margins of shrubberies, where, in company with Solomon's Seal, the grace of its pendent flowers will perhaps be even better appreciated. It thrives in almost any kind of soil, and is readily multiplied by separation of the bulbs. It is an excellent plant for the wild garden, and increases as rapidly as the common Daffodil under the same circumstances. Another Snowflake, a form of L. æstivum, known in gardens under the name of L. pulchellum, is L. Hernandezi, a native of Majorca and Minorca. This grows to about the same height as L. æstivum, but has narrower leaves, flowers only half the size, and usually not more than three on each stem; it also flowers nearly a month earlier. Being greatly inferior in its appearance to L. æstivum, it is not much cultivated. Those, however, who wish to grow it, can treat it in the same manner as L. æstivum. The latter is naturalised in some parts of England.

L. vernum *(Spring Snowflake)* bears flowers on stalks from 4 in. to 6 in. high. The fragrant drooping flower resembles that of a large Snowdrop, the tips of the petals being well marked with a green or yellowish spot. It is an excellent subject for rockwork, and no less valuable as a border plant, thriving in a light, rich, and well-drained soil. Imported bulbs of the spring Snowflake make little show for the first year or two, but after that time, when established in sandy loam and peat in a somewhat shady border, they flower very freely and regularly. Probably the vitality of the bulbs is impaired by the drying process to which all bulbs are subjected in going through a market, but this process has a beneficial effect on some species. L. carpaticum flowers a month or six weeks later than L. vernum, and, unlike that species, begins to expand its blossoms when the stalk and the leaves are scarcely above the ground; the petals, too, are tipped with yellow instead of green. Other cultivated Snowflakes are L. hyemale and L. roseum; both of these, however, are very rare and somewhat difficult to cultivate.

Leucophyta Brownei.—A New Holland plant, with slender hoary stems. It is now largely used in a small state for flat geometrical beds in summer. Propagated by cuttings in early spring. Compositæ.

Leuzea. — Coarse growing Composite plants. L. conifera and salina are the kinds in cultivation.

Levisticum. — Vigorous umbelliferous plants, cultivated in botanic gardens.

Lewisia rediviva *(Spatlum)*.—A most remarkable and beautiful rock garden plant, similar to the Portulaca. It is very dwarf, being only 1 in. or so high, and has a small tuft of narrow leaves, from the centre of which the flower-stalks arise. The blossoms are large for the size of the plant, being from 1 in. to 2½ in. across, and vary from deep rose to white. The roots are succulent, and possess the power of retaining life under the most unfavourable conditions. It sometimes fails to develop leaves annually, and hence is often considered to be dead and a difficult plant to deal with, though such is not the case. It should be grown in sunshine, and not in shade, or it cannot be flowered. The plant itself should be kept high and dry, but the roots should have plenty of moisture; a crevice in a rockery is the best situation for it. If grown in pots the plants should be on broken stones, and the roots in light sandy loam with peat. After flowering, it shrivels up and becomes a withered, twisted mass, like so many bits of string, but this is the nature of the plant, and hence its name. Oregon, Utah, and Rocky Mountains.

Liatris *(Snakeroot)*.—A genus of N. American Composites, characterised by having the flower-heads, generally purple, arranged in long dense spikes. When well grown some of them are effective border flowers and well repay good cultivation. L. elegans grows about 2 ft. high, and has pale purple spikes 1 ft. or more in length. L. pycnostachya, which grows from 2 ft. to 4 ft., has deep purple flower-spikes, produced from August to October. L. spicata is one of the handsomest and neatest in growth. It grows from 1 ft. to 2 ft. high, and the violet-purple spikes continue a long time in beauty. L. scariosa, squarrosa, cylindracea, elegans, and pumila much resemble the foregoing, and, like them, will succeed well in any rich light soil. Propagated by division in spring or by seed.

Libanotis.—Weedy umbelliferous plants.

Libertia.—Beautiful Iridaceous plants, of which some three or four are hardy enough for the open border. L. formosa is beautiful at all seasons, even in the depth of winter, owing to the colour of its foliage, which is as green as that of the Holly, and it produces spikes of flowers of snowy whiteness more like those of some delicate Orchid than of an outdoor plant. It is neat, dwarf, and compact, and has flowers, twice the size of those of the others, lying close together on the stem, and reminding one of the old double white Rocket. L. ixioides, a New Zealand plant, is also a handsome evergreen species with narrow, grassy foliage and small, white blossoms. L. magellanica is also attractive when in flower. All of these thrive in warm borders of light soil, and in the rougher parts of the rock garden. Increased by seed or by careful division in spring.

Ligularia. — Large - growing perennial Composites remarkable for their bold, handsome foliage. L. macrophylla is a vigorous perennial, with an erect stem nearly 3½ ft. high, and very large, glaucous, erect leaves. The flowers are yellow, and are borne in a dense long spike at the end of the stem. The most suitable soil for this plant is that which is free, moist, and somewhat peaty. Multiplied by careful division in autumn or in spring. It is a useful subject for grouping with fine-leaved herbaceous plants, but will seldom command a place in the select flower

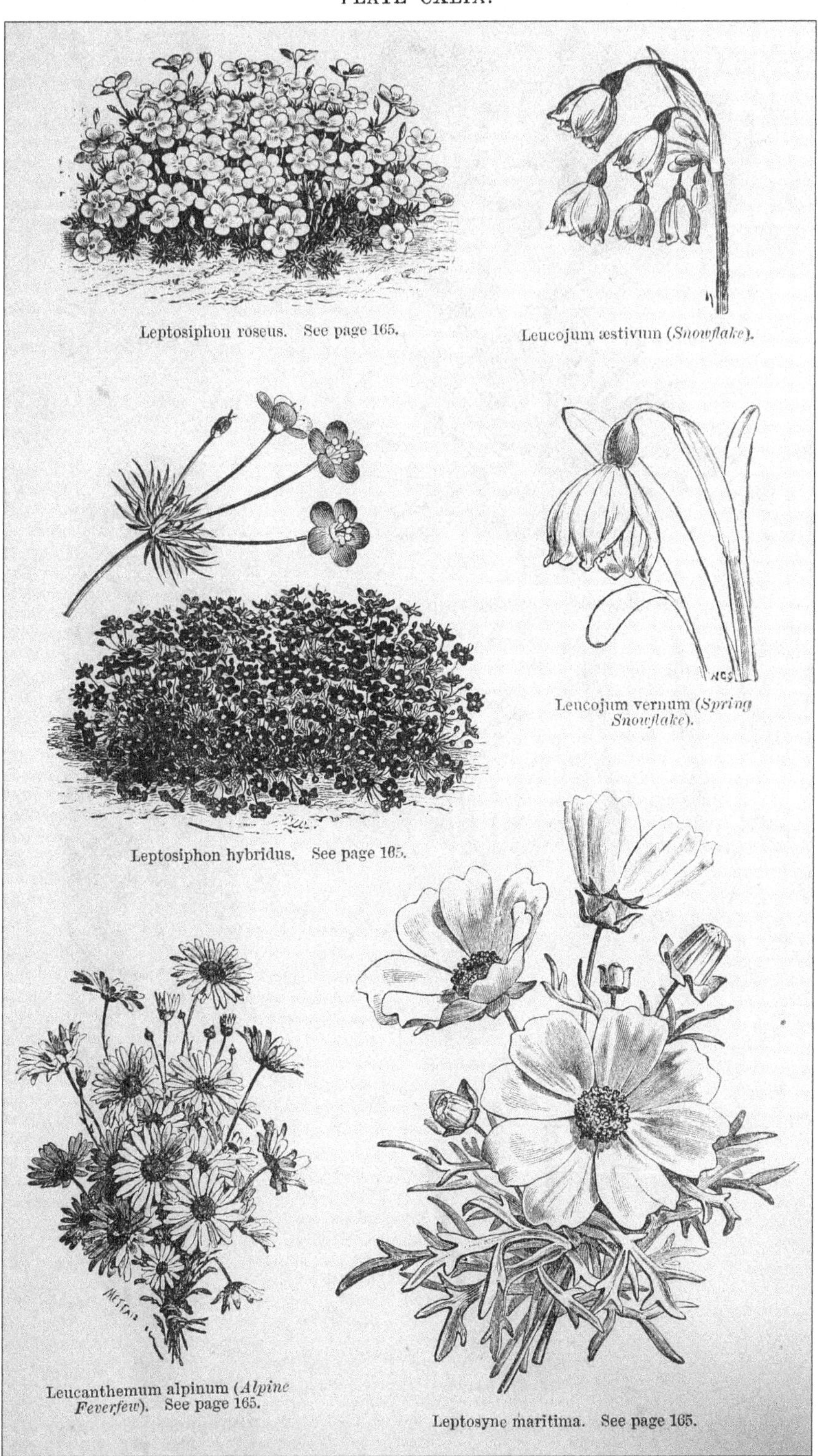

Leptosiphon roseus. See page 165.

Leucojum æstivum (*Snowflake*).

Leptosiphon hybridus. See page 165.

Leucojum vernum (*Spring Snowflake*).

Leucanthemum alpinum (*Alpine Feverfew*). See page 165.

Leptosyne maritima. See page 165.

[*To face page* **166.**

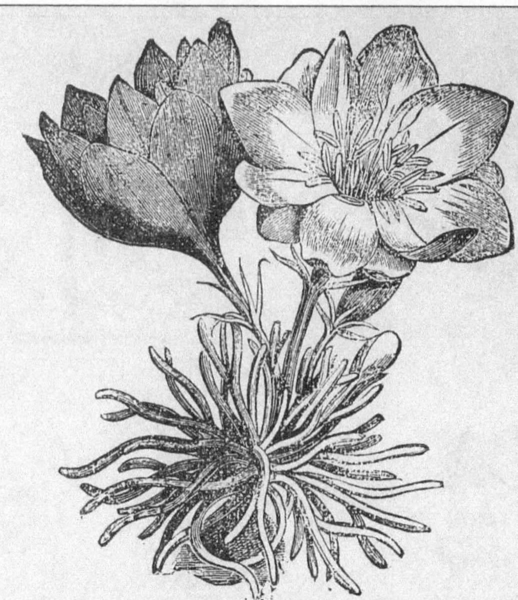

Lewisia rediviva (*Spatlum*). See page 166.

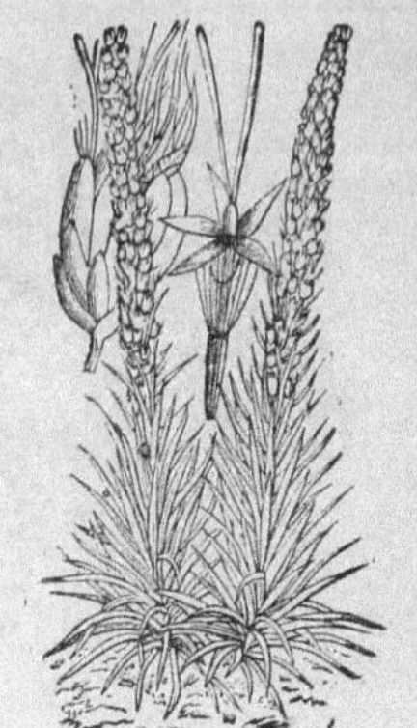

Liatris elegans (*Snakeroot*). See page 166.

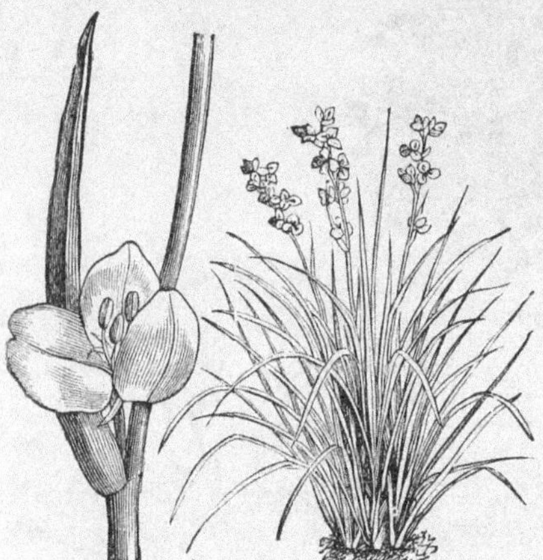

Libertia formosa. See page 166.

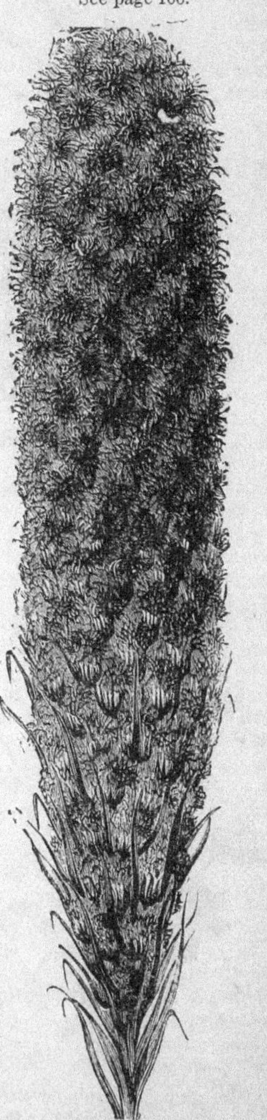

Liatris spicata. See page 166.

Liatris pycnostachya. See page 166.

garden. Caucasus. L. sibirica, Fischeri, and thyrsoidea are similarly fine-leaved plants, and worth growing with L. macrophylla for their foliage alone. The Japanese species, L. Kæmpferi and Hodgsoni, are better grown under glass, except in summer, when they may be used profitably among fine-leaved plants in the sub-tropical garden.

Ligusticum.—Plants of the Umbelliferæ; not useful in the flower garden.

Liliorhiza lanceolata (= *Fritillaria liliacea*).

Lilium (*Lily*).—The Lilies are such a beautiful class of plants, that no garden should be without a variety of them, combining, as they do, both stateliness and grace in habit with brilliant and delicate colouring in their flowers. The many kinds now in cultivation afford a rich choice for the garden. All are beautiful, but some better suited to particular localities than others. The habit and general character of the plants being so varied, their uses and position in the garden are likewise various, some being suited for the rock garden, others for the mixed border, many for the shrubbery—and especially the Rhododendron beds, while not a few are so robust that they are at home in the wild garden, holding their own against the native plants. The true place for the Lily, however, is the garden proper, and when its uses are understood and expressed the result will be a total change in the aspect of the flower garden. It is not only the culture that is important: the arrangement and grouping are even more so. A number of Lilies will grow in any ordinary soil; a good, rich loamy soil suits the largest number, while others want plenty of sand, so as to keep the soil free; and others, again, can be easily grown in ordinary soil if it is mixed with leaf mould or peat. It will thus be seen that there are no great difficulties in the way of growing a large number of kinds. In nearly all cases Lilies are more vigorous in growth, more brilliant in colour, where they receive a partial protection from severe frosts, and the flowers will last a longer time when they are sheltered from the scorching rays of the mid-day sun. The shrubbery border, among Rhododendrons (for those requiring peat), in the mixed border between shrubs and herbaceous plants, where the young shoots get a slight protection from the early frosts, are among the best situations. A very safe place is near the edge of a Rhododendron bed; soil that will grow Rhododendrons will also answer for most sorts of Lilies, and, in addition, protection will be afforded, for in some seasons, notably when cold and wet follow after drought, cultivators find that "blight and spot" greatly injure the growth and flowering of some species, even though the bulbs may be unhurt. It should be remembered that bulbs of nearly all Lilies occasionally lie dormant a whole season, and push out luxuriantly the following summer.

Manure should never be dug in with the bulbs, though they accept it gratefully as a top-dressing liberally applied after they have been established a year. The only manure to be dug in at the time of planting is rich peat and sand, one-third of the latter. This is advisedly called manure. L. auratum and some others in light soils are all the better for a top-dressing of clay put on dry and broken small. Though under each brief description of the chief kinds given below we have appended a word or two upon cultivation, it is, perhaps, advisable to give a few general remarks upon the subject. It should be borne in mind that however beautiful nearly all the known Lilies are, some of them are extremely fastidious; but there is a rare choice of beauty among those that are easily cultivated. Lilies for cultural purposes may be divided into three classes—first, those that are best grown in pots, such as neilgherrense, Wallichianum, philippinense, and nepalense, and, in many soils and climates, speciosum, auratum, and longiflorum; secondly, those that are best grown out of doors in loamy soil; thirdly, those that are best grown out of doors in peaty soil. On light soils the following kinds do remarkably well: All of the umbellatum, croceum, and elegans type; L. candidum, longiflorum and its varieties, chalcedonicum, excelsum, and the speciosum section; also tigrinum sinense. In deep loamy soil, L. auratum, Szovitzianum, Humboldti, the Tiger family, most of the Martagon group; while in an intermediate soil of leaf mould, loam, and sand, Buschianum, philadelphicum, pulchellum, Browni, giganteum, tenuifolium, Krameri, &c., should be planted. The North American forms require more peat and more moisture than the other groups. Lilies require, so far as their roots are concerned, a cool bottom, abundant moisture, and, for most kinds, a free drainage; for instance, the slope of a hill facing south-east or south-west, with water from above percolating through the sub-soil, so as always to afford a supply, yet without stagnation, would be an admirable site.

PROPAGATION.—This is generally and most readily done by means of the separation of the bulblets or offsets from the parent bulbs, and these, detached and grown in the same way as the parent, will in the course of a year or two make good flowering plants. The scales of the bulbs, too, afford a means of propagation; but this is a slower method. Raising plants from seed is a somewhat tedious process with the Lily, though many kinds in this country perfect seed in plenty, and in the case of such kinds as L. tenuifolium the seedlings will flower in the course of three or four years; others, again, will not flower for several years. The finest kinds, such as the Japanese and Californian Lilies, are now imported and sold so cheaply in large quantities, that it is scarcely necessary to propagate from home-grown plants. It will be well, however, if, by rapid increase or otherwise, these plants may become plentiful enough to adorn the smallest cottage gardens. As just stated, several kinds of Lilies, chiefly Japanese and Californian, are annually imported from their native habitats in large quantities. All bulbs as soon as received should be carefully examined, and any decaying matter removed. They should then be laid in soil, or, better still, Cocoa-nut fibre in a moderate condition of moisture, until the bulb recovers its usual plumpness and the roots are just on the point of starting from the base. Then they should be potted or planted out as required; but, before doing this, any decaying scales should be again removed, as a few of the outside ones are often bruised in transit, and after being in the soil a little time decay sets in, which if not then taken off may contaminate the whole bulb. Of those so imported, L. auratum and Krameri should, when potted, be surrounded with sand, while some do well without that precaution. L. Washingtonianum and rubescens are the most difficult to import among the N. American Lilies, suffering, as a rule, much more than the large solid bulbs of L. Humboldti or those of pardalinum, canadense, and superbum. These bulbs should be treated as above directed, but L. Washingtonianum, rubescens, and Humboldti should not be potted, as they never succeed in that way; indeed, the whole of the N. American Lilies do much better planted out. Those imported from Holland, such as the varieties of davuricum, elegans, and of speciosum, &c., arrive plump and sound, but even in their case it is much better to lay them in soil a little while before potting.

L. auratum.—One of the grandest of Lilies, now too well known to need description. It is a most valuable plant both as regards size and colour of the blossoms; some of the finest forms have flowers nearly 1 ft. across, with broad white petals copiously spotted with reddish-brown and with broad bands of golden yellow down the centre of each. The poorest forms have starry flowers and scarcely any markings. There are several named varieties that are particularly distinct, the chief of which are cruentum and rubro-vittatum, which have deep crimson instead of yellow bands down the petals; Wittei or virginale, the flowers of which have no colour but the golden bands; rubro-pictum, with a red stripe and spots; platyphyllum, with very large flowers and broad leaves; and Emperor, a grand flower, with reddish spots and centre. Besides these there are some very beautiful hybrids raised between this and some of the other species; for example, L. Parkmanni, between L. auratum and L. speciosum, the flowers of which are large, white, banded and spotted with carmine crimson. The bulbs of this Lily are now regularly imported in large quantities from Japan and sold at a cheap rate every autumn, so that it is readily procurable. Much has been written upon the culture of this Lily, but the matter is not so difficult as many suppose, provided a good start is made. It will thrive in ordinary garden soil, if well drained and fairly enriched. It grows freely in peat or loam, a mixture of both with a little road scraping best fulfilling its requirements. Where the soil is naturally of a poor, light, sandy description, it should be taken out to a depth of 18 in., and replaced either with the compost above mentioned, or with some fine, well-enriched mould. In this the bulbs should be planted, and should, as soon as growth commences in the spring, be mulched with some decomposed manure or short Grass. If the garden soil is fairly good, it will only need to be well stirred and manured, but the manure employed should be thoroughly decomposed. A sheltered situation should be chosen, screened, if possible, from the mid-day sun, and protected from westerly and southerly gales and heavy driving rains. Indeed, this is one of the principal conditions to observe, for this Lily is very susceptible to injury by cold draughts and cutting winds. Hence no better place can be chosen than a snug nook sheltered from the north and east by shrubs, but otherwise open to the sun. The best examples that have been grown of it were in a Rhododendron bed planted in a deep, moist, peaty soil, where they have been for years undisturbed. When planted among other things the young and tender uprising shoots in spring are greatly protected. As to propagation, there is scarcely any need to enlarge upon that subject, as the bulbs are imported so plentifully; but when it is desired to increase the home-grown plants, it is only necessary to separate the young bulbs and replant them in good soil. Those who increase this Lily from seed must be prepared to exercise a little patience, as the seed is long in germinating and the seedlings take several years before they flower. The seed should be sown, as soon as ripe, in a frame. The seedlings should be planted out as soon as the bulbs are of an appreciable size.

L. Browni is a fine Lily in the way of L. japonicum, but with larger flowers. It is readily distinguished from any other kind by the rich brownish-purple markings on the exterior of the blossoms, which in well-grown plants are sometimes 9 in. in length. It is a very satisfactory Lily to grow, as it is hardy and vigorous, and succeeds well without giving much trouble. In a soil and position such as suits L. auratum it flourishes well, and only needs to be lifted every few years and replanted in fresh, rich soil. It grows from 2 ft. to 4 ft. high, and has deep green foliage distinct from allied kinds. The variety Colchesteri is a handsome one.

L. bulbiferum is one of the handsomest of European Lilies, growing about 2 ft. high, and bearing large crimson flowers shading to orange. The variety umbellatum is a finer plant, of stronger growth, and bears its flowers in large umbelled clusters. This Lily is generally distinguished from its congeners by the bulblets on the axils of the leaves. It grows freely in any ordinary soil, and

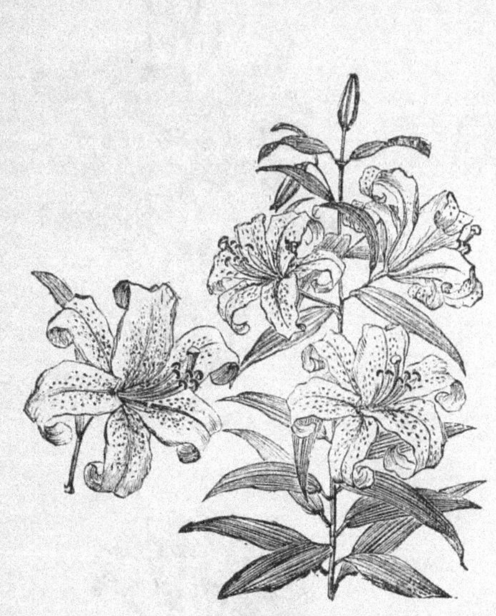

Lilium auratum.

Lilium Browni.

Lilium candidum in a border. See page 169.

[*To face page* 168.

PLATE CLII.

Lilium candidum (*White Lily*).

Lilium croceum (*Orange Lily*).

Lilium chalcedonicum (*Scarlet Martagon*).

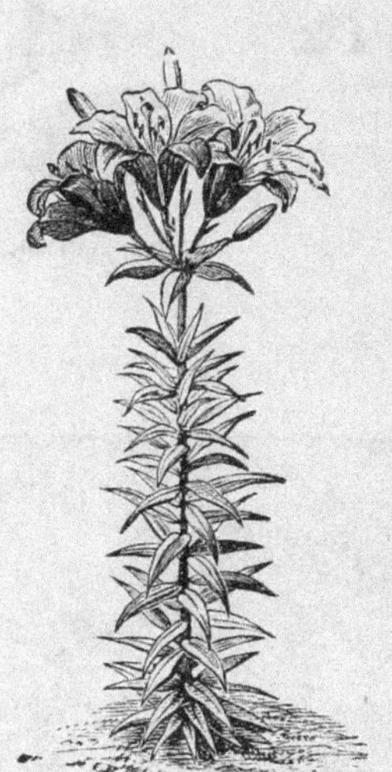

Lilium davuricum.

Lilium concolor.

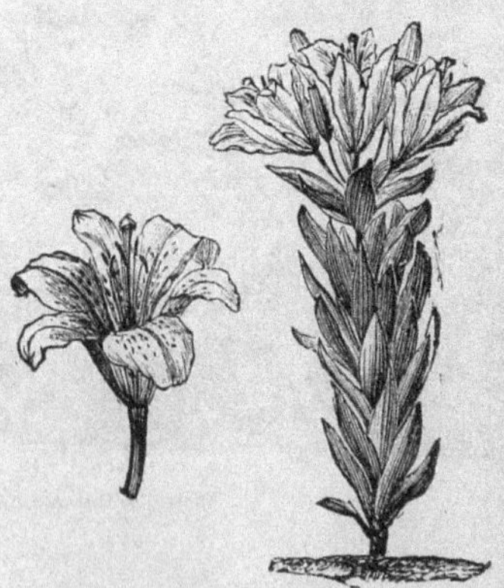

Lilium Thunbergianum (syn., elegans).

[*To face page* 169.

flowers in early summer. A capital plant for bold groups, thriving under partial shade or in the open.

L. canadense *(Canadian Lily)* is among the oldest of cultivated Lilies, and a beautiful one. It grows from 2 ft. to 4 ft. high, and bears, on slender stems, terminal clusters of drooping blossoms usually of an orange colour, and copiously spotted with deep brown. It occurs also with red flowers (rubrum) or with yellow flowers (flavum). L. parvum, L. nitidum, and L. columbianum are somewhat similar to the Canadian Lily, and require the same treatment—that is, a moist, deep, peaty soil enriched by decayed leaf mould, in a partially shaded position. The Canadian Lily flowers in late summer, and when in bold masses, such as seen often in the nurseries about London, is very attractive. Both it and allied forms will make elegant groups among choice shrubs such as Azaleas and Rhododendrons. By this plan we effect several improvements: get a second bloom and a variety of form from beds that formerly only had one blossoming season, and were poor and stiff in outline; prevent senseless digging of the beds once the groups are in place; keep the shrubs from growing into a solid, ugly mass, and grow and shelter our Lilies.

L. candidum.—One of the best known and loveliest of all the Lilies, seen in almost every cottage garden, and producing in summer its snow-white blooms. It is one of the kinds that dislike coddling or being meddled with, and thrives best when left undisturbed for years in good garden soil. Any attempt to deal with it in the same manner as with the more delicate ones generally results in failure. The best flowered plants are to be seen in old gardens, where the bulbs are allowed to run as they like with no attention whatever. Placed in bold masses, no plants can compare with the common white Lily when in bloom. It is so fair a flower, that there is scarcely a place which a good plant or well-grown group will not adorn. But it is the careful growth and proper placing of such lovely hardy plants as this that give the highest charm to the garden. For years it has been difficult to find even a miserable tuft in many "show" gardens, though nothing there displayed looked so well as a tall white Lily in a cottage garden. A moist loam seems to suit it generally, though, like other Lilies, it will grow on a variety of soils. The varieties peregrinum, striatum, and monstrosum are not so fine as the original, but the striped-leaved variety aureo-marginatis is valuable for the foliage in winter.

L. chalcedonicum *(Scarlet Martagon)* is a very old and handsome Lily, of tall, graceful growth and bearing several pendulous, vermilion-red, turban-shaped blossoms about the end of July. It is one of the easiest to cultivate, thrives in almost any kind of soil, and is best when well established and left undisturbed. There are a few varieties of it, majus being the largest and best. The others are græcum, rather taller than the type and having smaller flowers; pyrenaicum, which has yellow flowers; Heldreichi, tall and robust, flowering a week or two earlier; and maculatum, a very handsome form. Native of Greece and Ionian Isles. Similar to the scarlet Martagon is the Japanese L. callosum, a pretty Lily, from 1½ ft. to 3 ft. high, with slender stems bearing several brilliant scarlet blossoms in summer. L. carniolicum is likewise of a similar character, is from 1 ft. to 3 ft. high, and produces in early summer turban-shaped, nodding blossoms of a bright vermilion or yellow. Both these thrive under the same conditions as L. chalcedonicum.

L. concolor.—A pretty small-growing Lily from Japan, 1 ft. to 3 ft. high, bearing from three to six bright scarlet flowers spotted with black, star-shaped, and erect. There are some three or four varieties of this species—pulchellum, or Buschianum, an early-flowering variety from Siberia, growing from 1½ ft. to 2 ft. high, with crimson blossoms; Coridion, with flowers somewhat larger than the type and of a rich yellow, spotted with brown; sinicum, a Chinese form, with from four to six heavily-spotted crimson flowers, larger than concolor; and Partheneion, with scarlet flowers flushed with yellow. This charming Lily and its varieties are quite hardy, though they require some little attention in cultivating. They succeed in half-shady places in a soil composed of two parts peat, one of loam, and one of road scrapings; but the plant seems to require renewing every few years. This Lily and its allies are suited for grouping among the smaller and choicer evergreen shrubs where not grown in a special Lily bed.

L. croceum *(Orange Lily)* is one of the sturdiest and hardiest of Lilies, and therefore one of the commonest. It grows in almost any soil or position, and bears in early summer huge heads of large, rich orange flowers. In the mixed border it is an attractive object, but shows best on the margin of a shrubbery, where its stems just overtop the surrounding foliage. It is always best after some years' growth. A native of cold districts of mountains of Europe, this Lily is among those that may be naturalised, but it never grows so strongly as in rich garden ground. Lilies are said not to like manure, but we have never seen this Lily so fine as in well-manured ground after several years' growth. Indeed, we have planted it over a subsoil, so to say, of solid cow manure, and had bulbs and flowers of enormous size in two years. It is the Lily so much grown in the cottage gardens of the north of Ireland. A group in an open space among low evergreen shrubs is superb.

L. davuricum is a slender-growing European Lily with moderate-sized red flowers, spotted with black. Like L. elegans, it has several varieties, the chief being incomparable, Sappho, erectum, multiflorum, Don Juan, and Rubens. Being strong growers and free flowerers, these are fine plants for the mixed border, for margins of shrubberies, or for groups or masses, thriving in partial shade as well as in sunny places.

L. elegans.—One of the best of the early Lilies and one of the most generally grown, commonly under the name of Thunbergianum. It is a very variable species, there being about a dozen named varieties and others. The type grows about 1 ft. high, with stout, erect stems, furnished with numerous narrow leaves and terminated by a flower, 5 in. or 6 in. across, of a bright orange red A native of Japan, flowering with us about the beginning of July. Most of the varieties are so distinct that they merit a slight description. They are alutaceum, very dwarf, not more than 9 in. high, flower large, of a pale apricot colour, copiously spotted; armenaicum (venustum), about 1½ ft. high, with several moderate-sized flowers (in autumn) of a rich, glowing orange-red; atrosanguineum, about 1½ ft. high, flowers large, rich deep crimson; Batemanniæ, about 4 ft. high, with several moderate-sized flowers, produced in late summer, of a rich, unspotted apricot tint; bicolor, about 1 ft. high, flowers large, orange-red, flamed with a deeper hue; brevifolium, 1½ ft. high, flowers pale red and slightly spotted; citrinum, like armenaicum, but taller; fulgens, 1 ft. to 1½ ft. high, flowers large, four to six, of a deep red; sanguineum, 1 ft. to 1½ ft. high, with one or two large blood-red flowers; Van Houttei, 1½ ft. high, very deep crimson-red, spotted with black; Wallacei, 2½ ft. high, rich orange-red flowers, spotted with black; Wilsoni, 2 ft. high, with large apricot-tinted flowers—one of the latest to bloom. All the L. elegans group are perfectly hardy; they are vigorous growers in almost any kind of soil, but prefer a deep loamy one with an admixture of peat. They like an open position, and are suitable for planting around the margins of shrubberies. Small groups of them have a beautiful appearance in the open spaces that should exist in every shrubbery or Rhododendron bed. They are all excellent border plants, and the dwarf kinds may with propriety be introduced in suitable places in the rock garden. In all cases they must be placed in sunny situations.

L. giganteum.—A noble Lily, of huge growth and different in aspect to any other. Its bulb is large and conical, and develops handsome, shining, heart-shaped foliage in spreading tufts. The flower-stems are stout and erect, from 6 ft. to 10 ft. high, and terminated by a huge raceme, 1 ft. to 2 ft. in length, of about a dozen long, nodding, fragrant flowers, white, tinged with purple on the inside. It is one of the hardiest of Lilies, and succeeds well without much trouble. It flourishes best in a sheltered position, where there is an undergrowth of thin shrubs to protect the growth in spring. The soil must be deep, well drained, and consist of sandy peat and leaf mould, strengthened by a little rich loam. Years sometimes elapse before the tufts of foliage send up bloom. Native of Nepaul. L. cordifolium, a Japanese plant, is a similar, but inferior species, very rare in cultivation. It requires the same cultural treatment. A small group of three or four plants may be well placed in an open spot among shrubs growing in a free peaty soil. In such they do well, and when in flower the effect is all that could be desired.

L. Hansoni.—A handsome Japanese species growing about 4 ft. in height, having bright green leaves in whorls, and bearing about a dozen bright orange-yellow, brown-spotted flowers in a terminal spike. It flowers about the beginning of June, is quite hardy, and succeeds well in sheltered situations in a mixture of two parts of peat, one of loam, and one of road scrapings.

L. Humboldti is a very graceful Lily; the singular beauty of the blossoms and the elegant manner in which they droop from their slender stalks make it a most desirable plant, and its flowers, on account of their great substance, are more lasting than those of any other Californian Lily. The stems are stout and purplish, and attain a height of from 4 ft. to 8 ft. The leaves are in whorls of from ten to twenty each, and of a bright green. The flowers, which differ considerably with respect to colour and markings, are usually bright golden-yellow, richly spotted with crimson-purple. The variety ocellatum or Bloomerianum is dwarf, and has the petals tipped with brownish-crimson. It grows best in an open border of rich peaty or leafy soil of a good depth.

L. japonicum, or Krameri as it is more often called, possesses the most delicate beauty of any. The flowers are of the shape and nearly as large as those of L. auratum, and are either pure white or a delicate rosy pink—generally the latter. It grows from 1 ft. to 3 ft. high, and bears from one to five blooms—generally only one or two. It is somewhat difficult to grow owing to its delicate constitution, but the best specimens that have been produced in this country were grown under the same conditions as L. auratum and speciosum. On account of its beauty it is well deserving of the most careful attention. It is a lovely plant for a select spot between choice dwarf shrubs, growing in free peaty soil or deep sandy loam with some vegetable soil in it.

L. longiflorum (*White Trumpet Lily*).—This is among the most beautiful and most valuable of garden Lilies. The typical form grows from 1 foot to 3 feet high, the stems being terminated by long, tubular, waxy-white flowers, sweetly scented and produced in summer. There are several varieties of this Lily, the best being eximium, which, besides flowering a fortnight earlier than the type, bears larger and more numerous flowers, and is in every way superior to the ordinary longiflorum. Takesima is recognised by the purplish tint on the exterior of the blossoms and on the stem. Wilsoni, a grand Lily, grows nearly 4 ft. high, and has numerous

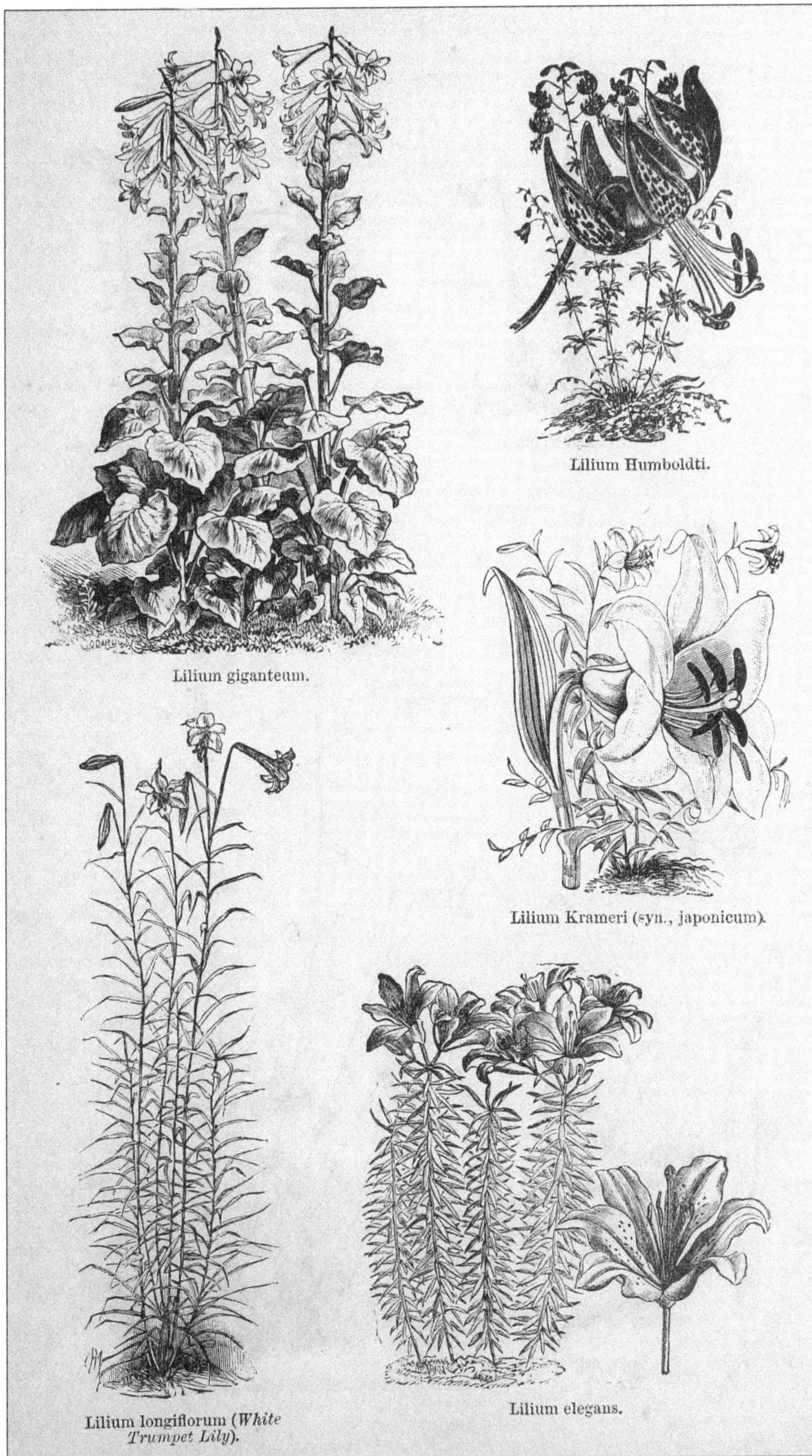

[*To face page* 170.

Lilium monadelphum.

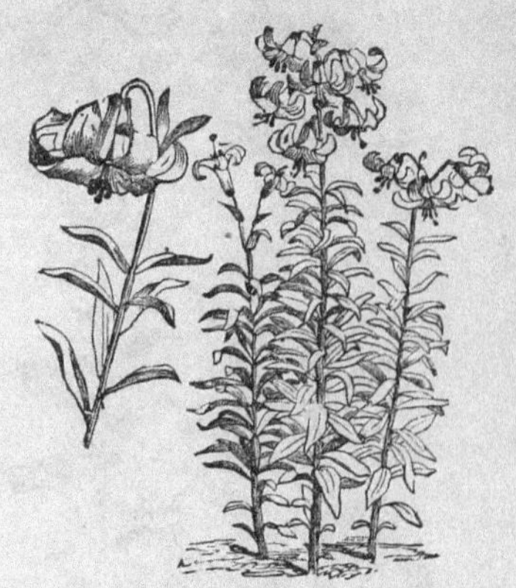

Lilium monadelphum var. Szovitzianum.

Lilium Martagon (*Turk's-cap Lily*).

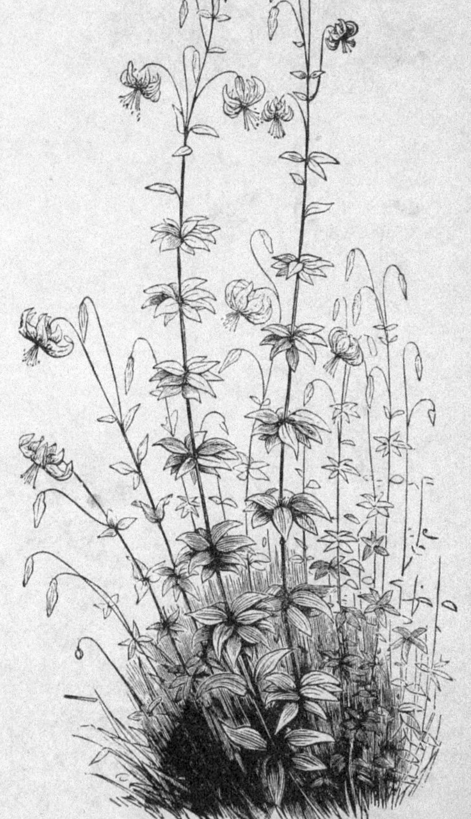

Lilium pardalinum (*Panther Lily*).

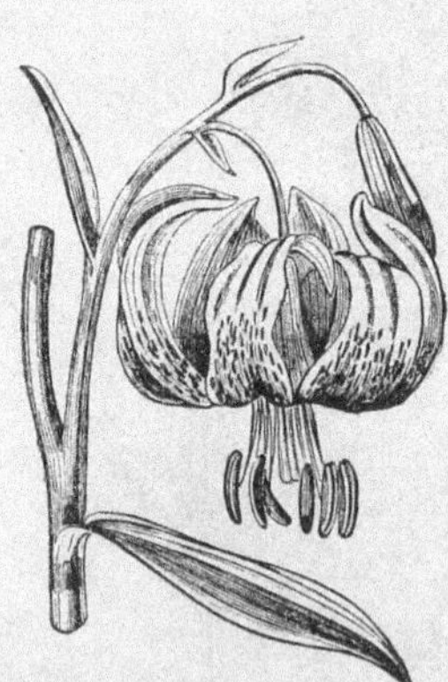

Lilium pomponium (*Little Turk's-cap*).

[*To face page* 171.

flowers about 9 in. in length. Madame Von Siebold is also a fine variety, which, though with smaller blooms than eximium, flowers a week or so earlier than that variety. The names Jama-Jura and Liukiu are native names for the varieties mentioned. The variegated-leaved form (albo-marginatum) is a desirable plant, as the variegation is distinct and constant. L. longiflorum and its varieties sometimes bloom very well in borders, but care should be taken that they are not injured by spring frosts. This Lily is such an early one, that, unless protected by the leaves of evergreens, its growth is apt to be checked. A well-drained, light loam, well enriched with leaf mould, suits it admirably. Wilsoni is benefited by a lighter soil than this, and by a little warmer and more sheltered position. When just pushing the growth in spring it is advisable to encircle the plants with a few dead branches if they are unprotected by shrubs. Where this fine species and its forms fail to grow in the ordinary soil of the garden, success is certain by making them a special soil composed of rotten manure or leaf mould or cocoa fibre. In such a mixture, so free and open that the hand could be pushed down below the bulb, we have seen them perfectly grown where the natural soil of the spot was too stiff and impervious. The hardier varieties of L. longiflorum are admirable plants for artistic gardening, their fine forms being very effective when tastefully grouped on the fringe of beds of choice bushes, touching and seeming to spring out of the Grass. They are also good in beds specially devoted to them alone or in combination with other plants. Similar to L. longiflorum are L. neilgherrense, philippinense, Wallichianum, and nepalense, but as neither is hardy and all are poor, unsatisfactory plants, even for the greenhouse, more need not be said of them here.

L. Martagon *(Turk's-cap Lily)*.—This is such a common Lily, that we need only mention its varieties. These are very fine, especially dalmaticum, which has flowers larger than those of the type and of a shining blackish-purple, a contrast to the loveliness of the pure white variety (album). Cattaniæ is only a form of dalmaticum and scarcely differs from it. Like the common Martagon Lily, the varieties thrive freely in a good loamy soil; they are perfectly hardy and rather partial to shade, growing freely in grassy places, open woods, or copses. Some of the finer varieties are good garden plants, and they should be seen in groups among hardy Azaleas or like flowering bushes, growing in the spaces between.

L. monadelphum is a magnificent Lily of noble growth. The stout flower-stems vary from 3 ft. to 5 ft. in height, and are terminated by a pyramidal cluster of from six to twenty turban-shaped flowers, ranging in colour from a rich canary-yellow to a pale lemon-yellow. Some forms have spotted flowers, and some are much larger than others. The varieties are known under the names of L. Szovitzianum, colchicum, and Loddigesianum. This Lily thrives best in a moist, deep, loamy soil, well enriched with good manure when planting; but it does not show its true character till after it has been planted several years. It rarely fails to do well, and is one of the least disappointing of all. It may be readily increased from root scales, a system that is taken advantage of by many cultivators, and is the only method of increasing and keeping pure any really good or marked variety. Seed, however, is the readiest way of acquiring a stock of this truly charming plant. The seeds are usually sown as soon as ripe in large shallow pans, where they generally remain for two years; by this time the bulbs have attained some considerable size; they are then planted in beds in rows 6 in. apart, and 4 in. bulb from bulb, re-planting when necessary. By this treatment flowers are frequently produced by seedling plants four or five years from the time of sowing.

L. Parryi is a new and distinct species from California, of elegant slender growth, and from 2 ft. to 4 ft. high. It bears graceful, trumpet-shaped flowers of a rich yellow, copiously spotted with chocolate-red, and delicately perfumed. The flowers are borne horizontally, and it is thus rendered very distinct. It grows in elevated districts in South California, in boggy ground. Not much is known of the culture of this Lily, but the finest plants have been produced where the soil chosen was two-thirds common peat and one-third loam, with plenty of coarse sand. A bed in a shady spot was selected, in which the bulbs were placed at a depth of 4 in., there being underneath them about 1 ft. of the soil. Here the strongest bulbs threw up stems 4 ft. in height, and the greatest number of blossoms on one stem for the first season was six.

L. pardalinum *(Panther Lily)*.—One of the handsomest of the Californian Lilies and one of the most valuable for English gardens, as it makes itself thoroughly at home in them, growing with a vigour equalling that which it acquires in its native habitat. It grows from 6 ft. to 8 ft. high, and has large drooping flowers of a bright orange, spotted with maroon. There are several varieties, the most distinct being—Bourgæi, one of the finest, producing stout stems 6 ft. to 7 ft. in height, bearing from twelve to twenty flowers of a bright crimson, shading to orange, and freely spotted with maroon—this blooms a fortnight later than any other; pallida, a dwarf variety, scarce reaching 5 ft. in height, flowers nearly double the size of the type, and paler in colour; californicum, a variety of more slender habit, growing from 3 ft. to 4 ft. in height, the brightest in colour; pallidifolium (puberulum), a small-growing form, with lighter coloured flowers; and Robinsoni, a robust variety, having stout stems 7 ft. to 8 ft. in height, and massive foliage; the flowers are large, of a bright vermilion shading to yellow, and freely spotted. This last is the noblest of all, and should be grown if possible. The Panther Lily is one of the most satisfactory of all Lilies to grow; it has a strong constitution, increases rapidly, soon becomes established, and rarely pines away, as many other kinds do. It likes a deep, light, good soil, enriched with plenty of decayed manure and leaf soil, where the roots can receive ample moisture. It should always be in a sheltered position, such as may be found on the sunny side of a bold group of shrubs or low trees. Where grown in a special bed the near shelter of hedges is desirable, though their roots are best kept away. Bare borders are not the places where this noble Lily does or looks best. There is no shelter or support in such places for plants that in their own country, where this species may be gathered in abundance, have for companions many shrubs and are sheltered by the finest trees of the northern world.

L. polyphyllum.—A rare and beautiful Lily, growing 2 ft. to 4 ft. high, and having large turban-shaped flowers of a waxy-white, copiously spotted and lined with purple. A native of North India. Mr. McIntosh, of Duneevan, Weybridge, who has been most successful with this Lily, writes thus to *The Garden:* "Sandy loam, peat or leaf-mould, sand, and charcoal, with a slight admixture of pulverised horse-droppings and good drainage under the bulbs, constitute all I have to tell, and I think early staking and tying may have something to do with many, but not all, growing taller than they otherwise might do."

L. pomponium.—This lovely Lily must not be confounded with the L. pomponium usually sold as such, this latter being simply the red variety of L. pyrenaicum. L. pomponium is elegant in growth, possesses a vigorous constitution, and blooms earlier than the numerous varieties of chalcedonicum and pyrenaicum to which it is related. It grows about 3 ft. in height, is of erect habit, and has long linear leaves. The flowers are produced in a lax raceme 1 ft. through, and a well-established plant will bear as many as twenty flowers. In rich loam it grows most luxuriantly both in sunshine and shade, and no difficulty is experienced with it in the case of either home-grown or imported roots. Native of the Maritime Alps. L. pyrenaicum, a similar, but smaller plant, with small yellow flowers, is a variety of L. pomponium, and there is also a red form of the Pyrenean Lily which is generally sold for the true L. pomponium, though it is a much inferior plant. These varieties require the same culture as L. pomponium. L. pomponium has an extremely offensive odour, and is not likely to be agreeable when cut for placing in the house.

L. speciosum, or lancifolium as it is erroneously called, is one of the most popular for pot culture, but it is none the less desirable for the open air, though it cannot be grown to such perfection as under glass, as it is of a somewhat delicate nature. It is well known and we need not describe it, but will mention the chief varieties. There is what is called the true speciosum, which has large deep rosy blossoms, richly spotted; vestale, pure white; album, white or faintly tinged with pink; rubrum, deep red; roseum, rosy-pink; punctatum, white spotted with pink; Krætzeri, very large white flowers with greenish stripe on the exterior; fasciatum album and fasciatum rubrum, two monstrous varieties bearing numerous flowers on flattened stems. Other fine varieties have originated in America, and among these Melpomene is very distinct. The beautiful hybrid, Mrs. A. Waterer, is large, white, and spotted with pink. All the varieties of L. speciosum require a sheltered situation, protected from winds and draughts, and a rich loamy soil mixed with peat and leaf manure. They flower for the most part in September, and last longer in bloom than many other Lilies. In good soils very happy use can be made of these handsome Lilies—in warm and sheltered places where their blooms may be fully developed.

L. superbum (*Swamp Lily*).—One of the stateliest of the N. American Lilies, with beautiful orange-red flowers, thickly spotted, produced late in summer. It may be at once distinguished from similar kinds by the purple-tinged stems, which rise from 5 ft. to 8 ft. high, and are very graceful, as they wave with the slightest breeze. The numerous flowers terminate the stems in a pyramidal cluster. It delights in a moist, deep soil consisting chiefly of peaty and decayed leaf manure, and is one of the best Lilies for growing in shady woods when the under-growth is not too rank. In the garden it should have snug glades and nooks protected by shrubs, and the soil must be of a moist, peaty description. L. carolinianum is a form of this species, but not so showy.

L. tenuifolium.—A most elegant Lily of dwarf growth, especially valuable on account of its earliness in flowering. It grows from 1 ft. to 1½ ft. high, and has narrow leaves on slender stems, furnished with a cluster of about a dozen turban-shaped flowers of a brilliant red colour, shining like sealing-wax. It succeeds well in open warm borders of light sandy loam, but it is all the better for a little protection, such as a hand-light or frame, as it flowers very early. Siberia and N. China. Similar to L. tenuifolium is L. callosum and its form stenophyllum, but these are less showy plants.

L. testaceum (*Nankeen Lily*).—This is such a distinct-coloured Lily that it should always be grown, especially as it is of easy culture and thrives in any ordinary soil, preferring, however, one that is peaty. It has the growth of the white L. candidum, but the flowers are a delicate apricot or nankeen colour. When well grown it is 6 ft. or 7 ft. high, and bears several flowers in a large head. Other names for this Lily are L. excelsum and isabellinum. It is one of the plants that grow freely in London.

PLATE CLV.

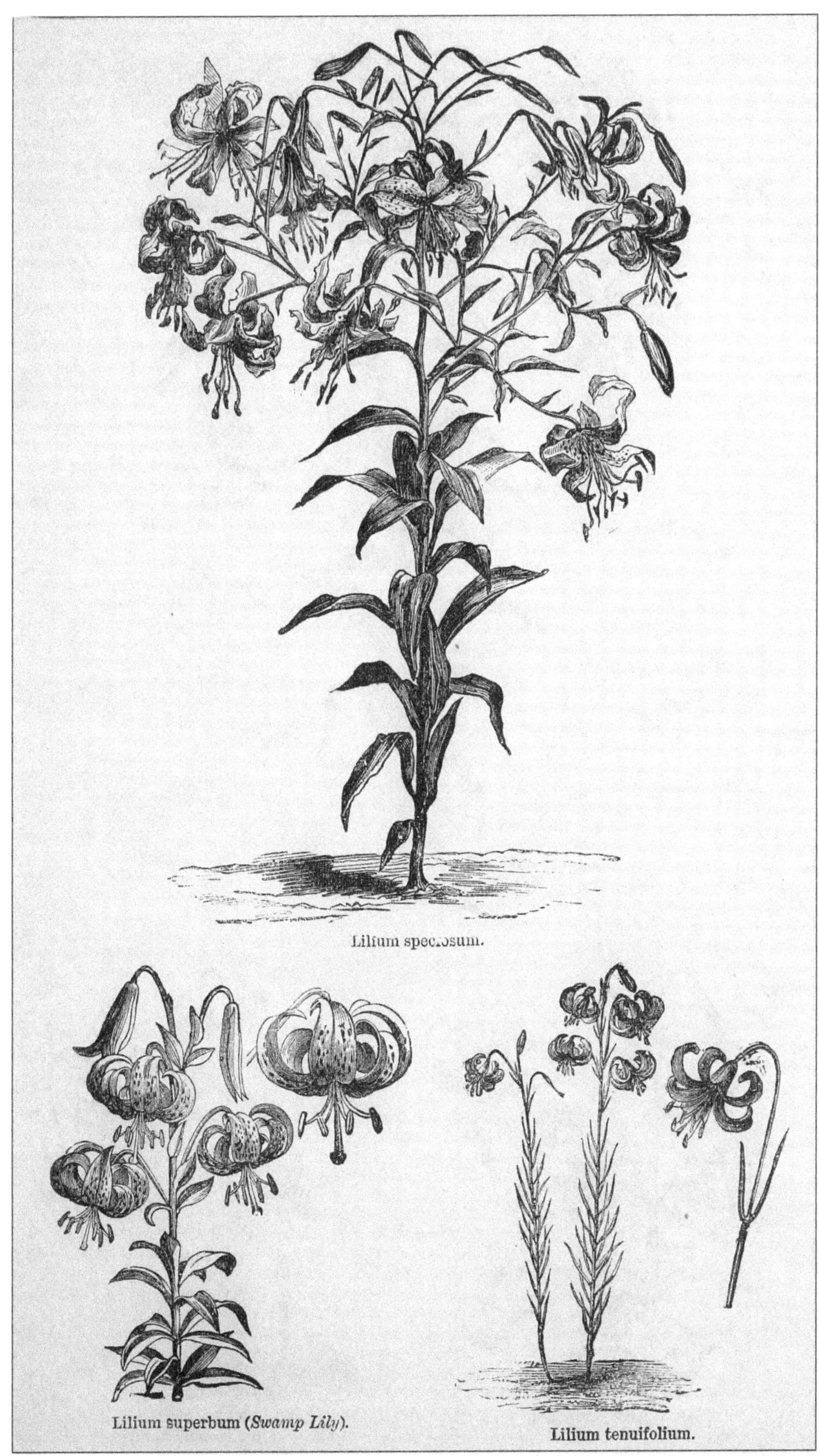

Lilium speciosum.

Lilium superbum (*Swamp Lily*).

Lilium tenuifolium.

[*To face page* 172.

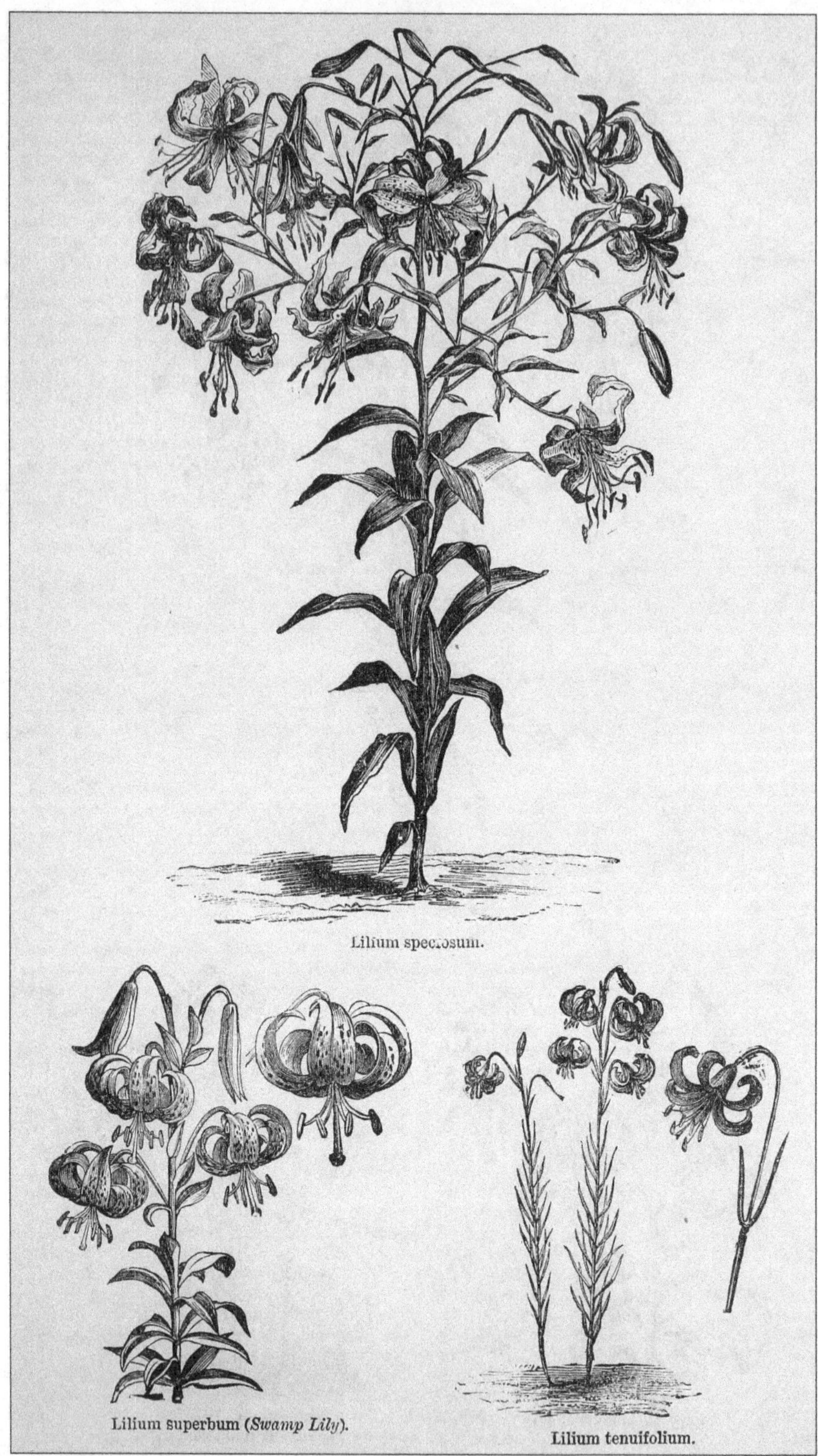

Lilium speciosum.

Lilium superbum (*Swamp Lily*).

Lilium tenuifolium.

[*To face page* 173.

L. tigrinum *(Tiger Lily).*—This is one of the commonest kinds, and is too well known to need description. No garden should be without it, for few plants are so attractive or have such stately growth as this Lily when grown well. The common kind is handsome, but the variety splendens is much finer, having larger flowers with larger spots; it is produced later in the season, and the plants often grow 7 ft. high. Fortunei is an early-flowering form and as desirable as splendens. The double-flowered variety (flore-pleno) is showy and a vigorous grower; erectum also is a distinct and desirable sort. There are other Lilies, referred to other species, which much resemble L. tigrinum. These are L. pseudo-tigrinum and varieties of Maximowiczi. The Tiger Lily is one very easy of cultivation, thriving best in deep sandy loam in an open, but sheltered position. The earliest varieties begin to flower at the end of August, and the latest last till the end of October. It may be quickly propagated by the bulblets, which form in the axils of the leaves.

L. Washingtonianum. — A lovely Californian Lily, growing from 2 ft. to 5 ft. high, bearing a cluster of large, white, purple-spotted flowers that become tinged with purple after expansion. Nearly allied to this, and by some considered a variety, is L. rubescens, which has smaller flowers (borne erect—not horizontal, as in the Washington Lily) of a pale lilac or nearly white. Neither of these is easy to grow, owing, probably, to their being but little understood at present. The best results have been obtained when they were grown in partially-shaded situations, in loose, peaty, well-drained, but moist soil. The wonderfully brilliant series of Lilies introduced to our gardens from N. America and Japan of recent years, which have given us wholly new aspects of vegetation in the flower garden, are accurately figured in colour and life size in the volumes of *The Garden*, and will be found enumerated in the general index to that work.

Lily *(Lilium).*

Lily of the Valley *(Convallaria majalis).*

Limnanthemum nymphæoides.—A pretty native water plant that may be sometimes seen in ponds or by the margins of slow-running streams. It has floating tufts of leaves, and bright yellow flowers 1 in. or more across. Where not plentiful in a wild state it is a desirable plant for introducing in lakes and other artificial water, or treating in the same manner as Limnocharis. Propagated by division or seed sown, as soon as gathered, in mud. Gentianaceæ. *Syn.*, Villarsia.

Limnanthes Douglasi.—A beautiful Californian hardy annual, particularly valuable because it flowers so early. The flowers are yellow and white, and there is also a pure white variety. Few annuals are more hardy than this, as the most severe winter does not injure it, and the seed germinates at a very low temperature. It is of a compact, but vigorous habit, a single plant covering, when in bloom, quite a square foot of soil; it should therefore be allowed quite that space for development. Unlike the generality of annuals, it neither requires a deep nor rich soil, but thrives admirably where the natural staple is poor and inclined to burn in the early summer months. This renders it valuable for planting on dry banks or similar situations, and it flowers more freely thus placed than when growing in rich soils. It generally sows itself on light soils, and gives no further trouble, but when wanted for a special purpose in spring the seed should be sown in autumn in boxes or in the open ground, or in spring for summer flowering. Tropæolaceæ.

Limnocharis Humboldti.—An interesting and pretty aquatic plant which in summer covers the surface of the water with broad, heart-shaped leaves, of a beautiful glistening green, and soft yellow flowers, which continue for several months. It commences, as a rule, to flower in August, and continues in blossom till quite late in the season; it will thrive either in running or still water, planted from 6 in. to 9 in. below the surface, and may also be grown successfully in tubs sunk in the ground. The tubs should be about 1½ ft. in depth, be half filled with loamy soil, and then filled up with water. In fountain-basins and clear, rather still, waters, where the plant will be fully exposed to the sun, it thrives and flowers freely during the summer months. It will not survive out of doors in winter, except in the mildest districts and placed at least 18 in. below the surface. It might, however, escape in water that does not freeze. Plants put out of a warm aquatic house in May soon begin to grow in the open air. It is, or used to be, most frequently seen in the aquatic stove. Division. Butomaceæ.

Linaria *(Toadflax).*—A very large genus of Figworts that includes some beautiful garden plants of annual and perennial duration, varying considerably in habit from very dwarf alpines to tall coarse plants. Among the alpine perennials the most desirable are those below mentioned.

L. alpina, from the Alps and Pyrenees, is found on moraines and débris of the mountains. It forms dense, dwarf, smooth and silvery tufts, covered with bluish-violet flowers, with two bosses of intense orange. Its habit is spreading, but neat and very dwarf, for it is rarely more than a few inches high. It is usually a biennial; but in favourable spots, both in a wild and cultivated state, becomes perennial. Its duration, however, is not of so much consequence, as it sows itself freely, and it is one of the most charming subjects that we can allow to "go wild" in sandy, gritty, and rather moist earth, or in chinks of rockwork. In moist districts it will sometimes even establish itself in the gravel walks. It is readily increased from seed, which should be sown in early spring in cold frames, or in the places it is destined to embellish out of doors. L. pilosa and pallida are very dwarf, creeping plants with conspicuous purple blossoms; both are very suitable for the rock garden, in moist spots where they can trail over the surface. L. origanifolia and its near ally, L. crassifolia, both of which are dwarf and have pretty purple flowers, are likewise suitable for the same purpose, as well as our common wild Ivy-leaved Toadflax (L. Cymbalaria), which drapes over many walls so gracefully. This last-named has a white and pretty variegated variety. The old plant itself would be fully described here were it not that it usually takes possession of old walls and other places suitable for its growth. The best of the taller kinds suitable for the border or for naturalising are

L. dalmatica, a really handsome plant, growing from 3 ft. to 5 ft. high, with very glaucous foliage. It is much branched, and bears in summer a profusion of large, showy, sulphur-yellow blossoms. It thrives best in a light, well-drained soil in warm places, and when once established can be eradicated but with difficulty. L. genistæfolia is a similar, but much inferior species, and also has yellow flowers.

L. triornithophora is a beautiful plant when well grown. When fully developed it is from 1 ft. to 1½ ft. high, and bears large, purple, long-spurred flowers in whorls of three. It is rather a delicate plant, and, though perennial, should be raised yearly from seed. L. triphylla is a similar plant.

L. purpurea, though not a showy plant, is well worth growing, as it thrives in dry places and is a capital subject for old walls. Even the common yellow Toadflax (L. vulgaris) is worth cultivating, particularly the singular and handsome Peloria variety, which has five spurs, and is remarkably curious as well as ornamental.

Some of the annual species are among the prettiest border flowers we have. All grow about 1 foot high, are profuse flowerers, very showy, and most effective when sown in broad masses. They should be sown in ordinary garden soil in early spring, and will then flower in July and August. The best kinds are L. reticulata, with small purple flowers; the variety aureo-purpurea is a charming plant, producing flowers in great variety, varying in colour from rose-purple to dark orange. L. bipartita, too, is a very variable species, the colours ranging from deep purple to white. Perezi has small yellow flowers; maroccána, violet to pink; and multipunctata, the dwarfest of the group, has black-spotted yellow flowers.

Lindelophia spectabilis.—A rather showy perennial Boragewort, growing about 1½ ft. high, bearing in early summer drooping clusters of deep purple-blue flowers. It is suitable for borders in sandy loam, and is quite hardy in well drained places, but it is not so valuable a plant as many others of the same order. Seed or division. North India. (= Cynoglossum longiflorum.)

Linnæa borealis *(Twin Flower)*.—A little creeping evergreen plant belonging to the Honeysuckle family, and as each slender, upright stalk bears two flowers, it has received the common name of Twin Flower. The flowers are white, often tinged with pink or purple, delicately fragrant, and droop with a modest air that is very charming. It is usually found in moist woods, where it forms a dense carpet. It is wrongly supposed to be rather difficult to cultivate. Little has to be done beyond planting healthy young plants in a moist, sandy border or rock garden, or a slightly raised bank. We have repeatedly seen the plant thriving freely in the open air—that is, in districts where the air is pure and the soil is fitted for it. It is an excellent plant for a moist rock garden, and when once established grows rapidly. Where it thrives it forms a charming fringe or carpet to a group or bed of the smallest alpine and rock shrubs, in cool borders or northern slopes of the rock garden. N. Europe, Asia, and America.

Linosyris vulgaris *(Goldilocks)*.—A showy native herb, growing from 1 ft. to 2 ft. high, and producing in late summer and autumn bright yellow flowers in terminal clusters. It grows in any ordinary soil, but is scarcely a garden plant. (= Chrysocoma Linosyris.) Compositæ.

Linum *(Flax)*.—The chief characteristic of this genus is elegance and lightness of growth; there is a striking sameness in most of the species, especially among the blue-flowered kinds, so that comparatively few are suitable garden plants. The most desirable are the following:—

L. campanulatum *(Yellow Herbaceous Flax)*.—A herbaceous plant with golden-yellow flowers on stems from 12 in. to 18 in. high, distinct from anything else in cultivation, and well worthy of a place in collections of alpine and herbaceous plants. A native of the south of Europe, flowering in summer, and flourishing freely in dry soil on the warm sides of banks or rockwork. Propagated by seed. Similar to this species is L. flavum, or tauricum, likewise a handsome and perfectly hardy yellow-flowered plant, but L. arboreum, a shrubby kind, also with yellow flowers, is not hardy in all districts, though a charming little evergreen rock-garden bush where it thrives.

L. grandiflorum is a showy hardy annual from Algeria, with deep red blossoms. It may be had in bloom from May till October by successional sowings. Seed sown in autumn will produce plants for early spring blooming, and sowings made from March till June will yield a display through the summer and autumn. Sown in pots in summer in good, rich soil, and plunged in a sunny border with plenty of water, plants may be obtained for the greenhouse or window during October and November. If protected from frost it is perennial.

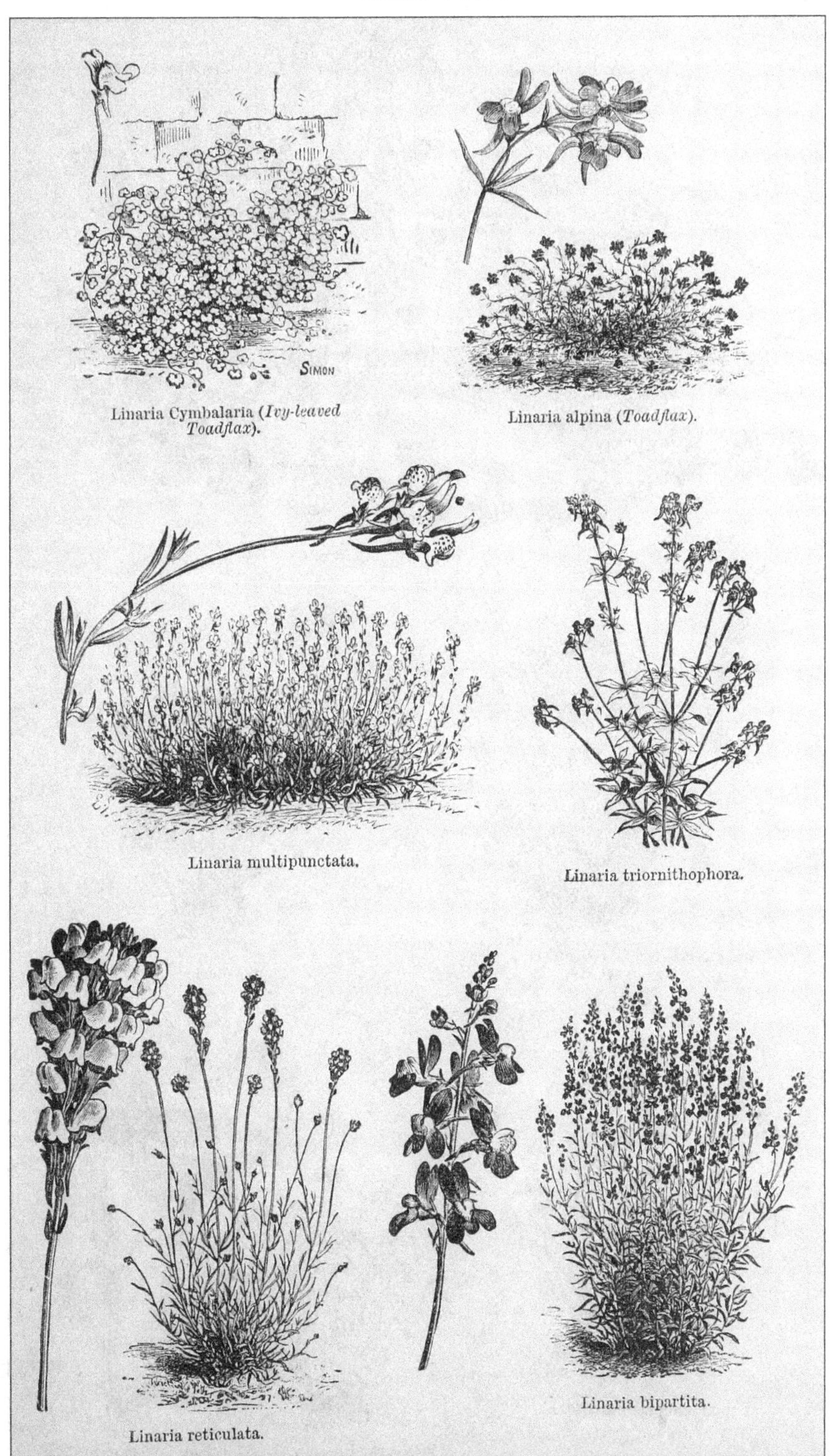

Linaria Cymbalaria (*Ivy-leaved Toadflax*).

Linaria alpina (*Toadflax*).

Linaria multipunctata.

Linaria triornithophora.

Linaria reticulata.

Linaria bipartita.

[*To face page* 174.

PLATE CLVIII.

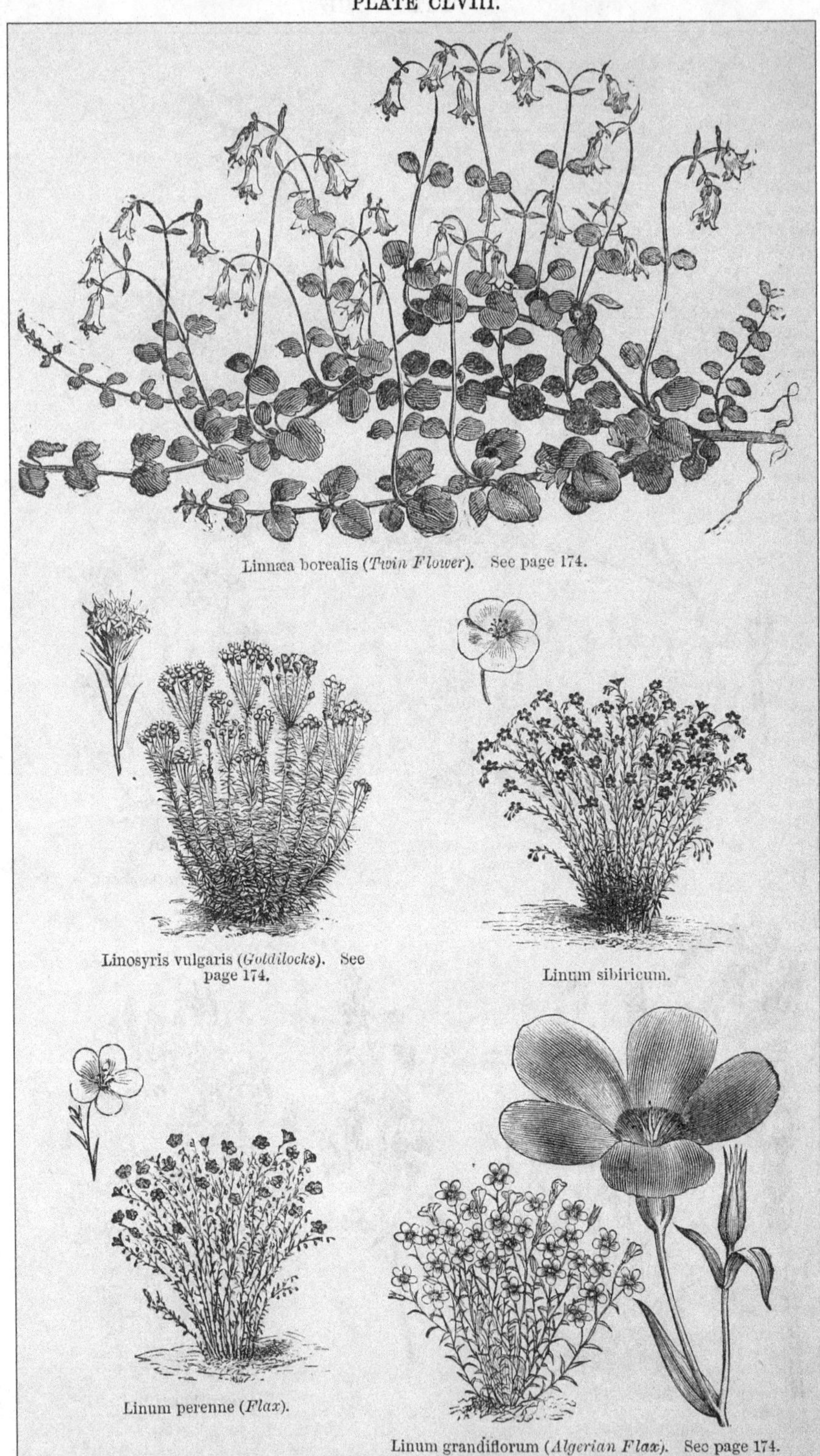

Linnæa borealis (*Twin Flower*). See page 174.

Linosyris vulgaris (*Goldilocks*). See page 174.

Linum sibiricum.

Linum perenne (*Flax*).

Linum grandiflorum (*Algerian Flax*). See page 174.

[*To face page* 175.

L. monogynum.—A beautiful kind with large pure white blossoms blooming in summer. It grows about 1½ ft high in good light soil, and its neat and slender habit renders it particularly pleasing and well adapted for rockwork borders or pot culture. It may readily be increased by seed or division; is hardy in the more temperate parts of England, but in the colder portions it is said to require some protection. L. candidissimum is a finer variety, and, moreover, hardier than the type. Both are natives of New Zealand.

L. narbonnense.—A beautiful and distinct sort, bearing during the summer months a profusion of large, light sky-blue flowers, with violet-blue veins. A fine ornament for borders, the flower garden, or the lower flanks of rockwork, on rich light soils, forming lovely masses of blue from 15 in. to 20 in. high. A native of Southern Europe. Other blue-flowered kinds similar to this, but inferior, are the common L. perenne, usitatissimum, alpinum, sibiricum, alpicola, collinum, austriacum, and crystallinum; all are European species, quite hardy, and make showy border plants, or do well for naturalising. There are white and rose-coloured varieties of L. perenne; both pretty plants.

L. salsoloides *(White Rock Flax)* is a procumbent half-shrubby species, essentially a rock garden plant. Its flowers are white, with a purplish eye, and, when abundant, the plant reminds one of some of our creeping white Phloxes. On the rockery, in a well-exposed sunny nook, it is perfectly hardy, and will trail freely over the surface of adjacent stones—conditions under which it will flower abundantly. It produces seeds but rarely, so that increase must be obtained by cuttings of the short shoots taken off with a heel about midsummer, just as they are hardening off; at that time they will be found to strike freely, and to make vigorous plants when potted off the following spring. Native of mountains of Europe. Another lose ally is L. viscosum, with pink flowers, but this is much less desirable.

Liparis—A genus of Orchids, for the most part inconspicuous and only suitable for botanical collections.

Lippia.—A genus of Verbenaceæ, of which the common Lemon Plant (L. citriodora), sometimes called Aloysia, is the most familiar example. L. nodiflora is a dwarf creeping perennial, bearing in summer pretty pink heads of blooms. It grows in any situation or soil, and is a capital plant for quickly covering bare spaces in the rock garden where choicer subjects will not thrive.

Listera.—Terrestrial Orchids, only of botanical interest.

Lithospermum *(Gromwell)*.—A few of these Borage-worts are pretty and worth growing. One of the finest is L. prostratum, a perfectly hardy little evergreen spreading plant, having rich and lovely blue flowers, with faint reddish-violet stripes, which are produced in great profusion when well grown. It is very hardy, and peculiarly valuable as a rock plant from its prostrate habit and the fine blue of its flowers—a blue scarcely surpassed by that of the Gentians. It may be planted so as to let its prostrate shoots fall down the sunny face of a rocky nook, or be allowed to spread into flat tufts on level parts of the rockwork. On dry and sandy soils it forms an excellent border plant, and where the soil is deep and good, as well as dry and sandy, it becomes a round, spreading mass, 1 ft. or more high. It is, in such soils, suited for the margins of beds of choice and dwarf shrubs, as an edging, as a single plant, or in groups. In heavy or wet soil it should be elevated on rockwork or banks and planted in sandy earth. It is sometimes grown as L. fruticosum, but the true L. fruticosum is a little bush, whereas this is prostrate. It flowers in early summer, and often continues a long time; the leaves are nearly oblong in outline, and covered with short bristle-like hairs. Easily propagated by cuttings. S. Europe.

L. petræum *(Rock Gromwell)*.—A neat and dressy dwarf shrub, somewhat like a Lilliputian Lavender bush, with small leaves of a greyish tone like those of Lavender. Late in May or early in June all the little grey shoots of the dwarf bush begin to exhibit a profusion of small, oblong purplish heads, and early in July the plant is in full blossom, the full-blown flowers being of a beautiful violet-blue. The best position for this plant is on the rockwork, somewhere near or on a level with the eye, on a well-drained, deep, but rather dryish sandy soil on the sunny side. Native of dry rocky places in Dalmatia and Southern Europe. Propagated by cuttings, or seeds if they can be obtained. L. purpureum-cœruleum, a British plant, L. Gastoni, and L. canescens are also worthy of culture.

Liver Leaf *(Hepatica)*.

Lizard Orchis *(Orchis hircina)*.

Lloydia serotina.—A small bulbous Liliaceous plant, most suitable for botanical collections. Alps.

Loasa.—All the species of the Loasa order are remarkable for the singular structure of their flowers and the stinging character of their foliage. There are but few in cultivation. L. hispida is a very ornamental plant, growing about 18 in. high, branching considerably when strong, and bearing deeply-cut foliage, with short stinging hairs. The flowers are 1 in. across, of a clear, bright lemon yellow, with the centre prettily variegated with green and white. It produces its blossoms for several weeks in succession during August and September. The other kinds in cultivation are—the beautiful new L. volcanica, with pure white flowers having red and white striped centres; L. lateritia, a twining species, with orange-red flowers; and L. triloba. All the species are natives of the cool regions of Peru and Brazil, and therefore can be grown in the open air in this country during summer. Treated as half-hardy annuals, and grown in a light, fertile soil, they make interesting plants for open borders; the climbing species, such as lateritia, require branches to twine among. All may be raised freely from seeds, which, in favourable seasons, are produced in abundance.

Lobelia.—A large genus containing, however, but few garden plants, but these few are of great beauty and indispensable. The species are annual and perennial. Of the perennials the best are

L. fulgens *(Cardinal Flower)*.—This is one of the handsomest of all open air flowers, for none can surpass it in the brilliancy of its rich vermilion flowers. Its bold erect habit and strikingly brilliant flowers adapt it for situations where bright colours are desirable. Planted in groups, backed up by, or in the near vicinity of, evergreens, it has a gorgeous effect. Its most congenial situation is amongst the lower shrubs in a border, as it thereby receives just that amount of shelter which is necessary to ensure its flowering in perfection. The reason why it is not so much in general cultivation is that in many localities it is very liable to rot off during the winter, while in a more than usually damp season it is not uncommon to hear complaints of the whole stock disappearing. The enemy of this Lobelia is a kind of rust, which fastens on the main fleshy roots when the plants are gone to rest, and, quickly eating them away, thoroughly rots them, and they eventually disappear. This disease, working as it does from the bottom at a time when growth is at a standstill, is not perceived in sufficient time for its ravages to be checked, and the whole plant, much to the grower's disappointment, will in the course of two or three weeks be completely ruined. The disease, which attacks the roots and ultimately destroys them, makes its appearance towards the latter end of October or the beginning of November, especially if the weather at that time should set in cold and wet. The plants should then be carefully taken up, preserving as much of the roots as possible, the soil shaken from them, and the roots well washed. If the disease be present it will be readily discovered, as it comes in the form of rusty-looking spots which eat their way into the roots. Wherever these are perceived they must be cut out with a sharp knife, as if only a small portion be left it will suffice to entirely destroy the vitality of the plants. When the plants are thus examined they may be either potted or laid in a frame in some free, sandy soil. Very fine specimens are obtained by potting and plunging in a slight bottom-heat, keeping the top quite cool. In about a fortnight they will have made fresh fibre, and this is the great object to be attained, as when once a new root is formed all danger of rotting is past. They may then be placed and kept in a cold frame during the winter, to be planted out where desired in spring; a very fine display may be thus secured. The bottom heat, however, is not at all indispensable; they will succeed very well indeed if carefully and but sparingly watered after potting. Similar to L. fulgens is L. cardinalis, splendens, and ignea. They differ chiefly in the depth of colour, but L. fulgens is the finest.

L. Erinus.—The dwarf section of Lobelia is one of the largest and the most important; hardly any flower garden is now considered complete unless one or more beds or borders is plentifully supplied with this beautiful Lobelia or some of its many varieties, and few features in the flower beds or borders are more thoroughly enjoyable and satisfactory than a perfect band or small bed of any of them from the middle of June till the frost clears the garden of its autumnal beauty. The chief points to start with in the successful culture of this Lobelia are good soil and well-grown established plants. The soil should be light and rich, and rest on a dry and perfectly drained bottom. It enjoys abundance of water when in robust and free growth, but nothing is more fatal to its well-being than stagnant water at the roots; if on a porous bottom it may be plentifully watered during a dry time in summer without fear of injuring the roots. The roots cannot make way nor can the plants thrive in a strong adhesive soil composed of clay or heavy loam, and if the compost be heavy, it must be lightened by a plentiful addition of leaf-mould, sand, or peat. This Lobelia thrives admirably in equal parts of rather sandy loam and leaf-mould with a fair admixture of sand to keep it open. Charcoal dust and peat also form capital additions to loam for their successful culture; likewise spent manure from Mushroom beds. A slight mulching of one-year-old sifted hotbed manure will be found a capital addition for beds or borders for keeping out the drought from and nourishing the roots of Lobelias through a dry season. One of the greatest difficulties, however, in carrying dwarf Lobelias in full beauty through the season is the freedom with which they produce seed. The moment the flowers fade they should be picked off, and so on persistently every week or ten days throughout the season. Of course the labour is great, but so is the reward. These dwarf Lobelias may be propagated by seeds or cuttings, or by lifting the plant, potting it, and placing it in a gentle bottom-heat until established; after that set it on a light, airy greenhouse or forcing-house shelf, when it may be increased to any extent by cuttings and root division in the spring. A stock should be planted on a piece of reserve ground for seed. This increase by cuttings, or conservation of the old plants by potting a few of them in the autumn, is also the best method of preserving and increasing the stock of special varieties. It strikes roots freely in a brisk heat in a moist propagating pit or frame in spring. The cuttings should be potted by the end of May in exactly the same way as seedlings sown in heat in September, October, or February. Those who want early Lobelias

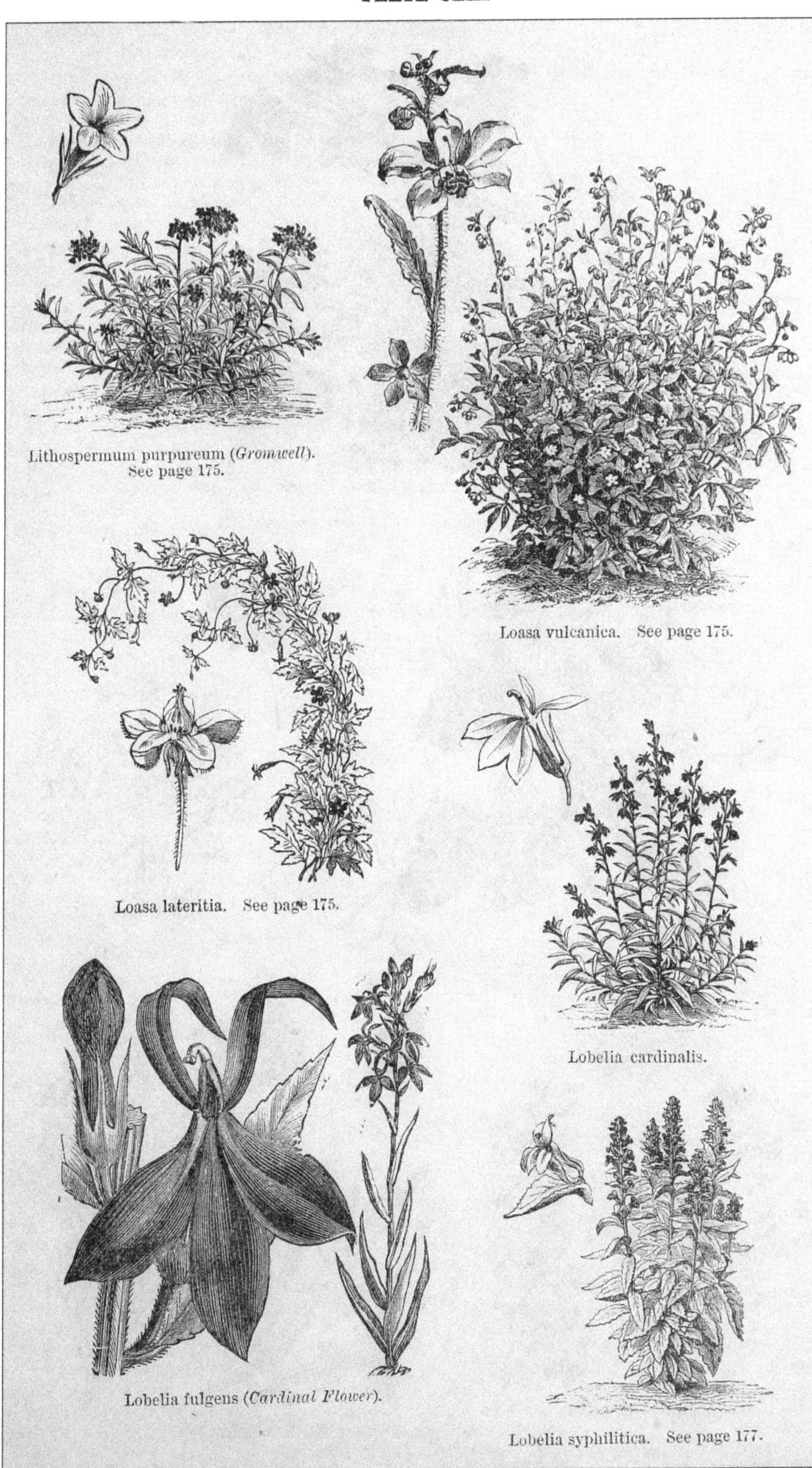
Lithospermum purpureum (*Gromwell*).
See page 175.
Loasa vulcanica. See page 175.
Loasa lateritia. See page 175.
Lobelia cardinalis.
Lobelia fulgens (*Cardinal Flower*).
Lobelia syphilitica. See page 177.

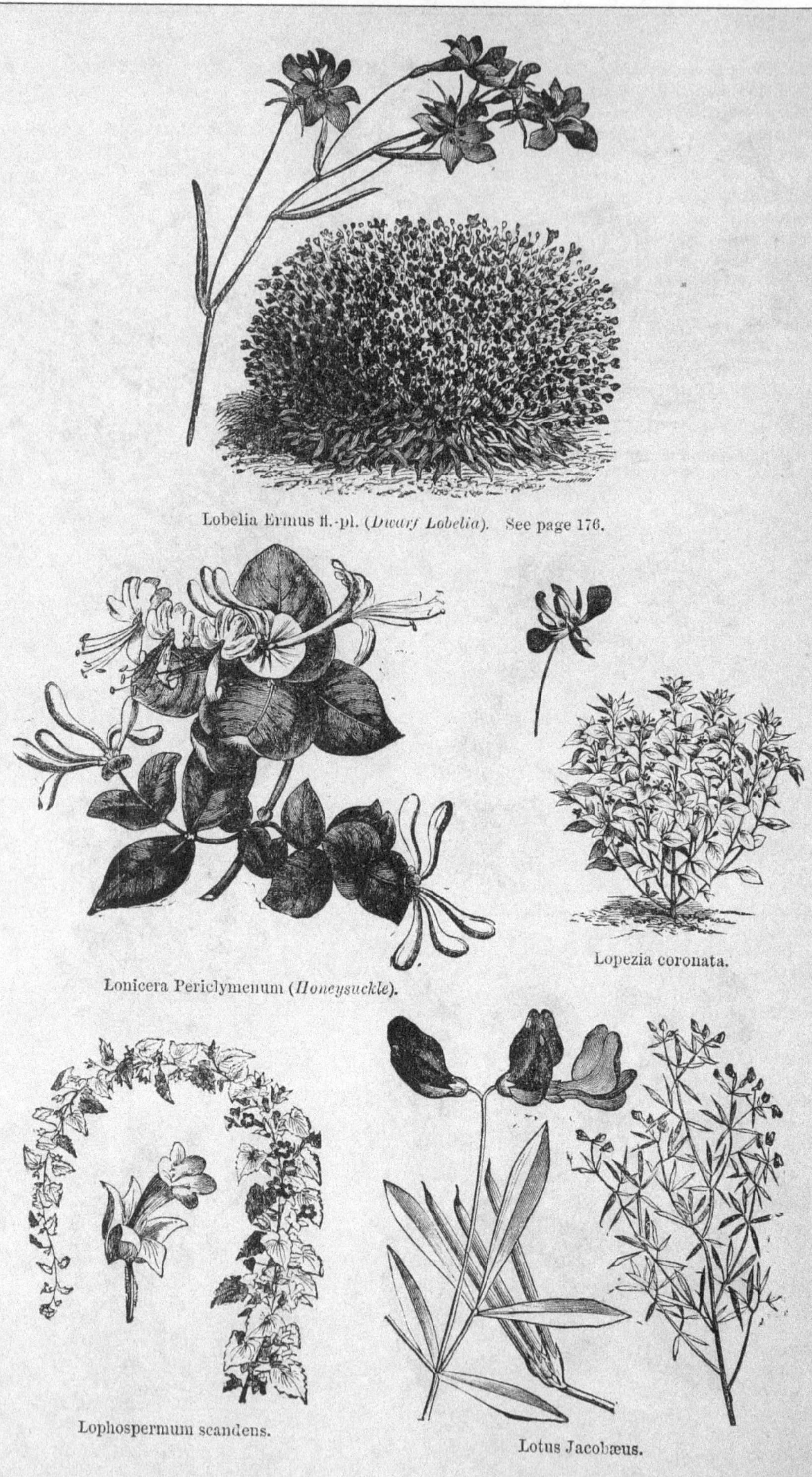

Lobelia Erinus fl.-pl. (*Dwarf Lobelia*). See page 176.

Lonicera Periclymenum (*Honeysuckle*).

Lopezia coronata.

Lophospermum scandens.

Lotus Jacobæus.

from seed should sow in the autumn, and prick the seedlings off in boxes or pans, or shift them into 2½-in. pots before winter; store them on shelves near the light, exposed to abundance of air, give another shift into 6-in. pots in March into equal parts of leaf-mould and loam, and such plants will be perfect for planting by the end of May. Spring-sown seedlings may go into smaller-sized pots, and be planted rather more closely, but will not flower so early nor so well. On the whole, therefore, autumnal propagation, either by means of cuttings or seeds, is preferable to the sowing of the seeds in spring. The varieties are very numerous, and it is a difficult matter to make a selection to suit every locality. L. Erinus is divided into five sections, viz., compacta, of which there is a white form; ramosoides; speciosa, of which the best are Blue Stone, Ebor, Blue Beauty, Emperor William, Blue King, Lustrous, Brilliant; pumila, of which grandiflora and magnifica are fine forms, as is also the pure white Mrs. Murphy; and Paxtoniana, which is of a lovely blue. The double variety is also beautiful where it succeeds well, but it is hardly to be depended upon in beds or borders. Sometimes it forms a complete sheet of bloom, and at others the shoots run up through it, as it were, and overpower and prevent it from blooming, thus giving it the appearance of tufts of Grass. Other dwarf Lobelias are ramosa, with large light-blue flowers, and coronopifolia, also with large blue flowers. Both are half-hardy annuals, requiring the same treatment as L. Erinus. L. lutea is a yellow-flowered perennial species, which, however, has not been sufficiently tested yet. It is very dwarf, flowers freely, but is not a showy plant. L. subnuda is a tender little species from Mexico, having handsomely marked foliage. It is not worth growing in a general way. L. ilicifolia is another dwarf, trailing species, a native of the Cape, and is best suited for growing in suspended pots in greenhouses, though in some localities it succeeds as a rock garden plant.

L. syphilitica is a hardy perennial species, growing about 2 ft. high, and bearing, from July to September, erect flower-stems, furnished thickly with light blue blossoms. It is not a very showy plant, but its varieties and the hybrids from it are ornamental. These occur in various shades —deep purple, blue, and light rose, and the hybrids Milleri and speciosa are also handsome. All these thrive best in moist borders, and so partial to moisture are they that they flourish well in the artificial bog. They are never so fine, however, as the forms of L. fulgens or L. ignea.

Lobel's Catchfly *(Silene Armeria).*

Loiseleuria procumbens.—A wiry trailing shrub that grows quite close to the ground, the plants occasionally forming a rather dense tuft that bears small reddish flowers in spring. It is not very attractive at any time in respect to bloom. It is very rarely seen in a thriving state under cultivation, and most of the plants transferred from the mountains to gardens usually perish. This sometimes occurs because the strongest-rooted and finest specimens are selected for transplantation, instead of the younger ones. Its true garden home is the rockwork, and it prefers deep, sandy peat. (= Azalea procumbens.) Ericaceæ.

Lomaria.—A genus of Ferns, for the most part tropical, and requiring artificial heat; but two or three are hardy enough to thrive in the open air in mild parts. L. alpina, a native of New Zealand, is dwarf and compact, and produces, from a creeping rhizome, abundance of dark shining green fronds, from 4 in. to 6 in. in height. It is specially adapted for the rock garden, and should receive similar treatment to that recommended for the Ceterach (to which it would form a charming companion), and should, like it, be associated with Sedums and such like alpine flowering plants. L. crenulata is similar, but not quite so hardy, though it will succeed in the mildest localities, as will also the Chili L. chilensis, a Tree Fern of noble growth. These Ferns should be placed in the snuggest quarters of the hardy fernery, and due care taken to protect them during severe cold.

London Pride *(Saxifraga umbrosa).*

Lonicera *(Honeysuckle)* — Although strictly shrubby plants, these are so graceful, fragrant, and beautiful, that they may be made a charming aid in the flower garden or pleasure ground. Wherever picturesque gardening is attempted they are beautiful when isolated and allowed to ramble as tufts in their own way, while in bold rock gardening or on banks, groups, or tufts they would occasionally be charming. In some Continental gardens it is the custom to train common Honeysuckles as low standards in borders, and well they look when in flower. The most desirable kinds are— L. Periclymenum, the common native Honeysuckle, and its varieties (serotina, belgica, and others), and L. etrusca, a pretty twining species with very sweet-scented blossoms. All of these are admirably adapted for the purpose recommended above.

Loosestrife *(Lysimachia vulgaris).*

Lopezia.—Mexican annuals of the Fuchsia family, not one of which is very showy. L. coronata grows from 1 ft. to 1½ ft. high, and produces plentifully small red flowers from July to September. L. racemosa is similar, but inferior. They thrive in ordinary soil. Seed should be sown in March in the open border.

Lophanthus.—Coarse growing plants of the Labiateæ.

Lophospermum scandens.—A tender climbing plant that has long slender stems, furnished with pale green, hairy leaves, and large pink flowers. It thrives well in the open air in summer, and is a beautiful plant for festooning old tree stumps, trellis arbours, and the like, or for trailing over dead branches placed against a warm south wall. It may be easily raised from seed in heat in early spring or autumn and kept through the winter, but the best plan is to lift the plants in autumn and winter in a greenhouse.

Lotus *(Bird's-foot Trefoil)*.—The majority of this genus is of a weedy nature, the only plant worth growing being the common L. corniculatus, which occurs in almost every lawn, meadow, or pasture, and forms tufts of bright yellow flowers, the upper part often red on the outside. Though so common, it is worthy of a place in the garden. The double-flowered variety is the best, as the flowers continue longest in perfection. L. creticus, maroccanus, sericeus, found in botanical gardens, are not nearly so showy. L. Jacobæus, a tender species with almost black flowers, succeeds in the open air in summer, and is all the better for planting out. The Lotus is best planted so that its shoots may fall in long and dense tufts over the face of rocks or stumps. It varies a good deal, but the common, low-tufted, spreading form, abundant in pastures and on sunny banks everywhere in Britain, is the most ornamental.

Love-in-a-mist *(Nigella damascena)*.

Love-lies-bleeding *(Amarantus caudatus)*.

Lunaria biennis *(Honesty)*.—This old-fashioned plant is beautiful when well grown, not only on account of its sweetly-scented purple blossoms, but also for the singular silvery flat seed-pods that succeed them. In borders, on the margins of shrubberies, and in half-shady situations, this plant is very effective in April and May, and it thrives well in any ordinary light garden soil. The Honesty is charming in a semi-wild state on chalky or dry banks and in open bushy places. Seeds should be sown every spring, and the plants thinned out during growth in order to make good ones for the next year. L. rediviva is a perennial kind similar to the Honesty, but with larger and more showy flowers. It grows 2 ft. or 3 ft. high and flowers in early summer. It thrives best in half-shady borders of good light soil. Division or seed. Mountain woods of Europe. Cruciferæ.

Lupinus *(Lupine)*.—A well known and beautiful genus of annual, biennial, and perennial plants, coming chiefly from N. America. The species in cultivation are few, though the names occurring in catalogues are numerous. The best of the perennial kinds are

L. arboreus *(Tree Lupine)*.—The large yellow Tree Lupine is a most precious plant. One can imagine nothing more beautiful than this for rough rocky banks or slopes. It flowers most profusely, and the scent of a single bush of it reminds one of that of a field of Beans. The purplish variety of the same plant, though good, is not nearly so valuable, and there are also inferior yellowish varieties. The best variety is the yellow. The bluish variety is not so good, because there are good blue perennial Lupines, whereas there is no other good yellow. It forms a roundish bush, from 2 ft. to 4 ft. high, and is easily raised from seed, though the true form is best increased from cuttings. It may be killed in severe winters, but is worth "trying again."

L. polyphyllus, one of the handsomest of all hardy plants. It grows from 3 ft. to 6 ft. high, has elegant foliage, and tall stately flower-spikes crowded with small blossoms, varying in colour from blue and purple to reddish purple and white. It flowers in summer, and continues long in beauty, and thrives in any kind of garden soil in open positions. It is a fine subject for naturalising, as it holds its own against other plants. The names of the principal varieties are argenteus, flexuosus, laxiflorus, Lachmanni, rivularis, and grandiflorus. L. nootkatensis is a dwarfer plant, with large spikes of blue and white blossoms. It flowers earlier than L. polyphyllus, and continues in bloom for a long time, but it is not a good perennial, and requires to be frequently raised from seeds.

The annual Lupines are among the most beautiful and popular of hardy annuals, being extremely varied in colour and all of the simplest culture. As they grow quickly, they need not be sown till about the middle of April; they thrive in any common soil. L. nanus is a handsome dwarf variety. L. subcarnosus is a beautiful ultramarine blue, and should always be grown. L. hybridus atrococcineus is the finest of all, having large handsome spikes of bloom; colour, bright crimson scarlet, with white tip; the flower-spikes are long and graceful. Other excellent sorts are mutabilis, Cruikshanki, Menziesi, luteus, superbus, pubescens, Hartwegi, and the varieties of Dunnetti. Many of the other sorts bear such a strong resemblance to each other, that they are not worth growing. The smaller annual Lupines are very pretty plants, and could be charmingly used to precede late-blooming and taller plants.

Luzuriaga radicans.—A small-growing Liliaceous evergreen plant from Chili. It is tolerably hardy in the mildest localities, though even in these it does not thrive so well as in a cool house. It is worthy of a trial in a cool bed of peat, on the north side of the rock garden, among the large alpine shrubs.

Lychnis *(Campion)*.—Plants of the Pink family, among which are a few well suited for the garden. All are of perennial duration.

L. chalcedonica.—This is an old border plant, from 1½ ft. to 4 ft. high, with large dense heads of brilliant scarlet flowers. It is of easy culture in any good ordinary soil. There is a handsome double scarlet variety, but the double white and single white kinds are not so desirable.

L. diurna.—The double deep purple-red sort of this common native is a very desirable plant, being very hardy, very showy, and never failing to produce a fine crop of bloom in early summer in any kind of garden soil. The double red variety, too, of L.

Lunaria biennis (*Honesty*).

Lunaria rediviva.

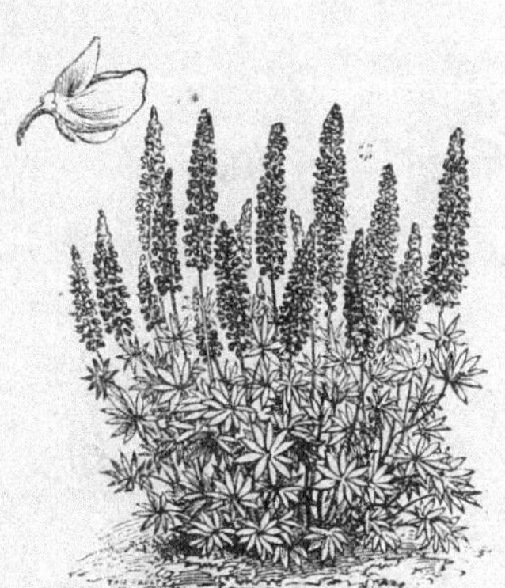
Lupinus polyphylius (*Lupine*).

Lupinus mutabilis.

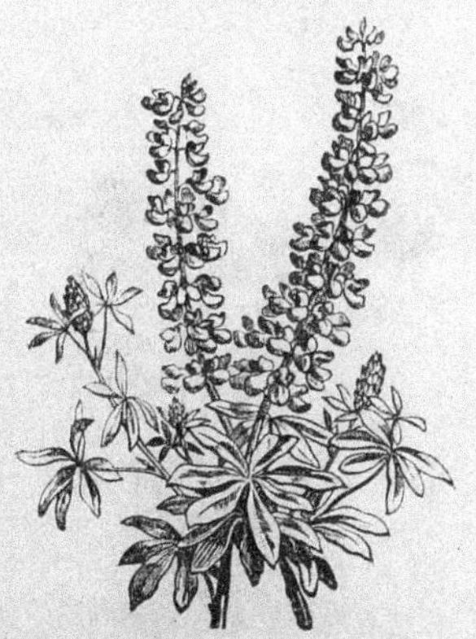
Lupinus pubescens.

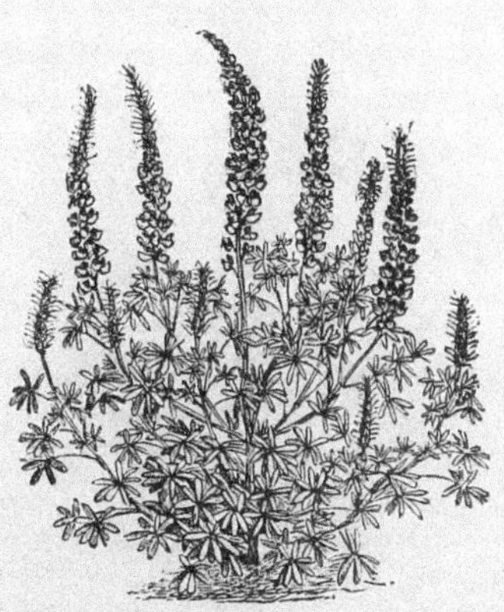
Lupinus Hartwegi (*Annual Lupine*).

[*To face page* 178.

Lychnis alpina.

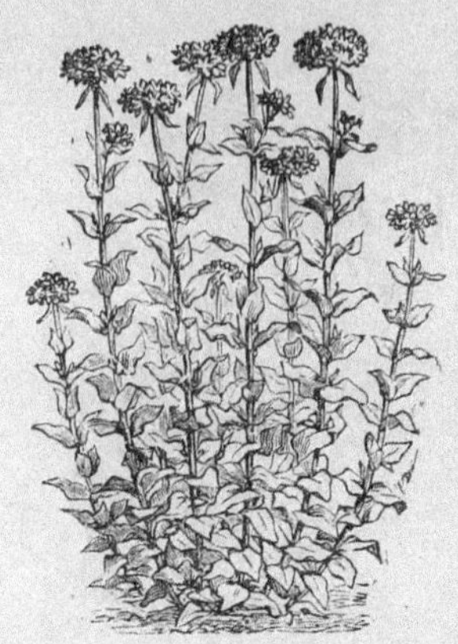

Lychnis chalcedonica (*Campion*).
See page 178.

Lychnis diurna fl -pl. See page 178.

Lychnis sylvestris fl.-pl.

Lychnis Preslii.

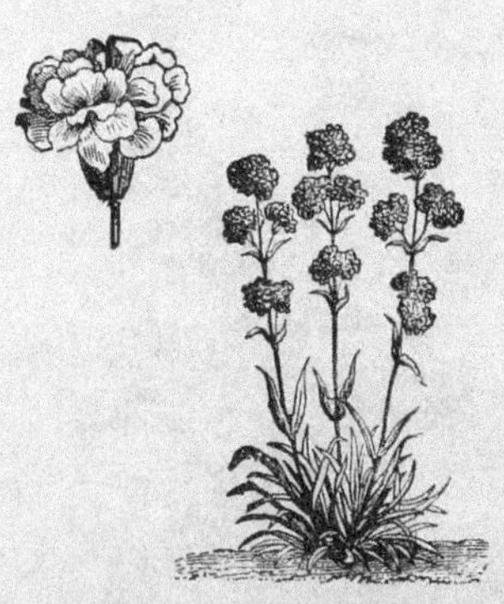

Lychnis Viscaria fl.-pl. (*German Catchfly*).

Flos-cuculi (Ragged Robin), another British plant, is a very ornamental border plant. Both should be propagated by division.

L. grandiflora.—This is a remarkably handsome plant, typical of the numerous varieties now in cultivation under the names of Bungeana, Haageana, &c. These all grow from 1 ft. to 2 ft. high, and bear in a cluster a dozen or so of flowers, each from 1 in. to 2 in. across, fringed at the edges and varying in colour from the most vivid scarlet to deep crimson, through pink to white. If fully exposed to the mid-day sun the colour of the flowers soon fades, but if grown in a partially shaded place they retain their true colour for a considerable time. They are all desirable border flowers, thriving best in warm, sheltered situations in good light soil, for though quite hardy the plants are apt to suffer from excessive moisture and cold combined. They are greatly benefited by frequent transplanting, say every other year. All the varieties may be raised by seeds or from cuttings. L. fulgens, a Siberian plant, is similar to the forms of L. grandiflora or coronata.

L. Lagascæ.—A lovely dwarf alpine plant, with a profusion of bright rose-coloured flowers, about ¾ in. across. It is peculiarly well suited for adorning fissures on the exposed faces of rocks, and should be associated with the smallest alpine flowers. It is easy of culture on rockwork in any free, sandy, or gritty soil. A thoroughly exposed position in the open air, however, should always be preferred, as the plant is very free in growth, as well as neat and hardy. The pale-flowered L. pyrenaica, which comes nearest to it, is not, since the introduction of the present subject, worthy of a place in any but a botanical or a very full collection. The flowers appear in early summer, and, when not drawn and weakened by shade, or by being placed in frames, it is in perfect condition at about 3 in. high, and is most readily increased by seed. *Syn.*, Petrocoptis Lagascæ.

L. Viscaria *(German Catchfly)*.—A British plant, having long, Grass-like leaves and showy panicles of rosy-red flowers, on stems from 10 in. to nearly 18 in. high, abundantly produced in June. The variety called splendens is the most worthy of garden cultivation, being of a bright colour. L. V. alba is a charming white variety worthy of a place, and so is the double variety, which has more rocket-like blooms. They are excellent plants for the rougher parts of rockwork, and as border plants on dry soils. The double variety is used with good effect as an edging plant about Paris. Any of the kinds are worthy of being naturalised on dryish slopes or rather open banks, on which they seem to form the largest, healthiest, and most enduring tufts. Easily propagated by seed or division.

L. alpina is a diminutive form of L. Viscaria, the tufts seldom being more than a few inches high, and not clammy. In cultivation it is a pretty and interesting, if not a brilliant, plant, and may be grown without difficulty on rockwork or in rather moist, sandy soil. Also a British plant.

Lycopodium dendroideum *(Ground Pine)*.—A Club Moss worthy of a place in the rock garden. The little stems, 6 in. to 9 in. high, are much branched and clothed with small, bright, shining green leaves. It flourishes best in a deep bed of moist peat in some part of the rock garden, where its distinct habit will prove attractive at all seasons. It is apparently difficult to increase, and as yet is exceedingly rare in this country. In attempting its culture the chief point to be observed is the selection of sound, well-rooted plants to begin with; small specimens may retain their verdure after the root has perished, and thus they often deceive. N. America, in moist woods.

Lygodium palmatum *(Climbing Fern)*.—An elegant North American Fern of twining growth, perfectly hardy if grown in a deep, peaty, moist soil in a sheltered and partially shady position. The stems are wiry, and are furnished with delicate green fronds cut in a fingered manner. It may be allowed to trail on the ground, but it likes best to twine around a rough branch or the branches of some choice shrub.

Lysimachia *(Loosestrife)*.—A small genus of the Primrose family of considerable diversity of growth. The most familiar example is the common creeping Jenny (L. Nummularia), than which there is no hardy flower more suitable for any position in which long-drooping, flower-laden shoots are desired, whether on points of the rockwork, or rootwork, or in rustic vases, or rapidly sloping banks. Creepers and trailers we have in abundance, but few that flower so profusely as this. Grows in any soil; in that which is moist and deep, the shoots will attain a length of nearly 3 ft., flowering the whole of their extent. Rarely or never seeds, but is as easily increased by division as the common Twitch. It flowers in early summer and often throughout the season, especially in the case of young plants. There is a golden-leaved variety which, though not quite so vigorous in constitution as the original type, is quite hardy. It retains its colour well, can be readily increased, is useful for rockeries or borders, and merits the name golden. The other kinds are of tall erect growth. L. vulgaris, thyrsiflora, lanceolata, ciliata, verticillata, punctata, and davurica all grow from 2 ft. to 3 ft., have spikes of yellow flowers, and delight in wet places; they are, therefore, suitable for planting by the sides of ponds, lakes, streams, and similar spots. Indeed, they grow almost anywhere, but if in a border they must have a place to themselves, as they spread so rapidly that they soon destroy weaker subjects. L. clethroides, a Japanese species, is a graceful and beautiful plant, from 2 ft. to 3 ft. high, bearing long, nodding, dense spikes of white blossoms, and the leaves when decaying in autumn display brilliant tints. L. Ephemerum is a similar plant, from S. Europe, but is scarcely so fine. Both are excellent subjects for naturalising, as, indeed, are all the others. There are some beautiful species, such as L. atropurpurea and lupinoides, which, though as yet rare, are very desirable. L. barystachya is a new species not much known.

Lythrum *(Purple Loosestrife)*.—The common waterside L. Salicaria is the most familiar plant of this genus and one of the showiest. It is well worthy of culture in localities where it is not plentiful. Pretty as is the ordinary wild kind, it is far surpassed by the varieties that have originated in gardens, of which superbum and roseum, both of which may be obtained in the hardy plant nurseries round London, are the finest. The colour of these is a much clearer rose than the common kind, and the spikes also are larger, particularly those of superbum, which, under good cultivation, rise 5 ft. or 6 ft. high. They are well worth growing by the sides of lakes or in marshy land, and are easily and rapidly increased by means of cuttings, which soon make good flowering specimens. Isolated plants in good soil make well-shaped bushes 3 ft. or 4 ft. high and as much through, and thus placed they look better than when planted closely in rows. L. virgatum, alatum, Græfferi, flexuosum, and diffusum are smaller plants, and not so showy, though they are not without beauty.

Machæranthera.—Biennial and annual Composite plants, native of N. America, now classed with Aster. M. canescens and tanacetifolia are in cultivation, but are scarcely worth growing.

Macleaya *(Bocconia)*.

Macrorhynchus.—A Composite of N. and S. America, allied to the Dandelion. M. grandiflorus, a Californian plant, is a rather handsome species with large yellow flower-heads, but it is rarely found in gardens.

Madaria elegans.—A hardy annual from California, having showy yellow and brown flowers. It succeeds well in half-shady places, and therefore is useful in certain positions. It requires the treatment of ordinary hardy annuals.

Madia.—A hardy annual Composite of no ornamental value.

Magydaris tomentosa.—A worthless umbelliferous plant.

Maianthemum bifolium.—A plant allied to the Lily of the Valley, and a native of our own country. Its habit and relationship make it interesting, and as it is easily grown in half-shady spots it is worth a place. Grows freely in any shady or half-shady place, and under or near Hollies or other bushes. It is not fitted for the border, and scarcely for the rock garden. *Syn.*, Convallaria bifolia.

Maiden-hair *(Adiantum)*.

Malaxis.—Terrestrial Orchids, of botanical interest mainly, and fitted for the bog garden.

Malcolmia *(Virginian Stock)*.—The old M. maritima needs no description; it is a charming, dwarf, hardy annual, growing in any soil and situation. The varieties are—the white (alba), alba nana, a dwarfer white than the other, and Crimson King (kermesina), a dwarf, deep red sort. These are all worthy of culture, and by some preferred to the common typical kind. The Virginian Stock, like many other annuals, does not show its full beauty from spring-sown seeds, though even those are pretty. Where it sows itself in the gravel it is often handsome. The plant, being easily raised and somewhat fleeting in bloom, is good when used as a surfacing plant in the spring or early summer garden, allowing bolder flowers to stand up from its pretty sheets of bloom. In flakes or masses, beds or lines, it is a pretty and effective plant.

Male Fern *(Aspidium Filix-mas)*.

Mallow *(Malva)*.

Malope grandiflora.—This, one of the most showy of hardy annuals, well deserves a place in every garden where a bold, showy crimson flower is desired. It grows from 18 in. to 24 in. high, and the better the plants are treated in regard to soil the finer they will bloom. If the Malope be sown in the open, the ground should be prepared by digging and manuring, and the seeds sown in some light, rich soil, ½ in. below the surface, and gently pressed down. Too often annuals only root superficially, and then when hot, dry weather sets in they soon go out of bloom. There is a white variety named M. g. alba, which is the counterpart of M. grandiflora, with the exception of the colour. The variety called M. g. rosea, white flushed with rose, is also very pretty and quite distinct. M. trifida is smaller in size in every part, but is also showy. These bold annuals are rarely used with good effect in gardens. For variety's sake, however, if for no other reason, they are worth a place occasionally. Like all annuals, they lend themselves to what we call rotation in the flower garden. If from any cause they get monotonous or uninteresting to us, it is worth while once in a way to try the effect of a batch of the best annuals, the more so if these are generally left out of our plans. The Malopes, being vigorous plants, are, as a rule, best in masses or groups. S. Europe. Malvaceæ.

Malva *(Mallow)*.—Of this genus there are a few ornamental garden plants, but the majority are of coarse and weedy growth. One of the most beautiful is the pure white flowered variety of the native Musk Mallow (M. moschata), which when in flower is highly attractive. It forms a branching, pyramidal bush, composed of numerous stems about 2 ft. high, and bears abundance of flowers from 1 in. to 1½ in. in diameter. It is a hardy perennial, and will grow in almost any soil or situation, but a hot, dry place suits it best. The whole plant is slightly Musk-scented. Among others, M. campanulata is a beautiful dwarf plant, but rare and difficult to grow well, as it is not quite hardy except in very mild districts. It is dwarf and

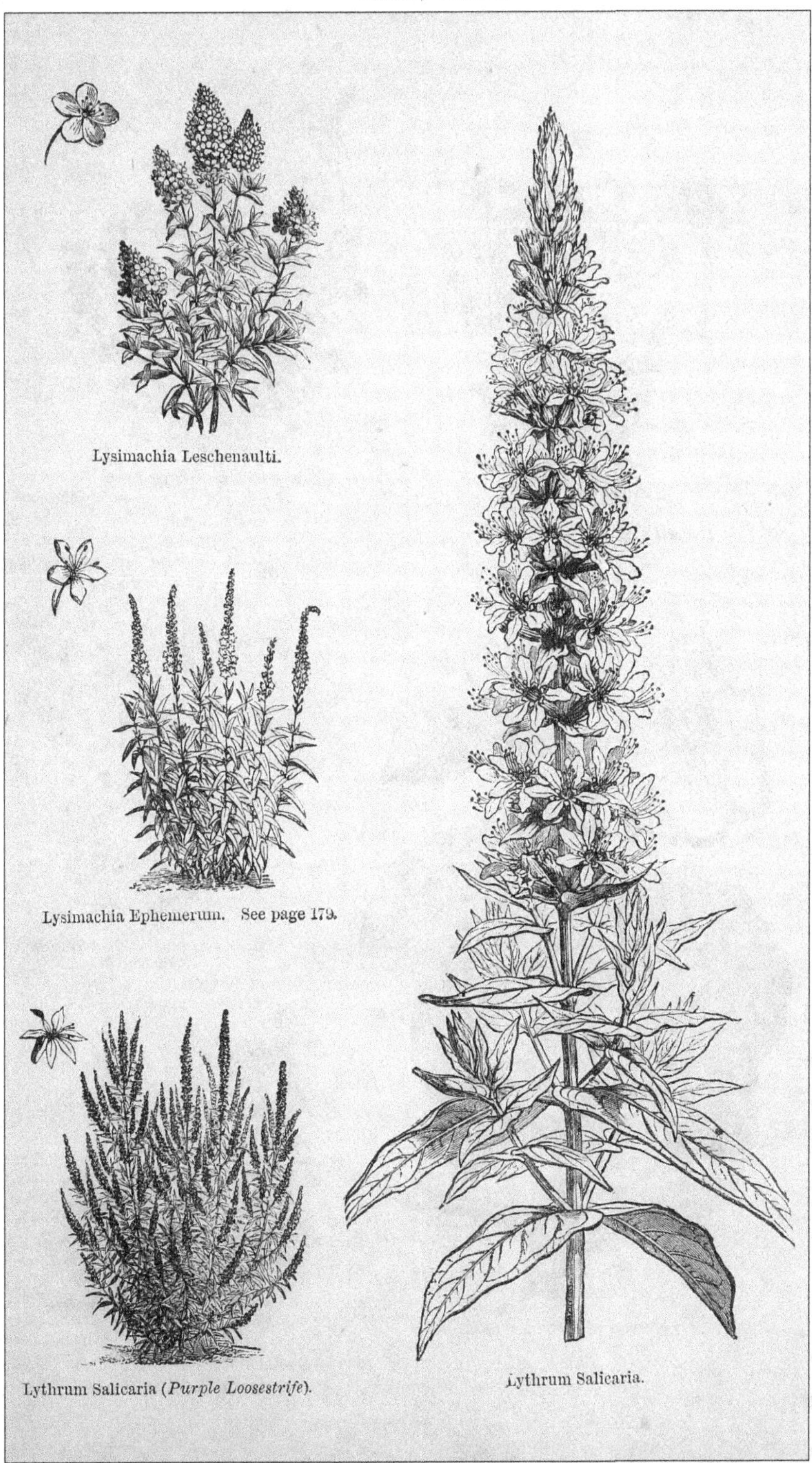

Lysimachia Leschenaulti.

Lysimachia Ephemerum. See page 179.

Lythrum Salicaria (*Purple Loosestrife*).

Lythrum Salicaria.

Madaria elegans. See page 180.

Malope trifida. See page 180.

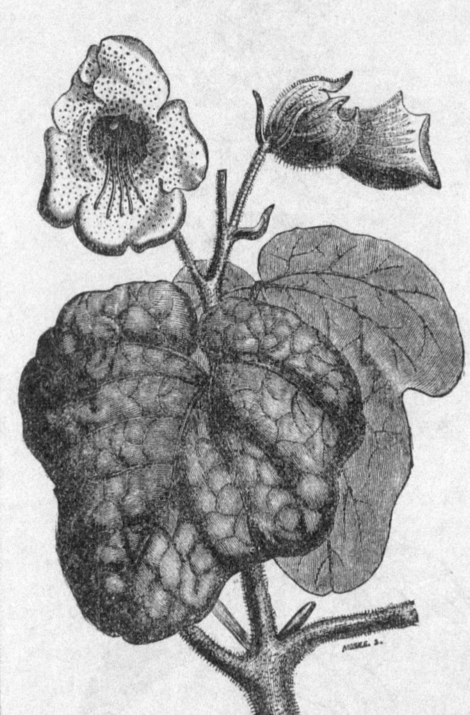

Flowers of Martynia lutea.

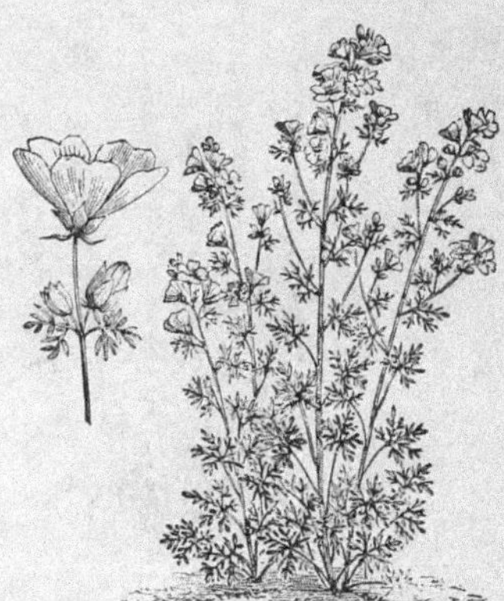

Malva moschata (*Musk Mallow*). See page 180.

Matricaria inodora (*Mayweed*).

[*To face page* 181.

spreading, and bears numerous lilac bell-shaped flowers. M. Alcea, Moreni, and mauritanica are worth growing when a full collection is desired, as is also the annual M. crispa, which grows from 3 ft. to 6 ft. high, forming an erect, pyramidal bush of broad leaves, with a curled or frizzled margin. Bushes of this are pretty in groups, beds, or borders. It may be sown in cool frames and put out early in May, by which means strong plants may be obtained early in the season.

Malvastrum.—These are similar to the preceding, but none of the species are thoroughly hardy, being natives of the warmer parts of North and South America. M. Munroanum is a dwarf plant with rather small orange-red flowers, and M. lateritium, a native of Buenos Ayres, is also dwarf, and has brick-red flowers. Sometimes these thrive in mild localities in light, warm soil on rock-work or well-drained borders.

Mandragora *(Mandrake).*— Curious plants of the Solanum family, suitable mainly for botanical collections. They are easily grown in warm, free soil, and enjoy borders at the foot of south walls.

Man Orchis *(Aceras anthropophora).*

Marguerite.—A name now commonly applied to Chrysanthemum frutescens, but the old French name for the Daisy.

Marrubium.—Unattractive plants of the Sage order, of which the common Horehound (M. vulgare) is the best known.

Marshallia cæspitosa.—This is an interesting Texan composite, from 6 in. to 9 in. high, each stem having a single Scabious-like white flower-head, about 1½ in. across. It is of perennial duration, and flowers the second season from seed. It forms a very neat border plant, blooming in June and July in light garden soil. Although not particularly showy, its flower-heads are so distinct that it is certainly well worth a place in the herbaceous border.

Marsh Fern *(Aspidium Thelypteris).*

Marsh Mallow *(Althœa officinalis).*

Marsh Marigold *(Caltha palustris).*

Martynia lutea.—A very pretty annual from Brazil, about 1½ ft. high, with roundish leaves and handsome yellow flowers in clusters. It requires a light, rich, cool soil, a warm position, and frequent watering in summer. Its large leaves and ornamental bloom make it a desirable subject for beds, groups, and borders. Increased by seed. M. fragrans is another species with sweet-scented flowers; it thrives in the open air in summer under similar conditions. It is best in rich borders, or among groups or beds of curious or distinct plants. M. proboscidea and others are suitable for the same purposes, but are less desirable.

Marvel of Peru *(Mirabilis Jalapa).*

Matricaria *(Mayweed).*— These are worthless weeds, except the double variety of M. inodora, which is a pretty plant with feathery foliage, somewhat like Fennel, and with large, perfectly double white flowers. It is creeping in growth, and requires considerable space to develop itself. If pegged down it forms a dense mass of growth, which, when in flower in autumn has a pretty effect. It is hardy, and quite perennial on most soils. It is easily propagated by cuttings or division in autumn or spring. (= Chrysanthemum inodorum fl.-pl.)

Matthiola *(Stock).*—From a few species of this genus have been obtained the numerous varieties of the garden Stocks, which have so long been such popular hardy flowers. The principal species from which the Stocks have been derived are M. incana, M. annua, and M. sinuata. The first named is a British plant growing on cliffs in the Isle of Wight, and is the origin of the Biennial, or Brompton and Queen Stocks; M. annua has yielded the Ten-week Stocks, and M. sinuata the others. These three primary divisions of the Stocks—the Ten-week, Intermediate, and Biennials—all require different treatment. These popular flowers are easily grown, fragrant, handsome, and have in all so many good qualities, that they deserve ever to have a place in our flower gardens.

TEN-WEEK STOCKS, if sown in spring, will flower continuously during the summer and autumn. The finest strain of the Ten-week Stock is that known as the large-flowering Pyramidal Ten-week, which are large, vigorous plants, branching freely, and producing a huge main spike of double flowers, with numerous branching spikes in succession. Where cut flowers are in request, a bed of these Stocks should be grown to assist in supplying them during the summer. The seed may be sown at any time from the middle of March onward, but it is always well to get Stocks from seed early. The seed can be sown thinly in pans or shallow boxes, in a gentle heat, and as soon as the plants can be handled without injury they should be transplanted into other pans or boxes and grown on quickly, care being taken at the same time not to draw them so as to make them weak and lanky. There are various places in most gardens where a bed or patch of Stocks might be grown with advantage; they should be given a good rich soil to grow in, and they will amply reward the cultivator. The German growers have a formidable list of kinds, many of which are more curious than showy. There are, however, sufficient leading colours among them, such as crimson, rose, purple, violet, and white, to yield distinct and pleasing hues. There is a strain of English-selected Stocks, known as Pyramidal, which are of tall growth, and remarkable for the large pyramids of fine flowers which they produce, but they are by no means so generally cultivated as they deserve to be. There is a very distinct type of Stocks known as Wallflower-leaved, which were introduced many years ago from the Grecian Archipelago, and which have shining, deep green leaves, not unlike those of a Wallflower. In all other respects the Wallflower-leaved type is like the ordinary German Stock. One of the finest varieties of this type, and at the same time one of the most beautiful Stocks in cultivation, bears the name of Mauve Beauty. It is a kind remarkable for its huge, compact heads of pale, lustrous, mauve-coloured flowers. The same treatment as that recommended for the Ten-week Stock will answer for this type of Stock. The autumn-flowering strain is very desirable, as the plants succeed the German varieties, and so prolong the season.

INTERMEDIATE STOCKS may either be sown in July or August, to stand the winter and flower early in spring, or in March to flower during the following autumn. In growth this strain is dwarf and bushy, and very free blooming. The varieties may be said to be confined to scarlet, purple, and white. There is a fine strain of this type grown in Scotland under the name of the East Lothian Intermediate Stock, and there it is much used for beds and borders, the climate appearing to suit it exactly. They are mainly used for late summer blooming, and are very effective. They are sown in the usual way about the end of March, planted out at the end of May when some 3 in. or 4 in. in height, and bloom finely through August and September, and even later, as they throw out numbers of side shoots that produce spikes of flowers. Thus, by using the autumn-sown Intermediate Stocks for early blooming, the ordinary large-flowering German Ten-week Stock for summer flowering, and the later East Lothian Intermediate Stocks for late summer, Stocks can be had in flower eight or nine months of the year without intermission.

BIENNIAL STOCKS comprise the Brompton and the Queen. They should be sown in June and July for flowering in the following spring or summer. The Queen and Brompton Stocks are closely allied, and are probably only varieties of the same kind; but it is curious to note that the seed of the white Brompton is pale in colour, that of the Queen quite dark. Old growers of the Stock distinguish the foliage of the Queen and Brompton Stocks in this manner. They assert that the under portion of the leaf of the Queen Stock is rough and woolly; that of the Brompton Stock is as smooth on the under part as on the upper. Of the Queen Stock there are three colours — purple, scarlet, and white; and of the Brompton Stock the same number, with the addition of a selected crimson variety of great beauty, but somewhat difficult to perpetuate. Both of these types are really biennials, and the seed should be sown at the end of July in beds and the plants transplanted to the open ground in the autumn. The difficulty of wintering the Brompton Stocks operates to deter many from attempting their cultivation. Even in the case of an unusually mild winter many will die. A well-drained subsoil and a porous surface soil suit them best. Shelter from hard frosts and nipping winds is of great service. A second transplantation of the seedlings has been tried with considerable success, the last one being made about December.

Maurandia Barclayana is an elegant twining plant from Mexico, grown in a greenhouse, but quite hardy enough to thrive in the open air in summer. It is admirably suited for covering trellises, fences, pillars, and the like. The flowers are deep violet and very showy; there are also white-flowered (alba), deep purple (atropurpurea), and rosy-purple (rosea), varieties. Easily raised from seed sown in early spring in heat; they will flower in the following summer if planted out in good soil in sheltered situations in May. Scrophulariaceæ.

May Apple *(Podophyllum peltatum)*.

Mazus Pumilio.—A distinct little New Zealand plant, vigorous in habit and creeping underground, so as rapidly to form wide and dense tufts, yet rarely more than ½ in. in height; the flowers, which are borne on very short stems, are pale violet; the leaves lie flat on the surface of the soil. It thrives in pots, cold frames, or in the open air, and is best placed in firm, open, bare spots on rockwork in free sandy soil in warm positions. It is not a showy, but an interesting plant, easily increased by division. Flowers in early summer. Figwort family.

Meadow Beauty *(Rhexia virginica)*.

Meadow Clary *(Salvia pratensis)*.

Meadow Rue *(Thalictrum)*.

Meadow Saffron *(Colchicum autumnale)*.

Meadow Saxifrage *(Saxifraga granulata)*.

Meconopsis.—The most familiar of these handsome Poppyworts is the common Welsh Poppy (M. cambrica); the others, about four in number, are all natives of the Himalayas. All are hardy, but only of biennial duration, and require some little attention to be grown well. They may be easily raised from seed sown in spring; indeed, the only way to ensure a good stock of strong plants is by annual sowings. The following is the most successful mode of cultivating them: A piece of ground is prepared by digging in good loam and well-rotted stable manure; a two-light frame is placed over it, and seedling plants are put in about March. As soon as the plants are fairly established the sashes are removed (unless the weather is frosty), and throughout the summer the plants are well supplied with water. During the following season, in April or May, they will have become large plants, often 2 ft. to 3 ft. in diameter, and are then removed to where they are wanted to flower. This may be readily done without checking them much, as they form such a large quantity of fibrous roots that usually a good ball of soil may be had with them. They are thus grown on as quickly as possible, treated like biennials. They should be planted out on well-drained rock-work in good soil, with a plentiful supply of

PLATE CLXV.

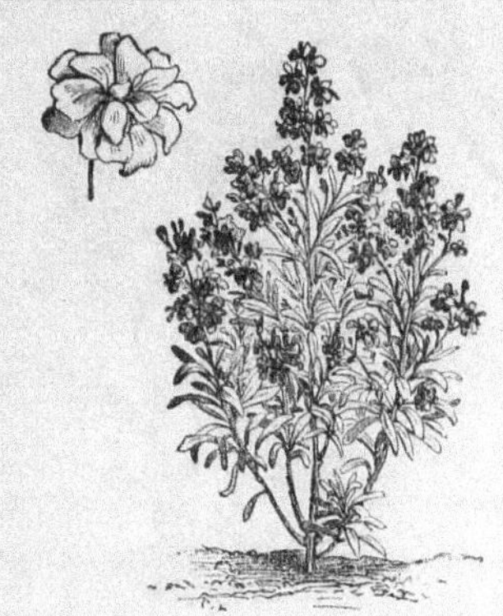

Mathiola incana (*Brompton Stock*).

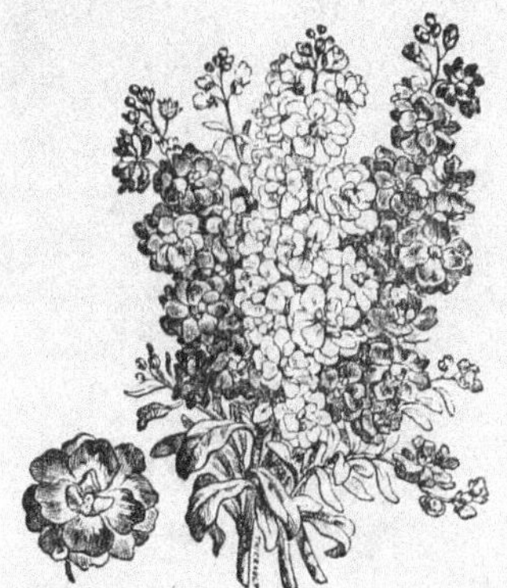

Mathiola annua fl.-pl. (*Ten-week Stock*).

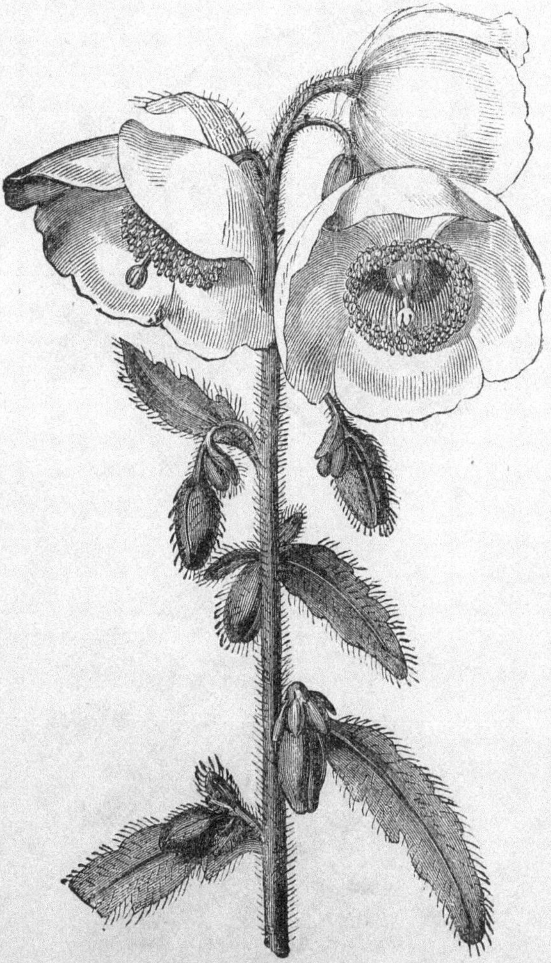

Meconopsis Wallichi (*Blue Himalayan Poppy*).

[*To face page* 182.

Maurandia Barclayana. See page 182.

Melittis Melissophyllum. See page 184.

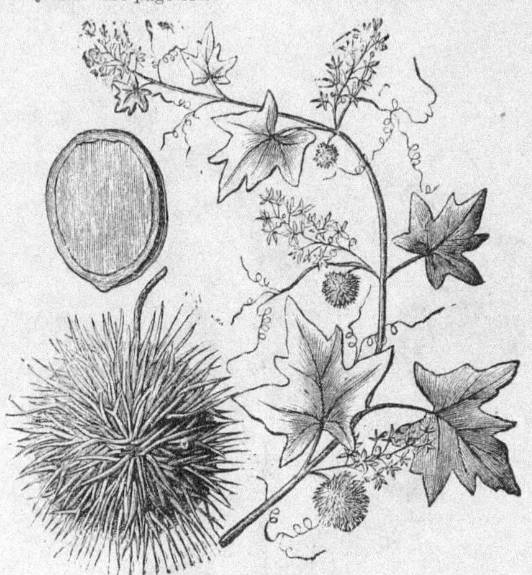

Megarrhiza californica (*Californian Big Root*).

Melianthus major.

[*To face page* 183.

water in summer, but they must be kept as dry as possible in winter, as excessive humidity in cold weather soon kills them. Pieces of sandstone broken finely should be placed under the leaves, so as to prevent them from coming in contact with the damp soil. A piece of glass placed over the leaves in a slanting position also protects them from too much moisture. Many plants require three or four years before flowering, and some may be kept in store pots for five or six years without showing any tendency to flower, but they are never nearly so fine when planted out after being cramped in this way. After flowering they all die.

M. aculeata is a singularly beautiful plant, with purple petals, like shot silk, which contrast charmingly with the numerous yellow stamens. The flowers are 2 in. across, on stems about 2 ft. high.

M. nepalensis has flower-stems from 3 ft. to 5 ft. high, and not much branched. The blossoms, which are produced freely, measure from 2 in. to 3½ in. across, nodding, and are of a pale golden-yellow colour.

M. simplicifolia produces a tuft of lance-shaped root leaves, from 3 in. to 5 in. long, very slightly toothed, and covered with a short, dense, brownish pubescence. The flower-stalk is unbranched, about 1 ft. high, and bears at its apex a single blossom, from 2 in. to 3 in. in diameter, of a violet-purple colour.

M. Wallichi is the finest of all, and a really handsome plant; it grows between 4 ft. and 5 ft. high, and forms an erect pyramid, the upper half of which is covered with pretty pale blue blossoms, which droop gracefully from the slender branchlets. It is a most conspicuous plant for the rockery, where it withstands the winter without the least injury. Well grown plants have leaves from 12 in. to 15 in. in length, bearing a great quantity of pale blue flowers, which open terminally. Separate flowers do not last long, but a few expand at one time. It takes fully a month before they are all expanded at the base, by which time the seeds of the first opened flowers are nearly ripe.

M. cambrica.—For the wild garden or wilderness the Welsh Poppy is one of the most suitable of plants, and at the same time one of the most charming. It is a cheerful plant to look at in all seasons; perched on some old dry wall, its masses of cut foliage are very refreshing in appearance, but when loaded with a profusion of large orange-yellow blossoms the plant, as a whole, is strikingly handsome; it is a determined coloniser, ready to hold its own under the most adverse circumstances. Its home is the wall, the rock, and the ruin. In many places it grows freely at the bottoms of walls, or even in gravel walks if allowed a chance. Where a plant is so easily grown naturally, so to say, there is no need to give it special care or place in the garden proper.

Medeola virginica.—A N. American Liliaceous plant of no particular garden value.

Medicago *(Medick)*.—This is a large genus, but it contains few, if any, good garden plants. One or two are useful for planting on banks or slopes, on which their wide-spreading masses may be seen to advantage, or on very rough rock or rootwork, so placed that the long shoots may fall over the brows of rocks. For this purpose the most suitable are M. falcata and elegans, both vigorous herbs with yellow flowers, thriving in almost any kind of soil. They are not suited for choice collections, but would yield good effects in certain special cases.

Megacarpæa polyandra.—A curious Cruciferous plant for botanical gardens.

Megarrhiza californica *(Californian Big Root)*.—A twining Bryony-like plant, having an enormous root. Only suitable for botanical gardens.

Megasea.—A section of the genus Saxifraga, mostly having large, broad foliage.

Melampyrum *(Cow Wheat)*.—Annual native plants not fit for garden culture.

Melanoselinum decipiens.—An umbelliferous shrub from Madeira, with a round simple stem, bare below, and large, spreading compound leaves. The flowers are white, and borne in umbels. Should be planted out in May. A useful subject for isolation on grass-plats, and requiring greenhouse or warm frame treatment in winter. Young plants are to be preferred for placing out. Multiplied by seed.

Melanthium triquetrum.—A bulbous-rooted plant little known to cultivation. It has long, round leaves, like those of some of the Alliums, and its flowers, which are small and have delicate mauve petals and a purple centre, are pretty, and are produced on spikes from 5 in. to 6 in. long. It is a half-hardy plant—hardy, perhaps, in favoured districts.

Melianthus major.—This is an effective half-hardy plant for the summer decoration of our flower gardens. Its finely-cut, large, glaucous leaves contrast effectively with the general types of vegetation, and as it is a plant of the easiest cultivation, the Melianthus has become a favourite in sub-tropical gardening. Propagation is by seed, from which it is freely produced, and plants raised from seed early in the season make good growth by planting-out time, and attain by midsummer a height of from 3 ft. to 4 ft. When it is desirable to have larger plants by planting-out time, it is advisable to sow the seeds in autumn and keep them growing throughout the winter, and a stronger and earlier development will be the result. The Melianthus is all but hardy when planted out upon a well-drained subsoil, in the south and western districts, in sheltered nooks, for although the stems may be cut down by frost, the roots will survive and push up in spring; but, like Cannas, Arundo Donax, and similar plants, they are not to be depended upon in wet, hard winters, even when carefully mulched and otherwise protected, frequently suffering from accumulated damp. It is far safer to lift the roots and store them away under the stage of a cool house or shed.

Melilotus *(Melilot)*.—A genus of the Pea family, of no garden interest.

Melissa officinalis *(Common Balm)*.—A well-known old garden plant, with a very grateful odour when bruised; 2 ft. to 3 ft. high. The variegated form is sometimes used as an edging plant, and the common kind might be naturalised in any position or soil by those who admire fragrant plants. Division. Europe.

Melittis Melissophyllum *(Bastard Balm)*.—A distinct-looking plant of the Salvia order, with from one to three flowers nearly or quite 1½ in. long. The peculiarly handsome purple lip reminds one of the flowers of some of the beautiful exotic Orchids rather than of those of a Labiate plant. M. grandiflora is merely a slight variety, differing in colour from the normal form. The plant is entirely distinct from any other in cultivation, and is well worthy of a position by shady wood and pleasure-ground walks, as it naturally inhabits woods. Woody spots near a fernery or rockwork suit it to perfection, and it grows very readily among shrubs, and also in the mixed border. It is found in a few localities in Southern England, and is widely distributed over Europe and Asia. Readily increased by seed or division, and flowers in May about London.

Mentha *(Mint)*.—There are but few ornamental plants of this genus. One is the variegated form of M. rotundifolia, common in most gardens, where it is useful for edgings or for clothing any bare, dry spots, as it will thrive almost anywhere and in any soil. Another is M. gibraltarica, a variety of the native M. Pulegium, now largely used for flat geometrical beds in summer on account of its dwarf, compact growth and deep green foliage, which retains its freshness throughout the season. It is one of the easiest plants to propagate, and may be increased with wonderful rapidity, as it will bear rapid forcing for cuttings early in spring. Inasmuch as its growth hugs the soil, and throws out roots at every joint, all one has to do is to keep cutting off little plants and potting them, or planting them in shallow boxes, and in a very short time they in their turn will bear cutting up in like manner. Being a native of S. Europe, it is somewhat tender, and is generally killed in winter. M. Requieni is a minute creeping plant like the Balearian Sandwort, and has a strong odour of Peppermint. It should be allowed to trail about among other similar plants in the rock garden. Native of Corsica.

Mentzelia.—These are lovely Californian plants, mostly of biennial duration, but no one has succeeded in growing them well in cultivation. They evidently require more attention than most half-hardy plants receive, but their beautiful bloom well repays the trouble. The most successful cultivator of these plants thus writes to *The Garden*: "I find it necessary to sow the seed as early in the season as possible, and to grow the seedlings on in a frame, giving liberal shifts, using a compost of fibry loam and a small quantity of leaf mould and sand. After the final shift they should be plunged in a sunny border until autumn, when they should be removed to a frame for wintering. In the spring they should be again plunged in the open air, and by assisting them with weak manure water occasionally, strong and healthy flowering specimens will be the result. When beginning to show flower they should be removed to a cool greenhouse or frame, as excessive humidity at this stage is injurious to them. They may be grown entirely in the open air, and with good results if the weather be favourable, yet in our climate the former mode is by far the most satisfactory." The following is a selection of the most showy of Mentzelias in cultivation. With regard to the others, the flowers are rather small, and, on the whole, scarcely worth the trouble of cultivating.

M. ornata is a biennial, growing from 2 ft. to 4 ft. in height. The flowers are from 2½ in. to 4 in. across, of a creamy white, and fragrant. It belongs to the vespertine section, or those in which the flowers fully expand only towards evening.

M. nuda, also 2 ft. to 4 ft. high, with flowers much resembling the last.

M. lævicaulis is another fine kind, with whitish stems, 1 ft. to 3 ft. in height, and with leaves much resembling the last, both covered with short and stout bristles. The flowers are of a rich, deep yellow colour, and open only in bright sunshine.

M. oligosperma is a perennial, from 1 ft. to 3 ft. high, with flowers 3 in. across, of a bright yellow colour, and opening in sunshine.

Menyanthes trifoliata *(Buckbean, or Marsh Trefoil)*.—A beautiful and fragrant plant, native of Britain—in shallow streams or pools, and very wet marshy ground or bogs. The plant forms strong, creeping, rooting stems which often float in deeper water. The flowers are borne on stout stalks, varying in length with the depth of the water, and are arranged in racemes, beautifully fringed and suffused with pink. It will be found easy to establish wherever the necessary conditions of its existence—shallow water or bog—are available, by introducing pieces of the stems, and securing them till, by the emission of the roots, they have secured themselves. In some moist soils it thrives in the ordinary mixed border.

Menziesia.—Dwarf shrubs, which, like the Heaths they resemble, are admirably suited for large rock gardens or anywhere where there is a moist peaty soil. They are all of neat growth, and bear pretty drooping flowers. M. cœrulea grows from 4 in. to 6 in. high, and has pinkish-lilac flowers. It is not

PLATE CLXVII.

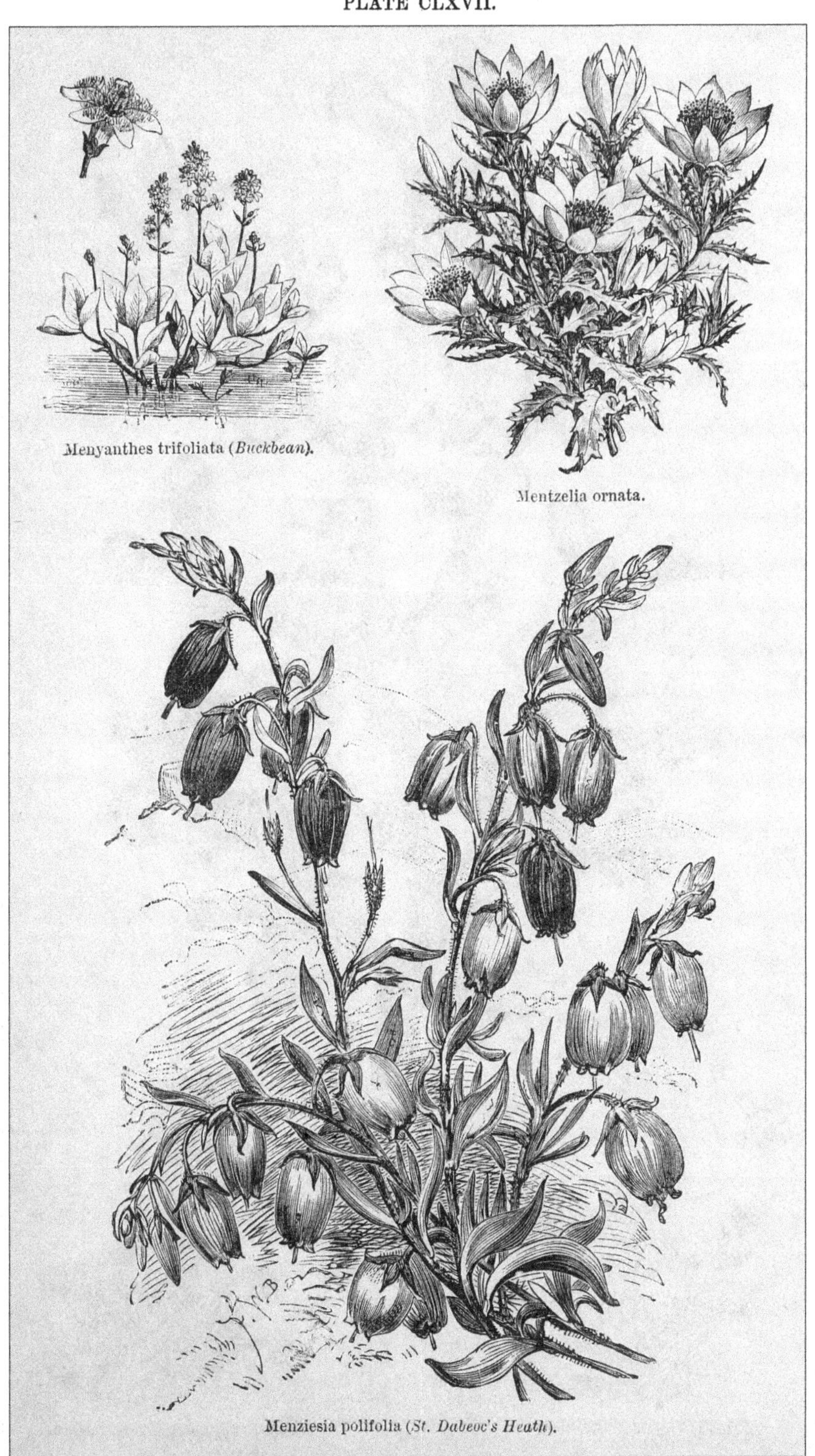

Menyanthes trifoliata (*Buckbean*).

Mentzelia ornata.

Menziesia polifolia (*St. Dabeoc's Heath*).

[*To face page* 184.

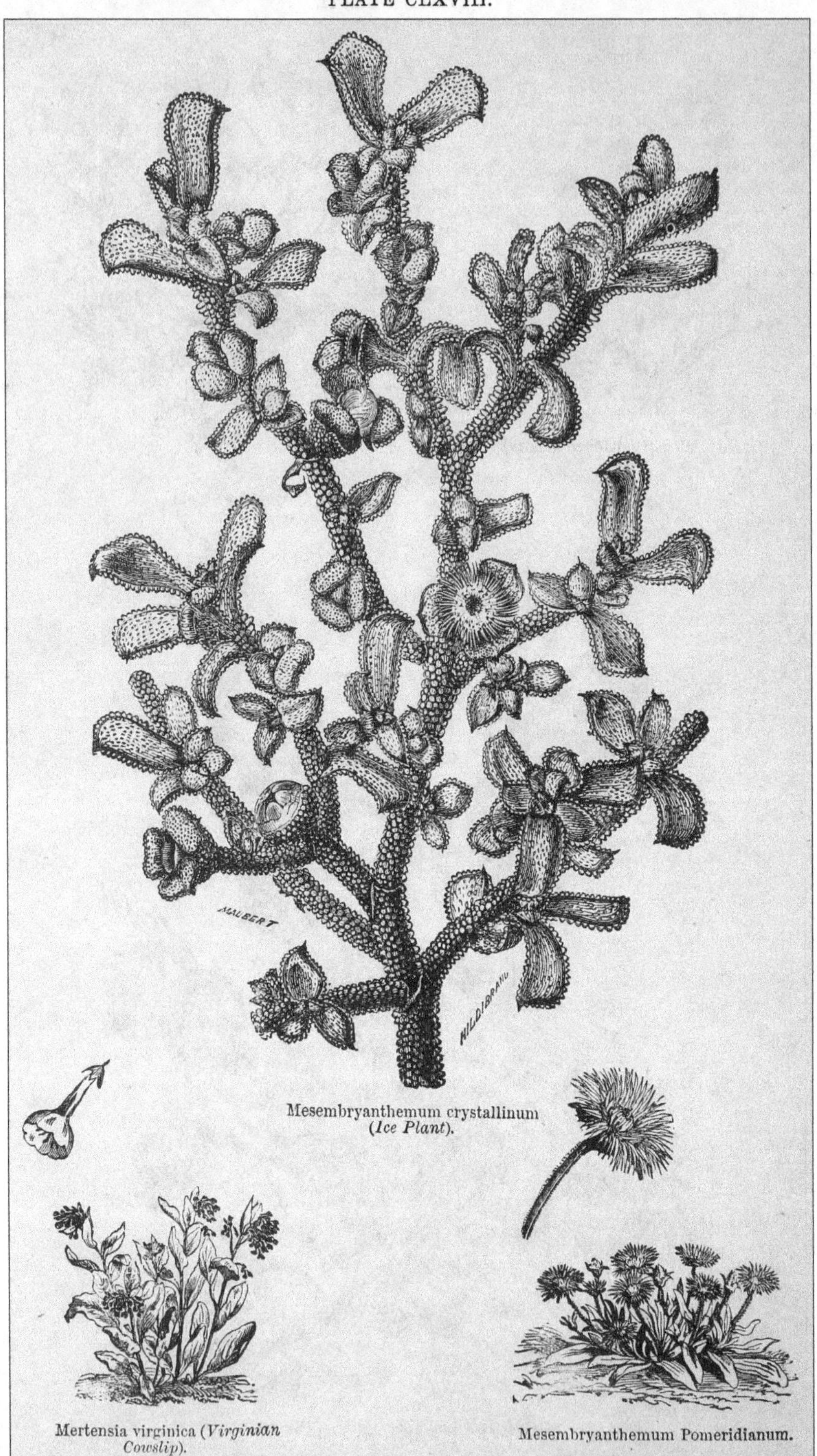

Mesembryanthemum crystallinum (*Ice Plant*).

Mertensia virginica (*Virginian Cowslip*).

Mesembryanthemum Pomeridianum.

[*To face page* 185.

so beautiful or brilliant in colour as M. empetriformis, but merits a place in full collections. Flowers rather late in summer and in autumn. Europe.

M. empetriformis.—A tiny shrub, producing numbers of rosy-purple bells in clusters on a dwarf Heath-like bush, seldom more than 6 in. high. This plant, very rarely seen in gardens, is one of the brightest of gems for the choice rock garden, thriving best in a rather moist sandy peat soil in fully-exposed positions. It flowers in summer, and is sometimes known as Phyllodoce empetriformis. This, unlike the rather tall and spreading M. polifolia, may be associated with the dwarfest alpine plants. N. America.

M. polifolia *(St. Dabeoc's Heath)* grows from 12 in. to 20 in.; the erect flowering-stems bear beautiful crimson-purple blooms in graceful, one-sided, drooping racemes. There is a pure white variety much less common and no less beautiful than the typical form, and also a pretty variety, with purple and white flowers, called bicolor. They flower in summer, and may be obtained from most nurserymen. The white variety is sometimes sold under the name of M. globosa and alba major. Abundant in some parts of Ireland, hence called Irish Heath.

Mercurialis *(Mercury)*.—Native weeds, and often troublesome garden pests.

Merendera Bulbocodium.—A bulbous plant, very like Bulbocodium vernum, but flowering in autumn. The flowers are large and handsome, and of a pale pinkish-lilac. Suitable for the rock garden and bulb garden, till plentiful enough to be used in borders. Increased by separation of the new bulbs and by seed. S. Europe.

Mertensia.—A genus of beautiful Borageworts, formerly known as Pulmonarias. There is something more beautiful in form about them, both in foliage and in stem, and also in the graceful way in which they rise into beautiful panicles of blue, than is to be found in almost any other family. There are in cultivation about half a dozen species, all of which are desirable garden plants.

M. alpina is a beautiful alpine kind, and should only be associated with the choicest of alpine plants. The leaves are bluish-green; the stem is only from 6 in. to 10 in. high, and has from one to three terminal drooping clusters of light blue flowers which appear in spring or early summer.

M. dahurica, although of a very slender habit and liable to be broken by high winds, is perfectly hardy. It grows from 6 in. to 12 in. high, with erect branching stems, and blooms in June, producing handsome, bright azure-blue, drooping flowers in racemose panicles. It is a very pretty plant, and suited either for the rock garden or borders, where it should be planted in a sheltered nook in a mixture of peat and loam. It is easily propagated by division or seed. (=Pulmonaria dahurica).

M. oblongifolia is another diminutive species with deep green, fleshy leaves. The stems are 6 in. to 9 in. high, and bear handsome clustered heads of brilliant blue flowers.

M. sibirica.—The peculiar value of this species is that it has the beauty of colour and grace of habit of the old Mertensia virginica, and at the same time grows and flowers for a long period in ordinary garden soil. The flowers are small and bell-shaped, and produced in loose drooping clusters that terminate in gracefully arching stems. The colour varies from a delicate pale purple-blue to a rosy pink in the young flowers. It is a more elegant plant than the Virginian Lungwort (M. virginica), an older and better known kind. It is a perfectly hardy perennial, and may be propagated by division.

M. virginica *(Virginian Cowslip)*.—One of the handsomest of all, producing in early spring drooping clusters of purple-blue blossoms on stems from 1 ft. to 1½ ft. high. The leaves are large and of a bluish-grey. The culture of the old Virginian Lungwort is interesting, because in many gardens one never sees it making the slightest progress. Some kind of sheltered, moist, peaty nook seems most suitable to it, and, from what one reads of its native habitat, this would seem to be a natural place for it. The finest plants are grown in moist, sandy, peat beds with shelter near, but even then its bloom is early and short, and it is not nearly so important a plant for our gardens as M. sibirica.

Mesembryanthemum *(Fig Marigold)*.—Of this very large genus there are several showy species well suited for the open air, though none can fairly lay claim to be perfectly hardy. The common Ice Plant (M. crystallinum) is grown in most large gardens for garnishing purposes; it is also sometimes used as a pot plant, but it is most effective when planted out on rockwork or on some old wall. It will, however, grow in any good soil in a sunny situation, and it might be used as a band to a sunny border. It will grow from 3 ft. to 4 ft. in a season, and on warm days has a cool and refreshing look. Its flowers are unimportant compared with the crystal-bespangled stems and foliage; and the richer the ground the greater the crystal development. Seeds of it should be sown in heat in March, and planted out from 6 in. to 8 in. apart. There are two varieties of this plant—one with red and the other with white flowers. M. cordifolium is a perennial, the variegated form of which is largely used in carpet gardening. M. Pomeridianum is a strong-growing species with broad foliage and large purple and rose-coloured flowers. It is not so common a kind as the last, but well deserves a place on a south border. M. tricolor is the most showy of the annual Mesembryanthemums. It is a neat plant with cylindrical foliage, and is not a creeper. It grows in neat, compact tufts from 4 in. to 6 in. in height; its flowers, which are abundant, are purple, rose, or white, the result being a good contrast. This species should be sown in a sandy soil in the open garden about the end of April; it dislikes transplantation, and it lasts longer in the ground than in a pot. Those who possess a large collection of this genus should turn the whole out on banks or rockwork and leave them there, taking cuttings off them yearly. They attain characters out-of-doors never seen in pots. The foliage of the whole family is singular and diversified, and the brilliant lustre of their gorgeous flowers—white, orange, rose, pink, crimson—is unequalled. They are children of the sun, and a rockery devoted solely to a collection of them in an open, sunny spot, is a sight worth seeing, and one that can only be compared to a pyrotechnic display of variously-coloured fire. A situation where the soil consists of little else than sand and gravel suits them perfectly. As these plants have been so little grown in the open it is scarcely known which are hardy and which are tender, but experiments in this matter would be interesting, for probably many would prove almost, if not quite, hardy in suitable situations.

Meum athamanticum.—A most elegant and graceful cut-leaved plant, dwarf and neat in habit. It is from 6 in. to 12 in. high, easy of growth in ordinary soils, and perfectly hardy and perennial. It is probable that in dry seasons it might wither too soon for association with autumn-flowering plants, but for rockwork, borders, or mixed arrangements of any sort it is invaluable. It is a British plant, easily increased by division.

Michauxia campanuloides.—A remarkable and highly ornamental plant of the Bellflower family, growing from 3 ft. to 8 ft. high. The flowers are white, tinged with purple, and are arranged in a pyramidal candelabra-like head, which makes it very striking and distinct. It is usually considered a biennial, but sometimes it flowers the third and even the fourth year. It should, however, be treated as a hardy biennial, and seedlings should be raised annually, so as always to have good flowering plants. It flourishes best in a moist and deep sandy loam. This is one of the plants only occasionally seen good, and therefore the enthusiastic cultivator will take all the greater pride in trying to grow it well. Its fine stately form and tall stature are very effective when the plant is tastefully placed in the mixed border or a nook to itself in a choice bed of evergreen shrubs. Warm and sheltered borders, and borders on the south side of walls seem to suit it best. Native of the Levant.

Microlepia anthriscifolia.—A most elegant Fern, growing from 6 in. to 12 in. high, that has proved perfectly hardy, even at York. It is deciduous, but is a charming plant in spring and summer, and is of easy culture. It thrives in the open as well as in the shade, and may be used with good effect as an edging to a sheltered border.

Microlonchus.—Composite plants allied to Centaurea, of no garden value.

Micromeria.—Labiate plants possessing a strong odour like Thyme, Mint, &c., but of no ornamental value.

Microsperma bartonioides *(Eucnide)*.

Mignonette *(Reseda odorata)*.

Mikania scandens *(German Ivy)*.—A slender twining perennial, with Ivy-like foliage and small flesh-coloured flowers. It is hardy in light, warm soils and situations, and is suitable for covering trellises and the like. N. America. Compositæ.

Milium *(Millet Grass)*.—A small genus of Grasses of which the native M. effusum is a plant worth cultivating for the sake of its elegant feathery plumes, which do well for associating with flowers in summer. Grows in any soil or position, but prefers moist places such as could be selected in most gardens. There are one or two other kinds worth growing.

Milk Thistle *(Silybum)*.

Milk Vetch *(Astragalus)*.

Milkwort *(Polygala)*.

Milla.—The bulbous plants, formerly known under this name, are now included with Brodiæa, under which name are placed the cultivated species. The only true Milla is said to be M. biflora, a beautiful plant with large snow-white blossoms deliciously scented. It is rather difficult to cultivate, but it is well worth any extra care. It is too choice to risk in the open border even if it be quite hardy, which is doubtful.

Mimulus *(Monkey Flower)*.—The cultivated species are valuable, showy, border flowers, for the most part native of California. These are all very partial to moisture, on which account they are suitable for growing in damp places, such as the artificial bog or in moist borders, and by the margins of artificial water and streams. The old M. cardinalis is a showy plant when well grown, and is deserving of a place in any garden. There are several varieties of it. The common Musk (M. moschatus) is very hardy and enduring in moist soil. It is worth growing in a corner in heavy or wet soil. M. luteus and its varieties, variegatus, cupreus, Tilingi, guttatus, and others are typical of the beautiful hybrid varieties now in gardens, which combine the dwarf habit and hardiness of the M. cupreus, with the larger richly spotted and blotched flowers of the other parent, the old M. variegatus. These hybrids, which are known as M. maculosus, also bear exposure to the sun better than the parents. There is also a strain with Hose-in-hose flowers, sometimes called double. These sorts should be grown in every garden, and a packet of seed will afford a wonderful variety. The seeds of the Mimulus should be merely sprinkled on the soil; if covered by it they may vegetate less quickly and abundantly. A little damp

PLATE CLXIX.

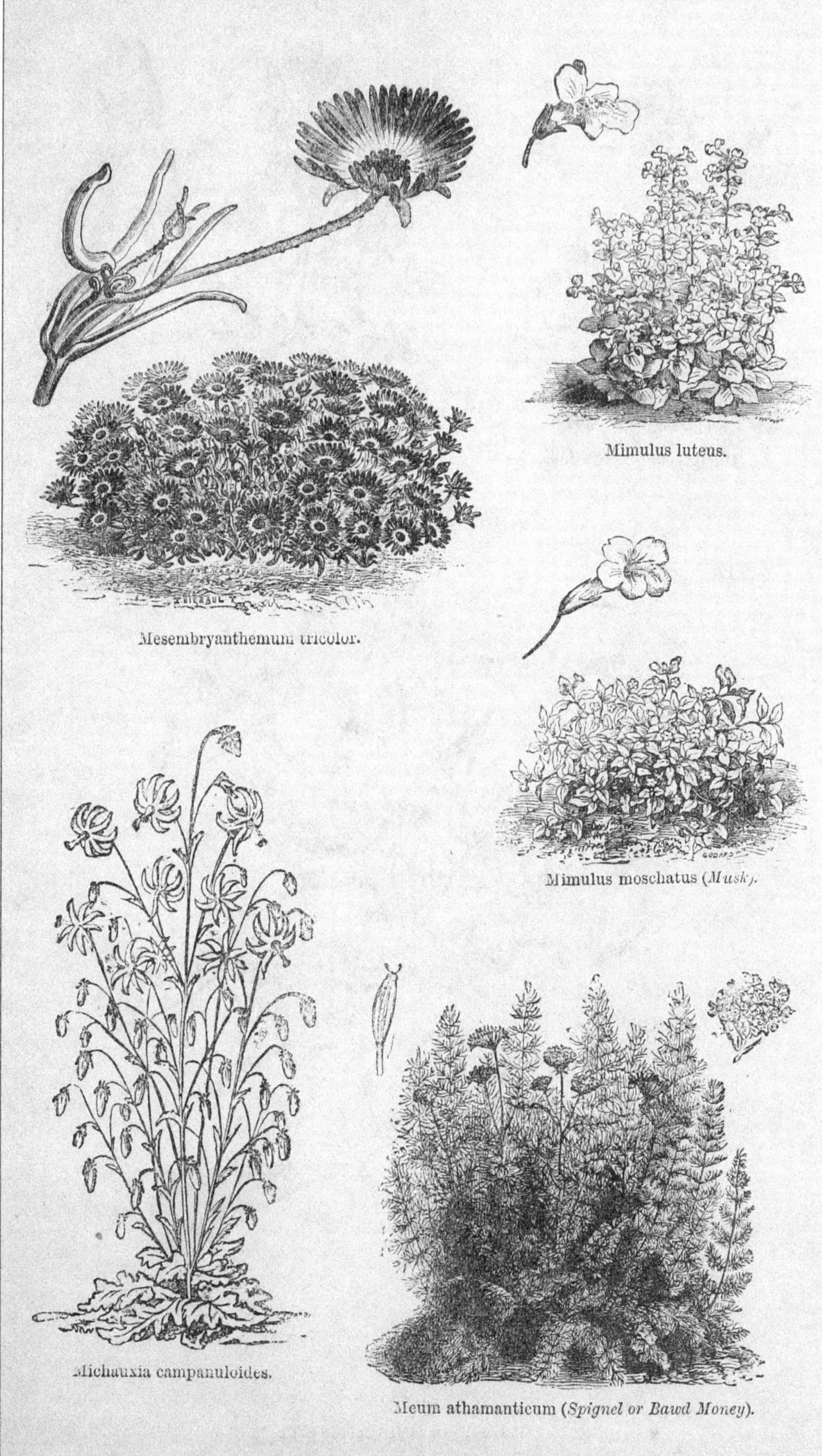

[*To face page* **186.**

PLATE CLXX.

Mimulus cupreus hybridus.

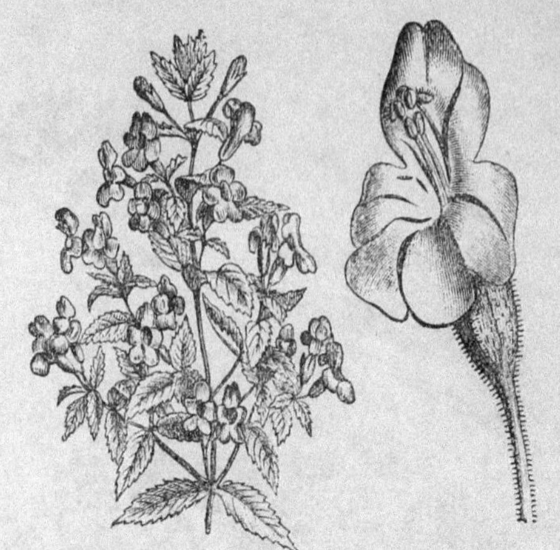

Mimulus cardinalis (*Monkey Flower*).

Mirabilis Jalapa (*Marvel of Peru*).

[*To face page* 187.

moss may, however, be laid over the surface, but this should be removed as soon as the seeds have germinated. Other cultivated species are M. primuloides, repens, &c., but they are only fit for botanical collections.

Mirabilis *(Marvel of Peru).*—Handsome herbaceous plants, the most familiar among which is M. Jalapa, which forms a dense, compact, round bush nearly 3 ft. across, and the same in height, covered with flowers. These are about 1 in. across, and their usual colours are white, rose, lilac, yellow, crimson (of various shades), and purple—striped, mottled, and selfs. The plants may be treated as half-hardy annuals, raised from seed in a warm frame, potted on, and planted out in May. They are perennials, however, producing a tapering black root, which must be lifted when the leaves are killed by frost and stored in sand during the winter. They should be started in pots in the spring and planted out as before. After the second year the roots become unwieldy, and are better discarded. The plants require a light, warm soil and all the sunshine they can get. The seeds ripen rapidly and readily, one seed only to each flower; and as they are large, they can be gathered from the ground beneath the plants. M. multiflora is a newly-introduced plant from California. It is somewhat similar to M. Jalapa, but dwarfer, and the flowers—a bright crimson-purple—are produced abundantly in large clusters, and expand in bright sunshine. It is quite a hardy perennial in light, warm soils, and is a most desirable border plant. M. longiflora, with long tubular flowers with carmine centres, is a capital plant for growing at the foot of a warm south wall. Native of Mexico.

Mitchella repens *(Variegated Partridge Berry).*—A very neat trailing, small, evergreen herb, 2 in. or 3 in. high. The flowers, which are white, are produced in summer, and succeeded by small, bright red berries. Flourishes best in shady spots near the rock garden or hardy fernery, in sandy peat and leaf mould. Division. North America.

Mitella *(Mitre-wort).*—North American plants of the Saxifrage family. M. diphylla is rather a pretty plant, with small white blossoms on spikes from 8 in. to 12 in. high. M. nuda and Breweri are suitable only for botanical gardens. They grow best in moist, peaty soil in half-shady situations.

Modiola.—The prettiest plant of this genus is M. geranioides, a hardy perennial growing about 6 in. high, and forming a dense, compact, trailing tuft of elegantly-cut foliage, which in summer is studded with a profusion of showy purplish crimson colour. It delights in a warm open place in the rock garden, where, if well drained, it is perfectly hardy. The best kind was figured in *The Garden* some time ago. Malvaceæ. North America.

Mœhringia muscosa.—A very dwarf, alpine, evergreen herb, 2 in. or 3 in. high, somewhat inconspicuous. It is best on the rock garden and borders, in fine very sandy loam. Division and seed.

Molinia cœrulea.—A native Grass, of which there is a variegated form not unfrequently used as an edging for flower beds, but there are other Grasses better adapted for this purpose.

Molopospermum cicutarium.—An ornamental umbelliferous plant, 5 ft. or more in height, with large, handsome leaves forming a dense irregular bush. It is hardy, and easily increased by seed or division, but it is, as yet, rarely met with. Loves a deep moist soil, but will thrive in any good garden soil. It a fine subject for isolation or grouping with other hardy and graceful-leaved umbelliferous plants. Carniola.

Monarda *(Horse Balm).*—All the Monardas are showy border flowers of the simplest culture, for they will thrive and flower finely in any position, and they are, moreover, equally indifferent as to soil; therefore, besides being admirably adapted for garden borders, they are excellent subjects for naturalisation in woods, shrubberies, and various wild places. All of them may be readily divided, and every division will grow. The perfume emitted by the leaves of many of them also tends to increase their value; and on this account some of them are called Balms, Bergamots, &c. There is a great variety of colour in the various kinds, the varieties of M. fistulosa alone representing more than half-a-dozen different shades. There are about half-a-dozen species in cultivation, but some of them are rarely met with. The red kinds scattered through the American woods in autumn are very handsome.

M. fistulosa *(Wild Bergamot)* is a very handsome, robust-growing perennial, from 2 ft. to 4 ft. in height. The flowers are very variable; the typical colour appears to be pale red, but it is sometimes purple-rose colour—in fact, every gradation to almost white may be found in the cultivated kinds. It is in full beauty during the summer months. M. didyma (Oswego Tea or Bee Balm) is another highly ornamental kind, and equally robust. It grows about 3 ft. high, and the flowers, which are borne in head-like whorls and continue in perfection for a considerable period during summer, are of a deep red colour. M. Kalmiana is a showy plant, taller and more robust in growth than the preceding, often attaining 4 ft. high. The flowers, which are somewhat downy and produced in dense whorls, are of a deep crimson colour. M. purpurea is somewhat similar in habit to the last, but differs in the size of the flowers, which are smaller and are invariably of a deep purplish-crimson. All are natives of N. America, and may be readily increased by division in spring, or by seed.

Monardella macrantha.—A charming little perennial plant from California. It grows from 4 in. to 6 in. high, and has slightly fragrant orange-scarlet blooms. It is not well known in this country, but is said to be worth cultivation. Labiatæ.

Monkey Flower (*Mimulus*).

Monkshood (*Aconitum Napellus*).

Monolopia major.—A half-hardy Californian annual, growing about 2 ft. high, and bearing pretty yellow flower-heads; but as there are so many perennials of a similar nature, it is scarcely worth a place.

Montbretia. — Iridaceous plants from the Cape of Good Hope, two of which are in cultivation. M. Pottsi, the best known species, grows from 2 ft. to 2½ ft. high, and bears abundance of bright green leaves, which resemble those of a Gladiolus both in form and in their singular plaited appearance. The flowers are about 1 in. long, and are borne on branching stems, each branch bearing as many as three dozen, closely arranged. They are of a bright orange-red with spots of a darker hue inside the tube, and these colours, when seen in profusion, render the plant highly attractive either out-of-doors or in the greenhouse. As regards cultivation, during the month of November or December, after the foliage has died down and the bulbs are matured for the season, a few may be taken up and separated and put in pots for conservatory decoration. Any bulbs out of the ground may then be planted and covered with 2 in. or 3 in. of equal parts of well mixed leaf-mould and sandy loam, to which may be added a third part of decomposed cow manure. In case of severe weather, a light covering of straw will prevent the bulbs from being uplifted by the frost. As soon as the season seems promising, the straw should be removed and a top dressing with a light covering of sand given. If the season is dry, an occasional watering with liquid manure would be advantageous, care being taken not to discolour the foliage with it. Another new kind is M. crocosmæflora, which has larger and deeper-coloured flowers, four or five times the size of—and borne on the spike in the same manner as—those of M. Pottsi. It is a hybrid, raised between Tritonia aurea and M. Pottsi, and requires the same treatment as the latter kind.

Morina longifolia.—A hardy perennial of handsome and singular appearance, with large, spiny leaves, resembling those of certain Thistles, and long spikes of whorled flowers, from 2 ft. to 3 ft. high. It grows well in ordinary, well-drained soil, preferring that which is mellow, deep, and moist; and is easily multiplied by sowing the seed as soon as it ripens in light, peaty, sandy soil. Being a fine-flowering plant, as well as remarkable for its leaves, it is excellent for every kind of mixed border, and also for grouping with the smaller and medium-sized perennials that have fine foliage or are singular in appearance. Nepaul. M. Wallichiana is probably a synonym or a slight variety. Dipsacaceæ.

Morna nitida.—A half-hardy annual from Swan River. It grows about 1 ft. high, and in late summer bears, in terminal clusters, bright yellow blossoms, nearly 1 in. across, similar to those of the Helichrysum. It is well worthy of cultivation, as it is an attractive plant. Seeds should be sown in early spring in a heated frame, and the seedlings transferred from small to larger pots before planting out, so as to make strong plants. Two or three may be placed in each pot. It thrives best in a sandy, peaty soil. (= M. elegans.) Compositæ.

Morning Glory (*Ipomæa*).

Morphixia. — A genus of Iridaceous plants, native of the Cape of Good Hope, and nearly allied to Sparaxis. There is but one cultivated species, M. paniculata, a plant growing about 1 ft. high. It has long narrow leaves and slender flower-spikes, bearing a profusion of showy buff-white blossoms with remarkably long tubes. The variety alba has white flowers with black centres; and rosea is of a rosy apricot hue. It flowers in April and June, and requires the same treatment as the Ixia and Sparaxis.

Moss Campion (*Silene acaulis*).

Moss Pink (*Phlox subulata*).

Moth Mullein (*Verbascum Blattaria*).

Mountain Avens (*Dryas octopetala*).

Mountain Cat's-foot (*Antennaria dioica*).

Mountain Fern (*Aspidium Oreopteris*).

Muhlenbeckia. — For covering trellis work or as isolated specimens on stumps, &c., these graceful and free-growing evergreen trailers are useful. The kinds in cultivation are all natives of New Zealand. The best known is M. complexa, which is a very rapid grower, with long, wiry, and entangled branches bearing minute oval-shaped leaves. The flowers are rather inconspicuous, being white and of a waxy substance, about ½ in. across. M. adpressa is larger in all its parts than the preceding, bearing heart-shaped leaves and long racemes of whitish flowers. M. varia is a small-growing kind, with fiddle-shaped leaves, and very distinct from either of the above. They are indifferent as to quality of the soil, but grow best in a rich compost. In severe winters it is advisable to give a little protection in the way of dried Fern or some such material, but in ordinary seasons this is not necessary. The Muhlenbeckia complexa is by no means difficult to propagate. It strikes freely in a cold frame, if the cuttings are put in in September, and the only point to be observed is that the cuttings should be selected from the thicker shoots that are generally sent up from the ground in the form of suckers, not from the twiggy shoots; the former strike freely and form snug little plants the first season; the latter, if they do strike, will be at least two years before they form well-established plants. This plant grows very freely in some mild districts, though near London it is not often seen out-of-doors.

Mulgedium Plumieri is a native of the Pyrenees, where it grows to a height of 4 ft. or 5 ft., but in our borders, and under the influence of a deep, strong clay, it is not

Mirabilis longiflora. See page 187.

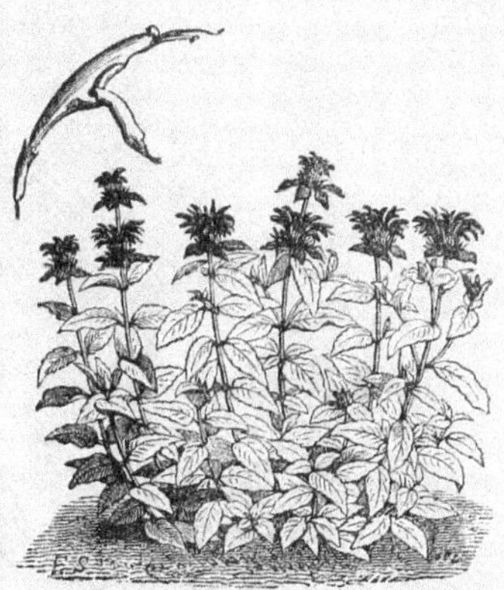

Monarda didyma (*Oswego Tea, or Bee Balm*).

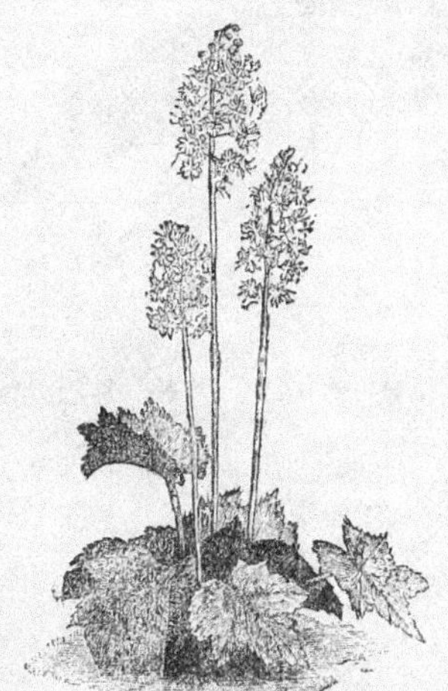

Mitella (*Mitre-wort*). See page 187.

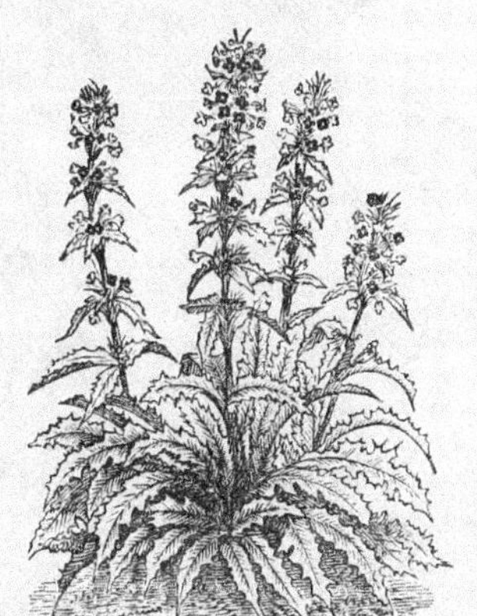

Morina longifolia.

Monolopia major.

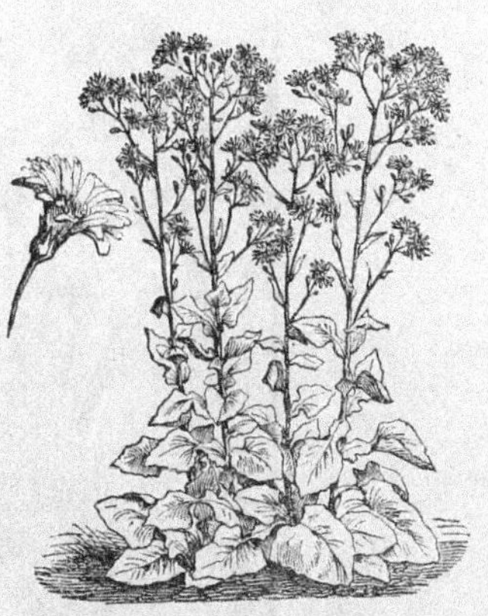

Mulgedium alpinum.

PLATE CLXXII.

Musa Ensete.

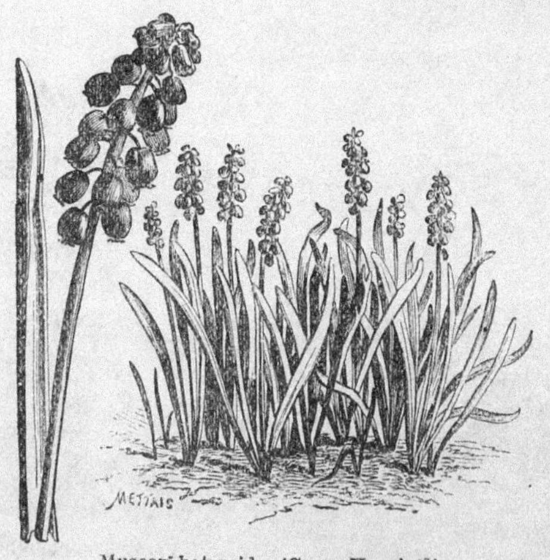

Muscari botryoides (*Grape Hyacinth*).

[*To face page* 189.

unusual to find it as much as 8 ft. or 9 ft. high. Its foliage is beautifully varied in outline. It should be planted in the rougher parts of a wild garden and left to take care of itself, as nothing seems to interfere with its rapid and vigorous growth. As an isolated plant on Grass it presents a bold and uncommon appearance, and when growing among other plants its remarkable leafage at once arrests observation; its blue flowers, too, are pretty. M. alpinum is a smaller and inferior plant. Both may be readily increased by seed or division.

Mullein (*Verbascum Thapsus*).

Muluccella lævis.—A singular plant of the Dead Nettle family. It is by no means showy, and the only recommendation it possesses as a garden plant is the singular form of the calyxes, which are shallowly bell-shaped and arranged densely on erect stems 1 ft. or so in height. It is a fine subject for skeletonising, and the stems, bracts, and calyxes may be done intact. For this purpose they should not be cut before autumn, when the plant is fully matured. It is a native of the Eastern Mediterranean, and should be treated as a half-hardy annual.

Musa.—These have of late years come to be extensively used during summer in our public parks, but in private gardens they are not so often met with, even where sub-tropical gardening is attempted. The cause of this is probably that it is generally supposed that great difficulties are experienced in preserving them through the winter. In the London parks Musas are often plunged in the ground in their pots during the summer, especially the smaller-sized plants, but the larger ones are usually planted out. When they are lifted in autumn, those in pots are stored away in houses, like other plants in pots, but the larger ones are lifted with comparatively small balls of earth and placed on shelves in houses in a temperature of not less than 45°. Here they are laid on their sides, their leaves being kept closely together, and thus they are allowed to remain throughout the winter, a mat being merely thrown over their roots. In February the roots are examined, and they are planted in trenches and subjected to an increased temperature, when they soon form new roots and begin to grow afresh. In June, after being gradually hardened by exposure to air, their leaves are tied up, the plants are lifted with as good balls as possible, and placed out in their summer quarters.

M. Ensete is the kind generally used in the open air, and it is one of the most noble plants used in what is termed the sub-tropical garden. Any one having a warm frame or greenhouse may grow it, and if planted out in June, in deep, warm, rich soil, it will grow well during summer in any sheltered position; at least, such is our experience of it in and near London and in the home counties.

Muscari (*Grape Hyacinth*).—These are pretty bulbous plants of the Lily family, all of the easiest culture and flowering in spring and early summer. Their proper position is either in the front row of the choice border or on rockwork, or they may be advantageously grown as window-plants in pots or boxes. In all cases they thrive best in rich, deep, sandy loam, and are easily multiplied by separation of the bulbs every third or fourth year. There is a large number of names, but there are few really distinct kinds.

M. armeniacum is one of the most handsome of all, and its beauty is considerably enhanced by its flowering at a period when most of the other kinds are past. Its flower-stems grow 8 in. high, and are terminated by a dense raceme, from 3 in. to 4 in. long, of bright dark blue flowers, with small whitish teeth. The foliage is much the same as that of the ordinary M. racemosum. Another beautiful kind is M. Szovitzianum, which comes early into bloom, and continues in blossom till the latest kinds have done flowering. In colour the blooms are a beautifully clear blue, the teeth of the corolla being white. The spike is of an oval shape, and larger than that of other species. These two, though not yet common, are among the finest species.

M. botryoides is a well known and deservedly favourite bulb, which has a distinctly dressy appearance from its little white teeth on its blue globose clusters. It grows about 9 in. in height, and is, therefore, very suitable for the front line of the border. The varieties pallidum and album are very distinct, and are even more beautiful; the former has pale sky-blue clusters. M. Heldreichi resembles M. botryoides, but is finer, by reason of its longer spike of flowers and its larger size, and it also flowers later.

M. comosum monstrosum (*Feather Hyacinth*) is quite distinct from any of the foregoing, growing 1 ft. or more in height; its flowers, of a beautiful mauve colour, bear a close resemblance to purple feathers, being cut into clusters of wavy filaments. Though, comparatively speaking, this species is now seldom seen, it is in every way qualified for a place in a collection of hardy flowers. M. moschatum has, in clusters, flowers of a dirty yellow hue and very inconspicuous, but it amply atones for its shortcomings in this respect by its delicious fragrance. Another sweet-smelling Muscari is M. luteum, the flowers of which fade by degrees from a dull purplish hue to one of a clear yellow.

M. racemosum, with its dark purple clusters and its strong smell of Plums, is a familiar old kind. Its leaves are long and weak, almost lying prostrate on the ground, whereas in M. botryoides and its varieties they stand boldly erect. It will hold its own anywhere, and, if permitted, will wander all over the mixed border, growing like a weed and in any soil. This plant has near relatives in M. commutatum (with blue flowers, darkening by degrees into purple) and M. neglectum—also a handsome kind. There are several other varieties offered in catalogues, but the best are mentioned above. Though coming chiefly from the south of Europe, they are all perfectly hardy, and will grow in any position in ordinary garden soil.

Musk Hyacinth *(Muscari moschatum).*

Musk Mallow *(Malva moschata).*

Mutisia decurrens.—Grown at the end of a glasshouse or against an open wall, this handsome composite plant, if planted in good soil, will make an attractive object during the summer. It is at present little grown in our gardens, although there is no reason why it should not be used, in conjunction with the smaller growing Nasturtiums or Convolvuluses, and similar plants, for covering fences, walls, or bare stones on rockwork. Its bright cinnamon-orange flowers are very showy, and will last in bloom for a considerable time. Native of Chili.

Myogalum (= Ornithogalum).

Myosotideum nobile.—A New Zealand plant about which very little is known in this country. In its native habitat it is a seaside plant, and is fond of damp sand, heavily manured with seaweed. It is said to be not difficult to grow, but is short-lived. In New Zealand it produces leaves measuring 18 in. across, and heads of blue and white flowers, 6 in. through. It is a gigantic Forget-me-not which has been introduced once or oftener, but has since been lost.

Myosotis. — A lovely genus of the Borage order, the cultivated ones being alpine or dwarf plants with blue flowers.

M. alpestris (Alpine Forget-me-not) is an exquisite plant, forming a low cushion of the loveliest blue. It is essentially a rock garden plant, which best enjoys moist, gritty soil and will not bear drought. It is so dwarf in habit that it should be surrounded only by half-buried pieces of sandstone or the like, which will prevent evaporation and preserve it from accident. It is easily raised from seed.

M. azorica *(Azorean Forget-me-not)* is a beautiful, but somewhat tender, kind, known at once by its deep blue blooms not having an "eye" of another colour in the centre. It is from 6 in. to 10 in. high, and the flowers, which are of rich purple when they first open and afterwards of an indigo blue, appear in summer. Being a native of the extreme western islands of the Azores, it is rather tender for outdoor culture, except in very exceptionally sheltered localities. The beauty of its dark flowers, of the richest indigo blue with a well-marked dash of purple, combined with the successional manner in which the flowering branches are produced, make it a most desirable border flower. It grows vigorously in good, light soil, and may be raised either from seed or cuttings. Impératrice Elizabeth is a nearly allied plant.

M. dissitiflora *(Early Forget-me-not)* is a very early-flowering beautiful plant, 6 in. to 12 in. high. The flowers are large, handsome, deep sky-blue, and numerous, and continue to bloom till the middle of summer. They resemble those of M. sylvatica more than any other, but stand further apart from each other on the spike. It is a beautiful border flower, and a worthy plant for the rock garden in open spots planted in broad masses. It is also invaluable as a spring bedding plant, or wherever spring flowers are much valued; and in the latter case it should be naturalised on all desirable spots. This will be an easy matter if the soil is tolerably good and moist, for it will then sow itself freely, but it is not easily done in dry gravelly ground in exposed places.

M. sylvatica *(Wood Forget-me-not)* is now popular in consequence of being used for flower beds in spring. There is a white, a rose-coloured, and a striped variety. Besides being grown in beds in the flower garden in spring, they should be found abundantly in a wild state by wood walks, in copses, &c. It sows itself freely in woods, and for garden use is best sown in beds in August every year.

M. palustris, although so common in wet ditches and by the sides of streams and canals throughout Britain, deserves a place in the garden among shrubs in peat beds. Used as a carpet beneath taller subjects, in small beds or borders in moist soil, or even as edgings. Division.

Myriophyllum. — Water plants of botanical interest mainly. They are graceful in form.

Myrrhis odorata *(Sweet Cicely).*—A graceful-looking native plant, with a peculiar, but grateful odour, and with sweet-tasted stems. It grows from 2 ft. to 3 ft. high. The flowers, which are produced in early summer, are white, in terminal compound umbels. It is suitable for naturalising near wood walks and in semi-wild places in any soil, and may be occasionally used among fine-leaved perennials. Division.

Napæa dioica.—A coarse-growing and unattractive plant of the Mallow family. Native of N. America.

Narcissus *(Daffodil).*—These are the hardiest and showiest of spring flowers. There are no plants more worthy of being well planted and cared for in our gardens than the various species of Daffodils, of which there are now a host of most beautiful varieties. They vary so much in size, colour, and form, that a most attractive garden may easily be made of Daffodils alone, and comprise such a wealth of beauty, that the commoner kinds ought to be planted by the thousand. There are but few species or varieties of Daffodils that will not succeed under ordinary circumstances in the open border. A few kinds—principally from Algiers, Spain, and the islands in the Mediterranean—flower mostly during the autumn and winter, and consequently are not adapted for outdoor cultivation, except in very favourable situations. Fortunately, however, there are, as we have said, but few

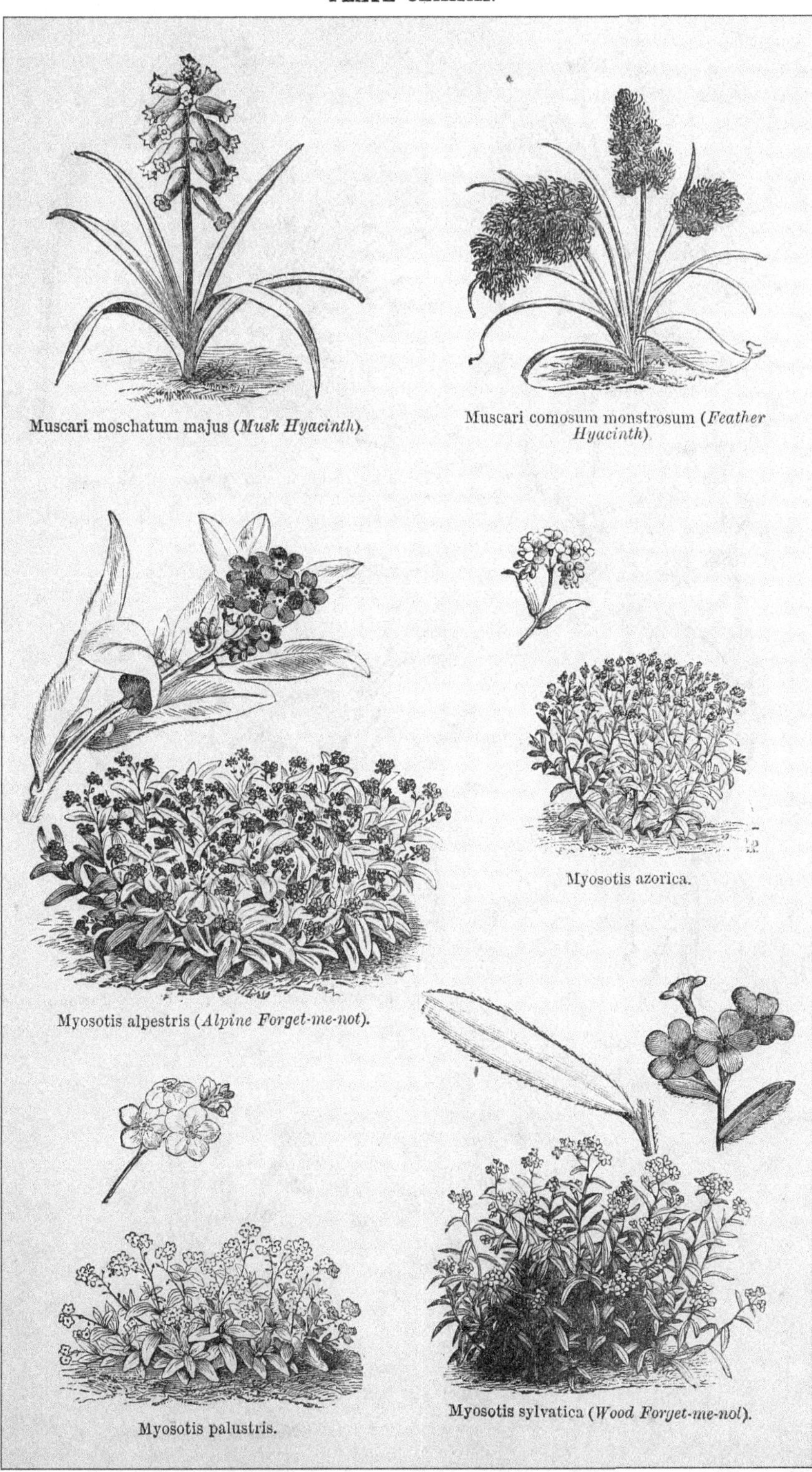

Muscari moschatum majus (*Musk Hyacinth*).

Muscari comosum monstrosum (*Feather Hyacinth*).

Myosotis alpestris (*Alpine Forget-me-not*).

Myosotis azorica.

Myosotis palustris.

Myosotis sylvatica (*Wood Forget-me-not*).

[*To face page* 190.

PLATE CLXXIV.

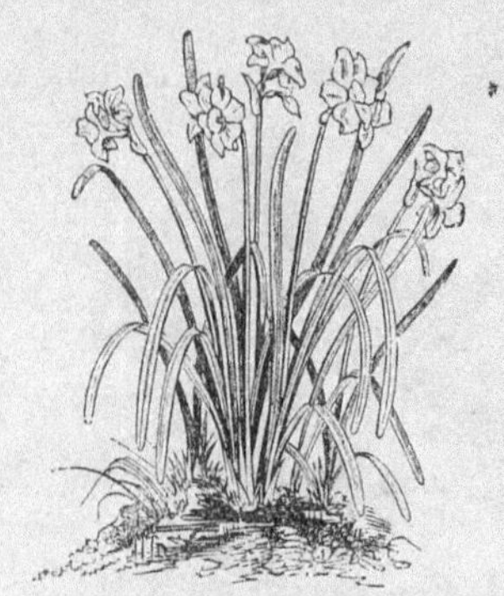

Narcissus biflorus.

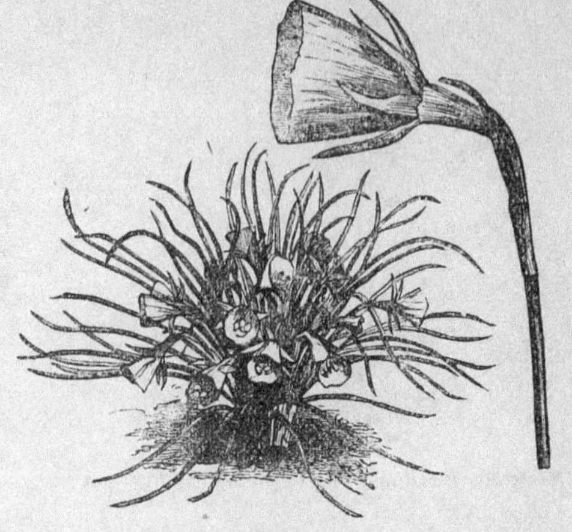

Narcissus Bulbocodium (*Hoop Petticoat Narcissus*).

Narcissus (monophyllus) Clusii.

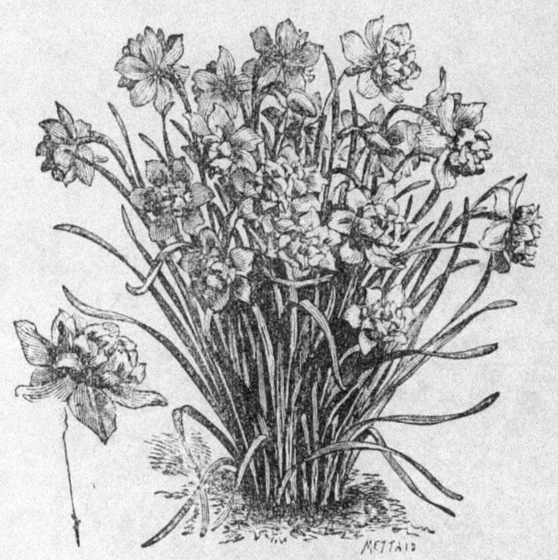

Narcissus incomparabilis fl.-pl. (*Double Peerless Daffodil*).

Narcissus gracilis.

Narcissus incomparabilis.

[*To face page* **191.**

of this character; the great bulk are not only perfectly hardy, but will grow in almost any soil or situation, and increase in numbers rapidly. Most of the Narcissi prefer a good rich, friable loam, annually enriched by a top-dressing of manure; but care must be taken to keep the manure away from the bulbs while in a fresh state. The small early-flowering kinds should be planted in sandy loam, and in rather drier positions than the more vigorous forms. When their fullest vigour is desired, all Narcissi must be replanted every three or four years, some even every two, but this depends a great deal upon the conditions under which the bulbs are placed. The best criterion as to when they require replanting is the production of small, weak foliage and a scanty proportion of flowers. When that occurs one may make sure that the bulbs have exhausted the soil, and that they have not sufficient room and strength to fully develop themselves. When this lifting process takes place, the whole of the soil should, before replanting, be well dressed with old manure and be thoroughly dug or trenched; if the latter, let the ground be moderately well trodden over to make it firm, as the roots of all plants—and more especially of Daffodils—like to fight their way against a pretty solid resistance. To change the roots to new ground is a better plan still. Imperfect or blind flowers are due to exhaustion of the soil, and may be easily guarded against by a winter's dressing of short rich manure. The margins of shrubberies are amongst the best places in the garden for the Daffodil; there are always little indentations—irregularities arising from the difference in growth of adjacent shrubs—where the mowing machine does not come, and in these, with admirable effect, may groups of Narcissi be placed. On the rougher parts of lawns, in open parks, and, indeed, in any grassy place, they may with great advantage be placed in bold groups. For cutting from the kinds best adapted for the purpose may be grown in beds. The few kinds, principally from Algiers, Spain, and the islands in the Mediterranean, consist of N. pachybulbos, viridiflorus, calathinus, Broussonetti, the various forms of Corbularia, and some others that flower mostly during the autumn and winter seasons.

N. bicolor, though strictly a variety of N. Pseudo-Narcissus, is a distinct type of Daffodil, distinguished from the Pseudo-Narcissus group in having the cup of the flower yellow and the petals white or primrose-yellow. N. bicolor itself is a handsome plant, but its varieties, such as Empress, Horsfieldi, and maximus, are much finer. These are extremely handsome, and should always have precedence. They are similar in appearance, but Horsfieldi is generally considered the finest. Other forms of bicolor worth growing are primulinus, sulphurescens, lorifolius, and breviflos. Horsfieldi and Empress flower the earliest. There are no double flowered forms of N. bicolor. All these forms are noble kinds, worthy of the best positions in the flower garden, and handsome as groups on grassy, sheltered banks in early spring. They are also lovely flowers for the house, and well worth growing for cutting for that purpose alone, if for no other.

N. biflorus is a native of Britain and N. Europe, and more hardy and robust than N. poeticus. The flowers are primrose-white, with a yellow crown. In Continental meadows this species and N. poeticus have given birth to spontaneous or natural hybrids. It is useful for wild garden or wood walks, and also for naturalising on grass or unmown lawns.

N. Bulbocodium *(Hoop Petticoat)*.—This and its varieties form quite a distinct section of the genus, and the group is also known under the name of Corbularia. In all of them the cup or corona of the flower is dilated like a crinoline. They are for the most part from S. Europe, and therefore are not so hardy as the majority of Narcissi. Warm, dry aspects should be selected for them, and they should be planted in sandy loam; a heavy, wet soil will not suit any of this group. The description of soil is of little importance, as they grow freely in almost pure sand, and quite as luxuriantly in stiff loam; but the latter should be in a well-drained situation and in a warm aspect. Where the soil and situation suit them, they soon form luxuriant masses of golden-yellow flowers, which appear at various times in different localities, but are generally produced in April and May. For edging small beds, for clumps in the borders, or for interspersing among alpine plants in the rock garden, the different varieties of Hoop Petticoat are charming. The chief varieties are tenuifolius, an early-flowering form; conspicuus, a handsome variety, with fine, large flowers; and monophyllus or Clusi. These are yellow-flowered kinds, except monophyllus, which has beautiful white blossoms, but is tender and difficult to cultivate, as it is a native of Algeria.

N. gracilis *(Yellow Rush-leaved N.)*.—From other Rush-leaved species this differs in having flowers similar to those of N. poeticus, but two to three on a stalk, each being of a soft sulphur-yellow colour instead of white. May be grown in pots or planted near a wall in rich sandy soil. Rather delicate, and not effective as a garden flower, although distinct and pleasing when cut for indoor uses. N. tenuior, often called the Silver Jonquil, is a slender and smaller form of this species; the flowers open yellow and die off nearly white.

N. incomparabilis *(Peerless Daffodil)*.—This, like N. Pseudo-Narcissus, is a common kind, affording numerous varieties, all very showy. The chief of these are the white and orange (albus and aurantius). There are two fine double sorts, the old Butter and

Eggs, or Sulphur Phœnix, and the Orange Phœnix. Many of the hybrid varieties range under this group, particularly those known as the Leedsi, Barri, and Nelsoni forms, which vary considerably both in size and in colour of the flowers. All these are vigorous growers, are from 1½ ft. to 2 ft. in height, and are useful for borders, margins of shrubberies, and for naturalising in woods. This species and its numerous varieties are free and bold flowerers—admirable as garden or wood plants.

N. Jonquilla *(Jonquil).*—A favourite old garden plant, commonly forced in pots for early spring flowering. It is of slender, Rush-like habit, and has three to eight golden flowers at the apex of a scape, 15 in. to 18 in. in height. It may be easily distinguished from N. odorus, which it most resembles, by its longer tubed flowers, the segments of the perianth being more slender and star-like. A double-flowered variety is grown, and so is a small greenish-flowered form with pointed segments and six-lobed cup. It is best when planted near a wall, although it is perfectly hardy on deep, light soils.

N. juncifolius *(Rush-leaved Daffodil).*—A sweet-scented, very dwarf, and pretty species, from the Pyrenees. The flowers, which are golden-yellow, are produced in early spring with the Primulas and Gentians in our gardens, but in April and May in its native habitats. It is one of the smallest and most beautiful Daffodils known, and one of the most valuable for rockwork, thriving freely in gritty or sandy earth, and being perfectly hardy; but as it flowers at a cold season, it is better planted on warm and slightly sheltered slopes on banks.

N. Macleayi.—A dwarf, sturdy plant, never very effective, except on the best of light, rich soils. It is easily recognised by its blunt, dense green foliage, and small white flowers having a cylindrical yellow crown nearly an inch long.

N. moschatus.—This is a beautiful species of dwarf growth, with flowers similar to the common Daffodil, but of a delicate creamy white. There are a few others similar to this, for example N. albicans, which has white blossoms with sulphur yellow cup, and N. cernuus, which has drooping creamy white flowers, of which there is also a double variety. N. tortuosus is another in a similar way, and all are worthy of culture. Another handsome species with white flowers is N. montanus or poculiformis, which is of taller growth than either of the preceding, has large, nodding flowers, and is more tender in constitution.

N. odorus *(Campernelle).*—Of this handsome species, which has rich yellow and deliciously-scented blossoms, there are three or four single varieties and a good double known as Queen Anne's Daffodil. All are worth cultivating, as they grow freely and give little trouble, and, being perfectly hardy in rich soils, are excellent for naturalising on the fringes of shrubberies or colonies on warm banks. We have never seen it effectively established in the wild garden, but have no doubt that it could be so grown. It is valuable for cutting.

N. poeticus *(Poet's Narcissus)* is a well-known and beautiful plant with white blossoms that have a small cup with a crimson or orange-red rim. Of this species there are many forms which prolong the season from March to May, and in the north to a still later period. The beautiful ornatus is the earliest, recurvus the next, and the double variety the latest. This last is a most valuable plant. In doubleness it varies in different soils, but it deserves to be extensively cultivated for the sake of its fragrant flowers, which open about the middle of May. As cut flowers they are very useful, and possess the advantage of lasting in good condition for a long time in water. Other handsome single forms are grandiflorus, poetarum, angustifolius, tripodalis, and spathulatus, all of which are worth growing, and, like the others, are suitable either for the border, margins of shady shrubberies, or for naturalising. The types known as Burbidgei are similar to poeticus, but have mostly yellow flowers. The finer types of the Poet's Narcissus should be grown for cutting in rich borders, the flowers being very beautiful; various kinds prolong the bloom for over six weeks.

N. Pseudo-Narcissus *(Common Daffodil or Lent Lily).*—This yields numerous varieties, all of which are beautiful garden plants, varying in size from the tiny N. minimus, which does not exceed 3 in. or 4 in. in height, to the handsome kinds that rise 2 ft. or 3 ft., such as maximus, Emperor, obvallaris, and lobularis. There is a great similarity about many of these forms, but they vary in time of flowering; thus cambricus, obvallaris, and maximus begin to bloom in February, and are closely followed by Telemonius, spurius, lobularis, princeps major, and rugilobus. Therefore it is important to have the sorts that have a distinct time of flowering. All the above, with several others less handsome, have wholly yellow flowers. The large double Daffodils are fine plants, the best being Telemonius plenus, lobularis plenus, grandiplenus (the largest of all), and the true double form of the common Daffodil, which is very rare. N. minor is a beautiful little Daffodil, a miniature of the common one, and in foliage and growth is not so vigorous as the Snowdrop when that is well grown. It is perfectly hardy and free, but being dwarfer than any mentioned above, care is required in the selection of a position, as it would scarcely do in long grass. A short bit of turf would do; but, better still, a bank covered with mossy Saxifrages, Spergula, or some dwarf plant through which the Narcissus could be scattered.

N. triandrus *(Cyclamen Daffodil)* is a most graceful kind, the divisions of the flower being turned back like those of the Persian Cyclamen or the American Cowslip.

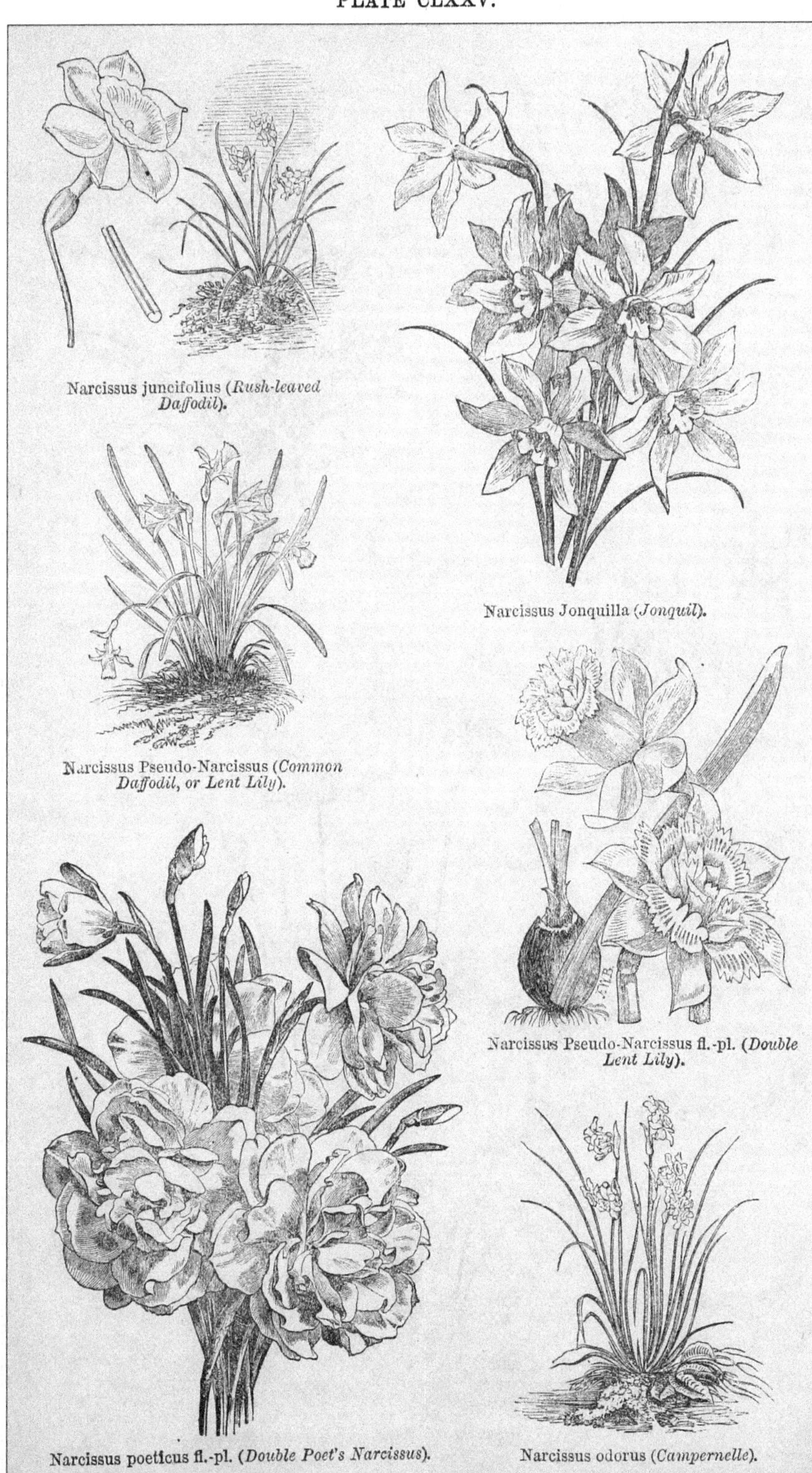

Narcissus juncifolius (*Rush-leaved Daffodil*).

Narcissus Jonquilla (*Jonquil*).

Narcissus Pseudo-Narcissus (*Common Daffodil, or Lent Lily*).

Narcissus Pseudo-Narcissus fl.-pl. (*Double Lent Lily*).

Narcissus poeticus fl.-pl. (*Double Poet's Narcissus*).

Narcissus odorus (*Campernelle*).

[*To face page* 192.

PLATE CLXXVI.

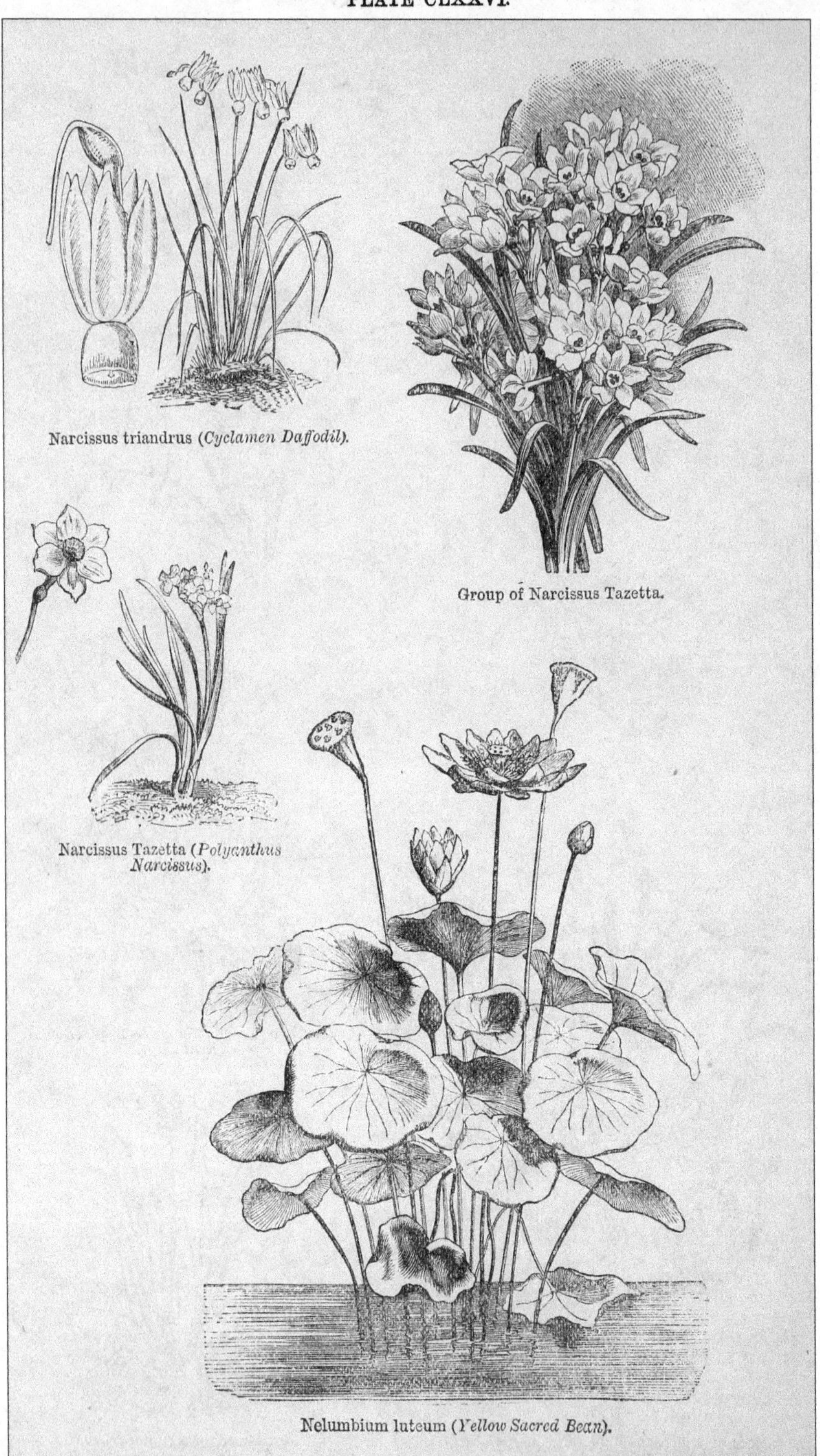

Narcissus triandrus (*Cyclamen Daffodil*).

Group of Narcissus Tazetta.

Narcissus Tazetta (*Polyanthus Narcissus*).

Nelumbium luteum (*Yellow Sacred Bean*).

[*To face page* 193.

The flowers are small, borne on slender stalks, and are pure white and pendulous. In the variety pulchellus the crown is white and the petals yellow; in cernuus the colour is pale yellow; and the other variety, nutans, is a stronger-growing plant than the type. This Narcissus is perfectly hardy, but does not increase so rapidly as others, and when planted out ought to have a carefully-prepared, light soil in a warm situation; it should not, like the stouter Daffodils, be placed among very coarse herbage.

N. Tazetta *(Polyanthus Narcissus)* is a common plant, and one of the most variable of all. It yields the numerous varieties so largely imported from Holland for pot culture in spring. All are fine border plants, and succeed well in any warm soil, more particularly at the foot of a warm wall. They flower somewhat late and are valuable on that account, and are especially valuable for cutting, as they have a delicious perfume. Among the varieties the charming little Paper-white (papyraceus) should never be omitted, and neither should aureus, floribundus, lacticolor, nor the large double form known as nobilissimus. The Dutch varieties are best adapted for pot culture, and, when done with in spring, should be planted out in groups in warm borders in any open place, where they will soon become established and produce a fine display every spring. There are several other species similar, but inferior, to N. Tazetta, such as pachybolbus, intermedius, elegans, bifrons, and Broussonetti. Being a South European and Italian plant, N. Tazetta does best in our extreme southern counties; elsewhere it and its varieties require warm, sandy loam and sheltered spots under south walls or among shrubs in sunny places.

HYBRID NARCISSI.—The species of Narcissus which best lend themselves to the hybridiser's purpose are N. poeticus, N. incomparabilis, N. Macleayi, N. montanus, and the forms of the common Daffodil (N. Pseudo-Narcissus). The type hybrids may be tabulated thus:—

N. Barri (incomparabilis and poeticus).—Smaller-flowered forms than N. Leedsi type, and more nearly approaching N. incomparabilis parent; varying from white to yellow, with yellow or orange cups.

N. Burbidgei (incomparabilis and poeticus).—Very near N. poeticus, but with larger and more star-shaped flowers; cups also larger, yellow or orange.

N. Humei (incomparabilis and Pseudo-Narcissus).—With flowers approaching the common Daffodil; the habit of the plant similar to that of N. incomparabilis.

N. Leedsi (incomparabilis and montanus).—Giving delicate white flowers of N. incomparabilis type, with pale sulphur cups.

N. Milneri (moschatus and incomparabilis).—Very near N. moschatus, or N. tortuosus parent;, Daffodil flowers of a pale primrose or sulphur crown; larger flowers and nearer to Daffodil type in habit than are the N. Humei forms.

N. Nelsoni (poeticus and Maçleayi).—Giving broad-petalled white flowers of firm texture, having yellow or orange cylindrical cups.

The best forms of each section are robust growers, and are distinct and effective as garden plants in warm, deep, sandy soil. Besides these, there is a race of Daffodils bearing several flowers on a stalk, of which N. tridymus (Bàckhouse), N. Mastersi, and N. bicolor dianthos are good examples. All these are fine, showy forms, well worth a place in every good garden.

Nardosmia *(Tussilago).*

Nardostachys Jatamansi.—A curious and interesting plant of the Valerian order, said to be the Spikenard of the ancients. It is hardy, and thrives in ordinary soil on a rockery, but is suitable only for botanical collections.

Narthecium ossifragum *(Bog Asphodel).*—A small native plant, in growth like an Iris, but with a spike of small yellow flowers. It is an interesting plant for the artificial bog.

Narthex asafœtida.—A rare economical plant of the umbelliferous family. It is of no ornamental value when there are so many fine Ferulas that surpass it in grace.

Nasturtium.—Cruciferous plants of no ornamental value. N. officinalis is the common Watercress.

Nasturtium —The popular name for Tropæolum.

Neja.—Small half-shrubby Composites, native of Brazil. These are strictly greenhouse plants, but N. gracilis thrives perfectly in the open air from spring to autumn, and it makes a pretty rock garden plant on account of its elegant feathery foliage and bright yellow, Daisy-like blossoms. The plants should be wintered in the greenhouse.

Nelumbium luteum *(Yellow Sacred Bean).*—This is one of the noblest of water plants, and reminds one of luxuriant tropical vegetation. The large round leaves, supported on tall stalks, push boldly out of the water for about 3 ft. The huge blossoms are pale yellow. This plant has been known to flower strongly and stand over 3 ft. above the water in the Garden of Plants at Paris, where it used to remain out all the winter in a fountain basin, the water of which was rather deep. It would probably flower out-of-doors in the south of England in a sunny and sheltered spot. It is rare, but may be procured in some nurseries or from America. The beautiful N. speciosum is another noble aquatic, and wherever there is a contrivance for heating the water in a small pond or tank in the open air, it would be well worth a trial

Nemastylus.—Bulbous plants of the Iris family. They are natives of the warm parts of N. America, and consequently none are quite hardy with us, though they may be grown successfully if treated in the same manner as recommended for the Ixia and

Sparaxis. There are three species introduced—N. geminiflora (N. acutus) and N. purpureus from Texas, and N. cœlestinus from Florida. All are of dwarf growth and bear showy blossoms, which, however, are fugacious, though produced continuously.

Nemesia floribunda.—A pretty hardy annual of the simplest culture. It grows about 1 ft. high, and bears in summer a profusion of Linaria-like, fragrant blossoms, white with yellow throats. N. versicolor is similar, but has blue, lilac, or yellow and white blossoms; its variety compacta, with blue and white flowers, is also a most desirable plant. If sown in masses in ordinary soil in early spring, and well thinned, the plants will have a pretty effect for several weeks after June. Cape of Good Hope. Scrophulariaceæ.

Nemophila.—These pretty Californian plants are among the most popular of hardy annuals, and from the many varieties enumerated in catalogues a good selection may be made. There are four species from which these varieties have been derived, viz., N. insignis, which has sky blue flowers, and of which grandiflora, alba, purpurea-rubra, and striata, all of which are worth growing, are varieties; N. atomaria, white flowers speckled with blue—its varieties are cœlestis, with sky-blue margin, oculata, pale blue and black centre, alba nigra, white and black centre; N. discoidalis, dark purple edged with white—its variety elegans is maroon margined with white; N. maculata, large, white, and blotched with violet, and the variety purpurea is of a mauve colour. These are all worth growing, thrive in any soil, and are of the simplest culture. Some pretty combinations may be effected in spring by arranging the masses in harmonising colours. All are particularly well adapted for edgings and for filling small beds, as they are compact in growth. The insignis section is especially desirable, and should always be preferred to the others. The periods of the year at which Nemophila seeds should be sown are early in August for spring flowering, and April for summer blooming. In order to secure a good display of flower, however, the best time is to sow in August, on light soil where the seed can germinate freely, and the plants will not acquire too robust a growth before winter sets in. If it be possible to sow the seed where the plants are to flower, the results will be most satisfactory; but if transplanting be necessary, it should be done early in the winter. A ball of earth should be attached to each plant, and to secure this thin sowing is indispensable. Hydrophyllaceæ.

Neottia.—Inconspicuous terrestrial Orchids.

Nepeta *(Cat Mint)*.—This is a numerous genus of Labiates, but contains few, if any, good garden plants. The ordinary Ground Ivy (N. Glechoma) is too common to cultivate, though it is an elegant plant in the rougher parts of the rock garden. N. macrantha has rather showy purple flowers, but the plant is too tall and coarse in growth for the border. N. Mussini is an old plant, and was once used a good deal for edgings to borders, a purpose for which its compact growth well suits it. The other species—about a dozen—are suitable only for botanical collections.

Nephrodium.—A genus of Ferns, the native species of which are alluded to under Aspidium. Besides these there are a few N. American species that are quite hardy, very handsome, and thrive perfectly under the same conditions as the native kinds. The chief are N. Goldieanum, intermedium, marginale, and noveboracense. There are also several Japanese and Chinese species that thrive in mild localities without protection, but these cannot be recommended for general culture. N. fragrans is a sweet-scented little form. It is somewhat delicate in constitution, but thrives well in a sheltered situation.

Nerine.—A beautiful genus of bulbous plants, mostly from South Africa. The hardiest is the Guernsey Lily (N. sarniensis), which for a long time has been a favourite object of culture in the Channel Islands, where it is quite hardy in light soil and thrives without any attention, though it is really a native of Japan. The flower-stems grow about 1 ft. high, and are surmounted by a large umbelled cluster of showy blossoms of a rich crimson. The bulbs are imported from the Channel Islands in large quantities. They require to be planted out in a light, warm soil in a sunny position, a border at the foot of a south wall being the best. The bulbs are usually sent over about August, and should be planted as soon as received. The plants will then flower in September or later. After the flower is past leaves will be produced, and these must be preserved from frosts either by lifting the plants and placing them under glass, or by preserving them by a handlight or other protective material. Protected in this way, the plants have in some localities withstood the winters for years. There are several other Nerines that, though not so hardy as the Guernsey Lily, succeed well in the open in summer, and, if protected, will flower in September or October, but they should not be left out during the winter except in extremely mild localities. The kinds that should be tried out are N. corusca, Fothergilli, pulchella, pudica, undulata, and their varieties.

Nertera depressa *(Fruiting Duckweed)*.—This is an extremely pretty plant when well grown and thickly studded with its tiny reddish-orange berries, and the minute round leaves are very suggestive of the Duckweed that infests the surface of our stagnant pools. It forms densely matted tufts, and when thriving soon spreads into a large mass. It is grown most successfully under glass, but flourishes in the open air, and is an excellent plant for level spots in the rock garden; it requires a little pro-

PLATE CLXXVII.

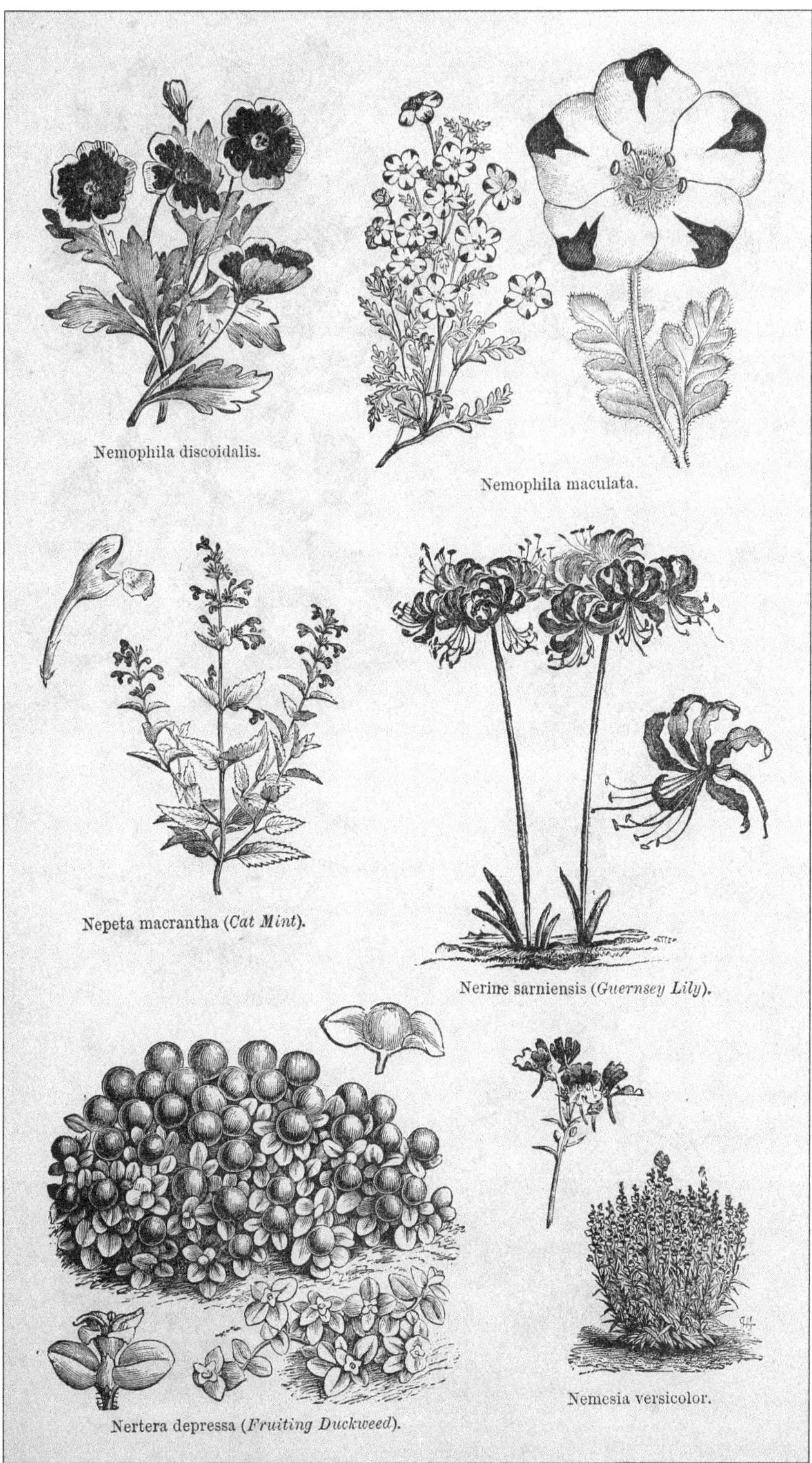

Nemophila discoidalis.

Nemophila maculata.

Nepeta macrantha (*Cat Mint*).

Nerine sarniensis (*Guernsey Lily*).

Nertera depressa (*Fruiting Duckweed*).

Nemesia versicolor.

[*To face page* 194.

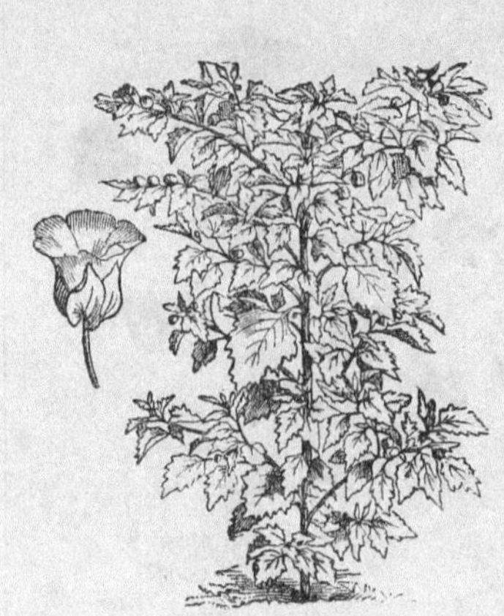

Nicandra physaloides.

Nicotiana macrophylla (*Tobacco*).

Nicotiana glauca.

Nigella hispanica.

Nierembergia frutescens.

Nigella damascena (*Fennel Flower*).

[*To face page* 195.

tection during the most severe cold. It may be readily propagated in the following manner: Take old plants and divide them into small portions, and then place them in small pots in a gentle heat for a time until started into growth; then remove into a cooler atmosphere. If a stock is required, the points of the young shoots should be taken off and inserted in pans filled with equal portions of leaf-mould and sand. When watered, a square of glass or bell-glass should be placed over the pan. Thus treated, the cuttings will take root in a few days; they should then be potted off into 2½-in. pots and again placed in a gentle heat until established, when they should be removed to a shelf in a greenhouse until the fruit is set. Rubiaceæ. New Zealand.

New Holland Daisy (*Vittadenia triloba*).

New Holland Violet (*Erpetion reniforme*).

New Zealand Flax (*Phormium tenax*).

Nicandra physaloides.—A pretty Peruvian half-hardy annual, about 2 ft. high. It is of stout growth, and bears in summer numerous showy, bell-like flowers, blue and white. It thrives in light soil and in an open position. Seed should be sown in early spring in heat, or in the open air about the end of March, and the seedlings transplanted in May. Being large, one plant is sufficient for a square yard. Solanaceæ.

Nicotiana (*Tobacco*).—These are noble half-hardy annuals of gigantic growth, excellent for associating with other stately subjects in the sub-tropical garden. There are several varieties, differing chiefly one from another in the stoutness and height of their stems, and size of leaf and flower. But these variations are also largely dependent on cultivation. N. macrophylla is the best. The deeper and richer the ground and more sheltered the position, the larger the Tobacco plants become in all their parts. Scarcely any plants equal the Tobaccos in rapidity of growth. They form noble groups of themselves, and they mix kindly and congruously with most other fine-foliaged plants. Place them, if possible, in a deep soil, rich in vegetable and animal remains, near the margin of a lake or stream, and note how they grow. They form capital backs to masses of Reeds, Pampas Grass, Bamboos, Rushes, or semi-aquatic vegetation of various other descriptions. They are tender, and must be sown in February in a warm house or frame. Prick off the plants as soon as they appear, and pot and grow them in a genial heat of, say, 60°. This will enable you to turn out, from 6-in. or 8-in. pots, fine plants about the end of May. They will start off at once, and will not cease growing until frost comes.

Nierembergia.—These are extremely pretty plants, but the only species that is quite hardy is N. rivularis (White Cup), one of the handsomest of all. The stems and foliage of this trail along the ground as dwarfly as those of the New Holland Violet, while from amongst them spring erect, open, cup-like flowers of a creamy-white tint, just barely pushed above the foliage. Sometimes the blossoms are faintly tinged with rose; they are usually nearly 2 in. across, with yellow centres, and continue blooming during the summer and autumn months. Their distant effect, suggesting Snowdrops at first, is very pleasing, and they are no less pretty when closely viewed. No collection of rock or herbaceous plants can be complete without this plant, while the tasteful flower gardener may well use it in his smaller designs. To grow it successfully two things are requisite, viz., a heavy soil and absolute firmness of the ground in which the plant grows. These, accompanied by a sunny aspect and culture on a level surface, will in all probability secure success. The tender kinds are N. frutescens, a sub-shrubby plant of erect growth, and N. filicaulis, or gracilis as it is called, which has slender drooping branches. Both have pretty white flowers pencilled with purple, and are suitable for the rock garden in summer or for drooping over the edges of vases. Propagated by cuttings in spring in heat.

Nigella (*Fennel Flower*).—Hardy annual plants of the Crowfoot family, all curious and pretty. They have feathery Fennel-like foliage and bluish or yellowish blossoms. N. sativa, orientalis, damascena, and hispanica are the kinds cultivated, the last being the most ornamental. It grows about 1 ft. high, and has showy blue flowers, blooming in July and onwards. There is a white variety and one with deep purple blossoms. All the Nigellas should be sown in the open border in March, in light warm soil, in the spot they are to occupy, as they do not succeed if transplanted. If sown in autumn the seedlings often survive the winter and flower early.

Night Scented Rocket (*Hesperis tristis*).

Nightshade (*Solanum*).

Nigritella.—Terrestrial Orchids that are not very attractive, but difficult to cultivate, and suitable only for botanical collections and for those that require a full collection of hardy Orchids. N. suaveolens, with pale rose flowers and sweetly scented, and N. coccinea, with red flowers, are the best species.

Nolana.—Small-growing hardy annuals from S. America, of which N. paradoxa, prostrata, and atriplicifolia are cultivated. They all have slender trailing stems, and flowers, similar to those of a Convolvulus, that are generally blue. The last named species is by far the best of the three. It has beautiful, very showy blue flowers with a white centre. There is a white variety of it (N. atriplicifolia alba). The Nolanas are suitable either for borders or the rock garden, as they thrive in any warm open situation in good light soil. As the seedlings do not transplant well, the seed should be sown in March in the open, and the plants well thinned out. Nolanaceæ.

Nolina georgiana.—A rare cultivated plant, allied to the Yucca.

Nonnea.—Plants of the Borage family, fit only for botanical collections, if grown at all.

Nothochlæna Marantæ.—An elegant New Holland Fern, the only species of the genus that is hardy with us. It has dark green erect fronds, from 6 in. to 12 in. high, proceeding from a creeping rhizome. It requires a moist peaty soil in a sheltered, shady spot in the rock garden or fernery.

Nothoscordum. — Liliaceous bulbous plants allied to the Allium. The three introduced species, N. bulgaricum, striatum, and fragrans, are unattractive, fit only for botanical collections or for naturalising in the wild garden, where they would soon become weeds.

Nuphar.—The most familiar plant in this genus is the common yellow Water Lily (N. lutea), which inhabits many of our lakes and slow-running rivers in abundance, and is too well known to need description. It has a very interesting miniature variety called pumila or minima, which is found wild in some of the Highland lakes of Scotland, and possesses the same vinous perfume as the type. N. advena is the North American representative of our yellow Water Lily; it nearly approaches it in general aspect, but may be at once distinguished by its larger size and by the leaves standing erect out of the water if it be shallow. It is altogether a much finer plant. N. Kalmiana, also a North American kind, much resembles the small variety of N. lutea, and is a very interesting plant to grow in company with it. The cultivation is the same as that of the Nymphæa.

Nuttallia (=Callirhoe).

Nycterinia.—Pretty half-hardy annuals from the Cape of Good Hope. N. selaginoides grows about 9 in. high, and forms a dense compact tuft of slender stems, which in late autumn are profusely set with small, white, orange-centred blossoms that are fragrant at night. N. capensis is about the same size and growth, but the flowers are larger, though not of such a pure white. These require to be sown early in heat, and transplanted in May in light, rich, sandy loam in warm borders. N. Lychnidea is a small, shrubby, perennial plant with yellowish-white blossoms. It thrives in warm borders in summer, and should be propagated either by cuttings in autumn or by seed in spring. Scrophulariaceæ.

Nymphæa *(Water Lily).*—No ornamental water should be without a few fine plants or groups of Water Lilies, even though the common N. alba may be met with in almost every lake and river. Water Lilies are seen to greatest advantage in small groups a few yards from the margin of the water, but isolated groups or single plants always look well no matter where they are placed. It should also be remembered that small groups and individual plants always produce finer foliage and flowers when thus isolated than when crowded together. In small gardens, even in the smallest, it is easy to grow Water Lilies in a tank sunk in the Grass, or even in a large tub, with very moist soil. Such, sunk in a little lawn with the margin of the receptable hidden, affords a very good way of growing Water Lilies, and one in which they look better than where there is a crowd of them spreading over an acre of pond. Their culture is of the simplest kind, for, if properly planted at first, they seldom give any trouble. Where it is convenient to drain off the water, the best mode of planting the larger kinds is to make a hillock of a compost consisting of good loam and a small quantity of well-decomposed manure and river sand; on the surface of this place some large stones to prevent the soil from being removed by the water. On this hillock place the plant so that the depth from its own crown to the surface of the water may not exceed 2 ft. If there be no means of lowering the water the best substitute is to put the plants into large baskets and to sink them to the proper depth. If the bottom be of a gravelly nature the plants will not spread much, but if otherwise they should be kept within bounds, or they will soon grow into a mass, which will considerably mar the effect as shown in the form of isolated patches. In the case of young plants and the small-growing kinds it is advisable to keep them in small baskets and in shallow water. The following are the hardy kinds:—

N. alba.—Of this, the common native Water Lily, there are three or four varieties, the finest by far being the rose coloured form (rosea) which has lately come into cultivation. It is exactly like the type except in the colour of the flowers, which is a beautiful delicate rose tint. This beautiful plant should in all cases be preferred to the common one, as it is simple to grow and is quite hardy. There are two other varieties of N. alba. The minor form is very interesting on account of its small size, the blossoms being but 1½ in. to 2 in. across, with the leaves small in proportion. The variety candida is a form intermediate in size between the last and the type.

N. odorata.—This N. American species is a near ally to N. alba, the most perceptible distinction between them being that N. odorata has larger blossoms, which measure from 6 in. to 9 in. across. The flowers are sweetly scented and have a decided tendency to assume a red colour, and the full development of this is admirably shown in the variety rosea, or minor, though the flowers are much smaller than those of the type. The variety maxima differs from the type only in having larger flowers. The variety reniformis has the lobes of the leaf much rounded, so as to assume a kidney shape, but there is no difference in the flower. It requires precisely the same treatment as N. alba, and will be found to be quite as hardy in the southern counties

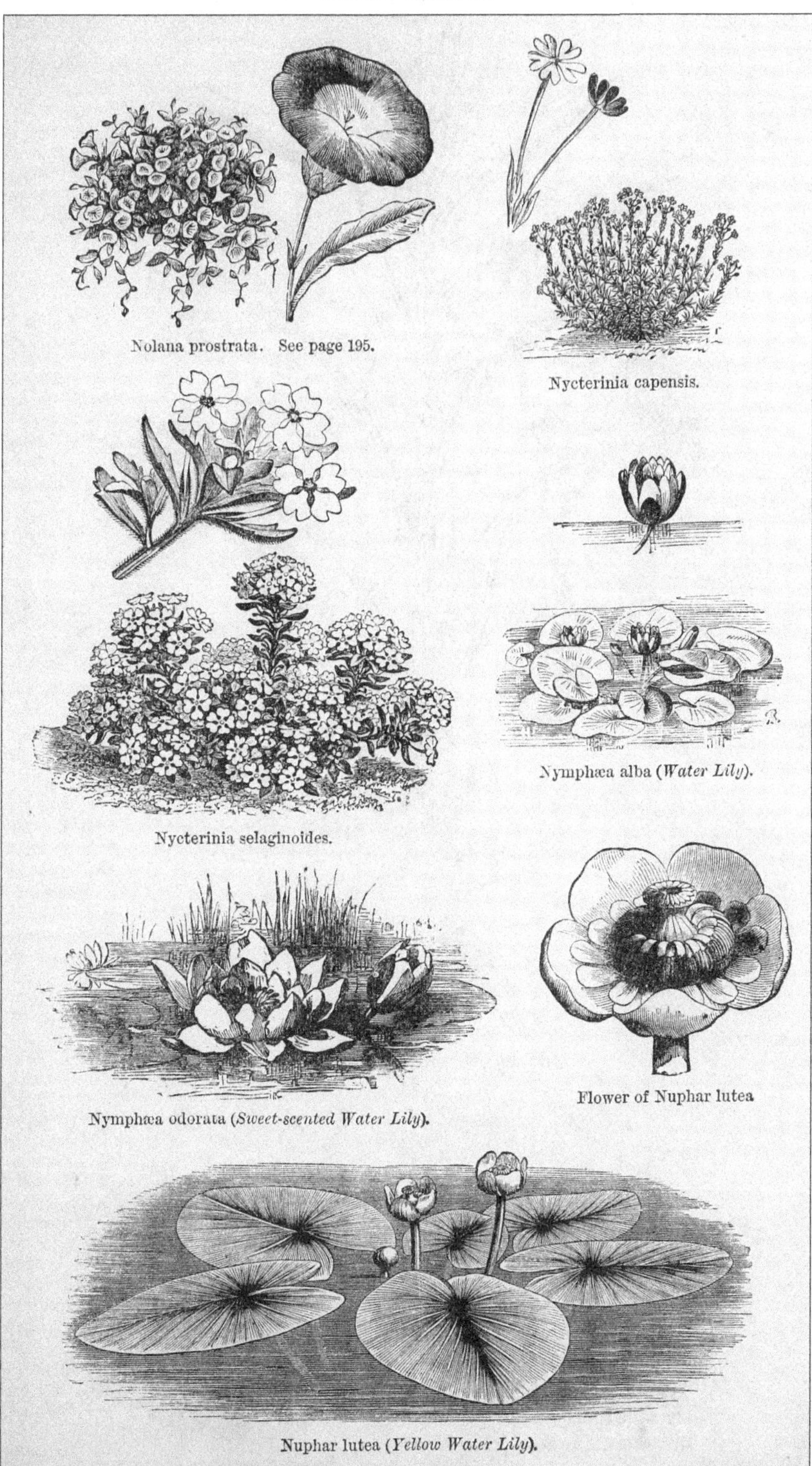

Nolana prostrata. See page 195.

Nycterinia capensis.

Nycterinia selaginoides.

Nymphæa alba (*Water Lily*).

Nymphæa odorata (*Sweet-scented Water Lily*).

Flower of Nuphar lutea

Nuphar lutea (*Yellow Water Lily*).

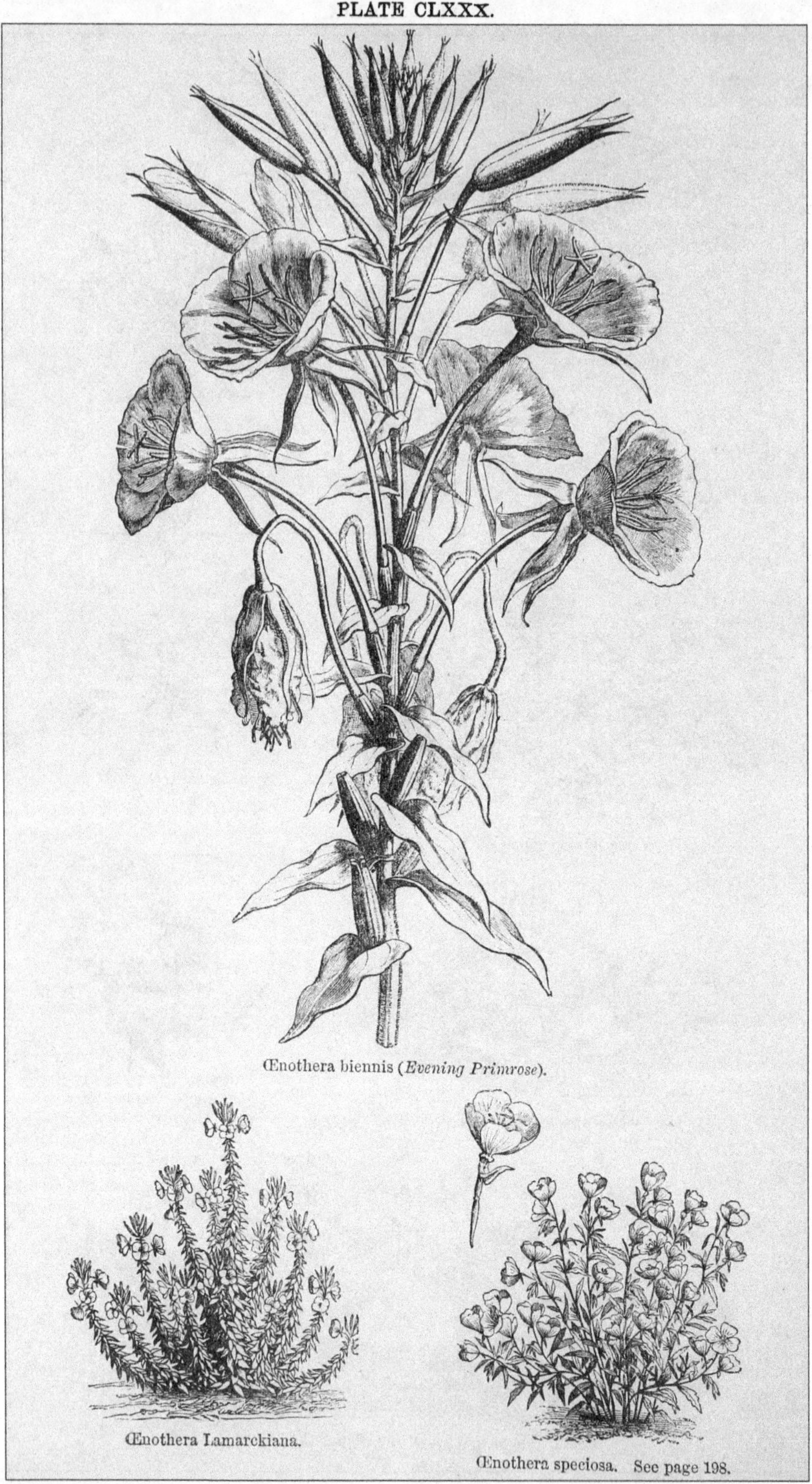

Œnothera biennis (*Evening Primrose*).

Œnothera Lamarckiana.

Œnothera speciosa. See page 198.

[*To face page* 197.

N. tuberosa is also a native of N. America. It much resembles N. alba, but differs from it principally in having tubers on the roots which spontaneously detach themselves from the plant, and so afford a ready means of propagation. The shining-leaved Water Lily (N. nitida) is also a near relative of N. alba, but has very shining leaves, and blossoms smaller and scentless. It inhabits the lakes and still waters of Siberia, and also the river Lena. The pigmy Water Lily (N. pygmæa) is a native of China and some parts of Siberia. It is the smallest of all, having leaves not more than 2 in. across, and very small flowers. One of the most interesting of all the Nymphæas is the yellow-flowered kind (N. flava), on account of its colour, as in no other sort, either tropical or temperate, is it found. It is in cultivation, but it is rare, and no one has experience in its culture. It is also a N. American species.

Oak Fern *(Polypodium Dryopteris).*

Obeliscaria *(Rudbeckia).*

Ocimum *(Basil).*—Labiate plants of no flower garden value.

Odontarrhena *(Alyssum).*

Œnothera *(Evening Primrose).*—These are amongst the most showy and hardy of late summer blooming plants, and are easily grown. From June onwards they are in their glory, but many species and varieties in late summer, from increased growth, become more full of flowers. They have large bright yellow or white flowers, which in many of the species are so freely and continuously produced as to make them of great value for the summer decoration of the outdoor garden. All are not equally fond of the night, as their name might suggest, for many species are open by day, as, for instance, Œ. linearis, speciosa, taraxacifolia, and trichocalyx. Many of our finest Evening Primroses are natives of west of the Mississippi States, as California, Utah, Missouri, and Texas. They all bloom the first season from early seedlings. Some of the true perennials, and particularly the prostrate-growing ones, are shy seeders, but the tall growers seed freely. Evening Primroses will grow in any good soil; the richer and deeper the better. The largest growers are very beautiful in any position. They, however, from their height and boldness, and the freedom with which they grow in almost any soil, are peculiarly suited for the wild garden, for shrubberies, copses, and the like, sowing themselves freely. In this case they are apt to become eventually somewhat too numerous and somewhat "starved," and they are best where confined to large groups in good ground. In any flower garden not confined to flat beds only, a small isolated bed of these Œnotheras looks well. Amongst them we have tall-growing sorts like Œ. Lamarckiana and fragrans, and decumbent carpets, as in trichocalyx and cæspitosa, white flowers, as in the last named two, coronopifolia and speciosa often changing with age to pink or rose. Few plants have more birlliant yellow blooms than those of missouriensis and Lamarckiana; besides, they are very large—4 in. to 6 in. across. Nearly all are more or less fragrant, cæspitosa, marginata, fragrans, and eximia being particularly so.

Œ. biennis, one of the commonest of all, may be seen in almost every garden. It is a handsome biennial, growing from 3 ft. to 5 ft. high, and having large bright yellow flowers. The variety grandiflora or Lamarckiana should always be preferred to the ordinary kind, as the flowers are larger and of finer colour. These are excellent border flowers; they have a bold effect in large masses, and are singularly well adapted for the wild garden, as they sow themselves and give no further trouble.

Œ. fruticosa *(Sundrops).*—This and its varieties are among the finest of hardy perennials, growing from 1 ft. to 3 ft. high, and bearing in profusion showy bright yellow blossoms. There are about half-a-dozen distinct varieties of this plant, the best being linearis or riparia, as it is most often called. This grows about 1½ ft. high, is bushy in growth, and bears an abundance of yellow blossoms, smaller than those of the type. It is one of the best of the yellow Evening Primroses for small beds, edgings, or as a groundwork for other plants. It becomes a mass of delicate flowers, is perfectly hardy, and goes on flowering even after the first frosts. It is always prudent to lift a few or strike a potful of cuttings in case of accident, though the old plants may be divided in spring to any extent. These plants thrive well in borders or margins of shrubberies in sandy loam. N. America.

Œ. glauca is another handsome N. American species similar to fruticosa. It is of sub-shrubby growth, becomes bushy, and bears a profusion of yellow flowers. The variety Fraseri is a finer plant, and where an attractive mass of yellow is desired through the summer, there are few hardy plants of easy cultivation that will give this so effectively as Fraser's Œnothera. In the case of large rockwork a few plants here and there would furnish colour. The yellow hue is a most pleasing one, and the plants bloom profusely and long.

Œ. marginata.—Although a dwarf plant—even when vigorous not more than 6 in. to 12 in. high—this blooms as nobly as any luxuriant native of the Tropics, the individual flowers being 4 in. to 5 in. across. They are white, changing, as the flower becomes older, to a very delicate rose, the blooms coming well above the toothed or jagged leaves as the evening approaches, and remaining in all their beauty during the night, emitting a Magnolia-like odour. This plant, tastefully arranged in the rock garden or in some quiet border, is one of the greatest charms of the garden, as welcome among the night bloomers as the nightingale among night singers. It is perennial, quite hardy, and increased by suckers, which are freely produced from the roots. Cuttings also root.

readily. Flowers in May. *Syn.*, Œ. cæspitosa. Œ. trichocalyx, another species of a similar nature, but probably only an annual, is a beautiful plant and well worth growing.

Œ. missouriensis.—A fine, yellow, herbaceous plant from N. America, with prostrate downy stems, and with rich clear golden-yellow flowers. These latter are from 4 in. to nearly 5 in. in diameter, and so freely produced that the plant may be said to cover the ground with tufts of gold. There is no more valuable border flower, and well placed on rockwork it is a handsome plant, especially when the luxuriant shoots are allowed to hang down. As the seed is but rarely perfected, it is better increased by careful division, or by cuttings made in April. When used as a border plant, it does not make such a free growth in cold clayey soils as it does in warm light ones. The blooms open best in the evening. *Syn.*, Œ. macrocarpa.

Œ. speciosa.—A very handsome plant, with an abundance of large flowers. These are at first white, but change to a delicate rose, and in these respects somewhat resemble Œ. taraxacifolia, but the plant is erect, with almost shrubby stems. It forms neat tufts, usually from 14 in. to 18 in. high, is a true perennial, and exceedingly valuable for borders, or the lower and rougher parts of rockwork. It is a native of North America, and is increased by division, cuttings, or seeds, the last, however, not being freely produced in this country. It flourishes vigorously in well-drained rich loam. It is best grown in a border of rich, light soil, where, if not disturbed, it will soon spread into a large tuft. In heavy soils it is apt to be injured during severe winters, so that to be successful with it the soil must be warm and of a light description.

Œ. taraxacifolia, a Chilian plant, is one of the finest of that section of Evening Primroses characterised by a low, trailing growth and large blossoms of a white-pinkish or yellow colour, expanding fullest towards evening. Œ. acaulis is a much inferior plant with smaller flowers, but is, possibly, only a variety of the other. Both are quite hardy and perennial in light soils, but often perish during winter on those that are wet and heavy. Œ. taraxacifolia has a fine effect when planted in a rich, deep soil in the rock garden, where its trailing stems can droop over the ledge of a block of stone. The flowers vary from 2½ in. to 3½ in. across, and are pure white when first opened, but afterwards gradually change to a delicate pink.

Œ. triloba is a handsome hardy annual species, of dwarf growth, and with large and showy yellow blossoms. It is also called Œ. rhizocarpa. Other showy annuals are Œ. sinuata and its variety maxima, macrantha, odorata, bistorta, Veitchiana, and Drummondi. These are all worthy of culture, require to be treated as half-hardy annuals, and should be grown in ordinary garden soil.

Omphalodes *(Navelwort).*—A small genus belonging to the Borage family. There are but three species in gardens; these are—

O. linifolia, a beautiful hardy annual from Portugal, which grows from 9 in. to 12 in. high, and is furnished with leaves of a smooth glaucous-green colour. The flowers, which are produced abundantly from June to August, are pure white, and resemble in form those of the common Forget-me-not, but are considerably larger. It may be grown in any ordinary garden soil. The seeds should be sown at intervals from April to June, or in September and October; the plant often sows itself.

O. Luciliæ is a charming rock plant, with flowers twice the size of O. verna, and of a peculiar lilac-blue colour, quite unlike that of any other flower. The foliage, too, is of a glaucous grey tint, which harmonises beautifully with the blossoms. It is perfectly hardy, and succeeds well when planted in a border or on a rockery, where its branches hang over the ledges, but the place must be thoroughly drained, for though it requires abundant waterings during the growing season, it is very impatient of stagnant moisture at the roots or about the crowns. Slugs are particularly fond of its foliage, and therefore the plants should be guarded against their attacks; for this purpose strips of perforated zinc about 3 in. wide, and bent so as to form a ring round the plant, will be found to be as effectual as any plan that can be adopted. It may be readily propagated by divison or by seeds, which in some seasons it produces freely. It grows quite freely in some soils, as, for example, in Wheeler's nursery at Warminster. Asia Minor.

O. verna *(Creeping Forget-me-not)* is a charming little plant, with handsome, deep, and clear blue flowers, with white throats. They are produced in early spring, and remind one of the finest Forget-me-nots, but are larger, of a more intense blue, and distinguished by the stems sending out runners from their base. It is indispensable for the rock and spring garden, and no plant is more worthy of naturalisation in half-wild places under shrubs or beneath Rhododendrons and the like. In cool, moist, thin woods it runs about as vigorously as any native plant, but it will thrive well by almost every wood-walk, and prove one of the prettiest plants in any position.

Onobrychis *(Sainfoin).*—A few graceful plants belong to this genus, but they are mostly of botanical interest.

Onoclea sensibilis.—A handsome hardy Fern, belonging to the group known as flowering Ferns, from the fact of the fertile frond being contracted, and thus having the appearance of an unopened spike of flowers. The fronds have a beautiful fresh light green colour, especially when they first make their appearance in spring. It is not very fastidious as regards soil and position, but still it succeeds best in a cool and moist situation, as, for instance, at the base of the rockwork, or

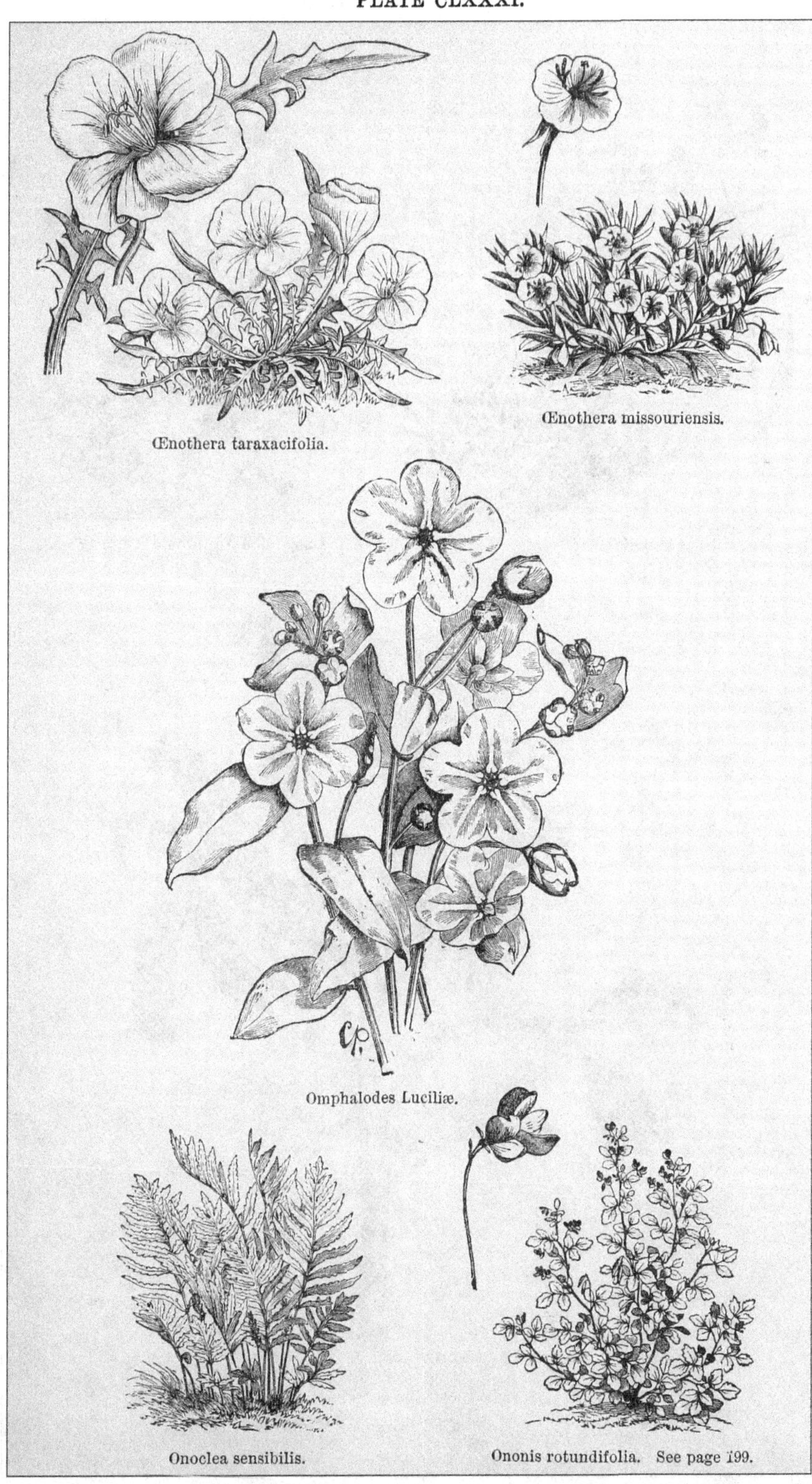

Œnothera taraxacifolia.

Œnothera missouriensis.

Omphalodes Luciliæ.

Onoclea sensibilis.

Ononis rotundifolia. See page 199.

PLATE CLXXXII.

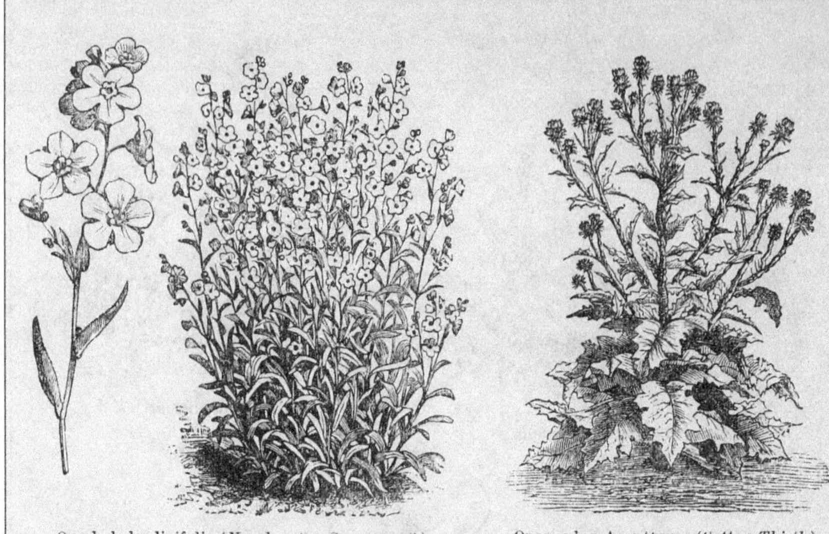

Omphalodes linifolia (*Navelwort*). See page 198.

Onopordon Acanthium (*Cotton Thistle*).

Onosma tauricum (*Golden Drop*).

among the occupants of the American garden, especially if sheltered a little by neighbouring plants. In such a position the pale green hue of the fronds contrasts charmingly with the dark foliage of the majority of its associates. One thing to be observed by those who are desirous of raising this class of Ferns from spores is this: If the fronds are allowed to remain on the plants until they appear to be ripe, it will be found that the spore cases are open and the spores shed, as they drop while the fronds look quite green; there ore, the best way is to cut off the frond as soon as indications of bursting are perceived, and to lay it in a sheet of paper for a few days, by which time all the spores will drop out. N. America.

Ononis *(Rest Harrow).*—A genus of the Pea family. The wild Liquorice (O. arvensis) is one of the prettiest of our wild plants, and well worthy of cultivation on banks and rough rockwork. It forms dense spreading tufts covered with racemes of pink flowers in summer. There is a white variety even more valuable for cultivation, and worthy of a better position and soil than the common form, which grows in any soil. No plants can be more readily increased from seed or by division. This plant is distinct from the spiny Ononis campestris, which forms stems nearly 2 ft. high, sometimes even more. O. rotundifolia is a distinct and pretty plant, hardy, and easily cultivated, flowering in May and June and through the summer. It attains a height of from 12 in. to 20 in., according to soil and position, increasing in height as the season advances. Suitable for the mixed border or rougher parts of rockwork; comes from the Pyrenees and Alps of Europe, and is easily propagated by seeds or division. These are the best of about half-a-dozen garden species, which include besides, O. fruticosa, Natrix and viscosa.

Onopordon *(Cotton Thistle).*—These noble plants, usually left to the botanic gardener, are valuable for their stately port, while their flowers are large and showy. They thrive as well in exposed places as in sheltered ones among shrubs, and may be effectively used in a variety of ways. Temperance in their use is, however, desirable, as in some situations they are apt to seed so freely that they require to be kept down judiciously. O. Acanthium is a native plant of bold habit and vigorous growth. It has stout, branching stems, often more than 5 ft. high and very large, covered with long, whitish, cobweb-like hairs; flowers purplish, in large terminal heads. O. illyricum has greener and more deeply-cut leaves, stiffer stems, a more branching habit, and leaves and stems much more spiny. O. arabicum grows to the height of nearly 8 ft., with an erect and very slightly branching habit, and has both sides of the leaves, as well as the stems, covered with a white down. O. græcum is also a handsome plant.

Onosma *(Golden Drop).*—A large genus of Borageworts, of which there are two species in cultivation. These are O. tauricum and O. echioides; the former was introduced about the beginning of the present century, but the latter has been in cultivation for nearly two hundred years. These two kinds much resemble each other, but O. tauricum is decidedly the finer, and should be grown in preference to O. echioides. There is no doubt, indeed, that this Taurian plant is among the finest of all the Onosmas, for none possess flowers of such a lovely citron-yellow colour. It is an evergreen perennial growing from 6 in. to 12 in. high, and forming a dense tuft of stems in a comparatively short period. It bears in summer drooping clusters of clear yellow, almond-scented blossoms which have gained for the plant the name of "Golden Drop." Native of the mountains of Greece and Tauria. The principal character that distinguishes O. echioides from the Taurian kind is its being a biennial, and hence it is of less value for the garden. The plant is similar in habit of growth to the preceding, but the flowers are pale yellow and not nearly so attractive. The best place for growing the perennial Golden Drop well is a properly constructed rock garden, thoroughly drained, in which provision is made for a good depth of soil, so that the plants may root strongly between the blocks of stone. The soil should be a good sandy loam, mixed with broken grit, and the plant placed between large blocks of stone, near which the roots ramify and are kept cool and moist. It may be propagated by seeds or cuttings. O. echioides requires similar treatment to the preceding, except that plants of it must be raised every year, and planted out in the rock garden to flower. It seeds freely enough in favourable seasons.

Onychium japonicum.—This elegant Japanese Fern is frequently grown in the greenhouse, and even in the stove. In most situations it is, however, perfectly hardy, but in severe winters some dead fronds of the common Brake may be thrown over it. The fronds are very finely divided, the segments being linear and of an intense dark green; the fronds vary from 1 foot to 2 feet in height, and are very useful for bouquet-making, or for placing loosely in vases with cut flowers.

Ophelia.—Plants of the Gentian order, and similar to Erythræa, but not much known to cultivators. They are probably not hardy.

Ophioglossum vulgatum *(Adder's-tongue)* is a common Fern, familiar to most people. It is generally found growing in moist meadows; therefore the best position for it in a garden would be in colonies, amongst groups of ornamental Grasses, in company with the Botrychiums and such flowering plants as Snowdrops and Anemones. O. lusitanicum, a dwarf variety, is an interesting plant, but so wayward and capricious in its likes and dislikes, that it is very difficult to cultivate with satisfaction.

Ophiopogon *(Snake's-beard).* — These are fibrous-rooted, herbaceous perennials, about 1½ ft. high. The flowers are usually small and lilac coloured, and appear numerously late in summer and in autumn in spikes, 2 in. to 5 in. long, rising from a deep green grassy tuft of evergreen foliage. They are suitable for borders or margins of shrubberies in sandy loam, but are scarcely ornamental. O. japonicus, Jaburan, spicatus, Muscari, and longifolius are the best known, but they are mainly suitable for botanical collections. All are propagated easily by division. In Italy these plants are used to make a green tuft in lieu of Grass, which perishes from the heat.

Ophrys.—These small terrestrial Orchids are singularly beautiful, and among the most curious of plants. There have been many in cultivation, but being chiefly from S. Europe and not hardy they must have protection, and then can be grown only with great attention. There are, however, a few native species that can be grown successfully. Of these one of the most singularly beautiful is the Bee Orchis (O. apifera). It varies from 6 in. to more than 1 ft. in height, with a few glaucous leaves near the ground; the lip of the flower is of a rich velvety brown with yellow markings, so that it bears a fanciful resemblance to a bee. It is usually considered very difficult to grow, but this is by no meansthe case, and it may be grown easily in rather warm and dry banks in the rock garden, planted in a deep little bed of calcareous soil, if that be convenient; if not, loam mixed with broken limestone may be used. It will be found to thrive best if the surface of the soil in which it grows be carpeted with the Lawn Pearlwort or some other very dwarf plant, and, failing these, with 1 in. or so of Cocoa fibre and sand, to keep the soil somewhat moist and compact about the plants. Flowers in early summer. Other interesting species to cultivate in a collection of hardy Orchids are O. muscifera, arachnites, aranifera, and Trolli.

Opuntia *(Prickly Fig).*—There are several species of Opuntia in cultivation, but few are hardy enough for growing in the open air. The hardiest are O vulgaris, missouriensis, humilis, brachyantha, and Rafinesquei; the last is the finest and one of the hardiest of all, and is a plant well worthy of culture. It bears in summer large showy yellow blossoms on its singular fleshy branches. There are many gardens in the south of England where good conditions for the growth of this plant exist. A little group of it in a sunny corner of the rock garden, with liberty to root into good, but not wet, soil, would thrive if sheltered by a few rocks or other surrounding objects from any passing danger to the stems. As it is perfectly evergreen and rather fragile, anything brushing against it would injure it; but it is easy, by the skilful placing around it of a few rough stones or little bushes, to prevent injury without in the least shading it. The ground from which it springs might, for the sake of obviating splashings, and also to secure a better effect, be surfaced with some dwarf mossy Saxifrage or Sandwort. Snails and slugs are extremely fond of this plant, and in the spring, and even in mild winters, will completely destroy it if not prevented. A liberal dressing of soot will keep these pests at a distance, and will impart luxuriance to the plants, for although the Cactus family object to manure in the soil, they fully appreciate a top-dressing of some concentrated manure, soot, or guano. The propagation of the hardy Cactus is easily effected. The cutting, consisting of a single joint, is placed in very sandy soil, and the pot containing it in a sunny, airy situation under glass; it should be watered very sparingly. In a short time it will form roots, and will then commence to push out several young shoots, which also in their turn may be taken off.

Orchis.—These terrestrial Orchids are singularly beautiful, and some are really showy plants and well worth attention on that account alone. Those who do not want a full collection will find the species mentioned below to be beautiful and easily grown plants, provided they are placed under the necessary conditions at the outset. Many of our native Orchises are very interesting and deserving of culture, but few cultivators succeed with them, chiefly because the plants are generally transplanted at a season when such an operation is most harmful. After success depends on transplanting. The usual plan is to do it just when the first or second flower has opened, so as to enjoy the remaining few in the garden, and, this done, the cultivation has in most cases begun and ended. This course of procedure is most mischievous, for at this period of growth the plant is forming a new tuber for the following year, and if this is in any way injured, it shrinks and dies. If, instead of this, the plants are marked when in flower and allowed to remain until August or September, when the newly-formed tuber will be thoroughly matured, the risk of transplanting it is considerably lessened, provided it be taken up with a good deep sod. It is advisable to take from the sod the roots of other plants, though some persons allow them to remain to spread over the Orchid border, thus imitating natural conditions by protecting the roots in winter and preventing evaporation in summer. In a garden, however, this presents an untidy appearance, and there is also the risk of cutting away the Orchids as well as the Grass, &c., in the flowering season. As substitutes for Grass such plants as the pretty Balearic Sandwort (Arenaria balearica), Lawn Pearlwort (Sagina glabra), the mossy and tufted section of Saxifrages, and many others, which are at all seasons green and suitable, may be used, and serve the purpose equally well. The situation for Orchises generally should be a partially shaded one, and the soil should consist of a deep, rich, fibry loam in a thoroughly drained border.

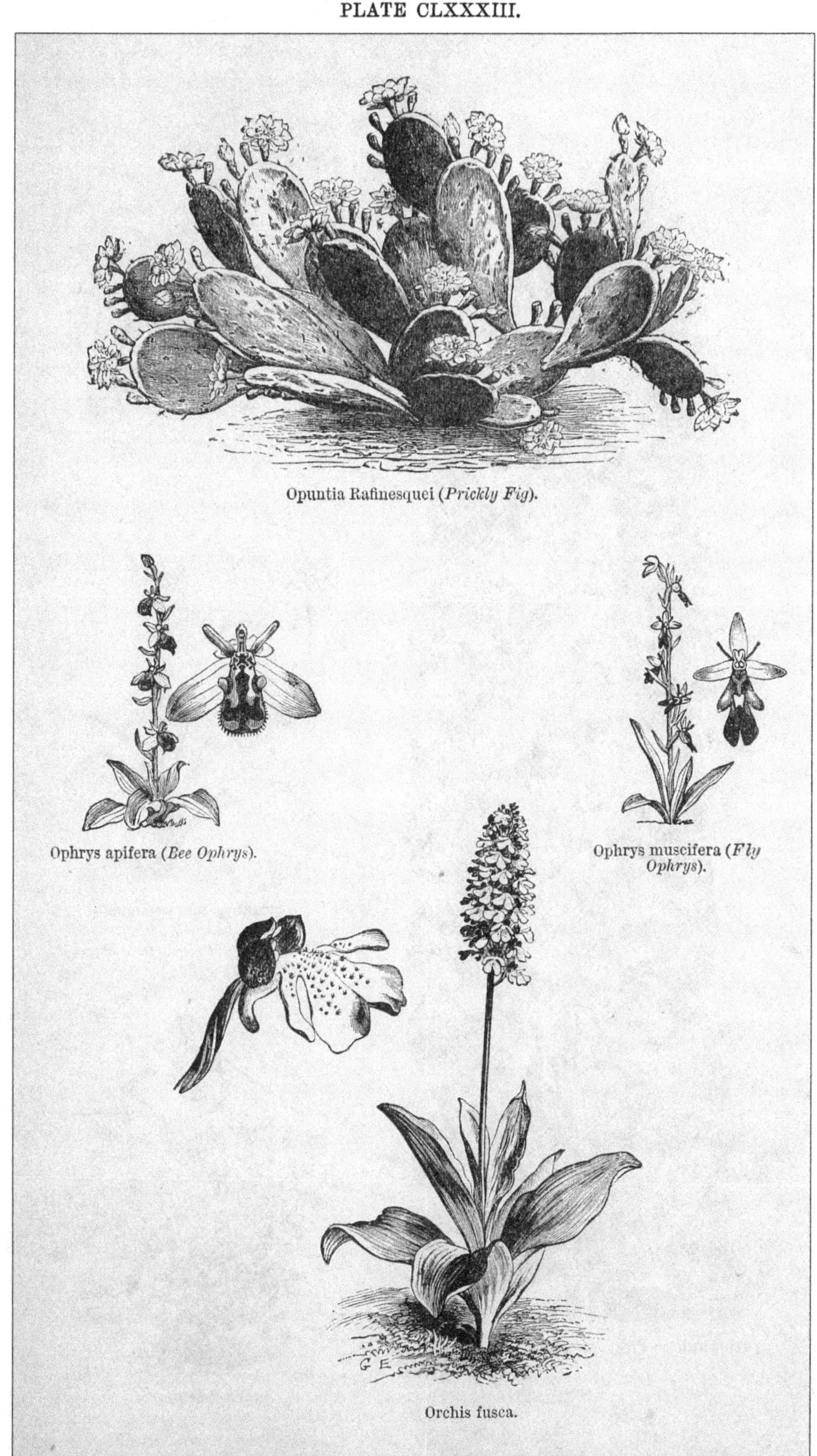

Opuntia Rafinesquei (*Prickly Fig*).

Ophrys apifera (*Bee Ophrys*).

Ophrys muscifera (*Fly Ophrys*).

Orchis fusca.

PLATE CLXXXIV.

Orchis latifolia (*Marsh Orchis*). Natural size.

Ornithogalum umbellatum.

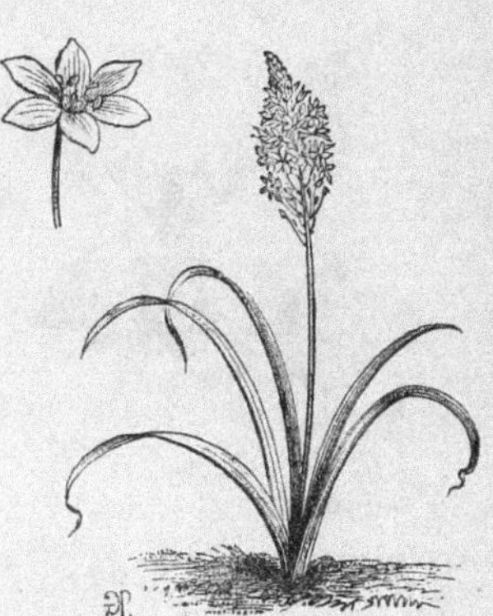

Ornithogalum pyramidale.

Orobus luteus.

[*To face page* **201.**

The following are the kinds most worthy of culture :—

O. foliosa.—This handsome terrestrial Orchid is one of the finest of the hardy kinds that thrive in our climate. It grows from 1 ft. to 2 ft. or more in height, and produces long dense spikes of rosy-purple blossoms, spotted with a darker hue. It begins to flower about the middle of May, and continues in full beauty for a considerable time. It delights in moist sheltered nooks at the base of the rock garden, or in some similar place, and it should be planted in deep, light soil. Native of Madeira.

O. latifolia *(Marsh Orchis)*.—A very ornamental native kind, 1 ft. to 1½ ft. high. The flowers, which are purple, are produced in early summer in long dense spikes. It is one of our commonest species, and easily grown. It forms luxuriant specimens in damp, boggy situations, in peat or leaf mould. There are several beautiful varieties of this Orchis, the best being præcox and sesquipedalis ; the latter is one of the finest of all hardy Orchids. It grows about 1½ ft. high, and the stem for fully a third of its length is furnished with densely-arranged flowers of large size and of a purplish-violet hue.

O. laxiflora is a handsome species, growing 1 ft. to 18 in. high, with flowers, opening in May and June, of a rich purplish-red, in long loose spikes. Native of Jersey and Guernsey, and suited for the rock garden in a moist spot, or the artificial bog ; it may also be naturalised. Division.

O. maculata.—This is one of the handsomest of British Orchises. It is usually prettily tinted in the poorest and driest soils, but is a very different object in a rich soil. If well grown in moist and rather stiff garden loam, it will surprise and please even those who know it well in a wild state. Obtain it at any season, and carefully plant twelve or twenty tubers in a patch in a half-shady and sheltered position in moist, deep soil. When the plants form dense patches two or three years afterwards they will prove more beautiful than any other Orchids from warmer climes. O. maculata flowers in summer, and may be associated with the Cypripediums, or planted in tufts in borders or on the margins of shrubberies. The variety superba should in all cases be preferred to the type, as it is a much finer plant.

Other beautiful kinds are O. papilionacea, purpurea, militaris, mascula, pyramidalis, spectabilis, tephrosanthos, and Robertiana, but all are more or less difficult to manage.

Oreocome Candollei is a very effective plant for the margins of shrubberies, or planted singly on a lawn. It makes a fine pyramid, 5 feet in height, furnished with large leaves that are as finely divided as those of a Todea, spread out horizontally, and recurve gracefully. They are of a fresh green colour, and the flowers, which rise well above the foliage, are in umbels, and pure white. It grows well in any ordinary garden soil, and is quite hardy. Himalayas. Umbelliferæ.

Oreoseris lanuginosa.—A composite plant of no garden value.

Origanum *(Marjoram)*.—The common O. vulgare is scarcely an ornamental plant, but another species, O. Dictamnus (the Dittany of Crete), is a pretty little plant, though somewhat tender, and more suitable for growing under glass than in the open air entirely. During mild winters, however, it survives unprotected. It has mottled foliage and small, purplish flowers in heads like the Hop ; hence it is sometimes called the Hop plant. O. Sipyleum is similar, and quite as pretty. If grown in the open these plants must have a warm spot in the rock garden in well-drained soil.

Orythia *(Tulipa)*.

Ornithogalum *(Star of Bethlehem)*.—Though this is a numerous genus of bulbous Liliaceous plants, there are not many good garden plants among those that are hardy. They all have white, or greenish-white, flowers, produced either in dwarf clusters or on tall spikes. The best are O. nutans, narbonnense, arabicum, umbellatum, and pyramidale. The last is a fine plant, with flower-stem, 3 ft. high, surmounted by a pyramidal cluster of white blossoms. These are all perfectly hardy in borders, and thrive in any good garden soil. They are seen to the best advantage in places where they have been left for years undisturbed in masses. Little colonies of them planted in Rhododendron beds, or other shrubberies where the shade is not too dense, produce a pretty effect in spring when in bloom. O. nutans, with its large spikes of flower — the outer part so delicately tinted with green—is a good early-flowering bulb, quite hardy and free in ordinary garden soil, as all the hardy kinds are.

Ornithopus *(Bird's-foot)*.— The commonest species is the little O. perpusillus, which often forms a dense carpet-like growth on lawns. It is, however, not worth cultivating.

Orobanche *(Broom Rape)*.—Curious leafless, parasitical plants often found growing on the roots of the Furze and Broom. They cannot be cultivated.

Orobus *(Bitter Vetch)*.—This genus is distinguished from Lathyrus chiefly by not being of a climbing habit of growth. All the species are perennial, and the majority are handsome when in flower ; this is usually in spring, and they are thus desirable plants. They are suitable for the mixed border, for the rougher parts of the rock garden, or for naturalising. There is a similarity between many of them, and therefore we mention the most distinct kinds only.

O. aurantius is a handsome plant from 18 in. to 24 in. high. The flowers are orange-yellow and appear early in summer. O. tauricus is a nearly allied species, also with orange-coloured flowers. Both require to be well established plants before they will bloom freely and be seen in perfection. They are suitable for borders in ordinary soil.

O. lathyroides is a lovely border plant. Its height, when growing vigorously, is from 18 in. to 24 in.; its flowers are bright blue, produced in dense racemes, and its general habit all that can be desired. It is increased freely by seeds, which it produces abundantly. It is a capital wild garden plant, as it flourishes in any soil.

O. vernus *(Spring Bitter Vetch).*—This is one of the most charming of the border flowers that usually begin to open at the end of April and the beginning of May. From black roots spring rich, healthy tufts of leaves with two or three pairs of shining leaflets; the flower-buds show soon after the leaves, and eventually almost cover the plants with beautiful blooms, purple and blue, with red veins, the keel of the flower tinted with green, and the whole changing to blue. Besides the typical form there are—tenuifolius, with very narrow leaflets and flowers very similar to the type, though the general habit of the plant is more lax; cyaneus, the most attractive of all, being larger and possessing a strange intermixture of colours, some a bright blue, others greenish-blue, and some with two or more distinct colours in the same flowers; flaccidus, a variety somewhat similar to tenuifolius, but brighter, denser in habit of growth, and with broader leaves. Then there is a double-flowered kind and a pure white variety, both desirable—the one on account of the lasting properties of the flowers, the other on account of its purity. All these are propagated very readily from seed, and also by division of the root, and attain their greatest development in deep, warm soils in sunny, sheltered positions. Some other species of Orobus useful for borders and rockwork are in cultivation, O. pubescens, O. canescens, O. varius, and O. Fischeri being among the best, but O. vernus and its forms are the handsomest of the species. All are of easy culture in ordinary garden soil, and are increased by seeds or division of the root.

Orontium aquaticum *(Golden Club).*—A handsome aquatic perennial of the Arum family, 12 in. to 18 in. high. The flowers—which are yellow, densely crowded all over the narrow spadix, and which emit a singular odour—are borne early in summer. The plant may be grown on the margins of ponds and fountain-basins, or in the very wettest part of the artificial bog. North America, in rivulets and bogs.

Orychophragmus sonchifolia.—A showy plant belonging to the Crucifer family, but scarcely hardy. It is an annual, though under certain conditions it is a biennial, or even perennial. It makes a showy plant, as its flowers are of a bright violet-blue colour, and under good culture it attains a height of 2 feet, the loose terminal racemes of flowers being about half that height. It is worth growing in mild localities as a half-hardy annual. (= Moricandia.) China.

Osmunda *(Royal Fern).*—This genus of so-called flowering Ferns is made familiar enough by the common native Royal Fern (O. regalis), which is to be found in many bogs and marshy woods. It is well worth cultivating, as it is the largest and most striking of our native Ferns (sometimes attaining a height of 8 ft.), and is one of the most ornamental subjects that can be grown in certain positions. It should be planted in moist peaty soil in half-shady places, on the banks of streams, or on the margins of pieces of water. It may also be planted in the water with good effect. It has been found to do well when exposed to the full sun, with its roots in a constantly moist, porous, moss-covered soil, in a position sheltered from strong winds. In shady positions it would be found to attain great stature if planted in deep, rich soil. The various North American Osmundas may be associated with this. There are several varieties to be met with in gardens. Popular and almost universally cultivated as Ferns are, the Royal Fern and several other Osmundas are rarely met with otherwise than in a shabby, or at best in a half-developed, condition. There are other hardy species not British. O. cinnamomea is an elegant N. American plant with pale green fronds; the variety angustata is smaller, and the fronds are not so inclined to droop. This species, like O. regalis, is deciduous. O. Claytoniana is a deciduous species, with vivid green fronds, that attains a height of from 2 ft. to 3 ft. O. interrupta is the same. O. gracilis is a native of Canada, and somewhat resembles a dwarf form of our native Royal Fern; the fronds are about 2 ft. high. O. spectabilis is a slender form of O. regalis; its fronds are smaller than those of the type, and the young ones always come up reddish-purple; this is its distinguishing feature. Native of N. America. These exotic species are of the simplest culture in any shady fernery, in moist peaty soil.

Ostrich Fern *(Struthiopteris).*

Oswego Tea *(Monarda didyma).*

Othonna cheirifolia *(Barbary Ragwort).*—A composite plant of distinct character, forming whitish-green tufts from 8 in. to 1 ft. high, or perhaps more on very rich soils. It has a spreading habit of growth, is evergreen, and flowers sparsely on heavy and cold soil, but on light soils it blooms often somewhat freely in May. The flowers are of a rich yellow, about 1½ in. across, but are not ornamental. It is chiefly useful for its distinct type of leaf and aspect on the rough rockwork or mixed border. A native of Barbary; propagated by cuttings. It perishes in severe winters, at least on clay soils.

Ourisia coccinea.—A bright little Chilian plant, of dwarf creeping growth, producing in early summer showy bright scarlet blossoms in slender clusters from 6 in. to 9 in. high. Though this plant is perfectly

PLATE CLXXXV.

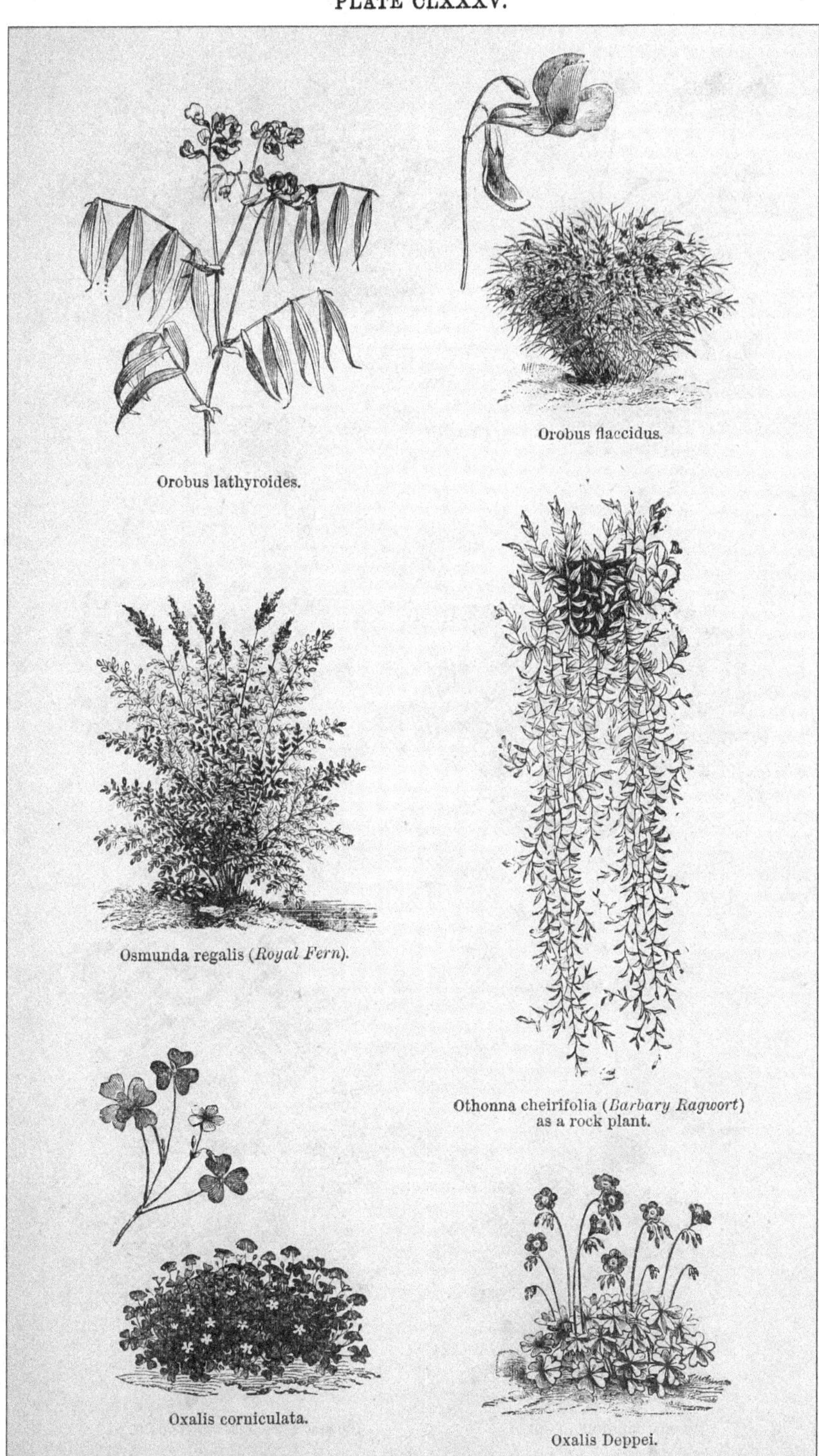

Orobus lathyroides.

Orobus flaccidus.

Osmunda regalis (*Royal Fern*).

Othonna cheirifolia (*Barbary Ragwort*) as a rock plant.

Oxalis corniculata.

Oxalis Deppei.

[*To face page* 202.

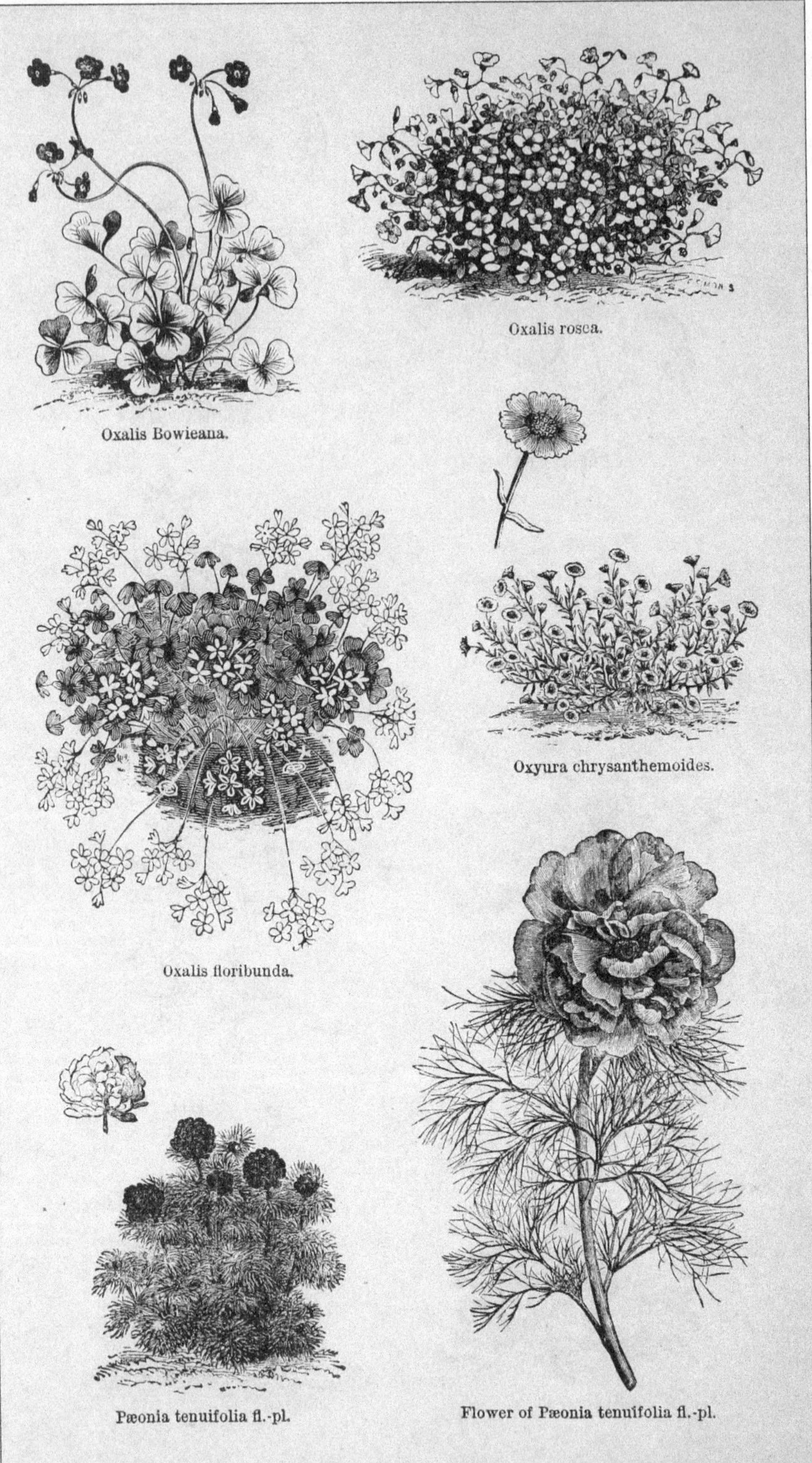

Oxalis Bowieana.

Oxalis rosea.

Oxyura chrysanthemoides.

Oxalis floribunda.

Pæonia tenuifolia fl.-pl.

Flower of Pæonia tenuifolia fl.-pl.

[*To face page* **203.**

hardy, it is reputed difficult to grow. To be successful with it, it must be placed hard against the side of a block of soft porous stone in a moist and partially-shaded place, such as the foot of a wall with an east aspect. In such a place its creeping stems will soon run along the face of the stone and will flourish and flower freely. A dry rockery is fatal to it. Scrophulariaceæ.

Oxalis.—This is a very numerous genus, consisting chiefly of plants requiring protection in their cultivation. A few, however, are quite hardy, and, when well grown, make extremely pretty plants, especially suitable, on account of their dwarf growth, for the rock garden. They all thrive best in sandy soil in the warmest and driest place in a garden. Where a full collection is not desired, the following are the most suitable :—

O. Bowieana is one of the finest. It is a robust-growing species, forming bold masses of leaves, 6 in. to 9 in. high. The flowers are dark rose, in large umbels, and are produced continuously throughout the summer. It is suitable only for warm borders or for the foot of a south wall. In cold soils this seldom or never flowers; on warm well-drained and very sandy ones it does so abundantly, and where this is the case it may be used with effect as an edging round beds of autumn-blooming plants. Division. Cape of Good Hope.

O. floribunda.—A free-flowering kind, quite hardy in all soils, and producing, for months in succession, numbers of rose-coloured flowers with dark veins. There is a white-flowered variety as free-flowering and in every way as valuable as the rose-coloured form. Both are very useful for rockwork and for the margins of borders, and are easily increased by division. This appears to be the commonest kind of Oxalis in cultivation. It is hardy enough to encourage one to attempt to naturalise it on any rocky place or about ruins. S. America.

O. lasiandra is one of the most distinct and beautiful kinds, producing very large dark green leaves and, in early summer, umbels of numerous flowers of a bright rose colour. Adapted to culture on warm rockwork. Mexico.

O. lobata.—This little gem is a stemless species with slender leaf-stalks and three deeply-lobed bright green leaflets. The blossoms, about ¾ in. across, are produced considerably above the leaves, are borne singly on slender, erect stalks, and are of a rich yellow colour, the centre being delicately pencilled with chocolate. It is very free flowering, and is a bright little plant during sunshine. It thrives in warm sandy loam on well-drained borders or rockwork. In mild winters it survives unprotected. Chili.

O. luteola is one of the most lovely of Oxalises, and, when planted out, forms a compact tuft. The flowers, in the bud state, are of a soft creamy-yellow, ½ in. in length, but when open they are pure white, shading to yellow towards the centre, and as large as a half-crown piece. O. luteola is not hardy, but if protected in winter will survive in light sandy soil.

There are other species of Wood Sorrel worthy of a place, especially on very dry sandy soils, and among them are O. Smithi (a beautiful species), rosea, Deppei, speciosa, arborea, violacea, versicolor, incarnata, tetraphylla, and venusta, and the British kinds O. Acetosella and corniculata. The rubra variety of the latter is sometimes used for bedding purposes, and should always be encouraged where there are old quarries and rough, rocky places, especially in a calcareous district, and their bareness might with advantage be hidden by this handsome plant, which speedily covers the most unpromising surfaces. In gardens, however, Wood Sorrels are apt to become troublesome weeds. If a collection be grown, it should be borne in mind that it is very difficult to preserve the correctness of the names, owing partly to the elasticity of the seed-pods, which permits the seeds to scatter in all directions, and also to the minute bulblets becoming mixed up with the earth.

Ox-eye (*Buphthalmum*).

Ox-eye Daisy (*Chrysanthemum Leucanthemum*).

Oxybaphus nyctagineus.—A North American plant similar to the Marvel of Peru. Of botanical interest only.

Oxyria.—Weedy Dock-like plants, of no ornamental value.

Oxytropis.—A large genus of the Pea family, nearly allied to the Astragalus. There are several species in cultivation, the best of which is O. pyrenaica, a very dwarf species, with pinnate leaves clothed with a short silky down. These leaves barely rise above the ground, as the short stems are nearly prostrate, and seldom exceed a few inches in height; the flowers, borne in heads of from four to fifteen, are of a purplish-lilac, barred with white. Not a showy, but withal it is a desirable little plant for those parts of rockwork devoted to very dwarf subjects. It is a native of the Pyrenees, rare in gardens, and increased by seed or division. It should be planted on well-exposed and bare parts of rockwork, in firm, sandy, or gravelly soil. Flowers in early summer. Another desirable plant is O. uralensis, a dwarf, compact species from the Ural Mountains; its flowers, which are produced in compact heads, rising about 4 in. high, are of a rosy-blue colour and exceedingly attractive. No collection should be without this plant. Other kinds are—O. montana, fœtida, and campestris and its several varieties; all of these are of dwarf growth and similar to those described, and, like them, thrive in open spots in the rock garden in sandy, loamy soil.

Oxyura chrysanthemoides. — A hardy annual composite, of neat growth and with showy yellow flowers. It should be sown in a broad mass in ordinary soil in autumn, and the seedlings well thinned out. It flowers from June to July. California.

Oyster Plant *(Mertensia maritima).*

Pachyphyton bracteosum *(Silver-bracts).*—A stout-growing succulent plant, commonly used for placing out in summer; usually in geometrical beds. The leaves are thick and fleshy and form rosettes, and the whole plant has a whitened tone, which gives it a distinct and ornamental character. Plants may be readily propagated by merely pulling off the leaves and placing them in pans in a dry, warm pit or stove.

Pachysandra procumbens *(Mountain Spurge).*—A small shrubby plant of dwarf, prostrate growth, only suitable for a botanical collection; though, like many other plants of no garden value, it is often named in trade catalogues.

Pæderota.—A genus of dwarf alpine heros belonging to the Figwort family and nearly allied to the Wulfenias. There are two species of Pæderota in gardens, and by far the more ornamental of these is P. Bonarota, which is found amongst rocks, &c., on the Alps in various parts of Austria and Carniola. It varies in height from 2 in. to 1 ft., which appears to be its maximum height. The flowers, which are rather small, and vary from pink to a bright violet colour, are borne in dense, erect, terminal racemes from 1 in. to 3 in. long, and are produced during May and June. Synonymous with P. cœrulea. P. Ageria, the other cultivated kind, is a much less showy plant. It grows from 1 ft. to 1½ ft. high; the blossoms are of a pale yellow colour, and produced in May and June. Native of the alpine districts of Carinthia, Carniola, and various parts of Italy.

Pæonia *(Pæony).*—The Pæonies are among the noblest and most beautiful of hardy flowers, and indispensable for the garden. They not only combine stateliness of growth with beauty of colouring, but in many the huge blossoms possess the delicious fragrance of a Tea Rose. Though there are several typical species to be found in botanical collections, by far the most important are the many hybrid varieties that have been obtained by intercrossing some half a dozen kinds. The Pæonies are divided into two groups—the tree or shrubby kinds, the varieties of P. Moutan; and the herbaceous kinds, of which the common P. officinalis is typical. The hybrid sorts have chiefly been obtained from P. officinalis and other European kinds, and albiflora, sinensis, and edulis, Chinese species, the forms of the latter being particularly fine. The European varieties flower early and the Chinese late, so that the flowering season is considerably prolonged.

HYBRID VARIETIES.—Among these there is an extensive variety of colours—white, pale yellow, salmon, flesh coloured, and a numerous intermediate series between pale pink and the brightest purple. Among the oldest varieties the following are the most remarkable, viz., grandiflora, double white; Louis Van Houtte, papaveriflora, rubra triumphans, sulphurea plenissima, rosea superba, Zoé, Mme. Calot, Gloria Patriæ, and Prince Troubetskoy. The most beautiful among those of a more recent date are: Arthémise, atrosanguinea, Virgo Maria, Mme. Lemoine, L'Espérance, Triomphe de l'Exposition de Lille, Jeanne d'Arc, Eugène Verdier, and Mme. Lemoinier; and among those most worthy of notice may be named Mme. Lebon, Marie Lemoine, Henri Laurent, Mme. Jules Elie, multicolor, Stanley, Charlemagne, Mme. Geissler, Bernard Palissy, and Van Dyck. Besides these there are many other commoner varieties—for example, the varieties of P. officinalis (such as anemonæflora, rubra, and Sabini), and the varieties of P. albiflora, peregrina, paradoxa, and especially the small growing P. tenuifolia, with feathery foliage and large deep red blossoms. There is also a double variety of this species. These, as well as the varieties, are perfectly hardy, and need no care in winter to preserve them from frost, however severe.

CULTURE.—A good moist loam, particularly when enriched by the addition of cow manure, is the soil best suited to them. They can be planted at any time of the year, but from October to April is the best. Take care to have the ground well prepared for their reception by manuring and trenching it to the depth of about 3 ft., and on no account should they be planted nearer than 4 ft. apart in each direction. They must not be expected to flower well before the second or third year after planting. An open position renders the plants robust, and they do not require to be shaded from the sun until they flower, when some slight shade prolongs and preserves their delicate tints, and also enables the flowers to become more thoroughly developed than they otherwise would be; as soon as the buds are well formed, watering the plants judiciously now and then with liquid manure also greatly benefits them. It is likewise, of course, necessary when the tufts have become very strong, and have impoverished the soil, to separate and transplant them on fresh ground.

POSITION.—In most gardens there are generally to be found spots so much shaded that scarcely any plants will thrive in them. In such places Pæonies would grow luxuriantly; the colour of their blooms would in many cases be more intense, and they would last much longer than flowers fully exposed to the sun. They may therefore be made useful as well as ornamental, even in small pleasure grounds, although the proper place for them is undoubtedly the fronts of shrubberies, plantations, or by the sides of carriage drives. Where distant effect is required no plants so admirably answer the end, as their size and brilliancy render them strikingly visible, even at long distances. Planted on either side of a Grass walk, the effect which they produce is admirable, especially in the morning and at or near sunset, and when

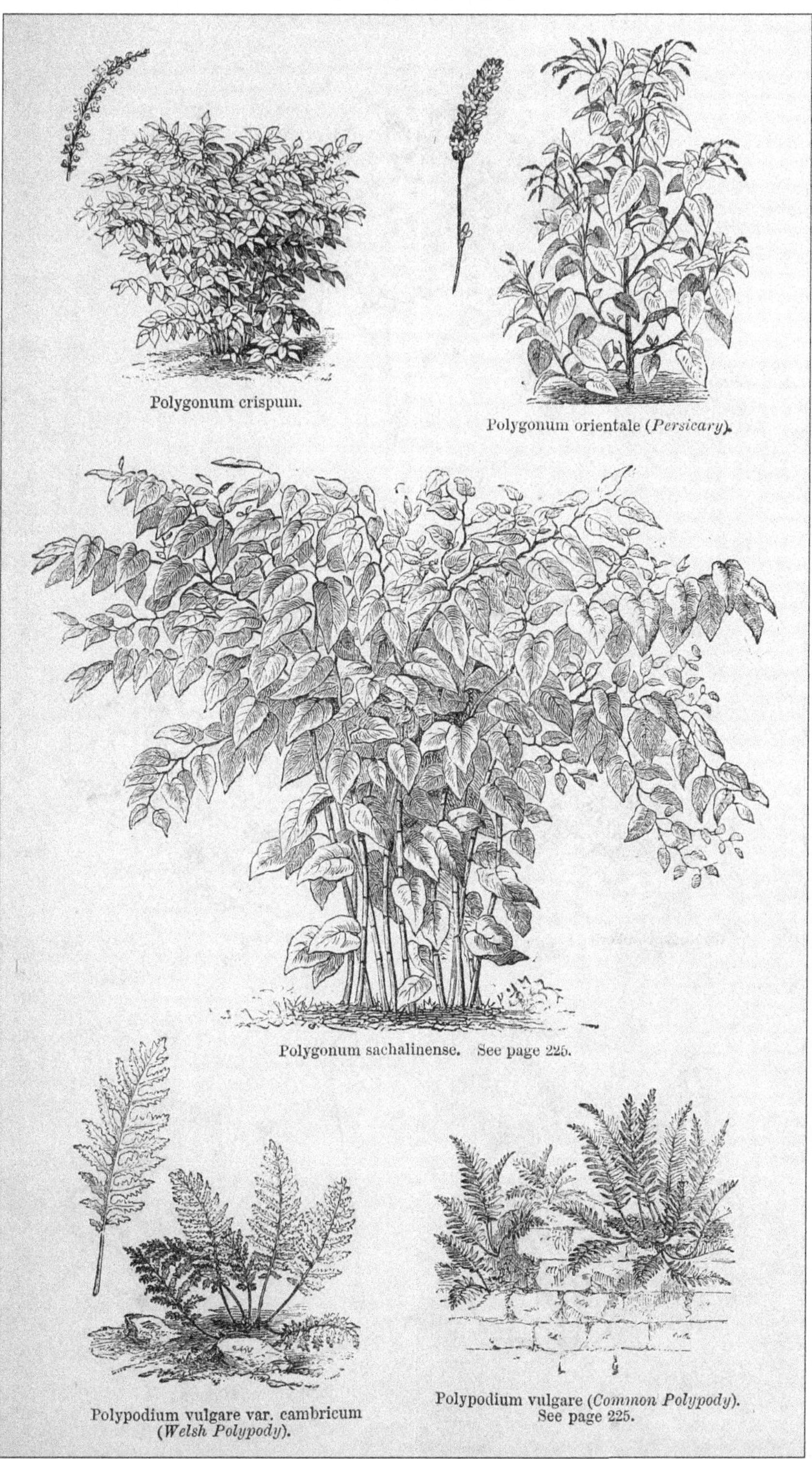

Polygonum crispum.

Polygonum orientale (*Persicary*).

Polygonum sachalinense. See page 225.

Polypodium vulgare var. cambricum (*Welsh Polypody*).

Polypodium vulgare (*Common Polypody*). See page 225.

Double flower of Pæonia Moutan.

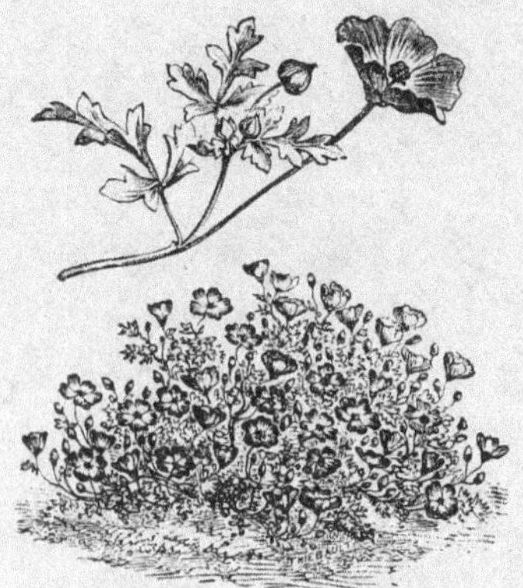

Palafoxia Hookeriana. See page 206.

Palava flexuosa (*Chilian Palava*). See page 206.

Pancratium illyricum. See page 206.

Pancratium maritimum. See page 206.

[*To face page* **205.**

planted in masses, as, for instance, in beds in pleasure grounds, they are invaluable for lighting up sombre nooks. If they are grown for their flowers only, or their buds, or for the purpose of increasing them, they may be placed in nursery lines in some rich part of the kitchen garden.

Besides being used for the garden proper, there are few plants more fitted for the wild garden than the Pæonies; and one of the boldest things we know of in wild gardening, and certainly the most brilliant, is the putting of a group of scarlet Pæonies in meadow Grass, where they have a singularly fine effect in early summer. It may be so managed that they come into the garden landscape, so to say, and be seen for a considerable distance from certain points of view. If so placed they could not be an eyesore or in the way when out of flower, as such plants sometimes are in the mixed border. In the future of our gardens there is a good deal to be done by the tasteful cultivator in considering well the positions suited for some kinds of plants. It is, for example, easy to arrange plants that are very handsome in spring and early summer, but do not continue in perfection very long, so that their effect when out of flower, or even their disappearance altogether, will not mar any "composition." This point is a very important one and well worth attending to. Pæonies are among the plants that deserve this kind of attention.

P. Moutan *(Tree Pæony).*—This is another noble type of plant from which we have derived a great wealth of beauty, for the varieties, like the herbaceous kinds, are very numerous. It is quite hardy, and, when properly planted, requires little care. It is invaluable for mixed borders, and especially well suited for forming single specimens on lawns; and if, during the winter months, its appearance is against it, this drawback is amply compensated for in summer, when it becomes covered with fine foliage and gorgeous blossoms. When, too, it breaks into growth in spring, its young leaves assume every shade of colour, from violet-crimson to green. Tree Pæonies are not very particular as to soil or position; they grow in nearly pure sand, flourishing as well as in a strong loam. They, however, prefer a good strong soil, and where it is too sandy it should be made a little better suited for them by adding decomposed manure and loam; or, where it is too clayey, it should be made lighter by means of the addition of manure, sand, and similar materials. Moutans are gross feeders, and amply repay for occasional top-dressings of half-decomposed cow manure. Of the scarcer and best varieties nurserymen generally send out one or two-year-old plants, grafted on the roots of Pæonia edulis. After a proper place has been selected in which to plant them, dig out a pit 1½ ft. deep and 2 ft. in diameter; put in a few inches of half-decomposed cow manure. and well mix it with the soil. The plants should be put in so that the graft may be buried a few inches under the ground, where it will, in time, throw out roots of its own. They do not flower well until the third year after planting, but after that they produce blossoms in profusion. Being naturally of what is termed slow growth, they are not propagated by division to any great extent, but chiefly by grafting upon the roots of the herbaceous varieties, which is done in August. The grafts are placed in frames, where they unite, and are transplanted the succeeding year in rows in the nursery.

September and October are the months best suited for their planting, but if in pots they may be put out in spring, when all danger of frost is over. Good plants set in autumn produce quantities of flowers the second or third year after planting. Each year they increase in size and beauty, and soon become the most showy and attractive features of the garden. They are the first of any of the varieties of Pæonies to flower, and put forth their blooms early in May. Until the second half of this century only such sorts as had white, rose, salmon, and lilac-coloured flowers were known; and we are indebted to Mr. Fortune for the introduction of his Chinese varieties, most of which have scarlet, violet, and magenta-coloured flowers. Von Siebold, too, introduced a number of Japanese varieties, which, however, form a different race, and are mostly single or semi-double. The following list contains some of the best varieties: Athlète, large, double, lilac; Bijou de Chusan, pure white; Carolina, bright salmon; Colonel Malcolm, violet; Comte de Flandres, very large, rose; Confucius, deep pink; Elisabeth, deep scarlet, very double; Farezzii, large, pale lilac striped with violet; Fragrans maxima fl.-pl., pale rose; Lambertiana, blush rose petals, tipped with violet; Louise Mouchelet, large double pink; Madame de Sainte-Rome, bright lilac-rose; Madame Stuart Low, bright salmon-red; Marie Ratier, large, rose; Odorata Maria, pale rose; Prince Troubetskoy, very large, double, deep lilac or violet; Purpurea, a deep amaranth, semi-double kind; Ranieri, bright amaranth; Rinzii, very large, bright rose; Rosini, a semi-double, brilliant rose-coloured variety; Rubra odorata plenissima, very large, double, lilac-rose; Souvenir de Madame Knorr, large, double blush; Triomphe de Malines, large, violet, a colour which deepens at the base of the petals; Triomphe de Vandermaelen, very large, and double violet-shaded rose; Vandermaeli, blush, almost white; Van Houttei, large, double, carmine; and Zenobia, white. Some of the more strikingly beautiful among them, such as Gloria Belgarum, Elisabeth, and Souvenir de Gand are well worth being put under glass—that is, they should have a sash or two put over them in spring to save them from late frosts and rainy weather. Of course plenty of air must be admitted, but, so treated, the flowers are gainers in an astonishing degree, both in size and colour.

Palafoxia Hookeriana.—This is a pretty dwarf annual composite plant, forming a dense tuft about 1 ft. high. The flower-heads are produced freely in loose clusters, and are of a pleasing rosy pink hue. It should be treated as a half-hardy annual, and grown in a warm border of sandy soil. Native of Mexico.

Palava.—A small genus of Mallow-like plants, native of Peru and Chili. The few cultivated species are all annual. P. rhombifolia grows about 1 ft. high, is somewhat procumbent, and has rosy purple blossoms, about 1 in. across, produced in July and August. P. moschata is similar, but of erect growth. P. flexuosa, a Chilian species, is a pretty plant, growing about 1½ ft. high, also with rosy purple blossoms. These all require to be sown in heat in early spring, and when the seedlings are large enough should be transplanted to the open border in ordinary garden soil, and in a sheltered situation.

Pampas Grass (*Gynerium argenteum*).

Pancratium.—The only really hardy species of this genus of bulbous plants is the South European P. illyricum, which grows from 1 ft. to 2 ft. high, and bears in summer a stout flower-stem terminated by numerous umbels of deliciously fragrant and attractive large white blossoms. It thrives best in a warm exposed border of sandy-loamy soil well drained. During winter the bulbs should be protected by a covering of litter or some such material. The plants are better for being transplanted about every third year; this operation should be done in autumn as soon as the leaves are decayed, and the bulbs should be replanted immediately. The Pancratiums may be propagated by offsets from the parent bulbs. The hardiest of the other species are P. parviflorum, maritimum, littorale, and rotatum, but these require more attention and only succeed well in exceptionally mild localities and on warm soils. They are best grown in a frame or cool greenhouse.

Panicum.—This is a very large genus of Grasses, chiefly tropical, though a few are hardy enough for out-door cultivation in our climate.

P. altissimum is a very handsome hardy perennial Grass, very like P. virgatum, but more elegant in habit. It forms dense, erect tufts, from 2 ft. to 6½ ft. high, according to climate, soil, and temperature. When in flower the plant presents a very attractive appearance, the flowers being of a dark chestnut-red tinge.

P. bulbosum.—A tall and strong species, with a free and beautiful inflorescence. It grows about 5 ft. high, and the flowers are very gracefully spread forth. It forms an elegant plant for the flower garden in which grace and variety are sought, and is suited for dotting about here and there, near the margins of shrubberies, &c., and for naturalisation.

P. capillare.—A hardy annual kind, growing in tufts from 16 in. to 20 in. high, and very ornamental when in full flower, the tufts being then covered with large, pyramidal panicles, which are borne both at the ends of the stems and in the axils of the stem-leaves. It grows in any soil or position, and sows itself, and is well suited for border beds or isolation, being one of the most graceful plants in cultivation.

P. virgatum.—A handsome, bold, hardy species from North America, growing, in good soil, to a height of nearly 3½ ft. It forms close, compact tufts of leaves, 1 ft. or more long, which, from July to the first frosts, are crowned with very large, dense, branching panicles. The general colour of the plant is a fine lively green, and its graceful habit renders it an admirable subject for the picturesque flower garden, the pleasure ground, &c., in isolated tufts. It is also fine for borders. The best mode of multiplying it is by division in the spring, when vegetation is just commencing.

Papaver (*Poppy*).—In this genus we have some of the most brilliant of hardy flowers, perfectly hardy, and of the simplest culture. Among the cultivated kinds there are a few good perennials, but the majority are annual and biennial. They range in stature from the tiny alpine Poppy to the stately growth of P. orientale and its varieties. The following is a selection of the best kinds for the garden:—

P. alpinum (*Alpine Poppy*).—This has large and beautiful white flowers, with yellow centres, and with smooth or hairy dissected leaves, cut into fine acute lobes. A native of the higher Alps of Europe, this plant may sometimes be seen in good condition in our gardens, but it is liable to perish as though not a true perennial. It varies a good deal as to colour, there being white, scarlet, and yellow forms in cultivation. The variety albiflorum has white flowers, spotted at the base, while the variety flaviflorum has showy orange flowers, grows 3 in. or 4 in. high, and is hairy. Easily raised from seed. P. pyrenaicum is similar to P. alpinum, but is of taller growth. It also occurs with white, yellow, and orange-red blossoms, which, however, do not always come true from seed.

P. nudicaule (*Iceland Poppy*).—A fine dwarf kind, with deeply lobed and cut leaves, and large rich yellow flowers on naked stems reaching from 12 in. to 15 in. high. It is a native of Siberia and the northern parts of America, and a handsome plant for borders or rockwork; it is easily raised from seed, and forms rich masses of cup-like flowers, but, like other dwarf Poppies, it does not seem to be permanent, and should be raised annually. There are several varieties with flowers white, yellow, and orange-red, one very large and bold and handsome.

P. orientale (*Oriental Poppy*) is the handsomest of all the Poppies, and among the noblest of hardy plants. The type is a fine plant, but the variety bracteatum is much

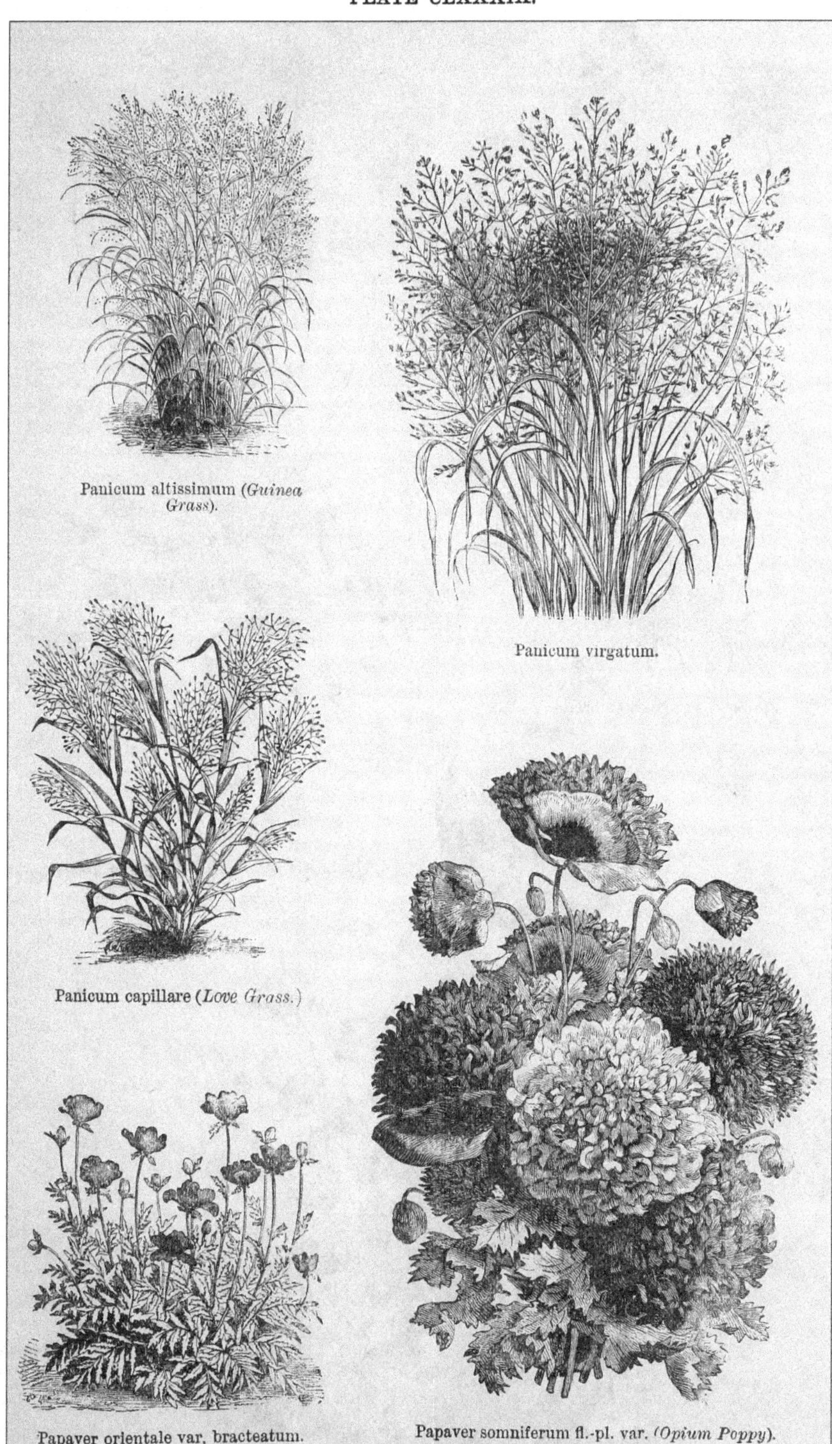

Panicum altissimum (*Guinea Grass*).

Panicum virgatum.

Panicum capillare (*Love Grass.*)

Papaver orientale var. bracteatum.

Papaver somniferum fl.-pl. var. (*Opium Poppy*).

[*To face page* 206.

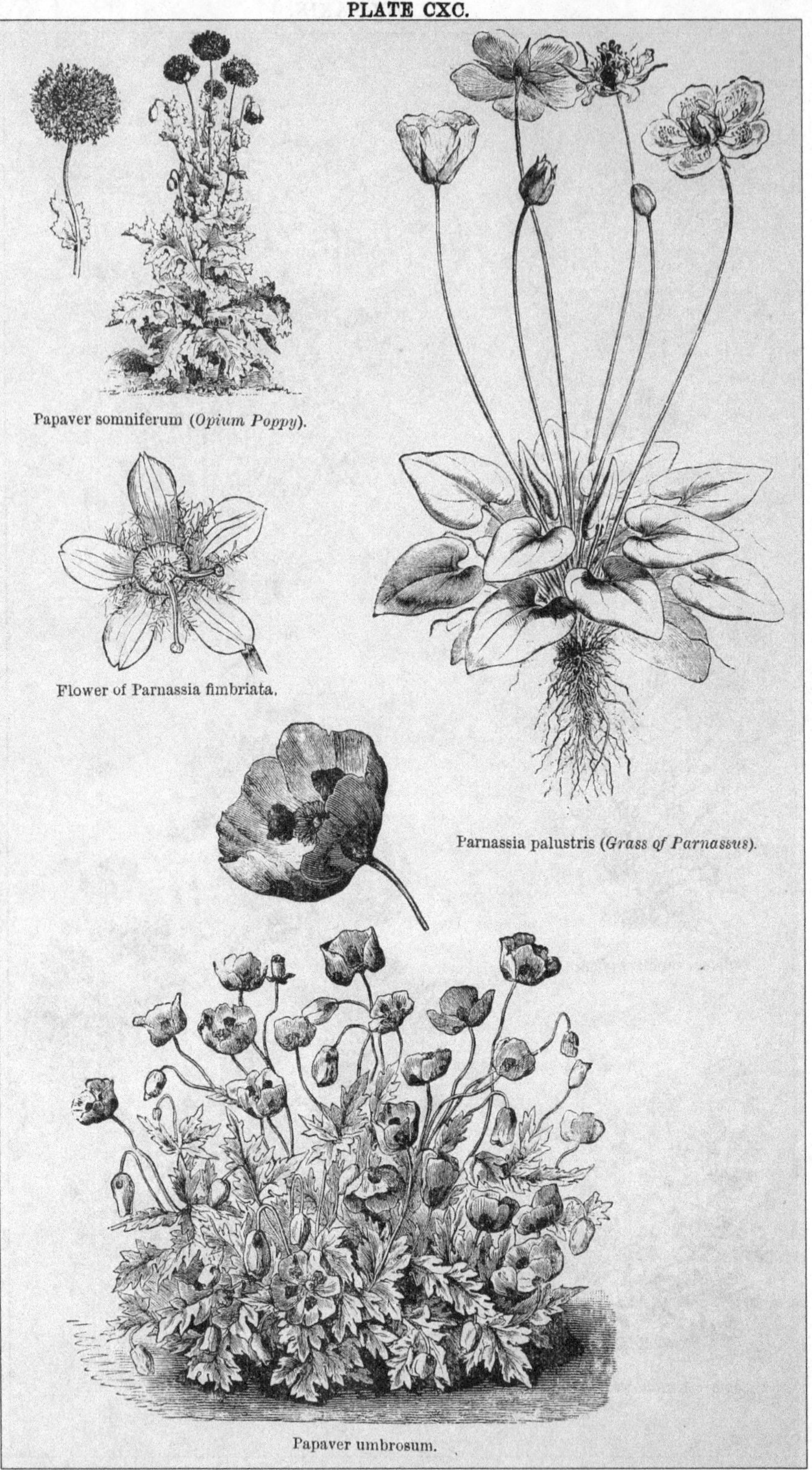

Papaver somniferum (*Opium Poppy*).

Flower of Parnassia fimbriata.

Parnassia palustris (*Grass of Parnassus*).

Papaver umbrosum.

[*To face page* 207.

superior in size and attractiveness. This variety forms huge masses of handsome foliage; the flowers are carried on stiff stalks, with leafy bracts at intervals, and one well-developed bract under each flower, which is 6 in. to 9 in. across, and brilliant scarlet. Each of the four petals has a deep, purple-black spot at its base inside, forming a black cross, and giving it a very distinct appearance. The original P. orientale has naked flower-stalks as a rule. The flowers are pure scarlet, but here again there is a variety, as a good many of them have a black spot. It seems as if the orientale has been crossed by bracteatum, for there are a good many hybrids in gardens. This Poppy's fault is its weaker stalk. It does not hold its large flowers erect like its rival, and is sooner over. There are several varieties of P. orientale besides the distinct bracteatum; for instance, some have no spots at the inner base of the petals, such as concolor; triumphans is of dwarfer habit, and very floriferous. These are all highly ornamental plants, valuable for borders, or for isolated masses on lawns, and also for the open parts of pleasure-grounds, as they flourish in almost any kind of well-drained soil. They prove most effective as isolated plants in the rougher parts of the pleasure-ground. A plant would also show to great advantage in a group of green-leaved subjects like the Ferulas.

P. Rhæas *(Common Corn Poppy)*.—What are known as the Carnation, Picotee, and Ranunculus Poppies are double-flowered forms of the common red Poppy. These varieties possess almost every shade of colour except blue and yellow; some are self-coloured while others are beautifully variegated. They are known, too, as French and German Poppies. Some are dwarfer than others, but all range between 2 ft. and 3 ft. in height. These Poppies, being hardy annuals, can be sown where they are to bloom, but they should be grown in good soil in order to bring out, in the fullest perfection, their size and colour. The seed, being very small, should be sown thinly, and the plants eventually thinned out to 6 in. or 8 in. apart; this enables the lateral shoots to develop themselves, and allows of space for the flowers. In semi-wild places, and by the sides of drives, they can be made conspicuous features; and their cultivation is recommended to lovers of hardy border flowers. Few annuals will afford such a brilliant display as the different kinds of Poppy do in outlying beds and borders during the summer months. Viewed from a distance even their colours are strikingly effective.

P. somniferum *(Opium Poppy)*.—This is a beautiful and most variable Poppy as regards the colour, and is, therefore, a valuable hardy annual. It generally grows about 2½ ft. in height, and varies in colour from white to deep crimson. The double scarlet, the double striped, and the double white are all varieties of this, and the flower-heads, being of great size, make a bold and striking effect when planted in masses. By selection, a type of Poppy called the Pæony-flowered has been obtained from the foregoing; it has large, but very double broad-petalled flowers of various colours, from pure white to dark crimson. This is known in some seed lists as P. pæoniæflorum, and is certainly distinct enough to represent a type. These require the same treatment as the preceding.

P. umbrosum is a strikingly brilliant hardy annual, about 2 ft. high, in habit like the common field Poppy. The colour of the flowers is a dazzling scarlet, with a jet-black blotch on the inner base of each petal, which is sometimes margined with ashy-grey. This black blotch is conspicuous also on the outer face of the petals, thereby rendering masses of the plant a grand sight. It flowers in early summer, and takes up, as it were, and perpetuates the waning glory of Anemone fulgens, which is scarcely more brilliant in colour. It is a native of the Caucasus, and therefore perfectly hardy in this country. Its seeds should be sown in autumn, so that strong plants may be ensured for flowering the following summer. P. arenarium is another showy annual species from the Caucasus. Other handsome Poppies, such as P. spicatum, pilosum, and lateritium, all perennial, and with orange-red blossoms, might be grown where a full collection is desired, and also the annual P. caucasicum and armeniacum. All are of the simplest culture.

Paradisia Liliastrum (=Anthericum).

Paranaphelius uniflorus.—A handsome composite perennial with large yellow flowers, but it is very rarely met with, and little is known of its hardiness or culture.

Pardanthus chinensis.—An Iris-like plant, from 3 ft. to 4 ft. high, bearing orange-coloured blossoms spotted with crimson. These, however, are very fugitive, on which account the plant is scarcely worth growing, particularly as it is not hardy. If grown it should be treated in the same manner as the Ixia and Sparaxis. Iridaceæ. China.

Parietaria.—Weedy plants of the Nettle family.

Paris quadrifolia *(Herb Paris)*.—A little native plant, growing about 1 ft. high, and with yellowish-green flowers. Though not strong, it is an interesting plant for naturalising in places where not plentiful—in quiet, shady, or moist bushy spots.

Parnassia *(Grass of Parnassus)*.—These are singularly interesting and pretty plants for the artificial bog or moist spots in the rock garden. In many of our moist heaths and bogs the Marsh Parnassia (P. palustris) is not unfrequently met with, and a very pretty plant it is—quite handsome enough to cultivate, particularly in gardens in which there is a suitable moist spot where it could grow as in its native haunts. There are, however, three other species, natives of North America, that are quite as showy as our native species, and the newest kind (P. fimbriata) is even more attractive, as it has larger flowers, with peculiar fringe-like appendages to the petals. It has kidney-shaped root leaves, resembling those of P. asarifolia, another hardy species, which grows about 9 in. high and bears white flowers similar, but without the singular fringes to the petals. P. caroliniana is the other species, and this only differs from P. asarifolia in the leaves being oval or heart-shaped; it also flowers about the same time, which is usually from the beginning of July till the end of August. These Parnassias, which are hardy, thrive best in a moist, peaty soil or spongy bog, such as exists by the sides of streams or pools.

Parochetus communis *(Shamrock Pea)*.—A beautiful little creeping perennial with Clover-like leaves. It grows from 2 in. to 3 in. high, and bears in spring attractive Pea-shaped blossoms of a beautiful deep blue. It is of easy culture in the rock garden and choice border, in warm positions, and where tender outdoors it may be grown in a cold frame. Propagated by division or seed. Native of Nepaul. Leguminosæ.

Paronychia.—Small-growing creeping plants of no ornamental value. P. serpyllifolia, on account of its dense turfy growth, might be made use of for clothing any dry bank where little else would thrive, or for covering any bare space in the rock garden.

Parsley Fern *(Allosorus crispus)*.

Partridge Berry *(Mitchella repens)*.

Pascalia glauca.—A Chilian plant of the Compositæ, related to Heliopsis. It is a perennial of erect growth, with glaucous leaves and yellow flower-heads. Thrives in ordinary soil, and is quite hardy.

Paspalum.—An enormous genus of Grasses, chiefly tropical and sub-tropical. P. elegans is a pretty hardy annual, and about the only cultivated kind that thrives in the open air and is worth attention.

Pasque Flower *(Anemone Pulsatilla)*.

Passerina nivalis *(Sparrow-wort)*.—An interesting dwarf alpine plant, nearly allied to the Daphne. It grows to about 1 ft. in height, and bears Mezereum-like blossoms. It is found at very high elevations on the Pyrenees.

Patersonia.—A genus of the Iridaceæ, of nature similar to the Libertia; but few, if any, of the species now exist in gardens. They are of doubtful merit and hardiness.

Patrinia.—A small genus of the Valerian family, not much known in cultivation. P. rupestris may be best described as a dwarf yellow Valerian. It succeeds in light soils, but must be preserved from excessive wet in winter. Siberia.

Peacock Iris *(Vieusseuxia glaucopis)*.

Pectis angustifolia.—A pretty dwarf annual composite from New Mexico, where it forms dense golden cushions 6 in. to 10 in. high. As there are many similar plants much hardier, this is scarcely worth attention except in botanical collections.

Pedicularis *(Lousewort)*.—Plants of the Figwort family; mostly parasitical and difficult to cultivate.

Peganum Harmala.—A South European plant allied to the Rue.

Pelargonium.—The numerous species of this genus are all, or nearly all, indigenous to the southern hemisphere, or have originated as hybrid or cross-bred varieties produced in this or other European countries. They are very often erroneously termed Geraniums—from which family of plants, although allied, they are totally distinct, Geraniums being chiefly indigenous to the northern half of the globe, some of them to England, and all of them hardy herbaceous plants. The genus Pelargonium contains many species, which botanists have divided into several sections. Probably all of them will grow and flower in the open air of this country during the summer months, without, however, being so hardy as to brave an ordinary British winter.

Of all varieties suited for outdoor decoration, the "zonals" are the most popular and the most useful. They are supposed to be descended from two distinct species, viz., Pelargonium zonale and P. inquinans. To the well-known modern Zonal Pelargonium, in its ever-increasing and improving varieties, allusion will in the first instance be made. As bedding-out plants they are certainly of the greatest importance; and, indeed, without them the bedding-out system, as it is called, would possibly never have become fashionable. The facility with which hybrid and improved varieties are produced among this class of plants has resulted in the introduction of numerous beautiful varieties, with flowers of nearly all shades of colour, from the purest white to the most intense scarlet; the richest purple and violet shades are also to be found, and these last almost appear to foretell the advent of even a blue-flowered Zonal Pelargonium. Great improvement has of late years been effected in many families of decorative plants, but in no instance has this advance been more marked and decided than in the case of Zonal Pelargoniums, which, in addition to their merits as decorative plants during the summer, either under glass or out-of-doors, are also found, with proper treatment, to flower throughout the entire winter. But it is as an out-of-doors plant that we are at present considering the Zonal Pelargonium, which, on account of the profusion and brilliancy of its bloom, has sometimes been too freely used in the flower garden and elsewhere, But these matters are now, generally speaking, better understood; and by the judicious use of plants with flowers of more subdued tints in connection with bright bloom, as well as the use of plants with highly ornamental foliage, anything like a glaring effect is generally avoided. The Pelargonium itself supplies this desideratum in an eminent degree, as among the zonal varieties there are many with variegated and exceedingly beautiful foliage. This is particularly the case in that section of the zonals known as the Tricolors, which, on account of their not succeeding so well in

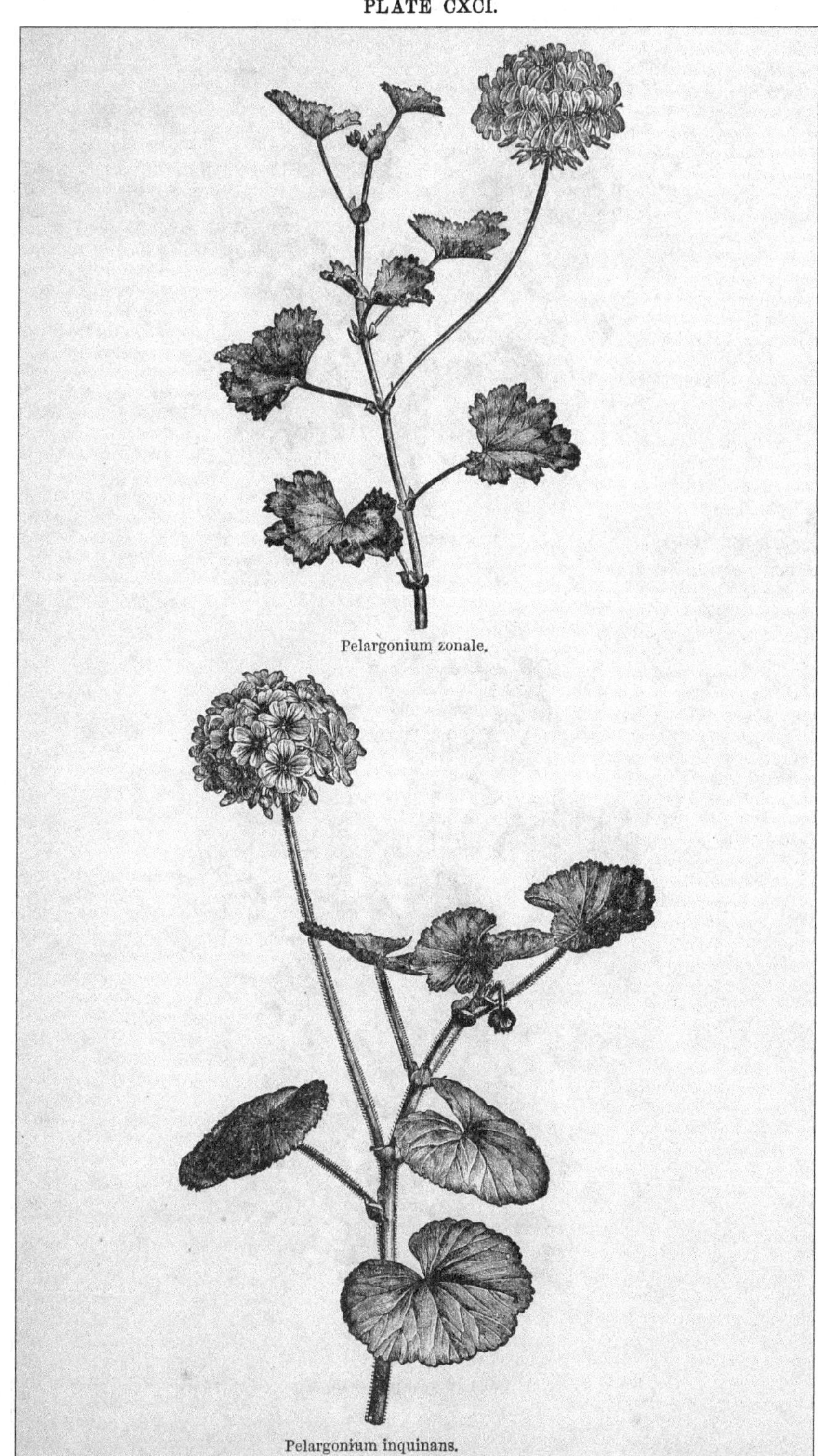

Pelargonium zonale.

Pelargonium inquinans.

[*To face page* 208.

Pelargonium zonale.

Group of zonal Pelargoniums.

[*To face page* 209.

some localities and in some kinds of soil as could be desired, are possibly less grown as outdoor plants than they deserve to be. Where they succeed, however—and this they will generally do with proper attention—they form very beautiful beds in the flower garden, more particularly when they are grown for the sake of their foliage alone, having the trusses of bloom removed from time to time as they appear. The varieties known as the "bronze zonals" are also very beautiful outdoor plants, free exposure to the open air adding greatly to the intensity of the rich tints of their coloured leaves.

There is possibly no family of decorative plants more easily increased than the different varieties of the Zonal Pelargonium. Cuttings may be inserted at any season when such can be obtained, and they will strike root freely in any ordinary light or sandy soil. This may be done in the open air during the summer and autumn, and of course under glass during winter and spring. The principal stock of these plants, however, is generally struck as cuttings inserted in pots, pans, or boxes in the open air during the early autumnal months, and such cuttings are generally considered to make the best plants. During the winter, the protection of a glass roof, and the exclusion of frost from under the same, are required. The Zonal Pelargoniums may also be successfully treated as annuals, and will be in bloom in even less than ten months from the time of sowing the seed. To secure this, the seed should be sown as soon as it is ripe, say during the month of August. Sufficient seed can always be had by retaining a few plants specially for the purpose in front of a greenhouse, or in any other light, airy, and yet warm situation. If the obtaining of new and improved varieties is one of the objects in view, recourse must be had to artificial fertilisation. But if the plants are merely required for the purpose of planting out or massing in the parterre, this need not be attempted; and should the plants used as seed-producers be all of one sort, the progeny may reasonably be expected to prove tolerably true, *i.e.*, the same variety as the parent plants. The seeds should be sown in pots of convenient dimensions, say 6 in. in diameter, filled to within ½ in. of the rims with light, turfy soil; they may be placed tolerably thick, gently pressed into the soil, and slightly covered with the same. Water with a fine rose, cover the surface of the pots with a piece of glass, and place them in a temperature of 65° more or less; the seeds will soon vegetate, when the piece of glass should be removed, and the seedling plants, when large enough, pricked into seed-pans and kept near to the glass in a reduced temperature during the winter. Early in the following March the plants should be potted singly into 3-in. pots and placed in a cold pit, or a similar structure where frost is excluded, until the time arrives when they may be safely planted in the flower garden or where they may be required. Such plants may, during the early part of the season, bloom less profusely than those propagated by cuttings; but their neat and compact habit, with their bright and healthy foliage, will amply compensate for any little paucity of bloom.

Many sorts remarkable for the beauty of their blooms, as regards form and brilliant tints, are, nevertheless, on account of the habit of growth of the plants themselves, unsuited for the purpose of bedding out; and few, if any, of the many beautiful varieties of zonals producing double flowers can be recommended as bedding plants or for outdoor culture unless it be in the form of standards, with clean stems, some 2½ ft. or 3 ft. in length. In this form they are in some situations very effective. When grown in this form, the plants are compelled to draw their sustenance through a single stem of comparatively small diameter, and this has the effect of inducing an exceedingly free-flowering habit in almost all the varieties. Of course it is necessary to assist the slender stems by neat, but strong stakes to enable them to support the heavy heads, the principal branches of which should be secured to strong circular wires or hoops. Single-flowered varieties may also with equal facility be formed into standards, and in their case seedling plants are likely to form specimens sooner than propagated plants. Standard plants of whatever kind should, in the autumn or before frost sets in, be well cut back, taken up, and carefully repotted in pots not larger than may be required to contain the roots, and after being staked should be placed for a time in a temperature not under 60° until they become fairly established. Treated thus annually, such plants are to be found in perfect health, even when not less than twelve or fourteen years old.

Next in importance to the zonals for outdoor culture are the Ivy-leaved species, or the several varieties of Pelargonium lateripes, more particularly now that a union has been effected between them and the zonal varieties, the result of which has been many highly interesting and beautiful sorts, all of which are useful for outdoor culture. As to treatment, suitable soil, method of propagation, &c., they may be said to be in nearly all respects identical with the zonal varieties, with the possible exception of being somewhat more tender, and requiring in a slight degree more warmth during winter.

Among the various other sections of the genus Pelargonium there are but few varieties found suitable for outdoor culture, or as useful bedding plants, their tendency being when planted out in the open air, in even light or poor soil, to become too luxuriant, and consequently to produce only a minimum of flowers. Each section of the family will now, however, be considered separately, and the various varieties, in each section, that have been found to succeed well when planted out will be mentioned. In such a section as that of the zonals, however, which contains so very many varieties, and from which older varieties are being annually discarded, to be replaced by others that are not always improvements upon them, it will be necessary to give only a short list of what are really known to be desirable sorts. At the same time it must be borne in mind that there are sorts found to succeed admirably in certain soils and situations which are by no means successful in others.

The following are a few of the many varieties of Zonal Pelargoniums suited for outdoor culture, or for massing in the flower garden or elsewhere: Anna Pfitzer, Ball of Fire, Corsair, Culford Rose, Distinction, Dr. Orton, Fire King, Harry Hieover, Henry Jacoby, Havelock, Jenny Dodds, John Gibbons, King of the Bedders, Master Christine, Mrs. Lancaster, Mrs. Turner, Mrs. Miles, Mulberry, Newland's Mary, Vanessa, Vesuvius and its salmon-coloured variety, Violet Hill Nosegay, White Perfection, White Princess, and White Vesuvius.

The bronze zonal varieties are as well adapted for bedding out as the green-leaved kinds, the constitutional vigour of the plants being in all respects equal to that of the latter. Their flowers vary in colour. The bright golden ground colour and rich leaf zones of some of them, however, show to greater advantage when the blooms are removed. The following are a few of those that may be considered the best bedders: Black Douglas, Bronze Beauty, Bronze Queen, Crown Prince, Gilt with Gold, Golden Harry Hieover.

There are also some useful bedding varieties with yellow zoneless leaves, such as Crystal Palace Gem, Golden Christine, and Robert Fish; while Happy Thought is a singular variety, each leaf having a large disc or centre of a creamy white colour, while the margins are green. This variety is inclined to grow rather too robust when planted out in rich soils.

As has already been said, the variegated zonals, or golden tricolors, are not found to succeed equally well in all kinds of soil as bedding plants. But with ordinary care the following varieties will generally be found to give satisfaction: Mrs. Pollock, Sophia Cussack, Sophia Dumaresque, Beautiful Star, Victoria Regina, Edward Richard Benyon, Macbeth, Lady Cullum, Peter Grieve, William Sandy, Prince of Wales, and Howarth Ashton.

The great drawback as regards the silver tricolor sorts, when planted out in the open air, is the circumstance of the central or green portion of the leaves expanding faster than the white or coloured margins of the same; consequently the centres of the leaves become somewhat puckered. The following are among the best of them for this purpose: Italia Unita, Lass o' Gowrie, Eva Fish, Maxwell Masters, Lady Dorothy Neville, and Miss Farren.

Among silver-margined zoneless sorts, Mangle's Variegated, although a very old variety, is still found to be useful, together with Silver Chain, Flower of Spring, Mrs. J. C. Mappin, Princess Alexandra, and Waltham Bride, the three last named having pure white flowers.

Of the Ivy-leaved sorts, together with their hybrid varieties, those named are useful as bedding plants: Album grandiflorum, Duke of Edinburgh, l'Elégante, Bridal Wreath, Willsi roseum, Dolly Varden, and Emperor.

Comparatively few of the Cape species of the Pelargonium, or their hybrid varieties, are of much use as bedding plants; a few of them, however, are sometimes used with pretty good effect. Of these are, Diadematum, Lady Mary Fox, Lady Plymouth, Pretty Polly, Prince of Orange, Rollisson's Unique, Crimson Unique, &c. Most of the sweet-scented sorts, when planted out-of-doors during the summer months, succeed admirably, and furnish abundance of fine, healthy, fragrant, flowering shoots, to arrange in connection with other cut flowers in glasses, vases, &c., as may be required.—P. G.

Pellæa atropurpurea.—An elegant dwarf North American Fern, quite hardy in sheltered localities if planted in a shady peat border that is well drained in winter. Some of the other North American species might prove hardy if tried.

Peltandra virginica is a handsome-leaved plant of the Arum family, native of North America. The leaves are broad and of a deep green, and when the plant is well grown it presents a handsome appearance. It grows from 2 ft. to 3 ft. high, is perfectly hardy, and is excellent for planting in shallow pools or lakes like the Callas. (= Caladium virginicum.)

Peltaria alliacea.—A Cruciferous plant growing from 1 ft. to 2 ft. high, and producing in spring numerous small white blossoms. It has a strong odour of Garlic, and is suited for botanical collections only.

Pennisetum longistylum.—One of the most elegant of Grasses. It grows from 1 ft. to 1½ ft. high, the flower-spikes being borne on slender stems. The spikes are from 4 in. to 6 in. long, have a singularly twisted appearance, and are enveloped in a feathery down of a purplish colour. It is admirably adapted for cutting purposes, as it lasts a long time in perfection. Unlike most of the ornamental Grasses, it is of perennial duration and perfectly hardy. P. fimbriatum is a similar species and equally desirable.

Penthorum sedoides (*Virginian Stone-crop*).—A Sedum-like plant from N. America, valuable for the botanic garden only.

Pentstemon.—Amongst genera peculiar to the northern portion of the New World, to which our herbaceous borders are indebted for their gaiety, there is, perhaps, none more important than the Pentstemon. The North American continent, amid the rocky ravines whence the mighty rivers of that part of the world derive their origin, is the home of the Pentstemons. Varied in colour, profusely floriferous, and possessing a graceful habit of growth, Pentstemons have a value for the decoration of our flower beds, rock gardens, and shrubberies that few other plants possess, the more so as their blooming season may be said to be distributed over a period of at least five months, commencing with the charming blue P. procerus in June, and finishing off with the endless varieties of P. Hartwegi, in all shades of rose, scarlet, and crimson, whose beauty holds its own even in the dull, dreary month of November long after the more fragile denizens of the flower garden have become things of the past. Within the past few years much has been done also in regard to the improvement of the Pentstemon by judicious selection of seminal varieties of the old P. Hartwegi and P. gentianoides, which, though they possess a wide variation in colour, lack the beautiful clear blue which we find in some of the species, and therefore have a somewhat monotonous effect. These garden varieties, or so-called hybrids, may be ranged under two series of colours—those from P. Hartwegi belonging to the red-flowered set, and those from P. gentianoides to the purple-flowered class; and, strange to remark, these series of colours will be found to run strictly parallel with other structural peculiarities. As regards culture, the species have unfortunately gained the reputation of being somewhat difficult to manage, as some, especially those of the shrubby section, show a remarkable tendency to suddenly die away when they are apparently in robust health. In order to ensure ordinary success, thorough drainage of the spot in which they are planted is a primary consideration, as it is a well-known fact that these, as well as a host of other Californian plants, suffer more from excessive moisture at the roots than from the coldness of our winters. The soil best suited for all Pentstemons consists of good friable loam with an admixture of well-decayed leaf mould and sharp sand. It is specially advisable in the case of Pentstemons to have, in addition to the border specimens, a few plants in cold frames, in order to be able to fill vacancies, should they occur. They may be propagated either by means of cuttings or seeds. The former mode applies chiefly to the shrubby kinds, which strike freely in spring; and in favourable seasons seeds are freely produced from those from which it is not practicable to obtain cuttings. Any attempt to multiply them by division of the tufts will be found to result for the most part in the entire loss of the plants. P. barbatus and P. procerus, however, endure this mode of propagation. Seed should be sown in February or March on a gentle hotbed under a frame, in seed-pans well drained with broken plaster and filled with a compost of peat soil and sand. In April the seedlings should be pricked out under a frame, and these, planted out in May, will, as a rule, usually come into flower by autumn of the same year. Another mode is to sow in May or June in the open air, in ground enriched with leaf-mould. The seed beds should be covered with chopped Moss to preserve a uniform coolness and humidity. In August the seedlings should be potted and removed to a greenhouse or conservatory for the winter. It is necessary to observe that the seed sown at either of these seasons frequently does not germinate until the following year.

The following is a descriptive list of some of the best cultivated species that are obtainable. Many are excluded, however; some on account of their rarity, and others, such as P. antirrhinioides, cordifolius, and Lobbianus, because they are not sufficiently hardy for border culture, though they succeed well enough if planted against a warm wall.

P. azureus is a well known and very effective dwarf kind, with numerous erect branches, bearing many blossoms in whorls. The blossoms, which are rather large, are of a clear violet-blue colour, begin to show towards the end of summer, and last a long time. From the Sacramento Valley, California.

P. barbatus.—A handsome and well-known plant, commonly known as Chelone barbata. It possesses, like P. procerus, a peculiarity of developing underground stems all round the parent, thus forming a readily divisible tuft or patch, from which arise several graceful stems, almost devoid of foliage, to a height of 3 ft. or more. The flowers are numerous, supported on pedicels of varying length, springing from the axils of the bracts; they are of a rosy scarlet, diminishing in brightness in the lower part of the tube. Were its flower-stems capable of standing without a stake, its value would be much enhanced, but it is well worthy of any little trouble that may be required in this process. The flowers are produced in succession, hence its beauty is retained for a considerable time. There is also a white variety of it. The variety Torreyi is a very fine, robust plant, differing chiefly in its greater height and stronger growth, as well as in the general absence of the beard on the lower lip of the flower, which characterises the species. Being a native of Colorado and Northern Mexico, it has a hardier constitution than the typical species, and is a very showy border perennial of easy culture.

P. campanulatus is an old inhabitant of our garden borders. It is a slender-growing plant, attaining a height of about 18 in., branching freely, and in the southern counties acquiring an almost shrubby character; the flowers are rose-coloured, and produced in one-sided racemes. It continues blooming for a long period, as the primary branches are succeeded by the secondary ones. It is a native of Mexico. P. pulchellus is a variety of this species.

P. Cobæa is one of the handsomest of the yet introduced kinds. It grows about 2½ ft. high, and has flowers nearly 2 in. in length, of a pale purple, distinctly pencilled with red streaks and delicately suffused with yellow, the base of the tube being of a creamy-white. They are freely produced in long, leafy racemes late in autumn. It thrives vigorously and generally quite unprotected, but it is very difficult to propagate in quantity. The name does service in trade lists for several spurious kinds, but it can be readily recognised when in flower by the above brief description. Native of the interior of Texas.

P. cyananthus.—This is a distinct and lovely kind, of erect growth, attaining 3 ft. to 4 ft. in height. The flowers appear in clusters on dense spikes, 1 ft. or more in length, composed of short stout stalks, and are of a bright azure-blue colour. It generally flowers during May and June in favourable seasons. Native of the Rocky Mountains. The newly-introduced variety Brandegei is a decided improvement on the type, as it is of a more robust habit, and its flowers are brighter coloured.

P. diffusus is a beautiful semi-shrubby kind, from 2 ft. to 4 ft. high, with freely-produced flowers, of a violet-purple colour, that form a large, loose, many-branched head. It flowers throughout the greater part of summer and autumn. Its near relative, P. Richardsoni, much resembles it, but is inferior in point of beauty, and P. Mackayanus and P. argutus are much the same. Though perfectly hardy, and capable of enduring the vicissitudes of climate and the extreme cold with which, in its native locality, it is associated, we find that it is liable to succumb to the damp of our early winters; or, should it survive this, to the keen incisive character of north-eastern blasts with which our climate is only too familiar, and which scorch the early spring growth as though it had been subjected to the lee-side of a blast furnace. Like all the Pentstemons, it is readily increased by means of cuttings, and possibly would come true from seed, but in this country the latter are rarely matured. Native of the rocky ravines of the tributaries of the Columbia River.

P. Digitalis is a large-leaved free-growing kind of erect habit, but is not very showy. The same remark applies to P. pubescens, lævigatus, perfoliatus, and glandulosus.

P. Fendleri.—This is a very pretty and distinct species, very glaucous, with a long, erect, one-sided raceme of flowers of a very pleasing light purple colour. In height it rarely exceeds 12 in. to 15 in. It is quite hardy in ordinary soils, and is one of the most distinct species in cultivation. P. Wrighti is a plant of a similar character with magenta-tinted blossoms, and the variety angustifolius is likewise a pretty plant. Both are worthy of culture.

P. Hartwegi is more generally known as P. gentianoides. Not only was the old species valuable as one of our very best autumn-flowering border plants, but in its progeny, called into existence by the skill of the florist, we have not only endless variety of colour, but increased size of bloom, the old, narrow, tubular flower acquiring the dimensions almost of a Foxglove. It was recorded as found by Humboldt and Bonpland, growing on lofty mountains in Mexico, at an altitude of nearly 11,000 ft., about the beginning of this century, but was not introduced into cultivation till 1828.

P. heterophyllus.—A handsome sub-shrubby kind of dwarf habit. It has showy flowers produced singly or in pairs from the axils of the upper leaves. They are of a pink-lilac colour in the type, but seminal varieties are very liable to vary. It is hardier than many species but often succumbs to severe winters; therefore reserve plants should be secured. Sacramento Valley, California.

P. humilis is a very distinct alpine species, of dwarf stature, rarely exceeding 6 in. or 8 in. in height. It forms compact tufts, and is remarkably free-flowering, and the blossoms, which are large for the size of the plant, present a very attractive appearance, on account of their pleasing blue colour, diffused with a reddish-purple hue. It should be planted in the rock garden in the most select and fully exposed spot. A compost of gritty loam and well-decomposed leaf-mould should be used, and during summer it should be copiously supplied with water. It blooms in the early part of June, and is a native of the Rocky Mountains, abundant about Pike's Peak.

P. hybridus—The lovely race of plants now in cultivation, which are among the most precious of all flowers, are generally supposed to have descended from P. gentianoides, a species that has disappeared from our garden, but there is no doubt that the majority have been derived from the pretty P. Hartwegi, a tolerably common plant in its simple state in most botanical collections. P. Cobæa, too, has probably been employed in hybridising, for some of the varieties bear a strong resemblance to that species. Whatever their correct parentage may be, they are a beautiful race of plants, and much should be made of them in the garden, as they are very valuable in autumn and carry their beauty into winter.

The varieties of Pentstemon generally succeed in any good soil, but in a good loam enriched with manure and leaf-soil they are certain to do well. They can be planted out singly or in groups in the mixed border or in beds, in which the various colours become charmingly blended. Amongst them we have a wonderful wealth of colour varying from pure white to glowing scarlet, the intermediate shades consisting of pink, rose, purple, carmine, and purplish-lilac. If good plants of Pentstemons be put out by the end of April, they will commence to bloom about the middle of June, and will continue to yield a succession of flowers until winter sets in. They require a deep, moderately rich soil, and a position fully exposed to the sun. They are increased both by means of cuttings and seeds. The former method must be resorted to in order to increase any particular variety.

PLATE CXCIII.

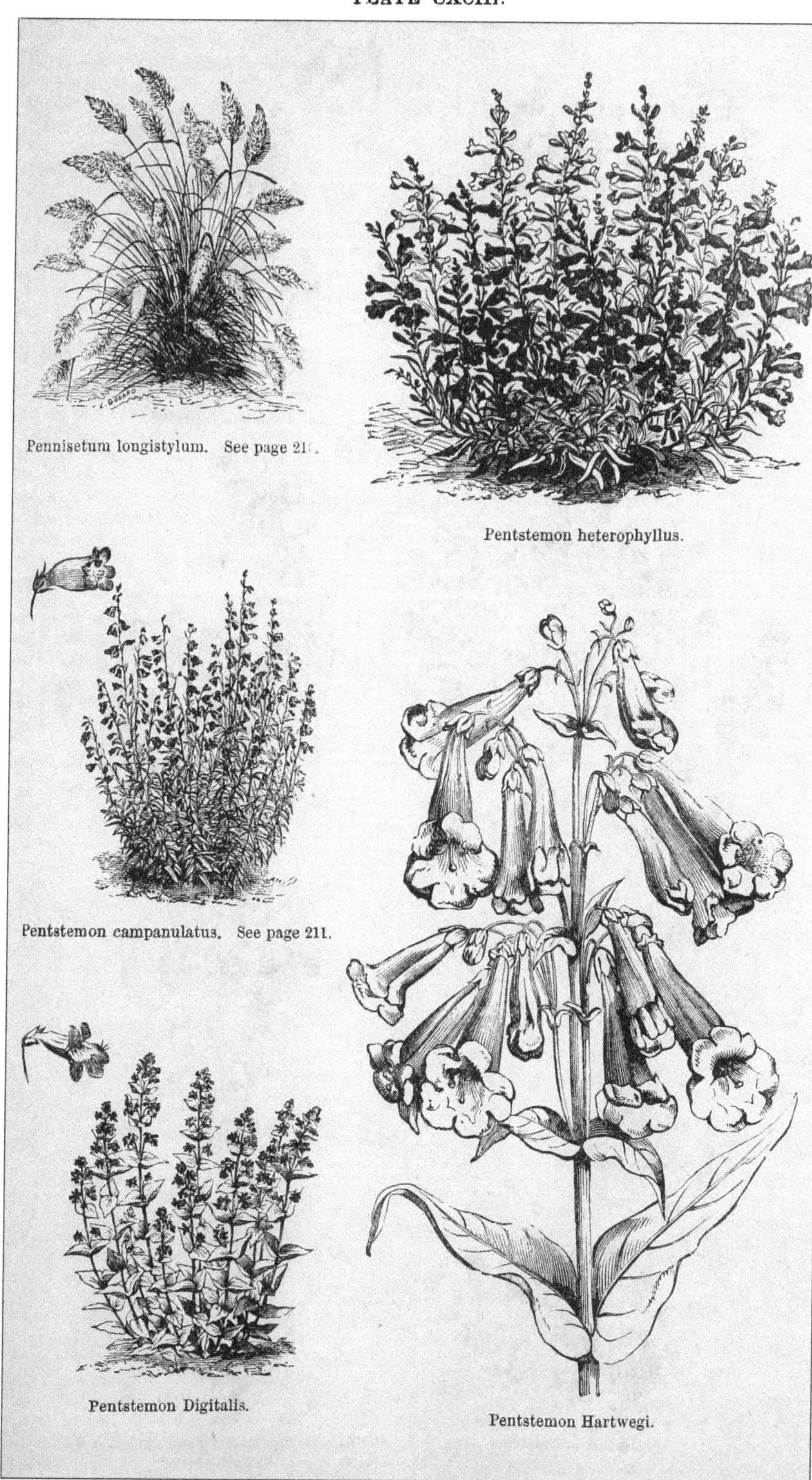

Pennisetum longistylum. See page 21[illegible].

Pentstemon heterophyllus.

Pentstemon campanulatus. See page 211.

Pentstemon Digitalis.

Pentstemon Hartwegi.

PLATE CXCIV.

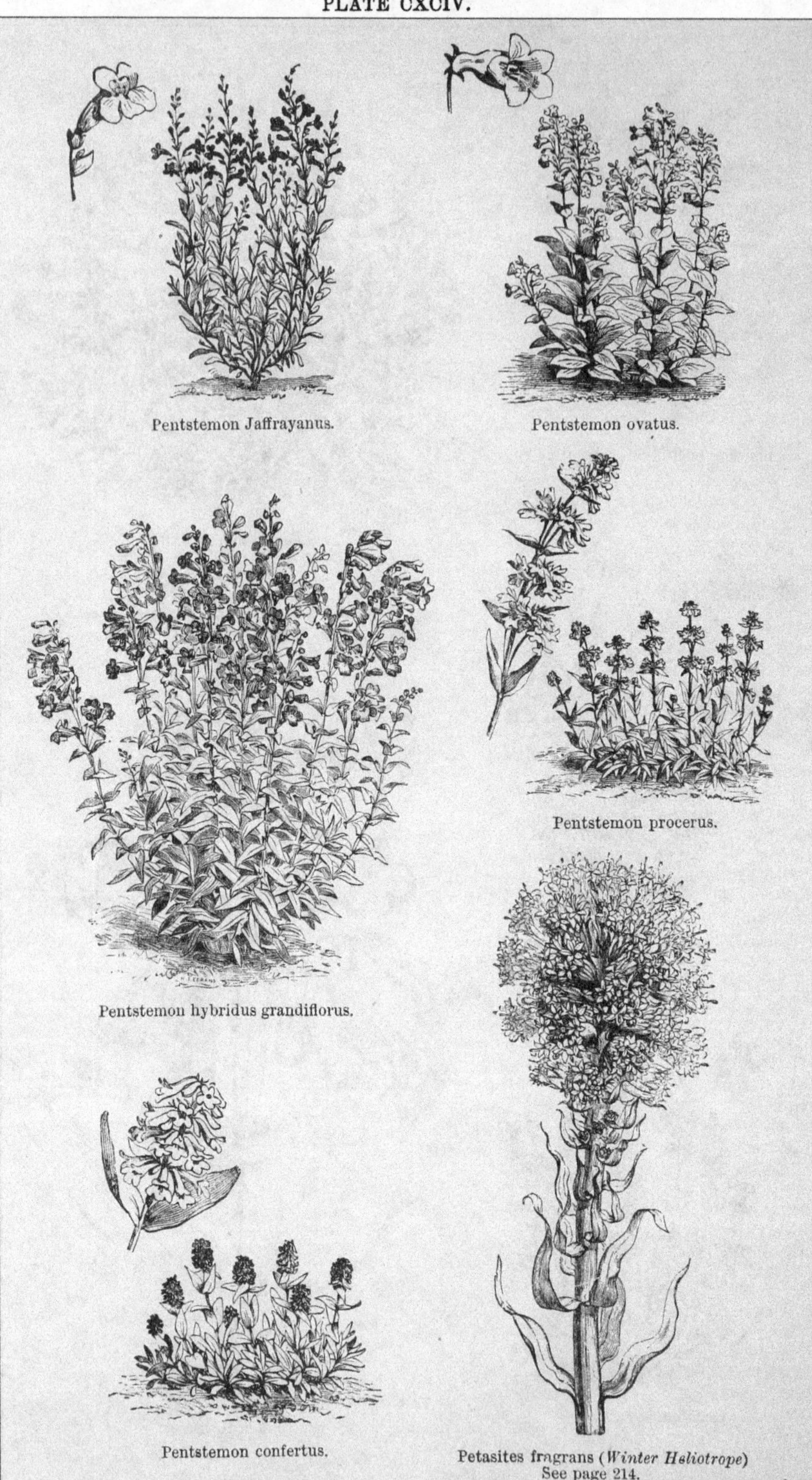

Pentstemon Jaffrayanus.

Pentstemon ovatus.

Pentstemon hybridus grandiflorus.

Pentstemon procerus.

Pentstemon confertus.

Petasites fragrans (*Winter Heliotrope*)
See page 214.

[*To face page* **213.**

Cuttings should be taken in August or early in September from the young growths thrown up round the main stem, and they should be put into a prepared sandy bed, on a shady border, under a hand-glass, or in boxes or pots placed in a cold frame. They root readily, and those in boxes or pots might be wintered in this way, and not transplanted till spring. Those struck on the border should be lifted and potted, or planted out in a cold frame for the winter, or transplanted to the open ground in a well prepared bed, and protected with a little litter or branches of evergreens during severe weather. Under general circumstances the young plants should not be planted out till the March or April following. When it is desired to increase the stock of any one or more varieties as rapidly as possible, the store pots of cuttings rooted in autumn should be taken into a gentle bottom-heat in spring, and induced to grow; and if the young growths be taken off as soon as they are 2 in. in length, and put in pans of sandy soil in the same temperature, they will quickly strike, and by May and June, if properly treated, will have grown into healthy plants. Seed of the best varieties only should be sown. The Pentstemon is a very free seeder, and there is no difficulty in obtaining some. In saving seed for sowing, only the very finest varieties should be selected for the purpose, and those showing novelty of character—for variation is always a most acceptable characteristic—and such flowers can scarcely fail to yield something well worthy of cultivation. The seed should be sown in February or early in March in a gentle heat. It will quickly germinate, and when the plants are large enough to handle they should be pricked off into shallow boxes, and, after a time, hardened off in a cold frame. Here the plants can remain till the end of May or later, according to their size, and then be planted out in well prepared beds. A generous soil will serve to bring out as fully as possible the quality of the seedling flowers. When they flower, which they will do by August and September, any varieties of extra good quality should be marked for propagation by cuttings or for seeding from, while the inferior ones will do for the mixed border. If the bed of seedlings be allowed to stand for another season (and it is always a good plan to do this), the seed-stalks should be cut away as soon as ripe, the bed cleaned, top dressed with leaf-soil and short manure in spring, and it will yield a plentiful harvest of flowers the following summer. During winter seedlings should be protected by a cold frame, and planted out early in April in a deep soil and in a sunny situation. The varieties change so frequently and so many are raised from time to time that we do not give the names of the varieties supposed to have most merit at present, but leave the selection to individual taste, assisted by the catalogues of the best growers.

P. Jaffrayanus is a remarkably showy kind, and the best of the blue-flowered class. The glaucous foliage contrasts finely with the bright, clear blue blossoms, which are borne in profusion during the greater part of the summer. It forms a handsome border plant of dwarf stature, and not being what may be called a good persistent perennial, should always be supplemented by a young stock of seedlings, which will bloom much more vigorously than old plants. Strong sunshine or wet injures the appearance of the flowers, and this is the only drawback to its cultivation. Native of North California.

P. lætus is a close ally of P. azureus and P. heterophyllus, and, like them, is of dwarf branching habit, with blue flowers in terminal raceme-like panicles. It grows about 1½ ft. high, and blooms in July and August. It is a native of California, and is probably as hardy as most of the species from that region.

P. Murrayanus.—A noble and distinct plant, and one of the most satisfactory to grow. When well grown it has stems from 4 ft. to 6 ft. high, furnished with tiers of brilliant scarlet flowers, and besides being specially attractive when in flower, it is very handsome in foliage, having broad and very glaucous lower leaves, and large, stem-piercing upper ones. It should be raised from seed annually, and the seedlings grown well for flowering the following summer. There are few plants more worthy of extra care and attention than this beautiful Pentstemon. It is a native of Texas, and loves a warm, sunny soil. P. centranthifolius is similar to P. Murrayanus, but is not such a handsome plant, though it is somewhat easier to grow, and is hardier.

P. ovatus is a fine, vigorous-growing plant, attaining a height of from 3 ft. to 4 ft. The flowers are comparatively small, but produced in densely arranged masses, and exhibit various shades of colour, from intense ultramarine to deep rosy-purple. The brilliancy of colour, rich profusion, and the handsome contour of the plant all unite to give it a special value. It should be considered a biennial, as it usually flowers so vigorously the second year that it makes little or no provision below for the following year's growth. Native of mountains of Columbia, and known also in gardens as P. glaucus.

P. Palmeri.—This handsome species is of robust habit, attaining in good soil a height of from 3 ft. to 5 ft. The flowers are peach-coloured, streaked with red, and are borne in a many-flowered naked panicle, from 18 in. to 2 ft. long; the corolla, which is peach-coloured, is remarkable for its short inflated tube and gaping mouth. It is quite hardy, will succeed in almost any well-drained soil, and flowers about midsummer. P. spectabilis is similar to this species.

P. procerus is a beautiful little plant, and about the hardiest of all the species, as it takes care of itself in any soil. It is of a creeping habit of growth, sending up from the tufted base numerous flowering-stems 6 in. to 12 in. high. The flowers are small, in dense spikes, and, being of a lovely amethyst-blue, it forms a charming plant, adapted alike for the border or rockery. It seeds abundantly, grows freely in any ordinary garden soil, and is the earliest to blossom of all the Pentstemons. P. nitidus and P. micranthus are synonymous with P. procerus, and P. confertus is somewhat similar. The latter has straggling stems, and though by no means showy it is a very distinct species.

P. Scouleri is a semi-shrubby plant of very distinct habit and of small twiggy growth. The flowers, which are of a slaty-bluish purple, are large, and arranged in short terminal racemes; they are not produced in any great abundance, but the whole plant, in its dwarf and compact habit of growth, and in its flowers, possesses charms sufficiently distinct to render it well worthy of general cultivation. It may be readily increased by cuttings in spring of the young shoots, which strike freely in a little bottom-heat, similar to that used for ordinary bedding plants. P. crassifolius bears in many respects a close relationship to the preceding species. It is of a dwarf, shrubby habit of growth; the flowers are produced in short, terminal racemes, and are of a charming light lavender colour. It is rare in cultivation, but is a most desirable plant, and admirably adapted for a dry knoll of the rockery, well exposed to the sun and erected on the basis of a good deep mass of bog soil or peat, from which, while the location of the plant is dry, the roots may find an abundant store. P. Menziesi resembles P. Scouleri, but has reddish-purple flowers. P. speciosus is a remarkably handsome kind, with stems 3 ft. or 4 ft. in height, bearing many flowered clusters of flowers, which are sky-blue, varying to a reddish hue. P. glaber is nearly related to the preceding, but is of dwarfer growth. The flowers, of various shades of purple, are borne in crowded spikes about 1 ft. in length, and are produced early in summer. On account of its dwarfness it is better suited for rockwork decoration than the majority of the other kinds. P. grandiflorus is an extremely handsome kind, allied to the two preceding. It grows about 3 ft. high, and the flowers, which are large and of a beautiful pink colour, are produced from July to September. Another similar species is P. secundiflorus, which is distinct in having its clear blue and violet blossoms in one-sided racemes. It is about 1½ ft. high when well grown. P. acuminatus is likewise a beautiful and similar kind. These all require to be raised from seed annually, and to be planted out the second year.

Perilla nankinensis.—A dark-leaved half-hardy annual, much used in summer bedding. It has dark vinous-purple foliage, and is generally used in lines, but a few plants in a group here and there in the mixed border and sub-tropical garden are effective. The seed should be sown in pans or boxes about the middle of February, in heat. The seedlings should be transplanted into boxes in soil not over-rich. After being gradually hardened off, they should be planted out about the end of May. For those who have not command of artificial heat in spring this is not a very suitable plant, as it requires heat to get it to a suitable size for planting in proper time. It has frequently been used with unrefined taste and with the worst results in bedding out.

Periwinkle (*Vinca*).

Persicary (*Polygonum Persicaria*).

Petalostemon.—Not a very attractive genus. Two species are occasionally met with, P. candidum and P. violaceum, but they are scarcely to be found outside botanic gardens. They are both perennials, with pinnate foliage and small purplish or whitish flowers. Not much is known about their culture.

Petasites fragrans (*Winter Heliotrope*).—A weedy looking plant, 4 in. to 12 in. high. The flowers appear in December and January unless the weather is severe; they are deliciously fragrant, of a pale dingy lilac colour, and are borne in a rather short racemose panicle. The plant is unfit for garden culture, as it runs very much at the root and becomes a perfect weed, but it may be planted in semi-wild places, lanes, and hedges, as it is very useful for bouquets in winter. It may also occupy and carpet, so to say, a small clump of shrubbery, where it can be conveniently gathered, and to which it should be confined. Another species, P. vulgaris (Common Butterbur), is a native herb, 2 ft. to 2½ ft. high, closely allied to the common Coltsfoot, but with great Rhubarb-like foliage. The flowers, which appear in spring before the leaves, are dull pinkish purple. Exotic plants with less effective leaves than this have been lately much used in gardens; it, however, should not be allowed to come nearer to the garden than the margins of some adjacent stream, or in a moist bottom among other large-leaved herbaceous plants. Division.

Petrocallis pyrenaica (*Rock Beauty*).—A small and beautiful alpine plant, forming a dense cushion 2 in. to 3 in. high, and when not in flower resembling a mossy Saxifrage. The flowers are pale lilac, faintly veined, are sweet-scented, and appear in April. The plant is only suited for careful culture on the well-made rockwork, for though perfectly hardy, it is of a fragile nature. It should be planted in sandy fibry loam, in rather level warm spots on rockwork, where it could root freely into the moist soil, and yet be near the congenial influences of the broken rocks and stones, down the buried sides of which it could send its roots. It should always have a sunny position. It may also be grown in pots plunged in sand in the open air and in frames in winter, but it becomes drawn and delicate under glass protection of any kind. Easily increased by careful division, and may also be raised from seed. Alps and Pyrenees.

Petunias.—Of late years these have not been so largely used for summer bedding as formerly, owing doubtless to the great in-

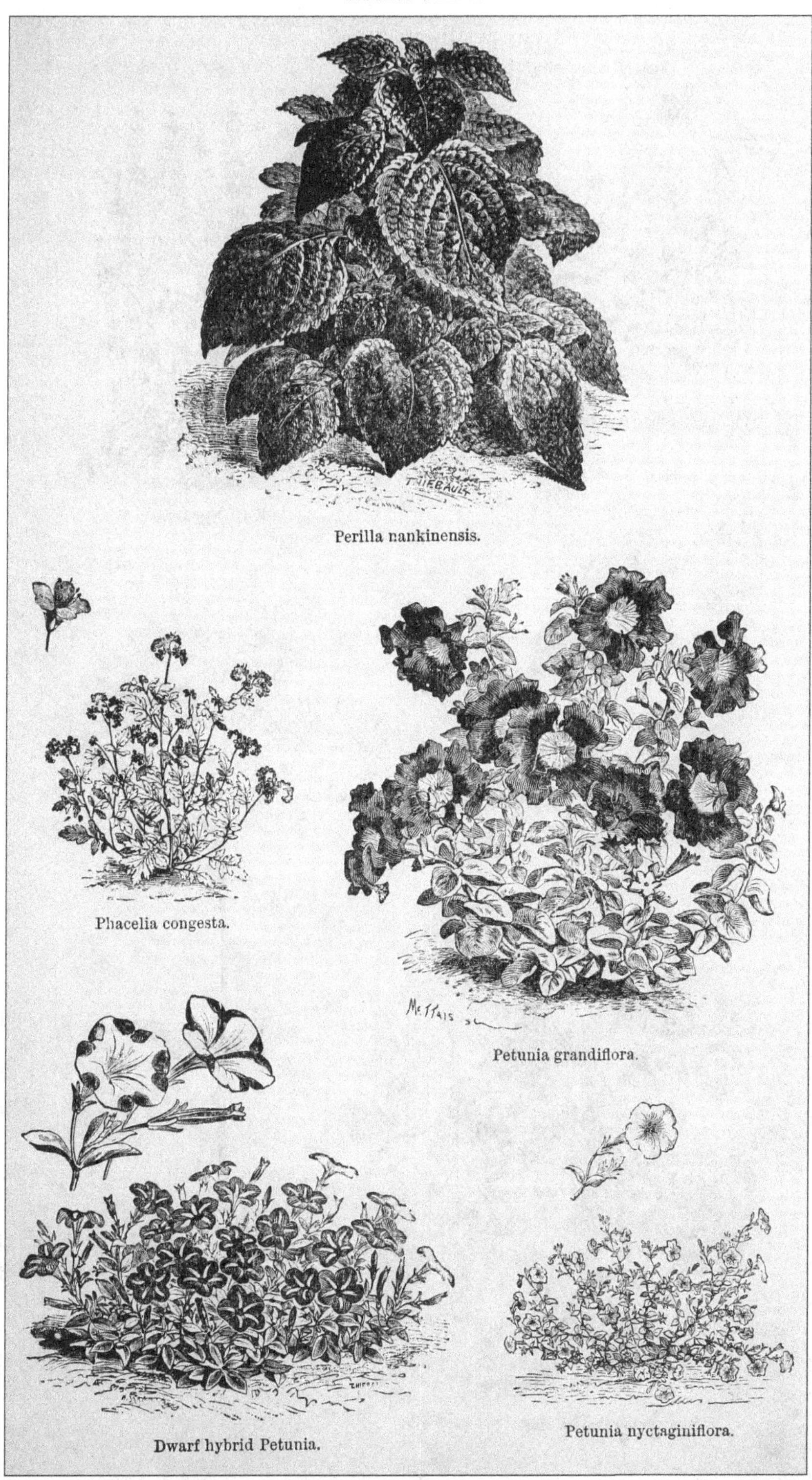

Perilla nankinensis.

Phacelia congesta.

Petunia grandiflora.

Dwarf hybrid Petunia.

Petunia nyctaginiflora.

[*To face page* **214.**

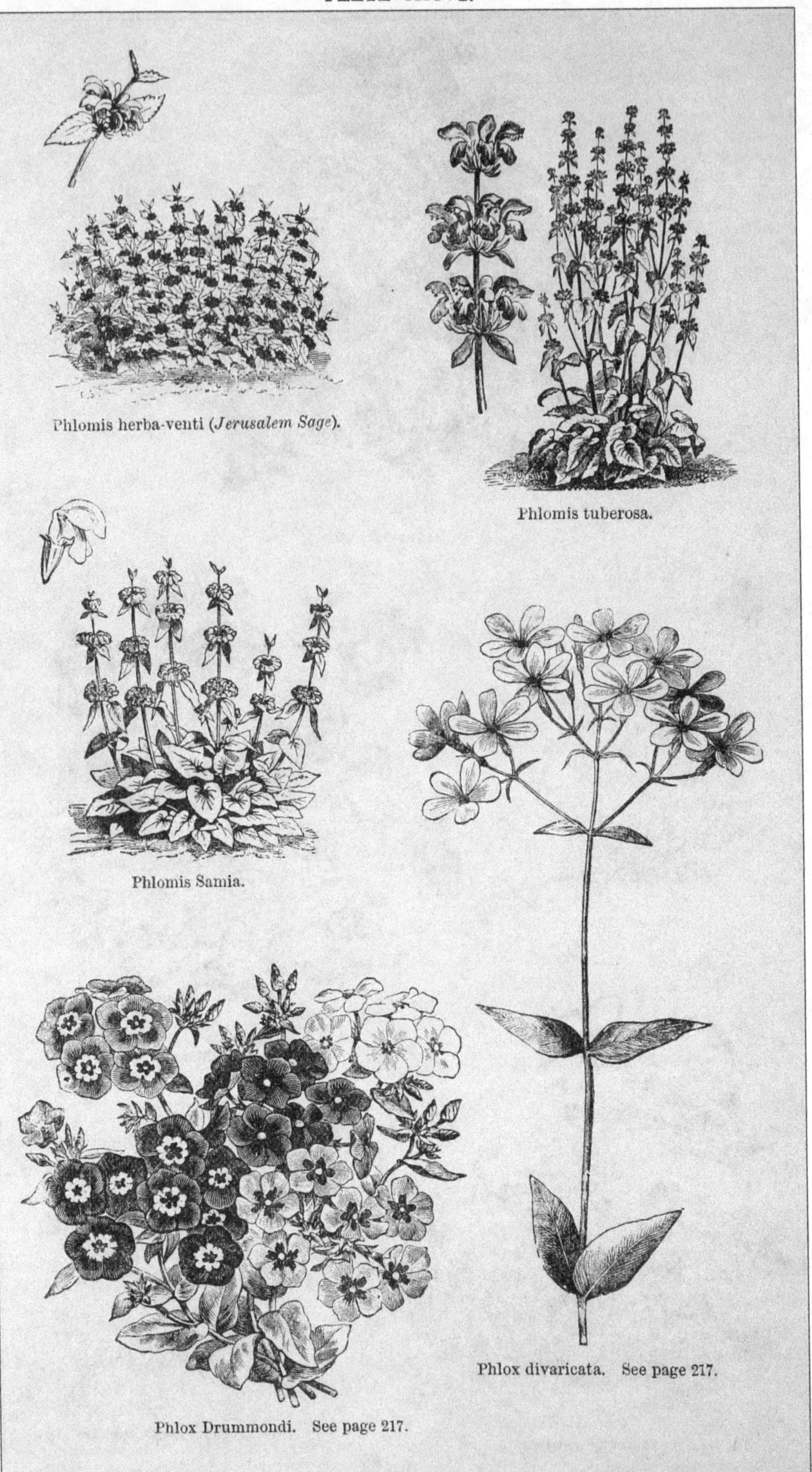

Phlomis herba-venti (*Jerusalem Sage*).

Phlomis tuberosa.

Phlomis Samia.

Phlox divaricata. See page 217.

Phlox Drummondi. See page 217.

crease that has taken place in bedding varieties of Pelargoniums, and to the more general use of fine-foliaged plants. They are, however, so extremely well suited for planting in large vases, in baskets of mixed plants, and also against low trellises, under windows and walls, that their culture is worthy of being continued and extended. In certain positions some of the varieties produce a charming effect when planted in masses; but it is necessary that the spot chosen for them should be open and sunny and the soil deep and rich. In low damp situations they mildew and canker as soon as the first cold nights of autumn set in. The best bedding varieties are Spitfire, dark purplish-crimson; Dr. Hogg, purple, with white throat; Miss Amy, crimson and white; Countess of Ellesmere, rosy crimson, with a lighter throat; and Delicata, white, striped with purple. Seedlings, too, now come so good that they are frequently planted in mixed borders for cutting. If sown in heat in February or March, good plants may be had for putting out at the end of May, and earlier than that it is not safe to plant them. The named kinds must be propagated from cuttings inserted in August; at that season they strike quickly when placed on a bed of leaves or other fermenting material where the temperature ranges from 70° to 75°, and where there is a top heat of 65°. As soon as rooted they should be taken out of the bottom heat and placed in cold frames till frosty nights set in; they should then be removed to an intermediate house and placed on shelves near the glass, there to remain in store pots till spring, when they should be potted off singly and grown on as sturdily as possible till planting-out time. The roots are so brittle that, however well they are rooted, the soil does not adhere to them; and this is a circumstance that renders it necessary to pot singly, for if put in pans or boxes, and transplanted from them to the beds, they suffer greatly, and are a long time in getting re-established.—W. W.

Phaca *(Bastard Vetch)* consists of a few dwarf plants, such as P. astragalina, australis, and frigida, which are now placed under Astragalus.

Phacelia.—A small genus of Californian plants, the cultivated kinds of which are all hardy annuals, but none very desirable. P. congesta is the best, as it is smaller than either P tanacetifolia or circinalis. They have small blue or violet flowers in dense heads. Hydrophyllaceæ.

Phalangium (= Anthericum).

Phalocallis plumbea. — A Mexican bulbous plant of the Iridaceæ. It has slender stems a yard or so high bearing flowers, about 3 in. across, lead coloured tinged with yellow. They are, however, very fugitive, lasting often but a few hours. It is not very hardy, but is an interesting plant to grow with tender bulbs such as Sparaxis and Ixia.

Pharbitis (= Ipomœa).

Phaseolus *(Kidney Bean)*.—A very large genus of the Pea family, of which the common Scarlet Runner (P. multiflorus) and the French Bean (P. vulgaris) are the most familiar examples. The other species that thrive in the open air are only suitable for botanical collections.

Pheasant's-eye *(Adonis autumnalis* and *Narcissus poeticus)*.

Phleum *(Cat's-tail)*.—A large genus of Grasses, several being natives. The variegated-leaved form of P. pratense is pretty and constant. Suitable for edgings in any soil. Division in spring.

Phlomis *(Jerusalem Sage)*.—These rank amongst the finest of hardy plants belonging to the Sage family. There are about a dozen species and varieties of the genus in cultivation, and amongst them great diversity of size and habit. Some, such as P. fruticosa, are shrubs, others noble herbaceous plants, while others again, such as P. armeniaca, are sufficiently alpine in character to allow of their being grown in the rock garden. The most desirable of the species to cultivate is P. fruticosa, a half shrubby plant growing from 3 ft. to 4 ft. high; its branches and leaves are covered with a rusty down, and the flowers are produced in dense whorls, clothing about half the length of the branches. Their colour is a rich yellow, and they are very attractive during June, July, and August. This species is perfectly hardy, thrives well in any common soil, and may be grown either in the mixed border or associated with shrubs, but it should always be placed in an open spot. P. ferruginea is similar, but not so hardy, and, moreover, not such an effective plant as the preceding. There are a few other shrubby kinds, but none of them are so fine as P. fruticosa. Of herbaceous kinds the best is P. Herba-venti, a strong growing plant, which forms an erect spreading mass from 1 ft. to 3 ft. high. Its flowers, which are borne in dense whorls, are a rich purplish violet. P. tuberosa and purpurea, both with purple flowers, are handsome plants when grown well, and, with P. Herba-venti, are excellent subjects for naturalising, as they flourish in any soil or situation. The best herbaceous kinds with yellow flowers are P. Russelliana and P. Samia, both of which grow about 3 ft. high and bear in summer a profusion of flowers in whorls. They are strong growers and do well for naturalising. P. armeniaca, a very dwarf species, has neat silvery leaves and reddish-purple flowers, and is very suitable for the rock garden. P. cashmeriana, a handsome species, has lately reappeared in cultivation. It is herbaceous and somewhat resembles P. Herba-venti, but the flower-heads are denser; the flowers, too, are larger and have a broad violet-purple lip. All the species are easily propagated—the shrubby kinds by cuttings and seed, the herbaceous sorts by division and seed.

Phlox.—The Phlox is one of the most popular genera with which our gardens are enriched from the woods, plains, and mountains of North America. It contains, for the most part, perennials, all of which are showy garden plants. The annual P. Drummondi alone has produced distinct varieties enough to furnish a garden with almost every shade in colour, while the perennial species are very numerous, and present such a variety in habit, that they may for garden purposes be regarded in three distinct groups. One set is properly alpine in habit; of this the beautiful P. subulata, or Moss Pink, is the best known, but there are many others in the Rocky Mountains and westward, some of them more truly alpine, and unknown to cultivators. Next to these are several that may be grouped as running or creeping Phloxes. These are perennial, but their principal stems are prostrate, though their flowering stems are erect. Lastly, there are the well-known tall garden Phloxes, which are generally called the perennial Phloxes, though all but P. Drummondi are also perennial. Perennial Phloxes have been so hybridised that the original species are quite lost sight of, and a vast number of garden forms of the greatest beauty and variety are the result. It is remarkable that one genus should have produced what may be regarded as the most widely popular annual and well-nigh the most useful perennials of our gardens, but its possibilities have not yet been exhausted. In the alpine section, while there are yet a number unknown to our gardens, which may be regarded as raw material in reserve, florists have been at work upon the one so long cultivated. The Moss Pink, P. subulata, varies so much in the wild state, that its forms have been described and named as species; this has of late been taken in hand by hybridists, and varieties of great beauty have been obtained.

AUTUMN OR LATE-FLOWERING PHLOXES have been obtained by hybridising and selecting from various North American species, but principally from P. paniculata and its varieties acuminata, decussata, and pyramidalis. This class is stronger in growth and taller than the early Phloxes, which they immediately succeed in flower, and thus prolong the season at least two months from the end of July. They are remarkable for their exceedingly bright and varied colours, which include all shades from rich vermilion to pure white, some also being beautifully striped. There are in this group endless varieties more or less distinct, but the following two dozen sorts will be found a good selection: Coccinea, David Syme, Gavin Greenshields, Jane Welsh, Jenny Grieve, Lothair, Matthew Miller, Mrs. Keynes, Monsieur Rafarin, Rêve d'Or, Robert Paterson, William Blackwood, Andrew Borrowman, Carnation, Henry Cannell, James Alexander, James Cocker, Madame Verlot, Major Molesworth, Miss Wallace, Mrs. Tennant, Thos. Chisholm, Triomphe du Parc de Neuilly, and William Veitch. For large beds, back lines of borders, or for park decoration, the following are the most effective varieties, and can be used according to the shades of colour required, viz: Coccinea, rich vermilion; Carnation, white and spotted with purple; James Alexander, rich crimson; Lothair, bright scarlet; Mrs. Keynes, pure white; Robert Paterson, rich crimson; William Blackwood, rosy salmon; Miss Wallace, pure white; and Major Molesworth, scarlet, with a crimson eye. When cultivated in beds or borders, the early and late sorts should be planted alternately, and arranged according to their heights and colours. Thus a mass of bloom, lasting for at least three months, would be produced, and the only labour required after planting would be to give each plant a good stake and tie up when necessary. In spring the number of shoots should be reduced according to the strength of the plant and nature of the variety. They are also much improved by a liberal top-dressing every spring of good, rich soil; and in very hot and dry seasons a good deluge of water will very much prolong the duration of bloom.

EARLY OR SUMMER-FLOWERING PHLOXES have chiefly been derived from P. suffruticosa. They include many varieties, varying principally in colour, and flowering during June and July. These plants will grow satisfactorily in any good border or bed, but if the ground be not naturally suited for it, it can, with very little labour or expense, be made so. If the subsoil be too wet, it must be drained, and about 9 in. of good Hazel loam should be laid on the surface and enriched with good old manure and a small quantity of broken bones. In the herbaceous border a pit can be taken out—say 12 in. square and 9 in. deep—and filled with this compost. This section is extremely useful in the herbaceous border in June and July, when it is in flower, coming as it does between the spring and autumn-flowering sorts. The following are the names of twenty-four of the finest: Beauty, Beauty of Edinburgh, Bridesmaid, Conqueror, Caller Ou, George Eyles, James Nicholson, Mrs. P. Guthrie, Mrs. Burton, Mrs. Gellatly, Philip Pollock, William Mitchell, Allan McLean, Dr. Robert Black, Duchess of Athole, Indian Chief, Mary Shaw, Mrs. Ritchie, President, Redbraes, Socrates, The Bouquet, The Deacon, and The Shah.

PROPAGATION may be effected by means of seed, cuttings of the stems and roots, and also by division. The seed should be gathered from the best sorts only. It should be sown at once in boxes or pans in good free loam, and kept in a greenhouse or warm pit close to the glass. The young seedlings will make their appearance above ground in February and March; and as soon as they are fit to handle, let them be pricked into boxes of good soil, and kept close and warm for a short time, when they can be grown along with the other plants intended for the decoration of the flower garden. The strongest will be fit to plant out in April and May, and they will flower the first season, but not strongly and well before the second year. Then all the best sorts ought to be marked and grown a third year, in order to test them with the best-named kinds. Cuttings can be taken at all seasons of the year; and in the case of new or rare sorts, a few might be treated like Dahlias in a hot pit, the cuttings being taken off as soon as they are 3 in. in length; or they might be cut into single eyes, when they will root in a short time in the bed of a propagating house. They can also be propagated by means of the roots, the oldest of which should be cut into pieces about ½ in. in length, sown, so to speak, in boxes, and treated like seedlings. The leaves also strike, but they are long in making shoots, and altogether this mode of propagation is too slow to be recommended. The mode of propagating by division consists simply in taking the old plant and cutting it into small pieces. As regards the properties of an exhibition Phlox, the form should be a perfect circle, quite smooth on the edge, and of good substance; the colours should be clear and well defined, the flowers arranged close on the spike and overlapping each other; their size should be about 1 in. in diameter. The spike of the early Phlox ought to be quite conical, and that of the late one a corymb. The habit of the plant should be strong and erect, with plenty of broad and healthy foliage, and it should not exceed 3 ft. or 4 ft. in height.

P. Carolina is a handsome plant, growing about 1 ft. high, with slender stems terminated by a cluster of large showy blossoms of a rich deep rose. P. ovata is an equally desirable form with broader leaves; and nitida is also a handsome form. P. glaberrima is of more slender growth, is not so attractive, and is more suitable for a botanical collection. These flower in summer, and grow well in any ordinary soil in an open position.

P. divaricata.—A very distinct and handsome plant, larger than both the Creeping Phlox and Moss Pink, attaining a height of about 1 ft., and bearing large lilac-purple blossoms; the leaves are rounded at the base, oblong egg-shaped or oblong lance-shaped in outline. The plant thrives very well on rockwork in good garden soil, and flowers in summer. A native of North America, and increased by division.

P. Drummondi.—This is deservedly one of the most popular of all half-hardy annuals, and nothing can excel its beauty and usefulness either as a border plant or for furnishing cut flowers. Its colours are varied and brilliant, and are not injured by bad weather, like those of many other flowers. Considering the easy culture of this half-hardy annual, and the fact that from a shilling packet of seed so many pleasing colours may be obtained, the wonder is that it is not more generally and extensively grown in gardens. It may be used in a variety of ways apart from border decoration. Beds of standard Roses carpeted with it are highly effective; and the fact that the plant does not in any way interfere with the growth and well-being of the standard Roses, the naked stems of which they both hide and ornament, should in itself be a sufficient inducement to plant this Phlox specially for that purpose. It is also a very suitable plant for the embellishment of rustic vases and boxes; but it is when grown in masses that its beauty and diversity of colour are seen to the best advantage. The seed should be sown about the first week in March in shallow pans or boxes, using a light rich soil, and placing the seeds in a warm and rather moist temperature. As soon as the plants can be handled without injury, it is a good plan to prick them off into boxes, or on a bed in a warm house, in a temperature of from 50° to 60°. They soon make growth, and as the weather gets warm out of doors the boxes should be placed out in the shade to harden the plants. Those growing in a bed should be again transplanted to a prepared bed in a cold frame, kept covered for a few days and hardened gradually. When the plants are from 3 in. to 4 in. high they should have the main shoot pinched out; this induces them to make a lateral and bushy growth, and greatly assists the prolongation of the period of flowering. The soil in which the plants are to grow should be rich; an abundance of decayed leaves suits this Phlox well. The bed should be in a position fully exposed to the sun, and as long as there is a good moist soil at the roots they will not be injured in the hottest weather. Although this Phlox is generally treated as an annual, it strikes freely from cuttings in autumn; these come in usefully for pot culture and early spring bloom in the conservatory or greenhouse. The varieties of this annual are endless, and a packet of mixed seed will suffice for ordinary purposes, though there are some very distinct named sorts that differ not only in colour, but in growth. Among these are Hortensiflora, Rose Chamois, Heynholdi cardinalis, Radowitzi, grandiflora, Victoria, and compacta. The two first named varieties are very dwarf and compact, and of all the strains there is an indescribable range of colour.

P. pilosa is a pretty species that grows to a height of 10 in. or 12 in., and flowers abundantly from June to August. The flowers are from ½ in. to ¾ in. in diameter, and are purple in colour. It is one of the rarest of the cultivated species of Phlox, though there is a spurious kind sometimes sold for it. The true plant reminds one at once of the annual P. Drummondi by its leaves, habit, size, and by the colour of the flowers, which are of a deep rose and borne in rather large flat clusters. Another rare species is the true P. bifida, a very elegant plant, the leaves of which are narrow and the corolla cut into very narrow segments; the colour is a bluish-purple.

P. reptans.—This is a beautiful little plant with the large flowers and richness of colour of the taller Phloxes. It mantles over the old border and rockwork with a healthy soft green about 1 in. or 2 in. thick, and then sends up numbers of stems from 4 in. to 6 in. at the end of April or beginning of May, each producing from five to eight deep rose flowers. It is by no means fastidious as to soil or situation, but will be found to thrive best in peat or light rich soils. As it creeps along the ground and gives off numbers of little rootlets from the joints, it is propagated with the greatest ease and facility. It is almost indispensable for rock or rootwork; makes very pretty edgings and tufts around the margins of beds of plants, &c.; and also capital tufts on the front edge of the mixed border. It is also useful for spring bedding, and good tufts of it might be employed for vases, or for the edges of raised beds. Everybody who cares about hardy spring flowers should grow it for some of these purposes. It is also known as P. verna and P. stolonifera. P. procumbens is similar, but taller, and not so desirable a plant.

P. subulata *(Moss Pink)*.—A Moss-like little evergreen, with stems from 4 in. to 1 ft. long, but always prostrate, so that the dense matted tufts are seldom more than 6 in. high except in very favourable rich and moist, but sandy and well-drained, soil, where, when the plant is fully exposed, the tufts attain a diameter of several feet, and a height of 1 ft. or more. The leaves are awl-shaped or pointed, and very numerous; the flowers, which are of pinkish purple or rose colour, with a dark centre, are so densely produced that the plants are completely hidden by them during the blooming season.

P. setacea is considered the same as P. subulata, but its leaves are longer and arranged further apart on its trailing stems, and the whole plant is less hard and rigid in its texture. The flowers, which are of a charming soft rosy pink, have delicate markings at the mouth of the tube. Besides the original, there is a handsome variety that originated in Scotland, and which is distinguished by a greater laxity of growth and a much deeper tint of colour, almost approaching a crimson; this is known as P. setacea violacea. Both are lovely plants and desirable for rockwork, where, with the roots deeply seated among the fissures in the enjoyment of coolness and moisture, the plants are enabled to sustain and thrive luxuriantly in any amount of sunshine. P. frondosa is another form of P. subulata. So vigorous is the growth of this that it makes an admirable plant for the front rank of a sunny herbaceous border, where in any ordinary light garden soil it will trail in a few years so as to cover almost a square yard of surface. Its trailing branches are rendered more massive-looking by the formation of a dense rosette of leaves in the axil of each of the older leaves. P. nivalis is of equal trailing habit to the preceding, but of smaller growth, and the leaves are shorter and more densely arranged; the flowers are of snow-white purity. P. Nelsoni is no doubt a hybrid between P. subulata and its forms, possessing an intermediate character as regards foliage. It has pure white flowers with a charming pink eye. Besides this, the late Mr. Nelson, of Aldborough, raised a large number of seedlings, as varied in hue as Phlox Drummondi, the most distinct being Vivid, Model, Grandiflora, Aldboroughensis, Bride, and Perfection. All of these are very lovely and almost indispensable to every good rock garden or border.

CULTURE.—So closely allied are the dwarf species, that the cultural remarks which apply to one will equally well apply to all. Well-drained ordinary garden soil, with a good sunny exposure, constitute the necessary elements of success. Though perfectly hardy and unaffected by even extreme frosts, they cannot endure the damp atmosphere of mild winters, which causes large patches of decay, if not the absolute death of the plants. None of these species seed freely, and their increase must be by means of cuttings or by layers; the former method requires very careful manipulation, as the cuttings are liable to break off just above the node or joint to which the leaves are attached, thus completely neutralising the chance of success. A sharp knife and a careful hand will soon remove the two or three pairs of leaves with their included buds without damaging either the slender stem or joint; and these, when taken off in July, when the branches are just commencing to harden, inserted in sandy soil in a frame where they can be shaded from full sunshine, and given the benefit of the night dews by the removal of the lights, will soon root and become good flowering plants the following season. Where large patches are growing, the readiest mode of increase is sprinkling sandy soil over the entire plant and working the same gently amongst the branches with the hand. If this be done during the summer or early in autumn, the trailing branches will be found to have formed roots, and may be readily removed and planted elsewhere the following season. By this means well-established plants are formed at once; but where numbers are required, increase by means of cuttings is preferable. There are few plants more valuable than these Phloxes for the decoration of the spring garden borders or rockwork, as it is hardy, dwarf, neat in habit, profuse in bloom, and forms gay cushions on the level ground, or pendent sheets from the tops of crags or from chinks on rockwork. It occurs in a wild state on rocky hills and sandy wastes in North America.

Phormium tenax *(New Zealand Flax)*.—A hardy plant, with something of the habit of a large Iris, forming tufts of broad, shining, leathery leaves from 5 ft. to 6½ ft. high, gracefully arching at the top. The flowers, of a lemon colour, are borne in erect loose spikes just above the foliage. Generally with us it will be found to enjoy greenhouse temperature, though in genial places in south and west of England and Ireland it does very well in the open air. Its best use is for the decoration of the garden in summer, a few specimens well grown and plunged in

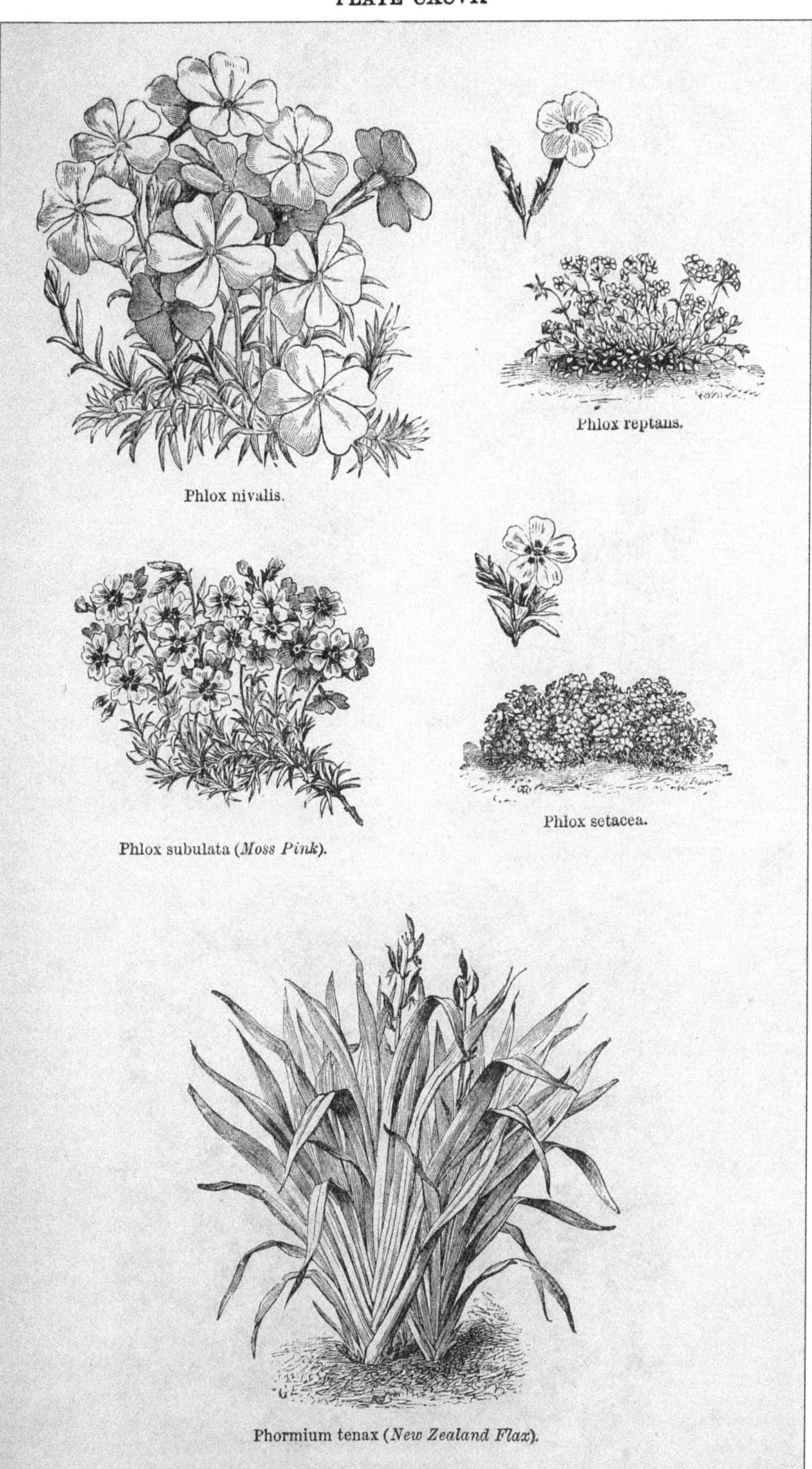

Phlox nivalis.

Phlox reptans.

Phlox subulata (*Moss Pink*).

Phlox setacea.

Phormium tenax (*New Zealand Flax*).

[*To face page* 218.

Phygelius capensis.

Physalis Alkekengi (*Winter Cherry*).

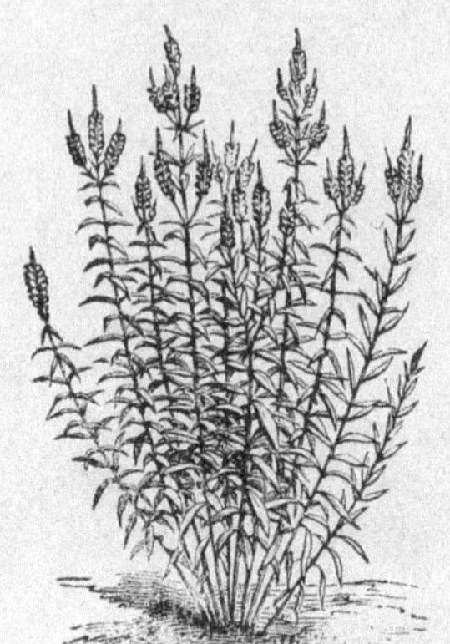

Physostegia virginiana.

Phyteuma orbiculare.

Phyteuma comosum.

[*To face page* 219.

the Grass or centre of a bed giving a most distinct aspect to the scene. The larger such plants are, the better, of course, will be the effect. The small ones will prove equally useful and effective in vases, to which they will add a grace that vases rarely now possess. This plant is pre-eminently useful from its being alike good for the house, conservatory, and hall-decoration in winter. It is multiplied by division of the tufts in summer, and thrives best in a light deep soil. When grown large in tubs it is very suitable for the large conservatory and for important positions in the flower garden. In the extreme south of England and Ireland the New Zealand Flax will thrive in the water as well as on dry land; and where this is the case it may of course be used with fine effect as an aquatic. The variegated variety is a capital plant to try in the open air in warm situations in the south and west of England and Ireland, where the green plant sometimes does so well. In any case it will do finely out-of-doors for the summer.

Phragmites communis (*Great Reed*).—This common, indigenous, waterside Grass is an elegant plant, and should always find a place by the margins of artificial water in localities where it is not plentiful.

Phryma leptostachya. — A North American plant of the Verbena family, but only valuable for botanical collections.

Phygelius capensis. — This highly ornamental plant appears to be nearly related to the Chelone and Pentstemon, and it bears a great resemblance to P. Torreyi with regard to the flowers, which are of a brilliant scarlet. It grows some 3 ft. or 4 ft. high, and has numerous semi-shrubby stems, each terminated by a long branching raceme of flowers. It is hardy in the neighbourhood of London, though it does not flourish so well in the open border as it does under the shelter of a wall. In this latter position it attains large proportions, and thrives vigorously. It is a most persistent flowerer, beginning in May or June, and producing blooms till far on in the autumn. The soil it prefers is that of a bright, rich character, but in warm seashore districts it is not at all fastidious in this respect. It may be readily increased by portions of the rootstock, the bases of the stems being invariably furnished with a few rootlets. S. Africa. Scrophulariaceæ.

Phyllostachys bambusoides. — A noble fine-foliaged plant, resembling and allied to the Bamboos. It grows from 10 ft. to 12 ft. high, and produces broad shining panicles, nearly 2 ft. in length. It requires the same treatment as the Bamboos, and, like them, is best suited for sheltered, sunny positions in the picturesque garden in deep, sandy, well-drained loam. It is quite hardy in the southern counties.

Physalis Alkekengi (*Winter Cherry*).—This is a singularly handsome plant, bearing in autumn and winter bright orange-red, bladder-like calyxes, which enclose the Cherry-like fruits. It grows from 1 ft. to 1½ ft. high, and is a hardy perennial in warm borders of loamy soil in a sunny situation. It is a good old plant, and worth a little care as to position. Propagated easily by division or seed. S. Europe. Solanaceæ.

Physochlaina.—Plants belonging to the Solanum family, of botanical interest. P. grandiflora, from Thibet, grows about 1 ft. high, and has yellow flowers, veined with purple, appearing in May. P. orientalis, of a similar habit, has dark purple flowers, and is a native of Iberia. Both are strong, hardy perennials, and thrive in any kind of light garden soil.

Physospermum.—Umbelliferous plants of no garden value.

Physostegia. — Handsome perennial plants of the Labiatæ, closely allied to Dracocephalum, useful for associating with the bolder types of hardy plants. There are three kinds, but all are much alike. The finest are — P. virginiana, which has erect stems, from 1 ft. to 4 ft. high, with flesh-coloured or purple flowers crowded in terminal racemes. P. imbricata is distinguished from P. virginiana by its stems being higher and more slender, its leaves broader, its calyx globular, and not egg-shaped, and its flowers of larger size and of a deeper colour. Texas. P. denticulata is a similar, but less showy and rarer kind than P. virginiana. All flower in summer, thrive in any ordinary soil, and may be readily propagated by division in spring. They are excellent subjects for naturalising in moist loam.

Phyteuma (*Rampion*).—Although the Rampions are not as showy as other members of the Hairbell family, they are pretty, neat, and interesting. Their flowers are produced in either globose or elongated heads, and make up their want as regards size by number. All enjoy a sunny position, and some make capital rock plants. P. orbiculare is a rare and very desirable native. It elevates its flower-heads 1 ft. to 2 ft. high, and should accordingly be classed amongst plants for the front rank of the border; but from a cultivator's point of view it is better grown among rock plants. It would then be free from the destructive effects produced by the use of the hoe and rake. It is extremely impatient of removal or division, and should be raised from seed sown in autumn in a cool frame. It flourishes in a dry position on the rock garden in a mixture of limestone grit, peat, and sand and loam. The flowers, which are violet-blue, are produced in July. P. Sieberi is a very desirable and neat plant for the moist, but sunny, parts of the rock garden ; it does exceedingly well in a mixture of leaf-mould, peat, and sand. It forms cushion-like tufts, and produces abundance of dark blue flower-heads, on stems from 4 in. to 6 in. long, in May and June. It is increased by division. P. humile is a capital rock plant, having a neat tufted habit; it requires a dry, sheltered position in winter, but should be liberally supplied with water in summer. It is not so neat as the last, but is a little larger in all its parts. The flowers, which are blue, are produced in June on stems 6 in. high. Increased by division. P. comosum is an extremely slow-growing plant, and requires to be carefully watched, or a slug will do considerable damage in a single night. It is very interesting and a genuine rock-plant, suitable for a fissure vertical or sloping to the sun. It does best amongst a mixture of peat, sand, or grit, with a little loam, where it can root to the depth of 2 ft. It has almost stalkless flower-heads of dark purple flowers, and dark, Holly-like leaves. It flowers in June and July, and is increased by means of seed. P. Charmeli and P. Scheuchzeri are too nearly allied for the pair to be in the same garden, except for purposes of comparison. The latter is dwarfest in habit; the flowers are borne on stems varying from 6 in. to 12 in. in height, and are of a pretty blue colour. It is evergreen, and must be increased by means of seed sown in autumn.

Phytolacca decandra *(Virginian Poke).*—A vigorous herbaceous perennial, from $5\frac{1}{2}$ ft. to nearly 10 ft. high, with stems of a reddish hue, the branches, leaf-stalks, veins of the leaves, and flower-stalks also red. The flowers are numerous, in cylindrical spikes, and are at first white, afterwards changing to a delicate rose colour. In autumn the leaves change to a uniform reddish tinge, which has a fine effect contrasted with the numerous pendent purple berries. This is a very hardy plant, requiring scarcely any attention and growing in almost any kind of soil. Multiplied either by seed or by division. It forms a very free and vigorous mass of vegetation, and though perhaps scarcely refined enough in leaf to justify its being recommended for flower garden use, no plant is more worthy of a place wherever a rich herbaceous vegetation is desired, whether near the rougher approaches of a hardy fernery, open glades near woodland walks, or any like positions. N. America.

P. icosandra grows from 2 ft. to 3 ft. high, and has a bushy habit, the leaves being similar to those of a Hydrangea. It bears longish spikes of rather inconspicuous creamy-white flowers, which are succeeded by clusters of fruit, similar in size and shape to Indian Corn, but composed, as it were, of ripe Blackberries. It requires the same treatment and position as P. decandra.

Picridium and **Picris.**—Weedy composites.

Picrorhiza Kurroa.—An interesting small perennial of the Figwort family, from the Himalayas. Suitable for botanical collections only.

Pigeon Berry *(Phytolacca decandra).*

Pimpernel *(Anagallis).*

Pimpernella.—Umbelliferous plants of no ornamental value.

Pine Barren Beauty *(Pyxidanthera barbulata).*

Pinguicula *(Butterwort).*—These are pretty and interesting bog plants very desirable for moist spots in the rock garden or the artificial bog. There are about half-a-dozen kinds, all of which are of dwarf growth and bear a resemblance to each other. All except P. vallisneriæfolia are natives. P. grandiflora (Irish Butterwort) is the finest of all. Its flowers are large, and uniformly of a decided blue-purple, and the leaves, which are broad and spreading, lie flat upon the rock or soil. It prefers the shady side of a moist, mossy rock where the face is steep and the narrow chinks are filled with rich loam. If planted in earth alone, where the drainage is imperfect, it usually perishes in winter. The blossoms are borne four or five from one crown, the crowns being usually solitary—a natural result of their dense, wide-spreading leaves. P. alpina differs from all the other species in having white flowers, with more or less lemon-yellow stains on the lip, and sometimes tinted with pale pink. It roots firmly, by means of strong woody fibres, in peaty soil mingled with shale or rough gravel, in shady humid positions, such as are afforded by high rockwork with a north aspect, or the shelter of a north wall. P. vulgaris is too well known to require description. It grows freely in any sunny position in rich moist peat or peaty loam. A small form, with leaves almost exactly resembling those of P. alpina, both in form and colour, occurs in alpine bogs in the north of England. P. lusitanica, found on the west coast of Scotland and in Ireland, is smaller than any of the preceding, and has pale yellowish flowers. It grows in peaty bogs exposed to the sun.

P. vallisneriæfolia.—This beautiful species differs from others of the genus in its clustered habit of growth, six, eight, ten, or even twelve crowns being sometimes densely massed together in one clump. Its leaves are of a pale yellowish-green, sometimes very thin in texture and almost pellucid, linear, produced erect in dense tufts, and undulated at the margins. Towards the end of the season they become much elongated, not unfrequently measuring 4 in. or 5 in., and occasionally even 6 in. or 7 in. long. The flowers are large, and of a soft purple or lilac-purple, with conspicuous white or pale centres. Dripping fissures and ledges of calcareous rocks (frequently in tufa) suit it perfectly. It requires very free drainage, continuous moisture, and a very humid atmosphere. Mountains of Spain.

Pink *(Dianthus).*

Pipsissewa *(Chimaphila).*

Piptatherum multiflorum is a large perennial Grass, worth growing for its elegant feathery panicles, which are useful for arranging with cut flowers. Grows vigorously in any soil, and is quite hardy.

Pisum maritimum *(Sea Pea).*—An interesting and ornamental native prostrate

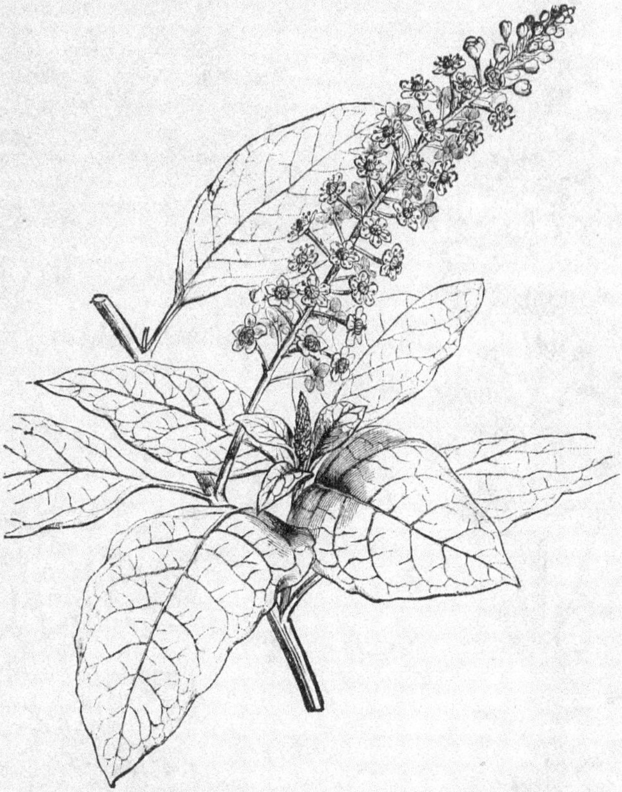

Flowering branch of Phytolacca decandra (*Virginian Poke*).

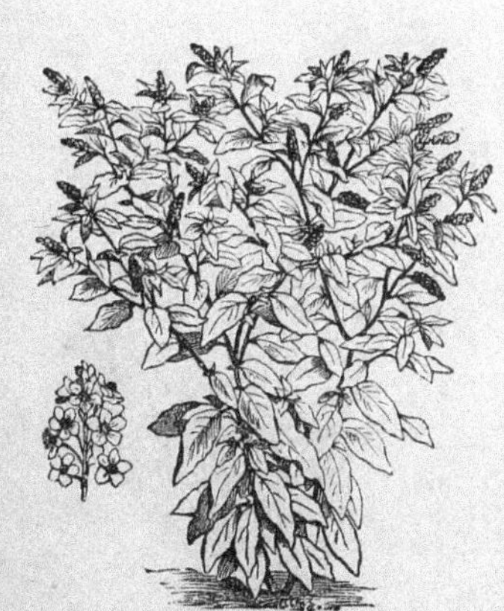

Phytolacca icosandra.

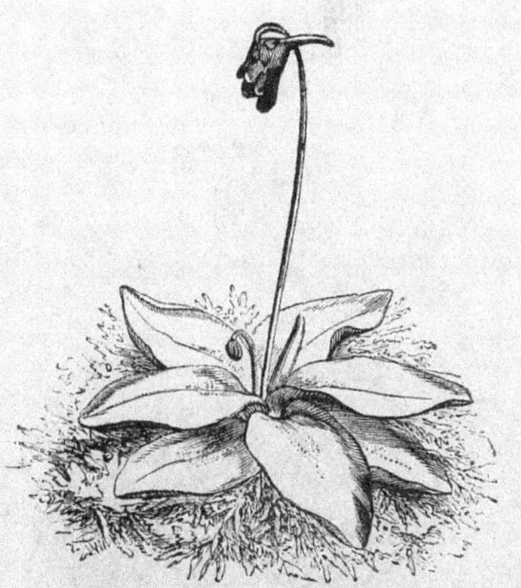

Pinguicula vulgaris (*Butterwort*).

[*To face page* **220.**

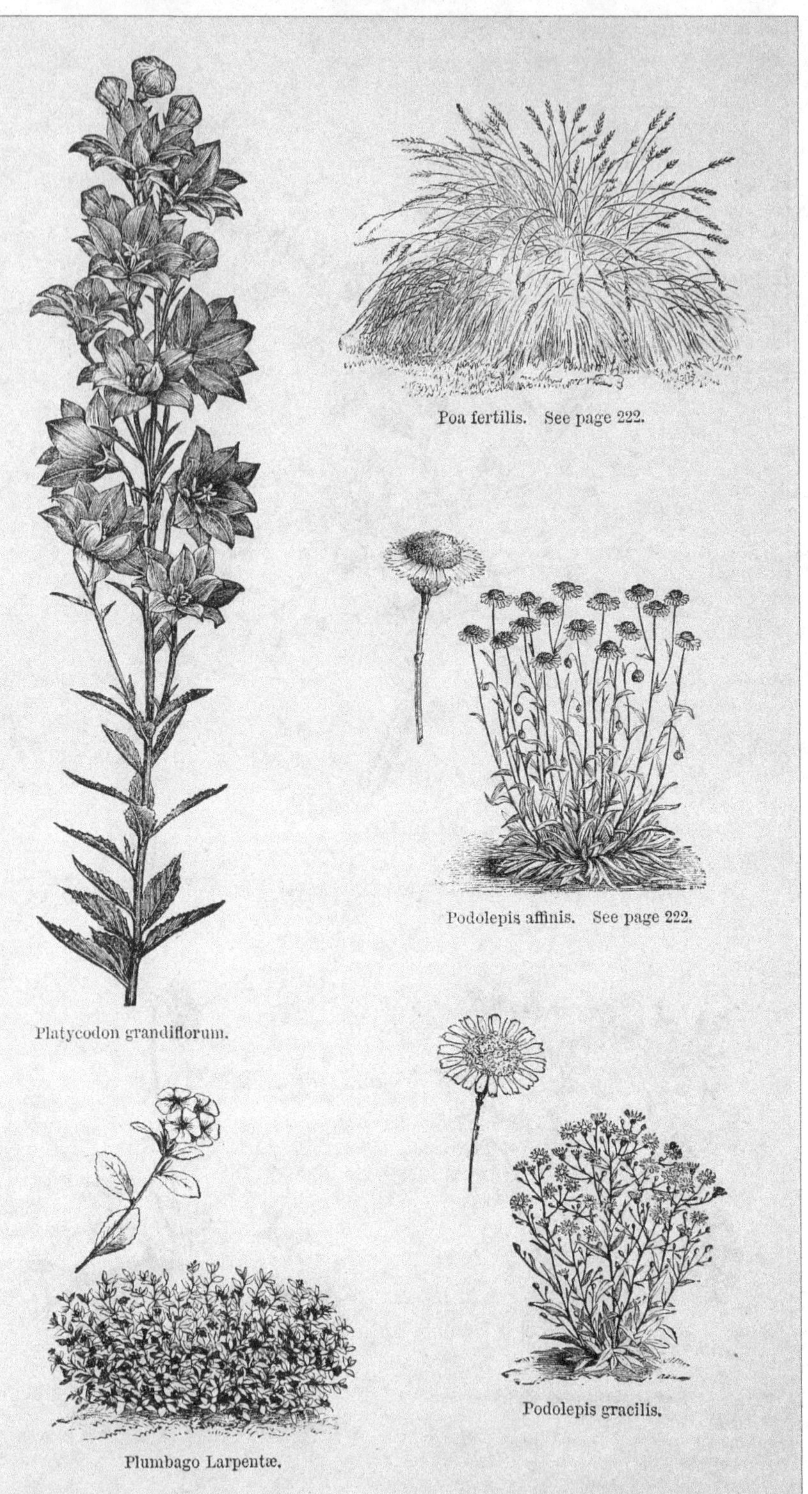

Poa fertilis. See page 222.

Podolepis affinis. See page 222.

Platycodon grandiflorum.

Plumbago Larpentæ.

Podolepis gracilis.

[*To face page* 221.

plant, with bluish-purple flowers produced in summer. It is suited for borders and fringes of shrubberies, in deep sandy soil. In the College Botanic Gardens, at Dublin, this forms a very free and handsome plant, perhaps from its proximity to the sea, and from its being rooted in a rich marine deposit. Division and seed.

Pityrosperma.—Japanese plants, usually met with in botanical gardens, but not of much ornamental value.

Plagius grandiflora (=Balsamita).

Plantago (*Plantain*). — All weedy plants.

Platanthera (=Habenaria).

Platycodon grandiflorum.—A handsome Campanula-like plant, a good perennial, perfectly hardy in light dry soils, but most impatient of damp and undrained situations, where its thick fleshy roots are sure to decay, sometimes from below upwards, but more generally from above downwards, rotting off just at the neck. The flowers are as much as 2 in. to 3 in. across, of a deep blue colour, with a slightly slaty shade therein. They are produced in clusters at the summit of each of the branches, which proceed from the old root-stock, rising to a height of about 18 in., and being very slender at the base. If neglected in the matter of supports in their early stage of growth, they are sure to fall to the ground, thus giving the plant, however beautiful its blossoms may be, an untidy appearance. If such a result occurs from early neglect, it is almost impossible to remedy it when the flowers are nearly developed, as branch after branch will break away, if made to assume the usually erect position, in the process of tying. When down, it will be better to leave them as they are, allowing the bloom to compensate, by its beauty, for any untidy appearance that the plant may present, merely taking the precaution to peg down the branches, else the sportive winds may whisk them round, and effect their total dislocation from the parent stock. Possibly the best position for such a plant would be overhanging a rock ledge in a sunny corner of the rock garden, where a negligent character of growth would be in keeping with its situation, and its flowers, produced on a level with the eye, would be shown to great advantage. Like most of the Campanulas, it has a tendency to sport in colour, and to revert from blue to white through various modifications. Equally pretty and acceptable as a garden plant is the white variety, though it is by no means so frequently met with as the blue type. A good rich loamy soil suits the plant best, but it should be well drained and the situation an open one. As regards propagation, the best mode is that of raising seedlings, as seed can be readily procured. The young shoots taken off when about 3 in. long, in spring, will strike, but not freely, if placed in a gentle bottom heat, and the plant is a bad one to divide—indeed, the attempt often results in failure, and, if done at all, must be carried out in May, when the growth has just commenced. There are a few varieties of this plant, the most distinct being P. autumnale, or chinense, from China and Japan. Compared with P. grandiflorum, it is both taller and more robust in growth, attaining a height of 3 ft. under favourable cirumstances; its leaves are narrower, but more densely arranged than those of P. grandiflorum, and its flowers, though smaller in size, are produced in greater quantities and are pretty evenly distributed along the upper half of the stems. From its taller stature, and also the fact that its young shoots are far more woody and vigorous in their growth than those of P. grandiflorum, its true position should be in the second or third rank of the herbaceous border. Besides a white variety, there is a tendency to become semi-double, by a sort of "hose-in-hose" re-duplication of the corolla, similar to what occurs in many of our Campanulas. The new dwarf variety, Mariesi, from Japan, is distinct and highly desirable. The typical P. grandiflorum is a native of Siberia, and flowers in July. It is known also as Campanula grandiflora.

Platyloma.—In some mild localities the elegant New Zealand Fern, P. rotundifolium, will thrive in the open, but it seldom does satisfactorily. P. atropurpureum is synonymous with Pellæa.

Platystemon californicus.—A pretty Californian hardy annual Poppywort. It is a dwarf trailer, grows freely, and forms a dense tuft which in summer is studded thickly with sulphur-yellow blossoms. It is of the simplest culture, merely requiring to be sown in ordinary soil in the open border either in autumn or spring; the seedlings should be well thinned out. P. leiocarpum is a similar kind.

Plectogyne variegata.—A Liliaceous plant with broad leaves, like the common Aspidistra lurida. It is a handsome-leaved plant and worthy of extended culture, but as yet it is not to be seen elsewhere than at Kew, where it is hardy in light soil and unprotected in winter. Japan.

Plectranthes.—Labiate plants, chiefly tropical. P. australis and glaucocalyx, which are grown in the open air, are of no use for the flower garden.

Plumbago Larpentæ.—A dwarf herbaceous plant, perfectly hardy, and a first-rate ornament for rockwork, banks, or sunny borders. Its numerous wiry stems form neat and full tufts from 6 in. to 10 in. high, according to soil and position. In September these become nearly covered with flowers, arranged in close trusses at the end of the shoots, and of a fine cobalt blue, afterwards changing to violet. The bloom usually lasts till the frosts. In all cases it is desirable to give it a warm sandy loam or other light soil and a sunny, warm position, as under these conditions the show of bloom is much finer. In consequence of its semi-prostrate habit, it is well suited for planting above the upper edges of vertical stones or slopes on rockwork, and it may also be used with good effect as a border plant, or as an edging plant in the flower garden, particularly in the case of slightly raised beds. A native of China. Very easily increased by division of the root during winter or early spring.

P. capensis, usually grown under glass, is a capital plant for planting out in summer, for it produces its lovely pale blue flowers profusely and continuously throughout the summer. For this purpose plants should be specially prepared, young ones being always the best for edgings, though taller ones may be used in certain positions. The plant is used with admirable effect in German gardens.

Poa.—A numerous genus of perennial and annual Grasses, few of which, however, are worth cultivating. P. fertilis is one of the most elegant of Grasses, forming dense tufts of long, soft, smooth, slender leaves, which arch in the most graceful manner on every side, and, in the flowering season, are surmounted by airy, diffuse, purplish or violet-tinged panicles, rising to a height of from 20 in. to 3 ft., the grassy tufts being usually about half that height. Isolated on lawns, it is most effective, and when once planted in good soil gives no further trouble. P. aquatica is a stout, rapidly increasing native Grass, growing from 4 ft. to 6 ft. high. It mostly occurs in wet ditches, by rivers, and in marshes. It is one of the boldest and handsomest of hardy Grasses for planting by the margins of pieces of artificial water or streams, associated with such plants as the Typhas, Acorus, Bulrush, and Water Dock.

Pocockia cretica. — An interesting little annual of the Pea family, but not worth attention except in a botanical collection.

Podolepis.—A pretty half-hardy annual from New Holland, belonging to the Compositæ. P. gracilis is of slender growth, about 1½ ft. high, and bears terminal flower-heads of rosy purple about 1 in. across. There is also a white variety. P. acuminata and chrysantha are similar, but have yellow flowers. They require to be sown in spring in light warm soils.

Podophyllum peltatum *(May Apple)*.—An interesting plant with glossy, green, wrinkled leaves, borne aloft, umbrella-like, on slender undulating wand-like stems, about 1 ft. high. Its Christmas Rose-like flowers of waxy whiteness, which are produced in May, are succeeded by green Crab-like fruit; hence the popular name. It is a capital plant for shady peat borders, or naturalised in woods, in moist vegetable soil, and in shady or half-shady positions.

P. Emodi is even handsomer, but is of similar aspect. The stem and leaves are suffused with a reddish tinge, which gives the plant a very distinct appearance. The fruits are 2 in. long, and of a coral-red colour. The plant succeeds perfectly in peaty soil if planted in warm sheltered spots. In such positions it is a capital subject for the margins of beds of American plants. Like the preceding, it may be increased by seed or division. Both plants are so distinct in habit, and so effective in spring and early summer, that they well deserve a fitting position in all good gardens.

Podospermum.—Composite plants of no ornamental value.

Poet's Daffodil *(Narcissus poeticus)*.

Pogogyne. — A small genus of the Labiatæ, chiefly valuable for the botanical garden.

Pogonia.—A small genus of terrestrial Orchids, chiefly of North America, but few, if any, have been grown well in this country. P. ophioglossoides is said to be a pretty plant, worth cultivating, and no doubt it will thrive under treatment similar to that given to other like plants.

Pokeweed *(Phytolacca)*.

Polanisia *(Cleome)*.

Polemonium *(Greek Valerian)*. — A small genus of Phloxworts containing a few good garden plants. The most distinct kinds are—

P. cœruleum *(Jacob's Ladder)*, with which most people are familiar, as it is a useful, though common, border plant. Besides the original blue-flowered species there is one with white blossoms, and another handsome form with variegated foliage, which is so desirable as an edging plant in the flower garden that its propagation and culture are of importance. On good garden soils generally it will be found almost as easy of culture as the common form. It is, however, apt to go off on a very wet clayey soil, but flourishes finely in deep, rich, but well-drained loam. As regards the propagation, it is effected by simply digging up well-established old plants, pulling them to pieces, and then planting them immediately in a nursery bed of good soil. This is best done in early autumn, so that they may be nicely established in the nursery beds before midwinter. Where plants are merely required for borders and rockwork, it is simply necessary that the old stools should be taken up, divided, and replanted, where desired, in the old-fashioned way of dealing with herbaceous plants. As the variegated variety is grown for its leaf beauty alone, the flower-stems should be removed when they appear. There are several other species in cultivation, but they are not of a sufficiently perennial or ornamental character. In the north, where its variegation is both more constant and more decided than in the south, this is an important garden plant.

P. confertum.—This is the latest addition to the cultivated kinds, and is one of the finest of all. It is a dwarf plant, has slender and deeply-cut leaves, and bears dense clusters of blossoms on stoutish stems about 6 in. high. The colour is a deep blue, very attractive, and the plant is quite distinct from any other. It requires a warm spot in the rock garden in a well-drained, deep loamy soil, rather stiff than otherwise

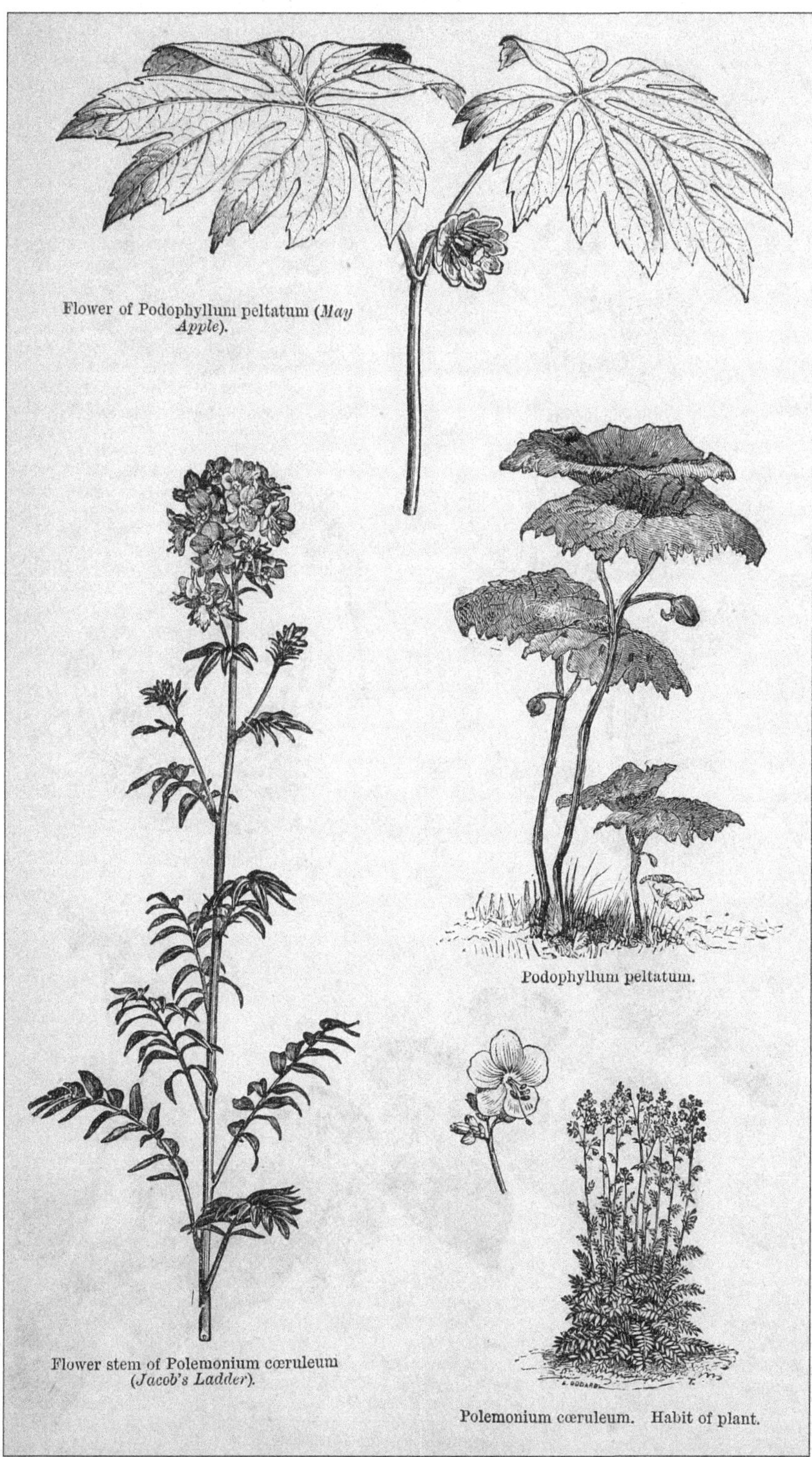

Flower of Podophyllum peltatum (*May Apple*).

Podophyllum peltatum.

Flower stem of Polemonium cœruleum (*Jacob's Ladder*).

Polemonium cœruleum. Habit of plant.

[*To face page* 222.

PLATE CCII.

Polemonium reptans.

Polygala Chamæbuxus (*Box-leaved Milkwort*).

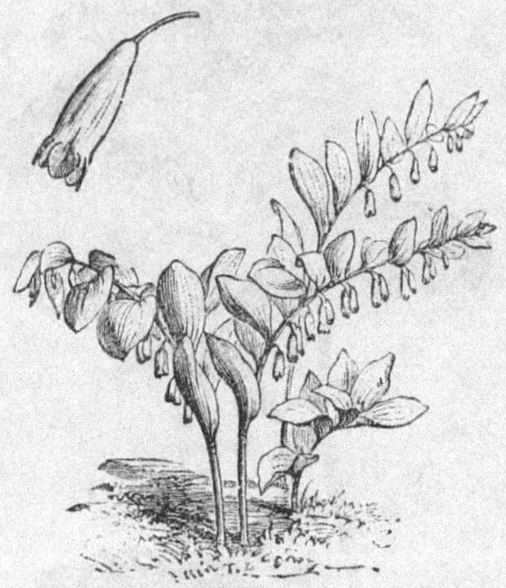

Polygonatum multiflorum (*Solomon's Seal*).

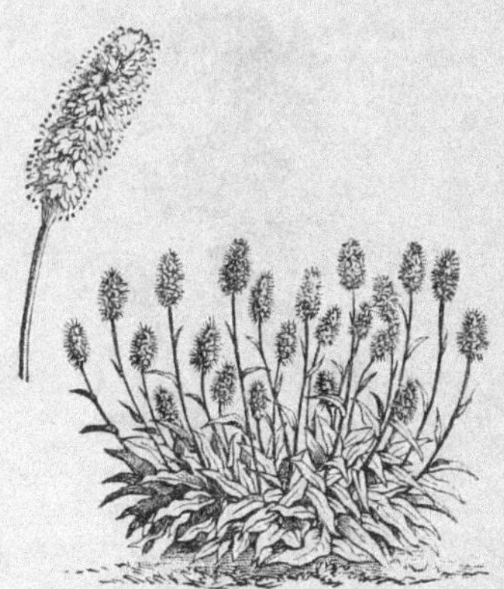

Polygonum Bistorta (*Snakeweed*).

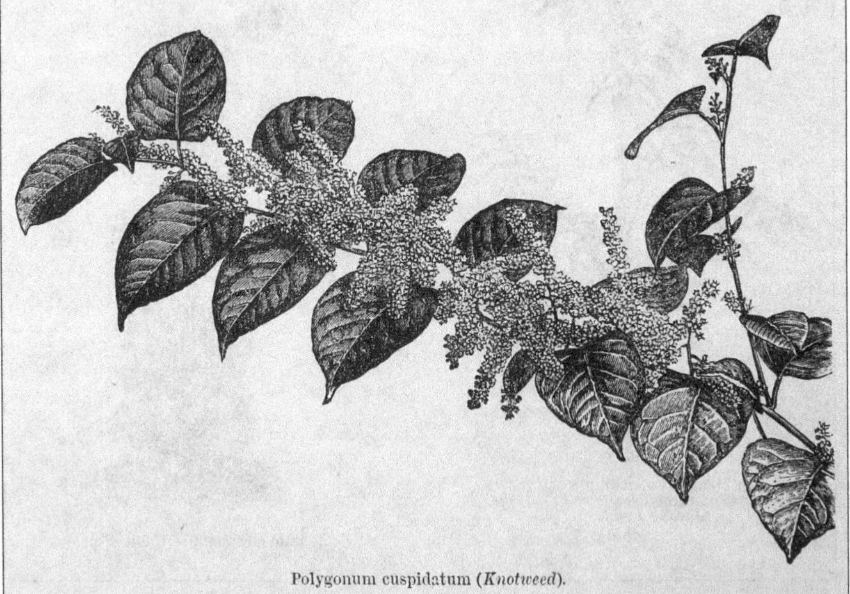

Polygonum cuspidatum (*Knotweed*).

[*To face page* **223.**

Though it requires plenty of moisture in summer, it is impatient of excessive dampness about the roots in winter, and hence precautions should be taken in this respect. It should be allowed to remain undisturbed for years after it has become established. It is a native of the Rocky Mountains, and is quite hardy.

P. humile is a plant of low stature, truly alpine in character. Its flowers are of a pale blue colour, and produced on stems a few inches high. Unlike P. reptans, this species loves a dry situation and a light sandy soil; under these conditions it is perfectly hardy, but if placed on a damp subsoil it is sure to die off in winter. N. America. P. mexicanum is a similar, but larger plant, and only of biennial duration; therefore it is scarcely worth cultivating generally.

P. reptans is an American species, and though, as regards beauty, far inferior to P. confertum, it is, nevertheless, an alpine plant well worthy of cultivation. Its stems are creeping; the flowers form a loose, drooping panicle, elevated some 6 in. or 8 in. above the ground, and are of a slate-blue colour. It is much infested by snails, which devour it ravenously, especially the scaly root-stocks during winter; therefore a watchful eye is always required to prevent their depredations. P. sibiricum, grandiflorum, and Richardsoni much resemble the common P. cœruleum, but are of larger growth, more vigorous, and bear larger flowers, and, therefore, they should be preferred for border decoration.

Polianthes tuberosa *(Tuberose).*—Though a native of the East Indies, strong imported bulbs of this deliciously fragrant plant will flower well in the open air during the month of August in warm soils. It must, however, be remembered that, like Hyacinths and similar bulbs, Tuberoses will not flower so well the second year as they do the first. Sufficient strength to flower well is only found in bulbs grown and matured under the sunny suns of Italy and France, from whence most of the Tuberoses grown in this country are imported. In the neighbourhood of London we have seen the Tuberose flowering freely in the open border. The plants were vigorous and healthy, and the flower-spikes thickly studded with buds and blossoms. They were growing in a light, sandy, well-drained soil, in which the bulbs remained all the winter, being slightly protected during severe weather by means of ashes or other dry material. It is better, however, to plant fresh sound bulbs every spring, as the result is more satisfactory.

Polygala *(Milkwort).*—The hardy species of Milkwort are neat dwarf plants, with flowers much resembling those of the Pea family.

P. Chamæbuxus *(Box-leaved Milkwort).*—A valuable little creeping shrub, a native of the Alps of Austria and Switzerland, where it often forms but very small plants; in our gardens, however, on peaty soil, and in some fine sandy loams, it spreads out into compact tufts covered with cream-coloured and yellow flowers. The variety purpurea is a much prettier plant, and deserves a place in every rock garden. The colour of the flowers is a lovely bright magenta-purple, with a clear yellow centre. It appears to have a greater tendency than P. Chamæbuxus to expand a portion of its blossoms during the autumnal months. It is a plant of easy culture, succeeding in any sandy, well-drained soil, and in almost any situation. It seems to do best in sandy peat, and slightly shaded from the mid-day sun, and is a most desirable addition to hardy spring flowering plants, being alike useful in the rock garden and ordinary flower border. Even when out of flower it is interesting, owing to its dwarf, compact habit, bright shining evergreen leaves, and olive-purplish stems. P. paucifolia is an interesting and handsome perennial 3 in. to 4 in. high. The flowers, which appear in summer, are generally rosy-purple, but sometimes white; they are large and handsome, about three-quarters of an inch long, and are borne one to three on stems springing from the slender prostrate shoots, which also bear concealed flowers. It is suited for the rock garden, in leaf-mould and sand, associated with such plants as Linnæa borealis, Trientalis, Mitchella, &c., in half-shady places. North America. Some of the British Milkworts are very pretty and worth cultivating, especially P. calcarea and vulgaris. These are very handsome and easily grown, and do very well on rockwork in sunny chinks, planted in calcareous soil. They form neat dressy tufts of blue, purple-pink, and white flowers, and bloom profusely in early summer. Seed may be gathered from wild plants and sown in sandy soil. Plants carefully taken up from their native positions have also been established in gardens.

Polygonatum *(Solomon's Seal).*—Of this genus there are in cultivation about a dozen species, the majority of which are similar to each other. They have tall, elegantly arching stems, thickly set with pretty pendent white blossoms. They grow from 1½ ft. to 3 ft. high, but P. giganteum, a lately introduced kind, attains as much as 6 ft. in height. The most distinct, besides our native species P. multiflorum, are P. pubescens, oppositifolium, latifolium, and officinale; there is also a double-flowered variety of P. multiflorum, and another with handsome variegated foliage. These thrive in any position in sandy loam, and are worthy of a place in the choicest border or group, but are, perhaps, seen to greatest advantage when leaning forth from beneath shrubs or low trees, on the margin of a shrubbery or grove, where their elegant growth bears a marked contrast to the surrounding vegetation. They should be abundantly grown as wild plants in woods. All are easily multiplied by division of the rootstocks. P. verticillatum and P. roseum are quite distinct in growth from the foregoing, as the stems are erect in growth and bear whorls of leaves and small blossoms which in P. roseum are of a purplish pink hue. These are best suited for border culture in good loamy soil.

Polygonum *(Knotweed)*.—Now that beauty of form is beginning to be appreciated, many plants of graceful proportions that were formerly discarded as worthless, either on account of their gross habit or inconspicuous flowers, are being brought into notice; and it is more than probable that these will form a prominent feature in the gardens of the future. The vast genus Polygonum, which comprises 150 species of world-wide distribution, the majority of which are insignificant weeds, nevertheless includes several noble plants, which claim a consideration that has hitherto been denied them. They are of the easiest culture, thriving in any ordinary garden soil, but being greatly improved by cultivation. All those of a bushy habit should be planted so as to have a clear space all round, in order to give the foliage all the air and light possible, as overcrowding is frequently the cause of naked stems and a straggling habit, to remedy which tying-in has to be resorted to. This detracts much from their natural appearance, as their beauty consists in the innumerable flower-spikes rising above a gracefully-developed mass of foliage continuous to the ground. Those of the P. cuspidatum type, which have few hardy representatives, invariably produce stems of sufficient strength to support their spreading crowns of foliage, though a tendency to droop at the points is not unusual; but this is by no means an objectionable feature. The annuals, unless grown as single specimens and in sheltered situations, will require support; but the dwarf perennials, most of which are evergreen, need very little attention beyond an occasional trimming. The stems of all the tall hardy species, being of annual duration, die off in the autumn, and the fact that the succeeding ones do not appear before April or May is a circumstance that must be taken into consideration when planting for effect. Propagation is effected by means of cuttings, division, or seeds, which, if of the P. orientale group, should not be sown in the open ground until the middle of April. Several attempts have been made to popularise this class of plants, but owing to neglect to keep them within bounds they have fallen into disrepute. To remedy this encroaching propensity, which is their chief drawback, their underground creeping rhizomes require to be cut back every spring. The selection described below includes most of the best sorts for garden purposes, and illustrates the various types of the genus.

P. affine, one of the Bistorta group of Knotweeds, is a native of the Himalayas, where it constitutes a very ornamental feature of the alpine vegetation. It inhabits the wet river banks and meadows, and hangs in rosy clumps from moist precipices. Cultivated plants of this species are from 6 in. to 8 in. high, with a few narrow leaves and rosy-red flowers in dense spikes. It flowers freely in the open border in September and October, and is an attractive plant. P. Brunonis is a similar plant, equally as desirable for the border or rock garden. The flowers are of a pale rose or flesh colour, and are borne in dense erect spikes nearly 18 in. high. It continues to bloom more or less throughout the summer months.

P. alpinum is a native of the Swiss Alps, and a very old inhabitant of our gardens. Its stems attain the height of 3 ft. to 4 ft., with ovate-lanceolate deep green leaves, with ciliated margins. Its flowers, which appear early and continue in bloom for several weeks, are arranged in snow-white panicles, and prove very serviceable where quantities of cut flowers are in request, but it is scarcely good enough for borders.

P. capitatum.—This is a charming little annual of a spreading habit, with oval greyish-green leaves, with a dark blotch in the centre of each, and numerous globose heads of pink flowers. When once established in light warm soils, it appears every year from self-sown seeds. Its neat habit and the delicacy of its flowers are qualities that never fail to attract admirers.

P. compactum is a species similar to P. cuspidatum. It has stems 1 ft. or 2 ft. high, forming a compact tuft. Its flowers, which are produced in great profusion late in the autumn, are white, and the leaves are similar to those of P. cuspidatum, though much smaller. Although not very showy, its late flowering season will recommend it to lovers of herbaceous plants.

P. cuspidatum, also known as P. Sieboldi, is a plant of sterling merit, now becoming quite common. It belongs to a section with a semi-arborescent habit and a peculiar curve of the stem, which brings nearly the whole of the foliage of each stem into the same plane. Its shoots are copious, speckled with purple; its broadly-ovate leaves, which are of a dark dull green, are frequently variegated with faint silvery blotches, and its creamy-white flowers are borne in great profusion. Its stately habit of growth and the luxuriance of its foliage are attractions of no ordinary character, which cannot fail to strike the most casual observer, more especially when the plant is in full bloom. It is a native of Japan, and is undoubtedly one of the finest herbaceous plants in cultivation. To do it justice it should be grown as an isolated specimen, either on the turf, or in some prominent position of the wild garden.

P. orientale *(Persicary)*.—A gigantic free-growing annual, growing to the height of 8 ft. to 10 ft. Its stems are very robust, and give off numerous lateral shoots, which are supplied with large, oblong, acute, rich green leaves, measuring 10 in. to 15 in. in length, of which those nearest the flowers are stem-clasping, and more inclined to be

PLATE CCIII.

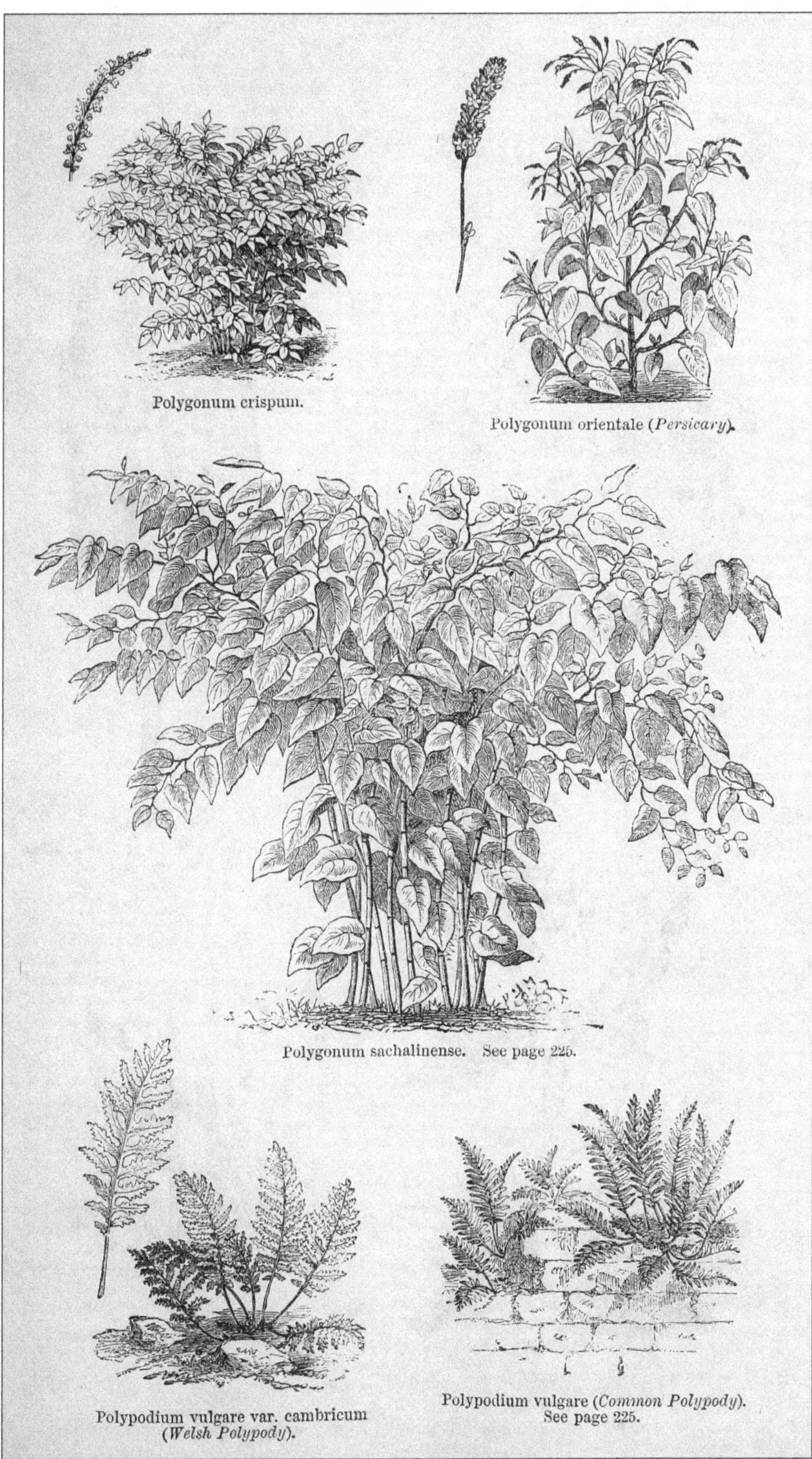

Polygonum crispum.

Polygonum orientale (*Persicary*).

Polygonum sachalinense. See page 225.

Polypodium vulgare var. cambricum (*Welsh Polypody*).

Polypodium vulgare (*Common Polypody*). See page 225.

Polymnia grandis.

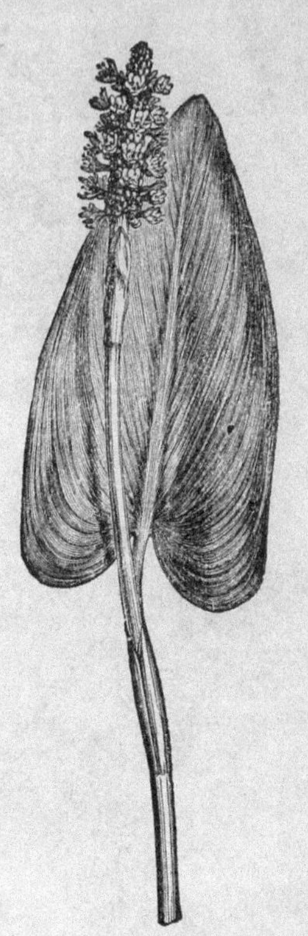

Flower-spike and leaf of Pontederia cordata. See page 226.

Portulaca grandiflora fl.-pl. (*Purslane*). See page 226.

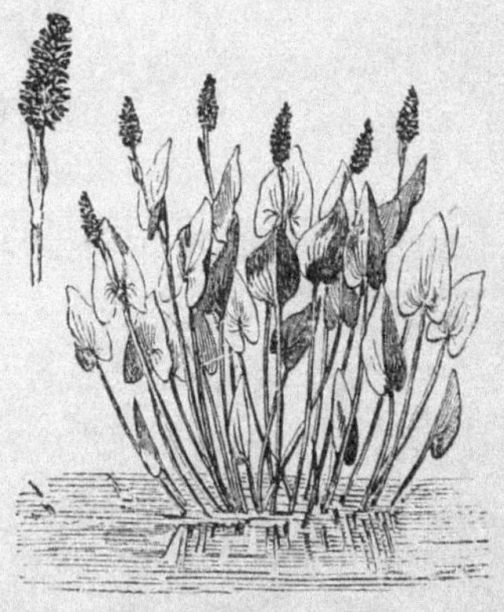

Pontederia cordata (*Pickerel Weed*).

[*To face page* 225.

sagittate. The inflorescence is both terminal and axillary, and consists of slender spikes of crimson flowers, which make their appearance in July and continue until the frosts. To obtain a good development of its foliage it should be grown as a single specimen and without shade. It will then bloom freely, and make one of the grandest of ornaments either for the sub-tropical garden or for some conspicuous place among its congeners in the wild garden. There is a variety of this with pure white flowers, though both are rarely met with. It is a native of Northern India.

P. sachalinense. — A native of the island of Sachalian, and of a similar habit to the preceding, but of much larger proportions. It often attains a height of 10 ft. to 12 ft., with broadly-oblong, bright green leaves upwards of 1 foot in length. Its flowers are rather inconspicuous, being of a greenish-white, and disposed in slender drooping racemes. It luxuriates in a moist subsoil near the margin of water, where it is very effective in company with grassy vegetation. It also makes a fine bold feature planted either on the turf or in a good position where it can develop its noble proportions. No better plant could be employed for naturalisation in semi-wild places, and also for association with vigorous herbaceous plants on the turf in the pleasure ground. For those whose gardens will not admit of such special places for this and similar plants, a tuft, well placed on the edge of a shrubbery, will well repay the planter. It would, however, be best not in the shrubbery, but just a few feet free of it.

P. vaccinifolium.—Few plants surpass this for rockwork. It differs widely from all its congeners at present in cultivation, unless P. crispum, once a favourite flower, resembling it in its twiggy habit, is still to be found in gardens. It is quite hardy, thrives in almost any moist soil, and is seen to the best advantage where its shoots can ramble over stones or tree stumps. Under favourable conditions it grows rapidly, and produces its Whortleberry-like leaves and rosy flower-spikes in profusion. This is also a native of the Himalayas.

Polymnia grandis.—This is second to no other plant for its dignified and finished effect in the flower garden. It forms a handsome shrub with large, opposite, much-divided, and elegantly-lobed leaves, which are often nearly 3 ft. long, presenting very striking and luxuriant masses of foliage. The stem and leaf-stalks are spotted with white, and the leaves when young are covered with a soft white down. Like most large soft-growing things in this way, it is best planted out in a young state, so as to ensure a fresh and unstinted growth. It is easily multiplied from cuttings, which plants freely produce if placed in heat in January. It is best planted out at the end of May, and should be in every collection. Mexico. (= Montagnea heracleifolia.)

Polypodium *(Polypody)*.—In this large genus of Ferns there are several valuable hardy kinds, the principal being the common P. vulgare, of which there are in cultivation a score or so of varieties differing from each other more or less widely. The most distinct and beautiful of these are cambricum, elegantissimum, omnilacerum, and pulcherrimum, and these are also the freest in growth. Like almost all Ferns, they prefer shade, but if well supplied with water at the root during summer they will thrive in perfection even when exposed to the full rays of the sun. They should be planted in a compost consisting of fibry loam and tough and fibry peat, also a liberal admixture of leaf-mould and well-decayed woody matter, and each autumn a thin top-dressing of similar material may with advantage be added; by this means a plentiful supply of natural food will always be available for its roots. The evergreen Polypodiums associate well with flowering bulbous and other plants that do not require frequent removing, or they might be made to cover bare spaces beneath trees, or to overrun stumps; and a beautiful effect, too, might be obtained by their use as a carpet or setting for some of the plants in the rock garden. They are also very useful for cutting purposes; and no garden, however small or extensive, should be without some good patches of them, planted especially for furnishing a good supply at all seasons of their beautiful fronds. Besides P. vulgare and its varieties, there are several deciduous species, such as P. Dryopteris (Oak Fern) and P. Phegopteris (Beech Fern), well known to all Fern lovers; they thrive best in a compost of peat, loam, and sharp sand, with the addition of some broken lumps of sandstone. They prefer a dry situation, and do well in the rock garden; indeed, they cannot be out of place in any situation provided it is not fully exposed to the sun. A slightly shaded spot should therefore be selected for them, where they might be planted in conjunction with flowering plants that are amenable to the same treatment, and which will afford the needed shelter. P. Robertianum (Limestone Polypody) is a beautiful deciduous species, but unfortunately somewhat difficult to manage; it will, however, endure sunshine, and therefore a dry, sheltered position should be afforded it. The soil it likes best is a mixture of sandy and fibry loam, with a plentiful supply of pounded limestone added thereto. P. alpestre is a distinct and handsome Fern, bearing a very close resemblance to the Lady Fern; the fronds are dark green and handsome, and sometimes exceed 2 ft. in length. It may with advantage be grouped with the Lady Ferns in our gardens, as it flourishes under similar treatment. P. hexagonopterum, a native of N. America, is quite hardy in sheltered positions, and produces elegant tapering dark green fronds about 1 ft. in height.

Polypogon *(Beard Grass)*.—Pretty Grasses, chiefly annual. P. monspeliensis and littoralis are native. P. affinis, a native of Chili, is an elegant annual worth cultivating. All should be sown in the open border in autumn or spring.

Polystichum *(Aspidium).*

Pontederia cordata *(Pickerel Weed).*—One of the handsomest water plants in cultivation, combining grace of habit and leaf with beauty of flower. It forms thick tufts of almost arrow-shaped, erect, long-stalked leaves from 1½ ft. to more than 2 ft. high, crowned with handsome blue flower-spikes, which issue from the leaf-stalks just below the base of the leaves. P. angustifolia has narrower leaves; both should be planted in shallow pools of water. Multiplied by division of the tufts at any season. N. America.

Poppy *(Papaver).*

Portulaca *(Purslane).*—The beautiful half-hardy annual Portulacas, now so much grown, are chiefly varieties of P. grandiflora, a native of Chili. The varieties are very numerous, and occur in great diversity of colour. The names of the principal are alba, aurea, aurea-striata, caryophylloides (rose streaked with crimson), rosea, splendens (deep crimson-purple), striatiflora (white and crimson striped), Thellusoni (crimson-scarlet) Thorborni (yellow). These are nearly all represented by full double flowers, which are the more desirable as they are more enduring. They grow from 6 in. to 9 in. high, and form a dense carpet of foliage, and the effect of the brilliant flowers in full sunshine is very fine. The seeds should be sown in heat in February or March, and the seedlings planted out in May in good soil in an open, exposed position, and ample space should be allowed for each plant to develop itself. It is best to pot off the seedlings singly in small pots, so that by planting out time they may be good strong plants. They begin to flower at the end of June, and continue for several weeks. Sometimes the plants sow themselves in autumn, but they are seldom so satisfactory as when treated as above.

Potamogeton *(Pondweed).* — Chiefly native aquatic plants of no ornamental value.

Potentilla *(Cinquefoil).*—An extensive genus of the Rose family, but containing comparatively few garden plants, the most important being the fine hybrid varieties that have been obtained of recent years by hybridising a few of the showy Himalayan species such as P. insignis and P. atrosanguinea. These two species, the former with clear yellow, and the latter with deep velvety-crimson flowers, are well worth growing, as is also the beautiful rosy-pink P. colorata, a plant that flowers throughout the summer; but these three are about the only typical species of tall growth that are worth cultivating. The double flowered kinds are most showy, and they posse s the additional advantage of lasting in perfection a longer time than the single sorts both on the plants and in a cut state. There are about three dozen distinct named varieties, and a great number of them have emanated from Continental sources, but all may now be obtained in any of the large hardy plant nurseries. These represent every type of shade, size, and colour that it is possible to obtain, though such a large number is not indispensable, as a good selection of the most dis tinct kinds would be found to embrace most of the qualities of the whole race. The culture of these plants is, as is the case with most hardy flowers, a simple matter; they luxuriate in a light deep soil. The more fully exposed the position that they occupy is, the better it appears to suit them; and they soon form vigorous specimens, and produce flowers in great profusion for many weeks in succession—in fact, from the beginning of summer till the middle of autumn.

The following is a good selection of double flowered sorts: M. Rouillard, reddish-crimson; Belzebuth, dark crimson; Chromatella, yellow; Dr. Andry, scarlet, margined with yellow; Escarboucle, crimson; Belisaire, reddish orange; Vase d'Or, yellow; Le Dante, orange, shaded with scarlet; Louis Van Houtte, crimson; Phœbus, rich yellow; Le Vesuve, crimson, margined with yellow; Versicolor, yellow suffused with brownish-crimson; Vulcan, scarlet, shaded with yellow; Variabilis flore-pleno, yellow, margined with scarlet; Eldorado, scarlet-crimson, margined with yellow; Perfecta plena, bright scarlet-crimson, slightly tinged with yellow; Imbricata plena, orange-scarlet; Etna, reddish-crimson; Panorama, yellow, heavily stained with scarlet; Nigra plena, dark crimson; Meteor, yellow, suffused and blotched with scarlet; Meirsschaerti flore-pleno, yellow, veined and striped with crimson; William Rollisson, deep orange-scarlet, yellow centre; Fenelon, orange and scarlet; Purpurea lutea plena, scarlet-crimson, slightly tipped with yellow.

Among the dwarf alpine species there are some very beautiful plants that are indispensable to the rock garden. Of these the following are the best:—

P. alba *(White Cinquefoil).*—A pretty species with the leaves quite silvery, with dense silky down on the lower sides. It is a very dwarf kind, neat and not rampant in habit, with white Strawberry-like flowers, nearly 1 in. across, with a dark orange ring at the base. A native of the Alps and Pyrenees, of the easiest culture in ordinary soil, and a fitting ornament for borders or rockwork. It flowers in early summer, and is easily increased by division.

P. alpestris *(Alpine Cinquefoil).*—A rare native plant, closely allied to the spring Potentilla (P. verna). When well grown it forms tufts nearly 1 ft. high, with flowers of a bright yellow, about 1 in. across. It is well worthy of a place on rockwork; it cares little how cold the position, and will enjoy a moist, deep soil. P. verna is also worthy of a place in the garden, and is of the easiest culture. It is not a very common plant, but is found

Portulaca grandiflora fl.-pl.

Potentilla atrosanguinea (*Himalayan Cinquefoil*).

Portulaca grandiflora (*Purslane*).

Group of Potentillas (*Cinquefoils*).

[*To face page* 226.

Flowers of Primula cortusoides amœna (Sieboldi). Natural size.

[To face page 227.

in several parts of the country on rocks and dry banks.

P. ambigua is a creeping plant of dwarf growth and compact habit, bearing in summer numerous large, clear yellow blossoms which just overtop the dense carpet of foliage. It is a plant of the simplest culture and perfectly hardy, requiring only a good, deep, but well-drained soil in an open position in the rock garden. Himalayas.

P. calabra (*Calabrian Cinquefoil*).—A very silvery species, particularly on the under sides of the leaves; the shoots are prostrate, with lemon-yellow flowers nearly 1 in. across. The species is chiefly valuable from the hue of its leaves; it flowers in May and June, and flourishes freely in sandy soil. It is worthy of a place in the rock garden and wherever dwarf Potentillas are grown. A native of Italy and Southern Europe.

P. nitida (*Shining Cinquefoil*).—A beautiful little plant, only a couple of inches high, with silky silvery leaves of three leaflets each, rarely more; the flowers are of a pretty and delicate rose. This native of the Alps is well worthy of a choice place in the rock garden, and is of the easiest culture and propagation.

P. pyrenaica (*Pyrenean Cinquefoil*).—A dwarf, but vigorous and showy species, with fine, large, deep golden-yellow flowers. It is a native of high valleys in the Central and Southern Pyrenees, and easily increased by division or seeds. It will grow, without any particular attention, on rockwork or in the mixed border. The shrubby kind, P. fruticosa, and its varieties are worth growing among small shrubs in rougher parts of the rock garden or naturalising.

Prangos fœniculacea.—An umbelliferous plant of no ornamental value.

Pratia angulata.—There are but few prettier subjects for the rock garden than this New Zealand plant. It creeps over the surface of the soil after the manner of the Fruiting Duckweed (Nertera depressa). The flowers, which resemble those of the common dwarf Lobelia, are borne very profusely, and are of a pure white colour. In autumn they are replaced by numerous violet-coloured berries about the size of large Peas. If placed on an exposed part of the rockery, with ample room to spread, an abundance of blossoms will be the result. It is perfectly hardy and a desirable little plant. Syn., Lobelia littoralis.

Prenanthes.—Composite plants allied to Lactuca, and of botanical interest only.

Prickly Poppy (*Argemone*).

Prickly Thrift (*Acantholimon*).

Primula (*Primrose*).—There is so much charm and beauty among Primroses that no garden can be complete without them, and there is scarcely a species that is not worth cultivating. There exists, too, such a diversity of habit and growth amongst them, that they may be put to a variety of uses. Some are at home on the sunny slopes of the rock garden, others in shade, while many make excellent border flowers, and a few of the exotic species are at home in the woodland with our common Primrose. In dealing with such a large family, which consists of nearly a hundred species and varieties, we have confined ourselves to those that are the most desirable of the cultivated kinds and the most distinct for general cultivation. There exists so much confusion among certain sections of the genus, particularly in the species which inhabit the European Alps and the Himalayas, that we have not attempted to deal with them exhaustively; while others, again, such as P. nivalis, are too little known in gardens to render it necessary that we should speak of their merits.

P. amœna.—An uncommon and beautiful species, allied to our own wild Primrose, but quite distinct. The flowers are large, of a pleasing purple hue, and come out before the snow has left the ground. It is so much earlier than the common Primrose, that while that species is in full flower, amœna has quite finished blooming, and sent up almost the same kind of strong tuft of leaves which the common Primrose does after its flowers are faded. It flourishes quite freely in common borders, and is one of the most valuable additions to the early spring garden and mixed border that has been made for many years. As the leaves are rather large, a sheltered and slightly shaded position will most tend to the perfect health and development of the plant. It is charming for rockwork or well-arranged borders, and, when plentiful enough, will, no doubt, be used in other ways. It is readily propagated by division of the root, and is a native of the Caucasus. The corolla is purplish-lilac in bud or when recently expanded, but turns bluer after a few days. The umbel is many-flowered, and the blooms, which are larger than those of P. denticulata, are borne about 6 in. or 7 in. high; the leaves are woolly beneath and toothed. There is a stemless variety, which would probably prove a great addition to our gardens. P. sibirica is somewhat similar to P. amœna, but is rare.

P. Auricula (*Common Auricula*).—This is, in a wild state, one of the many charming Primulas that rival the Gentians, Pinks, and Forget-me-nots in making the flora of alpine fields so exquisitely beautiful and interesting. Possessing a vigorous constitution, and sporting into a goodly number of varieties when raised from seed, it attracted early attention from lovers of flowers; its more striking variations were perpetuated and classified, and thus it became a "florists' flower." The varieties of cultivated Auriculas may be roughly thrown into two classes: 1st, self-coloured varieties, with the outer and larger portion of the flower of one colour or shaded, the centre or eye being white or yellow, and the flowers and other parts usually smooth, and not powdery; 2nd, those with flowers and stems thickly covered with a white powdery matter or "paste." The handsomest of the former kinds, known by the name of "alpines," to distinguish them from the florists' varieties, are the hardiest of all. The florists' favourites are always readily distinguished by the dense mealy matter with which the parts of the flower are covered. They are divided by florists into four sections—green-edged, grey-edged, white-edged, and selfs. In the green-edged varieties, the gorge or throat of the flower is usually yellow or yellowish; then comes a ring, varying in width, of white powdery matter, surrounded by another of some dark colour, and beyond this a green edge, which is sometimes ½ in. in width. The outer portion of the flower is really and palpably a monstrous development of the petal into a leaf-like substance, identical in texture with that of the leaves. The "grey-edged" have also the margin of a green, leafy texture, but so thickly covered with powder that this is not distinctly seen. This, too, is the case with the "white-edged," the differences being in the thickness and hue of the "paste" or powdery matter. In fact, the terms green-edged, grey-edged, and white-edged are simply used to express slight differences between flowers all having an abnormal development of the petals into leafy texture. It is a curious fact that between the white and the grey the line of demarcation is imaginary, and both these classes occasionally produce green-edged flowers. The "selfs" are really distinct in having the outer and larger portion of the corolla of the ordinary texture, a ring of powdery matter surrounding the eye.

The enumeration and classification of such slight differences merely tend to throw obstacles in the way of the flower being generally grown and enjoyed in gardens. By all means let the florists maintain them, but those who merely want to embellish their gardens with some of the prettier varieties need not trouble themselves with named sorts at all. One fact concerning the florists' kinds should, however, be borne in mind—they are the most delicate and difficult to cultivate. The curious developments of powdery matter, green margins, &c., have a tendency to enfeeble the constitution of the plant. They are, in fact, variations that, occurring in Nature, would have little or no chance of surviving in the struggle for life. The general grower will do well to select the free sorts—alpines, and good varieties of the common border kinds. An especial merit of these is that they may be grown in the open air on rockwork and borders, while the florists' kinds must be grown in frames.

The free-growing kinds are most likely to be enjoyed in all classes of gardens. Their culture is very simple, light vegetable soil and plenty of moisture during the growing season being the essentials. In many districts the moisture of our climate suits the Auricula to perfection, and in such may be seen great tufts of it grown in gardens without any attention. In others it must be protected against excessive drought by stones placed round the plants, and Cocoa-fibre and leaf-mould are also useful as a surfacing. However, none but good varieties of the "alpine" section would justify even this trouble; and, wherever practicable, we should prefer to place these on rockwork, on spots where they could root freely into rich, light soil, and have some shelter. They would cause little or no trouble, save taking up, dividing, and replanting every second or third year, or as often as they become too crowded or lanky. The very common kinds may be planted as edgings, or in beds in the spring garden. In all places where the plant naturally does freely, improved varieties should be substituted for the common old border kind.

Auriculas are easily propagated by division in spring or autumn—best in early autumn. They are also easily raised from seed, which ripens in July, the common practice being to sow it in the following January in a gentle heat. It should be sown in pans thinly. The plants need not be disturbed till they are big enough to prick into a bed of fine rich and light soil on a half-shady border in the open air. It is a most desirable practice to raise seedlings, as in this way many beautiful varieties may be obtained. When a desirable variety is noticed among the seedlings, it should be marked and placed under conditions best calculated to ensure its health and rapid increase, and propagated by division as fast as possible.

As to the florists' varieties, endless precise descriptions of the modes of culture considered necessary to success with these have been given by good cultivators; but the essential points may be summed up in a few words. They require protection in frames or pits during the winter and spring months; and they may be placed in the open air in summer and early autumn. Their suitable winter quarters are shallow pits, in which they are placed as near to the light as may be convenient, the lights being left off in mild weather, and air being given at all times, except in severe frosts. Air at night as well as by day is decidedly beneficial. The aspect of the pit or frame may be the usual one for the winter months; but as soon as the plants begin to show flower, they ought to be removed to one with a northern exposure, so that the bloom may be prolonged. Here, with abundance of air, they form objects of much interest and beauty through the month of April and the first weeks of May. After flowering they should be potted in May, and kept shaded till they have recovered. The potting is usually a process of carefully shaking away all the soil and putting the plant in fresh compost. The practice is well-founded, for it is the habit of this plant and its wild allies to put forth young roots higher up on the stem every year, and the encouragement of these young roots is sure to lead to a good result. The pots generally used (the 4-in. size) are quite large enough where the annual disrooting system is practised, one sucker of a kind being placed in the centre of each pot. The wisdom of applying this system of potting to every plant is, however, doubtful,

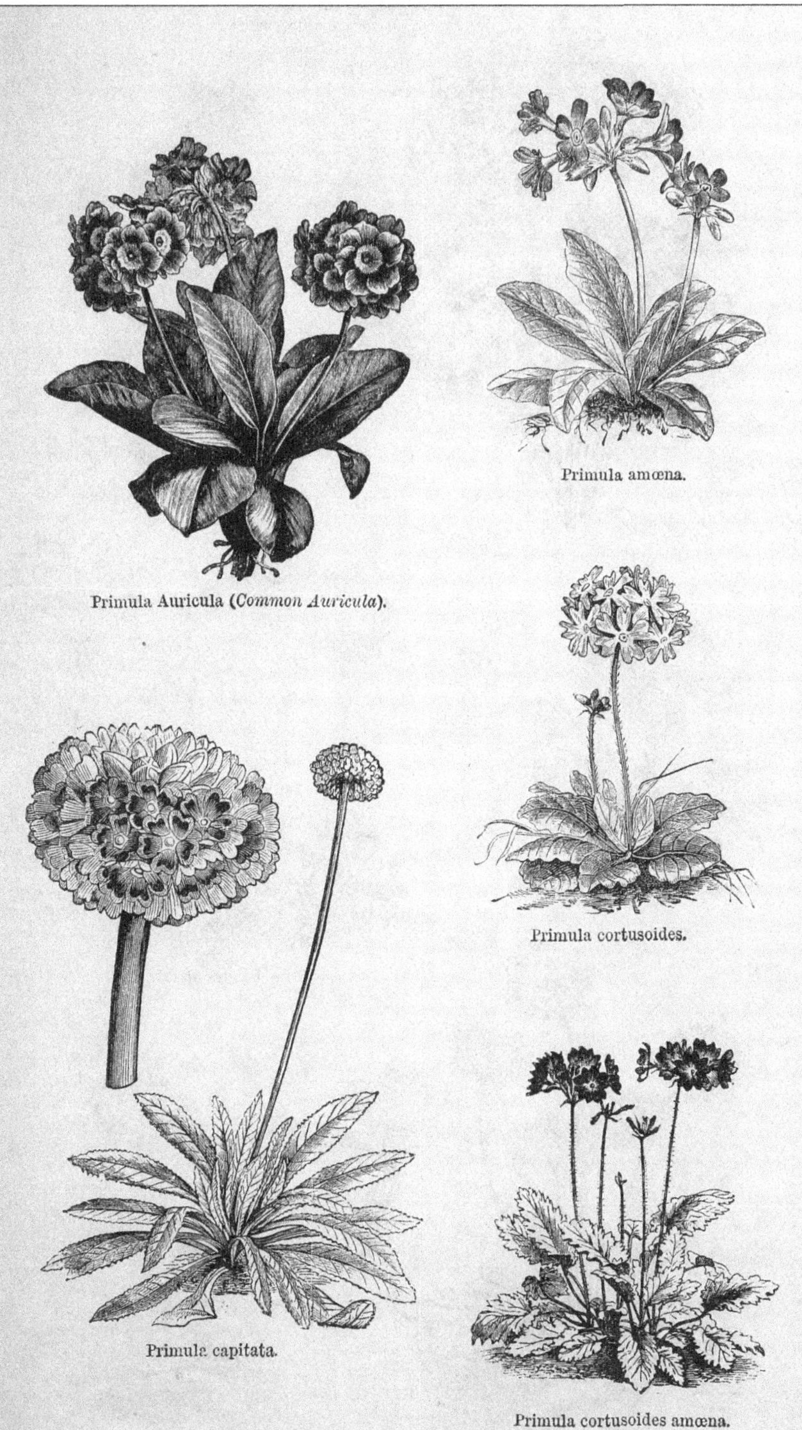

Primula Auricula (*Common Auricula*).

Primula amœna.

Primula cortusoides.

Primula capitata.

Primula cortusoides amœna.

[*To face page* 228.

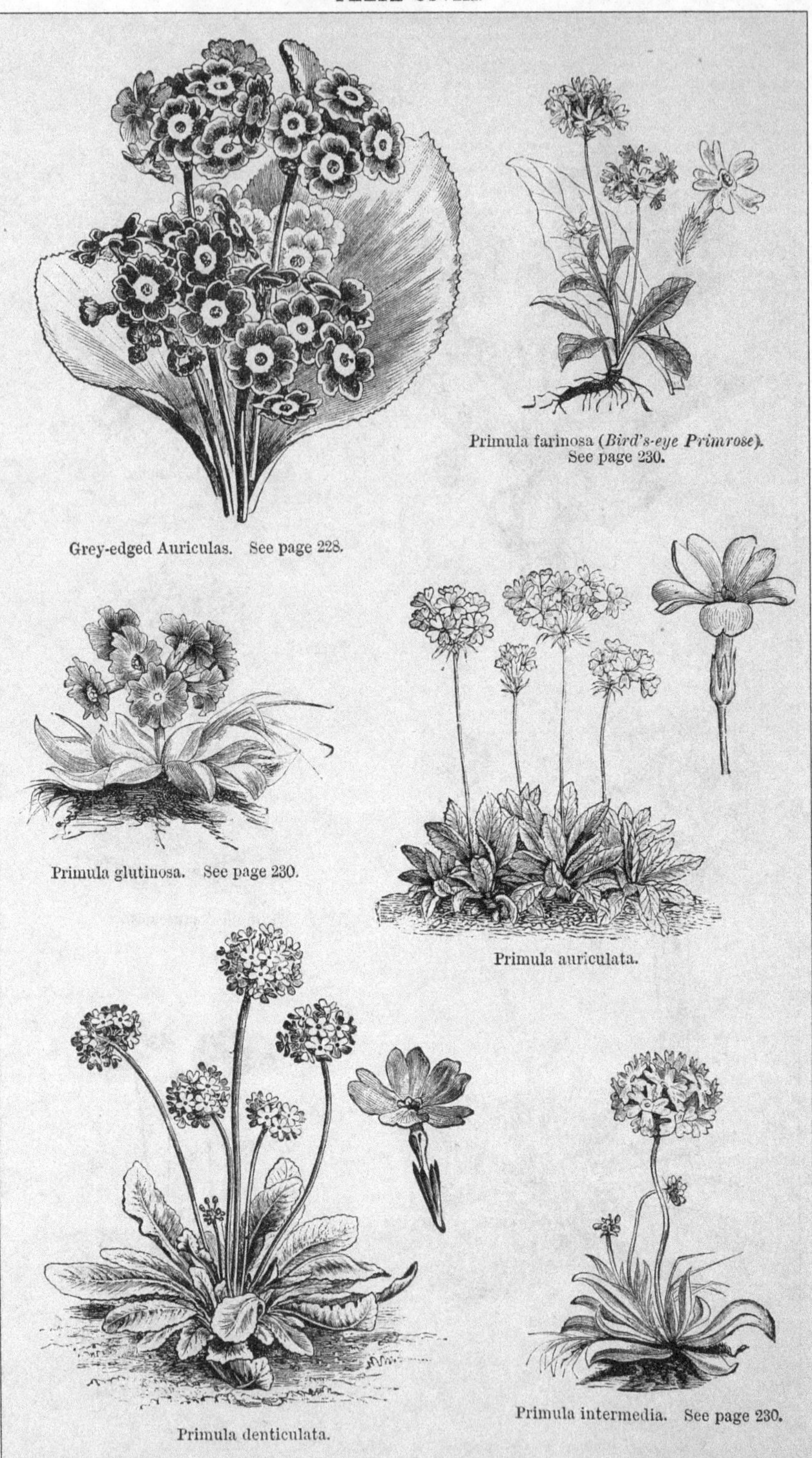

Grey-edged Auriculas. See page 228.

Primula farinosa (*Bird's-eye Primrose*). See page 230.

Primula glutinosa. See page 230.

Primula auriculata.

Primula denticulata.

Primula intermedia. See page 230.

[*To face page* **229.**

and it is better to select those that have sound roots, and are set firmly and low in the earth, and, disturbing the ball but little, give them a careful shift into a 5-in. pot. In case of growing the alpine kinds in pots—and they are quite as well worthy of it as the others—instead of confining themselves to one plant in a small pot, growers should put five or six of a kind in a 6-in. pot, one in the centre and four or five round the side, thus forming a handsome specimen. Or the same principle might be carried out in pans, and with the free-growing florists' varieties as well as the alpines. In summer all the plants should be placed in the open air, and on boards, slates, a bed of coal-ashes, or some substance that can prevent the entrance of worms into the pots. Some careful growers guard the plants from heavy rains; this is unnecessary if the pots are perfectly drained and everything else as it ought to be. The florists rarely plunge them; but, if plunged in a bed of clean sharp sand, or any like material placed on a well-drained bottom, and free from the earthworm, they would be in a safer and certainly less troublesome condition, because free from the vicissitudes that must attend all plants exposed in a fragile porous shell containing but a few inches of soil.

The perfect development of the choicest florists' kinds is secured by a simple mixture of one part good turfy loam, one part leaf-mould, and another composed of well-decayed cow manure and silver or sharp river sand. It should be observed that some pot their plants in August, but the best time is just after the flowering, as, if disrooted in the autumn, the plants have not that accumulated strength for flowering which they possess when the blooming time is preceded by a long period of undisturbed growth. Although we have gone so far in giving directions as to the culture of the florists' varieties, we again earnestly advise all who care about the flower to cultivate the free and hardy forms that thrive in the open air. It is a good plan to select bright or delicate self or other colours that please one. Such kinds should be increased, so that definite effects may be worked out with each colour.

P. capitata.—One of the finest of all Primroses. It is in the way of P. denticulata, but is very distinct from a garden standpoint. It has a tuft of sharply-toothed pale green leaves, not half the size of a fully-developed P. denticulata. It produces in autumn dense heads of flowers of the deepest Tyrian purple, enveloped in a white mealy powder, which shows off the colour to advantage. It is very variable as regards depth of colour, some forms being much superior to others. It is not such a vigorous grower as P. denticulata, though quite hardy, and it cannot be termed a good perennial in our climate, as it is apt to go off after it has flowered well; therefore it is advisable to raise seedlings, and as the plant produces these freely in most seasons, and they flower the second year, this is an easy matter. It thrives best in an open position, with a north aspect, in good loamy soil, well watered in dry weather.

P. cortusoides.—A distinct species, bearing deep rosy clusters of flowers on stalks from 6 in. to 10 in. high. In consequence of its taller and freer habit it is liable to be injured if placed in an exposed spot or open border; therefore it should be put in a sheltered position, in a sunny nook on rockwork, surrounded by low shrubs, &c., or in any position where it will not be exposed to cutting winds, and at the same time not be shaded to its injury. The soil should be light and rich, and a surfacing of Cocoa-fibre or leaf-mould would be beneficial in dry positions. It is one of the most beautiful and easily raised of the Primulas, being readily increased from seed and quite hardy, at least where any pains are taken to give it a well-drained and suitable position. It forms a charming ornament for the lower and less exposed parts of rockwork, for a sunny sheltered border near a wall or house, or for the margin of the choice shrubbery. A native of Siberia.

P. denticulata.—A pretty Himalayan Primrose, of robust growth, having large tufts of broad foliage, which develop in spring large, dense clusters of blossoms on stout, erect stems. It is a most variable plant, and consequently some of the more distinct forms have received garden names, of which these mentioned below are the principal. The type is paler in colour than any of the others, and the foliage and flower-stalks are without meal. It is from 8 in. to 10 in. high. P. pulcherrima is a great improvement on the original. It grows from 10 in. to 12 in. high, and has a more globular flower-truss, of a deeper lilac colour. Its stalks are olive-green, powdered slightly with meal, which is also scattered sparsely over the leaves. This is a very beautiful plant when in flower, and is quite hardy. P. Henryi is a very strong-growing variety, but does not otherwise differ from pulcherrima. It is, however, a very fine plant, often measuring 2 ft. across, and in Ireland it reaches even larger dimensions. P. cashmeriana is by far the finest variety. The flowers are of a lovely, dark lilac, closely set together in almost a perfect globe on stalks over 1 ft. high, and last from March till May. The foliage, too, is beautiful; both it and the stalk are of a bright pale green, thickly powdered with meal, like our own P. farinosa, which in many points it strongly resembles.

This species is perfectly hardy, though the foliage is liable to be injured by early spring frosts. It may be placed either in the rock garden or in an ordinary border, and grows vigorously in a deep, moist loamy soil, enriched by manure. It prefers a shady situation, but with a clear sky overhead, and delights in an abundance of moisture during warm summers. If grown in masses in beds the plants will, when in flower, be greatly benefited by a hand-light or frame placed over them, as this preserves the bloom. P. erosa is similar to P. denticulata, but is smaller, the flowers are paler and not so hardy, and altogether it is an inferior plant.

P. farinosa *(Bird's-eye Primrose)*.—A charming native species, with silvery leaves, in small rosettes, and flower-stems generally from 3 in. to 12 in. high, but sometimes more. The flowers, which are borne in a compact umbel in early summer, are lilac-purple with a yellow eye. In our gardens it loves a moist vegetable soil, and in moist and elevated parts of the country it flourishes on rockwork and slightly elevated beds without any attention; but in most districts a little care is necessary to ensure its perfect health. On rockwork, a moist, deep, and well-drained crevice, filled with peaty soil or fibry sandy loam, will suit it to perfection. When planted on the rock garden in the drier districts, it would be well to cover the soil with Cocoa-fibre or leaf-mould, which would protect the surface from being baked and from excessive evaporation; broken bits of sandstone would also do. It varies a little in the colour of the flower, there being pink, rose, and deep crimson shades. P. farinosa acaulis is a very diminutive variety of the preceding. The flowers nestle in the hearts of the leaves, and both flowers and leaves are very small. When a number of plants are grown together on one sod, or in one pan, they form a singularly charming little cushion of leaves and flowers not more than ½ in. high. Being so diminutive, greater care should be bestowed on it whether it is grown on rockwork or in pots. P. scotica is also a native plant similar to P. farinosa, and requires similar treatment in cultivation. The flowers, which are produced in April, are rich purple with a yellow eye, and are borne on stems a few inches high. Native of damp pastures in the northern counties of Scotland.

P. glutinosa.—A distinct little Primrose, rare in gardens, and flourishing in peaty soil at elevations of 7000 ft. to 8000 ft. on mountains near Gastein and Salzburg, in the Tyrol, and in Lower Austria. It grows from 3 in. to 5 in. high, and bears from one to five blossoms of a peculiar purplish-mauve, with the divisions rather deeply cleft. Suitable for rockwork or pots in moist peaty or very sandy soil. Similar to this species are P. tirolensis, Flœrkiana, Allioni, and others, all natives of the European Alps.

P. grandis.—A distinct species from the Caucasus, remarkable only for the large size of its foliage and the smallness of its flowers.

P. integrifolia.—A diminutive Primrose, easily recognised by its smooth, shining leaves, which lie quite close to the ground, and in spring, when in bloom, by its handsome rose flowers, which are borne one to three on a dwarf stem, and are often large enough to obscure the plant that bears them. There is no difficulty in growing this plant on flat, exposed parts of the rock garden, with the soil moist and free, but firm. The best way is to form a wide tuft of it, by dotting from six to a dozen plants over one spot, and, if in a dry district, scattering a little Cocoa-fibre mixed with sand between them to prevent evaporation. This, or stones, will help till the plants become established. It flowers in early summer, and is increased by division and by seeds. P. Candolleana is another name for this, and P. glaucescens, spectabilis, Clusiana, and Wulfeniana are of a like character. All are natives of the European Alps.

P. intermedia.—A charming plant, a hybrid between P. ciliata and P. Auricula. Its habit closely resembles that of some of the dwarf forms of alpine Auriculas, its flowers, purplish-crimson in colour, with a conspicuous yellow eye, being borne on stout, erect scapes. It is delicately fragrant, and well deserves culture on sheltered portions of the rock garden, where its richly-tinted blossoms are seen to advantage.

P. japonica.—One of the handsomest Primroses known, and now too common to need description. It has proved a first-rate border plant, is a good perennial, and is not in the least tender. In moist shady spots, in deep rich loam, it grows as vigorous as a Cabbage and throws up flower-stems 2 ft. or more high, unfolding tier after tier of its beautiful blossoms in continuous succession for several weeks. There are several forms of it differing in the colour of the blossoms; there is a white, a pale pink, and a rose, but the best is the original rich crimson form. It may be grown either in the rock garden or border, and is also an excellent wild garden plant, as it thrives almost anywhere and sows itself freely. It is said to be rabbit-proof. In raising it from seed it should be borne in mind that the seed remains some time dormant, unless sown as soon as gathered, and that it must on no account be sown in heat. A cool frame is the proper place for the seed-pan, and care must be taken to prevent or keep down the growth of Moss and Liverwort on the surface of the soil, till the seed has germinated.

P. latifolia.—A handsome and fragrant Primrose, with from two to twenty violet flowers in a head. It is less viscid, larger, and more robust than the better known P. viscosa of the Alps; the leaves sometimes attain a length of 4 in. and a breadth of nearly 2 in. It grows to a height of from 4 in. to 8 in., flowers in early summer, and in a pure air will thrive on sunny slopes of rockwork in sandy peat, with plenty of moisture during the dry season, and perfect drainage in winter. Like P. viscosa, it will bear frequent division, and it may also be well and easily grown in cold frames or pits. European Alps.

P. longiflora is related to our Bird's-eye Primrose, but is distinct from it, and considerably larger than the best varieties of that species, the lilac tube of the flower being more than 1 in. long. It is not at all difficult to cultivate on rockwork or in pots, and the treatment and position recommended for P.

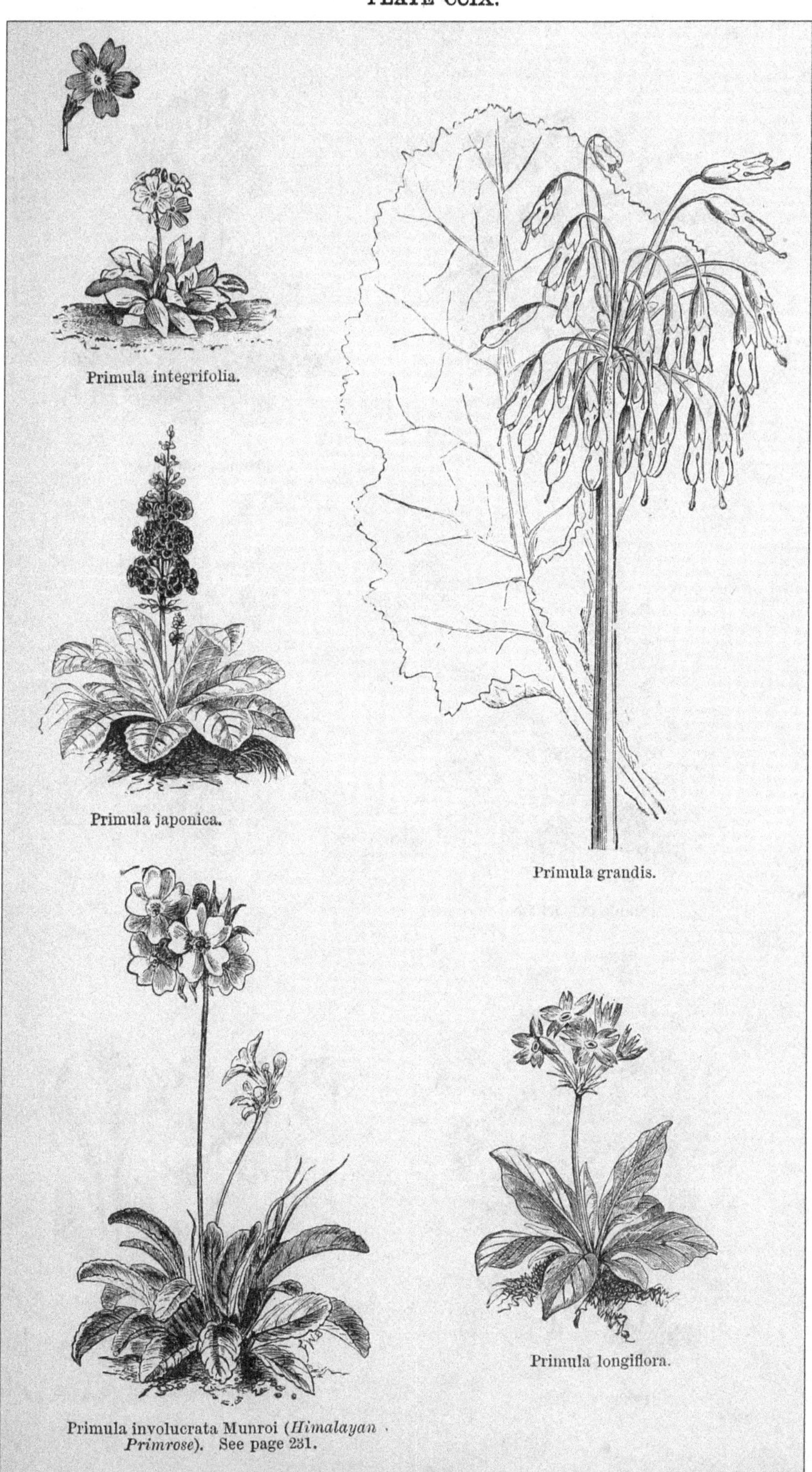

Primula integrifolia.

Primula japonica.

Primula grandis.

Primula involucrata Munroi (*Himalayan Primrose*). See page 231.

Primula longiflora.

Primula minima (*Fairy Primrose*).

Primula marginata.

Primula pulcherrima.

Primula Palinuri.

Primula Parryi.

Primula sikkimensis. See page 232.

[*To face page* **231.**

farinosa will suit it perfectly. In colour it is deeper than the Bird's-eye Primrose. Native of Austria.

P. luteola.—One of the handsomest of the yellow-flowered Primroses, and a noble plant when well grown, the flower-stems often attaining from 1½ ft. to 2 ft. in height; usually, however, they are under 1 ft. The flowers are like a Polyanthus or Auricula, but are borne in more compact heads; the stems sometimes become fasciated, and thus carry a huge cluster of flowers from 4 in. to 6 in. across. It is perfectly hardy, likes a moist situation in full exposure, and if put out in rich borders of rather moist soil, or on the lower banks of the rock garden, or in a copse with a good bed of leaf-soil, it will soon repay the planter. Native of the Caucasus. It has been well figured in *The Garden*, from plants that flowered in Oxfordshire, at Chipping Norton.

P. marginata.—One of the most attractive of the family, and readily distinguished by the silvery margin on its greyish leaves, and by its sweet, soft, violet-rose flowers, which appear in April and May. Even when not in flower the plant is pleasing from the tone of the margins and surfaces of the leaves. Our wet and mild winters are doubtless the cause of this and other kinds becoming rather lanky in the stems after being more than a year or so in one spot. When the stems become long, and emit roots above the surface, it is a good plan to divide the plants and insert each portion firmly down to the leaves. This will be all the more beneficial in dry districts, where the little roots that issue from the stems would be more likely to perish. It is a charming ornament for rockwork where it thrives freely. In the open ground a few bits of broken rock placed around each plant, or amongst the plants if they are in groups or tufts, will do good by preventing evaporation, and also by acting as a protection to the plant, which rarely exceeds a few inches in height. A native of the Alps.

P. minima *(Fairy Primrose)*.—A diminutive plant, one of the smallest of the European Primroses. It grows only an inch or so high, and though its leaves are so small its flowers are nearly 1 in. across, and almost cover the tiny rosettes of foliage. It blooms in early summer, and the stem rarely bears more than one flower, which is rose coloured, but sometimes white. Bare spots in firm open parts of the rockwork are the best places for it, and the soil should be very sandy peat and loam. It is peculiarly suited for association with the very dwarfest and choicest alpine plants. It may be propagated by division or by seed. Mountains of S. Europe. P. Flœrkiana is very like the Fairy Primrose, probably is a variety of it, and differs only in bearing two, three, or more flowers, instead of a single bloom. It is a native of Austria, and will be found to enjoy the same conditions and positions on rockwork as the preceding. Of both it is desirable to establish wide-spreading patches on firm bare spots, scattering half an inch of silver sand between the plants to keep the ground cool.

P. Munroi.—This has not the brilliancy or dwarfness of the Primulas of the high Alps, nor the vigour of our own wild kinds, but it has the merit of distinctness, and is of the easiest culture in any moist, boggy soil. It grows at very high elevations on the mountains of Northern India, in the vicinity of water, and bears creamy-white flowers, with a yellowish eye, more than an inch across, on stems 5 in. to 7 in. high. These latter spring from smooth green leaves, which are 2 in. long and nearly as much across, and which have a heart-shaped base. The flowers, which appear from March to May, are very sweet, and altogether it highly merits culture in a moist spot on the select rockwork or in a bog. P. involucrata is a very closely allied kind, from the same regions. It is, however, somewhat smaller, the leaves are not heart-shaped at the base, and the plant is not quite so ornamental. It thrives under the same conditions as its relative.

P. Palinuri.—This is quite removed from other cultivated Primroses, inasmuch as it seems to grow all to leaf and stem; whereas many of the other kinds often hide their leaves with flowers. In April the bright yellow flowers appear in a bunch at the top of a powdery stem, emit a Cowslip-like perfume, and are ornamental, though they rarely seem to fulfil the promise of the vigorous-looking plant. It flourishes in rich light soil as a border plant in various parts of these islands, so that nothing more need be said of its culture. Established plants are easily increased by division. It is well suited for some isolated nook on rockwork, where there is an unusually deep bed of soil. Southern Italy.

P. Parryi.—A pretty Primrose, bearing about a dozen large, bright, purple, yellow-eyed flowers, nearly 1 in. across, on stems about 1 ft. high. Though an undoubted alpine, growing on the margins of streams near the snow line, where its roots are constantly bathed in the ice-cold water, it has succeeded in the open border in moist deep soil of a loamy nature mingled with peat. It is perfectly hardy, requiring protection by partial shade from extreme heat rather than from cold. N. America.

P. purpurea.—A handsome Primrose, allied to P. denticulata, but far finer; the flowers, of an exquisite purple, are larger, and are borne in heads about 3 in. across; the leaves are entire, and it may thus be distinguished from its near relations. Sheltered and warm positions, but not very shady, on rockwork, or in the open parts of the hardy fernery, will best suit it, the soil being of a light, deep, sandy loam, well enriched with decomposed leaf-mould. It never thrives so well as when planted in nooks at the base of rocks, where it enjoys more heat than if exposed. This must not be confused with the common P. purpurea, which is but a variety of P. denticulata.

P. rosea is a charming little Primrose, with flowers, of the loveliest carmine-pink, produced in heads like the Polyanthus. Its pale green leaves form compact tufts, and the flower-stems, from 4 in. to 9 in. high, are produced in early spring, often as many as half a dozen from one plant. It is perfectly hardy, and though it has only recently been introduced from the Himalayas, it has become quite acclimatised and grows vigorously in almost any soil, preferring, however, one of deep rich loam in a moist, shady position in the rock garden. Where plentiful it should be tried in various positions and soils, as it has not yet been thoroughly tested with regard to its likes and dislikes.

P. Sieboldi.—Though this very handsome Primrose is considered as a variety of P. cortusoides, it is very distinct in many important particulars, for apart from the size of the flowers and the breadth of the foliage, the creeping root, the exclusively vernal habit of the plant, the pseudo-lobed or grooved seed-vessel, and the roundish flattened form of the seed, especially the two last features, warrant the belief in its distinctness from P. cortusoides from a garden point of view. It is at any rate one of the showiest and most charming of all, easy to grow and as hardy as many others. Since its introduction from Japan numerous beautiful varieties have been raised, some of the most distinct being clarkiæflora, lilacina, marginata, fimbriata oculata, vincæflora, cœrulea-alba, Mauve Beauty, Lavender Queen, laciniata, and maxima. These possess a great diversity of colour, and some have the petals beautifully fringed. One chief merit possessed by these Japanese Primroses is that they bloom early, coming in about the month of April, when flowering plants are not particularly plentiful; another is, that they are remarkably free bloomers, throwing up successional flower-stems, and lasting a long time in perfection. Their cultivation, also, is comparatively easy. The best soil for them is light, rich, free material, consisting of fibry loam, leaf-mould, pulverised manure, and some grit to keep it open. They are impatient of excessive moisture, and when planted in the open ground should be in thoroughly drained soil, or in raised positions on rockwork. The roots creep just below the surface and form eyes, by means of which any one variety can be easily propagated. P. Sieboldi is a hardy herbaceous perennial, which loses its leaves in autumn and winter, when it goes to rest, and breaks up again early in spring.

P. sikkimensis.—This is a robust growing species, deciduous or herbaceous in our climate, and quite distinct from all other known species. When well grown, it throws up strong flower-stems, from 15 in. to 24 in. high, bearing numerous bell-shaped flowers, pale yellow, without a spot of any other colour, the pedicels mealy, the blooms of an agreeable and peculiar perfume. Some of the stems bear a head of more than five dozen buds and flowers, and each individual flower is nearly 1 in. long and more than ½ in. across. It starts into growth quite late in the spring, in April or early in May, and should have a shady position when in bloom, as its delicate blossoms suffer from cutting winds and bright sunshine. It is perfectly hardy, and loves deep, well-drained, and moist ground; spots in the lower parts of rockwork near water, or in deep boggy places, suit it best. It begins to flower in May, remains in flower for many weeks, and is readily increased, either by seeds sown as soon as they are ripe in summer, or by careful division in autumn or spring. It is said to be the pride of all the Primroses of the mountains of India, inhabiting wet, boggy localities at elevations of from 12,000 feet to 17,000 feet, and covering acres of ground with its yellow flowers.

P. Stuarti *(Stuart's Primrose)*.—A noble and vigorous yellow Primrose, growing to about 16 in. high. It has leaves nearly 1 ft. long, and the umbels are many-flowered. Like P. denticulata, the place most suitable for this is some perfectly drained and sheltered spot on slightly elevated rockwork; if convenient, plant it against the base of rocks, which will shelter it from cutting winds, though, when sufficiently plentiful, this precaution may be dispensed with. A light deep soil, never allowed to get dry or arid in summer, will suit it well. Mountains of India.

P. villosa is a lovely little Primrose, though one of the oldest cultivated. It is known by its dark green obovate or suborbicular leaves with close-set teeth, covered with glandular hairs, and viscid on both sides; the flower-stems, which elevate the sweet blooms barely above the foliage, are also viscid. It is well adapted for rockwork, on which it may be grown in any position, in light peaty or spongy loam, with about one-half its bulk of fine sand, provided its roots are kept moist during the dry season. It is easily increased by division, and may also be raised from seed. Varieties are sometimes, but rarely, found with white flowers. It is sometimes grown under the name of P. viscosa. The variety nivea or nivalis is a beautiful plant, dwarf and neat in growth; it freely produces trusses of lovely white flowers, quite distinct in aspect from any other in cultivation. It is very easy of culture, and may be grown in pots or in the open ground. It deserves a select position on rockwork or in the border, a light free soil, and plenty of water during the warm season. It flowers in April and May, and is a native of the Alps. Similar to P. villosa are P. ciliata, Steini, hirsuta, pubescens, rhætica, pedemontana, œnensis, and Dinyana, charming little species from the European Alps, and very desirable for cultivation. All thrive under the same conditions as P. villosa.

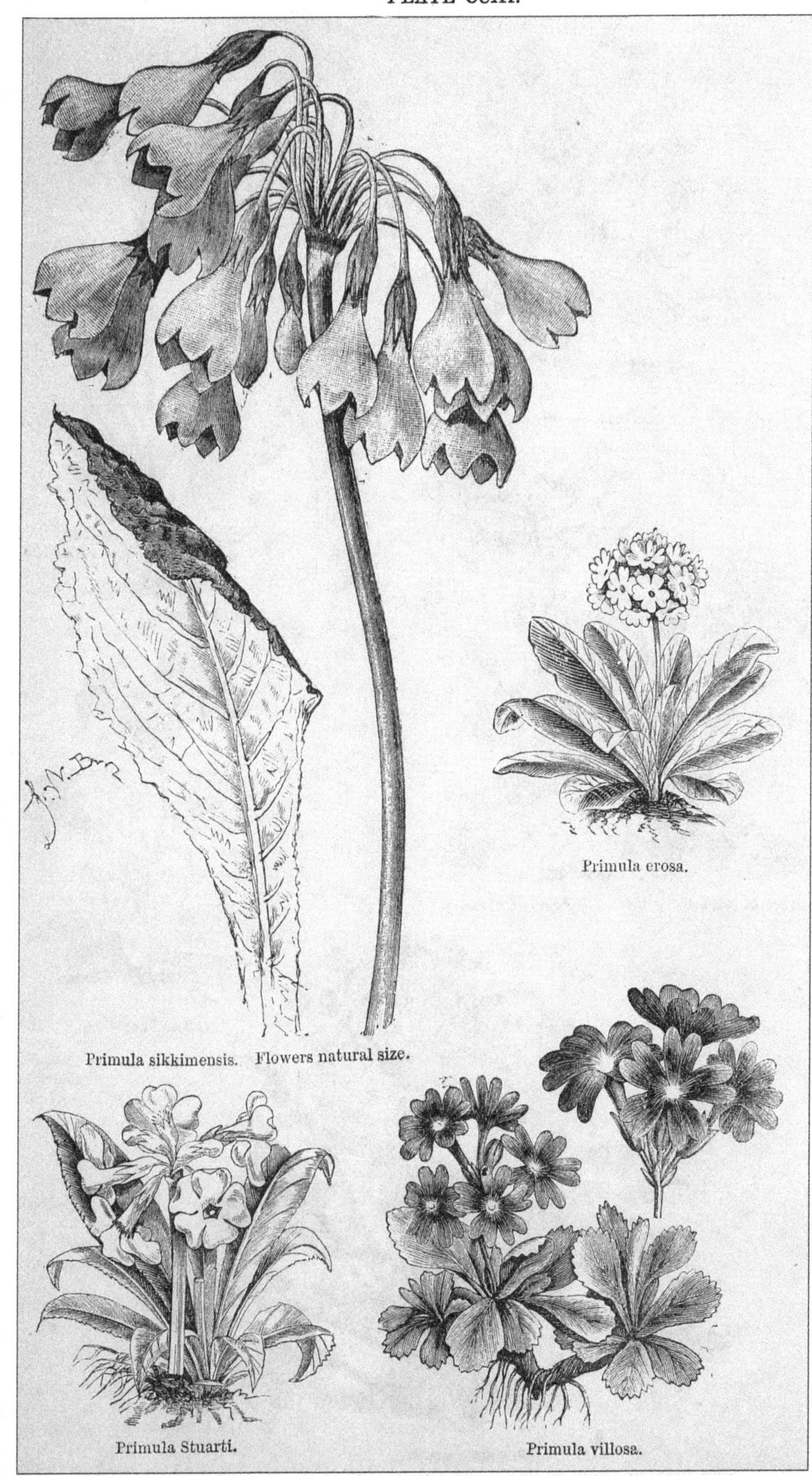

Primula sikkimensis. Flowers natural size.

Primula erosa.

Primula Stuarti.

Primula villosa.

[To face page 232.

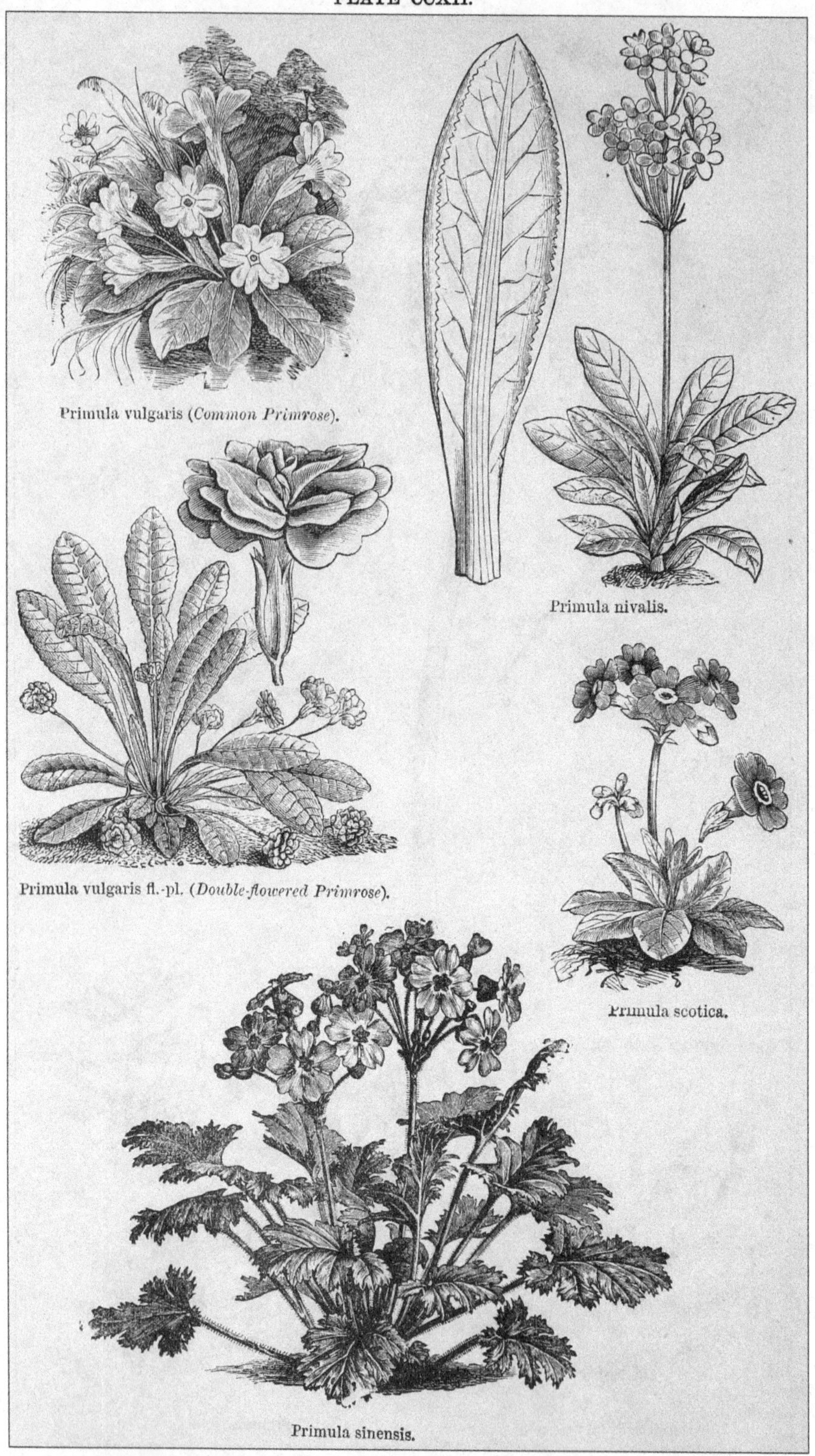

Primula vulgaris (*Common Primrose*).

Primula nivalis.

Primula vulgaris fl.-pl. (*Double-flowered Primrose*).

Primula scotica.

Primula sinensis.

[*To face page* 233.

P. vulgaris *(Common Primrose).*—Of all the Primula family none can excel our native Primroses in loveliness, and they are the earliest of all to flower. The Gentians and dwarf Primulas do not do more for the Alps than this charming wilding for the hedgebanks, groves, open woods, and borders of fields and streams of the British Isles. In some places it varies a good deal in colour, and some of the prettiest of the wild varieties are worthy of being introduced into shrubberies and semi-wild places, and also into gardens. Although it does not vary so much as the Cowslip, yet it does so in a degree that ought to make it much more valuable in the hands of flower gardeners than it is at present. All the varieties would perhaps find their most appropriate home in our woods and shrubberies; but so long as rich and lovely colour and fragrance are esteemed in the flower garden in spring, some of the more distinctly toned varieties should be sought after. Varied hues of yellow, red, rose, lilac, bluish-violet, lilac-rose, and white have already been raised, and no doubt many others will be raised in future, particularly if the good single varieties should become popular plants for the spring flower garden. Striking and desirable variations from the commoner types will then be much more likely to be preserved. For shrubberies and woodland walks these single varieties will always prove more useful and effective than the old double kinds, because more vigorous and easily increased. All the varieties are readily increased by division of the offsets, or by seeds, which are produced in abundance. Placed in woods and shrubberies, the plants will take care of themselves—a quality that adds to their charms. For the flower garden some system of culture must be pursued; a very simple one will secure the best results both as to the production of vigorous free-blooming plants and an abundant stock. After the summer occupants of the flower beds are faded and removed in autumn, the Primroses and other spring flowers are planted in the beds as the taste of the grower may direct. About the middle or the end of May it will be time to think of preparing the beds for their summer ornaments, and by that time the Primroses will have begun to fade after furnishing a long and abundant bloom. Then take them up, divide the offsets singly, doing this in a shed or shady position if the day be sunny. Tufts of new or scarce varieties, or those of which a large stock is required, may be divided into the smallest offsets; where much increase is not required, the plants should simply be parted sufficiently to allow of their healthy development. As soon as they are parted, they should be planted in the kitchen garden or some by-place. The more rich and moist the soil is, the better they will grow; and if the position be a half-shady one, it will be an improvement. The alleys between the Asparagus beds would do admirably for them, in case room was scarce and more convenient positions could not be found. If the weather be very bright, it would be desirable to shade them for a few days after planting, simply by spreading boughs or old garden mats over them, and it need hardly be added they should at this time be thoroughly watered. They should be planted in lines, 10 in. or 12 in. apart each way, if the plants be strong and regular in their development, and, if small offsets, closer in the lines in proportion to their size. By autumn they will make fine plants, and may then be taken up; as much of the root as will come up with ordinary care, but not of necessity any soil or ball, should be preserved, and they should be transferred to the beds in the flower garden or pleasure ground. The varieties of the single coloured Primroses are so numerous, that it seems a folly to name them; but a few of the most distinct that are propagated by division have received names. Amongst these may be mentioned Auriculæflora, one of the finest; Altaica, or grandiflora, also a beautiful sort; Rosy Morn, deep rosy-red; Gem of Roses, rosy-pink; Queen of Violets, deep purplish-violet; Crimson Banner, deep maroon-crimson; Violacea, pale purple; Fairy Queen, pure white, with good eye; Sulphurea, large sulphur colour; Virginia, pure white; Brilliant, rich vermilion-red; King of Crimsons, rich massive crimson; Violetta, a very beautiful violet-purple; Lustrous, very deep coloured crimson, with small perfect lemon eye; and Scott Wilson, a singular bluish-purple. The propagation of these kinds, as well as of all the perennial Primroses, is necessarily slow, except where they can be reproduced true from seed. A seedling plant may produce two others the first year after blooming; these may again produce six or eight the next year, and thus it will take several years to work up a hundred plants, a fact which shows that some patience must be exercised ere the newest forms can be circulated largely.

DOUBLE VARIETIES.—The forms of the plant most precious for the garden proper are the beautiful old double kinds. No sweeter or prettier flowers ever warmed into beauty under a northern sun than their richly and delicately-tinted little rosettes. Once they were grown in every garden; then the day came when they, in common with many hardy flowers, were cast aside to make way for gaudier things; but now people are beginning to grow them again, and are inquiring where they can get some old and half-lost kinds they used to know long ago. The best known and most distinctly marked kinds are the double lilac, double purple, double sulphur, double white, double crimson, and double red. These and several closely allied forms are occasionally honoured with Latin names descriptive of their shades of colour. In catalogues of the present day will be found the following: Primula vulgaris alba plena, lilacina plena, purpurea plena, rosea plena, rubra plena, sulphurea plena; but we had better speak of them in plain English and confine the Latin term to the species. The double kinds, more delicate and slower growing than the single ones, require more care, and in their case the development of healthy foliage after the flowering season should be the object of those who wish to succeed with them. It is a curious fact that the deeper the hue in the double kinds the less robust the plant, and both the rich crimsons and deep purples are most difficult to cultivate in ordinary summers; but in the extreme north, where the climate is at once temperate and moist, they grow almost with luxuriance. The climate of Ireland also favours them considerably, but in the south and midland districts it is found necessary to afford them shade and abundant moisture during summer, and the protection of glass generally from the effects of continued frosts and rains in the winter. The white, lilac, and sulphur kinds are, on the other hand, very hardy, and, if established, appear to stand our summers and winters with impunity.

Shelter and partial shade are the two conditions chiefly necessary to their successful culture. Open woods, copses, and half-shady places are the favourite haunts of the Primrose in a wild state. In them, in addition to the shade, it enjoys shelter not merely from tall objects around, but also from the long Grass and other herbaceous plants growing in close proximity; and we should also take into account the moisture consequent upon such companionship, and let these facts guide us in the culture of the double kinds. As will be readily seen, a plant exposed to the full sun on a naked border would be under a very different condition to one in a thin wood; the excessive evaporation and searing away of the leaves by the wind would be quite sufficient to account for the failure of the exposed plant.

It is therefore desirable, in the case of the beautiful double Primroses, to plant them in borders in slightly shaded and sheltered positions, using light rich vegetable soil, and, if convenient, keeping the earth from being too rapidly dried up by spreading Cocoa fibre or leaf-mould on it in summer. It would be better to plant them permanently in some favourite spot and leave them alone than to change them repeatedly from place to place. They may, however, be employed as bedding plants, and successfully treated in the manner recommended for the single varieties, but for this purpose they are not so useful or pretty as when seen in good single tufts. They are increased by division of the roots, and taking them up in order to divide these is the only disturbance they should suffer. The double Primroses well grown, and the same kinds barely existing, are such very different objects, that nobody will begrudge giving them the trifling attention necessary to their perfect development. Occasionally they may be seen flourishing by chance in some cottage or old country garden, where they find a home more congenial than the prim and bare fashionable flower garden.

THE POLYANTHUS.—Though the origin of this old-fashioned, yet beautiful, flower is somewhat involved in obscurity, it is considered a form of the common P. vulgaris with the stems developed. Polyanthuses present a wonderful array of beauty, and are not at all sufficiently appreciated. For rich and charmingly inlaid colouring they surpass all other flowers of our gardens in spring. It would require pages to describe even the good varieties. At one time the Polyanthus was highly esteemed as a florist's flower, and none in existence better deserved the attention and esteem of amateurs; but nearly all the choice old kinds are now lost, and florists who really pay the flower any attention are very few indeed. In consequence, however, of the great facility with which varieties are raised from seed, nobody need be without handsome kinds, and raising them will prove interesting amusement for the amateur. The laws of the florists are in this case of little more value than usual, and Maddock, in the following passage, describes a very beautiful type of the numerous variations of this flower: "The ground colour is most to be admired when shaded with dark rich crimson resembling velvet, with one mark or stripe in the centre of each division of the limb, bold and distinct from the edging down to the eye, where it should terminate in a fine point." He further says: "The pips should be large, quite flat, and as round as may be consistent with their peculiarly beautiful figure, which is circular, excepting those small indentures between each division of the limb, which divide it into five or six heart-like segments. The edging should resemble a bright gold lace, bold, clear, and distinct, and so nearly of the same colour as the eye and stripes as scarcely to be distinguished. In short, the Polyanthus should possess a graceful elegance of form, a richness of colouring, and symmetry of parts not to be found united in any other flower." Here, however, as in most similar cases, the general cultivator would do well to select the most diverse of the varieties that he may raise from seed or otherwise become possessed of, and not be tied by any conventional rules.

As to the capabilities of the various kinds of Polyanthus, it would be difficult to name a hardy flower so generally useful. The finer varieties are worthy of a place on rockwork amidst the choicest alpine plants; the showier ones do for bedding in the spring garden. Numbers of the vigorous varieties so easily raised from seed will form the most appropriate ornaments that can be massed alongside of shady walks in pleasure grounds. Some may be employed as edgings. The enthusiastic florist grows the finer ones in pots, though many varieties are worthy of being abundantly naturalised in pleasure grounds and along wood walks. They are perhaps scarcely so much to be recommended for using in masses in the spring garden as are the finer varieties of the Primrose—requiring, in fact, to be seen rather closely to be admired; but wherever flowers are placed for their individual beauty rather than for their mere effect as colouring agents, they are invaluable, and should be seen in strong tufts round every shrubbery border. The cultivation of Polyanthuses is, happily, almost as simple as that of meadow grass. They grow vigorously in almost any kind of garden soil, but best in that which is somewhat rich and moist. They thrive in the full sun, but enjoy best a partially shaded and sheltered position, and are somewhat impatient of heat and drought. When grown for bedding purposes, they are, like the Primroses, &c., removed from the flower garden to the kitchen garden or nursery in early summer, and again conveyed there when the summer bedding plants have passed away.

Some varieties, a good deal larger in their parts than the common type, have been raised of late, and these are also very easy of culture and very vigorous; there are very few, if any, double varieties. There are, however, some that are curious and interesting from the duplication of the calyx or corolla, and these are popularly known as "hose-in-hose" Polyanthuses; they grow with the same facility as the others. Where soil is prepared for the choicer varieties, any good loam with a free addition of decomposed leaf-mould and decomposed cow manure and sand will form an admirable compost. The Polyanthus may be raised with great facility from seed, which should be sown immediately after it is gathered from the plants, say about the end of June. It will grow with vigour if kept till the following spring, but by sowing it immediately nearly a year is gained. The amateur wishing to raise choice kinds had better sow the seed in rough wooden boxes or in pans, but for all ordinary purposes a bed of finely pulverised soil in the open air will answer to perfection. Sowings in early spring are better made in rough shallow boxes or pans placed in cold frames, as time will be gained thereby; but the best plan is not to lose the time by allowing the seed to remain idle in the drawer all the autumn and winter, but to sow it directly it is ripe, and by doing so have strong plants the following spring.

The Gold-laced Polyanthuses are much prized, and they are very beautiful. The best are those that were raised years ago, such as Cheshire Favourite, George the Fourth, Formosa, Duke of Wellington, Black Prince, Lancashire Hero, and others, and they are found mentioned in most florists' catalogues of hardy plants. The common Oxlip is a hybrid more or less intermediate between the Cowslip and the Primrose. It differs from the true or Bardfield Oxlip by having much larger and brighter-coloured flowers on longer footstalks, and by showing in the throat of the flower the five bosses characteristic of the Primrose and Cowslip. Some varieties approach the Cowslip, and others the Primrose in character. There are various forms, not usually so ornamental as the varieties of the Polyanthus, but more worthy of culture than the true Oxlip of the eastern counties. When cultivated in gardens, the positions and treatment that suit the Polyanthuses and Primroses will also suit the Oxlip. P. suaveolens is a variety of the Cowslip found in many parts of the Continent, and not sufficiently distinct or ornamental to merit cultivation. P. elatior is the true as distinguished from the common Oxlip. It is not a very ornamental species, the flowers being of a pale buff yellow, and it is readily distinguished by its funnel, and not saucer-shaped, corolla, which is also quite destitute of the bosses that are present in the Primrose and Cowslip. It is found in woods and meadows on clayey soils in the eastern counties of England, particularly in Essex, Suffolk, and Cambridgeshire. It is of easy culture, and most suitable for botanic gardens or collections of interesting plants, being neither distinct nor ornamental enough for very limited collections of ornamental kinds only. This plant is also known by the name of the Bardfield Oxlip. The blue Polyanthus (P. elatior cœrulea) is a singularly handsome variety, having flowers of a slaty-blue colour. It is now rare in gardens, but is well worth growing.

Prince's Feather *(Amarantus).*

Prosartes.—Liliaceous plants, allied to Polygonatum (Solomon's Seal), but the species at present in cultivation are scarcely worth attention.

Prunella grandiflora *(Large Self-heal).*—A handsome and vigorous plant, readily distinguished by its large flowers from the common British Self-heal (P. vulgaris), which is unworthy of cultivation. There is a white as well as a purple variety, both handsome plants that thrive in almost any ground, but prefer a moist and free soil and a position somewhat shaded. They are apt to go off in winter on the London clay, at least on the level ground, but are well suited for the mixed border, for banks, or for naturalisation in copses. A native of Continental Europe, flowering in summer. The variety laciniata has deeply-cut leaves. P. pyrenaica (Pyrenean Self-heal) is allied to the preceding species, and considered a variety of it. It is 8 in. to 1 ft. or more high, and the flowers, which are of a beautiful violet-purple, are larger than those of P. grandiflora. The same positions and treatment as recommended for the preceding kind are suitable. Labiatæ. (= Brunella.)

Psilostemon orientale (= Borago).

Psoralea.—Perennial plants of the Pea family, of which there are some half a dozen kinds in cultivation. None, however, are attractive enough for general cultivation, and they are suitable only for botanical collections. P. acaulis, bituminosa, macrostachya, orbicularis, physodes, Onobrychis, and strobilina are the species in cultivation at Kew.

Ptarmica (= Achillea).

Pteris *(Brake)*.—The common Bracken (P. aquilina) is the only thoroughly hardy species of this genus, and this is generally so common as not to merit cultivation. If anyone, however, wishes to introduce it where it is not plentiful, it should be borne in mind that to transplant it successfully large sods containing the strong creeping roots must be dug up and planted in light soil; if peaty, so much the better. In very mild localities, such species as P. cretica and the elegant P. scaberula, from New Zealand, will sometimes thrive in sheltered nooks.

Pterocephalus Parnassi.--A Scabious-like plant of dwarf, compact growth, forming a dense rounded mass of hoary foliage which in summer is studded with mauve-coloured flower-heads. It is a most desirable plant; thrives best in light, warm soils, and is adapted either for the rock garden or the ordinary border. (= Scabiosa pterocephala.) Greece.

Puccoon *(Sanguinaria canadensis)*.

Pulicaria *(Fleabane)*.—Composite plants of a weedy character.

Pulmonaria *(Lungwort)*.—These are very vigorous and hardy, thriving on any soil, and forming attractive beds in the spring garden. They are also worthy of being planted in semi-wild places, but do not deserve a place in the select borders except where more valuable plants are very scarce. Most of them grow well under the shade of trees, and all thrive best in shade. There are about half a dozen kinds, all of which bear a strong resemblance to each other. P. officinalis and angustifolia are native plants, the former with rose flowers turning to blue, the latter with blue flowers. P. mollis is intermediate between these two, and P. grandiflora is somewhat similar to P. officinalis, of which P. saccharata is a synonym. P. azurea has rich blue flowers. All are of compact growth, and form dense tufts of foliage, which is generally handsomely blotched and speckled with white. Chiefly natives of Europe. Boraginaceæ.

Pulsatilla (= Anemone).

Purple Coneflower *(Echinacea)*.

Purple Loosestrife *(Lythrum Salicaria)*.

Purple Sweet Sultan *(Centaurea moschata)*.

Puschkinia scilloides.—A charming bulbous plant, one of the most beautiful of spring flowers. In respect of growth it bears a resemblance to some of the Scillas, but the flowers, produced on spikes 4 in. or 5 in. high, are of a delicate blue, with each petal marked through the centre with a darker hue. There are two forms of the plant, differing only in the flower-clusters. In the ordinary form the cluster is looser and fewer-flowered than the other, which is called compacta on account of the blossoms being more numerous and densely arranged on the spike; hence this latter is the handsomer. P. scilloides is also known as P. libanotica and P. sicula. The culture of this plant is an easy matter in any garden. It is perfectly hardy, provided the position is a dry one and the soil thoroughly drained. It delights in a sunny border with a southern aspect near a wall, but if such is not available, an open border well drained and slightly raised above the ordinary ground level will suit it. Like all bulbous plants of a similar character, it will not thrive if mixed indiscriminately with other plants, if of coarse growth, for the shade and the consequent dampness have an injurious effect on the bulbs. The soil should be light and friable and about 1 ft. in depth; the bulbs should be planted about 4 in. deep. During winter a protective mulching is advisable, but it should be removed as soon as the severe cold is past. After the flowering season, which is late in spring, the surface of the soil should be entirely exposed, so as to allow it to become warm and dry, in order to well ripen the bulbs, a most important point with this and most bulbous plants. If these precautions are taken and the plant placed under proper conditions at the outset, there is no reason why it should not flourish to perfection in any garden. It grows naturally in shady situations in sub-alpine districts in Asia Minor.

Pycnanthemum.—Plants of the Labiatæ, for botanic gardens only.

Pyrethrum.—By far the most important of the numerous species of this genus is the Caucasian P. roseum, which has yielded the innumerable varieties, both single and double, that have now become such popular border flowers. They have much to recommend them; they are extremely showy, very hardy, easy to grow, and invaluable as cut flowers for several months in summer and autumn, as they last so persistently and are little affected by either sun or rain. The colour of the blossoms is continually becoming more varied and their shape more refined, and though they are in their fullest beauty in June, yet they are seldom altogether flowerless throughout the summer; and, by judicious stopping and thinning, a succession can easily be kept up. They are also invaluable for autumn decoration, for if cut down after flowering in June they flower again in autumn. They are easily propagated by division of the root and from seed, but the latter mode cannot be recommended, as the proportion of good varieties from the best of seed is remarkably small. The proper time for propagation is in spring. Take the plants up, shake all soil from them, and pull them to pieces, putting them in small pots, and placing them in a cold frame for a few weeks until they become established. Care should be taken not to keep them too close, as they are apt to damp. When established they may be planted out in their proper quarters. A good rich loam suits them best, although they will grow and flower freely in any good garden soil, and the more well-rotted manure there is incorporated with the soil the better the plants grow and the more luxuriantly they flower. Mulching, especially

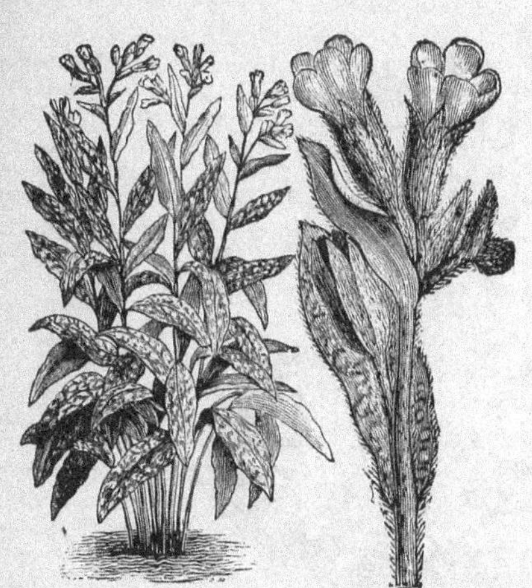

Pulmonaria saccharata (*Lungwort*).

Prunella grandiflora (*Large Self-heal*).
See page 235.

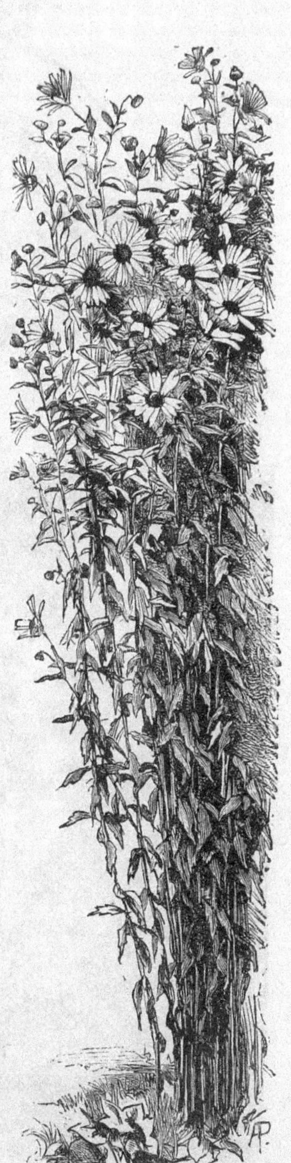

Pyrethrum uliginosum.

Pyrethrum Tchihatchewi.

[*To face page* 236.

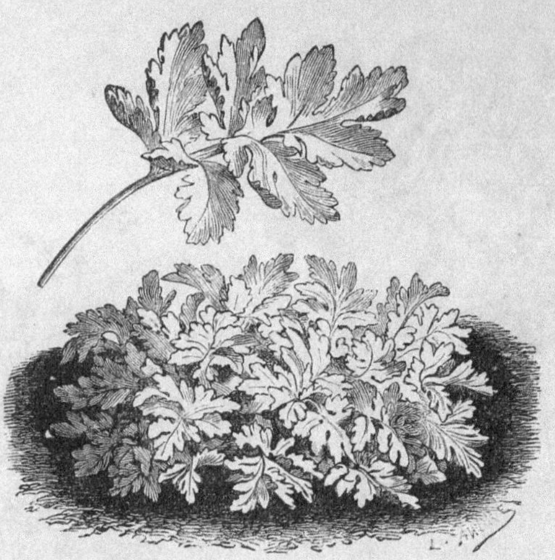

Pyrethrum Parthenium aureum (*Golden Feather Feverfew*).

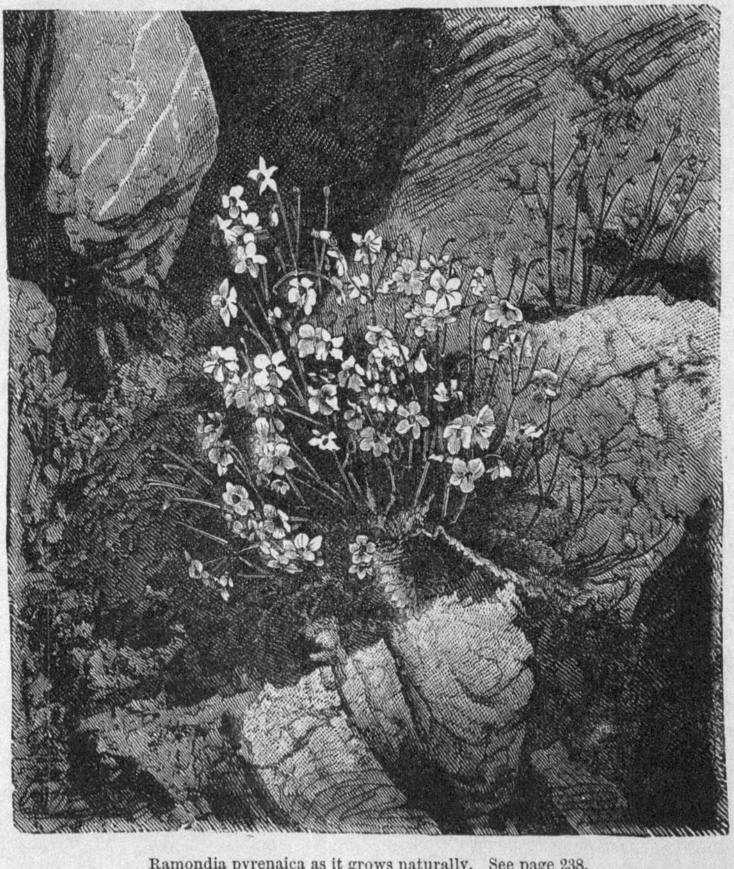

Ramondia pyrenaica as it grows naturally. See page 238.

[*To face page* **237.**

in dry soils, is very advantageous, as it keeps the ground moist and cool. The varieties are so numerous that it is a difficult matter to make a selection, and new sorts are continually being raised, but the following comprise some of the best: White and white-shaded —Boule de Neige, Delicatum, Madame Billiard, Nancy, Niveum plenum, Olivia, Argentine, Prince de Metternich, and Ne Plus Ultra. White with yellow centre—Bonamy, Impératrice Charlotte, La Belle Blonde, Virginale, and Voie lactée. Purple and red—Mrs. Dix, Rubrum plenum, Mons. Barral, Brilliant, and Wilhelm Kramper. Crimson—Michael Buckner, Miss Plinkie, Modèle, Multiflorum, Prince Teck, Progress, Emile Lemoine, and Marquis of Bute. Carmine and pink—Carmineum plenum, Charles Baltet, Floribundum plenum, Gloire de Stalle, Imbricatum plenum, Nemesis, Fulgens plenissimum, Haage et Schmidt, Iveryanum, J. N. Twerdy, and Rev. J. Dix. Yellow—Sulphureum plenum, Solfaterre. Lilac and rose—Comte de Montbrun, Delicatissimum, Dr. Livingstone, Gaiety, Galathée, Hermann Stenger, Lady Blanche, Lischen, Minerva, Uzziel, and Roseum plenum. These are mostly double-flowered sorts; there is also a great diversity of colour amongst the single kinds, and they are quite as beautiful as, and more suitable for vases than, the heavy-headed double flowers. Other species of Pyrethrum of garden value are:—

P. Parthenium *(Feverfew)*, the golden-leaved variety of which (called Golden Feather) is now common in every garden. Of this, too, there are several forms, one called laciniatum being very distinct from the older kind. These have their uses in geometrical borders where they have a bright effect. Their culture is of the simplest description, as they are raised annually from seed sown in spring in heat, the seedlings being pricked off in pans and transferred, when large enough, to the open borders. Here they are sufficiently hardy to withstand the winter unprotected. New plants raised every year are the best, as the plants flower the second year and lose their compact, neat growth.

P. Tchihatchewi *(Turfing Daisy)*.—A Caucasian Composite, chiefly remarkable for its power of resisting drought, its foliage retaining its verdure even in dry weather, and when planted on banks or slopes where but few perennials would flourish. Being of dwarf, creeping habit of growth, it quickly forms a carpet of green, which needs no alteration beyond that of removing the flower-stems thrown up, which, though not altogether devoid of interest, may yet in some situations interfere with the utility of the plant. The flower-heads themselves resemble closely those of the common Ox-eye Chrysanthemum Leucanthemum, having a white ray and a yellow disk, but the stems are much dwarfer, scarcely exceeding 1 ft. in height. It may be used with good effect for covering dry banks.

P. uliginosum is one of the noblest of all tall-growing herbaceous plants, forming dense tufts 5 ft. to 7 ft. in height, terminated by lax clusters of pure white flowers, each about twice the size of those of the Ox-eye Daisy. It is a stately plant for a moist, rich border, and is robust enough for naturalising in damp places. It is excellent for cutting purposes; its value is increased by the blossoms being produced late in autumn before the Chrysanthemums come in. It thrives best in a deep, moist, loamy soil, and may be readily propagated by division. *Syn.*, P. serotinum. Hungary.

Pyrola rotundifolia *(Larger Winter-green)*.—A rare native plant, inhabiting woods, shady, bushy, and reedy places; it has leathery leaves and erect stems, which bear long and handsome slightly-drooping racemes of pure white fragrant flowers, half an inch across, ten to twenty being borne on a stem from 6 in. to 12 in. high. Pyrola rotundifolia var. arenaria is another very graceful plant, found wild on sandy sea-shores, and differing from the preceding in being dwarfer, deep green, and smooth, and in generally having several empty bracts below the inflorescence. Both are beautiful plants for shady mossy flanks of rockwork, in free sandy and vegetable soil, and flourish more readily in cultivation than any species of their family. In America there are varieties of this plant with flesh-coloured and reddish flowers, but none of these are in cultivation with us. Pyrola uniflora, media, minor, and secunda are also interesting British plants, of which the first, a very rare one in our flora, is the most ornamental. P. elliptica, a native of North America, is also in our gardens, though rare. Any of these plants that can be obtained are worthy of a place in thin mossy copses on light, sandy, vegetable soil, or in moist and half-shady parts of the rockwork or fernery.

Pyrrhopappus carolinianus.—A North American Composite of no ornamental value.

Pyxidanthera barbulata *(Pine Barren Beauty)*.—A curious and minute American plant, plentiful in sandy dry "Pine barrens" from New Jersey to North Carolina. It is an evergreen shrub, yet smaller than many Mosses; the flowers are very numerous, rose-coloured in bud, and white when open. The effect of the rosy buds and white flowers on the dense dwarf cushions is singularly pretty. Generally found in low, but not wet, places, and usually on little mounds; it is a gem for the rock-work, on which it should be planted in pure sand and vegetable mould, and fully exposed to the sun. It flowers in early summer, is increased by division, and is as yet very scarce. Polemoniaceæ.

Quamash *(Camassia esculenta)*.

Quamoclit coccinea.—A pretty Convolvulus-like plant with fine scarlet flowers, and the profusion with which they are produced compensates for their rather small size. Its stems are very slender and of very rapid growth, attaining a height of from 6 ft. to 8 ft. in a few weeks, and it continues in bloom during the entire summer. It may be treated either as a half-hardy annual, in which case it should be sown under glass or on a hotbed in February or March, or as a hardy annual, the seeds being sown in April or May in the open ground. Q. hederæfolia is another pretty scarlet-flowered species, having lobed foliage and requiring the same treatment. Both are excellent plants for trellises in sheltered situations, as they produce an abundance of flowers from July to September.

Ramondia pyrenaica.—An interesting and ornamental Pyrenean plant, having leaves in rosettes spreading very close to the ground. The flowers, which are of a purple-violet colour with orange-yellow centre, are borne on stems from 2 in. to 6 in. long, and are from 1 in. to 1½ in. across. It does very well on rockwork, in mossy fissures filled with well-drained peaty earth, and is easily grown in cold frames in well-drained pots, well watered during the warm months. It flowers in spring and early summer, and is increased from seed. It is found in the valleys of the Pyrenees, growing on the face of steep and rather shady rocks. There is a pure white variety, but it is rare. It does well in borders of American shrubs in peat soil. Cyrtandraceæ.

Rampion *(Campanula Rapunculus)*.

Ranunculus *(Crowfoot)*.—The Crowfoots are a very numerous family. Many are useless weeds, while others, such as the Persian Ranunculus, are among the choicest garden flowers we have. Though there are so many weedy subjects among them, there are various kinds which should be found in every garden of hardy flowers, and will be found worthy ornaments of the border or rock garden. These are for the most part of the simplest culture, but R. asiaticus requires somewhat special treatment.

R. aconitifolius.—The double-flowered variety of this species, known as Fair Maids of France, is a very pretty plant that grows about 18 in. high, and forms a dense tuft profusely laden for several weeks in early summer with small rosette-like blossoms of snowy whiteness. Thrives best in a deep, moist loam.

R. acris.—The double form of this plant also is pretty, and is commonly known as Bachelors' Buttons. Like the last, it is a profuse flowerer, the blossoms being in button-like rosettes and Buttercup-yellow in colour. It is a useful border plant, and thrives well in a moist soil, fully exposed.

R. amplexicaulis is a lovely and most important garden plant. It grows about 1 ft. high, and the slender stems are furnished with glaucous grey foliage and a few pure white blossoms 1 in. across, with bright yellow centres. It blooms in April and May, and makes a pretty border flower, flourishing best in a deep, moist loam.

R. anemonoides is a new alpine species of very dwarf growth. It has, like some of the Anemones, finely divided foliage produced in a dense tuft. The flowers, which are borne on short stalks, and are 1 in. across, are white, tinged with purple, while the unexpanded buds are wholly tinged with rosy-purple. Not much is known of its culture at present, but doubtless it will thrive in any well constructed rock garden. R. alpestris is somewhat similar, but is a rare cultivated plant.

R. asiaticus.—It is scarcely necessary to speak of the beauties of this old-fashioned plant, with its neat, dressy, double flowers sporting every colour of the rainbow. The varieties are innumerable, and are divided into various sections, such as the Dutch, Scotch, Persian, and Turkish. Each of these represents a distinct race; but all are beautiful and well deserving of any amount of care and attention in their cultivation. They look well anywhere or in any position in the garden, but best where seen in bold masses. The culture of the Ranunculus is usually considered somewhat difficult, though it is simple if a few essential particulars are observed. The soil best suited for the Ranunculus is loam thoroughly mixed with a third of its bulk of good decayed stable manure. The situation should be open, but not exposed. The prepared soil should occupy about 15 in. in depth of the bed, and should be put in a month or so previous to planting. This takes place about the last half of February, but in some seasons it may be done in October, though such an early date is not advisable. In planting the roots, drills about 5 in. apart and 1½ in. deep should be made with a small hoe; the claws of the roots should be placed downwards and pressed firmly in the soil. Afterwards the soil should be raked over the roots and a top-dressing of about 2 in. of good loam given. If the surface soil is light, it may be gently beaten with a spade in order to obtain a firm surface, and this may be repeated just before the foliage appears, which is in about a month or six weeks after planting. As the Ranunculus delights in a moist soil, water should be plentifully supplied if there is a deficiency of rain, and in no case must the roots be allowed to become very dry. A light top-dressing of artificial manure or guano will be beneficial just as the foliage is developed. When the flowers are past and the leaves fade away, the roots must be taken up, dried, and stored in a cool place in sand till next planting season. If left in the ground after the foliage is decayed, the roots are injured by rains and are never strong. The Persian varieties are the finest as regards compactness and symmetry of growth as well as beauty of colouring, but the Turban varieties are of hardier constitution and freer growth than the edged and spotted kinds, and therefore are better adapted for beds, lines, and masses. The Scotch and the Dutch varieties are also fine for masses in beds, being all of highly effective colours. It is useless to

PLATE CCXV.

Ranunculus asiaticus fl.-pl.

Semi-double Persian Ranunculus.

Ranunculus aconitifolius (*Fair Maids of France*).

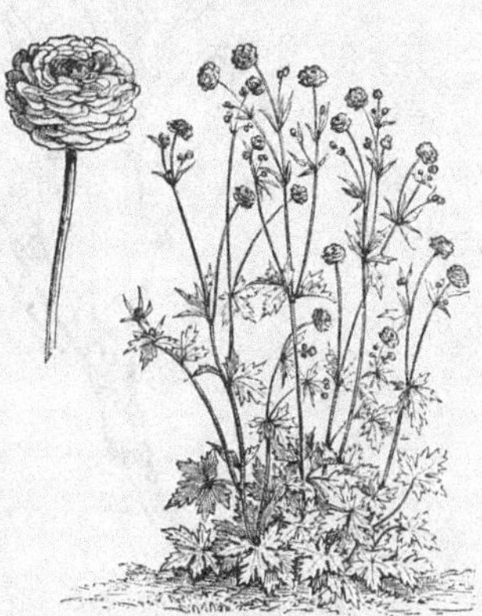

Ranunculus acris fl.-pl. (*Crowfoot, or Buttercup*).

Ranunculus parnassifolius (*Cyclamen Ranunculus*).

[*To face page* 238.

PLATE CCXVI.

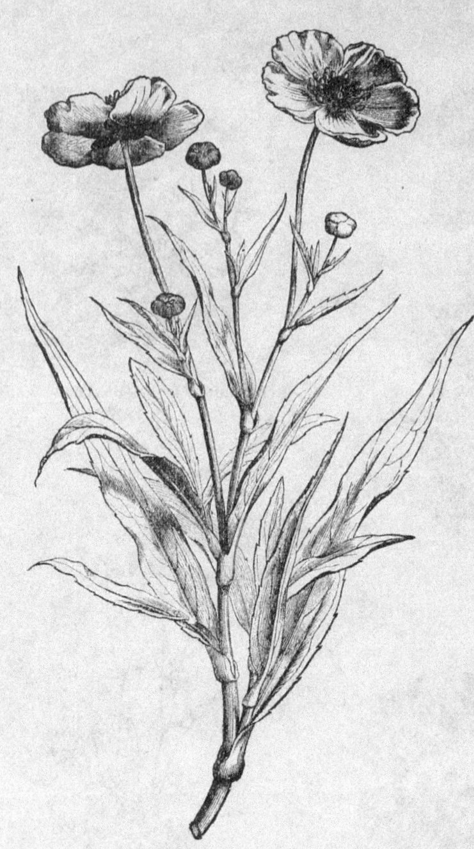

Ranunculus Lingua (*Great Spearwort*).

Ranunculus bulbosus fl.-pl.

Rehmannia glutinosa.

Rhaponticum scariosum (*Swiss Centaury*).

Reseda odorata (*Mignonette*).

[*To face page* **239.**

enumerate varieties, as they are usually sold in colours and nearly every bulb catalogue mentions them.

R. bulbosus fl.-pl. is a showy plant, about 1 ft. in height, producing in early summer numerous very double yellow blossoms. It grows well in any soil. Of R. repens there are two double varieties, one of neat growth, the other untidy in appearance.

R. bullatus is a fine border plant, growing only about 6 in. high. It has large, orange-yellow blossoms like those of the Marsh Marigold (Caltha palustris). It is not so hardy as the majority of Crowfoots, and therefore should be placed in warm, dry soil.

R. cortusæfolius, a handsome Madeiran plant, is scarcely hardy enough for open-air culture, though if given a little attention, it succeeds tolerably well in the most favoured localities. It has large foliage, and heads of bright yellow flowers. R. platanifolius is a similar plant, requiring the same treatment.

R. glacialis, an alpine species, does not grow more than 3 in. or 4 in. in height, and requires some little attention. It does best in light, gritty soil kept moist during the heat of summer. The same treatment serves for R. montanus, another alpine species, and also for such kinds as R. Traunfellneri.

R. Lingua, a native species, is a noble aquatic, well deserving of cultivation in localities where it does not occur naturally. It has large leaves that rise boldly out of the water, and the flowers are large, yellow in colour, and attractive. It requires to be grown immersed in water about 1 ft. The other aquatic species, such as R. aquaticus and fluitans, are pretty plants for artificial water, but they should be introduced cautiously, as when once established they spread rapidly and become too plentiful.

R. Lyalli.—A beautiful New Zealand species called the Water Lily of the Shepherds, but it is not, unfortunately, hardy enough for open air culture. It succeeds in a frame, and is well worthy of attention. It has recently been introduced, but is rare.

R. parnassifolius has Cyclamen-like leaves. It is a beautiful plant, but is a rather difficult subject to cultivate. To grow it well it must have the most select spot in the rock garden and a deal of attention; the same remark applies to a similar kind, R. Thora, which also has yellow blossoms.

R. rutæfolius has Rue-like leaves and white flowers with dark yellow centres. As it comes from the highest parts of the Alps, it requires the same treatment as the choicest alpine plants. It should be in a fully exposed spot in moist soil with plenty of grit in it.

R. speciosus is a very showy plant, with compact, rosette-like flowers of a bright yellow, produced in May. It succeeds in any light soil partially shaded. Where a fuller collection is required, such species as R. gramineus, chærophyllus, illyricus, fumariæfolius, and Flammula might be added.

Reaumeria hypericoides.—A rare and singular plant allied to the Hypericum, which it somewhat resembles. Valuable for botanical collections, as it illustrates the Reaumuriaceæ. Orient.

Red Campion *(Lychnis diurna).*

Red Hawkweed *(Boerhhausia rubra).*

Red Valerian *(Centranthus ruber).*

Reed Grass *(Arundo).*

Rehmannia chinensis.—A handsome perennial plant of the Gesnera family, of dwarf growth. It produces in summer large tubular flowers of a purplish colour, striped with a darker hue. It is not a thoroughly hardy plant, but succeeds well in warm, sheltered localities. The best plants we have seen of it grew in a moist peaty border under the shade of a wall running east and west. Generally it is best to winter the plants under glass. China.

Reineckia carnea.—A dwarf Liliaceous plant from China. It is of low tufted growth and has grassy foliage, which in the type is dark, shining green, but in the variegated form is handsomely striped with creamy-yellow and green. This variety is the most valuable plant, as it makes a pretty edging plant, is hardy, and of sturdy growth. It grows well in any ordinary soil in an open situation. The flowers are not remarkable.

Reseda *(Mignonette).*—The only species of Reseda worth attention is the common Mignonette, which is an admired plant everywhere on account of its delicious fragrance. Seed sown in the open ground in March or April will produce a few weeks later flowering plants, which continue to bloom till late in autumn, and an abundant supply of cut flowers may be thereby obtained. If, however, a few extra fine masses of plants be wished for, the seed should be sown in pans about the end of March, the seedlings placed singly in 3-in. pots, and ultimately planted out in good soil in an open position. A little attention should be given with regard to thinning out the weak shoots and stopping the more vigorous ones. Of the red-flowered sorts Victoria is the best, and of the light the tall spiral variety. During mild winters plants sown in autumn will survive and produce flowers in early summer. These are altogether finer than spring-sown plants.

Rest Harrow *(Ononis).*

Rhaponticum cynaroides *(Swiss Centaury).*—A hardy perennial from the Pyrenees, 3 ft. or more in height. It has a stout stem and large leaves, covered underneath with silvery down. The flowers are purple, and are borne in very large heads. It thrives in a deep, substantial, moist, but well-drained and free soil. It is worthy of a place in full collections of hardy fine-leaved plants, for borders, the margins of groups, and for isolation. R. pulchrum is another hardy perennial from the Caucasus, with stems 2 ft. or more in height. The leaves are ashy or sea-green in colour, and the flowers are small and purplish. It is a suitable subject for embellishing dry, arid, rocky, positions, and R. scariosum is useful for the same purpose. All are easily increased by division. Compositæ.

Rheum *(Rhubarb).* — The Rhubarbs, from their vigour and picturesqueness, are well worthy of cultivation among hardy, fine-leaved plants. They are so hardy that they may be planted in any soil, and afterwards left to take care of themselves. Their fine leaves and bold habit make them valuable ornaments for the margins of shrubberies and for semi-wild places where a very free and luxuriant type of vegetation is desired. Though not particular as to soil, they enjoy it when it is deep and rich, and the more it is made so the better they will grow. R. Emodi is a fine herbaceous and large-foliaged plant, and should be planted in groups in the turf of the pleasure ground in good soil. It grows about 5 ft. high, and is very imposing with its wrinkled leaves and large red veins. It is, however, surpassed in appearance by R. officinale, which should be noted by all who desire bold and striking effects from hardy plants. This is, as regards foliage, the most effective hardy plant that has been introduced for many years. All the old Rhubarbs are insignificant beside it. It produces a fine effect early in the year, and is of luxuriant growth, easily propagated, and perfectly hardy, bearing our northern winters with the greatest impunity. It is really a grand plant for placing near the shrubbery, on the turf, or for the wild garden, where its fine form would be very striking. It would be well in certain gardens to form a little colony of fine-leaved, hardy plants; in a small glade with rich soil a novel and fine effect could readily be produced by a good selection, embracing the Ferulas, Heracleums, Rhubarbs, Acanthuses, Yuccas, the common Artichoke, Gunnera scabra, and many other vigorous, hardy subjects. R. palmatum tanguticum is a very fine kind of Rhubarb that, as regards vigour of growth and size, seems distinct from the old Rheum palmatum, which is a smaller plant, and generally a very slow-growing one. R. tanguticum increases rapidly. The fine and boldly-incised foliage will be welcome to those who grow the other hardy species now in our gardens. R. nobile is a most remarkable species, distinct from all the rest, as it forms a dense pyramid of foliage. It is, however, one of the most difficult to cultivate, and the only place it has succeeded in Europe is in the Edinburgh Botanic Garden. It is a native of the Sikkim Mountains. R. Ribes is too rare and too delicate a species to recommend. The common Rhubarbs, said to have sprung chiefly from Rheum Rhaponticum and R. undulatum, are worth planting for ornamental purposes. They have been so planted in Hyde Park, but in masses—not the proper way to employ them. Kinds deserving of notice are R. australe, R. compactum, R. rugosum, R. hybridum, Victoria Rhubarb (a variety with very large leaves and long red stalks), Myatt's Linnæus, and Prince Albert. Scott's Monarch is the most imposing and ornamental of all garden varieties.

Rhexia virginica *(Meadow Beauty).*—A dwarf northern representative of a beautiful and brilliantly-coloured South American family, which has vivid rosy flowers, and grows 6 in. or 8 in. high. It grows in sandy swamps in New England and the Eastern States, and also as far west as Illinois and Wisconsin. It is rarely seen well cultivated; but if healthy roots are put in a boggy or marshy place in peaty or sandy soil, they will thrive. Rhexia Mariana is even scarcer in this country, and is hardly so brilliant as the preceding. It does not grow so far north in America as does R. virginica, but it is plentiful in the sandy fields in New Jersey, and would probably thrive in sandy heath soil. R. virginica, however, is far more worthy of attention. It should not be taken in hand by persons who do not possess a bit of suitable boggy ground, unshaded by trees, and not overrun by coarse plants. The Rhexias are plants that are impatient of too much division, and to give them a fair chance good tufts should be obtained from their native localities. If these are secured and planted in a sandy peat bed on a cool clay subsoil, the chances of success are almost as certain as are the chances to the contrary where the plants are broken up into small fragments. Melastomaceæ.

Rhinanthus.—Pretty native plants, but not suitable for gardens.

Rhinopetalum Karelini.—A bulbous plant of the Lily family, allied to, and sometimes classed with, Fritillaria. It grows from 4 in. to 5 in. high, with two or three stem-clasping broad leaves, and a terminal raceme of slightly-drooping bell-like flowers. These are about 1 inch across, of a pale purple colour, with darker veins and a few darker spots, and with a distinct yellowish-green pit at the base of each reflexed segment. It is a native of Central Asia, and flowers in late autumn or early winter; it is, therefore, valuable and interesting for a winter-flowering collection of outdoor plants. According to Dr. Regel's experience, this bulb must be kept in dry sand until November, and should not be growing or showing bloom before spring. If planted in November, growth is retarded, and the plant blooms in spring, as it ought to do; those flowering in autumn will invariably dwindle away, and will not produce another new bulb. It should be planted in light soil in well-drained borders having a warm exposure.

Rhodanthe. — Charming half-hardy annuals from Australia, valuable as border flowers and for winter bouquets, as they come under the class of everlasting flowers. They are all of slender growth, from 1 ft. to 1½ ft. high, with glaucous grey foliage and showy flowers. The original species, R. Manglesi, has fine rose-coloured blossoms with yellow centres, and there is a double variety. R. maculata differs in having a deep crimson

PLATE CCXVII.

[*To face page* **240.**

PLATE CCXVIII.

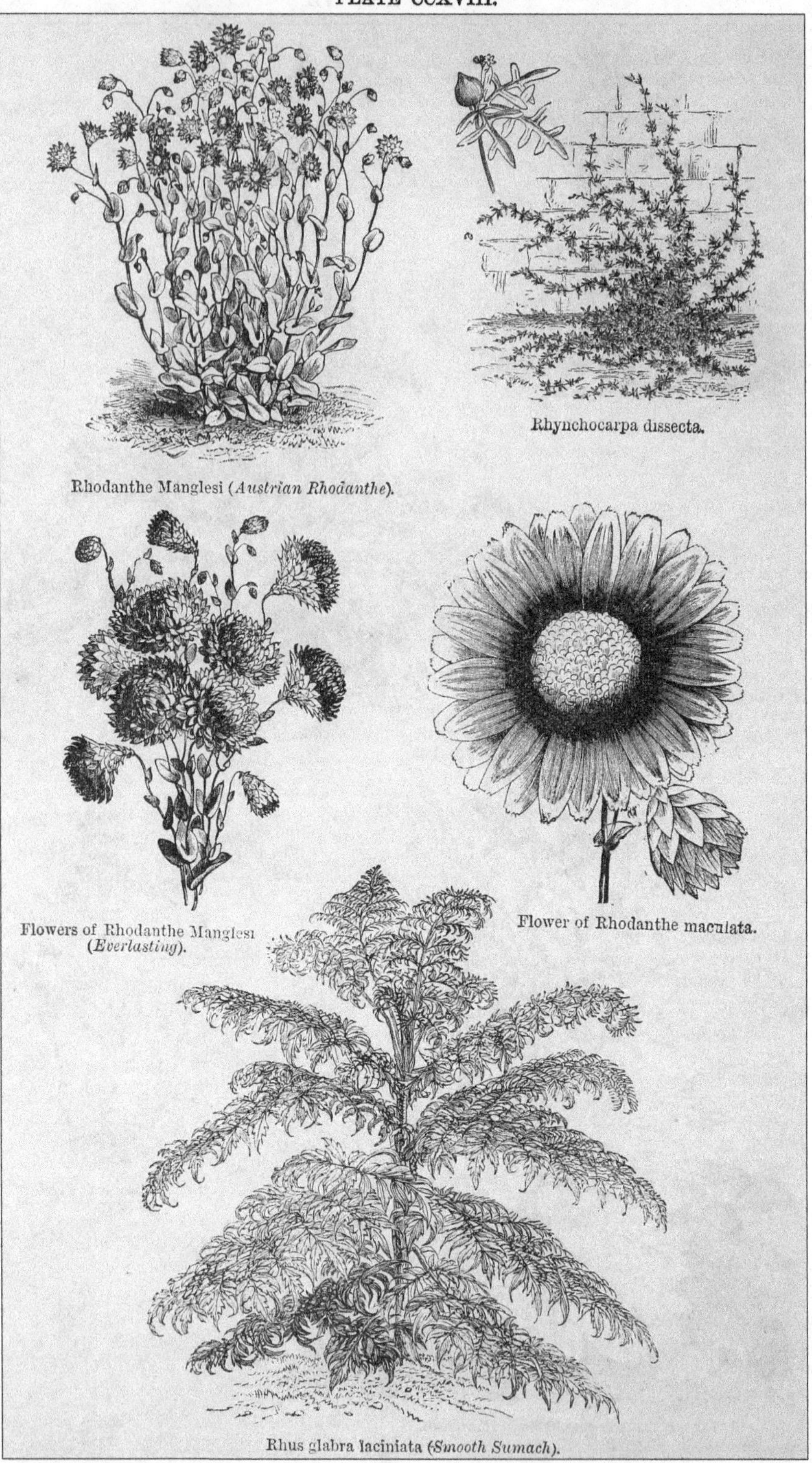

Rhodanthe Manglesi (*Austrian Rhodanthe*).

Rhynchocarpa dissecta.

Flowers of Rhodanthe Manglesi (*Everlasting*).

Flower of Rhodanthe maculata.

Rhus glabra laciniata (*Smooth Sumach*).

[*To face page* **241.**

ring encircling the eye; there is also a pure white variety of this kind. R. atrosanguinea differs considerably in habit from R. maculata, being not only dwarfer, but more branched, and consequently affording a larger number of blossoms. Its foliage is longer and more pointed, and of a bluish-green tint. The blossoms, of a bright magenta colour, are rather smaller than those of maculata, but average 1 in. in diameter. This beautiful species is rather less hardy than maculata and its white variety, but is sufficiently hardy for open-air cultivation. These should all be sown thinly in pots in February or March in heat, and the seedlings should be pricked while young, as they do not transplant well when large. They ought to be protected and attended to for a few days after transplanting; the position should be a warm, open one, and the soil well manured and of a light, friable nature—if peaty, the better. The seeds require to be freely watered in order to induce them to germinate quickly; the pot in which they are sown may even be plunged under water for a few seconds with advantage, and the soil being thus saturated will often require no further watering until the seeds are above ground. If the seeds are thus treated, and placed in a temperature of 65° to 70°, they will speedily give proof of their vitality, but if insufficiently watered, will remain dormant for several weeks. They flower from July to October. Compositæ.

Rhodea japonica.—A Liliaceous plant with thick, broad leaves of a deep green; in the variegated form they are elegantly striped with white and green. This latter is, therefore, the handsomer plant and the most desirable for general culture. R. japonica is hardy, but does not make vigorous growth, except in the most favoured districts, and it seldom or never flowers in the open. An ordinary garden soil suits it well, but the position should be a warm and sheltered one. Native of Japan.

Rhodiola. — Plants of the Crassula family, resembling some of the larger Stonecrops. They have fleshy leaves and heads of small flowers, which are not, however, very attractive. There are two cultivated kinds, R. rosea and R. sibirica, but neither can be recommended to the general cultivator, being better suited for the botanical garden.

Rhodochiton volubile.—This is a beautiful, slender, twining plant that is usually grown in greenhouses, but which, like Lophospermum scandens, succeeds well in the open air in sheltered parts. In warm and sheltered localities it is well worth a trial, for it is a beautiful plant for festooning arbours, trellises, &c. Strong young plants should be placed out in May in good soil.

Rhododendron.—Though these for the most part are shrubs of large growth, there are some comparatively dwarf kinds that may be associated with alpine plants in the rock garden—indeed, some are but a span or so high. One of the prettiest of these is R. Chamæcistus, a Liliputian species, which has tiny leaves, and produces in early summer exquisite flowers of a lively purple colour the size of those of Kalmia latifolia. This plant is very rarely seen in good health in gardens, and for its successful cultivation requires to be planted in limestone fissures, in peat, loam, and sand in about equal proportions. A native of calcareous rocks in the Tyrol, and one of the most precious of dwarf rock shrubs. The well-known R. ferrugineum and hirsutum, each of which bears the name of alpine Rose, and which often terminate the woody vegetation on the great mountain chains of Europe, are easily obtained in nurseries, and are well suited for large rock-work, in deep peat soil, where they attain a height of about 18 in. R. Wilsonianum, myrtifolium, amœnum, hybridum, dauricum-atrovirens, Gowenianum, odoratum, and Torlonianum are also comparatively dwarf kinds, which may be tastefully employed in the rock garden—the last two are very sweetly scented. They should not be too intimately associated with minute alpine plants.

Rhubarb *(Rheum).*

Rhus *(Sumach).* — Though these are shrubs, they are of value as fine-foliaged subjects in summer when they are in beautiful leaf. One of the finest for the purpose is R. glabra laciniata, a variety of the smooth or scarlet Sumach. It grows from 4 ft. to 7 ft. high, and has finely-cut and elegant leaves, the strongest being about 1 ft. long when the plant has been established a year or two. When seen on an established plant, these leaves combine the beauty of those of the finest Grevillea with that of a Fern frond, while the youngest and unfolding leaves remind one of the aspect of a finely-cut umbelliferous plant in spring. The variety observable in the shape, size, and aspect of the foliage makes the bush charming, while the midribs of the fully-grown leaves are red, and in autumn the whole glow off into bright colour after the fashion of American shrubs and trees. Its chief merit is that, in addition to being so elegant in foliage, it has a very dwarf habit and is hardy. R. vernicifera is distinct from the preceding. It has fine leaves, and is useful for grouping with the preceding or other hardy shrubs of like character.

Rhynchocarpa dissecta.—A South African plant, with perennial tuberous roots and incised foliage. The fruits are orange-red, terminated by a kind of beak. It should be planted in warm situations and light rich soil, and is more likely to succeed in the southern counties than in the northern. It may be grown under glass with more success. Under the climate of Paris it does well. Cucurbitaceæ.

Richardia æthiopica and **albo-maculata** *(Calla).*

Ricinus communis *(Castor-oil Plant).* —When well grown in the open air, there is not in the whole range of cultivated plants a more imposing subject than this. It may have been seen in the London parks of late years nearly 12 ft. high, and with leaves nearly a yard wide. The Ricinus is a good plant for making bold and noble beds near those of the more brilliant flowers, and tends to vary the flower garden finely. It is not well to associate it closely with bedding plants, in consequence of the strong growth and shading power of the leaves. It is a good plan to make a compact group of the plant in the centre of some wide circular bed and to surround it with a band of a dwarfer subject, say the Aralia or Caladium, and then finish with whatever arrangement of the flowering plants may be most admired. A bold and striking centre may be obtained, while the effect of the flowers is much enhanced, especially if the planting be nicely graduated and tastefully done. For such groups the varieties of the Castor-oil plant are not likely to be surpassed. It requires a bed of very rich, deep earth to make it attain great dimensions and beauty; but in all parts, and with ordinary attention, it grows well. In order to raise the plants a brisk hotbed is needed in February or March, in which to plunge the pots of rich light soil in which the seeds are sown. The pots should be well drained, and the soil be pressed down tolerably firm with a little finely sifted soil over the seeds. Afterwards a good watering should be given. When the plants are large enough, pot them singly into 4-in. pots in soil composed of sandy loam and leaf-mould or rotted manure; keep them in a warm, moist temperature, and also keep them well supplied with water at the roots. Remove, when the roots have reached the sides of the pots, into 6-in. or 8-in. pots. About the end of May gradually inure them to a cool temperature, and after a few weeks place them in a sheltered position out-of-doors. By the end of June they may be planted out permanently in the open ground; the more sheltered the situation the better. Dig out holes for them, in the bottom of which place a few forkfuls of manure; if from a warm manure bed, so much the better. Plant and water with soft rain water, and mulch the surface with manure. During hot weather liberal supplies of manure water will be of great benefit.

The best varieties are sanguineus, borboniensis, Gibsoni (a very fine dark variety), giganteus, Belot Desfougerès (a very tall and branching kind), viridis (of a uniform lively green colour), insignis, africanus, africanus albidus, minor, hybridus, microcarpus, macrophyllus, atropurpureus, and sanguinolentus, all of which are forms of R. communis, a native of the East Indies.

Rock Cress (*Arabis*).

Rocket (*Hesperis*).

Rocket Larkspur (*Delphinium Ajacis*).

Rock Jasmine (*Androsace Chamæjasme*).

Rockwood Lily (*Ranunculus Lyalli*).

Rodgersia podophylla.—A handsome leaved plant of the Saxifrage family. The leaves, which measure 1 ft. or more across, are supported on erect stalks from 2 ft. to 4 ft. high, and are cleft into five broad divisions. They are of a bronzy-green hue, which renders this very distinct from any other hardy plant. The flowers, produced on tall branching spikes, are inconspicuous. It is rather uncommon at present, but it certainly deserves attention from those who admire fine-foliaged plants. It likes a peaty soil and a shady situation, and is easily propagated by cutting the stoloniferous root-stock, from one of which as many as twenty plants can be made in one year. It is a native of Japan and perfectly hardy in our climate.

Romanzowia.—The cultivated species of this genus, R. sitchensis and R. unalaschcensis, both from Russian America, are small, dwarf plants that form in summer a dense tuft of foliage studded with tiny white flowers. Both are quite hardy, and grow well in open positions in the rock garden. They are interesting, but not attractive plants. Hydrophyllaceæ.

Romneya Coulteri.—This singular Californian Poppywort is found on the borders of streams near San Diego, and is a strong-growing, much-branching plant, with lobed, glaucous foliage. The flowers are pure white and large, nearly equalling in size those of the White Water Lily. It is a showy plant where the summer heat is sufficiently powerful to perfect its growth and fully develop its flowers before our early autumn frosts overtake it. A plant of it that bloomed in the open border at Glasnevin in October, 1876, was about 4 ft. high and 3 ft. through, with numerous lateral shoots, each of which, as well as the terminal branches, bore a flower-bud at the point. Mr. Thompson, of Ipswich, who introduced it, says: "My experience of it leads to the conclusion that, except in favoured localities in the south, it is but half hardy in England. It may, however, be easily wintered in a frame, and it pots up from the borders very well. It evidently loves a moist soil, but will grow fairly well in ordinary garden mould. It is not so free a bloomer as could be wished, at least in this latitude. I have never had more than a few flowers fully developed on each stem, but whether that is owing to deficient moisture, or heat, or both, is uncertain, but probably of warmth. It must be propagated by seeds or cuttings. It does not divide, so far as my experience has gone. The Romneya is certainly a true perennial, but I do not think it can be successfully treated as a perfectly hardy perennial."

Romulea.—Bulbous plants of the Iris family, of dwarf growth, with grassy foliage and showy blossoms. None, however, are perfectly hardy, and they require to be grown either in frames or in very warm, sheltered borders, in light soil. The best known are R. Bulbocodium, ramiflora, and Columnæ, natives of South Europe, and R. rosea and

PLATE CCXIX.

Ricinus communis (*Castor-oil Plant*).

Rodgersia podophylla.

[*To face page* **242.**

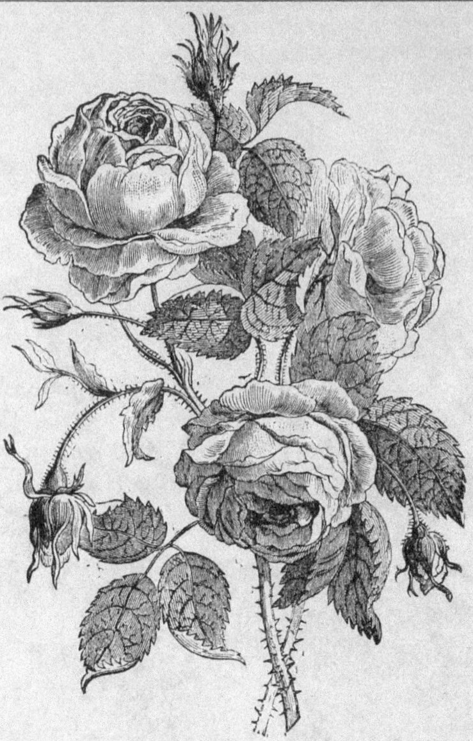

Rosa centifolia (*Provence Rose, Cabbage Rose*).

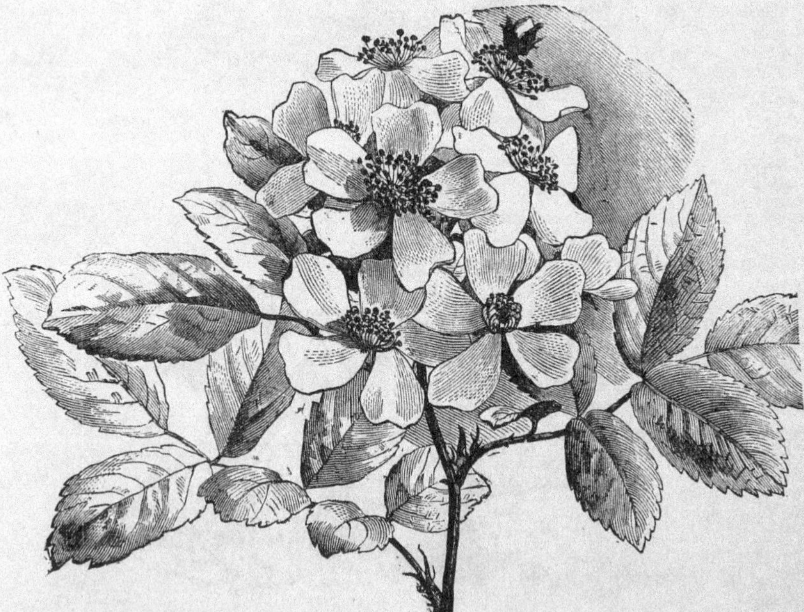

Rosa polyantha (*Bramble Rose*). See page 244.

R. Macowani from the Cape of Good Hope. The showy Crocus-like flowers of these open fullest in sunshine.

Rosa *(Rose)*.—That the queen of flowers should have a garden or plot to herself, where her majesty may not be incommoded, has long been insisted on by those who grow the Rose for the sake of exhibiting cut blooms, and, indeed, the idea of a rosery, as a separate and usually ugly pest, has long been held, but to those who desire rather to have a garden beautified by the presence of the Rose in its greatest luxuriance and beauty, as far as effect goes, there is yet considerable doubt as to the feasibility of such a thing, and difficulty in carrying out such a plan when begun. The Rose of modern gardens is such an artificial compound, so entirely the result of years of selection and interbreeding, that it will not show itself in full beauty except under the most favourable conditions, such as we cannot always give it in the mixed garden. It will, therefore, be well to consider how we may best arrange the various members of this favourite family, so that the original species may be found in our gardens as well as the beautiful hybrids and varieties raised therefrom.

Roses have many points of interest, some of which have been forgotten until lately. Besides the beauty of their flowers, some species have such beautiful hips that, were it for no other reason, they are worthy of a place in our gardens. As no garden is complete without the Rose, the Rose should be found in every variety of garden; and where can a more fit place for the natural species of Rose be found than on the banks that compose or surround the bolder kind of rock garden? The climbing varieties should grace our walls, shade our arcaded walks, festoon our trees, clamber up poles, or spray over the waste spaces. The Tea Rose should have its sunny bank or warm wall, and the Hybrid Perpetuals occupy the special plots provided for them. Besides, there is no better way of showing the beauty of a Rose than by growing it on its own roots in mixed borders, and should it be cut down by frost now and then it will spring up again with the herbaceous plants in spring. The effect of a good Rose or a group of Roses so grown is excellent.

CLIMBING ROSES.—If we look round in southern continental gardens where not one tithe of the labour and care lavished on English gardens has ever been bestowed, but where a warmer climate and more constant sun brings out luxuriant growth in many things, we see such arcades, bowers, pillars, and climbing masses of beautiful Roses on all sides as make one feel discontented with our beautiful individual blooms, and the absence from our gardens of these luxuriant masses that neither require nor obtain any special care whatever from one year's end to the other in the warmer parts of Europe. If, as is only too true, the varieties of the Rose that produce such glorious effects in foreign gardens are not hardy enough for us, why do we not try to raise new varieties that shall resist our cold and changeable seasons? Surely there is choice enough of species and varieties in a plant that ranges from Kamtschatka to India, among which we may find something that shall be the parent of hardy climbing varieties, as beautiful in our climate as the Noisette and indica major Roses are in the south of France and elsewhere. We have R. sempervirens and several garden varieties, such as Félicité Perpetuée and others, that will climb a pillar or shade an arcade. We have the Ayrshire Roses, R. arvensis and its varieties, such as Rampant, Ruga, &c., which are very charming in their way, but bloom only in summer, as do the varieties of the Boursault Rose (R. alpina). They are all quite hardy and of vigorous climbing growth, but they do not satisfy those who love the Hybrid Perpetual, the Noisette, or the Banksian Rose. We have the continuous blooming R. rugosa and the beautiful R. sinica, the parent of the so-called large white Banksian Rose Fortunei. There is, too, the semi-double yellow R. Fortunei, also from China. Cannot some hybrids be raised from these and the semperflorens, alpina, or arvensis species? or should we use the R. canina, the common Dog Rose, for a parent? Let us, however, make use of what we have already at hand, and plant in the wilder parts such hardy climbers as are already mentioned; make combinations of the excellent Gloire de Dijon and its progeny (none of which, unless it be Belle Lyonnaise, come up to their parent in abundance or sweetness of bloom) with such red climbing Roses as can be found hardy. Cheshunt Hybrid, where temperatures below zero are not to be found, is perhaps the best; and the old semi-double Bourbon Rose, Gloire des Rosomanes, will succeed in warm soils, and light up an arch with its bright red blooms in autumn as well as in summer. The many climbing varieties of H. P. Roses that have lately been raised are all good in their way, but they demand good soil and space for themselves. When it is a warm wall that needs clothing, then it is that the Banksian or various hybrids of Noisette and Tea Roses may be used, though all are liable to be cut down to the ground in cold situations and seasons. Of these, for general effect, the palm is borne away by Ophirie, whose vigour of growth is amazing and whose glossy sub-evergreen leaves and apricot clusters of bloom contrast exquisitely with the autumn Clematis. For sweetness, as well as continuity of bloom, Lamarque's clusters of lemon-white flowers must stand first; and Maréchal Niel is of course unrivalled for the splendour and sweetness of its golden blooms, but it is only a shy bloomer in autumn. Climbing Aimée Vibert, which is the only thoroughly hardy member of this section of the Rose family, should be found in every garden. Its blooms are individually small, but its white clusters are so continuously abundant and the foliage so persistent as to make it rank very high as a decorative garden Rose. Rêve d'Or, in a warm situation, is a delightful climber, and may be called a climbing Madame Falcot, so bright are its half-expanded buds. Many other charming Roses there are in this section, but, as a rule, they do not succeed in English gardens unless the situation be exceptionally good. If some hybrids were raised between the Rosa indica major, the Dog Rose, the single Banksian, and the Japanese R. rugosa, it is probable that a new race of perfectly hardy and decorative climbers would result.

HYBRID PERPETUALS.—These are so well known and grown that it seems superfluous to attempt to add anything to what has been admirably stated by others, but it should be stated that generous treatment is the only means of assuring a good show of bloom in summer and autumn. Lists of the various varieties for different purposes are given in every Rose catalogue.

THE SPECIES OF ROSE.—No one who has seen the dainty Cistus-like spotted blooms of the yellow Rosa berberifolia and its glossy leaves and dwarf growth will forget to give it the sunny and somewhat dry situation in the rock garden that suits it best. The single Macartney Rose (Rosa bracteata), with its large white single flowers and orange stamens, is, though less hardy, even more charming in foliage and more vigorous in growth, and deserves a wall when it can be given. Its Japanese variety—called R. Camellia—is equally attractive; but the palm for hardiness and decorativeness in exposed situations must be given to another Japanese Rose (Rosa rugosa), which is exceedingly hardy by the sea as well as inland, fears not the bitterest frost, and braves the keenest winds. It is now well known; both its white and purple-red flowered varieties are to be found in many a garden, and its large and handsome fruit is seen in autumn. Persia, which has sent us R. berberifolia, has also given us the Persian yellow; but it is from the Levant we obtain that most perfect of all Roses in colour and shape, Rosa sulphurea, the despair of many a cultivator. Its glaucous blue leaves and purely golden globes of bloom can never be forgotten by those who have seen it in full beauty. There is no doubt that there is greater chance of flowering this beautiful low-growing Rose where there is reflected heat and sunshine from rocks than in the Rose garden. Rosa lutea, the single yellow, and R. punicea, the brilliant Austrian Brier, are also suitable rock garden Roses. Rosa alpina is sometimes so richly coloured as to adorn any situation; it has very handsome hips. R. pyrenaica, a tiny form of this species, is a favourite already in many places, and has long, handsome fruits to succeed its short-lived flowers. The Scotch Rose (R. spinosissima) with its white flowers is not unworthy a place in the wilder rock garden, and its double varieties thrive in cold and shaded situations where no other Rose would exist. The Sweet Brier also deserves a place by it for the sake of its delicious fragrance, its bright hips, and its pretty flowers. Another Japanese Rose, but lately introduced, the Bramble Rose (R. polyantha), may be planted for its distinct habit and thyrsoid heads of white flowers; but it is much to be hoped that a greatly superior Chinese variety, with larger flowers and growth, and more golden-yellow stamens, may soon be introduced, as in its native country it thrives anywhere and stands severe frost. The little fairy Rose, R. Lawrenceana, is very pretty among rockwork, where it gains the needful warmth and protection. Rosa microphylla is very quaint in flower and fruit, and its double form will give a good effect in the wild Rose garden where there is a fair amount of shelter.

TEA ROSES.—These prefer a more open and sunny situation than is generally given them, and will thrive on dry banks with a southern exposure if watered and mulched the first season while they are establishing themselves. For future results it is best to plant out Tea Roses in early May, having first bloomed them in pots under glass, so that they have some firm foliage to help the root action when planted. For some time they will appear to be at a standstill, but the autumn bloom will be good, and the next year's growth and flower all that can be wished for.

BOURBON ROSES are amongst the most useful of garden Roses where they succeed; they seem, however, somewhat capricious, with the exception of Souvenir de la Malmaison, a Rose well known to the veriest tyro in Rose growing. Sir Joseph Paxton and Baron Gonella are two useful varieties.

MONTHLY ROSES, as they are commonly called, are not nearly as much grown nowadays as they deserve to be. When the soil is light and warm they will thrive anywhere, and charming beds may be made of these and hardy Fuchsias that are cut down each winter, the dwarfer varieties being placed at the edge. In sunny glades, where there are clumps of Pampas Grass, Tritomas, and other autumn-flowering plants, a most charming effect may be produced by the addition of such a combination.

SUMMER ROSES.—This section is disappearing from our gardens; but though a race of continuous blooming Roses is always a thing to be sought for, there are some old favourites, such as the Rosa centifolia (Provence, or Cabbage Rose), that should always find a place; and the beautiful Moss Rose must not be forgotten, though the season of its flower be not long. The Hybrid Chinas are most useful as pillar or climbing Roses; and Blairi No. 2, Charles Lawson, Fulgens, and Vivid are, in the season of their flowering, worthy representatives of the queen of flowers.

Rosa borbonica.

Dwarf Pompone Rose.

Rose Souvenir de la Malmaison (*Bourbon Rose*).

Rosa cristata (*Moss Rose*).

Rosa lutea (*Single Yellow Rose*).

[*To face page* 244.

Rosa bengalensis (*Bengal Rose*).

Tea Rose Madame Willermoz.

Climbing Rose Gloire de Dijon.

Rosa damascena (*Damask Rose*).

[*To face page* **245.**

PLANTING.—Heavy soil best suits Roses. With plenty of manure a sandy soil may produce a few good blooms the first season, but that will be all; consequently, sandy soil should be as little used as possible. Neither does very old worn-out soil grow them well, and where the beds in which they are to be planted is of this, a good deal of heavy loam should be added. Where Rose blooms of the finest "show" form are wanted, all the old soil should be taken out, and the bed filled up wholly with new loam. In all cases the new soil should be got in, and the beds thoroughly prepared, before planting. When the plants have to be put back in the same bed, they should be lifted out carefully and laid in stock until the bed is ready for them. It is destructive to allow them to become dry at the root through being out of the ground. The best way is to put in new soil first, making the bed up with this, and digging the manure in afterwards. The soil ought to be at least 2 ft. deep, but the manure need not be put down quite this depth, as doing so would only tend to draw the roots downwards—a bad result. If the manure is close to where the roots will be, it is in its proper place. It is generally known that Roses like manure, but when it is put in such quantities that it lies in lumps and heaps together under the surface it is injurious, and it is only when well mixed through the soil that the greatest benefits are to be had from it. The poorer the soil the greater the quantity of manure that must be given, and at all times it should be of the best quality. Half rotten straw and leaves are no good; heavy pig or cow manure is the best. Do not crowd the plants together. Individual plants well grown are more pleasing than a crowd and confusion. From 2 ft. to 3 ft. apart is close enough for any Roses, and some of them may have more space than this. In planting, make the hole big enough to let the roots in comfortably; putting them in in a bundle or one on the top of another is not the way. They ought to radiate from the centre and spread out well. The long thick fibreless roots of old plants should be cut carefully in to 8 in. or 10 in. from the base of the stem. Deep planting should also be avoided. Let the roots be covered with the soil about 3 in. and no more. When let down a foot or more to keep the wind from blowing them away, they will neither grow nor bloom well. Make the soil fine round and over the roots; standards and dwarfs alike should be tied to a firm stake, as it is important that the roots should not be moved after planting. A good coating of half-decayed manure should be placed over the surface of the beds.

PRUNING.—Even moderate-growing Roses may make several feet of wood a year, and need restraint unless a few only are to be grown. No doubt there is a strong tendency to plant Roses too closely together, and a good many gardens would look richer and better with fewer Roses allowed to grow larger than with a crowd of smaller Roses robbing each other. But while the present modes of growing Roses continue, a considerable amount of pruning will be needed to keep each plant in its place, and in due proportion to others. On most Rose shoots there are probably from six to a dozen blooms in embryo. Were all these allowed to remain, they must necessarily be smaller than if only one or two of them were allowed to develop into flowers. Pruning, so far as it reduces the number of flowers, concentrates the force of the plant, and thus heightens the colour and enlarges the size of the Roses. This is so obvious as to need no proof. The shoots of summer Roses, such as the Cabbage and Moss, are spurred back to two or three buds, or even less. Each of these produces one or more flowers of higher quality than if the entire shoot were left intact. The same principle is kept in view in the pruning of other Roses, though in some varieties feet or even yards of young wood may be left instead of eighths or quarters of an inch.

Pruning for form and vigour is absolutely necessary in gardens; in shrubberies, and in woods, and in the case of isolated Roses on turf the more the trees or bushes are left to themselves to wander freely as they list, the more artistic and beautiful the effect. But generally in gardens Roses must be pruned into form and kept in shape afterwards by the knife. No doubt many of them are over-pruned, and all the grace and not a little of the beauty cut out of them. Still it need not be so. Pruning may be made to heighten beauty as well as to mar it. And then we prune for vigour as well as form. By cutting out exhausted branches we cause young and more vigorous ones to spring forth from their base, and thus force the Rose to renew its youth. The removal of weakly, worthless, unsightly, and dead wood improves the appearance and health of our Roses. As soon as a branch fails it should be pruned out before it becomes diseased.—E. W.

ROSE HEDGES.—These are not uncommon in the south of France. Mr. J. C. Clarke tells us how to form them with success in England. "If the soil is naturally a good Rose soil, the work will be light. In that case mark out the position of the hedge 2 ft. wide, trench up that space 2 ft. deep, and incorporate with the soil at various depths a liberal quantity of well-rotted manure. Where there is any doubt about the staple being of the right sort, the whole of it should be removed, and its place supplied with a mixture consisting of three-parts loam and one of manure, but there are many gardens the soil of which, with the addition of one barrowful of loam to every yard length of hedge, and about half that quantity of manure, will grow Roses in a satisfactory manner. When practicable, the preparation of the ground should be done in dry weather, as there is some danger of the loam running together in heavy masses if moved about when it is wet. Plants on their own roots are indispensable for this purpose, and if they can be had from 2 ft. to 3 ft. high so much the better, as they will form a hedge the sooner. Those that have had one season's growth in pots and another in the open ground are what we have used; these have been two years planted, and some of them have this year made shoots 2 ft. long. Thoroughly hardened and established plants in pots may, of course, be used in the absence of the others. The time of planting must depend on the condition of the plants, but if only small plants in pots are to be had the planting should be done in April or May. In any case it should be done when the soil is moderately dry, and some finely-sifted mould should be prepared to place round the roots, about which the ground should be made moderately firm. Deep planting must be avoided. The crown should be about 2 in. under the surface, as the soil will afford it some protection during severe weather. As soon as the planting is done some support should be given to the branches; a neat stake and a strong tie will prevent them from being blown about by the wind. When this is done place a layer of short, rotten manure over the roots; this should be 3 in. thick and 1 ft. wide on each side. During the first two years very little pruning will be necessary. The second spring after planting any strong shoots that exceed 3 ft. in length should be cut back to that point. In the ground should be placed a few neat sticks, to which some of the lower branches should be tied to form the base of the hedge and bring it into shape. After the second year the growth will gain more vigour and increase in length. The strongest shoots should be cut down to 4 ft. the third year, and from that time the height should be allowed to increase slowly, so as to give the lower branches time to fill up the base. Some supports will be necessary to keep the growth in shape. The after-management consists in giving the roots a good dressing of rotten manure every winter. I find the best plan is to rake away the soil from over the roots, lay the manure on them, and then replace the soil. It seems hardly necessary to say that the plants will be greatly benefited by copious supplies of water, especially during the first two years after planting."

Rose Mallow *(Hibiscus).*

Rose of Heaven *(Agrostemma cœli-rosa).*

Rosewort *(Sedum Rhodiola).*

Rosmarinus officinalis *(Rosemary).*—This well-known fragrant shrub should always have a place in gardens. It is hardy and will grow in any ordinary soil In the embellishment of dry, warm rocky banks it is useful; all like its fragrance, and the flowers are pretty and plentiful on dry soils.

Rosy Rock Cress *(Arabis blepharophylla).*

Rubia.—This, for the most part, is a genus of weedy plants. Our native species, R. peregrina, is as handsome as any on account of its deep vinous purple foliage; its slender shoots are useful in winter for arranging with flowers. Rubiaceæ.

Rubus *(Bramble).*—The plants of this genus are mostly shrubby, though a few of the small growing kinds, such as R. arcticus (which grows a few inches high and bears numerous rosy pink blossoms), the Cloudberry, R. Chamæmorus (also dwarf and with white blossoms), and R. saxatilis, are very pretty. These are excellent plants for growing in moist, partially shaded spots in the rock garden in peaty soil.

Rudbeckia *(Coneflower).*—A genus of North American Composites, with showy yellow flower-heads, which usually have a dark raised cone-like centre. All, except R. bicolor, are perennials and hardy, but there is so much similarity among them that, outside of full or botanical collections, the best only should be grown. One of the best of all is R. speciosa, called also Obeliscaria speciosa and R. Newmani. The flowers are of a rich orange-yellow, with conspicuous, velvety maroon centres; individually they are from 3 in. to 4 in. in diameter, and are borne in dense masses in great profusion from early in August till late in October. The height of the plant is from 2 ft. to 2½ ft., and its habit is neat and compact. It is easily cultivated, flourishing in any ordinary garden soil, but succeeding best in that which is moderately rich, light, and well drained. It forms a very gay bed during summer and autumn. To make up for its autumn flowering, blue-branching Larkspur planted alternately in the same bed and pegged down has a good effect. The foliage of the Delphinium lightens up that of the Rudbeckia, and it comes early into flower. When both plants are in bloom they look well. The Rudbeckia may be easily and rapidly increased by division of the roots, and when fairly established, annual top-dressing is all that it requires, but, like all the herbaceous tribe, it enjoys good fresh soil, and when the beds become crowded the plants should be broken up and replanted. Among other showy kinds are R. hirta, subtomentosa, pinnata, and laciniata, but R. speciosa is better than any of these. The annual R. bicolor unites a dwarf branching habit of growth with the effective contrast of colour afforded by a bright yellow ray and blackish purple disc. It grows about 1½ ft. high; the flowers are about 3 in. in diameter, and resemble closely those of R. speciosa. It is of easy cultivation as a half-hardy annual in any soil, commencing to bloom in July, and continuing in flower till very late in the autumn. As a free-blooming and showy annual it can be recommended as well deserving of cultivation. It is a native of Texas. R. purpurea is synonymous with Echinacea purpurea.

Rue Anemone *(Thalictrum anemonoides).*

Rumex *(Dock).*—The only plant of this genus worth attention is our native Water Dock (R. Hydrolapathum). This is a very

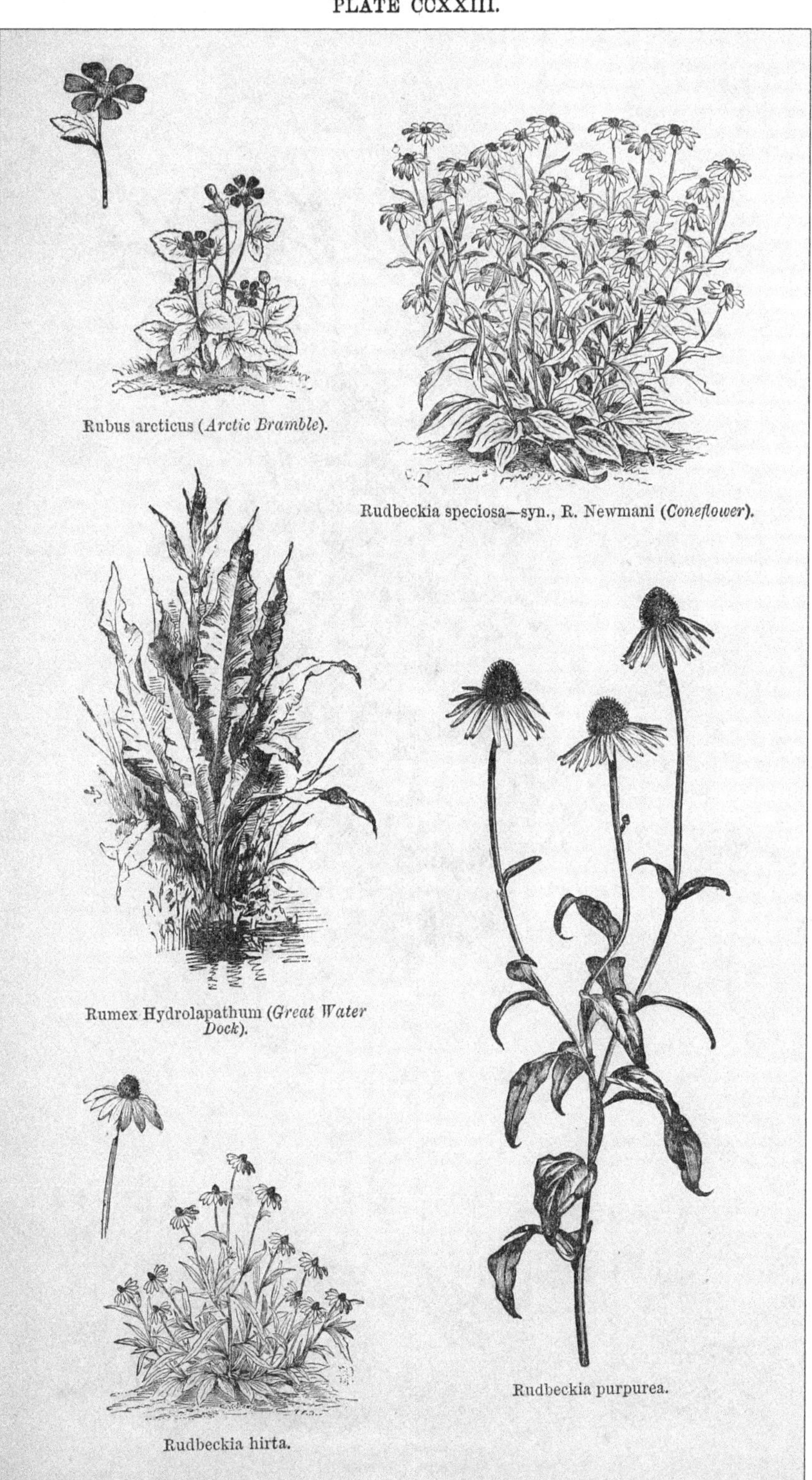

Rubus arcticus (*Arctic Bramble*).

Rudbeckia speciosa—syn., R. Newmani (*Coneflower*).

Rumex Hydrolapathum (*Great Water Dock*).

Rudbeckia hirta.

Rudbeckia purpurea.

[*To face page* **246.**

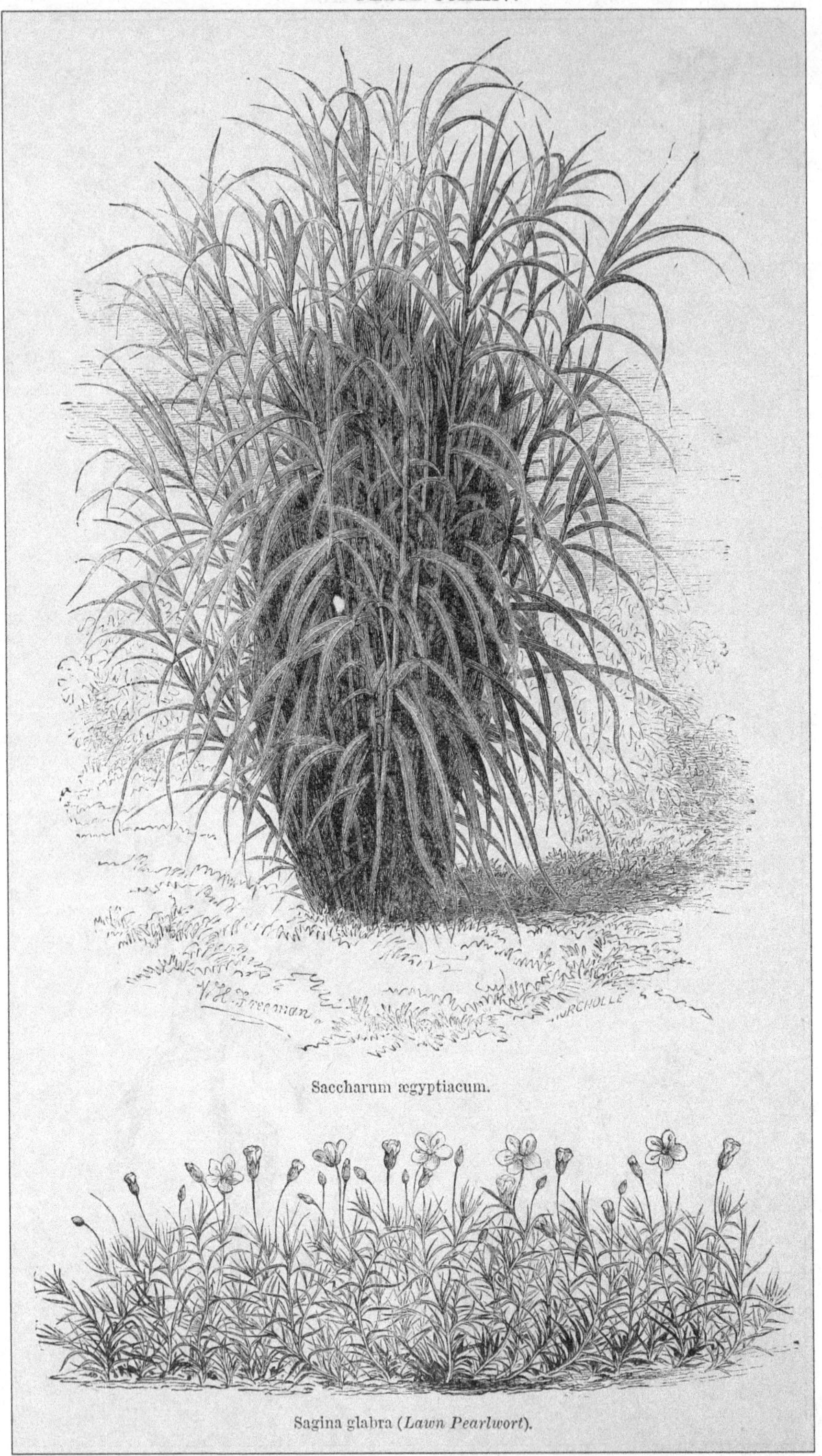

Saccharum ægyptiacum.

Sagina glabra (*Lawn Pearlwort*).

[*To face page* **247.**

large water plant of a size and habit sufficiently striking to entitle it to a place amongst ornamental subjects by the waterside. The leaves are sometimes 2 ft. or more in length, and form erect tufts of a very imposing character. The flowering stem is frequently 6 ft. in height, and bears a very large, dense, pyramidal panicle of a reddish or olive-fawn colour. The plant is always striking in appearance, but is most effective in autumn, when the leaves change to a lurid red, which colour they retain for some time. This plant will well repay the trifling trouble of procuring and planting it. A root or two deposited in the mud near the bank of a pond or slow stream will require no further attention.

Ruscus *(Butcher's Broom).*—The five known kinds of Butcher's Broom are ornamental, as well as curious, plants, and are well worth a place in gardens. With one exception, they are hardy, and in a state of nature are distributed throughout Europe, temperate Asia, and North Africa. All the hardy kinds are useful in planting under the drip and shade of trees where but few other evergreens would exist. The most ready method of propagation is by division of the roots. R. aculeatus (the Common Butcher's Broom) is the best known. It is a true native and is found in copses and woods. The small greenish flowers, which appear in April, are succeeded by bright red berries about the size of Peas. This dwarf, dense, much-branched shrub rarely attains a greater height than 2 ft.; its thick, white, twining roots strike deep into the ground, and when once established, even under apparently very adverse conditions, it grows freely. The Alexandrian Laurel (R. racemosus) is an elegant shrub with glossy, dark green leaves. The habit of this plant is not nearly so dense as that of the Butcher's Broom, but it grows to about the same height, and its stems are very valuable for cutting from in winter, or, in fact, at any season; it should be grown in quantity for this purpose. It flourishes best in deep loamy soil, in a partially shady situation. R. Hypophyllum is a very dwarf kind from the south of Europe, and R. Hypoglossum also comes from the same region as the preceding species, and is very similar in general appearance and size, but both are fit only for botanical collections.

Ruta *(Rue).*—The common Rue (R. graveolens) is not an ornamental plant, but there are two other species that deserve attention. R. albiflora is a graceful autumn-flowering plant which grows about 2 ft. high. In foliage it somewhat resembles the common Rue, but the leaves are more finely divided and more glaucous. The small pure white blossoms are borne profusely in large, terminal panicles, which droop very gracefully and last until sharp frosts curtail their beauty. In some localities it is quite hardy, but, generally speaking, it should, if not planted against a wall, have some slight protection in severe weather. It is a native of Nepaul, and also known as Bœnninghausenia albiflora. Another pretty plant is the Padua Rue (Ruta patavina), which grows about 4 in. to 6 in. high, and bears small golden yellow flowers, which, however, have the same odour as the common Rue. It is of about the same degree of hardiness as R. albiflora.

Sabbatia *(American Centaury).*—A lovely genus of N. American annuals and biennials of the Gentian family. There are three introduced species—S. chloroides, with large pink blossoms; S. campestris, light rose; and S. angularis, with purplish-red blossoms. The variety S. angularis alba has pure white blossoms, and is also at present in cultivation. The plants are not difficult to manage, provided the peculiar habitat of each species be imitated as nearly as possible. For instance, S. chloroides, being found in bogs, requires a very moist position; S. campestris, an open and drier place; S. angularis, a more sheltered situation than the others, and it must also have partial shade in imitation of that afforded by the surrounding vegetation amongst which it is found growing in a wild state. The soil should consist of equal parts of good fibry loam and finely sifted leaf-mould, with enough sand to make it open. The only means of propagation is by seeds, which should be sown in summer, and the seedlings potted off before they become in the least drawn. If this is not done they will make but weakly plants; they should be wintered in a cold, airy frame. In spring, repeatedly stopping the shoots will induce them to form bushy plants before flowering All the above kinds are but of biennial duration, and plants should be raised annually.

Saccharum ægyptiacum.—A vigorous perennial Grass, forming ample tufts of reed-like downy stems, 6 ft. to 10 ft. high and clothed with very graceful foliage. It is well adapted for ornamenting the margins of pieces of water, the slopes and other parts of pleasure grounds, &c., in a warm position. In our climate it does not flower, but even without its fine feathery plumes it is a pretty plant from its foliage and habit alone. It is easily and quickly multiplied by division in spring; the offsets should be started in a frame or pit. When established they may be planted out in May or June. N. Africa. S. Maddeni is a quick-growing hardy perennial, attaining a height of about 5 ft. It has handsome foliage, and is well worthy of culture for associating with other large-growing Grasses.

Sacred Bean *(Nelumbium).*

Sagina *(Pearlwort).*—The only species worthy of culture in this genus, which consists chiefly of British weeds, is the Lawn Pearlwort (S. glabra), a plant very generally known in consequence of being much talked of a few years since as a substitute for lawn Grass. Though it has not answered the expectations formed of it in that way, it is none the less a very beautiful and minute alpine plant, forming on level soil carpets almost as compact and smooth as velvet, which carpets are dotted and in early summer perfectly starred with numerous small, but pretty, white flowers. It is most readily multiplied by pulling the tufts into small pieces and replanting them at a few inches apart; they soon meet and form a carpet. It is also readily increased by seeds, but this mode is rarely worth resorting to, unless it is desired to propagate the plant largely for lawn making. Although it does not generally form a permanent and satisfactory turf, yet it is quite possible, by selecting a rather deep, sandy soil, and by keeping it perfectly clean and well rolled, to make a turf of it; but this is rarely worth attempting, except on a small scale, and when it begins to perish in flakes here and there, it should be taken up and replanted. It is very commonly grown in gardens under the name of Spergula pilifera. High mountains in Corsica.

Sagittaria *(Arrowhead).*— These are handsome aquatic plants belonging to the Water Plantain family (Alismaceæ). The most familiar species is our native Arrowhead, which is known by its arrow-shaped leaves and tall spikes of white blossoms. The double variety of this plant is very handsome; it is by far the most desirable, as the blossoms last much longer. S. heterophylla and S. obtusa are two N. American species, but scarcely so handsome as the British kind. The Arrowheads should always be introduced in artificial water; they grow best in water 1 ft. deep, with their tubers planted in mud.

St. Bruno's Lily *(Anthericum Liliastrum).*

St. Dabeoc's Heath *(Menziesia polifolia).*

St. John's Wort *(Hypericum).*

Salix *(Willow).*—Among the Willows there are some minute alpine species that are interesting for the rock garden or margins of beds of dwarf shrubs, though they are more adapted for the botanical than the general cultivator. The dwarf creeping kinds in gardens are S. herbacea, lanata, reticulata, retusa, and serpyllifolia, all natives of the northern parts of Europe and America. They grow well in any ordinary garden soil among stones.

Salpiglossis sinuata.—A beautiful plant and one of the finest of half-hardy annuals. It is of slender growth, with erect stem from 1 ft. to 2 ft. high, and with large funnel-shaped blossoms. These resemble Petunias, and are beautifully feathered, and veined with dark lines on a ground colour varying from pure white to deep crimson, yellow, orange, purple, and every intermediate shade. As the colour of the blossoms is so variable the plant is known as S. variabilis, and the varieties are described under Latin names according to the various tints. Where all are good and worth growing a selection of the varieties is difficult to make, but a packet of mixed seed will produce a wonderful variety of colours, and make a fine display from late summer till autumn. The plant thrives in a light, rich, sandy loam, and should be treated as a half-hardy annual. It is a native of Chili, and belongs to the Solanum family.

Salvia *(Sage).*—The Sages constitute one of the most natural and easily recognised genera in the vegetable kingdom, and yet they are exceedingly numerous, and diverse in duration, in the size and shape of the leaves and in the colouring of the flowers. They are found in almost all sub-tropical and temperate countries, but they find their greatest concentration in certain regions of the northern hemisphere, notably in the mountains of Tropical America and in the countries bordering the Mediterranean Sea. The best hardy ones are from the Mediterranean region, and the showiest from the mountains of Tropical America, especially Mexico. Few, if any, of the Mexican species are really hardy, but many of them are amongst the best ornaments of the conservatory and greenhouse during the autumn and winter months, and others are exceedingly effective in our beds and borders during the summer. Respecting the culture and propagation of Sages, little need be said, for few of them require special treatment. The herbaceous perennials may be rapidly propagated by division or from seed, and the half-shrubby species are as easily propagated from cuttings of the young soft shoots in heat. The hardy perennial species, or those requiring only a little care in the selection of a suitable soil and situation, are few compared with the large number of half-hardy kinds that we possess. One of the handsomest of the hardy sorts is S. pratensis, an indigenous species that sports into several varieties, which differ from each other in the colour of the flowers, there being alba, rubra, and bicolor. Handsomer even than the preceding is S. sylvestris, the long flower-spikes of which are deep purple and very showy. The well-known S. Sclarea and the variety bracteata make noble plants in a mixed border, and so does S. Forskohlei, a species similar in habit and colour of flowers to the preceding. The finest of all is S. hians, which, however, is rarely seen. Some of the forms of the common garden Sage (S. officinalis) are also very pretty, especially the variegated-leaved kind. The North American S. Pitcheri, a species with intensely blue flowers, is extremely ornamental, and so is its white-flowered variety. The pretty purple and red-topped Clary (S. Horminum) is a South European annual of easy culture. The tufts of coloured bracts terminating each stem render it very useful for cutting as well as a valuable decorative border plant. The silvery Clary (S. argentea) is also an excellent border plant. It has silvery leaves, from 6 in. to 12 in. long, which are very ornamental. When allowed to spread flat on the ground, seed may be sown every spring in patches on banks or borders of light, sandy

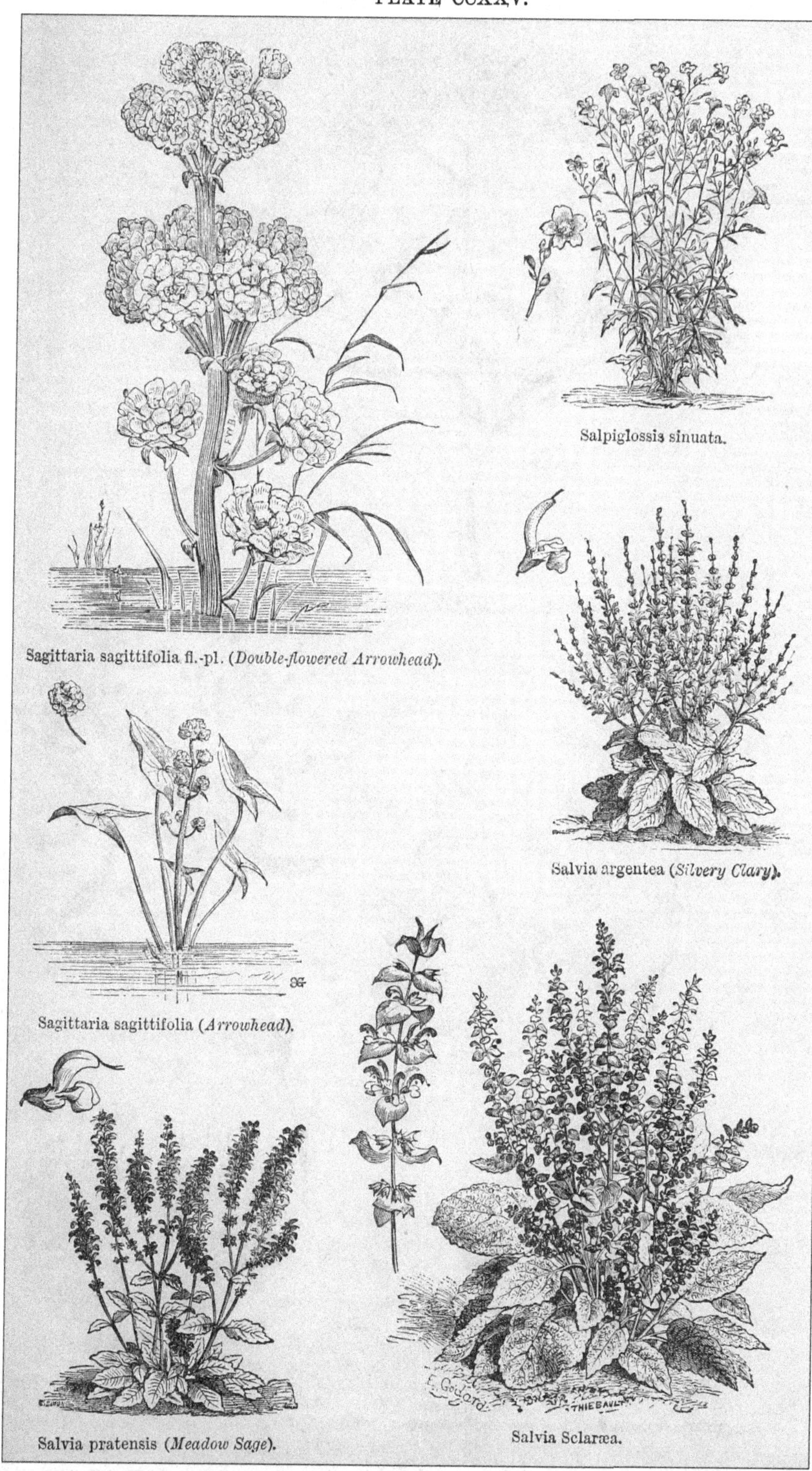

Sagittaria sagittifolia fl.-pl. (*Double-flowered Arrowhead*).

Salpiglossis sinuata.

Salvia argentea (*Silvery Clary*).

Sagittaria sagittifolia (*Arrowhead*).

Salvia pratensis (*Meadow Sage*).

Salvia Sclaræa.

[*To face page* **248.**

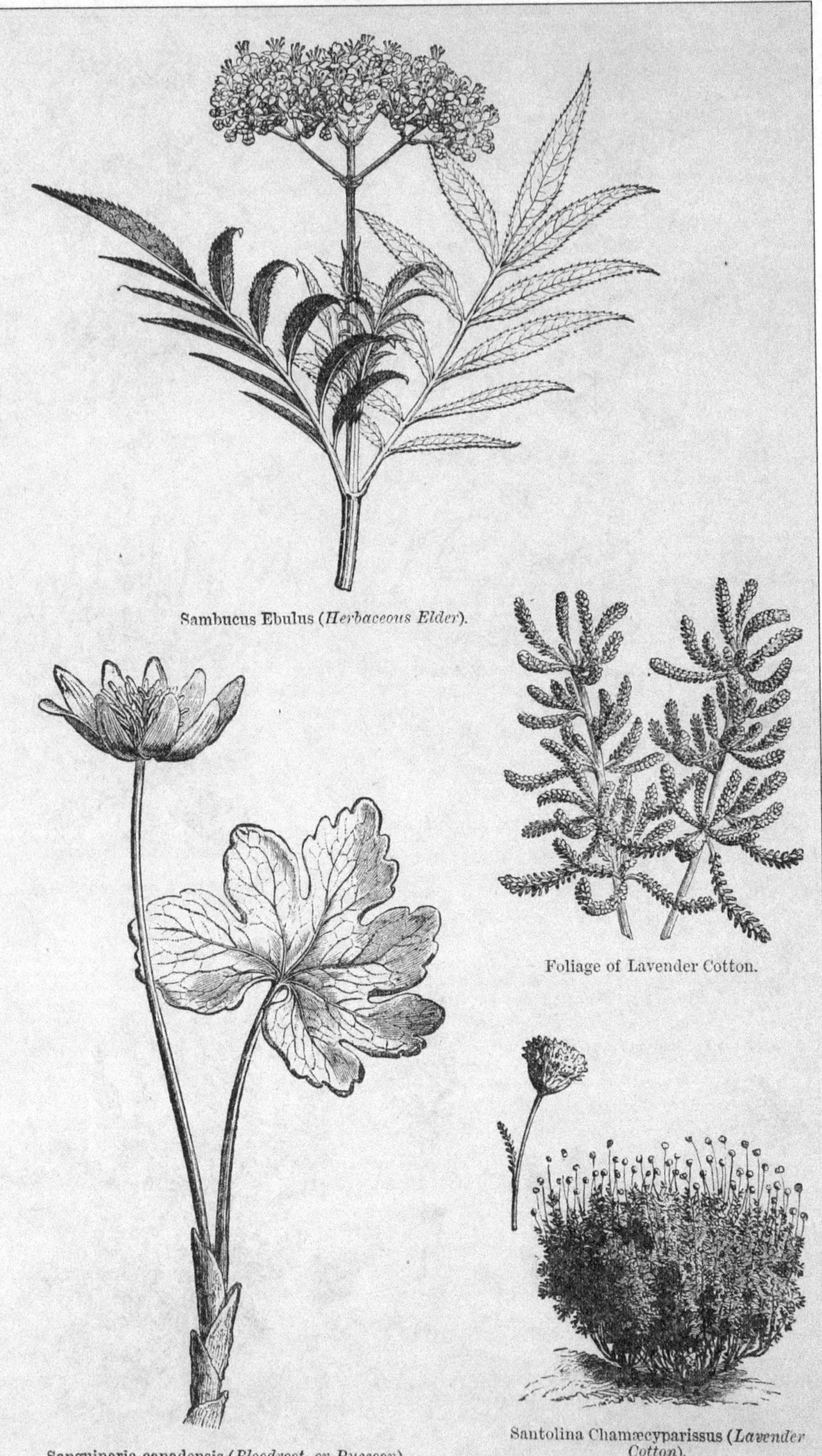

Sambucus Ebulus (*Herbaceous Elder*).

Foliage of Lavender Cotton.

Sanguinaria canadensis (*Bloodroot, or Puccoon*).

Santolina Chamæcyparissus (*Lavender Cotton*).

[*To face page* 249.

soil. S. candelabrum, a native of the south of Spain, is a half-shrubby species like the kitchen Sage, and has similar foliage, associated with ample panicles of rich violet and white flowers, borne on long stalks, clear of the leaves. S. taraxacifolia is a similarly handsome plant.

Of all the half hardy species, the S. patens is, doubtless, the most brilliant in colour, being surpassed by none and equalled by few flowers in cultivation. Although not hardy, it is easily preserved through the winter, and it is easily propagated from cuttings. In some districts, when in light warm soils, it survives an ordinary winter, but, as a rule, it is far safer to lift the plants. S. cacaliæfolia is a beautiful plant similar to S. patens, of a similar degree of hardiness to those mentioned below. S. porphyranthera is a dwarf, close-growing species with rich crimson flowers. It is very floriferous, and rarely fails to succeed in the open border. S. farinacea is a beautiful kind, the blossoms of which are of a light lavender with a white lip, quite different from that of porphyranthera. The flower-spike is covered with white powder. S. interrupta, a species from Morocco, is also very fine, with large light blue and white flowers. S. Grahami, another very old Mexican kind, has a very distinct habit and blossoms of a bright carmine colour. S. angustifolia, azurea, and Pitcheri are all pretty blue-flowered kinds, well worthy of open-air culture, as are also several of the kinds usually grown in greenhouses, such as S. Heeri, fulgens, gesneræfolia, elegans, tricolor, rutilans, splendens, and their varieties, for though some of them do not flower till autumn, they make better plants for indoor decoration by being planted out during summer.

Among these tender kinds there are several beautiful plants that flourish well in summer in the open border, where they have a fine effect. They are all easily propagated by cuttings in August and September in a close cold frame, and in spring in the same way as Heliotropes or Ageratums. When large plants are required, the old ones can be lifted and potted, or put close together in deep boxes. They should be cut down to within 6 in. of the soil when potted. Whether old plants or potted cuttings, they are easily wintered in any dry place where frost can be excluded. They are not particular as to soil, and thrive in any common garden soil moderately enriched.

Sambucus Ebulus.—A herbaceous kind of Elder, having handsome, spreading foliage, elegantly cut into leaflets. It is an admirable plant for associating with other fine-leaved plants, but is scarcely suitable for border culture. It will grow in any soil, is perfectly hardy, and may be readily increased by division. For dry banks in the open or shade it is a capital plant, and never becomes sickly through drought. The stems, being herbaceous, die down in autumn. As in most of the Elders, the flowers are inconspicuous. Native of Europe and the Caucasus. S. adnata and racemosa are similar plants, also suitable for the purposes above mentioned.

Samolus littoralis.—A pretty trailing plant, which has long, slender stems, furnished with small evergreen foliage and, in summer, with numerous pink blossoms. It is a desirable plant for the artificial bog, or for moist spots in the rock garden, as it delights in plenty of moisture at the roots. A peaty soil suits it best. A native of New Zealand, but quite hardy; it is still rare in cultivation. S. Valerandi, a British plant, is not of any garden value. Primulaceæ.

Sand Verbena *(Abronia)*.

Sandwort *(Arenaria)*.

Sanguinaria canadensis *(Bloodroot or Puccoon)*.—A singular and pretty plant, with thick creeping rootstocks, which send up kidney-shaped glaucous leaves some 6 in. high. The flowers, produced singly on stems as high as the leaves, are 1 in. across, white, with a tassel of yellow stamens. They appear in spring, and when in a good sized tuft a very pretty effect is produced. The variety grandiflora has larger blossoms than the type. It grows well in any border, but planted here and there under the branches of deciduous trees on lawns it soon spreads about, and without any attention becomes a charming naturalised plant. It prefers rather a moist soil. It may be increased in autumn by division, but the fleshy stems must not be long out of the ground. Papaveraceæ.

Santolina *(Lavender Cotton)*.—A genus of dwarf shrubby plants, remarkable only for their neat habit of growth, and for the pretty hoary foliage that most of them possess. One of the most distinct and useful is S. incana, a small silvery shrub, with numerous branches and narrow leaves. The whole plant is covered with dense white down, and forms neat prostrate tufts of edgings in the flower garden. The flowers are rather small, pale greenish-yellow, and in no way attractive, but the plant is popular from its neat habit and silvery hue. It grows readily in ordinary soil on the level border, and may be tastefully used on slopes of rockwork. It is considered a variety of the better known S. Chamæcyparissus, the Lavender Cotton; this and its other variety squarrosa are suitable for very large rockworks and banks, as they form spreading silvery bushes, 2 ft. high, in suitable soil, but are not suited for intimate association with very dwarf alpine plants. They are also used with good effect in mass bedding in toning down the glaring colours of other plants. Other species of Santolina suited for large rockworks are S. pectinata and S. viridis, which form bushes somewhat like the Lavender Cotton. S. alpina is of more alpine habit, forming dense mats quite close to the ground, from which spring yellow button-like flowers on long slender stems. It grows in any soil, and may be used on the less important parts of the rock garden. Cuttings of the shrubby species strike readily in spring or autumn, and S. alpina is easily increased by division.

Sanvitalia procumbens.—A pretty hardy annual from Mexico, having slender trailing branches and bright yellow flowers. In the single-flowered kind the blossoms have a dark purple centre, but in the double (S. procumbens fl.-pl.), which is by far the showiest, they are wholly of a bright yellow. It flowers from July till late in September. Owing to its dwarfness and compact growth, it is very useful for masses in beds or the front rows of borders, and even grown in suspended baskets it is very useful, as the slender branches droop gracefully on all sides of the basket. It may be sown in any ordinary garden soil in autumn for spring flowering, or in March and April for summer blooming. Compositæ.

Saponaria *(Soapwort)*.—In this genus of the Pink family there are a few very desirable garden plants.

S. cæspitosa is a neat little alpine plant from the higher regions of the Central and Eastern Pyrenees. There it flowers in the month of August, but in the lowlands its beautiful rose-coloured blossoms are developed towards the end of June. It is perennial, and forms rosettes of linear leaves, which are somewhat thick, glabrous, and of a beautiful green colour. The flowers, which form a thick cluster, are supported on short stout stems. This graceful little plant is very valuable for planting on rockwork. A sandy soil suits it best, and it will pass the winter perfectly in the open air without any particular care being bestowed on it.

S. calabrica is a pretty prostrate hardy annual, from 6 in. to 9 in. high. The slender stems, from June to September, are covered with a profusion of small blossoms, pink in the type and white in the variety alba. It is much used for mass beds and edgings. Seeds may be sown in the open border in April, or earlier, in heat, if bloom is required early in the season. It thrives best in a moderately rich sandy loam. S. Vaccaria, another annual, is not worth attention.

S. ocymoides is a beautiful trailing rock plant with prostrate stems and an abundance of rosy flowers, so densely produced as to completely cover the cushions of leaves and branches. It is easily raised from seed or from cuttings, thrives in almost any soil, and is one of the most valuable plants we have for clothing the most arid parts of rockwork, particularly in positions where a drooping plant is desired. The shoots fall profusely over the face of the rocks, and become masses of rosy bloom in early summer. The plant is also excellent for planting on ruins and old walls, on which the seed should be sown in mossy chinks in spots where a little soil is gathered. It is also valuable for the border, as it forms roundish, spreading cushions, with masses of flowers, and is well worthy of being naturalised in bare and rocky places. It is a native of Southern and Central Europe in stony and rocky places, and grows freely in poor soil, but when it is planted with the view of allowing it to fall freely over the face of the rock, a greater development will be secured by putting it in deep rich loam.

S. officinalis *(Common Soapwort)*.—This common native is a handsome plant when in flower. It grows about 2 ft. high and produces a profusion of large blossoms, varying from white to rose-pink. The double variety is the most desirable. It is a rambling plant, and soon spreads rapidly; therefore it should not be planted in select borders. It is a capital plant for rough places in the pleasure ground and wild garden, as it grows in any soil. Readily propagated by division.

Sarana kamtschatica (= Fritillaria).

Sarcodes sanguinea *(Californian Snow plant)*.—A curious parasitical plant, native of California, which probably cannot be grown in gardens. Ericaceæ.

Sarracenia purpurea *(Huntsman's Horn)*.—This is most singular, one of a family of Pitcher plants peculiar to North America. This species, which is the hardiest of all, is very handsome when well grown. The curious leaves, which are hollowed out like a horn, and are of a deep blood-red colour, form a compact tuft 1 ft. or more in breadth and the same in height. The flowers, too, are of singular shape, but are not very attractive. This is a good plant for growing in the artificial bog, or in damp spots in the rock garden, in open and fully exposed positions, in company with Cypripedium spectabile, the Parnassias, &c. The soil should consist of fibrous peat and Sphagnum Moss (which is common in most marshy places), mixed well together; there should also be placed around the plant a surface layer of living Moss, which tends to keep the plant moist. The soil must be perpetually moist, but not too much saturated. The plant is quite hardy if grown under these conditions, but precautions should be taken to prevent birds from disturbing the soil, so as to expose the roots. It is frequently imported in large quantities from America, and may be obtained cheaply. It is not easy to propagate, but sometimes may be raised from seed. S. flava is the next hardiest species, but is rarely satisfactory in the open air. Some of the hybrids between S. purpurea and others may, however, prove to be hardy.

Satureia.—Labiate plants of no ornamental value.

Satyrion *(Satyrium)*.

Satyrium.—Singular and beautiful terrestrial Orchids, natives of the Cape of Good Hope, and therefore scarcely hardy enough for open-air culture here. With care and attention, however, they succeed fairly well in sheltered and warm localities if protected well in winter, but, like the Disa, also from the Cape, they thrive best under glass pro-

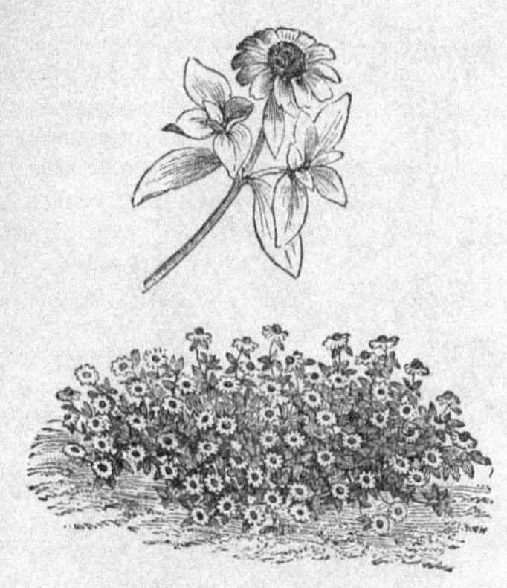

Sanvitalia procumbens.

Double flowers of Sanvitalia procumbens.

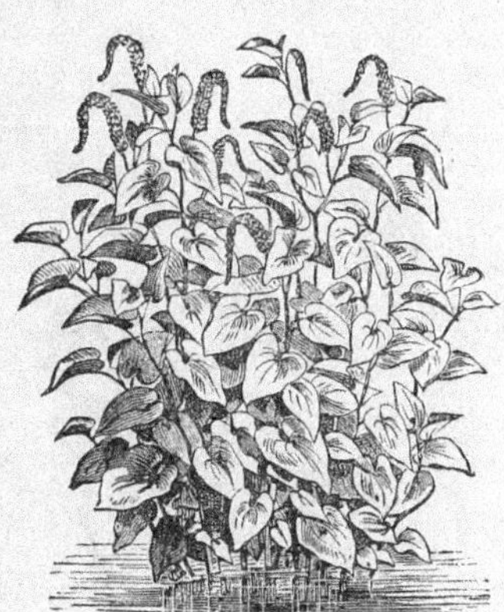

Saururus cernuus. See page 251.

Saponaria cæspitosa (*Soapwort*).

Saponaria officinalis fl.-pl. (*Double-flowered Soapwort*).

Saponaria ocymoides.

Saxifraga Burseriana. See page 252.

Saxifraga cordifolia.

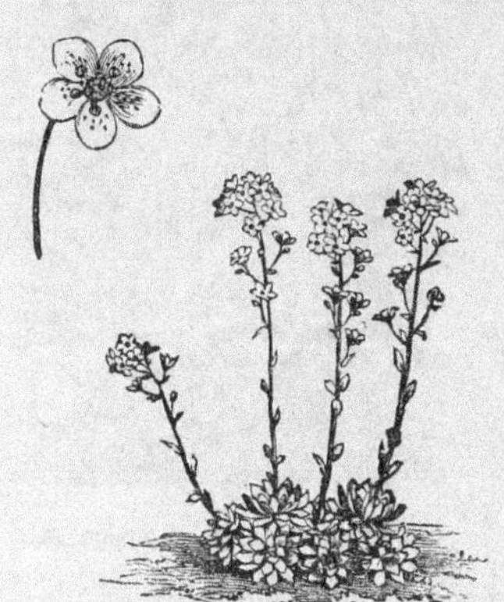

Saxifraga Aizoon.

Saxifraga cæspitosa (*Tufted Saxifrage*).

Saxifraga crassifolia. See page 253.

tection. S. carneum and aureum are in cultivation, and both are handsome.

Saururus.—Curious plants, suitable only for the botanic garden.

Saussurea. — Composite plants, of Thistle-like aspect, and suitable only for botanical or full collections. Some, such as S. pygmæa, grow only a few inches high, and may be cultivated in the rock garden. The leaves of S. discolor are white on the under-surface, and thus contrast well with the heads of light purple blossoms. S. alpina is a native species. All grow well in any soil.

Saxifraga *(Saxifrage)*—Perhaps this genus includes more true alpine flowers than any other. In the Arctic circle, in the highest alpine regions, on the arid mountains of Southern and Eastern Europe and Northern Africa, throughout the length and breadth of Europe and Northern Asia, they are found presenting many interesting varieties of form and colour. From their alpine habitats one might expect these to be as difficult of cultivation as other alpine plants, but they are the easiest to grow of all. Hence they were common in collections of alpine flowers when few other families were represented in them. Of late years many very pretty species have been introduced, and so great are the variety and merit of the family now, that a very interesting garden might be made of its members alone. For purposes of cultivation some kind of rough division of the members is convenient, as they are very different in aspect and uses. The most ordinary form is the Mossy or Hypnoides section, of which there are many kinds in cultivation. Their delicate moss-like spreading tuft of foliage so freshly green, especially in autumn and winter, when most plants show decay, and their countless white flowers springing from this carpet in spring, make them very precious inhabitants of gardens. These mossy kinds are especially suited for the tasteful practice of carpeting the bare ground beneath taller plants. They are admirable for their fresh green hue, with which they clothe rocks and banks in winter. They are indeed the most valuable winter "greens" in the alpine flora. Next to these we may place the very extensive silvery group as represented by such kinds as S. Aizoon and the great pyramidal-flowering S. Cotyledon of the Alps. Considering the freedom with which these grow in all cool climates, even on the level ground, and their beauty of flower and of foliage, they are perhaps the most precious group of alpine flowers we possess. Anybody with a cottage garden can grow them. The London Pride section is another of great beauty in its way. These thrive under ordinary circumstances in lowland gardens, &c., and soon naturalise themselves in lowland woods and copses. But the most brilliant Saxifrages, so far as flower is concerned, are the purple Saxifrage (S. oppositifolia) group and its near allies. Here we have tufts of splendid colour in spring with the same dwarfness and the most perfect hardiness. The large, leathery-leaved group, of which the Siberian S. crassifolia is best known, is also of much importance; the constitution of its members is such that they thrive in ordinary soil and on the level ground; there are various other minor groups. Such of the smaller and rarer alpine species as require any particular attention should be planted in moist sandy loam mingled with grit and broken stone, and made very firm. Very dwarf and rather slow-growing kinds, like S. cæsia and S. aretioides, should be surrounded by half-buried pieces of stone, so as to prevent their being trampled on or overrun, and stone will also help to preserve the ground in a moist, healthy condition in the dry season, when they are most likely to suffer. Very dry winds in spring sometimes have a bad effect on Saxifrages when such precautions are not taken. When in established tufts they are apt to throw out stem-roots into their own cushions, so to say—cushions frequently moist during the autumn and winter months. When the tufts are suddenly dried, the plants suffer if the ground roots be dried too.

The following are among the most important of the cultivated kinds, though of course the list excludes many species that are very difficult to grow or are scarcely procurable, and which are found only in very full collections.

S. aizoides.—A native plant, very abundant in Scotland, the north of England, and some parts of Ireland, in wet places, by the sides of mountain rills or streams. It produces at the end of summer or autumn an abundance of bright yellow flowers, ½ in. across, and dotted with red towards the base. It forms dense, dwarf, bright green masses of leaves, and has leafy branched flower-stems, by which it is distinguished from the other yellow Saxifrages. Although a moisture-loving mountain plant, it is quite easy to grow in lowland gardens, naturally doing best in moist ground. Wherever a small stream or rill is introduced to the rock garden or its neighbourhood, it may be most appropriately used, and planted so as to form wide-spreading masses, as it does on its native mountains. Easily propagated by division or by seed. S. autumnalis is similar.

S. Aizoon deserves to be cultivated as a rock, border, and edging plant. Plants of it established two or three years form grey-silvery tufts 1 ft. or more in diameter, and about 6 in. high, sometimes a few inches more; these great tufts do not flower so freely as the wild plants, but this need not be regretted, as it is the silvery mass, and not the flowers, that is sought. The plant is easily distinguished by its rather oblong and obtuse leaves, bordered with fine teeth, densely margined with encrusted pores, and stiffly ciliated at the base, and by the flower-stems, which grow from 6 in. to 15 in. high, are furnished with glandular hairs on the upper part, but are usually smooth on the lower. As to its culture, nothing can be easier; it is very often grown in pots, but flourishes as freely as any native plant, and is best perhaps when exposed to the full sun. There are several named varieties. S. pectinata, Hosti, intacta, rosularis minor, australis, cartilaginea, and various others are only slight variations from the typical S. Aizoon. Division in spring.

S. Andrewsi.—Among the green-leaved Saxifrages there is no better kind than this. Its flowers are freely produced, prettily spotted, and larger than those of S. umbrosa. It is more worthy of a position on a rockwork than the London Pride, but does quite freely on any border soil, merely requiring to be replanted occasionally when it spreads into very large tufts, or to have a dressing of fine light compost sprinkled over it annually. The variety Guthrieana is handsome and distinct.

S. aretioides.—A very gem of the encrusted section, forming cushions of little silvery rosettes, almost as small and dense as those of Androsace helvetica, and about ½ in. high. It has rich golden-yellow flowers, which appear in April on stems a little more than 1 inch high, and remind one of the flowers of Aretia Vitalliana. It is not difficult to grow, but requires a moist and well-drained soil, and being so dwarf and tiny, must be guarded from being overrun by coarser neighbours. A native of the Pyrenees; increased by seed and careful division.

S. aspera.—A small, grey, tufted, prostrate plant. The flowers are small and few, of a dull white colour, and are borne on stems about 3 in. high. S. bryoides, a variety of this, forms a dense tuft, with pale yellow flowers. Both are worthy of growing for their Moss-like character, and are natives of the Pyrenees—S. bryoides in the most elevated regions. They grow freely in the open air in London, but rarely flower.

S. Burseriana.—None of the Saxifrages surpass this in vernal beauty. It is dwarf, indeed, almost Moss-like in habit, forming broad patches and spreading rapidly over the earthy interstices of warm, moist sandstone rockwork, if planted where it does not suffer from stagnant moisture. The blossoms are borne singly on slender red stalks, which rise 2 in. or 3 in. above the general surface of the plant, and are pure white, the margins of the over-lapping petals being elegantly frilled or crisped. Interspersed among the fully-expanded flowers, the unopened buds (which are of a dullish crimson-brown colour) show themselves to excellent advantage, and enhance the pearly whiteness of the petals. The flowers appear freely in January and February before the Snowdrops, and even before the flowers are expanded, the brownish-scarlet buds, just emerging from compact silvery tufts of foliage, have a cheerful effect. It soon forms good-sized tufts in the open border or on rockwork, preferring a dry, sunny situation and calcareous soil. It is a native of the Austrian Alps, and a plant that all lovers of hardy spring flowers should possess. There are two or three distinct forms of this species which differ chiefly in habit of growth, one being much more tufted than the others. There is also a form with larger flowers than those of the type.

S. cæsia.—This resembles an Androsace in the dwarfness and neatness of its tufts. On the Alps it stains the rocks and stones like a silvery moss, and on level ground, where it has some depth of soil, it spreads into beautiful little cushions from 2 in. to 6 in. across. It is easily known, either in cultivation or in a wild state, by its exceeding dwarfness and neatness, and by its three-sided keeled leaves, regularly margined with white crustaceous dots. It bears pretty, white flowers, about the third of an inch in diameter, on thread-like, smooth stems, 1 in. to 3 in. high. A native of the high Alps and Pyrenees, it thrives perfectly well in our gardens in very firm, sandy soil, fully exposed, and abundantly supplied with water in summer. It may also be grown well in pots or pans in cold frames near the glass; but, being very minute, no matter where it is placed, the first consideration should be to keep it distinct from all coarse neighbours, and even the smallest weeds will injure or obscure it if allowed to grow. Flowers in summer, and is increased by seeds or careful division. Of a similar character to S. cæsia are S. calyciflora, luteo-viridis, Kotschyi, valdensis, squarrosa, and diapensoides, all of which are of pigmy growth, and, for the most part, extremely difficult to cultivate successfully, though the beauty they afford amply repays any trouble bestowed on them. They should be grown in the same way as S. cæsia.

S. cæspitosa.—One of the dwarf-growing kinds that form dense, carpet-like masses of foliage, arranged in neat tufts which in summer are studded with white blossoms. It is one of the most variable of all Saxifrages, and hence a host of names are found in gardens. The most distinct of the numerous varieties are palmata and grœnlandica, both of which, like the type, succeed in almost any garden soil, and in any situation, and are very valuable plants for margins to herbaceous borders, while they are beautiful coverings for moist banks in sun or shade.

S. ceratophylla *(Stag's-horn Saxifrage)*.—A fine and ornamental species of the mossy section, with very dark, finely-divided leaves. The flowers are pure white, and abundantly produced in loose panicles in early summer. It quickly forms strong tufts in any good garden soil, and is admirably adapted for covering rockwork of any description, either as wide level tufts on the flat portions, or pendent sheets from the brows of rocks. It may also be used with good effect in borders. A native of Spain; increased by seed or division. Similar to this species are numerous others, such as S. paniculata, ladanifera, Wilkommiana, geranioides, irrigua, ajugæfolia, aquatica, Schraderi, &c., all of which are desirable for a full collection.

S. ciliata.—One of the broad-leaved or Megasea section. It has large broad leaves (covered with soft hairs), produced from creeping stems, which, in well established plants, form a spreading mass. The flower-stems rise from 6 in. to 9 in. high, and bear numerous large, flesh-coloured flowers in spring. Being a native of North India, it is suitable for culture in the open air in the south of England only. It is, however, so handsome and distinct that wherever it may be grown it should be tried. A sheltered nook in the rock garden, partially shaded, suits it best.

S. cordifolia.—This is entirely different in aspect from the ordinary dwarf section of Saxifrages. It has very ample, roundish, heart-shaped leaves on long and thick stalks; the flowers are of a clear rose arranged in dense masses, and are half concealed among the great leaves in early spring, apparently hiding under them from the cutting breath of March. This Saxifrage and its varieties grow and flower in any soil and position, and are thoroughly hardy; but it is desirable to encourage their early-flowering habit by placing them in warm, sunny positions, where the fine flowers may be induced to open well. They are perhaps more fitted for association with the larger spring flowers and with herbaceous plants than with dwarf alpines, and are well worthy of being naturalised on bare sunny banks, in sunny wild parts of the pleasure-ground, or by wood walks. They may also be used with fine effect on rough rock or root work, near cascades, or on rocky margins to streams or artificial water, their fine evergreen, glossy foliage being quite distinct. They may, in fact, be called the fine-foliaged plants of the rocks. There are several handsome varieties of it, particularly one called purpurea, which is the finest of all the group. Native of Siberia.

S. Cotyledon *(Pyramidal Saxifrage)*.—A noble Saxifrage, which embellishes with its great silvery rosettes and elegant pyramids of white flowers many parts of the great mountain ranges of Europe, from the Pyrenees to Lapland. It is easily known by its rather broad leaves, margined with encrusted pores, and its fine handsome bloom. It is the largest of the cultivated Saxifrages, and also the finest except S. longifolia, the linear leaves of which it does not possess. The rosettes of the pyramidal Saxifrage differ a good deal in size. When grown in tufts, they are generally much smaller, from being crowded, than isolated specimens. The flower-stem varies from 6 in. to 30 in. high, and about London, in common soil, they will often attain a height of 20 in. In cultivation the plant usually attains a greater size than on its native rocks. A variety more pyramidal in growth and more robust than the type is known in gardens under the erroneous name of S. nepalensis—sometimes S. pyramidalis, which is much more appropriate. It is of easy culture, the chief point to be observed being to denude the parent plant of offsets as soon as they appear. Good specimens will thus be obtained. The plant may be grown either in the rock garden or in the ordinary border.

S. crassifolia.—A well-known species of the Megasea section with large broad leaves. Its flowers are produced in dense panicled cymes, rising from the terminal shoots in showy pendent masses; they are of a light rosy colour with the slightest lilac tint, and are produced in the months of March and April. In spite of the somewhat coarse appearance of the leaves, the plant forms a very gay and useful accessory to the beauty of the spring garden. It is suitable for the same purposes as S. cordifolia. The chief varieties are ovata, which throws its deep rose-coloured flowers well above the foliage; rubra, similar to the last, but with flowers of a deep tinge of rose; orbicularis, which is sometimes ranked as a species, but is nothing more than a small-growing form, with the leaves rather broader than those of ovata, and with a more branching habit; it is a free bloomer, producing an abundance of light rosy flowers, which are well elevated above the foliage; and media, a distinct and ornamental variety, with large, dark, shining green leaves and bright rosy-pink flowers, produced in very large clusters in great profusion on strong stems. There is also a variety with variegated foliage called aureo-marginata. Siberia.

S. Cymbalaria.—Quite distinct in aspect from any of the family, and one of the most useful of all, being a bright and continuous bloomer. Little tufts of it in early spring form masses of bright yellow flowers set on light green, glossy, small, Ivy-like leaves, the whole being not more than 3 in. high. These, instead of fading, preserve the same little rounded pyramid of golden flowers until autumn, when they are about 12 in. high. It is an annual or biennial plant, which sows itself abundantly, coming up in the same spot. It is peculiarly suitable for moist spots on or near rockwork, grows freely on the level ground, and might be readily naturalised on the margins of a rocky stream and in various other places in large pleasure grounds.

S. flagellaris is not only one of the most distinct of its race, but one of the freest growing. Like its near ally, S. Hirculus, it has large, bright yellow blossoms, borne, three together, on viscid stems 3 in. or 4 in. high. From each rosette slender thread-like or wire-like stolons are thrown off, which, rooting at the tips, quickly form new rosettes in moist, peaty, and gritty soil. One of the most arctic of plants.

S. Fortunei has large panicles of white blossoms rising in profusion from rosettes of dark green, roundish leaves. As it is not particular as to treatment, and flowers in autumn, it is a desirable plant.

S. Geum.—This is very like the London Pride in habit and flowers, and is useful for the same purposes and is cultivated with the same facility as that plant. It will grow freely in woods or borders, particularly in moist districts, and is worthy of naturalisation in the former. Like its neighbours, it is, of course, suitable for rockwork, but does not deserve that position so well as numbers of plants much more difficult to grow. S. hirsuta comes near this, and is probably a variety, the chief difference being that the leaves are longer than broad, are less heart-shaped, and more hairy; it is suitable for similar positions. Nearly allied also is S. cuneifolia, with its numerous varieties, all of which are desirable for a full collection.

S. granulata (*Meadow Saxifrage*).—A lowland plant, with several small scaly bulbs in a crown at the root, and numerous white flowers three quarters of an inch across. It is common in meadows and banks in England. The double form is very handsome and is useful in the spring garden as a border plant, or on rougher parts of rockwork.

S. Hirculus is a native plant found on mountains in a few localities in Britain. It has weak prostrate stems and flower-stems, about 6 in. high, bearing a solitary yellow flower. The plant is handsome under congenial circumstances—that is, in very moist peaty soil, or even in loam when wet—but is otherwise unworthy of culture for ornament. S. Hirculus major produces flowers nearly double the size of the original species; the foliage also is broader and of much greater substance, and this form is also more amenable to cultivation.

S. hypnoides.—A very variable plant in its stems, leaves, and flowers, but usually forming mossy tufts of the deepest and freshest green. No plant is more useful for forming carpets of the most refreshing green in winter in almost any soil. For this reason it is peculiarly well suited for planting in the low rocky borders often made in town and villa gardens, and should be largely used by those who desire to rest their eyes on glistening verdure during the winter months. It thrives either on rockwork or the level ground, in half-shady positions or fully exposed to the sun, forming the fullest and healthiest tufts in the latter case, and flowering profusely in early summer. Nothing can be easier to grow or to increase by division. It is also suitable for forming dwarf verdant carpets in the flower garden or on the rockwork with a view to placing one or more plants on the surface. Under this species may be grouped S. hirta, S. affinis, S. incurvifolia, S. platypetala, S. decipiens, and several others, all exhibiting differences which some think sufficient to mark them as species. They present considerable differences in appearance when grown together in a garden, and many no doubt think them worthy of a place; they all thrive with the same freedom as the Mossy Saxifrage, appearing to suffer from drought or very drying winds only. If, when first planted, a few largish stones are buried in the earth round each, the plants will soon lap over them, the stones will serve to preserve the moisture in each tuft, and the plants will be much less likely to suffer from drying winds. S. densa and Whitlavi are two of the very best free-growing species, and are suitable for a margin, being compact and always green.

S. juniperina (*Juniper Saxifrage*) is one of the most distinct and desirable kinds in cultivation, having spine-pointed leaves, densely set in cushioned masses, looking, if one may so speak, like Juniper bushes compressed into the size of small round pincushions, and with little seen but the prickly points of the leaves. The flowers are yellow, arranged in spikes on a leafy stem, and appear in summer. It thrives very well in moist, sandy, firm soil, and is well worthy of a place in the rock garden, and also in every collection of alpine plants grown in pots. A native of the Caucasus; propagated by seed and careful division.

S. lantoscana.—One of the finest of all the incrusted-leaved section. It is somewhat similar to the pyramidal variety of S. Cotyledon, but the plant is smaller, the leaves narrower and more crowded in the rosette, and the flower-spike, which is not borne erect, but slightly drooping, is more densely furnished with flowers. It is a plant easily grown in a fully exposed position of a well formed rockery, in a gritty soil well drained. It remains long in flower, and is one of the most satisfactory of rock garden plants.

S. ligulata is a Nepaul plant with broadly obovate leaves. The flowers are produced in small cymose panicles, white, with a rosy tint towards the margin of the petals; the anthers before expansion are of a deep crimson colour, which adds much to the beauty of the flowers. Coming from Nepaul, and with a tendency to very early spring growth, it is liable to suffer from frosts, and this form of injury occurring in three or four consecutive seasons so weakens the plant, as ultimately to kill it; care, therefore, should be taken to give it a nice sheltered situation, where it may also have the benefit of a bit of shade. The varieties rubra and speciosa, the latter particularly, are finer than the type in every way. S. ligulata may be associated with others of the Megasea section.

S. longifolia.—The single rosettes of this Pyrenean plant are often 6 in., 7 in., and 8 in. in diameter; it may well be termed the queen of the silvery section of Saxifrages, by which section are meant those that have their greyish leathery leaves margined with dots of white, so as to give the whole a silvery character. This is so beautifully marked in that way, that it is attractive at all seasons, while in early summer it pushes up massive foxbrush-like columns of flowers from 1 ft. to 2 ft. long, the stem covered with short, stiff, gland-tipped hairs, and bearing a multitude of pure white flowers. It is perfectly hardy in this country, not difficult of

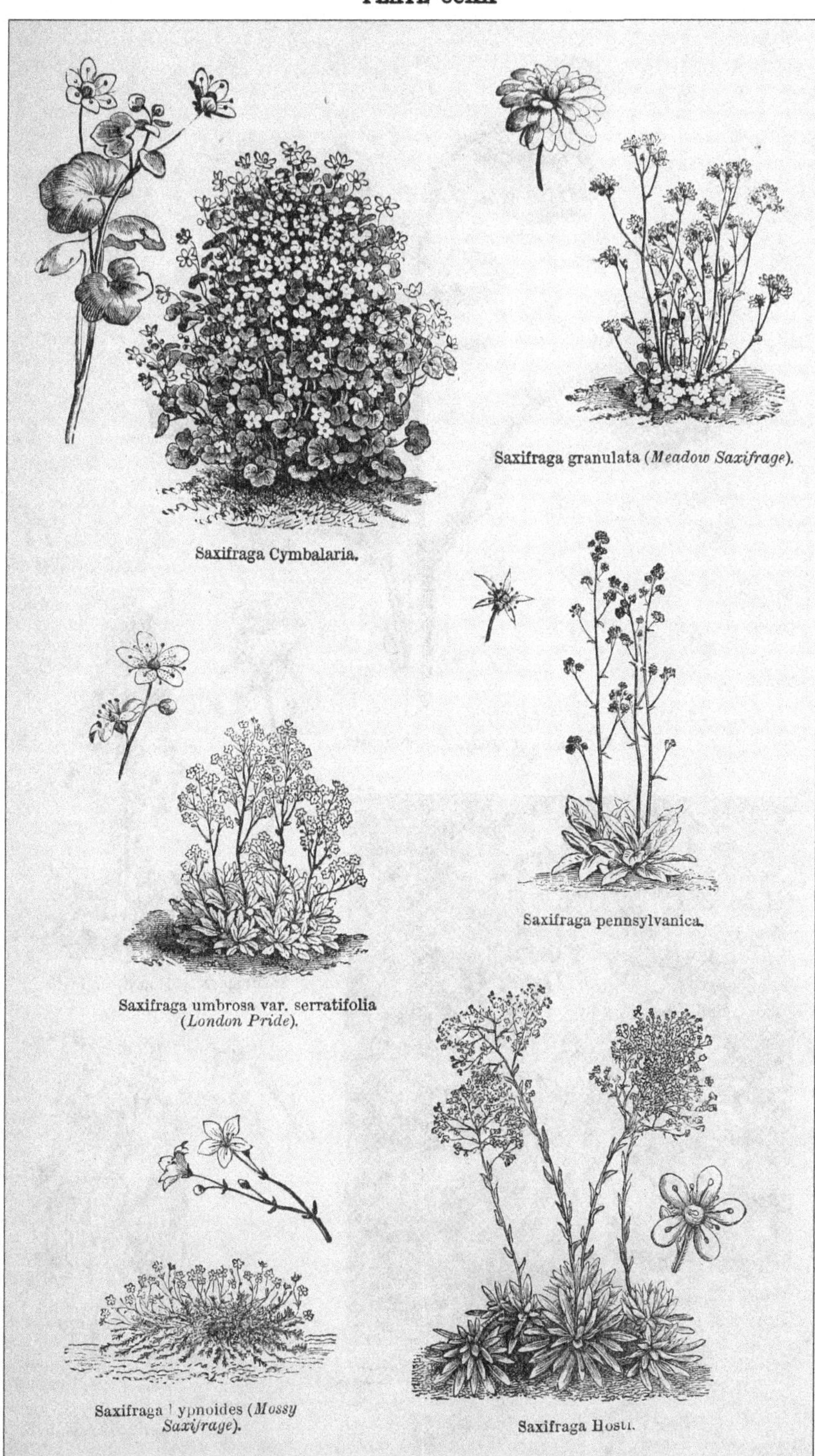

Saxifraga Cymbalaria.

Saxifraga granulata (*Meadow Saxifrage*).

Saxifraga umbrosa var. serratifolia (*London Pride*).

Saxifraga pennsylvanica.

Saxifraga hypnoides (*Mossy Saxifrage*).

Saxifraga Hostii.

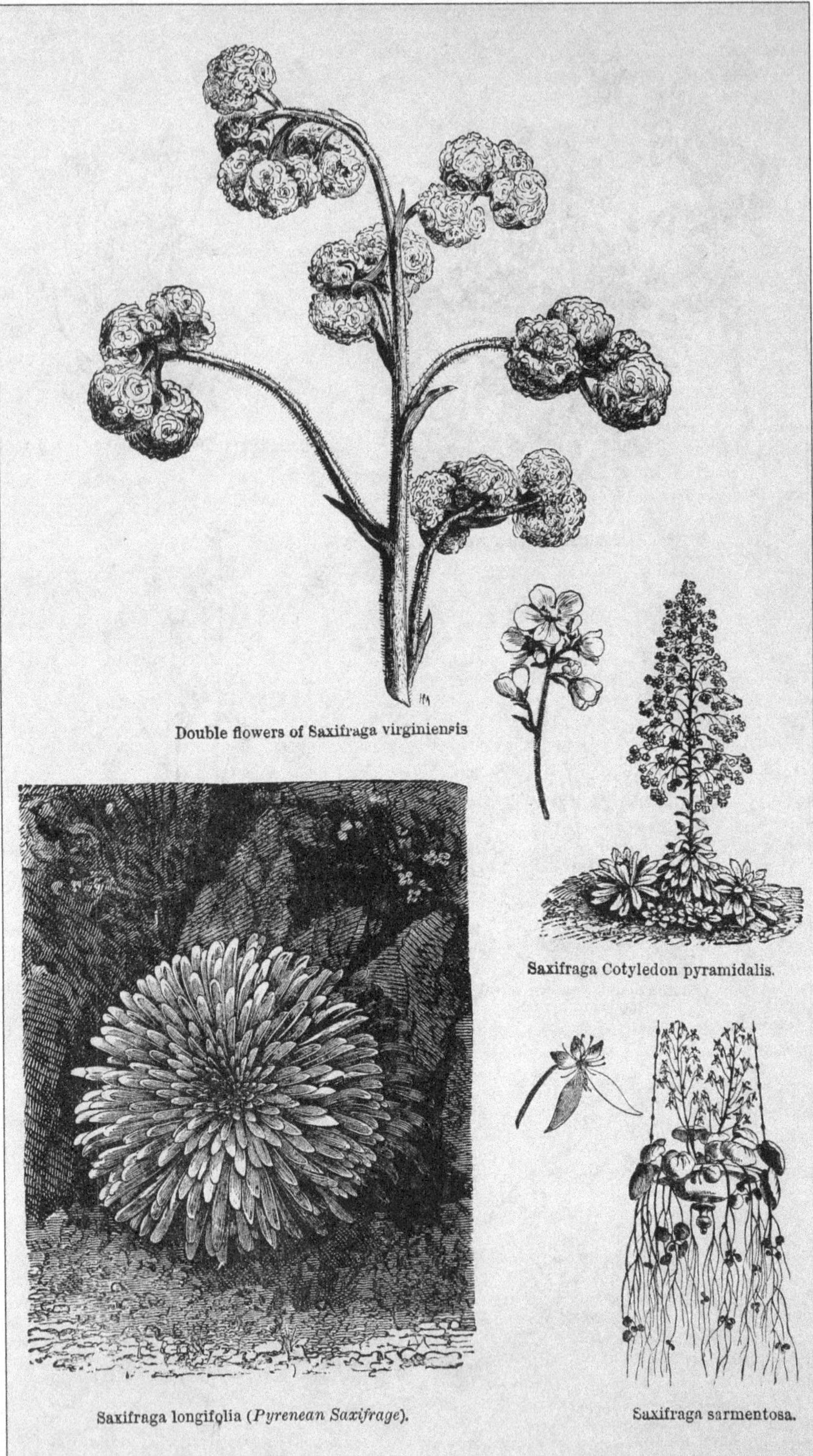

Double flowers of Saxifraga virginiensis

Saxifraga Cotyledon pyramidalis.

Saxifraga longifolia (*Pyrenean Saxifrage*).

Saxifraga sarmentosa.

[*To face page* 255.

culture, and may be grown in various ways. On some perpendicular chink in the face of a rockwork into which it can root deeply it is very striking when the long outer leaves of the rosette spread away from the densely-packed centre. It may also be grown on the face of an old wall, beginning with a very small plant, which should be carefully packed into a chink with a little soil. Here the stiff leaves will, when they roll out, adhere firmly to the wall, eventually forming a large silver star on its surface. It will thrive on a raised bed or border if surrounded by a few stones to prevent evaporation and to guard it from injury. It also thrives in a greenhouse or frame, and perhaps the readiest way of making a weakly young plant from the nursery develop into a sturdy rosette is to put it in a 6-in. pot, well drained and filled with a mixture of sandy loam and stable manure, and place it in a sunny pit or frame, giving it plenty of water in spring, summer, and autumn. It is propagated by seeds, which it produces freely. In gathering them it should be observed that they ripen gradually from the bottom of the stem upwards, so that the seed-vessels there should be cut off first, leaving the unripe capsules to mature; the plant should be visited every day or two to collect them as they ripen successively. S. lingulata is by some united with the preceding, from which it chiefly differs in having smaller flowers, in the leaves and stems being smooth and not glandular, in its shorter stems, and in the leaves in the rosette being shorter and very much fewer in number than in S. longifolia. This is also a charming rock plant, and will succeed with the same treatment and in the same positions as S. longifolia. S. crustata is considered a very small variety of S. longifolia; being much smaller, it will require more care in planting and to be associated with dwarfer plants.

S. Maweana is a very handsome species of the cæspitosa section. It is larger than any other both as regards the foliage and flowers. The latter are pure white, as large as a shilling, and produced so abundantly as to form in early summer dense masses of white. After flowering this species has the peculiarity of forming buds on the stems, which remain dormant till the following spring. It is of easy culture, but is rare. Similar to S. Maweana, but even finer, is a new kind called S. Wallacei. It is far more robust, earlier, freer as regards flowering, and very easily propagated. It differs from S. Maweana in not developing the small, round buds during the summer. It is of easy culture, and a most desirable plant for either the border or rock garden.

S. muscoides *(Mossy Saxifrage)*.—A beautiful little plant of small growth, forming a dense carpet of bright green foliage like S. hypnoides and cæspitosa. There are several forms of it, but the best is atropurpurea, which produces a dense mass of deep red-purple blossoms on stalks a few inches high. The varieties pygmæa and crocea are likewise pretty, as are also the allied S. exarata, pedemontana, Rhei, aromatica, and a few others, all of which are excellent for the rock garden, growing well in almost any soil.

S. mutata.—A yellow-flowered species bearing considerable similitude to S. lingulata; its flower panicle is about 18 in. high; it is rarely seen in cultivation, owing to the fact that it not unfrequently exhausts all its vigour in producing blooms, and rarely matures seeds in this country; further, it does not produce offsets, as most of this section do. It is a native of the Alps, but limited in its distribution. An allied species is S. florulenta, a beautiful plant of the Maritime Alps, but which no one has succeeded in cultivating in this country.

S. oppositifolia.—It is impossible to speak too highly of the beauties of this bright little mountaineer, so distinct in colour and in habit from the familiar members of its family. The moment the snow melts, its tiny herbage glows into solid sheets of purplish rose colour; the flowers are solitary, on short, erect little stems, and are often so thickly produced as to quite hide the leaves, which are small, opposite, and densely crowded. There are several varieties in cultivation; that known as splendens has flowers of much greater brilliancy, though slightly smaller, than those of the type. In bud especially the colour is almost carmine and is exquisitely beautiful. In density of bloom it approaches the typical form, but rarely quite equals it. This variety was obtained many years ago on the mountains of Scotland. S. oppositifolia major has flowers twice the size of the type with less of the purple tinge. Its colour is a clear rose, inclining to cherry. The petals are broad and well rounded, but when fully expanded they are less compact than those of the type. S. oppositifolia pyrenaica differs conspicuously in habit. Its shoots are ascending, and are twice as robust as those of any of the preceding. The flowers also are larger, but usually more sparse. Its finest form is S. oppositifolia pyrenaica maxima, which has blossoms as large as shillings, of a lovely light rose colour. S. oppositifolia alba has white flowers, which form a pleasing contrast to the other varieties. S. Rudolphiana has a more spreading habit of growth, and the rosy-purple flowers are borne sometimes singly, sometimes (though rarely) in pairs. Its alliance is with S. biflora, the habit of which is loose, but which is a beautiful dwarf species. The flowers are in clusters of from two to four, and vary in tints from bright rose to deep blood colour; the petals are narrow and wide apart. S. Kochi is similar in habit to the last, but it has rosy purple flowers, also borne in twos and fours together at the extremities of the shoots; it is good in form, the petals being broad and close. S. retusa has foliage very short, firm, dense, and compact. Its flowers are small, and in clusters at the extremity of erect stalks; the petals are narrow, pale rose, sometimes brighter. It blooms rather later than the varieties of S. oppositifolia. S. Wulfeniana is closely allied to the preceding, of which it is probably the Tyrolese form. S. oppositifolia and its varieties flourish in open, rich, loamy soil, either in the common garden border or in fissures of rockwork; but in either case the soil should be trenched or prepared fully 2 ft. deep, so that the roots may reach a level below that which is affected by summer drought. They are perhaps finest when placed in a fissure or on a ledge of rockwork, where the roots can ramble backward or down to any depth, the soil being rich, light loam, mixed with fragments of limestone or grit (small fragments of any rock will do) and a little river sand. They should always be placed so as to get sunshine; they grow, but will not flower freely in the shade. The same treatment, with a little admixture of peat or vegetable mould, suits S. retusa and S. Wulfeniana. The Tyrolese species (biflora, Rudolphiana, and Kochi) are less easy to please. They grow wild on the moraines of glaciers, where light vegetable soil, sand, and *débris* of every kind blend with massive rocks, now coating the surface, now filling the interstices; water dripping or oozing around, and frequently flowing in volume within 2 ft. or 3 ft., so as to soak the bases of the rocks on which their rosy carpet is spread. They will grow in pots, nevertheless, but rarely with the same freedom which characterises the varieties of S. oppositifolia.

S. peltata.—The shield-like form of the leaves of this plant is unique amongst Saxifrages. So distinct is S. peltata in this respect that some have referred it to a section under the name Peltiphyllum. It grows about a yard high and as much or even more across, and is a remarkably bold and handsome plant. Its leaves, which arise from a very thick and fleshy creeping rootstock, have stout erect stalks, on the apex of which the target-like blade of the leaf attains 1 ft. or more in diameter. The flowers, which are produced in spring, a little before the leaves make their appearance, are borne on stalks from 1 ft. to 2 ft. high, in loose clusters from 3 in. to 6 in. in diameter. They are individually about ½ in. across, of a white or very pale pink tint. This Saxifrage is a native of California, where it is found growing in the neighbourhood and in the beds of quick-running streamlets throughout the Sierra Nevada of California. In cultivation it succeeds best in a deep moist border, consisting chiefly of peaty soil. It may be easily propagated by means of division or seeds, which in some seasons are produced in abundance.

S. purpurascens is, perhaps, the very finest of the whole of the Magasea section. It is neatly edged with red, the same colour being also conspicuous in the mid-rib. The stem rises to a height of 10 in. or 12 in., and is considerably branched; the flowers are produced in pendent masses, of red and purple Though the plant is by no means a rapid grower, it possesses a good vigorous constitution, and is perfectly hardy. It succeeds best in a moist peaty soil in a rather sheltered spot. Native of high elevations about Sikkim.

S. Rocheliana.—A very compact and dwarf kind, forming dense silvery rosettes of tongue-shaped white-margined leaves, with distinctly impressed dots. It is distinguished among the dwarf silvery Saxifrages by producing large white flowers on sturdy little stems in spring. There is no more exquisite plant for rockwork, for culture in pans, or for small rocky or elevated borders. Any free, good, moist, loamy soil will suit it, and it thrives very well on borders in London. It should always be exposed to the full sun, and deserves to be associated with the choicest spring flowers and alpine plants. A native of Austria; increased by seeds or careful division. S. coriophylla is similar to Rochel's Saxifrage, but is not such a valuable plant.

S. sancta.—A beautiful species, lately come into cultivation, and therefore not much known yet. It promises, however, to become a valuable rock garden plant. It is of dwarf spreading growth, forming a dense carpet-like mass of deep green foliage, which in early spring is studded with numerous bright yellow blossoms on stems an inch or so high. It seems to be a free grower in any position in the rock garden.

S. sarmentosa.—A well-known old plant, with roundish leaves, mottled above and red beneath, with numbers of creeping, long, and slender runners, producing young plants Strawberry fashion. Striking and singular in leafage, it is also ornamental in bloom, and, growing freely in the dry air of a sitting-room, may be seen gracefully suspended in numerous cottage windows. It, perhaps, is most at home running wild on banks or rocks in the cool greenhouse or conservatory; however, it lives in the open air in mild parts of England, and where this is the case may be used in graceful association with Ferns and other creeping plants. It is a native of China, flowering in summer. Closely allied to S. sarmentosa is the delicate Dodder-like Saxifrage, S. cuscutæformis, so called from having thread-like runners like the stems of a Dodder, and distinguished by having much smaller leaves, and petals more equal in size than those of sarmentosa, in which the two outer ones are much larger than the others. It will serve for the same purposes as the creeping Saxifrage, but, being much more delicate and fragile in habit, will require a little more care. It is a beautiful plant for growing amongst Moss in a cool fernery, where it is perfectly at home and displays the delicate markings of its leaves to full advantage when contrasted with the green surroundings. The plants grown in gardens as S. japonica and S. tri-

PLATE CCXXXI.

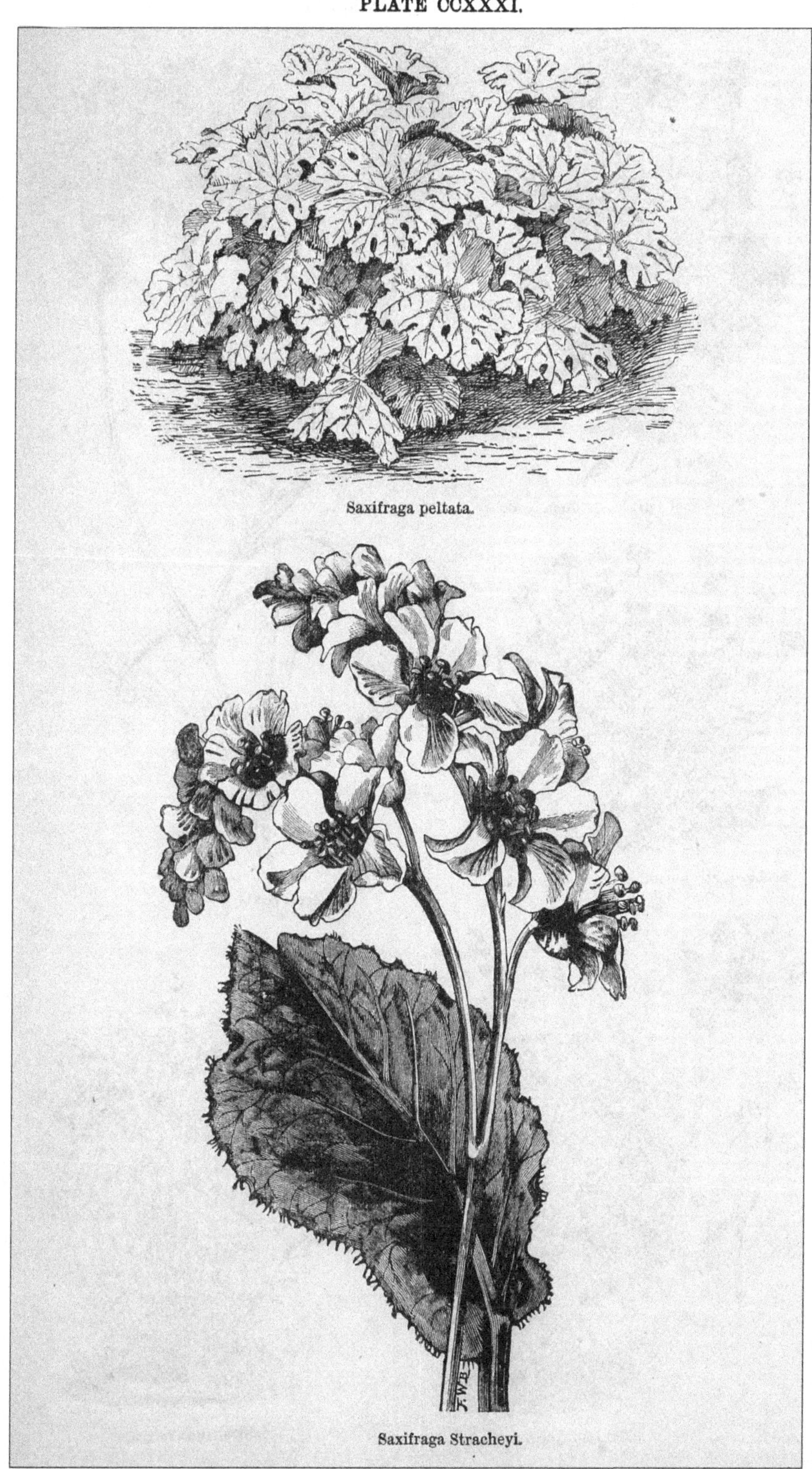

Saxifraga peltata.

Saxifraga Stracheyi.

[*To face page* 256.

PLATE CCXXXII.

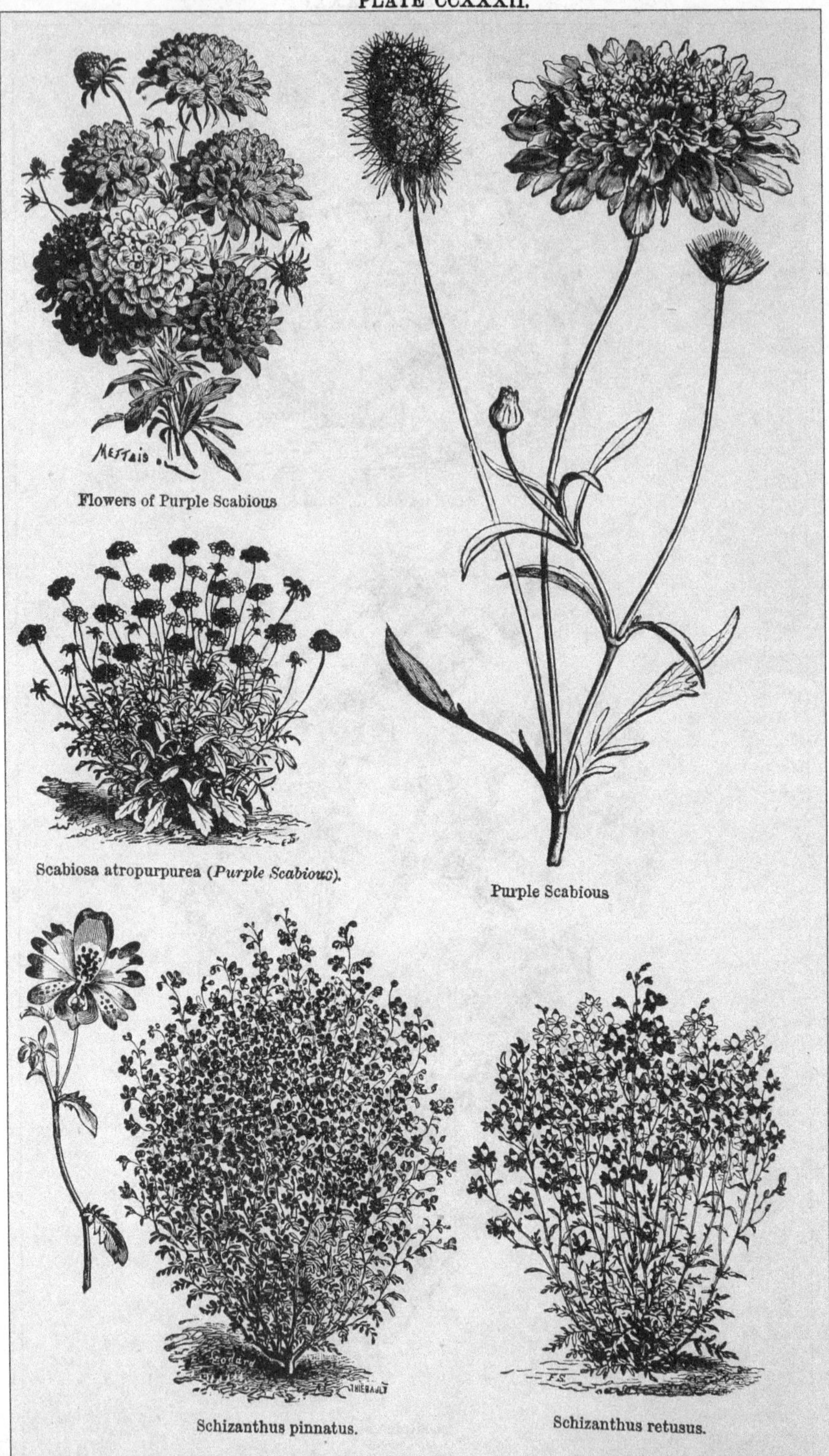

Flowers of Purple Scabious

Purple Scabious

Scabiosa atropurpurea (*Purple Scabious*).

Schizanthus pinnatus.

Schizanthus retusus.

[*To face page* **257.**

color are considered varieties of the creeping Saxifrage (S. sarmentosa).

S. squarrosa.—This pigmy alpine is the smallest of the Crustaceous section of Saxifrages in cultivation. When out of flower it looks like a Lichen if not closely observed, but in flower the hoary mass is studded with small white blossoms, which have a very pretty appearance. It is somewhat difficult to grow successfully; it requires the most select part of the rock garden in a fully exposed position among sandstone, but in deep, gritty, moist soil.

S. Stracheyi is a strong-growing plant, the leaves of which are nearly as broad as long. Its flowers are produced on broad branching panicles, of a light pinkish colour with a shade of lilac therein. Its closest ally is S. ciliata, than which, however, it is hardier. It blooms in March, and should have allotted to it in cultivation a special nook, sheltered either by shrubs in the border or by protecting masses in the rockery, otherwise, if exposed to the bleak winds with which the month of March is usually and unhappily associated, the plant would soon be shorn of the delicacy and beauty of its blossoms.

S. tenella.—A very handsome prostrate plant, forming tufts of delicate fine-leaved branches, about 4 in. or 5 in. high, which root as they grow. The flowers, which appear in summer, are numerous, whitish-yellow, and arranged in a loose panicle. Similar in gro th are S. aspera, bryoides, sedoides, Seguieri, Stelleriana, and tricuspidata, all of which are suitable for clothing the bare parts of rockwork and slopes, in moist soil, and in cool positions. Division in the end of summer or in spring.

S. umbrosa *(London Pride)*.—This almost universally cultivated plant grows abundantly on the mountains round Killarney, though it was much grown in our gardens before it was recognised as a native of Ireland. It is distinguished from S. Geum by having oval-oblong leaves, narrowed and not heart-shaped at the base; its flowers, too, are a little larger and more freely dotted with red. It is much used as an edging plant in old gardens, and, being such a pleasing evergreen, should be freely used for embellishing the rough parts of rockwork, the fringes of cascades, &c. It is naturalised in several parts of England, and grows freely among dwarf herbage, or in rocky ground in woods. There are several varieties, as, for example, S. punctata and S. serratifolia, which are distinct enough when grown side by side, but submit to the same culture. The variety Oglevieana is a most distinct form of this species, producing its pinkish blossoms in dense, dwarf panicles not over 6 in. high. S. rotundifolia and similar kinds are related to S. umbrosa, but are scarcely worth culture, except in botanical collections.

S. virginiensis.—The typical form of this plant is weedy, but the double-flowered variety is handsome, having flowers very double, and pure white, produced in erect spikes about 6 in. high, Grows in any soil in borders or rock garden.

Scabiosa *(Scabious)*.—A small genus of the Teazel family, consisting of perennial, biennial, and annual plants, some of which are valuable garden flowers. Of the perennials, which number about a dozen in cultivation, by far the finest is S. caucasica, which grows from 1½ ft. to 3 ft. high, and bears in summer large heads of pale lilac-blue flowers on long slender stalks. It is a large spreading plant, and requires plenty of room to develop itself. It grows freely in any ordinary soil, in open situations, and is an excellent plant for naturalising. Less desirable, though pretty plants, are S. graminifolia, about 18 in. high, with pale blue flowers, and S. Webbiana, a very dwarf tufted kind with white blossoms. Both thrive in ordinary garden soil. Among the biennial species the finest is S. atropurpurea, a very handsome plant, which for two or three centuries has been a favourite object of culture in English gardens. It is a plant that will not fail to give satisfaction, both as regards showiness and length of bloom, which continues from June to October; more, it will flower all through the winter if put through the necessary course of treatment. The typical plant grows about 3 ft. high, but there is now a dwarf variety (nana) that scarcely exceeds 1 ft., and this to some would be the most desirable, as it is neater and more compact. The normal colour is a deep rich maroon-crimson, but there is a pure white kind; another variety, with deep purple flowers margined with white; and still another (striata) with streaked and spotted flowers. The variety foliis aureis has yellow foliage, and is very distinct. The flowers have a sweet musky odour, and are peculiarly adapted for cutting purposes. Being biennial, S. atropurpurea requires to be raised annually from seeds, which should be sown in the reserve border in March or April in good soil, and when large enough the seedlings should be well thinned. In autumn they may be transplanted to their permanent places in the borders where they are intended to flower. Thus treated, they will become strong before winter sets in, and will flower early the following summer, and produce an abundance of seeds. By sowing under glass earlier in the year the plants will flower the same year. The dwarf Scabious is now used for pot culture in winter, for which purpose the seed is sown in summer, and the plants grown strongly for winter flowering. It is one of the most useful of all plants for furnishing cut blooms.

Scaly Fern *(Ceterach officinarum)*.

Scarlet Painted Cup *(Castilleja coccinea)*.

Schistostega pennata *(Iridescent Moss)*.—Mosses are small, but this is so very small, that it would hardly be noticed by the naked eye were it not for the iridescent gleams of beautiful colour which it displays when viewed in positions where it flourishes.

It is no exaggeration to state that some of the stones and sods on which it grows look as if sown with a mixture of gold and the material that goes to form the wings of green humming birds. It is almost startling to see this little gem for the first time, no "plant" being visible to the casual visitor, who wonders what produces the exquisite colour. It was supposed to require a particular kind of rock on which to grow; but lately its wonderful coruscations have been seen spreading over sods of turf and masses of peat, quite as well as on the chips of rock brought from its native place. Messrs. Backhouse have it in perfection both in the open air, in a quiet, deep gorge of rocks, where it obtains sufficient moisture without being washed by rains, and also in their underground fernery, constructed for the rarer filmy Ferns. This proves that there is no insurmountable difficulty in establishing it in such positions; and certainly the most graceful dwarf or biggest Tree Fern is not capable of adding to them a more decided charm than this diminutive moss.

Schivereckia podolica.—A small alpine plant of the Crucifer family, and nearly allied to Alyssum. It has hoary foliage, and produces, in early summer, a profusion of small white blossoms. It is suited for the rock garden or margins of borders, and will grow well in any ordinary soil; but it is not of the first merit. Native of South Russia, and quite hardy.

Schizanthus.—These Chilian plants are very pretty annuals of elegant growth, and produce in summer a profusion of showy and curiously shaped blossoms. There are three or four species in cultivation, and these have yielded numerous varieties. S. pinnatus is a hardy annual, from 1½ ft. to 3 ft. high, with rosy-purple and yellow blossoms, copiously spotted. Of this species the chief varieties are—papilionaceus (purple spotted), Priesti (white), atropurpureus (deep purple with dark eye), and Tom Thumb (a compact-growing, dwarf variety). S. porrigens is similar to pinnatus, but with larger flowers. These, being hardy, may be sown in the open border in spring or autumn in light sandy loam. The half-hardy kinds are S. retusus (deep rose and orange flowers with crimson tip), Grahami (lilac and orange), and Hookeri (pale rose and yellow). These, like the hardy kinds, are very beautiful, and worthy of being grown well. They should be treated as other half-hardy annuals, sown in spring in heat; or treated as biennials, sown in August, and the plants preserved in the greenhouse till May, when they should be planted out. These also like a good, rich, sandy loam to grow in.

Schizopetalon Walkeri.—A singular Cruciferous half-hardy annual from Chili. It grows about 1 ft. high, and bears, on slender stems, numerous white, almond-scented blossoms, which are elegantly fringed at the edges. It should be sown in the open border in April or May, in light, warm, rich soil, when it will flower in July and August; or it may be sown in pots, and the ball of earth transferred to the border without disturbance, as the plants do not well bear transplanting.

Schizostylis coccinea.—A handsome bulbous plant, with the habit of a Gladiolus. It is from 2 ft. to 3 ft. high. The flowers, which are produced late in autumn, are of a bright crimson, and resemble in form those of Tritonia aurea. They are borne in a one-sided spike, opening from below upwards. The plant is very useful, and should be grown in quantity wherever cut flowers are in request during the winter months. It is perfectly hardy, and will flower out-of-doors in a mild autumn; but to obtain the flower in its full beauty it should have some protection from inclement weather. If, when planted close to a wall or fence, some temporary protection from severe frosts can be given, a good row will yield a large quantity of spikes for cutting purposes. A well-drained soil of sandy peat and loam suits it best. Caffraria.

Schœnus nigricans.—A Sedge-like plant of no garden value.

Schrankia uncinata *(Sensitive Brier).*—A rare North American plant of the Rose family, suitable for the botanical garden.

Scilla *(Squill).*—The Squills comprise a numerous genus of bulbous plants, but few of them are suitable open-air garden flowers. These few, however, are indispensable to every garden, as they are all very beautiful and flower in spring; indeed, some bloom even before spring commences. They are of the simplest culture. They should be planted at the depth of a few inches when the bulbs are at rest—that is early in the autumn—in any good garden soil that is not too heavy; and if, when once established, they are not disturbed for years, except, perhaps, when given a slight annual top-dressing of manure, the cultivator will be annually rewarded with a display that will amply repay any trouble that he may have taken. Some kinds, especially the many-coloured varieties of the Spanish Squill (S. hispanica—syn., S. campanulata) and the English (S. nutans), are admirably adapted for introducing into the wild garden by the sides of woodland walks, margins of shrubberies, &c., situations in which they form an attractive feature in spring.

As regards propagation, offsets may be taken from established clumps during summer, and a very interesting, though somewhat slow, mode of procedure is that of raising them from seeds, which in some seasons are produced plentifully; by this mode many varieties of real merit, both as regards size and colour, have been obtained, and there is still ample room for further improvement in this direction. The following is a selection of the best kinds:—

S. amœna.—A distinct, early spring-flowering Squill, opening about three weeks

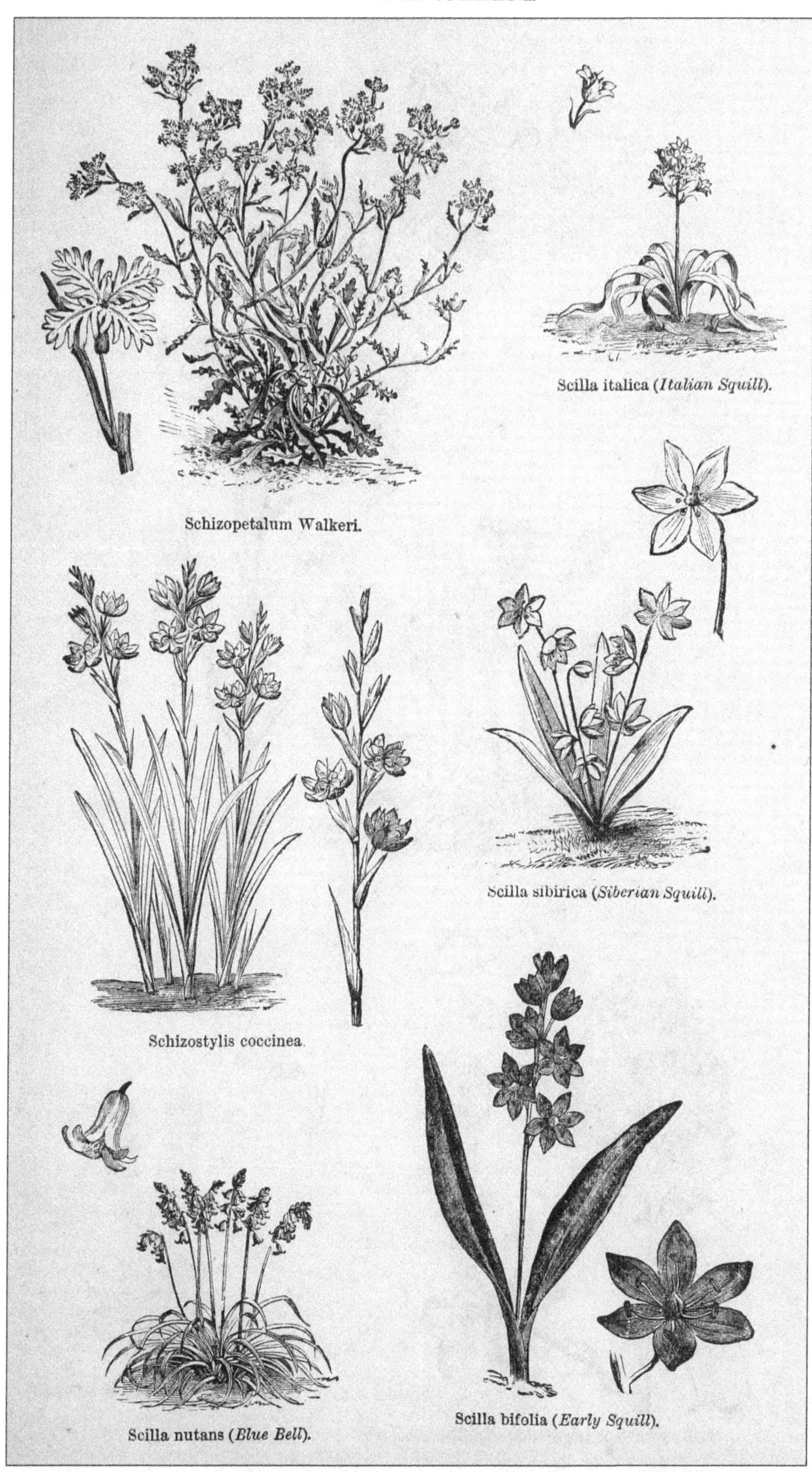

Schizopetalum Walkeri.

Scilla italica (*Italian Squill*).

Scilla sibirica (*Siberian Squill*).

Schizostylis coccinea.

Scilla bifolia (*Early Squill*).

Scilla nutans (*Blue Bell*).

[*To face page* 258.

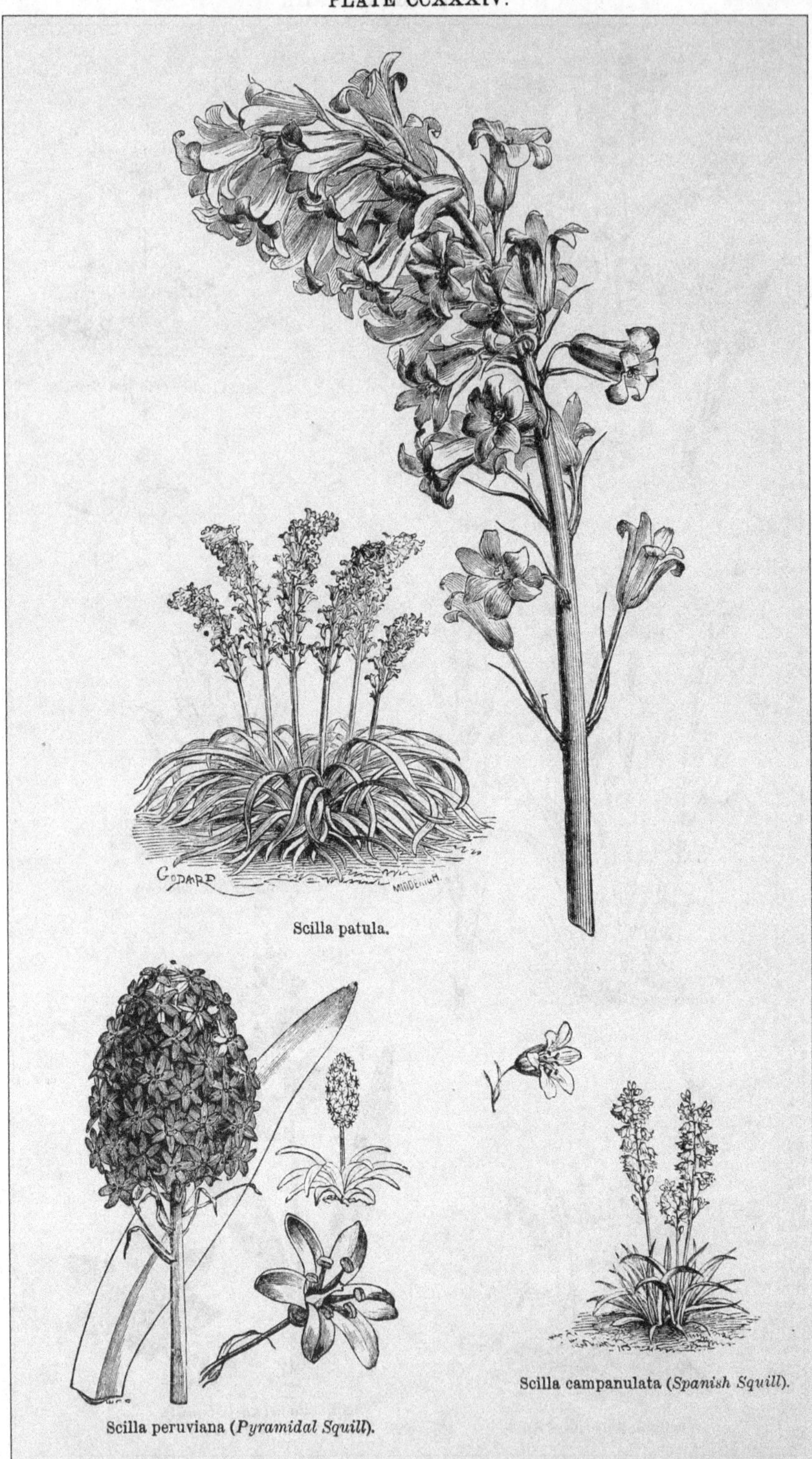

Scilla patula.

Scilla peruviana (*Pyramidal Squill*).

Scilla campanulata (*Spanish Squill*).

[*To face page* 259.

after S. sibirica, and readily known from any of its relatives by the large, yellowish-green ovary showing conspicuously in the centre of the dark indigo-blue flowers. Though sufficiently attractive to merit a place in borders and collections of hardy bulbs, it is less ornamental than any other kind here mentioned, the flowers being arranged in a somewhat sparse and rigid manner, and having none of the grace characteristic of S. campanulata and the varieties of S. nutans, or of the dwarfness and brilliancy of S. sibirica. The leaves, usually about half an inch across, attain a height of about 1 ft., and are very easily injured by cold or wind, so that a sheltered position is that best suited to its wants. It is not exactly suited for choice rockwork, though well worthy of a place in borders and of being naturalised on sunny banks in semi-wild spots. A native of the Tyrol; increased from seeds or by separation of the bulbs.

S bifolia *(Early Squill)*.—Although not nearly so well known or popular as S. sibirica, this is quite as worthy of cultivation, as it produces in the very dawn of spring, indeed often in winter, rich masses of dark blue flowers four to six on a spike, and forms very handsome tufts of vegetation from 6 in. to 10 in. high, according to the richness and lightness of the soil, and the warmth and shelter of the aspect. It thrives well in almost any position in ordinary garden soil, the lighter the better. Although it blooms earlier than S. sibirica, it does not withstand cold wintry and spring rains and storms nearly so well as that species, and therefore it would be well to place some tufts of it in warm, sunny spots either on the rockwork or sheltered borders. A native of Southern and Central Europe. The varieties are numerous and important, and the following are most distinct and most worthy of cultivation: alba, candida, carnea, compacta, maxima, metallica, rosea, pallida, præcox, and taurica. The name S. præcox, which occurs so often in gardens and in nurserymen's catalogues, does not really belong to a distinct species, and, when most properly applied, refers to the variety of S. bifolia, that usually flowers somewhat earlier than the common form. The variety taurica is a very handsome plant, differing from the type in being larger in size and more robust in habit; its flowers are also larger and more numerous, and it produces several leaves, thus departing from the bifoliate character which is generally observed by the type, but which, by the way, is not an absolutely reliable one.

S. campanulata *(Spanish Squill)*.—A vigorous species, long cultivated in England. It is one of the finest ornaments among early summer-flowering bulbs, and, though a more southern species than most of the others, it is the most robust of the family. It is easily known by its strong pyramidal raceme of pendent, short-stalked, large, bell-shaped flowers, usually of a clear light blue. A variety known as major is larger in all its parts, and a noble early summer flower; and the white and rose coloured varieties (alba and rosea) are also excellent. It is never seen to greater advantage than when peeping here and there from the fringes of shrubberies and beds of evergreens, the shelter it receives in such positions protecting its very large leaves from strong winds. It is, however, sturdy enough to thrive in any position. Comes from the south of Europe, attains a height of from 12 in. to 18 in., and deserves to be naturalised by the side of wood walks and in the semi-wild parts of every pleasure-ground. (= S. hispanica.)

S. italica *(Italian Squill)*.—This Squill, with its pale blue flowers, intensely blue stamens, and delicious odour, is one of the most interesting and distinct, if not the most brilliant, of cultivated kinds. It grows from 5 in. to 10 in. high, the leaves somewhat shorter, slightly keeled, and oblique; the flowers small, spreading in short conical racemes, and opening in May. It is perfectly hardy, living in almost any soil, but thriving best in sandy and warm ones. Increased by division, which should be performed only every three or four years, when the bulbs should be planted in fresh positions. It is worthy of a sheltered sunny spot on rockwork, particularly as it does not seem to thrive so freely in this country as some of the other species. Native of Italy and S. Europe generally.

S. nutans *(Bluebell)*.—A well-known and much admired native plant, abounding in almost every wood and copse; the flowers are always arranged in a gracefully drooping fashion on one side of the stem. The Bluebell is very common, but its several beautiful varieties are not so well known as they deserve to be, although fitted to be great ornaments of the early summer garden—particularly the white variety, alba; the rose-coloured one, rosea; the pale blue, cœrulea; and a pleasing "French white" variety. The variety bracteata is interesting on account of its abnormally long bracts, and cernua is a Portuguese form with reddish flowers. S. patula is closely allied to the Bluebell, with flowers of a pleasing violet-blue, not sweet, like those of that species, nor arranged on one side, but larger and more open, with narrow bracts. These are all highly suitable for planting here and there in tufts along the margins of shrubberies, near rockwork, for borders, the spring garden, and for naturalisation in woods, among the common blue kind.

S. peruviana *(Pyramidal Squill)*.—The Peruvian Squill, which, however, is not a native of Peru, is a very noble plant where it thrives, and it does so perfectly in many mild parts of these islands, though it suffers on cold soils. The flowers are of a fine blue, very numerous, arranged in a superb, regular, umbel-like pyramid, which lengthens during the flowering period. The white stamens contrast charmingly with the blue of the flowers. In all but the warmest parts of the country this fine plant should have a somewhat elevated, warm, and sheltered position, a deep, light, and well-drained soil, and the large Pear-shaped bulbs should be planted 6 in. under the surface, as they will thus be better enabled to withstand the cold. It is a native of Southern Europe and North Africa, grows from 6 in. to 18 in. high, flowers in May and June, and deserves a place in a sheltered, sunny nook on every rockwork, and on every warm raised bed or border devoted to choice hardy bulbs. There is a white variety, S. peruviana alba, which is not quite so beautiful as the ordinary form, and there are varieties with reddish flowers, such as sub-carnea and elegans, and others again with whitish and yellowish blossoms, such as pallidiflora, flaveola, and sub-albida. S. Hughi, from the island of Maretina, near Sicily, is a more robust form of S. peruviana, with the flower-stalks tinged with red. S. Cupani, another Sicilian Squill, is similar to the Peruvian, but is less showy. Tufts of the Peruvian Squill should be taken up, when at rest, every three or four years, the bulbs divided, and immediately replanted.

S. sibirica *(Siberian Squill).*—This minute gem among the flowers of earliest spring is so beautiful that no rockwork, spring garden, or garden of any kind can be complete without its striking and peculiar shade of porcelain blue which quite distinguishes it from the other species. It has had a great number of synonyms, but, unlike S. bifolia, has sported into few varieties, S. amœnula being the only one worth mentioning, and that is not really distinct, but is desirable, as it flowers a fortnight earlier than the type. Varieties with larger blossoms and with one instead of from two to five on a stem are preserved in herbariums and occasionally cultivated, but these are only trifling variations, often arising from the conditions in which they are placed. It is perfectly hardy in this country, and, like most other bulbs, thrives best in a good sandy soil. Bulbs of it that have been used for forcing should never be thrown away; if they are allowed to fully develop their leaves and go to rest in a pit or frame, and are afterwards planted out in open spots, in warm soil, they will thrive well. It is unnecessary to disturb the tufts, except every two or three years for the sake of dividing them, when they grow vigorously. It comes in flower in very early spring a little later than S. bifolia, but withstands the storms better than that plant, and remains much longer in bloom. In places where it does not thrive very freely, from the cold nature of the soil or other causes, it would be well, in placing tufts of it on rockwork or on borders, to put it in sheltered positions, so that the leaves may not be injured by the wind, and the plant thereby weakened. It may be used with good effect as an edging to beds of spring flowers or choice alpine shrubs.

Of other cultivated Squills, the British ones, S. verna and S. autumnalis, are certainly not worthy of cultivation except in botanical collections; the plant usually sold by the Dutch and by our seedsmen as S. hyacinthoides is generally S. campanulata, and occasionally S. patula. The true S. hyacinthoides of Southern Europe is scarcely worthy of cultivation; S. cernua is not sufficiently distinct from S. patula, and one or two southern species allied to S. peruviana have not been proved sufficiently hardy for general cultivation.

Scirpus.—Sedge-like plants of no value, except for fringing artificial lakes and ponds, which too often present a bare appearance. There are numerous native species that might be readily transplanted, S. triqueter, atrovirens, and lacustris being the best. S. lacustris (the Bullrush) grows from 3 ft. to 8 ft. high, and is very effective on the margins of lakes or streams, associated with other tall and imposing aquatic plants.

Scleranthus. — Weedy plants of no value.

Sclerolepis verticillata.—A North American bog plant of dwarf creeping growth, with small heads of pink flowers, produced in August. For botanic gardens.

Scoliopus Bigelowi.—A very singular Liliaceous plant from California, allied to Trillium. It is of dwarf growth, has a pair of broad oval leaves, and produces during the summer several dull-coloured inconspicuous flowers. It is only to be recommended for the curious garden. It grows well in shady and moist peat borders.

Scolopendrium vulgare *(Common Hart's-tongue)* is one of the most widely distributed and best known of all hardy evergreen British Ferns. It has broken from the type into almost innumerable forms and varieties, all interesting, and some very striking and beautiful. It has the merit, too, of being a wonderfully accommodating plant, one that will thrive in almost any position; and so variable, indeed, are the places in which it thrives, that good use may readily be made of it in gardens. Like nearly all Ferns, it prefers a shady situation, though in a wild state it may sometimes be met with on dry stone and brick walls, as well as on shady moist banks. Its favourite position is, however, by the side of a stream in a shady ravine, and fine plants of it have been seen between the joints of brickwork at the top of old wells, a position in which the fronds attain wonderful dimensions. The first point to be considered is the all-important one of providing suitable soil for it, for, because it is common, any sort of soil is too often given it, and failure is the result. A suitable compost for it should consist of about equal portions of fibrous peat and loam, with some good sharp sand added thereto, and also a rather liberal addition of broken oyster-shells or limestone. As to position, the Scolopendriums should be associated with the Lastreas, Polystichums, and Lady Ferns, or be placed in groups on the rock garden in conjunction with some flowering plant that will

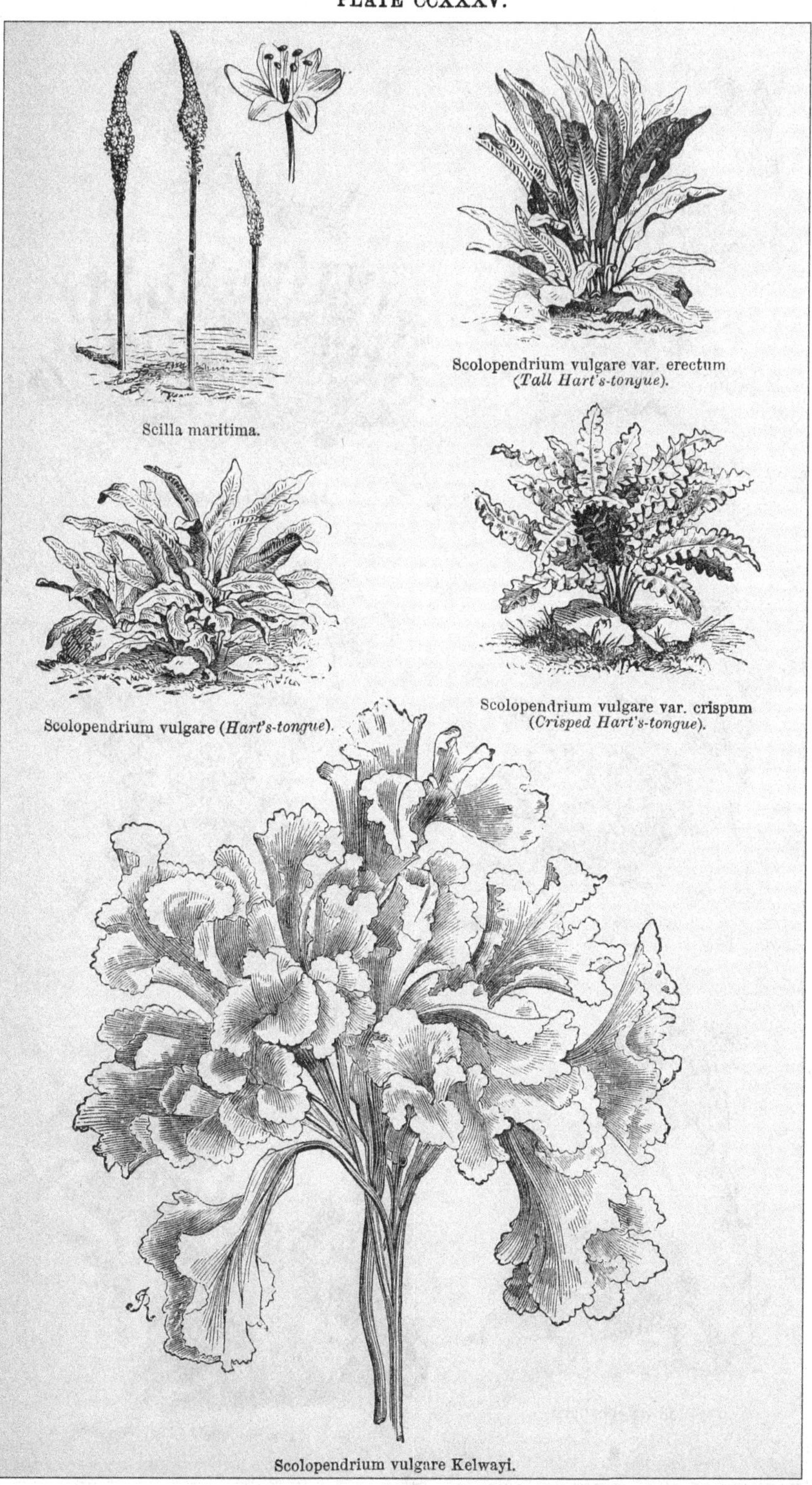

Scilla maritima.

Scolopendrium vulgare var. erectum (*Tall Hart's-tongue*).

Scolopendrium vulgare (*Hart's-tongue*).

Scolopendrium vulgare var. crispum (*Crisped Hart's-tongue*).

Scolopendrium vulgare Kelwayi.

[*To face page* 260.

PLATE CCXXXVI.

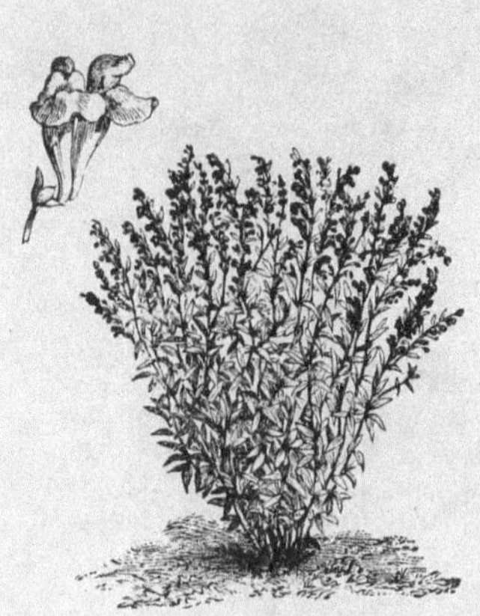

Scutellaria macrantha (*Siberian Skullcap*). See page 262.

Scutellaria lupulina bicolor.

Sedum acre (*Wall Pepper*). See page 262.

Sedum album.

Sedum dasyphyllum.

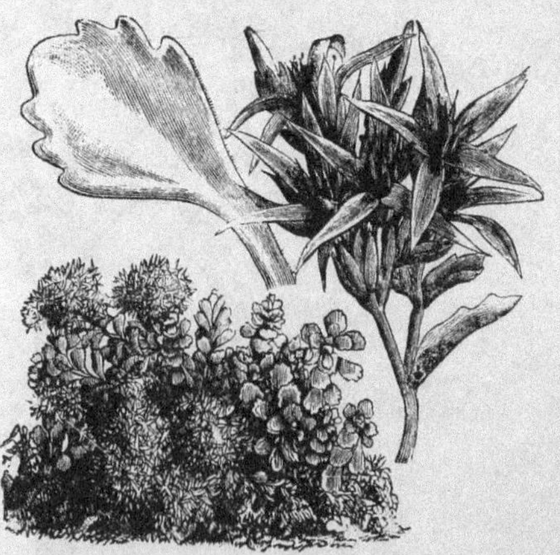

Sedum spurium splendens. See page 265.

thrive well in similar soil, and under conditions suitable to their mutual well-being. During hot, dusty weather in summer a daily syringing will much refresh and invigorate them; the best time in the day in which to do this is in the afternoon. All the Hart's-tongues are well adapted for culture in pots, and make useful plants for room and table decoration.

Some idea of the amount of variation to be found among the sports of the Hart's-tongue may be formed from the fact that no fewer than 400 varieties were described twenty years ago, and since then this number has been considerably augmented. The greater part of these, however, are such deformities —vegetable cripples, so to speak—that not more than one-fourth are really worth cultivation. A select few of the characteristic forms of each group might be used with advantage where collections of hardy Ferns are in course of formation, as, being evergreen, and possessed of great diversity as regards cutting, they supply a type of foliage that is uncommon among hardy Ferns. The following selection commences with the simple forms, and ends with those much divided. S. latifolium is a fine bold variety, with ovate or oblong fronds, 8 in. to 10 in. long, and 2 in. to 3 in. broad, with wavy margins and an obtuse point. Being of a spreading habit, its broad, bright green foliage, seen from above looks very fine. S. reniforme has fronds oblong, roundish, or kidney-shaped. In length they vary from 4 in. to 6 in., and in breadth from 1 in. to 2 in. S. cornutum is an interesting form with fronds similar to the preceding, with the midrib, which terminates a short distance before reaching the point of the frond, prolonged into a horn-like appendage. S. marginatum is a very distinct variety with a linear pointed frond, a distinctly crenated margin, and a prominent vein running down the frond on the under side just within the margin. Of S. pinnatifidum, some of the best forms are very attractive; the sides of the fronds are cut down nearly to the midrib into blunt or acute lobes. S. crispum is an old favourite, and by far the prettiest of the simple forms; its fronds retain the habit of the normal plant, but the margin is regularly undulated or frilled from the base to the point. A form of this called grandidens has its undulated margin deeply incised. Stansfieldi is a form with curled and twisted incisions, and Wrigleyi is a luxuriant form with erect fronds upwards of 3 ft. in length. S. laceratum has broad, flat fronds, which are deeply cut into lobes of variable length and breadth, some measuring as much as 3 in. in length. In S. sagittato-cristatum the fronds are three-lobed or sagittate, with the margin wavy, and each lobe surmounted with a crest. S. cervi-cornu is a singular form, with the lower part of the frond narrow, and the upper divided into two, three, or more divisions, each of which is again divided into slender, erect teeth. In S. contractum the blade is very narrow, with a toothed margin, and a terminal flat hemispherical crest of serrated lobes. The frond of S. acrocladon is narrow, slightly widened at the base, and divided at the upper end into several wedge-shaped divisions, with their upper margin deeply incised. S. sagittato-polycuspis is a robust form, with the basal lobes enlarged and the upper portion deeply cleft; the separate divisions are again cut into acuminate lobes, which are variously twisted. S. patulum is one of the ramose type, with fronds cut down near to the base into two or three divisions, each of which has a narrow wing and a terminal broad crest of flat-pointed lobes. S. digitatum is an extremely elegant plant, with fronds 1 ft. or more in length, divided in a digitate manner, the separate divisions of which are terminated by a multifid crest of forked and twisted lobes. S. tortuoso-cristatum is an interesting form allied to the preceding, with its divisions incised to the base. The terminal crest is contorted in various ways. The fronds collectively form a cushion-like mass, 1 ft. in breadth. S. Elworthi is a dwarf plant not more than 6 in. in height, with a profusion of fan-shaped fronds divided into from three to five triangular divisions, the margins of which are cut into roundish lobes and crenelles. The fronds of S. Gloveri have the central rib cut into four or five divisions, each of which is palmate, quite flat, and regularly crenate. In S. concavo-ramosum the fronds are divided into four or five distinctly-stalked divisions, with triangular blades, the lower half of which is laciniated at the edges, and the upper half deeply cleft and forked. S. Kelwayi is a handsome form, the fronds of which have a large terminal crest some 6 in. or 10 in. broad. S. ramosa-marginatum comes near Kelwayi in form of the frond, but differs in having a nearly flat crest, which is not unlike the tasselled-frond extremities of the maximum form of Pteris serrulata cristata. It also has a broad-winged stalk, which is wanting in Kelwayi. The base of the frond of S. corymbiferum is normal, but the upper half is cut into innumerable contorted and twisted incisions, with an appearance similar to the leaf ends of some of the ragged Kales. S. Coolingi is very similar to the last, but the incision forms an intricate mass of slender segments, curled and twisted in various ways so as to form a globular head. This kind of incision represents the extreme form of variation.—J. McSmith.

Scolymus *(Oyster Plant)*.—Composite plants of large coarse growth with Thistle-like foliage and bright yellow flowers. They are not suitable for the ordinary border, but may be grown with good effect in bold masses in the rougher parts of the garden. S. hispanicus, a perennial, is the best known, and S. grandiflorus is the handsomest. Both grow well in light sandy soil in an open position. S. maculatus is an annual with silvery blotched foliage. It is scarcely worth attention.

Scopolia.—The cultivated species, S. carniolica and lurida, are curious plants.

They are interesting for botanical collections, but not for general culture.

Scorpiurus.—Annual plants of the Pea family, but curious only.

Scorzonera.—Perennial plants of the Composite family, of botanical interest only. They are chiefly natives of S. Europe and the East.

Scotch Asphodel (*Tofieldia palustris*).

Scotch Thistle (*Onopordon Acanthium*).

Scrophularia (*Figwort*).—The only garden plant worth mentioning in this genus is the variegated-leaved variety of S. nodosa, one of our common native plants. It is useful as an edging to tall perennials, and in summer its green and white foliage is very handsome. It requires a good soil to grow well, and frequent pinching to prevent it from flowering. The flowers of all the species are inconspicuous. It may be easily propagated by cuttings in autumn or division in spring.

Scutellaria (*Skullcap*).—A genus of perennial plants of the Figwort family, several species of which are in cultivation, though but few are good garden plants. These few are handsome border flowers, or they may, on account of their dwarf, neat growth, be appropriately given a place in a large rock garden in any kind of soil in an open, sunny situation. They are all easily propagated by division or seeds. S. macrantha, a native of Siberia, is the finest of all the species of this genus. It forms a hard, woody rootstock, and is an excellent perennial alpine plant; it grows 9 in. high, producing an abundance of rich, velvety, dark blue flowers, much finer in colour than those of S. japonica, which, however, is also a handsome plant. The alpine Skullcap (S. alpina) is a spreading plant with all the vigour of the coarsest weeds of its Natural Order, but withal neat in habit and ornamental in flower. The stems are prostrate, but so abundantly produced that they rise into a full round tuft, 1 ft. high or more in the centre, falling low to the sides; the leaves are ovate-roundish or heart-shaped at the base, very shortly stalked and notched, and the flowers are borne in terminal heads, at first short, afterwards elongating. They are purplish, or have the lower lip white or yellow. The variety bicolor, with the upper lip purplish and lower pure white, is very pretty. S. lupulina is a very ornamental kind with yellow flowers. These are admirably suited for borders, the margins of shrubberies, and the rougher parts of rockwork. A native of the Pyrenees, Swiss and Tyrolese Alps, and many other parts of Europe and Asia; readily increased by division, and flowering freely in summer. Of other kinds of Scutellaria in cultivation, S. japonica, orientalis, scordiifolia, altaica, galericulata, peregrina, and the British S. minor, an interesting little plant for the artificial bog, are among the best, but it is doubtful if they are worth a place in any but a very large collection.

Scypanthus elegans (*Cup Flower*) is a beautiful half-hardy climbing annual from Chili, valuable for trailing over a trellis or against a wall. It is a slender plant from 5 ft. to 8 ft. high with forked stems. The leaves are deeply cut, and thus the graceful appearance of the plant is enhanced. The flowers, produced singly in the forks of the branches, are cup-like in shape and of a bright golden yellow, with fine red spots inside. It blooms profusely from August till October, and is easily cultivated in rich light soil. It should be treated as a half-hardy annual. Loasaceæ.

Sea Bindweed (*Convolvulus Soldanella*).

Sea Cottonweed (*Diotis maritima*).

Sea Heath (*Frankenia lævis*).

Sea Holly (*Eryngium maritimum*).

Sea Lavender (*Statice*).

Sea Pea (*Pisum maritimum*).

Sedum (*Stonecrop*).—The Stonecrops, like the Houseleeks and Saxifrages, are among the commonest of garden plants, and few plants are more accommodating, for they may be grown in the ordinary border, on rockwork, walls, ruins, or, indeed, in any place where the roots can obtain a foothold. Like the Saxifrages, too, the Stonecrops represent a great diversity of habit, some, such as S. acre, being humble and creeping; while others, such as S. spectabile, are stately border plants. There is a large number in cultivation, but we mention only the most desirable of the hardy kinds, nearly all of which are perennial and easily cultivated.

S. acre (*Wall Pepper*).—Growing on walls, thatched houses, rocks, and sandy places in almost all parts of Britain, this little plant, with its small, thick, bright green leaves and brilliant yellow flowers, is as well known as the common Houseleek. Like the Daisy, it is so very abundant in an uncultivated state that there is rarely occasion to introduce it to gardens, though it is one of the most brilliant and distinct of its very large family. Sheets of it in bloom look very gay, and it may well be used with dwarf alpine plants in forming carpets of living mosaic-work in gardens. The fact that it runs wild on comparatively new brick walls round London does away with the necessity of speaking of its cultivation or propagation. There is a variegated or yellow-tipped variety (aureum), the tips of the shoots of which become of a yellow tone in early spring, so that the tufts or flakes look quite showy at that season. It is suitable for use in the spring garden, on the rockwork, and for the same purposes as the ordinary form, but it is not so robust. The variety elegans has the tips and young leaves of a pale silvery tone, but it is not so effective as aureum and not so vigorous. These varieties are now largely used in geometrical bedding. The golden variety is a beautiful winter decorative plant. Its golden

PLATE CCXXXVII.

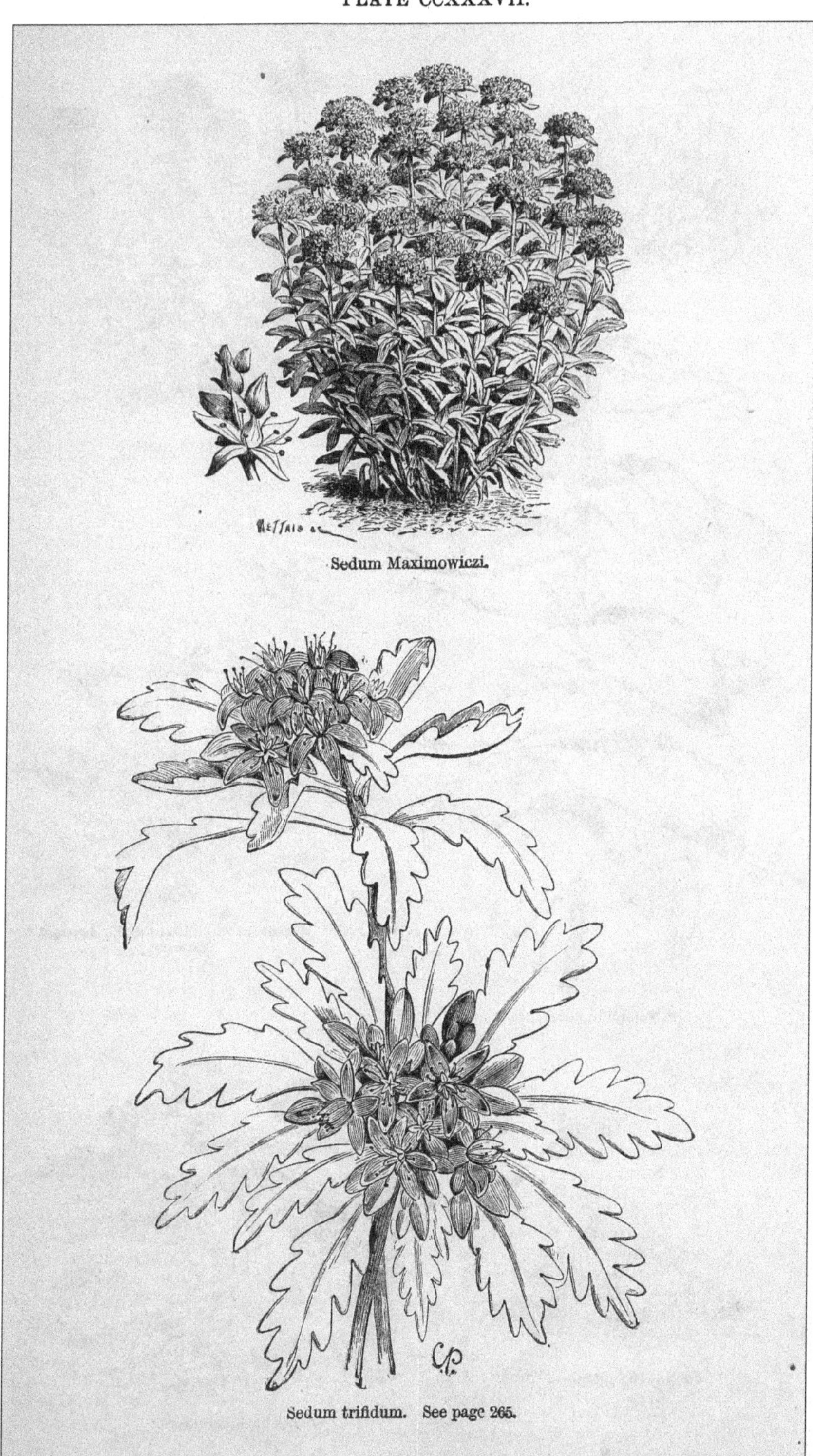

Sedum Maximowiczi.

Sedum trifidum. See page 265.

[*To face page* 262.

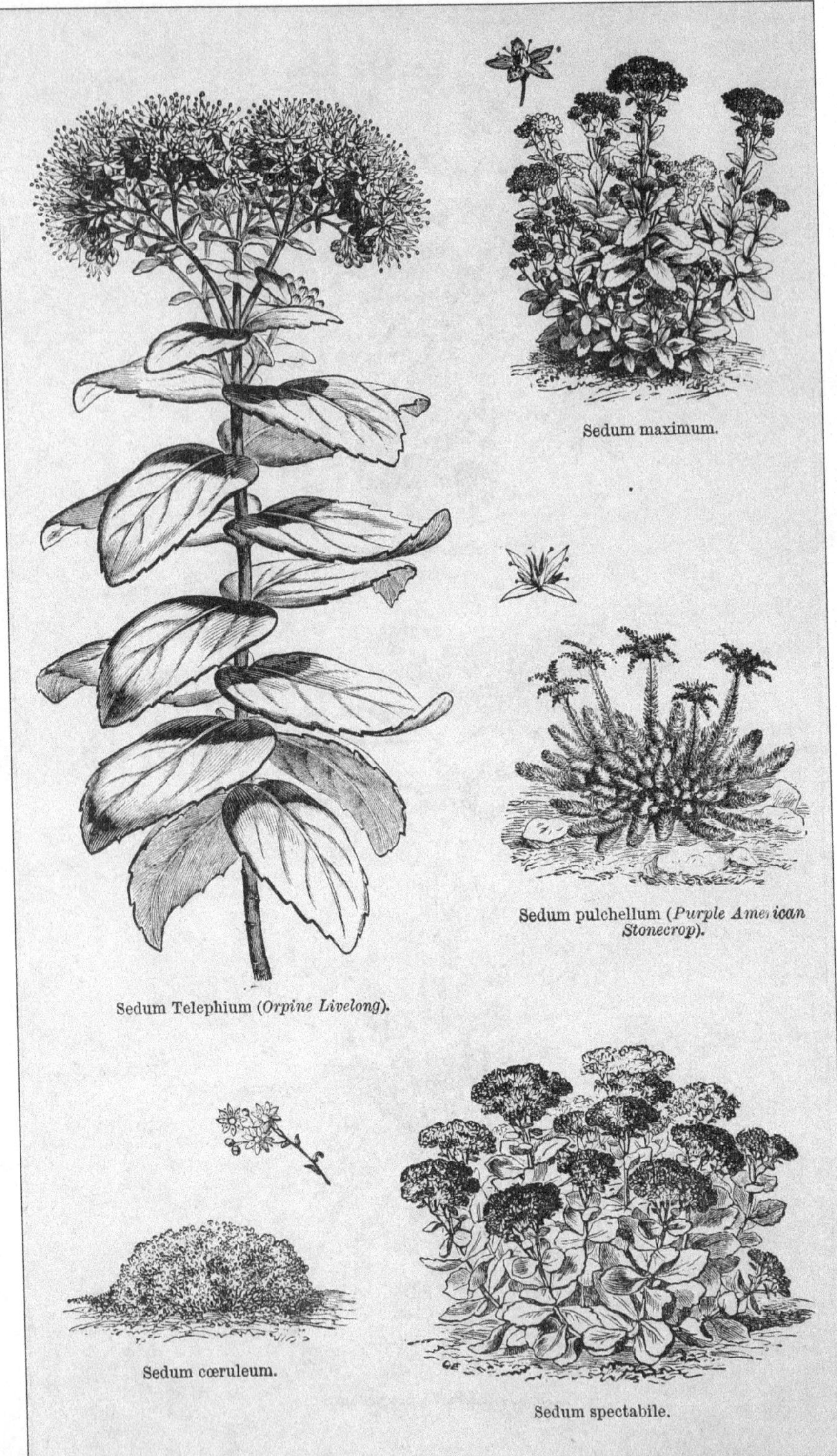

Sedum maximum.

Sedum pulchellum (*Purple American Stonecrop*).

Sedum Telephium (*Orpine Livelong*).

Sedum cœruleum.

Sedum spectabile.

[*To face page* 263.

tips peep out in November, are brilliant through the winter, and only vanish with the heat of May; still, except as a green carpet plant, it is not much used for summer bedding. The silver-tipped variety has its colour best when the weather is warm and dry. It grows freely, is close and dense, and forms a very pleasing groundwork; the variegation is most displayed on the side shoots that break out from established plants. It succeeds best when divided and transplanted in autumn, as it then gets well established in thick masses by the following spring, and is therefore better able to withstand the drought of summer. S. sexangulare is similar to S. acre and is also much used in carpet bedding.

S. Aizoon grows 1 ft. or more in height, and has erect stems terminated by dense clusters of yellow flowers. It is perfectly hardy, and is an old garden plant suitable for the border or large rock garden in open positions in light soil. Native of Siberia and Japan. S. Maximowiczi and S. Selskyanum are similar.

S. Anacampseros *(Evergreen Orpine).*—A species easily recognised by its very obtuse and entire glaucous leaves, closely arranged in pyramidal rosettes on prostrate branches that do not flower. The rose-coloured flowers, which are in corymbs, are not very ornamental, but the distinct aspect of the plant will secure it a place on the rockwork, or among very dwarf border plants. A native of the Alps, Pyrenees, and mountains of Dauphiny; flowers in summer, is easily propagated by division, and thrives in any soil.

S. Ewersi.—An exceedingly neat, diminutive species, with glaucous and broad leaves and purplish flowers in terminal corymbs. The whole plant is of a pleasing silvery tone and rather delicate appearance, but it is quite hardy. It is easily increased by division, and flowers in summer. It rarely rises above 2 in. or 3 in. high, and merits a place on every rockwork and in collections of the dwarfest hardy succulent plants. Siberia and N. India.

S. glaucum.—A minute species of a greyish tone, forming dense spreading tufts of short stems, densely clothed with thick leaves, and producing, rather sparsely, somewhat inconspicuous flowers. Its neat habit has caused it to become quite popular in our gardens of late years as a minute edging or surfacing plant, and for the former purpose it is, perhaps, better divided every spring; thin but regular lines planted at that season form neat, swelling, chubby-looking edgings which are 4 in. or 5 in. across in autumn. On the rockwork it may be used in any spot that is to spare, either to form a turf under other plants or for its own sake. Various other Sedums are very nearly allied to this, such as S. dasyphyllum, glanduliferum, farinosum, and brevifolium: but although hardy on walls and rocks, these have not the vigour and constitution of many of the other Stonecrops, and it is desirable to establish them on an old wall or dry stony part of rockwork, so as to secure a stock in case the plants perish in winter on low ground. S. hispanicum is a synonym of S. glaucum, of which there are also several named varieties. They are very suitable for association with such plants as the Cobweb Houseleek, and are interesting subjects for naturalisation on old ruins.

S. kamtschaticum *(Orange Stonecrop).*—A broad-leaved species, not unlike Sedum spurium in habit, but at once distinguished by its dark orange-yellow flowers. It is a prostrate plant, quite hardy, succeeding in almost any soil, but best in a warm, rich loam, and flowering profusely in summer. Highly suitable for the rougher parts of the rock garden, where it will take care of itself. It is a capital plant for the margin of the mixed border. S. hybridum is a similar species, and suitable for the same purpose.

S. Lydium.—A pretty little plant, similar to S. glaucum, but the tiny, crowded leaves are greenish and red tipped. Of plants adapted for edgings, or for covering slopes bordering footpaths, none excel this Sedum, which, when grown on rockwork, is not to be compared with what it is when grown where it can get plenty of moisture. When exposed to heat and drought it becomes almost red in colour; but when grown where there is an abundance of moisture, it spreads rapidly and assumes a rich, deep green hue. It roots on the surface with great rapidity, and may, therefore, be speedily propagated. Very small pieces put in the soil in spring soon form a perfect mass of green, scarcely exceeding an inch in height and as level as a piece of turf. Its rich verdure is pleasant to the eye, and it always looks neat, and needs but little attention to keep it in order. In spots where turf will not thrive this Sedum will do well. It is known also as S. lividum, pulchellum, and anglicum. Native of Asia Minor.

S. maximum.—One of the large-growing species, and, like S. Telephium, is extremely variable, there being no fewer than a dozen named varieties. Of these, by far the most important garden plant is hæmatodes or atropurpureum, as it is commonly called, the stems of which, as well as the large fleshy leaves, being of a deep vivid purple. Like the other forms, it grows from 1 ft. to 2 ft. high, but the flowers are not very showy. It is, however, a most telling plant, being of a bold, stately growth, and admirably adapted for massing. It should be planted in the poorest, stony, gravelly soil, and smoke will not injure it. On rockwork of a white calcareous nature it is all that can be desired.

S. pulchellum *(Purple American Stonecrop).*—A very pretty species, at once distinguished by its purplish flowers, which are arranged in several spreading and recurved branchlets, bird's-foot fashion, with numerous spreading stems densely clothed with alternate obtuse leaves. It is abundant in North America, and at present is very rarely seen in our gardens, though it is far more worthy of cultivation than many commonly grown. In France it is a good deal used as an edging plant, for which purpose, as well as for rockwork, it is well suited; and it is also a highly appropriate plant for the front margin of a mixed border, flowering in summer, growing in any soil, and being easily increased by division.

S. rupestre *(Rock Stonecrop)*.—A glaucous, densely-tufted native plant, with numerous spreading shoots generally rooting at the base, but quite erect at the apex. It has rather loose corymbs of yellow flowers, and is frequently grown as an edging and border plant in gardens, though it is not so ornamental as some rarely grown kinds. There are several kinds similar to this, such as S. pruinatum, a handsome glaucous-leaved kind, commonly known as S. elegans; Fosterianum, with light green leaves; S. reflexum, and its several varieties, among them the curious crested variety cristatum, known also as monstrosum and fasciatum. Another native kind, but of a different growth, is S. album, which has brownish-green leaves and white or pinkish flowers. Like the common Stonecrop, this occurs on old roofs and rocky places in many parts of Europe, and may be cultivated with the same facility as that well-known plant. These are all worthy of naturalisation on walls or old ruins, in places where they do not occur naturally, and also on the margins of the pathways or the less important surfaces of the rock garden.

S. sempervivoides *(Scarlet Stonecrop)*.—This is a beautiful and very distinct species with rosettes of leaves similar to that of the common Houseleek (Sempervivum tectorum). The leaves are thick, fleshy, and pale green, and blotched here and there with reddish-brown. The flowers are small, of a brilliant deep scarlet, and form a dense head similar to the well known Rochea falcata. This Stonecrop is strictly a biennial, flowering freely the second year from seed and then dying. It cannot be considered hardy, but it will grow very freely on a dry bank during summer, and stand a large amount of frost, provided it is dry, but frost and wet combined will kill it. Seeds of it germinate freely, and it is very easily managed. It should be sown in January in gentle heat; great care should be taken in watering, as the seed is very fine, and it would be better not to water at all, but to keep the pot plunged, so that the soil should not get dry. As soon as the young plants are large enough to handle they should be pricked off, kept near the glass, and by no means allowed to get dry. As soon as they are large enough they should be potted singly, and treated liberally with water while growing, and by the end of the summer they ought to be as large as a crown-piece. During autumn and winter water must be withheld, the plants being kept just sufficiently moist to prevent the leaves from shrivelling, and in spring they should be repotted and kept growing freely until the flower-spike is fully developed. The time of flowering depends a great deal upon time of sowing, treatment, &c., but under ordinary circumstances it is in July or August. The flowers will last in good condition for six or seven weeks. Native of Asia Minor and the mountains of the Caucasus. It is erroneously called Umbilicus Sempervivum, but the latter is a totally distinct plant.

S. Sieboldi.—A well-known and elegant species, frequently cultivated in pots. It has roundish leaves of a pleasing glaucous tone, bluntly toothed in their upper part, in whorls of three on the numerous stems, which in autumn bear the soft rosy flowers in small round bouquets. At first the boldly-ascending stems form neat tufts, but as they lengthen, they bend outwards with the weight of the buds and flowers at the points, thus making the plant a graceful object for pots, small baskets, or vases. It is hardy, and merits a place on the rockwork, especially in positions where its graceful habit may be seen to advantage—that is to say, where its branches may fall without touching the earth; but, except in favoured places, it does not make such a strong and satisfactory growth as most of the other Sedums, and it is, perhaps, seen to greatest advantage as a frame or greenhouse plant. A native of Japan; easily propagated by division. In late autumn the leaves often assume a lovely rosy-coral hue. There is a variegated variety, but it is not so good as the ordinary form and is much more tender.

S. spectabile.—This is the handsomest of the tall-growing species, and is worthy of a place in the choicest collection of hardy plants, being very distinct and beautiful. It is an erect-growing plant, with stout stems, from 1 ft. to 18 in. high, furnished with broad glaucous leaves. The flowers, which are rosy-purple, are produced in dense, broad corymbs. In the flower garden it may be usefully employed in many ways, either for breaking the uniformity of large flat surfaces, or for centres to small beds; and where the style is formal and severe its rigid aspect will be found to harmonise well with the surroundings. For this purpose the plant is best divided into single crowns annually early in spring. It will increase in three years from a single crown to about two hundred plants. Previous to its coming into flower, the glaucous foliage of this Sedum tends to give a pleasant relief to any high-coloured plant that may be near it. Its fine heads of rosy-purple flowers expand about the beginning of August, and remain for two months, and sometimes longer, in perfection. It is particularly effective when placed in the centres of beds, or in patches by itself in borders. Like others of its class, it withstands extreme cold, heat, or wet with impunity. Not many plants will grow, and fewer still will flower, in shaded places; but this Sedum doing both to perfection, should be generally cultivated in such situations. A rich soil suits this Stonecrop best, but it thrives in nearly any kind. Native of Japan. Known also as S. Fabaria. S. erythrostichum, or albo-roseum, is a similar plant, but is scarcely so handsome as S. spectabile.

S. spurium *(Purple Stonecrop).*—Several kinds of Sedum, with large, flat, crenate leaves, occur in our gardens, and of these this is much the best, its rosy-purple corymbs of flowers being handsome compared with the dull whitish flowers of allied kinds. It is a native of the Caucasus, and exceedingly well suited for forming edgings, the margin of a mixed border, or the rock garden. It is not sufficiently grown in gardens. It is of the easiest culture and propagation, and blooms late in summer, and often through the autumn, making a bright display. It is known also as S. stoloniferum, its correct name. Other species in this way, but not so ornamental, are S. oppositifolium, trifidum, dentatum, and ibericum, the latter having white flowers. S. denticulatum is the same as S. spurium.

S. Telephium.—None of the Stonecrops are so variable as this British species. No fewer than twenty forms have received names either as sub-species or varieties, but our native form is as showy as any of them, and is the most desirable for cultivating as a border flower. It grows from 1 ft. to 2 ft. high, and has stout, erect stems furnished with roundish fleshy leaves, and terminated by dense broad clusters of bloom which are usually of a bright rosy-purple hue, but sometimes white. It is not an uncommon plant, and may be found generally distributed about the country, usually growing in hedgerows and thickets, where late in summer and in autumn it produces its showy blossoms. In the garden this Stonecrop, and, indeed, all the other varieties and allied species of about the same size, are particularly useful for planting in places that would be too dry for other plants, such as on rough rockwork and dry borders. They prefer, however, to be treated as liberally as other plants, and well repay any attention by growing and flowering more vigorously. When cut, the flowers last a long time in perfection; the stems are so tenacious of life after they have been severed from the root, that they are often called Everlasting Livelongs.

The preceding are the most distinct and ornamental kinds in cultivation. The pretty S. cœruleum is an annual, and S. carneum variegatum not hardy enough to stand our winters. There are many other species and reputed species now in cultivation in this country, all of the easiest culture in any ordinary soil.

Selaginella.—Of this large genus of Lycopods there are a few hardy kinds which are valuable for carpeting the surface soil in the fernery, or shady spots in the rock garden. The kinds are S. denticulata, helvetica, and rupestris, all small trailing plants of a delicate green, mossy growth. S. Kraussiana, so common in plant houses generally under the name of S. denticulata, is quite hardy in many places, and in Ireland grows and thrives better than either of the others. A well-drained peaty soil should always be used for these plants, and the position should be a shaded and sheltered one.

Sempervivum *(Houseleek).* — The Houseleeks comprise a large genus of succulent plants, very many of which are hardy perennials, and of these the common Houseleek (S. tectorum), so often seen growing in patches on old roofs and walls, is the most familiar example. There is such a strong family likeness throughout all the hardy members of the genus that if one is known, all the others are easily recognisable. They form rosette-like tufts of fleshy leaves, some more compact and smaller than others, and differ more conspicuously in the colour of the foliage, some being deep red and others pale green. The flowers of most of them are pretty, and of a reddish tinge; several are yellow, though as regards their value in point of beauty they may be dispensed with. They may be used in a variety of ways in the garden. All are capital subjects for clothing dry sandy parts of the rock garden where few other alpines will thrive, for old walls, ruins, and the like, they are excellent, merely requiring to be placed at the outset in chinks with a little soil. The majority of them will thrive perfectly in any ordinary border, provided the soil be not too stiff and damp; but they prefer a dry elevated position, and in all cases they must, in order to do well, have full exposure to the sun. Propagation of nearly all is most easily effected by means of offsets, which, as a rule, are abundantly produced. Of late years some of the larger kinds, such as S. calcareum, have been used for geometrical beds. The following is a selection of about a hundred named kinds, grouped under well known types to which they are allied. All are natives of Europe and W. Asia:—

S. arachnoideum *(Cobweb Houseleek).*—One of the most singular of alpine plants, its tiny rosettes of fleshy leaves being covered at the top with a thick white down, which intertwines itself all over each plant like a spider's web. Widely distributed over the Alps and Pyrenees, this plant is perfectly hardy in our gardens, in which, however, it is rarely seen, except as a frame plant. It thrives in exposed spots, in sunny arid parts of rockwork, forming sheets of whitish rosettes, which look as if a thousand fine-spinning spiders had been at work upon them, and which send up pretty rose-coloured flowers in summer. About London it sometimes suffers from the sparrows plundering the "down." It should be on every rockwork; is easily increased by division, and thrives in moist sandy loam. Similar to this species are S. tortuosum (or Webbianum of gardens),

S. Fauconneti, heterotrichum, and Laggeri, all remarkable for the rosettes of leaves being united by a web of white threads.

S. arenarium *(Sand Houseleek)*, when grown in dense patches, has a lovely effect. It is much smaller than S. globiferum, to which it is allied, and is usually of a richly crimson colour. The leaves in the rosettes are not incurved, as in the latter species; the flowers are small, yellow, and pretty. S. Heuffelli, a similar species, is remarkable for the rich tint (almost chocolate-crimson) its foliage assumes in the autumn. The flowers are yellow. S. hirtum and Neilreichi are species of a similar character, as is also S. soboliferum, which is often confused with S. globiferum.

S. calcareum *(Glaucous Houseleek)*.—Than this species, now becoming very common in cultivation, under the garden name of S. californicum, no finer Houseleek has been introduced. It is as easily grown and as hardy as the common Houseleek (S. tectorum). Planted singly, the rosettes of S. calcareum sometimes attain a diameter of nearly 5 in., and as the leaves are of a decided glaucous tone, distinctly tipped at the points with chocolate, it is deservedly very popular for forming edgings in the flower garden. It is also admirable for the rockwork, is easily increased by division, and thrives in any soil. Other cultivated kinds in the same way are S. glaucum, Camollei, Lamottei, Verloti, and juratense, all desirable where a full collection is required.

S. fimbriatum *(Fringed Houseleek)*.—One of the most profusely blooming kinds. The flowers, which appear in summer, are of a dark rose colour, and are borne on stems 6 in. to 10 in. high. The leaves are in small rosettes, are smooth on both sides, strongly fringed on the edges, and marked with a large purple spot on the end, which terminates in a long point. S Funcki, Powelli, barbatulum, atlanticum, and piliferum are similar.

S. globiferum *(Hen-and-chicken Houseleek)*.—This is one of the neatest and most distinct in appearance of the family, being particularly distinguished by growing in firm, dense tufts, and throwing off little round offsets so abundantly that these are pushed clear above the tufts, and lie rootless, small, brownish green balls on the surface. The full-grown rosettes are of a peculiarly light green, and of a decided chocolate-brown at the tips of the under-side of the leaves for nearly one-third of their length. The small leaves of the young rosettes, all turning inward, appear of a purplish-brown colour. The rosettes are usually not more than 1½ in. in diameter. The plant has small yellow flowers, and is admirably suited for forming wide tufts on rockwork, on banks beneath the eye. It grows freely in any soil. It is also known as S. soboliferum. There are several other kinds in gardens similar to this, though not possessing its peculiar characteristics. Among these are S Wulfeni, Brauni, ruthenicum, Pittoni, Zellebori, grandiflorum, and albidum. All of these have yellow blossoms, and, being all distinct, are worth cultivation in a full collection.

S montanum *(Mountain Houseleek)*.—A dark green kind, smaller than the common Houseleek, with a very pleasing, almost geometrical, arrangement of leaves. These are pubescent and glandular on both sides, ciliated, and form neat rosettes, from which spring dull rosy flowers in summer. It is very suitable for forming edgings or for rockwork; like all the others, it grows in any soil, and, like all its fellows, is very easily propagated. A native of the Alps. Others similar to this are S. assimile and flagelliforme.

S. Reginæ Amaliæ is one of the newest and handsomest of all. It forms the largest rosettes of any; the leaves are broad and green tipped with reddish brown. It is not a very rapid grower and is a difficult kind to propagate, as it does not throw out offsets so freely as the others. Therefore it has not yet become very common. It belongs to the yellow-flowered section, but rarely blooms.

S. tectorum *(Common Houseleek)*.—This is a native of rocky places in the great mountain ranges of Europe and Asia, but having been cultivated from time immemorial on housetops and old walls, it is well known to everybody. Like some less known species, it may be used in flower gardening for forming dwarf borders, &c., though it would be better to give a position in gardens to somewhat rarer species. It varies somewhat, a glaucous form called rusticum being one of the most distinct. Other varieties in a similar way are Royeni, Rœgnerianum, Sequieri, calcaratum, and Greenei; the last is a new species, smaller in growth than the common Houseleek, but otherwise similar.

S triste is distinct from all the other Houseleeks on account of the rosettes of leaves being of a deep, dull red, which renders it a handsome plant. It grows about the size of the common S. tectorum, and is quite as vigorous and rapid a grower in any light, warm soil. As a contrast to other plants, this, by its singular colour, is valuable. At present, however, it is not much known.

Besides these there are several tender species that are now largely used for summer gardening. The most popular of these are—S. tabulæforme, a singularly handsome species, with broad rosettes of leaves that lie flat on the surface of the soil; S. Bolli, with leaves that form a dense, cup-like tuft from 3 in. to 6 in. across; and S. arboreum, a tall, straggling plant, with stout branches terminated by a rosette of foliage. These are generally propagated in quantity in heat in autumn or spring by means of offsets, and the plants are placed in the open border in May. Other species are hardy enough to be employed for a similar purpose, but these are the most useful.

Sempervivum tectorum (*Houseleek*).

Sempervivum arachnoideum (*Cobweb Houseleek*). See page 265.

[*To face page* **266.**

PLATE CCXL.

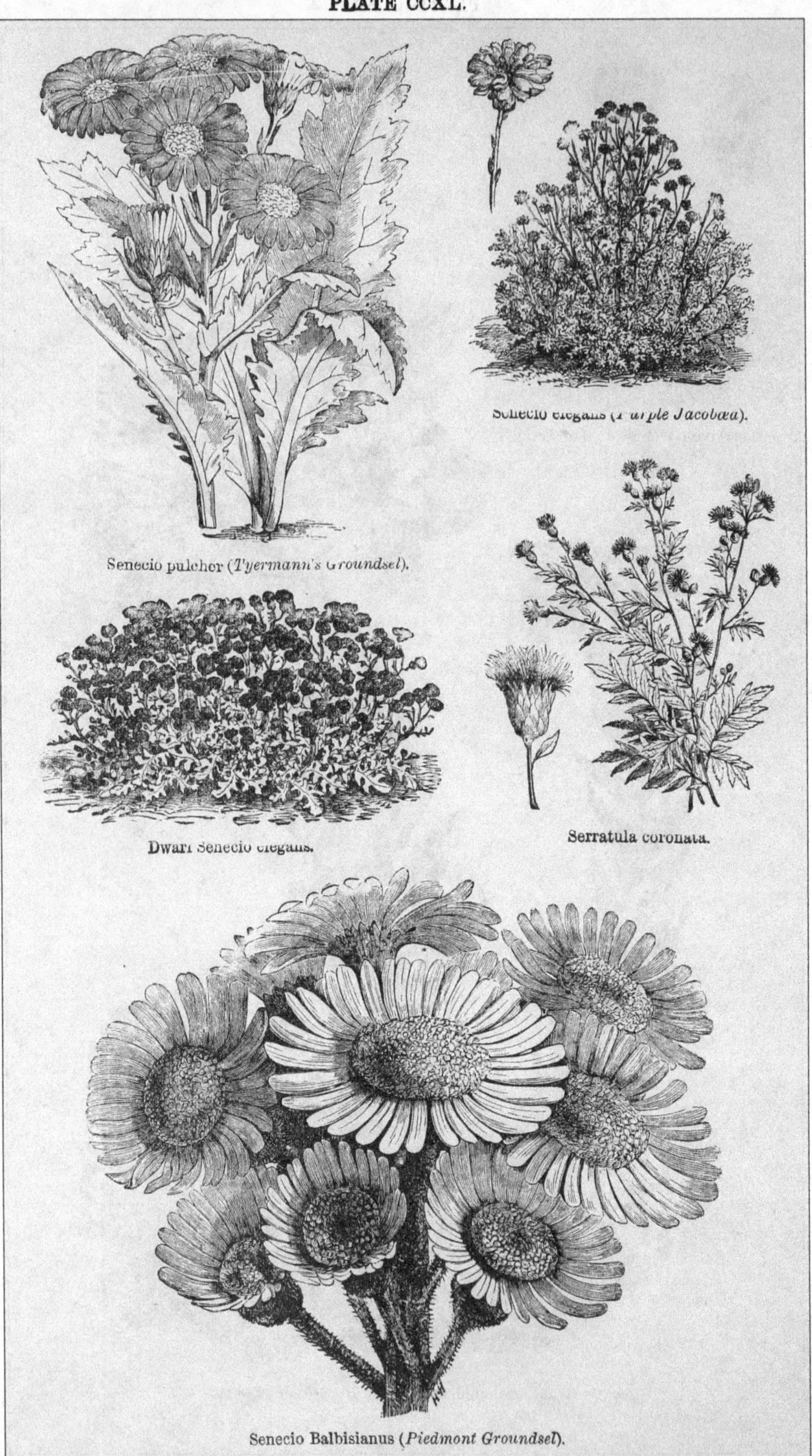

Senecio pulcher (*Tyermann's Groundsel*).

Senecio elegans (*Purple Jacobœa*).

Dwarf Senecio elegans.

Serratula coronata.

Senecio Balbisianus (*Piedmont Groundsel*).

[*To face page* 267.

Senecio (*Groundsel*). — The greater portion of this large genus of Composites are troublesome weeds, but a few are well worthy of cultivation in any garden. The following are among the most desirable:—

S. argenteus (*Silvery Groundsel*).—A sturdy, but minute silvery plant, almost like a miniature of the popular Centaurea ragusina. The leaves are quite silvery, and vary from ½ in. to 1½ in. long. The whole plant is not more than 2 in. high when fully developed and established; it stands any weather, and will live everywhere in sandy soil in well-drained borders. It is valuable for rockwork or borders, and well fitted to form beautiful dwarf edgings. The flower is not attractive, but, like the Centaurea and Cineraria maritima, the plant is valuable for the effect of its foliage. Similar to this, but inferior, are S. incanus, uniflorus, and carniolicus, all of which are good rock garden plants, and may be freely increased by division.

S. artemisiæfolius is a neat little perennial, with showy clear yellow flowers in broad clusters on stems from 1 ft. to 1½ ft. high. The leaves are deep green and finely divided, giving the plant an elegant feathery appearance. The foliage of S. abrotanifolius is similar to that of S. artemisiæfolius, but the flowers are larger, fewer in number, and of a good orange-yellow colour. Both are European plants and perfectly hardy, thriving well in ordinary soil in the rock garden or border.

S. Doronicum is among the showiest of the genus and one of the most useful. It grows from 1 ft. to 3 ft. high, and produces, on stout stalks in summer, numerous large, bright yellow flowers. It is of easy culture in any soil, is perfectly hardy, and may be increased by seed or division. Central Europe.

S. elegans (*Purple Jacobæa*). — This beautiful half-hardy annual from the Cape of Good Hope has for generations been a favourite garden plant, and is still very popular. It has numerous varieties both with single and double flowers, the latter being the showiest and most desirable. There is also a dwarf strain (nana), which grows about 1 ft. in height. The colours of these sorts vary from white to deep crimson. S. elegans should be treated in the manner recommended for half-hardy annuals. It grows best in rich sandy loam, and flowers from July to October, according to the time of sowing. These, like all other annuals, look best in good-sized masses.

S. japonicus.—This is one of the handsomest of the large species. It is of bold habit, growing about 5 ft. high, with leaves nearly 1 ft. across, divided into about nine divisions. The flower-stems are slightly branched, and the flower-heads are about 3 in. across, with the outer narrow florets of a rich orange colour. It is a native of Japan, but perfectly hardy. It is a moisture-loving plant, and should have an abundant supply of water in summer. It should be grown in a rich and moderately stiff loamy soil, by the margin of a lake or pond, so that its roots might be well supplied with moisture. *Syn.*, Erythrochæte palmatifida.

S. pulcher is one of the handsomest species, growing from 2 ft. to 3 ft. high, and producing in autumn several large flowers from 2 in. to 3 in. across, of a rich rosy-purple with bright yellow centres. The flower-stems are stout and branched, and the leaves are large and fleshy. It is perfectly hardy, but owing to its late season of flowering its beauty is somewhat impaired. It succeeds best in a good, deep, loamy soil, in a moist and somewhat shaded and sheltered position. It rarely ripens seed in this country, but it may be freely propagated in spring by cuttings of the roots, 1 inch long, sown like seeds in a pan of light sandy earth, placed in a cool frame or on a shelf in the greenhouse. Native of Buenos Ayres.

S. saracenicus grows wild in moist situations in some parts of the west of England, where it attains a height of from 4 ft. to 5 ft. It is a showy plant, and one that is suitable for planting on the margins of ponds or streams in semi-wild places, where it spreads rapidly, and when associated with the Willow Herb (Epilobium angustifolium) a beautiful contrast is the result, the habit of growth and colour of the flowers of each being distinct and effective. Other coarse growing species similar to this are S. Dorio and macrophyllus, also suitable for the wild garden, but not for border culture.

S. spathulæfolius is a rare species. It is both an interesting and pretty plant and is perfectly hardy. The flowers are a pleasing orange yellow, about the size of a shilling, and are borne in terminal clusters from 6 in. to 12 in. in height. Another species in a similar way is the Piedmont Groundsel (Senecio Balbisianus), which inhabits elevated districts in Northern Italy, particularly Piedmont. It grows from 3 in. to 9 in. high, and is furnished with hoary root leaves. The colour of the flowers is a golden-yellow, and, contrasted with the foliage, they have a remarkably bright appearance. Both species flourish on well-drained rockwork in light rubbly soil in an exposed dry situation.

Sensitive Fern (*Onoclea sensibilis*).

Serapias.—Terrestrial Orchids from South Europe, well worth growing in a hardy Orchid collection, as the flowers of all are singular, and some of them beautiful. The most desirable are S. cordigera, with large, showy flowers, chiefly of a blood-red colour; S. lingua, of a peculiar brownish purple, and S. longipetala, with large, rosy-red blossoms. These all grow from 9 in. to 12 in. high, and have broad and erect flower-stems, upon which the flowers are densely arranged. They succeed best in a soil composed of two parts of peat, one of loam, and one of sand and leaf-mould. The position should be partially shaded, and well sheltered from cold winds.

Serratula.—Perennial plants belonging to the Compositæ, but being all more or less of a Thistle-like aspect, they are not suitable for general culture. S. coronata and tinctoria are the best.

Seseli gummiferum *(Gum Seseli)*.—A handsome, silvery plant, 18 in. to 3 ft. high, with elegantly-divided leaves of a peculiarly pleasing glaucous or almost silvery tone. It is only a biennial, but it is so unique in its way that some persons might like to grow it, and if so the best position is on dry and sunny banks, or raised beds or borders. None of the other species are worthy of attention.

Sesleria.—Grasses not worth cultivating, though our native blue Grass, S. cœrulea, is pretty when in bloom.

Setaria.—Annual Grasses, not sufficiently ornamental to merit cultivation except in botanical gardens.

Shamrock Pea *(Parochetus communis)*.

Sheep's Scabious *(Jasione)*.

Sheffieldia repens.—A little New Zealand creeping plant, with small, slender stems and small leaves. Tiny white flowers are produced in summer. It is an interesting plant for the rock garden, and will grow in any good soil in a well drained spot; it is quite hardy. Primulaceæ.

Shortia galacifolia.—An interesting new and rare North American plant. It is not by any means showy, but is a welcome garden alpine. It is about 4 in. high, with Galax-like leaves, both as regards form and colour, and pure white bell flowers, borne singly on stems 3 in. high. Little or nothing is known of its culture in this country.

Sibbaldia.—Small growing plants of the Rose family, resembling some of the dwarf Potentillas. S. cuneata, one of the few species in cultivation, is a dwarf alpine with small yellow flowers, and is a suitable plant for open spots in the rock garden.

Sibthorpia europæa *(Moneywort)*.—A little creeping native plant with slender stems and small round leaves. In summer it forms a dense carpet on the surface of moist soil. It should always be grown in the artificial bog garden, as it is a capital plant to run among taller subjects. The variegated form is by far the prettiest plant, but unfortunately it seldom succeeds in the open, being much more delicate than the type. In a cool house or frame, however, no prettier foliaged plant could be grown. Shady banks and ditches will suit it. The flowers are inconspicuous. Scrophulariaceæ.

Sida.—A genus of Malvaceæ, the cultivated species of which, S. dioica and Napæa, are coarse-growing, unattractive plants, suitable only for naturalising in the wild garden.

Sidalcea.—Plants of the Mallow family, scarcely suitable for general culture. S. acerifolia, malvæflora, and oregana are the species in cultivation

Sideritis *(Ironwort)*.—Interesting, but not attractive, plants of the Sage family. S. hyssopifolia, taurica, and syriaca are the species in cultivation. All are perennial, of tufted growth, and bear yellow flowers. Desirable only for botanical collections.

Siegesbeckia orientalis.—An East Indian annual, belonging to the Compositæ, but of little value.

Sieversia *(Geum)*.

Silene *(Catchfly)*.—This is a genus of considerable extent, containing amongst its numerous annual and biennial species few showy plants, but among the perennials there are various species of great beauty. Southern and Central Europe is the home of the Silene, though a few species extend westward to America, a few eastward to Siberia, and a sprinkling of them will be found on the southern shores of the Mediterranean and in Asia Minor. The species showy enough for general cultivation are very few compared with the large number that exist, and of which there are crowds in botanic gardens. The following are all of dwarf growth, and can be recommended chiefly for the rock garden:—

S. acaulis *(Cushion Pink)*.—A very dwarf alpine herb tufted into light green masses like a wide-spreading Moss, but quite firm. In summer it becomes a mass of pink, rose, or crimson flowers barely peeping above the leaves. Many places on the mountains of Scotland, Northern Ireland, North Wales, and the mountains in the lake district of England are quite sheeted over with its firm, flat tufts of verdure, which are often several feet in diameter. It is in cultivation as beautiful and distinct as in a wild state, and grows freely in almost any soil on rockwork (not shaded) or in pots and pans. It is easily increased by division. There are several varieties: alba, the white one; exscapa, with the flower-stems even less developed than in the usual form; and muscoides, dwarfer still; but none of them are far removed from the typical form or of greater importance either from a horticultural or botanical point of view.

S. alpestris *(Alpine Catchfly)*.—This possesses every quality that renders an alpine plant worthy of extended garden culture, great beauty of bloom, perfect hardiness, very dwarf and compact habit, and a constitution that enables it to flourish in any soil. It grows from 4 in. to 6 in. high, and the flowers, which appear in May, are of a pure and polished whiteness, with the petals notched. It is one of those plants that should be used in abundance in every rock garden. A native of the Alps of Europe; very readily increased by seed or by division. Some forms of this species are quite sticky from viscid matter, and others are perfectly free from it. S. quadridentata and quadrifida, similar to S. alpestris, are also worth cultivating in a full collection. Both thrive under the same treatment. All are pro-

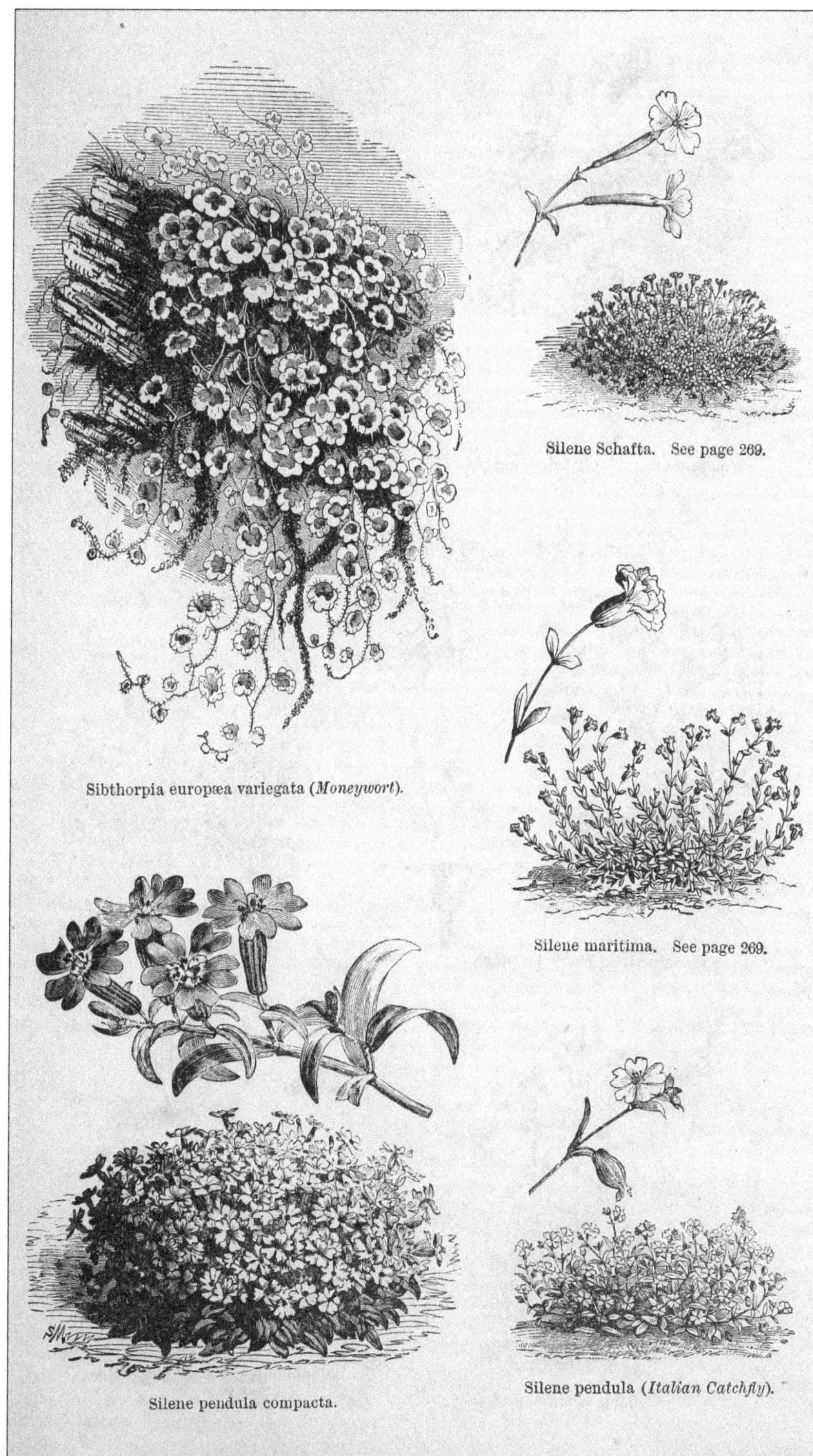

Sibthorpia europæa variegata (*Moneywort*).

Silene Schafta. See page 269.

Silene maritima. See page 269.

Silene pendula compacta.

Silene pendula (*Italian Catchfly*).

Silene pendula fl.-pl. (*Double flowers of Italian Catchfly*).

Silene Armeria compacta.

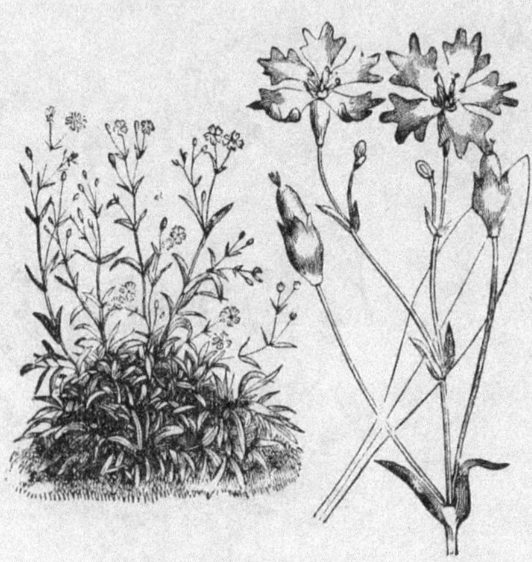

Silene alpestris (*Alpine Catchfly*).

Silene Saxifraga. see page 270.

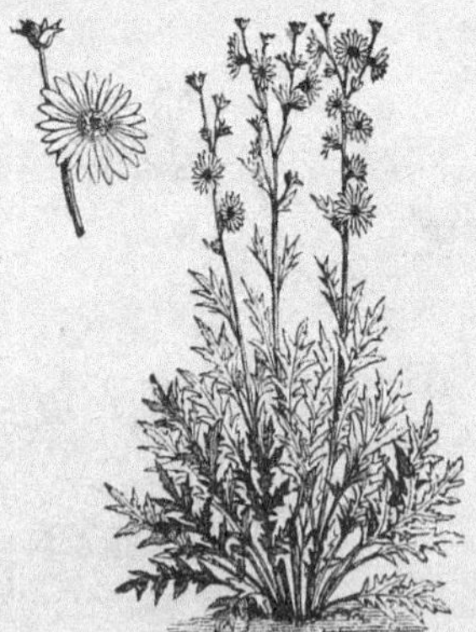

Silphium laciniatum (*Rosin Plant*).

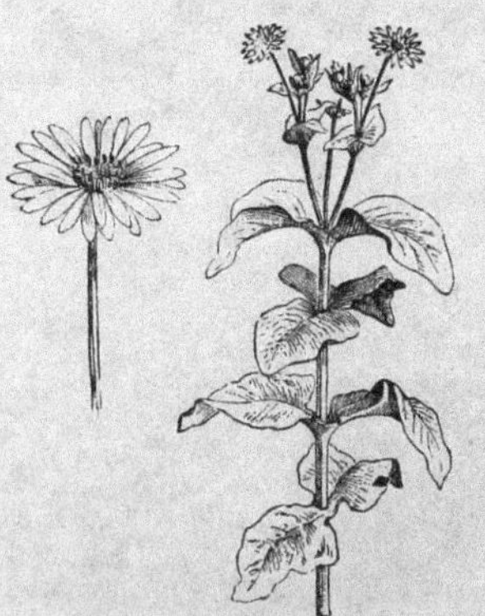

Silphium perfoliatum (*Cup Plant*). See page 270.

pagated either by division in spring or by seeds.

S. Elisabethæ.—A remarkably beautiful, and as yet very rare, alpine plant, quite distinct from all its congeners, the flowers looking more like those of some handsome but diminutive Clarkia than those of the commonly grown Silenes. They are very large, of a bright rose colour, and have the claws or bases of the petals white. From one to seven are borne on stems 3 in. or 4 in. high. This species is considered difficult to cultivate, but when once strong plants are secured, it will be found to be as easily grown as the Cushion Pink. It is rare in a wild state, occurring in the Tyrol and Italy, growing amidst shattered fragments of rock, and sometimes in flaky rocks without any soil. It grows freely enough in a warm nook on rockwork, in a mixture of about one-third good loam, one-third peat, and one-third broken stones. It should be planted in such a position that the roots can penetrate back, say from 18 in. to 24 in., into congenial soil. Flowers in summer, rather late, and is easily increased by seeds.

S. maritima.—A British plant, not uncommon on sand, shingle, or rocks by the sea, or on wet rocks on mountains, forming level carpets of smooth glaucous leaves, from which spring generally solitary flowers about 1 in. across, white, with purple inflated calices. The handsome double variety of this plant, S. maritima fl.-pl., is well worthy of culture, not only for its flowers, but for the dense, sea-green, spreading carpet of leaves that it forms, and which makes it particularly suitable for the margins of raised borders, for hanging over the faces of stones in the rougher parts of rockwork, or for the front edge of the mixed border. The flowers appear in June, and, in the case of the double variety, rarely rise more than a couple of inches above the leaves, which form a tuft about 2 in. deep. There is, likewise, a pretty rose coloured variety, which is less rambling than the type.

S. pendula, a handsome, rose-coloured, biennial plant, now much used in the spring flower garden, is a well known species. There are several varieties of it, notably compacta, compacta alba, Bonnetti, ruberrima, variegata, and Zulu King, all of which are decided improvements on the original. The compacta varieties are mostly used for spring bedding, as they are very dwarf (only about 4 in. high), and form compact rounded tufts. The other forms grow from 6 in. to 12 in. high. This species is perfectly hardy, and in order to obtain the finest plants for spring flowering the seeds should be sown in autumn in the reserve garden and then transplanted to their permanent beds. It flowers from May to August, according to the season of sowing. Native of Italy and Sicily.

S. pennsylvanica.—The wild Pink of the Americans is a dwarf and handsome plant with narrow, spoon-shaped, and nearly smooth root-leaves, those on the stems lance-shaped. It forms dense patches, and produces clusters of six or eight purplish-rose flowers, about 1 in. across, borne on stems from 4 in. to 7 in. high. A native of many parts of North America, in sandy gravelly places. Flowers from April to June. When strongly grown it forms an attractive border plant. It succeeds best in rather light sandy soil, but it is not particularly fastidious in its requirements. It will occasionally produce flowers the first year from seeds, but more generally it does not blossom till the second season. It may also be readily increased by cuttings.

S. Pumilio.—A beautiful species from the Tyrol, resembling the Cushion Pink of our own mountains in its dwarf firm tufts of shining green leaves, which are, however, a little more succulent and obtuse. It has also much larger and handsomer rose coloured flowers, which are taller than those of S. acaulis, and yet are scarcely more than 1 in. above the flat mass of leaves, so that the whole plant seldom attains a height of more than between 2 in. and 3 in. It will thrive as well on our rockworks as S. acaulis. It should be planted in deep sandy loam on a well-drained and thoroughly exposed spot, sufficiently moist in summer, facing the south, a few stones being placed round the neck of the young plant to keep it firm and prevent evaporation. Once it begins to spread, it will take care of itself. There is a white variety, but it is not in cultivation.

S. Schafta.—A much-branched plant, not compressed into hard cushions like the alpine, stemless, or dwarf Silenes, but withal forming very neat tufts, from 4 in. to 6 in. high, and becoming covered with large purplish-rose flowers from July to September, and even later. It comes from the Caucasus, is perfectly hardy, and is a fine ornament for the front margin of the mixed border, but it is particularly suitable for almost any position on rockwork. In planting, it may be as well to bear in mind its late flowering habit; it should not be used where a spring or early summer bloom is chiefly sought, but it may be employed in the summer flower garden in edgings to permanent beds, in the small circles round standard Roses, &c., with better effect than most alpine plants. It is easily raised from seed or increased by division of established tufts.

S. virginica *(Fire Pink).*—A brilliant perennial, with flowers of the richest and brightest scarlet, 2 in. across, and sometimes more. The stalks, which are somewhat slender, lie on the soil, and the blooms are borne a few inches above it. It succeeds well in the open air on a well-drained rockery, but it is one of those plants that require to be looked after, particularly in winter, when it is liable to "damp off" from excessive moisture. It is a native of open woods from New York southwards, and flowers there from June to August. It is increased by division or seed, the latter being preferable, as the strongest plants are obtained therefrom; besides, the plant does not bear dividing well. Similar in character to the Fire Pink are S. regia and S. rotundifolia, both rare cultivated plants from N. America, but well worthy of culture.

Having in cultivation such brilliant and distinct plants as the preceding Catchflies, we must consider S. Zawadski (dwarf, with white flowers), the woody S. arborescens (a dwarf, shrubby, evergreen species, with rose coloured flowers), the dirty white S. Saxifraga, S. Bolanderi, S. fimbriata, S. laciniata, and many others only worthy of a place in very large collections or in botanic gardens. S. rupestris, a sparkling-looking, dwarf, white species, little more than 3 in. high when in bloom, and reminding one of a dwarf S. alpestris, is better worthy of a place, and so is S. Hookeri, a beautiful dwarf Californian species, which, however, is very rarely cultivated.

Silkweed *(Asclepias).*

Silphium *(Rosin Plant).*—Coarse growing North American Composites. They are very showy, Sunflower-like plants, with a stately and large habit of growth. No border of the mixed type should contain such plants, which, however, are among those that first suggested the idea of the "wild garden," in which they will be found at home among the most vigorous growers, as they thrive and flower freely on the worst clay soils. The chief kinds are the following: S. laciniatum, a vigorous perennial with a stout, round stem, often upwards of 8 ft. in height. The leaves are large and deeply divided; the flowers are of a fine yellow, with a brownish centre, in large, horizontal or drooping heads, which have the peculiarity of facing the east. S. perfoliatum (Cup Plant) grows from 4 ft. to 8 ft. in height, with broad leaves 6 in. to 15 in. long, and yellow. Flower-heads about 2 in. across. S. terebinthinaceum (Prairie Dock) has stems from 4 ft. to 10 ft. high, panicled at the summit, and bearing many small heads of light yellow flowers. The leaves, which are borne on slender stalks, are ovate, thick and rough, and from 1 ft. to 2 ft. long. A variety (pinnatifidum) has the leaves deeply cut or pinnatifid. This species is remarkable for its strong turpentine odour. Other species are S. trifoliatum, integrifolium, and ternatum, all of which are of coarse growth like the others. If planted in numbers in bold masses, these plants produce a stately effect in the wild garden, particularly in autumn.

Silverbracts *(Pachyphitum bracteosum).*

Silybum marianum *(Milk Thistle).*—A very robust and vigorous growing native biennial, 5 ft. or more in height. It is strikingly handsome in appearance, and well deserves to be associated with other large, fine-foliaged plants. Its leaves are of great size, variously cut and undulated, tipped and margined with scattered spines; they are of a bright, glistening green colour, marbled and variegated with broad white veins. It is easily raised from seed, and thrives in almost any kind of well-drained soil. Additional vigour and development may be thrown into the foliage by pinching off the flower-stems on their first appearance. If a few plants are raised in the garden and planted out in rough and somewhat bare places or banks, &c., the Milk Thistle will soon establish itself permanently. Silybum eburneum is a more tender species, very closely resembling the above, but with spines that appear as if made of ivory. (=Carduus.)

Simethis.—A genus of Liliaceæ, of no ornamental value. S. bicolor is a rare British plant.

Sisymbrium.—Chiefly annual weeds belonging to the Cruciferæ. S. millefolium, a perennial species, has elegant feathery foliage of a whitish colour and small yellow flowers. It is rather tender, but worthy of cultivation in mild districts. It grows well in any light soil.

Sisyrinchium.—A genus of Iridaceous plants, of which there are about half-a-dozen species in gardens. Of these, however, only one is worthy of general culture. This is S. grandiflorum, a beautiful perennial species that flowers in early spring. The foliage is narrow and grass-like; the flowers, which are produced on slender stems 6 in. to 12 in. high, are bell-shaped and drooping, of a rich deep purple in the typical plant, and of a pure transparent whiteness in the variety album. Both are extremely graceful and pretty plants, and no garden should be without them. They are charming for the rock garden or borders, in light peaty soil or sandy loam, in warm positions. They may be multiplied by careful division in autumn. N.-W. America.

Skullcap *(Scutellaria).*

Smilacina *(False Solomon's Seal).*—Interesting perennials, allied to and resembling the Solomon's Seal. S. racemosa and stellata are the cultivated species; both grow about 2 ft. high and bear in early summer numerous small white flowers on slender stems. Although scarcely worthy of a position in borders, they are desirable for planting in moist shrubberies, or for naturalisation in open woods, copses, or on bushy banks. They thrive in ordinary soil, best, however, in a deep vegetable one, and will grow well under heavy shade, and also in dry places. Both North American plants.

Smilax *(Green Brier).*—There are a few hardy species of these evergreen climbing shrubby plants that are useful for garlanding bold rocks, and also for planting singly on a lawn or bank in a dry-bottomed garden. The kinds best suited for the purpose are S. aspera, from South Europe, and S. herbacea, rotundifolia, and tamnoides, natives of North America. All of these have slender, wiry stems and evergreen foliage, and their habit of growth admirably adapts them for the purposes above mentioned.

Smooth Lungwort *(Mertensia).*

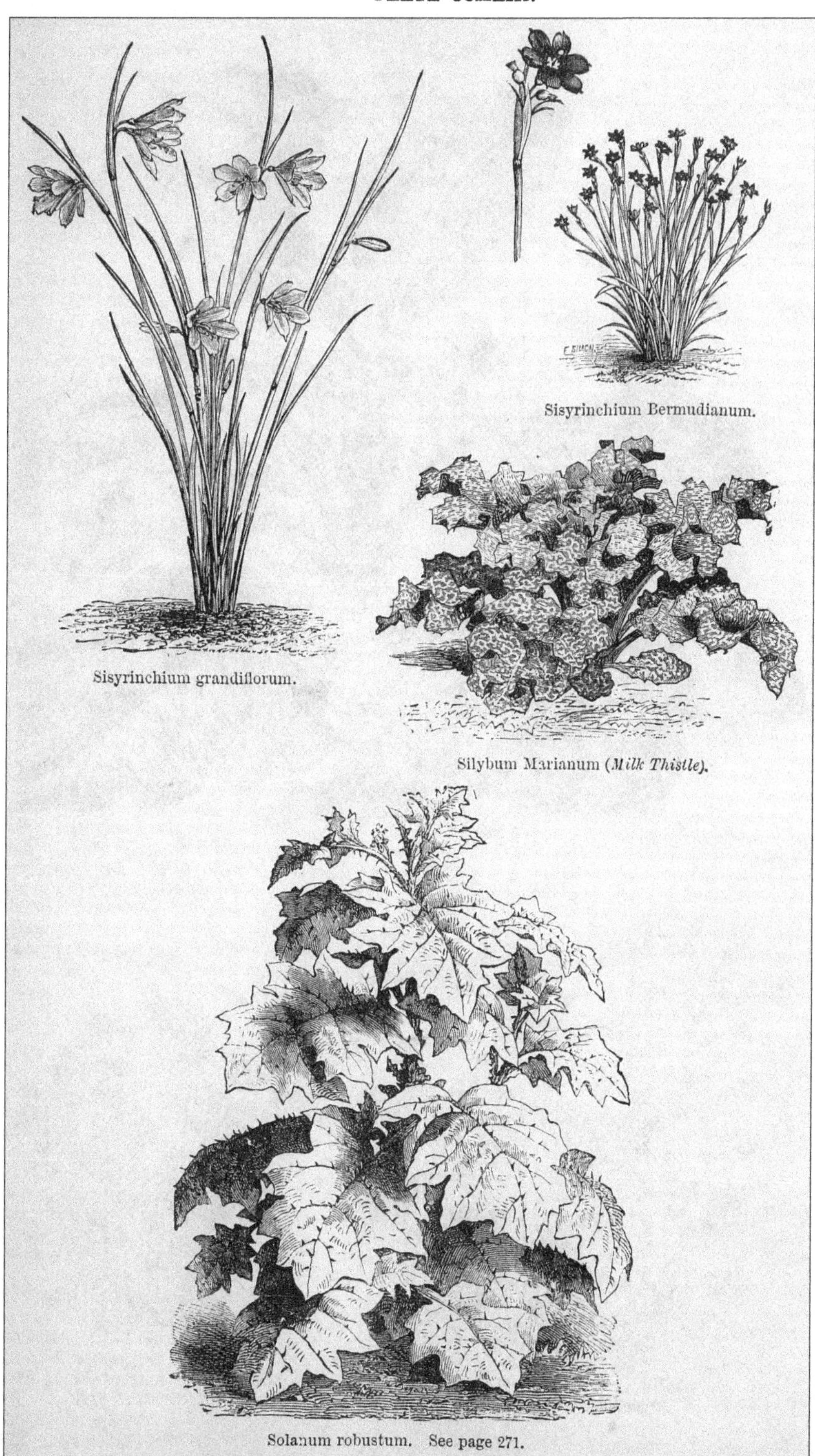

Sisyrinchium Bermudianum.

Sisyrinchium grandiflorum.

Silybum Marianum (*Milk Thistle*).

Solanum robustum. See page 271.

[*To face page* 270.

Solanum crinitum.

Solanum Warscewiczi.

Solanum laciniatum.

[*To face page* 271.

Smyrnium.—Plants of the Umbelliferæ. Of botanical value only.

Snapdragon (*Antirrhinum majus*).

Snowdrop (*Galanthus*).

Snowdrop Windflower (*Anemone sylvestris*).

Snowflake (*Leucojum*).

Soapwort (*Saponaria*).

Sogalina trilobata.—A rather coarse growing Mexican Composite.

Solanum.—This family, so wonderfully varied, affords numerous species that look graceful and imposing in leaf when in a young and free-growing state. In selecting examples we must be careful, as our climate is a shade too cold for some of the kinds grown on the Continent, and many of them are of too ragged an aspect to be tolerated in a tasteful garden. Half a dozen species or so are useful, but there is quite a crowd of narrow-leaved and ignoble ones which may well be passed over.

Most of these plants may be raised from seed, while they are also freely grown from cuttings, which, struck in February, will make good plants by May. All the kinds named are suitable for association with the larger leaved plants, though they do not as a rule attain such height and vigorous development as those of the first rank like the Ricinus. As a rule, temperate house treatment in winter is required, and they should be planted out about the middle or end of May in rich light soil, a warm position, and perfect shelter.

S. betaceum.—A small tree from South America, which in our climate attains a height of nearly 10 ft. if taken up in autumn and kept through the winter in a house. The stems are stout, smooth, and fleshy. The leaves, which resemble those of the Beet, are of an oval, pointed shape, and of a deep green colour, tinged with violet in the variety purpureum. The flowers are small, rose colour, in pendent cyme-like clusters, and are succeeded by fruit, of the shape and size of a fowl's egg, which become of a fine deep scarlet colour during the winter. Some varieties have flowers tinged with purple and fruit striped with brown. It may be placed to great advantage in groups in round beds with dwarfer plants or shrubs at the base, or with climbing plants ascending the stems, but it is much better isolated on slopes, &c. It is a vigorous grower, and should have rich soil.

S. crinitum.—A vigorous-growing species from Guiana, 5 ft. or more in height, with stout stems, set with short strong spines and dense long hairs. It has very large, roundish leaves, which in good soil attain a length of 2½ ft. The flowers are very large and white; the berries are roundish, villose, and twice or thrice as large as a Cherry. This attains a remarkable development in sheltered, warm spots in the south of England. It is fine in medium-sized groups.

S. Karsteni.—This, which is more commonly known as S. callicarpum, is a robust, slightly branching, arborescent shrub about 5 ft. high, covered with long hairs interspersed with spines. It is of a variable greyish-violet hue. The leaves are oval, broad, and 2 ft. or more in length. The flowers are large, of a fine delicate violet colour, and borne in crowded, almost one-sided, clusters. This plant is best isolated, as when placed in close groups the leaves of the associated subjects are apt to tear it. Venezuela.

S. lanceolatum.—This is the best kind for blooming qualities. The foliage, which is somewhat fluffy and Willow-like, possesses no marked character, but the mauve-coloured flowers are borne abundantly in clusters, each containing twenty or more blooms; the stamens, being of an orange colour, add to the effect. There are a dozen or more species that flower freely, but have little beauty of leaf; among the best of these is S. Rantonnettii, which has very pretty dark purple flowers, more than 1 in. across, with orange centres. It forms a neat bush, and flowers freely in the southern counties in warm sunny spots and on light soils. Mexico.

S. macranthum.—A fine species from Brazil, one of the best in cultivation, and somewhat resembling Polymnia grandis. It grows nearly 7 ft. high in one year, and has a stout, simple, spiny stem of a deep shining green, with grayish spots, and sparsely armed with very strong shortish spines. The leaves are elegant and deeply cut; some of them are over 2½ ft. long, and fall gracefully earthwards; they are of a light green on the upper surface, with red veinings, the under side having a reddish hue. The flowers, seldom seen in our climate, are of a fine violet colour, and grow in corymbs. It will not attain its full character and large dimensions in cold places, and should therefore have positions as warm as possible. Increased by cuttings, which, struck in February, are fit to plant out in May.

S. robustum.—A Brazilian species with a vigorous, much-branching stem, more than 3 ft. high, and furnished with very sharp and strong spines, and densely-set, long, reddish, viscous hairs. The leaves are very large; the flowers are white, with orange stamens, and are borne in unilateral clusters. The berries are round, of a brown colour, and the size of a small Cherry. As a foliage plant this is a subject of considerable merit, and one of those most suitable for our climate. It requires a warm, sunny aspect in a position that is at the same time both airy and sheltered from strong winds.

S. Warscewiczi.—A very fine and ornamental kind, resembling S. macranthum, but with a lower and more thickset habit, and branching more at the base. The flowers are white and small. The stem is armed with strong slightly recurved spines, and both the stems and the petioles of the leaves are covered with a very dense crop of short stiff brown hairs that scarcely rise above the skin.

This is one of the handsomest and best kinds grown.

Soldanella.—The few plants comprising this genus are all of minute growth and very beautiful. They bear a strong resemblance to each other, but are all worth growing, as they do not flower at the same period.

S. alpina is one of the most interesting of the plants that live near the snow line on many of the great mountain-chains of Europe. It is not brilliant, but withal it is beautiful in its pendent, pale bluish, open, bell-shaped flowers, which are cut into numerous, narrow, linear strips, three or four being borne on a stem. This latter is from 2 in. to 6 in. in height, and springs from a dwarf carpet of leathery, shining, roundish, or kidney-shaped leaves. It is comparatively rare in gardens, and considered difficult to cultivate, but if healthy young plants are placed out of doors on the rockwork or a raised border, in a little bed of deep and very sandy loam, they will be found to succeed perfectly well, especially in all moist districts. In dry ones also it will be easy to prevent evaporation by covering the ground near the young plants with some Cocoa fibre mixed with sand to give it weight. Its most suitable position is a level spot on the rockwork near the eye.

S. montana is very nearly allied to the preceding; in fact, except that it is usually somewhat larger in all its parts than alpina, and that the flowers are of a bluer purple, there is no great difference in its character. It also inhabits several of the great Continental chains, and will be found to thrive under the same treatment as the preceding. Both are readily increased by division, though, as they are usually starved and delicate from being confined in small worm-defiled pots, exposed to daily vicissitudes, they are rarely strong enough to be pulled to pieces.

S. pusilla has kidney-shaped leaves, heart-shaped at the base. The corolla is not deeply cut into fringes. The very small S. minima, with minute round leaves and one flower fringed only for a portion of its length, is also in cultivation, though rare. Both these plants will thrive under the conditions given for the others; but, being much smaller, especially the last, they require more care in planting, and should be associated with the most minute alpine plants, in a mixture of peat and good loam with plenty of sharp sand. They should get abundance of water in summer, especially in dry districts. S. Clusii and S. Wheeleri are kinds similar to those mentioned above.

Solidago (*Golden Rod*).—This is a large genus of North American composites, of which too many have found their way into cultivation, for scarcely one of them is fitted for garden culture. In borders they merely serve to exterminate much more valuable plants, and to give a coarse and ragged aspect to the garden, and they are such gross feeders that they soon impoverish any good border in which they are placed. They hold their own in any rough, open shrubbery or copse, among the coarsest indigenous vegetation, in shade or otherwise. If, however, a few are required for a full collection, the following will be found the best: S. altissima, canadensis, grandiflora, nutans, multiflora, rigida, and Virgaurea, all of which grow from 3 ft. to 5 ft. high, and have clusters of yellow flowers produced in autumn.

Solomon's Seal (*Polygonatum*).

Sonchus (*Sow Thistle*).—The Marsh Sow Thistle (S. palustris) is really a noble plant when well grown, as it attains as much as 6 ft. in height, with broad, deeply cut foliage, but with unattractive yellow blossoms. It is a desirable plant for damp places in the wild garden, as it takes care of itself, and soon develops into fine proportions. It is a native, and the only kind worthy of attention.

Sophora.—The perennial species of this genus are not important. One or two species used to be in cultivation at Kew.

Sorghum halepense.—A handsome hardy Grass from South Europe, North Africa, and Syria, with an erect stem 3½ ft. high, and broad flat leaves, more than 1 ft. long, chiefly collected around the base of the plant. It is most attractive when in flower in the end of summer, the inflorescence consisting of a dense panicle of purplish awned flowers. Suitable for isolation, groups, or borders.

Sowbread (*Cyclamen*).

Sparaxis.—Like the Ixias, these are charming bulbous plants from the Cape of Good Hope. There are in cultivation very many varieties, which have been derived chiefly from S. grandiflora and S. tricolor. These grow about 1 ft. high, are of slender growth, and bear large showy flowers, which vary from white to bright scarlet and deep crimson; the majority of the varieties have dark centres. They are very valuable plants for early summer flower, and well repay any trouble in cultivating. They should be treated in exactly the same way as the Ixias.

S. pulcherrima, a native of the Cape, is so remarkably distinct from everything that we have been in the habit of looking upon as a Sparaxis, that its claim to be a member of that genus has often been the subject of comment. It is a most lovely plant; its tall and graceful flower-stems rise to a height of from 5 ft. to 6 ft., and wave about when agitated by the wind, but are of so tough and wiry a character, that they are never broken in a storm (as the much stouter and stronger-looking, but in reality far more brittle, stems of the Gynerium argenteum, or Pampas Grass, so frequently are). It continues to produce its lovely deep rosy-purple Foxglove-shaped bells on almost invisibly fine wire-like lateral footstalks for from six to seven weeks. The usual colour of its bell-like blossoms is a rosy purple, but there are forms of purple-crimson, and almost every intermediate shade to

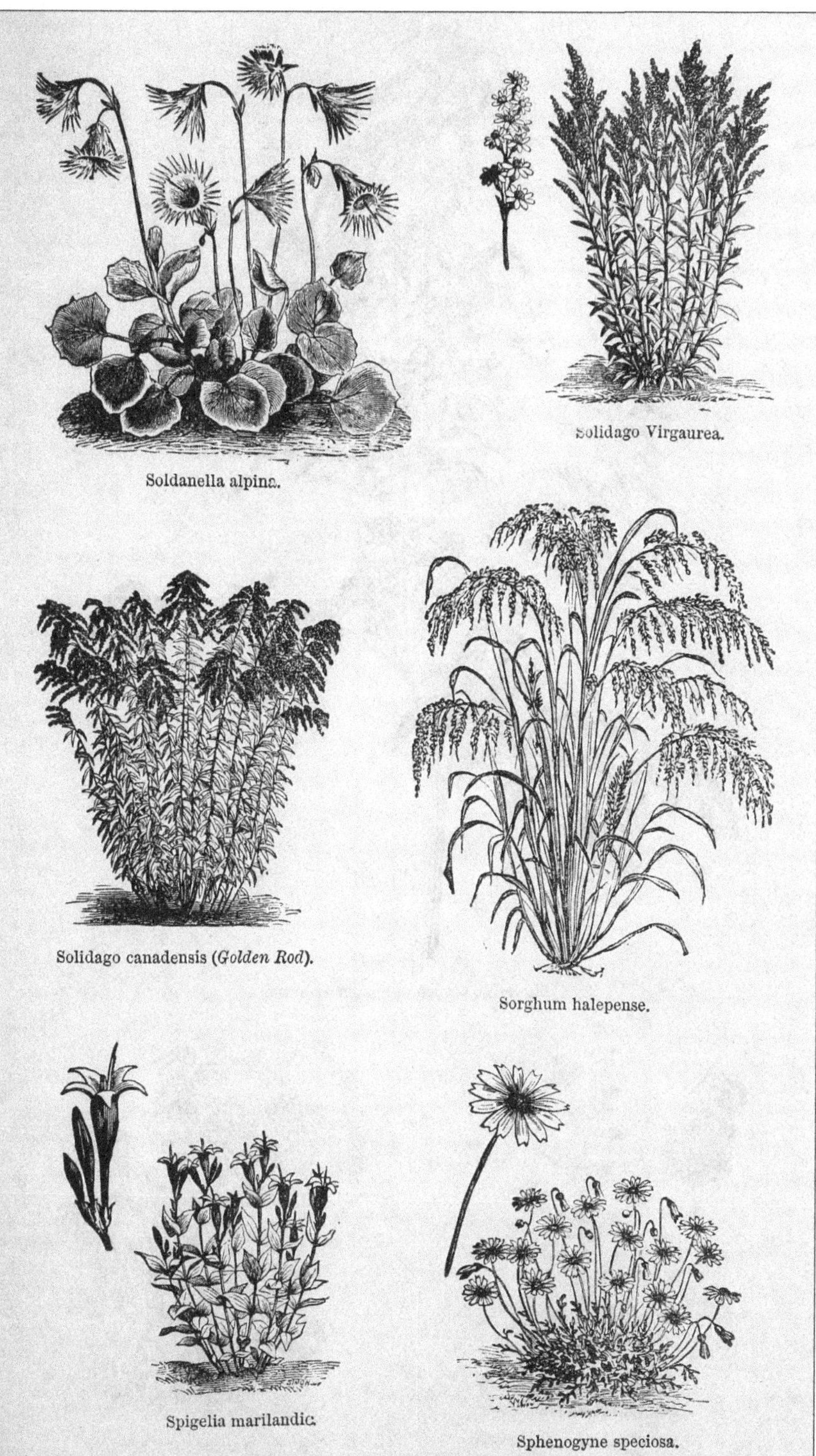

Soldanella alpina.

Solidago Virgaurea.

Solidago canadensis (*Golden Rod*).

Sorghum halepense.

Spigelia marilandica.

Sphenogyne speciosa.

[*To face page* 272.

PLATE CCXLVI.

Flowering branch of Spiræa Aruncus.

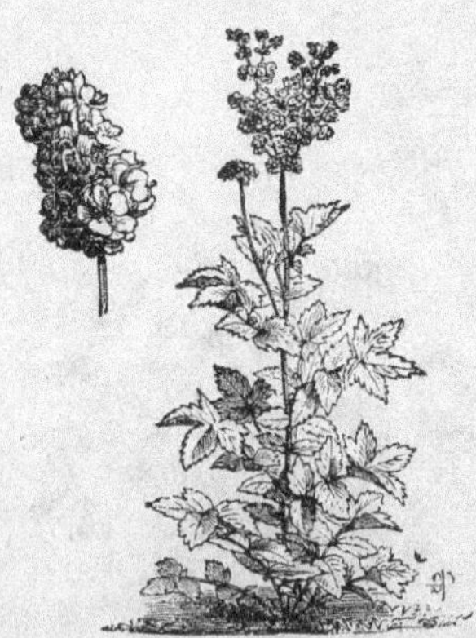

Spiræa Ulmaria fl.-pl. (*Double Meadow Sweet*).

Spiræa Aruncus (*Goat's-beard*).

[*To face page* 273.

nearly white, and some are beautifully striped. An allied species is S. Thunbergi, but S. pulcherrima is by far the finer and more elegant of the two. S. Thunbergi, although a very handsome plant, has nothing of the graceful waving beauty displayed by the tall, slender, arching flower-stems of S. pulcherrima. It has a stiffer habit and is a much dwarfer plant, with erect flower-stems seldom more than 2½ ft. high. The flowers, moreover, have very short stalks, and are not pendulous like those of S. pulcherrima.

The best position for Sparaxis pulcherrima is in clumps among shrubs, where its whip-like spikes rise up gracefully, and are seen to good advantage; the shrubs too, afford shelter. It is a plant that will well repay a little care the first year or two. In such a position Tritonia aurea might be associated with it, and as they flower at the same time they associate well together. It is about as hardy as the Tritonia, Montbretia Pottsi, and similar plants, though unfortunately more difficult to establish in the first instance. It has a great objection to removal, which, when it must be done, should be done as soon as the flowers begin to fade. Immediately after this stage stout fleshy roots are emitted from the bulbs, and if these are seriously broken or bruised much injury is the result. Its general appearance would lead to the supposition that it is a water-loving plant, but it succeeds well both in dry and damp positions, provided it has a rich friable soil or is stimulated with liquids when beginning to grow. It succeeds in and probably requires a much stronger soil than is found to answer for the Sparaxis grandiflora family. Amongst seedlings of S. pulcherrima there occur long and short flowers, some with the perianth slightly closing towards the mouth and others reflexed—H. C. S.

Sparganium.— British water plants, with the aspect of the common Flag, but suitable only for botanical collections, or for fringing artificial water in localities where not naturally plentiful. S. ramosum and S. simplex are both common natives. Reed Mace family.

Spartina.—Grasses of no ornamental value.

Specularia (*Venus' Looking-glass*).—These are so very similar to the Campanulas that they are very often placed with them, though they are distinct enough for garden purposes. S. Speculum, with numerous open bell-like flowers of a bright violet-purple, is one of our showiest annuals. Besides the large-flowered form called grandiflora, both in purple and white, there are the double-flowered kind, which comes true from seed, and a procumbent variety, dwarf and compact in habit, and bearing violet-blue flowers. S. pentagonia is another well-known favourite. It has flowers larger, but less profusely produced than the last. Its colour is purple, with the centre of a deep blue shade. This and the foregoing are particularly desirable, for, in addition to their showiness, they generally scatter their own seeds, which germinate year after year and give no further trouble; indeed it is necessary to prevent them from becoming too plentiful, therefore they are as useful as perennials. Both are perfectly hardy.

Speedwell (*Veronica*).

Spergula (*Spurry*).

Sphenogyne speciosa.—A beautiful Mexican half-hardy annual of the Compositæ. It is of slender, much-branched growth, and attains a height of about 1 ft. The flowers, produced from July to September, are yellow, with a brownish centre encircled by a conspicuous black ring. In the variety aurea the centre is orange. The flowers close at dusk. Though it will succeed if sown in the open in spring, it is much earlier and better if treated as a half-hardy annual and sown in early spring in heat. Any ordinary light soil suits it.

Spider Orchis (*Ophrys aranifera*).

Spider-wort (*Tradescantia*).

Spigelia marilandica (*Worm Grass*).—A beautiful native of North America, distinct from all other hardy plants. It forms a dense tuft of slender stems about 1 ft. high, each terminated by long tubular flowers which are deep red on the outside, and deep yellow within. These colours make a brilliant show. It is rarely found in gardens and is considered difficult to cultivate. In its native country it grows in sheltered situations, the roots finding their way deep down into a body of rich vegetable mould. Shelter from cutting winds, a free root-run in congenial soil, partial shade in summer, with abundance of moisture in hot weather, are the essential points in the culture of this beautiful plant. A position in which the natural conditions can be imitated should be chosen; and where the soil is not of the right description, it should be taken out to the extent of 2 ft. in depth and filled up with a well-sanded mixture of loam, leaf-mould, and peat. It is suitable for borders or the lower parts of the rock garden or for margins of beds of American plants.

Spignel (*Meum athamanticum*).

Spilanthes.—Weedy plants of the Composite family.

Spiræa (*Meadow Sweet*).—This extensive genus includes both shrubby and herbaceous species, but the latter are the only kinds we have to deal with here. They constitute a beautiful class of plants almost indispensable to every garden. There are in cultivation only about half-a-dozen species and their varieties, but all are worthy ornaments of any garden.

S. Aruncus (*Goat's-beard*) is a vigorous perennial that grows from 3 ft. to 5 ft. high and flowers in summer. It is a most valuable herbaceous plant, being beautiful in foliage and habit, as well as in its flowers, which are freely produced in large, gracefully-drooping plumes. It should, however, be noted that there are some inferior varieties of this plant not worth growing. In its best forms it is as good in midsummer as the Pampas Grass is in autumn. It is found in various parts of Europe, Asia, and America, and is a valuable subject for grouping with other fine-foliaged herbaceous plants. It thrives in ordinary soil, but best in a deep moist loam, and is easily multiplied by division of the tufts. S. astilboides, a recently introduced plant from Japan, is a variety of S. Aruncus and perfectly hardy, and bears snowy white feathery plumes.

S. cæspitosa is one of the smallest, if not the smallest, of Spiræas. The leaves, which are from ¼ in. to 1 in. long, narrow, and silky and glaucous on both surfaces, are arranged in dense rosette-like tufts springing from a woody root-stock; they are numerously produced, and form a spreading carpet-like tuft similar to the Stemless Catchfly (Silene acaulis). The flowers are from 1 in. to 3 in. high, terminated by a dense cone-like spike of flowers, which are very small and white. It is a native of the mountains of North America, from New Mexico to Northern Nevada, where it is found growing on rocks, &c. It seems to thrive well in an open position in the rock garden in good deep soil.

S. Filipendula *(Dropwort)*.—A rather common, but very pleasing British species. It grows from 1 ft. to 2 ft. high, and has yellowish-white flowers (often tipped with red) in loose terminal clusters. The leaves are mostly on the lower part of the stem, and when the flower-stems are pinched off, it forms a very effective edging plant, the Fern-like aspect of its foliage rendering it very distinct from many other plants that are used for this purpose. The double variety (S. Filipendula fl.-pl.) is very useful in the mixed border and for cuttings. It thrives in ordinary soil, and is multiplied by division of the tufts.

S. lobata *(Queen of the Prairie)* is one of the handsomest of the hardy Spiræas. It grows from 1½ ft. to 3 ft. high, and has deep rosy-carmine flowers in large terminal compound cymes. The leaves are pinnate, the leaflets pointed and irregularly toothed. It does best in sandy loam, and is valuable for the mixed border or for planting on the margins of shrubberies, or in beds among groups of the finer perennials. *Syn.*—S. venusta. Similar to S. lobata is S. Humboldti and S. digitata, two handsome plants well worthy of a place in every garden.

S. palmata is one of the most beautiful of the herbaceous species and among the finest of hardy plants. It has handsome palmate foliage and bears in late summer broad clusters of lovely rosy crimson blossoms. When well grown it attains 4 ft., but often it is seen less than half that height. It is generally considered tender, and is therefore grown largely in pots, but it is perfectly hardy, and thrives well in moist, deep loam, well enriched by decayed manure. It is a fine plant for many positions in large rock gardens, in borders, on the margins of shrubberies, or for naturalising, as it is quite vigorous enough to take care of itself. It looks best when grown in masses. The variety called elegans, said to be a hybrid production, has pale pinkish flowers, and is altogether an inferior plant to the best forms of S. palmata.

S. Ulmaria.—This common British Meadow Sweet would, no doubt, be considered a plant of high merit if it were an exotic. It is seldom seen in gardens, but there are often worse things to be found in our borders. It is too well known to need description here. It deserves a place, if only for the sake of variety, in the mixed border, on the margins of shrubberies, or in the rougher parts of pleasure grounds, where it may be advantageously planted with other subjects which do not require much looking after. Almost any soil will suit it; if moist, so much the better. The variegated leaved form of this plant is ornamental, the creamy-yellow and green variegation being effective.

Spiranthes.—A curious genus of terrestrial Orchids, generally with small inconspicuous flowers, white, or greenish white. They grow freely in damp, boggy places, but are suited only for botanical gardens or full collections of hardy Orchids.

Spotted Wintergreen *(Chimaphila maculata)*.

Spraguea umbellata.—A singular and pretty Californian plant, allied to the Claytonia. It grows from 6 in. to 9 in. high, has fleshy foliage, and spikes of showy pinkish blossoms. It has not been cultivated much here, but so far as is known it is an annual or biennial, according to the treatment that may be given it. Sown early in February in a warm frame, pricked out singly in small pots and planted out in May, the seedlings will bloom in August and September. If sown in May the plant will not flower till the following summer. In light soils it will resist an ordinary winter, but on the whole it is best protected in a frame. Like most tap-rooted plants, it does not bear transplantation well, except while small. If seeds are plentiful they may be sown in the open ground, and stronger plants will result, but, being usually scarce, sowing in pots in a moderate temperature is to be preferred.

Spring Bitter Vetch *(Orobus vernus)*.

Squill *(Scilla nutans)*.

Stachys *(Woundwort)*.—This is a large genus of the Labiateæ, but few species are worth cultivating. The common S. lanata, the woolly-leaved plant so much used as edgings, is the most familiar. It thrives in any kind of soil, and may be propagated easily in autumn or spring by division. It is also suited for naturalising. S. coccinea is rather a pretty perennial with red flowers in spikes. It grows about 1 ft. high, and succeeds in any soil in a partially shaded border.

Starflower *(Trientalis europœa)*.

Star of Bethlehem *(Ornithogalum)*.

Star Thistle *(Centaurea)*.

Stavesacre *(Delphinium Staphisagria)*.

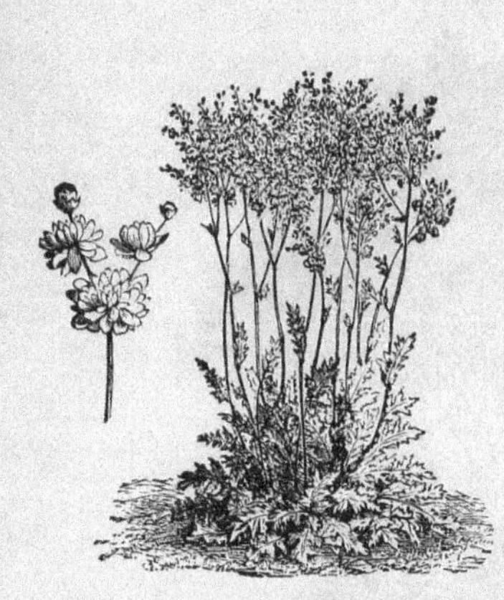

Spiræa Filipendula fl.-pl. (*Dropwort*).

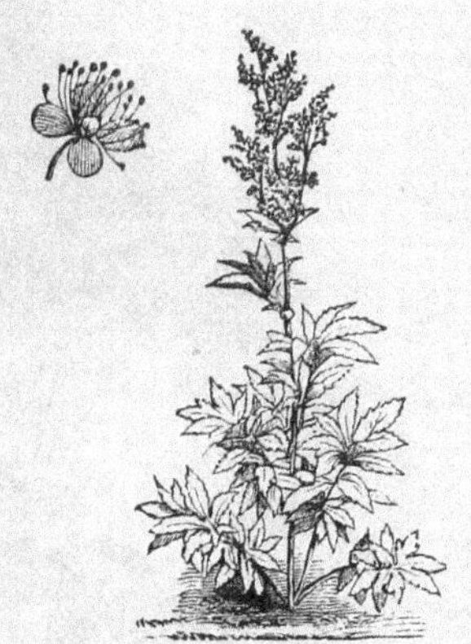

Spiræa lobata (*Queen of the Prairie*).

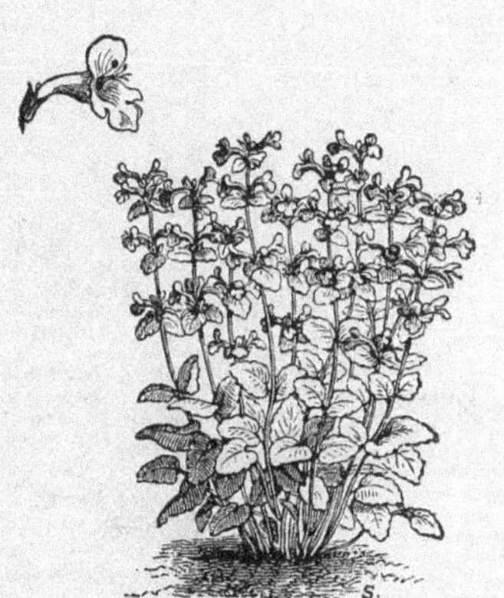

Stachys grandiflora (*Woundwort*).

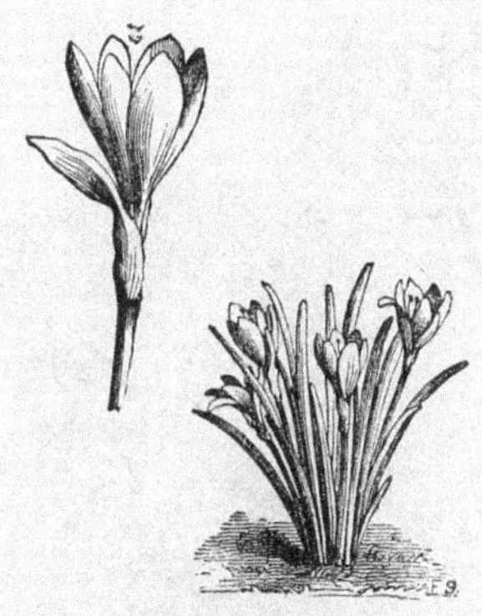

Sternbergia lutea.

Statice tatarica.

Statice eximia.

[*To face page* 274.

Statice latifolia (*Sea Lavender*).

Stevia purpurea.

Stipa pennata.

Stokesia cyanea.

Struthiopteris germanica (*Ostrich Fern*). See page 276.

[*To face page* 275.

Statice (Sea Lavender).—A large genus belonging to the Leadwort or Plumbago family. There are about a score of hardy or half-hardy species in gardens all of which are dwarf perennials and annuals, chiefly natives of seashore and mountainous districts. The majority bear large and very twiggy flower-stems covered with myriads of small flowers, and as these are for the most part dry and membraneous, they retain their colour for a long time after cut from the plants; they are, therefore, used to intermix with other everlasting flowers for vase decoration in winter. The larger growing species all thrive admirably without care if planted in an open, exposed bed of sandy soil, and many of them are admirable for growing in the rock garden. The best of the larger kinds are S. Limonium, of which there are several varieties; S. latifolia, the finest of all, producing wide spreading flower-stems carrying a profusion of small purplish blue flowers; and S. tatarica, a dwarfer species, with red flowers quite distinct from any other. There are other species similar to these, such as S. occidentalis, ovalifolia, spathulata, Gmelini, but are not sufficiently distinct for general culture. The smaller-growing species, such as S. minuta, minutiflora, caspia, eximia, are suitable rock plants. Among the half-hardy annuals and biennials the best are S. Bonduelli, yellow, a biennial if protected in winter; S. spicata, with small, rosy flowers in spikes; Thouini, violet, and very free flowering; and sinuata, purple and white, a very pretty plant and easy to grow. There are several varieties of S. sinuata hybrida possessing flowers of varied colours and make pretty border flowers. All the annual and biennial kinds require to be raised from seed in early spring, and planted out as soon as large enough. The half-hardy biennials need protection during winter, so that it is best not to plant them out until the following spring after raising.

Stellaria *(Chickweed).*—The golden-leaved form of the Grass-leaved Chickweed (S. graminea aurea) is now a favourite plant for geometrical bedding.

Stenactis (=Erigeron).

Stenanthium occidentale.—A Liliaceous plant from 6 in. to 24 in. high, bearing greenish purple flowers in a raceme. It grows naturally upon the mossy banks of mountain streams in Oregon, but is fit for the botanic garden only.

Stenosiphon virgatus.—This, a N. American perennial, is not an attractive plant.

Stenotaphrum americanum.—A dwarf, creeping Grass, growing best in damp, boggy places. It is, therefore, suitable as a carpeting plant in the artificial bog. There is a pretty variegated form. It is not a thoroughly hardy plant.

Sternbergia lutea.—A beautiful bulbous plant, with narrow, deep green leaves and showy, bright yellow flowers, produced in autumn, and often in winter. The whole plant is about 6 in. high, and forms a compact tuft. S. angustifolia is also a narrow-leaved variety, but has rather larger flowers. S. sicula is a miniature of S. lutea; S. crociflora and Clusiana are probably synonymous. These are plants of simple culture, and should be planted 6 in. deep, and a surface carpet of Sedum given, as such keeps the bulbs sheltered in winter, and protected from drought in summer. The soil should be good, deep, and fairly dry, and the position sheltered. Native of S. Europe. (=Oporanthus luteus and Amaryllis lutea.)

Stevia.—A genus of perennial Composites from Mexico, for the most part of shrubby growth and somewhat tender. They are, therefore, scarcely worth attention, except in botanical collections. They have Eupatorium-like heads of flowers. S. purpurea and S. serrata are the best known.

Stipa (*Feather Grass*).—The plants of this genus are all more or less interesting and ornamental, but not one of them approaches in elegance the South European S. pennata. It is only when gathered, and the awns collected in bundles, that its beauty is seen, bunches of it almost equalling in beauty the tail feathers of the bird of paradise. In May and June the plant, which at other times is hardly to be distinguished from a strong, stiff tuft of common Grass, presents a very different appearance, the tuft being then surmounted by numerous flower-stems nearly 2 ft. high, gracefully arching, and densely covered with long, twisted, feathery spikes. S. pennata loves a deep, sandy loam and may be used with fair effect in groups of small plants or isolated, but its flowers continue too short a time in bloom to make it very valuable away from borders. Division or seed. There are other elegant Feather Grasses worth growing in a full collection, such as S. calamagrostis, capillata, and elegantissima.

Stobæa purpurea.—A member of a numerous genus of South African Thistles, or rather Thistle-like plants, of which S. purpurea is perhaps the finest. It grows from 2 ft. to 3 ft. high, and bears in autumn numerous large bluish purple flowers. It is hardy and will grow in any kind of soil, but best in one that is deep and rich. It may be readily increased by detaching the running underground stems that push up in spring. Seeds are sparingly produced, except in very favourable seasons.

Stokesia cyanea.—A handsome hardy perennial of stout, free growth, from 1½ ft. to 2 ft. high, producing in September large, showy, blue flowers somewhat similar to a China Aster. It is a really fine autumn plant, and on this account it is now largely grown for supplying the market with blue flowers in autumn. It grows freely in good warm soils, but on account of its late flowering it sometimes does not expand its flowers well. In naturally damp localities, a hand-light placed over the plants at the flowering season will be found to be very beneficial, but it should be so arranged that it allows free admission of air on all sides while it protects from excessive moisture. It is also very useful for conservatory decoration in autumn and winter. It may be readily increased by division in spring. The slips, after being taken off, should be inserted in a warm border or frame, a few inches apart, in sharp sandy soil. As soon as they get well rooted and begin to grow afresh, they should be transplanted, mixing a little river sand and leaf-mould with the soil. Native of the Southern United States.

Stonecrop (*Sedum*).

Stork's-bill (*Erodium*).

Stratiotes aloides (*Water Soldier*).—An interesting native water plant, with leaves that form a compact vasiform tuft, from the centre of which arises in summer a spike of unattractive blossoms. If introduced to artificial lakes or ponds it will take care of itself. Hydrocharidaceæ.

Strawberry Blite (*Blitum*).

Streptopus.—A genus of Liliaceous plants, allied to Solomon's Seal; all natives of America. None of them are showy enough for general culture, though they are interesting botanically. S. amplexicaulis is a graceful plant, producing in autumn clusters of oval red berries. They succeed well in peaty soil among American shrubs.

Strumaria gemmata.—A pretty bulbousplant of the Amaryllis family. It is sometimes catalogued as a hardy plant, but it is not so, as it is a native of the Cape. If grown at all it should have frame protection, or should be treated as Ixias and Sparaxis.

Struthiopteris (*Ostrich Fern*).—This is a fine, hardy, exotic genus of Ferns, the fronds of which have somewhat the appearance of ostrich feathers. Each plant produces two kinds of fronds—fertile and sterile; the fertile are always grouped in the centre of the plant, and the sterile form, as it were, a cordon around them. They can be easily increased by division of the creeping underground stems, which run for some distance around well established plants. They require good peat and loam, well drained, to grow in, and should be planted in groups in bold, slightly-sheltered spots, where their noble appearance during their season of growth would not fail to be admired. Being deciduous, it would be advisable to plant amongst and around them some Polystichums or other evergreen robust-growing Ferns for winter effect, while for other seasons some of the finer varieties of Lilies that we now possess would form a useful mixture. The names of the two kinds of Struthiopteris adapted to our gardens are S. germanica and S. pennsylvanica. The former is a native of Germany, and one of the most elegant of hardy Ferns, with fronds nearly 3 ft. long. It is particularly suited for the embellishment of the slopes of pleasure grounds, cascades, grottoes, and rough rockwork, the margins of streams and pieces of water, and will thrive in the full sunshine or in the shade. S. pennsylvanica very closely resembles S. germanica, the chief point of difference being the narrowness of the fertile fronds of the former species. Both kinds will prove very effective in adding beauty of form to a garden, and should by no means be confined to the fernery proper.

Stylidium.—Interesting New Holland plants, sometimes cultivated in the open air and frames, but too tender for open-air culture generally.

Stylophorum diphyllum.—A handsome Poppywort, somewhat resembling the Celandine (Chelidonium majus), but a much finer plant. The foliage is greyish, and the flowers are large, bright yellow, and attractive, and plentifully produced in early summer. It grows from 1 ft. to 2 ft. high; in any ordinary garden soil it is quite hardy. N. America. S. ohioense and S. japonicum are the same.

Sullivantia ohioensis.—A little plant of the Saxifrage family, but only worthy of the attention of the botanical cultivator. N. America.

Summer Snowflake (*Leucojum æstivum*).

Sundew (*Drosera*).

Sundrops (*Œnothera fruticosa*).

Sunflower (*Helianthus*).

Sun Rose (*Helianthemum*).

Sweet Alyssum (*Koniga maritima*).

Sweet Cicely (*Myrrhis odorata*).

Sweet Flag (*Acorus Calamus*).

Sweet Pea (*Lathyrus odoratus*).

Sweet William (*Dianthus barbatus*).

Swertia perennis (*Marsh Swertia*).—An interesting and singular perennial, with slender, erect stems growing from 1 ft. to 3 ft. high, and terminated by erect spikes of flowers. These are greyish-purple spotted with black, and are produced in summer. It is not a showy plant, but is an interesting subject for the bog garden, for moist spots near the rock garden, or for naturalising in any damp places in peaty soil. Propagated by seed or division. S. speciosa and S. marginata are said to be handsome plants, but they are as yet not known in gardens.

Symphyandra. — Campanula - like plants, two of which are in cultivation. Both are showy and hardy perennials. S. pendula has branched pendulous stems and very large, cream - coloured, drooping, bell - like flowers, which are almost hidden amongst the leaves. It is a very hardy dwarf plant, rarely reaching 1 ft. high; it is a native of rocky places in the Caucasus, and easily increased by seed. In consequence of its pendulous habit, it is seen to best effect when elevated to the level of the eye in the rock garden, but it is also a first-rate border plant, and thrives in ordinary garden soil. S. Wanneri is also of dwarf habit, rarely exceeding 1 ft. in height. Its flowers are borne copiously on branching racemes, and are of a deep mauve colour. It differs from S. pendula in its corolla being very slightly lobed and in the colour of the blossoms. Like the other, it succeeds well in company with

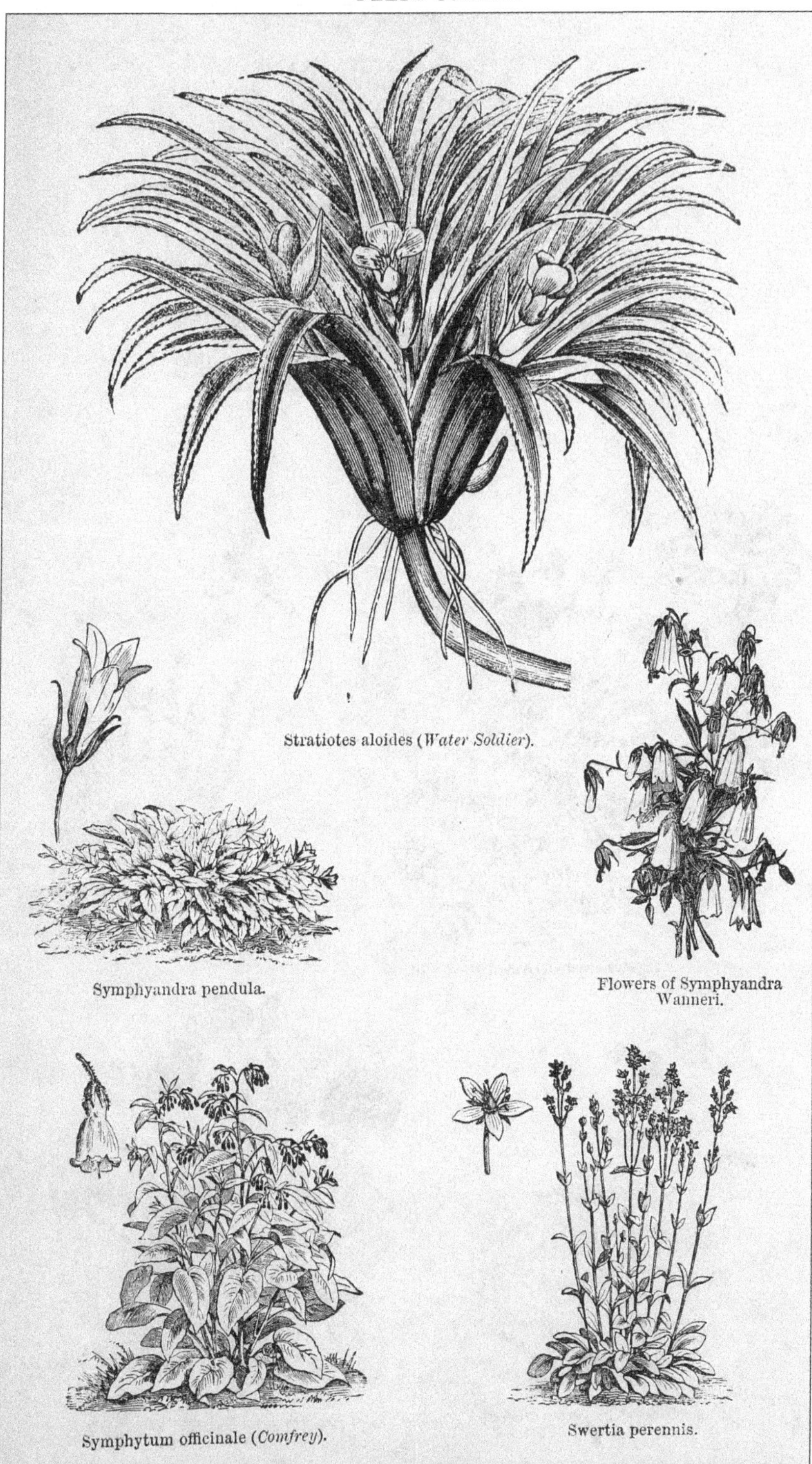

Stratiotes aloides (*Water Soldier*).

Symphyandra pendula.

Flowers of Symphyandra Wanneri.

Symphytum officinale (*Comfrey*).

Swertia perennis.

[*To face page* 276.

Tagetes tenuifolia (T. signata pumila).

Tagetes erecta (*African Marigold*).

Teucrium Chamædrys (*Wall Germander*).

Tanacetum vulgare var. crispum (*Crisped Tansy*).

Tagetes patula (*French Marigold*).

[*To face page* 277.

the majority of the Bellflowers, but prefers a light, warm, rich soil in a partially shady situation. Both plants are apt to be short lived, and therefore duplicates should be kept at hand. Native of Austria.

Symphytum *(Comfrey)*.—This genus of Borage-worts is composed chiefly of bold, but somewhat coarse growing plants scarcely suitable for the border, but admirably adapted for naturalising in rather open sunny places, for when well developed their bold foliage has a fine effect in masses. The largest growing kinds best suited for the wild garden are S. asperrimum and caucasicum. The Bohemian Comfrey (S. bohemicum) is suitable for the border. It is a very handsome and brilliantly coloured perennial, growing about 1 ft. to 1½ ft. high. The flowers are produced in early summer in erect twin racemes, and are of a brilliant reddish purple. The variegated leaved form of the common Comfrey (S. officinale) is very handsome, on account of its bold and striking variegation. In spring the leaves form rosettes close to the ground, and later in the season the stems and leaves rise and form a larger plant with numerous drooping blue flowers over all. The plant may be used with excellent effect in the garden of hardy flowers, although generally it is only seen in mixed collections of hardy variegated plants. Both it and S. bohemicum thrive in any ordinary garden soil in open sunny borders.

Symplocarpus fœtidus *(Skunk Cabbage)*.—A curious plant with bronzy-purple cowls, marbled with green, which are not without beauty when closely examined. Its leaves are very large. It is worthy of a place in the wild garden or botanical collection, and thrives in moist, deep soil. N. America. Arum family.

Tagetes. — A genus of Composites, familiar by the well-known French and African Marigolds, beautiful half-hardy annuals that have been favourite objects of culture for centuries. The genus consists of annuals and perennials, but of the latter none are hardy enough to make satisfactory garden plants, though one or two, such as T. lucida and Parryi, are desirable for a full collection. The annuals in cultivation are all natives of Mexico. The following are the most desirable :—

T. erecta *(African Marigold)* is easily distinguished by its stiff, erect habit, its pale green stalks and foliage, and its huge massive double orange and yellow and intermediate shaded blooms. It is one of the peculiarities of the African Marigold that seeds saved from the finest double flowers will always produce quite one-third of single flowers, but the rest are invariably good. Though occasionally grown for exhibition, the points required are not so difficult to acquire as is the case with the striped French kinds. Round, massive centres, well filled up and large, are the chief requirements. In garden borders, where a clump of three or four plants will produce telling masses of bloom from June till November, size is of less importance, but single flowers are ineffective and are best eradicated. It is wisest to sow seed under glass in April, for even without bottom heat it will then germinate freely. When the young plants are 3 in. in height, dibble them out again either into a frame or under handlights till very stout in the stems, as slugs are very partial to the young plants and will soon make a clearance of them. Where extra large blooms are desired the soil for the plants can hardly be too rich, and the buds on the branches should be thinned out. A good strain will give flowers thus treated nearly as large as Dahlia blooms. Though generally known as tender annuals, few summer flowers are hardier, for they need little special attention, and give a grand show of flowers.

T. patula *(French Marigold)* is a very charming garden summer annual. In it we have a flower that is highly favoured for exhibition, but the hard rules of the florists are such as to favour only the striped flowers, which are very beautiful, but, moreover, very monotonous. Colours, however, are not so limited as is the case with the African Marigold. We have yellow, orange, chestnut, mottled, striped, and various other hues. In habit, the older and coarser-growing forms of T. patula are spreading, yet tall, and make in good soil huge plants carrying scores of flowers. These are of medium size, and if good, are very double, rounded, sometimes partly reflexed, and invariably very pretty. The striped forms can be kept true only by growing them quite free from all other sorts, but even in the best strains the flowers will vary. Sometimes one plant will produce perfect blooms, and at other times self-yellow or maroon flowers. The unpleasant perfume emitted by the annual Marigolds renders them unsuited for cutting. Of more recent introduction are the compact habited forms of the French Marigold. These are bushy, stout and robust, usually growing to a height of from 12 in. to 15 in., and carrying dense heads of very perfect flowers. When on these are fully developed the beautiful striped qualities found in the older forms, these latter will probably be no longer cultivated. These dwarf forms make very effective masses and answer admirably as edgings of beds of the tall Africans. There is little to add to the directions for cultivation as given above. It is absolutely needful that the plants should be well hardened before being planted out, but when this is done it should be singly, and not in clumps.

T. tenuifolia is an allied kind to the French Marigolds, but is marked by much smaller flowers, some double and others single. The plant was formerly largely grown for the production of summer bedding effects. The very elegantly-cut leafage of this form is perhaps its most pleasing feature. As the plant needs a little starving to induce it to bloom freely when employed in beds and masses, the soil must be rather poor. This, as all other Marigolds, stands drought well. Having regard to the difficulty there is in getting any tender flowers to produce a good mass of yellow during the summer, perhaps this dwarf yellow Marigold, and particularly the variety pumila, is amongst the best available for that purpose. *Syn.*, T. signata.

Tamus communis *(Black Bryony)*.—A common climber in hedges and copses.

Tanacetum vulgare var. crispum.—A very elegant variety of the common Tansy, much dwarfer in stature, and with smaller emerald-green leaves, which are very elegantly cut, and have a crisped or frizzled appearance. It is quite hardy, and forms an effective ornament on the margins of shrubberies. It does best fully exposed, in deep but rather moist soil, but will grow anywhere. By thinning out the shoots in spring so that each remaining one shall have free room to suspend its graceful leaves, it looks much better than when the stems are crowded. The flowers should be pinched off before they open. It is increased by division in autumn, when every crown will make a plant. Britain.

Tansy *(Tanacetum vulgare)*.

Teasel *(Dipsacus Fullonum)*.

Tecophylæa cyanocrocus. — A beautiful spring-flowering bulbous plant from Chili, of dwarf growth, and bearing large, open flowers of an intensely deep blue. The variety Leichtlini has a white centre and a sweet perfume. It is not what can be termed a thoroughly hardy plant, except in very mild localities, but it succeeds well under frame culture. Flowering-sized bulbs should be procured and planted 3 in. deep in rich soil in a frame about August. If potted, a depth of 2 in. is sufficient, and after this the pots should be plunged and the plants treated like those outside. They should be kept cool, and as much air as possible admitted. The lights ought to be taken off when the weather becomes warm in February and March, and the pots allowed to remain exposed until the flowers begin to expand, when the plants may be transferred to the greenhouse. The bright colours of the deliciously-scented large flowers will always secure a first-rate position for this desirable bulb from "the Valley of Paradise."—M. L.

Tellima.—A North American genus of the Saxifrage family resembling the Heuchera. There are but two cultivated species. T. grandiflora has prettily coloured and veined leaves like Heuchera Richardsoni, and spikes of small, yellowish, bell-like flowers. T. parviflora is not such a desirable plant. Both grow in tufts and thrive well in moist, peaty soil with the Heucheras. Division in autumn.

Teucrium *(Germander)*.—A large genus of the Labiatæ, but containing few garden plants, and these are commendable only for their neat, dwarf growth. The following are the most desirable of about a score of species in cultivation. T. Chamædrys (Wall Germander), a compact perennial, 6 in. to 10 in. high, with shining foliage. The flowers are reddish-purple, and borne profusely in summer. The plant is found throughout Europe on walls, rocks, &c., and is suitable for borders and naturalisation on ruins, stony banks, &c., in any light soil; it is used also as an edging plant. T. Marum (Cat Thyme) is a small, grey, wiry-branched shrub with somewhat the habit of the common Thyme. The flowers, which are produced in summer, are bright red. Being a Spanish plant, it is only likely to prove hardy in the southern parts of these islands, and then only on ruins, old walls, or in dry chinks in chalk or gravel pits. If planted out, the soil should be of the driest and poorest description, brick rubbish, &c., with sand and a little poor dry loam. It should be placed in positions where cats cannot get at it, as those animals usually destroy it. Cuttings. T. Polium (Poly Germander), a curious dwarf whitish herb, 3 in. to 5 in. high. The flowers, which are small, pale yellow, and densely covered with short, yellow down, also appear in summer. The plant is suited for the rock garden in sunny spots and in light free soil. It is not hardy except in the milder southern districts and in favourable spots in the rock garden, where it grows freely. Seed, cuttings, and division. South Europe. T. pyrenaicum (Pyrenean Germander), a dwarf hardy perennial, 3 in. to 7 in. high. The flowers are borne in dense, terminal clusters, and are purplish and white. The leaves are thickly covered with soft down, as are also the stem and short branches. A very desirable plant for the rock garden and borders in ordinary soil. Seed, cuttings, or division. Other species that might be included in a full collection are T. hyrcanicum, lusitanicum, orientale, and multiflorum.

Thalia dealbata is one of the most stately of all hardy aquatics, quite different from the Cannas, to which, however, it is closely related. It is a native of S. Carolina, and its glaucous foliage and elegant panicles of purple flowers render it a most desirable plant for cool aquaria or for planting along the margins of shallow ponds or streams in the open air, inasmuch as it has proved to be hardy in sheltered positions in this country. The best mode of growing it is in pots or tubs pierced with holes, in a mixture of stiff peat and clayey soil, with a portion of river mud and sand. In winter these pots or tubs may be submerged to a greater depth, and the plants be thus effectually protected. It would not attain its greatest size out-of-doors, except in warm places in the southern counties, in which it might be planted out directly. Propagated easily by division.

Thalictrum *(Meadow Rue)*.—An extensive genus of Crowfoots, all of which have much-divided elegant foliage, but none have showy flowers. A few of the smaller-growing species possess foliage of a most graceful kind, rivalling in delicacy of form and colour some of the charming Maidenhair Ferns. They are valuable, therefore, for many ornamental purposes, both in association with flowering plants and with those of fine or characteristic foliage.

T. anemonoides *(Rue Anemone)*.—A delicate, diminutive, and interesting species, of dwarf habit, usually only a few inches high. The flowers are white, nearly 1 in. in diameter, and open in April and May. The plant is best adapted for cultivating on rockwork in deep, moist soil, and partial shade. A double-flowered variety is in cultivation, and may be preferred to the single one. Native of North America.

T. minus, one of the most désirable of all the species, forms compact tufts, from 1 ft. tó 18 in. high, very symmetrical, and of a slightly glaucous hue. It may be grown in any soil, and requires only that the slender flower-stems, which appear in May and June, should be pinched off. Not alone in its aspect, as a little bushy tuft, does it resemble the Maiden-hair Fern, for the leaves are almost pretty enough to pass, when mingled with flowers, for those of the Fern ; they are stiffer and more lasting than Fern fronds, and are well suited for mingling with vases of flowers, &c. There are probably several forms or varieties of this plant. It would look pretty isolated in large tufts as an edging, or in borders, or in groups of dwarf subjects. Easily increased by division. T. adiantifolium is a similar plant.

T. tuberosum grows about 9 in. high. Besides the graceful foliage which we find in all the dwarf forms of the genus Thalictrum, we have in this instance an additional beauty in the abundant mass of yellowish cream-coloured flowers which this plant produces. It is perfectly hardy, and thrives in a deep peat soil. S. Europe.

Besides these dwarf kinds there are about two dozen species in cultivation, all of which are of tall growth, ranging from 3 ft. to 6 ft. high. There is a great sameness amongst them, as all have elegant, finely cut foliage. One or two of them are well worthy of cultivating in a mixed border, as their elegant growth associates admirably with gay flowers, and for this purpose we should choose T. aquilegifolium, which is about 4 ft. in height, and a vigorous grower in any soil. There are two or three varieties of it, one (atropurpureum) with dark purplish stems and leaves. All of the Meadow Rues are admirable subjects for naturalising.

Thapsia. — Umbelliferous perennial plants, chiefly of botanical interest.

Thermopsis.—This genus is represented by only two species. Both are showy plants from 2 ft. to 3 ft. high, of slender erect growth, and bear long terminal spikes of attractive Lupine-like yellow blossoms. T. montana is the superior plant, and though it is said to be but a variety of the older species, T. fabacea or rhombifolia, it is quite distinct enough for garden purposes. It is of graceful growth, and, flowering at the same time as the perennial Lupine, it will be found useful for association with that and other border plants of the season. It grows best in good soil in an open situation, and as it does not divide well it should be propagated by seeds. It is a native of California and other parts of Western North America, while T. fabacea occurs farther north. T. barbata, a beautiful Himalayan species, with purple flowers, has apparently gone out of cultivation.

Theropogon pallidus.—A pretty Liliaceous plant growing from 6 in. to 1 ft. high. It has grassy foliage similar to that of an Ophiopogon, and its flowers much resemble those of the Lily of the Valley, but are rather larger and pink in colour. It flowers later than this Lily. It is presumably quite hardy in this country, as it inhabits high elevations in the Himalayas. Found growing on mossy rocks, the bases of old trees, &c., where the roots spread superficially in the loose soil.

Thladiantha dubia. — A handsome creeping perennial of the Gourd family. The long climbing stems bear a profusion of bright yellow flowers, together with heart-shaped leaves of an agreeable lively green colour ; and as the plant is very hardy, it may be effectively employed for covering trellises, arbours, &c. In the neighbourhood of Paris it survives the winter in the open ground. N. China and India.

Thlaspi latifolium.—A dwarf, but vigorous perennial, 6 in. to 12 in. high with large root-leaves, and flowers somewhat like those of Arabis albida, but larger. Suitable for borders, the spring garden, in beds, and naturalised in association with the dwarfer flowers of spring and early summer ; thrives in ordinary garden soil. Caucasia. Division and seed. T. alpestre is a biennial, scarcely worth the trouble of cultivating. T. rotundifolium is synonymous with Iberidella.

Thorn Apple *(Datura)*.

Thrift *(Armeria vulgaris)*.

Thunbergia alata.—These beautiful half-hardy annuals, so common in greenhouses, are elegant dwarf climbers of the easiest culture and thrive well in summer in the open air, where they are valuable for draping dwarf trellises or similar positions. The flowers of the typical kind are showy and of a yellowish-buff colour, but there are several varieties differing in colour. Alba is pure white ; aurantiaca, bright orange ; Fryeri, orange, with white eye ; Doddsi has variegated foliage ; and there are others with yellow and sulphur coloured flowers. They grow from 4 ft. to 5 ft. high, and their slender stems are copiously furnished with bloom from July to October. The seeds should be sown in heat in early spring, the seedlings potted separately as soon as large enough and planted out in May in good light soil. East Indies. Acanthaceæ.

Thymophylla aurea.—A neat little annual Composite from Colorado. It is of very dwarf habit, forming a branching tuft about 4 in. high and 6 in. to 9 in. in diameter. The flowers are in terminal heads, about ½ in. across, resembling a miniature single Marigold, with a bright yellow ray and disk, and the resemblance to a Tagetes is still more marked in the cup-like involucrum and tiny akenes, which yield a strong aromatic odour. It is of the easiest cultivation as a hardy annual, and prefers a rather dry soil. (= Lowellia aurea.)

Thymus *(Thyme)*.—There are numerous Thymes in cultivation, but a small number only are ornamental enough for general cultivation, though few plants are more suited for the most arid parts of rockwork, and for those places in which, from various causes, many other plants will not thrive, but they spread so quickly into wide dense cushions that they ought not to be placed near any delicate or very minute alpine plants. T. Serpyllum albus, the white variety of our common wild Thyme, is a lovely plant for a sunny bank. Nothing can be more charming than such a bank covered with a mixture of the common wild form and the white variety. T. lanuginosus, though usually considered a very woolly variety of T. Serpyllum, our common British Thyme, is a far more ornamental plant, pleasing at all seasons, and forming wide cushions in any soil, provided it be thoroughly exposed to the sun. Another very desirable plant is the variegated form of the Lemon-scented Thyme (T. citriodorus aureus). It is singular that the variegated form is more robust than the green-leaved kind. Apart from the beautiful character of the leafage of this Thyme, however, it retains its foliage all through the winter, and in this respect is much more useful for the flower garden than the common variety. Indeed, looking at the beauty of its foliage and the perennial character of the leaves of the Golden Thyme, it seems difficult to understand why the common kind should be grown at all. The golden kind grows from 6 in. to 9 in. high, and, being of dense, compact growth, is useful, and is much used as an edging plant. It may be readily increased by cuttings, which strike readily in September in hand-glasses or cold frames, and should be planted out in spring. In selecting cuttings choose those best variegated, otherwise there is a tendency to revert to the normal green type. Various other kinds of Thyme are worthy of a place on the dry, arid slopes of the large rock garden and on old ruins. The minute, creeping, and strongly Peppermint-scented T. corsicus, with flowers so small that they are almost invisible, should be planted on every rockwork, where it will soon become one of the welcome weeds. Other kinds in cultivation are T. azoricus, azureus, bracteosus, Zygis, thuriferus, Chamædrys, and Mastichina.

Tigridia Pavonia *(Tiger Flower)*, one of the most gorgeous of all flowers, is an Iridaceous plant, about 2 ft. high, with erect, rigid foliage and triangular-shaped flowers, about 6 in. across, of a brilliant scarlet, copiously spotted with crimson. The flowers last only one day, but are produced in quick succession from the same stem for at least six or eight weeks. Many seminal varieties have sprung from it, the most noteworthy being T. Wheeleri, a kind whose flowers are larger and a shade deeper in colour than those of the type. T. splendens and T. speciosa are also larger, and their colour is brighter; but description fails to convey an idea of the distinctive points of these varieties. The form grandiflora is superior to the type on account of its larger and richer-coloured blossoms, its small, fiddle-shaped petals being more copiously spotted; the deep green of the plaited, sword-like foliage much enhances the effect. T. Pavonia var. conchiflora differs from the type in being of less robust habit and in the blossoms being bright yellow and spotted with scarlet. That this is but a variety of Pavonia is evident from the fact that there are recorded instances of bulbs producing blossoms in which yellow predominated in the early part of the season, and later on the same bulbs bore scarlet flowers. Planted in groups this charming plant is a highly effective object associated with Lilies, Gladioli, and other noble subjects. As regards culture, about the end of March prepare a bed which will be best if in a partially shaded position; thoroughly drain it with rubble, &c., placed at a depth of 1½ ft. or 2 ft., and fill in with a compost of rich loam and well-decayed leaf-mould in equal quantities, with a sprinkling of sand, well mixed up together. Early in April plant the bulbs from 5 in. to 6 in. apart and 3 in. deep, placing a little sand under and around each. They will require no further care, except, if the weather be dry, slight waterings up to the time when they begin to expand their blossoms. When the foliage becomes decayed in autumn, the bulbs should be carefully lifted and tied in bunches of about twelve, according to their size, and hung in a cool, airy place until next planting season. Some cultivators allow the bulbs to remain undisturbed during the winter, and in warm localities and in light soils, perhaps, it is advisable for a year or two; but, as a general rule, the plan recommended is preferable, as it entails but little trouble, and the bulbs are at hand for the purposes of propagation. The latter is effected by means of offsets, which are freely produced, and if carefully removed and treated as above, they will flower the second year. Seeds afford a wholesale means of propagation. They should be sown either as soon as ripe or in the following spring in shallow pans, and placed in a heated frame or greenhouse. As soon as the seedlings are of sufficient size they should be pricked off into other pans, after which they should be treated as matured bulbs, and in the third and fourth seasons they will produce blossoms.

Toadflax *(Linaria)*.

Tofieldia.—A genus of Liliaceous plants suitable for botanical collections. T. palustris

Thalictrum aquilegifolium (*Meadow Rue*).

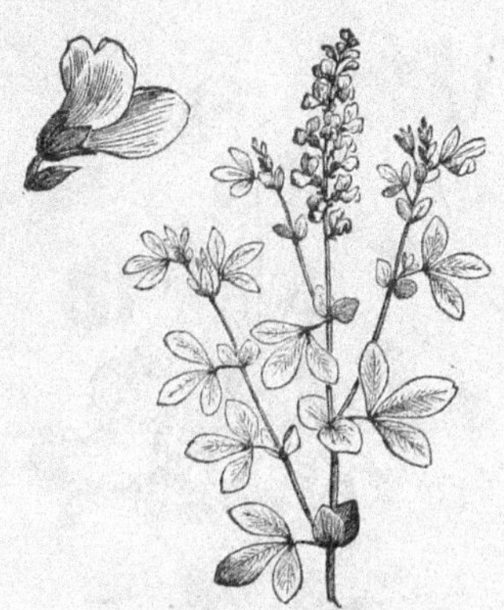

Thermopsis fabacea

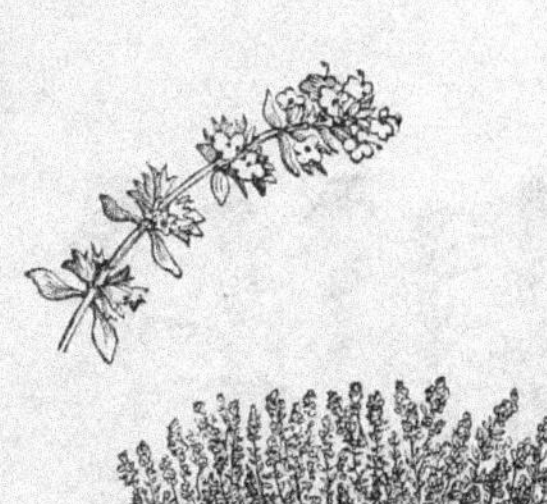

Thymus vulgaris (*Thyme*).

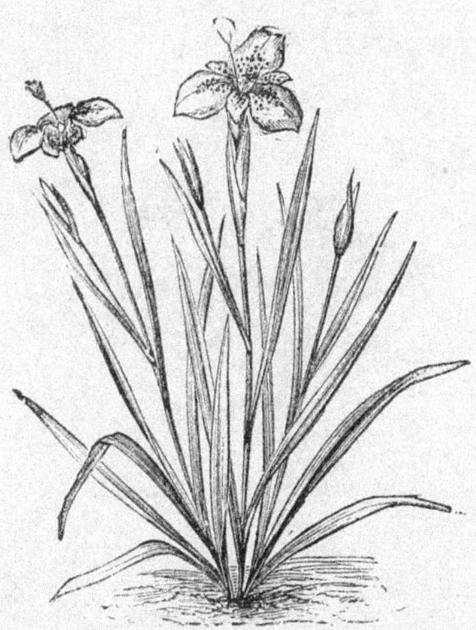

Tigridia Pavonia (*Tiger Flower*).

Tournefortia heliotropioides.

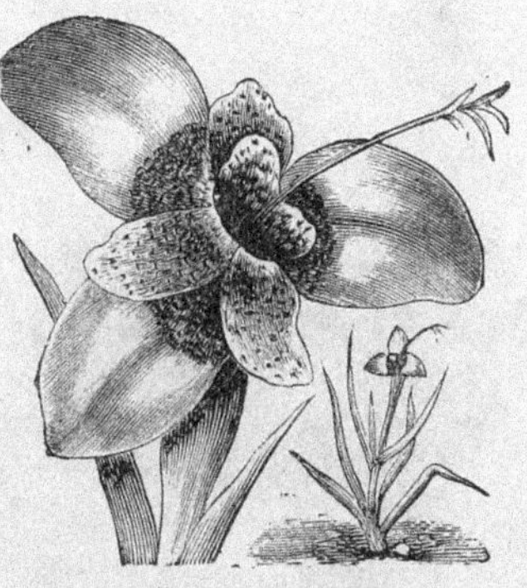

Flower of Tigridia Pavonia.

[*To face page* 280.

Trachelium cœruleum (*Blue Throatwort*).

Tradescantia virginica (*Virginian Spiderwort*).

Trapa natans (*Water Chestnut*).

Tricyrtis hirta.

Trifolium rubens.

Fruit of Trapa natans.

[*To face page* 281.

is a native perennial, found in marshy places.

Tolmiæa Menziesi is a curious rather than handsome Saxifrageous plant from Oregon, related to Tiarella. It propagates naturally and freely by adventitious buds, produced at the junction of the leaf-stalk with the blade in the manner of Begonia. For botanic gardens.

Tolpis barbata (*Yellow Hawkweed*).—A composite from S. Europe, with pale yellow flowers. It is rather a weedy plant, but a variety with white flowers is worth cultivating. It is a hardy annual, and grows in any soil.

Tommasinia verticillaris.—An umbelliferous plant for botanic collections.

Tooth Violet (*Dentaria bulbifera*).

Tovaria oleracea.—A handsome Liliaceous perennial plant, nearly related to Smilacina and similar to S. racemosa. The stems grow about 1½ ft. high and arch gracefully. The flowers are borne in broad racemes, about 4 in. across, at the apex of the stems. The colour is pure white, which contrasts with the purplish hue of the stems and flower-stalks. Native of Sikkim, and quite hardy.

Tournefortia heliotropioides.—A modest, half-hardy plant, from Buenos Ayres, much resembling the common Heliotrope, having pale lilac flowers, which are borne in clusters in much the same way. It should be treated as a half-hardy annual, sown in heat in early spring, and the seedlings planted out in May. It likes a good, deep, loamy soil and an open position in order to develop itself freely. As it is of spreading growth, the seedlings should not be placed too near each other.

Trachelium cœruleum (*Blue Throatwort*).—An attractive perennial from the Mediterranean region, belonging to the Campanula family. It is a sub-shrubby, much-branched plant, from 1 ft. to 2 ft. high, producing in summer broad clusters of small blossoms, which vary from blue in the type to white, and lilac in the varieties. It is not thoroughly hardy; therefore must be grown in only the warmest situations in dry borders, rocky banks, and old ruins or walls. It is an elegant plant for vases and such like purposes. It should be protected during severe cold. Propagated by seed or cuttings.

Trachymene cœrulea (*Didiscus*).

Tradescantia virginica (*Virginian Spiderwort*).—Some of the hardy Tradescantias are pretty plants, but this is by far the best, and with its varieties represents all the beauty of the others. It is a distinct and valuable perennial, growing from 1 ft. to 2½ ft. high, flowering abundantly during summer. Besides the type, which has showy purple-blue flowers, there are several varieties, one with double violet, one with single rose coloured, one with single lilac, and one with single white blossoms. All are well worth growing in mixed borders of plants. They will grow in any soil, from that found ordinarily in gardens to the wettest clays. They are suitable for the mixed border, margins of shrubberies, the rougher parts of extensive rock gardens, and for naturalising in the wild garden. They may be easily propagated by division in spring. Among the other kinds in cultivation are T. Barclayana, iridescens, and subaspera, all desirable where a full collection is required.

Tragopogon.—A genus of composites of botanical interest only.

Trapa natans (*Water Chestnut*).—An interesting aquatic plant, native of the south of Europe. Its hardiness is doubtful.

Trautvetteria palmata.—A plant of the Crowfoot family. For botanical collections.

Treacle Mustard (*Erysimum*).

Tree Mallow (*Lavatera arborea*).

Trichomanes.—Beautiful Filmy Ferns not adapted for open-air culture. They require a special place under glass, so that they can at all times be kept growing in an atmosphere heavily charged with moisture, and be kept shaded. In the mildest parts, perhaps, the Killarney Fern (T. radicans) and similar kinds would succeed.

Trichonema (= Romulea).

Trichosanthes (*Snake Gourd*). — Cucurbitaceous plants, producing singularly-shaped fruits like the Gourds, but better suited for culture in hothouses than in the open air.

Tricyrtis hirta. — A singularly interesting perennial from Japan. It grows about 3 ft. high, has slender, erect stems terminated by a few curiously-shaped blossoms, pinkish and copiously spotted with purplish-black. It is perfectly hardy, but flowers so late in autumn that it invariably becomes damaged by frosts. On this account the variety nigra is more desirable, as it flowers fully three weeks earlier, and the flowers, moreover, are distinct from, and more attractive than, those of the type. T. pilosa is dwarfer, but otherwise a similar, though rarer plant. Both thrive well in a moist peat border, partially shaded, and if somewhat protected, so much the better.

Trientalis europæus (*Star-flower*).—A delicate and graceful inhabitant of shady, woody, and mossy places, with erect, slender stems rarely more than 6 in. high, bearing from one to four slender flower-stems, each of which supports a star-shaped white or pink-tipped flower. With healthy well-rooted plants to begin with, it is not difficult to establish among bog shrubs in some half-shady part of the rock garden, or in the shade of Rhododendrons, &c., in peat soil. It is very suitable for association with the Linnæa, the Pyrolas, and Pinguiculas among mossy rocks. Flowers in early summer, and is increased by division of the creeping root-stocks. Native of Europe, Asia, and America.

Trifolium (*Trefoil*).— Though a large genus, there are very few garden plants among the Trefoils, and these few are not generally known. There are among them a few dwarf creeping alpines that are desirable for the rock garden. Of these the best are T. uniflorum, a neat trailing plant with pink and white flowers, borne singly and larger than those of any of the genus. They are studded profusely over the surface of the plant, rendering it a very striking object. Of the easiest culture, delighting in an exposed position on the rock garden and an open space on which to creep. T. alpinum is a stout spreading kind 3 in. to 6 in. high. The flowers, which appear in summer, are large, but not brilliant; the upper petal is flesh coloured and streaked with purple. Suitable for the rock garden and margins of borders. Another kind is T. repens purpureum. Among the taller kinds suitable for the mixed border is T. rubens, a stout perennial, about 1 ft. high, with large dense heads of carmine flowers produced in early summer. It grows almost anywhere, seeming to prefer dry, calcareous, marly or gravelly soil, and is therefore well adapted for naturalisation on arid declivities with a southern aspect. T. pannonicum, with creamy white flowers, is another ornamental plant. All may be readily propagated by division or seed.

Trigonella.—A genus of the Pea family, chiefly annuals, and of no ornamental value.

Trillium *(Wood Lily)*.—A small genus of perennial plants of low growth, all inhabiting the shady woods of N. America. There are several species in cultivation, the finest by far being T. grandiflorum (White Wood Lily) which is one of the most singular and beautiful of all hardy plants. It grows from 6 in. to 1 ft. high, and when in good health, each stem bears a lovely, white, three-petalled flower, fairer than the white Lily, and almost as large when the plant is strong; but much depends on the vigour of the specimens. It thrives under almost any kind of treatment, and becomes a free-growing herb of goodly size in a shady, peaty border in the open air. If placed in a sunny or exposed position, the large soft green leaves are not sufficiently developed, and consequently the plant fails to become strong. Depressed shady nooks in the rock garden or hardy fernery will suit it admirably. It is now sold cheaply. There are several other species in cultivation—T. atropurpureum, erythrocarpum, sessile, and pendulum, none of them equal to T. grandiflorum, but some are pretty, and all interesting.

Triteleia uniflora *(Spring Star-flower)*. —A delicately-coloured, free-flowering, hardy, bulbous plant, 4 in. to 6 in. high. The flowers are of an iridescent white with bluish reflections, and marked on the outside through the middle of the divisions with a violet streak, which is continued down the tube. They open with the morning sun, are conspicuously beautiful on bright days, and close in dull and sunless weather. The plant comes into flower with or before Scilla sibirica, and remains during the last days of April still in effective bloom, when the vivid blue of the Squill has been long replaced by green leaves. It flowers profusely in pots, and even when placed in clay in a most unfavourable position it will flower boldly. There are several forms which differ in the shade of their flowers. Associated with the best Scillas, Leucojum vernum, Iris reticulata, dwarf Daffodils, and the like, it forms a charming addition to the select spring garden, and is equally useful for rockwork, borders, or edgings. Native of Mendoza, South America. Triteleia (Leucocoryne) alliacea is a nearly allied plant, but scarcely so ornamental; it will thrive under similar circumstances. For other species see Brodiæa.

Triticum.—A genus of Grasses of no garden value.

Tritoma *(Kniphofia)*.

Tritonia aurea.—This beautiful bulbous plant from the Cape is so common as not to need description. Though usually grown as a greenhouse plant, it is a valuable open-air flower. The bulbs should be bought as soon as they can be got, and should be out of the ground for as short a time as possible. They succeed well as border plants in any soil except clay, but seem to like moist beds of peat soil best; they may be planted out in April or May. Sometimes the shoots are killed to the ground by frost in May without injury to the bulb, though of course the growth is thereby weakened and retarded. Treated as cold frame bulbs and planted out, they will flower in the south in August, and in the north of England a month later. In cold places it is better to bring them on for a longer time under glass and plant them out later, or to treat them altogether as greenhouse bulbs, for which they are well suited. Though tolerably hardy, there are two objections, besides the lateness of their flowering, to leaving them out all winter in the open ground—one, that they are liable to be killed in severe winters unless well protected with litter; the other, that, owing to their habit of straying, in which they much resemble Lilies of the Valley, they are apt to leave the place in which they were planted, and come up where they are not wanted. It is better, therefore, to lift the whole stock in autumn, say about the middle of November, and as, if necessary, many may be potted together, and separated when planting-out time comes, this takes little trouble or room. The lifting must be done carefully, for in September, although the tops are not dead, the shoots that form the new bulbs next year are already several inches long, sometimes even more than 1 ft. The old bulbs can readily be distinguished by having no stalks or shoots, and may be pulled off and thrown away. It will be understood after what has been written above that anything like drying off or storing the roots in a dry place is fatal. They had better not be left uncovered for a single day.

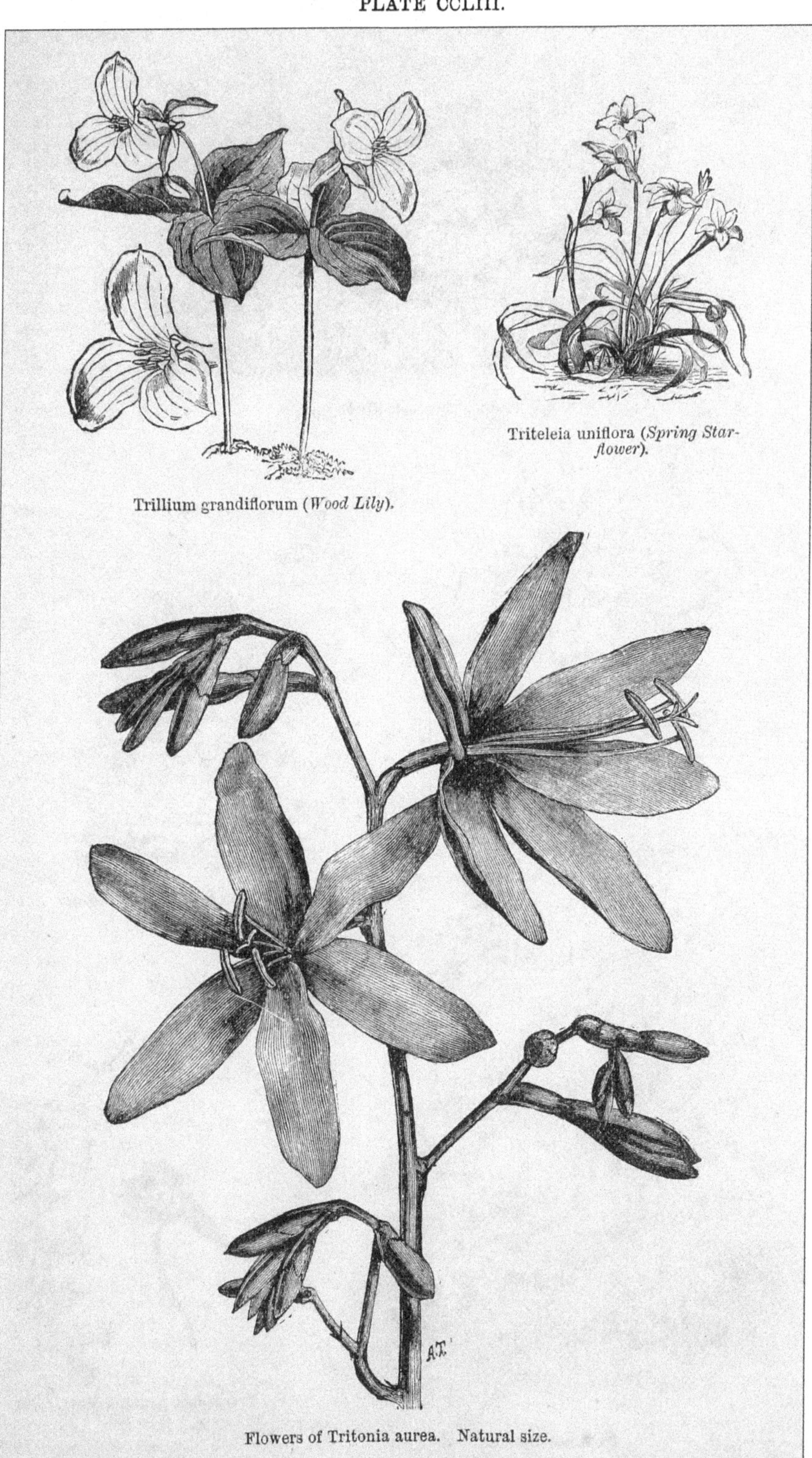

Trillium grandiflorum (*Wood Lily*).

Triteleia uniflora (*Spring Star-flower*).

Flowers of Tritonia aurea. Natural size.

[*To face page* 282.

Trollius europæus (*Globe-flower*).

Flower of Trollius asiaticus.

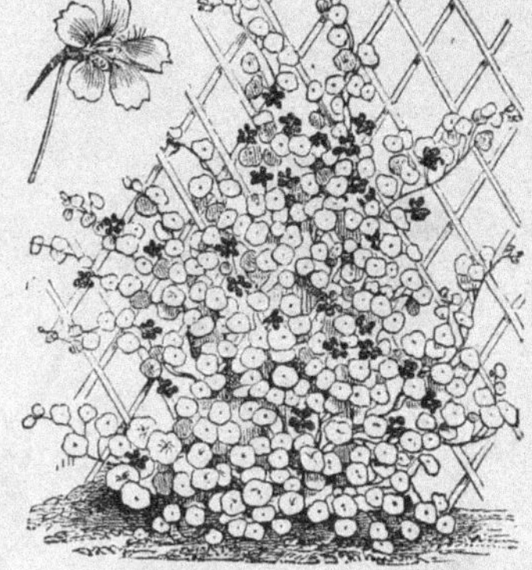

Tropæolum majus (*Nasturtium, Indian Cress*). See page 284.

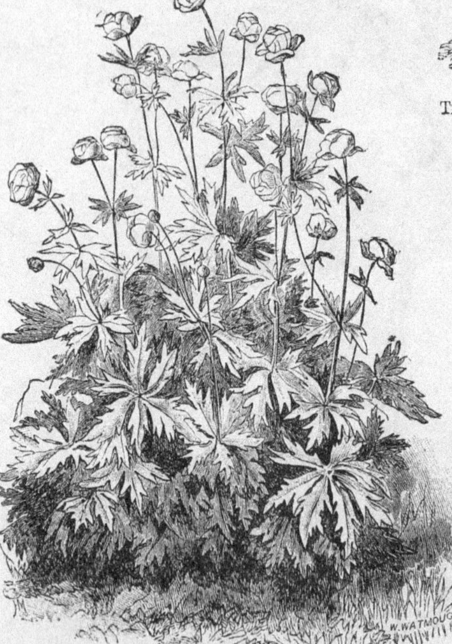

Trollius dauricus.

Tropæolum pentaphyllum. See page 285.

None of the other Tritonias are hardy enough for open-air culture. T. aurea is also known as Crocosma aurea.

Trollius (*Globe-flower*). — Few plants are more desirable for our gardens than Globe-flowers. All the species, though varying considerably in height, are of compact habit, and with them such things as stakes, or other modes of support, need never be resorted to. They may be grown in beds, borders, on lawns, by the side of ponds or streams, and any one not possessing a garden may grow them well enough in deep pots. All are of a dense habit of growth, the foliage and flowers rising from an underground crown, which does not possess the rambling proclivities that mar the value of many an otherwise good herbaceous plant. The roots are numerous and deep searching, especially in a border where perfect drainage removes the water-level to a considerable depth. The flowers present considerable variation in colour, from a pale yellow to deep golden, almost bordering on vermilion. All the species are spring or summer bloomers, and are at their best in April, May, and June. Occasionally in old-established plants a few autumnal flowers are developed in September and October, but these depend alike on the season and the strength of the plant itself. They may be most readily increased by division of the root, which operation should be performed either in September or in March; the former is to be preferred, as the plants have then an opportunity, during the remaining portion of the autumn, of making fresh roots and thoroughly repairing, before the dry early summer sets in, the damages unavoidable in the operation. When divided in March, a few dry days, accompanied by sunshine, will cause the foliage to be uncomfortably prostrate on the ground, and the blossoms are equally certain to be puny and short-lived. They may also be propagated by seeds, which in vigorous well-established plants are produced freely, and generally retain, with marked persistency, the specific characters. It is to be noted, however, that they rarely vegetate the same year that they are sown, but come up vigorously the following spring, and, if carefully attended to, will make fine flowering plants the second season after vegetating; they will not, however, attain their full development until the fourth year or even later. They grow freely in any soil, are partial to a good stiff loam overlying a cool moist subsoil, but, if cultivated in a dry situation, should have a good supply of manure, not only to act as a stimulant, but as a mechanical and moisture-retaining element in the soil; for it should be remembered that the mountain meadows they affect are almost invariably supplied with cool water springs below, which enable the plants to withstand the burning heat of an uninterrupted day's sunshine without showing, by flaccid leaves, any indication of exhaustion.

T. europæus (*Mountain Globe-flower*) is the native species with which we are all familiar. Its height is about 15 in.; the flowers are of a lemon-yellow colour, forming a perfectly globular flower from 1 in. to 2 in. in diameter. This plant is of common occurrence throughout the upland meadows of Europe, and is by no means an especial native of Britain alone. T. albus is considered by some as a mere variation of T. europæus; it is occasionally met with under the title of T. pumilus. The flowers are of a pale lemon colour, but not white, as the specific name would indicate. T. albus is dwarf in stature as well as small in development. The plant, which, whether known by the white or dwarf titles, is especially fond of peat soil—more so, perhaps, than any other species of the genus—is exceedingly compact, and distinct in the arrangement of its leaves. These, like its blossoms, are scarcely half the size of the common European species.

T. dauricus is distinguished from the foregoing by its gigantic growth, both in respect of plant development and bloom. Its leaves are large, much divided, of a deep olive-green, and supported on long foot-stalks. The flowers are large and lemon coloured, the sepals overlapping one another in a globose form. It is a native of Dahuria, and is a most desirable plant. It seeds freely, and may by this means be readily increased, perfectly true to its type. In heavy soil, which appears to be admirably adapted to the requirements of the genus, it attains the height of 3 ft. or more. It is identical with the species known by the names of T. giganteus and T. Demayanus.

T. asiaticus (*Asiatic Globe-flower*) has leaves that are much more divided and also distinguished by their bronzy green colour; the flowers are similar in size to those of the last species, but of a golden-yellow and not globular, the sepals, as they reach maturity, expanding. When growing vigorously it attains a height of 18 in.; it flowers in the early part of May, rarely perfects its seeds, but is readily increased by root division. It is a native of Siberia.

T. napellifolius is a handsome strong-growing species; the flower is a globose, deep yellow, bordering upon orange, and more than 2 in. in diameter. It is a native of Central Europe, where, on the slopes of the Carpathian Mountains, it forms a very conspicuous object, and in our herbaceous borders it carries off the palm as unquestionably the most showy of the genus.

T. sinensis is a very distinct plant, a native of Japan and China. The flower-stems attain a height of 2 ft. or 3 ft. The flowers are of a deep yellow, the sepals partially expanded. It blooms in the month of July, and in this respect bears some affinity to the American variety of the Asiatic Globe-flower, to which it is in some degree related.

Besides the foregoing, we have T. altaicus, T. caucasicus, T. intermedius, T. tauricus, and T. medius, all of which bear a very close relationship to either the European or Asiatic forms. So slight, indeed, is the difference, that in the species above enumerated the ordinary cultivator will find all that is required so far as the genus can supply towards the decoration of the early summer borders.

Tropæolum *(Nasturtium)*.—This genus comprises about forty species, all of which are almost confined to the mountainous region on the western side of South America, from New Granada to Chili. Very few of them descend into the tropical zone, and therefore they do not require great heat, which is, indeed, rather unfavourable to them. On the other hand, the first frost cuts most of them to the ground. They love a half-shaded situation, and all of them will succeed in the open air in the summer. It is quite unnecessary to dwell on their merits as ornamental plants, because three or four species and their hybrid varieties are familiar and fully appreciated. We would only call attention to the fact that there are many other species equally, or even more, attractive than the better-known ones. There are annual and perennial species, and the latter may be subdivided into two groups, the one with fibrous and the other with tuberous roots. The rapid growth of the annuals, T. majus, minus, &c., is proverbial, and their hardiness in a temperature above the freezing point, as well as their indifference to the nature and quality of the soil, recommends them where anything unsightly is to be hidden. All are not so easy of culture as the annual species, at least in our climate. Among the numerous species in cultivation the following are the most fitted for the open air:—

T. aduncum *(Canary Creeper)*. — Undoubtedly the favourite of the genus, a distinction it fully merits, as it is almost without a rival in lightness and elegance; indeed, one might say that it is peerless among yellow flowers. The yellow is of the most pleasing shade, having more of the green than the red element in it. Its precise home is uncertain, as it is now spread from Mexico all through the eastern side of South America to Chili; but it has doubtless spread from the Andes. This popular annual is a valuable garden plant, as it accommodates itself to any conditions, thriving in sun or shade, but best in a position with a north aspect. For festooning trellises, arbours, shrubs, &c., it is unsurpassed, as it rarely fails to succeed well and produce a profusion of its pretty yellow blossoms. Seeds should be sown in the open ground in April in sandy loam. It is known also as T. peregrinum and T. canariense.

T. Lobbianum.—This is a most beautiful annual species, of vigorous climbing growth, and easily recognised from the old T. majus by being more or less hairy. There are many varieties of it, differing chiefly in the colours of the blossoms, which are mostly yellow, scarlet, and crimson. Of named kinds there are no finer forms than those known in seed shops as Coronet, a large yellow sort; Perfection, rich crimson-scarlet; and Octoroon, deep mulberry. These are all great improvements on the old Nasturtiums, and those who have greenhouses will find the scarlet kind useful for winter decoration in pots. T. Lobbianum is an excellent plant for clothing unsightly spots or for providing temporary shelter during the summer time. It may be readily raised from seeds sown about the middle of April where the plants are required, and all the after-culture needed is merely the guiding (it can hardly be called training) of the leading shoots in the direction in which it is wished they should grow, the rest being left to Nature. This Tropæolum has a most pleasing effect when the seed is sown here and there amongst shrubs in the back part of the border. As the plants grow they attach themselves to the bushes, and, climbing over or through them, throw out in all directions wreaths of lovely blossoms which last in beauty until cut down by frost. When that takes place the haulm should be pulled up and removed, and thus any unsightliness is speedily obviated. Temporary floral fences may also be made with this plant with good effect. All that is required is a row of Pea stakes for the support of the shoots, and it may also be made to assume a pyramidal form by allowing it to overrun the dead tops of young Fir trees, which it will do most effectually, provided they are not too large. In short, there is no end to the uses to which these climbing Tropæolums may be put.

T. majus *(Large Indian Cress or Nasturtium)* is too well known to need description. It differs from T. minus in its larger growth, and from T. Lobbianum in the absence of hairiness. There are many beautiful varieties, principally the result of hybridising it with Lobb's Nasturtium and others. The principal named sorts in seed catalogues are atrosanguineum, dark orange-crimson; citrinum, or Shillingi, pale yellow; and Scheuermannianum, pale yellow with dark spots. There is a double variety, too, which is likewise very handsome. These climbing sorts are useful for the same purposes as T. Lobbianum, and require similar treatment. The most important varieties of this plant are the dwarf or Tom Thumb strain, of which there is a large number of sorts, the most valuable being what is called the compactum strain. Few annuals come into flower more quickly than these dwarf Nasturtiums, and few bloom more constantly or longer. Such rich, bold colours as are found in the compactum strains are marvellously effective when seen in masses, and if the soil be somewhat poor, the mass of bloom is so much the greater. The forms of compactum are so constant, that they are never without flowers from first to last. In this respect these outshine the

Tropæolum Lobbianum.

Dwarf Tropæolum majus.

[To face page 284.

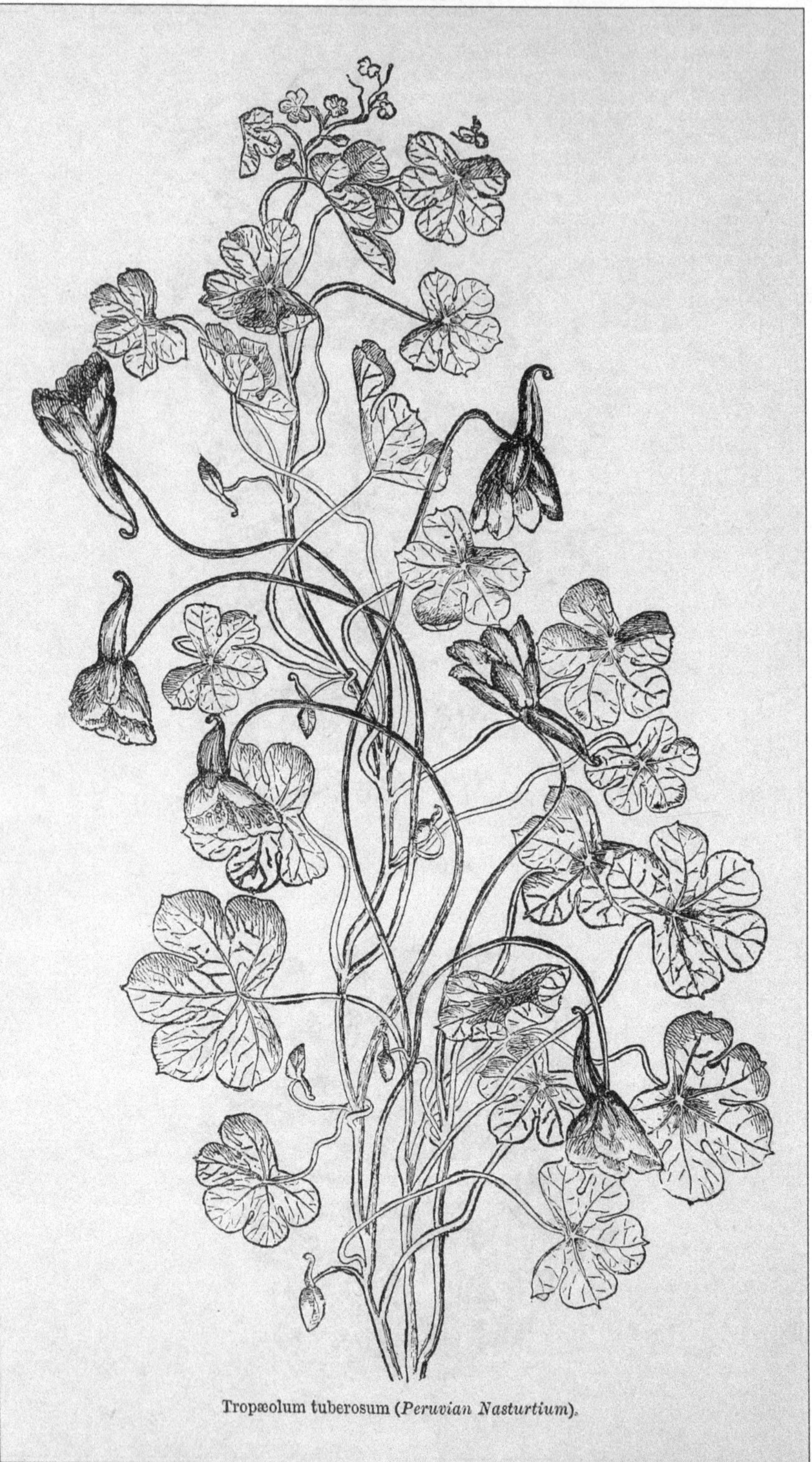

Tropæolum tuberosum (*Peruvian Nasturtium*).

[*To face page* 285.

Tom Thumb kinds, which have a short season of flower and then collapse. They can be had good from seed, and generally true, but it must be understood that actual truthfulness as regards varieties can only be ensured by means of propagation by cuttings. Fortunately, cuttings are furnished by the plants in great abundance, and these will strike freely anywhere under glass if put in about the middle of September. A few dozens put into store pots will yield a large number of cuttings in spring, and these spring-struck plants make the best for summer bedding. Plants from cuttings also make less leafage, furnish more bloom, and, curiously enough, produce little seed, but what is so produced may be relied upon to fairly reproduce the parents if no other kinds of Nasturtium are growing near. The Tom Thumb kinds may be sown in the open ground in spring, but the compactum race grow too freely when so raised, and are best sown under glass and then planted out. Treated thus, they bloom earlier, are more compact, and can be planted out to succeed spring bedding plants. All who love rich masses of colour will find these dwarf Nasturtiums or Tropæolums worth cultivation. The varieties are so numerous and at the same time so uniformly beautiful, that it is a difficult matter to make a selection, for the catalogues of the various seedsmen abound with names of particular and favourite sorts. Of T. minus also there are several varieties, including a double flowered one, but these are less desirable than T. majus.

T. pentaphyllum.—A rapidly-growing climber, 6 ft. to nearly 10 ft. high. The flowers, which are yellowish-red, are produced abundantly in summer. It is an admirable plant for covering pillars, walls, chains, bowers, &c., and does best in light and warm loams or calcareous soils. It revels in sunshine, and does splendidly on the south end of a greenhouse or other warm aspect. Division or seed. Chili.

T. polyphyllum.—This is one of the most valuable hardy plants ever introduced, not only for its freedom of growth and flower, and the readiness with which it may be grown, but also for its picturesque way of growth, for, while its foliage may form a dense carpet over a bank, the wreaths of flowers usually throw themselves into irregular windings and groupings. It is a very distinct-looking subject whether in or out of flower. The leaves are glaucous, almost Rue-like in tone, and cut into fine divisions or leaflets. When planted on a warm, sunny rockwork the stems creep about, snake-like, through the vegetation around, some to the extent of 3 ft. or 4 ft. The flowers are a deep yellow, and produced as freely as the leaves. It is a tuberous-rooted kind, quite hardy in dry situations on rockwork and sunny banks, where it should not be often disturbed. It springs up early, and dies down at the end of summer. Native of the Cordilleras of Chili.

T. speciosum *(Flame Nasturtium)*.—A splendid creeping plant with long and elegant annual shoots, gracefully clothed with six-lobed leaves, from the axils of which spring such brilliant vermilion flowers that a long shoot of the plant is startlingly effective, especially if seen wandering alone among Ivy leaves or among verdure of any kind. It has been introduced a considerable time from South America, but, notwithstanding its graceful beauty and perfect hardiness, it is but very little grown or known, especially in Southern England. It is impossible to find anything more worthy of a position in which its shoots may fall over or climb up the face of some high rock or bank in the rock garden; it is also suitable for an open spot in the hardy fernery, or for any other position in which its peculiar beauty may be seen to full advantage. It is very beautiful when clambering through evergreen shrubs, nailed against terrace walls. It enjoys a deep, rich, and rather moist soil, and apparently flourishes best in cool moist places, or in those near the sea. No pains should be spared to establish this plant in a vigorous condition. When a position is selected for it, the soil should be made light, and deep, and free by the addition of leaf-mould, peat, fibry loam, and sand, as the nature of the ground may require, and the surface should be mulched in summer with an inch or two of decomposed manure or leaf-mould to prevent excessive evaporation; but whatever kind of manure is used, it must be well decayed. The young plants are best planted in spring, the roots inserted 6 in. or 8 in. in the soil, and they should be well watered. They will also enjoy a deep bed of manure beneath the roots. It is best planted where the shoots may ramble among the spray of shrubs, or Ferns, or trailers, but as it must in the first instance be placed on a cleared spot, it is well to put a few branchlets over the roots so that the young shoots may crawl over them when they begin to grow. When established they may be allowed to take care of themselves, and it is much better to let them have their own wild way than to resort to any kind of staking or support, except that afforded by other subjects growing near. It may be propagated by division or seed. The latter should be sown as they are ripe, in light loam, leaf-mould, and sand in a pan or box; they must be placed in a pit or frame, and the soil kept moist, but not wet, until the plants make their appearance in spring. The careful division of the old roots is much the best and easiest method of propagating this Tropæolum.

T. tuberosum.—A distinct and beautiful climbing species with tuberous roots. The slender stems grow from 2 ft. to 4 ft. high, and bear in summer a profusion of showy scarlet and yellow flowers on slender stalks. It should be grown in open spots in the poorest of soils, and the branches may either be supported or allowed to trail along the ground. As it is not perfectly hardy, the tubers should be lifted in autumn, stored in a dry place through the winter, and planted in spring. Native of Peru.

Truffle Sunrose (*Helianthemum Tuberaria*).

Trumpet Weed (*Eupatorium purpureum*).

Tulipa (*Tulip*).—This is a large genus of bulbous plants, but comparatively few species are in general cultivation. This, no doubt, is owing to the fact that most of them are inferior in beauty to the numerous varieties that have so long been such favourite garden plants, and which have been derived principally from T. Gesneriana, T. suaveolens and T. præcox. These varieties of Tulip are so valuable that no garden should be without them, as indeed very few are, particularly as their culture is very simple. With regard to the early flowering kinds the bulbs are usually taken up and stored away in summer, but these, if left undisturbed in a warm, sunny border, increase rapidly and produce plenty of fine flowers. This cannot always be done, however, and therefore the best way is to have a reserve garden to which the Tulips can be removed when it is time to commence the summer planting. Here the bulbs will not only mature themselves, but throw off offsets which in two or three years make fine flowering bulbs. A good light and moist sandy soil is the kind of home best fitted for the Tulips during the summer. The blooming season is not nearly so short as is generally supposed, for between the earliest and the latest flowering kinds a considerable time intervenes. But beds of Tulips should be carpeted with small tufted or creeping plants, and there are many hardy flowering and foliage plants suited for the purpose. The White Rock Cress (Arabis albida), together with its variegated form, the Aubrietias, Hepaticas, Primroses, Cowslips, the drooping Catchfly (Silene pendula), Pansies, early-flowering Violas, Saxifrages, the white Iberis corifolia, Sedum acre aureum, the pretty creeping Ajuga reptans rubra, and many others make excellent carpets for beds of bulbs. When a collection of Tulips is sufficiently large to admit of its being done it is a good plan to rest the bulbs every third year by preventing them from blooming. They occupy but a small space in a reserve garden, and can be planted quite thickly. In order to grow Tulips to perfection, a light, rich, well-drained soil is required, but as their roots rarely extend more than 9 in. or 10 in. from the bulb, a position is easily made for them. They should be planted with from 3 in. to 4 in. of soil above the crown of the bulb; if planted nearer the surface and a severe winter follows they are liable to injury.

The late flowering Tulips are chiefly descendants from T. Gesneriana, itself a very handsome plant in the wild state, particularly its variety fulgens, which has very large cup-shaped flowers of a glossy deep crimson. For centuries this class of Tulips has been cultivated, and at one time and even still are classed among florists' flowers. They are now divided into four sections, viz., breeders, or self-flowers, bizarres, bybloemens, and roses. When a seedling Tulip flowers for the first time, it is usually a self, and in the course of a few years (but occasionally as long as thirty years) they will break into the flamed or feathered state. A feathered Tulip has the colour finely pencilled round the margin of the petals, the base of the flower being pure; in the bizarre it should be clear yellow, and in the rose and bybloemens white. In the flamed flower stripes of colour descend from the top of the petals towards the base of the cup. The colours in the bizarres are red, brownish-red, chestnut, and maroon; in the bybloemens, black and various shades of purple are the prevailing colours; and in the roses, they are rose of various shades and deep red or scarlet. The Tulip requires a rich soil and the careful hand of the gardener; the bulbs should be planted early in November, and although they are sufficiently hardy to withstand our most severe winters, they must be protected either by canvas or glass frames as soon as the flowers and buds are observed, otherwise they are sure to be injured by wet lodging in the axils of the leaves, followed by frosts. The succulent leaves are also easily broken by high winds. Some persons cover the beds entirely with glass; others have stout canvas fixed to rollers to run up and down. The object aimed at should be to expose the plants to light and air on every favourable opportunity, but to shelter them from high winds, heavy rains, or hail storms, and to cover the beds on frosty nights. In planting the bulbs it is usual to put a little clean sand around them. Although many varieties are of a tall habit, and the flowers are heavy, the stems are usually strong enough to support them without sticks. The time of lifting the bulbs should be fixed by the condition of the flower-stems; when these will bend without breaking, they may be taken up, dried, and stored away until planting time.

Among the wild types of Tulip there are some really beautiful plants quite distinct from the garden varieties, but a few only are worthy of mention here, though many others might well be included in a full collection. Many of the European and some of the Asiatic species succeed perfectly in the open border, but it is best to lift and replant the bulbs every year or two. The only mode by which many of the species can be propagated is by seed, as offsets are produced very sparingly, and in some cases hardly ever; the seed, if sown when ripe, germinates in spring, and produces full-sized bulbs in six or seven years. The finest and most distinct among the species for general culture are those mentioned below.

T. Celsiana.—A species whose bright yellow flowers, much smaller than those of the common bedding Tulips, sometimes, when in clumps, remind one of a yellow Crocus;

PLATE CCLVII.

Tropæolum aduncum—syn., T. canariense (*Canary Creeper*). See page 284.

Tulipa suaveolens.

Tulipa præcox.

[*To face page* 286.

Tulipa Gesneriana.

Tulipa Greigi (*Turkestan Tulip*).

Tulipa sylvestris (*Wild Tulip*).

[*To face page* 287.

the outside of the petals is tinted with reddish-brown and green. It begins to flower about the 1st of May, and usually attains a height of 6 in. to 8 in., and sometimes 12 in. The bulbs emit stolons after flowering. Comes from Southern Europe and the shores of the Mediterranean, and is well suited for rock-work or choice borders in well-drained sandy soil.

T. Clusiana is delicate in tone, humble in stature, and modestly pretty in appearance. The bulbs are very small, the stem reaching from 6 in. to 9 in. high, seldom more, and sometimes flowering when little more than 3 in. high. The flower is small, with a purplish spot at the base of each petal; the three outer divisions of the petals are stained with a pleasing rose; the three inner ones are of a pure transparent white. It is a native of the south of Europe, a little more delicate than most of its family, and requires to be planted in good light vegetable earth in a warm, sheltered, and well-drained position. Although so small, it will be the better of being planted rather deeply, say at from 6 in. to 9 in., and of being placed in some snug spot where it need not be disturbed too often. Readily known from other species by the peculiarity of its colouring, and well adapted for the rock garden or the collection of hardy bulbs. T. stellata, a North Indian species, is somewhat similar, and a very beautiful plant. Its large white blossoms appear in early summer. It is somewhat tender and needs protection.

T. Greigi.—Of all the known species of Tulip this is perhaps the most showy and desirable as a garden plant. It blooms freely in April or May, its large goblet-shaped flowers being generally of a vivid orange-scarlet colour; but there are also purple and yellow-flowered forms. The bulbs are so extremely hardy that they will withstand freezing and thawing with impunity, and even when the leaves are half-grown they will endure a temperature as low as zero without any protection. The plant is a vigorous grower, attaining a height of from 9 in. to 15 in., and bearing flowers from 4 in. to 6 in. in diameter when fully expanded, and three or four lance-shaped, glaucous leaves with undulated margins, the whole of the upper surface being boldly blotched with purple or chocolate-brown. Varieties occur without spots, and others with yellow and spotless flowers. It grows freely in any light rich soil in an open, sunny position, and rarely requires transplanting. A native of Turkestan.

T. Oculus-solis is a handsome plant, with brilliant scarlet blossoms that have black spots at the base of the interior. The varieties maleolens, præcox, and Strangwaysi are also very showy. Of a similar character are T. altaica, Fransoniana, undulatifolia, elegans, and bœtica.

T. pulchella is typical of a numerous race of beautiful dwarf-growing species, which, though rare at present, must eventually find their way into all good gardens, for they are so extremely beautiful when seen in full blossom in a warm, sunny border. T. pulchella has pretty purplish-pink blossoms, on stems some 2 in. or 3 in. high; T. saxatilis, pale magenta flowers, with bright yellow centres; T. biflora, white with yellow eye; T. sylvestris, our only indigenous Tulip, is a beautiful species, with showy blossoms of a clear yellow colour, produced in early summer; Biebersteiniana, also with yellow flowers; Orphanidea, salmon colour, shaded with purple; and tricolor, white on the inside, and yellow at the base. Besides these there is a host of others not generally known, but which, on account of the beautiful blossoms and simple culture, are very desirable for a garden.

Tunica Saxifraga.—A small plan with narrow leaves and a profusion of wiry stems that bear elegant rosy flowers, small, but very numerous. It thrives without particular care on most soils, and forms tufts a few inches high. It is a native of arid, stony places on the Pyrenees and Alps, often descending into the low country, where it is found on the tops of walls. There can be no doubt that it will grow in like positions in this country, and also on ruins, while it is a neat plant for the rock garden or the margin of the mixed border. It is not unlike a Gypsophila in appearance, is easily raised from seed, and thrives in poor soil.

Tupa. — Perennial plants from Chili, much resembling the Lobelia; indeed, so much alike are they, that some place them in the same genus. There are two kinds in gardens, but neither is perfectly hardy, and both must be treated as half-hardy subjects. T. Feuillei, the best known, grows from 2 ft. to 4 ft. high. The flowers appear in rather dense, erect spikes or racemes, terminating each stem; they are brilliant scarlet, and appear in September and October. The plant is also known as Lobelia Tupa. T. Bridgesi grows about 4 ft. high, with pink flowers in dense racemes, appearing in late summer. Both of these must have a warm and sheltered position in the open air, and the soil must be light and warm. In autumn the plants should be lifted, potted, and protected through the winter in a greenhouse. Both may be propagated by division.

Tussilago (*Coltsfoot*).—The variegated form of the common Coltsfoot (T. Farfara) is an ornamental plant, perfectly hardy. It increases itself, like a Nettle or Couch Grass, by running underground, is rather troublesome in spreading, and not easily eradicated once it gets a footing. It may be used with good effect in shady positions where other plants will not thrive. It does well as an edging to a clump of Ferns, or as a groundwork to plants with graceful foliage. Another desirable species is T. fragrans, named the Winter Heliotrope, on account of the delicious fragrance of its spikes of purplish flowers. It is of the easiest cultivation, as it spreads almost as freely as the irrepressible Coltsfoot, and is quite hardy. A waste corner cannot be better occupied than by this delicious flower. It grows well in the shade, but flowers best in rather sunny places. The best place is the home wood or wild shrubbery. It rarely blooms freely if overhung by evergreen trees or shrubs. In midwinter its flowers are useful for room adornment.

Twig Rush *(Cladium Mariscus).*

Twin-flower *(Linnæa borealis).*

Twining Fumitory *(Adlumia cirrhosa).*

Twin-leaf *(Jeffersonia diphylla).*

Typha latifolia *(Reed Mace).* — A native aquatic plant, growing in tufts of two-rowed flat leaves from 1½ ft. to 2 ft. long and 1 in. or 1½ in. wide. From the centre of each tuft springs a stem 6 ft. or 7 ft. high, which in the flowering season is terminated by a close cylindrical spike 9 in. long, of a dark olive colour, changing to a brownish-black as it ripens. This is one of the most striking and ornamental of our British water plants, and may be used with excellent effect grouped with such subjects as the Great Water Dock. T. angustifolia resembles the preceding in all respects except in the size of its leaves and spike. Of the two it is perhaps the more graceful in aspect. There are also one or two smaller forms, suited for collections of aquatic plants.

Uhdea bipinnatifida.—This sub-tropical plant has very handsome leaves, and is of a good habit. The leaves are of a slightly glaucous or silvery character. The plant is well suited for forming rich masses of foliage, and may be seen in the London parks planted in large beds along with the Polymnia, Ricinus, Cannas, &c. It has a fine effect when isolated on turf. It is easily propagated by cuttings taken from old plants, kept in a cool stove, greenhouse, or pit during the winter months, and placed in heat to afford cuttings freely in early spring. Under the ordinary treatment of cuttings on hotbeds or in a moist, warm propagating house, it grows as freely as could be desired, and may be planted out at the end of May or the beginning of June. U. pyramidata has been less cultivated in England than the preceding, from which it is distinct in appearance. It is of a lighter and fresher green, and inclined to grow larger in habit, having more of the aspect of a Malva in foliage. Useful for the same purposes as the preceding kind, but not so valuable. Both natives of Mexico.

Ullucus tuberosus.—A singular plant, which bears small tubers similar to Potatoes. It is interesting for botanical collections, but of no ornamental value. S. America. Portulacaceæ.

Umbilicus.—A small genus of succulent plants similar to the Houseleeks; few of them are hardy enough for open air culture. The best are here mentioned. U. spinosus is a very singular looking plant, with somewhat the appearance of a small Apicra or Haworthia. The leaves form a rosette like a Sempervivum, each leaf bearing a spine at the apex. The flowers, which appear early in summer, are yellow, and form a terminal cylindrical spike on the top of the flower-stem. It is a good plant for the rock garden, in dry, sunny spots, and is tolerably hardy, but its chief enemies are slugs, which destroy it whenever they have a chance. Native of Siberia, China, and Japan. U. chrysanthus is about 4 in. high, and resembles a small Houseleek; the flowers are yellowish, in short panicles. Suitable for the same positions as the preceding. U. Sempervivum strongly resembles some species of Sempervivum, forming a rosette-like tuft of succulent leaves, and producing in the second year of its growth a large cluster of pink flowers on a stem about 6 in. in height. It is useful for carpet bedding, and when used for this purpose the flower-stems must be pinched out. It is a native of Kurdistan, and is hardy in England if planted on rockwork or in well-drained soils.

Umbrella-leaf *(Diphylleia).*

Unicorn plant *(Martynia).*

Uniola latifolia.—A handsome perennial Grass from North America. It grows 2 ft. or 3 ft. high, and has a large, loose panicle, bearing large flattened spikelets. A clump placed in rich garden soil gathers strength from year to year, and when well established is a beautiful object. The loose, drooping panicles have a wonderfully graceful expression when living, and this is not lost when added to a winter bouquet.

Urginea maritima *(Scilla).*

Uropetalum serotinum *(Dipcadi).*

Urospermum Dalechampi. — A rather handsome composite from South Europe. It is of dwarf, tufted growth, and produces large heads of lemon-yellow blossoms. It thrives in any light soil in an open position, and is quite hardy.

Urtica *(Nettle).*—This genus contains no garden plants.

Uvularia.—A pretty genus of dwarf, slender plants, allied to the Solomon's Seal, but bearing yellow blossoms. There are four cultivated species, U. chinensis, grandiflora, puberula, and sessilifolia. Of these U. grandiflora is by far the finest plant; indeed, it is the only one worth growing generally. It attains a height of from 1 ft. to 2 ft., and the numerous slender stems form a dense, compact tuft. The flowers are long and clear yellow, gracefully drooping, and are very attractive in early summer. It is a capital peat border plant, and thrives best in a moist peaty soil, in a partially shaded place. It is a native of North America, as are all the others except U. chinensis.

Vaccinium Vitis-idæa *(Red Whortleberry).*—A dwarf British evergreen with Box-like foliage and clusters of small pale flowers. These appear in summer, and are followed by berries about the size of Red Currants, very like those of the Cranberry, on wiry stems from 3 in. to 9 in. high. It forms a

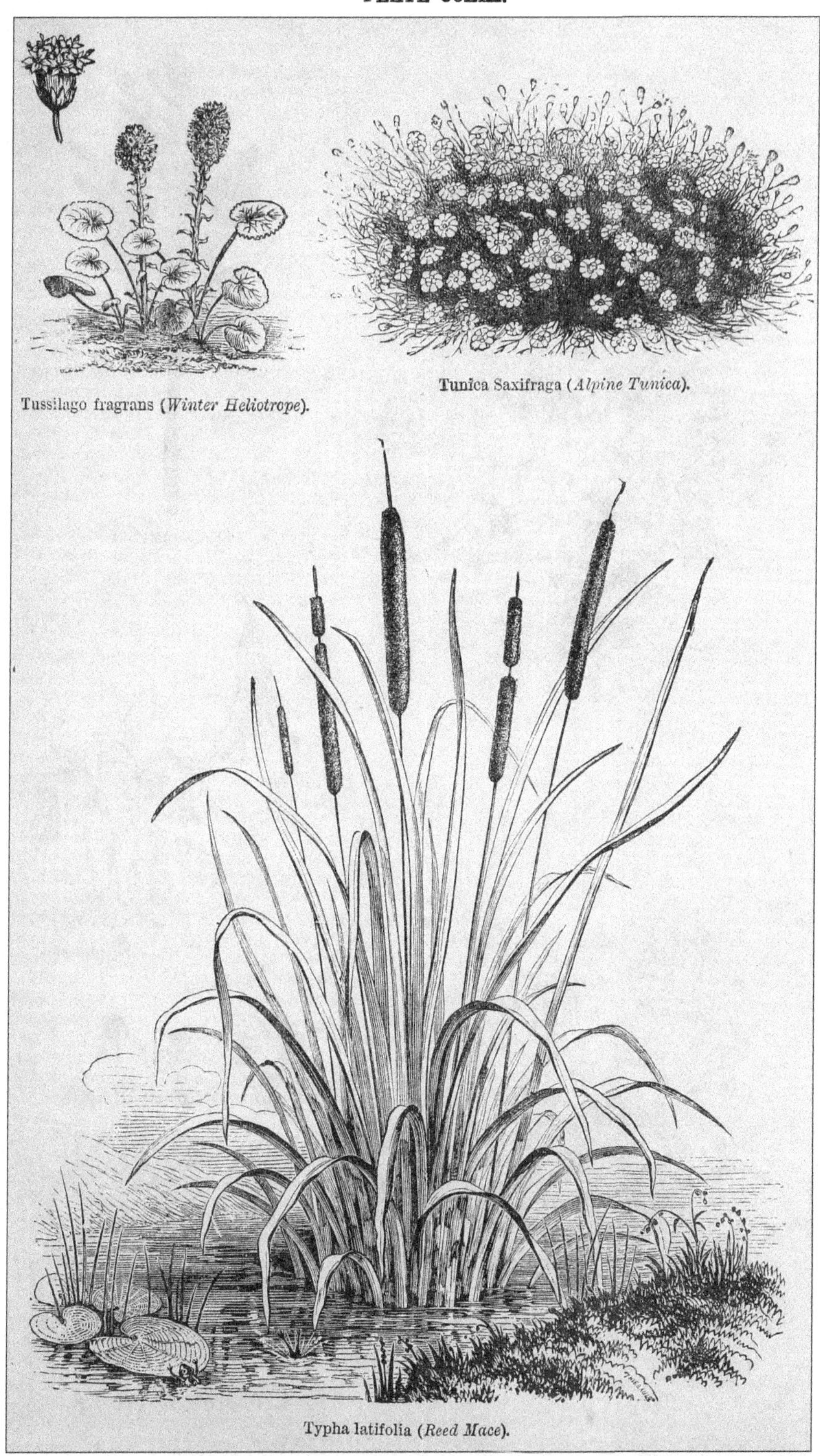

Tussilago fragrans (*Winter Heliotrope*).

Tunica Saxifraga (*Alpine Tunica*).

Typha latifolia (*Reed Mace*).

[*To face page* 288.

Uhdea bipinnatifida.

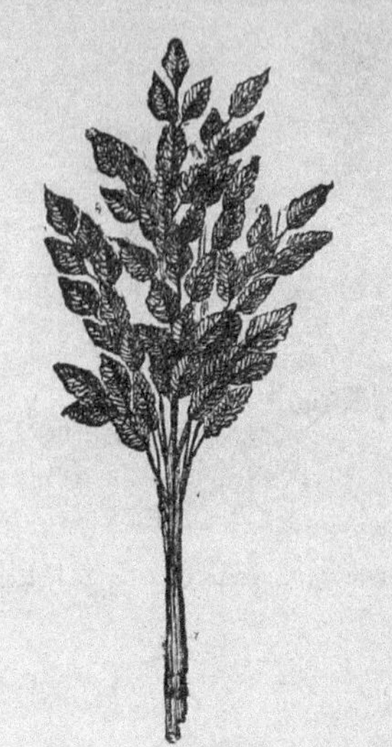

Uniola latifolia.

Uvularia grandiflora.

Umbilicus Sempervivum.

Valeriana officinalis (*Valerian*).

[*To face page* 289.

neat little bush on rockwork, or in beds in peat soil. The Marsh Cranberry (V. Oxycoccos), a native of wet bogs in Britain, with very slender creeping shoots and drooping dark rose flowers, requiring wetter soil than the preceding, is also worthy of a place where bog plants are admired. The American Cranberry (V. macrocarpum), a much larger plant, distinguished from the preceding by its oblong-obtuse leaves and very much larger fruit, is also worthy of a place in moist sandy peat, associated with bog shrubs. It fruits profusely in beds of peat soil.

Valeriana *(Valerian)*.—The only plant of this small genus worth cultivating in a general way is the golden-leaved variety of V. Phu, which is a highly effective plant, particularly in spring, when its foliage is newly developed. It is of tufted, neat habit, grows freely in any soil, and is perfectly hardy. There are a few dwarf alpine species sometimes grown in full collections, but none of them are very attractive, and the unpleasant odour of the flowers of most of them renders them still less desirable. Some of the larger species are pretty in rough places.

Valerianella *(Lamb's Lettuce)*.—Small plants that are not ornamental.

Valloradia plumbaginoides *(Plumbago Larpentæ)*.

Vallota purpurea *(Scarborough Lily)*.—This brilliantly-flowered bulbous plant, so well known in gardens, is hardy in mild climates if planted in warm situations in light soil, such as at the foot of a south wall. In such positions it often thrives better than when grown in pots under glass, but under such circumstances the bulbs must have protection during severe frosts. The outdoor culture of this bulb deserves more attention than it has hitherto had. Some flowers sent us by Mr. Kingsmill, grown in his flower garden, were superb. Propagated by off-sets detached from the parent bulbs. Native of the Cape of Good Hope.

Venidium calendulaceum. — A beautiful half-hardy perennial composite plant from the Cape of Good Hope. It is of dwarf spreading growth and bears in summer a profusion of showy Marigold-like yellow blossoms. If several plants are put out together on a warm, sunny border the effect is very good. Cuttings put in in August root freely, and may be potted and kept in the greenhouse through the winter till May, when they should be planted out. In winter the plants must not have much water, or they will damp off. It may be readily raised from seeds sown in a hot-bed in early spring; the seedlings should be planted out in May in friable soil in a warm exposure.

Venus's Looking-glass *(Specularia Speculum)*.

Veratrum album *(White Hellebore)*.—A handsome, erect perennial of pyramidal habit, 3½ ft. to 5 ft. high, with large plaited leaves. The flowers, of a yellowish-white colour, are borne in numerous dense spikes on the top of the stem, forming a large panicle. As the leaves are handsome, it is worth a place in full collections of fine-foliaged hardy herbaceous plants, and would look to best advantage in small groups in the rougher parts of the pleasure ground and by wood walks. Thrives best in peaty soil; it is best multiplied by division, as the seed is very slow and capricious in germinating, sometimes not starting until the second year, and it is some years before the seedlings are strong enough to flower. The root of this plant is exceedingly poisonous. V. nigrum differs from V. album in having more slender stems, narrower leaves, and blackish-purple flowers. V. viride resembles V. album, but its flowers are of a green colour. France.

Verbascum *(Mullein)*.—For the most part the Mulleins in cultivation are only of biennial duration, and are unsatisfactory plants to cultivate, though some of them are stately and handsome when well grown. Of the cultivated kinds, which number about a dozen, one of the finest is

V. Chaixi, or V. vernale, as it is also called. It is a true perennial species—at least, on warm soils—and in this respect quite unlike other Mulleins that are sometimes seen in our gardens. It also has the advantage of great height, attaining 10 ft., or even more. It has large green leaves, which come up rather early, and are extremely effective. The colour is good, and the panicle of flowers enormous. The quantity of yellow flowers with purplish filaments that are borne on one of these great branching panicles is something marvellous. The use of such a plant cannot be difficult to define, it being so good in form and so distinct in habit. For the back part of a mixed border, for grouping with other plants of remarkable size or form of foliage, or for placing here and there in open spaces among shrubs, it is well suited. A bold group of it arranged on the Grass by itself, in deep, light, and well-dressed soil, would be effective in a picturesque garden. Other tall-growing kinds with yellow flowers are V. phlomoides, nigrum, gnaphalodes, Thapsus, pulverulentum, Blattaria, and pyramidatum, which may be added to a full collection.

V. phœniceum is a very handsome border plant, distinct from all the others by the flowers being of various hues, but usually of a violet-blue, overlying a yellow ground striped with violet. It is of slender growth, from 2 ft. to 4 ft. high, and the blossoms, which are large and showy, are produced numerously in long spikes. The varieties are known as album, atroviolaceum, lilacinum, roseum, and violaceum. On warm, dry soil it is a hardy perennial, but in cold and wet localities it is only biennial. It may be easily raised by seed, and frequently sows itself. It likes a good, light soil and an open, warm situation. South Europe.

Verbena.—The beautiful race of garden varieties of the Verbena has long been a favourite class of plants for the flower garden, though of late years the flowers have not been so popular, probably on account of the disease that attacks them. They are very beautiful plants, and are cultivated with comparative ease; out-of-doors they bloom profusely, and stand well till quite late in the autumn; and if the lustre of the flowers happen to become dimmed by a storm, a burst of sunshine quickly restores their lost beauty. As in the case of other flowers, the present fine development of Verbenas is, doubtless, the result of years of patient industry. From the Continent have come many fine varieties, and we have also some excellent kinds, the produce of English-saved seed. The majority are unsuitable for open-air culture, and some of the older kinds are still the best, such as Celestial Blue, Crimson King, Edwin Day, Firefly, James Birkbeck, La Grande, Boule de Neige, Mrs. Mole, Oxonian, Purple King, Wonderful; and among the newer sorts are Gruss au Erfurt and Mdlle. Emilie Hulter. A pretty and effective bed may be secured by mixing a few good varieties together. When grown out-of-doors the Verbena should have liberal treatment; a dry, open border should be selected for it, and the ground should be trenched and well dressed with spent hotbed manure and leaf-soil. Under such circumstances the plants can be put out about the end of May, and as they make growth the shoots require to be pegged securely over the bed and be kept well thinned. The plants of the named sorts are propagated either in autumn or spring, the latter season being the best. Good plants of Verbenas cannot be had except good cuttings are procurable. The best means of ensuring these for spring propagation is to keep a few store plants in pots all the summer, cutting them pretty closely in the autumn, and giving them a shift into larger pots of rich soil. Soon afterwards they should be set in a cool house or pit from which frost is excluded, and in this way better cuttings will be ensured in spring than those from autumn-struck plants. Of late years Verbenas have been raised from seed like other half-hardy annuals with the greatest success, and they are valuable for garden adornment in summer, and, besides, such a plan avoids cutting the plants for propagation and the risks of disease and insects. Keeping the plants through the winter free from insects and disease is a troublesome point; but, beginning with seedlings under fair conditions, we should avoid that, and the seedlings would probably give us a degree of vigour that would get over the so-called disease. In any case the fact is an interesting one, that Verbenas in any quantity and in the greatest vigour may be raised from seed in the year in which they adorn a garden, or, in other words, may be treated as annual plants. The wonderful diversity and brilliancy of colour and the profusion with which the flowers are borne combine to render the Verbena grown in this way one of the most valuable plants we possess. The seeds should be sown about the middle of January in light soil in a warm frame or pit. The seedlings should be pricked out when a few weeks old in 2½-in. pots, and when fully established the plants should be placed near the glass in a pit kept well ventilated in order to induce stout, hardy growth. About the end of March the seedlings should be potted singly in 2½-in. pots, and a month later into 3-in. pots; they should then be planted out about the middle of May about 2 ft. apart in a sunny border, and in a short time the plants will be aglow with flower. Verbena seed is sold in colours—scarlet, blue, white, carnation flaked, and other forms, and all come remarkably true. The scarlet is just a fine reproduction of the old scarlet Defiance, that prince of scarlet bedding plants of some years since, and the growth and quantity of bloom produced is marvellous. The compact habited, purplish-red flowered kermesina is a very pleasing and effective kind.

V. venosa is a beautiful kind, growing from 1 ft. to 1½ ft. high, and producing heads of purple-violet blossoms. It is much hardier than ordinary Verbenas, and is not so liable to be injured by mildew nor to be damaged by bad weather. It looks all the brighter for drenching rains, and lasts very late in the season. When all the ordinary varieties of Verbenas fail this one is sure to give satisfaction. It is easily kept through the winter, and if its fleshy roots are stored thickly in boxes any number of plants may be propagated in spring from the young shoots that are abundantly thrown out. It should be planted rather thickly, and pegged down until the ground is covered, when it will continue to flower until the last of the summer flowers are removed or destroyed by the frost. It is easily raised from seed, which should be sown four months before the plants are wanted, as the seeds frequently take long to germinate. In cases in which it is necessary to lift the roots in autumn they may be placed at once in the boxes in which they are to be started, and kept in a cool place until the time arrives for putting them in heat. In herbaceous borders the roots may be allowed to remain for years, but they should be protected throughout the winter, lest they should be injured should the frost be severe. Being of a branching and wide-spreading habit, it has a grand effect in large patches, bound round with some contrasting colour; Mangles' variegated Pelargonium and it associate beautifully together.

Verbesina encelioides. — A half-hardy annual composite plant from 1 ft. to 2 ft. high, having broad clusters of golden-yellow blossoms. An interesting plant for a full collection. California, Texas, and Mexico. Among other species are the following: V. gigantea, an ornamental shrub from Jamaica, about 6½ ft. high, forming, when young, a very pleasing subject for decorative purposes, its round green stems being covered with large, winged, pinnate leaves of a glistening delicate green colour, and very elegant outline. Suitable for rich beds or

PLATE CCLXI.

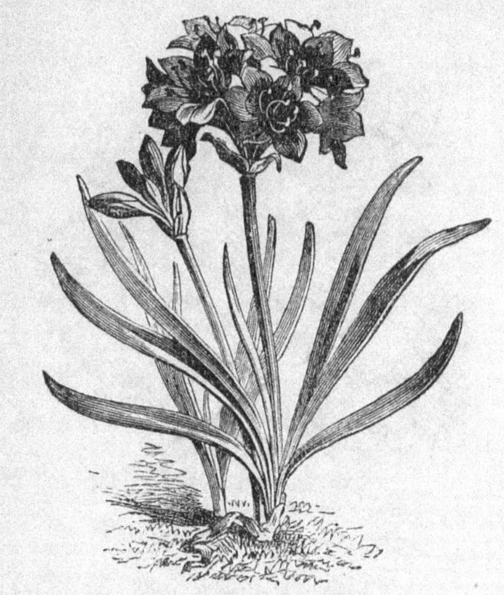

Vallota purpurea (*Scarborough Lily*).

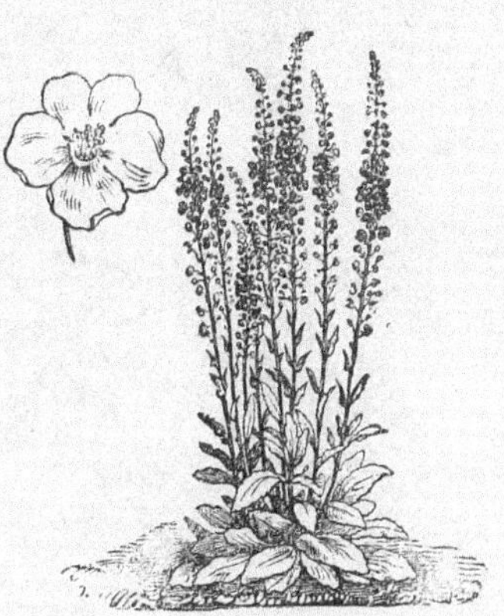

Verbascum phœniceum (*Mullein*).

Venidium calendulaceum.

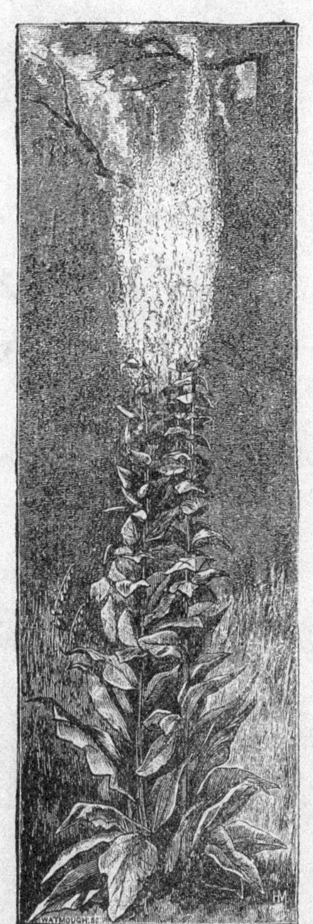

Verbascum Chaixi.

Veratrum nigrum.

[*To face page* 290.

PLATE CCLXII.

Verbena venosa.

Verbena (*Italian Hybrid*).

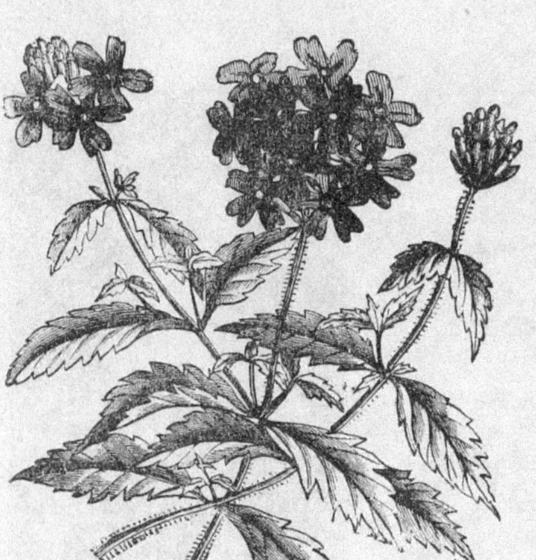

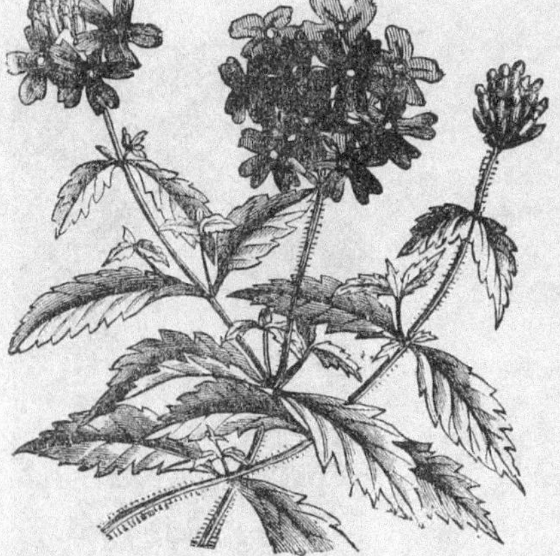

Verbena chamædrifolia.

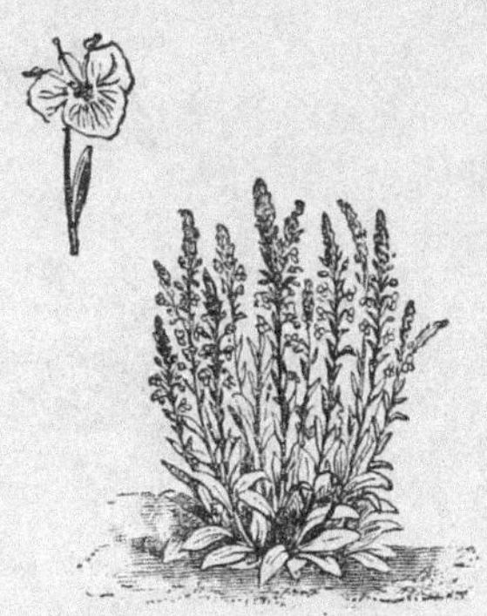

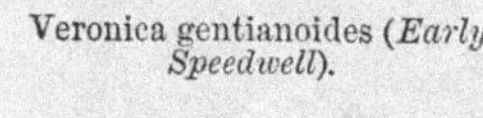

Veronica gentianoides (*Early Speedwell*).

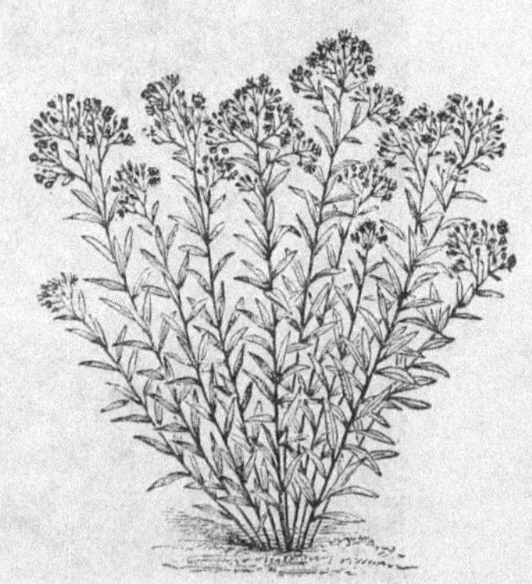

Vernonia præalta (*Ironweed*).

Veronica prostrata (*Dwarf Speedwell*).

[*To face page* 291.

groups, and should be planted out at the end of May or early in June. V. pinnatifida is a rough, half-shrubby species with a winged stem and woolly oval leaves with lobed or toothed margins; they are larger than those of the preceding species, growing 3 ft. long by 14 in. broad in the first year. Both species require hothouse treatment in winter, and are multiplied by cuttings in early spring. Young plants are to be preferred for effect, and will be much the better for as warm and sheltered a position and as rich and light a soil as can be conveniently given them.

Vernonia *(Ironweed).* — A genus of coarse-growing Compositæ from North America, of which some half a dozen species are in cultivation. They bloom so late with us, and so often suffer from our cold autumnal rains, that they are scarcely worth a place as border plants, but V. præalta at least is a fine plant for the wild garden, being so stately in habit that even if its flowers escape us, or are injured, it may still be considered worth room in a copse, or ditch, or open spot in a wood. V. noveboracensis is a vigorous growing perennial, producing rosy-purple flowers in abundance. Although the stems are somewhat naked, it might be used with good effect in shrubberies along with such Asters as Novæ-Angliæ, or among vigorous perennials in semi-wild situations in any soil. All may be easily propagated by division.

Veronica *(Speedwell).*—This genus is a very large one, and includes some very ornamental evergreen shrubs. These are mostly natives of New Zealand, and none of them are quite hardy in all parts of England. Of hardy perennial Speedwells, herbaceous or trailing, or carpeting in their habit of growth, that are suitable for cultivation in English gardens there are at least fifty species, and more than twice as many varieties. The normal colour of the flowers is generally some shade of blue, but, like many other blue flowers, their colour is often imperfect, and varies to rose or dull white. There is great variation of form in plants of the same species, and some seem to differ far more from one another than from plants of different species. It is, therefore, not surprising to find that the naming of the Speedwell is difficult and perplexing. We cannot do more than mention a few of the more distinct and ornamental forms. All the species mentioned below are quite hardy, and will grow in any soil and almost any situation. Propagation by division of the roots of nearly all Speedwells is so easy, that little need be said about it. The herbaceous kinds root from every shoot, and the creeping kinds root as they creep. They are so hardy, that they may be divided or transplanted at all seasons. Such kinds as V. longifolia require frequent division to prevent the shoots becoming too crowded. Most of them ripen abundance of seed, and seedlings come up round them which vary in colour and form.

V. amethystina grows from 1 ft. to 1½ ft. high, is of good colour, but rather too diffuse in habit. It should be cut down in autumn, as it trails in an untidy way if allowed to continue its growth. It produces its terminal racemes of blossoms abundantly in June.

V. Chamædrys *(Germander Speedwell)* has been recommended for covering the surface of beds in which late-flowering bulbs are grown. A curious variety of it, named pedunculata, is quite different from the type in appearance, and is a neat plant.

V. gentianoides is one of the earliest of the Speedwells, flowering in May. Three forms of it are common—the type with grey flowers, a variety with white flowers and bright glossy leaves like those of the Gentianella, and another with handsome variegated leaves; all are worth growing.

V. incana, also called candida, is a dwarf plant with silvery leaves and dark, rich purple flowers. It grows 6 in. high, and seems to flourish anywhere. It is used with good effect in geometrical bedding, its grey leaves being in contrast to most other foliage. It is easily propagated in autumn or spring by division. V. neglecta is similar to this, but inferior.

V. longifolia, the commonest garden species, is generally sold as V. spicata in four distinct varieties—blue, white, rose coloured, and purple, with variegated leaves. The variegation of the leaves is uncertain and irregular, but the habit of the plant is very good. By the rich colour of the flower, the length of the flower-spike, and its sturdy and compact growth, this species and its varieties make handsome border plants. They grow well in any ordinary soil.

V. pectinata, with elegant serrated downy leaves and blue or rose-coloured flowers, is a pretty trailing kind, admirably adapted for clothing dry spots in the rock garden, or for the margins of borders and other places. It grows in almost any kind of soil or position.

V. prostrata.—A very dwarf species, forming spreading tufts and bearing deep blue flowers. There are several varieties with rose-coloured and white blooms, appearing in early summer. It is a hardy and pretty plant, flowering so freely that when in full perfection the leaves are often quite obscured by the flowers. A native of France, Central and Southern Europe, occurring on stony hills and in dry grassy places, and in cultivation succeeding perfectly in sandy soil. It is an admirable plant for rockwork, its prostrate habit fitting it best for sloping positions or fissures on vertical faces of rocks.

V. repens is a dense, close-growing creeper, covering the soil as it proceeds with a perfect soft carpet of bright green foliage. It is covered with pale bluish flowers in spring, and thrives well on soil that is moderately dry. It delights, however, in moist corners on rockeries, and is an admirable minute rock plant.

V. rupestris is a very good rockery plant, trailing neatly and closely, and flowering abundantly in June. Those who have seen large masses of it on rockeries will want no further recommendation. There are several nearly allied alpine species.

V. satureiæfolia is one of the best Speedwells; the flowers are about the size of V. saxatilis, of the same intense blue, and produced in abundant upright racemes. It is a somewhat rare plant, but when once seen growing in perfection it is not to be forgotten.

V. saxatilis.—A native of alpine rocks in various parts of Europe, and also in a few places in the highlands of Scotland, forming neat tufts 6 in. or 8 in. high. The flowers are a little more than ½ in. across, and of a pretty blue, striped with violet, with a narrow, but decided ring of crimson near the bottom of the cup, the base of which is pure white. It is also an excellent rock garden plant.

V. spicata is a dwarf native plant, not more than 5 in. or 6 in. high. It is useful for bare corners of rockeries; it seldom flowers before the end of July. V. corymbosa is a name given to varieties of two or three species. The best seems to be a form of V. spicata; it is a profuse and continuous flowerer, and one of the best for rockeries. V. hybrida is generally classed as a variety of V. spicata, though it seems quite distinct, being far more robust in habit, with flowers varying in colour from dark purple to lavender and light rose. It grows wild in profusion on mountain limestone hills near Llandudno and other parts of the north-western counties. Both these species increase much in size under cultivation, especially when raised from seed.

V. subsessilis is botanically considered to be but a variety of V. longifolia, with stalkless leaves, but for garden purposes it is very distinct. It is without doubt the handsomest of all the hardy Veronicas, and one of the choicest acquisitions recently added to the mixed flower border. It comes from Japan, but its constitution is certainly superior to that of most Japanese plants, which require a warm spring and a hot summer for their successful development. This Speedwell grows on and flourishes in spite of spring frosts and cold summers. Its large dense spikes of the deepest purple-blue blossoms render it a most effective plant, and it should always have a position among the choicest hardy flowers. It succeeds best in a good deep loamy soil in an open situation. It is easily propagated by division or seed.

V. taurica.—A very dwarf, wiry, and almost woody species, forming neat dark green tufts, under 3 in. high; the leaves are crowded, the upper ones distinctly toothed; the flowers, which are abundantly produced, are of a fine gentian-blue. It is, perhaps, the neatest of all rock Veronicas for forming spreading tufts in level spots, or tufts drooping from chinks, and it is admirable also for the margin of the mixed border, being thoroughly hardy, growing in ordinary well-drained garden soil, and flowering in early summer. Suitable for association with the dwarfest alpine plants and mountain shrubs, being itself indeed a tiny, compact, prostrate shrub. A native of Tauria. Increased by division or by cuttings.

V. Teucrium.—A Continental plant, the stems of which form spreading masses, from 8 in. to 1 ft. high, covered in early summer with flowers of an intense blue. These are at first in dense racemes, which afterwards become much longer, the fine blue corolla having oval segments, the three lower ones pointed. It is an excellent plant either for the rockwork or borders, is easily increased by seeds or division, and grows freely in ordinary garden soil.

V. virginica and other tall species grow from 3 ft. to 4 ft. high, and flower in July, but are deficient in colour.

Amongst dwarfer kinds may be added to our list V. verbenacea, fruticulosa, alpina, aphylla, Nummularia, Guthrieana, austriaca, incisa, bellidioides, and Dabneyi, all species well worth cultivation, of good, dwarf habit, and admirably adapted for a rockery. Nor should the pink variety of V. officinalis be omitted—forming, as it does when established, dense patches of pink-coloured blossoms, some elevated 3 in. above the surface of the ground.

Vesicaria.—This genus numbers about twenty species, but about half a dozen kinds only are showy enough for cultivation, the others being straggling and weedy in appearance. V. græca is the handsomest of the genus, and for several weeks is a really showy plant when in flower in the early part of the summer. It bears a strong resemblance to the better known V. utriculata, long cultivated in gardens, but is more showy, and begins to flower about a fortnight earlier. It flowers profusely, the blossoms opening in continuous succession for several inches on each stem. It is a hardy evergreen perennial, inhabiting some rocky districts in Dalmatia and other places in South Europe. It may be propagated, like V. utriculata, by means of cuttings placed in soil under a hand-glass, but the best mode is by seeds, which in favourable seasons are produced plentifully. Both flourish on dry, sunny parts of the rock garden in dryish soil.

Vicia *(Vetch)*.—A large genus of perennial and annual plants, several of which are natives. For the full collection, V. Cracca, Orobus, sylvatica, Sepium, and argentea would be the most desirable. The last named has elegant silvery leaves, but is rare in cultivation. All grow freely in almost any soil, and are propagated by seeds.

Vieusseuxia glaucopis *(Blue-eyed Peacock Iris)*.—An exquisitely beautiful bulbous plant, 9 in. to 15 in. high. The flowers, which appear in early summer, are about 2 in. across, pure white, with a beautiful porcelain-blue stain nearly ⅓ in. broad at the

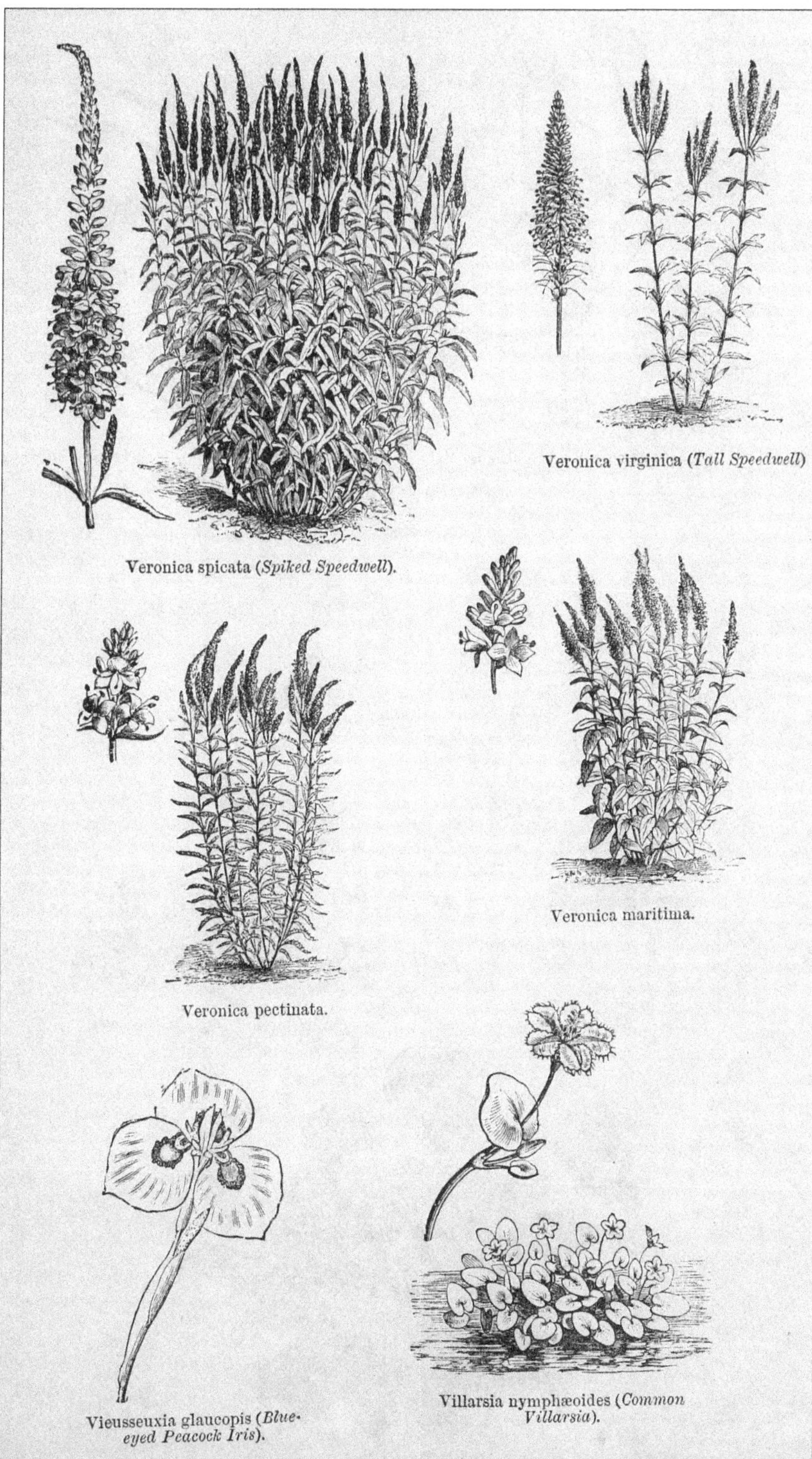

Veronica virginica (*Tall Speedwell*)

Veronica spicata (*Spiked Speedwell*).

Veronica maritima.

Veronica pectinata.

Vieusseuxia glaucopis (*Blue-eyed Peacock Iris*).

Villarsia nymphæoides (*Common Villarsia*).

[*To face page* **292.**

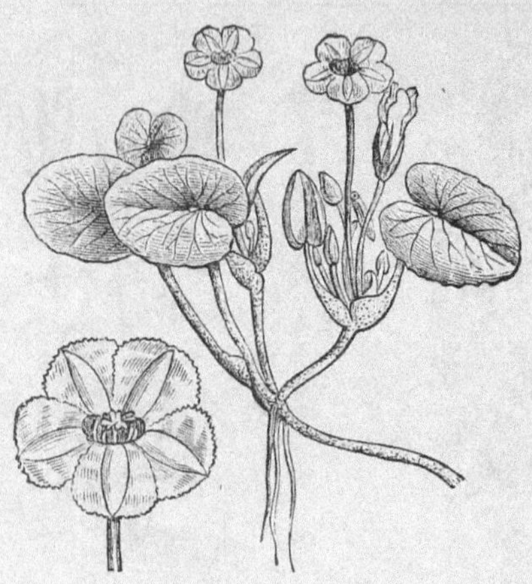

Flowers of Villarsia nymphæoides.

Vinca herbacea.

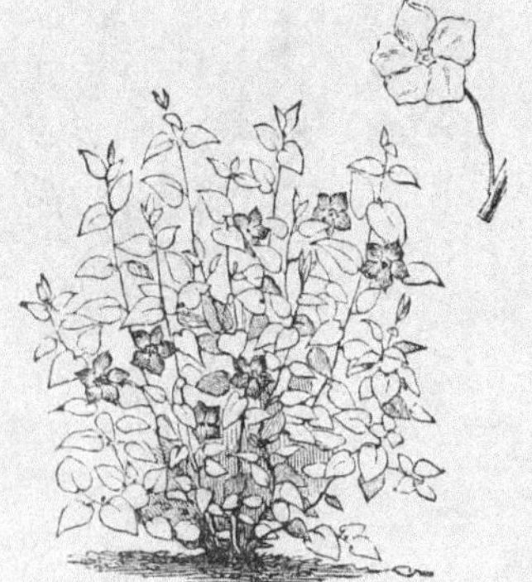

Vinca major (*Greater Periwinkle*).

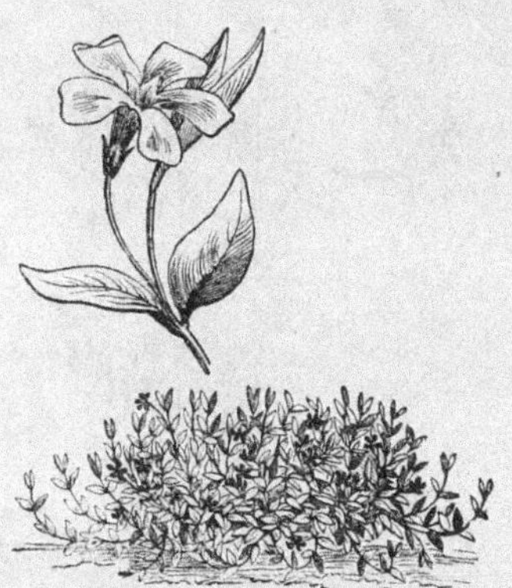

Vinca minor (*Lesser Periwinkle*).

Viola tricolor (*Winter Heartsease*).

[*To face page* **293.**

base of each of the three larger divisions; the spot is margined with dark purple teeth, and is of a fine deep violet at the base. It should be grown on warm, sheltered borders in sandy peat or sandy loam and leaf-mould; the bulbs should be planted rather deep, say with 4 in. of firm soil above the crest. Propagation is effected by separation of the bulbs in autumn. Cape of Good Hope. (= Iris Pavonia.)

Viguiera.—A genus of North American composites to which Harpalium rigidum is said to belong.

Vilfa.—Grasses of no garden value.

Villarsia nymphæoides *(Common Villarsia).*—An attractive British aquatic, with leaves like those of a Water Lily, but smaller; they float on the surface of the water. The flowers are yellow, borne singly on stalks as long as those of the leaf, and are produced in summer. Where not naturally plentiful it should be encouraged in lakes, ponds, and quiet bays in streams. Division.

Vinca *(Periwinkle).*—A small genus of herbaceous perennials of trailing growth. They are useful for a variety of purposes, as they are hardy and vigorous in almost any soil or position. The well-known V. major (the common Periwinkle) is very useful on large rockwork, on masses of rootwork, near cascades, &c., and also in rocky places or banks in the wilder parts of pleasure grounds or by wood walks. A variety called elegantissima is finely blotched and variegated with creamy white, and there are several other variegated varieties, including a golden-leaved kind, all exceedingly useful in the positions above named. The lesser Periwinkle (V. minor), a much smaller plant than the preceding, is also useful for like positions; there are several varieties of it well worthy of cultivation, a white-flowered one (V. minor alba), one with reddish flowers, one or two double varieties, and also, as of the larger, several variegated forms. V. herbacea is a plant much less frequent than our common Periwinkles, and more worthy of culture on rocks, as it is not rampant in habit. It is a native of Hungary, and flowers in spring and early summer, the stems dying down every year unlike those of its more familiar relatives; it thrives best in an open position. V. acutiloba is a distinct and elegant Periwinkle, valuable because it flowers late in the autumn and in winter, and also on account of the delicate mauve colour of the blossoms. It is not a variety of either of the old and long-cultivated Vincas, but is a species from the south of Europe. It is particularly suitable for the embellishment of sunny banks and slopes and for warm borders.

Viola *(Violet).*—A beautiful and well-known family, many of which are alpine, and some among the most beautiful ornaments with which the alpine turf is bedecked. The common Violet, indeed, may almost be claimed as a true alpine plant itself, for it wanders along the hedgerow, and hillside, and copses, and thin woods all the way to Sweden. And it is only a mountaineer in constitution that can, in spite of the bitterly cold winds that, every spring, sweep over the woods and plains of Europe and N. Asia stain the carpet of dead leaves in the copse with purple, and fill the air with perfume before a green bud has opened. Among the Violets, as among many other families, it is impossible to divide the true alpine species from those that haunt the plains, heaths, woods, copses and hedges, bogs, hills, meadows, and sandy shores. From all, the wild world of flowers and our gardens derive a precious treasure of beauty and delicate fragrance. There are, it is true, some inconspicuous species without distinct attraction for the garden, but, on the other hand, there is no family that has given anything more precious to our gardens than the numerous races of Pansies and large showy, as well as sweet-scented, Violets. Far above the faint blue carpets of our various scentless wild Violets, in woods and heaths, and thickets and bogs, the miniature Pansies that find their home among our lowland field weeds; the larger Pansy-like Violas (varieties of V. lutea) which flower so richly in the mountain pastures, and even on the tops of the stone walls in Northern England; above the large and free-growing Violets of the American heaths and thickets we have the true alpine Violets, like the yellow two-flowered Violet (V. biflora), and large blue Violets like V. calcarata and V. cornuta. It would be difficult to exaggerate the beauty of these alpine Violets. They grow in a turf of high alpine plants not more than an inch or so high. The leaves do not show above this densely-matted alpine turf, but the flowers start up, waving everywhere thousands of little banners. Violets of all kinds are of the easiest culture in our gardens; even the highest alpine kinds thrive with little care, and V. cornuta and calcarata of the Alps and Pyrenees thrive not only freely, but much more so than in their native uplands, the growth of the foliage and stems being much stronger in our gardens. Some of the many varieties of the Sweet Violet raised of late years might be naturalised with advantage, being of stronger growth. They and the stronger-growing American kinds are likely to prove very useful in the wild garden. Slow-growing compact kinds, like the American Bird's-foot Violet, enjoy, from their stature and comparative slowness of growth, a position in the rock garden or choice border, in which they are of easy culture in moist sandy soil.

The following are among the most desirable and distinct for general cultivation:—

V. biflora *(Two-flowered Yellow Violet).*—This bright little Violet is a lovely ornament on the Alps, in many parts of which every chink between the moist rocks is densely clothed with it. It even crawls far under the great boulders and rocks, and lines shallow caves with its fresh verdure and little golden stars. It is readily known from any other cultivated species by its very small, but bright yellow flowers, the lips of which are streaked with black, being usually borne in pairs, and by its kidney or heart-shaped leaves. It will be found especially useful in large rock gardens, where rude flights of stone are constructed to give one or more winding pathways over the mass, as it will run through every chink between the steps, and tend to make them, as well as the most select spots, replete with life and interest. If obtained in a small or weakly condition, it may seem difficult to establish, but this is not by any means the case; once fairly started in a moist and half-shady spot, it soon begins to creep about rapidly, and may then be readily increased by division. When well established on suitable rockwork, it is able to take care of itself. Native of Europe, Asia, and America.

V. calcarata *(Spurred Violet)*.—This plant comes very near the well-known Viola cornuta in the flower and spur, but is easily known by its habit of increasing by runners under the earth, somewhat after the manner of Campanula pulla, instead of forming strong leafy tufts like V. cornuta. It is a very pretty plant on the Alps, usually in very high situations, amidst very dwarf flowers, and is sometimes so plentiful that its large purple flowers form sheets of colour, the leaves being scarcely seen amidst the other dwarf plants that form the turf. It is as charming a plant in the rock garden as in its native wilds. There is a yellow variety, named flava, which is the same as V. Zoysi.

V. cornuta *(Horned Pansy)*. — This Pyrenean and Alpine Violet is now to be seen in almost every garden, its pale blue or mauve-coloured and sweet-scented flowers, so abundantly produced, making it very valuable in lines, borders, and mixtures. It has been cultivated for ages in our gardens as a rockwork and border plant, but its value as a continuous bloomer, and consequent capacity for bedding, only came into prominent notice a few years ago. Generally speaking, it does poorly on dry soils and in warm districts, and exceedingly well in wet places. In many cold and stormy districts the blue Lobelia, so fine in the south, grows quite to Grass instead of flower; that which spoils the Lobelia will highly improve this Violet. It is largely used in all kinds of border decoration, and the numerous varieties of it now existing afford a wide range of colour. It is quite easily propagated by division, cuttings, or seeds.

V. cucullata is unusually floriferous, and bears some resemblance to the common Violet, but has not its delicious scent. It, however, flowers much more freely than the sweet-scented Violet, and its bold foliage, which is sometimes variegated, makes it a valuable acquisition. It came originally from North America, and belongs to a section in which there are some fine varieties, such as V. primulæfolia, semperflorens, blanda, obliqua, sagittata, palmata, delphinifolia, canadensis, pubescens, striata, and others. All of these are worthy of culture where a full collection is desired.

V. gracilis is a remarkably pretty species of dwarf growth, producing in spring an abundance of deep purple blossoms in dense tufts. It is quite hardy in light soil, and never fails to produce annually a profusion of bloom. Native of Mount Olympus.

V. lutea *(Mountain Violet)*.—In cultivation the yellow form of this Violet is a very neat and compact plant, rising from 2 in. to 6 in. high, and flowering abundantly from the month of April onwards. The flowers are of a peculiarly rich and handsome yellow, the three lower petals striped with thin lines of rich black. It possesses first-class qualities as a bedding plant, and is less uncertain in its growth on the majority of soils than V. cornuta, while it is dwarf, neat, and of a colour much to be desired for the flower garden. Tasteful use may be found for it in the margining of choice beds, in forming low and rich mixtures with bright-leaved plants like Amarantus or Coleus, kept very dwarf, and in other ways.

V. Munbyana.—This Algerian species, one of the prettiest of Violets, is an abundant flowerer, a free and robust grower, and quite hardy. It generally begins to flower about the end of February, but attains its greatest beauty in May. The flowers resemble those of V. cornuta, and are of a deep purple-blue. There is also a yellow variety.

V. odorata *(Sweet Violet)*.—This well-known plant is in a wild state widely spread over Europe and Russian Asia, and is common in various parts of Britain, but it is best known because it is found in almost every garden, and because enormous quantities of it are sold in London, Paris, and many other cities. It need not be described; besides, its delicious odour distinguishes it immediately from the numerous other Violets. It is too well known to require praise, but it is very seldom used in the best way. The Sweet Violet and most of its varieties may be used in many places where few other things but weeds succeed; it will form carpets for open groves or the fringes of woods, or in open parts of copses or on hedgebanks, demanding in such positions no care, and rewarding the planter by filling the cold March air with unrivalled sweetness. In the garden, instead of confining it to a solitary bed for cutting from, as is often done, it should be permitted to fringe the margins of shrubberies, or the margins of rockwork or ferneries, or any like places where it may be allowed to exist and take care of itself. It will grow in almost any soil, but succeeds best in free sandy loams, and should be put in such when there is any choice. It is well to naturalise the plant on sunny banks, and fringes of woods, and on the warmer sides of bushy places, to encourage a very early bloom.

The cultivation of the Sweet Violet in gardens is of great importance, not only for the

PLATE CCLXV

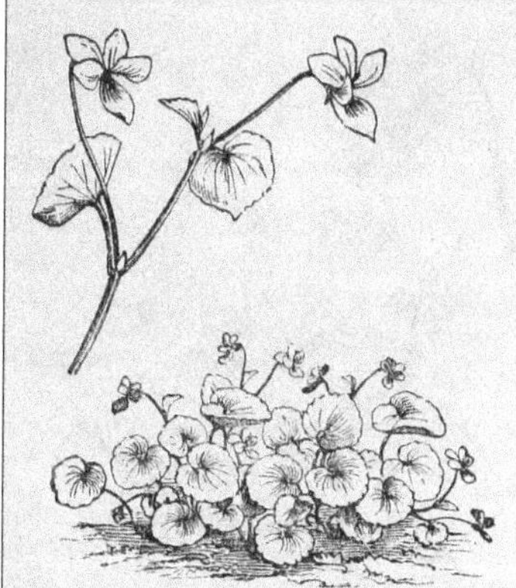

Viola biflora (*Twin-flowered yellow Violet*).

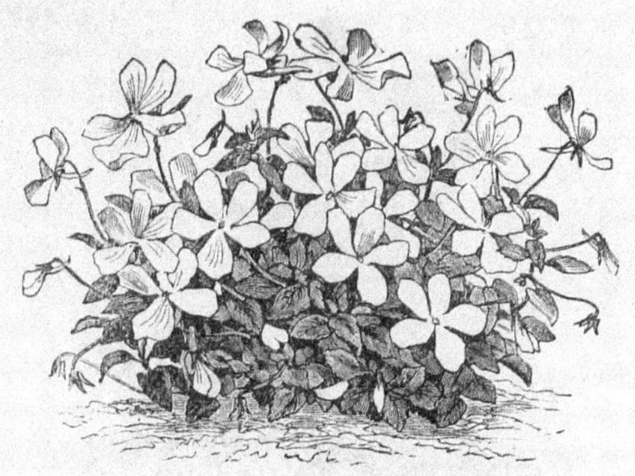

Viola cornuta var. alba (*White Horned Pansy*).

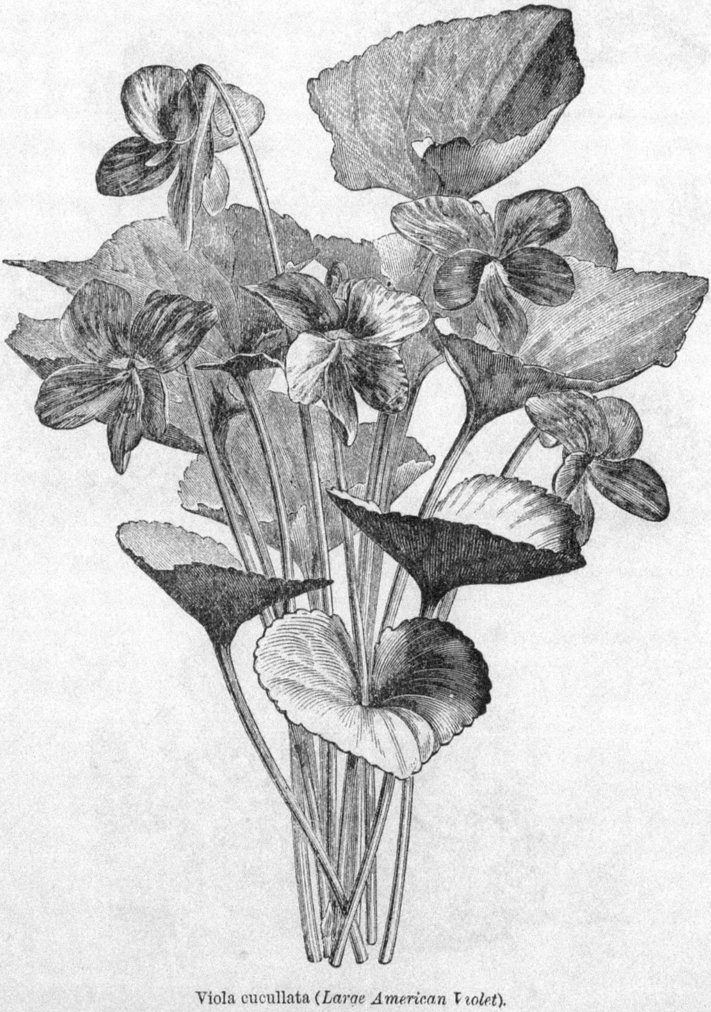

Viola cucullata (*Large American Violet*).

[*To face page* **294.**

PLATE CCLXVI.

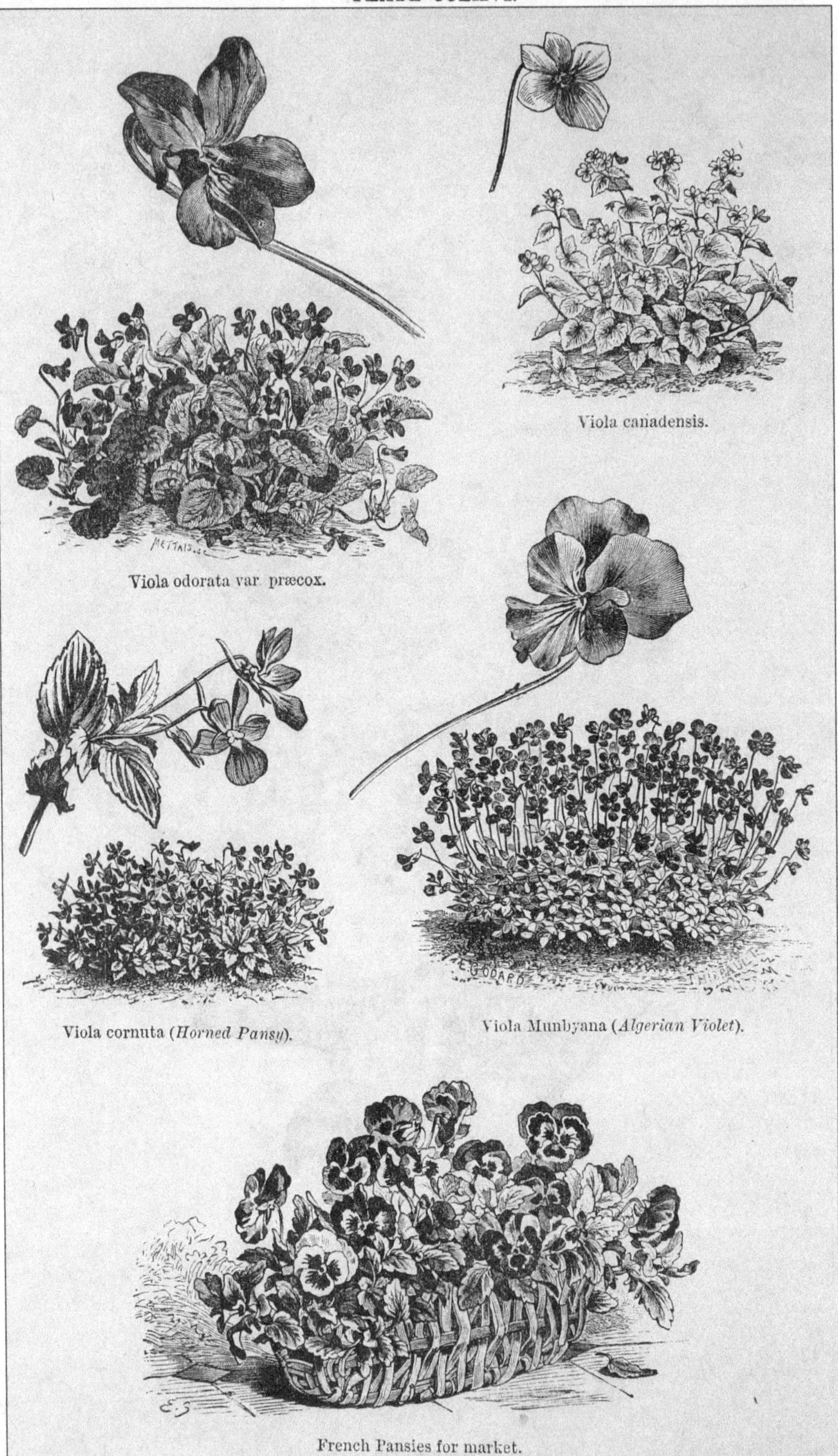

Viola canadensis.

Viola odorata var. præcox.

Viola cornuta (*Horned Pansy*).

Viola Munbyana (*Algerian Violet*).

French Pansies for market.

[*To face page* **295.**

supply of private wants, but also for the vast supply required in large cities.

About Paris the cultivation of the Violet for the markets is carried on to a great extent, and in some places near that city three or four acres may be seen covered with them, the ground being well exposed to the midday sun and of a rich, free, and warm nature, the plantations being made in spring, and those required during winter being grown in frames. It is almost needless to say that they may be propagated to any extent by division, but strong, healthy, free-flowering plants are also easily raised from seed, which should be sown as soon after gathering as is convenient. In cold, dry parts and in gardens where Violets do not succeed well, and where they are required in midwinter, it is better to raise a number of healthy plants every year and put them in a light frame in a sunny position in autumn. With but little trouble one may, however, have Violets from January until May, when they bloom in the open ground. Plants, obtained by setting out runners in spring in rich soil, and giving all the water they need in dry weather, may be set early in autumn in a common cold frame. They should be allowed to grow until winter comes on, the frames should be filled with leaves, the sashes put on, and over these a shutter. Of course the plants must be treated like others under glass, and have abundance of air on mild days, and water as needed. A frame of three sashes, with boards to separate it into three parts, may be uncovered, one sash at a time, at intervals of two or three weeks, and thus a succession of flowers will be kept up. Violets do not like forcing, neither do they need it if their crowns are ripened early, and generally tempted, by the protection of glass, to open out genially and exhibit their fragrant blossoms.

In the open border they thrive well on a moderately heavy, rich soil; if it happens to be light and gravelly, some stiff material and plenty of manure must be added to it, and if poor and hard clay, it will be benefited by the addition of some sharp, gritty matter and abundance of rotten manure. Violets require shelter, but not that of a wall. Their natural shelter is a hedgerow, in which they get currents of pure air, which are essential for keeping down red spider, and for maintaining the foliage in a healthy state. In town gardens, and in gardens surrounded by high walls, they are seldom healthy. They grow well on the north or north-east side of a Hornbeam hedge, provided it is somewhat naked at bottom, so as to allow the sun to shine on their leaves early in spring, and also affords a partial shade in summer. When, however, the soil is deep and rich, they will bear a considerable amount of sunshine without injury. It is well to have a few plants in different positions, so as to ensure a succession of bloom. On south borders they dwindle and die; but a few roots on sunny banks will produce some early pickings. Violets of all kinds are easily increased by means of cuttings made from stout, short runners. All that are wiry and hard should be rejected, and runners should not be taken from plants that have grown in pots or under glass. The cuttings should be taken off the first week in April if they are intended to bloom next year, put under hand-lights on a shady border, and kept close until they have begun to grow, when the lights may be tilted a little, gradually increasing the space, until at last they may be wholly dispensed with. By September they will be ready for transplanting, when they may be placed in beds 4 ft. wide, three rows in a bed, 1 ft. apart. This will afford space to hoe between the rows while they are growing. They will soon spread and fill the beds, but must not be allowed to remain more than two, or, at the most, three years in the same place, else the flowers will become small and short stemmed. If they are permitted to remain more than two years on the same piece of ground, they must receive either liberal top-dressings of rotten manure, or copious applications of manure water. Another mode of propagation, which is perhaps attended with less trouble, is, as soon as they have done blooming, to get a few large plants and tear them into as many pieces as possible with a little bit of root attached to each. Little pieces without roots may be placed under hand-lights and treated the same as cuttings.

The insects that trouble the Violet most are green-fly and red spider. The first is generally the result of a close, unhealthy atmosphere, and is most easily got rid of by gentle smokings. Red spider is caused by strong sun and by dryness at the roots; hand-dusting with sulphur is the best remedy, but it is easy to prevent its occurrence by maintaining a damp atmosphere by syringing the plants and surroundings.

The varieties of the Violet are very numerous; thus we have the single white and single rose, double white and double rose, the small Russian, the Czar (a very large and sweet variety), the Queen of Violets (with flowers almost as large as those of the double white Cherry), and the perpetual blooming Violet, well known in France as the Violette des Quatre Saisons. The last differs but slightly from the common Sweet Violet, but is valuable for flowering long and continuously in autumn, winter, and spring. It is the variety used by the cultivators round Paris. The double white or, as it becomes in the open air, rosy-white Belle de Chatenay has a robust habit, and is a favourite kind either for open beds or for frames. Though not so pure as the old double white kind, it blooms more freely, and is not so loose in its growth. Marie Louise is a fine kind, and a great advance upon the Neapolitan; the flowers are larger and rather deeper coloured, and more freely produced. The old double blue has, perhaps, the fullest and neatest flowers of all the kinds, but the stems are short. It is, however, very beautiful when grown in frames or in beds in the open ground, where the thick growth keeps the flowers well up from the soil. Blandyana, another double, is of a better habit and a somewhat freer bloomer; the flowers are rather larger, but not quite so dark or so neatly formed. The Neapolitan will doubtless ever be a favourite, in spite of other and newer kinds, but it is not quite so hardy, and needs a frame to protect it in severe weather.

V. pedata *(Bird's-foot Violet)*.—The most beautiful of the American Violets, with handsome flowers, 1 in. across, pale or deep lilac, purple or blue, the two upper petals sometimes deep violet, and velvety like a Pansy; the leaves are deeply divided, like the foot of a bird, and the plant is very dwarf and compact in habit. The variety bicolor is a much prettier plant than the ordinary form. The flowers are larger, and the petals arranged flat like those of a Pansy; the two upper ones are a rich velvety purple, and the three lower a delicate blush tint. In the typical form the flowers are produced plentifully, and the plants are free in growth in a light rich soil in partial shade, but the variety bicolor is somewhat fastidious, and succeeds only in certain localities, and then needs special attention; but any extra care is well bestowed on such a beautiful plant. It is rare even in its native locality, but the type is plentiful in the sandy or gravelly soil in the Northern States of America, Flowers in summer, and is increased by seeds or division. It is best adapted for the choice rock garden, but may also be grown in borders where the soil is peaty, sandy, and moist. It does freely in pots where alpines are grown in cold frames, and should be amongst those that are grown for exhibition.

V. rothomagensis *(Rouen Violet)*.—This is a distinct and handsome plant belonging to the tricolor group. It is of dwarf growth with low, creeping stems, from which arise in spring numerous purple and white blossoms. It is a free grower, but being a native of Sicily it is not so hardy as some of the other Violets, and should be grown in a light soil and warm border.

V. tricolor *(Heartsease)*.—The Pansy is usually included under the head of V. tricolor, though it is more likely to have descended from V. altaica; in any case a good many kinds seem very nearly allied to that species. But the kinds are so numerous, so varied, and, withal, so distinct from any really wild species of Violet in cultivation, that little can be traced of their origin. Of one thing we may be certain: the parents of this precious race were true mountaineers. Alpines only could give birth to such rich and brilliant colour and noble amplitude of bloom. Its season never ends; it blooms often cheerfully enough at Christmas, and is sheeted with delightful gold and purple when the Hawthorn is whitened with blossoms. Such a flower must not be ignored on our rock gardens, even though it thrives in almost any soil and position. It may be treated as an annual, biennial, or perennial, according to climate, position, and soil. Belonging by nature to a more northerly region than the south of England, the wild Viola tricolor is one of the commonest weeds in Scotland; it may, however, be grown in the south, provided it is sheltered from the midday sun. A north or, better still, a north-east exposure suits it capitally if it is sheltered by tall trees or buildings, so that it may get the cool sun of the early morning hours only.

For border decoration the best way is to grow the plants from seed, and the best variety for the tyro to begin with is the Belgian or fancy Pansies, which are remarkable for the strange and almost gorgeous variety of their colours, and many of the blooms are of unusual size. They are more hardy as seedlings and more robust as plants than the other kinds, and have the additional advantage of yielding a greater variety of colours. The seed should be sown in July or August in pans of light leafy soil, such as sand, leaf-mould, and mould from rotted turves, and placed in a cool, shady place. When a mixed packet of seed is sown, it is important to sow each seed separately, at distances of 1½ in. or so, that the first seeds which germinate may be removed as soon as they have made three pairs of leaves without disturbing the weaker and more backward ones, for amongst those seeds which are the last to germinate will be found the greatest proportion of finely-coloured flowers. It follows from this that it is important to sow the seed when fresh.

It is rarely convenient to plant the seedlings at once in their blooming places. They must therefore be placed in pots plunged in a cool place in the open ground, and be shifted to their final places in time to get well established before winter sets in. They stand the winter well, the only danger being in weather when heavy rain or sleet is immediately succeeded by a sharp frost. A pot inverted over each plant to protect the soil about it from too much wet would, in such cases, be sufficient. It is not advisable to move Pansies in the spring unless they have been kept in pots during the winter, when they may be planted with as little root disturbance as possible.

Pansies are divided into two sections—the show or English kinds, and the fancy or Belgian kinds. The first comprises five divisions, as under: white, yellow, and purple selfs and white, and yellow ground belted Pansies. The self kinds must be of clear, decided colours, and a black and well defined blotch under the eye; the belted kinds have a ground of white or yellow, also with centre blotch, and a broad margin of bronzy-red, chestnut, purple, crimson, or some other hue, and the colours must be in all cases dense, and the margins very distinctly defined. All these flowers should be rounded, of good form, and stout of petal; also of good size, but great size is of less importance in an ex-

Double White Violet, Belle de Chatenay.

Viola tricolor (*Heartsease*).

Large-flowered Show Pansies.

[*To face page* 296.

PLATE CCLXVIII.

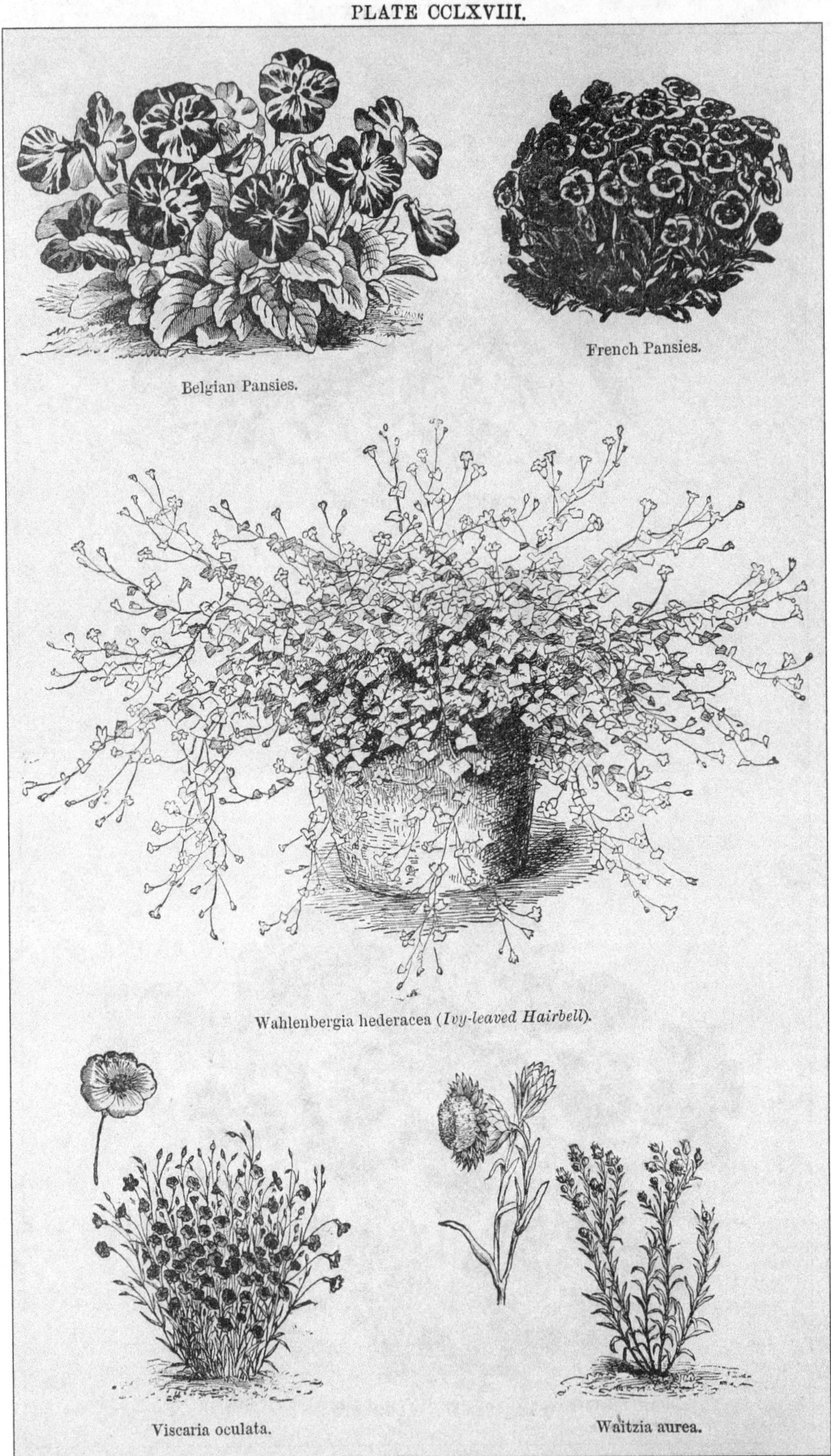

Belgian Pansies.

French Pansies.

Wahlenbergia hederacea (*Ivy-leaved Hairbell*).

Viscaria oculata.

Waitzia aurea.

[*To face page* 297.

hibition flower than perfection or quality in the markings. A correct fancy Pansy should have a very large, deep-coloured blotch, covering nearly the whole of the bottom petal and portions of the side petals; and the other portion of the flower may be white, yellow, buff, red, maroon, purple, crimson, and various other shades, but none so dense as the black hue of the centre blotch; some also are flaked or parti-coloured, but all good ones are singularly showy and beautiful beyond the imagination of those who know Pansies only by what they have seen of common strains. Some of the bedding Pansies are very free-flowering—Cliveden Yellow, Cliveden Blue, King of Blacks, and a host of other named sorts mentioned in catalogues are very effective, although a little less robust, and not so quick in growth as the Belgians. In using seedling plants from mixed seed, the arrangement of colours must be left to chance, as even the colour of the foliage is no guide to that of the flowers. Named Pansies come fairly true from seed, but the only way to secure a stock of any particular variety is to make cuttings. When any plant or plants show flowers such as are desirable to perpetuate, the best way is to sacrifice the bloom for the year, pinch the bloom-buds off as fast as they show, feed the plant well with dressings of leaf-mould pricked in about the roots, peg the first shoots down so as to leave the crown of the plant exposed; fresh, healthy shoots will rise from that. A few of these should be taken off when they have made three or four pairs of leaves, and be planted in light soil, sand, and leaf-mould, under a hand-glass, and kept moist and shaded. The pegged-down stems will produce shoots which may be taken off in the same way; when well rooted they may be treated as seedlings.

The best plants are those that combine fine flowers in profusion with a dwarf, short-stemmed, stocky habit, so that the plant, when in flower, is a round, green bush, with the flowers carried just ½ in. or so clear of the leaves. It is useless to attempt to save seed until a stock of first-class plants is obtained. July is early enough to sow the seed in the south of England, but it may be sown earlier further north, until in Scotland it should be sown in the spring.

Violet Cress (*Ionopsidium acaule*).

Viper's Bugloss (*Echium vulgare*).

Virginian Cowslip (*Mertensia virginica*).

Virginian Spiderwort (*Tradescantia virginica*).

Viscaria oculata.—This is a showy and beautiful hardy annual, well suited to the hardy flower border. The seed should be sown in autumn or spring, and the seedlings thinned out when large enough. It grows about 6 in. or 8 in. high, and bears a profusion of rose coloured blossoms having a dark centre. There are numerous varieties, such as cardinalis (bright crimson-purple), cœrulea (bluish), alba (white), Dunnetti (rose), splendens (scarlet), picta elegans (crimson-purple, edged with white), and a dwarf variety, nana, about 9 in. high, which is very desirable. V. cœli rosea is likewise a beautiful hardy annual, sometimes called Agrostemma. If it is sown in autumn in warm sandy soil, and the plants are duly thinned out, it will produce compact cushions of rosy-blue coloured blossoms. It may be sown in spring, but the results will not be so good as those from autumn sowing. V. oculata is a native of Algeria, V. cœli rosea of S. Europe. Pink family.

Vittadenia triloba and **australis** (= Erigeron mucronatum).

Wachendorfia. — A small genus of bulbous plants. They are natives of the Cape of Good Hope, and therefore not hardy; they require precisely the same treatment as the Cape Ixias and Sparaxis. There are only a few species in cultivation, W. paniculata, hirsuta, and thyrsiflora being the best known. These are similar to each other, and have tall, slender branching flower-stems, which bear in early spring numerous large yellow blossoms that are very showy and attractive.

Wahlenbergia hederacea (*Ivy-leaved Hairbell*).—A pretty and elegant little native, closely allied to Campanula, under which name it is often associated. It is a weakly, creeping thing, with almost thread-like branches, which bear small delicate leaves and flowers of a faint bluish purple, less tnan ½ in. long and drooping in the bud. There is, however, an interest and peculiar grace about it that we do not find in more robust members of the same family. It creeps over bare spots by the side of rills and on moist banks, and wherever there is a moist, boggy spct near the rockwork or by the side of a streamlet or in an artificial bog it will be found worthy of a place as an interesting native. It occurs chiefly in Ireland and Western England, and less abundantly in the east. Increased by division. The various species of Edraianthus are known also as Wahlenbergias.

Waitzia.—A genus of half-hardy annuals of the Composite family, all natives of Australia. There are four kinds in cultivation, and all are valuable for their pretty flowers which, like those of Helichrysum, are termed Everlastings, and are very useful for winter bouquets. Of W. acuminata there are two varieties with purple and yellow flowers. W. aurea is bright yellow, W. corymbosa has flowers white and purple, and W. grandiflora has flowers like aurea, but finer. All grow about 1 ft. high, and require to be treated as other tender annuals, such as Rhodanthe. They succeed best in sandy peat in an open position, but the seedlings should be shifted into different-sized pots before planting out in May. They flower in August and September.

Waldsteinia.—Dwarf Rosaceous plants, three of which are in cultivation—W. geoides, fragarioides, and trifolia. The last is the most attractive plant, but not one of them is ornamental enough for border culture. They are good only for naturalising on dry banks and such places, as they grow in any soil. Their flowers are yellow and produced in spring.

Wallflower (*Cheiranthus Cheiri*).

Wall Pennywort (*Cotyledon Umbilicus*).

Wall Rue (*Asplenium Ruta-muraria*).

Water Avens (*Geum rivale*).

Water Dock (*Rumex Hydrolapathum*).

Water Flag (*Iris Pseud-acorus*).

Water Leaf (*Hydrophyllum*).

Water Pennywort (*Hydrocotyle*).

Water Plantain (*Alisma Plantago*).

Water Soldier (*Stratiotes aloides*).

Water Violet (*Hottonia palustris*).

Watsonia.—Bulbous plants representing some of the most beautiful to be found in the large Iridaceous family. Several of the species, and these the finest, were favourites long ago in gardens. They are too tender to be included among what are called hardy plants, but some succeed perfectly well in open borders in the southern counties. The genus is not a large one, as it numbers but a dozen species and about as many varieties, half of which are variations from W. Meriana. All the known species are natives of South Africa, but their headquarters are at the Cape. What with the hybrids and the many seminal varieties, there is a great diversity of colours, and some of the trade lists even advertise a "mixed" selection representing "all colours." The commonest known species seem to be W. Meriana with its native varieties; W. coccinea, W. iridifolia, W. rosea alba, W. humilis, W. angusta (likewise known as W. fulgida), and W. aletroides. All these belong to the true Watsonias, which have much more showy flowers than the other sections of the genus. The white-flowered Watsonia (W. alba) is a lovely plant, which flowers in early summer. With regard to the culture of these plants, treatment similar to that recommended for the early-flowering Gladioli will suit them. Where there are means of growing these and similar Cape bulbous plants in frames, a good deal of trouble is saved, and, moreover, the plants so grown would produce finer flowers, as the young growths would be protected at a time when they most needed it. As a general rule, however, growing the plants in open borders of light rich soil in warm situations will be found satisfactory. Among the numerous varieties offered in trade lists, the following may be taken as a representative set, viz., W. coccinea, fulgens, Meriana, alba, humilis, marginata, rosea, speciosa, fulgida, brevifolia, angustifolia, Grootvorst, Louis XVI., Wreede, Duchess, George IV., Chilea, Duc de Berri, and Blucher.

Welsh Poppy (*Meconopsis cambrica*).

White Cup (*Nierembergia rivularis*).

White Wood Lily (*Trillium grandiflorum*).

Whitlavia grandiflora.—A hardy Californian annual, allied to the Nemophila, and of remarkable beauty. It attains a height of about 1 ft., is of branched growth, and produces an abundance of showy bell-shaped blossoms of a rich, deep blue colour. There is a variety with white blossoms, and another handsome variety called gloxinioides with white and blue flowers. This annual is quite hardy, and may be sown either in autumn or spring in the open border in good friable soil. It is best used as a carpet or ground-work plant. Hydrophyllaceæ.

Wigandia.—A genus of noble-leaved plants, native of the Tropics, but succeeding well in the open air in summer. The best known is W. caracasana, a native of the mountainous regions of New Granada, which, from the nobility of its port and the size and form of its leaves, is entitled to hold a place among the finest plants of our gardens. Under the climate of London it has made leaves as remarkable for size as for strong veining and texture. It will be found to succeed in the midland and southern counties of England, though too much care cannot be taken to secure for it a warm, sheltered position and free good soil. It may be used with superb effect either in a mass or as a single plant. It is frequently propagated by cuttings of the roots, and grown in a moist and genial temperature through the spring months, kept near the light so as to preserve it in a dwarf and well-clothed condition; and, like all the other plants in this class, it should be very carefully hardened off previous to planting out at the end of May. It is, however, much better raised from cuttings of the shoots, if these are to be had. It may also be raised from seed. W. macrophylla, from Mexico, has stems covered with short stinging hairs, bearing brownish viscid drops, which adhere to the hand like oil when the stem is touched. W. imperialis, a new variety, is said to excel the others in its growth. W. Vigieri is another fine kind of quick and vigorous growth, and remarkable habit. It often has leaves 3 ft. 9 in. long, including the leaf-stalk, and 22 in. across, and the stem, nearly 7 ft. high and 3 in. in diameter, bears a column of such leaves. It is distinguished from either W. caracasana or macrophylla by the leaves and the stems being covered in a greater degree with glossy, slender, stinging bodies. These are so thickly produced as to give the stems a glistening appearance. W. urens is another species often planted, but decidedly inferior to either of the foregoing, except in power of stinging, in which way it is not likely to be surpassed. All have blue or violet blossoms produced in clusters, but they are not often borne in the open air with us. In their native habitats they range from 3 ft. to 12 ft. high, W. caracasana being the tallest.

Wilbrandia drastica.—An interesting Cucurbit from Brazil. It is of graceful growth, and bears dense clusters of small

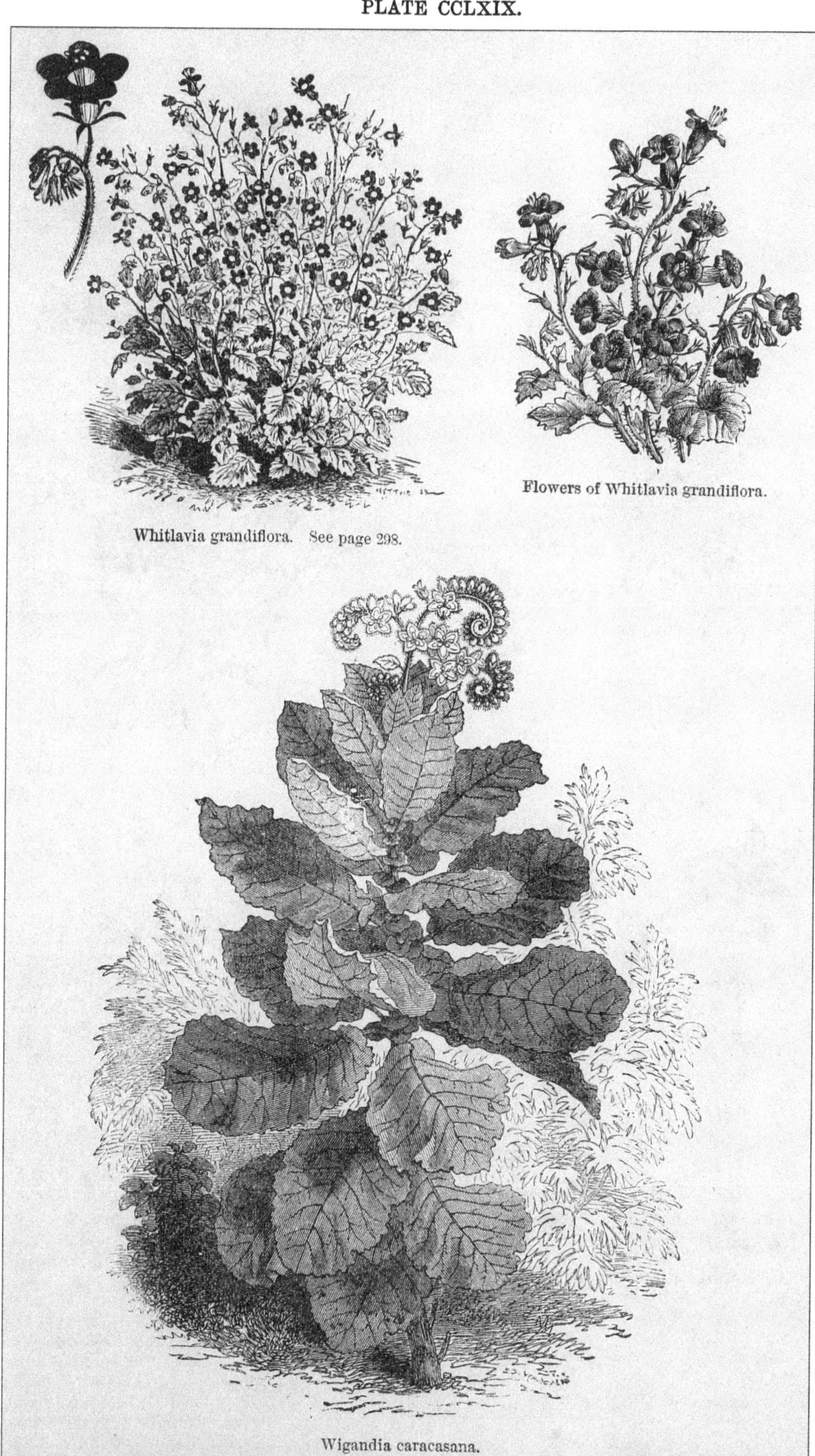

Whitlavia grandiflora. See page 298.

Flowers of Whitlavia grandiflora.

Wigandia caracasana.

PLATE CCLXX.

Xerophyllum asphodeloides.

Flowers of Xeranthemum annuum.

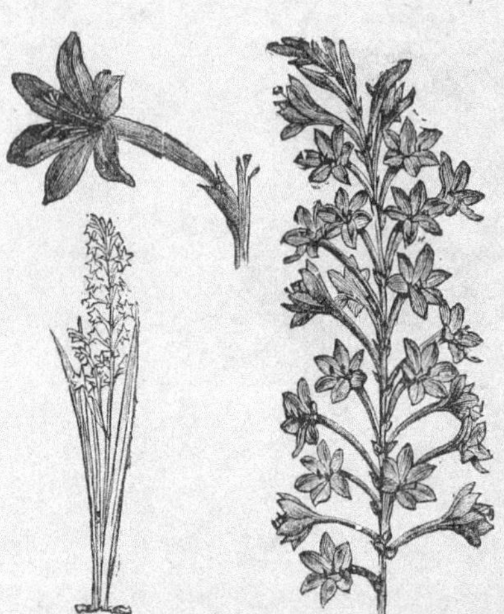

Watsonia Meriana. See page 298.

Xeranthemum annuum.

fruits. It should be treated as a half-hardy annual like the Gourds.

Wild Bergamot (*Monarda fistulosa*).

Wild Indigo (*Baptisia tinctoria*).

Wild Liquorice (*Ononis arvensis*).

Wild Senna (*Cassia marilandica*).

Willow Herb (*Epilobium*).

Windflower (*Anemone*).

Winter Aconite (*Eranthis hyemalis*).

Winter Cherry (*Physalis Alkekengi*).

Winter Cress (*Barbarea*).

Winter Green (*Pyrola*).

Wood Lily (*Trillium*).

Woodruff (*Asperula*).

Woodsia. — These pretty deciduous hardy Ferns are admirably adapted for a northern position in the alpine or rock garden. As they are impatient of sunshine, the drainage of the spot on which they are planted should receive special attention, and they should be placed in a mixture of fibry peat and loam, with some sandstone broken up and mixed with it. It is a good plan to place them between little blocks of sandstone just peeping out of the soil, a position for which they have a great partiality. These blocks of stone could, if necessary, be covered with Sedums and other suitable flowering rock plants. W. ilvensis makes a very pretty pot plant well drained. The best hardy species are W. ilvensis and W. alpina; there is also a very beautiful North American kind named W. obtusa.

Wood Sorrel (*Oxalis*).

Woodwardia.—Of this noble genus of Ferns there are a few hardy species; all are handsome plants, having broad and beautifully arching fronds, which are very ornamental, especially if seen a little above the level of the eye. They thrive well under the ordinary conditions of the hardy fernery, and in any place in a shady position and a light, peaty soil, that is moist in summer. The principal hardy kinds are W. areolata (angustifolia) and W. virginica, both from N. America; W. japonica and W. orientalis, from Japan; and W. radicans, from Madeira This last is the tenderest, and requires a more sheltered position, and perhaps protection in severe cold.

Woolly Thistle (*Carduus eriophorus*).

Worm Grass (*Spigelia marilandica*).

Wormwood (*Artemisia*).

Woundwort (*Anthyllis Vulneraria*).

Wulfenia carinthiaca. — A remarkably dwarf, almost stemless evergreen herb, 12 in. to 18 in. high, bearing in summer showy spikes of purplish-blue drooping flowers. Found only on one or two mountains in Carinthia. A very ornamental plant for rockwork or borders, in light, moist, sandy loam. W. Amherstiana is another species from the Himalayas. It is a similar plant to the Carinthian species, but showier. It is rare, and we have only seen it in Kew Gardens. It grows freely in any position in the rock garden, but prefers a shady spot and light, rich soil; it is perfectly hardy. Scrophulariaceæ.

Wyethia.—A genus of North American plants of the Compositæ; for botanical gardens.

Xanthium.—Curious annual plants of the Compositæ; of botanical interest only.

Xanthosoma sagittifolium. — A West Indian plant of the Arum family with very much the habit and appearance of Caladium esculentum, but not so valuable. It has arrow-shaped leaves of a dark green colour, supported on rather slender stalks. Another equally handsome and large species is X. violaceum, the leaves and leaf-stalks of which are suffused with a delicate violet hue, slightly inclining to hoariness. The positions and treatment should be similar to those recommended for Caladium esculentum. The plants should be tried only in the warmer parts of the country, and should not be placed in the open air till the beginning of June.

Xeranthemum annuum.—This hardy annual is one of the prettiest of what are known as Everlasting flowers. It grows about 2 ft. high, and if sown in patches yields abundant masses of white, purple, and yellow, double and single and semi-double blossoms. A packet of mixed seed sown in March in any ordinary garden soil will give a variety of colour, and if it is preferred to keep the colours distinct, then the different kinds must be obtained. The principal ones are —Album, white; imperiale, dark violet-purple; plenissimum, dark purple, double; superbissimum, double, Globe-flowered; and Tom Thumb, a compact dwarf variety. The flowers are excellent for cutting, and if dried in autumn are useful for winter decoration. S. Europe. Compositæ.

Xerophyllum asphodeloides. — A tuberous-rooted plant with the aspect of an Asphodel, very interesting and beautiful. It forms a spreading tuft of grassy leaves when well grown, and bears a flower-stem from 1 ft. to 4 ft. high, terminated by a compact raceme of numerous white blossoms. It grows well in a moist sandy peaty border, or in the drier parts of the artificial bog. A common plant in the Pine barrens in North America.

Ximenesia encelioides (=Verbesina encelioides).

Yarrow (*Achillea*).

Yellow Hawkweed (*Tolpis barbata*).

Yellow Pond Lily (*Nuphar*).

Yellow Sweet Sultan (*Amberboa odorata*).

Youngia.—A genus of the Compositæ, of which one or two species are in cultivation, and these are suitable only for botanical collections.

Yucca (*Adam's Needle*).—The Yucca has no rival among hardy plants in its own peculiar habit and style of growth. Though the stiffest of all our hardy plants, it yet has a grace and elegance peculiarly its own; and this elegance shows itself under all conditions, provided the plant is not cramped for room, It seems equally fitted to stand alone on a lawn, in the centre of a bed, or in numbers grouped with other plants or forming a bed by themselves. The Yuccas look especially well on rockwork in any part, either at the bottom, the sides, or the top. They are not very particular about soil, growing in good rich mould, in stiff clay, or any well-drained soil; but they do not flourish so well in sand, chalk, or peat. Their complete hardiness is another great recommendation; all of those mentioned below are so hardy that it is almost impossible to kill them. The suckers are apt to die down to the ground when first planted if they have not been very carefully taken from the parent plant; but if left alone, they will in a few months renew their growth. There are several species hardy and well suited for flower garden purposes, and, more advantageous still, distinct from each other. The effect afforded by them, when well developed, is equal to that of any hothouse plant that we can venture to place in the open air for the summer, while they are green and ornamental at all seasons. The free flowering kinds, filamentosa and flaccida, may be associated with any of our nobler autumn flowering plants, from the Gladiolus to the great Statice latifolia. The species that do not flower so often, like pendula and gloriosa, are simply magnificent in their effect when grown in the full sun and planted in good soil. They are mostly easily increased by division of the stem and rhizome; and should in all cases be planted well and singly, beginning with healthy young plants, so as to secure perfectly developed specimens.

Y. aloifolia.—A fine and distinct species, with a stem when fully developed as thick as a man's arm, and rising to a height of from 6 ft. to 18 ft. The leaves are numerous, rigidly ascending, dark green, with a slight glaucous bloom; they are 18 in. to 21 in. long and broad at the middle, with the horny margin rolled in for 2 in. or 3 in. below the point, and finely toothed in the remaining portion. The flowers are almost pure white, in a vast pyramidal panicle. This plant is hardy, but the fact is not generally known. It should be tried on well-drained slopes in good sandy loam. There are some varieties, and quadricolor and versicolor having the leaves variously edged with green, yellow, and red are the finest. These variegated varieties are also very hardy, but as they are as yet far from common, it will be best to utilise them in the greenhouse or conservatory, or to place them in the open air during summer. They look very pretty isolated on the Grass, the pots plunged to the rim. S. America and W. Indies.

Y. angustifolia.—The smallest of all, the whole plant, when in flower, not being more than 2 ft. or 3 ft. high. The leaves are thick and rigid in texture, from 15 in. to 18 in. long and about $\frac{1}{4}$ in. broad, of a pale sea-green colour, with numerous white filaments at the edges. The inflorescence is a simple raceme of white flowers slightly tinged with yellow. Till more plentiful this had better be grown in warm borders, in well drained sandy loam. It is excellent for rockwork. N. America.

Y. canaliculata.—The leaves of this species are entire, *i.e.*, neither toothed nor filamentose at the margin, and form a dense rosette on a stem that rises 1 ft. or 2 ft. above the level of the ground. Each leaf is from 20 in. to 24 in. long, and 2 in. to $2\frac{1}{4}$ in. broad at the middle, very strong and rigid, and deeply concave on the face. The flowers are of a creamy white, in a large panicle 4 ft. to 5 ft. high. Fine for isolation or groups. Till more plentiful, should be encouraged in favourable positions and on warm soils. Mexico.

Y. filamentosa.—A very common and well-known species, with apple-green leaves and a much-branched panicle, 4 ft. to 6 ft. high. It varies very much when raised from seed. One variety (concava) has short, strong, broad leaves, with the face more concave than in the type; another variety (maxima) has leaves nearly 2 ft. long by $2\frac{1}{2}$ in. broad, with a panicle 7 ft. to 8 ft. in height. This species flowers with much vigour and beauty, and is well worth cultivating in every garden, not only in the flower garden or pleasure ground, but also on therough rockwork, or any spot requiring a distinct type of hardy vegetation; its fine, though delicate, variegated variety is suitable for like purposes. All the varieties thrive best and flower most abundantly in peaty or fine sandy soil. N. America.

Y. flaccida.—A stemless species, somewhat resembling Y. filamentosa, but smaller, with a downy branching panicle 3 ft. to 4 ft. high. The foliage is in close rosettes of leaves, $1\frac{1}{2}$ ft. to 2 ft. long, by about $1\frac{1}{2}$ in. broad at the middle, often fringed with filaments on the edges, the young ones nearly erect, the old ones abruptly reflexed at the middle, almost appearing as if broken. This gives such an irregular aspect to the tufts, that this kind is at once distinguished from any of the varieties of Y. filamentosa. It also flowers more regularly and abundantly than its relative, and is exceedingly well suited for groups of the finer hardy plants, for borders, or for being planted in large isolated tufts. N. America.

Y. glaucescens.—A very free-flowering kind, with a panicle 3 ft. to 4 ft. high, the branches of which are short and very downy. Leaves sea-green, about 18 in. long, with a few filaments on the margins. The flowers are of a greenish-yellow colour, and when in bud are tinged with pink, which tends to give the whole inflorescence a peculiarly pleasing tone. A very useful and ornamental sort—fine for groups, borders, isolation, or placing among low shrubs. N. America.

Y. gloriosa.—A species of large and imposing proportions, with a distinct habit

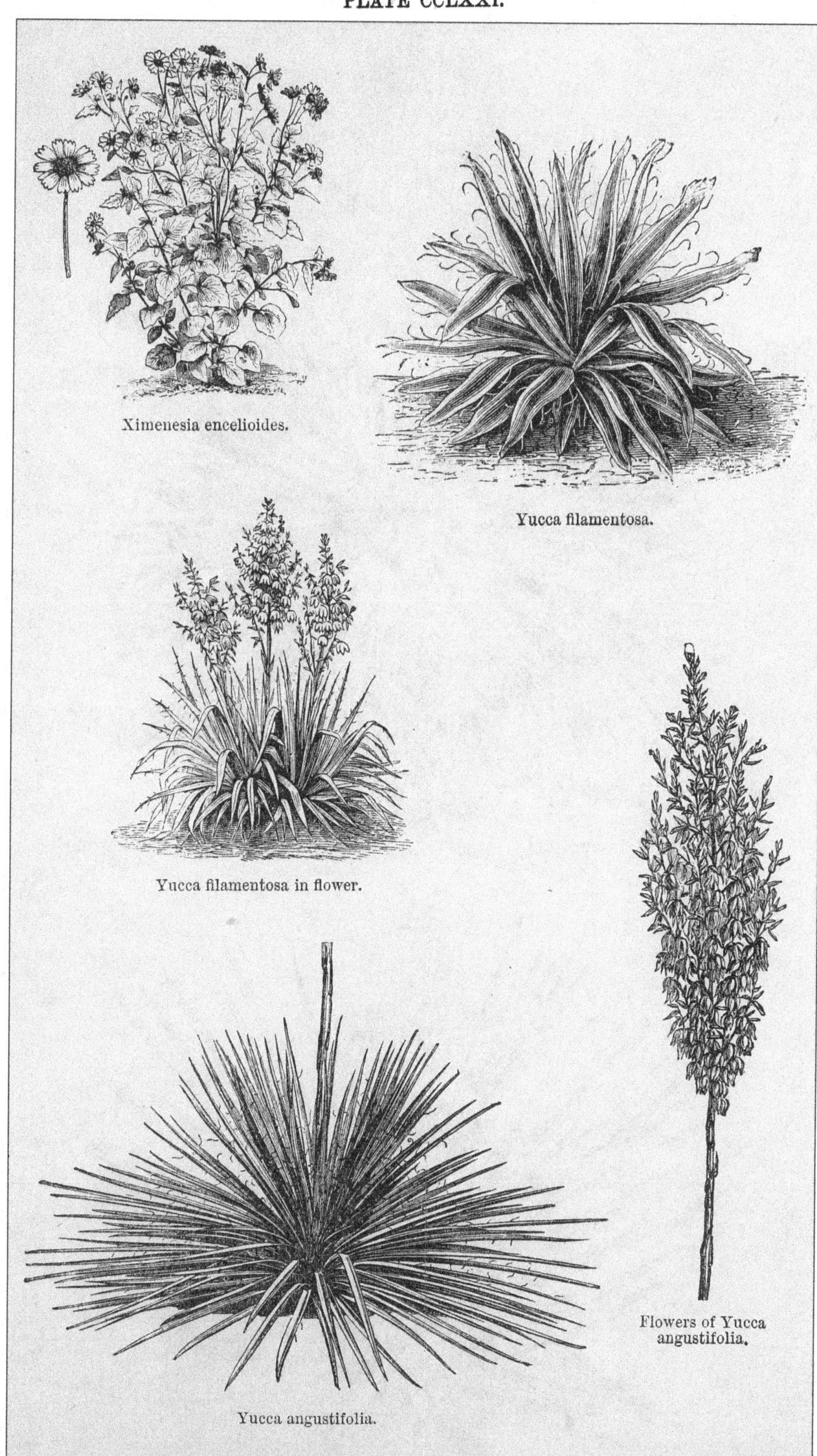

Ximenesia encelioides.

Yucca filamentosa.

Yucca filamentosa in flower.

Flowers of Yucca angustifolia.

Yucca angustifolia.

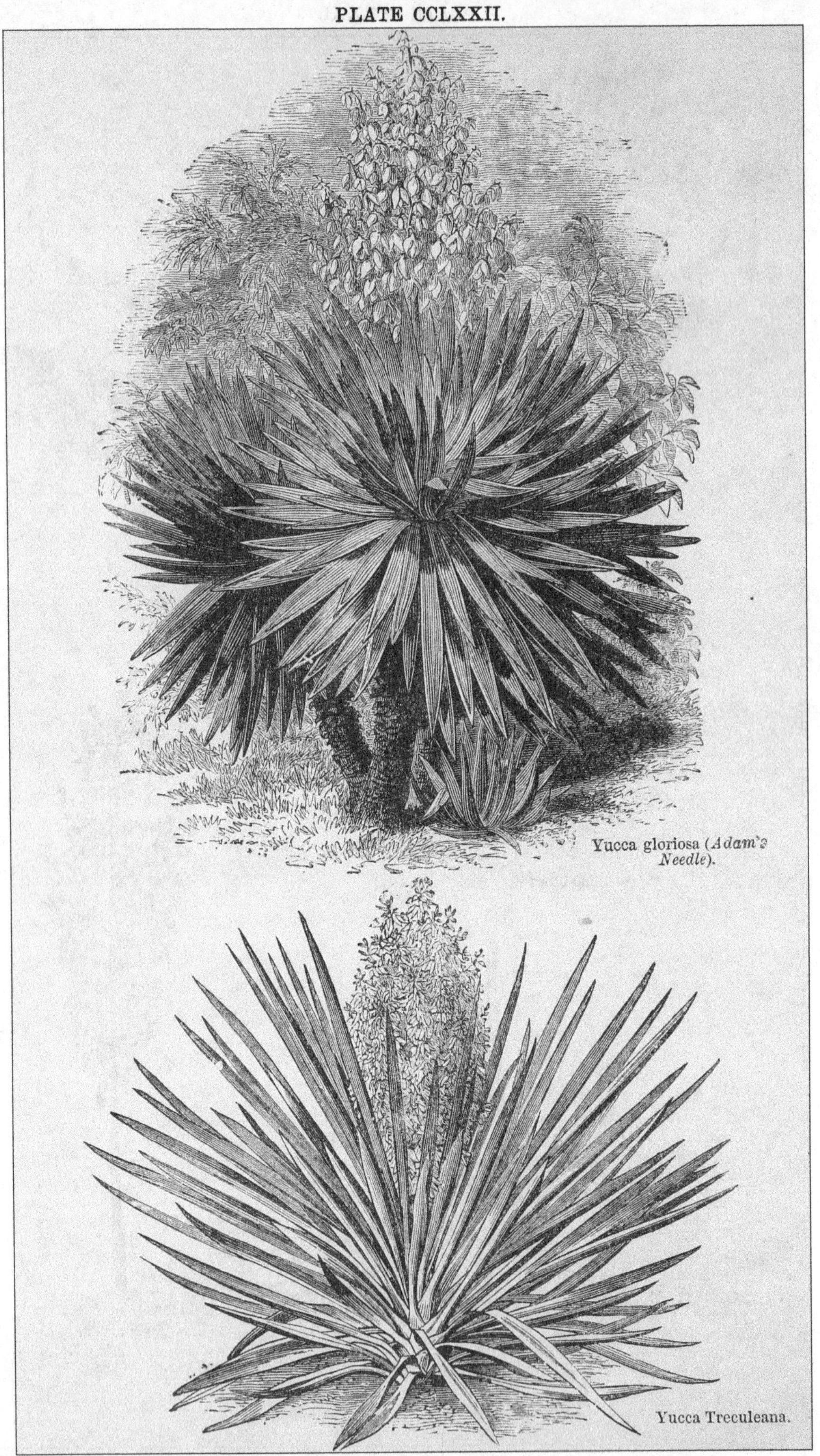

Yucca gloriosa (*Adam's Needle*).

Yucca Treculeana.

[*To face page* 301.

and somewhat rigid aspect. The flower-stem is over 7 ft. high, much branched, and bears an immense pyramidal panicle of large, almost pure white flowers. Its leaves are numerous, stiff, and pointed. It is one of the noblest plants in our gardens, and suitable for use in almost any position. It varies very much when grown from seed, and this is a good recommendation, as the greater variety of fine form we have the better. The chief varieties in cultivation are Y. g. longifolia, plicata, maculata, glaucescens, and minor. The soil for this plant should be a rich, deep loam. N. America.

Y. pendula.—The best species, perhaps, considering its graceful habit, which is invaluable. It grows about 6½ ft. high, the leaves being at first erect and of a sea-green colour, afterwards becoming reflexed and changing to a deep green. Old and well-established plants of it standing alone on the Grass are pictures of grace and symmetry, from the lower leaves which sweep the ground to the central ones that point up as straight as a needle. It is amusing to think of people putting tender plants in the open air, and running with sheets to protect them from the cold and rain of early summer and autumn, while perhaps not a good specimen of this fine thing is to be seen in the place. There is no plant more suited for planting between and associating with flower-beds. N. America. (= Y. recurva.)

Y. Treculeana.—This species is one of the most remarkable of the genus from its habit, and especially from the dimensions to which its foliage attains. Like many plants of its family, young specimens differ considerably from those which have reached maturity. Thus, while the former have their leaves bent, generally inflected, the full-grown plants exhibit them erect, rigid, very long, and very straight. The stem of this plant is stout, about 10 in. in diameter, furnished on all sides with leaves about 4 feet long, straight, thick, deeply channelled, acuminate for a considerable length, very finely toothed on the edges (which are of a brownish red and scarious), and ending in a stiff, very sharp point. The flower-stalk is very stout, about 4 ft. long, much branched; the branches erect, from 1 ft. to 1 ft. 8 in. long, bearing throughout their entire length flowers with long and narrow petals of a yellowish white, shining, and, as it were, glazed. It is a hardy and very vigorous plant. On the Continent specimens of more than 6½ ft. in diameter may frequently be seen. It is fine for banks and knolls, placed singly, or for the boldest groups. It comes from Texas. Any person wishing to have a distinct collection of Yuccas would find the above suit his purpose, though there are still several other species all more or less desirable. The dead flower-stems of Yuccas make capital supports for delicate creepers.

Zacintha verrucosa. — An annual composite of no ornamental value.

Zapania nodiflora *(Creeping Vervain)*.—A pretty and modest-looking, compactly-spreading, trailing plant, with prostrate stems 2 ft. or 3 ft. or more in length. It blooms late in summer, producing small purplish flowers in small roundish heads. Very suitable for the rougher parts of the rock garden or for borders or edgings in any rather warm soil. Asia and America. (= Lippia nodiflora.) Verbenaceæ.

Zauschneria californica.—This is a very distinct and bright perennial from California, and quite hardy in sheltered places in warm soils; in cold localities it requires in winter a little protection, such as a covering of coal ashes. It grows 12 in. to 18 in. high, and yields abundance of gracefully drooping, bright vermilion flowers. It flourishes in sandy loam on a rockery, and grows capitally on an old wall. On heavy and moist soils we have never observed it thrive. It can be easily propagated in spring by division of the roots, or raised from seeds sown on a gentle hotbed in spring. It flowers during summer and autumn. Onagraceæ.

Zea Mays *(Indian Corn)*.—This is one of the noblest of the Grasses that thrive in our climate, and is almost an indispensable adornment to our gardens, where it has a fine appearance either in isolated masses or associated with other fine foliage subjects. Cuzko and Caragua are the largest and finest of the green varieties, and gracillima the smallest and most graceful. The variegated or Japanese Maize is a very handsome variety. Its beautiful variegation is reproduced true from seed. It is particularly useful for intermingling with arrangements of ordinary bedding plants for vases, the outer margins of beds of sub-tropical plants, and like positions where its variegation may be well seen, and where its graceful leaves will prove effective among subjects of dumpy habit. It should in all cases have light, rich, warm soil. It has a habit of breaking into shoots rather freely near the base of the central stem, and where it grows very freely this should recommend it for planting in an isolated manner, or in groups of three or five on the turf. The seeds of the Maize should be sown in a gentle hotbed in April, although occasionally it will succeed if sown out-of-doors. Gradually harden off the plants before they have made more than three or four little leaves, keeping them in a cool frame very near the glass, so as to keep them sturdy, and finally exposing them in the same position by taking the lights quite off. This course is perhaps the more desirable in the case of the variegated Maize, which does not grow so vigorously as the green kinds. In neither case should the plants be drawn up long in heat, for if thus treated, they will not thrive so well. The first few leaves the variegated kind makes are green, but they soon begin to manifest the striping. They should be planted out about the middle of May.

Zephyranthes *(Zephyr Flower)*.—This beautiful genus, which has been termed the Crocus of America, includes about fourteen species. They are low-growing, bulbous plants, with grassy leaves, which appear in spring with or before the blossoms, and white or rosy-pink Crocus-like flowers, mostly large and handsome. They require rest during the winter, and at that season are best kept dry, being planted out in the full sun in spring in very sandy soil. They also do well in the greenhouse, planted four or six in a pot. They are increased by means of offsets. The following are the principal cultivated species:—

Z. Atamasco *(Atamasco Lily)*.—This handsome plant is a native of North America, where it is a conspicuous ornament of damp places in woods and fields, presenting, when in flower, a very beautiful appearance. It has glossy leaves, appearing at the same time as the blossoms, which they slightly exceed in height; the blossoms, which are white, striped with rose, are about 3 in. long, borne singly upon a scape 6 in. high. This species flowers from May to July, usually during the last named month. It grows well in the open border, and increases rapidly by offsets, which should be removed and divided in the spring of every third or fourth year. Z. candida is a similar species, but not so hardy.

Z. carinata.—This lovely plant has narrow leaves and a flower-stem about 6 in. high, bearing a delicate rosy flower 2 in. or 3 in. in length. It blossoms freely in the open border if kept dry in winter, and should be grown in a light sandy loam. This species is widely distributed in South America. Z. rosea, a similarly beautiful species, with flowers of a bright rose colour, is a native of the mountains of Cuba.

Z. tubispatha.—A handsome plant, bearing a white, slightly fragrant flower, from 2 in. to 3 in. long. It is a native of Antigua and the Blue Mountains of Jamaica, and is properly a stove plant, but in very mild localities, if well protected in winter, it would thrive and flower well. A pretty pink hybrid from this species crossed by Z. carinata is sometimes met with in cultivation under the name of Z. Spofforthiana. Z. Treatiæ, a new species in the way of Z. Atamasco, is yet too rare to speak of its culture or hardiness.

Zietenia lavandulæfolia.—An evergreen, spreading, half-shrubby perennial of a greyish hue, with a stem 6 in. to 12 in. high. The flowers appear in summer, and are purple, in whorls, forming a spike about 6 in. long, with a very slender, downy stalk. Suitable for the margins of borders and the rougher parts of the rock garden, or naturalised in any soil, wet or dry. Division. Caucasus. Labiatæ.

Zinnia.—Few half-hardy annuals are more satisfactory than these when well cared for. The flowers exhibit great brilliancy of colour, and, whether planted in beds, rows, or singly, they are amongst the most effective of summer blooming plants. Another point in their favour is that they flower well up to the autumn, and that the blooms do not easily become injured by inclement weather, but retain all their freshness and gay colouring when many other bright flowering subjects present but a sorry appearance. In mixed borders, or in beds among sub-tropical plants, these are always attractive when well grown. They require a deep loamy soil and a warm, open situation. The seed should be sown in gentle warmth, but nothing is gained by sowing before the middle or latter end of March, as the young plants are apt to become somewhat root-bound and stinted for nutriment when they have to stand for some considerable time before planting out, in which case they lose something of that fresh, free growth which should at all times be maintained until they come into flower. Once the tissues harden to the extent of bringing the young plants to a standstill, there is but little chance of rapid progress being made when they are set out in the open ground. Neither is it advisable to plant out much before the end of the first week in June, as the Zinnia is very susceptible to atmospheric changes, and is completely ruined by a few degrees of frost. Plant in well-stirred, fairly enriched soil in full exposure, for the Zinnia loves to bask in the sun's fiercest rays, merely demanding a surface covering, to protect the roots, and plenty of moisture at all times. Planted in a bed by themselves, they would be greatly improved if the soil could be thrown out, and a good depth of fermenting manure be trodden well in, the soil being afterwards replaced. Both the single and double forms of Zinnia make fine garden plants when well grown, displaying a diversity and brilliancy of colour equalled by few plants. The double forms have of late been most in request, although both the double and single varieties have been greatly improved. There is one good characteristic about the double Zinnias—they are not in some instances of such rank and unwieldy growth as the single types, and the process of dwarfing has gone hand-in-hand with that of multiplying the petals in the flowers. Careful selection also has done something in the way of inducing a better habit of growth; and it will be observed that it often happens that particular types of flower get improved both in habit and blooms at the same time. Some of the single Zinnias are very beautiful, such as the yellow, carmine, rosy-purple, scarlet, crimson, and orange. Z. elegans is the principal species from which the numerous varieties mentioned in seed catalogues have been derived, both single and double-flowered dwarf (pumila) as well as tall kinds. Z. Darwini is a beautiful hybrid variety with very double flowers of various colours. Z. Haageana, known also as Z. mexicana, has a very neat habit of growth and rich orange-yellow blossoms. It

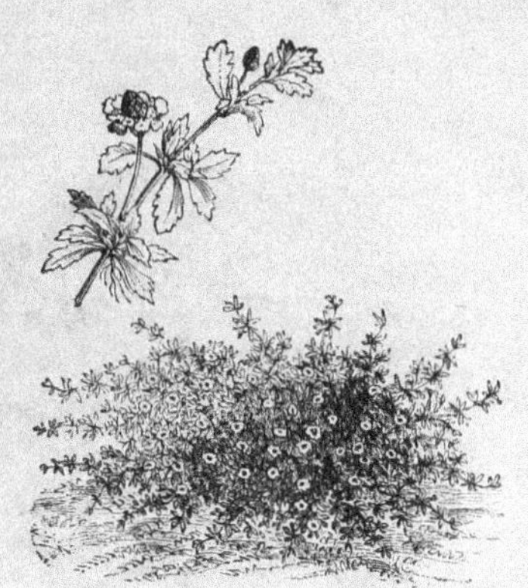

Zapania nodiflora (*Creeping Vervain*).
See page 301.

Zauschneria californica. See page 301.

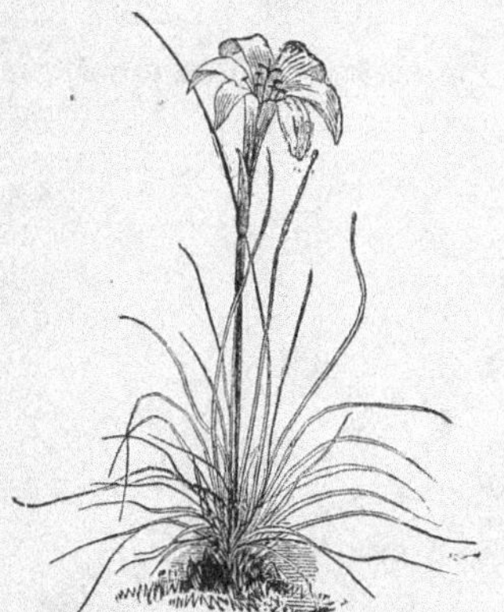

Zephyranthes Atamasco (*Atamasco Lily*).

Zea Mays (*Indian Corn*).

Flower of Zinnia Haageana.

PLATE CCLXXIV.

Zinnia elegans fl.-pl.

Double-flowered Zinnia Darwini

Zinnia Haageana fl.-pl.

[*To face page* **303.**

also occurs with double flowers. All are natives of Mexico.

Zozimia.—Plants of the Umbelliferæ; for botanical collection.

Zygadenus.—A small genus of the Lily family (synonymous with Anticlea). Not of much value as ornamental plants, as the flowers are all greenish-yellow, but on account of their distinct growth they are worth cultivating in a botanical or full collection. There are three species in cultivation, all from California. They are slender bulbous plants, with narrow, grassy foliage, and tall, branching flower-stems, ranging from 1 ft. to 4 ft. high. Z. Fremonti (also known as Z. glaberrimus, Z. chloranthus, and Z. Douglasi) is the largest flowered species. Z. venenosus (Z. Nuttalli) and Z. paniculatus are the two other kinds. They thrive best in a moist peaty border in a shady position protected from cold winds.

Zygophyllum Fabago.—An interesting South European plant of botanical interest only.

Printed in the United States
By Bookmasters